Lecture Notes in Artificial Intelligence 5782

Edited by R. Goebel, J. Siekmann, and W. Wahlster

Subseries of Lecture Notes in Computer Science

Wray Buntine Marko Grobelnik
Dunja Mladenić John Shawe-Taylor (Eds.)

Machine Learning and Knowledge Discovery in Databases

European Conference, ECML PKDD 2009
Bled, Slovenia, September 7-11, 2009
Proceedings, Part II

Springer

Series Editors

Randy Goebel, University of Alberta, Edmonton, Canada
Jörg Siekmann, University of Saarland, Saarbrücken, Germany
Wolfgang Wahlster, DFKI and University of Saarland, Saarbrücken, Germany

Volume Editors

Wray Buntine
NICTA
Locked Bag 8001, Canberra, 2601, Australia
and
Helsinki Institute of IT
Finland
E-mail: wray.buntine@nicta.com.au

Marko Grobelnik
Dunja Mladenić
Jožef Stefan Institute
Jamova 39, 1000 Ljubljana, Slovenia
E-mail: {marko.grobelnik,dunja.mladenic}@ijs.si

John Shawe-Taylor
University College London
Gower St., London, WC1E 6BT, UK
E-mail: jst@cs.ucl.ac.uk

Library of Congress Control Number: 2009933615

CR Subject Classification (1998): I.2, H.2.8, H.3, G.3, H.5, G.2, I.7

LNCS Sublibrary: SL 7 – Artificial Intelligence

ISSN 0302-9743
ISBN-10 3-642-04173-6 Springer Berlin Heidelberg New York
ISBN-13 978-3-642-04173-0 Springer Berlin Heidelberg New York

springer.com

Printed in Germany

Typesetting: Camera-ready by author, data conversion by Scientific Publishing Services, Chennai, India
Printed on acid-free paper SPIN: 12754656 06/3180 5 4 3 2 1 0

Preface

The year 2008 was the first year that the previously separate European Conferences on Machine Learning (ECML) and the Conference on Principles and Practice of Knowledge Discovery in Databases (PKDD) were merged into a unified event. This is a natural evolution after eight consecutive years of their being collocated after the first joint conference in Freiburg in 2001. The European Conference on Machine Learning (ECML) traces its origins to 1986, when the first European Working Session on Learning was held in Orsay, France followed by the second European Working Session on Learning held in Bled, the location of this year's ECML PKDD 2009 conference. The European Conference on Principles and Practice of Knowledge Discovery in Databases (PKDD) was first held in 1997 in Trondheim, Norway. Over the years, the ECML/PKDD series has evolved into one of the largest and most selective international conferences in machine learning and data mining, the only one that provides a common forum for the two closely related fields. In 2009, ECML PKDD conference was held during September 7–11 in Bled, Slovenia.

The conference used a hierarchical reviewing process. We nominated 26 Area Chairs, each of them responsible for one sub-field or several closely related research topics. Suitable areas were selected on the basis of the submission statistics for ECML PKDD 2008 and from last year's International Conference on Machine Learning (ICML 2008) and International Conference on Knowledge Discovery and Data Mining (KDD 2008) to ensure a proper load balance among the Area Chairs. A joint Program Committee (PC) was nominated consisting of some 300 renowned researchers, mostly proposed by the Area Chairs. In order to make best use of the reviewing capabilities we initially only requested that two reviews be sought. However, in the event of an inconsistency between the two assessments a third review was requested. Papers receiving two very positive reviews were considered for inclusion in the two special issues of *Machine Learning* and *Data Mining and Knowledge Discovery* appearing in time for the conference. A further review was also sought for these papers in order to assess their suitability to appear in journal form. Aleksander Kolcz was the Best Papers Chair responsible for overseeing the selection of papers for these special issues.

ECML PKDD 2009 received 679 abstract submissions resulting in a final total of 422 papers that were submitted and not withdrawn during the reviewing process. Based on the reviews, and on discussions among the reviewers, the Area Chairs provided a recommendation for each paper with a ranking of the borderline papers. The three Program Chairs made the final program decisions after merging the opinions of the 26 Area Chairs.

All accepted papers were of equal status with an oral presentation, poster presentation and 16 pages in the proceedings, with the exception of those accepted for the special issues of journals that were only allocated a single page abstract in the proceedings. We have selected a total of 106 papers of which 14 were be equally divided between the two special issues. The acceptance rate for all papers is therefore 25%, in line with the high-quality standards of the conference series. It is inevitable with such a low acceptance rate that some good papers were rejected and we hope that authors of these papers were not discouraged by their disappointment. We are, however, confident that the accepted papers are of a high quality, making a very exciting and stimulating conference. In addition to research papers, 15 demo papers were accepted, each having 4 pages in the proceedings and demo of the system during the poster session. In addition to the paper and poster/demo sessions, ECML PKDD 2009 also featured five invited talks, ten workshops, six tutorials, and the ECML PKDD discovery challenge and industrial track. The selection of Invited Speakers covered a broad range from theoretical to leading application-orientated research. Together they made a very strong addition to the conference program. We are grateful to Shai Ben-David (University of Waterloo, Canada), Nello Cristianini (University of Bristol, UK), Mark Greaves (Vulcan Inc.), Rosie Jones (Yahoo! Research), Ralf Steinberger (European Commission - Joint Research Centre) for their participation in ECML PKDD 2009. The abstracts of their presentations are included in this volume.

This year we continued to promote an Industrial Track chaired by Marko Grobelnik (Jožef Stefan Institute, Slovenia) and Nataša Milić-Frayling (Microsoft Research, Cambridge, UK) consisting of selected talks with a strong industrial component presenting research from the area covered by the ECML PKDD conference. We have also included a Demonstration Track chaired by Alejandro Jaimes Larrarte, providing a venue for exciting exemplars of applications of novel technologies.

As in recent years, the conference proceedings were available on-line to conference participants during the conference. We are grateful to Springer for accommodating this access channel for the proceedings.

As in previous years we will continue with the recently established tradition of videorecording the event, ensuring an enduring record of the event made accessible at http://videolectures.net/. Mitja Jermol is the Video Chair overseeing this aspect of the organization.

This year's Discovery Challenge was coordinated by Andreas Hotho together with Folke Eisterlehner and Robert Jäschke. It involved three tasks in the area of tag recommendation.

We are all indebted to the Area Chairs, Program Committee members and external reviewers for their commitment and hard work that resulted in a rich but selective scientific program for ECML PKDD 2009. We are particularly grateful to those reviewers who helped with additional reviews at a very short notice to assist us in a small number of difficult decisions. We further thank the Workshop and Tutorial Chairs Ravid Ghani and Cédric Archambeau for selecting and coordinating the ten workshops and six tutorials that accompany the conference; the workshop organizers, tutorial presenters, and the organizers of the discovery challenge, the Industrial and Demonstration Tracks; the Video Chair;

the Publicity Chair David Hardoon; and Richard van de Stadt and CyberChair-PRO for highly competent and flexible support when confronted by novel features in our handling of the papers. Special thanks are due to the Local Chair, Tina Anžič, for the many hours spent ensuring the success of the conference. Finally, we are grateful to the Steering Committee and the ECML PKDD community that entrusted us with the organization of the ECML PKDD 2009.

Most of all, however, we would like to thank all the authors who trusted us with their submissions, thereby contributing to the main yearly European-focussed international event in the life of our expanding research community.

June 2009

Dunja Mladenić
Wray Buntine
Marko Grobelnik
John Shawe-Taylor

Organization

General Chair

Dunja Mladenić	Jožef Stefan Institute, Slovenia

Program Chairs

Wray Buntine	Helsinki Institute of IT, Finland; NICTA, Australia
Marko Grobelnik	Jožef Stefan Institute, Slovenia
John Shawe-Taylor	University College London, UK

Local Chair

Tina Anžic	Jožef Stefan Institute, Slovenia

Tutorial Chair

Cédric Archambeau	University College London, UK

Workshop Chair

Rayid Ghani	Accenture Technology Labs, USA

Discovery Challenge Chairs

Robert Jäschke	University of Kassel, Germany
Andreas Hotho	University of Kassel, Germany
Folke Eisterlehner	University of Kassel, Germany

Industrial Track Chairs

Marko Grobelnik	Jožef Stefan Institute, Slovenia
Nataša Milić-Frayling	Microsoft Research Cambridge, UK

Demo Chair

Alejandro Jaimes Larrarte	Telefonica Research, Spain

Best Paper Chair

Aleksander Kolcz — Microsoft Live Labs, USA

Publicity Chair

David Hardoon — University College London, UK

Video Chair

Mitja Jermol — Jožef Stefan Institute, Slovenia

Steering Committee

Walter Daelemans
Katharina Morik
Joost N. Kok
Dunja Mladenić
Andrzej Skowron
Bart Goethals
Johannes Fürnkranz
Stan Matwin
Tobias Scheffer
Myra Spiliopoulou

Area Chairs

Francis Bach
Francesco Bonchi
Toon Calders
Walter Daelemans
Johannes Fürnkranz
João Gama
Eamonn Keogh
Alek Kolcz
Stan Matwin
Claire Nedellec
David Silver
Gerd Stumme
Michael Witbrock
Hendrik Blockeel
Pavel Brazdil
Nitesh Chawla
Tijl De Bie
Thomas Gärtner
Bart Goethals
Joost Kok
Jure Leskovec
Taneli Mielikainen
Martin Scholz
Steffen Staab
Luis Torgo
Stefan Wrobel

Program Committee

Ameen Abu-Hanna
Osman Abul
Lada Adamic
Abdullah Al Mueen
Enrique Alfonseca
Erick Alphonse
Carlos Alzate
Massih-Reza Amini
Gennady Andrienko
Annalisa Appice
Hiroki Arimura
Andrew Arnold
Sitaram Asur
Martin Atzmueller
Nathalie Aussenac-Gilles
Paulo Azevedo
Lars Backstrom
Tony Bagnall
Roberto Basili
Vladimir Batagelj
Ron Bekkerman
Marc Bellemare
Paul Bennett
Bettina Berendt
Tanya Berger-Wolf
Michael Berthold
Sourangshu Bhattacharya
Concha Bielza
Misha Bilenko
Stephan Bloehdorn
Christian Bockermann
Mario Boley
Christian Borgelt
Karsten Borgwardt
Henrik Bostrom
Guillaume Bouchard
Jean-François Boulicaut
Janez Brank
Ulf Brefeld
Bjorn Bringmann
Paul Buitelaar
Rui Camacho
Stephane Canu
Olivier Cappe
Andre Carvalho
Carlos Castillo
Ciro Cattuto
Vineet Chaoji
Sanjay Chawla
David Cieslak
Philipp Cimiano
Lucchese Claudio
Vincent Claveau
Fabrice Colas
Antoine Cornuejols
Christophe Costa Florencio
Fabrizio Costa
Bruno Cremilleux
Padraig Cunningham
Alfredo Cuzzocrea
Florence D'Alche-Buc
Claudia d'Amato
Gautam Das
Kamalika Das
Jesse Davis
Alneu de Andrade Lopes
Jeroen de Bruin
Marco de Gemmis
Jeroen De Knijf
Thomas Degris-Dard
Jose del Campo-Avila
Krzysztof Dembczynski
Laura Dietz
Carlos Diuk
Kurt Driessens
Pierre Dupont
Jennifer Dy
Saso Dzeroski
Charles Elkan
Tapio Elomaa
Damien Ernst
Floriana Esposito
Fazel Famili
Nicola Fanizzi
Amir-massoud Farahmand
Ad Feelders

Xiaoli Fern
Daan Fierens
Ilias Flaounas
George Forman
Blaz Fortuna
Eibe Frank
Jordan Frank
Mohamed Gaber
Dragan Gamberger
Gemma Garriga
Gilles Gasso
Eric Gaussier
Ricard Gavalda
Floris Geerts
Peter Geibel
Lise Getoor
Olivier Gevaert
Rayid Ghani
Fosca Gianotti
Melanie Gnasa
Henrik Grosskreutz
Amit Gruber
Vincent Guigue
Robert Gwadera
Larry Hall
Zaid Harchaoui
Hannes Heikinheimo
Iris Hendrickx
Mark Herbster
Tom Heskes
Melanie Hilario
Alexander Hinneburg
Susanne Hoche
Frank Höppner
Geoff Holmes
Tamas Horvath
Bettina Hoser
Veronique Hoste
Andreas Hotho
Eyke Hüllermeier
Inaki Inza
Mariya Ishteva
Robert Jaeschke
Longin Jan Latecki
Nathalie Japkowicz
Szymon Jaroszewicz
Thorsten Joachims
Alipio Jorge
Felix Jungermann
Matti Kaariainen
Alexandros Kalousis
Murat Kantarcioglu
Samuel Kaski
Philip Kegelmeyer
Kristian Kersting
Svetlana Kiritchenko
Igor Kononenko
Anna Koop
Walter Kosters
Wojciech Kotlowski
Stefan Kramer
Andreas Krause
Yuval Krymolowski
Miroslav Kubat
Ravi Kumar
James Kwok
Bill Lampos
Niels Landwehr
Mark Last
Nada Lavrac
Lihong Li
Jessica Lin
Charles Ling
Huan Liu
XiaoHui Liu
Eneldo Loza Mencia
Peter Lucas
Elliot Ludvig
Yiming Ma
Sofus Macskassy
Michael Madden
Donato Malerba
Bradley Malin
Lluis Marquez
Michael May
Prem Melville
Rosa Meo
Pauli Miettinen
Roser Morante
Fabian Morchen

Katharina Morik
Flavia Moser
Fabien Moutarde
Klaus-Robert Müller
Ion Muslea
Amedeo Napoli
Olfa Nasraoui
Vivi Nastase
James Neufeld
Alexandru Niculescu-Mizil
Siegfried Nijssen
Joakim Nivre
Blaž Novak
Ann Nowe
Alexandros Ntoulas
Andreas Nuernberger
Guillaume Obozinski
Arlindo Oliveira
Martijn van Otterlo
Gerhard Paass
Cosmin Paduraru
Georgios Paliouras
Themis Palpanas
Junfeng Pan
Rong Pan
Andrea Passerini
Mykola Pechenizkiy
Dmitry Pechyony
Dino Pedreschi
Kristiaan Pelckmans
Jose-Maria Pena
Ruggero Pensa
Raffaele Perego
Bernhard Pfahringer
Christian Plagemann
Barnabas Poczos
Doina Precup
Philippe Preux
Kai Puolamaki
Peter van der Putten
Sampo Pyysalo
Predrag Radivojac
Troy Raeder Davood Rafiei
Chedy Raissi
Shyam Rajaram
Alain Rakotomamonjy
Liva Ralaivola
Jan Ramon
Chotirat Ratanamahatana
Chandan Reddy
Ehud Reiter
Elisa Ricci
Martin Riedmiller
Celine Robardet
Marko Robnik-Sikonja
Pedro Rodrigues
Teemu Roos
Fabrice Rossi
Volker Roth
Celine Rouveirol
Ulrich Rueckert
Stefan Rüping
Yvan Saeys
Lorenza Saitta
Scott Sanner
Vitor Santos Costa
Craig Saunders
Yucel Saygin
Lars Schmidt-Thieme
Jouni Seppanen
Shashi Shekhar
Jin Shieh
Stefan Siersdorfer
Tomi Silander
Ricardo Silva
Ozgur Simsek
Ajit Singh
Sergej Sizov
Carlos Soares
Maarten van Someren
Yang Song
Elaine Sousa
Myra Spiliopoulou
Karsten Steinhaeuser
David Stern
Jan Struyf
Jiang Su
Masashi Sugiyama
Johan Suykens
Vojtech Svatek

Sandor Szedmak
Nikolaj Tatti
Evimaria Terzi
Gerald Tesauro
Hanghang Tong
Volker Tresp
Koji Tsuda
Ville Tuulos
Rasmus Ulslev Pedersen
Dries Van Dyck
Stijn Vanderlooy
Sergei Vassilvitskii
Cor Veenman
Paola Velardi
Shankar Vembu
Celine Vens
Jean-Philippe Vert
Ricardo Vilalta
Michalis Vlachos
Christel Vrain
Jilles Vreeken
Christian Walder
Xiaoyue Wang
Markus Weimer
David Wingate
Michael Wurst
Dragomir Yankov
Lexiang Ye
Jie Yin
François Yvon
Menno van Zaanen
Bianca Zadrozny
Osmar Zaiane
Mikhail Zaslavskiy
Gerson Zaverucha
Filip Zelezny
Justin Zhan
Bin Zhang
Zhi-Hua Zhou
Qiang Zhu
Xiaojin Zhu
Albrecht Zimmermann

Additional Reviewers

Dima Alberg
Anelia Angelova
Mohammad Aziz
Michele Berlingerio
Marenglen Biba
Alexander Binder
Zoran Bosnić
Christos Boutsidis
Fabian Buchwald
Markus Bundschus
Wray Buntine
Lijuan Cai
Michelangelo Ceci
Weiwei Cheng
Joaquim Costa
Dave DeBarr
Marcos Domingues
Jun Du
Charles Elkan
Daan Fierens
Nuno A. Fonseca
Dmitriy Fradkin
Thomas Gabel
Zeno Gantner
Robby Goetschalckx
Habiba
Niina Haiminen
Katja Hansen
Andreas Hapfelmeier
Jingrui He
Raymond Heatherly
Thibault Helleputte
James Henderson
Yi Huang
Ali Inan
Tsuyoshi Ide
Szymon Jaroszewicz
Takafumi Kanamori
Alexandros Karatzoglou
Hisashi Kashima
Tsuyoshi Kato
Maarten Keijzer

Evan Kirshenbaum
Arto Klami
Thoralf Klein
Marius Kloft
Arne Koopman
Da Kuang
Mayank Lahiri
Tobias Lang
Lieven De Lathauwer
Gayle Leen
Jens Lehmann
Guy Lever
Biao Li
Nan Li
Shuyan Li
Yu-Feng Li
Yuan Li
Grigorios Loukides
Jose A. Lozano
Juan Luo
Hamid Reza Maei
Michael Mampaey
Alain-Pierre Manine
Leandro Marinho
Eneldo Loza Mencia
João Mendes-Moreira
Olana Missura
Matteo Mordacchini
Marianne Mueller
Eileen A. Ni
Martijn van Otterlo
Aline Paes
Indranil Palit
Sang-Hyeun Park
Aritz Perez
Claudia Perlich
Georgios Petasis
Dimitrios Pierrakos
Fabio Pinelli
Cristiano Pitangui
Troy Raeder
Alain Rakotomamonjy
Steffen Rendle
Achim Rettinger
François Rioult
Jan Rupnik
Jarkko Salojärvi
Leander Schietgat
Jana Schmidt
Armin Shmilovici
David Silver
Karsten Steinhäuser
Erik Štrumbelj
Jan Struyf
Ilija Subašić
Taiji Suzuki
Takashi Takenouchi
Nicola Tonellotto
Grigorios Tsoumakas
Mikalai Tsytsarau
Stijn Vanderlooy
Guy Van den Broeck
Sicco Verwer
Ricardo Vilalta
Dimitrios Vogiatzis
Petar Vračar
Chris Watkins
Joerg Wicker
Eric Wiewiora
Fuxiao Xin
Xingwei Yan
Xingwei Yang
Monika Zakova
De-Chuan Zhan
Indre Zliobaite

Sponsors

We wish to express our gratitude to the sponsors of ECML PKDD 2009 for their essential contribution to the conference. We wish to thank Jožef Stefan Institute, Slovenia, for providing financial and organizational means for the conference; the European Office of Aerospace Research and Development, a detachment of U.S. Air Force Office of Scientific Research (EOARD) for generous financial support; Pascal European Network of Excellence (PASCAL2) for sponsoring the Invited Speaker program and the videorecording of the conference; Slovenia Research Agency (ARRS); Google for supporting a Poster Reception; Microsoft Research Ltd., Cambridge, UK for supporting the Industrial track; Yahoo! Research, Quintelligence, Hewlett-Packard Labs, and ACTIVE European Integrated project for their financial support; the *Machine Learning* Journal for supporting the Student Best Paper Award; the *Data Mining and Knowledge Discovery* Journal for supporting the Student Best Paper Award; Nokia for sponsoring the Discovery Challenge Awards and the Best Demo Award.

Table of Contents – Part II

Regular Papers

Demo Papers

Table of Contents – Part I

Invited Talks (Abstracts)

Machine Learning Journal Abstracts

Data Mining and Knowledge Discovery Journal Abstracts

Regular Papers

Decomposition Algorithms for Training Large-Scale Semiparametric Support Vector Machines

Sangkyun Lee and Stephen J. Wright

Computer Sciences Department, University of Wisconsin-Madison,
1210 W. Dayton St., Madison, WI 53706, USA
{sklee,swright}@cs.wisc.edu

Abstract. We describe a method for solving large-scale semiparametric support vector machines (SVMs) for regression problems. Most of the approaches proposed to date for large-scale SVMs cannot accommodate the multiple equality constraints that appear in semiparametric problems. Our approach uses a decomposition framework, with a primal-dual algorithm to find an approximate saddle point for the min-max formulation of each subproblem. We compare our method with algorithms previously proposed for semiparametric SVMs, and show that it scales well as the number of training examples grows.

Keywords: semiparametric SVM, regression, decomposition, primal-dual gradient projection.

1 Introduction

Support Vector Machines (SVMs) are the most widely used nonparametric methods in machine learning, which aims to find a function that performs well in classifying or fitting given data. The power of SVM lies in the fact that it does not require the user to define the class of functions from which the observations might have been generated. In a sense, this is also a weakness, in that prior knowledge of the function class is often available for use. *Semiparametric* SVM formulations introduce parametric components into the model of the classifying / regression function, alongside the nonparametric contribution. The basis functions in the parametric part of the model can be chosen to embed prior knowledge and can be used for analyzing the effects of certain covariates, thus giving semiparametric SVM the potential advantages of both parametric and nonparametric methods.

Despite the benefits, semiparametric models have not drawn much attention from the machine learning community, possibly in part because the optimization problems arising from semiparametric SVMs are harder to solve than those generated by standard SVMs. This paper describes an efficient approach for finding solutions to large-scale semiparametric SVM problems. We focus on the formulation of semiparametric SVM regression first introduced in [1], which gives rise to

W. Buntine et al. (Eds.): ECML PKDD 2009, Part II, LNAI 5782, pp. 1–14, 2009.

a dual problem which is a convex quadratic program (QP) with several equality constraints as well as bound constraints.

To motivate our description of solvers for semiparametric SVMs, we discuss first the state of the art for solvers that tackle the standard SVM dual formulation, which is

$$\min_{\mathbf{x}} \frac{1}{2}\mathbf{x}^T \boldsymbol{Q}\mathbf{x} + \boldsymbol{p}^T\mathbf{x} \qquad \text{s.t.} \quad \mathbf{y}^T\mathbf{x} = 0, \qquad \mathbf{0} \le \mathbf{x} \le C\mathbf{1} \ , \tag{1}$$

where $\mathbf{x}$, $\mathbf{y}$, and $\mathbf{1} := (1, 1, \ldots, 1)$ are column vectors of length n. Many effective algorithms for this problem solve a sequence of subproblems, each of which updates some subvector of $\mathbf{x}$ while leaving the remaining elements unchanged. These algorithms can be categorized into two distinct groups. In the first group, the subvector is very short, typically containing just two components. Since the subproblem can be solved analytically for such a small number of variables, no numerical solver is needed. The subproblems are inexpensive, but many iterations are usually needed to reach a solution with acceptable quality. Sequential Minimal Optimization (SMO) [2] and its variants such as LIBSVM [3] fall into this category. In the second group of solvers, the subvectors are longer, requiring the subproblems to be solved with a QP solver that exploits the structure of the application. Although we face the burden of designing an efficient, robust QP solver, methods in the second group often show faster convergence than those in the first group. Successful instances of methods in the second group include SVM^{light}[4] and GPDT [5,6]. The QP solvers used in the second group can be applied to the full problem, thus solving it in one "outer" iteration, though this approach is not usually effective for large data sets.

In general, the methods in both groups discussed above are specialized to handle the single equality constraint in (1) along with the bound constraints. The analytic subproblem solution in SMO can be acquired only when the subproblem has up to one (or two in case of the modified SMO [7]) equality constraint. The subproblem selection algorithm of SVM^{light} strongly depends upon the existence of a single equality constraint; the same is true of GPDT, which uses a projection algorithm from [8]. Semiparametric SVMs, however, require solution of the following generalization of (1):

$$\min_{\mathbf{x}} F(\mathbf{x}) := \frac{1}{2}\mathbf{x}^T \boldsymbol{Q}\mathbf{x} + \boldsymbol{p}^T\mathbf{x} \qquad \text{s.t.} \quad \boldsymbol{A}\mathbf{x} = \boldsymbol{b}, \qquad \mathbf{0} \le \mathbf{x} \le C\mathbf{1}, \tag{2}$$

where $\boldsymbol{A} \in \mathbb{R}^{K \times n}$ and $\boldsymbol{b} \in \mathbb{R}^K$, where $K \ge 1$ is the number of parametric basis functions that we wish to include in the model. For semiparametric SVM regression, Smola, Frieß, and Schölkopf [1] proposed to apply a primal-dual interior point method based on the code LOQO. The size of problems that can be handled is thus limited by the need to perform a full evaluation of the matrix $\boldsymbol{Q}$ and the need for repeated factorizations of matrices of about this size. (The approach could however be used as the inner loop of a decomposition method in the second group discussed above.) Kienzle and Schölkopf [9] suggested a *Minimal Primal Dual* (MPD) algorithm. This algorithm use a variant of the method of multipliers to formulate a sequence of convex quadratic programs of dimension n with bound constraints only (no equalities), which are solved by a method that selects a single component for updating at each iteration. (In this sense, it is akin to the methods in the first group described above.) We give further details on MPD as we introduce our methods below. This approach does not scale well as the size n of the problem grows, but its performance can be improved by embedding it in a decomposition framework, as described below. We include both MPD and its decomposition variants in our computational tests of Sect. 5.

In this paper, we propose an approach that is related to MPD but that differs in several ways. First, it is a primal-dual approach; we alternate between steps in a subvector of $\mathbf{x}$ and steps in the Lagrange multipliers for the constraints $\boldsymbol{A}\mathbf{x} = \boldsymbol{b}$. Second, subvectors of $\mathbf{x}$ with more than 1 element are allowed. Third, two-metric gradient projection techniques are used in taking steps in the $\mathbf{x}$ components. Throughout, we take account of the fact that n may be very large, that $\boldsymbol{Q}$ cannot practically be computed and stored in its entirety, and that operations involving even modest-sized submatrices of $\boldsymbol{Q}$ are expensive.

We compare our approach computationally with MPD as stand-alone solvers, and also in a decomposition framework.

The remainder of the paper is structured as follows. In the next section, we define the semiparametric SVM regression problem and show that its dual has the form (2). Section 3 outlines the decomposition framework, while Sect. 4 describes the primal-dual method that we propose for solving the subproblems that arise from decomposition. Section 5 presents some computational results.

2 Semiparametric SVM Regression

We consider a regression problem for data $\{(\boldsymbol{t}_i, \boldsymbol{y}_i)\}_{i=1}^M$ where $\boldsymbol{t}_i \in \mathbb{R}^N$ are feature vectors and $\boldsymbol{y}_i \in \mathbb{R}$ are outcomes. We wish to find a function h that minimizes ϵ-insensitive loss function $\ell_\epsilon(h; \boldsymbol{t}, \boldsymbol{y}) := \max\{0, |\boldsymbol{y} - h(\boldsymbol{t})| - \epsilon\}$, while maximizing the margin as in [10]. Following [1,9], we formulate the semiparametric SVM regression problem as follows:

$$\min_{\boldsymbol{w},\boldsymbol{\beta},\boldsymbol{\xi},\boldsymbol{\xi}^*} \quad \frac{1}{2}\boldsymbol{w}^T\boldsymbol{w} + C\sum_{i=1}^{M}(\boldsymbol{\xi}_i + \boldsymbol{\xi}_i^*) \tag{3a}$$

$$\text{s.t.} \quad \boldsymbol{y}_i - \langle \boldsymbol{w}, \phi(\boldsymbol{t}_i)\rangle - \sum_{j=1}^{K} \boldsymbol{\beta}_j \psi_j(\boldsymbol{t}_i) \le \epsilon + \boldsymbol{\xi}_i \qquad \text{for } i = 1, \ldots, M \tag{3b}$$

$$\langle \boldsymbol{w}, \phi(\boldsymbol{t}_i)\rangle + \sum_{j=1}^{K} \boldsymbol{\beta}_j \psi_j(\boldsymbol{t}_i) - \boldsymbol{y}_i \le \epsilon + \boldsymbol{\xi}_i^* \qquad \text{for } i = 1, \ldots, M \tag{3c}$$

$$\boldsymbol{\xi} \ge \mathbf{0},\ \boldsymbol{\xi}^* \ge \mathbf{0}\ . \tag{3d}$$

where ϕ is a feature mapping function which defines a positive semidefinite kernel $\kappa(\boldsymbol{t}_i, \boldsymbol{t}_j) := \langle \phi(\boldsymbol{t}_i), \phi(\boldsymbol{t}_j)\rangle$, for all $i, j \in \{1, \ldots, M\}$, while $\{\psi_j\}_{j=1}^K$ are the basis functions for the parametric part of the model function. The model function is defined as an extended linear model of parametric and nonparametric parts, that is, $h(\boldsymbol{t}) = \langle \boldsymbol{w}, \phi(\boldsymbol{t}) \rangle + \sum_{j=1}^{K} \boldsymbol{\beta}_j \psi_j(\boldsymbol{t})$. We typically have $K \ll M$. If $K = 1$ and ψ_1 is a constant function, we recover the standard SVM regression problem.

The Wolfe-dual of (3) has the form (2), where

$$\mathbf{x} = \begin{bmatrix} \boldsymbol{\alpha} \\ \boldsymbol{\alpha}^* \end{bmatrix} \in \mathbb{R}^{2M} \text{ for the dual vectors } \boldsymbol{\alpha} \text{ and } \boldsymbol{\alpha}^* \text{ of (3b) and (3c), resp.,}$$

$$\boldsymbol{p} = [\epsilon - \boldsymbol{y}_1, \ldots, \epsilon - \boldsymbol{y}_M, \epsilon + \boldsymbol{y}_1, \ldots, \epsilon + \boldsymbol{y}_M]^T \in \mathbb{R}^{2M} \ ,$$

$$\boldsymbol{Q}_{ij} = \begin{cases} \boldsymbol{y}_i \boldsymbol{y}_j \kappa(\boldsymbol{t}_i, \boldsymbol{t}_j) & \text{if } 1 \le i,j \le M, \text{ or } M+1 \le i,j \le 2M \\ -\boldsymbol{y}_i \boldsymbol{y}_j \kappa(\boldsymbol{t}_i, \boldsymbol{t}_j) & \text{otherwise} \end{cases} ,$$

$$\boldsymbol{b} = \boldsymbol{0} \ ,$$

and

$$\boldsymbol{A} = \begin{bmatrix} \psi_1(\boldsymbol{t}_1) & \cdots & \psi_1(\boldsymbol{t}_M) & -\psi_1(\boldsymbol{t}_1) & \cdots & -\psi_1(\boldsymbol{t}_M) \\ \psi_2(\boldsymbol{t}_1) & \cdots & \psi_2(\boldsymbol{t}_M) & -\psi_2(\boldsymbol{t}_1) & \cdots & -\psi_2(\boldsymbol{t}_M) \\ \vdots & \ddots & \vdots & \vdots & \ddots & \vdots \\ \psi_K(\boldsymbol{t}_1) & \cdots & \psi_K(\boldsymbol{t}_M) & -\psi_K(\boldsymbol{t}_1) & \cdots & -\psi_K(\boldsymbol{t}_M) \end{bmatrix} \in \mathbb{R}^{K \times 2M} \ .$$

Introducing $\boldsymbol{\eta}$ as the Lagrange multipliers for the constraints $\boldsymbol{A}\mathbf{x} = \boldsymbol{b}$ in (2), the Karush-Kuhn-Tucker (KKT) optimality conditions for (2), stated here for later reference, are as follows:

$$\left(\boldsymbol{Q}\mathbf{x} + \boldsymbol{p} + \boldsymbol{A}^T\boldsymbol{\eta}\right)_i \ge 0 \qquad \text{if } \mathbf{x}_i = 0 \tag{4a}$$

$$\left(\boldsymbol{Q}\mathbf{x} + \boldsymbol{p} + \boldsymbol{A}^T\boldsymbol{\eta}\right)_i \le 0 \qquad \text{if } \mathbf{x}_i = C \tag{4b}$$

$$\left(\boldsymbol{Q}\mathbf{x} + \boldsymbol{p} + \boldsymbol{A}^T\boldsymbol{\eta}\right)_i = 0 \qquad \text{if } \mathbf{x}_i \in (0, C) \tag{4c}$$

$$\boldsymbol{A}\mathbf{x} = \boldsymbol{b} \tag{4d}$$

$$\mathbf{0} \le \mathbf{x} \le C\mathbf{1} \ . \tag{4e}$$

If the kernel function κ is positive semidefinite, the Hessian matrix $\boldsymbol{Q}$ of (2) is also positive semidefinite, by definition. Therefore the objective function $F(\cdot)$ of (2) is convex, and as we only have linear constraints, the dual objective of (2) is a concave function in terms of the dual variable $\boldsymbol{\eta}$. Therefore the primal-dual pair $(\mathbf{x}, \boldsymbol{\eta})$ satisfying the conditions in (4) is the saddle point of (2). Moreover, $\boldsymbol{\eta}$ agrees with $\boldsymbol{\beta}$ in (3) since $\boldsymbol{\eta}$ is the double dual variable of $\boldsymbol{\beta}$ (refer [11] for details.) As our primal-dual solver discussed in Sect. 4 provides the optimal value of $\boldsymbol{\eta}$, there is no need to compute $\boldsymbol{\beta}$ separately.

3 Decomposition Framework

In this section we outline the decomposition strategy, giving details of two key aspects.

3.1 Subproblem Definition

The convex quadratic program (2) becomes harder to solve as the number of variables $n := 2M$ grows (where M is the number of data points), as the Hessian $\boldsymbol{Q}$ in (2) is dense and poorly conditioned for typical choices of the kernel function κ. The decomposition framework can alleviate these difficulties by working with a subset $\mathbf{x}_{\mathcal{B}}, \mathcal{B} \subset \{1, 2, \ldots, n\}$ of the variables at a time, fixing the other variables $\mathbf{x}_{\mathcal{N}}, \mathcal{N} = \{1, 2, \ldots, n\} \setminus \mathcal{B}$ at their current values. We usually choose the number of elements $n_{\mathcal{B}}$ in $\mathcal{B}$ to be much smaller than n. By partitioning the data objects $\boldsymbol{p}$, $\boldsymbol{A}$, and $\boldsymbol{Q}$ in the obvious way, we obtain the following subproblem at outer iteration k:

$$\begin{aligned} \min_{\mathbf{x}_{\mathcal{B}}} \quad & f(\mathbf{x}_{\mathcal{B}}) := \frac{1}{2}\mathbf{x}_{\mathcal{B}}^T \boldsymbol{Q}_{\mathcal{B}\mathcal{B}} \mathbf{x}_{\mathcal{B}} + (\boldsymbol{Q}_{\mathcal{B}\mathcal{N}} \mathbf{x}_{\mathcal{N}}^k + \boldsymbol{p}_{\mathcal{B}})^T \mathbf{x}_{\mathcal{B}} \\ \text{s.t.} \quad & \boldsymbol{A}_{\mathcal{B}} \mathbf{x}_{\mathcal{B}} = -\boldsymbol{A}_{\mathcal{N}} \mathbf{x}_{\mathcal{N}}^k + \boldsymbol{b}, \qquad 0 \le \mathbf{x}_{\mathcal{B}} \le C\mathbf{1}, \end{aligned} \tag{5}$$

where $\mathbf{x}_{\mathcal{N}}^k$ contains the current values of the $\mathcal{N}$ components. This problem has the same form as (2); we discuss solution methods in Sect. 4.

Since our emphasis in this paper is computational, we leave a convergence theory for this decomposition framework for future work. Suffice for the present to make a few remarks. If $\mathcal{B}$ is chosen so that the columns of $\boldsymbol{A}_{\mathcal{B}}$ corresponding to components of $\mathbf{x}_{\mathcal{B}}$ that are away from their bounds in (5) form a full-row-rank matrix, and if appropriate two-sided projections of $\boldsymbol{Q}_{\mathcal{B}\mathcal{B}}$ are positive definite, then (5) has a primal-dual solution $(\mathbf{x}_{\mathcal{B}}^*, \boldsymbol{\eta}^*)$ that corresponds to a solution $(\mathbf{x}^*, \boldsymbol{\eta}^*) = (\mathbf{x}_{\mathcal{B}}^*, \mathbf{x}_{\mathcal{N}}^*, \boldsymbol{\eta}^*)$ of (2), when $\mathbf{x}_{\mathcal{N}}^k = \mathbf{x}_{\mathcal{N}}^*$. Perturbation results can be used to derive a local convergence theory, and it may be possible to derive a global theory from appropriate generalizations of the results in [12].

3.2 Working Set Selection

The selection of working set $\mathcal{B}$ at each outer iteration is inspired by the approach of Joachims [4], later improved by Serafini and Zanni [6]. The size of the working set is fixed at some value $n_{\mathcal{B}}$, of which up to n_{c} are allowed to be "fresh" indices while the remainder are carried over from the current working set. Given the current primal-dual iterate $(\mathbf{x}^{k+1}, \boldsymbol{\eta}^{k+1})$, we find the indices corresponding to the nonzero components $\boldsymbol{d}_i$ obtained from the following problem:

$$\begin{aligned} \min_{\boldsymbol{d}} \quad & \left(\nabla F(\mathbf{x}^{k+1}) + (\boldsymbol{\eta}^{k+1})^T \boldsymbol{A}\right)^T \boldsymbol{d} \\ \text{s.t.} \quad & \begin{array}{ll} 0 \le \boldsymbol{d}_i \le 1 & \text{if } \mathbf{x}_i^{k+1} = 0, \\ -1 \le \boldsymbol{d}_i \le 0 & \text{if } \mathbf{x}_i^{k+1} = C, \\ -1 \le \boldsymbol{d}_i \le 1 & \text{if } \mathbf{x}_i^{k+1} \in (0, C), \\ \#\{\boldsymbol{d}_i | \boldsymbol{d}_i \ne 0\} \le n_{\mathrm{c}}. & \end{array} \end{aligned} \tag{6}$$

Note that the objective function of (6) is a linearization of the Lagrangian function of F at the current primal-dual pair $(\mathbf{x}^{k+1}, \boldsymbol{\eta}^{k+1})$. Our approach is motivated by the KKT conditions (4), and indeed can be solved by simply sorting the violations of these conditions. It contrasts with previous methods [4,6,12], in which the equality constraints are enforced explicitly in the working set selection subproblem. Our approach has no requirements on the size of n_c, yet it is still effective when $\boldsymbol{\eta}^{k+1}$ is close to the optimal value $\boldsymbol{\eta}^*$.

Earlier analysis of decomposition algorithms based on working set selection schemes has been performed by Lin [13], who shows linear convergence for the case of a single constraint, under positive definiteness assumptions on $\boldsymbol{Q}$. Tseng and Yun [12] proposed a decomposition framework for a formulation similar to (2) that includes multiple equality constraints. They present a convergence analysis which assumes that the subproblems at each step of decomposition are solved exactly, although they do not discuss techniques for solving the subproblem. Their working set selection algorithm requires relatively high complexity ($\mathcal{O}(K^3n^2)$) in general, compared with the $\mathcal{O}(n \log n)$ complexity of our approach.

The (up to) n_c new components from (6) are augmented to a total of $n_{\mathcal{B}}$ entries by adding indices from the previous working set $\mathcal{B}$ according to a certain priority. We choose the indices of the off-bounds components ($0 < \mathbf{x}_i^{k+1} < C$) first, and then those of lower and upper bounds. We reduce n_c as the change between two consecutive working sets decreases, as in [6]. We observe that adaptive reduction of n_c provides better convergence of the Lagrange multiplier $\boldsymbol{\eta}^k$, and helps avoid zigzagging between two working sets without making further progress. Adaptive reduction also helps not to degrade the benefit of optimizing many new components in a single decomposition step.

Our decomposition framework is summarized in Algorithm 1.

Algorithm 1. Decomposition Framework

1. **Initialization.** Choose an initial point $\mathbf{x}^1$ of (2) (possibly infeasible), initial guess of the Lagrange multiplier $\boldsymbol{\eta}^1$, positive integers $n_{\mathcal{B}} \geq K$ and $0 < n_c < n_{\mathcal{B}}$, and convergence tolerance `tolD`. Choose an initial working set $\mathcal{B}$ and set $k \leftarrow 1$.

2. **Subproblem.** Solve the subproblem (5) for the current working set $\mathcal{B}$, to obtain solution $\mathbf{x}_{\mathcal{B}}^{k+1}$ together with Lagrange multiplier $\boldsymbol{\eta}^{k+1}$ of the equality constraints. Set $\mathbf{x}^{k+1} = (\mathbf{x}_{\mathcal{B}}^{k+1}, \mathbf{x}_{\mathcal{N}}^{k})$.

3. **Gradient Update.** Evaluate the gradient of the Lagrangian of (2), by incrementally updating ∇F, as indicated here:

$$\nabla F(\mathbf{x}^{k+1}) + (\boldsymbol{\eta}^{k+1})^T \boldsymbol{A} = \nabla F(\mathbf{x}^k) + \begin{bmatrix} \boldsymbol{Q}_{\mathcal{B}\mathcal{B}} \\ \boldsymbol{Q}_{\mathcal{N}\mathcal{B}} \end{bmatrix} (\mathbf{x}_{\mathcal{B}}^{k+1} - \mathbf{x}_{\mathcal{B}}^{k}) + (\boldsymbol{\eta}^{k+1})^T \boldsymbol{A} \ .$$

4. **Convergence Check.** If the maximal violation of the KKT conditions (4) falls below `tolD`, terminate with the primal-dual solution $(\mathbf{x}^{k+1}, \boldsymbol{\eta}^{k+1})$.

5. **Working Set Update.** Find a new working set $\mathcal{B}$ as described in Sect. 3.2.
6. Set $k \leftarrow k+1$ and go to step 2.

---

4 Subproblem Solver

Recalling that the decomposition framework requires both a primal solution $\mathbf{x}_\mathcal{B}$ and Lagrange multipliers $\boldsymbol{\eta}$ to be obtained for the subproblem (5), we consider the following min-max formulation of (5):

$$\max_{\boldsymbol{\eta}} \min_{\mathbf{x}_\mathcal{B} \in \Omega} \quad L(\mathbf{x}_\mathcal{B}, \boldsymbol{\eta}) \ , \tag{7}$$

where $\Omega = \{\mathbf{x} \in \mathbb{R}^{n_\mathcal{B}} | \mathbf{0} \leq \mathbf{x} \leq C\mathbf{1}\}$ and

$$L(\mathbf{x}_\mathcal{B}, \boldsymbol{\eta}) := f(\mathbf{x}_\mathcal{B}) + \boldsymbol{\eta}^T(\boldsymbol{A}_\mathcal{B}\mathbf{x}_\mathcal{B} + \boldsymbol{A}_\mathcal{N}\mathbf{x}_\mathcal{N}^k) \ .$$

In this section we describe a primal-dual approach for solving (7), in which steps are taken in $\mathbf{x}_\mathcal{B}$ and $\boldsymbol{\eta}$ in an alternating fashion. Scalings that include second-order information are applied to both primal and dual steps. We call the approach PDSG (for "Primal-Dual Scaled Gradient").

Our approach can be viewed as an extreme variant of the method of multipliers [14], in which we do not attempt to minimize the augmented Lagrangian between updates of the Lagrange multiplier estimates, but rather take a single step along a partial, scaled, and projected gradient direction in the primal space. In describing the general form of each iteration, we use superscripts ℓ to denote iteration counts, bearing in mind that they refer to the *inner* iterations of the decomposition framework (and hence are distinct from the superscripts k of the previous section, which denote outer iterations).

$$\mathbf{x}_\mathcal{B}^{\ell+1} \leftarrow \mathbf{x}_\mathcal{B}^\ell + s(\mathbf{x}_\mathcal{B}^\ell, \boldsymbol{\eta}^\ell) \tag{8a}$$

$$\boldsymbol{\eta}^{\ell+1} \leftarrow \boldsymbol{\eta}^\ell + t(\mathbf{x}_\mathcal{B}^{\ell+1}, \boldsymbol{\eta}^\ell) \ , \tag{8b}$$

where $s(\cdot,\cdot)$ and $t(\cdot,\cdot)$ are steps, defined below. In computational testing, we found PDSG to be superior to methods more like traditional method-of-multiplier approaches, which would take multiple steps in $\mathbf{x}_\mathcal{B}$ in between successive steps in $\boldsymbol{\eta}$.

Primal Step. In the ℓ-th iteration of the subproblem solver, we choose a small sub-working set $\mathcal{W}^\ell \subset \mathcal{B}$ containing at most $n_\mathcal{W}$ elements (where $n_\mathcal{W}$ is a user-defined parameter), containing those indices in $\mathcal{B}$ that are among the $n_\mathcal{W}$ most-violated KKT conditions (4a)-(4c) for the subproblem (5). We define the further subset $\bar{\mathcal{W}}^\ell$ by selecting those indices $i \in \mathcal{W}^\ell$ that are *not* at one of their bounds 0 and C. We then construct the block-diagonal $n_\mathcal{B} \times n_\mathcal{B}$ matrix $\boldsymbol{H}^\ell$, as follows:

$$\boldsymbol{H}_{ij}^\ell = \begin{cases} \boldsymbol{Q}_{ij} + \tau\delta_{ij} & \text{if } i \in \bar{\mathcal{W}}^\ell \text{ and } j \in \bar{\mathcal{W}}^\ell \\ \boldsymbol{Q}_{ii} & \text{if } i = j \text{ and } i \in \mathcal{W}^\ell \setminus \bar{\mathcal{W}}^\ell \\ \infty & \text{if } i = j \text{ and } i \notin \mathcal{W}^\ell \\ 0 & \text{otherwise,} \end{cases} \tag{9}$$

where $\delta_{ij} = 1$ if $i = j$ and 0 otherwise, while τ is a small positive parameter (we use $\tau = 10^{-8}$) chosen to ensure that the "block" part of $\boldsymbol{H}^\ell$ is numerically non-singular. Since we apply the inverse of this matrix to the gradient in computing the step, the components of the matrix-vector product that correspond to the ∞ entries will evaluate to zero. Specifically, we obtain the search direction as follows:

$$d^\ell := \mathbf{x}_\mathcal{B}^\ell - \mathbb{P}_\Omega \left(\mathbf{x}_\mathcal{B}^\ell - \left(\boldsymbol{H}^\ell\right)^{-1} \nabla_{\mathbf{x}_\mathcal{B}} L(\mathbf{x}_\mathcal{B}^\ell, \boldsymbol{\eta}^\ell) \right) \tag{10}$$

where $\mathbb{P}_\Omega(\cdot)$ is a projection operator to the set Ω, which is trivial to compute since this set is defined by simple bounds. This is essentially the two-metric gradient projection search direction [15] applied to the subvector defined by $\mathcal{W}^\ell$. Given this direction, the primal step s from (8a) is defined to be

$$s(\mathbf{x}_\mathcal{B}^\ell, \boldsymbol{\eta}^\ell) = \alpha_\ell \boldsymbol{d}^\ell \ , \tag{11}$$

where $\alpha_\ell \in \mathbb{R}$ is the unconstrained minimizer of $L(\cdot, \boldsymbol{\eta}^\ell)$ along the line segment connecting $\mathbf{x}_\mathcal{B}^\ell$ to $\mathbf{x}_\mathcal{B}^\ell + \boldsymbol{d}^\ell$.

Dual Update. The step in the dual variable $\boldsymbol{\eta}$ is a Newton-like step in the dual objective function for (5), which is

$$g(\boldsymbol{\eta}) := \min_{\mathbf{x}_\mathcal{B} \in \Omega} L(\mathbf{x}_\mathcal{B}, \boldsymbol{\eta}).$$

This is a piecewise quadratic concave function. Since its second derivative does not exist, we cannot take a true Newton step. However, we use a slight modification of the procedure in Kienzle and Schölkopf [9] to form a diagonal approximation $\boldsymbol{G}$ to this matrix. Their procedure progressively updates $\boldsymbol{G}$ by applying one step of Gauss-Jacobi-like procedure at each iteration of the MPD optimization scheme. Unlike MPD, our modification estimates $\boldsymbol{G}$ both internally and externally to the optimization loop. The external estimation ensures us to have an approximation with a certain quality before performing any dual updates. We refer the reader to [9] for additional details. The dual step t in (8b) is thus simply

$$t(\mathbf{x}_\mathcal{B}^{\ell+1}, \boldsymbol{\eta}^\ell) = -\boldsymbol{G}^{-1} \nabla_{\boldsymbol{\eta}} L(\mathbf{x}_\mathcal{B}^{\ell+1}, \boldsymbol{\eta}^\ell). \tag{12}$$

Our subproblem algorithm is summarized in Algorithm 2.

Algorithm 2. Subproblem solver: PDSG

---

1. **Initialization.** Given a index set $\mathcal{B}$, choose initial points $\mathbf{x}_\mathcal{B}^1$ and $\boldsymbol{\eta}^1$. Choose $n_\mathcal{W}$ such that $1 \leq n_\mathcal{W} \leq n_\mathcal{B}$. Choose small positive convergence tolerance `tolS`. Set $\ell \leftarrow 1$.

2. **Sub-Working Set Selection.** Construct $\mathcal{W}^\ell$ (with at most $n_\mathcal{W}$ elements) and $\bar{\mathcal{W}}^\ell$ as described above.

3. **Primal-Dual Update.** Take the primal step according to (8a) and (11), then the dual step according to (8b) and (12).

4. **Convergence Check.** If the maximal KKT violation of the current primal-dual pair $(\mathbf{x}_\mathcal{B}^{\ell+1}, \boldsymbol{\eta}^{\ell+1})$ is less than `tolS`, exit. Otherwise, go to step 2.

---

5 Experiments

We report on computational experiments that show the intrinsic benefits of the PDSG approach, as well as the benefits of the decomposition strategy, when applied to a simple semiparametric SVM regression problem. We compare PDSG with the MPD algorithm of Kienzle and Schölkopf [9], which has slightly better performance and lower memory requirement than the interior-point-based approach used in [1]. We also show the advantage of semiparametric modeling on a real world problem.

Implementations. We implemented both the decomposition framework (Algorithm 1) and the PDSG subproblem solver (Algorithm 2) in C++. The code was developed by modifying the GPDT code of Serafini, Zanghirati, and Zanni [5][1], and retains many features of this code. Our code caches once-computed kernel entries for reuse, with the least-recently-used (LRU) replacement strategy. For efficiency, our subproblem solver exploits warm starting; the most recent values of the primal and dual variables are used as the starting points in the next invocation of the subproblem solver. We also implemented the MPD solver [9] in C++, again basing the implementation on GPDT. Our codes can be invoked either with the decomposition framework, or in "stand-alone" mode, in which the solver is applied directly to the stated problem.

5.1 Toy Problem

For the semiparametric regression test problem, we choose the modified Mexican hat function studied in [1,9]:

$$\omega(t) = \sin(t) + \operatorname{sinc}\left(2\pi(t-5)\right) \ .$$

To generate data, we sample the function ω at uniform random points $t_i \in \mathbb{R}$ in the interval $[0, 10]$, making M samples in total. The observations y_i's are corrupted with additive Gaussian noise ζ_i with mean 0 and standard deviation 0.2, that is, $y_i = \omega(t_i) + \zeta_i$. In the training process, we use Gaussian kernel $\kappa(x, y) = \exp(-\gamma||x - y||^2)$ with $\gamma = 0.25$, and set the insensitivity width ϵ of the loss function to $\epsilon = 0.05$, as in [1]. The optimal tradeoff parameter value of $C = 0.5$ is found by 10-fold cross validation (CV) in [1] using very small samples ($M = 50$). Since we are interested in the convergence behavior of algorithms with larger samples, we performed computational experiments with $C = 0.1$, $C = 1$, and $C = 10$. Our model is $h(t) = \langle \boldsymbol{w}, \phi(t)\rangle + \sum_{j=1}^{K} \boldsymbol{\beta}_j \psi_j(t)$, with two basis functions $\psi_1(t) = \sin(t)$ and $\psi_2(t) = \operatorname{sinc}\left(2\pi(t-5)\right)$ as in [9].

The size of the sample dataset M is varied from 500 to 100000. The subproblem size $n_{\mathcal{B}}$ and the maximum number of new components in each subproblem n_c are fixed to 500 and 100, respectively, as these values gave good performance on the largest data set. Similarly, we fix the sub-working set size $n_{\mathcal{W}}$ to 2. (We tried

[1] GPDT is available at `http://mloss.org/software/view/54/`

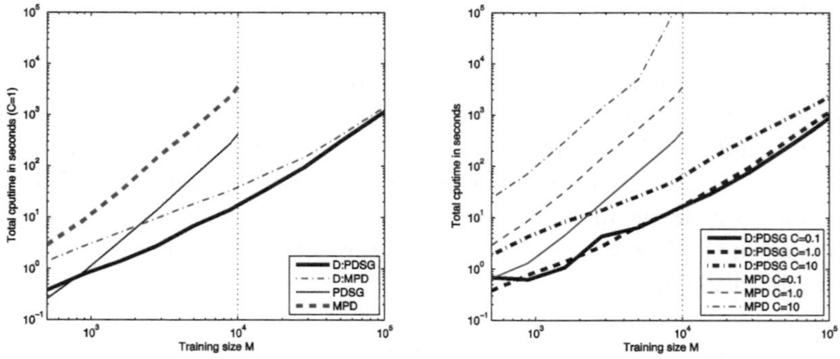

Fig. 1. Left plot shows total runtimes using solvers PDSG and MPD in stand-alone mode and inside of the decomposition framework (D:PDSG and D:MPD) with $C = 1$. Right plot shows the total runtimes of D:PDSG (our proposed method) and MPD with different C values. For larger number of training examples M, updating of the full gradient in Step 3 of Algorithm 1 dominates the computation, blurring the distinction between PDSG and MPD as subproblem solvers (left plot). D:PDSG outperforms MPD for all C values tried (right plot). Stand-alone algorithms are run only for training-set size up to 10000 because of their high computational cost.

various other values between 1 and 25, but 2 was slightly better than several alternatives.) In each setting, we use a kernel cache of 400MB in size.

Growth of the total runtime of the algorithms with increasing size of the data set is shown in Fig. 1. When the decomposition framework is used, the stopping threshold values are set to `tolD` $= 0.001$ and `tolS` $= 0.0005$. In stand-alone settings, we set `tolS` $= 0.001$. We impose a slightly tighter threshold on subproblem solvers inside the decomposition framework to reduce the number of decomposition steps. Outer iterations in the decomposition framework become more costly as the number of variables increases, mainly because the full gradient update in Step 3 of Algorithm 1 becomes more expensive. The benefit of using decomposition framework becomes larger as the dataset size grows. For instance, D:PDSG is about 100 times faster than MPD when $M = 10000$. In decomposition settings, using PDSG as the inner solver found the solution two to three times faster than using MPD as the inner solver on average. Our proposed method D:PDSG shows quite stable scaling behavior for different values of C.

Convergence and Complexity. The different convergence behavior of PDSG and MPD is illustrated in Fig. 2. Here both solvers are asked to solve a semiparametric regression problem discussed above with 1000 samples, in stand-alone mode. In the top and middle plots, the dual and primal infeasibility, respectively, are more rapidly reduced with PDSG than with MPD. (Note that since we project the iterates $\mathbf{x}^k$ to the bound constraints set, the KKT condition (4e) is always satisfied.) The bottom plot of Fig. 2 shows the changes of the first Lagrange multiplier (the coefficient of the first basis function). In that, MPD is showing the typical behavior of the method of multipliers: sudden changes are made,

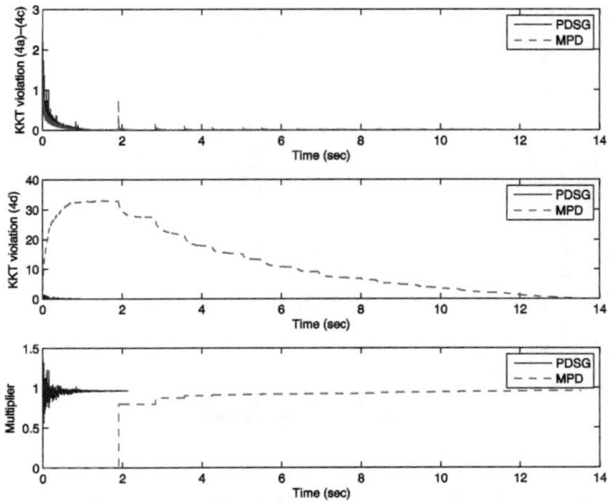

Fig. 2. Convergence of PDSG and MPD in stand-alone mode (Mexican hat, dataset size M=1000). PDSG requires about 2 seconds to reach convergence, whereas MPD takes about 14 seconds. (Top) maximum violation of the dual feasibility conditions (4a), (4b), (4c). (Middle) maximum violation of the primal equality constraints (4d). (Bottom) convergence of the first Lagrange multiplier to its optimal value of 1. The horizontal axis represents elapsed CPU time.

but time gaps between such changes are rather large. In contrast, PDSG keeps making changes to the multiplier, resulting in a faster approach to the optimal value.

When the sub-working-set size $n_{\mathcal{W}}$ is smaller than the working-set size $n_{\mathcal{B}}$ of the subproblem (5), PDSG has computational complexity $\mathcal{O}(Kn_{\mathcal{B}})$, the same as MPD, where K is the number of equality constraints in (2). Dual updates in Algorithm 2 requires $\mathcal{O}(Kn_{\mathcal{B}})$ operations; all primal updates are done in $\mathcal{O}(n_{\mathcal{B}})$. The effect of increasing K on the total time taken by D:PDSG is shown in Fig. 3. We use the basis functions

$$\psi_j(t) = \begin{cases} \cos(j\pi t) & j = 0, 2, 4, \ldots \\ \sin(j\pi t) & j = 1, 3, 5, \ldots \end{cases}$$

and datasets of size $M = 1000$ randomly sampled from the Mexican hat function. Other settings are the same as the previous experiment. As expected, we observe linear scaling of total runtime with K.

5.2 Milan Respiratory Illness Dataset

We consider a dataset[2] from the study on the effect of air pollution on respiratory illness in Milan, Italy, during 1980–89 [16]. This dataset consists of daily records of environmental conditions and the number of deaths due to respiratory diseases

[2] Available at `http://www.uow.edu.au/~mwand/webspr/data.html`

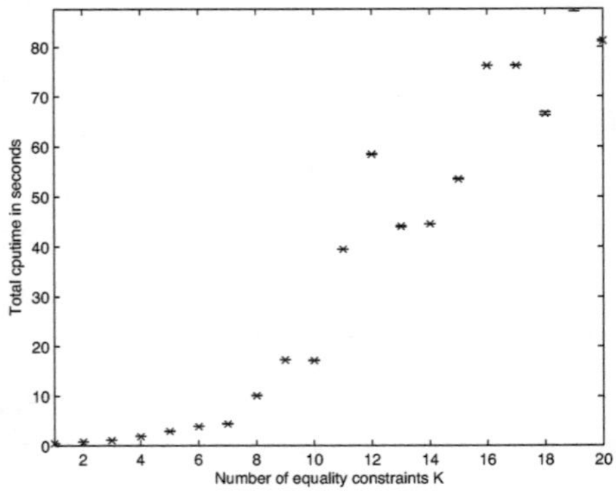

Fig. 3. Total solution time for D:PDSG with increasing number of equality constraints K. Measurements are averaged over 10 repetitions with different random datasets (M=1000) sampled from the Mexican hat function, and error bars (hardly visible) show the standard deviations. The time complexity of D:PDSG is $\mathcal{O}(uKn_{\mathcal{B}})$ where u is the number of outer iterations. Solver time appears to increase linearly with K.

(total 3652 records, 9 features). All features are scaled linearly to the range [0, 1]. We construct a test set by holding out 20% of randomly chosen records from the dataset, using the remaining records for training.

We hypothesize a simple semiparametric model to predict the number of respiratory deaths, inspired by [16]:

$$h_{\text{sp}}(\boldsymbol{t}) = \langle \boldsymbol{w}, \phi(\boldsymbol{t}) \rangle + \boldsymbol{\beta}_1(\boldsymbol{t}_{\text{temp}}) + \boldsymbol{\beta}_2(\boldsymbol{t}_{\text{SO}_2}) + \boldsymbol{\beta}_3(\boldsymbol{t}_{\text{temp}})^2 + \boldsymbol{\beta}_4(\boldsymbol{t}_{\text{SO}_2})^2 + \boldsymbol{\beta}_5 \ ,$$

where the features $\boldsymbol{t}_{\text{temp}}$ and $\boldsymbol{t}_{\text{SO}_2}$ correspond to mean temperature and SO_2 level of the day, respectively. Our purpose is to study how those two elements affect the respiratory illness.

We fit our semiparametric model to the training data, and compare its prediction performance on the test set to that of a nonparametric model

$$h_{\text{np}}(\boldsymbol{t}) = \langle \boldsymbol{w}, \phi(\boldsymbol{t}) \rangle + \boldsymbol{\beta}_1 \ .$$

With Gaussian kernel ($\gamma = 25.0$) and ϵ-insensitive loss function ($\epsilon = 0.01$), we perform 10-fold CV on the training set to determine the best balancing parameter C for each of semiparametric and nonparametric models independently.

The results are shown in Table 1. The semiparametric model attained smaller prediction error on the test set than the nonparametric model, indicating that the embedding of prior knowledge in h_{sp} while retaining the power of nonparametric approaches is beneficial. Moreover, the parametric components in the trained semiparametric model

$$h_{\text{sp}}(\boldsymbol{t}) = \langle \boldsymbol{w}^*, \phi(\boldsymbol{t}) \rangle - 0.30(\boldsymbol{t}_{\text{temp}}) + 0.26(\boldsymbol{t}_{\text{SO}_2}) + 0.22(\boldsymbol{t}_{\text{temp}})^2 - 0.07(\boldsymbol{t}_{\text{SO}_2})^2 + 0.22 \ .$$

Table 1. Nonparametric and semiparametric regression on Milan dataset. The loss penalty parameter C is determined by cross validation. Comparing the prediction performance on the test set by mean square error (MSE) values, the semiparametric model performed better than the nonparametric model by 2.8%. No significant difference of the number of support vectors (SVs) was found between the two methods.

Model	C	Fraction of SVs	Training Time (s)	Test Error (MSE)
Nonparametric ($h_{\rm np}$)	0.025	46.7%	1.17	0.019368
Semiparametric ($h_{\rm sp}$)	0.01	46.9%	5.35	0.018828

reveal that (i) deaths are lower in the middle of the temperature range, and (ii) there is an almost linear increase of death rate with SO_2 level. These results broadly agree with the outcomes of [16], which were acquired from completely different statistical analysis techniques. It is difficult to perform model interpretation of this type with nonparametric approaches.

6 Conclusions

We have presented a new method for semiparametric SVM regression problems, which extends a number of previous approaches in being able to handle multiple equality constraints. Our method combines a decomposition framework with a primal-dual scaled gradient solver for the subproblems. Computational tests indicate that the approach improves on previously proposed methods.

Future work includes reducing the cost of the full gradient update by using a randomized sampling procedure for the components of the gradient, as has been tried in a different context in [17]. While the concept is simple, it is not straightforward to implement this technique in conjunction with caching of kernel entries, which is so important to efficient implementation of SVM solvers based on QP formulations. Other research topics include devising a more effective update strategy for the dual variables $\boldsymbol{\eta}$ in the subproblem solver, and theoretical analyses both of the decomposition framework (including the working set selection technique) and the subproblem solver.

Acknowledgements

The authors acknowledge the support of NSF Grants CCF-0430504, DMS-0427689, CNS-0540147, and DMS-0914524. The first author was supported in part by Samsung Scholarship from the Samsung Foundation of Culture.

References

1. Smola, A.J., Frieß, T.T., Schölkopf, B.: Semiparametric support vector and linear programming machines. In: Advances in Neural Information Processing Systems 11, pp. 585–591. MIT Press, Cambridge (1999)

2. Platt, J.C.: Fast training of support vector machines using sequential minimal optimization. In: Schölkopf, B., Burges, C., Smola, A. (eds.) Advances in Kernel Methods - Support Vector Learning, pp. 185–208. MIT Press, Cambridge (1999)
3. Chang, C.C., Lin, C.J.: LIBSVM: a library for support vector machines (April 2009) version 2.89, `http://www.csie.ntu.edu.tw/~cjlin/libsvm`
4. Joachims, T.: Making large-scale support vector machine learning practical. In: Schölkopf, B., Burges, C., Smola, A. (eds.) Advances in Kernel Methods - Support Vector Learning, pp. 169–184. MIT Press, Cambridge (1999)
5. Serafini, T., Zanghirati, G., Zanni, L.: Gradient projection methods for large quadratic programs and applications in training support vector machines. Optimization Methods and Software 20(2–3), 353–378 (2004)
6. Serafini, T., Zanni, L.: On the working set selection in gradient projection-based decomposition techniques for support vector machines. Optimization Methods and Software 20, 583–596 (2005)
7. Keerthi, S.S., Gilbert, E.G.: Convergence of a generalized smo algorithm for svm classifier design. Machine Learning 46(1-3), 351–360 (2002)
8. Dai, Y.H., Fletcher, R.: New algorithms for singly linearly constrained quadratic programs subject to lower and upper bounds. Mathematical Programming, Series A 106, 403–421 (2006)
9. Kienzle, W., Schölkopf, B.: Training support vector machines with multiple equality constraints. In: Gama, J., Camacho, R., Brazdil, P.B., Jorge, A.M., Torgo, L. (eds.) ECML 2005. LNCS (LNAI), vol. 3720, pp. 182–193. Springer, Heidelberg (2005)
10. Boser, B.E., Guyon, I.M., Vapnik, V.N.: A training algorithm for optimal margin classifiers. In: COLT 1992: Proceedings of the fifth annual workshop on Computational learning theory, pp. 144–152. ACM, New York (1992)
11. Scholkopf, B., Smola, A.J.: Learning with Kernels: Support Vector Machines, Regularization, Optimization, and Beyond. MIT Press, Cambridge (2001)
12. Tseng, P., Yun, S.: A coordinate gradient descent method for linearly constrained smooth optimization and support vector machines training. Published online in Computational Optimization and Applications (October 2008)
13. Lin, C.J.: Linear convergence of a decomposition method for support vector machines. Technical report, Department of Computer Science and Information Engineering, National Taiwan University (2001)
14. Bertsekas, D.P.: Nonlinear Programming, 2nd edn. Athena Scientific, Belmont (1999)
15. Gafni, E.M., Bertsekas, D.P.: Two-metric projection methods for constrained optimization. SIAM Journal on Control and Optimization 22, 936–964 (1984)
16. Vigotti, M.A., Rossi, G., Bisanti, L., Zanobetti, A., Schwartz, J.: Short term effects of urban air pollution on respiratory health in Milan, Italy, 1980-1989. Journal of Epidemiology Community Health 50, s71–s75 (1996)
17. Shi, W., Wahba, G., Wright, S.J., Lee, K., Klein, R., Klein, B.: LASSO-Patternsearch algorithm with application to opthalmology data. Statistics and its Interface 1, 137–153 (2008)

A Convex Method for Locating Regions of Interest with Multi-instance Learning

Yu-Feng Li[1], James T. Kwok[2], Ivor W. Tsang[3], and Zhi-Hua Zhou[1]

[1] National Key Laboratory for Novel Software Technology
Nanjing University, Nanjing 210093, China
{liyf,zhouzh}@lamda.nju.edu.cn
[2] Department of Computer Science and Engineering
Hong Kong University of Science and Technology, Hong Kong, China
jamesk@cse.ust.hk
[3] School of Computer Engineering
Nanyang Technological University, Singapore 639798
IvorTsang@ntu.edu.sg

Abstract. In content-based image retrieval (CBIR) and image screening, it is often desirable to locate the regions of interest (ROI) in the images automatically. This can be accomplished with multi-instance learning techniques by treating each image as a bag of instances (regions). Many SVM-based methods are successful in predicting the bag labels, however, few of them can locate the ROIs. Moreover, they are often based on either local search or an EM-style strategy, and may get stuck in local minima easily. In this paper, we propose two convex optimization methods which maximize the margin of concepts via key instance generation at the instance-level and bag-level, respectively. Our formulation can be solved efficiently with a cutting plane algorithm. Experiments show that the proposed methods can effectively locate ROIs, and they also achieve performances competitive with state-of-the-art algorithms on benchmark data sets.

1 Introduction

With the rapid expansion of digital image collections, content-based image retrieval (CBIR) has attracted more and more interest. The main difficulty of CBIR lies in the gap between the high-level image semantics and the low-level image features. Much endeavor has been devoted to bridging this gap, it remains unsolved yet. Generally, the user first poses in the query and relevance feedback process several labeled images that are relevant/irrelevant to an underlying target concept. Then the CBIR system attempts to retrieve all images from the database that are relevant to the concept. It is noteworthy that although the user feeds whole images to the system, usually s/he is only interested in some regions, i.e., *regions of interest* (ROIs), in the images.

For medical and military applications which require a fast scanning of huge amount of images to detect suspect areas, it is very desirable if ROIs can be identified and exhibited when suspected images are presented to the examiner. Even in common CBIR scenarios, considering that the system usually returns a lot of images, the explicit identification of ROIs may help the user in recognizing images s/he really wants more quickly.

W. Buntine et al. (Eds.): ECML PKDD 2009, Part II, LNAI 5782, pp. 15–30, 2009.

In multi-instance learning [6], the training examples are *bags* each containing many instances. A bag is positively labeled if it contains at least one positive instance, and negatively labeled otherwise. The task is to learn a model from the training bags for correctly labeling unseen bags. Multi-instance learning is difficult because that, unlike conventional supervised learning tasks where all the training instances are labeled, here the labels of the individual instances are unknown. It is obvious that if a whole image is regarded as a bag with its regions being regarded as instances, the problem of determining whether an image is relevant to a target concept or not can be viewed as a multi-instance problem. So, it is not surprising that multi-instance learning has been found very useful in tasks involving image analysis.

In general, three kinds of multi-instance learning approaches can be used to locate the ROIs. The first is the Diverse Density (DD) algorithm [15] and its variants, e.g., EM-DD [26] and multi-instance logistic regression [19]. These methods apply gradient search with multiple restarts to identify an instance which maximizes the *diverse density*, that is, an instance close to every positive bags while far from negative bags. The instance is then regarded as the prototype of the target concept. It is obvious that DD can be applied to locate ROIs. A serious problem with this kind of methods is the huge time cost, since they have to perform gradient search starting from every instance in every positive bag.

The second approach is the C*k*NN-ROI algorithm [29], which is a variant of Citation-*k*NN [23]. This approach uses Citation-*k*NN to predict whether a bag is positive or not. It takes the minimum distance between the nearest pair of instances from two bags as the distance between bags, and then utilizes *citers* of the neighbors to improve performance. Subsequently, each instance in a positive bag is regarded as a bag and a score is calculated by considering its distance to other bags, from which the key instance can be decided. The time complexity of C*k*NN-ROI is mainly dominated by the calculation of neighbors, and is much more efficient than DD. However, this algorithm is based on heuristics and the theoretical justification has not been established yet.

The third approach is MI-SVM [1]. While many SVM-based multi-instance learning methods have been developed [1,3,4], to the best of our knowledge, MI-SVM is the only one that can locate the ROIs. The MI-SVM locates ROI (also referred to as the key instance) with an EM-style procedure. It first starts with a SVM using some multi-instance kernel [8] and picks the key instances according to the SVM prediction, and the SVM is then retrained with respect to the key instance assignment; the procedure is repeated until convergence. Empirical study shows that MI-SVM is efficient and works well on many multi-instance data sets. In fact, MI-SVM can be viewed as a *constrained concave-convex programming* (CCCP) method whose convergence has been well-studied [5]. Each MI-SVM iteration only involves the solving of a convex optimization problem, however, the optimization problem as a whole is still non-convex and suffers from local minima.

In this paper, we focus on SVM-based methods and propose the KI-SVM (key-instance support vector machine) algorithm. We formulate the problem as a convex optimization problem. At each iteration, KI-SVM generates a violated key instance assignment and then combines them via efficient multiple kernel learning. It is noteworthy that it involves a series of standard SVM subproblems that can be solved with various state-of-the-art SVM implementations in a scalable and efficient manner, such as SVM-*perf* [10], LIBSVM [7], LIBLINEAR [9] and CVM [21]. Two variants of the KI-SVM, namely, Ins-KI-SVM and Bag-KI-SVM, are proposed for locating the key instances at the instance-level and bag-level, respectively.

The rest of the paper is organized as follows. Section 2 briefly introduces MI-SVM. Section 3 proposes our KI-SVM method. Experimental results are reported in Section 4. The last section concludes the paper.

2 Multi-instance Support Vector Machines

In the sequel, we denote the transpose of a vector/matrix (in both the input and feature spaces) by the superscript $'$. The zero vector and the vector of all ones are denoted as $\mathbf{0}, \mathbf{1} \in \mathbb{R}^n$, respectively. Moreover, the inequality $\mathbf{v} = [v_1, \cdots, v_k]' \geq \mathbf{0}$ means that $v_i \geq 0$ for $i = 1, \cdots, k$.

In multi-instance classification, we are given a set of training bags $\{(B_1, y_1), \cdots, (B_m, y_m)\}$, where $B_i = \{\mathbf{x}_{i,1}, \mathbf{x}_{i,2}, \cdots, \mathbf{x}_{i,m_i}\}$ is the ith bag containing instances $\mathbf{x}_{i,j}$'s, m_i is the size of bag B_i, and $y_i \in \{\pm 1\}$ is its bag label. Suppose the decision function is denoted as $f(\mathbf{x})$. As is common in the traditional MI setting, we take $f(B_i) = \max_{1 \leq j \leq m_i} f(\mathbf{x}_{i,j})$. Furthermore, $\mathbf{x}_{i,l} = \arg\max_{\mathbf{x}_{i,j}} f(\mathbf{x}_{i,j})$ is viewed as the key instance of a positive bag B_i. For simplification, we assume that the decision function is a linear model, i.e., $f(\mathbf{x}) = \mathbf{w}'\phi(\mathbf{x})$, where ϕ is the feature map induced by some kernel k.

The goal is to find f that minimizes the structural risk functional

$$\Omega(\|\mathbf{w}\|_p) + C \sum\nolimits_{i=1}^{m} \ell \left(-y_i \max_{1 \leq j \leq m_i} \mathbf{w}'\phi(\mathbf{x}_{i,j}) \right), \tag{1}$$

where Ω can be any strictly monotonically increasing function, $\ell(\cdot)$ is a monotonically increasing loss function, and C is a regularization parameter that balances the empirical risk functional and the model complexity. In this paper, we focus on $\Omega(\|\mathbf{w}\|_p) = \frac{1}{2}||\mathbf{w}||^2$ and the squared hinge loss. So, (1) becomes:

$$\min_{\mathbf{w},\rho,\boldsymbol{\xi}} \ \frac{1}{2}||\mathbf{w}||_2^2 - \rho + \frac{C}{2} \sum_{i=1}^{m} \xi_i^2 \tag{2}$$

$$\text{s.t } y_i \max_{1 \leq j \leq m_i} \mathbf{w}'\phi(\mathbf{x}_{i,j}) \geq \rho - \xi_i, \ i = 1 \cdots, m, \tag{3}$$

where $\boldsymbol{\xi} = [\xi_1, \cdots, \xi_m]'$. This, however, is a non-convex problem because of the max operator for positive bags.

Andrews *et al.* [1] proposed two heuristic extensions of the support vector machines, namely, the mi-SVM and MI-SVM, for this multi-instance learning problem. The mi-SVM treats the MI learning problem in a supervised learning manner, while the MI-SVM focuses on finding the key instance in each bag. Later, Cheung and Kwok [5] proposed the use of the *constrained concave-convex programming* (CCCP) method, which has well-studied convergence properties, for this optimization problem. However, while each iteration only involves the solving of a convex optimization problem, the optimization problem as a whole is non-convex and so still suffers from the problem of local minima.

3 KI-SVM

In this section, we propose two versions of KI-SVM, namely the Ins-KI-SVM (instance-level KI-SVM) and Bag-KI-SVM (bag-level KI-SVM).

3.1 Mathematical Formulation

Let p be the number of positive bags. Without loss of generality, we assume that the positive bags are ordered before negative bags, i.e., $y_i = 1$ for all $1 \leq i \leq p$ and -1 otherwise. Moreover, let $J_i = \sum_{t=1}^{i} m_t$.

For a positive bag B_i, we use a binary vector $\mathbf{d}_i = [d_{i,1}, \cdots, d_{i,m_i}]' \in \{0,1\}^{m_i}$ to indicate which instance in B_i is its key instance. Here, followed the traditional multi-instance setup, we assume that each positive bag has only one key instance, so $\sum_{j=1}^{m_i} d_{i,j} = 1$[1]. In the following, let $\mathbf{d} = [\mathbf{d}_1, \cdots, \mathbf{d}_p]$, and Δ be its domain. Moreover, note that $\max_{1 \leq j \leq m_i} \mathbf{w}'\phi(\mathbf{x}_{i,j})$ in (3) can be written as $\max_{\mathbf{d}_i} \sum_{j=1}^{m_i} d_{i,j}\mathbf{w}'\phi(\mathbf{x}_{i,j})$ in this case.

For a negative bag B_i, all its instances are negative and the corresponding constraint (3) can be replaced by $-\mathbf{w}'\phi(\mathbf{x}_{i,j}) \geq \rho - \xi_i$ for every instance in B_i. Moreover, we relax the problem by allowing the slack variable ξ_i to be different for different instances of bag B_i. This leads to a set of slack variables $\{\xi_{s(i,j)}\}_{i=1,\cdots,m; j=1,\cdots,m_i}$, where $s(i,j) = J_{i-1} - J_p + j + p$ is the indexing function that numbers these slack variables from $p+1$ to $N = J_m - J_p + p$.

Combining all these together, (2) can be rewritten as:

$$\text{(Ins-KI-SVM)} \quad \min_{\mathbf{w},\rho,\boldsymbol{\xi},\mathbf{d}} \frac{1}{2}||\mathbf{w}||_2^2 - \rho + \frac{C}{2}\sum_{i=1}^{p}\xi_i^2 + \frac{\lambda C}{2}\sum_{i=p+1}^{m}\sum_{j=1}^{m_i}\xi_{s(i,j)}^2$$

$$\text{s.t.} \quad \sum_{j=1}^{m_i} \mathbf{w}' d_{i,j}\phi(\mathbf{x}_{i,j}) \geq \rho - \xi_i, \quad i = 1, \cdots, p,$$

$$-\mathbf{w}'\phi(\mathbf{x}_{i,j}) \geq \rho - \xi_{s(i,j)}, \quad i = p+1, \cdots, m, \quad j = 1, \cdots, m_i, \tag{4}$$

where λ balances the slack variables from the positive and negative bags.

Note that each instance in a negative bag leads to a constraint in (4). Potentially, this may result in a large number of constraints in optimization. Here, we consider another variant that simply represents each negative bag in the constraint by the mean of its instances. It has been shown that this representation is reasonable and effective in many cases [8,24]. Thus, we have the following optimization problem:

---

[1] In many cases the standard assumption of multi-instance learning, that is, the positive label is triggered by a key instance, does not hold. Instead, the positive label may be triggered by more than one key instances [22,24,30]. Suppose the number of key instances is v, we can simply set $\sum_{j=1}^{m_i} d_{i,j} = v$, and thus our proposal can also handle this situation with a known v.

$$\text{(Bag-KI-SVM)} \min_{\mathbf{w},\rho,\boldsymbol{\xi},\mathbf{d}} \frac{1}{2}||\mathbf{w}||_2^2 - \rho + \frac{C}{2}\sum_{i=1}^{p}\xi_i^2 + \frac{\lambda C}{2}\sum_{i=p+1}^{m}\xi_i^2$$

$$\text{s.t.} \sum_{j=1}^{m_i}\mathbf{w}'d_{i,j}\phi(\mathbf{x}_{i,j}) \geq \rho - \xi_i, \quad i = 1, \cdots, p,$$

$$-\mathbf{w}'\frac{\sum_{j=1}^{m_i}\phi(\mathbf{x}_{i,j})}{m_i} \geq \rho - \xi_i, \quad i = p+1, \cdots, m. \tag{5}$$

Hence, instead of a total of $\sum_{i=p+1}^{m} m_i$ constraints for the negative bags in (4), there are now only $m - p$ corresponding constraints in (5). As (4) considers each *instance* (in a negative bag) as one constraint, while (5) only represents the whole negative *bag* as a constraint. Therefore, we will refer to the formulations in (4) and (5) as the instance-level KI-SVM (Ins-KI-SVM) and bag-level KI-SVM (Bag-KI-SVM), respectively.

As (4) and (5) are similar in form, we consider in the following a more general optimization problem for easier exposition:

$$\min_{\mathbf{w},\rho,\boldsymbol{\xi},\mathbf{d}} \frac{1}{2}||\mathbf{w}||_2^2 - \rho + \frac{C}{2}\sum_{i=1}^{p}\xi_i^2 + \frac{\lambda C}{2}\sum_{i=p+1}^{r}\xi_i^2$$

$$\text{s.t.} \sum_{j=1}^{m_i}\mathbf{w}'d_{i,j}\phi(\mathbf{x}_{i,j}) \geq \rho - \xi_i, \quad i = 1, \cdots, p,$$

$$-\mathbf{w}'\psi(\hat{\mathbf{x}}_i) \geq \rho - \xi_i, \quad i = p+1, \cdots, r. \tag{6}$$

It is easy to see that both the Ins-KI-SVM and Bag-KI-SVM are special cases of (6). Specifically, when $r = N$, and $\psi(\hat{\mathbf{x}}_{s(i,j)}) = \phi(\mathbf{x}_{i,j})$ for the second constraint, (6) reduces to the Ins-KI-SVM. Alternatively, when $r = m$, and $\psi(\hat{\mathbf{x}}_i) = \frac{\sum_{j=1}^{m_i}\phi(\mathbf{x}_{i,j})}{m_i}$ for the second constraint, then (6) becomes the Bag-KI-SVM.

By using the method of Lagrange multipliers, the Lagrangian can be obtained as:

$$\mathcal{L}(\mathbf{w}, \rho, \boldsymbol{\xi}, \mathbf{d}, \boldsymbol{\alpha})$$
$$= \frac{1}{2}||\mathbf{w}||_2^2 - \rho + \frac{C}{2}\sum_{i=1}^{p}\xi_i^2 + \frac{\lambda C}{2}\sum_{i=p+1}^{r}\xi_i^2 - \sum_{i=1}^{p}\alpha_i(\sum_{j=1}^{m_i}\mathbf{w}'d_{i,j}\phi(\mathbf{x}_{i,j}) - \rho + \xi_i)$$
$$- \sum_{i=p+1}^{r}\alpha_i(-\mathbf{w}'\psi(\hat{\mathbf{x}}_i) - \rho + \xi_i).$$

By setting the partial derivatives with respect to the $\mathbf{w}, \rho, \boldsymbol{\xi}$ to zeros, we have

$$\frac{\partial L}{\partial \mathbf{w}} = \mathbf{w} - \sum_{i=1}^{p}\alpha_i\sum_{j=1}^{m_i}d_{i,j}\phi(\mathbf{x}_{i,j}) + \sum_{i=p+1}^{r}\alpha_i\psi(\hat{\mathbf{x}}_i) = 0,$$

$$\frac{\partial L}{\partial \rho} = -1 + \sum_{i=1}^{r}\alpha_i = 0,$$

$$\frac{\partial L}{\partial \xi_i} = C\xi_i - \alpha_i = 0, \forall i = 1, \cdots, p,$$
$$\frac{\partial L}{\partial \xi_i} = \lambda C\xi_i - \alpha_i = 0, \forall i = p+1, \cdots, r.$$

Then, the dual of (6) can be obtained as

$$\min_{\mathbf{d}\in\Delta} \max_{\boldsymbol{\alpha}\in\mathcal{A}} -\tfrac{1}{2}(\boldsymbol{\alpha}\odot\hat{\mathbf{y}})'\Big(\mathbf{K}^{\mathbf{d}} + \mathbf{E}\Big)(\boldsymbol{\alpha}\odot\hat{\mathbf{y}}), \tag{7}$$

where $\boldsymbol{\alpha} = [\alpha_1, \cdots, \alpha_r]' \in \mathbb{R}^r$ is the vector of Lagrange multipliers, $\mathcal{A} = \{\boldsymbol{\alpha} \mid \sum_{i=1}^r \alpha_i = 1, \alpha_i \geq 0\}$, $\hat{\mathbf{y}} = [\mathbf{1}_p, -\mathbf{1}_{r-p}] \in \mathbb{R}^r$, $\odot$ denotes the element-wise product of two matrices, $\mathbf{E} \in \mathbb{R}^{r\times r}$ is a diagonal matrix with diagonal entries

$$E_{i,i} = \begin{cases} \frac{1}{C} & i = 1, \cdots, p, \\ \frac{1}{\lambda C} & \text{otherwise}, \end{cases}$$

and $\mathbf{K}^{\mathbf{d}} \in \mathbb{R}^{r\times r}$ is the kernel matrix where $\mathbf{K}^{\mathbf{d}}_{ij} = (\boldsymbol{\psi}^{\mathbf{d}}_i)'(\boldsymbol{\psi}^{\mathbf{d}}_j)$ with

$$\boldsymbol{\psi}^{\mathbf{d}}_i = \begin{cases} \sum_{j=1}^{m_i} d_{i,j}\phi(\mathbf{x}_{i,j})' & i = 1, \cdots, p, \\ \psi(\hat{\mathbf{x}}_i) & i = p+1, \cdots, r. \end{cases} \tag{8}$$

Note that (7) is a mixed-integer programming problem, and so is computationally intractable in general.

3.2 Convex Relaxation

The main difficulty of (7) lies in the variables $\mathbf{d}$ which is hard to optimize in general. But once the $\mathbf{d}$ is given, the inner problem of (7) will become a standard SVM which could be solved in an efficient manner. This simple observation motivates us to avoid optimizing $\mathbf{d}$, alternatively, to learn the optimal combination of some $\mathbf{d}$'s. Further observed that each $\mathbf{d}$ corresponds to a kernel $\mathbf{K}^{\mathbf{d}}$, learning the optimal convex combination will become multiple kernel learning (MKL) [13] which is convex and efficient in general.

In detail, we consider a minimax relaxation [14] by exchanging the order of $\min_{\mathbf{d}}$ and $\max_{\boldsymbol{\alpha}}$. According to the minimax inequality [12], (7) can be lower-bounded by

$$\begin{aligned} &\max_{\boldsymbol{\alpha}\in\mathcal{A}} \min_{\mathbf{d}\in\Delta} \quad -\frac{1}{2}(\boldsymbol{\alpha}\odot\hat{\mathbf{y}})'\Big(\mathbf{K}^{\mathbf{d}} + \mathbf{E}\Big)(\boldsymbol{\alpha}\odot\hat{\mathbf{y}}) \\ &= \max_{\boldsymbol{\alpha}\in\mathcal{A}} \Big\{ \max_{\theta} -\theta \\ &\qquad\qquad \text{s.t.} \quad \theta \geq \frac{1}{2}(\boldsymbol{\alpha}\odot\hat{\mathbf{y}})'\Big(\mathbf{K}^{\mathbf{d}_t} + \mathbf{E}\Big)(\boldsymbol{\alpha}\odot\hat{\mathbf{y}}), \ \forall \mathbf{d}_t \in \Delta\Big\}. \end{aligned} \tag{9}$$

By introducing the dual variable $\mu_t \geq 0$ for each constraint, then its Lagrangian is

$$-\theta + \sum_{t:\mathbf{d}_t\in\Delta} \mu_t\Big(\theta - \frac{1}{2}(\boldsymbol{\alpha}\odot\hat{\mathbf{y}})'\Big(\mathbf{K}^{\mathbf{d}_t} + \mathbf{E}\Big)(\boldsymbol{\alpha}\odot\hat{\mathbf{y}})\Big).$$

Algorithm 1. Cutting plane algorithm for KI-SVM

1: Initialize $\mathbf{d}$ to $\mathbf{d}_0$, and set $\mathcal{C} = \{\mathbf{d}_0\}$.
2: Run MKL for the subset of kernel matrices selected in $\mathcal{C}$ and obtain $\boldsymbol{\alpha}$ from (10). Let o_1 be the objective value obtained.
3: Find a constraint (indexed by $\hat{\mathbf{d}}$) violated by the current solution and set $\mathcal{C} = \hat{\mathbf{d}} \bigcup \mathcal{C}$.
4: Set $o_2 = o_1$. Run MKL for the subset of kernel matrices selected in $\mathcal{C}$ and obtain $\boldsymbol{\alpha}$ from (10). Let o_1 be the objective value obtained.
5: Repeat steps 3-4 until $|\frac{o_2 - o_1}{o_2}| < \epsilon$.

It can be further noted that $\sum \mu_t = 1$ by setting the derivative w.r.t. θ to zero. Let $\boldsymbol{\mu}$ be the vector of μ_t's, and $\mathcal{M}$ be the simplex $\{\boldsymbol{\mu} \mid \sum \mu_t = 1, \mu_t \geq 0\}$. Then (9) becomes

$$\max_{\boldsymbol{\alpha} \in \mathcal{A}} \min_{\boldsymbol{\mu} \in \mathcal{M}} -\frac{1}{2}(\boldsymbol{\alpha} \odot \hat{\mathbf{y}})' \Big(\sum_{t:\mathbf{d}_t \in \Delta} \mu_t \mathbf{K}^{\mathbf{d}_t} + \mathbf{E} \Big) (\boldsymbol{\alpha} \odot \hat{\mathbf{y}}) \tag{10}$$

$$= \min_{\boldsymbol{\mu} \in \mathcal{M}} \max_{\boldsymbol{\alpha} \in \mathcal{A}} -\frac{1}{2}(\boldsymbol{\alpha} \odot \hat{\mathbf{y}})' \Big(\sum_{t:\mathbf{d}_t \in \Delta} \mu_t \mathbf{K}^{\mathbf{d}_t} + \mathbf{E} \Big) (\boldsymbol{\alpha} \odot \hat{\mathbf{y}}). \tag{11}$$

Here, we can interchange the order of the max and min operators as the objective in (10) is concave in $\boldsymbol{\alpha}$ and convex in $\boldsymbol{\mu}$ [13]. It is noteworthy that (11) can be regarded as multiple kernel learning (MKL) [13], where the kernel matrix to be learned is a convex combination of the base kernel matrices $\{\mathbf{K}^{\mathbf{d}_t} : \mathbf{d}_t \in \Delta\}$. However, the number of feasible vectors $\mathbf{d}_t \in \Delta$ is exponential, the set of base kernels is also exponential in size and so direct MKL is still computationally intractable.

In this paper, we apply the cutting plane method [11] to handle this exponential number of constraints. The cutting plane algorithm is described in Algorithm 1. First, as in [1], we initialize $\mathbf{d}_0$ as the average value, i.e., $\{d_{i,j} = 1/m_i, i = 1, \cdots, p; j = 1, \cdots, m_i\}$ and initialize the working set $\mathcal{C}$ to $\{\mathbf{d}_0\}$. Since the size of $\mathcal{C}$ (and thus the number of base kernel matrices) is no longer exponential, one can perform MKL with the subset of kernel matrices in $\mathcal{C}$, obtain $\boldsymbol{\alpha}$ from (10) and record the objective value o_1 in step 2. In step 3, an inequality constraint in (9) (which is indexed by a particular $\hat{\mathbf{d}}$) that is violated by the current solution is then added to $\mathcal{C}$. In step 4, we first set $o_2 = o_1$, then we perform MKL again and record the new objective value o_1. We repeat step 3 and step 4 until the gap between o_1 and o_2 is small enough. ϵ is simply set as 0.001 in our experiments.

Two important issues need to be addressed in the cutting plane algorithm, i.e., how to efficiently solve the MKL problem in Steps 2 and 4 and how to efficiently find the a violated constraint in Step 3? These will be addressed in Sections 3.3 and 3.4, respectively.

3.3 MKL on Subset of Kernel Matrices in $\mathcal{C}$

In recent years, a number of MKL methods have been developed in the literature [2,13,17,18,20,25]. In this paper, an adaptation of the SimpleMKL algorithm [18] is used to solve the MKL problem in Algorithm 1.

Specifically, suppose that the current $\mathcal{C} = \{\mathbf{d}_1, \cdots, \mathbf{d}_T\}$. Recall that the feature map induced by the base kernel matrix $\mathbf{K}^{\mathbf{d}_t}$ is given in (8). As in the derivation of the SimpleMKL algorithm, we consider the following optimization problem that corresponds to the MKL problem in (11).

$$\begin{aligned}
\min_{\boldsymbol{\mu}\in\mathcal{M},\mathbf{w},\boldsymbol{\xi}} \quad & \frac{1}{2}\sum_{t=1}^{T}\frac{||\mathbf{w}_t||^2}{\mu_t} - \rho + \frac{C}{2}\sum_{i=1}^{p}\xi_i^2 + \frac{\lambda C}{2}\sum_{i=p+1}^{r}\xi_i^2 \\
\text{s.t.} \quad & \sum_{t=1}^{T}\left(\sum_{j=1}^{m_i}\mathbf{w}_t' d_{i,j}^t \phi(\mathbf{x}_{i,j})\right) \geq \rho - \xi_i,\ i = 1, \cdots, p, \\
& -\sum_{t=1}^{T}\mathbf{w}_t'\psi(\hat{\mathbf{x}}_i) \geq \rho - \xi_i, \quad i = p+1, \cdots, r.
\end{aligned} \tag{12}$$

It is easy to verify that its dual is

$$\begin{aligned}
\max_{\boldsymbol{\alpha}\in\mathcal{A},\theta} \quad & -\frac{1}{2}\boldsymbol{\alpha}'\mathbf{E}\boldsymbol{\alpha} - \theta \\
\text{s.t.} \quad & \theta \geq \frac{1}{2}(\boldsymbol{\alpha}\odot\hat{\mathbf{y}})'\mathbf{K}^{\mathbf{d}_t}(\boldsymbol{\alpha}\odot\hat{\mathbf{y}})\ \ t = 1, \ldots, T,
\end{aligned}$$

which is the same as (9). Following SimpleMKL, we solve (11) (or, equivalently, (12)) iteratively. First, by fixing the mixing coefficients $\boldsymbol{\mu} = [\mu_1, \cdots, \mu_T]'$ of the base kernel matrices and we solve the SVM's dual

$$\max_{\boldsymbol{\alpha}\in\mathcal{A}} -\tfrac{1}{2}(\boldsymbol{\alpha}\odot\hat{\mathbf{y}})'\Big(\textstyle\sum_{t=1}^{T}\mu_t\mathbf{K}^{\mathbf{d}_t} + \mathbf{E}\Big)(\boldsymbol{\alpha}\odot\hat{\mathbf{y}}).$$

Then, by fixing $\boldsymbol{\alpha}$, we use the reduced gradient method to update $\boldsymbol{\mu}$. These two steps are iterated until convergence.

3.4 Finding a Violated Constraint

While the cutting plane algorithm only needs to find a violated constraint in each iteration, it is customary to find the most violated constraint. In the context of (9), we then have to find the $\hat{\mathbf{d}}$ that maximizes

$$\max_{\mathbf{d}\in\Delta}\sum\nolimits_{i,j=1}^{r}\alpha_i\alpha_j\hat{y}_i\hat{y}_j(\boldsymbol{\psi}_i^{\mathbf{d}})'(\boldsymbol{\psi}_j^{\mathbf{d}}). \tag{13}$$

However, this is a concave QP and so can not be solved efficiently. Note, however, that while the use of the most violated constraint may lead to faster convergence, the cutting plane method only requires a violated constraint at each iteration. Hence, we propose in the following a simple and efficient method for finding a good approximation of the most violated $\hat{\mathbf{d}}$.

First, note that maximizing (13) could be rewritten as $\|\sum_{i=1}^{r}\alpha_i\,\hat{y}_i\boldsymbol{\psi}_i^{\mathbf{d}}\|_2$. Using the definition of $\boldsymbol{\psi}_i^{\mathbf{d}}$ in (8), this can be rewritten as

$$\max_{\mathbf{d}\in\Delta}\left\|\sum_{i=1}^{p}\alpha_i\sum_{j=1}^{m_i}d_{i,j}\phi(\mathbf{x}_{i,j}) - \sum_{i=p+1}^{r}\alpha_i\psi(\hat{\mathbf{x}}_i)\right\|_2. \tag{14}$$

The key is to replace the ℓ_2-norm above with the infinity-norm. For simplicity, let $\phi(\mathbf{x}) = [x^{(1)}, x^{(2)}, \cdots, x^{(g)}]'$ and $\psi(\hat{\mathbf{x}}) = [\hat{x}^{(1)}, \hat{x}^{(2)}, \cdots, \hat{x}^{(g)}]'$, where g is the dimensionality of $\phi(\mathbf{x})$ and $\psi(\hat{\mathbf{x}})$. Then, we have

$$\max_{\mathbf{d}\in\Delta}\left\|\sum_{i=1}^{p}\alpha_i\sum_{j=1}^{m_i}d_{i,j}\psi(\hat{\mathbf{x}}_{i,j}) - \sum_{i=p+1}^{r}\alpha_i\psi(\hat{\mathbf{x}}_i)\right\|_\infty$$

$$= \max_{l=1,\cdots,g}\max_{\mathbf{d}\in\Delta}\left|\sum_{i=1}^{p}\alpha_i\sum_{j=1}^{m_i}d_{i,j}x_{i,j}^{(l)} - \sum_{i=p+1}^{r}\alpha_i\hat{x}_i^{(l)}\right|. \tag{15}$$

The absolute sign for each inner subproblem (defined on the lth feature)

$$\max_{\mathbf{d}\in\Delta}\left|\sum_{i=1}^{p}\alpha_i\sum_{j=1}^{m_i}d_{i,j}x_{i,j}^{(l)} - \sum_{i=p+1}^{r}\alpha_i\hat{x}_i^{(l)}\right|. \tag{16}$$

can be removed by writing as the maximum of:

$$\max_{\mathbf{d}\in\Delta}\sum_{i=1}^{p}\alpha_i\sum_{j=1}^{m_i}d_{i,j}x_{i,j}^{(l)} - \sum_{i=p+1}^{r}\alpha_i\hat{x}_i^{(l)}, \tag{17}$$

and

$$\max_{\mathbf{d}\in\Delta} -\sum_{i=1}^{p}\alpha_i\sum_{j=1}^{m_i}d_{i,j}x_{i,j}^{(l)} + \sum_{i=p+1}^{r}\alpha_i\hat{x}_i^{(l)}. \tag{18}$$

Recall that each $d_{i,j} \in \{0, 1\}$. Hence, by setting the key instance of (the positive) bag B_i to be the one corresponding to $\arg\max_{1\le j\le m_i} x_{i,j}^{(l)}$, i.e.,

$$d_{i,j} = \begin{cases} 1 & j = \arg\max_{1\le j'\le m_i} x_{i,j'}^{(l)}, \\ 0 & \text{otherwise}, \end{cases}$$

the maximum in (17) can be obtained as

$$\sum_{i=1}^{p}\alpha_i\max_{1\le j\le m_i}x_{i,j}^{(l)} - \sum_{i=p+1}^{r}\alpha_i\hat{x}_i^{(l)}. \tag{19}$$

Similarly, for (18), we set the key instance of (the positive) bag B_i to be the one corresponding to $\arg\min_{1\le j\le m_i} x_{i,j}^{(l)}$, i.e.,

$$d_{i,j} = \begin{cases} 1 & j = \arg\min_{1\le j'\le m_i} x_{i,j'}^{(l)}, \\ 0 & \text{otherwise}, \end{cases}$$

then the maximum in (18) is obtained as

$$-\sum_{i=1}^{p}\alpha_i\min_{1\le j\le m_i}x_{i,j}^{(l)} + \sum_{i=p+1}^{r}\alpha_i\hat{x}_i^{(l)}. \tag{20}$$

Algorithm 2. Local search for $\mathbf{d}$. Here, $obj(\mathbf{d})$ is the objective value in (13).

Initialize $\mathbf{d} = \arg\max_{\mathbf{d}\in\{\mathbf{d}_1,\cdots,\mathbf{d}_T,\hat{\mathbf{d}}\}} obj(\mathbf{d})$, $v = obj(\mathbf{d})$.
if $\mathbf{d} = \hat{\mathbf{d}}$ **then**
  return $\mathbf{d}$;
end if
for $i = 1 : p$ **do**
  $\mathbf{d}'_l = \mathbf{d}_l, \forall l \neq i$.
  for $j = 1 : m_i$ **do**
    Set $d'_{i,j} = 1$, $d'_{i,q} = 0\ \forall q \neq j$
    if $obj(\mathbf{d}') > v$ **then**
      $\mathbf{d} = \mathbf{d}'$ and $v = obj(\mathbf{d}')$.
    end if
  end for
end for
return $\mathbf{d}$;

These two candidate values (i.e., (19) and (20)) are then compared, and the larger value is the solution of the lth subproblem in (16). With g features, there are thus a total of $2g$ candidates for $\hat{\mathbf{d}}$. By evaluating the objective values for these $2g$ candidates, we can obtain the solution of (15) and thus the key instance assignment $\hat{\mathbf{d}}$.

Note that for all the positive bags, both $\max_{1\leq j\leq m_i} x_{i,j}^{(l)}$ and $\min_{1\leq j\leq m_i} x_{i,j}^{(l)}$ can be pre-computed. Moreover, this pre-processing takes $O(gJ_p)$ time and space only. When a new $\boldsymbol{\alpha}$ is obtained by SimpleMKL, the processing above takes $O(2gr)$ time. Therefore, $\hat{\mathbf{d}}$ can be solved efficiently without the use of any numeric optimization solver.

However, a deficiency of this infinity-norm approximation is that the $\hat{\mathbf{d}}$ obtained may not always correspond to a violated constraint. As the cutting plane algorithm only requires the addition of a violated constraint at each iteration, a simple local search is used to refine the $\hat{\mathbf{d}}$ solution (Algorithm 2). Specifically, we iteratively update the key instance assignment for each positive bag, while keeping the key instance assignments for all the other positive bags fixed. Finally, the $\mathbf{d}$ that leads to the largest objective value in (13) will be reported.

3.5 Prediction

On prediction, each instance $\mathbf{x}$ can be treated as a bag, and its output from the KI-SVM is given by $f(\mathbf{x}) = \sum_{t=1}^{T} \mu_t \sum_{i=1}^{N} \alpha_i \hat{y}_i (\psi_i^{\mathbf{d}_t})'\phi(\mathbf{x})$.

4 Experiments

In this section, we evaluate the proposed methods on both CBIR image data and benchmark data sets of multi-instance learning.

Table 1. Some statistics of the image data set

concept	#images	average #ROIs per image
castle	100	19.39
firework	100	27.23
mountain	100	24.93
sunset	100	2.32
waterfall	100	13.89

4.1 Locating ROI in Each Image

We employ the image database that has been used by Zhou *et al.* [29] in studying the ROI detection performance of multi-instance learning methods. This database consists of 500 COREL images from five image categories: *castle*, *firework*, *mountain*, *sunset* and *waterfall*. Each category corresponds to a target concept to be retrieved. Moreover, each image is of size 160×160, and is converted to the multi-instance feature representation by using the bag generator SBN [16]. Each region (instance) in the image (bag) is of size 20×20. Some of these regions are labeled manually as ROIs. A summary of the data set is shown in Table 1.

The one-vs-rest strategy is used. In particular, a training set of 50 images is created by randomly sampling 10 images from each of the five categories. The remaining 450 images constitute a test set. The training/test partition is randomly generated 30 times, and the average performance is recorded.

The proposed KI-SVMs are compared with the MI-SVM [1] and two other SVM-based methods in multi-instance learning, namely the mi-SVM [1] and the SVM with a multi-instance kernel (MI-Kernel) [8]. Moreover, we further compare with three state-of-art methods on locating the ROIs, namely, Diverse Density (DD) [15], EM-DD [26] and C*k*NN-ROI [29]. For the MI-SVM, mi-SVM, MI-Kernel and KI-SVMs, the RBF kernel is used and the parameters are selected using cross-validation on the training sets. Experiments are performed on a PC with 2GHz Intel Xeon(R)2-Duo running Windows XP with 4GB memory.

Following [29], we evaluate the *success rate*, i.e., the ratio of the number of successes divided by the total number of relevant images. For each relevant image in the database, if the ROI returned by the algorithm is a real ROI, then it is counted as a success. For a fair comparison, all the SVM-based methods are only allowed to identify one ROI, which is the region in the image with maximum prediction value.

Table 2 shows the success rates (with standard deviations) of the various methods. Besides, we also show the rank of each method in terms of its success rate. As can be seen, among all the SVM-based methods, Ins-KI-SVM achieves the best performance on all five concepts. As for its performance comparison with the other non-SVM type methods, Ins-KI-SVM is still always better than DD and C*k*NN-ROI, and is comparable to EM-DD. In particular, EM-DD achieves the best performance on two out of five categories, while Ins-KI-SVM achieves the best performance on the other three. As can be seen, the proposed Bag-KI-SVM also achieves

Table 2. Success rate (%) in locating ROIs. The number in parentheses is the relative rank of the algorithm on the corresponding data set (the smaller the rank, the better the performance).

Method		*castle*	*firework*	*mountain*	*sunset*	*waterfall*	total rank
SVM methods	Ins-KI-SVM	64.74 (2) ±6.64	**83.70 (1)** ±15.43	76.78 (2) ±5.46	**66.85 (1)** ±6.03	**63.41 (1)** ±10.56	7
	Bag-KI-SVM	60.63 (3) ±7.53	54.00 (4) ±22.13	72.70 (3) ±7.66	47.78 (4) ±13.25	45.04 (2) ±21.53	16
	MI-SVM	56.63 (4) ±5.06	58.04 (3) ±20.31	67.63 (5) ±8.43	33.30 (6) ±2.67	33.30 (5) ±8.98	23
	mi-SVM	51.44 (6) ±4.93	40.74 (6) ±4.24	67.37 (6) ±4.48	32.19 (7) ±1.66	22.04 (7) ±4.97	32
	MI-Kernel	50.52 (7) ±4.46	36.37 (8) ±7.92	65.67 (7) ±5.18	32.15 (8) ±1.67	19.93 (8) ±4.65	38
non-SVM methods	DD	35.89 (8) ±15.23	38.67 (7) ±30.67	68.11 (4) ±7.54	57.00 (2) ±18.40	37.78 (4) ±29.61	25
	EM-DD	**76.00 (1)** ±4.63	79.89 (2) ±19.25	**77.22 (1)** ±13.29	53.56 (3) ±16.81	44.33 (3) ±15.13	10
	C*k*NN-ROI	51.48 (5) ±4.59	43.63 (5) ±12.40	60.59 (8) ±4.38	34.59 (5) ±2.57	30.48 (6) ±6.34	29

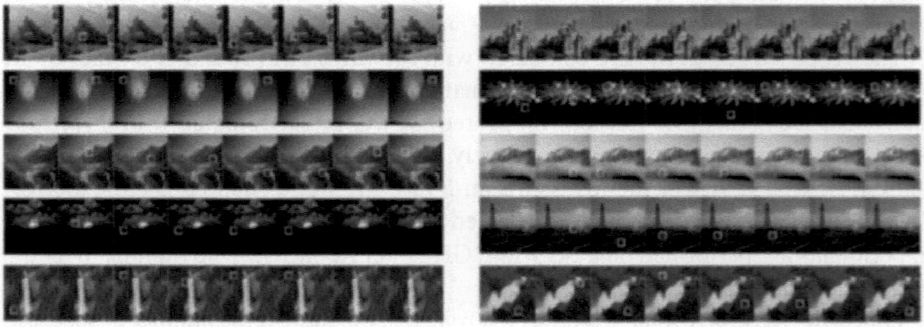

Fig. 1. ROIs located by (from left to right) DD, EM-DD, C*k*NN-ROI, MI-SVM, mi-SVM, MI-Kernel, Ins-KI-SVM and Bag-KI-SVM. Each row shows one category (top to bottom: *castle*, *firework*, *mountain*, *sunset* and *waterfall*).

highly competitive performance with the other state-of-the-art multi-instance learning methods. Fig. 1 shows some example images with the located ROIs. It can be observed that Ins-KI-SVM can correctly identify more ROIs than the other methods.

Each multi-instance algorithm typically has higher confidences (i.e., higher prediction value on the predicted ROI) on some bags than in others. In the next experiment, instead of reporting one ROI in each image, we vary a threshold on the confidence so

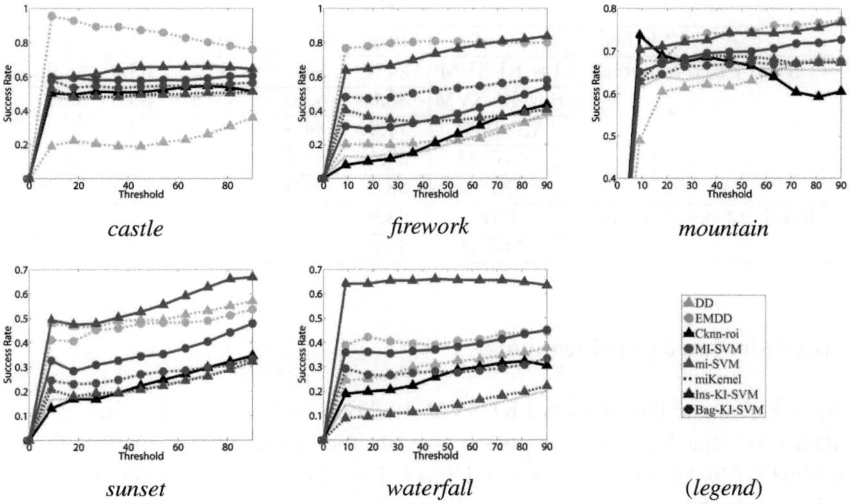

Fig. 2. Success rates when different number of top-confident bags are considered

Table 3. Average wall clock time per query (in seconds)

non-SVM-based methods			SVM-based methods				
DD	EM-DD	CkNN-ROI	MI-SVM	mi-SVM	MI-Kernel	Ins-KI-SVM	Bag-KI-SVM
155.02	15.91	0.003	6.03	6.39	3.04	19.47	5.57

that more than one ROIs can be detected. Fig. 2 shows how the success rate varies when different number of top-confident bags are considered. As can be seen, the proposed Ins-KI-SVM and Bag-KI-SVM achieve highly competitive performance. In particular, Ins-KI-SVM is consistently better than all the other SVM-based methods across all the settings.

Table 3 compares the average query time for the various methods. As can be seen, DD is the slowest since it has to perform gradient descent with multiple restarts. EM-DD is about ten times faster than DD as it involves a much smaller DD optimization at each step. Moreover, note that both the Ins-KI-SVM and mi-SVM work at the instance level while MI-SVM and Bag-KI-SVM work at the bag level. Therefore, MI-SVM and Bag-KI-SVM are in general faster than Ins-KI-SVM and mi-SVM. On the other hand, CkNN-ROI is very efficient as it pre-computes the distances and only needs to compute the citer and reference information when locating ROIs. Moreover, unlike CkNN-ROI which uses the standard Euclidean distance, MI-Kernel needs to compute a small kernel matrix. Therefore, MI-Kernel is slower than CkNN-ROI but is still faster than the other SVM methods in that it only needs to solve the SVM once. However, although CkNN-ROI and MI-Kernel are fast, their performance is much inferior to those of the proposed KI-SVMs, as shown in Table 2.

Table 4. Testing accuracy (%) on the multi-instance classification benchmark data sets

Methods		*Musk1*	*Musk2*	*Elephant*	*Fox*	*Tiger*
SVM-based Methods	Ins-KI-SVM	84.0	84.4	83.5	**63.4**	82.9
	Bag-KI-SVM	**88.0**	82.0	**84.5**	60.5	**85.0**
	MI-SVM	77.9	84.3	81.4	59.4	84.0
	mi-SVM	87.4	83.6	82.0	58.2	78.9
	MI-Kernel	**88.0**	**89.3**	84.3	60.3	84.2
Non-SVM-based Methods	DD	**88.0**	84.0	N/A	N/A	N/A
	EM-DD	84.8	84.9	78.3	56.1	72.1

4.2 Multi-instance Classification

Finally, we evaluate the proposed KI-SVM methods on five multi-instance classification data sets[2] that have been popularly used in the literature [1,5,6,8,28]. These include *Musk1*, *Musk2*, *Elephant*, *Fox* and *Tiger*. The *Musk1* data set contains 47 positive and 45 negative bags, *Musk2* contains 39 positive and 63 negative bags, and each of the remaining three data sets contains 100 positive and 100 negative bags. Details of these data sets can be found in [1,6]. The RBF kernel is used and the parameters are determined by cross-validation on the training set. Comparison is made with the MI-SVM [1], mi-SVM [1], SVM with MI-Kernel [8], DD [15] and EM-DD [26]. Ten-fold cross-validation is used to measure the performance[3]. The average test accuracies of the various methods are shown in Table 4. As can be seen, the performance of KI-SVMs are competitive with all these state-of-the-art methods.

5 Conclusion

Locating ROI is an important problem in many real-world image involved applications. In this paper, we focus on SVM-based methods, and propose two convex optimization methods, Ins-KI-SVM and Bag-KI-SVM, for locating ROIs in images. The KI-SVMs are efficient and based on convex relaxation of the multi-instance SVM. They maximize the margin via generating the most violated key instance step by step, and then combines them via efficient multiple kernel learning. Experiments show that KI-SVMs achieve excellent performance in locating ROIs. The performance of KI-SVMs on multi-instance classification is also competitive with other state-of-the-art methods.

The current work assumes that the bag labels are triggered by single key instances. However, it is very likely that some labels are triggered by several instances together instead of a single key instance. Moreover, some recent studies disclosed that in multi-instance learning the instances should not be treated as i.i.d. samples [27,28]. To identify key instances or key instance groups under these considerations will be studied in the future.

[2] http://www.cs.columbia.edu/∼andrews/mil/datasets.html

[3] The accuracies of these methods were taken from their corresponding literatures. All of them were obtained by ten-fold cross-validation.

Acknowledgements

This research was supported by the National Science Foundation of China (60635030, 60721002), the National High Technology Research and Development Program of China (2007AA01Z169), the Jiangsu Science Foundation (BK2008018), Jiangsu 333 High-Level Talent Cultivation Program, the Research Grants Council of the Hong Kong Special Administrative Region (614907), and the Singapore NTU AcRF Tier-1 Research Grant (RG15/08).

References

1. Andrews, S., Tsochantaridis, I., Hofmann, T.: Support vector machines for multiple-instance learning. In: Becker, S., Thrun, S., Obermayer, K. (eds.) Advances in Neural Information Processing Systems 15, pp. 577–584. MIT Press, Cambridge (2003)
2. Bach, F.R., Lanckriet, G.R.G., Jordan, M.I.: Multiple kernel learning, conic duality, and the SMO algorithm. In: Proceedings of the 21st International Conference on Machine Learning, Banff, Canada, pp. 41–48 (2004)
3. Bi, J., Chen, Y., Wang, J.Z.: A sparse support vector machine approach to region-based image categorization. In: Proceedings of the IEEE Computer Society Conference on Computer Vision and Pattern Recognition, San Diego, CA, pp. 1121–1128 (2005)
4. Chen, Y., Wang, J.Z.: Image categorization by learning and reasoning with regions. Journal of Machine Learning Research 5, 913–939 (2004)
5. Cheung, P.M., Kwok, J.T.: A regularization framework for multiple-instance learning. In: Proceedings of the 23rd International Conference on Machine Learning, Pittsburgh, PA, pp. 193–200 (2006)
6. Dietterich, T.G., Lathrop, R.H., Lozano-Pérez, T.: Solving the multiple instance problem with axis-parallel rectangles. Artificial Intelligence 89(1-2), 31–71 (1997)
7. Fan, R.E., Chen, P.H., Lin, C.J.: Working set selection using second order information for training support vector machines. Journal of Machine Learning Research 6, 1889–1918 (2005)
8. Gärtner, T., Flach, P.A., Kowalczyk, A., Smola, A.J.: Multi-instance kernels. In: Proceedings of the 19th International Conference on Machine Learning, Sydney, Australia, pp. 179–186 (2002)
9. Hsieh, C.J., Chang, K.W., Lin, C.J., Keerthi, S.S., Sundararajan, S.: A dual coordinate descent method for large-scale linear SVM. In: Proceedings of the 25th International Conference on Machine Learning, Helsinki, Finland, pp. 408–415 (2008)
10. Joachims, T.: Training linear SVMs in linear time. In: Proceedings to the 12th ACM SIGKDD Conference on Knowledge Discovery and Data Mining, Philadelphia, PA, pp. 217–226 (2006)
11. Kelley, J.E.: The cutting plane method for solving convex programs. Journal of the SIAM 8(4), 703–712 (1960)
12. Kim, S.-J., Boyd, S.: A minimax theorem with applications to machine learning, signal processing, and finance. SIAM Journal on Optimization 19(3), 1344–1367 (2008)
13. Lanckriet, G.R.G., Cristianini, N., Bartlett, P., El Ghaoui, L., Jordan, M.I.: Learning the kernel matrix with semidefinite programming. Journal of Machine Learning Research 5, 27–72 (2004)
14. Li, Y.-F., Tsang, I.W., Kwok, J.T., Zhou, Z.-H.: Tighter and convex maximum margin clustering. In: Proceeding of the 12th International Conference on Artificial Intelligence and Statistics, Clearwater Beach, FL, pp. 344–351 (2009)

15. Maron, O., Lozano-Pérez, T.: A framework for multiple-instance learning. In: Jordan, M.I., Kearns, M.J., Solla, S.A. (eds.) Advances in Neural Information Processing Systems 10, pp. 570–576. MIT Press, Cambridge (1998)
16. Maron, O., Ratan, A.L.: Multiple-instance learning for natural scene classification. In: Proceedings of the 15th International Conference on Machine Learning, Madison, WI, pp. 341–349 (1998)
17. Rakotomamonjy, A., Bach, F., Canu, S., Grandvalet, Y.: More efficiency in multiple kernel learning. In: Proceedings of the 24th International Conference on Machine Learning, Corvalis, OR, pp. 775–782 (2007)
18. Rakotomamonjy, A., Bach, F.R., Canu, S., Grandvalet, Y.: SimpleMKL. Journal of Machine Learning Research 9, 2491–2521 (2008)
19. Ray, S., Craven, M.: Supervised versus multiple instance learning: an empirical comparison. In: Proceedings of the 22nd International Conference on Machine Learning, Bonn, Germany, pp. 697–704 (2005)
20. Sonnenburg, S., Rätsch, G., Schäfer, C., Schölkopf, B.: Large scale multiple kernel learning. Journal of Machine Learning Research 7, 1531–1565 (2006)
21. Tsang, I.W., Kwok, J.T., Cheung, P.: Core vector machines: fast SVM training on very large data sets. Journal of Machine Learning Research 6, 363–392 (2006)
22. Wang, H.Y., Yang, Q., Zha, H.: Adaptive p-posterior mixture-model kernels for multiple instance learning. In: Proceedings of the 25th International Conference on Machine Learning, Helsinki, Finland, pp. 1136–1143 (2008)
23. Wang, J., Zucker, J.D.: Solving the multiple-instance problem: A lazy learning approach. In: Proceedings of the 17th International Conference on Machine Learning, Stanford, CA, pp. 1119–1125 (2000)
24. Xu, X., Frank, E.: Logistic regression and boosting for labeled bags of instances. In: Dai, H., Srikant, R., Zhang, C. (eds.) PAKDD 2004. LNCS (LNAI), vol. 3056, pp. 272–281. Springer, Heidelberg (2004)
25. Xu, Z., Jin, R., King, I., Lyu, M.R.: An extended level method for efficient multiple kernel learning. In: Koller, D., Schuurmans, D., Bengio, Y., Bottou, L. (eds.) Advances in Neural Information Processing Systems 21, pp. 1825–1832. MIT Press, Cambridge (2009)
26. Zhang, Q., Goldman, S.A.: EM-DD: An improved multiple-instance learning technique. In: Dietterich, T.G., Becker, S., Ghahramani, Z. (eds.) Advances in Neural Information Processing Systems 14, pp. 1073–1080. MIT Press, Cambridge (2002)
27. Zhou, Z.-H., Sun, Y.-Y., Li, Y.-F.: Multi-instance learning by treating instances as non-i.i.d. samples. In: Proceedings of the 26th International Conference on Machine Learning, Montreal, Canada (2009)
28. Zhou, Z.-H., Xu, J.-M.: On the relation between multi-instance learning and semi-supervised learning. In: Proceedings of the 24th International Conference on Machine Learning, Corvalis, OR, pp. 1167–1174 (2007)
29. Zhou, Z.-H., Xue, X.-B., Jiang, Y.: Locating regions of interest in CBIR with multi-instance learning techniques. In: Proceedings of the 18th Australian Joint Conference on Artificial Intelligence, Sydney, Australia, pp. 92–101 (2005)
30. Zhou, Z.-H., Zhang, M.-L.: Multi-instance multi-label learning with application to scene classification. In: Schölkopf, B., Platt, J., Hofmann, T. (eds.) Advances in Neural Information Processing Systems 19, pp. 1609–1616. MIT Press, Cambridge (2007)

Active Learning for Reward Estimation in Inverse Reinforcement Learning*

Manuel Lopes[1], Francisco Melo[2], and Luis Montesano[3]

[1] Instituto de Sistemas e Robótica - Instituto Superior Técnico
Lisboa, Portugal
macl@isr.ist.utl.pt
[2] Carnegie Mellon University
Pittsburgh, PA, USA
fmelo@cs.cmu.edu
[3] Universidad de Zaragoza
Zaragoza, Spain
lmontesa@unizar.es

Abstract. Inverse reinforcement learning addresses the general problem of recovering a reward function from samples of a policy provided by an expert/demonstrator. In this paper, we introduce *active learning* for inverse reinforcement learning. We propose an algorithm that allows the agent to *query* the demonstrator for samples at specific states, instead of relying only on samples provided at "arbitrary" states. The purpose of our algorithm is to estimate the reward function with similar accuracy as other methods from the literature while reducing the amount of policy samples required from the expert. We also discuss the use of our algorithm in higher dimensional problems, using both Monte Carlo and gradient methods. We present illustrative results of our algorithm in several simulated examples of different complexities.

1 Introduction

We address the general problem of *learning from demonstration*. In this class of problems, an agent is given a set of sample situation-action pairs by a demonstrator, from which it must recover the overall demonstrated behavior and/or corresponding task description. In this paper we are particularly interested in recovering the task description. In other words, the agent infers the underlying task that the demonstrator is trying to solve. From this task description, the agent can then construct its own policy to solve the recovered task. One interesting aspect of this approach is that it can accommodate for differences between the demonstrator and the learner [1]. The learner is not just replicating the observed trajectory, but is inferring the "reason" behind such behavior.

* Work partially supported by the ICTI and FCT, under the CMU-Portugal Program, the (POS_C) program that includes FEDER funds and the projects PTDC/EEA-ACR/70174/2006, (FP6-IST-004370) RobotCub and (FP7-231640) Handle.

W. Buntine et al. (Eds.): ECML PKDD 2009, Part II, LNAI 5782, pp. 31–46, 2009.

We formalize our problem using Markov decision processes (MDP). Within this formalism, the demonstration consists of a set of state-action pairs and the compact task representation takes the form of a *reward function*. Learning from demonstration in MDPs has been explored in different ways in the literature [2, 3, 4], and is usually known as *inverse reinforcement learning*. The seminal paper [3] gives the first formal treatment of inverse reinforcement learning as well as several algorithms to compute a reward description from a demonstration. This problem has since been addressed in several other works [4, 5, 6, 7, 8].

The general IRL problem poses several interesting challenges to be dealt with. On one hand, the process of searching for the "right" reward function typically requires the underlying MDP to be solved multiple times, making this process potentially computationally expensive in large problems. Furthermore, it is unreasonable to assume that the desired policy is completely specified, as this is impractical in problems with more than a few dozen states, or that there is no noise in the demonstration. Finally, the IRL problem is *ill-posed*, in the sense that there is not a single reward function that renders a given policy optimal and also there are usually multiple optimal policies for the same reward function [9]. This means that, even if the desired policy is completely specified to the learner, the problem remains ill-posed, as additional criteria are necessary to disambiguate between multiple rewards yielding the same optimal policy.

Probabilistic sample-based approaches to the IRL problem [4, 5, 6] partly address these issues, alleviating the requirement for complete and correct demonstration while restricting the set of possible solution rewards. These approaches allow the solution to IRL to be "better conditioned" by increasing the size of the demonstration and are robust to suboptimal actions in the demonstration[1]. However, this will typically require a large amount of data (samples) for a good estimate of the reward function to be recovered.

In this paper we propose the use of *active learning* to partly mitigate the need for large amounts of data during learning. We adopt a Bayesian approach to IRL, following [6]. The idea behind active learning is to reduce the data requirements of learning algorithms by actively selecting potentially informative samples, in contrast with random sampling from a predefined distribution [10]. In our case we use this idea to reduce the number of samples required from the expert, and only ask the expert to demonstrate the desired behavior at the most informative states. We compute the posterior distribution over possible reward functions and use this information to *actively select* the states whose action the expert should provide. Experimental results show that our approach generally reduces the amount of data required to learn the reward function. Also, it is more adequate in terms of interaction with the expert, as it requires the expert to illustrate the desired behavior on fewer instances.

---

[1] By considering demonstrations in which suboptimal actions can also be sampled, the learner is provided with a *ranking* of actions instead of just an indication of the optimal actions. Demonstrations are thus "more informative", enforcing a more constrained set of possible reward functions than in the general case where only optimal policies are provided. This, in turn, simplifies the search problem.

2 Background

In this section we review some background material on MDPs and inverse reinforcement learning.

2.1 Markov Decision Processes

A *Markov decision process* (MDP) is a tuple $(\mathcal{X}, \mathcal{A}, \mathsf{P}, r, \gamma)$, where $\mathcal{X}$ represents the finite state-space, $\mathcal{A}$ the finite action space, P the transition probabilities, r the reward function and is γ a discount factor. $\mathsf{P}_a(x, y)$ denotes the probability of transitioning from state x to state y when action a is taken. The purpose of the agent is to choose the action sequence $\{A_t\}$ maximizing

$$V(x) = \mathbb{E}\left[\sum_{t=0}^{\infty} \gamma^t r(X_t, A_t) \mid X_0 = x\right].$$

A *policy* is a mapping $\pi : \mathcal{X} \times \mathcal{A} \to [0, 1]$, where $\pi(x, a)$ is the probability of choosing action $a \in \mathcal{A}$ in state $x \in \mathcal{X}$. Associated with any such policy there is a *value-function* V^π, $V^\pi(x) = \mathbb{E}_\pi\left[\sum_{t=0}^{\infty} \gamma^t r(X_t, A_t) \mid X_0 = x\right]$,where the expectation is now taken with respect to policy π. For any given MDP there exists at least one policy π^* such that

$$V^{\pi^*}(x) \geq V^\pi(x).$$

Any such policy is an *optimal policy* for that MDP and the corresponding value function is denoted by V^*.

Given any policy π, the following recursion holds

$$V^\pi(x) = r_\pi(x) + \gamma \sum_{y \in \mathcal{X}} \mathsf{P}_\pi(x, y) V^\pi(y)$$

where $\mathsf{P}_\pi(x, y) = \sum_{a \in \mathcal{A}} \pi(x, a) \mathsf{P}_a(x, y)$ and $r_\pi(x) = \sum_{a \in \mathcal{A}} \pi(x, a) r(x, a)$. For the particular case of the optimal policy π^*, the above recursion becomes

$$V^*(x) = \max_{a \in \mathcal{A}} \left[r(x, a) + \gamma \sum_{y \in \mathcal{X}} \mathsf{P}_a(x, y) V^*(y) \right].$$

We also define the Q-function associated with a policy π as

$$Q^\pi(x, a) = r(x, a) + \gamma \sum_{y \in \mathcal{X}} \mathsf{P}_a(x, y) V^\pi(y)$$

Sometimes, it will be convenient to write the above expressions using vector notation, leading to the expressions

$$\mathbf{V}^\pi = \mathbf{r}_\pi + \gamma \mathsf{P}_\pi \mathbf{V}^\pi \qquad \mathbf{Q}^\pi_a = \mathbf{r}_a + \gamma \mathsf{P}_a \mathbf{V}^\pi \tag{1a}$$

$$\mathbf{V}^* = \max_{a \in \mathcal{A}} [\mathbf{r}_a + \gamma \mathsf{P}_a \mathbf{V}^*] \qquad \mathbf{Q}^*_a = \mathbf{r}_a + \gamma \mathsf{P}_a \mathbf{V}^*, \tag{1b}$$

where $\mathbf{r}_a$ and $\mathbf{Q}_a$ denote the ath columns of matrices $\mathbf{r}$ and $\mathbf{Q}$, respectively.

2.2 Bayesian Inverse Reinforcement Learning

As seen above, an MDP describes a sequential decision making problem in which an agent must choose its actions so as to maximize the total discounted reward. In this sense, the reward function in an MDP encodes the *task* of the agent.

Inverse reinforcement learning (IRL) deals with the problem of recovering the task representation (*i.e.*, the reward function) given a demonstration of the task to be performed (*i.e.*, the desired policy). In this paper, similarly to [6], IRL is cast as an *inference problem*, in which the agent is provided with a noisy sample of the desired policy from which it must estimate a reward function explaining the policy.

Our working assumption is that there is one reward function, r_{target}, that the demonstrator wants the agent to maximize. We denote the corresponding optimal Q-function by Q^*_{target}. Given this reward function, the demonstrator will choose an action $a \in \mathcal{A}$ in state $x \in \mathcal{X}$ with probability

$$\mathbb{P}\left[A_{\text{demo}} = a \mid X_{\text{demo}} = x, r_{\text{target}}\right] = \frac{e^{\eta Q^*_{\text{target}}(x,a)}}{\sum_{b \in \mathcal{A}} e^{\eta Q^*_{\text{target}}(x,b)}}, \tag{2}$$

where η is a non-negative constant.

We consider the demonstration as a sequence $\mathcal{D}$ of state-action pairs,

$$\mathcal{D} = \{(x_1, a_1), (x_2, a_2), \ldots, (x_n, a_n)\},$$

From (2), for any given r-function, the *likelihood* of a pair $(x, a) \in \mathcal{X} \times \mathcal{A}$ is given by

$$L_r(x, a) = \mathbb{P}\left[(x, a) \mid r\right] = \frac{e^{\eta Q^*_r(x,a)}}{\sum_{b \in A} e^{\eta Q^*_r(x,b)}},$$

where we denoted by $Q^*_r(x, a)$ the optimal Q-function associated with reward r. The constant η can be seen as a *confidence parameter* that translates the confidence of the agent on the demonstration. Note that, according to the above likelihood model, evaluating the likelihood of a state-action pair given a reward r requires the computation of Q^*_r. This can be done, for example, using dynamic programming, which requires knowledge of the transition probabilities P. In the remainder of the paper, we assume these transtion probabilities are known.

Assuming independence between the state-action pairs in the demonstration, the likelihood of the demonstration $\mathcal{D}$ is

$$L_r(\mathcal{D}) = \prod_{(x_i, a_i) \in \mathcal{D}} L_r(x_i, a_i).$$

Given a prior distribution $\mathbb{P}[r]$ over the space of possible reward functions, we have

$$\mathbb{P}\left[r \mid \mathcal{D}\right] \propto L_r(\mathcal{D}) \mathbb{P}\left[r\right].$$

The posterior distribution takes into account the demonstration and prior information and will provide the information used to actively select samples to be included in the demonstration. From this distribution we can extract several policies and rewards, for instance the mean policy $\pi_{\mathcal{D}}$:

$$\pi_{\mathcal{D}}(x, a) = \int \pi_r(x, a)\mathbb{P}\left[r \mid \mathcal{D}\right] dr, \tag{3}$$

or the *maximum a posteriori*

$$r^* = \max_r \mathbb{P}\left[r \mid \mathcal{D}\right], \tag{4}$$

We conclude this section by describing two methods used to address the IRL problem within this Bayesian framework.

2.3 Two Methods for Bayesian IRL

So far, we cast the IRL problem as an *inference problem*. In the continuation we describe two methods to address this inference problem, one that directly approximates the maximum given in (4) and another that estimates the complete posterior distribution $\mathbb{P}\left[r \mid \mathcal{D}\right]$.

Gradient-based IRL. The first approach considers a *uniform prior* over the space of possible rewards. This means that maximizing $\mathbb{P}\left[r \mid \mathcal{D}\right]$ is equivalent to maximizing the likelihood $L_r(\mathcal{D})$. To this purpose, we implement a gradient-ascent algorithm on the space of rewards, taking advantage of the structure of the underlying MDP. A similar approach has been adopted in [4].

We start by writing the log-likelihood of the demonstration given r:

$$\Lambda_r(\mathcal{D}) = \sum_{(x_i, a_i) \in \mathcal{D}} \log(L_r(x_i, a_i)). \tag{5}$$

We can now write

$$\left[\nabla_r \Lambda_r(\mathcal{D})\right]_{xa} = \sum_{(x_i, a_i) \in \mathcal{D}} \frac{1}{L_r(x_i, a_i)} \frac{\partial L_r(x_i, a_i)}{\partial r_{xa}}. \tag{6}$$

To compute $\nabla_r L_r(x, a)$, we observe that

$$\nabla_r L_r(x, a) = \frac{dL_r}{dQ^*}(x, a)\frac{dQ^*}{dr}(x, a). \tag{7}$$

Computing the derivative of L_r with respect to each component of Q^* yields

$$\frac{dL_r}{dQ^*_{yb}}(x, a) = \eta L_r(x, a)\big(\delta_{yb}(x, a) - L_r(y, b)\delta_y(x)\big),$$

with $x, y \in \mathcal{X}$ and $a, b \in \mathcal{A}$. In the above expression, $\delta_u(v)$ denotes the Kronecker delta function.

To compute $\frac{dQ^*}{dr}$, we recall that $\mathbf{Q}_a^* = \mathbf{r}_a + \gamma \mathsf{P}_a(\mathbf{I} - \gamma \mathsf{P}_{\pi^*})^{-1}\mathbf{r}_{\pi^*}$. We also note that, except for those points in reward space where the policy is not differentiable with respect to r — corresponding to situations in which a small change in a particular component of the reward function induces a change in a component of the policy, — *the policy remains unchanged under a small variation in the reward function.* We thus consider the approximation $\frac{dQ^*}{dr_{zu}}(x,a) \approx \frac{\partial Q^*}{\partial r_{zu}}(x,a)$ that ignores the dependence of the policy on r. The gradient estimate thus obtained corresponds to the actual gradient except near those reward functions on which the policy is not differentiable[2]. Considering the above approximation, and letting $\mathbf{T} = \mathbf{I} - \gamma \mathsf{P}_{\pi^*}$, we have

$$\frac{\partial Q^*}{\partial r_{zu}}(x,a) = \delta_{zu}(x,a) + \gamma \sum_{y \in \mathcal{X}} \mathsf{P}_a(x,y)\mathbf{T}^{-1}(y,z)\pi^*(z,u), \tag{8}$$

with $x, y, z \in \mathcal{X}$ and $a, u \in \mathcal{A}$.

Putting everything together, the method essentially proceeds by considering some initial estimate r_0 and then use the gradient computation outlined above to perform the update

$$\mathbf{r}_{t+1} = \mathbf{r}_t + \alpha_t \nabla_r \Lambda_{r_t}(\mathcal{D})$$

MCMC IRL. The second approach, proposed in [6], uses the Monte-Carlo Markov chain (MCMC) algorithm to approximate the posterior $\mathbb{P}[r \mid \mathcal{D}]$. The MCMC algorithm thus generates a set of sample reward functions, $\{r_1, \ldots, r_N\}$, distributed according to the target distribution, $\mathbb{P}[r \mid \mathcal{D}]$. Then,

$$\mathbb{P}[r \mid \mathcal{D}] \approx \frac{1}{N} \sum_{i=1}^{N} \delta(r, r_i). \tag{9}$$

In the MCMC algorithm, these samples correspond to a sample trajectory of a Markov chain designed so that its invariant distribution matches the target distribution, $\mathbb{P}[r \mid \mathcal{D}]$ [11].

We implement PolicyWalk, an MCMC algorithm described in [6]. In this particular variation, the reward space is discretized into a uniform grid and the MCMC samples jump between neighboring nodes in this grid (see Fig. 1). In other words, a new sample is obtained from the current sample to one of the neighboring nodes in the grid. The new sample is accepted according to the ratio between the posterior probabilities of the current and new samples. Reward functions with higher (posterior) probability are thus selected more often than those with lower probability, and the method is guaranteed to sample according to the true posterior distribution.

A problem with this method is that, for large-dimensional problems, it generally requires a large number of sample rewards to ensure that the estimate of $\mathbb{P}[r \mid \mathcal{D}]$ is accurately represented by the sample set. We refer to the result

[2] It is noteworthy that Rademacher's theorem guarantees that the set of such reward functions is null-measured. We refer to [4] for further details.

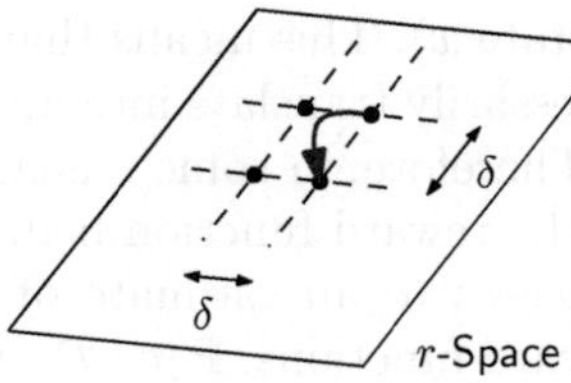

Fig. 1. Representation of the PolicyWalk variation of MCMC. We refer to [6] for details.

in [6], in which the number of samples N required to ensure an estimation error bounded by ε must be $O(M^2 \log(1/\varepsilon))$, where M is the dimension of the reward function. This means that the number of samples grows (roughly) quadratically with the dimension of the reward space. Furthermore, this result assumes that $\|r\|_\infty \leq 1/M$. As noted in [6], this condition on r can be ensured by rescaling the reward function, which does not affect the optimal policy. It does, however, affect the likelihood function and, consequently, $\mathbb{P}[r \mid \mathcal{D}]$. Whenever such rescaling is not possible or convenient and the rewards can only be bounded by some value C, the previous lower-bound on the sample size roughly deteriorates to $O(M^6 e^{2M} \log(1/\varepsilon))$, which quickly becomes prohibitive for large M.

3 Active Learning for Reward Estimation

In the previous section we discussed two possible (Bayesian) approaches to the IRL problem. In these approaches, the agent is provided with a demonstration $\mathcal{D}$, consisting of pairs (x_i, a_i) of states and corresponding actions. From this demonstration the agent must identify the underlying target task.

In the active learning setting, we now assume that, after some initial batch of data $\mathcal{D}$, the agent has the possibility to *query* the expert for the action at particular states chosen by the agent. In this section we propose a criterion to select such states and discuss how this can be used within the IRL framework. We also discuss on-line versions of the methods in the previous section that are able to cope with the successive additional data provided by the expert as a result of the agent's queries.

3.1 Active Sampling

The active learning strategies presented below rely on the uncertainty about the parameter to be estimated to select new data points. As discussed in Section 2, the parameter to be estimated in the IRL setting is the task description, *i.e.*, the reward function. Unfortunately, the relation between rewards and policies is not one-to-one, making active learning in this setting more complex than in other settings.

This is easily seen by thinking of a scenario in which *all* possible reward functions give rise to the same policy in a given state x (this can be the case if there is only one action available in state x). This means that a large uncertainty in the reward function does not necessarily translate into uncertainty in terms of which action to choose in state x. Therefore, in some scenarios it may not be possible to completely disambiguate the reward function behind the demonstration.

In our setting, we have access to an estimate of the posterior distribution over the space of possible reward functions, $\mathbb{P}[r \mid \mathcal{D}]$. From this distribution, the agent should be able to choose a state and query the expert about the corresponding action in such a way that the additional sample is as useful/informative as possible. Notice that the posterior distribution, $\mathbb{P}[r \mid \mathcal{D}]$, does not differentiate between states (it is a distribution over functions) and, as such, a standard variance criterion cannot be used directly.

We are interested in finding a criterion to choose the states to query the demonstrator so as to recover the correct reward (or, at least, the optimal target behavior) while requiring significantly less data than if the agent was provided with randomly chosen state-action pairs. To this purpose, we define the set $\mathcal{R}_{xa}(p)$ as the set of reward functions r such that $\pi_r(x,a) = p$. Now for each pair $(x,a) \in \mathcal{X} \times \mathcal{A}$, the distribution $\mathbb{P}[r \mid \mathcal{D}]$ in turn induces a distribution over the possible values p for $\pi(x,a)$. This distribution can be characterized by means of the following density

$$\bar{\mu}_{xa}(p) = \mathbb{P}[\pi(x,a) = p \mid \mathcal{D}] = \mathbb{P}[r \in \mathcal{R}_{xa}(p) \mid \mathcal{D}]. \tag{10}$$

Using the above distribution, the agent can now query the demonstrator about the correct action in states where the uncertainty on the policy is larger, *i.e.*, in states where $\bar{\mu}_{xa}$ exhibits larger "spread".

One possibility is to rely on some measure of *entropy* associated with $\bar{\mu}_{xa}$. Given that $\bar{\mu}_{xa}$ corresponds to a continuous distribution, the appropriate concept is that of *differential entropy*. Unfortunately, as is well-known, differential entropy as a measure of uncertainty does not exhibit the same appealing properties as its discrete counterpart. To tackle this difficulty, we simply consider a partition of the interval $I = [0,1]$ into K subintervals I_k, with $I_k = (\frac{k}{K}, \frac{k+1}{K}]$, $k = 0, \ldots, K-1$, and $I_0 = [0, 1/K]$. We can now define a new *discrete* probability distribution

$$\mu_{xa}(k) = \mathbb{P}[\pi(x,a) \in I_k \mid \mathcal{D}] = \int_{I_k} \bar{\mu}_{xa}(p)dp, \quad k = 1, \ldots, K.$$

The distribution μ_{xa} thus defined is a discretized version of the density in (10), for which we can compute the associated (Shannon) entropy, $H(\mu_{xa})$. As such, for each state $x \in \mathcal{X}$, we define the *mean entropy* as

$$\bar{H}(x) = \frac{1}{|\mathcal{A}|} \sum_a H(\mu_{xa}) = -\frac{1}{|\mathcal{A}|} \sum_{a,k} \mu_{xa}(k) \log \mu_{xa}(k)$$

and let the agent query the expert about the action to be taken at the state x^* given by

$$x^* = \arg\max_{x \in \mathcal{X}} \bar{H}(x),$$

with ties broken arbitrarily. Given the estimate (9) for $\mathbb{P}[r \mid \mathcal{D}]$, this yields

$$\mu_{xa}(k) \approx \frac{1}{N} \sum_i \mathbb{I}_{I_k}(\pi_i(x,a)), \tag{11}$$

where π_i is the policy associated with the ith reward sampled in the MC method and $\mathbb{I}_{I_k}$ is the indicator function for the set I_k. This finally yields

$$\bar{H}(x) \approx -\frac{1}{|\mathcal{A}|N} \sum_{i,a,k} \mathbb{I}_{I_k}(\pi_i(x,a)) \log \frac{\sum_i \mathbb{I}_{I_k}(\pi_i(x,a))}{N}.$$

It is worth mentioning that, in the context of IRL, there are two main sources of uncertainty in recovering the reward function. One depends on the natural ambiguity of the problem: for any particular policy, there are typically multiple reward functions that solve the IRL problem. This type of ambiguity appear even with perfect knowledge of the policy, and is therefore independent of the particular process by which states are sampled. The other source of uncertainty arises from the fact that the policy is not accurately specified in certain states. This class of ambiguity can be addressed by sampling these states until the policy is properly specified. Our entropy-based criterion does precisely this.

3.2 Active IRL

We conclude this section by describing how the active sampling strategy above can be combined with the IRL methods in Section 2.3.

Algorithm 1. General active IRL algorithm

Require: Initial demo $\mathcal{D}$
1: Estimate $\mathbb{P}[r \mid \mathcal{D}]$ using general MC algorithm
2: **for all** $x \in \mathcal{X}$ **do**
3: Compute $\bar{H}(x)$
4: **end for**
5: Query action for $x^* = \arg\max_x \bar{H}(x)$
6: Add new sample to $\mathcal{D}$
7: Return to 1

The fundamental idea is simply to use the data from an initial demonstration to compute a first posterior $\mathbb{P}[r \mid \mathcal{D}]$, use this distribution to query further states, recompute $\mathbb{P}[r \mid \mathcal{D}]$, and so on. This yields the general algorithm summarized in Algorithm 1. We note that running MCMC and recompute $\mathbb{P}[r \mid \mathcal{D}]$ at each iteration is very time consuming, even more so in large-dimensional problems. For efficiency, step 1 can take advantage of several optimizations of MCMC such as sequential and hybrid monte-carlo.

In very large dimensional spaces, however, the MC-based approach becomes computationally too expensive. We thus propose an approximation to the general Algorithm 1 that uses the gradient-based algorithm in Section 2.3. The idea

Algorithm 2. Active gradient-based IRL algorithm

Require: Initial demo $\mathcal{D}$
1: Compute r^* as in (4)
2: Estimate $\mathbb{P}[r \mid \mathcal{D}]$ in a neighborhood of r^*
3: **for all** $x \in \mathcal{X}$ **do**
4: Compute $\bar{H}(x)$
5: **end for**
6: Query action for $x^* = \arg\max_x \bar{H}(x)$
7: Add new sample to $\mathcal{D}$
8: Return to 1

behind this method is to replace step 1 in Algorithm 1 by two steps, as seen in Algorithm 2. The algorithm thus proceeds by computing the maximum-likelihood estimate r^* as described in Section 2.3. It then uses Monte-Carlo sampling to approximate $\mathbb{P}[r \mid \mathcal{D}]$ *in a neighborhood* $B_\varepsilon(r^*)$ *of* r^* and uses this estimate to compute $H(x)$ as in Algorithm 1. The principle behind this second approach is that the policy π_{r^*} should provide a reasonable approximation to the target policy. Therefore, the algorithm can focus on estimating $\mathbb{P}[r \mid \mathcal{D}]$ only in a neighborhood of r^*. As expected, this significantly reduces the computational requirements of the algorithm.

It is worth mentioning that the first method, relying on standard MC sampling, eventually converges to the true posterior distribution as the number of samples goes to infinity. The second method first reaches a local maximum of the posterior and then only estimates the posterior around that point. If the posterior is unimodal, it is expectable that the second method brings significant advantages in computational terms; however, if the posterior is multimodal, this local approach might not be able properly represent the posterior distribution. In any case, as discussed in Section 2.3, in high-dimensional problems, the MCMC method requires a prohibitive amount of samples to provide accurate estimates, rendering such approach inviable.

4 Simulations

We now illustrate the application of the proposed algorithms in several problems of varying complexity.

4.1 Finding the Maximum of a Quadratic Function

We start by a simple problem of finding the maximum of a quadratic function. Such a problem can be described by the MDP in Fig. 2, where each state corresponds to a discrete value between −1 and 1. The state-space thus consists of 21 states and 2 actions that we denote a_l and a_r. Each action moves the agent deterministically to the contiguous state in the corresponding direction (a_l to the left, a_r to the right). For simplicity, we consider a reward function parameterized using a two-dimensional parameter vector $\boldsymbol{\theta}$, yielding

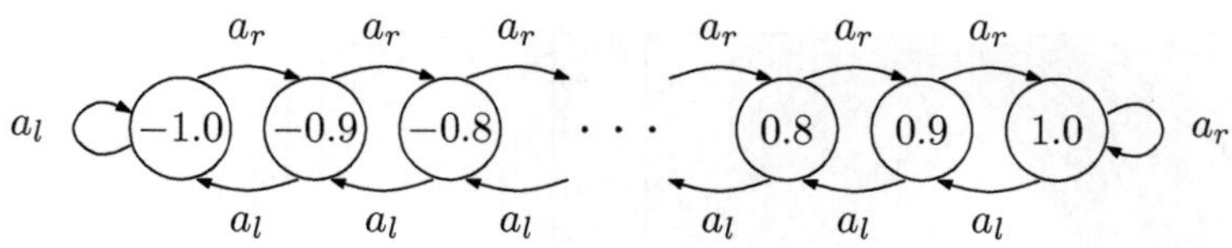

Fig. 2. Simple MDP where the agent must find the maximum of a function

$$r(x) = \theta_1(x - \theta_2)^2,$$

corresponding to a quadratic function with a (double) zero at θ_2 and concavity given by θ_1. For the MDP thus defined, the optimal policy either moves the agent toward the state in which the maximum is attained (if $\theta_1 < 0$) or toward one of the states ± 1 (if $\theta_1 > 0$).

For our IRL problem, we consider the reward function, $r(x) = -(x - 0.15)^2$, for which the agent should learn the parameter $\boldsymbol{\theta}$ from a demonstration. The initial demonstration consisted on the optimal actions for the extreme states:

$$\mathcal{D} = \{(-1.0, a_r), (-0.9, a_r), (-0.8, a_r), (0.8, a_l), (0.9, a_l), (1.0, a_l)\}$$

and immediately establishes that $\theta_1 < 0$.

Figure 3 presents the results obtained using Algorithm 1, with the confidence parameter in the likelyhood function set to $\eta = 500$ and $N = 400$ in the MCMC estimation. The plots on the left represent the reward functions sampled in the MCMC step of the algorithm and the plots on the right the corresponding average policy $\pi_{\mathcal{D}}$. In the depicted run, the queried states were $x = 0$ at iteration 1, $x = 0.3$ at iteration 2, $x = 0.1$ at iteration 3, and $x = 0.2$ at iteration 4. It is clear from the first iteration that the initial demonstration only allows the agent to place θ_2 somewhere in the interval $[-0.8, 0.8]$ (note the spread in the sampled reward functions). Subsequent iterations show the distribution to concentrate on the true value of θ_2 (visible in the fact that the sampled rewards all exhibit a peak around the true value). Also, the policy clearly converges to the optimal policy in iteration 5 and the corresponding variance decreases to 0.

We conclude by noting that our algorithm is roughly implementing the *bisection method*, known to be an efficient method to determine the maximum of a function. This toy example provides a first illustration of our active IRL algorithm at work and the evolution of the posterior distributions over r along the iterations of the algorithm.

4.2 Puddle World

We now illustrate the application of our algorithm in a more complex problem. This problem is known in the reinforcement learning literature as the *puddle world*. The puddle world consists in a continuous-state MDP in which an agentmust reach a goal region while avoiding a penalty region (the "puddle"), as

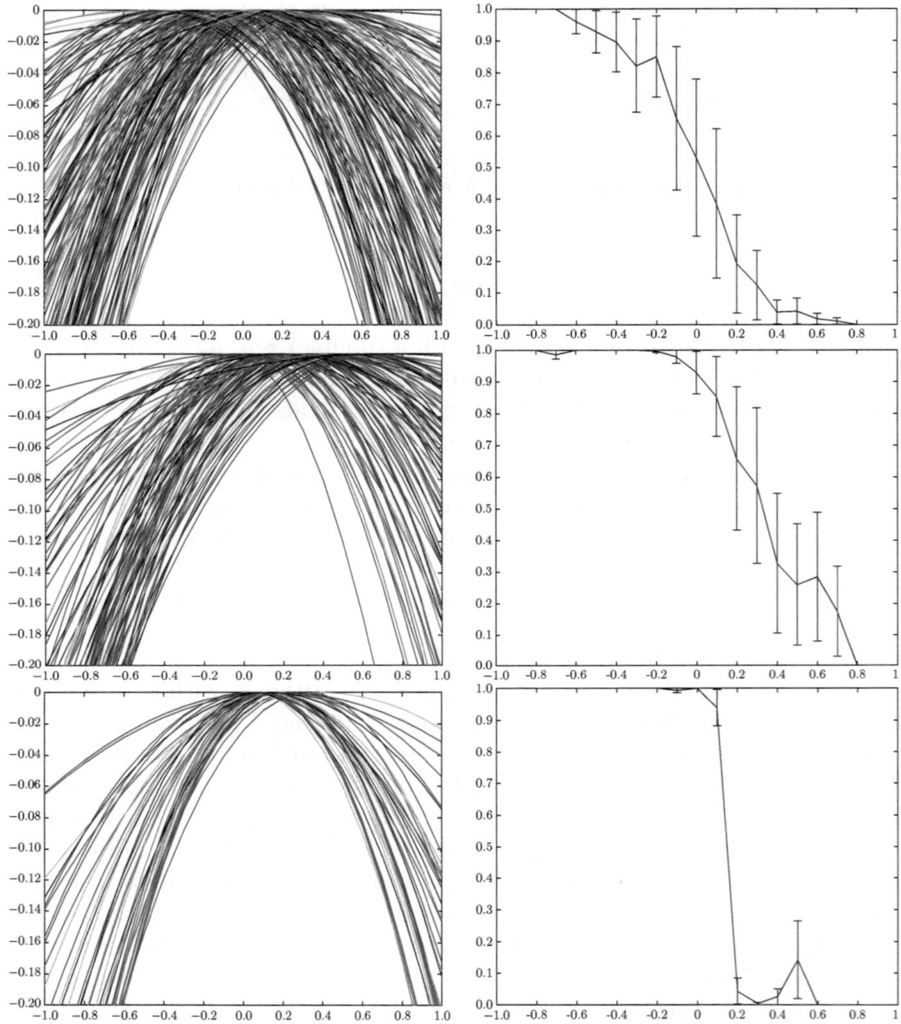

Fig. 3. Sample iterations (1st, 2nd and 5th) of Algorithm 1 in the problem of Fig. 2. On the left are samples obtained from $\mathbb{P}[r \mid \mathcal{D}]$ and on the right the corresponding π_D, showing the mean the and variance of the optimal action for each state (1- move right, 0- move left).

depicted in Fig. 4. This example illustrates our active IRL algorithm at work in a more complex problem that can still visualized.

The MDP has a continuous state-space, consisting of the unit square, and a discrete action-space that includes the four actions N (north), S (south), E (east), and W (west). Each action moves the agent 0.05 in the corresponding direction. Since the MDP has a continuous state-space, exact solution methods are not available. We adopt a batch approximate RL method known as *fitted*

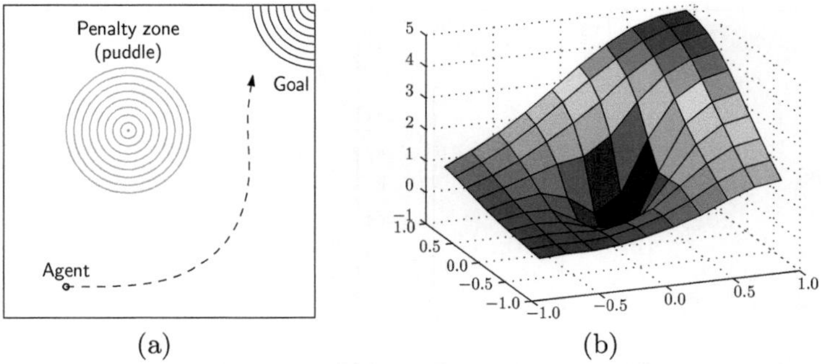

Fig. 4. Representation of the puddle world and a possible value function

Q-iteration that essentially samples the underlying MDP and uses regression to approximate the optimal Q-function [12]. The fact that we must resort to function approximation implies that the exact optimal policy cannot be recovered but only an approximation thereof. This will somewhat impact the ability of our algorithm to properly estimate the posterior $\mathbb{P}[r \mid \mathcal{D}]$.

In the puddle world, the reward function can be represented as

$$r(x) = r_{\text{goal}} \exp\big((\mathbf{x} - \boldsymbol{\mu}_{\text{goal}})^2/\alpha\big) + r_{\text{puddle}} \exp\big((\mathbf{x} - \boldsymbol{\mu}_{\text{puddle}})^2/\alpha\big),$$

where r_{goal} and r_{puddle} represent the reward and maximum penalty received in the goal position and in the center of the puddle, respectively. The parameters $\boldsymbol{\mu}_{\text{goal}}$ and $\boldsymbol{\mu}_{\text{puddle}}$ define the location of the goal and puddle, respectively. The parameter α is fixed *a priori* and roughly defines the width of both regions. For our IRL problem, the agent should learn the parameters $\boldsymbol{\mu}_{\text{goal}}, \boldsymbol{\mu}_{\text{puddle}}, r_{\text{goal}}$, and r_{puddle} from a demonstration.

Figure 5 presents two sample iterations of Algorithm 1. To solve the MDP we ran fitted Q-iteration with a batch of 3,200 sample transitions. We ran MCMC with $N = 800$. Notice that after the first iteration (using the initial demonstration), the MCMC samples are already spread around the true parameters. At each iteration, the algorithm is allowed to query the expert in 10 states. In the depicted run, the algorithm queried states around the goal region — to pinpoint the goal region — and around the puddle — to pinpoint the puddle region.

4.3 Random Scenarios

We now illustrate the application of our approach in random scenarios with different complexity. We also discuss the scalability of our algorithm and statistical significance of the results.

These general MDPs in this section consist of squared grid-worlds with varying number of states. At each state, the agent has 4 actions available (N, S, E, W), that moves the agent in the corresponding direction. We divide our results in two classes, corresponding to *parameterized rewards* and general rewards.

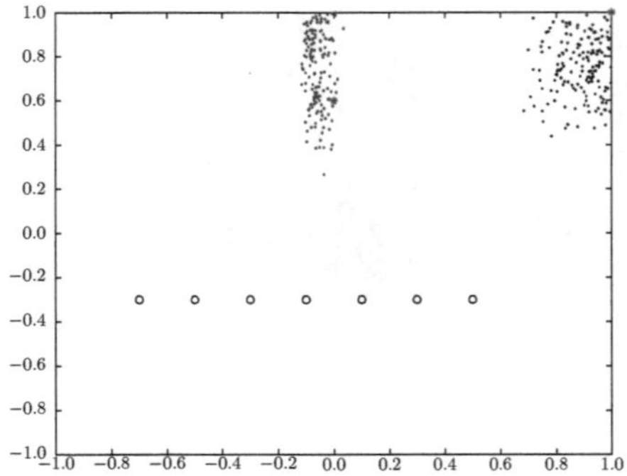
(a) Iteration 1.

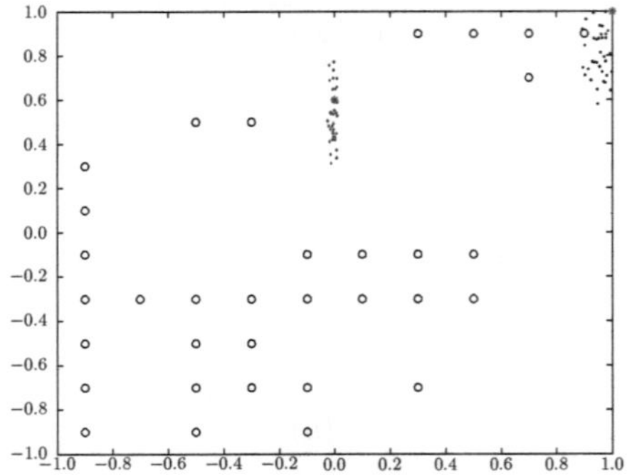
(b) Iteration 3.

Fig. 5. Two sample iterations of Algorithm 1. The red stars (*) represent the target values for the parameters $\boldsymbol{\mu}_{\text{goal}}$ and $\boldsymbol{\mu}_{\text{puddle}}$. The green and blue dots ($\cdot$) represents the sampled posterior distribution over possible value of these parameters. The circles ($\circ$) denotes the states included in the demonstration.

Parameterized Rewards. We start by considering a simple parameterization of the reward function of the form $\delta_{x^*}(x)$. Therefore, the only parameter to be learnt is the position of the goal state x^* in the grid.

We applied Algorithm 2 to a 15×15 grid-world. The estimation in step 2 of the algorithm uses $N = 15$. At each iteration, the agent is allowed to query the expert in 10 states. Figure 6(b) shows the error between the estimated policy and the target policy as a function of the size of the demonstration, averaged over 50 independent trials. Our approach clearly outperforms random sampling, attaining the same error while requiring about 1/3 of the samples.

We conclude by noting that we chose to run Algorithm 2 in this scenarion since (as discussed in Section 2.3, the MCMC component in Algorithm 1 does not scale well with the number of states. Indeed, for a similar scenario with 100 states, the MCMC-based algorithm required around 12,000 MC samples, for each of which an MDP must be solved. In that same 100-state scenarion, Algorithm 2 required around 50 gradient steps and then 20 MC samples to compute the local approximation of the posterior, thus requiring a total of 70 MDPs to be solved.

Non-parameterized reward. We now consider a more general situation, in which the reward function is a vector $\mathbf{r}$ in the $|\mathcal{X}|$-dimensional unit square. In this case, the reward value is merely a real-valued function $r : \mathcal{X} \to [0; 1]$, and the problem is significantly more complex than in the previous case.

We applied Algorithm 2 to a 10×10 grid-world. The estimation in step 2 of the algorithm uses $N = 40$. At each iteration, the agent is allowed to query the expert in 2 states. Figure 6(a) shows the error between the estimated policy and the target policy as a function of the size of the demonstration, averaged over 50 independent trials. In this case, it is clear that there is no apparent advantage in using the active learning approach. Whatever small advantage there may be is clearly outweighed by the added computational cost.

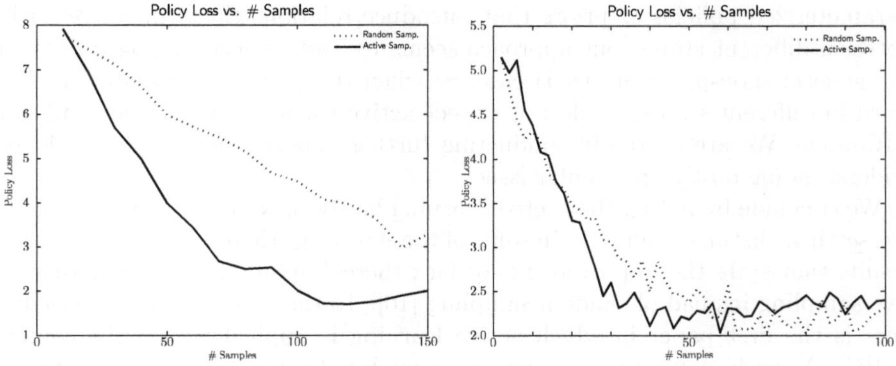

(a) (b)

Fig. 6. Performance of Algorithm 2 comparing active sampling vs. random sampling as a function of the demonstration size. (a) Results with parameterized rewards in a 15×15 grid-world. (b) Results with general (non-parameterized) rewards in a 10×10 grid-world.

These results illustrate, in a sense, some of the issues already discussed in Section 3.1. When considering a non-parameterized form for the reward function and a prior over possible rewards that is state-wise independent, there is not enough structure in the problem to generalize the observed policy from observed states to non-observed states. In fact, the space of general (non-parameterized) reward functions has enough degrees of freedom to yield any possible policy. In this case, any sampling criterion will, at best, provide only a mild advantage over uniform sampling. On the other hand, when using parameterized rewards or a prior that weights positively ties between the reward in different states (*e.g.*, an Ising prior [6]), the policy in some states restricts the possible policies on other states. In this case, sampling certain states can certainly contribute to disambiguate the policy in other states, bringing significant advantages to an active sampling approach over a uniform sampling approach.

5 Conclusions

In this paper we introduced the first active learning algorithm explicitly designed to estimate rewards from a noisy and sampled demonstration of an unknown optimal policy. We used a full Bayesian approach and estimate the posterior probability of each action in each state, given the demonstration. By measuring the state-wise entropy in this distribution, the algorithm is able to select the potentially most informative state to be queried to the expert. This is particularly important when the cost of providing a demonstration is high.

As discussed in Section 4, our results indicate that the effectiveness of active learning in the described IRL setting may greatly depend on the prior knowledge about the reward function or the policy. In particular, when considering parameterized policies or priors that introduce relations (in terms of rewards) between different states, our approach seems to lead to encouraging results. In the general (non-parameterized) case, or when the prior "decorrelates" the reward in different states, we do not expect active learning to bring a significant advantage. We are currently conducting further experiments to gain a clearer understanding on this particular issue.

We conclude by noting that active learning has been widely applied to numerous settings distinct from IRL. In some of these settings there are even theoretical results that state the improvements or lack thereof arising from considering active sampling instead of random sampling [10]. To the extent of our knowledge, ours is the first paper in which active learning is applied within the context of IRL. As such, many new avenues of research naturally appear. In particular, even if the ambiguities inherent to IRL problems make it somewhat distinct from other settings, we believe that it should be possible (at least in some problems) to theoretically asses the usefulness of active learning in IRL.

References

1. Lopes, M., Melo, F., Kenward, B., Santos-Victor, J.: A computational model of social-learning mechanisms. Adaptive Behavior (to appear, 2009)
2. Melo, F., Lopes, M., Santos-Victor, J., Ribeiro, M.: A unified framework for imitation-like behaviors. In: Proc. 4th Int. Symp. Imitation in Animals and Artifacts (2007)
3. Ng, A., Russell, S.: Algorithms for inverse reinforcement learning. In: Proc. 17th Int. Conf. Machine Learning, pp. 663–670 (2000)
4. Neu, G., Szepesvári, C.: Apprenticeship learning using inverse reinforcement learning and gradient methods. In: Proc. 23rd Conf. Uncertainty in Artificial Intelligence, pp. 295–302 (2007)
5. Abbeel, P., Ng, A.: Apprenticeship learning via inverse reinforcement learning. In: Proc. 21st Int. Conf. Machine Learning, pp. 1–8 (2004)
6. Ramachandran, D., Amir, E.: Bayesian inverse reinforcement learning. In: Proc. 20th Int. Joint Conf. Artificial Intelligence, pp. 2586–2591 (2007)
7. Syed, U., Schapire, R., Bowling, M.: Apprenticeship learning using linear programming. In: Proc. 25th Int. Conf. Machine Learning, pp. 1032–1039 (2008)
8. Ziebart, B., Maas, A., Bagnell, J., Dey, A.: Maximum entropy inverse reinforcement learning. In: Proc. 23rd AAAI Conf. Artificial Intelligence, pp. 1433–1438 (2008)
9. Ng, A.Y., Harada, D., Russell, S.: Policy invariance under reward transformations: Theory and application to reward shaping. In: Proc. 16th Int. Conf. Machine Learning, pp. 278–287 (1999)
10. Settles, B.: Active learning literature survey. CS Tech. Rep. 1648, Univ. Wisconsin-Madison (2009)
11. Andrieu, C., de Freitas, N., Doucet, A., Jordan, M.: An introduction to MCMC for machine learning. Machine Learning 50, 5–43 (2003)
12. Timmer, S., Riedmiller, M.: Fitted Q-iteration with CMACs. In: Int. Symp. Approximate Dynamic Programming and Reinforcement Learning (2007)

Simulated Iterative Classification A New Learning Procedure for Graph Labeling

Francis Maes, Stéphane Peters, Ludovic Denoyer, and Patrick Gallinari

LIP6 - University Pierre et Marie Curie
104 avenue du Président Kennedy, Paris, France

Abstract. Collective classification refers to the classification of interlinked and relational objects described as nodes in a graph. The Iterative Classification Algorithm (ICA) is a simple, efficient and widely used method to solve this problem. It is representative of a family of methods for which inference proceeds as an iterative process: at each step, nodes of the graph are classified according to the *current predicted labels* of their neighbors. We show that learning in this class of models suffers from a training bias. We propose a new family of methods, called Simulated ICA, which helps reducing this training bias by simulating inference during learning. Several variants of the method are introduced. They are both simple, efficient and scale well. Experiments performed on a series of 7 datasets show that the proposed methods outperform representative state-of-the-art algorithms while keeping a low complexity.

1 Introduction

A fundamental assumption that underlies most existing work in machine learning is that data is *independently and identically distributed* (*i.i.d.*). Web pages classification, WebSpam detection, community identification in social networks and peer-to-peer files analysis are typical applications where data is naturally organized according to a graph structure. In these applications, the elements to classify (Web pages or users of files for example) are interdependent: the label of one element may have a direct influence on other labels in the graph. Problems involving the classification of graph nodes are generally known as *graph labeling* problems [5] or as *collective classification* problems [11]. Due to the irrelevancy of the *i.i.d.* assumption, new models have been proposed recently to perform machine learning on such networked data.

Different variants of the graph labeling problem have been investigated. For *inductive graph labeling*, training and test are performed in distinct steps. The goal here is to learn classifiers able to label any node in new graphs or subgraphs. This is an extension of the classical *supervised classification task* to interdependent data. Typical applications include email classification, region or object labeling in images or sequence labeling. For *transductive graph labeling*, node labeling is performed in a single step where both labeled and unlabeled data are considered simultaneously. This corresponds to applications like WebSpam detection or social network analysis. Note that some problems like web

W. Buntine et al. (Eds.): ECML PKDD 2009, Part II, LNAI 5782, pp. 47–62, 2009.

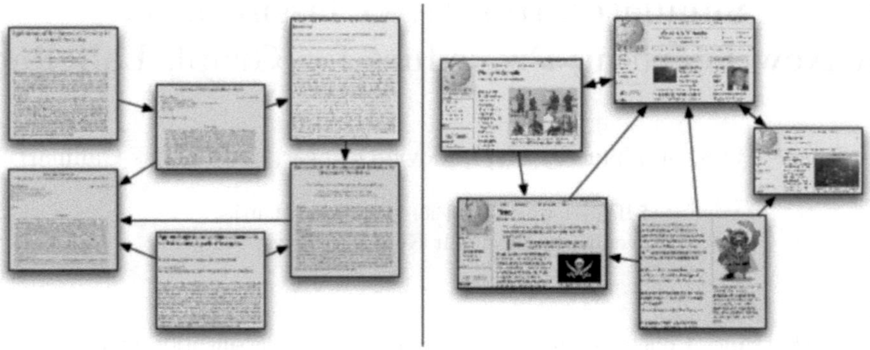

Fig. 1. Two examples of graph labeling problems. Left: categorization of scientific articles related by citation links. Right: classification of Web pages into relevant and non-relevant pages for a given query.

page classification may be handled under either the inductive or the transductive settings. In this article, we focus on the inductive graph labeling problem and we will use the name *collective classification* for this setting.

Graph labeling is often a hard problem and performing exact inference is generally prohibitive. Practical algorithms then rely on approximate inference methods. Many collective classification algorithms make use of local classifiers. Inference then amounts at iteratively labeling nodes: each iteration takes into account labels predicted at preceding steps. Several such local classifier techniques have been proposed [16,15]. Among them, the Iterative Classification Algorithm (ICA) has received a growing interest in the past years. It has been shown to be more robust and accurate than most alternative solutions, it is simple to understand and scales well *w.r.t.* the size of data, making it possible to label graphs containing thousands to millions of nodes.

Like most local collective classification methods, training and inference for ICA are performed differently. For the former, a local classifier is trained classically, using as inputs some node characteristics and its *correct* neighboring labels. Inference, on the opposite, is an iterative process, where the node labels are repeatedly re-estimated, given the *current estimated* labels of neighbors. Prediction errors on labels may then propagate during the iterations and the classifier will then have difficulties to generalize correctly. This is mainly caused by the bias between training – which assume perfect labels for neighboring nodes – and inference – which may iterate over wrong labels. In this paper, we build on this idea and introduce methods for reducing this training bias. The *Simulated Iterative Classification* (SICA) models proposed here introduce different ways to simulate the inference process during learning. The local classifier being trained on situations close to test ones, this will allow reducing the bias of classical ICA. We present different variants of this SICA algorithm. We also introduce the SICA$^+$ algorithm with *pass-dependent classifiers*, which relies on the idea of using a different local classifier for each iteration of the algorithm.

The underlying idea of this last model is similar to the one developed for the *Stacked Learning* method [8], however it proceeds differently, replacing the cross validation steps used for stacked learning by inference simulation. This family of techniques provides computationally efficient algorithms which outperform many other methods and which can be used for large scale graph labeling.

The contributions of this paper are twofold. Firstly, we propose a new collective classification algorithm. Its inference procedure is similar to ICA. Its learning procedure incorporates *inference simulation* avoiding the training bias of ICA. The SICA model has the same low inference complexity than ICA but provides higher performance. Several variants of the SICA method are introduced and compared. Secondly, we present a set of experiments on seven graph labeling datasets showing the efficiency of our approach.

The paper is structured as follows. Section 2 is an overview of related work. Section 3 defines the graph labeling problem formally and describes ICA. Section 4 introduces our contribution: *Simulated ICA*. We demonstrate the efficiency of the proposed approach with several experiments and comparisons with state-of-the-art models in Section 5. Finally, we conclude and discuss the generality of the proposed approach in Section 6.

2 Related Work

Two main directions have been developed independently for learning to label graphs.

The first one has been proposed for semi-supervised learning and is sometimes called *learning on data manifolds* [20]. The graph here reflects the intrinsic structure of the data and describes local constraints among data points [19]. Graph labeling is formalized as the minimization of an objective function over both labeled and unlabeled nodes. This objective function aims at finding a classifier corresponding to a good balance between an error minimization term defined on the labeled nodes and a local smoothness constraint which imposes close scores for connected nodes. All these methods rely on transductive learning. Different type of models have been developed along this idea:

- Models based on *label propogation* [21,2] only operate on the node labels without considering any other node feature. The labels are propagated iteratively from the labeled nodes to the rest of the graph. The models mainly differ by the regularization function they rely on.
- Models taking into account both the structure of the graph and the node features. Belkin et al. [3] developed a general framework allowing the use of both local node features and weighted links. Other models have been proposed by Zhang et al. [18] for web page classification. They have been adapted by Abernethy et al. [1] for the WebSpam detection task.

The second direction sometimes called collective or relational classification directly attacks the problem of graph labeling. It makes use of different models like ICA, Gibbs sampling, Relaxation Labeling and Loopy Belief Propagation (see

[16,15] for a review and a comparison of these methods). There are two main groups of methods:

- *Local classifier based methods* make use of a local classifier for classifying a node knowing both its content and the labels of its neighbors. For example, the *Iterative Classification* model [16] iteratively uses the base classifier during a fixed number of iterations. Gibbs sampling [9] aims at finding the best set of labels, by sampling each node label iteratively according to the probability distribution given by the local classifier.
- *Global models* try to optimize a global function over the graph. Since this is NP-hard, these methods propose different approximation algorithms in order to reduce the complexity of the optimization. The more popular methods are Loopy Belief Propagation [13] and Relaxation Labeling [10].

Note that methods of this second family are generally used in an inductive setting. They also suffer from the same training label bias as ICA since training and test operate differently. Finally, it is worth emphasizing Stacked Graphical Learning [8,12], a collective classification method developed independently. It makes use of a chain of classifiers, used iteratively to label the nodes. Each classifier uses as input the output of the preceding one. The originality of this algorithm comes from the way it learns the stacked classifiers using a cross-validation based approach. In the context of graph labeling, this algorithm was successfully used for WebSpam classification [6] or for layout document structuring [7].

Graph labeling problem is an instance of the more general framework of *structured prediction* or *structured output classification* [17]. In principle, any general structured prediction methods could be applicable to solve the graph labeling problem. However they have a high computational complexity and are not used for practical applications, especially on a large scale.

3 Supervised Graph Labeling

We use the following notations in the remainder of the paper. A directed graph is a couple $G = (X, E)$ where $X = (x_1, \ldots, x_N)$ is a set of nodes and $E = \{(i,j)\}_{i,j \in [1..N]^2}$ is a set of directed edges. We denote (G, Y) the labeled graph G where $Y = (y_1, \ldots, y_N)$ is the set of labels, with y_i the label corresponding to node x_i. Nodes in X are described through feature vectors $x_i \in \mathbb{R}^d$ where d is the number of features. We consider here the multiclass single-label problem: labels belong to a predefined set of possible labels $\mathcal{L} = (l_1, ..., l_L)$ where L is the number of possible labels[1].

For example, in a document classification task, nodes x_i may be documents described with vectors of word frequencies, edges (i, j) may correspond to citation links and labels y_i may be document categories.

We consider the inductive graph labeling problem. Given a training labeled directed graph (G, Y), the aim is to learn a model that is able to label the nodes

[1] Depending on the communities, labels may also be called *classes* or *categories* in the literature.

Input: A labeled graph (G, Y)
Input: A multiclass learning algorithm $\mathcal{A}$
Output: A base classifier $P(y_i|x_i, \mathcal{N}(x_i))$
foreach $x_i \in X$ **do**
| submit training example $((x_i, \mathcal{N}(x_i)), y_i)$ to $\mathcal{A}$
end
return the classifier trained by $\mathcal{A}$

Algorithm 1. ICA Learning algorithm

of any new graph G'. Note that this framework is rather general, since undirected graph can be seen as particular cases of directed graph and since G may contain multiple disconnected components, *i.e.* multiple different training graphs.

3.1 Iterative Classification

ICA is a graph labeling method based on iterative classification of graph nodes given both their local features and the labels of neighboring nodes. Let $\mathcal{N}(x_i)$ be the set of labels of neighbors of x_i. Typically, neighboring nodes are those directly connected to x_i, i.e:

$$\mathcal{N}(x_i) = \{y_j \mid (i,j) \in E\} \cup \{y_j \mid (j,i) \in E\}$$

ICA relies on the assumption that the probability $P(y_i = l|x_i, G)$ of a label can be approximated by the local probability $P(y_i|x_i, \mathcal{N}(x_i))$, which only depends on the associated content x_i and on the set of neighboring labels. Note that, since the number of neighboring labels $\mathcal{N}(x_i)$ is variable and depends on node x_i, neighbors information needs to be encoded as a fixed-length vector to enable the use of usual classification techniques. This is detailed and illustrated in Section 5.

Learning. The learning procedure of ICA is given in Algorithm 1. $P(y_i|x_i, \mathcal{N}(x_i))$ is estimated by a multiclass classifier. Possible multiclass base classifiers include neural networks, decision trees, naive Bayes or support vector machines. The base classifier is learned thanks to a learning algorithm $\mathcal{A}$ given a labeled training graph (G, Y). Learning the base classifier simply consists in creating one classification example per node in the graph and then running $\mathcal{A}$. In these classification examples, inputs are pairs $(x_i, \mathcal{N}(x_i))$ and outputs are correct labels $y_i \in \mathcal{L}$. Note that both batch learning algorithms and online learning algorithms may be used in conjunction with ICA.

Inference. ICA performs inference in two steps, as illustrated in Algorithm 2. The first step, *bootstrapping*, predicts an initial label for all the nodes of the graph. This step may either be performed by the base classifier – by removing neighboring labels information – or by another classifier trained to predict labels given the local features only. The second step is the *iterative classification* process

Input: A unlabeled graph $G = (V, E)$
Input: A classifier $P(y_i|x_i, \mathcal{N}(x_i))$
Output: The set of predicted labels Y

```
// Bootstrapping
```
foreach $x_i \in X$ **do**
  $y_i \leftarrow \text{argmax}_{l \in \mathcal{L}} P(y_i = l|x_i)$
end

```
// Iterative Classification
```
repeat
  Generate ordering $\mathcal{O}$ over nodes of G.
  foreach $i \in \mathcal{O}$ **do**
    $y_i \leftarrow \text{argmax}_{l \in \mathcal{L}} P(y_i = l|x_i, \mathcal{N}(x_i))$
  end
until *all labels have stabilized or a threshold number of iterations have elapsed*
return Y

Algorithm 2. ICA Inference algorithm

itself, which re-estimates the labels of each node several times, picking them in a random order and using the base classifier. ICA inference may converge exactly; if none of the labels change during an iteration, the algorithm stops. Otherwise, inference is usually stopped after a given number of iterations.

The main advantage of ICA is that both training and inference have linear complexities *w.r.t.* the number of nodes of the graphs. Furthermore, even if it is not guaranteed to converge, ICA has shown to behave well in practice and to give nice performance on a wide range of real-world problems [15].

4 Simulated ICA

When using ICA to infer the labels of a directed graph G, the base classifier is used repeatedly to predict the label of a node given its content and neighboring labels. Since it is very rare to reach perfect classification, the base classifier of ICA often makes some prediction errors during inference. Since prediction errors become inputs for later classification problems, ICA raises an important bias between training and inference; in the former case, neighboring labels are always correct, while they may be noisy in the latter case. This training/inference bias is illustrated in Figure 2.

The training/inference bias raised by ICA corresponds to a general problem that appears as soon as *predictions become inputs for later classification problems.* In the context of collective classification, the authors of Stacked Learning [8] identified the same training/inference bias. Both the Simulated ICA approach detailed below and Stacked Learning are motivated by the same concern: learning with intermediary predictions that are adapted towards the natural bias of the classifier.

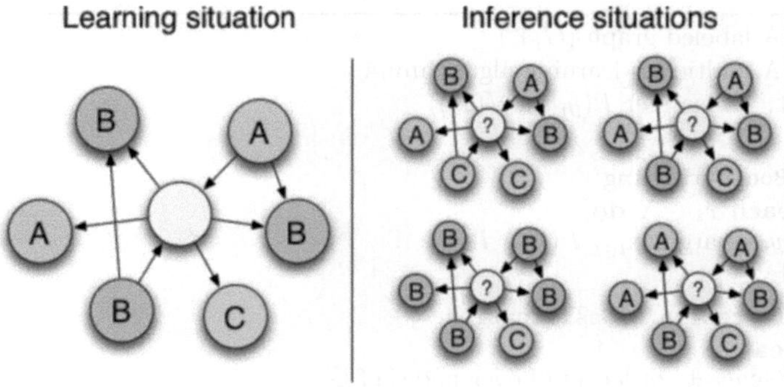

Fig. 2. Illustration of the learning/inference bias of ICA. Left: during training, only perfect neighboring labels are considered. Right: during inference, several neighboring labels may be wrong due to classification errors.

4.1 Learning Algorithm

In order to remove the learning/inference bias of ICA, the base classifier should be trained on situations representative of what happens during inference. In other words, the base classifier should be trained with corrupted neighboring labels. Furthermore, these labels should be biased towards the natural bias of inference: errors that are more frequent during inference should appear more during learning. Simulated ICA (SICA) relies on a simple, but powerful, idea to make training examples representative of inference behavior: *simulation*. We propose to simulate inference during learning, in order to create the training examples representative of the real use of the base classifier.

The general scheme of the Simulated ICA learning procedure is given by Algorithm 3. SICA is an iterative learning algorithm which repeatedly runs ICA inference, using labels which are *sampled* based on the currently learned classifier $P(y_i|x_i, \mathcal{N}(x_i))$. This sampling operation can be performed in different ways, which are detailed in Section 4.2. At each inference step, one training example is submitted to the learning algorithm $\mathcal{A}$, to train the base classifier for predicting the correct label y_i given the node x_i and the current neighboring labels. The key property of training examples in SICA is that they rely on *currently predicted neighboring labels* instead of assuming perfectly labeled neighbors, as it is the case of ICA.

Similarly to classical ICA, both batch and stochastic learning algorithms may be used inside SICA. In our experiments, we used stochastic descent learning algorithm to update the parameters of the base classifier.

4.2 Sampling Methods

When learning with SICA, the aim of simulation is to generate a maximum number of situations that are representative of ICA's inference behavior. SICA therefore relies on a *sampling* operation, whose role is to select the labels y_i used during learning. We propose three sampling strategies:

```
Input: A labeled graph (G, Y)
Input: A multiclass learning algorithm 𝒜
Output: A classifier P(y_i|x_i, 𝒩(x_i))
repeat
    // Bootstrapping
    foreach x_i ∈ X do
        y_i ← argmax_{l∈ℒ} P(y_i = l|x_i)
    end
    // Iterative Classification
    repeat
        Generate ordering 𝒪 over nodes of G.
        foreach i ∈ 𝒪 do
            sample y_i given P(y_i = l|x_i, 𝒩(x_i))
            submit training example ((x_i, 𝒩(x_i)), y_i) to 𝒜
        end
    until iterative classification terminates
until learning has converged or a threshold number of iterations have elapsed
return the classifier trained by 𝒜
```

Algorithm 3. Simulated ICA Learning algorithm

- SICA-DET (Deterministic). The simplest way to perform sampling in SICA consist in selecting the labels y_i that are *predicted* by the current classifier $P(y_i|x_i, \mathcal{N}(x_i))$. Formally, the sampling operation used in SICA-DET is defined as follows:

$$y_i = \operatorname*{argmax}_{l \in \mathcal{L}} P(y_i = l|x_i, \mathcal{N}(x_i))$$

- SICA-UNI (Uniform Noise). In order to increase the range of generated inference situations, one simple variant of SICA called SICA-UNI, consists in introducing stochasticity into the inference process by selecting labels randomly with a small probability. Formally, the sampling operation used in SICA-UNI is defined as follows:

$$y_i = \begin{cases} \text{a random label } l \in \mathcal{L} & \text{with probability } \epsilon \\ \operatorname{argmax}_{l \in \mathcal{L}} P(y_i = l|x_i, \mathcal{N}(x_i)) & \text{with probability } 1 - \epsilon \end{cases}$$

  where $\epsilon \in [0, 1]$ is a parameter controlling the tradeoff between random sampling and SICA-DET style sampling.
- SICA-PROB (Probabilistic). Instead of using uniform noise, a more natural alternative consists in sampling labels from the $P(y_i|x_i, \mathcal{N}(x_i))$ distribution:

$$y_i = \text{sample } l \in \mathcal{L} \text{ from } P(y_i = l|x_i, \mathcal{N}(x_i))$$

The advantages of SICA-PROB and SICA-UNI over SICA-DET are twofold. Firstly, stochasticity enables to generate a wider range of inference situations and thus creates more training examples for the base classifier. This may thus lead to more robust classifiers, especially when using few training nodes. Secondly, stochasticity may contribute to make training conditions closer to test conditions. In this sense, simulating the inference procedure with additional noise is an alternative to the cross-validation approach of Stacked Learning. In both cases, the aim is to learn with intermediary predictions that correctly reflect the behavior of the model on testing data. Stacked Learning creates the intermediary predictions by cross-validation, while we propose to directly modify the predicted labels on training data. We show in Section 5 that SICA-PROB and SICA-UNI frequently outperform Stacked Learning experimentally.

4.3 One Classifier Per Pass

ICA operates in several passes, where each node of the graph is re-estimated during one pass. Since the problem of first estimating the labels may slightly differ from the problem of re-estimating the labels at the second pass or at the third pass, the current pass number may have a direct influence on the base classifier. Instead of learning a unique classifier $P(y_i|x_i, \mathcal{N}(x_i))$, SICA can be modified to take the current pass number, t, into account, by learning a classifier $P(y_i|x_i, \mathcal{N}(x_i), t)$.

SICA-DET, SICA-UNI and SICA-PROB can be modified by learning *one distinct classifier per pass.* This leads to three new variants of SICA that are denoted SICA$^+$-DET, SICA$^+$-UNI and SICA$^+$-PROB in the remainder of this paper. As an example, our experiments use SICA with 5 maximum inference passes, which leads to a set of 5 slightly different classifiers, each one specialized for a given inference pass. Concretely, using one classifier per pass makes it possible to learn finer inference-dependent behavior. This idea is similar to stacking: in both SICA$^+$ and Stacked Learning, there is one classifier per inference pass and each of these classifiers is trained to compensate the errors of previous classifiers.

SICA and SICA$^+$ both perform learning and inference in the same way. The only difference concerns the number of parameters which is multiplied by the number of inference passes by using one disctinct classifier per pass.

5 Experiments

In this section, we describe experimental comparisons between Simulated ICA and various state-of-the-art models for inductive graph labeling.

5.1 Datasets

We performed experiments on four datasets of webpages and on three datasets of scientific articles, whose characteristics are summarized in Table 1. In the former datasets, nodes correspond to webpages, links represent web hyperlinks and the aim is to predict the category of each webpage. In the latter datasets, nodes are scientific articles, links are citation links and the aim is to predict the

Table 1. This table shows different charasteristics (numbers of nodes, links, features and classes) of the seven different datasets used in our experiments. The four first datasets are small scale graphs from WEBKB, the two following ones are medium scale graphs and the last one is our large scale dataset.

Dataset	Nodes	Links	Features	Classes
		Small scale – WEBKB		
CORNELL	195	304	1703	5
TEXAS	187	328	1703	5
WASHINGTON	230	446	1703	5
WISCONSIN	265	530	1703	5
		Medium scale		
CITESEER	3312	4715	3703	6
CORA-I	2708	5429	1433	7
		Large scale		
CORA-II	36954	136024	11816	11

category of each article. All nodes are described by feature vectors, with features that correspond to the most frequent words and that indicate whether or not the associated words appear. CORA-II was introduced by A. McCallum and was preprocessed by keeping only the words appearing in at least 10 documents[2]. The other datasets were preprocessed by Getoor et al[3].

5.2 Experimental Setup

State-of-the-art models. In order to evaluate Simulated ICA, we have implemented three state-of-the-art graph labeling models: Iterative Classification (ICA), Gibbs Sampling (GS) and Stacked Learning. Our implementation of Stacked Learning uses 5-fold cross-validation during learning to create the intermediary predictions. STACK2 is the simplest Stacked Learning model, where the labels are first estimated based on the content only and are then re-estimated by taking both the initial predictions and the content into account. STACK3 is a Stacked Learning model with three stacks: labels are first estimated using STACK2 and are then re-estimated with a classifier that uses both the content and the predictions of STACK2. Similarly, STACK4 and STACK5 are Stacked Learning models that respectively rely on STACK3 and STACK4 to provide intermediary predictions.

Baselines. We have also compared Simulated ICA with two baselines named *Content-Only* (CO) and *Optimistic* (OPT). The former is a classifier ignoring the graph structure and taking only the content of nodes into account during classification. The latter is a classifier which assumes the availability of perfect neighboring labels for each node, during both *training and inference*. Note that

[2] CORA-II is available at `http://www.cs.umass.edu/~mccallum/code-data.html`

[3] Datasets available at `http://www.cs.umd.edu/~sen/lbc-proj/LBC.html`, see [14] for more details.

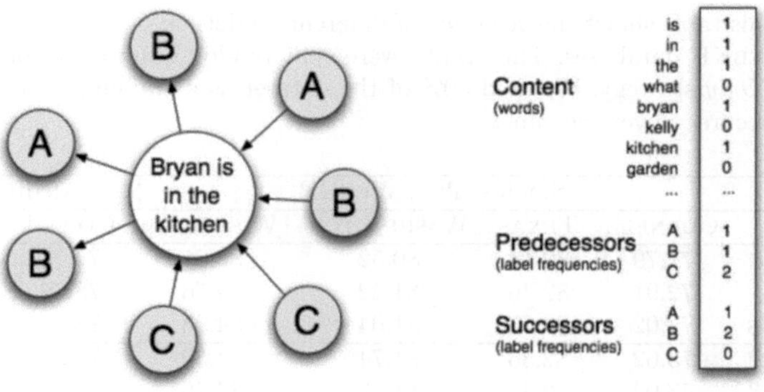

Fig. 3. A feature function to jointly describe content of nodes and neighboring labels. Feature vectors contain three parts. The first one describes the content of the node and the other two describe the labels of the node neighbors. For each label, there is a feature which counts the number of predecessor (*resp.* successors) with this label.

due to this dependency, the OPT baseline is not a "true model" able to generalize to new unlabeled graphs.

Base classifier. In order to make the comparison fair, we used the *same base classifier*, the *same learning algorithm* and the *same tuning procedure* for all models. The base classifiers are L2-regularized maximum-entropy classifiers [4] learned with stochastic gradient descent. For each model, we tried ten different regularizer values and kept the best-performing ones.

In order to take simultaneously the content of nodes and their neighboring labels, the base classifiers rely on a feature function that jointly maps contents and associated neighboring labels to scalar vectors. The feature function used in our experiments is illustrated in Figure 3.

Data splitting. In order to evaluate the generalization abilities of the models, we have split each dataset into one training graph and one testing graph. As [16], we have used a random sampling strategy: both the training and the testing graphs are random subsets of the whole graphs. When splitting graphs randomly, it is often the case that some links connect train nodes to test nodes. The simplest way to deal with these *train-test links* is simply to ignore them; this approach is called *Test Links Only*. Since many links may be discarded with the *Test Links Only* approach, we have also adopted the alternative approach proposed by [14] called *Complete Links* in which no link is suppressed and in which we assume, during inference, to know the correct labels of training nodes connected to testing nodes. For the purpose of comparison with [14], we used 10-folds cross-validation with the *Complete Links* strategy. Note that, if we used 10-folds with *Test Links Only*, 90% of the links involving testing nodes would be discarded. In order to make the average number of links connected to testing nodes and to training nodes equal, we used *Test Links Only* with two-fold cross-validation.

Table 2. This table shows the accuracy of different models on CORA-I, CITESEER and the four WEBKB databases. The graphs were split randomly into two folds by using *Test Links Only* strategy. We used 50% of the dataset as a training set and took the mean test accuracy over ten runs.

	Small scale – WEBKB				Medium scale	
	CORNELL	TEXAS	WASHINGTON	WISCONSIN	CORA-I	CITESEER
CO	71.79	82.14	80.52	85.21	74.73	71.73
ICA	72.91	82.46	81.22	84.76	78.69	73.14
GS	73.02	82.67	81.04	84.98	78.67	72.8
STACK2	73.02	82.35	81.74	84.38	78.85	73.25
STACK3	73.02	82.46	81.13	84.76	79.73	73.34
STACK4	73.02	82.67	81.3	84.83	79.53	73.25
STACK5	72.91	82.35	81.13	84.76	79.81	73.22
SICA-DET	73.53	82.24	81.83	85.21	78.77	72.92
SICA-UNI	**74.24**	83.21	82.35	85.36	79.29	73.21
SICA-PROB	74.04	**83.42**	**82.43**	**85.74**	79.32	73.47
SICA$^+$-DET	71.48	80.32	80	83.32	79.18	73.59
SICA$^+$-UNI	72.71	82.35	81.13	84.45	79.59	73.70
SICA$^+$-PROB	73.42	82.35	81.91	85.06	**80.01**	**74.02**
OPT	72.5	82.78	82.17	84.46	83.16	76.18

5.3 Results

In our experiments, the parameters was the same as these in [16] for the baselines (CO, OPT, ICA, GS and STACK): the maximum number of training iterations was set to 100, for Iterative Classification we used a maximum of 100 inference passes and for Gibbs Sampling we performed 1000 samples of each node label. For the SICA inference, we used a maximum of 5 passes. Moreover, we fixed the uniform noise percentage in SICA-UNI and SICA$^+$-UNI to 10%.

Comparisons with state-of-the-art models. We compared the six variants of SICA described in Section 4 with the baselines described previously on the small and medium scale datasets. Firstly, we have split each dataset into two halves by using the *Test Links Only* strategy. Each experiment was performed 10 times with different random splits and we report the mean test accuracies in Table 2. As expected, SICA models outperform ICA on all datasets. More interestingly, our models also outperform Gibbs Sampling and Stacked Learning in nearly all cases. In particular, the SICA-UNI and SICA-PROB models that rely on stochasticity outperform Stacked Learning on five datasets out of six, whereas this tendency is less clear for SICA-DET. As discussed in Section 4, adding a perturbation to the predicted labels proves to be a good alternative to the heavy cross-validation based prediction process adopted by Stacked Learning. Among the two perturbation approaches, SICA-PROB most-of-time reaches better results than SICA-UNI. A deeper comparison of these sampling approaches in given below. Using one classifier per pass (SICA$^+$-DET, SICA$^+$-UNI and SICA$^+$-PROB)

Table 3. This table shows the accuracy of different models on CORA-I, CITESEER and the four WEBKB databases. Here, the graphs were split randomly into ten folds by using *Complete Links* strategy. Then, we did a 10-folds cross-validation and took the mean test accuracy.

	Small scale – WEBKB				Medium scale	
	CORNELL	TEXAS	WASHINGTON	WISCONSIN	CORA-I	CITESEER
CO	79.5	86.64	84.35	89.4	77.43	72.98
ICA	79.47	87.22	85.65	89.79	88.52	77.63
GS	80.5	87.22	85.22	89.79	88.18	77.47
STACK2	78.92	**89.91**	87.39	89.42	88.07	76.72
STACK3	78.42	88.27	**88.26**	89.03	88.18	77.35
STACK4	78.97	88.3	86.96	89.8	88.4	77.23
STACK5	78.45	88.83	87.83	89.42	88.4	77.08
SICA-DET	**81.55**	87.75	85.65	89.79	88.37	76.27
SICA-UNI	**81.55**	88.27	85.65	89.79	88.26	76.48
SICA-PROB	81.53	87.75	86.52	**90.16**	88.37	76.33
SICA$^+$-DET	79.5	86.7	84.78	89.42	**88.74**	77.75
SICA$^+$-UNI	80.03	86.70	86.52	89.42	88.63	77.93
SICA$^+$-PROB	81.05	87.22	85.65	89.79	88.66	**78.02**
OPT	79.97	87.75	86.09	89.77	88.85	78.08

improves over the basic versions of SICA on the two medium scale datasets (+ 0.69% on CORA-I and + 0.55% on CITESEER). On the small datasets, using one classifier per pass slightly deteriorates the results. We believe that this is due to estimation problems related to the large number of parameters on these approaches.

The second set of experiments aims at comparing our results to those of [16]. Here, each dataset was split into ten folds by using the *Complete Links* strategy. Table 3 gives the 10-fold cross-validation accuracies (90% training nodes and 10% testing nodes) for all models and all datasets. As previously, our models most-of-the-time outperform ICA and GS. On small datasets, Stacked Learning is competitive with our models. However, these results should be taken with a grain of salt, since, when using 90% training nodes with *Complete Links*, a large majority of testing-node neighbors are in the training set. Consequently, the various methods that take wrong labels into account during learning (SICA and Stacked Learning) have a more limited interest in this case.

Impact of the uniform noise percentage. Next, we evaluated the impact of uniform noise percentage on SICA-UNI. Figure 4 compares the three sampling methods on our two medium scale datasets. With a reasonable noise percentage (below 40%), the accuracy of SICA-UNI is between that of SICA-DET and that of SICA-PROB. Once more, this confirms the contribution of stochasticity in the simulation process of SICA.

In most cases, SICA-PROB outperforms SICA-UNI. SICA-PROB has another key advantage over the latter: it does not rely on additional hyper-parameter,

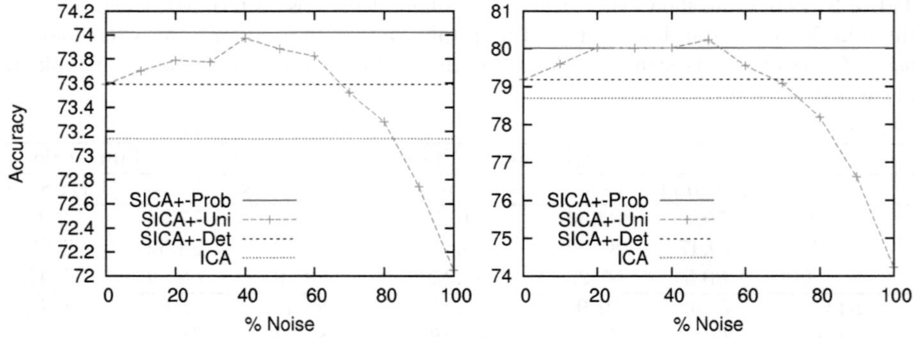

Fig. 4. Accuracy for varying percentage of uniform noise in SICA-UNI on CITESEER (on the left) and CORA-I (on the right)

Table 4. Test accuracies and training/inference time on CORA-II. The first two columns give the mean test accuracy of the models with the *Test Links Only* and *Complete Links* splitting strategies. The last two columns show approximate training time and inference time. The results for STACK5 are incomplete due to the excessive time requirement of this method (since we average our results over 10 runs and try 10 different regularizer values, the experiment would need more than three months to complete).

	Test Links Only	Complete Links	Training time	Inference time
CO	49	49	2 min	300 ms
ICA	51.9	58.05	6 min	20 s
GS	44.43	55.42	6 min	10 min
STACK2	54.46	56.28	13 min	1 s
STACK3	56.07	57.99	1 h	1.5 s
STACK4	**56.67**	58.52	4.5 h	2 s
STACK5	-	-	20 h	2.5 s
SICA-DET	56.02	58.52	5 min	4 s
SICA-UNI	56.20	**59.57**	3 min	4 s
SICA-PROB	56.3	59.14	3 min	4 s
SICA$^+$-DET	55.5	58.87	5 min	4 s
SICA$^+$-UNI	56.25	59.56	3 min	4 s
SICA$^+$-PROB	56.2	58.91	2 min	4 s
OPT	66.40	67.71	3.5 min	600 ms

which makes its tuning much easier. With a too high noise percentage, information in neighboring labels becomes irrelevant and thus accuracy drops down to the one of a *Content-Only* classifier.

Large scale. In order to show that our model can deal with large-scale graphs, we performed a set of experiments on the CORA-II database. For each model, we performed 10 runs with 20% training nodes and 80% training nodes selected randomly. The mean test accuracies are reported in Table 4. The first two columns give scores respectively for *Test Links Only* and *Complete Links*. The last two columns give the CPU time needed to train a single model and the time needed to fully label the test graph[4]. Experiments were performed on standard 3.2 Ghz computer.

Our approaches clearly outperform ICA and GS on CORA-II (up to +3% improvement) and behave similarly to Stacked Learning with a much lower training time. Indeed, since each stack involves making 5 folds and learning a model on each sub-fold recursively, STACK models have a training time which is exponential *w.r.t.* the number of stacks. Instead, all our models – that use 5 inference passes – were learned in a few minutes.

6 Conclusion

In this paper, we have introduced the Simulated Iterative Classification Algorithm (SICA), a new learning algorithm for ICA. The core idea of SICA is to simulate ICA's inference during learning. We argued that simulation is a simple and efficient way to create training examples that are representative of *real inference situations*. We have shown that the proposed approach outperforms state-of-the-art models (Iterative Classification, Gibbs Sampling and Stacked Learning) on a wide range of datasets. Furthermore, we have shown that the model scales well, which makes it possible to label graphs containing thousands to millions of nodes.

Our future work will primarily focus on generalizing the idea of simulation during learning to semi-supervised graph labeling problems. We believe that one promising approach is to develop (fast and scalable) incremental inference algorithms, that takes both the labeled and the unlabeled nodes into account, and to learn them using simulation.

One key characteristic of ICA is that it relies on a classifier whose inputs depend on its previous predictions. Although this paper is focused on supervised graph labeling problems, we believe that the proposed idea of simulating inference during learning is relevant to a wider class of problems where predictions become inputs for later classification problems. In order to tackle error propagation problems in such algorithms involving classifier chains, simulation is a key solution, which appears to be both simple and efficient.

Acknowledgement

We would like to thank the French National Research Agency (ANR) for financial support under the project *MADSPAM 2.0*.

[4] Note that SICA models were carefully optimized in our implementation, which explains their relative low training times.

References

1. Abernethy, J., Chapelle, O., Castillo, C.: Witch: A new approach to web spam detection. Technical report, Yahoo! Research (2008)
2. Agarwal, S.: Ranking on graph data. In: ICML 2006, pp. 25–32. ACM, New York (2006)
3. Belkin, M., Niyogi, P., Sindhwani, V.: Manifold regularization: A geometric framework for learning from labeled and unlabeled examples. Journal of Machine Learning Research 7, 2399–2434 (2006)
4. Berger, A.L., Pietra, S.D., Della Pietra, V.J.: A maximum entropy approach to natural language processing. Computational Linguistics 22(1), 39–71 (1996)
5. Castillo, C., Davison, B.D., Denoyer, L., Gallinari, P. (eds.): Proceedings of the Graph Labelling Workshop and Web Spam Challenge (2007)
6. Castillo, C., Donato, D., Gionis, A., Murdock, V., Silvestri, F.: Know your neighbors: web spam detection using the web topology. In: SIGIR 2007, pp. 423–430. ACM, New York (2007)
7. Chidlovskii, B., Lecerf, L.: Stacked dependency networks for layout document structuring. In: SAC, pp. 424–428 (2008)
8. Cohen, W.W., de Carvalho, V.R.: Stacked sequential learning. In: IJCAI, pp. 671–676 (2005)
9. Hastings, W.K.: Monte carlo sampling methods using markov chains and their applications. Biometrika 57(1), 97–109 (1970)
10. Hummel, R.A., Zucker, S.W.: On the foundations of relaxation labeling processes, pp. 585–605 (1987)
11. Jensen, D., Neville, J., Gallagher, B.: Why collective inference improves relational classification. In: ACM SIGKDD 2004, pp. 593–598. ACM, New York (2004)
12. Kou, Z., Cohen, W.W.: Stacked graphical models for efficient inference in markov random fields. In: SDM (2007)
13. Kschischang, F.R., Frey, B.J.: Iterative decoding of compound codes by probability propagation in graphical models. IEEE Journal on Selected Areas in Communications 16, 219–230 (1998)
14. Lu, Q., Getoor, L.: Link-based classification using labeled and unlabeled data. In: ICML: Workshop from Labeled to Unlabeled Data (2003)
15. Macskassy, S.A., Provost, F.: Classification in networked data: A toolkit and a univariate case study. J. Mach. Learn. Res. 8, 935–983 (2007)
16. Sen, P., Namata, G.M., Bilgic, M., Getoor, L., Gallagher, B., Eliassi-Rad, T.: Collective classification in network data. Technical Report CS-TR-4905, University of Maryland, College Park (2008)
17. Taskar, B., Chatalbashev, V., Koller, D., Guestrin, C.: Learning structured prediction models: A large margin approach. In: ICML 2005, Bonn, Germany (2005)
18. Zhang, T., Popescul, A., Dom, B.: Linear prediction models with graph regularization for web-page categorization. In: KDD 2006: Proceedings of the 12th ACM SIGKDD, pp. 821–826. ACM, New York (2006)
19. Zhou, D., Schölkopf, B.: Regularization on discrete spaces. In: Kropatsch, W.G., Sablatnig, R., Hanbury, A. (eds.) DAGM 2005. LNCS, vol. 3663, pp. 361–368. Springer, Heidelberg (2005)
20. Zhou, D., Schölkopf, B., Hofmann, T.: Semi-supervised learning on directed graphs. In: NIPS, pp. 1633–1640. MIT Press, Cambridge (2005)
21. Zhu, X., Ghahramani, Z.: Learning from labeled and unlabeled data with label propagation. Technical report (2002)

Graph-Based Discrete Differential Geometry for Critical Instance Filtering

Elena Marchiori

Department of Computer Science, Radboud University, Nijmegen, The Netherlands

Abstract. Graph theory has been shown to provide a powerful tool for representing and tackling machine learning problems, such as clustering, semi-supervised learning, and feature ranking. This paper proposes a graph-based discrete differential operator for detecting and eliminating competence-critical instances and class label noise from a training set in order to improve classification performance. Results of extensive experiments on artificial and real-life classification problems substantiate the effectiveness of the proposed approach.

1 Introduction

In graph-based data analysis, a dataset is represented as a graph, where the vertices are the instances of the dataset and the edges encode a pairwise relationship between instances. For instance, the nearest neighbor relation between points of a finite set in the Euclidean space can be described by the popular nearest neighbor (proximity) graph [3,27]. Concepts and methods from graph theory are then used for extracting knowledge from such a representation. In particular, the graph Laplacian provides a natural interpretation to the geometric structure of datasets. It has been used in machine learning for tackling diverse tasks such as dimensionality reduction and clustering, e.g., [4,29], feature selection, e.g., [19,34], and semi-supervised learning, e.g., [35,36].

This paper shows how the graph Laplacian operator can be directly used for filtering competence-critical instances and class label noise from a training set, in order to improve test accuracy.

Research on instance selection focusses mainly on three types of filtering techniques [8]: competence preservation, competence enhancement, and hybrid approaches. *Competence preservation* algorithms, e.g., [1,14], remove irrelevant points, that is, that do not affect the classification accuracy of the training set. *Competence enhancement* methods, e.g., [23,26,28,31], remove noisy points, such as those with a wrong class label, as well as points close to the decision boundary, yielding to smoother decision boundaries, in order to increase classifier accuracy. *Hybrid* methods, e.g., [8,20,24,25,32], aim at finding a subset of the training set that is both noise free and does not contain irrelevant points. Alternative methods use prototypes instead of instances of the training set, see for instance [21].

W. Buntine et al. (Eds.): ECML PKDD 2009, Part II, LNAI 5782, pp. 63–78, 2009.

The algorithm proposed here belongs to the so-called *competence enhancement* methods. It differs from previous methods for this task in the way it extracts information from the neighborhood of an instance in order to measure its relevance. Indeed, while previous methods are based on 'static' measures, such as being correctly classified, the proposed method uses a 'differential' measure, defined by means of a graph Laplacian operator. Specifically, we consider the graph-based representation of two class dependent K nearest neighbor (KNN) relations defined over pairs of instances: the within- and between- class KNN. The between-class KNN graph is used to define a graph Laplacian operator. The within-class KNN graph is used to define the within-class degree function, mapping vertices to their degree.

The application of the Laplacian operator to such function, called *Laplace scoring*, provides such differential measure of instance relevance. It measures the flow of the within-class degree function at each vertex of the between-class KNN graph. Vertices with negative Laplace score are either close to the KNN decision boundary or are outliers. This motivates the introduction of a simple Laplace-based instance filtering algorithm, which removes instances having negative Laplace score.

To the best of our knowledge, this work presents the first attempt to perform class noise instance filtering using a graph-based differential approach.

In order to test comparatively the effectiveness of this approach, extensive experiments on artificial and real-life data sets are conducted. We consider three classifiers: the KNN classifier without instance filtering, with the popular Wilson's editing [31], and with Laplace filtering. Results of the experiments indicate best test accuracy performance of Laplace filtering over the other methods, as well as superior robustness with respect to the presence of class noise in the training set.

Furthermore, comparison of Laplacian filtering with state-of-the-art editing algorithms indicate similar or improved generalization performance of the `1-NN`.

Finally, we investigate the use of Laplacian filtering for improving the performance of classifiers other than `1-NN`. We consider SVMs with RBF kernels. These are related to NN methods, because each RBF measures the distance of a test instance to one of the training instances. SVM training keeps certain training instances as support vectors, and discards others. In this way, SVM/RBF may also be viewed as a competence-enhancement filtering method. We investigate the effect of Laplacian filtering as pre-processing step by performing experiments on datasets with different levels of added class noise. Results of these experiments show that at all considered levels of class noise, Laplacian filtering has no significant positive effect on the generalization performance of SVMs with RBF kernel.

In general, the results substantiate the effectiveness of graph Laplacian operators for tackling the class noise filtering problem. Therefore this contribution adds yet another successful application of such powerful graph-based framework in machine learning.

We begin by setting the stage with the notation and main concepts used in this study.

2 Background

In this paper we use X to denote a dataset of n instances $X = \{x_1, \ldots, x_n\}$, where x_i is a real-valued vector of dimension m. Let C denote the set of class labels of X, and let $l : X \to C$ the function mapping each instance x_i to its class label $l(x_i)$.

A graph $G = (V, E)$ consists of a finite set V and a subset $E \subset V \times V$. The elements of V are the *vertices* of the graph and those of E are the edges of the graph. In this work we consider undirected graphs, that is, such that for each edge $(u, v) \in E$ we have also $(v, u) \in E$. We say that u and v are *adjacent* vertices, denoted by $u \sim v$, if $(u, v) \in E$. The degree function d of a graph is defined by $d(u) = |\{v \mid u \sim v\}|$, where $|S|$ denotes the cardinality of a set S.

The graph normalized Laplacian can be defined as follows. Suppose $|V| = n$. Consider the $n \times n$ matrix L, defined as

$$L(u,v) = \begin{cases} d(u) & \text{if } u = v, \\ -1 & \text{if } (u,v) \in E, \\ 0 & \text{otherwise.} \end{cases}$$

The (normalized) Laplacian of G is the $n \times n$ matrix

$$\mathcal{L}(u,v) = \begin{cases} 1 & \text{if } u = v \text{ and } d(u) \neq 0, \\ \frac{-1}{\sqrt{d(u)d(v)}} & \text{if } (u,v) \in E, \\ 0 & \text{otherwise.} \end{cases}$$

The graph Laplacian operator $\mathcal{L}$ maps real-valued functions on vertices to real-valued functions on vertices, defined by

$$\mathcal{L}(f)(u) = \frac{1}{\sqrt{d(u)}} \sum_{u \sim v} \left(\frac{f(u)}{\sqrt{d(u)}} - \frac{f(v)}{\sqrt{d(v)}} \right).$$

3 Laplacian Instance Filtering

Denote by l a generic element of C. Let X_l be the subset of X consisting of those instances having class label equal to l. Define $KNN(x, l)$ to be the set of K nearest neighbors of x computed among those instance in X_l, excluding x.

Define the following two graphs. The *within-class KNN graph*, denoted by $G_{wc} = (V, E_{wc})$, such that $V = X$, and

$$E_{wc} = \{(x_i, x_j) \mid x_j \in \texttt{KNN}(x_i, l(x_i)) \text{ or } x_i \in \texttt{KNN}(x_j, l(x_j))\}.$$

G_{wc} represents the (symmetric) nearest neighbor relation between points of the same class in the training set.

Analogously, define the *between-class KNN graph*, denoted by $G_{bc} = (V, E_{bc})$, such that $V = X$,

$$E_{bc} = \{(x_i, x_j) \mid (x_j \in \texttt{KNN}(x_i, l) \text{ and } l \neq l(x_i)) \text{ or } (x_i \in \texttt{KNN}(x_j, l) \text{ and } l \neq l(x_j))\}.$$

G_{bc} represents the (symmetric) KNN relation between points of each pair of different classes in the training set. Note that this relation differs from the nearest unlike neighbor relation (NUN) [15] because it considers all pairs of different classes, while NUN considers one class versus the union of all the other classes. Clearly, for binary classification problems the two relations coincide.

The within- and between-class graphs with $K = 1$ for a toy binary classification problem are shown in Figure 1.

Let $\mathcal{L}$ be the Laplacian operator of G_{bc}. Let g denote the within-class degree function, mapping vertex i to its degree in G_{wc}. $g(i)$ can be viewed as an estimate of the density of points of class label $l(x_i)$ around x_i, since the more instances with label $l(x_i)$ are close to x_i, the larger the $g(i)$ will be [34].

We define the *Laplace score*, denoted by $Score$, to be the real-valued function on vertices such that

$$Score(u) = \mathcal{L}(g)(u).$$

This function assigns a small score to an instance whose neighbors from different classes (that is, its adjacent vertices in G_{bc}) are in a region containing many points of their own class, and few points of classes different from their one.

The within-class degree and Laplace score of instances for the considered toy classification example are shown in Figure 1. Observe that in this example points with negative Laplace score are close to the one nearest neighbor class decision boundary. This motivates the introduction of the simple Algorithm 1 for class noise filtering, which removes from the training set those instances with negative Laplace score.

The time complexity of this algorithm is dominated by the cost of building the between- and within-class graphs, which is quadratic in the size of the training set. However, this bound can be reduced to $O(n \log(n))$ (for small input dimension) by using metric trees or other spatial data structures, as shown for

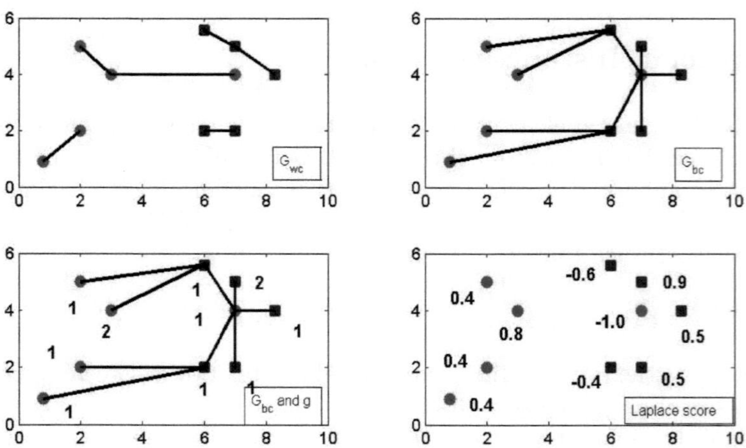

Fig. 1. Graphs and Laplace score with $K = 1$ of a training set for a toy binary classification problem in the real plane

Algorithm 1. Laplace instance filtering

Input: training data X of size n
number K of nearest neighbors
Output: subset S of X
G_{bc} = between-class KNN graph of X
G_{wc} = within-class KNN graph of X
g = degree function of G_{wc}
for $i = 1$ **to** n **do**
  $Score(i) = \mathcal{L}(g)(i)$
end for
$S = \{x_i \in X \mid Score(i) \geq 0\}$

example in [18], as well as structures optimized for a data and query distribution [5,9].

Application of Laplace instance filtering to an instance of the XOR classification problem is illustrated in Figure 2. Points filtered out by the algorithm are highlighted with circles.

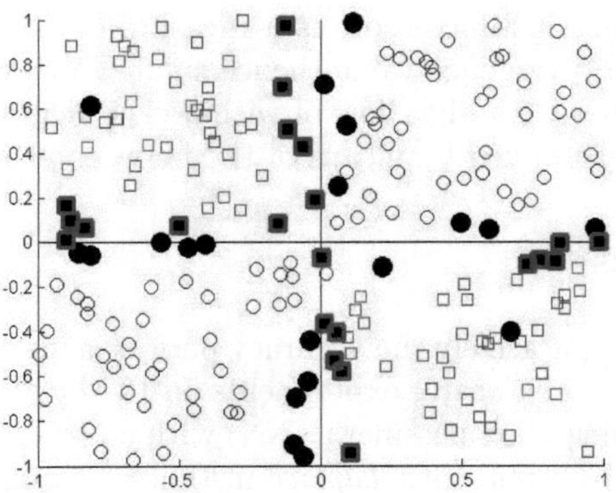

Fig. 2. Application of Laplace filtering with $K = 1$ to an instance, with class noise examples, of the XOR classification problem in the real plane. The points removed have filled markings.

3.1 Justification

The score of an instance can be interpreted as a discrete divergence measure. This can be shown by using discrete analogous of differential operators. To this end, we use the results contained in [36].

Indeed, consider the graph gradient operator, mapping real-valued functions of the vertices into real-valued functions on edges.

$$(\nabla\phi)(u, v) = \frac{\phi(v)}{d(v)} - \frac{\phi(u)}{d(u)}.$$

Observe that in this definition, before computing the variation of ϕ between two adjacent vertices, the function value is split at each vertex along its adjacent edges.

The graph gradient can also be defined at vertex v, as

$$\nabla\phi(u) = \{(\nabla\phi)(u,v) \mid (u,v) \in E\}.$$

The graph divergence maps real-valued functions of the edges into real-valued functions on vertices.

$$(\texttt{div}\ \psi)(u) = \sum_{u\sim v} \frac{1}{\sqrt{d(v)}}(\psi(u,v) - \psi(v,u)).$$

The divergence measures the net outflow of function ψ at each vertex.

The following equality relates the graph divergence and Laplacian operators:

$$\mathcal{L}(\phi) = -\frac{1}{2}\texttt{div}\ (\nabla\phi).$$

By instantiating the above formula with the graph Laplacian of G_{bc} and the within-class degree function g we obtain

$$Score = -\frac{1}{2}\texttt{div}\ (\nabla g).$$

Therefore, the Laplace instance score is a measure of negative divergence. An instance having high divergence value (hence small Score value) can be considered critical, since there is a high flow of within-class degree at that instance in a neighborhood characterized by means of the between-class graph.

4 Experiments

In order to assess comparatively the accuracy performance of the proposed filtering method, we conduct extensive experiments on 19 Machine Learning datasets, using the K-nearest neighbor classifier (KNN) with no training set pre-processing [13], here called *No-filtering*, with *Laplace* instance filtering, and with the popular *Wilson*'s filtering algorithm. The latter one removes those instances that do not agree with the majority of its K nearest neighbors. We consider three instances of each algorithm obtained by setting the number K of neighbors to $1, 3, 5$, respectively, resulting in a total of nine classifiers.

Cross validation is applied to each dataset. Specifically, for each partition of the dataset, each filtering algorithm is applied to the training set X from which a subset S is returned. The KNN classifier that uses only points of S is applied to the test set.

4.1 Datasets

We consider 3 artificial datasets (Banana, g50c, g10n) and 16 real-life ones, with different characteristics as shown in Table 1. These datasets have been used in previous studies on model selection for (semi)supervised learning.

Specifically, Raetsch's binary classification benchmark datasets have been used in [22]: they consists of 1 artificial and 12 real-life datasets from the UCI,

Table 1. Datasets used in the experiments. CL = number of classes, TR = training set, TE = test set, VA = number of variables, Cl.Inst. = number of instances in each class.

DATASET	CL	VA	TR	CL.INST.	TE	CL.INST.
1 BANANA	2	2	400	212-188	4900	2712-2188
2 B.CANCER	2	9	200	140-60	77	56-21
3 DIABETES	2	8	468	300-168	300	200-100
4 GERMAN	2	20	700	478-222	300	222-78
5 HEART	2	13	170	93-77	100	57-43
6 IMAGE	2	18	1300	560-740	1010	430-580
7 RINGNORM	2	20	400	196-204	7000	3540-3460
8 F.SOLAR	2	9	666	293-373	400	184-216
9 SPLICE	2	60	1000	525-475	2175	1123-1052
10 THYROID	2	5	140	97-43	75	53-22
11 TITANIC	2	3	150	104-46	2051	1386-66
12 TWONORM	2	20	400	186-214	7000	3511-3489
13 WAVEFORM	2	21	400	279-121	4600	3074-1526
14 IRIS	3	4	120	40-40-40	30	10-10-10
15 BREAST-W	2	9	546	353-193	137	91-46
16 BUPA	2	6	276	119-157	69	26-43
17 PIMA	2	8	615	398-217	153	102-51
18 G50	2	50	550	252-248	50	23-27
19 G10N	2	10	550	245-255	50	29-21

DELVE and STATLOG benchmark repositories. For each experiment, the 100 (20 for `Splice` and `Image`) partitions of each dataset into training and test set available in the repository are used here.

Two artificial binary classification problems from Chapelle's benchmark datasets [11], `g50c` and `g10n`, are generated from two standard normal multivariate Gaussians. In `g50c`, the labels correspond to the Gaussians, and the means are located in a 50-dimensional space such that the Bayes' error is 5%. In contrast, `g10n` is a deterministic problem in 10 dimensions, where the decision function traverses the centers of the Gaussians, and depends on only two of the input dimensions. For each experiment, the 10 partitions of each dataset into training and test set available in the repository are used.

Finally, four standard benchmark datasets from the UCI Machine Learning repository are used: `Iris`, `Bupa`, `Pima`, and `Breast-W`. For each experiment, 100 partitions of each dataset into training and test set are used. Each partition randomly divides the dataset into training and test set, equal to 80% and 20% of the data, respectively.

4.2 Results

Results of the experiments are summarized in Table 2 (also plotted in the first row of Figure 3). The table contains average accuracy results of the

Table 2. A(M) = average (median) results over datasets. S = +/- number of times Laplace average accuracy is significantly better (+) or significantly worse (-) than the other algorithm, according to a paired t-test at 0.01 significance level. W = a '+' indicates Laplace significantly better than the other algorithm at a 0.01 significance level according to a Wilcoxon test for paired samples.

	Laplace			Wilson			No-Filtering		
	1	3	5	1	3	5	1	3	5
1 Banana	88.5	88.9	88.7	87.8	88.2	88.3	86.4	87.9	88.3
2 B.Cancer	70.6	73.0	74.2	69.4	73.4	73.6	67.3	68.5	71.2
3 Diabetes	73.9	74.0	73.8	72.7	73.2	73.6	69.9	72.4	72.6
4 German	74.0	74.0	73.1	73.0	73.9	73.9	70.5	73.1	74.1
5 Heart	81.6	82.6	82.9	80.6	82.0	82.7	76.8	80.5	81.8
6 Image	94.9	92.3	90.8	95.8	94.6	94.1	96.6	95.7	95.1
7 Ringnorm	67.3	63.2	61.0	54.8	51.2	50.6	64.9	59.5	56.7
8 F.Solar	64.0	64.6	64.7	61.4	62.8	62.7	60.7	62.3	62.2
9 Splice	73.3	76.4	77.1	68.4	68.2	66.7	71.1	72.6	73.3
10 Thyroid	94.3	91.9	89.5	94.0	91.9	89.5	95.6	93.8	92.6
11 Titanic	77.2	77.0	77.2	67.3	72.5	74.5	66.9	72.3	74.0
12 Twonorm	95.5	96.6	96.9	94.1	95.9	96.4	93.3	95.5	96.2
13 Waveform	86.2	87.4	87.3	85.4	86.9	87.5	84.1	86.3	87.3
14 Iris	95.2	95.1	94.8	96.1	95.5	95.8	95.6	95.1	95.8
15 Breast-W	97.1	97.3	97.1	96.9	97.2	96.9	96.2	97.1	97.4
16 Bupa	65.8	69.2	68.4	63.5	67.0	67.4	61.2	64.3	66.5
17 Pima	72.5	74.2	75.0	69.6	73.1	73.8	67.3	69.9	72.0
18 g50	85.6	89.8	91.2	82.2	87.2	92.4	79.6	88.4	92.0
19 g10	74.6	79.0	80.8	74.0	79.2	80.0	75.2	78.4	78.2
A	80.6	81.4	81.3	78.3	79.7	80.0	77.9	79.7	80.4
M	77.3	79.0	80.8	74.0	79.2	80.0	75.2	78.4	78.2
S	N/A	N/A	N/A	14/2	15/3	11/2	13/2	7/3	11/5
W	N/A	N/A	N/A	+	+	+	+	+	+

algorithms on each classification task, their average and median, the outcome of a paired t-test on the results of each classification task, and the outcome of a paired Wilcoxon test on the (average) results of the entire set of classification tasks.

Results of a paired t-test at a 0.01 significance level shows improved accuracy performance of Laplace (see row 'S' in Table 2) on the majority of the datasets. Application of the non parametric Wilcoxon test for paired samples at a 0.01 significance level to the average results on the entire set of classification tasks, indicates that KNN with Laplace filtering outperforms the other algorithms.

In summary, the experimental analysis indicates effectiveness of Laplace-based instance filtering and robustness with respect to the presence of high number of variables, training examples, noise and irrelevant variables.

We turn now to the experimental analysis of classifier robustness with respect to the presence of class noise.

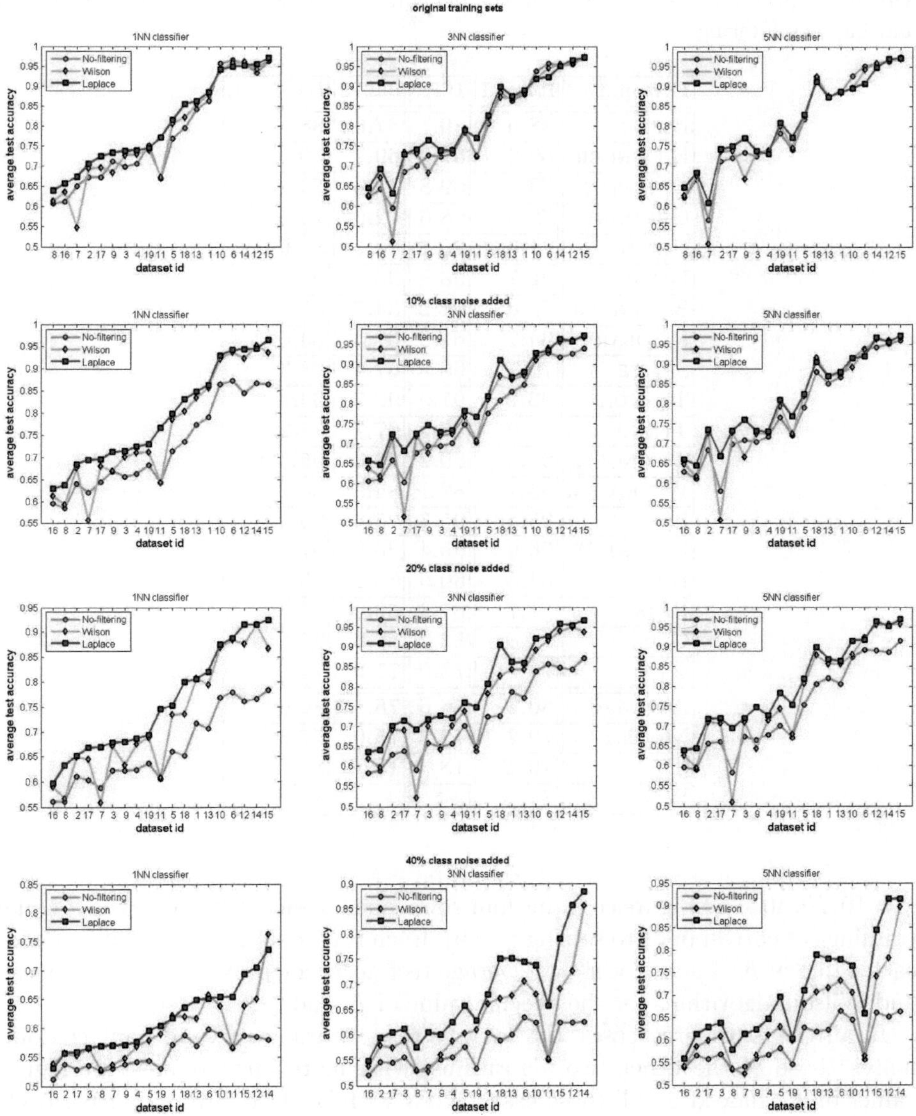

Fig. 3. Average test accuracy performance of the methods

4.3 Robustness to Class Noise

In order to analyze experimentally the robustness of the methods with respect to the presence of class noise in the training set, all experiments are repeated with modified training sets. The new training sets are obtained by changing the class labels of a given percentage γ of randomly selected instances.

Figure 3 shows plots of the average accuracy of the nine KNN classifiers using the original datasets and those obtained by adding $\gamma\%$ class noise, with

Table 3. Results of experiments on ML benchmark datasets of HMN-EI, ICF, DROP3, and Laplace filtering

Dataset	HMN-EI	ICF	DROP3	LAPLACE
BANANA	88.6	86.1	87.6	88.5
B.CANCER	69.2	67.0	69.7	70.6
DIABETES	73.5	69.8	72.3	73.9
GERMAN	72.9	68.6	72.0	74.0
HEART	81.6	76.7	80.2	81.6
IMAGE	92.7	93.8	95.1	94.9
RINGNORM	65.6	61.2	54.7	67.3
F.SOLAR	64.7	61.0	61.4	64.0
SPLICE	70.7	66.3	67.6	73.3
THYROID	93.2	91.9	92.7	94.3
TITANIC	76.0	67.5	67.7	77.2
TWONORM	95.9	89.2	94.3	95.5
WAVEFORM	85.4	82.1	84.9	86.2
IRIS	95.4	95.3	95.8	95.2
BREAST-W	96.9	95.4	96.8	97.1
BUPA	64.5	60.9	63.1	65.8
PIMA	71.7	67.9	69.4	72.5
G50	86.8	82.2	82.8	85.6
G10	79.2	73.0	75.0	74.6
Average	80.2	76.6	78.1	80.6
Median	79.2	73.0	75.0	77.3
S	10/2	18/0	13/0	N/A
W	~	+	+	N/A

$\gamma = 10, 20, 40$. The Figure contains four rows, one for each value of γ (the original training set corresponds to setting $\gamma = 0$). Each row contains three plots, one for each value of K. Each plot shows average test accuracy of No-filtering, Laplace, and Wilson algorithms for the specific value of K and γ.

In all the considered cases, the average test accuracy curve of Laplace dominates those of the other two algorithms, with more improvement for higher values of K. Indeed, in all these cases, KNN with Laplace filtering outperforms significantly the other classifiers.

These results substantiate robustness of the Laplace-based instance filtering approach for KNN with respect to the presence of class noise.

5 Comparison with Other Methods

5.1 Editing Algorithms

In order to compare the performance of the proposed method with that of state-of-the-art editing algorithms, we report in Figure 3 the test accuracy results of the `1-NN` classifier, achieved by the state-of-the-art instance editing algorithms recently investigated in [20]: Iterative Case Filtering (`ICF`) [7], Decremental Reduction Optimization (`DROP3`) [32,33], and Hit Miss Network Editing (`HMN-EI`) [20].

`ICF` first applies `E-NN` noise reduction iteratively until it cannot remove any point, and next iteratively removes points. At each iteration all points for which the so-called *reachability* set is smaller than the *coverage* one are deleted. The reachability of a point x consists of the points inside the largest hyper-sphere containing only points of the same class as x. The *coverage* of x is defined as the set of points that contain x in their reachability set.

`DROP3` first applies a pre-processing step which discards points of X misclassified by their K nearest neighbors, and then removes a point x from X if the accuracy of the KNN rule on the set of its associates does not decrease. Each point has a list of K nearest neighbors and a list of associates, which are updated each time a point is removed from X. A point y is an *associate* of x if x belongs to the set of K nearest neighbors of y. If x is removed then the list of K nearest neighbors of each of its associates y is updated by adding a new neighbor point z, and y is added to the list of associates of z. The removal rule is applied to the

Table 4. Results of SVM/RBF with Laplace pre-processing (LAPLACE) and without (SVM-RBF) at different levels of class noise γ

	$\gamma = 0$		$\gamma = 20$		$\gamma = 40$	
	SVM-RBF	LAPLACE	SVM-RBF	LAPLACE	SVM-RBF	LAPLACE
1	89.3	89.3	87.3	87.4	65.7	66.0
2	73.1	72.9	71.1	71.1	62.5	63.1
3	76.5	76.5	74.4	74.4	64.9	64.5
4	76.2	76.4	73.3	73.3	65.7	66.1
5	83.9	84.1	81.1	81.6	64.5	65.4
6	96.5	96.5	92.9	92.9	78.7	78.9
7	98.2	98.1	96.5	96.2	81.3	79.0
8	66.8	66.8	65.0	65.1	57.9	58.4
9	88.8	88.6	83.3	83.3	68.0	68.2
10	95.2	95.0	91.5	91.9	76.4	76.4
11	77.3	77.2	76.1	75.9	67.1	68.0
12	97.5	97.5	96.5	96.9	88.0	89.5
13	89.8	89.8	87.1	87.3	74.2	75.9
14	95.8	95.8	95.1	95.6	90.6	91.0
15	94.2	95.0	96.2	96.2	90.2	90.8
16	70.5	70.7	64.2	64.6	54.9	54.9
17	75.6	75.8	72.9	73.0	64.7	64.7
18	95.8	95.8	93.4	92.6	78.6	81.6
19	94.2	95.0	88.4	89.0	69.8	68.2
A	86.1	86.2	83.5	83.6	71.8	72.1
M	89.3	89.3	87.1	87.3	68.0	68.2
S	0/0	0/0	0/0	0/0	0/0	2/0
W	~	~	~	~	~	~

points sorted in decreasing order of distance from their nearest neighbor from the other classes (nearest enemy).

`HMN-EI` is an iterative heuristic algorithm based on a directed graph-based representation of the training set, called hit miss network. Topological properties of such network are used for designing an iterative algorithm that removes points considered irrelevant or harmful to the `1-NN` generalization performance. The empirical error on the training set is used as criterion for terminating the iterative process.

Results in Figure 3 show that test accuracy of Laplace filtering is similar or better than the one of these state-of-the-art methods. However, observe that editing algorithms also reduce storage, while Laplace filtering is specifically designed for removing critical instances. In order to reduce also storage reduction, one could use Laplace instance filtering as pre-processing step, followed by the application of a competence preservation algorithm, such as [1].

5.2 SVM with RBF Kernels and Optimized Parameters

A family of classifiers different from KNN, whose training process results in the selection of a subset of the training set, are the Support Vector Machines (SVMs). They map training points x into a higher (possibly infinite) dimensional feature space by the function θ. Then SVM finds a linear separating hyperplane with the maximal margin in this higher dimensional feature space. Given a training set of real-valued instance-label pairs $(x_i, l(x_i)), i = 1, \ldots, n$ the support vector machines (SVM) [6,12] require the solution of the following optimization problem:

$$\min_{w,b,\xi} w^T w + C \sum_{i=1}^{n} \xi_i$$

$$\text{such that } l(x_i)(w^T \xi(x_i) + b) \geq 1 - \xi_i \text{ and } \xi_i \geq 0.$$

$C > 0$ is the penalty parameter of the empirical error term. Furthermore, $K(x, y) = \xi(x)^T \xi(y)$ is called the kernel function. In particular, SVMs with Radial Basis Function (RBF) kernel use $K(x, y) = e^{-\sigma||x-y||^2}$. The set of points with $\xi_i > 0$ are called support vectors. They uniquely identify the separating hyperplane.

It is interesting to investigate whether the use of Laplacian filtering as preprocessing step improves the performance of SVMs with RBF kernel and optimized parameters.

To this aim, the experimental evaluation described in the previous section is used. Specifically, first cross-validation is applied to search for the given training set for the optimal values of the soft-margin C parameter and the RBF parameter σ[1]. Next, Laplacian filtering with $K = 1$ and Euclidean distance is applied for discarding critical instances from the training set. Finally, a SVM with RBF

[1] In the experiments we use the Matlab functions implemented by S. Hashemi of LIBSVM's library [10].

kernel is trained on the selected instances, using the given optimal values for σ and C.

Results of experiments are reported in Table 4. The new training sets are obtained by changing the class labels of a given percentage γ of randomly selected instances. We consider $\gamma = 0, 20, 40$. Laplace filtering does not appear to affect significantly the test accuracy at the considered levels of class noise. This result is not very surprising, since SVMs with RBF kernel and 'optimized' parameters selection have in general good generalization accuracy in the presence of noise.

6 Conclusions and Future Work

This paper introduced a graph differential operator for scoring instances of a training set. We showed how this scoring is related to the flow of the within-class density in the between-class KNN graph, and observed empirically that instances with negative score are close to the class decision boundary or are outliers. This observation motivated the design of a simple algorithm for class noise instance filtering which removes instances having negative score.

We performed extensive experiments on artificial and real-life datasets and analyzed the test accuracy of KNN classifier without filtering, with a traditional filtering algorithm, and with Laplace filtering. The results indicated superior performance of Laplace filtering over the other algorithms. Experiments with modified training sets obtained by permuting the class label of a percentage of their instances were conducted, to investigate robustness of the approach to the presence of class noise. Results of the experiments substantiated the robustness of Laplacian filtering, which achieved significantly better test accuracy performance than the other algorithms, at each of the considered levels of class noise. Comparison of Laplacian filtering with state-of-the-art editing algorithms indicated similar or improved generalization performance of the `1-NN`.

Finally, we investigated whether the use of Laplacian filtering as preprocessing step improves the performance of classifiers other than KNN. We considered SVMs with RBF kernels. These are related to NN methods, because each RBF measures the distance of a test instance to one of the training instances. SVM training keeps certain training instances as support vectors, and discards others. In this way, SVM/RBF may be view as a competence-enhancement filtering method. Results of extensive experiments seemed to indicate no significant effect of Laplacian filtering on the generalization performance of SVM with RBF kernel. The benefits of noise reduction are much more apparent for kNN because it does not really have an induction step and uses examples directly for classification. SVMs with RBF kernel and equipped with cross validation for selecting optimal values of their parameters, provide a rather powerful tool for selecting the centers and parameters of the RBF's, which is robust to the presence of noise.

In summary, these results show that Laplacian instance filtering provides a simple yet effective tool for improving accuracy performance of nearest neighbor classifiers.

We conclude with a discussion of issues and future research directions.

The Laplacian filtering algorithm does not takes into account the effect of removing one instances on the remaining ones. An adaptive approach, consisting in removing the instance having the largest negative score, and then updating the score of the remaining instances, and so on, could possibly improve the effectiveness of the algorithm. However, such an approach would increase the algorithmic complexity of the algorithm.

In the present algorithm, instances with negative Laplacian score are considered critical. Replacing such "rule of the thumb" with an incremental procedure for selecting a cutoff value will possibly have a beneficial effect. Such a procedure could be based on the leave-one-out error of the original training set, using the KNN classifier with actual set of instances incrementally constructed starting from a core subset consisting of instances with high score.

In those cases where the underlying metric is corrupted (e.g., due to irrelevant features), instance selection methods that directly depend on the underlying similarity measure, such as Laplacian filtering, may possibly fail to improve the classification performance of the KNN classifier. In such cases hybridization with metric learning techniques (cf. e.g., [17,16]), could help to overcome this drawback. In the metric learning approach the goal is typically to change the metric in order to repair the KNN classifier. We are investigating an hybridization of Laplacian filtering with Weinberger's et al. method [30], for effective repairing of the metric and removal of critical instances.

Another important issue in instance filtering is scalability. Recently, an instance selection method based on distributed computing has been proposed for speeding up execution of the algorithm without affecting training set accuracy [2]. It is interesting to investigate whether this approach can be used also to speed up execution of Laplace filtering, in order to allow its applicability to very large datasets.

References

1. Angiulli, F.: Fast condensed nearest neighbor rule. In: ICML 2005: Proceedings of the 22nd international conference on Machine learning, pp. 25–32. ACM, New York (2005)
2. Angiulli, F., Folino, G.: Distributed nearest neighbor-based condensation of very large data sets. IEEE Trans. on Knowl. and Data Eng. 19(12), 1593–1606 (2007)
3. Barnett, V.: The ordering of multivariate data. J. Roy. Statist. Soc., Ser. A 139(3), 318–355 (1976)
4. Belkin, M., Niyogi, P.: Laplacian eigenmaps and spectral techniques for embedding and clustering. In: Advances in Neural Information Processing Systems 14, pp. 585–591. MIT Press, Cambridge (2002)
5. Beygelzimer, A., Kakade, S., Langford, J.: Cover trees for nearest neighbor. In: ICML 2006: Proceedings of the 23rd international conference on Machine learning, pp. 97–104. ACM, New York (2006)
6. Boser, B.E., Guyon, I.M., Vapnik, V.N.: A training algorithm for optimal margin classifiers. In: COLT 1992: Proceedings of the fifth annual workshop on Computational learning theory, pp. 144–152. ACM, New York (1992)

7. Brighton, H., Mellish, C.: On the consistency of information filters for lazy learning algorithms. In: Żytkow, J.M., Rauch, J. (eds.) PKDD 1999. LNCS (LNAI), vol. 1704, pp. 283–288. Springer, Heidelberg (1999)
8. Brighton, H., Mellish, C.: Advances in instance selection for instance-based learning algorithms. Data Mining and Knowledge Discovery (6), 153–172 (2002)
9. Cayton, L., Dasgupta, S.: A learning framework for nearest neighbor search. In: NIPS, vol. 20 (2007)
10. Chang, C., Lin, C.: LIBSVM: a library for support vector machines (2001), `http://www.csie.ntu.edu.tw/~cjlin/libsvm`
11. Chapelle, O., Zien, A.: Semi-supervised classification by low density separation. In: Proceedings of the Tenth International Workshop on Artificial Intelligence and Statistics, pp. 57–64 (2005)
12. Cortes, C., Vapnik, V.: Support-vector networks. Mach. Learn. 20(3), 273–297 (1995)
13. Cover, T., Hart, P.: Nearest neighbor pattern classification. IEEE Transactions on Information Theory 13, 21–27 (1967)
14. Dasarathy, B.V.: Minimal consistent set (mcs) identification for optimal nearest neighbor decision systems design. IEEE Transactions on Systems, Man, and Cybernetics 24(3), 511–517 (1994)
15. Dasarathy, B.V.: Nearest unlike neighbor (nun): An aid to decision confidence estimation. Opt. Eng. 34(9), 2785–2792 (1995)
16. Domeniconi, C., Gunopulos, D., Peng, J.: Large margin nearest neighbor classifiers. IEEE Transactions on Neural Networks 16(4), 899–909 (2005)
17. Domeniconi, C., Peng, J., Gunopulos, D.: Locally adaptive metric nearest neighbor classification. IEEE Transactions on Pattern Analysis and Machine Intelligence 24(9), 1281–1285 (2002)
18. Grother, P.J., Candela, G.T., Blue, J.L.: Fast implementation of nearest neighbor classifiers. Pattern Recognition 30, 459–465 (1997)
19. He, X., Cai, D., Niyogi, P.: Laplacian score for feature selection. In: Advances in Neural Information Processing Systems 18 (2005)
20. Marchiori, E.: Hit miss networks with applications to instance selection. Journal of Machine Learning Research 9, 997–1017 (2008)
21. Pekalska, E., Duin, R.P.W., Paclík, P.: Prototype selection for dissimilarity-based classifiers. Pattern Recognition 39(2), 189–208 (2006)
22. Rätsch, G., Onoda, T., Müller, K.-R.: Soft margins for AdaBoost. Machine Learning 42(3), 287–320 (2001)
23. Sánchez, J.S., Pla, F., Ferri, F.J.: Prototype selection for the nearest neighbour rule through proximity graphs. Pattern Recognition Letters 18, 507–513 (1997)
24. Sebban, M., Nock, R., Lallich, S.: Boosting neighborhood-based classifiers. In: ICML, pp. 505–512 (2001)
25. Sebban, M., Nock, R., Lallich, S.: Stopping criterion for boosting-based data reduction techniques: from binary to multiclass problem. Journal of Machine Learning Research 3, 863–885 (2002)
26. Shin, H., Cho, S.: Neighborhood property based pattern selection for support vector machines. Neural Computation (19), 816–855 (2007)
27. Toussaint, G.T.: Proximity graphs for nearest neighbor decision rules: recent progress. In: Interface 2002, 34th Symposium on Computing and Statistics, pp. 83–106 (2002)

28. Vezhnevets, A., Barinova, O.: Avoiding boosting overfitting by removing "confusing" samples. In: Kok, J.N., Koronacki, J., Lopez de Mantaras, R., Matwin, S., Mladenič, D., Skowron, A. (eds.) ECML 2007. LNCS (LNAI), vol. 4701, pp. 430–441. Springer, Heidelberg (2007)
29. von Luxburg, U.: A tutorial on spectral clustering. Statistics and Computing 17(4), 395–416 (2007)
30. Weinberger, K.Q., Saul, L.K.: Distance metric learning for large margin nearest neighbor classification. Journal of Machine Learning Research 10, 207–244 (2009)
31. Wilson, D.L.: Asymptotic properties of nearest neighbor rules using edited data. IEEE Transactions on Systems, Man and Cybernetics (2), 408–420 (1972)
32. Randall Wilson, D., Martinez, T.R.: Instance pruning techniques. In: Proc. 14th International Conference on Machine Learning, pp. 403–411. Morgan Kaufmann, San Francisco (1997)
33. Wilson, D.R., Martinez, T.R.: Reduction techniques for instance-based learning algorithms. Machine Learning 38(3), 257–286 (2000)
34. Zhao, Z., Liu, H.: Spectral feature selection for supervised and unsupervised learning. In: ICML 2007: Proceedings of the 24th international conference on Machine learning, pp. 1151–1157. ACM Press, New York (2007)
35. Zhou, D., Huang, J., Schölkopf, B.: Learning from labeled and unlabeled data on a directed graph. In: ICML 2005: Proceedings of the 22nd international conference on Machine learning, pp. 1036–1043. ACM Press, New York (2005)
36. Zhou, D., Schölkopf, B.: Regularization on discrete spaces. In: Kropatsch, W.G., Sablatnig, R., Hanbury, A. (eds.) DAGM 2005. LNCS, vol. 3663, pp. 361–368. Springer, Heidelberg (2005)

Integrating Novel Class Detection with Classification for Concept-Drifting Data Streams

Mohammad M. Masud[1], Jing Gao[2],
Latifur Khan[1], Jiawei Han[2], and Bhavani Thuraisingham[1]

[1] University of Texas, Dallas
[2] University of Illinois, Urbana Champaign
mehedy@utdallas.edu, jinggao3@uiuc.edu,
lkhan@utdallas.edu, hanj@cs.uiuc.edu, bhavani.thuraisingham@utdallas.edu

Abstract. In a typical data stream classification task, it is assumed that the total number of classes are fixed. This assumption may not be valid in a real streaming environment, where new classes may evolve. Traditional data stream classification techniques are not capable of recognizing novel class instances until the appearance of the novel class is manually identified, and labeled instances of that class are presented to the learning algorithm for training. The problem becomes more challenging in the presence of concept-drift, when the underlying data distribution changes over time. We propose a novel and efficient technique that can automatically detect the emergence of a novel class in the presence of concept-drift by quantifying cohesion among unlabeled test instances, and separation of the test instances from training instances. Our approach is non-parametric, meaning, it does not assume any underlying distributions of data. Comparison with the state-of-the-art stream classification techniques prove the superiority of our approach.

1 Introduction

It is a major challenge to data mining community to mine the ever-growing streaming data. There are three major problems related to stream data classification. First, it is impractical to store and use all the historical data for training, since it would require infinite storage and running time. Second, there may be concept-drift in the data, meaning, the underlying concept of the data may change over time. Third, novel classes may evolve in the stream. There are many existing solutions in literature that solve the first two problems, such as single model incremental learning algorithms [1,2,11], and ensemble classifiers [3,5,9]. However, most of the existing techniques are not capable of detecting novel classes in the stream. On the other hand, our approach can handle both concept-drift, and detect novel classes at the same time.

Traditional classifiers can only correctly classify instances of those classes with which they have been trained. When a new class appears in the stream, all instances belonging to that class will be misclassified until the new class

W. Buntine et al. (Eds.): ECML PKDD 2009, Part II, LNAI 5782, pp. 79–94, 2009.

has been manually identified by some experts and a new model is trained with the labeled instances of that class. Our approach provides a solution to this problem by incorporating a novel class detector within a traditional classifier so that the emergence of a novel class can be identified without any manual intervention. The proposed novel class detection technique can benefit many applications in various domains, such as network intrusion detection and credit card fraud detection. For example, in the problem of intrusion detection, when a new kind of intrusion occurs, we should not only be able to detect that it is an intrusion, but also that it is a new kind of intrusion. With the intrusion type information, human experts would be able to analyze the intrusion more intensely, find a cure, set an alarm in advance and make the system more secure.

We propose an innovative approach to detect novel classes. It is different from traditional novelty (or anomaly/outlier) detection techniques in several ways. First, traditional novelty detection techniques [4,6,10] work by assuming or building a model of normal data, and simply identifying data points as outliers/anomalies that deviate from the "normal" points. But our goal is not only to detect whether a single data point deviates from the normality, but also to discover whether a group of outliers have any strong bond among themselves. Second, traditional novelty detectors can be considered as a "one-classs" model, which simply distinguish between normal and anomalous data, but cannot distinguish between two different kinds of anomalies. But our model is a "multi-class" model, meaning, it can distinguish among different classes of data and at the same time can detect presence of a novel class data, which is a unique combination of a traditional classifier with a novelty detector.

Our technique handles concept-drift by adapting an ensemble classification approach, which maintains an ensemble of M classifiers for classifying unlabeled data. The data stream is divided into equal-sized chunks, so that each chunk can be accommodated in memory and processed online. We train a classification model from each chunk as soon as it is labeled. The newly trained model replaces one of the existing models in the ensemble, if necessary. Thus, the ensemble evolves, reflecting the most up-to-date concept in the stream.

The central concept of our novel class detection technique is that each class must have an important property: the data points belonging to the same class should be closer to each other (cohesion) and should be far apart from the data points belonging to other classes (separation). Every time a new data chunk appears, we first detect the test instances that are *well-separated* from the training data (i.e. outliers). Then filtering is applied to remove the outliers that possibly appear as a result of concept-drift. Finally, if we find strong cohesion among those filtered outliers, we declare a novel class. When the true labels of the novel class(es) arrive and a new model is trained with the labeled instances, the existing ensemble is updated with that model. Therefore, the ensemble of models is continuously enriched with new classes.

We have several contributions. First, we provide a detailed understanding of the characteristic of a novel class, and propose a new technique that can detect novel classes in the presence of concept-drift in data streams. Second, we establish a framework for incorporating novel class detection mechanism into a traditional classifier. Finally, we apply our technique on both synthetic and real-world data and obtain much better results than state-of the art stream classification algorithms.

The rest of the paper is organized as follows. Section 2 discusses related work. Section 3 provides an overview of our approach and Section 4 discusses our approach in detail. Section 5 then describes the datasets and experimental evaluation of our technique. Section 6 concludes with discussion and suggestions for future work.

2 Related Work

Our work is related to both stream classification and novelty detection. There have been many works in stream data classification. There are two main approaches - single model classification, and ensemble classification. Some single-model techniques have been proposed to accommodate concept drift [1,2,11]. However, Our technique follows the ensemble approach. Several ensemble techniques for stream data mining have been proposed [3,5,9]. These ensemble approaches require simple operations to update the current concept, and they are found to be robust in handling concept-drift. Although these techniques can efficiently handle concept-drift, none of them can detect novel classes in the data stream. On the other hand, our technique is not only capable of handling concept-drift, but also able to detect novel classes in data streams. In this light, our technique is also related to novelty detection techniques.

A comprehensive study on novelty detection has been discussed in [4]. The authors categorize novelty detection techniques into two categories: statistical and neural network based. Our technique is related to the statistical approach. Statistical approaches are of two types: parametric, and non-parametric. Parametric approaches assume that data distributions are known (e.g. Gaussian), and try to estimate the parameters (e.g. mean and variance) of the distribution. If any test data falls outside the normal parameters of the model, it is declared as novel [6]. Our technique is a non-parametric approach. Non-parametric approaches like parzen window method [10] estimate the density of training data and reject patterns whose density is beyond a certain threshold. K-nearest neighbor (K-NN) based approaches for novelty detection are also non-parametric [12]. All of these techniques for novelty detection only consider whether a test instance is sufficiently close (or far) from the training data based on some appropriate metric (e.g., distance, density etc.). Our approach is different from these approaches in that we not only consider separation from normal data but also consider cohesion among the outliers. Besides, our model assimilates a novel class into the existing model, which enables it to distinguish future instances of that class from other classes. On the other hand, novelty detection techniques just remember the "normal" trend, and do not care about the similarities or dissimilarities among the anomalous instances.

A recent work in data stream mining domain [7] describes a clustering approach that can detect both concept-drift and novel class. This approach assumes that there is only one 'normal' class and all other classes are novel. Thus, it may not work well if more than one classes are to be considered as 'normal' or 'non-novel', but our approach can handle any number of existing classes. This makes our approach more effective in detecting novel classes than [7], which is justified by the experimental results.

3 Overview

Algorithm 1 outlines a summary of our technique. The data stream is divided into equal sized chunks. The latest chunk, which is unlabeled, is provided to the algorithm as input. At first it detects if there is any novel class in the chunk (line 1). The term "novel class" will be defined shortly. If a novel class is found, we detect the instances that belong to the class(es) (line 2). Then we use the ensemble $L = \{L_1, ..., L_M\}$ to classify the instances that do not belong to the novel class(es). When the data chunk becomes labeled, a new classifier L' trained using the chunk. Then the existing ensemble is updated by choosing the best M classifiers from the $M + 1$ classifiers $L \cup \{L'\}$ based on their accuracies on the latest labeled data chunk. Our algorithm will be mentioned henceforth as

Algorithm 1. MineClass

Input: D_n: the latest data chunk
  L: Current ensemble of best M classifiers
Output: Updated ensemble L

```
1: found ← DetectNovelClass(D_n,L) (algorithm 2, section 4.3)
2: if found then Y ← Novel_instances(D_n), X ← D_n − Y else X ← D_n
3: for each instance x ∈ X do Classify(L,x)
4: /*Assuming that D_n is now labeled*/
5: L' ← Train-and-create-inventory(D_n) (section 4.1)
6: L ← Update(L,L',D_n)
```

"MineClass", which stands for Mining novel Classes in data streams. MineClass should be applicable to any base learner. The only operation that is specific to a learning algorithm is *Train-and-create-inventory*. We will illustrate this operation for two base learners.

3.1 Classifiers Used

We apply our novelty detection technique on two different classifiers: decision tree, and K-NN. We keep M classification models in the ensemble. For decision tree classifier, each model is a decision tree. For K-NN, each model is usually the set of training data itself. However, storing all the raw training data is memory-inefficient and using them to classify unlabeled data is time-inefficient. We reduce both the time and memory requirement by building K clusters with the training data, saving the cluster summaries as classification models, and discarding the raw data. This process is explained in details in [5]. The cluster summaries are mentioned henceforth as "pseudopoint"s. Since we store and use only K pseudopoints, both the time and memory requirements become functions of K (a constant number). The clustering approach followed here is a constraint-based K-means clustering where the constraint is to minimize cluster impurity while minimizing the intra-cluster dispersion. A cluster is considered pure if it contains instances from only one class. The summary of each cluster consists of the centroid, and the frequencies of data points of each class in the cluster. Classification is done by finding the nearest cluster centroid from the test point, and assigning the class, that has the highest frequency, to the test point.

3.2 Assumptions

We begin with the definition of "novel" and "existing" class.

Definition 1 (Existing class and Novel class). *Let L be the current ensemble of classification models. A class c is an existing class if at least one of the models $L_i \in L$ has been trained with the instances of class c. Otherwise, c is a novel class.*

We assume that any class has the following essential property:

Property 1. *A data point should be closer to the data points of its own class (cohesion) and farther apart from the data points of other classes (separation).*

Our main assumption is that the instances belonging to a class c is generated by a an underlying generative model θ_c, and the instances in each class are independently identically distributed. With this assumption, we can reasonably argue that the instances which are close together are supposed to be generated

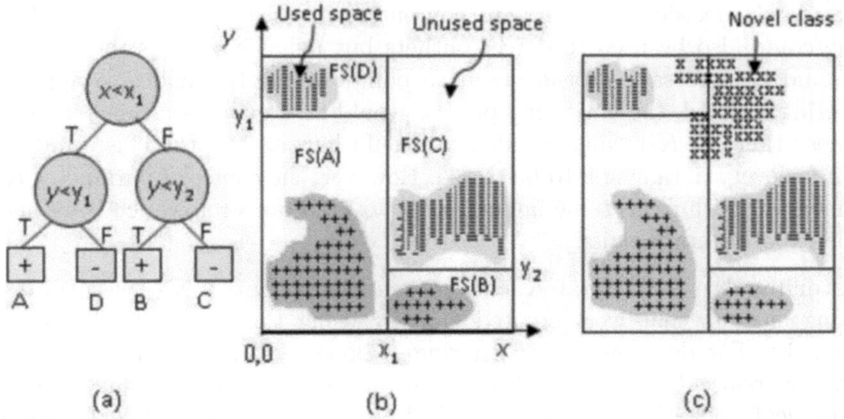

Fig. 1. (a) A decision tree and (b) corresponding feature space partitioning. FS(X) denotes the feature space defined by a leaf node X. The shaded areas show the used spaces of each partition. (c) A Novel class (denoted by x) arrives in the unused space.

by the same model, i.e., belong to the same class. We now show the basic idea of novel class detection using decision tree in figure 1. We introduce the notion of ***used space*** to denote a feature space occupied by any instance, and ***unused space*** to denote a feature space unused by an instance. According to property 1(separation), a novel class must arrive in the unused spaces. Besides, there must be strong cohesion (e.g. closeness) among the instances of the novel class. Thus, the two basic principles followed by our approach are: keeping track of the used spaces of each leaf node in a decision tree, and finding strong cohesion among the test instances that fall into the unused spaces.

4 Novel Class Detection

We follow two basic steps for novel class detection. First, the classifier is trained such that an inventory of the used spaces (described in section 3.2) is created and saved. This is done by clustering and and saving the cluster summary as "pseudopoint" (to be explained shortly). Secondly, these pseudopoints are used to detect outliers in the test data, and declare a novel class if there is strong cohesion among the outliers.

4.1 Saving the Inventory of Used Spaces During Training

The general idea of creating the inventory is to cluster the training data, and save the cluster centroids and other useful information as pseudopoints. These pseudopoints keep track of the use spaces. The way how this clustering is done may be specific to each base learner. For example, for decision tree, clustering is done at each leaf node of the tree, since we need to keep track of the used spaces for each leaf node separately. For the K-NN classifier discussed in section 3.1, already existing pseudopoints are utilized to store the inventory.

It should be noted here that K-means clustering appears to be the best choice for saving the decision boundary and computing the outliers. Density-based clustering could also be used to detect outliers but it has several problems. First, we would have to save all the raw data points at the leaf nodes to apply the clustering. Second, the clustering process would take quadratic time, compared to linear time for K-means. Finally, we would have to run the clustering algorithm for every data chunk to be tested. However, the choice of parameter K in K-means algorithm has some impact on the overall outcome, which is discussed in the experimental results.

Clustering: We build total K clusters per chunk. For K-NN, we utilize the existing clusters that were created globally using the approach discussed in section 3.1. For decision tree, clustering is done locally at each leaf node as follows. Suppose S is the chunk-size. During decision tree training, when we reach a leaf node l_i, we build $k_i = (t_i/S) * K$ clusters in that leaf, where t_i denotes the number of training instances that ended up in leaf node l_i.

Storing the cluster summary information: For each cluster, we store the following summary information in memory: i) ***Weight***, w: Defined as the total number of points in the cluster. ii) ***Centroid***, ζ. iii) ***Radius***, $\mathcal{R}$: Defined as the maximum distance between the centroid and the data points belonging to the cluster. iv) ***Mean distance***, μ_d: The mean distance from each point to the cluster centroid. The cluster summary of a cluster $\mathcal{H}_i$ will be referred to henceforth as a "pseudopoint" ψ_i. So, $w(\psi_i)$ denotes the weight of pseudopoint ψ_i. After computing the cluster summaries, the raw data are discarded. Let Ψ_j be the set of all pseudopoints stored in memory for a classifier L_j.

4.2 Outlier Detection and Filtering

Each pseudopoint ψ_i corresponds to a hypersphere in the feature space having center $\zeta(\psi_i)$ and radius $\mathcal{R}(\psi_i)$. Thus, the pseudopoints 'memorize' the used spaces. Let us denote the portion of feature space covered by a pseudopoint ψ_i as the "region" of ψ_i or $RE(\psi_i)$. So, the union of the regions covered by all the pseudopoints is the union of all the used spaces, which forms a decision boundary $\mathcal{B}(L_j) = \cup_{\psi_i \in \Psi_j} RE(\psi_i)$, for a classifier L_j. Now, we are ready to define outliers.

Definition 2 (Routlier). *Let x be a test point and ψ_{min} be the pseudopoint whose centroid is nearest to x. Then x is an Routlier (i.e., raw outlier) if it is outside $RE(\psi_{min})$, i.e., its distance from $\zeta(\psi_{min})$ is greater than $\mathcal{R}(\psi_{min})$.*

In other words, any point x outside the decision boundary $\mathcal{B}(L_j)$ is an *Routlier* for the classifier L_j. For K-NN, *Routlier*s are detected globally by testing x against all the psuedopoints. For decision tree, x is tested against only the psueodpoints stored at the leaf node where x belongs.

Filtering: According to definition 2, a test instance may be erroneously considered as an *Routlier* because of one or more of the following reasons: i) The test instance belongs to an existing class but it is a noise. ii) There has been a concept-drift and as a result, the decision boundary of an existing class has been shifted. iii) The decision tree has been trained with insufficient data. So, the predicted decision boundary is not the same as the actual one.

Due to these reasons, the outliers are filtered to ensure that any outlier that belongs to the existing classes does not end up in being declared as a new class instance. The filtering is done as follows: if a test instance is an *Routlier* to *all* the classifiers in the ensemble, then it is considered as a filtered outlier. All other *Routlier*s are filtered out.

Definition 3 (Foutlier). *A test instance is an Foutlier (i.e., filtered outlier) if it is an Routlier to all the classifiers L_i in the ensemble L.*

Intiuitively, being an $Foutlier$ is a necessary condition for being in a new class. Because, suppose an instance x is not an *Routlier* to some classifier L_i in the ensemble. Then x must be inside the decision boundary $\mathcal{B}(L_i)$. So, it violates property 1 (separation), and therefore, it cannot belong to a new class. Although being an $Foutlier$ is a necessary condition, it is not sufficient for being in a new class, since it does not guarantee the property 1 (cohesion). So, we proceed to the next step to verify whether the $Foutlier$s satisfy both cohesion and separation.

4.3 Detecting Novel Class

We perform several computations on the *Foutliers* to detect the arrival of a new class. First, we discuss the general concepts of these computations and later we describe how these computations are carried out efficiently. For every *Foutlier*, we define a λ_c-neighborhood as follows:

Definition 4 (λ_c-neighborhood). *The λ_c-neighborhood of an Foutlier x is the set of $\mathcal{N}$-nearest neighbors of x belonging to class c.*

Here $\mathcal{N}$ is a user defined parameter. For brevity, we denote the λ_c-neighborhood of an *Foutlier* x as $\lambda_c(x)$. Thus, $\lambda_+(x)$ of an *Foutlier* x is the set of $\mathcal{N}$ instances of class c_+, that are closest to the outlier x. Similarly, $\lambda_o(x)$ refers to the set of $\mathcal{N}$ *Foutliers* that are closest to x. This is illustrated in figure 2, where the *Foutliers* are shown as black dots, and the instances of class c_+ and class c_- are shown with the corresponding symbols. $\lambda_+(x)$ of the *Foutlier* x is the set of $\mathcal{N}$ $(= 3)$ instances belonging to class c_+ that are nearest to x (inside the circle), and so on. Next, we define the $\mathcal{N}$-neighborhood silhouette coefficient, ($\mathcal{N}$-NSC).

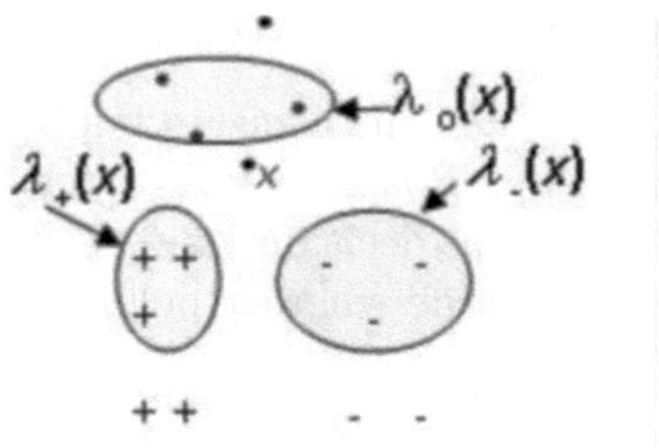

Fig. 2. λ_c-neighborhood with $\mathcal{N}$=3

Definition 5 ($\mathcal{N}$-NSC). *Let $a(x)$ be the average distance from an Foutlier x to the instances in $\lambda_o(x)$, and $b_c(x)$ be the average distance from x to the instances in $\lambda_c(x)$ (where c is an existing class). Let $b_{min}(x)$ be the minimum among all $b_c(x)$. Then $\mathcal{N}$-NSC of x is given by:*

$$\mathcal{N}\text{-}NSC(x) = \frac{b_{min}(x) - a(x)}{max(b_{min}(x), a(x))} \tag{1}$$

According to the definition, the value of $\mathcal{N}$-NSC is between -1 and +1. It is actually a unified measure of cohesion and separation. A negative value indicates that x is closer to the other classes (less separation) and farther away from its own class (less cohesion). We declare a *new class* if there are at least N' $(> \mathcal{N})$ *Foutliers*, whose $\mathcal{N}$-NSC is positive. In fact, we prove that this is a *necessary and sufficient condition* for a new class. This proof is omitted here due to space limitation, but can be obtained from [8].

It should be noted that the larger the value of $\mathcal{N}$, the greater the confidence with which we can decide whether a novel class has arrived. However, if $\mathcal{N}$ is too large, then we may also fail to detect a new class if the total number of instances belonging to the novel class in the corresponding data chunk is $\leq \mathcal{N}$. We experimentally find an optimal value of $\mathcal{N}$, which is explained in section 5.

Computing the set of novel class instances: Once we detect the presence of a novel class, the next step is to find those instances, and separate them from the existing class data. According to the *necessary and sufficient condition*, a set of *Foutlier* instances belong to a novel class if following three conditions satisfy: i) all the *Foutliers* in the set have positive $\mathcal{N}$-NSC, ii) all the *Foutliers* in the set have $\lambda_o(x)$ within the set, and iii) cardinality of the set $\geq \mathcal{N}$. Let $\mathcal{G}$ be such a set. Note that finding the exact set $\mathcal{G}$ is computationally expensive, so we follow an approximation. Let $\mathcal{G}'$ be the set of all *Foutliers* that have positive $\mathcal{N}$-NSC. If $|\mathcal{G}'| \geq \mathcal{N}$, then $\mathcal{G}'$ is an approximation of $\mathcal{G}$. It is possible that some of the data points in $\mathcal{G}'$ may not actually be a novel class instance or vice versa. However, in our experiments, we found that this approximation works well.

Speeding up the computation: Computing $\mathcal{N}$-NSC for every *Foutlier* instance x takes quadratic time in the number of *Foutliers*. In order to make the computation faster, we also create K_o pseudopoints from *Foutliers* using K-means clustering and perform the computations on the pseudopoints (referred to as *Fpseudopoints*), where $K_o = (N_o/S) * K$. Here S is the chunk size and N_o is the number of *Foutliers*. Thus, the time complexity to compute the $\mathcal{N}$-NSC of all of the *Fpseudopoints* is $O(K_o * (K_o + K))$, which is constant, since both K_o and K are independent of the input size. Note that $\mathcal{N}$-NSC of a *Fpseudopoint* is actually an approximate average of the $\mathcal{N}$-NSC of each *Foutlier* in that *Fpseudopoint*. By using this approximation, although we gain speed, we also lose some precision. However, this drop in precision is negligible when we keep sufficient number of pseudopoints, as shown in the exprimental results. The novel class detection process is summarized in algorithm 2 (DetectNovelClass).

This algorithm can detect one or more novel classes concurrently (i.e., in the same chunk) as long as each novel class follows property 1 and contains at least $\mathcal{N}$ instances. This is true even if the class distributions are skewed. However, if more than one such novel classes appear concurrently, our algorithm will identify the instances belonging to those classes as novel, without imposing any distinction among dissimilar novel class instances (i.e., it will treat them simply as "novel"). But the distinction will be learned by our model as soon those instances are labeled, and a classifier is trained with them.

Time complexity: Lines 1-3 of algorithm 2 requires $O(KSL)$ time where S is the chunk size. Line 4 (clustering) requires $O(KS)$ time, and the last for loop (5-10) requires $O(K^2L)$ time. Thus, the overall time complexity of algorithm 2 is $O(KS + KSL + K^2L) = O(K(S + SL + KL))$. Assuming that $S >> KL$, the complexity becomes $O(KS)$, which is linear in S. Thus, the overall time complexity (per chunk) of MineClass algorithm (algorithm 1) is $O(KS + f_c(LS) + f_t(S))$, where $f_c(n)$ is the time required to classify n instances and $f_t(n)$ is the time required to train a classifier with n training instances.

Algorithm 2. DetectNovelClass(D,L)

Input: D: An unlabeled data chunk
L: Current ensemble of best M classifiers
Output: true, if novel class is found; false, otherwise

for each instance $x \in D$ **do**
  if x is an *Routlier* to all classifiers $L_i \in L$
  then $FList \leftarrow FList \cup \{x\}$ /* x is an $Foutlier$*/
end for
Make K_o=($K * |FList|/|D|$) clusters with the instances in $FList$ using K-means clustering, and create *Fpseudopoints*
for each classifier $L_i \in L$ **do**
  Compute $\mathcal{N}$-NSC(ψ_j) for each *Fpseudopoint* ψ_j
  $\Psi_p \leftarrow$ the set of *Fpseudopoints* having positive $\mathcal{N}$-NSC(.).
  $w(\Psi_p) \leftarrow$ sum of $w(.)$ of all *Fpseudopoints* in Ψ_p.
  if $w(\Psi_p) > \mathcal{N}$ **then** NewClassVote++
end for
return NewClassVote > M - $NewClassVote$ /*Majority voting*/

Impact of evolving class labels on ensemble classification: As the reader might have realized already, arrival of novel classes in the stream causes the classifiers in the ensemble to have different sets of class labels. For example, suppose an older (earlier) classifier L_i in the ensemble has been trained with classes c_0 and c_1, and a newer (later) classifier L_j has been trained with classes c_1, and c_2, where c_2 is a new class that appeared after L_i had been trained. This puts a negative effect on voting decision, since the older classifier mis-classifies instances of c_2. So, rather than counting votes from each classifier, we selectively count their votes as follows: if a newer classifier L_j classifies a test instance x as class c, but an older classifier L_i does not have the class label c in its model, then the vote of L_i will be ignored if x is found to be an outlier for L_i. An opposite scenario occurs when the oldest classifier L_i is trained with some class c', but none of the later classifiers are trained with that class. This means class c' has been outdated, and, in that case, we remove L_i from the ensemble. In this way we ensure that older classifiers have less impact in the voting process. If class c' later re-appears in the stream, it will be automatically detected again as a novel class (see definition 1).

5 Experiments

We evaluate our proposed method on a number of synthetic and real datasets, but due to space limitations, we report results on four datasets.

5.1 Data Sets

Specific details of the data sets can be obtained from [8].

Synthetic data generation: There are two types of synthetic data: synthetic data with *only concept-drift (SynC)* and synthetic data with *concept-drift and novel-class (SynCN)*. SynC is generated using moving hyperplane, which contains 2 classes and 10 numeric attributes. SynCN is generated using Gaussian distribution, which contains 10 classes and 20 numeric attributes.

Real datasets: The two real datasets used in the experiments are the 10% version of the *KDDCup 99 network intrusion detection*, and *Forest Cover* dataset from UCI repository. We have used the 10% version of the KDDcup dataset, where novel classes appear more frequently than the full version, hence it is more challenging. KDDcup dataset contains around 490,000 instances, 23 classes, and 34 numeric attributes. Forest Cover dataset contains 7 classes, 54 attributes and around 581,000 instances. We arrange the Forest Cover dataset so that in any chunk at most 3 and at least 2 classes co-occur, and new classes appear randomly. All datasets are normalized to have attribute values within [0,1].

5.2 Experimental Setup

We implement our algorithm in Java. The code for decision tree has been adapted from the Weka machine learning open source repository (`http://www.cs.waikato.ac.nz/ml/weka/`). The experiments were run on an Intel P-IV machine with 2GB memory and 3GHz dual processor CPU. Our parameter settings are as follows, unless mentioned otherwise: i) K (number of pseudopoints per chunk) = 50, ii) $\mathcal{N}$ = 50, iii) M (ensemble size) = 6, iv) Chunk-size = 1,000 for synthetic datasets, and 4,000 for real datasets. These values of parameters are tuned to achieve an overall satisfactory performance.

Baseline method: To the best of our knowledge, there is no approach that can classify data streams *and* detect novel class. So, we compare MineClass with a combination of two baseline techniques: $OLINDDA$ [7], and Weighted Classifier Ensemble (WCE) [9], where the former works as novel class detector, and the latter performs classification. For each chunk, we first detect the novel class instances using $OLINDDA$. All other instances in the chunk are assumed to be in the existing classes, and they are classified using WCE. We use $OLINDDA$ as the novelty detector, since it is a recently proposed algorithm that is shown to have outperformed other novelty detection techniques in data streams [7].

However, $OLINDDA$ assumes that there is only one "normal" class, and all other classes are "novel". So, it is not directly applicable to the multi-class novelty detection problem, where any combination of classes can be considered as the "existing" classes. We propose two alternative solutions. First, we build parallel $OLINDDA$ models, one for each class, which evolve simultaneously. Whenever the instances of a novel class appear, we create a new $OLINDDA$ model for that class. A test instance is declared as novel, if *all the existing class models* identify this instance as novel. We will refer to this baseline method as WCE-OLINDDA_PARALLEL. Second, we initially build an $OLINDDA$ model with all the available classes. Whenever a novel class is found, the class is absorbed into the existing $OLINDDA$ model. Thus, only one "normal" model is maintained throughout the stream. This will be referred to as WCE-OLINDDA_SINGLE. In all experiments, the ensemble size and chunk-size are kept the same for both these techniques. Besides, the same base learner is used for WCE and MC. The parameter settings for $OLINDDA$ are: i) number of data points per cluster (N_{excl}) = 15, ii) least number of normal instances needed to update the existing model = 100, iii) least number of instances needed to build the initial model = 30. These parameters are chosen either according to the default values used in [7] or by trial and error to get an overall satisfactory performance. *We will henceforth use the acronyms **MC for MineClass, W-OP for WCE-OLINDDA_PARALLEL** and **W-OS for WCE-OLINDDA_SINGLE**.*

5.3 Performance Study

Evaluation approach: We use the following performance metrics for evaluation: M_{new} = % of novel class instances Misclassified as existing class, F_{new} = % of existing class instances Falsely identified as novel class, ERR = Total misclassification error (%)(including M_{new} and F_{new}). We build the initial models in each method with the first M chunks. From the $M{+}1^{st}$ chunk onward, we first evaluate the performances of each method on that chunk, then use that chunk to update the existing model. The performance metrics for each chunk for each method are saved and averaged for producing the summary result.

Results: Figures 3(a)-(d) show the ERR for decision tree classifier of each approach up to a certain point in the stream in different datasets. K-NN classifier also has similar results. For example, at X axis = 100, the Y values show the average ERR of each approach from the beginning of the stream to chunk 100. At this point, the ERR of MC, W-OP, and W-OS are 1.7%, 11.6% and 8.7%, respectively, for the KDD dataset (figure 3(c)). The arrival of novel a class in each dataset is marked by a cross (x) on the top border in each graph at the corresponding chunk. For example, on the SynCN dataset (figure 3(a)), W-OP and W-OS misses most of the novel class instances, which results in the spikes in their curves at the respective chunks (e.g. at chunks 12, 24, 37 etc.). W-OS misses almost 99% of the novel class instances. Similar spikes are observed for both W-OP and W-OS at the chunks where novel classes appear for KDD and Forest Cover datasets. For example, many novel classes appear between chunks 9-14 in KDD, most of which are missed by both W-OP and W-OS. Note that there is no novel class for SynC dataset. MC correctly detects most of these novel classes. Thus, MC outperforms both W-OP and W-OS in all datasets.

Table 1 summarizes the error metrics for each of the techniques in each dataset for decision tree, and K-NN. The columns headed by ERR, M_{new} and F_{new} report the average of the corresponding metric on an entire dataset. For example, while using decision tree in the SynC dataset, MC, W-OP and W-OS have almost the same ERR, which are 11.6%, 13.0%, and 12.5%, respectively. This is because SynC simulates only concept-drift, and both MC and WCE handle concept-drift in a similar manner. In SynCN dataset with decision tree, MC, W-OP, and W-OS have 0%, 89.4%, and 99.7% M_{new}, respectively. Thus, W-OS misses almost all of the novel class instances, whereas W-OP detects only 11% of them. MC correctly

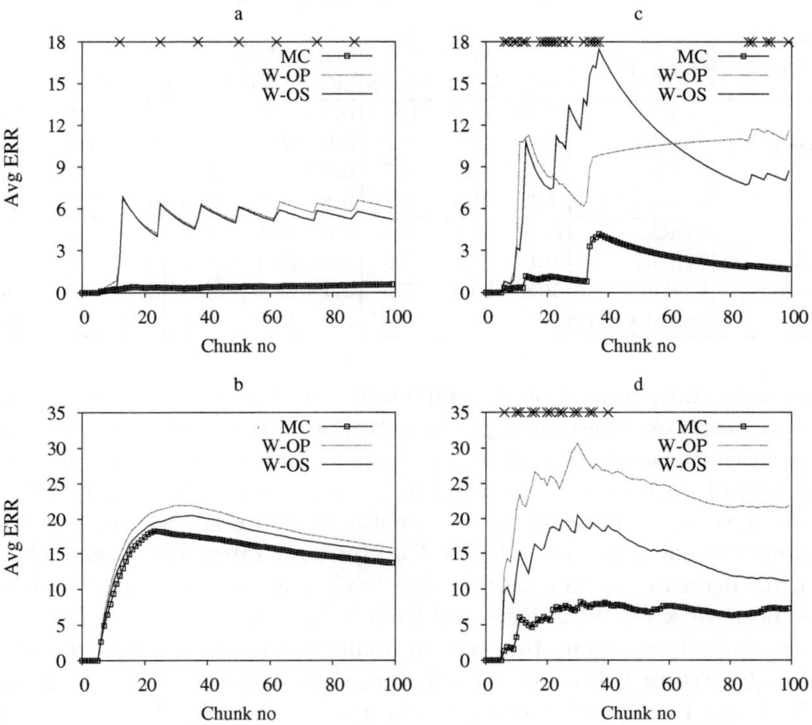

Fig. 3. Error comparison on (a) SynCN, (b) SynC, (c) KDD and (d) Forest Cover

detects all of the novel class instances. It is interesting that all approaches have lower error rates in SynCN than SynC. This is because SynCN is generated using Gaussian distribution, which is naturally easier for the classifiers to learn. W-OS miss-predicts almost all of the novel class instances in all datasets. The comparatively better ERR rate for W-OS over W-OP can be attributed to the lower false positive rate of W-OS, which occurs since almost all instances are identified as "normal" by W-OS. Again, the overall error (ERR) of MC is much lower than other methods in all datasets and for all classifiers. K-NN also has similar results for all datasets.

Figures 4(a)-(d) illustrate how the error rates of MC change for different parameter settings on KDD dataset and decision tree classifier. These parameters have similar effects on other datasets, and K-NN classifier. Figure 4(a) shows the effect of chunk size on ERR, F_{new}, and M_{new} rates for default values of other parameters. M_{new} reduces when chunk size is increased. This is desirable, because larger chunks reduce the risk of missing a novel class. But F_{new} rate slightly increases since the risk of identifying an existing class instance as novel also rises a little. These changes stabilizes from chunk size 4,000 (for Synthetic dataset, it is 1,000). That is why we use these values in our experiments. Figure 4(b) shows the effect of number of clusters (K) on error. Increasing K generally

Table 1. Performance comparison

Classifier	Dataset	ERR			M_{new}			F_{new}		
		MC	W-OP	W-OS	MC	W-OP	W-OS	MC	W-OP	W-OS
Decision tree	SynC	**11.6**	13.0	12.5	0.0	0.0	0.0	**0.0**	1.0	0.6
	SynCN	**0.6**	6.1	5.2	**0.0**	89.4	99.7	0.0	0.6	0.0
	KDD	**1.7**	11.6	8.7	**0.7**	26.7	99.4	1.5	7.0	**0.0**
	Forest Cover	**7.3**	21.8	8.7	**9.8**	18.5	99.4	1.7	15.0	**0.0**
K-NN	SynC	**11.7**	13.1	12.6	0.0	0.0	0.0	**0.0**	1.0	0.6
	SynCN	**0.8**	5.8	5.6	**0**	90.1	99.7	0.9	0.6	**0.0**
	KDD	**2.3**	10.0	7.0	**2.7**	29.0	99.4	2.2	7.1	**0.0**
	Forest Cover	**5.4**	19.2	8.9	**1.0**	18.5	94.0	4.5	15.0	**0.3**

reduces error rates, because outliers are more correctly detected, and as a result, M_{new} rate decreases. However, F_{new} rate also starts increasing slowly, since more test instances are becoming outliers (although they are not). The combined effect is that overall error keeps decreasing up to a certain value (e.g. K=50), and then becomes almost flat. This is why we use K=50 in our experiments. Figure 4(c) shows the effect of ensemble size (M) on error rates. We observe that the error rates decrease up to a certain size (=6), and become stable since then. This is because when M is increased from a low value (e.g., 2), classification error naturally decreases up to a certain point because of the reduction of error variance [9]. Figure 4(d) shows the effect of $\mathcal{N}$ on error rates. The x-axis in this chart is drawn in a logarithmic scale. Naturally, increasing $\mathcal{N}$ up to a certain point (e.g. 20) helps reducing error, since we know that a higher value of $\mathcal{N}$ gives us a greater confidence in declaring a new class (see section 4.3). But a too large value of $\mathcal{N}$ increases M_{new} and ERR rates, since a new class is missed by the algorithm if it has less than $\mathcal{N}$ instances in a data chunk. We have found that any value between 20 to 100 is the best choice for $\mathcal{N}$.

Running time: Table 2 compares the running times of MC, W-OP, and W-OS on each dataset for decision tree. K-NN also shows similar performances. The columns headed by "Time (sec)/chunk " show the average running times (train and test) in seconds per chunk, the columns headed by "Points/sec" show how many points have been processed (train and test) per second on average, and the columns headed by "speed gain" shows the ratio of the speed of MC to that of W-OP and W-OS, respectively. For example, MC is 2,095, and 105 times faster than W-OP on KDD dataset, and Forest Cover dataset, respectively. Also, MC is 203 and 27 times faster than W-OP and W-OS, respectively, on the SynCN dataset. W-OP and W-OS are slower on SynCN than on SynC dataset because SynCN dataset has more attributes (20 vs 10) and classes (10 vs 2). W-OP is relatively slower than W-OS since W-OP maintains C parallel models, where C is the number of existing classes, whereas W-OS maintains only one model. Both W-OP and W-OS are relatively faster on Forest Cover than KDD since Forest Cover has less number of classes, and relatively less evolution than KDD. The main reason for this extremely slow processing of W-OP and W-OS is that the number of clusters for each OLINDDA model keeps increasing linearly with

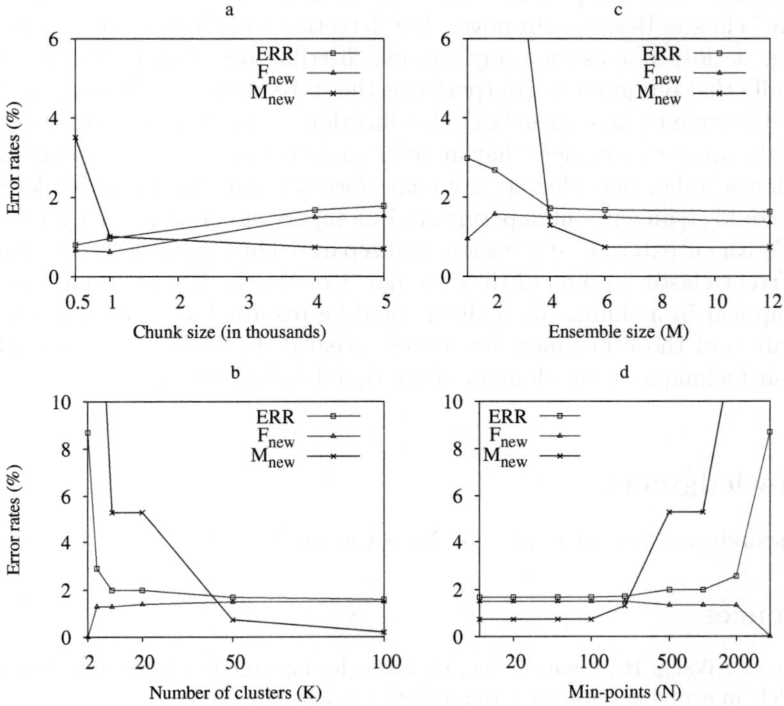

Fig. 4. Sensitivity to different parameters

Table 2. Running time comparison in all datasets

Dataset	Time(sec)/chunk			Points/sec			Speed gain	
	MC	W-OP	W-OS	MC	W-OP	W-OS	MC over W-OP	MC over W-OS
SynC	**0.18**	0.81	0.19	**5,446**	1,227	5,102	**4**	**1**
SynCN	**0.27**	52.9	7.34	**3,656**	18	135	**203**	**27**
KDD	**0.95**	1369.5	222.8	**4,190**	2	17	**2,095**	**246**
Forest Cover	**2.11**	213.1	10.79	**1,899**	18	370	**105**	**5**

the size of the data stream, causing both the memory requirement and running time to increases linearly. But the running time and memory requirement of MC remains the same over the entire length of the stream.

6 Conclusion

We have presented a novel technique to detect new classes in concept-drifting data streams. Most of the novelty detection techniques either assume that there is no concept-drift, or build a model for a single "normal" class and consider all other classes as novel. But our approach is capable of detecting novel classes in the presence of concept-drift, and even when the model consists of multiple "existing" classes. Besides, our novel class detection technique is non-parametric, meaning, it does not assume any specific distribution of data. We also show empirically that our approach outperforms the state-of-the art data stream based novelty detection techniques in both classification accuracy and processing speed.

It might appear to readers that in order to detect novel classes we are in fact examining whether new clusters are being formed, and therefore, the detection process could go on without supervision. But supervision is necessary for classification. Without external supervision, two separate clusters could be regarded as two different classes, although they are not. Conversely, if more than one novel classes appear in a chunk, all of them could be regarded as a single novel class if the labels of those instances are never revealed. In future, we would like to apply our technique in the domain of multiple-label instances.

Acknowledgment

This research was funded in part by NASA grant NNX08AC35A.

References

1. Chen, S., Wang, H., Zhou, S., Yu, P.: Stop chasing trends: Discovering high order models in evolving data. In: Proc. ICDE, pp. 923–932 (2008)
2. Hulten, G., Spencer, L., Domingos, P.: Mining time-changing data streams. In: Proc. ACM SIGKDD, pp. 97–106 (2001)
3. Kolter, J., Maloof, M.: Using additive expert ensembles to cope with concept drift. In: Proc. ICML, pp. 449–456 (2005)
4. Markou, M., Singh, S.: Novelty detection: A review-part 1: Statistical approaches, part 2: Neural network based approaches. Signal Processing 83 (2003)
5. Masud, M.M., Gao, J., Khan, L., Han, J., Thuraisingham, B.: A practical approach to classify evolving data streams: Training with limited amount of labeled data. In: Proc. ICDM, pp. 929–934 (2008)
6. Roberts, S.J.: Extreme value statistics for novelty detection in biomedical signal processing. In: Proc. Int. Conf. on Advances in Medical Signal and Information Processing, pp. 166–172 (2000)
7. Spinosa, E.J., de Leon, A.P., de Carvalho, F., Gama, J.: Olindda: a cluster-based approach for detecting novelty and concept drift in data streams. In: Proc. 2007 ACM symposium on Applied computing, pp. 448–452 (2007)
8. University of Texas at Dallas Technical report UTDCS-13-09 (June 2009), `http://www.utdallas.edu/~mmm058000/reports/UTDCS-13-09.pdf`
9. Wang, H., Fan, W., Yu, P., Han, J.: Mining concept-drifting data streams using ensemble classifiers. In: Proc. ACM SIGKDD, pp. 226–235 (2003)
10. yan Yeung, D., Chow, C.: Parzen-window network intrusion detectors. In: Proc. International Conference on Pattern Recognition, pp. 385–388 (2002)
11. Yang, Y., Wu, X., Zhu, X.: Combining proactive and reactive predictions for data streams. In: Proc. ACM SIGKDD, pp. 710–715 (2005)
12. Yang, Y., Zhang, J., Carbonell, J., Jin, C.: Topic-conditioned novelty detection. In: Proc. ACM SIGKDD, pp. 688–693 (2002)

Neural Networks for State Evaluation in General Game Playing

Daniel Michulke and Michael Thielscher

Department of Computer Science
Dresden University of Technology
{daniel.michulke,mit}@inf.tu-dresden.de

Abstract. Unlike traditional game playing, General Game Playing is concerned with agents capable of playing classes of games. Given the rules of an unknown game, the agent is supposed to play well without human intervention. For this purpose, agent systems that use deterministic game tree search need to automatically construct a state value function to guide search. Successful systems of this type use evaluation functions derived solely from the game rules, thus neglecting further improvements by experience. In addition, these functions are fixed in their form and do not necessarily capture the game's real state value function. In this work we present an approach for obtaining evaluation functions on the basis of neural networks that overcomes the aforementioned problems. A network initialization extracted from the game rules ensures reasonable behavior without the need for prior training. Later training, however, can lead to significant improvements in evaluation quality, as our results indicate.

1 Introduction

Developing an agent for a specific game allows to include game-specific knowledge as data structures (e.g. as in [15]) or manually designed features (like in today's chess programs [10]). These specializations, however, restrict the agent in that it cannot adapt to game modifications, let alone play completely different games.

In contrast, General Game Playing (GGP) is concerned with the development of systems that understand the rules of previously unknown games and learn to play these games well without human intervention. As such systems cannot benefit from prior knowledge, they have to be endowed with high-level cognitive abilities such as general strategic thinking and abstract reasoning. This makes GGP a good example of a challenging problem which encompasses a variety of AI research areas, including knowledge representation and reasoning, heuristic search, planning and learning.

To allow for comparison of GGP agent systems, the Game Description Language (GDL) has become standard. It allows to describe arbitrary deterministic n-player games with complete information by giving a formal axiomatization of their rules. Progress can be seen at the GGP competition held annually at the

W. Buntine et al. (Eds.): ECML PKDD 2009, Part II, LNAI 5782, pp. 95–110, 2009.

AAAI Conference [8], where the following two different approaches have been recently established in the development of general game playing agents: *simulation-based* approaches [6, 7] employ probabilistic look-ahead search in Monte-Carlo fashion, while *knowledge-based* systems refine domain knowledge (the game rules) in order to extract an evaluation function (or state value function) for assessing the leaf nodes in a depth-limited search tree [11, 2, 13]. Evaluation functions used in the latter type of systems share, however, two major disadvantages:

- They are fixed in their form and do not necessarily capture the real state value function of the game.
- They are determined solely on the basis of the game rules; later experience is ignored.

While the former problem can be solved by using any general function approximator, the latter one can be addressed by employing learning algorithms. Though neural networks represent a solution to the two issues, their training is known to be both time consuming and not converging in general. However, an initialization as done in [16, 3] allows for a sound state evaluation without prior training. In this paper we present an algorithm that, on the basis of $C\text{-}IL^2P$ [3], transforms the game rules to a set of neural networks that can be employed for state evaluation and need not be trained. Learning abilities, however, are retained, and experiments indicate a significant increase in evaluation quality in this case.

The rest paper is organized as follows. In section 2, we give a brief introduction on the fields encompassed by our approach. This is followed, in section 3, by a presentation of the transformation process of domain knowledge to a state value function. In section 4, we present experimental results. These are discussed in section 5. We conclude in section 6.

2 Background

2.1 Running Example: Pentago

For illustration purposes we use the game Pentago[1] as an example throughout this paper. Pentago is a two-player game played on a 6x6 board. The players take turn in first marking a cell that is blank, followed by rotating (either clockwise or counterclockwise) one of the four 3x3 quadrants of the board. The player wins who first achieves a line (horizontal, vertical, or diagonal) with five of his marks. If no blank cell is left on the board and nobody has won, the game ends in a draw. Figure 1 shows a final board position which is won for the white player. The four quadrants can be distinguished by their background color.

2.2 GDL

The Game Description Language (GDL) [12] has become the standard language for GGP, allowing for the definition of deterministic n-player games with

[1] The complete set of rules for this game, and for all others mentioned in this paper, can be found at `http://www.general-game-playing.de/game_db/doku.php`

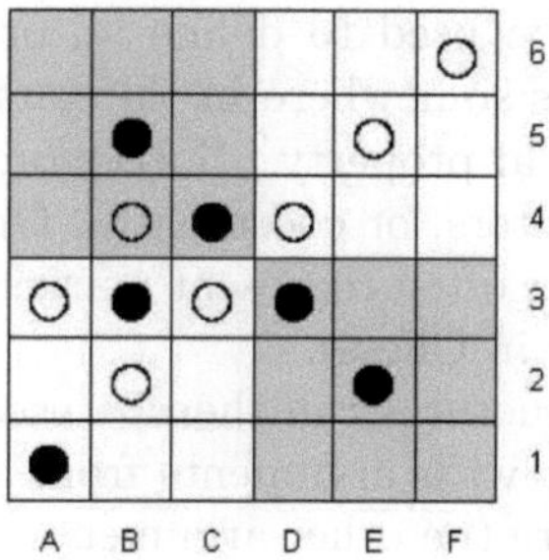

Fig. 1. White wins a match of Pentago

complete information. Examples are popular games like Tic-Tac-Toe, Chess, or Checkers.

GDL allows to give modular, logic-based specifications of games which are interpreted as finite state machines subject to certain restrictions. Each player has a specific role in a game. An actual match corresponds to traversing the finite state machine until a terminal state is reached. A goal function is defined on the set of terminal states that determines the outcome for each player. For the formal specification of the goal, the language uses the keyword `goal` in axioms of the form

```
(<= (goal ?role ?value) goalcondition)
```

where `?role` is the role the player occupies and `?value` gives the goal value (in the interval [0, 100] by convention) for this player if `goalcondition` is fulfilled. `(<= (goal white 100) (line white))` in Pentago thus exemplifies a goal condition where "white" (the player with the role "white") wins 100 points in case there is a line of white marks in the current state. While the roles in a game are defined as ground facts, the auxiliary relation `line` can be resolved down to facts holding in a state. These facts, called fluents, have the form `(fluent arg_1 ... arg_n)` and can be queried with the keyword `true`. We refer to [12] for the complete specification of GDL.

2.3 Knowledge-Based State Value Functions

Existing knowledge-based GGP agents, such as [11, 2, 13], employ different approaches to construct a state value function that gives a degree of preference of non-terminal states to guide search in an instance of the game (a match). For this purpose they analyze the game description and extract position features to be used in the value function. The automatic construction of an evaluation function takes place in three phases as follows.

In the first phase, game-specific features are identified. To begin with, all of the above mentioned systems attempt to detect static structures (that is, which do not depend on the current state of a game) such as successor relations. E.g., `(succ 2 3)` may define the constant `3` as the successor of `2` in a specific ordering, which then may be used to define an ordering over possible values for a variable which appears somewhere in the game rules. Note that this is a syntactic rather than a lexical property.[2] Successor relations are typically used to represent quantities, counters, or coordinates. On the other hand, constants without any underlying order often represent markers (like in Pentago) or pieces (unique markers as the king in Chess).

These identified static structures can then be used to detect high-level structures. For example, a fluent with arguments from a successor relation may be interpreted as a board, where the other arguments of this fluent either describe the contents of the individual cells (e.g., markers or quantities) or identify a specific instance of the board. The Pentago board, for example, is represented by `(cellholds ?q ?x ?y ?player)`, where `?x` and `?y` appear in a successor relation and where an instance of this fluent holds for every pair (`?x`, `?y`) in every state. Thus, `?x` and `?y` can be identified as coordinates. Because specific instances `?x` and `?y` can occur together with several instances of `?q`, the latter must be part of the coordinates as well. But `?q` does not occur in an underlying order, so that this argument is identified as marking different instances of the board. The remaining argument `?player`, then, must describe the contents of the board cell. Thus, with the coordinates `(?x ?y)`, with the four possible values for `?q`, and with `?player` describing the board content, as a high-level structure of Pentago one can identify four 3x3 boards, one for each quadrant.

In a similar fashion, quantities and other structural elements can be detected. The identified structures can be combined arithmetically in order to construct features out of them. In this way, a knowledge-based GGP system can measure, for example, the distance of a pawn to the promotion rank in chess-like games, or the difference in the number of pieces each player has still on the board in the game of checkers, etc.

Other, game-independent features may be added, such as "mobility," which is a measure of how much control a player has in a given game state by counting the number of available moves. Another feature may be "material," characterizing an advantage in the number of markers or pieces on a board (if applicable).

In the second phase of the automatic construction of an evaluation function, useful features are selected. Common conditions for selecting a feature are stability (e.g., the absence of wild oscillation over its domain) [2], an estimation for the effort it takes to compute a high-level feature, correlation with winning or losing terminal states, and the appearance in the goal formulas of a game [13].

In the third and last phase, the selected features are combined in a single state value function. Either the features are combined with fixed or estimated weights according to some predefined scheme, or the goal function of the game is fuzzified [13]. In the latter case, comparison relations that occur in the goal function of the game can be substituted by the features detected in the first phase. In this way, a boolean winning condition of having more markers than

[2] This can be identified by checking whether the relation in question is a singleton for each constant in any argument position, and whether its graph representation is acyclic.

the opponent, for example, can be mapped onto an arithmetic condition. This allows the player to prefer a lead with a higher margin to one with a smaller margin.

2.4 Neuro-Symbolic Integration

Symbolic knowledge offers the advantage to draw precise conclusions, provided there is a sound and sufficiently large knowledge base. Often, however, the knowledge base is insufficient, incorrect, or simply unavailable. Learning systems, on the other hand, can adapt to various domains, but require prior training and are prone to errors due to indiscriminate or too few learning examples as well as bad initialization. Neuro-Symbolic Integration attempts to unify both views to maintain their advantages while eliminating drawbacks by exploiting synergies. KBANN (Knowledge-Based Artificial Neural Network [16]) was the first approach to convincingly utilize both paradigms. Initialized with an "approximately correct" propositional domain theory, the network was automatically generated and trained and eventually led to a correction of the initial theory. Furthermore, KBANN learned much faster than randomly initialized nets and was able to qualitatively outperform any other algorithm during training as well as afterwards. While KBANN allowed the application of a standard learning algorithm, the algorithm still had some weaknesses that made it more complex than necessary and limited its application to domain theories with only a small number of rules and antecedents per rule.

In [9] it was shown that three-layer recurrent neural networks can represent propositional logic programs without any restrictions regarding the number of rules or antecedents. Learning via backpropagation, however, was not intended. Though it was possible to enhance this result to some extent to first-order logic (see [1] for an overview), it remains an open question whether this purely theoretical algorithm can be put into practice for a problem domain that is as challenging as GGP.

A combined approach was finally presented in [4], allowing an intuitive one-to-one transformation of propositional logic to neural nets while maintaining learning capabilities. We will use a more general version [3] by the same author.

3 From Domain Knowledge to State Values

Using the standard approaches explained above to identify and select features for state evaluation, we pay particular attention to the construction of the value function. As described, existing GGP systems use evaluation functions which are mathematical combinations of individual features. Though weights or similar measures of influence are determined during the analysis of the game rules, the functions are structurally fixed prior to the analysis and only cover a small part of the space of possible state value functions, which may not include useful (let alone optimal) value functions for the game at hand. Moreover, these functions remain static and do not adapt to playing experience. This is partly due to the one-time challenge situation enforced by the GGP championship rules that propose games as unique and non-repeating within the contest [8]. As a consequence, learning over matches is not applicable within this context. Nevertheless, the ability to adapt would allow for constant refinement of the evaluation function if the underlying formulas for the heuristics were designed in a dynamic way.

For this purpose, we will use neural networks as an established method to approximate functions. The idea is that by applying a neuro-symbolic translation algorithm, we can transform a ground-instantiated goal condition encoded in GDL to a bipolar neural net with a sigmoidal activation function that exactly captures its behavior and need not be trained. In this way, we would obtain a baseline state value function that is correct for all terminal states and provides a measure of preference for all non-terminal states by calculating fluent-wise their similarity to goal states.

3.1 Goal Conditions as Propositional Proof Trees

Consider the set of rules of a specific game defined in GDL. The given goal formulas define for every terminal state and for every role `?p` in the game a goal value `?gv`. For any specific rule of the form `(<= (goal ?p ?gv) conditions)`, the given conditions can be evaluated for any current (terminal or non-terminal) state. We can ground-instantiate a goal clause by iteratively substituting every free variable by the ground terms obtained from the signature of the game description. Algorithm *prop*(*argument*) (see Table 1) transforms a ground-instantiated goal condition to a propositional proof tree.

As some of the facts are static (that is, do not depend on the current state such as successor relations), they can instantly be proved to be false or true and subsequently be replaced by a propositional *false* or *true*. Naturally, this may lead to other simplifications, which can easily be implemented and hence shall not be explained further.

By calling *prop* on a ground-instantiated goal condition, we can thus obtain a propositional proof tree whose edges are either identities or negations, whose leafs are queries of facts in the current state, and whose non-leaf nodes are either conjunctions or disjunctions.

Table 1. Transforming a single goal condition to a propositional proof tree

if argument is		then
a fact	`(true fact)`	return $fact$
a negated argument	`(not argument)`	return $\neg prop(argument)$
the head of a clause	`(<= h b_1)` ... `(<= h b_n)`	return $\bigvee_{1 \le i \le n} prop(b_i)$
the body of a clause	`(c_1, ..., c_n)`	return $\bigwedge_{1 \le i \le n} prop(c_i)$
an inequality	`(distinct x1 x2)`	return $false$ if $x1 = x2$, else $true$

3.2 Propositional Proof Trees as Neural Networks

As has been shown in [16, 9], it is possible to encode propositional logic as a neural network. The two algorithms, however, were developed under different considerations. While [16] was mainly concerned with eliminating disadvantages of neural nets by the use of logic (namely, indifferent initialization, convergence problems, and long training time), [9] translated logic programs to neural nets so as to benefit from advantages of the different representation. An algorithm that got rid of the problems of each of the two approaches is described in [3] and will be used here with some modifications. The original algorithm was used for logic programs where literals in the body of a clause were interpreted as their conjunctions, while clauses with the same head were interpreted as disjunctions of their bodies. Therefore, we use the rules for encoding multiple clauses with the same head for disjunctions, and literals within a clause as rules for conjunctions. Furthermore, the net was built up as a recurrent net to implement the immediate consequent operator of the logic program. As this stepwise fashion is not needed, we build up a simpler feed-forward architecture that is similar to the propositional proof tree.

Consider a layered feed-forward neural net with the following properties: if neuron i is a successor of neuron j (e.g. there is a directed connection from neuron j to neuron i), then the output of neuron i is defined as $o_i = h(\sum_j w_{ij} o_j)$, where $h(x)$ is the bipolar activation function $h(x) = \frac{2}{1+e^{-\beta x}} - 1$. We define the interval $(A_{min}, 1]$ to denote truth and $[-1, A_{max})$ to denote falsity. Without loss of generality we set $A_{min} = -A_{max}$. Then we can translate every node of the propositional proof tree to a neuron by calling the following algorithm on the root of the tree and an "empty" neural net:

```
1 prop_to_net(node, net) {
2   if (count_children(node) > 0) {
3     for each (c in children(node)) {
4       if (is_positive(c))
5         add_conn(net, node, c, W);
6       else
7         add_conn(net, node, c, -W);
8       prop_to_net(c, net);
9     }
10    add_conn(net, node, bias, threshold_weight(node));
11  }
12  else
13    mark_as_input_node(net, node);
14 }
```

Basically, the algorithm does the following: for each non-leaf node (line 2) in the propositional tree an edge is added from the node to each of its children with weight W if the child is positive (that is, not negated; line 5) and $-W$ otherwise (line 7). The algorithm is called recursively for each child of the current node (line 8), and a connection to the bias unit (whose output is always 1) is added so as to function as a threshold. Its weight is determined by the type of the node (conjunction or disjunction) and the number n of its children:

$$threshold_{disj}(n) = \frac{(1+A_{min})*(n-1)}{2} * W \quad (1)$$

$$threshold_{conj}(n) = \frac{(1+A_{min})*(1-n)}{2} * W \quad (2)$$

Leafs of the tree are marked as input nodes (line 13) and used to evaluate the current state s in the match: if $fact$ holds in s, the node returns 1, else -1.

A_{min} / A_{max} and W are subject to some restrictions to ensure logically sound behavior:

$$A_{min} > \frac{\max(disj, conj) - 1}{\max(disj, conj) + 1} \quad (3)$$

$$W \geq \frac{2}{\beta} * \frac{ln(1 + A_{min}) - ln(1 - A_{min})}{\max(disj, conj)(A_{min} - 1) + A_{min} + 1} \quad (4)$$

Here, $disj$ is the largest number of children that a disjunction node in the tree has, while $conj$ is the largest number of children a conjunction node has. Parameter β controls the steepness of the activation function $h(x)$ and is usually set to 1. Note that the absolute weight w_{ij} of each connection has to be increased by a small random float to avoid symmetry effects when learning.

Encoding the goal conditions as neural network thus maintains the logical sound behavior of an automated proof procedure (for a proof see again [3]) while enabling it to perform "soft computing" and to adapt to experience via a standard backpropagation algorithm.

Practical Aspects of *C-IL²P*. Though the algorithm is logically sound, its application in the GGP domain is not as straightforward. Recall that the interval $[-1, A_{max})$ denotes falsity and $(A_{min}, 1]$ truth of a fact. A neuron's output representing a conjunction is thus higher than A_{min} iff the output of all positive preceding neurons is higher than A_{min} and of its negative preceding neurons is lower than A_{max}. Due to the monotonicity of the activation function, we know the output of a neuron to be lower if fewer of its antecedents are fulfilled. A problem, however, arises if no or few antecedents are fulfilled: in these cases the absolute value of the neuronal activation is high and, because of the squashing of the activation function, the output of the neuron stays approximately the same. In other words, for conjunction neurons with few fulfilled antecedents the derivative $h'(x) = 1 - h(x)^2$ is near zero and small changes in the input (e.g. one preceding neuron changing from -1 to $+1$) therefore have only little impact on the output of the neuron. While these differences of the output are small but still recognizable in the case of just one neuron, they eventually become smaller than machine precision after passing through several neurons. As a consequence, the neural network loses its ability to effectively guide search as it erroneously evaluates several states to the same value although there is a difference in their state value.

Regarding the network parameters, there are two reasons for this behavior. The first is the restriction of A_{min} (equation 3): while for a small number of rule antecedents k, A_{min} may be set to a small value as well, higher k fix it near 1 with the result that the intervals for truth and falsity become too small to, e.g., distinguish one neuronal state representing *false* from another. In the same way one can argue that with a high A_{min} the "forbidden zone" $[A_{max}, A_{min}]$ (the interval of output values "between *false* and *true*") becomes very large, resulting in a loss of output space, i.e., output resolution of the neuron.

Another reason is the restriction of the weight W (equation 4): for high values of W the space of activation values is larger, the absolute neuronal input is more likely far away from zero and thus its derivative close to zero. High absolute weights therefore contribute to the problem.

We are, however, not interested in a strict propositional evaluation of a goal condition, but in a degree of preference between states. Specifically, the following two conditions have to be satisfied:

- A terminal state fulfilling the underlying goal function will yield the greatest possible network output.
- Any state will yield a smaller value than another one if it constitutes a worse matching of the corresponding goal function, that is, it is fluent-wise less similar to the state pattern that corresponds to the goal function.

Therefore, we calculate the lower bound $l_A = \frac{\max(disj,conj)-1}{\max(disj,conj)+1}$ for A_{min} and set it such that it enables the smallest possible weight. With $A_{min} \in (l, 1)$, obviously the following condition is satisfied:

$$A_{min} = (l_A - 1) * \alpha + 1 \tag{5}$$

$$\alpha \in (0, 1) \tag{6}$$

We numerically determined W to be minimal for $0.15 \leq \alpha \leq 0.3$ and set it to $\alpha = 0.2$. As this is not enough, we additionally ignore the weight condition (equation 4) and set it to a fixed value $W = 1$.

3.3 From Neural Networks to State Values

To finally obtain a state value, we map the output of the neural networks $o_{player,gv} \in [-1, 1]$ to values in $[0, 1]$ and multiply these with the corresponding goal value. In this way, we obtain the share $v_{player,gv}(s)$ that each pair of network and goal value contributes to the state value. The normalized sum of these shares then forms our value function in the interval $[0, 100]$.

$$v_{player,gv}(s) = \frac{o_{player,gv}(s)+1}{2} * gv^p \tag{7}$$

$$V_{player}(s) = \frac{\sum_{gv \in GV} v_{player,gv}(s)}{\sum_{gv \in GV} gv^p} * 100 \tag{8}$$

With the selection of $p \in [1, \infty)$ we can choose the degree to which higher goal values are more influential on the state value function. For $p = 1$ every goal value is considered linearly and for $p \to \infty$ only the highest goal value is considered, resulting in a riskier behavior.

As we have to ensure that any calculated state value is better than a real loss (where $gv = 0$) and worse than a real win ($gv = 100$), the final state value is set to $V(s)' = V(s) * 0.98 + 1$.

3.4 Increasing Flexibility

The resulting set of neural networks can be enhanced in several ways. Any feature obtained by domain analysis or designed manually can be introduced in the networks by adding an input node for it, normalizing the output of the node to $[-1, 1]$, and connecting it to the output nodes of all networks for a player. In much the same way, spatial or quantitative comparisons in the goal conditions can be substituted by a more detailed feature to increase expressiveness.

3.5 Learning

As result of the transformation we get a set of neural networks to which back-propagation learning can be applied. To train these neural networks, we use terminal states s_{term} from past matches which can be obtained by self-play or are a by-product of probabilistic look-ahead searches as applied in Monte-Carlo search [6, 7]. The goal value of the terminal state indicates the corresponding network which then can be trained by presenting value 1 as signal for the output neuron. All other networks are trained with -1.

We can further increase the amount of available examples by applying the TD(λ) algorithm as used, e.g., in TD-Gammon [15]. By assuming the predecessors of a terminal state to likely leading to the goal value of the terminal state, we can use these predecessors as weaker but still valid training examples. We therefore discount the training signal relative to the distance to the terminal state in order to weaken the impact of the predecessor state on the neural network.

The training signal $t(s)$ for the k-th predecessor of the terminal state s_{term-k} is calculated thus:

$$t(s_{term-k}) = t(s_{term}) * \lambda^k \qquad (9)$$

The parameter $\lambda \in [0, 1]$ controls the speed of decay of the training signal. The network corresponding to the terminal state is thus trained with the signals $1, \lambda, \lambda^2, \ldots$ for the terminal state, its predecessor, the predecessor of the predecessor, and so on.

We further enhance TD(λ) to include domain knowledge by checking the predecessors of the terminal state subject to learning whether they inevitably lead to a terminal state with the same goal value. This is done by employing a full-depth game tree search and, in case the search yields a positive answer, substituting a potentially flawed training signal by a provably correct and more expressive one. Note that the required tree search was already performed during the match and can be reused here.

Assuming s_{term} leads to a goal value $gv(s_{term})$ we can thus train the corresponding neural network with the following signal:

$$t(s_{term-k}) = \left\{ \begin{array}{l} 1: \ minimax(s_{term-k}) = gv(s_{term}) \\ \lambda^k : else \end{array} \right\} \qquad (10)$$

Altogether, the network training over several states of one match has two major benefits:

- The network prefers stable features over wildly oscillating ones. The importance of this concept has been discussed in [2] and is implicitly included here.
- The learned patterns match states preceding the terminal states. Along with feedforward search during matches the algorithm thus exhibits a behavior comparable to bidirectional search.

3.6 Transformation of Pentago Rules

For better understanding we will explain the approach in the domain of the game Pentago by showing the result of the transformation for the winning goal condition for the player with the role "white". The goal condition in this game is given as `(<= (goal white 100) (line white))`. A white line, according to the rules, can be a white row, column, or diagonal. As the three cases are analogous, we will just concentrate on white winning with a row. The definition of auxiliary predicate `(row ?player)`, after substituting `?player` by `white`, is:

```
(<=  (row white)
     (role white)
     (true (cellholds ?q1 ?x1 ?y1 white))
     (globalindex ?q1 ?x1 ?y1 ?x1g ?yg)
     (succ ?x1g ?x2g)
     ...
     (succ ?x4g ?x5g)
     ...
     (true (cellholds ?q5 ?x5 ?y5 white))
     (globalindex ?q5 ?x5 ?y5 ?x5g ?yg))
```

Hence, a white row is implied by a conjunction of the following facts. The expression `cellholds` is a fluent which describes the contents of individual cells in a game state. Relation `globalindex` is essentially a function which translates the local quadrant coordinates `?q1`, `?x1`, and `?y1` to the global coordinates `?x1g` and `?yg`. Predicate `succ` describes a successor relation that connects five instances of `cellholds` such that the global x-coordinate of the second cell is a successor of the global x-coordinate of the first cell, and so on. All `cellholds` fluents need to refer to the same y-coordinate `?yg`. The head of the above clause, `row(white)`, is implied if there exists an instantiation for the 21 free variables such that each of the atoms in the body holds.

To translate this goal condition to a neural network, we have to ground-instantiate the rule. A naive ground instantiation would yield at least 3^{21} instances, as each domain of the variables has a size greater or equal to 3. While instantiations of this size cannot be handled efficiently, the inherent structure of the rule can be easily exploited. Because successor relations are functional, each variable `?xg2, ..., ?xg5` is in fact a function of `?xg1`. The domain of `?xg1` shrinks accordingly, indicating that the starting point of a row must have x-coordinate 1 or 2; if it started at 3, the penultimate cell in the row would have no successor.

The `globalindex` relation is functional and injective as well, allowing to bring down the number of different ground instances of the rule to 12, where all variables are functions of the global coordinates A1, A2, B1, ..., F2. With these twelve possible rows on the board we end up with a proof tree consisting of a disjunction of the rows with each of them being a conjunction of five adjacent cells. The `role` relation can be statically evaluated to *true* and thus immediately omitted in the conjunctions.

The resulting net consists of an input layer with 36 nodes of the form `(true (cellholds ?q ?x ?y white))`, a first hidden layer representing the 12 instances of the `row` rule as a conjunction of five adjacent cells, and a second hidden layer representing the generalized `row` rule as a disjunction of the twelve ground-instantiated ones. A similar structure is added for the possibility of a vertical or a diagonal white line, resulting in a total of three neurons in the third hidden layer. The output node would then again be a disjunction of these three cases.

4 Experiments

For testing purposes we implemented a standard General Game Player according to the approach described above. To evaluate its quality and efficiency, we let it play against "Fluxplayer" [13], the best non-simulation based system, ranked at least third in the three recent AAAI GGP World Championships 2006-2008. We refer to the agents with "Neuro" and "Fluxplayer" respectively throughout the experiments.

4.1 Experimental Setup

The tests were run on a dual-processor system at 3.16 GHz with 768 MB RAM available for each agent. The roles were switched after each match, thus forming pairs of matches in which each agent had the first move once. Both agents used α-β-search. For learning, we used all "solved" states that could be proven to lead to a terminal state with the same goal value. If the number of these states was less than 10% of the states occurring in the match, "unsolved" states were used in addition. In this way, at least 10% of the match states were used for training. To distinguish solved from unsolved states, we reused the values calculated in the match. Unsolved states were discounted by the factor of $\lambda = 0.8$ and the learning rate was set to 0.001.

Table 2. Average Goal Value achieved

	#Matches	Pentago		3D-Tic-Tac-Toe	
		Fluxplayer	Neuro	Fluxplayer	Neuro
Init 1-ply	300	57.33	42.67	50	50
Learning 1-ply	5x500	25.72	74.28	50	50
Init Real-Time	300	48.36	51.64	38.67	61.33
Learning Real-Time	5x500	45.46	54.54	45.5	54.5

The tests were run in two different games: Pentago and a 3D version of Tic-Tac-Toe where the first player to achieve a line of four marks in a 4x4x4 cube wins. The results can be seen in Table 2. For each of the two games we examined the two dimensions *Init vs Learning* and *1-ply vs Real-Time*:

Init 1-ply. The quality of the initialized evaluation function was determined in 300 matches with a search depth of 1 and newly initialized networks in each match.

Learning 1-ply. The improvement of evaluation accuracy through learning was determined by running 500 consecutive matches, starting with a newly initialized network and searching with depth 1. The test was run 5 times to minimize possible random effects.

Init Real-Time. The quality of the initialized evaluation function in a real-time scenario was determined in 300 matches with newly initialized networks in each match. Each system had 100 seconds before a match started and 5 seconds for each move.

Learning Real-Time. The improvement of playing strength through learning in a real-time scenario was determined by running 500 consecutive matches starting with a newly initialized network. Each system had 100 seconds before a match started and 5 seconds for each move. The test was run 5 times to minimize possible random effects.

4.2 Results

As can be seen, the 1-ply performance differs in the two games. In Pentago, Neuro wins 43% in the initial scenario, but reaches 74% when learning, showing that the initialized networks are near a local optimum.

In 3D-Tic-Tac-Toe, however, the player who has the first move has such a big advantage that it can enforce a win, resulting in a 50% win rate each. As no examples for how to win as non-starting player are available, learning has no impact on the win rate.

The real-time performance of Neuro in both games is at least equal to that of Fluxplayer. In Pentago, the initial real-time performance lies at roughly 52%. This can be partly attributed to a state evaluation rate approximately twice as high as that of Fluxplayer. The result is improved by another 3% with learning.

In 3D-Tic-Tac-Toe, the initial real-time performance with a 61% win rate is significantly higher than that of Fluxplayer. For this game, Neuro examines

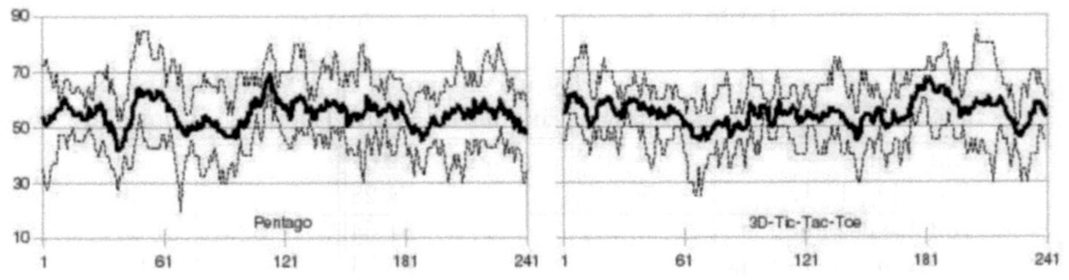

Fig. 2. Evolution of Win Rate in the Learning Real-Time Scenario

about 30% less states than Fluxplayer. The results show that learning decreases the win rate, which might indicate overfitting.

In general, it can be seen that learning has a positive effect if done with caution. Figure 2 depicts the evolution of the win rate in each game by plotting the moving maximum, average and minimum of the recent 10 pairs of matches. Obviously, learning did not converge. However, Neuro achieved in each test run in both games at least a 70% win rate over at least 20 matches at some point throughout the experiments, indicating its potential.

5 Discussion

As can be seen from our initial experiments, the neural approach to General Game Playing can prove advantageous when compared to the currently best non-simulation based system. The relative evaluation quality and, in particular, the real-time performance both depend on the type of the game. Given that the system works in principle while still being in an early development stage regarding possible optimizations, it is likely to outperform the current version of Fluxplayer after thorough parameter tuning and full exploitation of its learning and extension capabilities.

We have seen that learning leads to a significant boost in evaluation quality in the 1-ply scenario in Pentago. On the other hand, it only had a small effect in the real-time scenarios, indicating an overfitting of the network. The drop in win-rate in Pentago when switching from "learning 1-ply" to "learning real-time" as well as the lack of convergence of the win rates support this explanation.

While the initial results look promising and should be put on a broader empirical basis, there are some further theoretical problems related to the ideas presented in this paper. The major disadvantage of the approach is that it requires the use of ground-instantiated fluents. Depending on the complexity of the game, the number of ground instances is subject to combinatorial explosion and can thus easily become too large to be handled by a neural network in an efficient manner. Although there are possibilities to encode first-order logic in neural networks to some degree, current results are not convincing, as they are bound by machine precision for real numbers [14] or have not yet been applied to non-toy examples.

For this reason, we employ several techniques to limit or work around the problem for the time being.

- By exploiting domain constraints, we limit the number or the domain of free variables as demonstrated in the Pentago example.
- Clauses like `(<= c (a ?x) (b ?x))`, where variable bindings hold over more than just one antecedent, or clauses with too many possible ground instances, are redirected to the automated theorem proving system. In this case, we do not transform the clause, but add it as an input node.

Other problems are directly related to learning and can be seen as a consequence of the induction principle. For example, we have to assume the training states to be distributed among the game tree and opponents to play near optimal to avoid overfitting or biasing of the networks.

6 Conclusions

We have presented a method that overcomes the major restrictions of today's knowledge-based GGP systems. It allows to derive a state value function from the goal definition for a game, provided it can be ground-instantiated with reasonable effort. The goal conditions are used to initialize neural networks such that they need not be trained. The evaluation function can then benefit from training with states extracted from past matches, as our experimental results have shown.

6.1 Future Work

The overall evaluation quality is primarily based on the initial evaluation quality and its increase through training, making both the main areas of possible improvement. As none of the parameters used in the experiments were optimized, an analytical parameter determination of the $C\text{-}IL^2P$ parameters (A_{min}, W) and an autonomous adaptation of the learning parameters (learning rate, discount factor, ...) to specific games could further improve evaluation quality.

Furthermore, we intend to exploit the newly gained flexibility by adding new connections or hidden layer nodes to allow for the emergence of new features or by integrating features generated by other means. In fact, the connection weights of the networks offer a utility feedback for those features that could give rise to feature generation algorithms like in [5].

For the future, we intend to conduct further experiments in other games, address the above issues, and implement a complete GGP system on the basis of the approach.

References

[1] Bader, S., Hitzler, P.: Dimensions of neural-symbolic integration — a structured survey. In: We Will Show Them: Essays in Honour of Dov Gabbay. Kings College Publications, pp. 167–194 (2005)

[2] Clune, J.: Heuristic evaluation functions for general game playing. In: Proceedings of the AAAI National Conference on Artificial Intelligence, Vancouver, pp. 1134–1139. AAAI Press, Menlo Park (2007)
[3] d'Avila Garcez, A., Broda, K., Gabbay, D.: Neural Symbolic Learning Systems. Springer, Heidelberg (2002)
[4] d'Avila Garcez, A., Zaverucha, G., de Carvalho, L.: Logical inference and inductive learning in artificial neural networks. In: Proceedings of the ECAI Workshop on Neural Networks and Structured Knowledge (1996)
[5] Fawcett, T.: Feature Discovery for Problem Solving Systems. PhD thesis, University of Massachusetts, Amherst (1993)
[6] Finnsson, H.: Cadia-player: A general game playing agent. Master's thesis, School of Computer Science, Reykjavík University (2007)
[7] Finnsson, H., Björnsson, Y.: Simulation-based approach to general game playing. In: Proceedings of the AAAI National Conference on Artificial Intelligence, Chicago, pp. 259–264. AAAI Press, Menlo Park (2008)
[8] Genesereth, M., Love, N., Pell, B.: General game playing: Overview of the AAAI competition. AI Magazine 26, 62–72 (2005)
[9] Hölldobler, S., Kalinke, Y.: Towards a massively parallel computational model for logic programming. In: Proceedings of the ECAI Workshop on Combining Symbolic and Connectionist Processing, pp. 68–77 (1994)
[10] Hsu, F.: Behind Deep Blue: Building the Computer that Defeated the World Chess Champion. Princeton University Press, Princeton (2002)
[11] Kuhlmann, G., Dresner, K., Stone, P.: Automatic Heuristic Construction in a Complete General Game Player. In: Proceedings of the AAAI National Conference on Artificial Intelligence, Boston, pp. 1457–1462. AAAI Press, Menlo Park (2008)
[12] Love, N., Hinrichs, T., Haley, D., Schkufza, E., Genesereth, M.: General Game Playing: Game Description Language Specification. Technical Report LG–2006–01, Stanford Logic Group, Computer Science Department, Stanford University, 353 Serra Mall, Stanford, CA 94305 (2006), `http://games.stanford.edu`
[13] Schiffel, S., Thielscher, M.: Fluxplayer: A successful general game player. In: Proceedings of the AAAI National Conference on Artificial Intelligence, Vancouver, pp. 1191–1196. AAAI Press, Menlo Park (2007)
[14] Hitzler, P., Bader, S., Witzel, A.: Towards a massively parallel computational model for logic programming. In: Proceedings of the IJCAI Workshop on Neural-Symbolic Learning and Reasoning (2005)
[15] Tesauro, G.: Temporal difference learning and TD-gammon. Communications of the ACM 38, 58–68 (1995)
[16] Towell, G., Shavlik, J., Noordenier, M.: Refinement of approximate domain theories by knowledge based neural network. In: Proceedings of the AAAI National Conference on Artificial Intelligence, pp. 861–866 (1990)

Learning to Disambiguate Search Queries from Short Sessions

Lilyana Mihalkova and Raymond Mooney

The University of Texas, Austin
Department of Computer Sciences
1 University Station C0500
Austin, Texas 78712-0233
{lilyanam,mooney}@cs.utexas.edu

Abstract. Web searches tend to be short and ambiguous. It is therefore not surprising that Web query disambiguation is an actively researched topic. To provide a personalized experience for a user, most existing work relies on search engine log data in which the search activities of *that particular user*, as well as other users, are recorded over *long* periods of time. Such approaches may raise privacy concerns and may be difficult to implement for pragmatic reasons. We present an approach to Web query disambiguation that bases its predictions only on a short glimpse of user search activity, captured in a brief session of 4–6 previous searches on average. Our method exploits the relations of the current search session to previous similarly *short* sessions of other users in order to predict the user's intentions and is based on Markov logic, a statistical relational learning model that has been successfully applied to challenging language problems in the past. We present empirical results that demonstrate the effectiveness of our proposed approach on data collected from a commercial general-purpose search engine.

1 Introduction

Personalizing a user's Web search experience has become a vibrant area of research in recent years. One of the most actively researched topics in this area is Web query disambiguation, or automatically determining the intentions and goals of a user who enters an ambiguous query. This is not surprising, given the frequency of ambiguous searches and the unwillingness of users to enter long and descriptive queries. For example, Jansen and Spink [1] found that about 30% of search queries, submitted to several engines, consisted of a single word. Furthermore, Sanderson [2] reports that anywhere between roughly 7% and 23% of the queries frequently occurring in the logs of two search engines are ambiguous, with the average length of ambiguous queries being close to one.

Ambiguity exists not only in cases such as the all-too-familiar "jaguar" example (which can be a cat, car, or operating system), but also in searches that do not appear ambiguous on the surface. Queries that are commonly considered unambiguous often become ambiguous as a result of the wealth of Web sources,

W. Buntine et al. (Eds.): ECML PKDD 2009, Part II, LNAI 5782, pp. 111–127, 2009.

which examine different aspects of a given topic. For example, as we observed in our data, a search for "texas"[1] may be prompted by at least two different kinds of intentions. In one session, a user who had first searched for "george w. bush" proceeded to search for "texas" and selected `www.tea.state.tx.us`, thus indicating an interest in Texas government agencies. In another session, the user intended to learn about travel to Texas because repeated searches for "georgia travel" were followed by a search for "texas" and a click to `www.tourtexas.com`. This indicates that even a query, such as "texas" that normally refers to a single entity, may become ambiguous.

Most approaches to Web query disambiguation leverage a user's previous interactions with the search engine to predict her intentions when entering an ambiguous query. Typically, the actions of each user are logged over long periods of time, e.g., [3,4,5]. While techniques that assume the availability of long search histories *for each user* are applicable in some situations, in many cases such approaches may raise privacy concerns and may be difficult to implement for pragmatic reasons. After the release of AOL query log data allowed journalists to identify one user based on her searches [6], many people have become especially wary of having their search histories recorded. To address such concerns, we present an approach that bases its predictions only on short glimpses of user search activity, captured in a brief search session. Our approach relates the current search session to previous *short* sessions of *other* users based on the search activity in these sessions. Crucially, our approach does *not* assume the availability of user identifiers of any sort (i.e. IP addresses, login names, etc.) and thus such information, which could allow user searches to be tracked over long periods of time, does not need to be recorded when our approach is used.

As an example, consider the query "scrubs," which could refer either to the popular television show or to a type of medical uniform. Table 1 juxtaposes the users' actions in two sessions. The sessions are short, with each containing only two searches preceding the ambiguous query; nevertheless, this short glimpse of the users' actions is sufficient to provide an accurate idea of the users' intentions because by examining historical data, one may discover that people who search for radio stations are probably "ordinary" users and would therefore be interested in the television show. On the other hand, by relating Session 2 to sessions of other users who searched for medical-related items, we may be able to predict that the second user has more specialized interests.

Our proposed approach is appealing also from a pragmatic standpoint because it does not require search engines to store, manage, and protect long user-specific histories. Identifying users across search sessions is another difficulty arising from methods based on long user-specific search histories. One possibility, to require users to log in before providing personalized search, may be cumbersome. The alternative of using as an identifier the IP address of the computer from which the search was initiated is also unsatisfactory, especially in cases when entire organizations share the same IP address or when all members of a household

[1] We write these queries in lower-case because this is how they were typed by the searchers in our data set.

Table 1. Two sessions in which the users searched for the query "scrubs"

—Search Session 1—		—Search Session 2—	
98.7 fm	→ www.star987.com/main.html	huntsville hospital	→ www.huntsvillehospital.org
kroq	→ www.kroq.com/	ebay.com	→ ebay.com
scrubs	→ scrubs-tv.com	scrubs	→ www.scrubs.com

search from the same computer. Disambiguation techniques that explicitly do not use such identifiers and instead rely only on information from brief sessions avoid such difficulties.

When so little is known about a searcher, the problem of query disambiguation becomes very challenging. In fact, it has previously been argued that "it is difficult to build an appropriate user profile even when the user history is rich" [5]. We develop an approach that successfully leverages the small amount of information about a user captured in a short search session to improve the ranking of the returned search results. Our approach is based on statistical relational learning (SRL) [8] and exploits the relations between the session in which the ambiguous query is issued and previous sessions.

SRL addresses the problem of learning from multi-relational data models that support probabilistic reasoning. SRL is appealing for our problem because, first, the data is inherently relational—there are several types of entities: queries, clicked URLs, and sessions, which relate to each other in a variety of ways, e.g., two sessions may be related by virtue of containing clicks to the same URLs; queries may be related by sharing words. Second, data recording human interactions with a search engine is likely to be noisy. SRL models allow for probabilistic inference, helpful when reasoning from noisy data.

We used Markov logic networks (MLNs) [9]. An MLN consists of a set of weighted formulae in first-order logic and defines a Markov network when provided with a set of constants. The probability of a possible world decreases exponentially in the weight of formulae it fails to satisfy. We chose MLNs because of their generality, their successful application to other language-related tasks, e.g., [11,12,13], and the availability of a well-maintained code base [14].

2 Related Work

Personalized search is an important problem that has been studied under many settings and assumptions. We review some of this research and draw distinctions between previous work and ours.

Several authors have proposed techniques addressing the case where, for each particular user, a relatively long history of that user's interactions with the search engine is available. Sugiyama *et al.* [3] present a personalization method that builds a user preference model by modeling separately the long-term and "today's" user interests. In addition to relying on long-term records of user activity, their approach also uses the content of browsed pages. In contrast, we are interested in a more light-weight approach that does not necessarily use page content. Sun *et al.* [4] use spectral methods to perform personalization by organizing the data into a three-dimensional tensor comprised of users, queries, and clicked pages. These tensor-based methods are unlikely to be effective in our case because of data sparsity. A comprehensive empirical study of Web search personalization techniques is presented by Dou *et al.* [5]. These techniques also use longer-term histories (up to 12 days) of the same user. The authors find that the best-performing methods are based on the intuition that the Web pages most relevant to a user are those clicked frequently in the past by that user or by related users, where user similarity is measured by estimating user membership in a pre-defined set of categories. Such a strategy is unlikely to work in our setting because the sessions in our data represent one-time interactions that usually do not contain repeated clicks to the same URL. Joachims [15] and Radlinski and Joachims [16] use a clever method for deriving constraints about user preferences by observing whether or not the user clicked on or skipped over particular search results. These preferences are then used to train a system for ranking search results. All work discussed in this paragraph assumes that long-term information about *each user* is available. In contrast, we study the setting where personalization is performed based on records of very short interactions with the search engine.

To the best of our knowledge, the only previous work that targets query disambiguation from short sessions is that of Almeida and Almeida [17] in which users are identified as belonging to a set of communities in order to determine their interests. The authors experimented with data from online bookstore search sites for computer science literature, and their approach is tailored for situations when user interests fall into a small set of categories, organizing users into 10 communities. While in a more restricted application of search, such as specialized book search, this small number of communities may be sufficient to model different aspects of user interests, if, as in our case, the goal is to disambiguate queries in a general-purpose search engine, a small number of communities is likely to be insufficient to effectively model the variety of user interests, and allowing for more communities may be prohibitively costly. Privacy-aware Web personalization has been addressed by Krause and Horvitz [18], whose method considers the privacy cost of a given piece of user information and explicitly models the improvement in personalization versus the cost of the used information. While the ability to trade off performance with cost is highly desirable, their method relies on more information about the user than is available to us.

Query disambiguation is also related to determining user goals and intentions, as done by the TaskPredictor [19], which learns to predict the current task of a user based on the properties of the currently open window, or of an arriving e-mail message. Because training this system requires potentially sensitive information, it is intended to be run on the user's local machine. Another project [20] that relies on sensitive user information studies ways for personalizing Web search by constructing a user profile from long-term observations on the user's activities, ranging from browsing history to e-mail.

An orthogonal issue is producing a diverse set of documents for a given query. Recent work includes that of Chen and Karger [21], whose technique ranks results so as to cover as many different aspects of interest as possible, and that of Yue and Joachims [22] whose technique is based on structural SVMs. A related area is that of clustering search results in groups of common topics. Wang and Zhai [23] use search log data to learn useful aspects of queries in order to cluster them. The ability to disambiguate user intent complements these contributions because it would allow the most relevant cluster, or the most relevant results from a diverse set, to be placed ahead of all others on the search page.

Collaborative filtering, where the goal is to suggest items that would be of interest to a user, based on that and other users' previous preferences, is also related. Early comparative studies of collaborative filtering algorithms include [24,25]. More recently, Popescul *et al.* [26] and Melville *et al.* [27] proposed approaches that combine collaborative and content-based information in forming recommendations. These were not applied to personalizing Web search.

3 Background

This section provides some necessary background on first-order logic and MLNs.

First-order logic uses 4 types of symbols—constants, variables, predicates, and functions [28]. **Constants** describe the objects in the environment, e.g., `www.ecmlpkdd2009.net` and `1acadc00158440d9` are constants representing a url and a sessionId. **Predicates** represent relations, such as `ClickOn`, and can be thought of as functions that evaluate to `true` or `false`. A **term** is a constant, a variable, or a function applied to terms. Ground terms contain no variables. An **atom** is a predicate applied to terms. A positive (negative) **literal** is a (negated) atom. For example, `ClickOn(www.ecmlpkdd2009.net, 1acadc00158440d9)` is a ground positive literal. Its value is `true` iff `www.ecmlpkdd2009.net` is clicked in session `1acadc00158440d9`. A possible world is a truth assignment to all possible ground literals in an environment. A first-order formula uses conjunction ($\wedge$) and disjunction ($\vee$) to combine positive and negative literals into a logical statement. A **grounding** is a ground formula or literal.

A Markov logic network (MLN) [9] consists of a set of first-order formulae, each of which has an associated weight. MLNs can be viewed as relational analogs to Markov networks whose features are expressed in first-order logic. In this way MLNs combine the expressivity of first-order logic with the ability of probabilistic graphical models to reason under uncertainty.

Let $\mathbf{X}$ be the set of all possible ground literals in the environment, $\mathcal{F}$ be the set of all first-order formulae in the MLN, and w_i be the weight of formula $f_i \in \mathcal{F}$. Then, the probability of a particular truth assignment $\mathbf{x}$ to $\mathbf{X}$ is given by $P(\mathbf{X} = \mathbf{x}) = \frac{\exp\left(\sum_{f_i \in \mathcal{F}} w_i n_i(\mathbf{x})\right)}{\sum_{\mathbf{x}'} \exp\left(\sum_{f_i \in \mathcal{F}} w_i n_i(\mathbf{x}')\right)}$ [9], where $n_i(\mathbf{x})$ is the number of groundings of f_i that are true given the truth assignment $\mathbf{x}$ to $\mathbf{X}$. Intuitively w_i determines how much less likely a world is in which a grounding of f_i is not satisfied than one in which it is satisfied. The first-order formulae are called the *structure*. By grounding the formulae of an MLN with the constants in the environment, one defines a Markov network, over which inference can be performed to determine the probability that each of a set of unknown ground literals is true, given the truth values of a set of evidence ground literals. In our case, the evidence literals, which we define in Section 4, provide information on the user activity in the current session and how it relates to previous search sessions, and the goal is to predict the probability that each grounding of the `clickOn` predicate is true. Several algorithms are available to perform inference over a ground MLN. We used MC-SAT [10], which has been demonstrated to give good performance.

4 Proposed Approach

Our general approach follows that of previous applications of MLNs to specific problems, e.g., [12]: we hand-coded the structure of the model as a set of first-order formulae and learned weights for these formulae from the data. The key idea behind our approach is to relate the current, *active*, session A in which an ambiguous query Q is issued to previous, *background*, sessions from historical data, where it is assumed that both the active session and the background sessions are short. Sessions are related by sharing various types of information. We define the following predicates to capture these relationships. Since every training/testing example refers to a single (Q,A) pair, A and Q are implicit in the example and do not need to appear as arguments of the predicates.

-`result(r)`: `r` is a search result for Q.
-`choseResult(s, r)`: Background session `s` clicked on `r` after searching for Q.
-`clickOn(r)`: User in session A clicks on result `r` in response to the search for Q.
-`sharesClick(s, d)`: Sessions `s` and A share a click to URL with hostname `d`.
-`sharesKeywordBtwnClicks(s, k)`: Background session `s` and A share a keyword `k`, found in the hostnames of clicked URLs in each of the sessions.
-`sharesKeywordBtwnClickAndSearch(s, k)`: Background session `s` and A share a keyword `k`, found in the hostname of a clicked URL in A and a search in `s`.
-`sharesKeywordBtwnSearchAndClick(s, k)`: Background session `s` and A share a keyword `k`, found in a search in A and the hostname of a clicked URL in `s`.
-`sharesKeywordBtwnSearches(s, k)`: Sessions `s` and A share a keyword `k` that appeared in searches in both sessions.
-`clicksShareKeyword(r, d, k)`: Keyword k appears in the hostname of both result r and previous click d from session A.
-`clickAndSearchShareKeyword(r, s, k)`: Keyword k appears in the hostname of result r and in previous search query s from session A.

Fig. 1 illustrates the predicates used to relate two sessions. The last two predicates capture information local to the active session. In the active session A, only the clicks and searches temporally *preceding* Q are used. For the predicates in which a keyword relates two sessions, we used only keywords that appeared at least 100 times (i.e., we removed keywords appearing less than 0.00083% of the time) and at most 10,000 times (i.e., we removed the top 61 most popular

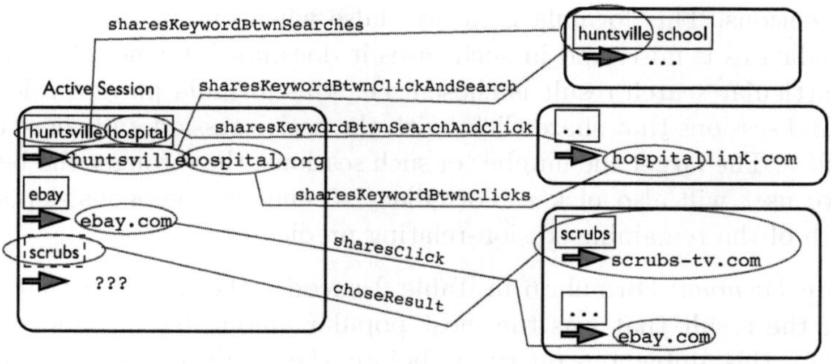

Fig. 1. An illustration of predicates that relate sessions. Tokens in boxes represent queries, whereas tokens preceded by an arrow represent the clicked result. The *active* session, on the left, is related to some of the *background* sessions, on the right, by shared clicks or keywords. Not all possible relations are drawn.

keywords) over our training data. This was done to avoid rare or misspelled keywords and to make the size of the data more manageable.

The goal is to predict the `clickOn(r)` predicate, given as evidence the values of the remaining ones. The search results available for a given query are then ranked by the predicted probability that the user will click on each of them.

4.1 Model Structure

This section describes the formulae used in our MLN models.

Collaborative Formulae: The collaborative formulae, shown in lines 1-5 of Table 2, draw inferences about the interests of the active user based on the choices made by related users from background sessions. For example, formula 1 establishes a relationship between the event that the active user chooses result r and the event that the user in a previous session s, related to the active session by sharing a click to a URL with hostname d, chose result r after searching for the current ambiguous query. This formula exploits one type of relation between the active session and background sessions to provide evidence of the active

Table 2. Formulae included in the model

1: result(r) ∧ sharesClick(s, d) ∧ choseResult(s, r) ∧ clickOn(r)
2: result(r) ∧ sharesKeywordBtwnClicks(s, k) ∧ choseResult(s, r) ∧ clickOn(r)
3: result(r) ∧ sharesKeywordBtwnClickAndSearch(s, k) ∧ choseResult(s, r) ∧ clickOn(r)
4: result(r) ∧ sharesKeywordBtwnSearchAndClick(s, k) ∧ choseResult(s, r) ∧ clickOn(r)
5: result(r) ∧ sharesKeywordBtwnSearches(s, k) ∧ choseResult(s, r) ∧ clickOn(r)
6: result(r) ∧ choseResult(s, r) ∧ clickOn(r)
7: result(r) ∧ clicksShareKeyword(r, d, k) ∧ clickOn(r)
8: result(r) ∧ clickAndSearchShareKeyword(r, s, k) ∧ clickOn(r)
9: result(r1) ∧ result(r2) ∧ r1 ≠ r2 ∧ clickOn(r1) ⇒ ¬clickOn(r2)

user's intentions. This formula is always false when one of the first three evidence predicates is false, and in such cases it does not influence the probability that a particular search result is chosen; i.e., this formula plays a role only for background sessions that share clicks with the active session and chose a particular result r. The larger the number of such sessions, the stronger the belief that the active user will also pick r. Formulae 2-5 encode analogous dependencies using each of the remaining session-relating predicates.

Popularity Formula: Formula 6 in Table 2 encodes the intuition that the user will click the result that was the most popular among background users that searched for this ambiguous query. As before, the result for which there are the largest number of clicks in background data, and thus the largest number of groundings of this formula that are not falsified by the evidence, will have the largest probability of being clicked.

Local Formulae: Formulae 7-8 in Table 2 use information local to the active session to predict the user's preferences. Formula 7 (8) states that the user will click a result that shares keywords with a previous result (search) from the active session. We clarify that keywords were *not* extracted from the pages to which a URL points, but only from the URL itself because we are interested in developing a light-weight re-ranker. Because in our setting sessions are very short, we do not expect the local formulae to contribute much to the overall model performance. We include them in order to verify this.

Balance Formula: Formula 9 in Table 2 sets up a competition among the possible results by stating that if the user clicks one of the results, the user will not click another one. This formula prevents all possible results from obtaining a very high probability of being clicked. This makes the model more discriminating and allows the same set of weights to perform well even as the number of groundings of the other formulae varies widely across active sessions.

These formulae encode "rules of thumb" and useful features, which we expect will hold in general, but may sometimes be violated, e.g., the balance formula is violated when a user clicks more than one result for a query. The ability of MLNs to combine such varied sources of information effectively and in a principled way is one of the main considerations that motivated our choice of model. Using these formulae, we defined three MLNs:

MLN 1 – Purely Collaborative: Contains only the collaborative formulae (1-5) and the balance formula (9).
MLN 2 – Collaborative and Popularity: Contains formulae 1-6 and the balance formula (9).
MLN 3 – Collaborative, Popularity, and Local: Contains all formulae. It can be viewed as a mixed collaborative-content-based model, e.g., [26,27].

4.2 Weight Learning

To learn weights for the structures defined above, we used the contrastive divergence algorithm (CD) described by Lowd and Domingos [29]. CD can be viewed as a voted-perceptron-like gradient descent algorithm in which the gradient for updating the weight of formula C_i is computed as the difference between the number of true groundings of C_i in the data and the expected number of true groundings of C_i, where the expectation is computed by carrying out a small number of MCMC steps over the model using the currently learned weights. Like Lowd and Domingos [29], we computed the expectations with MC-SAT [10]. We used the implementations of these algorithms in the Alchemy package [14], except that we adapted the existing implementation of CD so that learning can proceed in an online fashion, considering examples of sessions containing ambiguous queries one by one. This was done because otherwise our data was too large to fit in memory. We set the learning rate to 0.001 and the initial weight of formulae to 0.1 and kept all other parameters at their default values. Parameter values were selected on a validation set, strictly disjoint from our test set.

5 Data and Methodology

We used data provided by Microsoft Research containing anonymized query-log records collected from MSN Search in May 2006. The data consists of time-stamped records for individual short sessions, the queries issued in them, the URLs clicked for each query, the number of results available for each query and the position of each result in the ranked results. We removed queries for which nothing was clicked. The average number of clicked results per session, over all sessions in the data, is 3.28. The data does not specify what criteria were used to organize a set of user interactions into a session; e.g., we do not know how multiple open tabs in a browser were treated. Although some of the sessions may belong to the same users, the data excludes this information through the lack of user-specific identifiers. This dataset therefore perfectly mirrors the scenario of disambiguating user intent from short interactions that we address in this research. Because there is a one-to-one correspondence between users and sessions, we will use these two terms interchangeably.

The data has two main limitations. First, it does not state which search queries are ambiguous. Automatically detecting ambiguity from user behavior is an interesting research question but is not the focus of this work. We therefore employed a simple heuristic to obtain a (possibly noisy) set of ambiguous queries, using DMOZ (`www.dmoz.org`): a query string is considered ambiguous if, over all URLs clicked after searching for this string, at least two fall in different top-level DMOZ categories. This heuristic does not require human effort beyond that already invested in constructing DMOZ. We did not include DMOZ category information into our models because many Web pages are not classified in the hierarchy. We limited ourselves to strings containing up to two words, thus obtaining 6,360 distinct ambiguous query strings. Limiting the length of potentially ambiguous queries to two was motivated by the fact that most ambiguity occurs in short queries. For example Sanderson [2] found that the average length of ambiguous queries in two search log datasets ranges from 1.02 to 1.26 words. Queries of length at most two constituted 43.7% of all queries in our data. Of these queries, using the above method, we identified 2.4% as ambiguous, which agrees with the statistics reported by Sanderson, who found that between 0.8% and 3.9% of all queries are ambiguous [2].[2]

Another limitation is that our data does not list all URLs presented to the user after a search but just the clicked ones. To overcome this, we assumed that the set of all URLs clicked after searching for a particular ambiguous query string, over the entire dataset, was the set of results presented to the user. Our approach contrasts with that used in previous work, e.g., that of Dou *et al.* [5], in which missing possible results lists are generated by separately querying the MSN search engine (on which data was collected) for each query. Although the queries were performed less than a month after the data was collected, the authors found that 676 queries from 4,639 "lost the clicked web pages in downloaded search results." Because in our case almost 3 years have passed since the MSN06 data was collected, we preferred the simpler approach based on the available data. With this method, the average number of possible results for an ambiguous query string was 9.10. Figure 2 (a) shows the distribution over the number of ambiguous queries for which we have a particular number of possible results. Although this heuristic is imperfect, it is likely to bias the results *against* our proposed solution—since every possible result was found to be relevant by at least one user, our systems cannot get high scores by simply separating the useful results from the totally irrelevant ones.

Figure 2 (b) shows the distribution over the number of clicks preceding an ambiguous query in our test data. As can be seen, our test sessions, are indeed very short. Several of the predicates we define use keywords. To generate a list of keywords, we performed a pass over all training sessions. Any token separated by spaces was considered a keyword. As mentioned in Section 4, we then kept keywords that appeared at least 100 times and at most 10,000 times. To determine which keywords occur in a given hostname, we first use the non-alphanumeric characters in the hostname to break it down into pieces and then match each piece with keywords such that as much of the piece is covered as possible, using the smallest number of keywords.

To ensure a fair evaluation, the data was split into training and testing periods. The training period was used for training, validation, keyword generation, and *idf* [30] calculations (*idf*s were used by one of the baselines) and consisted of the first 25 days of data. The remaining 6 days were reserved for testing. Sessions that started in the training period and ended in the test period were discarded to avoid contaminating the test data. As validation/testing examples we used sessions that contained an ambiguous query from the training/testing periods respectively. To decrease the amount of random noise in the results, we removed from the test set sessions that contained no relational evidence, i.e., we removed the sessions that contain no true groundings of the `sharesKeyword/Click`

[2] In the Introduction, we cited Sanderson's findings for *frequently occurring* queries, whereas here we refer to his findings over *all* queries.

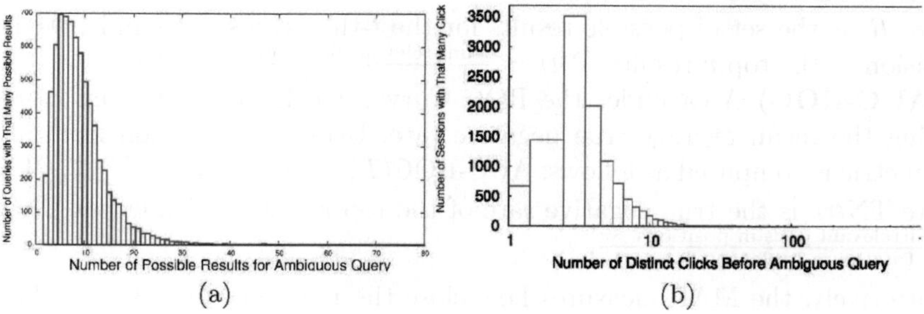

Fig. 2. Histograms showing (a) the distribution over the number of possible results available for an ambiguous query and (b) the distribution over the number of clicks preceding an ambiguous query in the test data. The X axis in (b) is drawn in log-scale.

predicates introduced in Section 4. In this way we obtained 11,234 test sessions, which constitutes 72% of the available test sessions. The distribution over the number of previous clicks in these sessions is shown in Figure 2 (b). As can be seen, the peak is at 3 distinct clicks before the ambiguous query.

During testing, only the information *preceding* the ambiguous query in the active test session is provided. The set of possible results for this ambiguous query string is given, and the goal is to rank these results based on how likely it is that they represent the intent of the user. The user may click more than one result after searching for a string. This behavior might be indicative of at least two possible scenarios: either the user is performing an exploratory search and all clicked results were relevant, or the user was dissatisfied with the results and kept clicking until finding a useful one. Since the data does not indicate which of these scenarios was the case, we treated all results clicked by the user after searching for the ambiguous query as relevant to his or her intentions. This presents yet another source of noise, and in the future we plan to explore approaches similar to the implicit feedback techniques described by Radlinski and Joachims [16] to disentangle these possibilities, although the exact method introduced by these authors would not be applicable to our data because it requires the availability of an *ordered* list of the results returned to the user by the search engine.

Learning was performed as described in Section 4.2. To evaluate the learned models, we used Alchemy's implementation [14] of the MC-SAT algorithm [10] for inference. During inference, we ran for 1,000 burn-in steps and 10,000 sampling steps. All other inference parameters were kept at their Alchemy defaults.

Evaluation Metrics: For evaluation purposes, query disambiguation can be viewed as an information retrieval problem: rank the set of possible results so that the URLs reflecting the user's intentions appear as close to the top as possible. Thus, we used standard information retrieval metrics to evaluate the performance of our system [30] (Chapter 8):

(MAP) Area under the precision-recall curve, which is identical to the Mean Average Precision metric, commonly used in IR. The MAP score is computed over a set of test instances T as follows: $\text{MAP}(T) = \frac{1}{|T|} \sum_{t \in \text{T}} \frac{1}{|R_t|} \sum_{r \in R_t} \text{P}@r$, where R_t is the set of possible results for the t-th test instance and P@r is the precision of the top r results: P@$r = \frac{\text{Num relevant docs among the top } r}{r}$.

(AUC-ROC) Area under the ROC Curve, which can be viewed as representing the mean average true negative rate. Using the notation from above, this metric is computed as follows: AUC-ROC$(T) = \frac{1}{|T|} \sum_{t \in \text{T}} \frac{1}{|R_t|} \sum_{r \in R_t}$ TN@r, where TN@r is the true negative rate of the top r results, defined as TN@$r = \frac{\text{Num irrelevant docs in positions} > r}{\text{Total num irrelevant docs}}$.

Intuitively, the MAP measures how close the relevant URLs are to the top. One disadvantage of this metric in our case is that it is insensitive to the number of results to be ranked. For example, ranking a relevant result in the second position obtains the same score both when the number of possibilities is 2 and when it is 100, even though in the second case the task is clearly more difficult. Assuming that the user starts scanning the page of returned results from top to bottom and does not consider any results appearing after the relevant ones, the AUC-ROC intuitively represents the percentage of irrelevant results that were *not* seen by the user. Thus, a random ranker would obtain an AUC-ROC of 0.5. Another useful characteristic of this measure is that unlike the MAP, it is sensitive to the number of possible results that are to be ranked.

A final issue is how to break ties when a relevant result has the same score as some irrelevant results. We report the **average case** in which the relevant result is placed in the middle position within the group of results with equal scores. For the most interesting systems, we also report the **worst case** in which the relevant result is placed last within the group of results that share scores. This is motivated by the goal of performing effective personalization *consistently*. The best case is not interesting because for it perfect performance can be obtained by giving all results the same score.

Systems Compared: We compared the MLNs from Section 4 to:

Random: Ranks the possible results randomly.

Collaborative-Pearson: Implements a standard collaborative filtering algorithm [25] that weights each previous user based on the Pearson correlation between the preferences (i.e. clicks) of that user and the active user. We considered a clicked result to have rating 1, and an unclicked result that was clicked by another user for the same query to have rating 0, and all other results to be unrated. The n closest neighbors are chosen (we used $n = 30$ following [25]), and the prediction that a given result is selected is formed as a weighted average of the deviations from the mean of each neighbor.

Collaborative-Cosine: Identical to **Collaborative-Pearson** except that it computes the similarity between the active user and a previous user as the cosine similarity between the *idf*-weighted vectors of their clicked results.

Popularity: Ranks each result according to the number of previous sessions that searched for the ambiguous query and chose it.

6 Results

Table 3 (a) presents the performance when ties among results with the same score are broken as in the average case. The **Collaborative-Pearson** baseline performs no better than **Random** on AUC-ROC and only slightly better than **Random** on MAP. Switching to cosine similarity in **Collaborative-Cosine** gives modest (but significant at the 99.996% level according to a paired t-test) improvements. The **Popularity** baseline is very strong and outperforms the other baselines, as well as **MLN 1**. However, combining popularity with relational information in **MLN 2** leads to significant gains in performance, and **MLN 2** achieves a significantly higher AUC-ROC score. **MLN 2**, our strongest model, highlights the main advantage of using MLNs: we were able to significantly improve **MLN 1** by incorporating a reliable source of information simply by adding the popularity formula to the model. Finally, as expected, we observe that adding local formulae in **MLN 3** does not improve performance. This demonstrates that the interactions of the active user prior to the ambiguous query are not directly helpful for determining intent and occurs as a result of the brevity of sessions in our data (cf. Figure 2 (b)). The inefficacy of local formulae may also be due to the fact that a session may continue when the user is dissatisfied with the results obtained so far. It is interesting to contrast this result with the findings of Dou *et al.* [5] who experimented with much longer

Table 3. **(a)** Results over all test sessions that contain an ambiguous query when ties in ranking are broken as in the **(a) average case** and **(b) worst case.** Numbers in bold present significant improvements over all preceding systems at the 99.996% level with a paired t-test. Additional significant differences are: in (a) **MLN 1** is a significant improvement over all baselines except **Popularity**, and **MLN 2** improves significantly over all preceding systems except for **Popularity** also in terms of MAP; in (a), there is no significant difference between the MAP scores of **Popularity** and **MLN 2**; in (a) and (b) the MAP score of **Popularity** is significantly higher than that of **MLN 1**.

System	MAP	AUC-ROC
Random	0.317	0.502
Collaborative-Pearson	0.333	0.502
Collaborative-Cosine	**0.360**	**0.521**
Popularity	**0.389**	**0.575**
MLN 1	0.375	0.563
MLN 2	0.386	**0.587**
MLN 3	0.366	0.583

(a)

System	MAP	AUC-ROC
Popularity	0.380	0.525
MLN 1	0.373	**0.563**
MLN 2	**0.385**	**0.586**
MLN 3	0.355	0.572

(b)

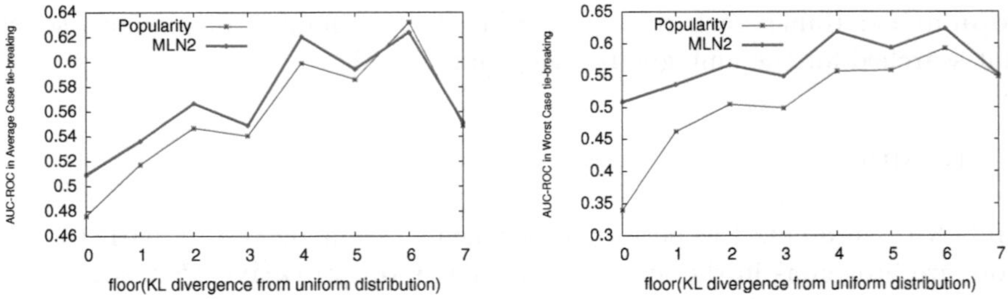

(a) (b)

Fig. 3. AUC-ROC when ranking ties are broken so as to simulate the **(a) average case** and **(b) worst case** for different bins of KL divergence of the distribution over possible results to uniform

sessions (up to 12 days) and reported that the previous interactions of the active user presented a very strong signal for personalization purposes. This emphasizes a fundamental difference in our assumptions about the data compared to previous research: because in our case user-specific session information is so limited, we cannot rely on only using the past preferences of the active user and must instead exploit relations to other, historical, users.

Next, we analyze in more detail the performance of the MLN systems to that of **Popularity**, which is the strongest baseline. Table 3 (b) presents the performance over all test sessions when ties in ranking are broken as in the worst case. As can be seen, **Popularity**'s AUC-ROC score decreases sharply, whereas the MLN models maintain their performance to almost the same level as in the average case. This behavior is observed partly because **Popularity** introduces many more ties among the scores of possible results than do the MLN models. In particular, averaged over all test sessions, the ratio between the number of possible results and the number of distinct scores for **Popularity** was 1.8, whereas for **MLN2** it was just 1.02. These results indicate that **Popularity**'s behavior is erratic and can, for the same user and the same query, lead to rankings that vary highly in quality. This kind of behavior can give the perception of poor quality to a frequent user. On the other hand, the MLN models are consistent, maintaining the quality of their rankings in the worst case.

Finally, we compare the performance of **Popularity** to that of **MLN 2** while varying the degree to which some of the possible results for an ambiguous query dominate in popularity over the rest. We formalized this as follows. Let q_Q be the empirical distribution over the results clicked for an ambiguous query Q. This distribution was measured empirically on the training data, i.e., for every ambiguous query, we determined from the training sessions the proportion of time each potential search result was clicked. We then separated the test examples into bins, such that bin i contains all test sessions s for which $\lfloor KL_{q_Q||\text{uniform}} \rfloor = i$, where Q is the ambiguous query in session s and $KL_{q_Q||\text{uniform}}$ is the KL divergence of q_Q to the uniform distribution. In other words, bin 0 contains the sessions in which the possible results for the ambiguous query were all chosen with roughly the same frequency. Higher-numbered bins contain sessions in which one of the search results strongly dominates in popularity over the other possibilities. When this is the case, predicting just based on the popularity of a result gives good performance. The more challenging scenario occurs in the lower-numbered bins where the preferences over possible results are more uniformly distributed. Figure 3 compares **Popularity** to **MLN 2** when ties in ranking are broken for the average and worst cases. **MLN 2** maintains a lead over **Popularity** until the last two bins in which the distribution over possible results is furthest from uniform. As we expect, the difference between the performance of the two systems shrinks as we move to higher-numbered bins, and **MLN 2** has a greater advantage over **Popularity** in the lower-numbered bins in which the need to disambiguate is more pressing. The sharp drop in accuracy observed in bin 7 is due to the fact that one of the ambiguous queries occurring in sessions in this bin was overwhelmingly followed by clicks to what seems to be a newly appearing Web page during the test period. That page was selected only 3 times in the training period while the most popular page in the training period was selected more than 2000 times. As a final but important note, inference over the learned models was very efficient and completed in the order of a second.

7 Conclusions and Future Work

We addressed Web query disambiguation in the challenging setting when the only information available about any particular user is that captured in a short search session of 4–6 previous searches on average. Using the language of MLNs, we developed an approach that draws heavily on different types of relations between search sessions and demonstrated that our approach significantly outperforms several natural baselines by successfully combining the inferences of collaborative and popularity formulae. In this way, we provided evidence that despite the sparseness and noise inherently present in a short search session, it is possible to output meaningful predictions about a searcher's underlying interests.

Here our goal was a light-weight approach to Web query disambiguation. In the future, we would like to experiment with richer sources of information, such as the actual content of clicked pages. A second avenue for future work involves improving supervision by discovering ways to decrease the amount of noise in the data and developing learning algorithms that are more tolerant to noise.

Acknowledgment

We thank Tuyen Huynh, Joe Reisinger, and the anonymous reviewers for their helpful comments. This research is supported by a gift from Microsoft Research and by ARO grant W911NF-08-1-0242. Experiments were run on the Mastodon Cluster, provided by NSF Grant EIA-0303609.

References

1. Jansen, B.J., Spink, A.: How are we searching the World Wide Web? A comparison of nine search engine transaction logs. Information Processing and Management 42, 248–263 (2006)
2. Sanderson, M.: Ambiguous queries: Test collections need more sense. In: SIGIR 2008 (2008)
3. Sugiyama, K., Hatano, K., Yoshikawa, M.: Adaptive web search based on user profile constructed without any effort from users. In: WWW 2004 (2004)
4. Sun, J., Zeng, H., Liu, H., Lu, Y., Chen, Z.: CubeSVD: A novel approach to personalized web search. In: WWW 2005 (2005)
5. Dou, Z., Song, R., Wen, J.: A large-scale evaluation and analysis of personalized search strategies. In: WWW 2007 (2007)
6. Barbaro, M., Zeller, T.: A face is exposed for AOL searcher no. 4417749. New York Times (August 2006), `http://www.nytimes.com/2006/08/09/technology/09aol.html?ex=1312776000` (Accessed on October 16, 2008)
7. Conti, G.: Googling considered harmful. In: New Security Paradigms Workshop, Dagstuhl, Germany (September 2006)
8. Getoor, L., Taskar, B. (eds.): Introduction to Statistical Relational Learning. MIT Press, Cambridge (2007)
9. Richardson, M., Domingos, P.: Markov logic networks. Machine Learning 62 (2006)
10. Poon, H., Domingos, P.: Sound and efficient inference with probabilistic and deterministic dependencies. In: AAAI 2006 (2006)
11. Singla, P., Domingos, P.: Entity resolution with Markov logic. In: ICDM 2006 (2006)
12. Poon, H., Domingos, P.: Joint inference in information extraction. In: AAAI 2007 (2007)
13. Wu, F., Weld, D.: Automatically refining the Wikipedia infobox ontology. In: WWW 2008 (2008)
14. Kok, S., Singla, P., Richardson, M., Domingos, P.: The Alchemy system for statistical relational AI. Technical report, Department of Computer Science and Engineering, University of Washington (2005), `http://www.cs.washington.edu/ai/alchemy`
15. Joachims, T.: Optimizing search engines using clickthrough data. In: KDD 2002 (2002)
16. Radlinski, F., Joachims, T.: Query chains: Learning to rank from implicit feedback. In: KDD 2005 (2005)
17. Almeida, R.B., Almeida, V.A.F.: A community-aware search engine. In: WWW 2004 (2004)
18. Krause, A., Horvitz, E.: A utility-theoretic approach to privacy and personalization. In: AAAI 2008 (2008)
19. Shen, J., Li, L., Dietterich, T.G., Herlocker, J.L.: A hybrid learning system for recognizing user tasks from desktop activities and email messages. In: IUI 2006 (2006)
20. Teevan, J., Dumais, S.T., Horvitz, E.: Personalizing search via automated analysis of interests and activities. In: SIGIR 2005 (2005)
21. Chen, H., Karger, D.R.: Less is more: Probabilistic models for retrieving fewer relevant documents. In: SIGIR 2006 (2006)

22. Yue, Y., Joachims, T.: Predicting diverse subsets using structural SVMs. In: ICML 2008 (2008)
23. Wang, X., Zhai, C.: Learn from web search logs to organize search results. In: SIGIR 2007 (2007)
24. Breese, J.S., Heckerman, D., Kadie, C.: Empirical analysis of predictive algorithms for collaborative filtering. Technical Report MSR-TR-98-12, Microsoft Research (1998)
25. Herlocker, J., Konstan, J., Borchers, A., Riedl, J.: An algorithmic framework for performing collaborative filtering. In: SIGIR 1999 (1999)
26. Popescul, A., Ungar, L.H., Pennock, D.M., Lawrence, S.: Probabilistic models for unified collaborative and content-based recommendation in sparse-data environments. In: UAI 2001 (2001)
27. Melville, P., Mooney, R.J., Nagarajan, R.: Content-boosted collaborative filtering for improved recommendations. In: AAAI 2002 (2002)
28. Russell, S., Norvig, P.: Artificial Intelligence: A Modern Approach, 2nd edn. Prentice Hall, Upper Saddle River (2003)
29. Lowd, D., Domingos, P.: Efficient weight learning for Markov logic networks. In: Kok, J.N., Koronacki, J., Lopez de Mantaras, R., Matwin, S., Mladenič, D., Skowron, A. (eds.) PKDD 2007. LNCS (LNAI), vol. 4702, pp. 200–211. Springer, Heidelberg (2007)
30. Manning, C.D., Raghavan, P., Schutze, H.: Introduction to Information Retrieval. Cambridge University Press, Cambridge (2008)

Dynamic Factor Graphs for Time Series Modeling

Piotr Mirowski and Yann LeCun

Courant Institute of Mathematical Sciences, New York University,
719 Broadway, New York, NY 10003 USA
{mirowski,yann}@cs.nyu.edu
http://cs.nyu.edu/~mirowski/

Abstract. This article presents a method for training Dynamic Factor Graphs (DFG) with continuous latent state variables. A DFG includes factors modeling joint probabilities between hidden and observed variables, and factors modeling dynamical constraints on hidden variables. The DFG assigns a scalar energy to each configuration of hidden and observed variables. A gradient-based inference procedure finds the minimum-energy state sequence for a given observation sequence. Because the factors are designed to ensure a constant partition function, they can be trained by minimizing the expected energy over training sequences with respect to the factors' parameters. These alternated inference and parameter updates can be seen as a deterministic EM-like procedure. Using smoothing regularizers, DFGs are shown to reconstruct chaotic attractors and to separate a mixture of independent oscillatory sources perfectly. DFGs outperform the best known algorithm on the CATS competition benchmark for time series prediction. DFGs also successfully reconstruct missing motion capture data.

Keywords: factor graphs, time series, dynamic Bayesian networks, recurrent networks, expectation-maximization.

1 Introduction

1.1 Background

Time series collected from real-world phenomena are often an incomplete picture of a complex underlying dynamical process with a high-dimensional state that cannot be directly observed. For example, human motion capture data gives the positions of a few markers that are the reflection of a large number of joint angles with complex kinematic and dynamical constraints. The aim of this article is to deal with situations in which the hidden state is continuous and high-dimensional, and the underlying dynamical process is highly non-linear, but essentially deterministic. It also deals with situations in which the observations have lower dimension than the state, and the relationship between states and observations may be non-linear. The situation occurs in numerous problems in speech and audio processing, financial data, and instrumentation data, for such

W. Buntine et al. (Eds.): ECML PKDD 2009, Part II, LNAI 5782, pp. 128–143, 2009.

tasks as prediction and source separation. It applies in particular to univariate chaotic time series which are often the projection of a multidimensional attractor generated by a multivariate system of nonlinear equations.

The simplest approach to modeling time series relies on time-delay embedding: the model learns to predict one sample from a number of past samples with a limited temporal span. This method can use linear auto-regressive models, as well as non-linear ones based on kernel methods (e.g. support-vector regression [13,14]), neural networks (including convolutional networks such as time delay neural networks [7,18]), and other non-linear regression models. Unfortunately, these approaches have a hard time capturing hidden dynamics with long-term dependency because the state information is only accessible indirectly (if at all) through a (possibly very long) sequence of observations [2].

To capture long-term dynamical dependencies, the model must have an internal state with dynamical constraints that predict the state at a given time from the states and observations at previous times (e.g. a state-space model). In general, the dependencies between state and observation variables can be expressed in the form of a *Factor Graph* [5] for sequential data, in which a graph motif is replicated at every time step. An example of such a representation of a state-space model is shown in Figure 1a. Groups of variables (circles) are connected to a factor (square) if a dependency exists between them. The factor can be expressed in the negative log domain: each factor computes an energy value that can be interpreted as the negative log likelihood of the configuration of the variables it connects with. The total energy of the system is the sum of the factors' energies, so that the maximum likelihood configuration of variables can be obtained by minimizing the total energy.

Figure 1a shows the structure used in Hidden Markov Models (HMM) and Kalman Filters, including Extended Kalman Filters (EKF) which can model non-linear dynamics. HMMs can capture long-term dependencies, but they are limited to discrete state spaces. Discretizing the state space of a high-dimensional continuous dynamical process to make it fit into the HMM framework is often impractical. Conversely, EKFs deal with continuous state spaces with non-linear dynamics, but much of the machinery for inference and for training the

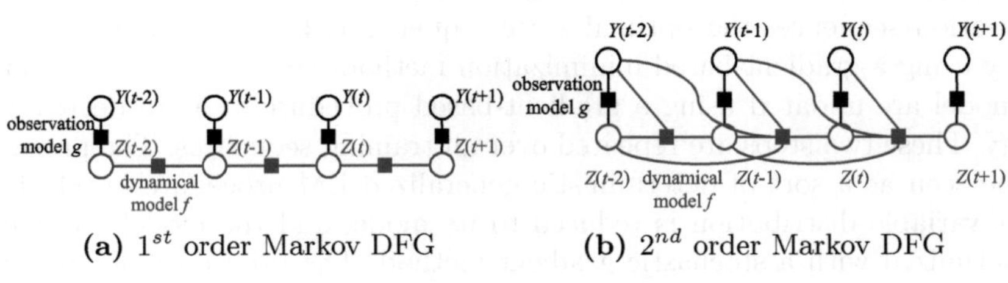

(a) 1st order Markov DFG **(b)** 2nd order Markov DFG

Fig. 1. (a) A simple Dynamical Factor Graph with a 1st order Markovian property, as used in HMMs and state-space models such as Kalman Filters. (b) A Dynamic Factor Graph where dynamics depend on the past two values of both latent state Z and observed variables Y.

parameters is linked to the problem of marginalizing over hidden state distributions and to propagating and estimating the covariances of the state distributions. This has lead several authors to limit the discussion to dynamics and observation functions that are linear, radial-basis functions networks [19,4] or single-hidden layer perceptrons [6]. More recently, Gaussian Processes with dynamics on latent variables have been introduced [20], but they suffer from a quadratic dependence on the number of training samples.

1.2 Dynamical Factor Graphs

By contrast with current state-space methods, our primary interest is to model processes whose underlying dynamics are essentially deterministic, but can be highly complex and non-linear. Hence our model will allow the use of complex functions to predict the state and observations, and will sacrifice the probabilistic nature of the inference. Instead, our inference process (including during learning) will produce the most likely (minimum energy) sequence of states given the observations. We call this method *Dynamic Factor Graph* (DFG), a natural extension of Factor Graphs specifically tuned for sequential data.

To model complex dynamics, the proposed model allows the state at a given time to depend on the states and observations over several past time steps. The corresponding DFG is depicted in Figure 1b. The graph structure is somewhat similar to that of Taylor and Hinton's Conditional Restricted Boltzmann Machine [17]. Ideally, training a CRBM would consist in minimizing the negative log-likelihood of the data under the model. But computing the gradient of the log partition function with respect to the parameters is intractable, hence Taylor and Hinton propose to use a form of the contrastive divergence procedure, which relies on Monte-Carlo sampling. To avoid costly sampling procedures, we design the factors in such a way that the partition function is constant, hence the likelihood of the data under the model can be maximized by simply minimizing the average energy with respect to the parameters for the optimal state sequences. To achieve this, the factors are designed so that the conditional distributions of state $Z(t)$ given previous states and observation[1], and the conditional distribution of the observation $Y(t)$ given the state $Z(t)$ are both Gaussians with a fixed covariance.

In a nutshell, the proposed training method is as follows. Given a training observation sequence, the optimal state sequence is found by minimizing the energy using a gradient-based minimization method. Second, the parameters of the model are updated using a gradient-based procedure so as to decrease the energy. These two steps are repeated over all training sequences. The procedure can be seen as a sort of deterministic generalized EM procedure in which the latent variable distribution is reduced to its mode, and the model parameters are optimized with a stochastic gradient method. The procedure assumes that the factors are differentiable with respect to their input variables and their parameters. This simple procedure will allow us to use sophisticated non-linear

[1] Throughout the article, $Y(t)$ denotes the value at time t of multivariate time series Y, and $\mathbf{Y}_{t-p}^{t-1} \equiv \{Y(t-t), Y(t-2), \ldots, Y(t-p)\}$ a time window of p samples preceding the current sample. $Z(t)$ denotes the hidden state at time t.

models for the dynamical and observation factors, such as stacks of non-linear filter banks (temporal convolutional networks). It is important to note that *the inference procedure operates at the sequence level*, and produces the most likely state sequence that best explains the entire observation. In other words, future observations may influence previous states.

In the DFG shown in Figure 1a, the dynamical factors compute an energy term of the form $E_d(t) = ||Z(t) - f\left(X(t), Z(t-1)\right)||^2$, which can seen as modeling the state $Z(t)$ as $f(X(t), Z(t-1))$ plus some Gaussian noise variable with a fixed variance $\epsilon(t)$ (inputs $X(t)$ are not used in experiments in this article). Similarly, the observation factors compute the energy $E_o(t) = ||Y(t) - g(Z(t))||^2$, which can be interpreted as $Y(t) = g\left(Z(t)\right) + \omega(t)$, where $\omega(t)$ is a Gaussian random variable with fixed variance.

Our article is organized in three additional sections. First, we explain the gradient-based approximate algorithm for parameter learning and deterministic latent state inference in the DFG model (2). We then evaluate DFGs on toy, benchmark and real-world datasets (3). Finally, we compare DFGs to previous methods for deterministic nonlinear dynamical systems and to training algorithms for Recurrent Neural Networks (4).

2 Methods

The following subsections detail the deterministic nonlinear (neural networks-based) or linear architectures of the proposed Dynamic Factor Graph (2.1) and define the EM-like, gradient-based inference (2.2) and learning (2.4) algorithms, as well as how DFGs are used for time-series prediction (2.3).

2.1 A Dynamic Factor Graph

Similarly to Hidden Markov Models, our proposed Dynamic Factor Graph contains an observation and a dynamical factors/models (see Figure 1a), with corresponding observed outputs and latent variables.

The *observation model* g links latent variable $Z(t)$ (an m-dimensional vector) to the observed variable $Y(t)$ (an n-dimensional vector) at time t under Gaussian noise model $\omega(t)$ (because the quadratic observation error is minimized). g can be nonlinear, but we considered in this article linear observation models, i.e. an $n \times m$ matrix parameterized by a weight vector $\mathbf{W}_o$. This model can be simplified even further by imposing each observed variable $y_i(t)$ of the multivariate time series Y to be the sum of k latent variables, with $m = k \times n$, and each latent variable contributing to only one observed variable. In the general case, the generative output is defined as:

$$Y(t) = Y^*(t) + \omega(t), \text{ where } Y^*(t) \equiv g\left(\mathbf{W}_o, Z(t)\right) \tag{1}$$

In its simplest form, the linear or nonlinear *dynamical model* f establishes a causal relationship between a sequence of p latent variables $\mathbf{Z}_{t-p}^{t-1}$ and latent variable $Z(t)$, under Gaussian noise model $\epsilon(t)$ (because the quadratic dynamic

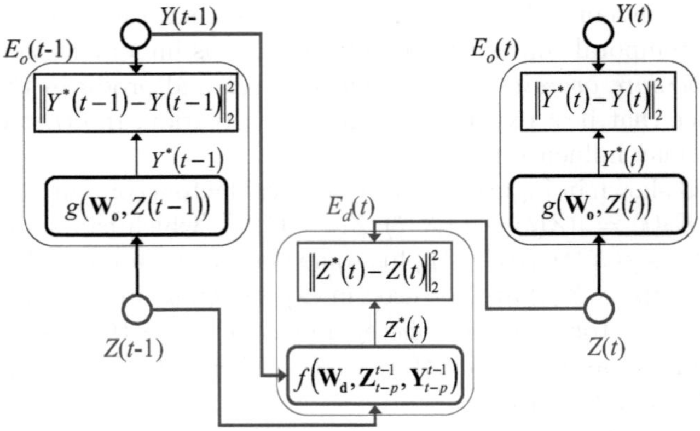

Fig. 2. Energy-based graph of a DFG with a 1st order Markovian architecture and additional dynamical dependencies on past observations. Observations $Y(t)$ are inferred as $Y^*(t)$ from latent variables $Z(t)$ using the observation model parameterized by $\mathbf{W}_o$. The (non)linear dynamical model parameterized by $\mathbf{W}_d$ produces transitions from a sequence of latent variables $\mathbf{Z}_{t-p}^{t-1}$ and observed output variables $\mathbf{Y}_{t-p}^{t-1}$ to $Z(t)$ (here $p = 1$). The total energy of the configuration of parameters and latent variables is the sum of the observation $E_o(.)$ and dynamic $E_d(.)$ errors.

error is minimized). (2) thus defines p^{th} order Markovian dynamics (see Figure 1a where $p = 1$). The dynamical model is parameterized by vector $\mathbf{W_d}$.

$$Z(t) = Z^*(t) + \epsilon(t), \text{ where } Z^*(t) \equiv f\left(\mathbf{W}_d, \mathbf{Z}_{t-p}^{t-1}\right) \quad (2)$$

Typically, one can use simple multivariate autoregressive linear functions to map the state variables, or can also resort to nonlinear dynamics modeled by a Convolutional Network [8] with convolutions (FIR filters) across time, as in Time-Delay Neural Networks [7,18].

Other dynamical models, different from the Hidden Markov Model, are also possible. For instance, latent variables $Z(t)$ can depend on a sequence of p past latent variables $\mathbf{Z}_{t-p}^{t-1}$ and p past observations $\mathbf{Y}_{t-p}^{t-1}$, using the same error term $\epsilon(t)$, as explained in (3) and illustrated on Figure 1b.

$$Z(t) = Z^*(t) + \epsilon(t), \text{ where } Z^*(t) \equiv f\left(\mathbf{W}_d, \mathbf{Z}_{t-p}^{t-1}, \mathbf{Y}_{t-p}^{t-1}\right) \quad (3)$$

Figure 2 displays the interaction between the observation (1) and dynamical (3) models, the observed Y and latent Z variables, and the quadratic error terms.

2.2 Inference in Dynamic Factor Graphs

Let us define the following *total* (4), *dynamical* (5) and *observation* (6) energies (quadratic errors) on a given time interval $[t_a, \ldots, t_b]$, where respective weight coefficients α, β are positive constants (in this article, $\alpha = \beta = 0.5$):

$$E\left(\mathbf{W}_d, \mathbf{W}_o, \mathbf{Y}_{t_a}^{t_b}\right) = \sum_{t=t_a}^{t_b} \left[\alpha E_d(t) + \beta E_o(t)\right] \tag{4}$$

$$E_d(t) \equiv \min_Z E_d\left(\mathbf{W}_d, \mathbf{Z}_{t-p}^{t-1}, Z(t)\right) \tag{5}$$

$$E_o(t) \equiv \min_Z E_o\left(\mathbf{W}_o, Z(t), Y(t)\right) \tag{6}$$

Inferring the sequence of latent variables $\{Z(t)\}_t$ in (4) and (5) is equivalent to simultaneous minimization of the sum of dynamical and observation energies at all times t:

$$E_d\left(\mathbf{W}_d, \mathbf{Z}_{t-p}^{t-1}, Z(t)\right) = ||Z^*(t) - Z(t)||_2^2 \tag{7}$$

$$E_o\left(\mathbf{W}_o, Z(t), Y(t)\right) = ||Y^*(t) - Y(t)||_2^2 \tag{8}$$

Observation and dynamical errors are expressed separately, either as Normalized Mean Square Errors (NMSE) or Signal-to-Noise Ratio (SNR).

2.3 Prediction in Dynamic Factor Graphs

Assuming fixed parameters $\mathbf{W}$ of the DFG, two modalities are possible for the prediction of unknown observed variables Y.

- *Closed-loop (iterated) prediction*: when the continuation of the time series is unknown, the only relevant information comes from the past. One uses the dynamical model to predict $Z^*(t)$ from $\mathbf{Y}_{t-p}^{t-1}$ and inferred $\mathbf{Z}_{t-p}^{t-1}$, set $Z(t) = Z^*(t)$, use the observation model to compute prediction $Y^*(t)$ from $Z(t)$, and iterate as long as necessary. If the dynamics depend on past observations, one also needs to rely on predictions $Y^*(t)$ in (3).
- *Prediction as inference*: this is the case when only some elements of Y are unknown (e.g. estimation of missing motion-capture data). First, one infers latent variables through gradient descent, and simply does not backpropagate errors from unknown observations. Then, missing values $y_i^*(t)$ are predicted from corresponding latent variables $Z(t)$.

2.4 Training of Dynamic Factor Graphs

Learning in an DFG consists in adjusting the parameters $\mathbf{W} = \left[\mathbf{W}_d^T, \mathbf{W}_o^T\right]$ in order to minimize the loss $L(\mathbf{W}, \mathbf{Y}, \tilde{Z})$:

$$L(\mathbf{W}, Y, Z) = E\left(\mathbf{W}, Y\right) + R_z(Z) + R\left(\mathbf{W}\right) \tag{9}$$

$$\tilde{Z} = argmin_Z L(\tilde{\mathbf{W}}, \mathbf{Y}, Z) \tag{10}$$

$$\tilde{\mathbf{W}} = argmin_{\mathbf{W}} L(\mathbf{W}, \mathbf{Y}, \tilde{Z}) \tag{11}$$

where $R(\mathbf{W})$ is a regularization term on the weights $\mathbf{W}_d$ and $\mathbf{W}_o$, and $R_z(Z)$ represents additional constraints on the latent variables further detailed. Minimization of this loss is done iteratively in an Expectation-Maximization-like fashion in which the states Z play the role of auxiliary variables. During inference, values of the model parameters are clamped and the hidden variables are relaxed to minimize the energy. The inference described in part (2.2) and equation (10) can be considered as the *E-step* (state update) of a gradient-based version of the EM algorithm. During learning, model parameters $\mathbf{W}$ are optimized to give lower energy to the current configuration of hidden and observed variables. The parameter-adjusting *M-step* (weight update) described by (11) is also gradient-based.

In its current implementation, the E-step inference is done by gradient descent on Z, with learning rate η_z typically equal to 0.5. The convergence criterion is when energy (4) stops decreasing. The M-step parameter learning is implemented as a stochastic gradient descent (diagonal Levenberg-Marquard) [9] with individual learning rates per weight (re-evaluated every 10000 weight updates) and global learning rate η_w typically equal to 0.01. These parameters were found by trial and error on a grid of possible values.

The state inference is not done on the full sequence at once, but on *mini-batches* (typically 20 to 100 samples), and the weights get updated once after each mini-batch inference, similarly to the Generalized EM algorithm. During one epoch of training, the batches are selected randomly and overlap in such a way that each state variable $Z(t)$ is re-inferred at least a dozen times in different mini-batches. This learning approximation echoes the one in regular stochastic gradient with no latent variables and enables to speed up the learning of the weight parameters.

The learning algorithm turns out to be particularly simple and flexible. The hidden state inference is however under-constrained, because of the higher dimensionality of the latent states and despite the dynamical model. For this reason, this article proposes to (in)directly regularize the hidden states in several ways.

First, one can add to the loss function an L_1 regularization term $R(\mathbf{W})$ on the weight parameters. This way, the dynamical model becomes "sparse" in terms of its inputs, e.g. the latent states. Regarding the term $R_z(\mathbf{Z})$, an L_2 norm on the hidden states $Z(t)$ limits their overall magnitude, and an L_1 norm enforces their sparsity both in time and across dimensions. Respective regularization coefficients λ_w and λ_z typically range from 0 to 0.1.

Pseudo-code of the EM-like Learning and Inference in DFGs

```
for each epoch k
  for each subsequence I in [1, T]
    repeat
      for each time index t within I
        forward-propagate Z(t) through f to get Z*(t)
        forward-propagate Z(t) through g to get Y*(t)
        back-propagate errors from ||Z(t)-Z*(t)||, add to dZ(t)
        back-propagate errors from ||Y(t)-Y*(t)||, add to dZ(t)
      update latent states Z(I) using gradients dZ(I)
```

```
until convergence, when energy E(I) stops decreasing
for each time index t within I
  back-propagate errors from ||Z(t)-Z*(t)||, add to dW
  back-propagate errors from ||Y(t)-Y*(t)||, add to dW
update parameters W using gradients dW
```

2.5 Smoothness Penalty on Latent Variables

The second type of constraints on the latent variables is the *smoothness penalty.* In an apparent contradiction with the dynamical model (2), this penalty forces two consecutive variables $z_i(t)$ and $z_i(t+1)$ to be similar. One can view it as an attempt at inferring slowly varying hidden states and at reducing noise in the states (which is particularly relevant when observation Y is sampled at a high frequency). By consequence, the dynamics of the latent states are smoother and simpler to learn. Constraint (12) is easy to derivate w.r.t. a state $z_i(t)$ and to integrate into the gradient descent optimization (10):

$$R_z\left(\mathbf{Z}_t^{t+1}\right) = \sum_i (z_i(t) - z_i(t+1))^2 \tag{12}$$

In addition to the smoothness penalty, we have investigated the decorrelation of multivariate latent variables $Z(t) = (z_1(t), z_2(t), \ldots, z_m(t))$. The justification was to impose to each component $\mathbf{z}_i$ to be independent, so that it followed its own dynamics, but we have not obtained satisfactory results yet. As reported in the next section, the interaction of the dynamical model, weight sparsification and smoothness penalty already enables the separation of latent variables.

3 Experimental Evaluation

First, working on toy problems, we investigate the latent variables that are inferred from an observed time series. We show that using smoothing regularizers, DFGs are able to perfectly separate a mixture of independent oscillatory sources (3.1), as well as to reconstruct the Lorenz chaotic attractor in the inferred state space (3.2). Secondly, we apply DFGs to two time series prediction and modeling problems. Subsection (3.3) details how DFGs outperform the best known algorithm on the CATS competition benchmark for time series prediction. In (3.4) we reconstruct realistic missing human motion capture marker data in a walk sequence.

3.1 Asynchronous Superimposed Sine Waves

The goal is to model a time series constituted by a sum of 5 asynchronous sinusoids: $y(t) = \sum_{j=1}^{5} \sin(\lambda_j t)$ (see Fig. 3a). Each component $x_j(t)$ can be considered as a "source", and $y(t)$ is a mixture. This problem has previously been tackled by employing Long-Short Term Memory (LSTM), a special architecture

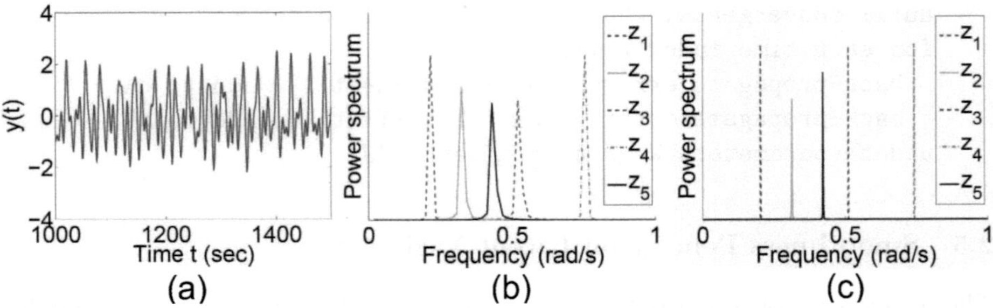

Fig. 3. (a) Superposition of five asynchronous sinusoids: $y(t) = \sum_{j=1}^{5} \sin(\lambda_j t)$ where $\lambda_1 = 0.2$, $\lambda_2 = 0.311$, $\lambda_3 = 0.42$, $\lambda_4 = 0.51$ and $\lambda_5 = 0.74$. Spectrum analysis shows that after learning and inference, each reconstructed state z_i isolates only one of the original sources x_j, both on the training (b) and testing (c) datasets.

of Recurrent Neural Networks that needs to be trained by genetic optimization [21].

After EM training and inference of hidden variables $Z(t)$ of dimension $m = 5$, frequency analysis of the inferred states on the training (Fig. 3b) and testing (Fig. 3c) datasets showed that each latent state $z_i(t)$ reconstructed one individual sinusoid. In other words, the 5 original sources from the observation mixture $y(t)$ were inferred on the 5 latent states. The observation SNR of 64dB, and the dynamical SNR of 54dB, on both the training and testing datasets, proved both that the dynamics of the original time series $y(t)$ were almost perfectly reconstructed. DFGs outperformed LSTMs on that task since the multi-step iterated (closed-loop) prediction of DFG did not decrease in SNR even after thousands of iterations, contrary to [21] where a reduction in SNR was already observed after around 700 iterations.

As architecture for the dynamical model, 5 independent Finite Impulse Response (FIR) filters of order 25 were chosen to model the state transitions: each of them acts as a band-pass filter and models an oscillator at a given frequency. One can hypothesize that the smoothness penalty (12), weighted by a small coefficient of 0.01 in the state regularization term $R_z(\mathbf{Z})$ helped shape the hidden states into perfect sinusoids. Note that the states or sources were made independent by employing five independent dynamical models for each state. This specific usage of DFG can be likened to Blind Source Separation from an unique source, and the use of independent filters for the latent states (or sources) echoes the approach of BSS using linear predictability and adaptive band-pass filters.

3.2 Lorenz Chaotic Data

As a second application, we considered the 3-variable (x_1, x_2, x_3) Lorenz dynamical system [12] generated by parameters $\rho = 16, b = 4, r = 45.92$ as in [13] (see Fig. 4a). Observations consisted in one-dimensional time series $y(t) = \sum_{j=1}^{3} x_j(t)$.

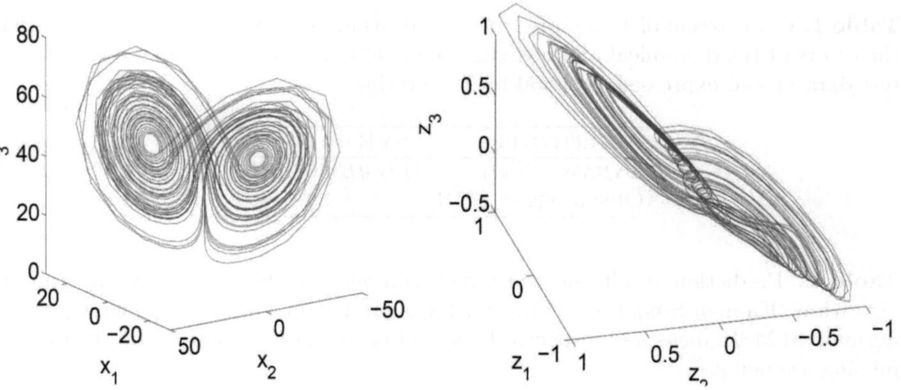

Fig. 4. Lorenz chaotic attractor (left) and the reconstructed chaotic attractor from the latent variables $Z(t) = \{z_1(t), z_2(t), z_3(t)\}$ after inference on the testing dataset (right)

The DFG was trained on 50s (2000 samples) and evaluated on the following 40s (1600 samples) of y. Latent variables $Z(t) = (z_1(t), z_2(t), z_3(t))$ had dimension $m = 3$, as it was greater than the attractor correlation dimension of 2.06 and equal to the number of explicit variables (sources). The dynamical model was implemented as a 3-layered convolutional network. The first layer contained 12 convolutional filters covering 3 time steps and one latent component, replicated on all latent components and every 2 time samples. The second layer contained 12 filters covering 3 time steps and all previous hidden units, and the last layer was fully connected to the previous 12 hidden units and 3 time steps. The dynamical model was autoregressive on $p = 11$ past values of Z, with a total of 571 unique parameters. "Smooth" consecutive states were enforced (12), thanks to the state regularization term $R_z(Z)$ weighted by a small coefficient of 0.01. After training the parameters of DFG, latent variables Z were inferred on the full length of the training and testing dataset, and plotted in 3D values of triplets $(z_1(t), z_2(t), z_3(t))$ (see Fig. 4b).

The 1-step dynamical SNR obtained with a training set of 2000 samples was higher than the 1-step prediction SNR reported for Support Vector Regression (SVR) [13] (see Table 1). According to the Takens theorem [16], it is possible to reconstruct an unknown (hidden) chaotic attractor from an adequately long window of observed variables, using time-delay embedding on $y(t)$, but we managed to reconstruct this attractor on the latent states $(z_1(t), z_2(t), z_3(t))$ inferred both from the training or testing datasets (Fig. 4). Although one of the "wings" of the reconstructed butterfly-shaped attractor is slightly twisted, one can clearly distinguish two basins of attraction and a chaotic orbit switching between one and the other. The reconstructed latent attractor has correlation dimensions 1.89 (training dataset) and 1.88 (test dataset).

Table 1. Comparison of 1-step prediction error using Support Vector Regression, with the errors of the dynamical and observation models of DFGs, measured on the Lorenz test dataset and expressed as signal-to-noise ratios

Architecture	SVR	DFG
Dynamic SNR	41.6 *dB*	46.2 *dB*
Observation SNR	-	31.6 *dB*

Table 2. Prediction results on the CATS competition dataset comparing the best algorithm (Kalman Smoothers [15]) and Dynamic Factor Graphs. E_1 and E_2 are unnormalized MSE, measured respectively on all five missing segments or on the first four missing segments.

Architecture	Kalman smoother	DFG
E1 (5 segments)	408	390
E2 (4 segments)	346	288

3.3 CATS Time Series Competition

Dynamic Factor Graphs were evaluated on time series prediction problems using the CATS benchmark dataset [10]. The goal of the competition was the prediction of 100 missing values divided into five groups of 20, the last group being at the end of the provided time series. The dataset presented a noisy and chaotic behaviour commonly observed in financial time series such as stock market prices.

In order to predict the missing values, the DFG was trained for 10 epochs on the known data (5 chunks of 980 points each). 5-dimensional latent states on the full 5000 point test time series were then inferred in one E-step, as described in section 2.3. The dynamical factor was the same as in section 3.2. As shown in Table 2, the DFG outperformed the best results obtained at the time of the competition, using a Kalman Smoother [15], and managed to approximate the behavior of the time series in the missing segments.

3.4 Estimation of Missing Motion Capture Data

Finally, DFGs were applied to the problem of estimating missing motion capture data. Such situations can arise when "the motion capture process [is] adversely affected by lighting and environmental effects, as well as noise during recording" [17]. Motion capture data[2] Y consisted of three 49-dimensional time series representing joint angles derived from 17 markers and coccyx, acquired on a subject walking and turning, and downsampled to 30Hz. Two sequences of 438 and 3128 samples were used for training, and one sequence of 260 samples for testing.

[2] We used motion capture data from the MIT database as well as sample Matlab code for motion playback and conversion, developped or adapted by Taylor, Hinton and Roweis, available at: `http://www.cs.toronto.edu/~gwtaylor/`

Table 3. Reconstruction error (NMSE) for 4 sets of missing joint angles from motion capture data (two blocks of 65 consecutive frames, about 2s, on either the left leg or entire upper body). DFGs are compared to standard nearest neighbors matching.

Method	Nearest Neighb.	DFG
Missing leg 1	0.77	0.59
Missing leg 2	0.47	0.39
Missing upper body 1	1.24	0.9
Missing upper body 2	0.8	0.48

We reproduced the experiments from [17], where Conditional Restricted Boltzman Machines (CRBM) were utilized. On the test sequence, two different sets of joint angles were erased, either the left leg (1) or the entire upper body (2). After training the DFG on the training sequences, missing joint angles $y_i(t)$ were inferred through the E-step inference. The DFG was the same as in sections 3.2 and 3.3, but with 147 hidden variables (3 per observed variable) and no smoothing. Table 3 shows that DFGs significantly outperformed nearest neighbor interpolation (detailed in [17]), by taking advantage of the motion dynamics modeled through dynamics on latent variables. Contrary to nearest neighbors matching, DFGs managed to infer smooth and realistic leg or upper body motion. Videos comparing the original walking motion sequence, and the DFG- and nearest neighbor-based reconstructions are available at `http://cs.nyu.edu/~mirowski/pub/mocap/`. Figure 5 illustrates the DFG-based reconstruction (we did not include nearest neighbor interpolation resuts because the reconstructed motion was significantly more "hashed" and discontinuous).

4 Discussion

In this section, we establish a comparison with other nonlinear dynamical systems with latent variables (4.1) and suggest that DFGs could be seen as an alternative method for training Recurrent Neural Networks (4.2).

4.1 Comparison with Other Nonlinear Dynamical Systems with Latent States

An earlier model of nonlinear dynamical system with hidden states is the Hidden Control Neural Network [11], where latent variables $Z(t)$ are added as an additional input to the dynamical model on the observations. Although the dynamical model is stationary, the latent variable $Z(t)$ modulates its dynamics, enabling a behavior more complex than in pure autoregressive systems. The training algorithm iteratively optimizes the weights $\mathbf{W}$ of the Time-Delay Neural Network (TDNN) and latent variables Z, inferred as
$\tilde{Z} \equiv argmin_Z \sum_t ||Y(t) - f_{\tilde{\mathbf{W}}}(Y(t-1), Z)||^2$.

The latter algorithm is likened to approximate maximum likelihood estimation, and iteratively finds a sequence of dynamic-modulating latent variables and

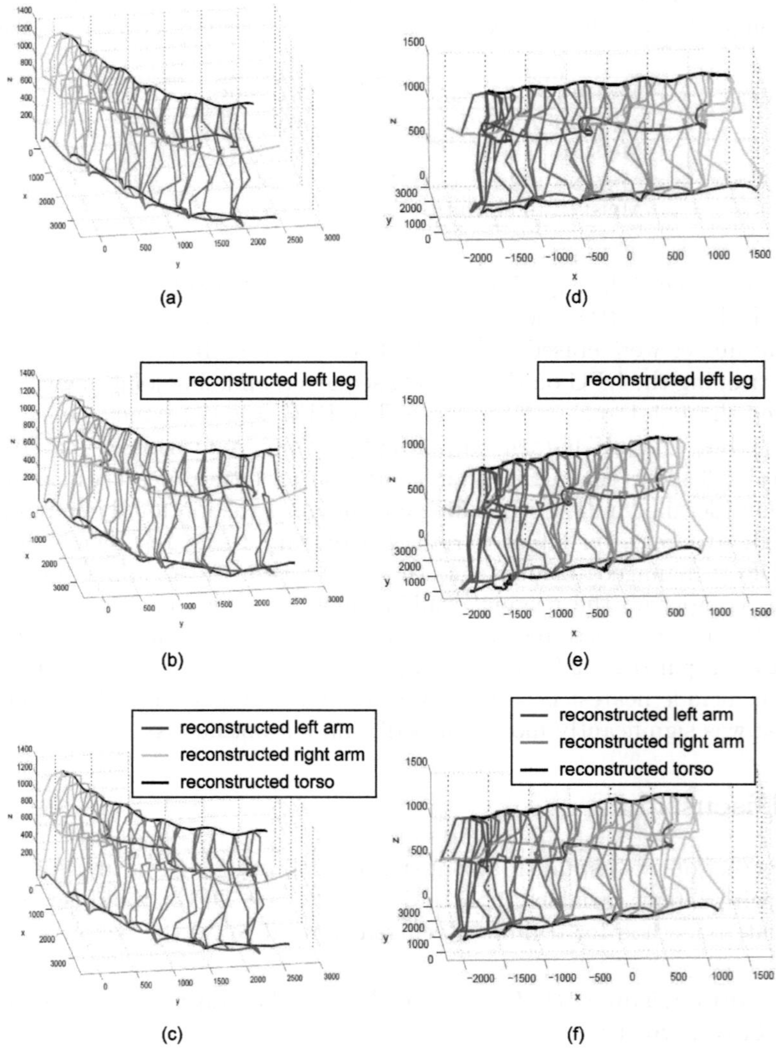

Fig. 5. Application of a DFG for the reconstruction of missing joint angles from motion capture marker data (1 test sequence of 260 frames at 30Hz). 4 sets of joint angles were alternatively "missing" (erased from the test data): 2 sequences of 65 frames, of either left leg or the entire upper body. (a) Subsequence of 65 frames at the beginning of the test data. (b) Reconstruction result after erasing the left leg markers from (a). (c) Reconstruction results after erasing the entire upper body markers from (a). (d) Subsequence of 65 frames towards the end of the test data. (e) Reconstruction result after erasing the left leg markers from (d). (f) Reconstruction results after erasing the entire upper body markers from (d).

learns dynamics on observed variables. DFGs are more general, as they allow the latent variables $Z(t)$ not only to modulate the dynamics of observed variables, but also to generate the observations $Y(t)$, as in DBNs. Moreover, [11] does not introduce dynamics between the latent variables themselves, whereas DFGs model complex nonlinear dynamics where hidden states $Z(t)$ depend on past states $\mathbf{Y}_{t-p}^{t-1}$ and observations $\mathbf{Z}_{t-p}^{t-1}$. Because our method benefits from highly complex non-linear dynamical factors, implemented as multi-stage temporal convolutional networks, it differs from other latent states and parameters estimation techniques, which generally rely on radial-basis functions [19,4].

The DFG introduced in this article also differs from another, more recent, model of DBN with deterministic nonlinear dynamics and explicit inference of latent variables. In [1], the hidden state inference is done by message passing in the forward direction only, whereas our method suggests hidden state inference as an iterative relaxation, i.e. a forward-backward message passing until "equilibrium".

In a limit case, DFGs could be restricted to a deterministic latent variable generation process like in [1]. One can indeed interpret the dynamical factor as hard constraints, rather than as an energy function. This can be done by setting the dynamical weight α to be much larger than the observation weight β in (4).

4.2 An Alternative Inference and Learning for Recurrent Neural Networks

An alternative way to model long-term dependencies is to use recurrent neural networks (RNN). The main difference with the proposed DFG model is that RNN use fully deterministic noiseless mappings for the state dynamics and the observations. Hence, there is no other inference procedure than running the network forward in time. Unlike with DFG, the state at time t is fully determined by the previous observations and states, and does not depend on future observations.

Exact gradient descent learning algorithms for Recurrent Neural Networks (RNN), such as Backpropagation Through Time (BPTT) or Real-Time Recurrent Learning (RTRL) [22], have limitations. The well-known problem of vanishing gradients is responsible for RNN to forget, during training, outputs or activations that are more than a dozen time steps back in time [2]. This is not an issue for DFG because the inference algorithm effectively computes "virtual targets" for the function f at every time step.

The faster of the two algorithms, BPTT, requires $O(T|\mathbf{W}|)$ weight updates per training epoch, where $|\mathbf{W}|$ is the number of parameters and T the length of the training sequence. The proposed EM-like procedure, which is dominated by the E-step, requires $O(aT|\mathbf{W}|)$ operations per training epoch, where a is the average number of E-step gradient descent steps before convergence (a few to a few dozens if the state learning rate is set properly).

Moreover, because the E-step optimization of hidden variables is done on mini-batches, longer sequences T simply provide with more training examples and thus facilitate learning; the increase in computational complexity is linear with T.

5 Conclusion

This article introduces a new method for learning deterministic nonlinear dynamical systems with highly complex dynamics. Our approximate training method is gradient-based and can be likened to Generalized Expectation-Maximization.

We have shown that with proper smoothness constraints on the inferred latent variables, Dynamical Factor Graphs manage to perfectly reconstruct multiple oscillatory sources or a multivariate chaotic attractor from an observed one-dimensional time series. DFGs also outperform Kalman Smoothers and other neural network techniques on a chaotic time series prediction tasks, the CATS competition benchmark. Finally, DFGs can be used for the estimation of missing motion capture data. Proper regularization such as smoothness or a sparsity penalty on the parameters enable to avoid trivial solutions for high-dimensional latent variables. We are now investigating the applicability of DFG to learning genetic regulatory networks from protein expression levels with missing values.

Acknowledgments. The authors wish to thank Marc'Aurelio Ranzato for fruitful discussions and Graham Williams for his feedback and dataset.

References

1. Barber, D.: Dynamic bayesian networks with deterministic latent tables. In: Advances in Neural Information Processing Systems NIPS 2003, pp. 729–736 (2003)
2. Bengio, Y., Simard, P., Frasconi, P.: Learning long-term dependencies with gradient descent is difficult. IEEE Transactions on Neural Networks 5, 157–166 (1994)
3. Dempster, A., Laird, N., Rubin, D.: Maximum likelihood from incomplete data via the EM algorithm. Journal of the Royal Statistical Society B 39, 1–38 (1977)
4. Ghahramani, Z., Roweis, S.: Learning nonlinear dynamical systems using an EM algorithm. In: Advances in Neural Information Processing Systems NIPS 1999 (1999)
5. Kschischang, F., Frey, B., Loeliger, H.-A.: Factor graphs and the sum-product algorithm. IEEE Transactions on Information Theory 47, 498–519 (2001)
6. Ilin, A., Valpola, H., Oja, E.: Nonlinear dynamical Factor Analysis for State Change Detection. IEEE Transactions on Neural Networks 15(3), 559–575 (2004)
7. Lang, K., Hinton, G.: The development of the time-delay neural network architecture for speech recognition. Technical Report CMU-CS-88-152, Carnegie-Mellon University (1988)
8. LeCun, Y., Bottou, L., Bengio, Y., Haffner, P.: Gradient-based learning applied to document recognition. Proceedings of the IEEE 86, 2278–2324 (1998a)
9. LeCun, Y., Bottou, L., Orr, G., Muller, K.: Efficient backprop. In: Orr, G.B., Müller, K.-R. (eds.) NIPS-WS 1996. LNCS, vol. 1524, p. 9. Springer, Heidelberg (1998b)
10. Lendasse, A., Oja, E., Simula, O.: Time series prediction competition: The CATS benchmark. In: Proceedings of IEEE International Joint Conference on Neural Networks IJCNN, pp. 1615–1620 (2004)

11. Levin, E.: Hidden control neural architecture modeling of nonlinear time-varying systems and its applications. IEEE Transactions on Neural Networks 4, 109–116 (1993)
12. Lorenz, E.: Deterministic nonperiodic flow. Journal of Atmospheric Sciences 20, 130–141 (1963)
13. Mattera, D., Haykin, S.: Support vector machines for dynamic reconstruction of a chaotic system. In: Scholkopf, B., Burges, C.J.C., Smola, A.J. (eds.) Advances in Kernel Methods: Support Vector Learning, pp. 212–239. MIT Press, Cambridge (1999)
14. Muller, K., Smola, A., Ratsch, G., Scholkopf, B., Kohlmorgen, J., Vapnik, V.: Using support vector machines for time-series prediction. In: Scholkopf, B., Burges, C.J.C., Smola, A.J. (eds.) Advances in Kernel Methods: Support Vector Learning, pp. 212–239. MIT Press, Cambridge (1999)
15. Sarkka, S., Vehtari, A., Lampinen, J.: Time series prediction by kalman smoother with crossvalidated noise density. In: Proceedings of IEEE International Joint Conference on Neural Networks IJCNN, pp. 1653–1657 (2004)
16. Takens, F.: Detecting strange attractors in turbulence. Lecture Notes in Mathematics, vol. 898, pp. 336–381 (1981)
17. Taylor, G., Hinton, G., Roweis, S.: Modeling human motion using binary latent variables. In: Advances in Neural Information Processing Systems NIPS 2006 (2006)
18. Wan, E.: Time series prediction by using a connectionist network with internal delay lines. In: Weigend, A.S., Gershenfeld, N.A. (eds.) Time Series Prediction: Forecasting the Future and Understanding the Past, pp. 195–217. Addison-Wesley, Reading (1993)
19. Wan, E., Nelson, A.: Dual kalman filtering methods for nonlinear prediction, estimation, and smoothing. In: Advances in Neural Information Processing Systems (1996)
20. Wang, J., Fleet, D., Hertzmann, A.: Gaussian process dynamical models. In: Advances in Neural Information Processing Systems, NIPS 2006 (2006)
21. Wierstra, D., Gomez, F., Schmidhuber, J.: Modeling systems with internal state using Evolino. In: Proceedings of the 2005 Conference on Genetic and Evolutionary Computation, pp. 1795–1802 (2005)
22. Williams, R., Zipser, D.: Gradient-based learning algorithms for recurrent networks and their computational complexity. In: Backpropagation: Theory, Architectures and Applications, pp. 433–486. Lawrence Erlbaum Associates, Mahwah (1995)

On Feature Selection, Bias-Variance, and Bagging

M. Arthur Munson[1] and Rich Caruana[2]

[1] Cornell University, Ithaca NY 14850, USA
mmunson@cs.cornell.edu
[2] Microsoft Corporation
rcaruana@microsoft.com

Abstract. We examine the mechanism by which feature selection improves the accuracy of supervised learning. An empirical bias/variance analysis as feature selection progresses indicates that the most accurate feature set corresponds to the best bias-variance trade-off point for the learning algorithm. Often, this is *not* the point separating relevant from irrelevant features, but where increasing variance outweighs the gains from adding more (weakly) relevant features. In other words, feature selection can be viewed as a variance reduction method that trades off the benefits of decreased variance (from the reduction in dimensionality) with the harm of increased bias (from eliminating some of the relevant features). If a variance reduction method like bagging is used, more (weakly) relevant features can be exploited and the most accurate feature set is usually larger. In many cases, the best performance is obtained by using all available features.

1 Introduction

In a collaboration with ecologists, we were faced with the following challenge: learn accurate models for the presence and absence of bird species from noisy observational data collected by volunteers watching bird feeders. Trying many different supervised learning algorithms (SVMs, boosted trees, neural nets, ...), we found that bagged decision trees yielded the best performance for the task. The resulting models were large, complicated, and used almost all of the 200 features available to the learning algorithm. Since the ultimate goal was to gain ecological understanding about avian population dynamics, we ran forward stepwise feature selection to find the smallest feature set yielding excellent performance.

To our surprise, after 30 steps of feature selection performance was still inferior to the performance when using all features and the gap was closing slowly as more features were added (see Figure 1). Unlike most learning algorithms, bagging appeared to perform remarkably well with many noisy features.

In this paper we examine how feature selection improves the accuracy of supervised learning through the lens of bias/variance analysis. We run feature selection for nineteen data sets and compare the bias-variance decompositions of single and bagged decision trees for many different feature subset sizes. The results

W. Buntine et al. (Eds.): ECML PKDD 2009, Part II, LNAI 5782, pp. 144–159, 2009.

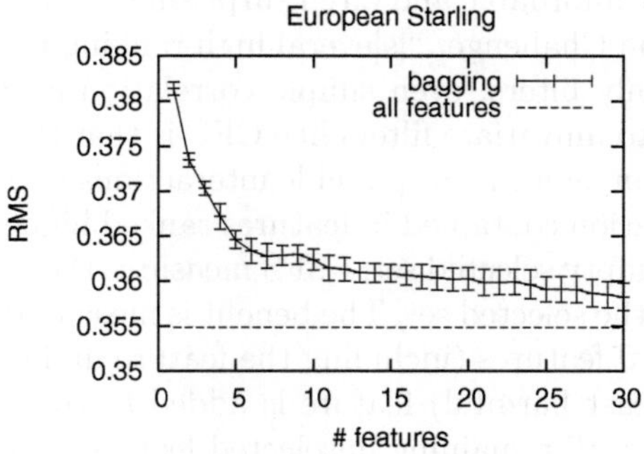

Fig. 1. Bagging performance with forward stepwise feature selection. The *all features* line shows performance of bagging with all 200 features.

show that the most accurate feature sets correspond to the best bias-variance trade-off point, and this depends on the learning algorithm. Particularly with high variance algorithms such as decision trees, this is usually *not* the separating point between relevant and irrelevant features. With too many variables, the increase in variance outweighs the potential gains of adding (weakly) relevant features. When bagging is used, however, the increases in variance are small, which makes the reduction in bias beneficial for many more features. In many cases, the best bagging performance is obtained by using all available features.

While it is known that ensemble methods improve the base learner's ability to ignore irrelevant features [1,2], little is known about their effects on weak/noisy features. To explore this, we generate synthetic data and randomly damage varying percentages of the feature values. The results show that bagging dramatically improves the ability of decision trees to profitably use noisy features.

2 Background

This section reviews feature selection and bagging, and situates the current paper in the context of prior work.

2.1 Feature Selection

Four reasons are traditionally given to motivate feature selection [3]: better predictive performance; computational efficiency from working with fewer inputs; cost savings from having to measure fewer features; and simpler, more intelligible models. Different types of feature selection exist to satisfy varying balances of these competing goals under a variety of data regimes. This work focuses on forward stepwise feature selection (FSFS) [4] and correlation-based feature filtering (CFF). FSFS is preferred when getting the best performance from the smallest feature set possible is important — as long as it is computationally feasible. For large data sets with hundreds or thousands of features, simple filter methods like CFF are affordable and often surprisingly competitive. In the NIPS 2003 Feature Selection Challenge, "[s]everal high ranking participants obtain[ed] good results using only filters, even simple correlation coefficients" ([5], p. 6). The main drawback to univariate filters like CFF is that they estimate the value of a feature in isolation, ignoring a) possible interactions with other features, and b) redundant information contained in features ranked higher (already selected).

Starting from an empty selected set, FSFS measures the benefit of adding each individual feature to the selected set. The benefit is measured by training a model using only the selected features (including the feature under consideration). The most beneficial (or least harmful) feature is added to the selected set, and the process is repeated for all remaining unselected features. The search stops after a fixed number of steps, once performance has stopped improving, or after all features have been selected. If feasible, the learning algorithm used in wrapper-based feature selection usually is the same algorithm to be used with the reduced feature set.

It is important for the search process to measure performance using data withheld during training to ensure good performance estimates. Additionally, the search process itself can potentially overfit this withheld data, so a third data set should be used to get an unbiased estimate of the selected subsets' performance [6]. The FSFS experiments below use a validation set to decide which feature to add, and a test set to measure the final performance.

CFF ranks the set of features by their *individual* correlation with the class label. Our experiments with large data sets use the magnitude of Pearson's correlation coefficient as the ranking criterion. The absolute correlation of feature $x_{.j}$ with the label y is:

$$r_j = \frac{|\sum_i (x_{ij} - \overline{x}_{.j})(y_i - \overline{y})|}{\sqrt{\sum_i (x_{ij} - \overline{x}_{.j})^2 \sum_i (y_i - \overline{y})^2}}$$

where i indexes over examples and $\overline{x}_{.j}$ and $\overline{y}$ are the respective means of $x_{.j}$ and y.[1] Features above a *cutoff* point are retained, while the others are discarded. Common strategies for selecting cutoff points include statistical tests of significance and cross-validated model performance at different ranks. We are interested in the performance at varying rank-levels, so we do not need to choose a cutoff.

Some researchers have looked at bias-variance estimates in the context of feature selection, but typically only for the final feature set selected (e.g. [7]). Van der Putten and van Someran [8] use bias-variance analysis to understand the wide performance spread of contestants in the 2000 CoIL challenge. They compare the bias-variance decompositions of a single subset (the top 7 features) against the original feature set, and find that feature selection is important for their problem (the decrease in variance outweighs the increase in bias).

[1] The high dimensional data sets we use are all binary classification problems with binary and/or continuous features, so Pearson's correlation coefficient is reasonable. Spearman's rank correlation would be a reasonable alternative for non-binary problems or nominal-valued features.

2.2 Bagging

Bagging [9] is a meta-learning algorithm that repeatedly creates sub-samples of the training data and trains a model (e.g. decision tree) on each sample. To make predictions, the bagged model averages the predictions of the constituent models. Bagging frequently improves the performance of a learning algorithm, and rarely hurts it. Bauer and Kohavi [10] showed empirically that the main benefit from bagging is a reduction in the variance of the underlying models. Bagging works best when models have good performance and make uncorrelated mistakes [11].

Several pieces of work exist that address features and bagging. We mention them here to avoid confusion and clarify the differences. (These techniques are not used in the experiments below.) First, *feature bagging* generates diverse simple models by training individual models with random samples of the features instead of (or in addition to) random samples of training examples [12,13], and is particularly useful for building ensembles with simple learners that are inherently stable [2]. In *ensemble feature selection* [14] multiple good feature sets are sought such that a) a good simple model can be built from each set, and b) the simple models are maximally diverse from each other. Finally, *feature selection using ensembles* [15] uses statistics derived from tree ensembles to rank features. More generally, ensembles have been used in feature selection to find more stable feature subsets [16,17].

3 Methodology

3.1 Learning Algorithms

To handle the wide range of data sizes, we used two different decision tree packages. In all cases bagging used 25 trees per ensemble, and training samples were drawn with replacement.

For data sets with small to medium dimensionality (< 200 features), we used minimum message length (MML) decision trees implemented in Buntine's IND package [18]. IND's MML trees use a Bayesian splitting rule with Wallace's MML tree prior [19] and use a small amount of pre-pruning to limit node expansions unwarranted by the tree's posterior probability. Predictions are smoothed by getting a prediction from the leaf and each of its ancestors on the path to the root; these fine- to coarse-grained predictions are combined in a weighted average. See Buntine [20] for full details.

We selected MML trees because the Bayesian smoothing makes them relatively low variance, so in our experience the individual trees perform well and seem to be resilient to spurious and noisy features. Thus, they are less likely to require feature selection to achieve good performance (vs. a less sophisticated decision tree like ID3), making them a strong baseline method. At the same time, they are not aggressively pruned and are large trees, making them good candidates for bagging.[2]

[2] Experiments with other very different tree methods such as C4.5 yield similar results.

For the high-dimensionality data sets, we used FEST[3], a decision tree package optimized for sparse data. To prevent overfitting, we tuned the maximum tree depth parameter in FEST to maximize performance of a single tree, using *all* features, on the validation fold of each data set. We tried depths of 1 through 10, and then the powers of 2 from 16 through 1024. The best performing depth was used for both single and bagged tree models.

A single FEST tree makes predictions from negative to positive infinity. We calibrated the predictions by fitting a sigmoid to convert them to probabilities [21]. Validation data was used to fit the calibrating sigmoid.

3.2 Performance Metrics

Model performance was measured using zero-one loss and squared error. Note that the models described above predict a probability distribution for an example, indicating the likelihood of the example belonging to each class. When the model needs to pick a single class (i.e. for zero-one loss), the class with the largest probability is chosen. A loss of zero represents perfect prediction for these measures.

Zero-one loss is the percentage of predictions that do not predict the correct class. It equals 1 − accuracy, and is often simply called the error rate for a classification model.

Mean squared error (MSE) is the average squared difference between the true prediction and the model's prediction. Let x denote an example, and let $p(x_k)$ and $q(x_k)$ be the true and predicted probability, respectively, that x is class k. Then:

$$\mathrm{MSE} \equiv \frac{1}{nK} \sum_x \sum_k (p(x_k) - q(x_k))^2$$

where K is the number of classes for the task.[4]

Zero-one loss frequently has high variance, so MSE was used as the performance metric during FSFS when deciding which feature to add.

3.3 Data Sets

We used 19 classification tasks in our experiments: American Goldfinch presence/absence at bird feeders (AMEGFI), Lark Bunting presence/absence in the plains east of the Rocky Mountains (BUNTING), forest cover-type (COVTYPE), Pima Indians Diabetes (PIMA), letter recognition (LETTERS), mushroom identification (MUSHROOM), land classification from satellite images (SATIMAGE, Statlog dataset), sonar classification (SONAR), soybean disease classification (SOYBEAN), spam detection (SPAMBASE and SPAMTREC[5]), cardiac abnormalities (SPECTF),

[3] `http://www.cs.cornell.edu/~nk/fest/`

[4] Normalizing by K is not strictly necessary, but places MSE on the same scale regardless of the number of classes.

[5] Created from TREC 2005 Spam Public Corpora. Nikos Karampatziakis, personal communication.

Table 1. Summary of datasets

Data Set	# Samples	# Features	# Classes	Max Depth
amegfi	23948	195	2	
bunting	20998	175	2	
covtype †‡	30,000	54	7	
letters †	20,000	16	26	
medis	14199	63	2	
mg5	22157	100	2	
mushroom †	8124	22	2	
pima †	768	8	2	
satimage †	6535	36	6	
sonar †	208	60	2	
soybean †	683	35	19	
spambase †	4601	57	2	
spectf †	266	44	2	
thyroid †	3772	27	5	
cryst	5,498	1341	2	4
digits	70,000	779	2	16–1024
real-sim	72,309	20,958	2	256–1024
spamtrec	87,688	405,915	2	256–1024
tis	13,375	927	2	5–6

†: Available from UCI Machine Learning Repository [22].
‡: First 30,000 examples from full data set.

hyper-thyroid conditions (THYROID), two medical prediction tasks (MEDIS and MG5), protein crystallography diffraction pattern analysis (CRYST[6]), hand-written digit recognition (DIGITS[7]), real vs. simulated auto racing and aviation text categorization (REAL-SIM[8]), and finding translation initiation sites (TIS[9].).

Table 1 summarizes the data sets; high-dimensional data sets are listed below the line along with the maximum tree depth(s) chosen during parameter tuning. Data sets were chosen to cover a range of sizes (number of examples) and dimensionalities (number of features). We used the first 30,000 points from the COVTYPE data set to make the experiments more affordable.

Each data set was divided into five folds. For FSFS, three folds were used for training, one for validation (to pick which feature to add), and one for testing final performance. For CFF, feature ranks were computed from the three training folds. The description for SATIMAGE warns against using cross-validation; for that data set we used the given train/test split instead of cross-validation and

[6] `http://ajbcentral.com/CrySis/dataset.html`, unscaled version

[7] MNIST data set converted to binary classification (class 0 = digits 5 or below; class 1 = rest). Original available from `http://yann.lecun.com/exdb/mnist/`

[8] `http://csie.ntu.edu.tw/~cjlin/libsvmtools/datasets`

[9] `http://datam.i2r.a-star.edu.sg/datasets/krbd/` (Kent Ridge Biomedical Data Repository)

pulled 435 examples from the train set to use as a validation set (about 10% of training).

3.4 Bias-Variance Decomposition

The bias-variance decomposition (BVD) of loss is a useful tool for understanding the performance characteristics of a learning algorithm. The squared error for a single example x can be decomposed into the sum of noise, bias, and variance [23], all non-negative. The noise is the intrinsic error / uncertainty for x's correct prediction, regardless of learning algorithm. Bias measures how closely the learning algorithm's average prediction (considering all possible training sets of a fixed size) matches the optimal prediction (the Bayes rate prediction). Finally, the variance of an algorithm is how much the algorithm's prediction fluctuates over different possible training sets of a given size.

We adopt the notation and definitions from Domingos [24] to formally express these quantities. Let $L(t, y)$ denote the squared loss of the prediction y for test example x which has the true value t. Further let $E_D[]$ be the expectation over the distribution of possible data sets of a fixed size; similarly, $E_t[]$ is the expectation over the distribution of possible true values for x (in a stochastic domain), and $E_{D,t}[]$ is over the joint distribution of D and t. Then expected squared loss for x can be decomposed as:

$$E_{D,t}[L(t,y)] = N(x) + B(x) + V(x), \qquad \begin{aligned} N(x) &= E_t[L(t,y_*)] \\ B(x) &= L(y_*, y_m) \\ V(x) &= E_D[L(y_m, y)] \end{aligned}$$

where y is the prediction from a model trained on data drawn from D, y_* is the *optimal prediction* that minimizes $E_t[L(t, y_*)]$, and y_m is the *main prediction* that minimizes $E_D[L(y, y_m)]$. For squared loss y_m is the mean prediction of the algorithm across possible training data sets. The expected bias and variance are computed by averaging over multiple test examples.

To estimate bias and variance on real data sets, we follow the same basic sampling procedure used by Bauer and Kohavi [10], since Bouckaert [25] shows that bootstrap sampling results in less reliable bias-variance estimates. The train and validation sets are pooled to create D. Twenty samples of size $|D|/2$ are drawn from D without replacement. Each sample is used to train a model that makes predictions on the test set. This empirical distribution of the algorithm's predictions is used to compute expected bias and variance. To improve the estimates of bias and variance, we repeat this process for each fold and average the estimates.

In practice, we cannot know y_* for real data so we follow previous authors [10,24] in using $y_* = t$. As a result, the bias and noise cannot be separated and are combined in one term for our estimates.

There are multiple proposals for the bias-variance decomposition of zero-one loss [26,24]. In the results below we focus on the decomposition for squared error because feature selection hill climbing used MSE. We did, however, compute the bias and variance of zero-one loss; the results were qualitatively identical to those obtained using the squared error decomposition.

4 Bias-Variance of Feature Selection

We estimated the bias-variance decomposition for all the data sets in Table 1 at multiple feature set sizes, for both single and bagged decision trees. Feature subset orderings were found using forward stepwise feature selection (FSFS, top of table) and correlation coefficients (CFF, bottom of table). FSFS evaluated performance using single trees or bagged trees, to match the algorithm used in the final comparison. After establishing subset orderings (and tuning the maximum tree depth for the high dimensional data sets), the training and validation sets were pooled as described in Sect. 3.4.[10] Bias-variance estimates were made at several points along the subset ordering sequence. The entire experiment was repeated across 5-folds and the 5 estimates averaged.[11]

The results cluster into two categories. Figure 2 shows representative results for two of the data sets. The *total* height of each bar is the error for the number of features on the x-axis. The pair of bars for each number of features correspond to using a single tree (left in pair) and using bagged trees (right in pair). Each bar is subdivided into portions that are due to a) the variance of the algorithm, and b) the bias of the algorithm. The bias portion also contains any noise inherent in the domain. For comparison's sake, results are shown for both mean squared error (left column) and zero-one loss (right column). For the moment we focus on patterns in the total error. Detailed observations about bias and variance are below.

Feature selection does not improve the performance of single or bagged trees on data sets in category one. Consider the graphs for COVTYPE (top row of Figure 2). Both bagging and the single tree perform as well (or better) using all features (right side of graph) than when using a subset (interior of graph). The graphs in Figure 3 show qualitatively similar results: feature selection does not improve the accuracy of single or bagged trees. (The results for zero-one loss are qualitatively the same as for squared error, and are omitted for most of the data sets.)

The second category, however, contains data sets on which feature selection improves single tree accuracy but does not improve bagging's accuracy. Looking at MEDIS (bottom row of Figure 2), the single tree achieves the minimum loss between five and ten features. Bagging, on the other hand, first reaches its minimum loss around 50 features, at which point the loss flattens out and stays roughly constant. The graphs in Figure 4 (*except* BUNTING — see discussion

[10] When validation data was needed to calibrate predictions, we set aside 10% of the training sample drawn from the pooled data. Thus, the calibration data varied with each training sample.

[11] The PIMA, SONAR, SPECTF, and THYROID data sets exhibited substantial variance in the results, so we repeated the 5-fold cross-validation five times using different seeds to divide the data into folds.

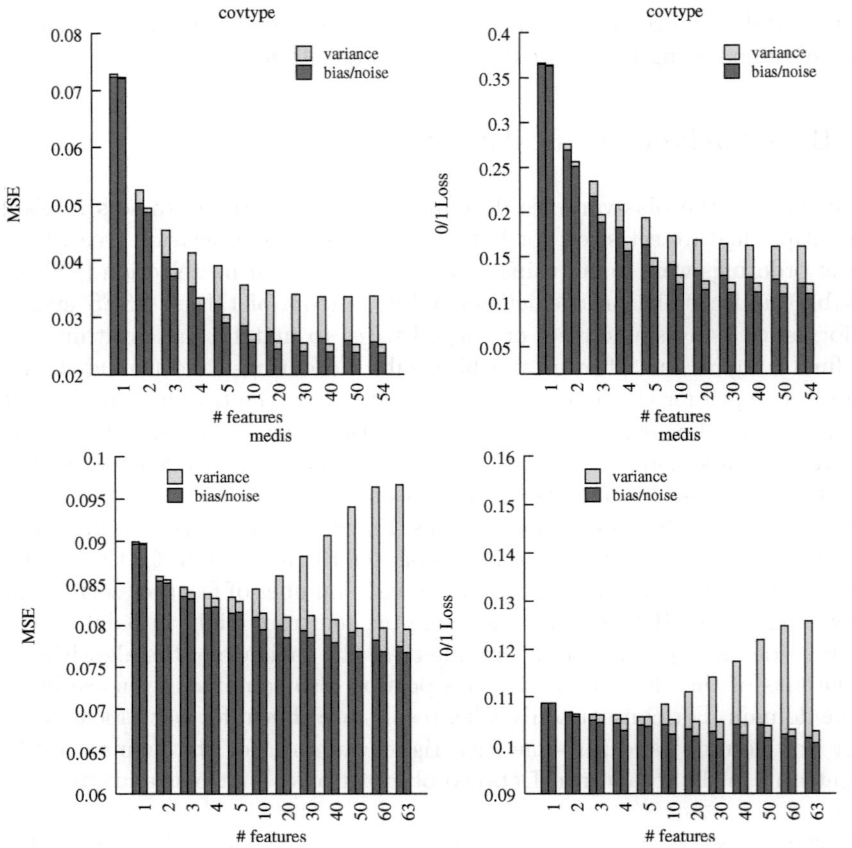

Fig. 2. Bias-variance decomposition of squared error and zero-one error for typical data sets. Left bar in pair: *single tree*; right bar: *bagging*. To better show interesting parts of graphs, the y-axes do not start at 0.

below) contain similar results. While single trees eventually lose performance as more features are added, bagging maintains or improves its performance with more features.

It is worth noting that for data sets in both categories (CRYST, LETTERS, MEDIS, PIMA, SATIMAGE, SONAR, SPAMBASE, SPECTF, TIS), bagging performance continues to improve as more features are added after the performance of single trees has plateaued (category 1) or peaked (category 2). In other words, bagging performance flattens further to the right in the graphs. Bagging seems to be capable of extracting information from noisy features as well as ignoring irrelevant ones. Section 5 explores this issue further.

For all the data sets, bias decreases as more features are added. This makes intuitive sense since extra features can be thought of as extra degrees of freedom. The decrease is largest for the first few features; after that, the bias levels off as the algorithms become sufficiently flexible. Although the bias error is very

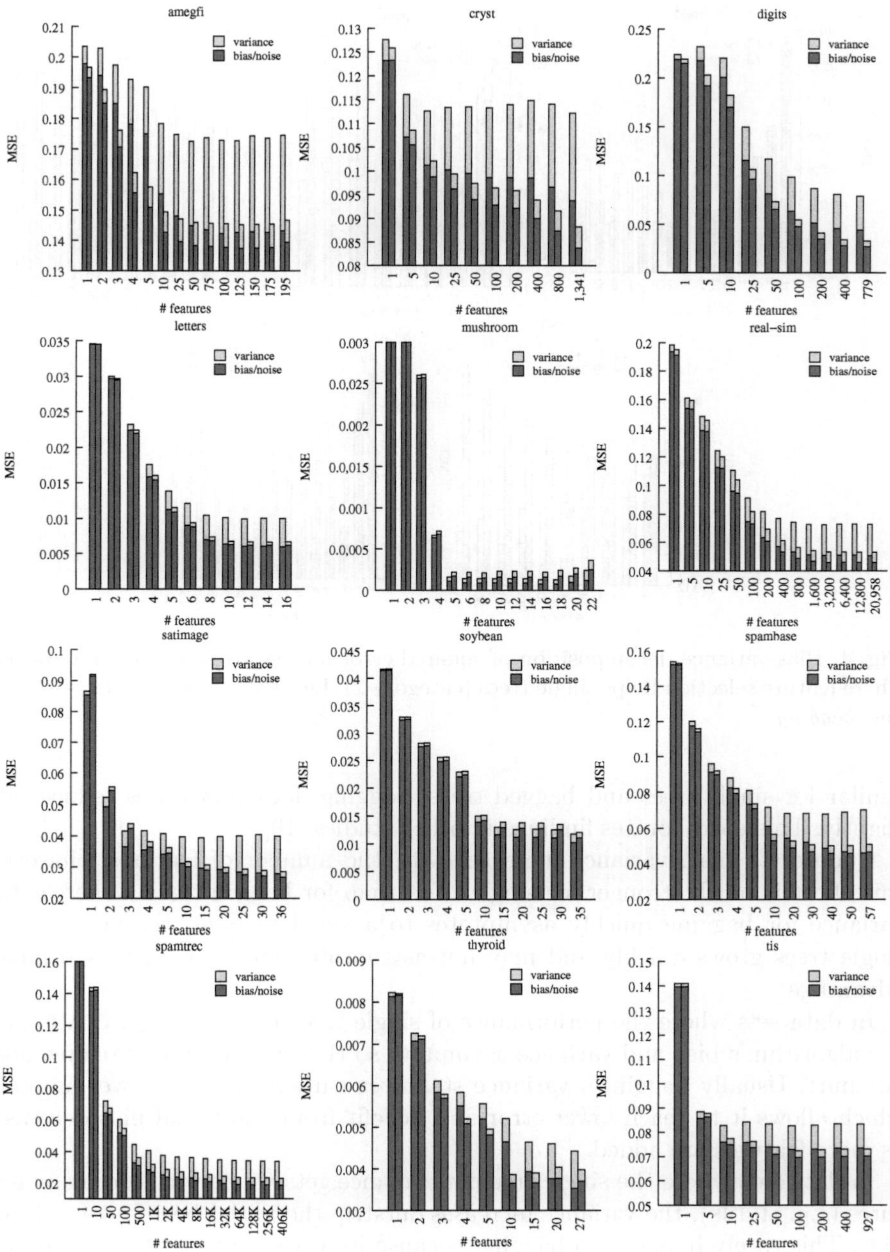

Fig. 3. Bias-variance decomposition of squared error for feature selection on data sets where feature selection does not improve performance (category 1). Left bar in pair: *single tree*; right bar: *bagging.*

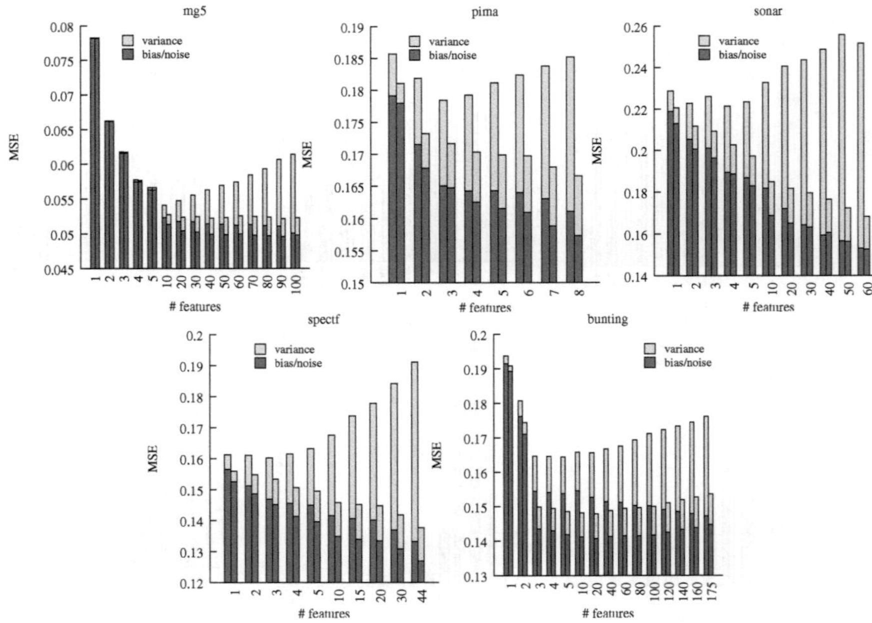

Fig. 4. Bias-variance decomposition of squared error for feature selection on data sets where feature selection helps single trees (category 2). Left bar in pair: *single tree*; right bar: *bagging.*

similar for single trees and bagged trees, bagging does sometimes reduce bias slightly. This corroborates findings in other studies [10].

Counter to bias, variance increases with the number of features. However, this effect is much stronger for single trees than for bagged trees. Whereas the variance for bagging quickly asymptotes to a small amount, the variance for single trees grows quickly and may not asymptote. This is bagging's primary advantage.

In data sets where the performance of single trees levels off (e.g. COVTYPE), the algorithm's bias and variance asymptote so that adding more features does not hurt. Usually bagging's variance stabilizes earlier and to a lower amount, which allows it to reach lower error and benefit from additional bias decreases as more features are added.

In data sets where the single tree performance gets worse with too many features (e.g. MEDIS), the variance increases outstrip the initial benefits of reduced bias. This rarely happens to bagging because its variance typically asymptotes to a small amount of error.

Figures 3 and 4 contain three anomalies, one large and two small. The most important anomaly is the graph for BUNTING, which does not fit into either category described above. On this data set, both single trees and bagging hit peak performance between 5 and 20 features, after which their performance degrades. Thinking that perhaps this domain was just extremely noisy and that more averaging would eliminate bagging's overfitting, we re-ran feature selection on one fold using 100 trees instead of 25. With 100 trees, bagging's performance was slightly better at all points along the x-axis (compared to bagging with 25 trees), but still overfit past 20 features and to the same degree. Further investigation revealed that this data set contains several features that can be combined to create semi-unique identifiers for individual examples in the training set. All the trees in the ensemble effectively memorize these identifiers and then do poorly on the validation and test data. This can be seen in the bias-variance decomposition. Although the variance has asymptoted, the bias for bagging stops decreasing and begins increasing. With the extra features, the trees in the ensemble more consistently construct unique identifiers for training examples and lose diversity in their incorrect predictions.[12]

The other two anomalies are that single decision trees perform (slightly) better than the bagged tree ensemble for the MUSHROOM and SOYBEAN data sets. For MUSHROOM, the single tree is extremely confident in the class probabilities it assigns, and always picks the right class (zero-one loss is 0%). Bagging also always picks the right class, but the randomization from sub-sampling the training data plus averaging results in *slightly* less confident class probabilities (probability mass is pushed away from the extremes). This small bias away from extreme values has a small effect on squared error. SOYBEAN has a different problem. This small data set has 19 classes. Cross-validation and bagging sampling reduces the number of cases for some classes in the training samples so that probability estimates become less reliable and MML pre-pruning prevents leaf expansions, yielding trees that are too small.

Throughout this section (and most of the paper), noise and bias have been conflated since we do not have a way to separate them on real data. We hypothesize that the large decreases in bias—coinciding with adding the first few features—is partly due to decreases in noise. Intuitively, the Bayes optimal error rate, given only a single feature as an information source, may be quite bad (effectively high noise). As more information becomes available (more features), the Bayes rate should improve as uncertainty decreases.

To summarize this section, these graphs show that bagging is resilient to noisy features. *Feature selection usually is unnecessary to get good performance from bagged models.* Further, picking the best subset size (using cross-validation, for

[12] A more detailed explanation follows. The task in BUNTING is to predict the presence or absence of a Lark Bunting. Data are collected at multiple sites; in particular, repeat observations are made at sites over time. Identifying the site is incredibly useful for predicting presence or absence, but is not ecologically interesting. Thus, the five data folds were partitioned by site (i.e. all examples from a site appear in a single fold). Most features are tied to location (e.g. habitat), so the decision trees can easily learn to map inputs to sites in the training set using only a few features. Trees that do this make bad predictions on the validation and test folds. If the folds are created by assigning examples to folds instead of sites (spreading sites across train/valid/test), bagging does not overfit while a single tree does.

example) *is not* equivalent to choosing the informative features and discarding irrelevant features. Rather, a discarded feature may be weakly informative (or correlated with a feature already selected) but cause too much variance when selected for the extra information to improve accuracy. The fact that discarded features are sometimes informative was previously noted and exploited to improve model accuracy by using discarded features as extra model outputs during training [27].

5 Noisy Informative Features

In the experiments above, bagging's performance continues to improve after the single tree's performance peaks or plateaus. This suggests that ensemble methods are not only resilient to irrelevant features [1], but also better able to take advantage of features containing useful but noisy information.

We generated synthetic data to study whether bagging improves the base learner's ability to use weak features. A binary classification problem was derived from the equation:

$$v = X_1 + X_2X_3 + X_4^2 + \text{sign}(X_5 + X_6)$$

The class label is 1 when $v \geq 0$, and 0 otherwise. Each X_i is a univariate Gaussian variable with 0 mean and unit variance. The sign(z) function returns 1 if $z > 0$ and -1 otherwise. This function was chosen to be challenging for decision tree learning algorithms.

We generated 5,000 examples using the above function, randomly corrupted some of the inputs to generate weak features, split the data set into 5 folds, and ran a bias-variance analysis using the procedure outlined in Sect. 3.4. A feature was corrupted by permuting a fraction of its values, chosen randomly among the examples. For example, at the 0.1 corruption level, 10% of the values in corrupted features are shuffled. This was repeated 20 times, creating 20 noisy versions of each corrupted feature. Half of the X_i features were corrupted, independently of each other, while the other X_i were left intact. Single decision trees and bagged decision trees were trained using the intact features and the noisy duplicates, but not the original versions of the corrupted features. To avoid experimental bias, this process was repeated for all $\binom{6}{3}$ combinations of choosing 3 features to corrupt, and the results averaged.

Figure 5 shows the results for different corruption levels. The far left column, *core*, is the error obtained when training using only the unblemished 6 original features, and shows the best performance obtainable on this data set for these algorithms (i.e. when the ideal feature set is used). The 0.0 column shows the performance obtained using only the 3 intact features, without any corrupted features. Performances that beat this baseline indicate an algorithm is learning something useful from noisy features. Finally, the far right column (1.0) shows the performance when the corrupted features are pure noise (irrelevant features).

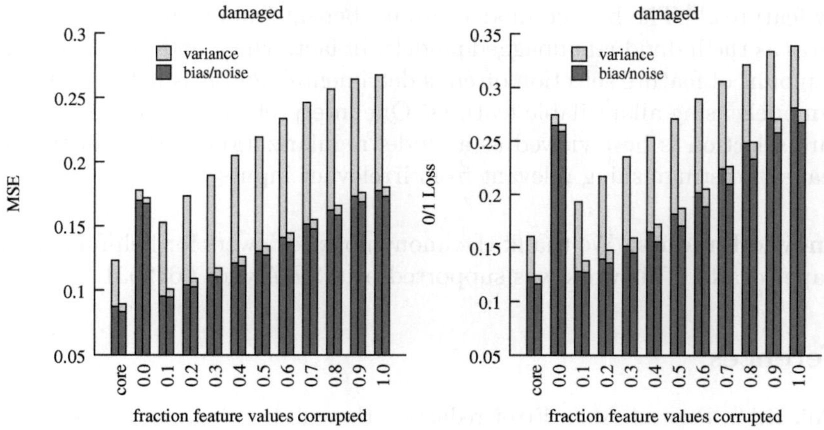

Fig. 5. Bias-variance decompositions for DAMAGED data sets with corrupted feature values. Left bar in pair: *single tree*; right bar: *bagging*. Note that the y-axes do not start at 0.

We make the following observations. First, at low corruption levels both single and bagged trees learn something useful from the noisy features. For bagged trees, performance is close to that of using the ideal feature set. Second, noisy features increase the bias (because noise is lumped in with bias in our empirical decomposition) of both single and bagged trees (vs. core), and increase the variance of single trees. Third, the main effect of increasing the corruption level is to increase the bias/noise component. Finally, the extra variance in the single trees means that the benefits of noisy features are quickly lost as the corruption level increases. At least for this synthetic task, the problem is more pronounced for squared error. In contrast, the bagged trees are remarkably resilient to damaging the feature values, and are able to extract useful information when as much as 80% of the values are corrupted.

6 Conclusions

Our experiments show that feature selection finds the feature set that represents the best trade-off between the bias of having too few features and the variance of having too many features. Because of this, most feature selection algorithms are not reliable methods for determining which features are relevant and irrelevant to a given problem: the threshold for feature inclusion/exclusion depends on the learning algorithm. Ultimately this limits the utility of feature selection for discovering which factors are important and unimportant in problems such as the avian analysis that originally motivated this work.

A by-product of our analysis is the discovery that when feature selection is too expensive to be feasible or effective, bagging provides a viable alternative to protect from the overfitting that can occur when models are trained with too many features.[13] The bagged models always benefit from using at least as many features as the individual unbagged models. In fact, when models will be bagged, any amount of feature selection often is detrimental, and it is better to train the base models using all available features. One interpretation of our results is that feature selection is best viewed as a model regularization method instead of as a means of distinguishing relevant from irrelevant inputs.

Acknowledgments. We thank the anonymous reviewers for helpful comments on paper drafts. This work was supported by NSF Award 0612031.

References

1. Ali, K.M., Pazzani, M.J.: Error reduction through learning multiple descriptions. Machine Learning 24(3), 173–202 (1996)
2. Bay, S.D.: Combining nearest neighbor classifiers through multiple feature subsets. In: ICML 1998: Proceedings of the 15th International Conference on Machine Learning, pp. 37–45. Morgan Kaufmann Publishers Inc., San Francisco (1998)
3. Guyon, I., Elisseeff, A.: An introduction to variable and feature selection. Journal of Machine Learning Research 3, 1157–1182 (2003)
4. Kohavi, R., John, G.H.: Wrappers for feature subset selection. Artificial Intelligence 97(1-2), 273–324 (1997)
5. Guyon, I., Gunn, S., Ben-Hur, A., Dror, G.: Result analysis of the NIPS 2003 feature selection challenge. In: Advances in Neural Information Processing Systems 17, pp. 545–552. MIT Press, Cambridge (2005)
6. Reunanen, J.: Overfitting in making comparisons between variable selection methods. Journal of Machine Learning Research 3, 1371–1382 (2003)
7. Loughrey, J., Cunningham, P.: Using early-stopping to avoid overfitting in wrapper-based feature selection employing stochastic search. Technical Report TCD-CS-2005-37, Trinity College Dublin, Department of Computer Science (May 2005)
8. van der Putten, P., van Someren, M.: A bias-variance analysis of a real world learning problem: The CoIL challenge 2000. Machine Learning 57(1-2), 177–195 (2004)
9. Breiman, L.: Bagging predictors. Machine Learning 24(2), 123–140 (1996)
10. Bauer, E., Kohavi, R.: An empirical comparison of voting classification algorithms: Bagging, boosting, and variants. Machine Learning 36(1-2), 105–139 (1999)
11. Dietterich, T.G.: Ensemble methods in machine learning. In: Kittler, J., Roli, F. (eds.) MCS 2000. LNCS, vol. 1857, pp. 1–15. Springer, Heidelberg (2000)
12. Ho, T.K.: The random subspace method for constructing decision forests. IEEE Transactions on Pattern Analysis and Machine Intelligence 20, 832–844 (1998)
13. Bryll, R., Gutierrez-Osuna, R., Quek, F.: Attribute bagging: Improving accuracy of classifier ensembles by using random feature subsets. Pattern Recognition 36(6), 1291–1302 (2003)

---

[13] Of course, feature rankers like CFF are less expensive than training a single bagged tree ensemble. Even with a ranker, however, there is the problem of choosing a cutoff threshold for which features to include — which typically requires training models for multiple candidate threshold levels. Thus, carefully choosing a threshold could easily cause a feature ranker to be more computationally expensive than training a single ensemble.

14. Opitz, D.W.: Feature selection for ensembles. In: AAAI 1999: Proceedings of the 16th National Conference on Artificial Intelligence, pp. 379–384. American Association for Artificial Intelligence, Menlo Park (1999)
15. Tuv, E., Borisov, A., Torkkola, K.: Feature selection using ensemble based ranking against artificial contrasts. In: International Joint Conference on Neural Networks, pp. 2181–2186 (2006)
16. Saeys, Y., Abeel, T., Peer, Y.: Robust feature selection using ensemble feature selection techniques. In: Daelemans, W., Goethals, B., Morik, K. (eds.) ECML PKDD 2008, Part II. LNCS (LNAI), vol. 5212, pp. 313–325. Springer, Heidelberg (2008)
17. Tuv, E.: Ensemble learning. In: Guyon, I., Gunn, S., Nikravesh, M., Zadeh, L.A. (eds.) Feature Extraction: Foundations, and Applications. Studies in Fuzziness and Soft Computing, vol. 207, pp. 187–204. Springer, Heidelberg (2006)
18. Buntine, W., Caruana, R.: Introduction to IND and recursive partitioning. Technical Report FIA-91-28, NASA Ames Research Center (October 1991)
19. Wallace, C.S., Patrick, J.D.: Coding decision trees. Machine Learning 11(1), 7–22 (1993)
20. Buntine, W.: Learning classification trees. Statistics and Computing 2(2), 63–73 (1992)
21. Platt, J.C.: Probabilistic outputs for support vector machines and comparison to regularized likelihood methods. In: Smola, A.J., Bartlett, P.J., Schoelköpf, B., Schuurmans, D. (eds.) Advances in Large Margin Classifiers, pp. 61–74. MIT Press, Cambridge (2000)
22. Asuncion, A., Newman, D.: UCI machine learning repository (2007)
23. Geman, S., Bienenstock, E., Doursat, R.: Neural networks and the bias/variance dilemma. Neural Computation 4(1), 1–58 (1992)
24. Domingos, P.: A unified bias-variance decomposition and its applications. In: Proceedings of the 17th International Conference on Machine Learning, pp. 231–238. Morgan Kaufmann, San Francisco (2000)
25. Bouckaert, R.R.: Practical bias variance decomposition. In: Wobcke, W., Zhang, M. (eds.) AI 2008. LNCS (LNAI), vol. 5360, pp. 247–257. Springer, Heidelberg (2008)
26. Kohavi, R., Wolpert, D.H.: Bias plus variance decomposition for zero-one loss functions. In: Proceedings of the Thirteenth International Conference on Machine Learning. Morgan Kaufmann, San Francisco (1996)
27. Caruana, R., de Sa, V.R.: Benefitting from the variables that variable selection discards. Journal of Machine Learning Research 3, 1245–1264 (2003)

Efficient Pruning Schemes for Distance-Based Outlier Detection

Nguyen Hoang Vu and Vivekanand Gopalkrishnan

Nanyang Technological University, 50 Nanyang Avenue, Singapore
ng0001vu@ntu.edu.sg, asvivek@ntu.edu.sg

Abstract. Outlier detection finds many applications, especially in domains that have scope for abnormal behavior. In this paper, we present a new technique for detecting distance-based outliers, aimed at reducing execution time associated with the detection process. Our approach operates in two phases and employs three pruning rules. In the first phase, we partition the data into clusters, and make an early estimate on the lower bound of outlier scores. Based on this lower bound, the second phase then processes relevant clusters using the traditional block nested-loop algorithm. Here two efficient pruning rules are utilized to quickly discard more non-outliers and reduce the search space. Detailed analysis of our approach shows that the additional overhead of the first phase is offset by the reduction in cost of the second phase. We also demonstrate the superiority of our approach over existing distance-based outlier detection methods by extensive empirical studies on real datasets.

1 Introduction

The problem of detecting abnormal events, also called outliers, has been widely studied in different research communities as rare classes mining [1], exception mining [2], outlier detection [3,4], etc. Researchers have developed several *supervised* and *unsupervised* techniques to mine outliers in *static databases* and also recently in *data streams* [9]. Unsupervised outlier detection can be further classified as distance-based [5,6,4,7], density-based [3,8,9] and deviation-based [10]. In this paper, we focus on distance-based outliers which have been popularly defined as: (a) data points from which there are fewer than p points that are within distance r [4], (b) top n data points whose distance to their corresponding k^{th} nearest neighbor are largest [7], and (c) top n data points whose total distance to their corresponding k nearest neighbors are largest [6]. As these definitions indicate, a significant amount of distance computations need to be performed in order to verify whether a data point is an outlier or not. This leads to high execution times and has motivated many attempts to produce efficient algorithms to mine outliers. Among them, outstanding work by Bay and Schwabacher [11] and Ghoting et al. [12] aim to reduce execution time by utilizing a simple pruning nested-loop algorithm.

Reducing time complexity of outlier detection techniques in general generates many benefits for various applications where the speed of detecting deviations

W. Buntine et al. (Eds.): ECML PKDD 2009, Part II, LNAI 5782, pp. 160–175, 2009.

plays a critical role (e.g., fraud detection, intrusion detection). To illustrate our point, let us consider a system in which data arrives in batches and each batch of data is stored in buffer memory. It may be assumed that the buffer size is large enough to accommodate each batch but if many batches are stored at the same time, buffer will overflow. Such scenario is very popular in applications dealing with data streams [13,9]. The task of the system is to identify abnormal records in each batch. The buffer is automatically flushed when this monitoring process is done. However, if the speed of the detection technique is slower than the speed of arrival of batches, we may lose data because of the problem of buffer overflows. Therefore, developing a fast detection algorithm becomes a necessity since it leads to higher throughput for the system. Additionally, the higher throughput will also yield higher detection accuracy since data loss is avoided.

Motivated by this issue, we focus on reducing the execution time and present a two-phased *MultI-Rule Outlier* (MIRO) detection approach. Based on the definition [6], we develop an outlier scoring criterion. Then in the first phase, we partition the data into clusters, and make an early estimate on the lower bound of outlier scores. This phase prunes clusters that cannot have outliers, and the second phase then processes the remaining clusters using the traditional block nested-loop algorithm. Here two pruning rules are utilized: a) first triangular inequality on the data point's outlier score is used, and then b) the outlier score is compared with the minimum score required to be an outlier. The second check is similar to that of ORCA [11]. However, while ORCA starts with a cutoff of 0, in MIRO the initial cutoff is obtained from the first phase, and hence converges faster. Though the pruning rules seem simple, their combined effect is strong and efficiently reduces the search space. The main contributions of this work can be summarized as follows:

- We analyze the problem of outlier detection from the outlier score perspective and introduce the concepts of global and local outlier score functions. This gives a summary classification of all existing detection techniques.
- We demonstrate a huge improvement in execution time by using multiple pruning rules in two phases, compared with outstanding existing nested-loop distance-based methods, ORCA [11] and RBRP [12]. Since ORCA, RBRP and MIRO use the same notion of outlier (Section 2), outliers identified by the three techniques are exactly the same.
- We illustrate the effectiveness of our pruning rules on the overall detection process and give a detailed theoretical analysis on how those rules lead to the superior performance of MIRO. With extremely low CPU cost, MIRO is very suitable for detecting outliers in streaming environments as well as other real-time applications.

The rest of this paper is organized as follows. We compare related work, and describe the problem formally in the next section. Then we present our MIRO approach in Section 3, and theoretically analyze its complexity in Section 4. Then we empirically compare our approach with other current-best approaches using real-world datasets in Section 5. Finally, we conclude in Section 6 with directions for future work.

2 Literature Review

2.1 Background

Consider a dataset DS with N data points in dim dimensions. While most of these data points are normal, some are abnormal (outlier), and our task is to mine these outliers. Assume a metric distance function D exists, using which we can measure the dissimilarity in dim space between two arbitrary data points. A general approach that has been used by most of the existing outlier detection methods [5,4,3] is to assign an *outlier score* (based on the distance function) to each individual data point, and then design the detection process based on this score. The use of the outlier score is analogous to the mapping of the multi-dimensional dataset to $\mathbb{R}$ space (the set of real numbers). In other words, we can define the outlier score function (F_{out}) which maps each data point in DS to a unique value in $\mathbb{R}$.

Among existing approaches to outlier detection problem, we can classify F_{out} into *global* and *local* score functions. An outlier score function is called *global* when the the value it assigns to a data point $p \in DS$, can be used to compare globally with other data points. More specifically, for two arbitrary data points p_1 and p_2 in DS, $F_{out}(p_1)$ and $F_{out}(p_2)$ can be compared with each other, and if $F_{out}(p_1) > F_{out}(p_2)$, p_1 has a larger possibility than p_2 to be an outlier. The definitions proposed by Angiulli et al. [6], Breunig et al. [3], and Ramaswamy et al. [7] straightforwardly adhere to this category. On the other hand, the definition of Ng and Knorr [4] can be converted to this category by taking the inverse of the number of neighbors within distance r of each data point. In contrast, a *local* outlier score function assigns to each data point p, a score that can only be used to compare within some local neighborhood. An example of such function was proposed in [8], where the local comparison space is the set of data points lying within the *circle* centered by p and the *radius* is user-defined. The choice of a global or local outlier score function clearly affects later stages of the algorithm design process. In this work, we employ a global outlier function based on [6], although the ideas employed in MIRO can also be adapted to use other functions. The intuition and quality of detection results of the chosen outlier definition are based on solid foundations as shown by prior work [6,11]. This definition is also employed in other popular techniques on outlier detection [12]. Therefore, in this paper we do not again demonstrate how well MIRO does in terms of actually discovering abnormalities in real data. Instead, we focus on showing its superiority in terms of CPU cost.

Let us denote the set of k nearest neighbors of a data point p in DS as kNN_p. We can now define F_{out} as follows.

Definition 1. [OUTLIER SCORE FUNCTION]. *The dissimilarity of a point p with respect to its k nearest neighbors is known by its cumulative neighborhood distance. This is defined as the total distance from p to its k nearest neighbors in DS. In other words, we have:* $F_{out}(p) = \sum_{m \in kNN_p} D(p, m)$.

Table 1. Definitions of symbols

Symbol	Definition
DS	The dataset
N	Number of points in the dataset
dim	Dimensionality of the data space
$D(p_1, p_2)$	Distance function between points p_1 and p_2
kNN_p	set of k nearest neighbors of a data point p
n	Number of outliers to be mined
F_{out}	Outlier Score Function

This definition has been proven by Angiulli et al. [6] to be more intuitive than the definition used by Ramaswamy et al. [7]. Given two positive integers k and n, our task is to mine top n outliers that have the largest outlier scores based on the chosen F_{out}. For ease of reference, symbols used in the definitions are presented in Table 1.

2.2 Related Work

Work in distance-based outlier detection was pioneered by Knorr and Ng in 1998 [4]. According to their proposal, outliers are points from which there are fewer than p other points within distance r. In order to detect such outliers, they introduced a nested-loop and a cell-based algorithm. The nested-loop algorithm has time complexity $O(N^2)$ and hence is usually not suitable for applications with a large dataset. On the other hand, the cell-based algorithm has time complexity linear with N, but exponential with the number of dimensions dim. In practice, this can only work efficiently when $dim \leq 4$, so it is not suitable in applications on high-dimensional datasets.

Ramaswamy et al. [7] had a different view of the problem. Instead of counting the r-neighborhood of a data point, their technique only takes the data point's distance to its kth nearest neighbor into account. They proposed three algorithms: nested-loop with $O(N^2)$ time complexity, and index-based and partition-based algorithms. The most efficient among these - the partition-based algorithm, partitions the dataset, and computes the upper and lower bounds of outlier scores for each partition[1]. Keeping track of the minimum lower bound computed so far, the algorithm terminates bound computations of partitions whose upper bound score is lower than this minimum bound. This effectively reduces the search space, and then index-based or nested-loop algorithms can be used on the remaining partitions to detect outliers. In Section 4, we prove that the theoretical complexity of the partition-based strategy is also quadratic to the dataset size. In general, early distance-based approaches were usually involved in time-consuming computations of nearest neighbors. Later techniques aim to reduce this time complexity by various means. Among these, approaches for pruning the outlier search space and distance computation reduction tech-

[1] Alternatively called clusters, micro-clusters.

niques are dominant. Computation reduction approaches [7,12,11,6] usually fix the desired number of outliers to a certain value (e.g., top n outliers), and deploy data structures similar to those used in Ramaswamy's index-based algorithm.

Bay and Schwabacher [11] provide detailed analysis for this type of algorithm, and discover that, in average case, the time complexity becomes linear with the data set size. However, their proposed technique, ORCA, depends on some assumptions: the data is in random order and the values of the data points are independent. The analysis provided also depends on the outlier score cutoff c which is initialized to 0. However, domain knowledge or a training phase can help to achieve a better cutoff. More specifically, the authors suggest that by training a subset of the original data set, an initial cutoff threshold can be obtained. During testing phase, the training set is placed at the top of the data set so that the cutoff threshold calculated during training phase can be retrieved very soon, and hence the pruning occurs at the very first stage of the detection process. The linear time complexity presented in [11] can only be obtained if the cutoff threshold c converges to $O(\sqrt{N})$ quickly [12]. However this occurs only when the dataset contains many outliers. Recognizing this limitation, Ghoting et al. [12] proposes RBRP, an algorithm which finds *approximate* nearest neighbors for every normal data point but exact ones for outliers. By avoiding expensive computations to find the exact nearest neighbors for normal records, RBRP works in $O(N \cdot lgN)$ time. The approach first clusters the dataset, and then searches for a data point's approximate nearest neighbors in its own cluster and neighboring clusters.

While the above mentioned techniques attempt to reduce execution time of the detection process, Tao et al. [14] aims at reducing I/O cost without any heuristic to minimize the CPU cost. Furthermore, it uses the notion of outliers introduced in [4], which has been shown to be difficult to apply in practice [7]. Hence, we choose not to compare our technique against the one in [14].

3 The MIRO Detection Approach

Our approach operates in two phases and employs three pruning rules. In the first phase, we partition DS into clusters, and compute upper and lower bounds of the outlier score for each cluster. Based on these bounds, some clusters are pruned, and the remaining candidates are sent for final processing in the traditional block nested-loop algorithm. Here two pruning rules are utilized: a) first triangular inequality on the data point's outlier score is used (R_1), and then b) the outlier score is compared with the minimum score required to be an outlier (R_2). The second check is similar to that of ORCA [11], however in MIRO the initial cutoff is obtained from the first phase (instead of using 0 as in [11]), and hence converges faster. The additional overhead of the first phase is offset by the reduction in cost of the second phase. While preprocessing by clustering has been proposed in RBRP, our preprocessing phase incorporates the pruning of unnecessary clusters while RBRP's does not. Additionally, the use of the simple triangular inequality in the second phase and the precomputation of the initial

Algorithm 1. Cluster

Input: M: the number of clusters, it: the number of iterations, DS: the dataset to be clustered
Output: B: the set of clusters

Set $Y = KMeans(M, it, DS)$
foreach *cluster* $y \in Y$ **do**
  if $\frac{|y|}{n_c} > M$ **then**
    $Cluster(M, it, y)$
  else if $\frac{|y|}{n_c} > 1$ **then**
    Set $Y' = KMeans(\lfloor \frac{|y|}{n_c} \rfloor, it, y)$
    foreach *cluster* $y' \in Y'$ **do**
      Add y' to B
  else
    Add y to B

cutoff of outlier score before this phase commences, generates the distinct advantages of MIRO's nested-loop compared to that of ORCA. The detailed process is described below.

3.1 Cluster Based Pruning

In this phase, we first cluster the dataset DS (using Algorithm 1) and subsequently identify upper and lower bounds of the outlier score for each resultant cluster (using Algorithm 2). Algorithm 1 is in fact based on the clustering algorithm of RBRP [12], however we have made some modifications. We denote the *expected number of data points per cluster* as n_c. By changing n_c, we can control the degree of homogeneity of clusters, i.e., points that are close to each other in space are likely assigned to the same cluster. It is noted that in our approach, n_c has the same role as the parameter $BinSize$ of RBRP. Compared to the original algorithm [12], the cost of clustering is saved for those resultant clusters y having $1 < |y|/nc \leq M$, since a) they are re-clustered only once with the number of clusters being $\lfloor \frac{|y|}{n_c} \rfloor \leq M$ and b) the time complexity of K-Means algorithm is proportional to the number of clusters produced. Hence, our clustering algorithm takes less time than that of RBRP.

Let C be the set of clusters obtained as a result of applying Algorithm 1 on DS with predetermined values of M and it. For each cluster $C_i \in C$, let $|C_i|$ denote its cardinality (or the number of data points allocated to C_i), o_{C_i} its centroid, and r_{C_i} its radius. l_{C_i}, u_{C_i} are the estimated lower and upper bounds of the outlier scores of all data points in C_i respectively. These bounds are only estimations since the true bounds can only be known when the true scores of member data points are identified. A data point p by itself is also a cluster C_i with $o_{C_i} = p$, $r_{C_i} = 0$, $l_{C_i} = u_{C_i} = F_{out}(p)$.

Definition 2. [DISTANCE BETWEEN CLUSTERS]
The minimum distance between clusters C_i *and* C_j *is*

$$minDis(C_i, C_j) = max\{D(o_{C_i}, o_{C_j}) - r_{C_i} - r_{C_j}, 0\},$$

and maximum distance between clusters C_i *and* C_j *is*

$$maxDis(C_i, C_j) = D(o_{C_i}, o_{C_j}) + r_{C_i} + r_{C_j}.$$

Given a cluster $C_i \in C$, we now need to find clusters that potentially contain k nearest neighbors for every point in C_i. So we first find a set of clusters, Min_{C_i}, *closest* to C_i in terms of $minDis()$, containing at least k data points, i.e., $Min_{C_i} \subseteq C \backslash C_i$, s.t. $minDis(C_j, C_i) \leq minDis(C_k, C_i)\ \forall C_j \in Min_{C_i}, C_k \in C \setminus \{C_i \bigcup Min_{C_i}\}$, the total number of data points in $Min_{C_i} \geq k$.

Similarly, we identify a set of clusters, Max_{C_i}, *closest* to C_i in terms of $maxDis()$, which also contains at least k data points in total.

Consider a data point $p \in C_i$. To compute the lower bound of its outlier score, we have to find the *closest* clusters to p in terms of $minDis()$. In order to do this we consider all clusters closest to C_i as well as other data points in C_i (as clusters). So we choose $Min_p = Min_{C_i} \bigcup C_i \setminus p$. In order to estimate the cumulative distance from p to its k nearest neighbors, we order Min_p and choose the top z clusters $M_1 \ldots M_z$ s.t. $\sum_{i=1}^{z-1} M_i < k \leq \sum_{i=1}^{z} M_i$. Now the lower bound of the outlier score of p can be computed as $l_p = \sum_{i=1}^{z-1} |M_i| \cdot minDis(p, M_i) + (k - \sum_{i=1}^{z-1} |M_i|) \cdot minDis(p, M_z)$.

Similarly we can compute the upper bound of p's outlier score, $u_p = \sum_{i=1}^{z-1} |M_i| \cdot maxDis(p, M_i) + (k - \sum_{i=1}^{z-1} |M_i|) \cdot maxDis(p, M_z)$, where $\{M_1 \ldots M_z\}$ are the top z clusters in Max_p defined as $Max_{C_i} \bigcup C_i \setminus p$.

Definition 3. [BOUNDS OF A CLUSTER'S OUTLIER SCORE]. *The upper and lower bounds of a cluster's outlier score in terms of its contained points are given as:* $u_{C_i} = max\{u_p, p \in C_i\}$ *and* $l_{C_i} = min\{l_p, p \in C_i\}$, *respectively.*

We now use a simple heuristic to prune clusters that do not contain outliers: pick clusters with the largest lower bounds of outlier scores, until we have a total of at least n data points. Let the last cluster picked be C_o. Clusters whose upper bounds of outlier scores are smaller than l_{C_o} cannot contain outliers, and are therefore pruned. This heuristic constitutes the first pruning phase and is presented in Algorithm 2. The value l_{C_o} is passed as an initial seed to the second pruning phase for faster pruning. While the above heuristic correctly prunes clusters containing data points which are all non-outliers, it may allow clusters containing some non-outliers. This happens for all clusters C_i, where $l_{C_i} \leq l_{C_o} \leq u_{C_i}$. This is undesirable, since not all data points in these clusters are potential outliers. In order to resolve this issue, we propose another heuristic called P_{points} which prunes all points $p \in C_i, u_p < l_{C_o}$. Time complexity of MIRO with and without P_{points} is discussed in Section 4.1.

Algorithm 2. PRUNECLUSTERS

```
lC_i, uC_i ← estimateBounds ∀_i C_i ∈ C
Identify C_o, lC_o
Prune C_i | uC_i < lC_o
Return lC_o, C
```

3.2 Nested-Loop Algorithm

After the lower bound on the outlier score is obtained from the first phase, we process the remaining clusters using the traditional nested-loop algorithm similar to ORCA [11]. In the second phase of MIRO (Algorithm 3) we employ two pruning rules (R_1 in line 9 and R_2 in line 13 of Algorithm 3). Similar to [11], we check if the outlier score of the data point is smaller than the current cutoff c on the outlier score (rule R_2). However, while ORCA initializes c as 0, in our second phase, we converge faster by choosing c from the first clustering phase (with or without P_{points}).

Let us consider an arbitrary data point q. If $c > kD(p,q) + F_{out}(q)$, then by our definition of outlier score and using triangular inequality, we can show that $c > \sum_{m \in kNN_q} D(p,m) \geq F_{out}(p)$, i.e., $c > F_{out}(p)$. Therefore p is not an outlier and can be pruned. Despite its simplicity, this pruning rule is extremely efficient in the final processing phase as shown in Section 5. By using the combination of two pruning rules, the execution time is further reduced, creating a huge advantage over ORCA and RBRP [12]. It is also noted that by reserving Min_{C_i} and Max_{C_i} for each remaining cluster C_i, we are able to limit the search space for each data point $p \in C_i$. More specifically, to process p, in the worst case we only have to scan $C_i \cup \{\bigcup_{C_1 \in Min_{C_i}} C_1\} \cup \{\bigcup_{C_2 \in Max_{C_i}} C_2\}$. The search space is therefore much smaller than the original dataset DS.

4 Theoretical Analysis

In addition to the notations stated in Table 1, we define the following new terms for analysis: (a) p_1 is the probability that a cluster will be pruned during the first phase, and (b) p_2 is the probability that a data point will be pruned by rule R_1 before it is scanned with the $(k+1)^{th}$ data point among the remaining ones. It is also noted that in practice, $n_c \leq k$ and $n \ll N$. In the following discussion, we present detailed time and space complexity analysis for MIRO.

4.1 Time Complexity of MIRO

The execution time cost of the first phase without P_{points} includes (a) the cost of clustering ($S_{cluster}$), (b) the cost of computing upper and lower bounds outlier score for all clusters (S_{bounds}), and (c) the pruning cost ($S_{pruning}$). The expected clustering cost is $O(N \cdot logN)$ according to [12]. Now, for a cluster C_i, we need to identify Min_{C_i} and Max_{C_i}. Since the mean size of each cluster is n_c, on average

Algorithm 3. FinalProcessing

```
Set c, C ← PruneClusters()
Set TopOut ← ∅
foreach remaining cluster C_i ∈ C do
    Set A ← C_i ∪ {⋃_{C_1 ∈ Min_{C_i}} C_1} ∪ {⋃_{C_2 ∈ Max_{C_i}} C_2}
    foreach data point p ∈ C_i do
        foreach cluster C_j ∈ A do
            foreach data point q ∈ C_j do
                if q ≠ p then
                    if (c − F_out(q))/k > D(p, q) then
                        Mark p as non-outlier
                        Process next data point in C_i
                    Update p's k nearest neighbors using q
                    if F_out(p) < c then
                        Mark p as non-outlier
                        Process next data point in C_i
        if p is outlier then
            Update TopOut with p
            if Min(TopOut) > c then
                Set c ← Min(TopOut)
```

we have $|Min_{C_i}| = |Max_{C_i}| = \lceil k/n_c \rceil$. A naïve approach sorts all clusters and extracts $\lceil k/n_c \rceil$ clusters for Min_{C_i}/Max_{C_i}, at a cost of $O(\frac{N}{n_c} \cdot log(\frac{N}{n_c}))$. However, we note that only $\lceil k/n_c \rceil$ clusters need to be reserved for Min_{C_i} as well as Max_{C_i}. Therefore a better approach is that for each cluster C_j, we compute the minimum/maximum distance from C_j to C_i and insert the result into the corresponding set. This approach leads to an total cost of $O(\frac{1}{2} \cdot \lceil \frac{k}{n_c} \rceil \cdot \frac{N}{n_c} \cdot (\frac{N}{n_c} - 1))$ over all clusters, which can be simplified to $O(\frac{N^2}{{n_c}^2})$. To estimate the cost of computing upper and lower bounds of the outlier score for each cluster C_i, we compute the cost of measuring the same bounds for each individual data point $p \in C_i$. To obtain p's bounds, we also need to extract $n_c + \lceil \frac{k}{n_c} - 1 \rceil$ clusters (including zero-radius ones) from a set of $n_c + \lceil \frac{k}{n_c} \rceil$ clusters. Since the number of items extracted is nearly no different from the total set of items, we apply the naïve sorting approach discussed above. As a consequence, the total cost incurred is $O((n_c + \lceil \frac{k}{n_c} \rceil) \cdot log(n_c + \lceil \frac{k}{n_c} \rceil))$, i.e., $O(n_c \cdot log(n_c))$. Hence, the cost of computing C_i's bounds $= O({n_c}^2 \cdot log(n_c))$. Therefore, $S_{bounds} = O(\frac{N}{n_c} \cdot {n_c}^2 \cdot log(n_c)) + O(\frac{N^2}{{n_c}^2}) = O(N \cdot n_c \cdot log(n_c)) + O(\frac{N^2}{{n_c}^2})$. To prune the clusters, we need to compute l_{C_o} and scan the whole set of clusters to check their corresponding upper bounds. To compute l_{C_o}, we need to extract $\lceil n/n_c \rceil$ clusters with largest lower bounds from a set of N/n_c clusters. In other words, $S_{pruning} = O(\lceil \frac{n}{n_c} \rceil \cdot \frac{N}{n_c}) + O(\frac{N}{n_c})$. Overall, the approximate overhead incurred by the first phase is:

$S_{phase1} = S_{cluster} + S_{bounds} + S_{pruning} = O(N \cdot logN) + O(N \cdot n_c \cdot log(n_c)) + O(\frac{N^2}{{n_c}^2}) + O(\lceil \frac{n}{n_c} \rceil \cdot \frac{N}{n_c}) + O(\frac{N}{n_c}) = O(N \cdot logN) + O(N \cdot n_c \cdot log(n_c)) + O(\frac{N^2}{{n_c}^2}) + O((\lceil \frac{n}{n_c} \rceil + 1) \cdot \frac{N}{n_c})$.

After the first phase, the number of remaining clusters is $(1 - p_1) \cdot \frac{N}{n_c}$, which implies that the total number of remaining data points is $n_c \cdot (1 - p_1) \cdot \frac{N}{n_c} = (1 - p_1) \cdot N$. Among them, the total number of data points pruned out by the rule R_1 with no more than k distance computations is $p_2 \cdot (1 - p_1) \cdot N$. On the other hand, for each of the data points left, we need to scan the entire cluster C_i as well as Min_{C_i} and Max_{C_i} in the worst case, i.e., the corresponding cost is $O(n_c + 2 \cdot n_c \cdot \lceil k/n_c \rceil)$, which simplifies to $O(3 \cdot n_c + 2 \cdot k)$. Hence the execution time of the second phase in the worst case can be expressed as:

$$S_{phase2} = O(k \cdot p_2 \cdot (1 - p_1) \cdot N + (3 \cdot n_c + 2 \cdot k) \cdot (1 - p_2) \cdot (1 - p_1) \cdot N)$$
$$= O((3 \cdot n_c \cdot (1 - p_2) + k \cdot (2 - p_2)) \cdot (1 - p_1) \cdot N).$$

Hence, the approximate cost of the whole algorithm is:

$$S_{phase1} + S_{phase2} = O(N \cdot logN) + O(N \cdot n_c \cdot log(n_c)) + O(\frac{N^2}{{n_c}^2}) + O((\lceil \frac{n}{n_c} \rceil + 1) \cdot \frac{N}{n_c}) + O((3 \cdot n_c \cdot (1 - p_2) + k \cdot (2 - p_2)) \cdot (1 - p_1) \cdot N).$$

We can also reclassify the whole detection process into a more detailed sequence of operations: (a) clustering, (b) identifying neighboring clusters for all clusters, (c) computing the bounds for clusters (we consider the process for each cluster as a operation, so we have N/n_c operations), (d) pruning clusters (N/n_c operations on average) and (e) final processing step ($(1 - p_1) \cdot N$ operations on average). Among them, the cost of the operations (a) and (b) are loglinear and quadratic w.r.t. N, respectively. On the other hand, each of the remaining operations incurs costs independent of N. Furthermore, when p_1 has large values, the execution time of the second phase becomes very small which compensates the overhead incurred by the first phase. In addition, when p_2 receives a large value, a larger portion of the remaining data points after the first phase require no more than k distance computations to be identified as normal records, and a smaller number of these remaining points require more than k distance computations. This fact leads to another reduction of execution time. Besides, the pre-computation of cutoff c helps contribute to further reduction of the execution time. Therefore, practically each of the operations performed in item (e) takes nearly constant time. By applying the accounting method of amortized analysis, we expect the expensive cost of operations (a) and (b) would be compensated by the remaining inexpensive ones, i.e., the *amortized running time* of each individual operation is inexpensive and non-quadratic w.r.t. N. In the experiments carried out in Section 5, we always have $\max(p_1, p_2) \geq 0.7$ which leads to the practical linear execution time w.r.t N. It is also noted that based on our analysis, this quadratic overhead w.r.t. N is common for techniques that utilize similar partition-based strategy such as [7], which though using less pruning rules than MIRO, still reports linear execution time performance w.r.t N.

Time complexity with P_{points}. In the above analysis, we assume that the P_{points} heuristic (c.f., Section 3.1) is not used for the first phase. In contrast, if this heuristic is considered, we prune all points whose upper bound of outlier score is less than the cutoff obtained by the clustering phase, so $S_{pruning}$ has to be recomputed. Particularly, after applying l_{C_o} for pruning out clusters, we perform an additional scan on the set of clusters left. The mean number of clusters to scan is therefore $(1 - p_1) \cdot \frac{N}{n_c}$, and the expected cost for scanning each cluster is n_c. Consequently, the additional cost is $O((1 - p_1) \cdot \frac{N}{n_c} \cdot n_c) = O((1 - p_1) \cdot N)$.

From the above analysis, it can be seen that the cost of S_{phase1} does not change theoretically whether P_{points} is used or not. But P_{points} is only effective if it does indeed help to prune out more data points after the first phase. We will examine that in Section 5.

4.2 Space Complexity of MIRO

As mentioned earlier, minimizing I/O cost is neither a focus of techniques in [11,12,6] nor of MIRO. Hence, in general MIRO uses space for: (a) storing the data points, and (b) storing the clusters created. Furthermore, the spatial cost for storing each cluster C_i can be simplified to the cost of storing its major components which include: (a) its member data points, and (b) Min_{C_i} as well as Max_{C_i}. This is simplified by space-efficient hash indexes, therefore each C_i takes $O(n_c + 2 \cdot \lceil \frac{k}{n_c} \rceil)$ space on average. Hence, the space complexity of MIRO is $O(N) + O(\frac{N}{n_c} \cdot (n_c + 2 \cdot \lceil \frac{k}{n_c} \rceil))$, which can be simplified to $O(N)$.

4.3 Analysis of Parameters Used

Cluster size. For a fixed dataset size, as the average cluster size n_c decreases, the total number of clusters will increase. Since the size of each cluster C_i becomes smaller, in order to compute the bounds of C_i, we need to include more clusters in Min_{C_i} as well as Max_{C_i}. In other words, more clusters are required for computing C_i's bounds. That increases S_{bounds} and leads to the increase in the overall execution time of our algorithm. In the extreme case, when $n_c = 1$, the first phase degrades to scanning the entire dataset, i.e., the total execution time becomes a normal nested-loop algorithm and the execution time saved during the second phase becomes insufficient to compensate this overhead. On the other hand, as n_c increases, there are less clusters than before. Since the size of each cluster becomes larger, we need to consider fewer clusters in the process of computing clusters' bounds on the outlier score. But that does not directly lead to a decrease in cost of computing bounds since we need to process more data points per cluster. Furthermore, as n_c increases and exceeds k, the lower bound score l_{C_o} becomes smaller since we only need to use data points in a cluster C_i to compute its bounds (the assumption here is that in general a cluster contains data that are relatively homogeneous). That means less clusters are pruned after the first phase hence the execution time will increase. Overall, we should choose a reasonable value of n_c such that the average number of data points per cluster is neither too small nor too large compared to k. More specifically, we need to identify a threshold for n_c such that as n_c increases above as well as decreases below this threshold, the execution time of MIRO will increase. Consequently, picking this threshold to be n_c will be a wise choice. From the above analysis, we conclude that the impact of n_c over the overall performance of MIRO is complex and identification of reasonable values for n_c by analytical methods is practically infeasible. Through empirical study carried out in Section 5, we show that $k/5$ is a possible candidate value.

Number of nearest neighbors. As the number of nearest neighbors taken into account for the computation of outlier score, k, increases, the value that F_{out} assigns to each individual data point p in DS will increase correspondingly. This in turn leads to an increase in the lower bound l_{C_o}, and hence more clusters may be pruned by the first phase of MIRO. However, as demonstrated before, an increase of k results in having to consider more clusters when computing outlier score bounds for an arbitrary cluster. Therefore, the cost of computing cluster's bounds will increase. The increase of k creates a two-fold effect: (a) a decrease in execution time since more data points are pruned, and (b) an increase in execution time due to the increase in the cost of computing clusters' bounds. Our experimental result in Section 5 shows that MIRO's execution time increases as k increases, i.e., the latter factor outperforms the former one.

5 Empirical Results and Analyses

In order to assess the effectiveness of our proposed technique, we performed extensive experiments on four real and high-dimensional datasets CorelHistogram, Covertype, Server[2] and Landsat[3]. All of these are original datasets except for Server which is extracted from KDD Cup 1999 data, using the procedure provided in [14]. For each set of input parameters that affect the performance of the corresponding algorithm, we ran the experiment ten times. The results presented are from average outcomes obtained from multiple runs. It is noted that we set $M = 10$ and $it = 5$ throughout all experiments. Through the empirical studies, we demonstrate:

- The efficiency of MIRO in reducing the execution time of the traditional nested-loop algorithm. We measure the scalability of MIRO's execution time against the dataset size (N) as well as the number of nearest neighbors (k) used. In the latter case, we present MIRO's performance with and without P_{points}. The result is then compared with ORCA [11] and RBRP [12] to highlight the merit of our method.
- The pruning power of MIRO, in both phases of processing, with and without P_{points}. In addition, we also assess the effect of k on the pruning quality. The sensitivity of MIRO's execution time with respect to the cluster size (n_c) is also presented.

[2] `http://www.ics.uci.edu/~mlearn/MLRepository.html`

[3] `http://vision.ece.ucsb.edu`

Execution time v/s. N: First we evaluate the scalability of execution time of three distance-based outlier detection techniques MIRO, RBRP and ORCA w.r.t the dataset size N. In this experiment, we chose the number of outliers mined $n = 30$, number of nearest neighbors $k = 50$, set the size of each cluster $n_c = 20$, and varied N. We chose the implementation of MIRO without P_{points} since the efficiency of P_{points} is highlighted in a later part of this section. We observe from the result (Figure 1) that MIRO scales better than RBRP and ORCA on all datasets, although its theoretical asymptotic time complexity is quadratic in N. This agrees with the amortized analysis in Section 4.1. In order to analyze the cause of MIRO's efficiency, we also compare the execution time with and without the first phase.

Execution time and MIRO's pruning power v/s. k: We now analyze the effect of the number of nearest neighbors (k) on execution time. This experiment is conducted on the entire datasets, and $n = 30$, $n_c = 20$ as in the previous case. The results (Figure 2) show that the execution time for every technique increases with k, but MIRO scales better (with and without P_{points}) compared to RBRP and ORCA. The reason is once again attributed to the effective pruning power of MIRO in both phases of processing. It is also clear that by using P_{points}, we are able to obtain better or equal performance in term of execution time. This observation is further analyzed later when we discuss the effect of k on MIRO's pruning power.

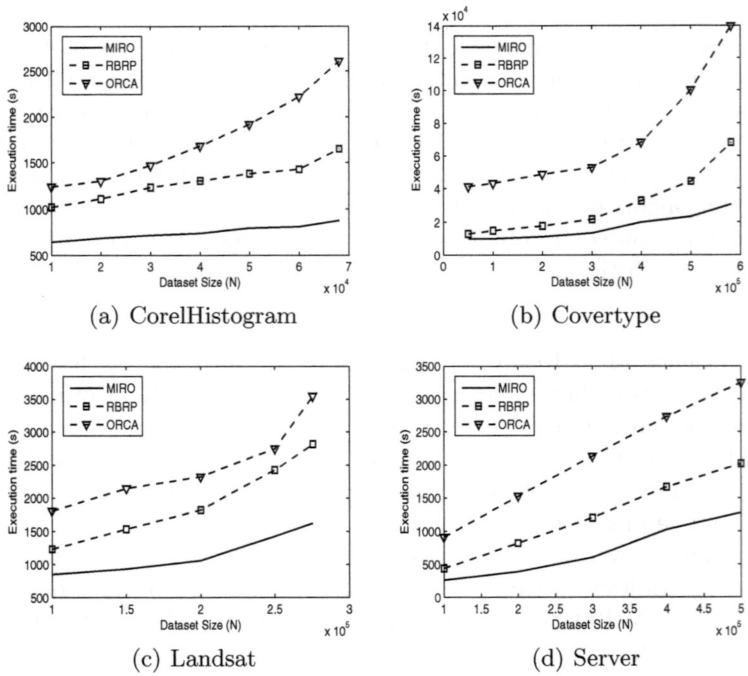

(a) CorelHistogram (b) Covertype

(c) Landsat (d) Server

Fig. 1. Execution time vs. the dataset size N

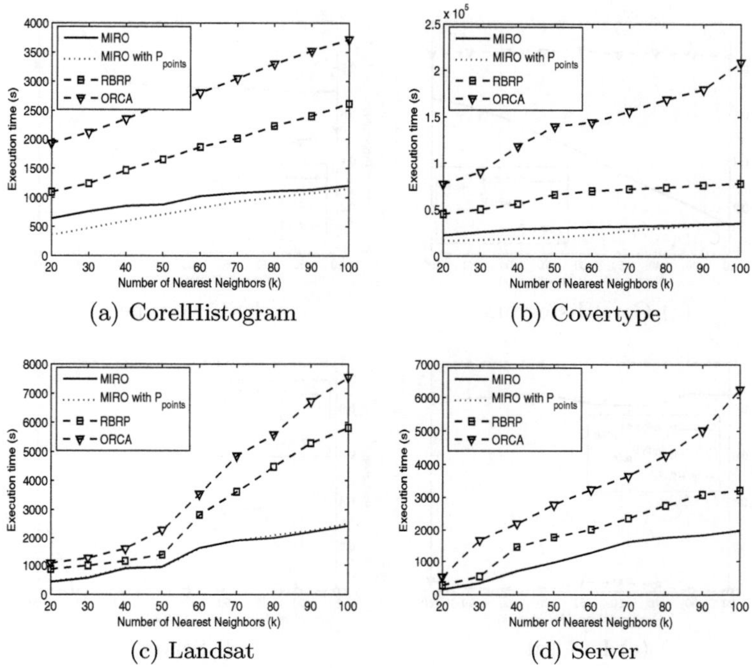

(a) CorelHistogram (b) Covertype

(c) Landsat (d) Server

Fig. 2. Execution time vs. the number of nearest neighbors k

Figure 3 presents two pruning probabilities in one plot for each dataset: the probability of pruning a cluster in the first phase (p_1), and the probability that a data point will be pruned out by rule R_1 before it is scanned with the $(k+1)^{th}$ data point among the remaining ones (p_2), as the number of nearest neighbors is varied. In all cases, very high values of p_1 and/or p_2 are achieved, with p_1 increasing when P_{points} is utilized. While we do not obtain high values for both p_1 and p_2 at the same time, we observe that in every case at least one of them receives a value greater than 0.7. This reflects a very high efficiency in pruning and explains why MIRO takes less execution time compared to RBRP and ORCA. In addition, the value of p_1 tends to increase as k increases (except in the case of Landsat dataset), which means more clusters will be pruned after the first phase when k receives higher value. This agrees with the discussion in Section 4.3. Furthermore, when p_1 without P_{points} already has relatively large value, applying P_{points} does not help much in increasing the pruning power of the first phase. This point is reflected by the tendency of p_1 with and without P_{points} to converge towards each other as p_1 increases. We also observe that when the pruning effect without using P_{points} is low, i.e., when p_1 is low, there will be a significant improvement in execution time if P_{points} is employed instead. This can be attributed to the fact that adjoining clusters' lower and upper outlier score bounds are too interleaved with each other which creates redundancy if we include the whole of each candidate cluster in the final processing step. In

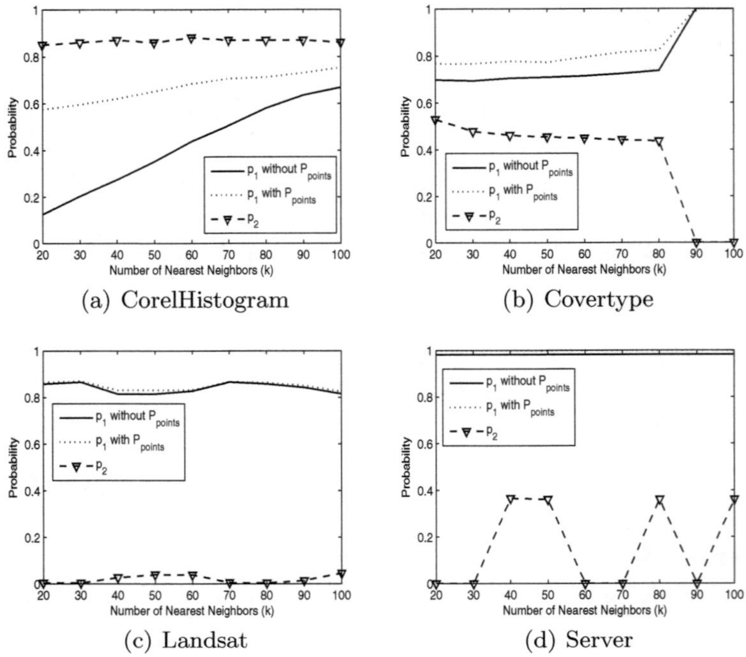

(a) CorelHistogram (b) Covertype

(c) Landsat (d) Server

Fig. 3. MIRO's pruning power vs. the number of nearest neighbors k

contrast, if the value of p_1 is already high, which means l_{C_o} has been identified wisely, using P_{points} may not improve MIRO's performance by much, although the pruning effect obtained is still equal or better. The reason is that increase in pruning power in such cases is not enough to compensate the additional time spent to run P_{points}. However, it is noted that when p_1 receives a higher value, the cost of executing P_{points}, which is $O((1 - p_1) \cdot N)$, becomes lower. Therefore, it can be concluded that applying P_{points} does not degrade performance by much, but may lead to significantly better performance.

Execution time v/s. n_c: For studying the effect of the average cluster size (n_c) on the execution time of MIRO, we set $n = 30$ while varying k. For each value of k, we run MIRO with $n_c \geq 1$ and $\leq k$ and note the value of n_c which yields smallest CPU cost. The result obtained suggests that n_c should be $k/5$. A good selection of n_c helps to balance the tradeoff between the time spent on computing clusters' bounds, as well as the pruning effect of the first phase of MIRO. In practice, we can also determine n_c by performing a training process on a subset of the original dataset with $n_c = k/5$ as the initial seed.

6 Conclusions

This work contributes to outlier detection research by proposing a new combination of several pruning strategies to produce an efficient distance-based outlier detection technique. The proposed technique, MIRO, consists of two pruning phases of processing which lead to amortized efficiency. During the first phase, a partition-based technique is employed to extract candidate clusters for the later processing step. Furthermore, an additional benefit of the first phase is that we are able to compute an initial value of the outlier cutoff threshold which is utilized in the nested-loop phase. In the second phase of MIRO, two pruning rules are employed to further reduce the overall temporal cost. In future work, we are considering to extend our analysis on more large and high-dimensional datasets to better study the full benefits of MIRO. We are also examining the possibility of applying the partition-based strategy to outlier detection problems where a *local* outlier score function is utilized. This will help us in building a general framework for creating faster detection techniques regardless of whether a *local* or *global* score function is employed.

References

1. Joshi, M.V., Agarwal, R.C., Kumar, V.: Mining needle in a haystack: Classifying rare classes via two-phase rule induction. In: SIGMOD Conference, pp. 91–102 (2001)
2. Suzuki, E., Zytkow, J.M.: Unified algorithm for undirected discovery of exception rules. In: Zighed, D.A., Komorowski, J., Żytkow, J.M. (eds.) PKDD 2000. LNCS (LNAI), vol. 1910, pp. 169–180. Springer, Heidelberg (2000)
3. Breunig, M.M., Kriegel, H.P., Ng, R.T., Sander, J.: LOF: Identifying density-based local outliers. In: SIGMOD Conference, pp. 93–104 (2000)
4. Knorr, E.M., Ng, R.T.: Algorithms for mining distance-based outliers in large datasets. In: VLDB, pp. 392–403 (1998)
5. Aggarwal, C.C., Yu, P.S.: An effective and efficient algorithm for high-dimensional outlier detection. VLDB Journal 14(2), 211–221 (2005)
6. Angiulli, F., Pizzuti, C.: Outlier mining in large high-dimensional data sets. IEEE Transactions on Knowledge and Data Engineering 17(2), 203–215 (2005)
7. Ramaswamy, S., Rastogi, R., Shim, K.: Efficient algorithms for mining outliers from large data sets. In: SIGMOD Conference, pp. 427–438 (2000)
8. Papadimitriou, S., Kitagawa, H., Gibbons, P.B., Faloutsos, C.: LOCI: Fast outlier detection using the local correlation integral. In: ICDE, pp. 315–324 (2003)
9. Nguyen, H.V., Vivekanand, G., Praneeth, N.: Online Outlier Detection Based on Relative Neighbourhood Dissimilarity. In: Bailey, J., Maier, D., Schewe, K.-D., Thalheim, B., Wang, X.S. (eds.) WISE 2008. LNCS, vol. 5175, pp. 50–61. Springer, Heidelberg (2008)
10. Arning, A., Agrawal, R., Raghavan, P.: A linear method for deviation detection in large databases. In: KDD, pp. 164–169 (1996)
11. Bay, S.D., Schwabacher, M.: Mining distance-based outliers in near linear time with randomization and a simple pruning rule. In: KDD, pp. 29–38 (2003)
12. Ghoting, A., Parthasarathy, S., Otey, M.E.: Fast mining of distance-based outliers in high dimensional datasets. In: SDM (2006)
13. Guha, S., Meyerson, A., Mishra, N., Motwani, R., O'Callaghan, L.: Clustering data streams: Theory and practice. IEEE Transactions on Knowledge and Data Engineering 15(3), 515–528 (2003)
14. Tao, Y., Xiao, X., Zhou, S.: Mining distance-based outliers from large databases in any metric space. In: KDD, pp. 394–403 (2006)

The Sensitivity of Latent Dirichlet Allocation for Information Retrieval

Laurence A.F. Park and Kotagiri Ramamohanarao

Department of Computer Science and Software Engineering,
The University of Melbourne, 3010, Australia

Abstract. It has been shown that the use of topic models for Information retrieval provides an increase in precision when used in the appropriate form. Latent Dirichlet Allocation (LDA) is a generative topic model that allows us to model documents using a Dirichlet prior. Using this topic model, we are able to obtain a fitted Dirichlet parameter that provides the maximum likelihood for the document set. In this article, we examine the sensitivity of LDA with respect to the Dirichlet parameter when used for Information retrieval. We compare the topic model computation times, storage requirements and retrieval precision of fitted LDA to LDA with a uniform Dirichlet prior. The results show there there is no significant benefit of using fitted LDA over the LDA with a constant Dirichlet parameter, hence showing that LDA is insensitive with respect to the Dirichlet parameter when used for Information retrieval.

Keywords: latent Dirichlet allocation, probabilistic latent semantic analysis, query expansion, thesaurus.

1 Introduction

Topic models allow us to represent documents as collections of topics, rather than collections of words. It has been shown that the use of topic models for Information retrieval on large documents provides a significant increase in precision when used in an appropriate form [1,2].

Latent Dirichlet Allocation (LDA) [3] is a generative topic model that allows us to model documents using a Dirichlet prior. By changing the Dirichlet parameter, we are able to control the number of topics that the model assigns to each word and document. By setting a small Dirichlet parameter, a small number of topics are assigned to each word; by increasing the parameter, we increase the distribution of topics to each word.

The original LDA model [3] provided a means of fitting the Dirichlet parameter, when given a document set. This fitting processes requires that the document models be recomputed until the maximum likelihood Dirichlet parameter is found. Later LDA models [4,5] have avoided fitting the Dirichlet parameter by simply providing an estimate of the parameter and computing the document models once.

In this article, we investigate the effect of fitting the Dirichlet parameter for Information retrieval. We examine the change in storage and retrieval precision

W. Buntine et al. (Eds.): ECML PKDD 2009, Part II, LNAI 5782, pp. 176–188, 2009.

obtained by changing the Dirichlet parameter, and hence observe the sensitivity of the LDA topic models with respect to the Dirichlet parameter when used for Information retrieval. We make the following contributions:

- a method of computing LDA term-term relationships, allowing us to use LDA in an efficient and effective Information retrieval setting in Section 2.3, and
- a detailed comparison of document retrieval using both fitted and unfitted LDA in Section 4.

The article will proceed as follows: Section 2 describes the LDA topic model and presents a derivation of probabilistic term relationships using LDA. Section 3 presents the information retrieval model used for efficient and effective retrieval using topic models. Section 4 provides a comprehensive set of experiments comparing the effectiveness of the fitted and unfitted LDA query expansion models.

2 Probabilistic Topic Models

The language modelling method of information retrieval assumes that each document is generated from a statistical document model. To create the document, a set number of terms are sampled from the document model.

The simplest document model is a multinomial distribution across all terms, where each term has a non-zero probability of being sampled. Terms that are related to the document will have higher probabilities, while those that are not related will have low probabilities.

In this section, we will examine two popular document modelling methods, called probabilistic latent semantic analysis and latent Dirichlet allocation, that assume that the models are dependent on a set of underlying topics.

2.1 Probabilistic Latent Semantic Analysis

The simple document model mentioned assumes that each term is independent. Using this model will result in over-fitting the document set and lead to poor generalisation.

Probabilistic latent semantic analysis (PLSA) [6] introduces a set of hidden topics into the document model, so rather than directly estimating the term distribution for each document, we estimate the topic distribution over all documents and the probabilistic relationship of each document and term to each topic. PLSA uses the following document generation model:

1. To set the number of words in the document (document length), take a sample N, where $N \sim \text{Poisson}(\xi)$, and ξ is th average document length. We now have a document with N empty word slots.
2. For each of the N word slots w_i:
   (a) Select a topic z_i by sampling the distribution Multinomial(ϕ_n), conditioned on the document d_n where $P(z_i|d_n) = \phi_{i,n}$.

(b) Select a term t_x by sampling the distribution Multinomial(β_i) conditioned on the topic z_i, where $P(t_x|z_i) = \beta_{x,i}$.

Using this method of document construction, we obtain the probability of sampling term t_x in document d_n as:

$$P(t_x|d_n) = \sum_i P(t_x|z_i)P(z_i|d_n)$$

where $P(z_i|d_n) = \phi_{i,n}$ and $P(t_x|z_i) = \beta_{x,i}$. From this we can compute the set of term-term relationships [1]:

$$P(t_x|t_y) = \frac{\sum_i P(t_x|z_i)P(t_y|z_i)P(z_i)}{\sum_j P(t_y|z_j)P(z_j)} \tag{1}$$

and use these relationships for query term expansion.

2.2 Latent Dirichlet Allocation

A criticism of the PLSA approach to document modelling is that we need to compute the multinomial distribution ϕ_j which provides the probability of sampling a topic for a given document ($P(z_i|d)$). If a new document appears, we are unable to include it, since we do not have a ϕ_j distribution for it.

Latent Dirichlet allocation [3] avoids this problem by conditioning the topic selection on a Dirichlet distribution, rather than on the specific document. Latent Dirichlet Allocation (LDA) uses the following document model:

1. To set the number of words in the document (document length), take a sample N, where $N \sim$ Poisson(ξ). We now have a document with N empty word slots.
2. Given the set $\alpha = \{\alpha_1, \ldots, \alpha_k\}$, where $\alpha_i > 0$ for all i, take the sample $\theta = \{\theta_1, \ldots, \theta_k\}$, where $\theta \sim$ Dirichlet(α).
3. For each of the N word slots w_i:
   (a) Select a topic z_i by sampling the distribution Multinomial(θ).
   (b) Select a term t_x by sampling the distribution Multinomial(β_i) conditioned on the topic z_i, where $P(t_x|z_i) = \beta_{x,i}$.

We can see that rather than the topic being sampled based on a document dependent multinomial distribution, LDA samples the topic from a multinomial distribution (θ) generated from a Dirichlet space.

This gives us the likelihood function for a document:

$$P(d|\alpha, \beta) = \int \left(\prod_{n=1}^{N} \sum_i P(w_n = t_x|z_i)P(z_i|\theta) \right) P(\theta|\alpha)d\theta$$

Note that $P(w_n = t_x|z_i) = P(t_x|z_i)$, since each document is considered a bag of words. By multiplying the likelihood of each document, we obtain the likelihood of the collection. Therefore, given a document collection, we are able to compute the α and β that are most likely used to generate the document collection by maximising the likelihood function.

2.3 LDA Probabilistic Term Relationships

It has been shown that direct use of topic models for Information retrieval on large document sets lead to huge index storage requirements and slow query times [1]. A more effective and efficient method of Information retrieval using topic models is to use the knowledge of each topic in the form of a query expansion. Therefore, to use LDA in this form, we must first compute the probabilistic relationships of each term pair, given the set of topics.

Once we obtain the LDA α and β for the document set, we are able to compute the LDA relationship between each term using the following:

$$\begin{aligned} P(t_x|t_y,\alpha) &= \sum_i P(t_x, z_i|t_y,\alpha) \\ &= \sum_i P(t_x|z_i,t_y,\alpha)P(z_i|t_y,\alpha) \\ &= \sum_i P(t_x|z_i,\alpha)P(z_i|t_y,\alpha) \end{aligned} \tag{2}$$

where $P(t_x|z_i,t_y) = P(t_x|z_i)$, assuming that t_x and t_y are conditionally independent given z_i. and:

$$P(t_x|z_i,\alpha) = \beta_{x,i} \tag{3}$$

The posterior distribution is computed using Bayes' theorem:

$$\begin{aligned} P(z_i|t_y,\alpha) &= \frac{P(t_y|z_i,\alpha)P(z_i|\alpha)}{P(t_y|\alpha)} \\ &= \frac{P(t_y|z_i,\alpha)P(z_i|\alpha)}{\sum_j P(t_y,z_j|\alpha)} \\ &= \frac{P(t_y|z_i,\alpha)P(z_i|\alpha)}{\sum_j P(t_y|z_j,\alpha)P(z_j|\alpha)} \end{aligned} \tag{4}$$

From this equation, we can obtain $P(t_y|z_i,\alpha)$ from β, but we need to derive an equation for $P(z_j|\alpha)$. We know that z_i is a sample from the multinomial distribution with parameter θ, therefore we can show:

$$\begin{aligned} P(z_i|\alpha) &= \int P(z_i,\theta|\alpha)d\theta \\ &= \int P(z_i|\theta,\alpha)P(\theta|\alpha)d\theta \\ &= \int P(z_i|\theta)P(\theta|\alpha)d\theta \end{aligned} \tag{5}$$

where $P(z_i|\theta,\alpha) = P(z_i|\theta)$, assuming that z_i and α are conditionally independent given θ.

Since $P(z_i|\theta) = \theta_i$, we can simply replace the probability function in equation 5 with a function of $g_i(\theta)$ that simply returns the ith element of θ:

$$\begin{aligned} P(z_i|\alpha) &= \int g_i(\theta)P(\theta|\alpha)d\theta \\ &= \mathbb{E}_{P(\theta|\alpha)}[g_i(\theta)] \\ &= \mathbb{E}_{P(\theta|\alpha)}[\theta_i] \end{aligned}$$

We know that $\theta \sim \text{Dirichlet}(\alpha)$, therefore:

$$\mathbb{E}_{P(\theta|\alpha)}[\theta_i] = \frac{\alpha_i}{\sum_k \alpha_k} \tag{6}$$

By substituting equation 6 into equation 4, we obtain:

$$\begin{aligned} P(z_i|t_y,\alpha) &= \frac{P(t_y|z_i,\alpha)\frac{\alpha_i}{\sum_k \alpha_k}}{\sum_j P(t_y|z_j,\alpha)\frac{\alpha_j}{\sum_k \alpha_k}} \\ &= \frac{P(t_y|z_i,\alpha)\alpha_i}{\sum_j P(t_y|z_j,\alpha)\alpha_j} \end{aligned}$$

and finally we substitute equation 3 to obtain:

$$P(z_i|t_y,\alpha) = \frac{\beta_{y,i}\alpha_i}{\sum_j \beta_{y,j}\alpha_j} \tag{7}$$

By substituting equation 7 into 2, we can obtain a closed solution for our term relationships:

$$\begin{aligned} P(t_x|t_y,\alpha) &= \sum_i P(t_x|z_i,\alpha)P(z_i|t_y,\alpha) \\ &= \sum_i \beta_{x,i}\frac{\beta_{y,i}\alpha_i}{\sum_j \beta_{y,j}\alpha_j} \\ &= \frac{\sum_i \beta_{x,i}\beta_{y,i}\alpha_i}{\sum_j \beta_{y,j}\alpha_j} \end{aligned} \tag{8}$$

Therefore, we are able to obtain the probabilistic relationships between each term using the α and β values computed by fitting the LDA model to our document collection. It is also easy to see the equivalence of equation 8 and the PLSA term relationships in equation 1.

2.4 Exchangeable and Uniform Dirichlet Distribution

An exchangeable Dirichlet prior implies that any of the Dirichlet prior parameters can be exchanged without affecting the distribution. To achieve this, an exchangeable Dirichlet prior must contain all equal elements.

When using an exchangeable Dirichlet prior ($\alpha_i = a$ for all i), the Dirichlet distribution simplifies to:

$$\begin{aligned}
P(\theta|\alpha) &= \frac{\Gamma(\sum_{i=1}^{k}\alpha_i)}{\prod_{i=1}^{k}\Gamma(\alpha_i)}\prod_{i=1}^{k}\theta_i^{\alpha_i-1} \\
&= \frac{\Gamma(\sum_{i=1}^{k}a)}{\prod_{i=1}^{k}\Gamma(a)}\prod_{i=1}^{k}\theta_i^{a-1} \\
&= \frac{\Gamma(ka)}{\Gamma(a)^k}\prod_{i=1}^{k}\theta_i^{a-1} \\
&= \frac{\Gamma(ka)}{\Gamma(a)^k}\left(\prod_{i=1}^{k}\theta_i\right)^{a-1}
\end{aligned}$$

A further specialisation of the Dirichlet prior is the Uniform Dirichlet prior ($\alpha_i = 1$ for all i), which is an exchangeable Dirichlet with $a = 1$:

$$\begin{aligned}
P(\theta|\alpha) &= \frac{\Gamma(k)}{\Gamma(1)^k}\left(\prod_{i=1}^{k}\theta_i\right)^{0} \\
&= \frac{\Gamma(k)}{1^k} \\
&= \Gamma(k) \\
&= (k-1)!
\end{aligned}$$

We can see that the probability is independent of the value of θ and therefore uniform across all values of θ.

To examine the term relationships produced when using an exchangeable Dirichlet distribution, we add the constraint that $\alpha_i = a$ for all i. By doing so, then we obtain the term relationship probabilities:

$$\begin{aligned}
P(t_x|t_y,\alpha) &= \frac{\sum_i \beta_{x,i}\beta_{y,i}\alpha_i}{\sum_j \beta_{y,j}\alpha_j} \\
&= \frac{\sum_i \beta_{x,i}\beta_{y,i}a}{\sum_j \beta_{y,j}a} \\
&= \frac{\sum_i \beta_{x,i}\beta_{y,i}}{\sum_j \beta_{y,j}} \qquad (9)
\end{aligned}$$

Therefore, when using an exchangeable Dirichlet prior for LDA, the term relationship probabilities become independent of α. When examining the term relationship probabilities using a uniform Dirichlet prior, we obtain the same equation. Note that although α does not appear in equation 9, the values of β are still affected by the choice of α, therefore we would not obtain the same term relationships for when choosing either an exchangeable or uniform Dirichlet prior.

When using a uniform Dirichlet prior, the topic distribution obtains a uniform prior and can be modelled using a multinomial distribution. It has been shown that under this condition, LDA is equivalent to probabilistic latent semantic analysis (PLSA) [7].

3 Topic Based Query Expansion

Information retrieval systems obtain their fast query times by building a fast query term lookup index over a sparse set of term frequencies. When a query is provided, the index lists corresponding to the query terms are extracted and combined to obtain the document scores. If term frequencies are stored, the index becomes sparse, and hence we will be able to process the index lists efficiently.

When using topic models such as LDA, every term has a probabilistic relationship to every document, and hence the data is not sparse. If we were to store these values in the document index, we would obtain a large dense index containing long index lists that would take substantial time to process.

Rather than store the topic model probabilities in the document index, it was found that it is more efficient to store these values as term relationships in a thesaurus and store the term frequencies in the document index [1,8,9,10]. By doing so, we obtain a compact document index for fast retrieval and control over the query time by using the topic models in the form of of a thesaurus for query expansion.

By using the index-thesaurus combination, we compute each document score using:

$$s_d(Q) = \mu \mathrm{S}_d(E) + (1 - \mu)\mathrm{S}_d(Q)$$

where Q is the set of query terms, E is the set of weighted expansion terms computed by applying the query terms to the topic model thesaurus (as shown in [1,2]), $\mathrm{S}_d()$ is the document scoring function for document d and $\mu \in [0, 1]$ is the query mixing parameter to combine the original query terms to the expansion terms.

In the following experiments, we chose $\mathrm{S}_d()$ to be the state-of-the-art BM25 probabilistic document scoring function [11].

4 Comparison of Fitted and Unfitted LDA Term Expansion

Now that we have shown how to compute the LDA term relationships using the LDA topic models, we will proceed to investigate their effectiveness. In this section, we will examine the precision obtained by LDA and examine if there is a statistically significant increase when fitting the Dirichlet parameter.

To build the LDA thesaurus, we first compute the set of $\beta_{x,i}$ values for the given document set using the LDA-C software[1] (which assumes an exchangeable

[1] http://www.cs.princeton.edu/~blei/lda-c/

Table 1. Statistics related to the LDA thesauruses used to store the probabilistic term relationships. Note that the Dirichlet parameter was set to the value of $\alpha = 1$, making the unfitted form of LDA equivalent to PLSA.

Document set	File size		Build time		Dirichlet α	
	Fitted	Unfitted	Fitted	Unfitted	Fitted	Unfitted
AP2	112 MB	87 MB	2 days	57 minutes	0.020	1
FR2	30 MB	29 MB	1 day	17 minutes	0.021	1
WSJ2	107 MB	90 MB	2.5 days	61 minutes	0.022	1
ZIFF2	52 MB	42 MB	2 days	32 minutes	0.033	1

Dirichlet prior) and then compute the thesaurus values using equation 9. Our experiments will compare the LDA with the Dirichlet parameter fitted to the document set (fitted LDA) to LDA with a uniform Dirichlet prior ($\alpha = 1$) representing the unfitted LDA.

Experiment 1: Effect of the number of topics and terms on Average Precision. The LDA thesaurus build time is dependent on the number of terms and the number of topics. Therefore our first set of experiments will examine the effect of changing the number of terms and topics on the retrieval average precision. Initial experiments compared the retrieval results using LDA term expansion when using 1) all terms and 100 topics, 2) 100 topics and terms that appear in at least 50 documents, and 3) 300 topics and terms that appear in at least 50 documents. Average precision (AP) results were obtained using the ZIFF2 document set from TREC[2] disk 2, with queries 51 to 200 (used with disk 2 for TRECs 1, 2 and 3).

We generated AP scores from the 150 queries for mix values of 0.0 to 1.0 in intervals of 0.1 and term expansions of 10, 20, 50, 100, 200, 500 and 1000. By applying the two-sided Wilcoxon Signed Rank test to our AP results, we found that there was no significant difference in the AP of any of the three forms of LDA term expansion. These results are consistent with the findings when using PLSA term expansion [1]. Based on these results, we will now only consider LDA term expansion using 100 topics and only terms that appear in at least 50 documents.

Experiment 2: Thesaurus storage and build time. Before performing any retrieval experiments, we must first construct the thesaurus containing the probabilistic term relationships.

Table 1 shows the thesaurus statistics corresponding to the fitted and unfitted forms of LDA term relationships for the AP2, FR2, WSJ2 and ZIFF2 document sets from TREC disk 2. We can see from this table that the fitted thesaurus file sizes are larger than the corresponding unfitted thesaurus, and that the term

[2] http://trec.nist.gov

Table 2. Mean average precision on FR2 using BM25 with fitted LDA term expansion. Single and double superscript daggers († and ††) show a significant increase in precision over unfitted LDA, using the same mix and expansion size, at the 0.1 and 0.05 levels respectively. Single and double subscript stars (∗ and ∗∗) show a significant increase in precision over BM25. Scores in *italics* show the greatest score for the given expansion size.

Mix	Expansion size						
	10	20	50	100	200	500	1000
0.0	0.197	0.197	0.197	0.197	0.197	0.197	0.197
0.1	0.198_{**}	0.198_{**}	0.198_{**}	0.198_{**}	0.199_{**}	0.203_{**}	0.203_{**}
0.2	0.198_{**}	0.198_{**}	0.203_{**}	0.203_{**}	0.204_{**}	0.204_{**}	0.204_{**}
0.3	0.198_{**}	0.199_{**}	0.204_{**}	0.204_{**}	0.205_{**}	0.206_{**}	0.206_{**}
0.4	0.202_{**}	0.204_{**}	0.204_{**}	0.206_{**}	0.209_{**}	0.210_{**}	0.210_{**}
0.5	0.203_{**}	0.204_{**}	0.207_{**}	0.209_{**}	0.212_{**}	0.214_{**}	0.216_{**}
0.6	0.204_{**}	0.205_{**}	0.209_{**}	0.212_{**}	0.215_{**}	0.218_{**}	0.220_{**}
0.7	0.207_{**}	0.210_{**}	0.213_{**}	0.215_{**}	0.221_{**}	0.222_{**}	0.227_{**}
0.8	*0.209*$_{**}$	*0.212*$_{**}$	*0.215*$_{**}$	*0.223*$_{**}$	*0.226*$_{**}$	*0.228*$_{**}$	*0.233*$_{**}$
0.9	0.207_{**}	0.205_{**}	0.211_{**}	0.212_{*}	0.213	0.219_{*}	0.219_{*}
1.0	0.018	0.023	0.031	0.037	0.036	0.037	0.043

Table 3. Mean average precision on ZIFF2 using BM25 with fitted LDA term expansion. Single and double superscript daggers († and ††) show a significant increase in precision over unfitted LDA, using the same mix and expansion size, at the 0.1 and 0.05 levels respectively. Single and double subscript stars (∗ and ∗∗) show a significant increase in precision over BM25. Scores in *italics* show the greatest score for the given expansion size.

Mix	Expansion size						
	10	20	50	100	200	500	1000
0.0	0.269	*0.269*	0.269	0.269	0.269	0.269	0.269
0.1	0.269_{*}	*0.269*$_{*}$	0.269_{**}	0.269_{**}	0.269_{**}	0.269_{**}	0.269_{**}
0.2	0.269_{**}	*0.269*$_{**}$	0.269_{*}	0.269	0.269	0.269_{*}	0.269
0.3	$0.269^{\dagger}$	$0.268^{\dagger}$	0.268	0.268	0.271	0.270	0.270
0.4	0.269	0.268	0.268	0.271	0.270	0.270	0.270
0.5	0.268	0.268	*0.271*	*0.272*	0.272	0.272	0.272
0.6	*0.270*	0.268	0.270	0.271	0.272	0.273	$0.273^{\dagger}$
0.7	0.269	0.269	0.270	0.270	0.273	0.272	*0.280*††
0.8	0.267	$0.268^{\dagger}$	$0.270^{\dagger}$	$0.271^{\dagger\dagger}$	*0.278*†	*0.278*†	$0.278^{\dagger}$
0.9	$0.257^{\dagger\dagger}$	$0.257^{\dagger\dagger}$	$0.259^{\dagger\dagger}$	$0.264^{\dagger\dagger}$	$0.256^{\dagger\dagger}$	$0.264^{\dagger\dagger}$	$0.253^{\dagger\dagger}$
1.0	0.014	0.016	0.015	0.015	$0.014^{\dagger}$	0.015	0.015

Table 4. Mean average precision on AP2 using BM25 with fitted LDA term expansion. Single and double superscript daggers († and ††) show a significant increase in precision over unfitted LDA, using the same mix and expansion size, at the 0.1 and 0.05 levels respectively. Single and double subscript stars (∗ and ∗∗) show a significant increase in precision over BM25. Scores in *italics* show the greatest score for the given expansion size.

Mix	Expansion size						
	10	20	50	100	200	500	1000
0.0	0.271	0.271	0.271	0.271	0.271	0.271	0.271
0.1	0.272_{**}	0.272_{**}	0.272_{**}	0.272_{**}	0.272_{**}	0.272_{**}	0.272_{**}
0.2	0.272_{**}	0.272_{**}	0.273_{**}	0.273_{**}	0.273_{**}	0.273_{**}	0.273_{**}
0.3	0.273_{**}	0.273_{**}	0.274_{**}	0.275_{**}	0.275_{**}	0.275_{**}	0.275_{**}
0.4	0.274_{**}	0.274_{**}	0.275_{**}	0.275_{**}	0.276_{**}	0.276_{**}	0.276_{**}
0.5	0.274_{**}	0.275_{**}	0.277_{**}	0.277_{**}	0.278_{**}	0.278_{**}	0.278_{**}
0.6	0.275_{**}	0.276_{**}	0.278_{**}	0.279_{**}	0.279_{**}	0.280_{**}	0.280_{**}
0.7	0.276_{**}	0.277_{**}	0.280_{**}	0.280_{**}	0.281_{**}	0.282_{**}	0.282_{**}
0.8	0.277_{**}	0.279_{**}	$\mathit{0.282}_{**}$	$\mathit{0.283}_{**}$	$\mathit{0.284}_{**}$	$\mathit{0.284}_{**}$	$\mathit{0.284}_{**}$
0.9	$\mathit{0.279}^{\dagger\dagger}_{**}$	$\mathit{0.281}^{\dagger}_{**}$	$0.281^{\dagger\dagger}_{**}$	$0.282^{\dagger\dagger}_{**}$	$0.282^{\dagger\dagger}_{**}$	$0.283^{\dagger\dagger}_{**}$	$0.282^{\dagger\dagger}_{**}$
1.0	0.030	0.031	0.037	0.039	0.039	0.038	0.037

relationships take substantially longer to compute (due to the fitting process). The table also shows the fitted α value, demonstrating that the computed fitted LDA term relationships are different from the computed unfitted LDA term relationships.

Experiment 3: Comparison of Average Precision when using fitted and unfitted LDA query expansion. Our next set of experiments compare the fitted LDA term expansion precision to unfitted LDA term expansion and the baseline BM25 (no term expansion). Again, we will use mixing values of 0.0 to 1.0 in intervals of 0.1 and term expansions of 10, 20, 50, 100, 200, 500 and 1000 for both fitted and unfitted LDA. For this set of experiments we will use the AP2, FR2, WSJ2 and ZIFF2 document sets from TREC disk 2 with queries 51-200. For each of the document sets, thesauruses were built for fitted and unfitted LDA using 100 topics and only those terms that appear in at least 50 documents. The mean average precision results are shown in Tables 2, 3, 4 and 5 for the documents sets FR2, ZIFF2, AP2 and WSJ2 respectively.

We can see from table 2 that fitted LDA did not provide a significant increase in precision over unfitted LDA for any values of mix or expansion size on FR2. Also, table 3 shows that fitted LDA did not provide a significant increase in precision over BM25 for most mix values and expansion sizes on ZIFF2.

The AP2 and WSJ2 results in tables 4 and 5 show that the fitted LDA term expansion provides the greatest mean average precision (MAP) at $\mu = 0.8$. We can also see that the MAP increases as the query expansion size is increased.

Table 5. Mean average precision on WSJ2 using BM25 with fitted LDA term expansion. Single and double superscript daggers († and ††) show a significant increase in precision over unfitted LDA, using the same mix and expansion size, at the 0.1 and 0.05 levels respectively. Single and double subscript stars (∗ and ∗∗) show a significant increase in precision over BM25. Scores in *italics* show the greatest score for the given expansion size.

Mix	Expansion size						
	10	20	50	100	200	500	1000
0.0	0.257	0.257	0.257	0.257	0.257	0.257	0.257
0.1	0.257_{**}	0.257_{**}	0.258_{**}	0.258_{**}	0.258_{**}	0.258_{**}	0.258_{**}
0.2	0.258_{**}	0.258_{**}	0.258_{**}	0.258_{**}	0.258_{**}	0.258_{**}	0.259_{**}
0.3	0.258_{**}	0.258_{**}	0.258_{**}	0.259_{**}	0.259_{**}	0.259_{**}	0.259_{**}
0.4	0.258_{**}	0.259_{**}	0.259_{**}	0.259_{**}	0.260_{**}	0.262_{**}	0.262_{**}
0.5	0.259_{**}	0.259_{**}	0.259_{**}	0.259_{**}	0.262_{**}	0.263_{**}	0.263_{**}
0.6	0.259_{**}	0.259_{**}	0.260_{**}	0.263_{**}	0.264_{**}	0.265_{**}	0.265_{**}
0.7	0.262_{**}	0.260_{**}	0.261_{**}	0.264_{**}	0.266_{**}	0.267_{**}	0.268_{**}
0.8	$\mathit{0.262}_{**}$	$\mathit{0.262}_{**}$	$\mathit{0.264}_{**}$	$\mathit{0.267}_{**}$	$\mathit{0.268}_{**}$	$\mathit{0.270}_{**}$	$\mathit{0.270}_{**}$
0.9	$0.260^{\dagger}$	$0.259^{\dagger}$	$0.260^{\dagger}_{*}$	$0.262^{\dagger\dagger}_{**}$	$0.263^{\dagger\dagger}_{**}$	$0.260^{\dagger\dagger}_{**}$	$0.260^{\dagger\dagger}_{**}$
1.0	0.025	0.025	0.029	0.031	0.033	0.032	0.032

Note that a mix of $\mu = 0$ implies that there are no expansion terms used and hence the score is simply the BM25 score.

Each of the tables contain information about statistical significance tests. For each run, we have reported statistically significant increases in Average Precision (AP) using the one sided Wilcoxon signed rank test at the 0.1 and 0.05 levels of fitted LDA over each of unfitted LDA (using the associated expansion size and mix) and BM25. We can see from both the AP2 and WSJ2 results that the fitted LDA thesaurus provides a significant increase in average precision over BM25 for most results where the mix values are between $\mu = 0.1$ and 0.9. We can also see that fitted LDA provides a significant increase over unfitted LDA for a mix of $\mu = 0.9$ only for both AP2 and WSJ2.

Given that fitted LDA provides the greatest mean average precision with a mix of $\mu = 0.8$, but does not provide a significant increase in precision over unfitted LDA for both of the AP2 and WSJ2 document set, we can deduce that we have not gained any precision advantages from fitting the Dirichlet parameter.

Table 6 contains the mix-expansion pair that provide the greatest MAP for each method on each data set. Also provided are the precision at 10 and mean reciprocal rank scores. This table shows little difference between the retrieval performance of fitted and unfitted LDA.

From this evidence we have obtained, we deduce that the extra computation required to compute the fitted LDA topic relationships do not provide any benefit over the unfitted LDA topic relationships for the all of the document sets examined.

Table 6. A comparison of the mean average precision (MAP), precision at 10 (Prec10) and mean reciprocal rank (MRR) of each system on each data set. The mix-expansion combination shown are those that produce the greatest MAP.

Data	Method	Mix	Expansion	MAP	Prec10	MRR
AP2	Fitted LDA	0.8	200	0.284	0.380	0.562
	Unfitted LDA	0.8	200	0.282	0.382	0.563
	BM25	N/A	N/A	0.271	0.355	0.537
FR2	Fitted LDA	0.8	1000	0.233	0.141	0.368
	Unfitted LDA	0.8	1000	0.234	0.141	0.367
	BM25	N/A	N/A	0.197	0.117	0.314
WSJ2	Fitted LDA	0.8	500	0.270	0.370	0.625
	Unfitted LDA	0.8	500	0.269	0.369	0.625
	BM25	N/A	N/A	0.257	0.353	0.572
ZIFF2	Fitted LDA	0.7	1000	0.280	0.169	0.427
	Unfitted LDA	0.8	500	0.280	0.165	0.460
	BM25	N/A	N/A	0.269	0.158	0.415

5 Conclusion

Latent Dirichlet allocation (LDA) is a generative topic model that allocates topics using the Dirichlet distribution. To compute the topic distribution, we need to obtain the Dirichlet parameter for the document set, which can be either fitted using maximum likelihood, or estimated.

In this article, we examined the effect of fitting the Dirichlet parameter with the LDA topic model on the precision for Information retrieval. To do so, we derived an expression for LDA term-term probabilistic relationships for use as a query expansion.

We compared the effectiveness of query expansion using fitted and unfitted LDA on several document sets and found that using fitted LDA did not provide a significant increase in Average Precision when operating under its peak settings (mix=0.8). We showed that by not fitting the Dirichlet parameter, we obtained a 50 to 90 times gain in computational efficiency when computing the topic models. Considering that the fitted LDA term relationships consume more storage and time to build, we conclude that fitting the Dirichlet parameter provides no advantage when using LDA for an Information retrieval task, hence showing that LDA is insensitive with respect to the Dirichlet parameter for Information retrieval.

References

1. Park, L.A.F., Ramamohanarao, K.: Efficient storage and retrieval of probabilistic latent semantic information for information retrieval. The International Journal on Very Large Data Bases 18(1), 141–156 (2009)

2. Park, L.A.F., Ramamohanarao, K.: An analysis of latent semantic term self- correlation. ACM Transactions on Information Systems 27(2), 1–35 (2009)
3. Blei, D.M., Ng, A.Y., Jordan, M.I.: Latent dirichlet allocation. Journal of Machine Learning Research 3, 993–1022 (2003)
4. Griffiths, T.L., Steyvers, M.: Finding scientific topics. Proceedings of the National Academy of Sciences of the United States of America 101(suppl. 1), 5228–5235 (2004)
5. Wei, X., Croft, W.B.: LDA-based document models for ad-hoc retrieval. In: SIGIR 2006: Proceedings of the 29th annual international ACM SIGIR conference on Research and development in information retrieval, pp. 178–185. ACM, New York (2006)
6. Hofmann, T.: Probabilistic latent semantic indexing. In: Proceedings of the 22nd annual international ACM SIGIR conference on Research and development in information retrieval, pp. 50–57. ACM Press, New York (1999)
7. Girolami, M., Kaban, A.: On an equivalence between PLSI and LDA. In: SIGIR 2003: Proceedings of the 26th annual international ACM SIGIR conference on Research and development in informaion retrieval, pp. 433–434. ACM Press, New York (2003)
8. Park, L.A.F., Ramamohanarao, K.: Query expansion using a collection dependent probabilistic latent semantic thesaurus. In: Zhou, Z.-H., Li, H., Yang, Q. (eds.) PAKDD 2007. LNCS (LNAI), vol. 4426, pp. 224–235. Springer, Heidelberg (2007)
9. Park, L.A.F., Ramamohanarao, K.: Hybrid pre-query term expansion using latent semantic analysis. In: Rastogi, R., Morik, K., Bramer, M., Wu, X. (eds.) The Fourth IEEE International Conference on Data Mining, pp. 178–185. IEEE Computer Society, Los Alamitos (2004)
10. Park, L.A.F., Ramaohanarao, K.: The effect of weighted term frequencies on probabilistic latent semantic term relationships. In: Amir, A., Turpin, A., Moffat, A. (eds.) SPIRE 2008. LNCS, vol. 5280, pp. 63–74. Springer, Heidelberg (2008)
11. Jones, K.S., Walker, S., Robertson, S.E.: A probabilistic model of information retrieval: development and comparative experiments, part 2. Information Processing and Management 36(6), 809–840 (2000)

Efficient Decoding of Ternary Error-Correcting Output Codes for Multiclass Classification

Sang-Hyeun Park and Johannes Fürnkranz

TU Darmstadt, Knowledge Engineering Group,
D-64289 Darmstadt, Germany
{park,juffi}@ke.tu-darmstadt.de

Abstract. We present an adaptive decoding algorithm for ternary ECOC matrices which reduces the number of needed classifier evaluations for multiclass classification. The resulting predictions are guaranteed to be equivalent with the original decoding strategy except for ambiguous final predictions. The technique works for Hamming Decoding and several commonly used alternative decoding strategies. We show its effectiveness in an extensive empirical evaluation considering various code design types: Nearly in all cases, a considerable reduction is possible. We also show that the performance gain depends on the sparsity and the dimension of the ECOC coding matrix.

1 Introduction

Error-correcting output codes (ECOCs) [6] are a well-known technique for handling multiclass classification problems, i.e., for problems where the target attribute is a categorical variable with $k > 2$ values. Their key idea is to reduce the k-class classification problem to a series of n binary problems, which can be handled by a 2-class classification algorithm, such as a SVM or a rule learning algorithm. Conventional ECOCs always use the entire dataset for training the binary classifier. Ternary ECOCs [1] are a generalization of the basic idea, which allows to ignore some examples in the trainingset of the corresponding binary classifier. For example, pairwise classification [8,9], which trains a classifier for each pair of classifiers, is a special case of this framework.

For many common general encoding techniques, the number of binary classifiers may exceed the number of classes by several orders of magnitude. This allows for greater distances between the code words, so that the mapping to the closest code word is not compromised by individual mistakes of a few classifiers. For example, for pairwise classification, the number of binary classifiers is quadratic in the number of classes. Thus, the increase in predictive accuracy comes with a corresponding increase in computational demands at classification time. In previous work [12], we recently proposed the QWEIGHTED algorithm, a fast decoding method for pairwise classifiers using a voting aggregation. Our experimental results showed that the quadratic complexity of the decoding step could be reduced to $O(k \log k)$ in practice. In this paper, we present QUICKECOC, a generalization of the above-mentioned algorithm to allow for quick decoding of arbitrary ternary ECOC ensembles with various decoding techniques. The resulting predictions are guaranteed to be equivalent to the original decoding strategy except for ambiguous final predictions.

W. Buntine et al. (Eds.): ECML PKDD 2009, Part II, LNAI 5782, pp. 189–204, 2009.

2 ECOC

Error-Correcting Codes (ECC) are a well-known topic in the field of Coding and Information Theory [11]. Their main purpose is to detect and correct errors in noisy physical communication channels. Dietterich and Bakiri [6] adapted this concept for multiclass classification and named it in this context as *Error Correcting Output Codes (ECOC)*. They consider classifier predictions as information signals which ideally describe the correct class for a given instance. Due to external influences (such as, e.g., a too small sample size) these signals are sometimes wrong, and such errors have to be detected and corrected. Formally, each class c_i $(i = 1 \ldots k)$ is associated with a so-called *code word* $\boldsymbol{cw}_i \in \{-1, 1\}^n$ of length n. In the context of ECOC, all relevant information is summarized in a so-called *coding matrix* $(m_{i,j}) = M \in \{-1, 1\}^{k \times n}$, whose i-th row describes code word $\boldsymbol{cw}_i$, whereas the j-th column represents a classifier f_j. Furthermore, the coding matrix implicitly describes a decomposition scheme of the original multiclass problem. In each column j the rows contain a (1) for all classes whose training examples are used as positive examples, and (-1) for all negative examples for the corresponding classifier f_j. For the classification of a test instance x, all binary classifiers are evaluated and their predictions, which form a *prediction vector* $\boldsymbol{p} = [f_1(x), \ldots, f_n(x)]$, are compared to the code words. The class c^* whose associated code word $\boldsymbol{cw}_{c^*}$ is "nearest" to $\boldsymbol{p}$ according to some distance measure $d(.)$ (such as the Hamming distance) is returned as the overall prediction, i.e. $c^* = \text{argmin}_c\ d(\boldsymbol{cw}_c, \boldsymbol{p})$.

Later, Allwein et al. [1] extended the ECOC approach to the ternary case, where code words are now of the form $\boldsymbol{cw}_i \in \{-1, 0, 1\}^n$. The additional code $m_{i,j} = 0$ denotes that examples of class c_i are ignored for training classifier f_j. We will say a classifier f_j is *incident* to a class c_i, if the examples of c_i are either positive or negative examples for f_j, i.e., if $m_{i,j} \neq 0$. This extension increases the expressive power of ECOCs, so that now nearly all common multiclass binarization methods can be modelled. This includes pairwise classification, which could not be modeled previously.

2.1 Code Design

Since the introduction of ECOC, a considerable amount of research has been devoted to code design (see, e.g., [5,14]), but without reaching a clear conclusion. We want to emphasize that our work does not contribute to this discussion, because we will not be concerned with comparing the predictive quality of different coding schemes. Our goal is to show that, irrespective of the selected coding scheme, we can achieve a substantial reduction in prediction time, without changing the predicted outcome.

Nevertheless, we will briefly review common coding schemes, because we will later demonstrate that our algorithm is applicable to different types of coding schemes. Essentially, one can distinguish between four code families, which we will discuss in the following four sections.

Exhaustive Ternary Codes. Exhaustive ternary codes cover all possible classifiers involving a given number of classes l. More formally, a (k, l)-exhaustive ternary code defines a ternary coding matrix M, for which every column j contains exactly l values, i.e., $\sum_{i \in K} |m_{i,j}| = l$. Obviously, in the context of multiclass classification, only columns with at least one positive $(+1)$ and one negative (-1) class are useful.

The number of classifiers for a (k, l) exhaustive ternary code is $\binom{k}{l}(2^{l-1} - 1)$, since the number of *binary* exhaustive codes is $2^{l-1} - 1$ and the number of combinations to select l row positions from k rows is $\binom{k}{l}$. These codes are a straightforward generalization of the exhaustive binary codes, which were considered in the first works on ECOC [6], to the ternary case. Note that $(k, 2)$-exhaustive codes correspond to pairwise classification.

In addition, we define a *cumulative* version of exhaustive ternary codes, which subsumes all (k, i)-codes with $i = 2 \ldots l$ up to a specific level l. In this case, we speak of (k, l)-cumulative exhaustive codes, which generate a total of $\sum_{i=2}^{l} \binom{k}{i}(2^{i-1} - 1)$ columns. For a dataset with k classes, (k, k)-cumulative exhaustive codes represent the set of all possible binary classifiers.

Random Codes. We consider two types of randomly generated codes. The first variant allows to control the probability distribution of the set of possible symbols $\{-1, 0, 1\}$ from which random columns are drawn. By specifying a parameter $r \in [0, 1]$, the probability for the zero symbol is set to $p(\{0\}) = r$, whereas the remainder is equally subdivided to the other symbols: $p(\{1\}) = p(\{-1\}) = \frac{1-p}{2}$. This type of code allows to control the *sparsity* of the coding matrix, which will be useful for evaluating which factors determine the performance of the QUICKECOC algorithm.

The second random code generation method selects randomly a subset from the set of all possible classifiers N_{all}. This corresponds to the cumulative ternary code matrix with $l = k$. Obviously, this variant guarantees that no duplicate classifiers are generated, whereas it can occur in the other variant. We do not enforce this, because we wanted to model and evaluate two interpretations of randomly generated codes: randomly filled matrices and randomly selected classifiers.

Coding Theory, BCH-Codes. Many different code types were developed within coding theory. We pick the so-called BCH Codes [3] as a representative, because they have been studied in depth and have properties which are favourable in practical applications. For example, the desired minimum Hamming distance of M can be specified, and fast decoding methods are available. Note, however, that efficient decoding in coding theory has the goal to minimize the complexity of finding the nearest code word given the received *full* code word, while we are interested in minimizing the classifier evaluations needed for finding the nearest code word respectively class. Although some concepts of efficient decoding in coding theory seem to be transferable to our setting, they lack, contrary to QUICKECOC, the capability to be a general purpose decoding method for arbitrary coding matrices.

A detailed description of this code family is beyond the scope of this paper, but we refer to [3,11] for a detailed description and further information regarding BCH-Codes. In our evaluation, we considered binary BCH codes of lengths $7, 15, 31, 63, 127$ and 255. Similarly to [6], we randomly selected k code words from the set of codes, if the number of classes is k.

The above techniques are general in the sense that they are applicable to every possible dataset. Often, it is possible to project data-specific relationships or expert knowledge explicitly to the coding matrix (see, e.g., [4]). We did not consider these types of codes in this paper, but note that our algorithm is applicable to all code types.

2.2 Decoding

The traditional ECOC framework is accompanied with the Hamming Distance. After receiving an ensemble of base predictions, the class with the shortest Hamming Distance is selected as the output. In the meantime, several decoding strategies have been proposed. Along with the generalisation of ECOCs to the ternary case, Allwein et al [1] proposed a loss-based strategy. Escalera et al. [7] discussed the shortcomings of traditional Hamming distance for ternary ECOCs and presented two novel decoding strategies, which should be more appropriate for dealing with the zero symbol. We considered all these decoding strategies in our work, and summarize them below.

In the following, let $\boldsymbol{cw_i} = (m_{i,1}, \ldots, m_{i,n})$ a code word from a ternary ECOC matrix and $\boldsymbol{p} = (p_1, \ldots, p_n)$ be the prediction vector.

Hamming Distance: Describes the number of bit positions in which $\boldsymbol{cw_i}$ and $\boldsymbol{p}$ differ. Zero symbols ($m_{i,j} = 0$) increase the distance by $\frac{1}{2}$. Note that the prediction vector is considered as a set of binary predictions which can only predict either -1 or 1.

$$d_H(\boldsymbol{cw_i}, \boldsymbol{p}) = \sum_{j=1}^{n} \frac{|m_{i,j} - p_j|}{2}$$

Euclidian Distance: Computes the distance of the two n-dimensional vectors in Euclidian space.

$$d_E(\boldsymbol{cw_i}, \boldsymbol{p}) = ||\boldsymbol{cw_i} - \boldsymbol{p}||_2 = \sqrt{\sum_{j=1}^{n} (m_{i,j} - p_j)^2}$$

Attenuated Euclidian/Hamming Distance: These measures simply ignore the zero symbols for computing the distance.

$$d_{AE}(\boldsymbol{cw_i}, \boldsymbol{p}) = \sqrt{\sum_{j=1}^{n} |m_{i,j}|(m_{i,j} - p_j)^2}$$

$$d_{AH}(\boldsymbol{cw_i}, \boldsymbol{p}) = \sum_{j=1}^{n} |m_{i,j}| \frac{|m_{i,j} - p_j|}{2}$$

Loss based: In loss based decoding we assume that we have given a score-based classifier $f(.)$.

$$d_L(\boldsymbol{cw_i}, \boldsymbol{p}) = \sum_{j=1}^{n} l(m_{i,j} \cdot f_j)$$

where $l(.)$ is the loss function. Typical functions are $l(s) = -s$ and $l(s) = e^{-s}$.

Laplace Strategy

$$d_{LA}(\boldsymbol{cw_i}, \boldsymbol{p}) = \frac{E + 1}{E + C + T} = \frac{d_{AH}(\boldsymbol{cw_i}, \boldsymbol{p}) + 1}{\sum_{j=1}^{n} |m_{i,j}| + T}$$

where C is the number of bit positions in which they are equal and E in which they differ. T is the number of involved classes, in our case $T = 2$, since we employ binary classifiers. Thus, the default value of $d_{LA}(.)$ is $\frac{1}{2}$.

Beta Density Distribution Pessimistic Strategy: This measure assumes that the distance is a Beta-distributed random variable parametrized by C and E of two code words. It can be seen as a probabilistic version of the Laplace strategy, because its expected value equals the one from the Laplace strategy. Please refer to [13,7] for a detailed description.

3 Efficient Decoding for ECOC

In this section, we will introduce the QUICKECOC algorithm for efficiently determining the predicted class without the need to evaluate all binary classifiers. It builds upon the QWEIGHTED algorithm [12], which is tailored to the special case of pairwise classification with voting aggregation as a decoding technique. We will first briefly recapitulate this algorithm in Section 3.1, and then discuss the three key modifications that have to be made: first, Hamming decoding has to be reduced to a voting process (Section 3.2), second, the heuristic for selecting the next classifier has to be adapted to the case where multiple classifiers can be incident with a pair of classes (Section 3.3), and finally the stopping criterion can be improved to take multiple incidences into account (Section 3.4). We will then present the generalized QUICKECOC algorithm for Hamming decoding in Section 3.5. Finally, we will discuss how QUICKECOC can be adapted to different decoding techniques (Section 3.6).

3.1 QWeighted for Pairwise Classification

Pairwise classification [8] tackles the problem of multiclass classification by decomposing the main problem into a set of binary problems, one problem for each pair of classes. At prediction time, all binary classifiers are queried, and each classifier emits a vote for one of its two classes. The class which receives the maximum amount of votes is eventually predicted.

Though it can be shown that the training time of pairwise classification is smaller than in the one-against-all case [9], a quadratic number of classifiers still has to be evaluated at classification time. The QWEIGHTED algorithm [12] addresses this problem by exploiting the fact that usually not all evaluations are necessary to compute the class with the maximum votes. If one class has received more votes than every other class can possibly achieve in their remaining evaluations, this class can be safely predicted. The QWEIGHTED algorithm tries to enforce this situation by always focusing on the class that has lost the least amount of voting mass. Experiments showed that QWEIGHTED uses an average runtime of $O(k \log k)$ instead of the $O(k^2)$ that would be required for computing the same prediction with all evaluations.

3.2 Reducing Hamming Distances to Voting

Obviously, pairwise classification may be considered as a special case of ternary ECOCs, where each column of the coding matrix contains exactly one positive $(+1)$, one negative (-1), and $k-2$ ignore values (0). Thus, it is natural to ask the the question whether the QWEIGHTED algorithm can be generalized to arbitrary ternary ECOCs.

To do so, we first have to consider that ECOCs typically use Hamming distance for decoding, whereas pairwise classification typically uses a simple voting procedure.[1] In

[1] Other choices for decoding pairwise classifiers are possible (cf., e.g., [16]), but voting is surprisingly stable. For example, one can show that weighted voting, where each binary vote is split according to the probability distribution estimated by the binary classifier, minimizes the Spearman rank correlation with the correct ranking of classes, provided that the classifier provides good probability estimates [10].

voting aggregation, the class that receives the most votes from the binary classifiers is predicted, i.e.,

$$\tilde{c} := \underset{i \in K}{\operatorname{argmax}} \sum_{j \neq i, j \in K} f_{i,j}$$

where $f_{i,j}$ is the prediction of the pairwise classifier that discriminates between classes c_i and c_j.

Traditional ECOC with Hamming decoding predicts the class c^* whose code word $\boldsymbol{cw}_{c^*}$ has the minimal Hamming Distance $d_H(\boldsymbol{cw}_{c^*}, \boldsymbol{p})$ to the prediction vector $\boldsymbol{p} = (p_1, \ldots, p_n)$. The following lemma allows to reduce minimization of Hamming distances to voting aggregation:

Lemma 1. *Let* $v_{i,j} := \left(1 - \frac{|m_{i,j} - p_j|}{2}\right)$ *be a voting procedure for classifier* j *for class* c_i *then*

$$\underset{i=1 \ldots n}{\operatorname{argmin}} \, d_H(\boldsymbol{cw_i}, \boldsymbol{p}) = \underset{i=1 \ldots n}{\operatorname{argmax}} \sum_{j \in N} v_{i,j}$$

Proof. Recall that

$$d_H(\boldsymbol{cw_i}, \boldsymbol{p}) = \sum_{a=1}^{n} \frac{|cw_{i_a} - p_a|}{2} = \sum_{a=1}^{n} \frac{|m_{i,a} - p_a|}{2}$$

and let $b_{i,a} := \frac{|m_{i,a} - p_a|}{2}$. Since $b_{i,a} \in \{0, 0.5, 1\}$ and

$$\min_{i \in 1 \ldots k} \sum_{a=1}^{n} b_{i,a} \rightarrow \max_{i \in 1 \ldots k} \sum_{a=1}^{n} 1 - b_{i,a} = \max_{i \in 1 \ldots k} \sum_{a=1}^{n} v_{i,a}$$

holds, we obtain the statement. □

To be clear, the above used definition of $v_{i,j}$ formalizes a voting procedure, for which class c_i receives one vote ($+1$), if the prediction p_j of classifier j equals the corresponding encoding bit $m_{i,j}$ and an half vote ($+0.5$) for the case $m_{i,j} = 0$ where the classifier was not trained with instances from c_i.

This voting schemes differs slightly from the commonly known voting aggregation. The exact voting aggregation procedure described within the ECOC framework would be

$$v_{i,j} = |m_{i,j}| \cdot \left(1 - \frac{|m_{i,j} - p_j|}{2}\right)$$

which ignores the zero symbols and is not equivalent with Hamming decoding for arbitrary ternary coding matrices (but for e.g. pairwise codes w.r.t final prediction). Nevertheless, it is easy to see, that voting aggregation is equivalent to ECOC decoding using the Attenuated Hamming distance.

3.3 Next Classifier Selection

The QWEIGHTED algorithm always pairs the current favorite (the class with the least amount of voting loss) with its strongest competitor (the class that has the least amount of voting loss among all classes with which it has not yet been paired), and evaluates the resulting classifier. The rationale behind this approach is that the current favorite can emerge as a winner as quickly as possible. In pairwise classification, the choice of a classifier for a given pair of classes is deterministic because, obviously, there is only one classifier that is incident with any given pair of classes.

General ECOC coding matrices, on the other hand, can involve more than two classes, and, conversely, a pair of classes may be incident to multiple binary classifiers. This has the consequence that the selection of the next classifier to evaluate has gained an additional degree of freedom. For example, assume a 4-class problem (A, B, C, D) using 3-level ternary exhaustive codes, and classes A and B have currently the greatest vote amount, we could select one of four different classifiers that discriminate the classes A and B, namely $A|BC$, $A|BD$, $AC|B$ and $AD|B$.

QUICKECOC uses a selection process which conforms to the key idea of QWEIGHTED: Given the current favorite class c_{i_0}, we select all incident classifiers N_{i_0}. Let K_j denote the set of classes, which are involved in the binary classifier f_j, but with a different sign than c_{i_0}. In other words, it contains all rows i of column j in the coding matrix M, for which holds: $m_{i,j} \neq m_{i_0,j} \wedge m_{i,j} \neq 0$. We then compute a score

$$s(j) = \sum_{i \in K_j} k - r(i)$$

for every classifier $c_j \in N_{i_0}$, where $r(i)$ is a function which returns the position of class c_i in a ranking, where all classes are increasingly ordered by their current votings respectively ordered decreasingly by distances. Finally, we select the classifier f_{j_0} with the maximal score $s(j_0)$. Roughly speaking, this relates to selecting the classifier which discriminates c_{i_0} to the greatest number of currently highly ranked classes.

We experienced that this simple score based selection was superior among other tested methods, whose presentation and evaluation we omit here. One point to note is, that for the special case of pairwise codes, this scheme is identical to the one used by QWEIGHTED.

3.4 Stopping Criterion

The key idea of the algorithm is to stop the evaluation of binary classifiers as soon as it is clear which class will be predicted, irrespective of the outcome of all other classifiers. Thus, the QUICKECOC algorithm has to check whether c_{i_0}, the current class with the minimal Hamming distance to $\boldsymbol{p}$, can be caught up by other classes at the current state. If not, c_{i_0} can be safely predicted.

A straight-forward adaptation of the QWEIGHTED algorithm for pairwise classification would simply compute the maximal possible Hamming distance for c_{i_0} and compare this distance to the current Hamming distances l_i of all other classes $c_i \in K \backslash \{c_{i_0}\}$. The maximal possible Hamming distance for c_{i_0} can be estimated by assuming that all outstanding evaluations involving c_{i_0} will increase its Hamming distance. Thus, we simply add the number of remaining incident classifiers of c_{i_0} to its current distance l_{i_0}.

However, this simple method makes the assumption that all binary classifiers only increase the Hamming distance of c_{i_0}, but not of the other classes. This is unnecessarily pessimistic, because each classifier will always increase the Hamming distance of *all* (or none) of the incident classifiers that have the same sign (positive or negative). Thus, we can refine the above procedure by computing a separate upper bound of l_{i_0} for each class c_i. This bound does not assume that all remaining incident classifiers will increase the distance for c_{i_0}, but only those where c_i and c_{i_0} are on different sides of the training set. For the cases where c_i was ignored in the training phase, $\frac{1}{2}$ is added to the distance, according to the definition of the Hamming distance for ternary code words. If there exist no class which can overtake c_{i_0}, the algorithm returns c_{i_0} as the prediction.

Note that the stopping criterion can only test whether no class can surpass the current favorite class. However, there may be other classes with the same Hamming distance. As the QUICKECOC algorithm will always return the first class that cannot be surpassed by other classes, this may not be the same class that is returned by the full ECOC ensemble. Thus, in the case, where the decoding is not unique, QUICKECOC may return a different prediction. However, in all cases where the code word minimal Hamming distance is unique, QUICKECOC will return exactly the same prediction.

3.5 Quick ECOC Algorithm

Algorithm 1 shows the pseudocode of the QUICKECOC algorithm. The algorithm maintains a vector $\boldsymbol{l} = (l_1, \ldots, l_k) \in \mathbb{R}^k$, where l_i indicates the current accumulated Hamming distance of the associated code word $\boldsymbol{cw}_i$ of class c_i to the currently evaluated prediction bits $\boldsymbol{p}$. The l_i can be seen as lower bounds of the distances $d_H(\boldsymbol{cw}_i, \boldsymbol{p})$, which are updated incrementally in a loop which essentially consists of four steps:

(1) Selection of the Next Classifier
(2) Classifier Evaluation and Update of Bounds $\boldsymbol{l}$
(3) First Stopping Criterion
(4) Second Stopping Criterion

(1): First, the next classifier is selected. Depending on the current Hamming distance values, the routine SELECTNEXTCLASSIFIER returns a classifier that pairs the current favorite $i_0 = \text{argmin}_i l_i$ with another class that is selected as described in Section 3.3. In the beginning all values l_i are zero, so that SELECTNEXTCLASSFIER returns an arbitrary classifier f_j.
(2): After the evaluation of f_j, $\boldsymbol{l}$ is updated using the Hamming distance projected to this classifier (as described in Section 3.2) and f_j is removed from the set of possible classifiers.
(3): In line 10, the first stopping criterion is checked. It checks whether the current favorite class i_0 can already be safely determined as the class with the maximum number of votes, as described in Section 3.4.
(4): At line 17, the algorithm stops when all incident classifiers of c_{i_0} have been evaluated (this criterion is actually a special case of **(3)** but it will be useful later). In this case, since it holds that $l_{i_0} \leq l_i$ for all classes c_i with l_{i_0} fixed and considering that l_i can only increase monotonically, we can safely ignore all remaining evaluations.

Algorithm 1. QuickECOC

Require: ECOC Matrix $M = (m_{i,j}) \in \{-1,0,1\}^{k \times n}$, binary classifiers $f_1, \ldots, f_n$, testing instance $\boldsymbol{x} \in X$

1: $\boldsymbol{l} \in \mathbb{R}^k \Leftarrow 0$ # Hamming distance vector
2: $c^* \leftarrow$ NULL , $N \leftarrow \{1, \ldots, n\}$
3: **while** $c^* =$ NULL **do**
4: $j \leftarrow$ SELECTNEXTCLASSIFIER($\boldsymbol{M}, \boldsymbol{l}$)
5: $p \leftarrow f_j(\boldsymbol{x})$ # Evaluate classifier
6: **for each** $i \in K$ **do**
7: $l_i \leftarrow l_i + \frac{|m_{i,j} - p|}{2}$
8: $M \leftarrow M \backslash M_j$, $N \leftarrow N \backslash \{j\}$
9: $i_0 = \underset{i \in K}{\operatorname{argmin}}\, l_i$
10: # First stop Criterion
11: abort $\leftarrow$ **true**
12: **for each** $i \in K \backslash \{i_0\}$ **do**
13: $e_{Full} \leftarrow |\{j \in N | m_{i,j} \times m_{i_0,j} = -1\}|$
14: $e_{Half} \leftarrow |\{j \in N | m_{i,j} \neq 0$ **and** $m_{i_0,j} = 0\}|$
15: **if** $l_{i_0} + e_{Full} + \frac{1}{2} e_{Half} > l_i$ **then**
16: abort $\leftarrow$ **false**
17: # Second stop Criterion
18: **if** abort **or** $\forall j \in N. m_{i_0,j} = 0$ **then**
19: $c^* \leftarrow c_{i_0}$
20: **return** c^*

3.6 Decoding Adaptions

All decoding methods that we discussed in section 2.2 are compatible with QUICK-ECOC by applying small modifications. In general, there are two locations where adaptations are needed. First, the statistics update step and the first stopping criteria have to be adapted according to the used distance measure. Second, some decoding strategies require a special treatment of the zero symbol, which can, in general, be modeled as a preprocessing step. We will briefly describe the modifications for all considered decoding strategies:

Euclidian Distance: For minimizing the Euclidian distance we can ignore the root operation and simply substitute the update statement of the pseudocode (line 7) with: $l_i \leftarrow l_i + (m_{i,j} - p)^2$. The factor for e_{Half} is changed to 1 and the one for e_{Full} to 4.

Att. Euclidian Distance: Similar to the above modifications we change line 7 with: $l_i \leftarrow l_i + |m_{i,j}|(m_{i,j} - p)^2$ and set the factor of e_{Full} to 4 and remove the occurrences of e_{Half}.

Loss based linear: For both loss based versions, we assume that we have given a normalizing function $w(.)$ which projects $f_i(x)$ from $[-\infty : \infty]$ to $[-1, 1]$, e.g.,

$$w(x) = \begin{cases} \frac{x}{\max f(x)} & x \geq 0 \\ \frac{x}{|\min f(x)|} & x < 0 \end{cases}$$

We substitute line 6 with: $p \leftarrow w(f_j(x))$ and the update procedure with: $l_i \leftarrow l_i + \frac{1-p\cdot m_{i,j}}{2}$ and remove the occurrences of e_{Half}.[2]

Loss based exponential: For the exponential loss, we have to change line 6 as above and the update step with $l_i \leftarrow l_i + e^{-p\cdot m_{i,j}}$. In addition, the factor of e_{Full} is set to e^1 and e_{Half} to e^{-1}.

Laplace Strategy: This strategy can be used by incorporating a class- respectively row-based incrementer. Note that each error bit between a code word $\boldsymbol{cw}$ and the prediction vector $\boldsymbol{p}$ amounts $\frac{1}{b+T}$ towards the total distance $d_{LA}(\boldsymbol{cw}, \boldsymbol{p})$, where b is the number of non-zero bits of $\boldsymbol{cw}$. This incrementer denoted by I_i for class c_i can be computed as a preprocessing step from the given ECOC Matrix. So, the update step has to be changed to $l_i \leftarrow l_i + I_i$ and the factor of e_{Full} changes to I_i. Besides, e_{Half} can be removed.

Beta Density Distribution Pessimistic Strategy: Here, we use an approximation of the original strategy. First, similar to the Laplace Strategy, an incrementer is used to determine $Z_i = \frac{E}{E+C}$. And second, instead of using a numerical integration to determine Z_i+a_i, its standard deviation is added, which is in compliance with the intended semantic of this overall strategy to incorporate the uncertainty. The incrementer I_i is again set during a preprocessing step and we change the update step to $l_i \leftarrow l_i+\min(1, (I_i+\sigma_i))$. The factor for e_{Full} has to be changed to I_i and e_{Half} has to be removed.[3]

In general, a distance measure is compatible to QUICKECOC if the distance can be determined bit-wise or incremental, and the iterative estimate of l_i has to be monotonically increasing, but must never over-estimate the true distance.

4 Experimental Evaluation

In this section, we evaluate the performance of QUICKECOC for a variety of different codes. In addition, we were interested to see if it works for all decoding methods and whether we can gain insights on which factors determine its performance.

4.1 Experimental Setup

All experiments were performed within the WEKA [15] framework using the decision tree learner J48 with default parameters as a base learner. All evaluations were performed using 10-fold stratified cross-validation. Our setup consisted of *5 encoding strategies* (BCH Codes and two versions each of exhaustive and random codes), *7 decoding methods* (Hamming, Euclidian, Att. Euclidian, linear loss-based, exponential loss-based, Laplacian Strategy and Beta Density Probabilistic Pessimistic) and *7 multi-class datasets* selected from the UCI Machine Repository [2].

[2] Note that we did not use such a normalizing function in our actual evaluation since we used a decision tree learner as our base learner. Although the normalization of score based functions, such as SVMs, is not a trivial task, the sketched function $w(.)$ could be possibly determined by estimating $\min f(x)$ and $\max f(x)$ during training time (e.g. saving the largest distances between instances to the hyperplane for each classifier).

[3] Note that this approximation yielded in all our evaluations the same prediction as the original strategy.

For the encoding strategies, we also tried several different parameters. Regarding the exhaustive codes, we evaluated all (k, l) codes ranging from $l = 2$ to $l = k$ per dataset and analogously for the cumulative version. For the generation of the first type of random codes the zero symbol probability was parametrized by $r = 0.2, 0.4, 0.6, 0.8$ and the dimension of the coding matrix was fixed to 50 % of the maximum possible dimension with respect to the number of classes. The second type of random codes was generated by randomly selecting $20\,\%, 40\,\%, 60\,\%$ and $80\,\%$ from the set of all valid classifiers respectively columns (all columns of an (k, k) cumulative ternary coding matrix) without repetition. Regarding BCH Codes, we generated $7, 15, 31, 63, 127$ and 255-bit BCH codes and randomly selected n rows matching the class count of the currently evaluated dataset. For the datasets *machine* and *ecoli* where the number of classes is greater than 7, we excluded the evaluation with 7-bit BCH codes.

The datasets were selected to have a rather low number of different classes. The main reason for this limitation was that for some considered code types the number of classifiers grows exponentially. Especially for the datasets with the maximum number of eight classes (*machine* and *ecoli*), the cumulative ternary exhaustive codes generates up to 3025 classifiers. In addition, we evaluated all possible combinations of decoding methods, code types with various parameters, which we can not present here completely (in total 1246 experiments) because of lack of space. Nevertheless, we want to stress that our technique is applicable to larger number of classes (with reasonable codes), and, as our results will show, the expected gain increases with the number of classes.

Because of the high number of experiments, we cannot present all results in detail, but will try to focus on the most interesting aspects. In addition to assess the general performance of QUICKECOC, we will analyze the influence of the sparsity of the code matrix, of the code length, and of different decoding strategies.

4.2 Reduction in Number of Evaluations

Table 1 shows the reduction in the number of classifier evaluations with QUICKECOC on all evaluated datasets with Hamming decoding and ternary exhaustive codes. In every column the average number of classifier evaluations is stated with its corresponding ratio to the number of generated classifiers in italics (the lower the better). The datasets are ordered from left to right by ascending class-count. As the level parameter l is bounded by the class-count k, some of the cells are empty.

Table 1. QUICKECOC performance using Hamming decoding and exhaustive ternary codes

l	vehicle	derm.	auto	glass	zoo	ecoli	machine
2	3.82 *0.637*	7.12 *0.475*	7.95 *0.379*	9.99 *0.476*	9.48 *0.451*	11.75 *0.420*	11.60 *0.414*
3	7.91 *0.659*	26.05 *0.434*	42.86 *0.408*	43.47 *0.414*	41.64 *0.397*	58.85 *0.350*	57.90 *0.345*
4	5.65 *0.808*	46.30 *0.441*	115.22 *0.470*	116.45 *0.475*	107.03 *0.437*	199.31 *0.407*	194.81 *0.398*
5		43.11 *0.479*	163.67 *0.520*	163.98 *0.521*	148.50 *0.471*	369.06 *0.439*	355.23 *0.423*
6		16.54 *0.534*	114.87 *0.529*	116.77 *0.538*	102.41 *0.472*	394.25 *0.454*	369.19 *0.425*
7			34.24 *0.543*	37.84 *0.601*	31.52 *0.500*	234.80 *0.466*	218.09 *0.433*
8						62.17 *0.490*	57.27 *0.451*

Table 2. QUICKECOC performance on BCH codes

	vehicle	derm.	auto	glass	zoo	ecoli	machine
7	0.764	0.774	0.851	0.880	0.834	-	-
15	0.646	0.656	0.699	0.717	0.659	0.670	0.648
31	0.571	0.564	0.607	0.662	0.581	0.602	0.558
63	0.519	0.506	0.567	0.616	0.517	0.540	0.509
127	0.489	0.447	0.522	0.565	0.477	0.493	0.459
255	0.410	0.380	0.450	0.467	0.397	0.417	0.388

One can clearly see that QUICKECOC is able to reduce the number of classifier evaluations for all datasets. The percentage of needed evaluations ranges from about 81 % ($vehicle$, $l = 4$) to only 35 % ($machine$, $l = 3$). Furthermore, one can observe a general trend of higher reduction by increasing class-count. This is particularly obvious, if we compare the reduction on the exhaustive codes (the last line of each column, where $l = k$), but can also be observed for individual code sizes (e.g., for $l = 3$). Although we have not performed a full evaluation on datasets with a larger amount of classes because of the exponential growth in the number of classifiers, a few informal and quick tests supported the trend: the higher the class-count, the higher the reduction.

Another interesting observation is that except for dataset *vehicle* the exhaustive ternary codes for level $l = 3$ consistently lead to the best QUICKECOC performance over all datasets. A possible explanation based on a "combinatorial trade-off" can be found in [13], which was omitted here because of space restrictions.

For BCH Codes, we can report also that in all cases a reduction was possible, as one can see in Table 2. Note that all coding matrices in this case are dense, i.e., no coding matrix contains a (0). Even in this case, we see that there was no situation, where all classifiers were needed for multiclass classification. And again, we observe that for higher dimensions (increasing the BCH bit code) higher reductions can be observed.

We do not show a detailed table of results for random codes, but they will be used in the following sections.

4.3 Sparsity of Coding Matrices

We define the sparsity of the ECOC matrix as the fraction of (0)-values it contains. Random codes provide a direct control over the matrix sparsity (as described in section 2.1), and are thus suitable for analyzing the influence of the sparsity degree of the ECOC matrix for QUICKECOC. Note, however, that the observed influences regarding sparsity and dimension of the matrix on the QUICKECOC performance can also be seen in the evaluations of the other code types, but not as clearly or structured as here.

Figure 1 shows QUICKECOC applied to random codes with varying matrix sparsity. A clear trend can be observed that the higher the sparsity of the coding matrix the better the reduction for all datasets. Keep in mind that the baseline performance (evaluating all binary classifiers) is a parallel to the x-axis with the y-value of 1.0. Note that the absolute reduction tends to be minimal over all considered datasets at datasets with higher class-counts i.e *machine* at 80 % sparsity, and the lowest reduction can be seen for the dataset *vehicle* with the smallest number of classes $n = 4$ at 20 % sparsity.

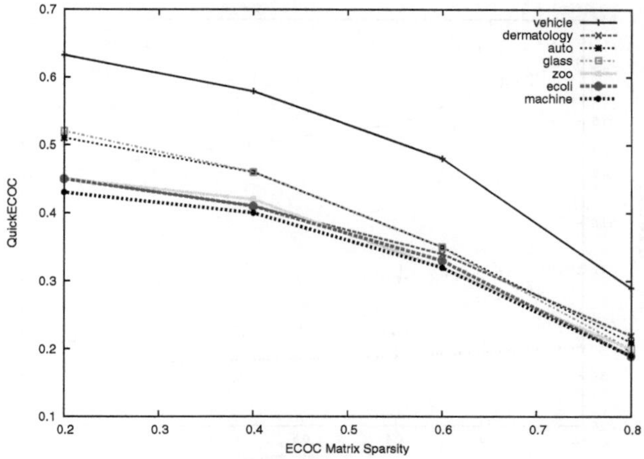

Fig. 1. QUICKECOC performance of random codes in dependence of sparsity

The main effect of an increase of sparsity on the coding matrices is that for each class the number of incident classifiers decreases. For sparsity 0, all classes are involved in all classifiers, for sparsity 0.5, each class is (on average) involved in only half of the classifiers. This will clearly affect the performance of the QUICKECOC algorithm. In particular, the second stopping criterion essentially specifies that the true class is found if all incident classifiers for the favorite class i_0 have been evaluated. Clearly, the algorithm will terminate faster for higher sparsity levels (ignoring, for the moment, the possibility that the first stopping criterion may lead to even faster termination).

4.4 Code Length

The second type of random codes, which were generated by randomly selecting a fixed number from the set of all possible binary classifiers can be seen in Fig. 2. All coding matrices for a k-class dataset have nearly the same sparsity, which relates to the average sparsity of (k, k) cumulative exhaustive codes and differ only in the length of the coding matrix (in percent of the total number of possible binary classifiers). This allows us to observe the effect of different numbers of classifiers on the QUICKECOC performance. Here, we can also see an consistent relationship, that higher dimensions lead to better performance, but the differences are not as remarkable as for sparse matrices.

For a possible explanation, assume a coding matrix with fixed sparsity and we vary the dimension. For a higher dimension the ratio of number of classifiers per class increases. Thus, on average, the number of incident classifiers for each class also increases. If we now assume that this increase is uniform for all classes, this has the effect that the distance vector $\boldsymbol{l}$ is multiplied by a positive factor $x > 1$, i.e., $\boldsymbol{l}^+ = \boldsymbol{l} * x$. This alone would not change the QUICKECOC performance, but if we consider that classifiers are not always perfect, we can expect that for higher number of classifiers, the variance of the overall prediction will be smaller. This smaller variance will lead to a more reliable voting vectors, which can, in turn, lead to earlier stopping. It also

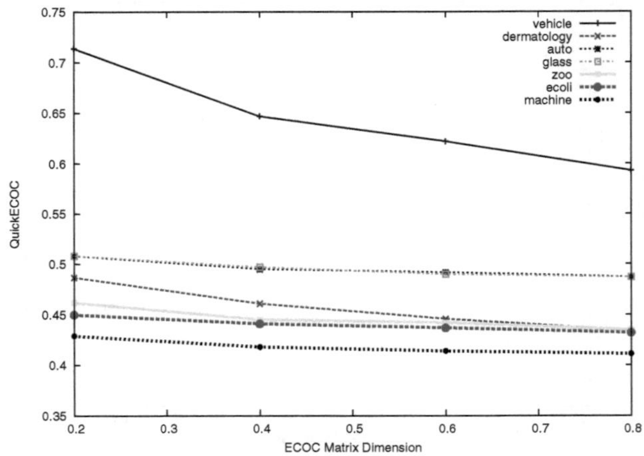

Fig. 2. QUICKECOC performance of random codes in dependence of code length

Table 3. QUICKECOC performance on *ecoli* with all decoding methods and cumulative exhaustive ternary codes

	Hamming	Euclidian	A. Euclidian	LBL	LBE	Laplace	BDDP	$\|N\|$
$l = 2$	*0.420*	*0.420*	*0.420*	*0.399*	*0.398*	*0.406*	*0.426*	28
$l = 3$	*0.331*	*0.331*	*0.331*	*0.335*	*0.350*	*0.332*	*0.333*	196
$l = 4$	*0.377*	*0.377*	*0.377*	*0.383*	*0.402*	*0.374*	*0.375*	686
$l = 5$	*0.400*	*0.400*	*0.400*	*0.414*	*0.439*	*0.399*	*0.401*	1526
$l = 6$	*0.421*	*0.421*	*0.421*	*0.437*	*0.466*	*0.419*	*0.418*	2394
$l = 7$	*0.427*	*0.427*	*0.427*	*0.444*	*0.475*	*0.426*	*0.425*	2898
$l = 8$	*0.428*	*0.428*	*0.428*	*0.446*	*0.477*	*0.427*	*0.426*	3025

seems reasonable that this effect will not have such a strong impact as the sparsity of the coding matrix, which we discussed in the previous section.

4.5 Different Decoding Strategies

As previously stated, because of the large number of experiments, we can not give a complete account of all results. We evaluated all combinations of experiments, that includes also all mentioned decoding methods. All the previously shown results were based on Hamming decoding, since it is still one of the commonly used decoding strategies even for ternary ECOC matrices. However, we emphasize, that all observations on this small subset of results can also be found in the experiments on the other decoding strategies. As an exemplary data point, Table 3 shows an overview of the QUICKECOC performance for all decoding strategies for the dataset *ecoli* using cumulative exhaustive ternary codes. It can be seen that the performance is quite comparable on all datasets. Even the optimal reduction for $l = 3$ can be found in the results of all decoding strategies.

5 Discussion

At first glance, the results may not be striking, because a saving of a little less than $40\,\%$ does not appear to be such a large gain. However, one must put these results in perspective. For example, for the *vehicle* dataset with a $(4, 3)$-exhaustive code, QUICKECOC evaluated $65.9\,\%$ of all classifiers (cf. Table 1). A $(4, 3)$-exhaustive code has 12 classifiers, and each individual class is involved in $75\,\%$ of these classifiers (cf. the example in section 2.1). Thus, on average, QUICKECOC did not even evaluate all the classifiers that involve the winning class before this class was predicted. Similarly, for (k, k)-exhaustive codes all classes are involved in all binary classifiers, but nevertheless, considerable savings are possible. It would be interesting to derive a lower bound on the possible optimal performance, and to relate these empirical results to such a bound.

One could also argue that in applications where the classification time is crucial, a parallel approach could be applied much more effectively. Since each classifier defined by a column of the ECOC matrix can be evaluated independently, the implementation could be done very easily. QUICKECOC loses this advantage because the choice of the next classifier to evaluate depends on the results of the previous evaluations. However, QUICKECOC can still be parallelized on the instance level instead of the classifier level. Given n processors or n threads we want to utilize, we select n incoming test instances and apply QUICKECOC for each of them. With this method a higher speed up can be expected as with a straight-forward parallelization of ECOC.

Another point is that the gains obtained by QUICKECOC are negligible in comparison to what can be gained by more efficient coding techniques. While this is true, we note that QUICKECOC can obtain gains independent of the used coding technique, and can thus be combined with any coding technique. In particular in time-critical applications, where classifiers are trained once in batch and then need to classify on-line on a stream of in-coming examples, the obtained savings can be decisive.

6 Conclusions

We have shown a general algorithm for reducing the number of classifier evaluations for ternary ECOC matrices without compromising the overall prediction. It is based on a similar algorithm that was tailored to pairwise classification. Since ternary ECOCs subsume nearly all possible binary decomposition schemes, the reduction applies now to a broader spectrum of applications. For example, data-specific optimal codes can now also take advantage of reduced classifier evaluations. Regardless of the used code, QUICKECOC improves the overall prediction efficiency. At first sight, the amount of improvement may not seem to be as striking as for the pairwise case, where we could report a reduction from k^2 to $k \log k$ [12], but one must keep in mind that in ECOC codings, each class has a much larger number of incident classifiers, and thus a higher number of evaluations must be expected to determine the winning class. We observed that the performance gain increases with higher sparsity of the coding matrix, again putting pairwise classification at the more efficient end of the spectrum. We also noted an increase in the performance gain with increasing code lengths of the chosen code.

There might still be some potential for improving our results with better heuristics for the selection of the next classifier, we have not yet thoroughly explored this parameter. For example, one could try to adapt ideas from active learning for this process. Furthermore, we consider an in-depth analysis of existing fast decoding methods in Coding Theory and the investigation of the transferability to the multiclass classification setting, because they seem to share some similarities.

Acknowledgments. We would like to thank Eyke Hüllermeier and Lorenz Weizsäcker for helpful suggestions and discussions. This work was supported by the *German Science Foundation (DFG).*

References

1. Allwein, E.L., Schapire, R.E., Singer, Y.: Reducing multiclass to binary: A unifying approach for margin classifiers. Journal of Machine Learning Research 1, 113–141 (2000)
2. Asuncion, A., Newman, D.: UCI machine learning repository (2007)
3. Bose, R.C., Ray-Chaudhuri, D.K.: On a class of error correcting binary group codes. Information and Control 3(1), 68–79 (1960)
4. Cardoso, J.S., da Costa, J.F.P.: Learning to classify ordinal data: The data replication method. Journal of Machine Learning Research 8, 1393–1429 (2007)
5. Crammer, K., Singer, Y.: On the learnability and design of output codes for multiclass problems. Machine Learning 47(2-3), 201–233 (2002)
6. Dietterich, T.G., Bakiri, G.: Solving multiclass learning problems via error-correcting output codes. Journal of Artificial Intelligence Research 2, 263–286 (1995)
7. Escalera, S., Pujol, O., Radeva, P.: Decoding of ternary error correcting output codes. In: Martínez-Trinidad, J.F., Carrasco Ochoa, J.A., Kittler, J. (eds.) CIARP 2006. LNCS, vol. 4225, pp. 753–763. Springer, Heidelberg (2006)
8. Friedman, J.H.: Another approach to polychotomous classification. Technical report, Department of Statistics, Stanford University, Stanford, CA (1996)
9. Fürnkranz, J.: Round robin classification. Journal of Machine Learning Research 2, 721–747 (2002)
10. Hüllermeier, E., Fürnkranz, J., Cheng, W., Brinker, K.: Label ranking by learning pairwise preferences. Artificial Intelligence 172, 1897–1916 (2008)
11. Macwilliams, F.J., Sloane, N.J.A.: The Theory of Error-Correcting Codes. North-Holland Mathematical Library. North Holland, Amsterdam (1983)
12. Park, S.-H., Fürnkranz, J.: Efficient pairwise classification. In: Kok, J.N., Koronacki, J., Lopez de Mantaras, R., Matwin, S., Mladenič, D., Skowron, A. (eds.) ECML 2007. LNCS (LNAI), vol. 4701, pp. 658–665. Springer, Heidelberg (2007)
13. Park, S.-H., Fürnkranz, J.: Efficient decoding of ternary error-correcting output codes for multiclass classification. Technical Report TUD-KE-2009-01, TU Darmstadt, Knowledge Engineering Group (2009)
14. Pimenta, E., Gama, J., de Leon Ferreira de Carvalho, A.C.P.: The dimension of ECOCs for multiclass classification problems. International Journal on Artificial Intelligence Tools 17(3), 433–447 (2008)
15. Witten, I.H., Frank, E.: Data Mining: Practical machine learning tools and techniques, 2nd edn. Morgan Kaufmann, San Francisco (2005)
16. Wu, T.-F., Lin, C.-J., Weng, R.C.: Probability estimates for multi-class classification by pairwise coupling. Journal of Machine Learning Research 5, 975–1005 (2004)

The Model of Most Informative Patterns and Its Application to Knowledge Extraction from Graph Databases

Frédéric Pennerath[1,2] and Amedeo Napoli[2]

[1] Supélec, Campus de Metz, 2 rue Édouard Belin 57070 Metz, France
frederic.pennerath@supelec.fr
[2] Orpailleur team, LORIA, BP 239, 54506 Vandoeuvre-lès-Nancy Cedex, France
amedeo.napoli@loria.fr

Abstract. This article introduces the class of Most Informative Patterns (MIPs) for characterizing a given dataset. MIPs form a reduced subset of non redundant closed patterns that are extracted from data thanks to a scoring function depending on domain knowledge. Accordingly, MIPs are designed for providing experts good insights on the content of datasets during data analysis. The article presents the model of MIPs and their formal properties wrt other kinds of patterns. Then, two algorithms for extracting MIPs are detailed: the first directly searches for MIPs in a dataset while the second screens MIPs from frequent patterns. The efficiencies of both algorithms are compared when applied to reference datasets. Finally the application of MIPs to labelled graphs, here molecular graphs, is discussed.

1 Introduction

Given a dataset describing objects by attributes (or items), a frequent itemset is a subset of attributes such that the number, also called support or frequency, of objects presenting all of these attributes is not less than some threshold. Since the first frequent itemset mining algorithm was proposed [1], frequent itemsets have become a major and prolific model in data-mining that has served many different applications and has been generalized to many different classes of patterns, like sequences, trees, or connected graphs (see for instance the Gaston algorithm [2] later used in Sect. 4.2). However searching frequent patterns is not an ultimate objective. Frequent patterns (of any type, even graphs) are generally considered as the result of an intermediate processing step, usually followed either by the extraction of frequent association rules, or by the extraction of a set of patterns of interest wrt some application specific criteria. In any case, resulting rules or patterns are usually sorted in decreasing order of some score so that only the head of the sorted list, whose members are sometimes called top-k patterns (like area-scored top-k patterns [3] later referred in Sect. 2.1), is considered.

For association rules, many scores are available like confidence or lift. For frequent patterns, scoring often serves supervised classification problems. Scores

W. Buntine et al. (Eds.): ECML PKDD 2009, Part II, LNAI 5782, pp. 205–220, 2009.

like p-value or information gain are then used to assess the discriminative power of patterns relatively to two sets of positive and negative examples. Whereas a direct scoring of patterns may make sense in the framework of machine learning problems, practical relevance of pattern scoring might be discussed in the framework of knowledge discovery, where selected rules or patterns are directly analyzed by experts. In that case two problems occur when providing experts lists of patterns sorted by decreasing order of score.

First finding a good qualitative scoring function is not an easy task in the context of knowledge discovery as scoring must predict interest of experts for patterns. This interest is typically the amount of novel information a pattern brings to experts relatively to their current state of knowledge but this information is obviously hardly assessable. Frequency is an example of a "bad" qualitative scoring function. Because of the anti-monotonic property of frequency, most frequent patterns tend to be the smallest and thus the least informative as well. An extreme example is the empty itemset that carries no information but has the largest possible frequency. However a good scoring function must somehow integrate frequency as the latter reflects likelihood of patterns, from highly improbable to very common. In many applications, the interest of a pattern thus balances between its frequency and the amount of information contained in its structure. Such a balance refers to the notion of data representativeness. The Minimal Description Length principle (MDL) provides a theoretical foundation to assess representativeness. This principle states the better a model helps to encode data with a reversible compression scheme, the more this model is representative of data. This principle has already been used to identify patterns representative of data. Data compression then consists in replacing every occurrence of these representative patterns by new attributes in datasets of attributes [4] or new vertices in datasets of graphs [5]. However MDL-based patterns are limited somehow as they do not take easily into account what experts know and want to know. A better solution is to provide a flexible model that accepts a large family of scoring functions tunable to experts' needs.

The second problem is information redundancy among extracted patterns: Since usual scoring functions are continuous, similar patterns are likely to have similar scores. Consequently top-k patterns gets saturated by patterns similar to the pattern of highest score, especially when patterns like graphs exhibit a high combinatorial power. In practice experts experience difficulties to distinguish patterns providing them new elements of information as they are flooded with redundant copies of already analyzed patterns. One way of reducing the number of useless frequent patterns to consider might consist in introducing additional constraints that patterns have to meet [6]. A common example of pattern constraints is provided by closed patterns: a pattern P is *closed* if the frequency of every pattern containing P is strictly smaller than the frequency of P. However, although constraints might reduce the number of patterns, they remain insensitive to pattern redundancy.

In this paper we propose to solve both previous problems by a pattern selection process that outputs a family of patterns we have called *Most Informative*

Patterns or MIPs. Intuitively MIPs are defined as local maxima of a scoring function. This function is only required to satisfy few conditions in order to assess pattern representativeness. The objective of MIP model is that every MIP reveals one independent element of interest for experts. In practice MIPs appear in limited number and are not structurally redundant compared to other pattern families so that experts can directly analyse them. The idea underlying the MIP model was initially motivated by a selective extraction of patterns from chemical reaction databases [7]. Contributions of this article are the generalization of this idea into a broad and formal model, the derivation of properties from the model, and the introduction, comparison, and application of two methods to extract frequent MIPs from itemset and graph datasets. To this end, the MIP model and its properties are introduced in Sect. 2, the MIP extraction methods in Sect. 3, and experiments in Sect. 4.

2 Introduction of Most Informative Patterns

2.1 An Example

In order to illustrate the redundancy problem, let consider the simple example of a dataset containing seven objects described by four attributes from a to d and whose descriptions are respectively a, b, ab, cd, abc, abd, and $abcd$. Let assume experts decide to score itemsets with the product of their length and their frequency (a MDL-related score sometimes called area function [3]). Figure 1 displays resulting frequency and score of every pattern inside the order diagram of itemsets ordered by subset inclusion. The list of itemsets sorted in decreasing order of score is: ab (score of 8); abc and abd (6); a and b (5); $abcd$, ac, bc, ad, bd, and cd (4); acd, bcd, c, and d (3); $\emptyset$ (0). When picking patterns from this list in that order, experts might ignore abc, abd, a, and b as these patterns are structurally similar to ab but with a lower score. For the same reason of redundancy, experts might ignore $abcd$, ac, bc not as interesting as abc, then ad and

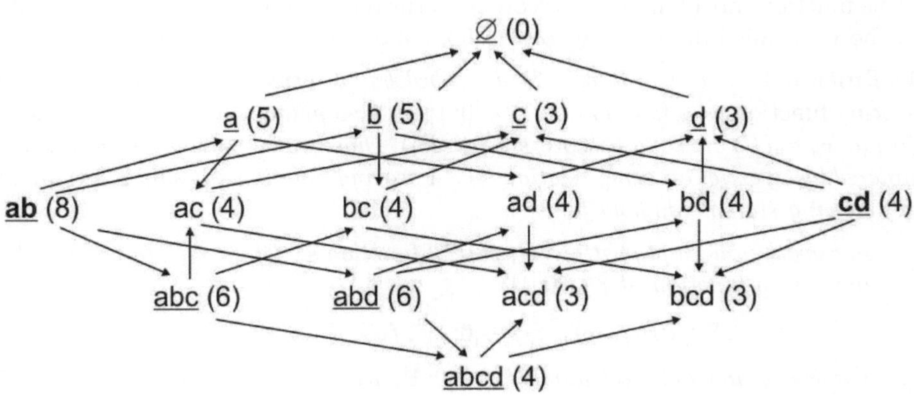

Fig. 1. Diagram order of itemsets. Every itemset is labeled with $(s = f \times l)$ where s, f and l are resp. its score, frequency and length. Closed patterns are underlined.

bd not as good as *abd*. However experts might consider next pattern *cd* that has a higher score than those of all similar patterns *acd*, *bcd*, *c* or *d*. Finally all remaining patterns are ignored as they are similar to patterns with better scores. The fact that *cd* is retained whereas its score is lower than those of many ignored patterns illustrates that scoring by itself is a limited approach. Introducing constraints may focus the analysis on a limited number of patterns of a particular type but does not remove pattern redundancy and may discard interesting patterns as well. For instance considering only closed patterns (underlined on Fig. 1) keeps redundant patterns like *a*, *b*, *ab*, *abc*, *abd*, *abcd*, whereas keeping only patterns containing item *a* removes the interesting pattern *cd*. MIPs formalize the screening process described on the previous example.

2.2 MIP Definition

Formally let consider a set $\mathcal{P}$ of patterns, ordered by a partial ordering relation $\leq_{\mathcal{P}}$. A *dataset* $\mathcal{D}$ of *objects* is then described by a function $d : \mathcal{D} \to \mathcal{P}$ mapping every object $o \in \mathcal{D}$ to its *description* $d(o) \in \mathcal{P}$. A pattern $P \in \mathcal{P}$ is said to *describe* an object o if $P \leq_{\mathcal{P}} d(o)$. The *support* or *frequency* of a pattern M is then the number $\sigma(P)$ of objects of $\mathcal{D}$ described by P whereas the *relative frequency* $\sigma_r(P)$ is the fraction of $\sigma(P)$ over the size $|\mathcal{D}|$ of the dataset. Support and relative frequency are non-increasing functions in the *pattern order* $(\mathcal{P}, \leq_{\mathcal{P}})$: the smaller a pattern is, the more objects it describes. In addition, the pattern order is assumed to contain a smallest pattern, called the *empty pattern* and denoted $\emptyset_{\mathcal{P}}$. One of the simplest examples of pattern order is the power set $\mathcal{P} = \mathfrak{P}(\mathcal{A})$ of a set $\mathcal{A}$ of attributes ordered by the subset inclusion relation $\leq_{\mathcal{P}} = \subseteq$, the empty pattern being the empty set. Another example of pattern order is the set of non-isomorphic connected graphs whose vertices and edges are tagged by labels taken from an arbitrary set $\mathcal{L}$. The ordering relation is then the isomorphic subgraph relation and the empty pattern is the empty graph.

As mentioned previously, the model of most informative patterns integrates a scoring function to assess the interest or relevance of a pattern. However only some functions are of interest to score patterns representative of data. The family of those so-called *informative scoring functions* is defined as follows.

Definition 1. *Given a dataset $\mathcal{D}$ described by patterns from order $(\mathcal{P}, \leq_{\mathcal{P}})$, a* scoring function *is a function $s : \mathcal{P} \times [0; 1] \to \$$ mapping a pattern P of relative frequency $\sigma_r(P)$ in $\mathcal{D}$ to a* score *$s(P, \sigma_r(P))$ whose value is taken from a set $\$$ ordered by a partial ordering relation $\leq_{\$}$. A scoring function s is said* informative *if following statements hold for s:*

1. *For every non-empty pattern P, partial function $s^P : f \mapsto s(P, f)$ is a strictly increasing function of $f \in [0; 1]$:*

$$\forall P \in \mathcal{P} \setminus \{\emptyset_{\mathcal{P}}\}, \forall (f_1, f_2) \in [0; 1]^2, f_1 < f_2 \Rightarrow s^P(f_1) <_{\$} s^P(f_2)$$

2. *For every non-null real number $f \in]0; 1]$, partial function $s^f : P \mapsto s(P, f)$ is a strictly increasing function of $P \in \mathcal{P}$:*

$$\forall f \in]0; 1], \forall (P_1, P_2) \in \mathcal{P}^2, P_1 <_{\mathcal{P}} P_2 \Rightarrow s^f(P_1) <_{\$} s^f(P_2)$$

3. *A pattern of zero frequency can never get a higher score than a pattern of non-zero frequency:*

$$\forall(P_1, P_2) \in \mathcal{P}^2, \nexists f > 0, s(P_1, f) <_{\$} s(P_2, 0)$$

The already used *area function* $s_a : (P, f) \mapsto |P| \cdot f$ is an example meeting all requirements of an informative function. This function may be interpreted wrt the MDL principle as an estimation of the amount of compressed space when replacing every occurrence of P by a new special symbol (attribute or vertex) [5]. In section 4, we propose to extend this area function by weighting attributes of an itemset or vertex/edge labels of a graph pattern with variable gains of information. The definition of the resulting scoring function is given for graphs (itemsets being equivalent to a graph whose isolated vertices have attributes as labels):

Definition 2. *The* information function s_i *is defined as:*

$$s_i : (g, \sigma_r) \mapsto I(g) \cdot \sigma_r$$

where the factor $I(g)$ of information related to graph pattern g is the sum of information carried by every vertex $v \in V(g)$ of label $l_v(v)$ and every edge $e \in E(g)$ of label $l_e(e)$:

$$I(g) = \sum_{v \in V(g)} i(l_v(v)) + \sum_{e \in E(g)} i(l_e(e))$$

Quantity of information associated to a vertex or edge label is in turn:

$$i(l) = -\log_2 \left(\frac{n(l)}{\sum_{l' \in \mathcal{L}} n(l')} \right)$$

where $n(l)$ is the number of vertices or edges in $\mathcal{D}$ carrying label l.

However many other informative functions can be considered here. In particular experts can complement or replace the previous factor $I(g)$ by other terms that grow with the pattern: number of vertices, edges and cycles of a given type, number of subgraphs isomorphic to some specific patterns, maximal degree, maximal length of paths or cycles.

Now the definition of MIPs formalizes the selection process described in the introductory example:

Definition 3. *Given a pattern order $(\mathcal{P}, \leq_{\mathcal{P}})$, a dataset $\mathcal{D}$ described by the previous set of patterns and an informative scoring function s defined on top of $\mathcal{D}$ and of scoring order $(\$, \leq_{\$})$,*

– *A pattern P' is a* neighbour *of pattern P if P' is an immediate predecessor or successor of P wrt pattern order $(\mathcal{P}, \leq_{\mathcal{P}})$, i.e. P and P' are comparable and no other pattern exists between P and P'.*

- *A pattern* $P' \in \mathcal{P}$ dominates *pattern* $P \in \mathcal{P}$ *if* P' *is a neighbour of* P *in* $(\mathcal{P}, \leq_{\mathcal{P}})$ *and scores of* P *and* P' *are comparable and verify* $s(P', \sigma_r(P')) >_{\$} s(P, \sigma_r(P))$.
- *A pattern* P *is a MIP if frequency* $\sigma_r(P)$ *of* P *is not null and if no pattern dominates* P.

Figure 2 represents diagram of Fig. 1 whose edges have been oriented according to the dominance relation: an arc drawn from m_1 to m_2 means m_1 dominates m_2 (rel. to s_a). Itemset *abc* is thus dominated by *ab* and dominates *ac*, *bc*, and

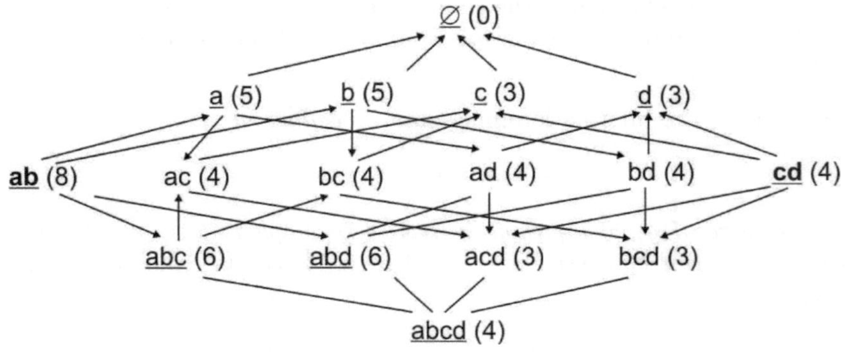

Fig. 2. Dominance relation between patterns from example of Fig.1. MIPs are bold.

abcd. Most informative patterns, in bold, are those pointed by no arc: they are as expected *ab* of score 8 and *cd* of score 4. The extraction of frequent MIPs consists in finding in dataset $\mathcal{D}$ scores and frequencies of all MIPs whose frequency is not less than some threshold σ_{min}. It is noticeable that a frequent pattern P may not be dominated by any immediate predecessor and any frequent immediate successor while being dominated by a non-frequent immediate successor. For instance, patterns c and d of Fig. 1 are frequent for $\sigma_{min} = 3$, but are not MIPs as they are dominated by the non-frequent pattern *cd*. Now some properties can be inferred from definitions of informative scoring functions and MIPs.

2.3 Properties

Let first assume the considered pattern order verifies the so-called "finiteness hypothesis": for every finite and non-empty dataset, the number of patterns of non-null frequency is finite and non-null. This is true for standard pattern orders like the sets of finite itemsets or finite graphs. This hypothesis allows to prove the following property:

Property 1. *The subset of most informative patterns of a finite non-empty dataset is not empty.*

Proof. This can be proved by contradiction. If every pattern is dominated by at least one pattern, it is possible to build recursively a sequence of patterns $(P_i)_{i\geq 1}$ from a pattern P_1 of positive frequency, such that for every index $i \geq 1$, P_{i+1} dominates P_i. Thanks to the third statement of def. 1, all those patterns have a positive frequency and thus, according to the finiteness hypothesis, build a subset of a finite set of patterns. Sequence (P_i) is thus finite and contains a cycle, contradicting the fact $(s(P_i))_{i\geq 1}$ is a strictly increasing sequence of scores.

However the extraction of frequent MIPs may produce no patterns if the threshold σ_{min} is too high. Another important property is related to closed patterns:

Property 2. *Every most informative pattern is a closed pattern.*

Proof. Let P' be a MIP relative to an informative scoring function s. If P' is not closed, there exists an immediate successor P'' of P' such that $\sigma_r(P'') = \sigma_r(P')$. Since $f = \sigma_r(P') \neq 0$, the second statement of def. 1 applies so that function $s^f : P \mapsto s(P, f)$ is strictly increasing. Since $P' <_{\mathcal{P}} P''$, $s^f(P') <_{\$} s^f(P'')$, and thus $s(P') <_{\$} s^f(P'')$. Because $\sigma_r(P'') = \sigma_r(P')$, $s^f(P'') = s(P'')$ and $s(P') <_{\$} s(P'')$. Domination of P'' over P' would contradict the hypothesis P' is a MIP.

On the example of Fig. 2, MIPs *abc* and *cd* appear to be closed. Conversely closed patterns can be seen as a particular case of MIPs:

Property 3. *Closed patterns are the most informative patterns relative to the informative scoring function equal to the identity* $Id : (P, f) \mapsto (P, f)$ *and the scoring order equal to the product order* $(\mathcal{P}, \leq_{\mathcal{P}}) \times ([0;1], \leq)$[1].

Proof. Given a MIP P, let assume P is not closed. At least one immediate successor P' would have the same frequency as P and since by definition $P <_{\mathcal{P}} P'$, P' would dominate P according to the definition of product order, contradicting the initial hypothesis. Conversely a closed pattern has a higher frequency than every immediate successor and is larger (rel. to $<_{\mathcal{P}}$) than every immediate predecessor, so that it cannot be dominated and thus is a MIP.

Both properties 2 and 3 prove together that closed patterns build the least restrictive family of most informative patterns (and thus the largest as well) among every possible choice of informative scoring functions.

3 Extraction of Frequent Most Informative Patterns

We propose two distinct approaches to extract frequent MIPs. The first one is a one-step extraction of MIPs from datasets, while the second is a two-step process that screens frequent MIPs from frequent patterns.

[1] The product order $(E_1 \times E_2, \leq_{12})$ of two orders $(E_1, \leq_1)$ and $(E_2, \leq_2)$ is defined by $(x_1, x_2) \leq_{12} (y_1, y_2)$ iff $x_1 \leq_1 y_1$ and $x_2 \leq_2 y_2$.

3.1 Direct Extraction Method

As seen in the previous section, every arc $P_1 \rightarrow P_2$ of the diagram order of $(\mathcal{P}, \leq_{\mathcal{P}})$ connects a pattern P_1 to an immediate successor P_2 of P_1. Since every arc defines a possible relation of dominance, an algorithm extracting frequent MIPs must potentially look at every arc whose origin P_1 is frequent. Consequently the direct extraction method explores the pattern order in a DFS manner and when crossing an arc $P_1 \rightarrow P_2$, compares scores of P_1 and P_2 and if these scores are comparable and different, withdraw one of the two patterns from the set of valid MIP candidates. In order to remember which patterns are still valid candidates, it is required to maintain a *mip* flag for every frequent pattern, initialized to true. To this end, a pattern dictionary $\mathcal{T}$ is used to map a pattern P to an entry $\mathcal{T}(P)$ containing the *mip* flag along with frequency and score of P. This dictionary uses a trie structure for storing canonical encoding of patterns. In case of itemsets, this encoding is simply the list of attributes sorted in some arbitrary order. In case of labeled connected graphs, encoding first assumes to compute a canonical ordering of vertices of this graph thanks to some state-of-the-art algorithm like Nauty [8], and then encodes the resulting canonical graph as a sequence of symbols for accessing the trie. The DFS exploration is performed thanks to a recursive function detailed on Fig. 3. This function `develop` takes a current pattern P and its entry e in $\mathcal{T}$ as arguments. Line 1 then computes in one single pass over $\mathcal{D}$, frequencies of the set S of all immediate successors of P occurring in $\mathcal{D}$ (i.e. of non-null frequency). This operation can be done efficiently by storing in memory all embeddings of the current pattern in dataset

Function `develop`(*pattern* P, *entry* e)

Data: Dataset $\mathcal{D}$, threshold σ_{min}, scoring function s and order $(\mathbb{S}, \leq_{\mathbb{S}})$
Result: List of frequent MIPs with their scores and frequencies

1 Extract set $S = \{(P', \sigma_r(P'))\}$ of all imm. succ. P' of P occur. in $\mathcal{D}$;
foreach $(P', \sigma'_r) \in S$ **do**
    if $\sigma'_r \geq \sigma_{min}$ **then**
        Search for entry e' mapped to P' in $\mathcal{T}$;
        if e' *does not exist* **then**
            Create entry e' such that $e'.score \leftarrow s(P', \sigma'_r)$, $e'.freq \leftarrow \sigma'_r$, and $e'.mpi \leftarrow$ true and map P' to e' in $\mathcal{T}$;
2   Call `develop` (P', e')
3   **if** $e.score <_{\mathbb{S}} e'.score$ **then**
            $e.mpi \leftarrow$ false
4   **else if** $e.score >_{\mathbb{S}} e'.score$ **then**
            $e'.mpi \leftarrow$ false
5   **else if** $e.score <_{\mathbb{S}} s(P', \sigma'_r)$ **then**
        $e.mpi \leftarrow$ false

Fig. 3. Recursive procedure for a direct extraction of frequent MIPs (using data structures like tid-lists [9] for itemsets or occurrence lists [10] for connected graphs). Then line 2 calls recursively the function in order to further develop every frequent immediate successor P' of P that has not been explored yet (i.e. that has not already been inserted in $\mathcal{T}$). In any case, scores of P and P' are compared (lines 3, 4, and 5) to discard dominated patterns from the set of MIP candidates. At the end of recursion started with the empty pattern as argument, the algorithm outputs frequent MIPs as patterns contained in $\mathcal{T}$ with a true flag, along with their scores and frequencies.

3.2 Frequent Pattern Screening Method

Another solution is to screen frequent MIPs from frequent patterns produced by an existing frequent pattern mining algorithm. This screening processes frequent patterns level by level as a level-wise algorithm like `Apriori` [1]: level of order n is the set of frequent patterns with the same length equal to n (i.e. number of attributes for itemsets and number of edges for graphs). More exactly the algorithm compares the score of every frequent pattern of level n with scores of their immediate predecessors of level $n-1$ for every non-empty level n. Comparison of scores allows to rule out i) MIP candidates of level n that are dominated by at least one immediate predecessor and ii) MIP candidates of level $n-1$ that are dominated by at least one **frequent** immediate successor. This processus is called the *primary screening* as it does not exactly produce the set of frequent MIPs but only the superset of frequent MIP candidates that are not dominated by any of their immediate predecessors and frequent successors. A *secondary screening* is required to rule out MIP candidates that are dominated by at least one non-frequent immediate successor. The method is summarized on Fig. 4. It takes as input the set $\mathcal{F}$ of frequent patterns wrt threshold σ_{min} and returns the list $\mathcal{I}$ of frequent MIPs. The idea is that lists L_{-1}, L'_{-1}, L''_{-1}, and L'_0 contain successive copies of level $l-1$ (for the three first lists) and of level l (for L'_0), where each pattern is tagged by its *mip* flag, score and frequency. The *mip* tag

Data: Dataset $\mathcal{D}$, threshold σ_{min}, scoring function s, order $(\mathbb{S}, \leq_{\mathbb{S}})$, and list $\mathcal{F}$ of frequent patterns with their frequencies
Result: List $\mathcal{I}$ of frequent MIPs with scores and frequencies

Partition $\mathcal{F}$ into levels $(\mathcal{F}_l)_{0 \leq l \leq k}$ of the same length l ; Load $\mathcal{F}_0$ into list L_{-1} ;
for *l from 1 to* $k+1$ **do**
    (Clear lists L'_{-1}, L'_0, and L''_{-1}) ;
    if $l \leq k$ **then**
        Primary screening between L_{-1} (lev. $l-1$) and $\mathcal{F}_l$ (lev. l) producing resp. MIP candidates in lists L'_{-1} (lev. $l-1$) and L'_0 (lev. l) ;
        Rename L'_0 in L_{-1}
    Secondary screening of L'_{-1} producing MIPs in list L''_{-1} ;
    Append L''_{-1} to $\mathcal{I}$

Fig. 4. Algorithm computing frequent MIPs by screening frequent patterns initialized to true, may get false during primary filtering at iteration l (from $\mathcal{F}_l$ to L'_0), during primary filtering at iter. $l+1$ (from $L'_0 = L_{-1}$ to L'_{-1}) or finally during secondary filtering at iter. $l+1$ (from L'_{-1} to L''_{-1}). At that stage members of L''_{-1} are necessarily MIPs and are added to $\mathcal{I}$.

Primary filtering consists i) first in loading L_{-1} into a pattern dictionary $\mathcal{T}$ identical to the one used by the first algorithm (i.e. a pattern is mapped to its *mip* flag, score and frequency) ii) then for every pattern P of $\mathcal{F}_l$ in computing every immediate predecessor P' of P and retrieving the entry of P' from $\mathcal{T}$ (that necessarily exists as P' is necessarily frequent) iii) in comparing scores and updating accordingly *mip* flags of P and P'. In case of itemsets, computing immediate predecessors of P consists in withdrawing any attribute of P but in case of connected graphs, this requires not only to withdraw any edge of P but also to ensure the resulting graph is still connected. In other words, only edges that are not bridges may be withdrawn. Bridge edges can be identified thanks to a DFS algorithm [11] of complexity linear in the number of edges of P.

Finally secondary filtering consists given any MIP candidate P (i.e. any pattern of L'_{-1} with a true *mip* flag), in computing in one pass over the dataset $\mathcal{D}$ the frequencies of all immediate successors of P occurring at least once in $\mathcal{D}$. Scores of these successors are then computed one by one until one of these scores is larger than score of P, otherwise P is output as a MIP.

4 Experiments

Experiments aim at answering two issues. The first is the comparison of algorithm performances on reference itemset datasets, while the second is a practical and qualitative assessment of MIP relevance on a reference graph dataset.

4.1 Performance Comparison

Both algorithms can be proved to be sound and complete so that they can be distinguished only by their performance and scalability. A theoretical comparison of algorithm complexity does not allow to draw conclusions as theoretical bounds mostly rely on non-assessable measures specific to datasets (like distribution of frequent patterns over levels, number of MIP candidates that have to be processed by the secondary filtering...). For this reason, this section proposes an experimental comparison of both algorithms. In order to ease comparison, tests have been performed on reference itemset datasets. Compared to other more complex pattern families like graphs, itemsets have the advantage to be simple to process so that the risk is reduced to bias performance measurement by differences of implementation quality.

For a fair comparison, time spent for searching frequent itemset has been included in the processing time of the screening process. To this end, a version[2] of

[2] The used implementation has been written by Bart Goethals and can be downloaded from http://www.adrem.ua.ac.be/~goethals/software/

Dataset	MUSHROOM	VOTE	RETAIL	CHESS
Object number	8124	435	88162	3196
Attribute number	119	17	16470	75
Rel./abs. threshold σ_{min}	4 % / 325	0.2 % / 1	0.01 % / 9	40 % / 1279
Total time for direct extraction (s)	901	1.5	576	6810
Total time for screening MIPs (s)	747	6.8	86	1304
includ. time used by FP-Growth (s)	34	0.5	21	84
N. of frequent patterns	3.957.084	44,073	322,924	6,472,981
N. of freq. closed patterns	15.463	6478	229,303	1,366,834
N. of freq. MIPs	21	7	1045	2

Fig. 5. Test results for various datasets using the area function s_a

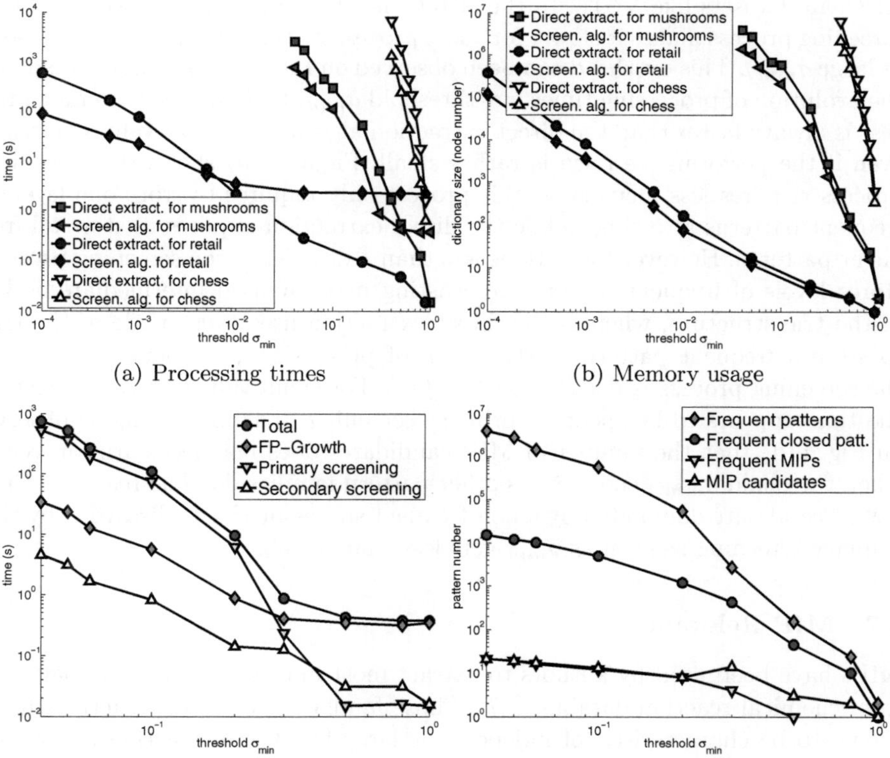

(a) Processing times

(b) Memory usage

(c) Proc. times of screening proc. steps

(d) Pattern numbers

Fig. 6. Details of the test results on log-log scales. Figures (c) and (d) focus on the MUSHROOM dataset.

`FP-growth` [12] has been used as one of the most efficient itemset mining algorithms. Tests have been run on a standard laptop (single thread on Intel Core 2, 1.8 GHz) on four datasets contrasting with each other by their size, density and purpose. All datasets are from the UCI repository, except RETAIL that is provided by Tom Brijs [13]. Table of Fig. 5 compares processing times of both algorithms and summarizes numbers of patterns when using the area function s_a. The table shows frequent MIPs are very few compared to frequent patterns (ratio from 2 to 5 decades) and closed frequent patterns (2 to 3 decades) even for sparse datasets like MUSHROOM. In case of CHESS, almost no MIPs are found as the dataset describes uncorrelated objects (i.e. winning or loosing chessboard configurations) that do not share enough common patterns to make some MIPs emerge. In comparison, datasets MUSHROOM, VOTE or RETAIL describe set of objects (resp. mushrooms, senators and customers) that are likely to build families sharing common attributes and thus to provide MIPs. Concerning efficiency, the screening process appears always faster than the direct extraction, expect for small datasets like VOTE. In the latter case, the larger time overhead of the screening process makes it slower for short processing time (i.e. for small datasets or large σ_{min}). This overhead can been observed on Fig. 6(a). The figure displays the evolution of processing times wrt threshold σ_{min}. It shows the screening process is always faster than the direct extraction algorithm for low values of σ_{min}, even if the performance ratio is rather small. Figure 6(b) shows the screening process requires less memory as this process only requires to store one level of frequent patterns at a time, while the direct extraction requires to store all frequent patterns. However the ratio is less than a decade, as the screening process stores levels of frequent patterns by wasting many unused intermediate nodes in the trie structure, whereas the direct extraction uses every node of the trie to store a frequent pattern. Distribution of processing time between steps of the screening process is detailed on Fig. 6(c). For small values of σ_{min}, most of the time appears to be spent on primary screening. It is interesting to observe on Fig. 6(d) that the number of MIP candidates does not necessarily increase when threshold σ_{min} decreases, as other pattern families do. The reason is that a MIP candidate dominated by a non-frequent successor P gets discarded by the primary screening as soon as σ_{min} gets less than $\sigma_r(P)$.

4.2 MIP Relevance

MIPs have been used by authors to extract most informative reaction patterns from chemical reaction databases [7]. Those families of chemical reactions have shown to be characteristic of independent large families of reactions. However chemical reaction processing requires to describe too many details so that we propose a somewhat simpler application that consists in extracting MIPs from 1408 molecular graphs contained in NCI DTP AIDS antiviral active and moderately active datasets (cf dtp.nci.nih.gov/docs/aids/aids_data.html) without taking into account negative examples (i.e. the inactive dataset). MIPs are extracted in two-steps, first by mining frequent subgraphs by `Gaston` [2], one of the most efficient algorithms to perform this task, and then by applying the

L_{mips} rank	L_{fcps} rank	L_{fps} rank	Score $s_i(P)$	Freq. $\sigma(P)$	Comment
1	1	1	76.4	888	Phenyl group
2	217	237	32.3	298	Sulfonyl + phenyl groups
3	224	244	31.8	401	First fragment of carbon skeleton
7	314	365	28	101	Signif. fragm. of AIDS active mol.
15	632	1344	24.2	106	Other significant fragment
53	1615	6765	22.3	116	Azo benzene group
74	2681	11528	21.7	216	Polycyclic aromatic hydrocarbon
80	3775	15046	21.4	107	Double aromatic amine
82	3837	15778	21.3	161	Sulfonic acid + phenyl groups
95	11799	38918	20.1	174	Diol group
111	37083	123812	17.5	249	Ether group
142	45806	211961	13.2	786	Carbonyl group
145	45950	213207	13.1	167	Phenyl + amide groups
152	47109	221915	12.1	270	Amide group
169	50985	237114	8.72	271	Alkene group
176	53288	241210	3.53	107	Sulfide group
177	53329	241261	2.2	211	Imine group
178	53333	241269	1.34	116	Sodium
179	53334	241270	1.27	117	Ammonium group

Fig. 7. The 19 frequent interesting MIPs and their ranks in L_{mips}, L_{fcps}, and L_{fps}

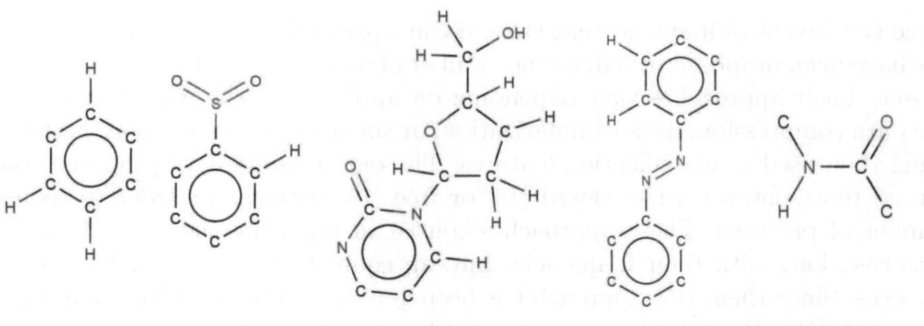

(a) 1st MIP (b) 2nd MIP (c) 7th MIP (d) 53rd MIP (e) 152nd MIP

Fig. 8. Some of the 19 frequent interesting MIPs

screening process to frequent subgraphs wrt to function s_i. For a threshold $\sigma_{min} = 100$ (7 %), the number of frequent patterns, frequent closed and MIPs are respectively 262728, 53335, and 179. Test has then consisted in reproducing the same visual analysis for the three pattern families. To this end, the three set of frequent patterns, frequent closed patterns and frequent MIPs have been sorted in decreasing order of scores in three lists L_{fps}, L_{fcps} and L_{mips}. For each list, the 179 first patterns (i.e. L_{mips} length) have been visually analyzed: a pattern has been considered as interesting only if it brings some new obvious pieces of chemical information (mostly defined in term of functional groups and cycle configurations) compared to previous patterns of higher scores. Whereas 19 MIPs have been identified as interesting in L_{mips}, all patterns from rank 2 to 179 in L_{fcps} and L_{fps} appear to be structural variations of the pattern of rank 1 (that is also the 1st MIP). This pattern is the phenyl ring shown on Fig. 8(a) that is common in molecules and is unrelated with the AIDS antiviral application. This shows how much frequent and even closed patterns are structurally redundant and not adapted to experts' visual analysis. In comparison the 179 MIPs provide 19 non-redundant interesting patterns described on Fig. 7. The increasing gap between the ranks of successive MIPs in L_{fcps} or L_{fps} gives an idea of the number of redudant patterns in those lists. Conversely a further analysis shows that each of the 1000 first closed patterns appears similar to one of the interesting MIPs. In other words, no important information appears to be lost when considering only the 19 MIPs. Some of these 19 interesting MIPs are represented on Fig. 8. The 7th MIP (cf Fig. 8(c)) is particularly interesting as it includes very specific chemical information but still appears in 71 active molecules and 30 moderately active molecules. Other MIPs appear to have various size, frequencies, and types of atoms or bonds. In particular some MIPs appear to represent well-known functional groups, e.g. amide group on Fig. 8(e).

5 Related Work

Since the advent of frequent itemsets and the Apriori algorithm [1], many methods have been proposed to reduce the number of frequent patterns to a restricted subset. Their approaches vary depending on applications these methods serve, like data compression, data summarization, or supervised classification, patterns being then used as classification features. The oldest works have proposed condensed representations like closed [14] or free [15] patterns in order to reduce number of patterns. These approaches consist in replacing the set of frequent patterns along with their frequencies into an equivalent and reduced subset of patterns. Since then, this approach has been generalized to other functions than frequency [16]. However in many practical applications, the compression gain appears insufficient to allow a direct interpretation of condensed representations by experts, especially when datasets are dense. As their direct analysis is impossible, methods have proposed to summarize set of frequent patterns by clustering frequent itemsets [17] or even graphs [18]. Other approaches have recently proposed to link pattern mining to constraint programming so that user-defined constraints can easily be injected into the mining process [6]. Whereas experts may this way focus on patterns with specific structures, pattern constraints are generally insensitive to pattern redundancy.

Recent works have been proposed to address specifically the problem of reducing pattern redundancy [4,19,20,21]. Most of these approaches aim at find a reduced set of patterns that covers (i.e. subsumes) the whole dataset: for instance Siebes et al. [4] use the MDL principle to encode transactions as unions of itemsets, whereas Bringmann et al. [21] find a basis of patterns, possibly graphs, whose the various combinations (as conjunctions of patterns) may match every transaction, one by one. Similarly Hasan et al. [20] proposes to extract from a graph dataset a basis of orthogonal (i.e. non-redundant) graph patterns with a large covering of data. In order to achieve coverage of transactions, all those methods produce patterns that are not defined on an individual basis but all together as a set of interdependent patterns. This set is generally defined as an optimum relative to some global scoring function. Optimizing such a global criterion requires a large amount of processing as the search space (i.e. the power set of the set of patterns!) is huge. For this reason, a greedy heuristic algorithm is generally used to select the next best pattern to add to the set under construction. In comparison, the MIP model contrasts on several points: first the purpose of MIPs is not covering all transactions but finding significant patterns relatively to user expectation (through a scoring function). Second the MIP model addresses the redundancy problem with considerations purely based on pattern space, not on transaction coverage. Third MIPs are defined on an individual basis, and for this reason, a complete extraction without heuristics is possible for reasonable frequency thresholds.

6 Conclusion

MIPs provide experts a very reduced set of patterns that are representative of a dataset and are not redundant compared to other families of patterns like closed patterns. In addition the model accepts a large choice of scoring functions in order to reflect representativeness wrt to expert knowledge. The method consisting in screening frequent MIPs from frequent patterns appears more efficient and more scalable than a direct extraction even if the gain varies from significant to slight levels, depending on datasets. The model has been tested on datasets made of itemsets but also of molecular graphs, and chemical reactions. In the two latter cases, MIPs have shown to provide significant patterns (i.e. molecule fragments or reaction patterns) characteristic of distinct families of objects (i.e. families of molecules or chemical reactions). However some MIPs still appear redundant for low level of scores because of noisy variations in the scoring function. Therefore we plan as a perspective, to screen more severely patterns according to their score in order to remove these artefacts.

References

1. Agrawal, R., Imielinski, T., Swami, A.N.: Mining association rules between sets of items in large databases. In: Buneman, P., Jajodia, S. (eds.) Proceedings of the 1993 ACM SIGMOD International Conference on Management of Data, Washington, D.C., pp. 207–216. ACM Press, New York (1993)
2. Nijssen, S., Kok, J.N.: A quickstart in frequent structure mining can make a difference. In: Proceedings of the Tenth ACM SIGKDD International Conference on Knowledge Discovery and Data Mining, Seattle, Washington, USA, pp. 647–652. ACM Press, New York (2004)
3. Soulet, A., Crémilleux, B.: Extraction des top-k motifs par approximer-et-pousser. In: Noirhomme-Fraiture, M., Venturini, G. (eds.) Extraction et gestion des connaissances (EGC 2007), Actes des cinquièmes journées Extraction et Gestion des Connaissances, Namur, Belgique. Volume RNTI-E-9 of Revue des Nouvelles Technologies de l'Information, Cépaduès-Éditions, January 23-26, vol. 2, pp. 271–282 (2007)
4. Siebes, A., Vreeken, J., van Leeuwen, M.: Item sets that compress. In: Ghosh, J., Lambert, D., Skillicorn, D.B., Srivastava, J. (eds.) SDM. SIAM, Philadelphia (2006)

5. Cook, D.J., Holder, L.B.: Substructure discovery using minimum description length and background knowledge. J. of Art. Intell. Res. 1, 231–255 (1994)
6. Raedt, L.D., Guns, T., Nijssen, S.: Constraint programming for itemset mining. In: Li, Y., Liu, B., Sarawagi, S. (eds.) KDD, pp. 204–212. ACM, New York (2008)
7. Pennerath, F., Napoli, A.: La famille des motifs les plus informatifs. application à l'extraction de graphes en chimie organique. Revue I3 8(2), 252 (2008)
8. McKay, B.D.: Practical graph isomorphism. Congr. Numer. 30, 45–87 (1981)
9. Zaki, M.J.: Scalable algorithms for association mining. IEEE T. Knowl. Data. En. 12(3), 372–390 (2000)
10. Borgelt, C., Berthold, M.R.: Mining molecular fragments: Finding relevant substructures of molecules. In: Proceedings of the 2002 IEEE International Conference on Data Mining (ICDM 2002), Maebashi City, Japan, vol. 51. IEEE Computer Society, Los Alamitos (2002)
11. Tarjan, R.E.: Depth-first search and linear graph algorithms. SIAM J. Comput. 1(2), 146–160 (1972)
12. Han, J., Pei, J., Yin, Y., Mao, R.: Mining frequent patterns without candidate generation: A frequent-pattern tree approach. Data Min. Knowl. Discov. 8(1), 53–87 (2004)
13. Brijs, T., Swinnen, G., Vanhoof, K., Wets, G.: Using association rules for product assortment decisions: A case study. In: KDD, pp. 254–260 (1999)
14. Pasquier, N., Bastide, Y., Taouil, R., Lakhal, L.: Efficient mining of association rules using closed itemset lattices. International Journal of Information Systems 24(1), 25–46 (1999)
15. Boulicaut, J.F., Bykowski, A., Rigotti, C.: Free-sets: A condensed representation of boolean data for the approximation of frequency queries. Data Min. Knowl. Discov. 7(1), 5–22 (2003)
16. Soulet, A., Crémilleux, B.: Adequate condensed representations of patterns. Data Min. Knowl. Discov. 17(1), 94–110 (2008)
17. Yan, X., Cheng, H., Han, J., Xin, D.: Summarizing itemset patterns: a profile-based approach. In: Grossman, R., Bayardo, R.J., Bennett, K.P. (eds.) KDD, pp. 314–323. ACM, New York (2005)
18. Chen, C., Lin, C.X., Yan, X., Han, J.: On effective presentation of graph patterns: a structural representative approach. In: Shanahan, J.G., Amer-Yahia, S., Manolescu, I., Zhang, Y., Evans, D.A., Kolcz, A., Choi, K.S., Chowdhury, A. (eds.) CIKM, pp. 299–308. ACM, New York (2008)
19. Gallo, A., Bie, T.D., Cristianini, N.: Mini: Mining informative non-redundant itemsets. In: Kok, J.N., Koronacki, J., Lopez de Mantaras, R., Matwin, S., Mladenič, D., Skowron, A. (eds.) PKDD 2007. LNCS (LNAI), vol. 4702, pp. 438–445. Springer, Heidelberg (2007)
20. Hasan, M.A., Chaoji, V., Salem, S., Besson, J., Zaki, M.J.: Origami: Mining representative orthogonal graph patterns. In: Proceedings of the 7th IEEE International Conference on Data Mining (ICDM 2007), pp. 153–162. IEEE Computer Society, Los Alamitos (2007)
21. Bringmann, B., Zimmermann, A.: One in a million: picking the right patterns. Knowledge and Information Systems (2008)

On Discriminative Parameter Learning of Bayesian Network Classifiers

Franz Pernkopf and Michael Wohlmayr

Graz University of Technology, Inffeldgasse 16c, A-8010 Graz, Austria
pernkopf@tugraz.at, michael.wohlmayr@tugraz.at

Abstract. We introduce three discriminative parameter learning algorithms for Bayesian network classifiers based on optimizing either the conditional likelihood (CL) or a lower-bound surrogate of the CL. One training procedure is based on the extended Baum-Welch (EBW) algorithm. Similarly, the remaining two approaches iteratively optimize the parameters (initialized to ML) with a 2-step algorithm. In the first step, either the class posterior probabilities or class assignments are determined based on current parameter estimates. Based on these posteriors (class assignment, respectively), the parameters are updated in the second step. We show that one of these algorithms is strongly related to EBW. Additionally, we compare all algorithms to conjugate gradient conditional likelihood (CGCL) parameter optimization [1].

We present classification results for frame- and segment-based phonetic classification and handwritten digit recognition. Discriminative parameter learning shows a significant improvement over generative ML estimation for naive Bayes (NB) and tree augmented naive Bayes (TAN) structures on all data sets. In general, the performance improvement of discriminative parameter learning is large for simple Bayesian network structures which are not optimized for classification.

1 Introduction

There are two avenues for learning statistical classifiers: Generative and discriminative approaches [2,3,4]. In generative classifiers we learn a model of the joint probability of the features and the corresponding class label and perform predictions by using Bayes rule to determine the class posterior probability. ML estimation is usually used to learn a generative classifier. Discriminative approaches model the class posterior probability directly. Hence, the class conditional probability is optimized when we learn the classifier which is most important for classification accuracy. There are several reasons for using discriminative rather than generative classifiers, one of which is that the classification problem should be solved most simply and directly, and never via a more general problem such as the intermediate step of estimating the joint distribution [5]. However, there are also a number of reasons why in certain contexts a generative model is preferred including: parameter tying and domain knowledge-based hierarchical decomposition is facilitated; it is easy to work with structured data; and it is easy to work with missing features by marginalizing over the unknown variables.

W. Buntine et al. (Eds.): ECML PKDD 2009, Part II, LNAI 5782, pp. 221–237, 2009.

The expectation-maximization (EM) algorithm is commonly used for generatively learning hidden variable models, e.g. Gaussian mixtures. It optimizes a globally valid lower bound of the likelihood function [6] which therefore guarantees an increase of the likelihood itself. The straightforward application of the EM algorithm to optimize the CL is not possible since we have to optimize a rational function and the constructed lower bounds are only locally valid, i.e. for current parameter estimates [7]. Due to this fact, convergence is not stringent. In [4] and [7] a global lower bound for *EM-like* CL optimization algorithms has been proposed.

A sufficient (but not necessary) condition for optimal classification is for the conditional likelihood (CL) to be optimized. Unfortunately, the CL function for Bayesian networks does not decompose and there is no closed-form solution for determining its parameters. In current approaches, the structure and/or the parameters are learned in a discriminative manner by maximizing CL[1]. Greiner et al. [1] express general Bayesian networks as standard logistic regression – they optimize parameters with respect to the conditional likelihood using a conjugate gradient method. Similarly, Roos et al. [8] provide conditions for general Bayesian networks under which the correspondence to logistic regression holds. An empirical and theoretical comparison of discriminative and generative classifiers (logistic regression and the NB classifier) is given in [9]. It is shown that for small sample sizes the generative NB classifier can outperform the discriminatively trained model. An experimental comparison of discriminative and generative parameter training on both discriminatively and generatively structured Bayesian network classifiers has been performed in [10].

The CL function is closely related to the maximum mutual information (MMI) criterion which is popular in the speech community [2,11]. It was proposed for hidden Markov models (HMM) [2] and attempts to maximize the posterior probability of the transcriptions given the utterances. In this context, the extended Baum-Welch (EBW) algorithm [12,13] has been introduced to discriminatively optimize HMMs. In [14], it has been applied to optimize Gaussian mixture models.

In this paper, we present three discriminative *EM-like* parameter learning methods for Bayesian network classifiers and compare them to conjugate gradient CL optimization [1]. As first method, we introduce EBW for Bayesian networks. The two remaining approaches are based on optimizing either the CL or a lower-bound surrogate of it. The algorithms are abbreviated as ECL (exact CL decomposition) and ACL (approximate CL decomposition), respectively. Our ACL algorithm has been formulated for HMMs in [11]. In fact, we can show that ECL with Laplace smoothing is closely related to EBW. Both algorithms (i.e. ECL and ACL) iteratively optimize the parameters (initialized to ML) with a 2-step algorithm similar as in [11]. In the first step, the class posterior probabilities for ECL (class assignments for ACL, respectively) are calculated based

---

[1] By "discriminative structure learning", we mean that the aim of optimization is to maximize a cost function that is suitable for reducing classification errors, such as conditional likelihood or classification rate.

on current parameter estimates. Based on these posteriors (class assignments, respectively), the parameters are updated in the second step. Additionally, we have to introduce mechanisms to avoid negative values for probabilities. However, both algorithms do not show a monotone improvement of the objective function as we have for the conjugate gradient approach [1] and the EBW method. Nevertheless, we obtain excellent results at low computational costs for two phonetic classification tasks using the TIMIT speech corpus [15] and for handwritten digit recognition using the MNIST and USPS data sets. Discriminative parameter learning significantly outperforms ML parameter estimation for NB and TAN structures. In general, the performance improvement is larger for simple structures which are not optimized for classification, i.e. for structures which are not optimized with respect to the CL or the classification rate (CR) [1].

The paper is organized as follows: In Section 2, we introduce our notation and briefly review Bayesian networks, ML parameter learning, NB, TAN, and 2-tree structures. In Section 3, we introduce all discriminative parameter learning algorithms, i.e. ECL, ACL, EBW, and CGCL. In Section 4, we present experimental results. Section 5 concludes the paper.

2 Bayesian Network Classifiers

A Bayesian network [16] $\mathcal{B} = \langle \mathcal{G}, \Theta \rangle$ is a directed acyclic graph $\mathcal{G} = (\mathbf{Z}, \mathbf{E})$ consisting of a set of nodes $\mathbf{Z}$ and a set of directed edges connecting the nodes. This graph represents factorization properties of the distribution of a set of random variables $\mathbf{Z} = \{Z_1, \ldots, Z_{N+1}\}$. Each variable in $\mathbf{Z}$ has values denoted by lower case letters $\{z_1, \ldots, z_{N+1}\}$. We use boldface capital letters, e.g. $\mathbf{Z}$, to denote a set of random variables and correspondingly boldface lower case letters denote a set of instantiations (values). Without loss of generality, in Bayesian network classifiers the random variable Z_1 represents the class variable $C \in \{1, \ldots, |C|\}$, $|C|$ is the cardinality of C or equivalently the number of classes, $\mathbf{X}_{1:N} = \{X_1, \ldots, X_N\} = \{Z_2, \ldots, Z_{N+1}\}$ denote the set of random variables of the N attributes of the classifier. In a Bayesian network each node is independent of its non-descendants given its parents [16]. Conditional independence reduces computation for exact inference on such a graph. The set of parameters which quantify the network are represented by Θ. Each node Z_j is represented as a local conditional probability distribution given its parents Z_{Π_j}. We use $\theta^j_{i|h}$ to denote a specific conditional probability table entry (assuming discrete variables), the probability that variable Z_j takes on its ith value assignment given that its parents Z_{Π_j} take their h^{th} (lexicographically ordered) assignment. That is, $\theta^j_{i|h} = P_\Theta \left(z_j = i | z_{\Pi_j} = h\right) = \prod_{i=1}^{|Z_j|} \prod_h \left(\theta^j_{i|h}\right)^{u^j_{i|h}}$, where

$$u^j_{i|h} = \begin{cases} 1, \text{ if } z_j = i \text{ and } z_{\Pi_j} = h \\ 0, \text{ otherwise} \end{cases}. \tag{1}$$

Hence, h contains the parent configuration assuming that the first element of h, i.e. h_1, relates to the conditioning class and the remaining elements $h \backslash h_1$ denote the conditioning on parent attribute values. The training data consists of M independent and identically distributed samples $\mathcal{S} = \{\mathbf{z}^m\}_{m=1}^M = \{(c^m, \mathbf{x}_{1:N}^m)\}_{m=1}^M$. The joint probability distribution of the network is determined by the local conditional probability distributions as

$$P_\Theta(\mathbf{Z} = \mathbf{z}^m) = \prod_{j=1}^{N+1} P_\Theta\left(Z_j = z_j^m | Z_{\Pi_j} = z_{\Pi_j}^m\right) = \prod_{j=1}^{N+1}\prod_{i=1}^{|Z_j|}\prod_h \left(\theta_{i|h}^j\right)^{u_{i|h}^{j,m}}, \quad (2)$$

where $u_{i|h}^{j,m}$ is the obvious extension of $u_{i|h}^j$ to the mth sample.

2.1 Generative ML Parameter Learning

The parameters of the generative model are learned by maximizing the log likelihood of the data which leads to the ML estimation of $\theta_{i|h}^j$. The log likelihood function of a fixed structure of $\mathcal{B}$ is

$$LL(\mathcal{B}|\mathcal{S}) = \sum_{m=1}^M \log P_\Theta(\mathbf{Z} = \mathbf{z}^m) = \sum_{m=1}^M \sum_{j=1}^{N+1}\sum_{i=1}^{|Z_j|}\sum_h u_{i|h}^{j,m} \log\left(\theta_{i|h}^j\right).$$

It is easy to show that the ML estimate of the parameters is

$$\theta_{i|h}^j = \frac{\sum_{m=1}^M u_{i|h}^{j,m}}{\sum_{m=1}^M \sum_{l=1}^{|Z_j|} u_{l|h}^{j,m}}, \quad (3)$$

using Lagrange multipliers to constrain the parameters to a valid normalized probability distribution, i.e. $\sum_{i=1}^{|Z_j|} \theta_{i|h}^j = 1$.

2.2 Structures

In this paper, we restrict our experiments to NB, TAN, and 2-tree classifier structures. The NB network assumes that all the attributes are conditionally independent given the class label. This means that, given C, any subset of $\mathbf{X}$ is independent of any other disjoint subset of $\mathbf{X}$. As reported in the literature [17], the performance of the NB classifier is surprisingly good even if the conditional independence assumption between attributes is unrealistic or even wrong for most of the data.

In order to correct some of the limitations of the NB classifier, Friedman et al. [17] introduced the TAN classifier. A TAN is based on structural augmentations of the NB network, where additional edges are added between attributes in order to relax some of the most flagrant conditional independence properties of NB. Each attribute may have at most one other attribute as an additional parent which means that the tree-width of the attribute induced sub-graph is unity, i.e. we have to learn a 1-tree over the attributes. The maximum number of edges added to relax the independence assumption between the attributes is $N - 1$.

A TAN network is typically initialized as a NB network and additional edges between attributes are determined through structure learning. An extension of the TAN network is the k-tree, where each attribute can have a maximum of k attribute nodes as parents. In this work, TAN and k-tree structures are restricted such that the class node remains parent-less, i.e. $C_{\Pi} = \emptyset$.

In the experiments, we apply one generative structure learning algorithm introduced in Friedman et al. [17] for constructing a TAN structure using the conditional mutual information (CMI) [18]. Additionally, we use two discriminative structure learning algorithms: a simple greedy heuristic [19] and an order-based greedy algorithm [20]. Both methods consider CR as scoring function.

3 Discriminative CL Parameter Learning

Optimizing CL is tightly connected to good classification performance. Hence, we want to learn parameters so that CL is maximized. Unfortunately, CL does not decompose as ML does. Consequently, there is no closed-form solution. The objective function of the conditional log likelihood (CLL) is

$$CLL(\mathcal{B}|\mathcal{S}) = \log \prod_{m=1}^{M} P_{\Theta}\left(C = c^m | \mathbf{X}_{1:N} = \mathbf{x}_{1:N}^m\right) = \\ \sum_{m=1}^{M} \left[\log P_{\Theta}\left(C = c^m, \mathbf{X}_{1:N} = \mathbf{x}_{1:N}^m\right) - \log \sum_{c=1}^{|C|} P_{\Theta}\left(C = c, \mathbf{X}_{1:N} = \mathbf{x}_{1:N}^m\right) \right]. \tag{4}$$

Greiner et al. [1] use a conjugate gradient descent algorithm with line-search to optimize $CLL(\mathcal{B}|\mathcal{S})$. In contrast, we aim to optimize the CL directly using an iterative *EM-like* procedure. In the following subsections, we introduce EBW, ECL, and ACL parameter learning for Bayesian networks. Additionally, we shortly review CGCL [1]. For the sake of brevity, we only notate instantiations of the random variables in the probabilities.

3.1 ECL Algorithm

We want to optimize $CLL(\mathcal{B}|\mathcal{S})$ (see Eq. (4)) under the constraints

$$\sum_{i=1}^{|Z_j|} \theta_{i|h}^{j} = 1 \qquad \forall h, j \tag{5}$$

using Lagrange multipliers ω_h^j. The Lagrangian function is given according to

$$L(\Theta, \omega) = \sum_{m=1}^{M} \left[\log P_{\Theta}\left(c^m, \mathbf{x}_{1:N}^m\right) - \log \sum_{c=1}^{|C|} P_{\Theta}\left(c, \mathbf{x}_{1:N}^m\right) \right] - \\ \sum_{j=1}^{N+1} \sum_{h} \omega_h^j \left(1 - \sum_{i=1}^{|Z_j|} \theta_{i|h}^j \right).$$

The derivative of the Lagrangian function is

$$\frac{\partial L(\Theta,\omega)}{\partial \theta^j_{i|h}} = \sum_{m=1}^{M} \left[\frac{\partial}{\partial \theta^j_{i|h}} \log P_\Theta\left(c^m, \mathbf{x}^m_{1:N}\right) - \frac{\frac{\partial}{\partial \theta^j_{i|h}} \sum_{c=1}^{|C|} P_\Theta\left(c, \mathbf{x}^m_{1:N}\right)}{\sum_{c=1}^{|C|} P_\Theta\left(c, \mathbf{x}^m_{1:N}\right)} \right] - \omega^j_h. \quad (6)$$

For TAN, NB, or 2-tree structures each parameter $\theta^j_{i|h}$ involves the class node value, either $C = i$ for $j = 1$ or $C = h_1$ for $j > 1$ where h_1 denotes the class instantiation $h_1 \in h$. Due to this fact, only one summand remains nonzero in $\frac{\partial}{\partial \theta^j_{i|h}} \sum_{c=1}^{|C|} P_\Theta\left(c, \mathbf{x}^m_{1:N}\right)$ of Eq. (6). We distinguish two cases for deriving the Lagrangian: class variable ($j = 1$) and attribute variables ($j > 1$).

Case 1. For the class variable, i.e. $j = 1$ and $h = \emptyset$, we get

$$\frac{\partial L(\Theta,\omega)}{\partial \theta^1_i} = \sum_{m=1}^{M} \left[\frac{\partial}{\partial \theta^1_i} \sum_{i=1}^{|C|} u^{1,m}_i \log\left(\theta^1_i\right) \right.$$

$$\left. - \frac{P_\Theta\left(i, \mathbf{x}^m_{1:N}\right)}{\sum_{c=1}^{|C|} P_\Theta\left(c, \mathbf{x}^m_{1:N}\right)} \frac{\partial}{\partial \theta^1_i} \log P_\Theta\left(i, \mathbf{x}^m_{1:N}\right) \right] - \omega^j_h = \sum_{m=1}^{M} \left[\frac{u^{1,m}_i}{\theta^1_i} - \frac{W^m_i}{\theta^1_i} \right] - \omega^j_h = 0, \quad (7)$$

where we use Eq. (2) for deriving the first term (omitting the sum over j and h) and we introduced the class posterior $W^m_i = P_\Theta\left(i | \mathbf{x}^m_{1:N}\right)$ as

$$W^m_i = \frac{P_\Theta\left(i, \mathbf{x}^m_{1:N}\right)}{\sum_{c=1}^{|C|} P_\Theta\left(c, \mathbf{x}^m_{1:N}\right)}.$$

Multiplying Eq. (7) by θ^1_i and summing over i, we can determine ω^j_h as

$$\omega^j_h = \sum_{m=1}^{M} \sum_{i=1}^{|C|} \left[u^{1,m}_i - W^m_i \right],$$

using the constraint of Eq. (5). Finally, we get for the parameters θ^1_i

$$\theta^1_i = \frac{\sum_{m=1}^{M} \left[u^{1,m}_i - \lambda W^m_i \right]}{\sum_{m=1}^{M} \sum_{i=1}^{|C|} \left[u^{1,m}_i - \lambda W^m_i \right]}, \quad (8)$$

where we introduced λ to weight the posterior W^m_i (to be described in the sequel).

Case 2. For the attribute variables, i.e. $j > 1$, we derive correspondingly and have

$$\frac{\partial L(\Theta,\omega)}{\partial \theta^j_{i|h}} = \sum_{m=1}^{M} \left[\frac{u^{j,m}_{i|h}}{\theta^j_{i|h}} - W^m_{h_1} \frac{v^{j,m}_{i|h\backslash h_1}}{\theta^j_{i|h}} \right] - \omega^j_h = 0,$$

where $W^m_{h_1} = P_\Theta\left(h_1|\mathbf{x}^m_{1:N}\right)$ is the posterior for class h_1 and sample m, and

$$v^{j,m}_{i|h\backslash h_1} = \begin{cases} 1, \text{ if } z^m_j = i \text{ and } z^m_{\Pi_j} = h\backslash h_1 \\ 0, \text{ otherwise} \end{cases}.$$

Employing the constraint from Eq. (5) we obtain the parameters $\theta^j_{i|h}$ as

$$\theta^j_{i|h} = \frac{\sum\limits_{m=1}^{M} \left[u^{j,m}_{i|h} - \lambda W^m_{h_1} v^{j,m}_{i|h\backslash h_1} \right]}{\sum\limits_{m=1}^{M} \sum\limits_{i=1}^{|Z_j|} \left[u^{j,m}_{i|h} - \lambda W^m_{h_1} v^{j,m}_{i|h\backslash h_1} \right]}. \tag{9}$$

Again, we introduced λ. Its value is in the range of $0 \leq \lambda \leq 1$. If we set λ to zero the second part of both equations vanishes and we obtain ML parameter learning (see Section 2.1). In the *discriminative* parameter learning case (i.e. $\lambda > 0$), parameters $\theta^j_{i|h}$ are affected by the samples m which have a large absolute value for the quantity $u^{j,m}_{i|h} - \lambda W^m_{h_1} v^{j,m}_{i|h\backslash h_1}$ (resp. $u^{1,m}_i - \lambda W^m_i$). These are the training samples belonging to class h_1 (resp. i for Eq. (8)) which have a low probability of being classified as h_1 (resp. i) under the current Θ, or the samples which are not in class h_1 (resp. i for Eq. (8)) but which have a large probability of being classified as h_1 (resp. i) [21,3]. Discriminative learning is concerned with establishing the optimal classification boundary with those samples which might be easily misclassified. Data samples which are simple to classify do not contribute much for discriminative parameter learning. In contrast, generative ML parameter learning optimizes the distribution of the samples belonging to a certain class irrespective of the samples from other classes.

A solution for Eq. (8) and (9) can be determined by the following iterative two step algorithm:

1. Estimate the posterior W^m_i ($W^m_{h_1}$ respectively) using the old parameters Θ.
2. Given W^m_i and $W^m_{h_1}$ from step 1, update the parameters according to Eq. (8) and (9).

The parameters Θ can be initialized randomly, however, empirical results showed that initialization of Θ to the ML estimates leads to better performance. Both steps are repeated iteratively until a specified number of iterations is reached. ECL parameter learning is performed once the structure of the Bayesian network is determined. For large values of λ the numerator/denominator in Eq. (8) and (9) might become negative. To tackle this, we introduce the following two strategies:

– Laplace-like smoothing: We introduce a discounting value D_h^j for variable j and conditioning parent values h for Eq. (9) (similarly for Eq. (8)) and obtain the parameters according to

$$\theta_{i|h}^j = \frac{-D_h^j + \sum_{m=1}^{M} \left[u_{i|h}^{j,m} - \lambda W_{h_1}^m v_{i|h\backslash h_1}^{j,m} \right]}{-|Z_j| D_h^j + \sum_{m=1}^{M} \sum_{i=1}^{|Z_j|} \left[u_{i|h}^{j,m} - \lambda W_{h_1}^m v_{i|h\backslash h_1}^{j,m} \right]}, \tag{10}$$

where

$$D_h^j = \min_{i \in |Z_j|} \left\{ 0, \sum_{m=1}^{M} \left[u_{i|h}^{j,m} - \lambda W_{h_1}^m v_{i|h\backslash h_1}^{j,m} \right] \right\}$$

– ML fallback: We set $\theta_{i|h}^j$ to the ML estimates of Eq. (3) for all i in case $\min_{i \in |Z_j|} \left\{ \sum_{m=1}^{M} \left[u_{i|h}^{j,m} - \lambda W_{h_1}^m v_{i|h\backslash h_1}^{j,m} \right] \right\} < 0$.

3.2 ACL Algorithm

In the approximated CLL optimization method we can find for the second term of the CL in Eq. (4), i.e. for the marginalization over C, a lower and an upper bound according to

$$\log \left(|C| \max_i F_i \right) \geq \log \sum_i F_i \geq \log \left(\max_i F_i \right).$$

Using the first inequality, we obtain a lower bound of the objective function

$$CLL\left(\mathcal{B}|\mathcal{S}\right) \geq \sum_{m=1}^{M} \left[\log P_\Theta \left(c^m, \mathbf{x}_{1:N}^m\right) - \log \max_i P_\Theta \left(i, \mathbf{x}_{1:N}^m\right) \right] - \log |C|,$$

where the last term $\log |C|$ is constant and can be neglected. Further,

$$CLL\left(\mathcal{B}|\mathcal{S}\right) \geq \sum_{m=1}^{M} \log P_\Theta \left(c^m, \mathbf{x}_{1:N}^m\right) - \sum_{c=1}^{|C|} \sum_{v \in B_c} \log P_\Theta \left(c, \mathbf{x}_{1:N}^v\right), \tag{11}$$

where the set B_c contains the indices of samples recognized as class c ($B_c \equiv \{m | c^m = \arg\max_i P_\Theta \left(i, \mathbf{x}_{1:N}^m\right)\}$). Equivalently, we introduce a modified training data set $\mathcal{S}_W$ by changing the class label of each sample m to the most probable class under the current Θ, i.e. $\mathcal{S}_W = \{(\arg\max_i P_\Theta \left(C = i, \mathbf{x}_{1:N}^m\right), \mathbf{x}_{1:N}^m)\}_{m=1}^M$. Thus, we can rewrite the lower bound of the objective function in Eq. (11) as

$$J\left(\Theta\right) = \sum_{m=1}^{M} \left[\log P_\Theta \left(c^{m,\mathcal{S}}, \mathbf{x}_{1:N}^{m,\mathcal{S}} \right) - \lambda \log P_\Theta \left(c^{m,\mathcal{S}_W}, \mathbf{x}_{1:N}^{m,\mathcal{S}_W} \right) \right],$$

where $0 \leq \lambda \leq 1$ determines the influence of the *discriminative* part of the objective function and the superscripts $\mathcal{S}$ and $\mathcal{S}_W$ refer to the corresponding data set. If $\lambda = 0$ the ML objective function is obtained (see Section 2.1). Similar as in Section 3.1, we can use Lagrange multipliers to include the parameter constraint (see Eq. (5)). The derivative of the Lagrangian leads to the expression for the parameters $\theta^j_{i|h}$ according to

$$\theta^j_{i|h} = \frac{\sum_{m=1}^{M} \left[u^{j,m}_{i|h} - \lambda w^{j,m}_{i|h} \right]}{\sum_{l=1}^{|Z_j|} \sum_{m=1}^{M} \left[u^{j,m}_{l|h} - \lambda w^{j,m}_{l|h} \right]}, \tag{12}$$

where $w^{j,m}_{i|h}$ for the class variable ($j = 1$) is

$$w^{1,m}_i = \begin{cases} 1, \text{ if } z^m_1 = i = \arg\max_{k \in C} P_\Theta\left(k, \mathbf{x}^m_{1:N}\right) \\ 0, \text{ otherwise} \end{cases}$$

and for the remaining variables ($j > 1$) we get

$$w^{j,m}_{i|h} = \begin{cases} 1, \text{ if } z^m_j = i \text{ and } z^m_{\Pi_j} = \{h^*_1 \cup h \backslash h_1\} \\ 0, \text{ otherwise} \end{cases},$$

where we replaced the original class label (i.e. h_1) by the most likely class (i.e. $h^*_1 = \arg\max_{k \in C} P_\Theta\left(k, \mathbf{x}^m_{1:N}\right)$) using the current parameter estimates. The parameters $\theta^j_{i|h}$ can be optimized in a discriminative manner using an algorithm analog to the iterative method introduced above (see Section 3.1), i.e. we have the following two steps per iteration:

1. Classify the data set $\mathcal{S}$ to establish $\mathcal{S}_W$ (which directly relates to $w^{1,m}_i$ and $w^{j,m}_{i|h}$) according to $\mathcal{S}_W = \{(\arg\max_i P_\Theta\left(C = i, \mathbf{X}_{1:N} = \mathbf{x}^m_{1:N}\right), \mathbf{x}^m_{1:N})\}^M_{m=1}$.
2. Compute new estimates of the parameters $\theta^j_{i|h}$ for all i, h, and j by optimizing $J\left(\Theta\right)$ according to Eq. (12).

3.3 CGCL Algorithm

The objective function of the conditional log likelihood is given in Eq. (4). As in [1] we use a conjugate gradient algorithm with line-search which requires both the objective function and its derivative. In particular, the *Polak-Ribiere* method is used [22]. Similar as in Section 3.1, we distinguish two cases for deriving $\frac{\partial CLL(\mathcal{B}|\mathcal{S})}{\partial \theta^j_{i|h}}$. For TAN, NB, or 2-tree structures each parameter $\theta^j_{i|h}$ involves the class node value, either $C = i$ for $j = 1$ (Case A) or $C = h_1$ for $j > 1$ (Case B) where h_1 denotes the class instantiation $h_1 \in h$. For case A and B we have the derivatives

$$\frac{\partial CLL\left(\mathcal{B}|\mathcal{S}\right)}{\partial \theta^1_i} = \sum_{m=1}^{M} \left[\frac{u^{1,m}_i}{\theta^1_i} - \frac{W^m_i}{\theta^1_i} \right], \text{and} \tag{13}$$

$$\frac{\partial CLL\left(\mathcal{B}|\mathcal{S}\right)}{\partial\theta^j_{i|h}} = \sum_{m=1}^{M}\left[\frac{u^{j,m}_{i|h}}{\theta^j_{i|h}} - W^m_{h_1}\frac{v^{j,m}_{i|h\backslash h_1}}{\theta^j_{i|h}}\right], \tag{14}$$

respectively. The probability $\theta^j_{i|h}$ is constrained to $\theta^j_{i|h} \geq 0$ and $\sum_{i=1}^{|Z_j|}\theta^j_{i|h} = 1$. We re-parameterize the problem to incorporate the constraints of $\theta^j_{i|h}$ in the conjugate gradient algorithm. Thus, we use different parameters $\beta^j_{i|h} \in \mathbb{R}$ according to

$$\theta^j_{i|h} = \frac{\exp\left(\beta^j_{i|h}\right)}{\sum_{l=1}^{|Z_j|}\exp\left(\beta^j_{l|h}\right)}.$$

This requires the gradient $\frac{\partial CLL(\mathcal{B}|\mathcal{S})}{\partial\beta^j_{i|h}}$ which is computed using the chain rule as

$$\frac{\partial CLL\left(\mathcal{B}|\mathcal{S}\right)}{\partial\beta^j_{i|h}} = \sum_{k=1}^{|Z_j|}\frac{\partial CLL\left(\mathcal{B}|\mathcal{S}\right)}{\partial\theta^j_{k|h}}\frac{\partial\theta^j_{k|h}}{\partial\beta^j_{i|h}} = \sum_{m=1}^{M}\left[u^{1,m}_i - W^m_i\right] - \theta^1_i\sum_{m=1}^{M}\sum_{c=1}^{|C|}\left[u^{1,m}_c - W^m_c\right]$$

for Case A and similarly for Case B we get the gradient

$$\frac{\partial CLL\left(\mathcal{B}|\mathcal{S}\right)}{\partial\beta^j_{i|h}} = \sum_{m=1}^{M}\left[u^{j,m}_{i|h} - W^m_{h_1}v^{j,m}_{i|h\backslash h_1}\right] - \theta^j_{i|h}\sum_{m=1}^{M}\sum_{l=1}^{|Z_j|}\left[u^{j,m}_{l|h} - W^m_{h_1}v^{j,m}_{l|h\backslash h_1}\right].$$

3.4 EBW Algorithm

The EBW algorithm uses the re-estimation equation [12,13] of the form

$$\theta^j_{i|h} \leftarrow \frac{\theta^j_{i|h}\left(\frac{\partial CLL(\mathcal{B}|\mathcal{S})}{\partial\theta^j_{i|h}} + D\right)}{\sum_{l}^{|Z_j|}\theta^j_{l|h}\left(\frac{\partial CLL(\mathcal{B}|\mathcal{S})}{\partial\theta^j_{l|h}} + D\right)}. \tag{15}$$

Considering the derivative in Eq. (14) (similar for Eq. (13)) we obtain

$$\theta^j_{i|h} \leftarrow \frac{\sum_{m=1}^{M}\left[u^{j,m}_{i|h} - W^m_{h_1}v^{j,m}_{i|h\backslash h_1}\right] + \theta^j_{i|h}D}{\sum_{m=1}^{M}\sum_{i=1}^{|Z_j|}\left[u^{j,m}_{i|h} - \lambda W^m_{h_1}v^{j,m}_{i|h\backslash h_1}\right] + D}.$$

In fact, this equation is related to the parameter estimation equation of ECL using Laplace smoothing (see Eq. (10)), i.e. no λ is needed and the value of D is set globally.

The EBW algorithm converges to a local optimum of $CLL\left(\mathcal{B}|\mathcal{S}\right)$ providing a sufficiently large value for D. Indeed, setting the constant D is not trivial. If it is chosen too large then training is slow and if it is too small the update may fail to increase the objective function. In practical implementations heuristics have been suggested [13,14]. The derivatives (Eq. (13) and (14)) are sensitive to small parameter values. Therefore, we use a more robust approximation for the derivative of Case B (similarly for Case A) as suggested in [23,13]

$$\frac{\partial CLL\left(\mathcal{B}|\mathcal{S}\right)}{\partial \theta^j_{i|h}} \approx \frac{\sum_{m=1}^{M} u^{j,m}_{i|h}}{\sum_{l=1}^{|Z_j|}\sum_{m=1}^{M} u^{j,m}_{l|h}} - \frac{\sum_{m=1}^{M} W^m_{h_1} v^{j,m}_{i|h\backslash h_1}}{\sum_{l=1}^{|Z_j|}\sum_{m=1}^{M} W^m_{h_1} v^{j,m}_{l|h\backslash h_1}}.$$

Setting D according to $D = 1 + \left|\min_{i,h,j} \frac{\partial CLL(\mathcal{B}|\mathcal{S})}{\partial \theta^j_{i|h}}\right|$ shows good performance in our experiments.

4 Experiments

We present results for frame- and segment-based phonetic classification using the TIMIT speech corpus [15] and for handwritten digit recognition using the MNIST [24] and the USPS data. In the following, we list the used structure learning algorithms for TAN and 2-trees [25]:

- TAN-CMI: Generative TAN structure learning using CMI [17].
- TAN-CR: Discriminative TAN structure learning using naive greedy heuristic [19].
- TAN-OMI-CR: Discriminative TAN structure learning using the efficient order-based heuristic [20].
- 2-tree-OMI-CR: Discriminative 2-tree structure learning using the order-based heuristic.

Once the structure has been determined discriminative parameter optimization is performed. In particular, we use the ECL and ACL parameter optimization (see Section 3.1 and 3.2). Additionally, EBW (see Section 3.4) and CGCL (see Section 3.3) parameter learning have been applied. The parameters are initialized to the ML estimates for all discriminative parameter learning methods (empirical results showed it performed better than random initialization). Similar as in [1] we use *cross tuning* to estimate the optimal number of iterations for CGCL. For ECL and ACL we perform 5 iterations — however, the best values are mostly found after the first iteration in case of ML parameter initialization. The λ for the best classification rate is empirically obtained in the range of 0.1 to 0.9. EBW uses 200 iterations. As mentioned earlier, this number depends on the choice of D. Continuous features were discretized using recursive minimal entropy partitioning [26] where the quantization intervals were determined using only the training data. Zero probabilities in the conditional probability tables are replaced with small values ($\varepsilon = 0.00001$). Further, we used the same data set partitioning for various learning algorithms.

4.1 Data Characteristics

TIMIT-4/6 Data: This data set is extracted from the TIMIT speech corpus using the dialect speaking region 4 which consists of 320 utterances from 16 male and 16 female speakers. Speech frames are classified into either four or six classes using 110134 and 121629 samples, respectively. Each sample is represented by 20 MFCC and wavelet-based features. We perform classification experiments on data of male speakers (Ma), female speakers (Fe), and both genders (Ma+Fe), all in all resulting in 6 distinct data sets (i.e., Ma, Fe, Ma+Fe $\times$ 4 and 6 classes). We use 70% of the data for training and the remaining 30% for testing. More details are given in [27].

TIMIT-39 Data: The difference to TIMIT-4/6 is as follows: The phonetic transcription boundaries specify a set of frames belonging to a particular phoneme. From this set of frames – the phonetic segment – a single feature vector is derived. In accordance with [28] 61 phonetic labels are combined into 39 classes, ignoring glottal stops. For training, 462 speakers from the standard NIST training set have been used. For testing the remaining 168 speakers from the overall 630 speakers were employed. We derive from each phonetic segment 66 features, i.e. MFCC's, Derivatives, and log duration. All together we have 140173 training samples and 50735 testing samples. Further information is given in [25].

MNIST Data: We present results for the handwritten digit MNIST data [24] which contains 60000 samples for training and 10000 digits for testing. We downsample the gray-level images by a factor of two which results in a resolution of 14×14 pixels, i.e. 196 features.

USPS Data: This data set contains 11000 uniformly distributed handwritten digit images from zip codes of mail envelopes. The data set is split into 8000 images for training and 3000 for testing. Each digit is represented as a 16×16 grayscale image, where each pixel is considered as feature.

4.2 Results

Tables 1, 2, 3, and 4 show the classification rates for TIMIT-39, MNIST, USPS, and the 6 TIMIT-4/6 data sets for various learning methods[2]. Additionally, we provide classification performances for TIMIT-4/6 employing support vector machines (SVMs) using a radial basis function (RBF) kernel[3].

Discriminative parameter learning using ECL, ACL, EBW and CGCL produces mostly a significantly better classification performance than ML parameter

[2] The average CR over the 6 TIMIT-4/6 data sets is determined by weighting the CR of each data set with the number of samples in the test set. These values are accumulated and normalized by the total amount of samples in all test sets.

[3] The SVM uses two parameters, namely C^* and σ, where C^* is the penalty parameter for the errors of the non-separable case and σ is the variance parameter for the RBF kernel. We set the values for these parameters to $C^* = 1$ and $\sigma = 0.05$. The optimal choice of the parameters and kernel function has been established during extensive experiments.

Table 1. Classification results in [%] for TIMIT-39 data with standard deviation. Best parameter learning results for each structure are emphasized using bold font.

	Parameter Learning				
Classifier	ML	ECL	ACL	EBW	CGCL
NB	61.70 ± 0.89	65.70 ± 0.87	65.33 ± 0.87	**70.35** ± 0.83	70.33 ± 0.83
TAN-CMI	65.40 ± 0.87	**66.33** ± 0.86	66.08 ± 0.86	65.38 ± 0.87	66.31 ± 0.86
TAN-OMI-CR	66.61 ± 0.86	66.94 ± 0.86	**67.41** ± 0.86	66.35 ± 0.86	66.87 ± 0.86
TAN-CR	66.78 ± 0.86	67.23 ± 0.86	**67.66** ± 0.85	66.73 ± 0.86	67.23 ± 0.86
2-tree-OMI-CR	66.94 ± 0.86	67.54 ± 0.85	**67.94** ± 0.85	66.86 ± 0.86	67.06 ± 0.86

Table 2. Classification results in [%] for MNIST data with standard deviation. Best parameter learning results for each structure are emphasized using bold font.

	Parameter Learning				
Classifier	ML	ECL	ACL	EBW	CGCL
NB	83.73 ± 0.37	87.75 ± 0.33	87.73 ± 0.33	91.65 ± 0.28	**91.70** ± 0.28
TAN-CMI	91.28 ± 0.28	92.74 ± 0.26	92.77 ± 0.26	93.21 ± 0.25	**93.80** ± 0.24
TAN-OMI-CR	92.01 ± 0.27	**93.62** ± 0.24	93.46 ± 0.25	**93.62** ± 0.24	93.39 ± 0.25
TAN-CR	92.58 ± 0.26	93.86 ± 0.24	93.69 ± 0.24	93.86 ± 0.24	**93.94** ± 0.24
2-tree-OMI-CR	92.69 ± 0.26	**93.11** ± 0.25	92.98 ± 0.26	92.93 ± 0.26	93.09 ± 0.25

Table 3. Classification results in [%] for USPS data with standard deviation. Best parameter learning results for each structure are emphasized using bold font.

	Parameter Learning				
Classifier	ML	ECL	ACL	EBW	CGCL
NB	87.10 ± 0.61	91.77 ± 0.50	91.83 ± 0.50	**94.03** ± 0.43	93.67 ± 0.44
TAN-CMI	91.90 ± 0.50	93.50 ± 0.45	93.60 ± 0.45	92.83 ± 0.47	**94.87** ± 0.40
TAN-OMI-CR	92.40 ± 0.48	94.27 ± 0.42	94.07 ± 0.43	93.73 ± 0.44	**94.90** ± 0.40
TAN-CR	92.57 ± 0.48	93.93 ± 0.44	94.13 ± 0.43	94.23 ± 0.43	**95.83** ± 0.36
2-tree-OMI-CR	94.03 ± 0.43	94.50 ± 0.42	94.60 ± 0.41	94.10 ± 0.43	**94.77** ± 0.41

learning on the same classifier structure. Especially, for cases where the structure of the underlying model is not optimized for classification [1] — the average improvement of discriminative parameter learning over ML estimation on NB and generative TAN-CMI structures is large.

In particular, for NB structures the convergent EBW and CGCL methods are superior compared to ECL and ACL. Analyzing the results of EBW and CGCL on TAN and 2-tree structures reveal that EBW may overfit the data and *cross tuning* in CGCL is too restrictive concerning the number of iterations. Hence, an alternative regularization method is required. This also explains the good performance of ECL and ACL on those structures even though both algorithms do not converge to a local optimum. *Cross tuning* in CGCL and the selection of D in EBW is time-consuming. Especially, the choice of D strongly influences the convergence rate of EBW. For the ECL and ACL methods we have to select $\lambda \in [0, \dots, 1]$. The best classification rates are mostly obtained after only 1 iteration. This renders ECL and ACL to be computationally less demanding than EBW and CGCL. For the discriminative 2-tree, discriminatively learned parameters do not help to outperform ML estimation using TIMIT-4/6 (contrary on the remaining data). Again, we suspect overfitting effects since the performance of

Table 4. Classification results in [%] for TIMIT-4/6 data with standard deviation. Best parameter learning results for each structure are emphasized using bold font.

Data set	Ma+Fe	Ma	Fe	Ma+Fe	Ma	Fe	
Number of Classes	4	4	4	6	6	6	
Classifier							Average
NB-ML	87.90 ± 0.18	88.69 ± 0.25	87.67 ± 0.25	81.82 ± 0.20	82.26 ± 0.28	81.93 ± 0.28	84.85
NB-ECL	91.36 ± 0.15	92.13 ± 0.21	90.84 ± 0.22	84.07 ± 0.19	84.75 ± 0.27	83.60 ± 0.27	87.59
NB-ACL	91.19 ± 0.16	91.81 ± 0.21	90.60 ± 0.23	83.67 ± 0.19	85.05 ± 0.26	83.48 ± 0.27	87.40
NB-EBW	91.61 ± 0.15	92.50 ± 0.21	91.15 ± 0.22	85.03 ± 0.19	86.01 ± 0.26	84.54 ± 0.27	88.27
NB-CGCL	**92.12** ± 0.15	**92.81** ± 0.20	**91.57** ± 0.22	**85.41** ± 0.18	**86.28** ± 0.26	**85.12** ± 0.26	**88.69**
TAN-CMI-ML	89.83 ± 0.17	90.20 ± 0.23	90.36 ± 0.23	82.23 ± 0.20	83.20 ± 0.28	82.99 ± 0.28	86.18
TAN-CMI-ECL	90.98 ± 0.16	91.32 ± 0.22	**91.00** ± 0.22	**83.98** ± 0.19	84.79 ± 0.27	83.71 ± 0.27	**87.42**
TAN-CMI-ACL	91.00 ± 0.16	91.36 ± 0.22	90.93 ± 0.22	83.70 ± 0.19	84.46 ± 0.27	83.52 ± 0.27	87.28
TAN-CMI-EBW	**91.35** ± 0.15	**92.01** ± 0.21	90.98 ± 0.22	83.39 ± 0.19	84.45 ± 0.27	83.38 ± 0.27	87.34
TAN-CMI-CGCL	90.96 ± 0.16	91.39 ± 0.22	90.92 ± 0.22	83.06 ± 0.20	**84.85** ± 0.27	**84.05** ± 0.27	87.22
TAN-OMI-CR-ML	91.19 ± 0.16	92.15 ± 0.21	90.51 ± 0.23	84.07 ± 0.19	84.68 ± 0.27	83.71 ± 0.27	87.52
TAN-OMI-CR-ECL	**91.72** ± 0.15	92.45 ± 0.21	**90.88** ± 0.22	**84.54** ± 0.19	85.01 ± 0.27	**84.26** ± 0.27	**87.96**
TAN-OMI-CR-ACL	91.67 ± 0.15	**92.49** ± 0.21	90.77 ± 0.22	84.37 ± 0.19	**85.09** ± 0.26	84.08 ± 0.27	87.88
TAN-OMI-CR-EBW	91.17 ± 0.16	92.48 ± 0.21	90.79 ± 0.22	83.07 ± 0.20	84.15 ± 0.27	83.33 ± 0.28	87.20
TAN-OMI-CR-CGCL	91.37 ± 0.15	92.28 ± 0.21	90.51 ± 0.23	84.00 ± 0.19	84.49 ± 0.27	83.75 ± 0.27	87.54
TAN-CR-ML	91.29 ± 0.16	91.81 ± 0.21	90.52 ± 0.23	84.35 ± 0.19	84.80 ± 0.27	83.93 ± 0.27	87.62
TAN-CR-ECL	**91.69** ± 0.15	92.17 ± 0.21	90.98 ± 0.22	**84.63** ± 0.19	**85.26** ± 0.26	**84.14** ± 0.27	**87.97**
TAN-CR-ACL	91.52 ± 0.15	92.07 ± 0.21	90.14 ± 0.23	84.59 ± 0.19	85.01 ± 0.27	83.84 ± 0.27	87.74
TAN-CR-EBW	91.03 ± 0.16	**92.32** ± 0.21	**91.07** ± 0.22	83.33 ± 0.20	84.73 ± 0.27	82.90 ± 0.28	87.27
TAN-CR-CGCL	91.29 ± 0.16	92.04 ± 0.21	90.52 ± 0.23	83.69 ± 0.19	84.83 ± 0.27	83.91 ± 0.27	87.48
2-Tree-OMI-CR-ML	**91.68** ± 0.15	**92.28** ± 0.21	91.03 ± 0.22	**84.52** ± 0.19	**85.43** ± 0.26	**84.31** ± 0.27	**88.01**
2-Tree-OMI-CR-ECL	91.53 ± 0.15	92.18 ± 0.21	90.98 ± 0.22	84.13 ± 0.19	85.00 ± 0.27	83.81 ± 0.27	87.73
2-Tree-OMI-CR-ACL	91.57 ± 0.15	92.06 ± 0.21	90.95 ± 0.22	84.34 ± 0.19	85.16 ± 0.26	84.07 ± 0.27	87.83
2-Tree-OMI-CR-EBW	91.63 ± 0.15	**92.28** ± 0.21	**91.06** ± 0.22	84.26 ± 0.19	85.22 ± 0.26	84.13 ± 0.27	87.88
2-Tree-OMI-CR-CGCL	91.28 ± 0.16	91.79 ± 0.21	90.53 ± 0.23	83.46 ± 0.19	84.48 ± 0.27	83.42 ± 0.27	87.27
SVM-1-5	**92.49** ± 0.14	**93.30** ± 0.20	**92.14** ± 0.21	**86.24** ± 0.18	**87.19** ± 0.25	**86.19** ± 0.25	**89.38**

discriminative parameter learning still improves on the training data. However, the best classification performances for TIMIT-4/6 are achieved with SVMs. One reason might be that SVMs are applied to the continuous feature domain. In contrast to SVMs, a Bayesian network is a generative model. It might be preferred since it is easy to work with missing features, parameter tying and knowledge-based hierarchical decomposition is facilitated, and it is easy to work with structured data.

Figure 1 shows the classification rate of ECL and ACL parameter learning depending on λ for TIMIT-39 on both the training and test data. A NB structure is used. The best result on the training data was obtained for $\lambda = 0.4$, whereby the corresponding CR on the test set is 65.70% and 65.33% for ECL and ACL, respectively. For large values of λ the CR drops. For ECL we use Laplace-like

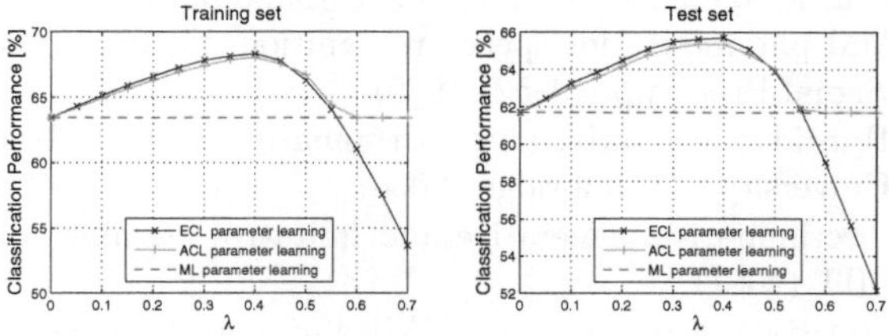

Fig. 1. Classification rate of TIMIT-39 versus λ using a NB structure

smoothing for cases where the numerator/denominator in Eq. (8) and (9) is negative, whereas for ACL we use the ML fallback. In general, there is no clear winner between ECL and ACL parameter optimization. To prevent negative parameter values during ECL and ACL optimization Laplace-like smoothing and a ML fallback scheme have been proposed. Since both strategies work best on different cases we present the best classification rate from either strategy.

5 Conclusion

We present three *EM-like* discriminative parameter learning algorithms for Bayesian network classifiers. As first method, we introduce the extended Baum-Welch algorithm. The two remaining approaches are based on iteratively optimizing either the CL or a lower-bound surrogate of the conditional likelihood. Both algorithms do not show a monotonously increasing objective function unlike the extended Baum-Welch approach. Experiments on various phonetic and handwritten digit classification tasks show that for NB and generatively and discriminatively learned TAN structures discriminative parameter optimization algorithms lead to significant improvements compared to the generative ML parameter estimation. In general, the benefit of discriminative parameter training is large for simple network structures which are not optimized for classification.

Acknowledgments

The authors thank the anonymous reviewers for many useful comments. This work was supported by the Austrian Science Fund (Grant number P19737-N15 and S10604-N13). Thanks to Jeff Bilmes for discussions and support in writing this paper.

References

1. Greiner, R., Su, X., Shen, S., Zhou, W.: Structural extension to logistic regression: Discriminative parameter learning of belief net classifiers. Machine Learning 59, 297–322 (2005)

2. Bahl, L., Brown, P., de Souza, P., Mercer, R.: Maximum Mutual Information estimation of HMM parameters for speech recognition. In: IEEE Conf. on Acoustics, Speech, and Signal Proc., pp. 49–52 (1986)
3. Lasserre, J.: Hybrid of generative and discriminative methods for machine learning. PhD thesis, University of Cambridge (2008)
4. Jebara, T.: Discriminative, generative and imitative learning. PhD thesis, Media Laboratory, MIT (2001)
5. Vapnik, V.: Statistical learning theory. Wiley & Sons, Chichester (1998)
6. Bishop, C.: Pattern recognition and machine learning. Springer, Heidelberg (2006)
7. Salojärvi, J., Puolamäki, K., Kaski, S.: Expectation maximization algorithms for conditional likelihoods. In: Inter. Conf. on Machine Learning (ICML), pp. 753–760 (2005)
8. Roos, T., Wettig, H., Grünwald, P., Myllymäki, P., Tirri, H.: On discriminative Bayesian network classifiers and logistic regression. Machine Learning 59, 267–296 (2005)
9. Ng, A., Jordan, M.: On discriminative vs. generative classifiers: A comparison of logistic regression and naive bayes. In: NIPS 14 (2002)
10. Pernkopf, F., Bilmes, J.: Discriminative versus generative parameter and structure learning of Bayesian network classifiers. In: Inter. Conf. on Machine Learning (ICML), pp. 657–664 (2005)
11. Ben-Yishai, A., Burshtein, D.: A discriminative training algorithm for hidden Markov models. IEEE Trans. on Speech and Audio Proc. 12(3), 204–217 (2004)
12. Gopalakrishnan, O., Kanevsky, D., Nàdas, A., Nahamoo, D.: An inequality for rational functions with applications to some statistical estimation problems. IEEE Transactions on Information Theory 37(1), 107–113 (1991)
13. Woodland, P., Povey, D.: Large scale discriminative training of hidden Markov models for speech recognition. Computer Speech and Language 16, 25–47 (2002)
14. Klautau, A., Jevtić, N., Orlitsky, A.: Discriminative Gaussian mixture models: A comparison with kernel classifiers. In: Inter. Conf. on Machine Learning (ICML), pp. 353–360 (2003)
15. Lamel, L., Kassel, R., Seneff, S.: Speech database development: Design and analysis of the acoustic-phonetic corpus. In: DARPA Speech Recognition Workshop, Report No. SAIC-86/1546 (1986)
16. Pearl, J.: Probabilistic reasoning in intelligent systems: Networks of plausible inference. Morgan Kaufmann, San Francisco (1988)
17. Friedman, N., Geiger, D., Goldszmidt, M.: Bayesian network classifiers. Machine Learning 29, 131–163 (1997)
18. Cover, T., Thomas, J.: Elements of information theory. John Wiley & Sons, Chichester (1991)
19. Keogh, E., Pazzani, M.: Learning augmented Bayesian classifiers: A comparison of distribution-based and classification-based approaches. In: Workshop on Artificial Intelligence and Statistics, pp. 225–230 (1999)
20. Pernkopf, F., Bilmes, J.: Order-based discriminative structure learning for Bayesian network classifiers. In: International Symposium on Artificial Intelligence and Mathematics (2008)
21. Bouchard, G., Triggs, B.: The trade-off between generative and discriminative classifiers. In: Intern. Conf. on Computational Statistics, pp. 721–728 (2004)
22. Bishop, C.: Neural networks for pattern recognition. Oxford University Press, Oxford (1995)

23. Merialdo, B.: Phonetic recognition using hidden Markov models and maximum mutual information training. In: IEEE International Conference on Acoustics, Speech, and Signal Processing (ICASSP), pp. 111–114 (1988)
24. LeCun, Y., Bottou, L., Bengio, Y., Haffner, P.: Gradient-based learning applied to document recognition. Proceedings fo the IEEE 86(11), 2278–2324 (1998)
25. Pernkopf, F., Bilmes, J.: Efficient heuristics for discriminative structure learning of Bayesian network classifiers. Technical report, Laboratory of Signal Processing and Speech Communication, Graz University of Technology (2009)
26. Fayyad, U., Irani, K.: Multi-interval discretizaton of continuous-valued attributes for classification learning. In: Joint Conf. on Artificial Intelligence, pp. 1022–1027 (1993)
27. Pernkopf, F., Van Pham, T., Bilmes, J.: Broad phonetic classification using discriminative Bayesian networks. Speech Communication 143(1), 123–138 (2008)
28. Halberstadt, A., Glass, J.: Heterogeneous measurements for phonetic classification. In: Proceedings of EUROSPEECH, pp. 401–404 (1997)

Mining Spatial Co-location Patterns with Dynamic Neighborhood Constraint

Feng Qian, Qinming He, and Jiangfeng He

College of Computer Science and
Technology, Zhejiang University, Hangzhou, China
{qfeng,hqm,jerson_hjf2003}@zju.edu.cn

Abstract. Spatial co-location pattern mining is an interesting and important issue in spatial data mining area which discovers the subsets of features whose events are frequently located together in geographic space. However, previous research literatures for mining co-location patterns assume a static neighborhood constraint that apparently introduces many drawbacks. In this paper, we conclude the preferences that algorithms rely on when making decisions for mining co-location patterns with dynamic neighborhood constraint. Based on this, we define the mining task as an optimization problem and propose a greedy algorithm for mining co-location patterns with dynamic neighborhood constraint. The experimental evaluation on a real world data set shows that our algorithm has a better capability than the previous approach on finding co-location patterns together with the consideration of the distribution of data set.

Keywords: Spatial data mining, Spatial co-location patterns, Spatial neighborhood constraint.

1 Introduction

Spatial co-location pattern mining [1] is an interesting and important issue in spatial data mining area which discovers the subsets of features (co-locations) whose events are frequently located together in geographic space. Its application domains include M-commerce, earth science, biology, public health, transportation, etc [2]. Take M-commerce as an example, in mobile computing, to provide location based services, service providers may tend to find services requested frequently located in spatial proximity.

Previous research literatures [1,2,3,4,5,6,7,8,9] for mining spatial co-location patterns require two constraints as prerequisite: the neighborhood constraint for spatial events and the prevalence constraint for spatial feature sets. As a result, two corresponding thresholds should be given from domain experts to help describe the constraints. A typical mining approach is carried out by two steps: identify the neighbor relations of spatial events with the neighborhood constraint; calculate the values of prevalence measure for different co-location candidates based on the neighbor relations of their events and find the winners whose values meet the prevalence constraint satisfaction. Apparently, some drawbacks are introduced by fixing on the neighbor relations early.

W. Buntine et al. (Eds.): ECML PKDD 2009, Part II, LNAI 5782, pp. 238–253, 2009.

1. It's hard to choose a definitely appropriate neighborhood threshold, even for domain experts. Usually, some repetitive trials should be put into practice to investigate the data sets.
2. The relationship between the neighbor relations and the prevalence measure is simply split.
3. Event types have different distributions in different areas of the global space. A static neighborhood threshold is too simple to evaluate their neighbor relations.

Hence, we suggest in this paper a heuristic approach which postpones the determination of neighbor relations to the process of the second step. A greedy algorithm is proposed to carefully select neighbor relations in stead of determining them through a uniform comparison with a simple distance based threshold. The selection strategy is based on the preferences of finding nontrivial co-locations early and the dynamic configuration of the spatial framework. Therefore, in our study of mining spatial co-location patterns, the neighborhood constraint is dynamic.

Related Work. The first general framework of mining spatial co-location patterns was proposed by Huang et al. [1] which adopts the aforementioned two-step approach. After that, different algorithms were proposed to improve the efficiency of the mining process, such as partial-join algorithm [3] and join-less algorithm [2] proposed by Yoo et al., and density based algorithm [4] proposed by Xiao et al.. In these literatures, the neighborhood constraint is described by a distance threshold which is the maximal distance allowed for two events to be neighbors. The work was extended to complex spatial co-location patterns by Munro et al. [5] and Verhein et al [6]. All these researches are limited to Huang's framework and approach.

Other researches deal with the interest measure for the prevalence of spatial co-location patterns. Huang et al. [7] adjusted the measure to treat the case with rare events. There was no change with the neighborhood constraint. Another algorithm was also given by Huang [8] that uses density ratio of different features to describe the neighborhood constraint together with a clustering approach. A buffer based model was used to describe the neighborhood constraint by Xiong et al. [9] that deal with extended spatial object such as lines and polygons. However, in the above algorithms, once the neighborhood constraint is given by user, the neighbor relations are determined and never change during the later process of evaluating the prevalence of co-locations.

On the other hand, in the field of spatial statistics, hypothesis testing methods are used to identify the correlation patterns. In Salmenkivi's work [10], the neighbor relations and prevalence measure were bound together. However, the work is not cost effective and furthermore can not handle the case of a co-location with more than two features. Sheng et al. [11] introduced the definition of influence function based on gaussian kernel to describe the neighborhood constraint. However, the algorithm assumed a distribution of features on the global space and it is expensive to compute the value of influence function.

Contributions. In this paper, we make the following contributions:

1. We conclude the preferences for selecting neighbor relations that are qualified for finding co-location patterns on early stage.
2. We define the mining task as an optimization problem and propose a greedy algorithm to mine co-location patterns with dynamic neighborhood constraint.
3. We also experimentally evaluate the algorithm on a real world data set and compare it with the previous approach.

The remainder of the paper is organized as follows: In Section 2, the relationship between neighborhood constraint and prevalence constraint is investigated together with several observations of the preferences for determining qualified neighbor relations for finding co-locations early. Section 3 define the co-location mining task as an optimization problem. A greedy algorithm for mining co-location patterns with dynamic neighborhood constraint is proposed in Section 4. Section 5 presents the experimental evaluation. And conclusion is discussed in Section 6.

2 Relationship between Neighborhood Constraint and Prevalence Constraint

In this section, we investigate the relationship between neighborhood constraint and prevalence constraint based on Huang's framework [1].The qualification of neighbor relations for detecting co-locations and the preferences for selecting qualified neighbor relations are studied. We first present some basic concepts.

Given a set of spatial features F, a set of their events E, and a neighborhood constraint, the objective of spatial co-location pattern mining is to find co-location rules in the form of $C_1 \rightarrow C_2$ (p, cp), where C_1, $C_2 \subseteq F$, and $C_1 \cap C_2 = \emptyset$ [2]. We denote the **co-location** as C=$\{C_1 \cup C_2\}$. The interest of a co-location rule can be measured by the constraints of prevalence (p) and conditional probability (cp).

The prevalence constraint is described as **participation index** (Pi) in the framework which is $Pi(C)$=$min_{f_i \in C}\{Pr(C, f_i)\}$ where $f_i \in F$ and Pr is **participation ratio** defined as $Pr(C,f_i)$=$\frac{Number\ of\ distinct\ events\ of\ f_i\ in\ instances\ of\ C}{Number\ of\ events\ of\ f_i}$ [2]. Since the conditional probability is similar defined and can be straightly measured based on prevalence constraint, we skip the discussion of it in the paper. We call an instance of C as a **clique instance** ($CI(C)$) [1] that all events in it are neighbors to each other based on the neighborhood constraint. In Huang's framework, the neighborhood constraint is described by a distance threshold which is the maximal distance allowed for two events to be neighbors.

To investigate the relationship between neighborhood constraint and prevalence constraint, we study an US National Transportation Atlas Database with Intermodal Terminal Facilities (NTAD-ITF) [12]. Every event in the data set has a location information with latitude and longitude, and is a facility with

Table 1. Description of NTAD-ITF with attributes

No.	Attribute	Description
1	ID	Event id.
2	TYPE	Name of the function of the primary function of the facility.
3	LATITUDE	Latitude for the location of the facility.
4	LONGITUDE	Longitude for the location of the facility.

Table 2. Description of NTAD-ITF with event number of facility types

No.	Facility Type	Abbreviation	Number of events
1	RAIL	R	2157
2	AIRPORT	A	398
3	TRACK	T	443
4	PORT	P	172
5	INTER PORT	I	87

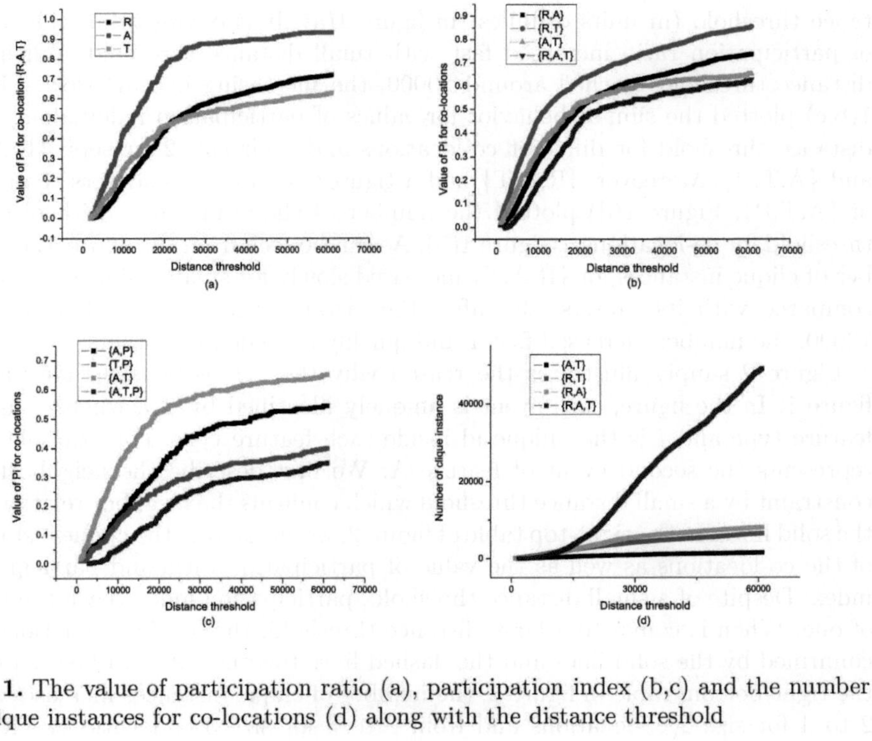

Fig. 1. The value of participation ratio (a), participation index (b,c) and the number of clique instances for co-locations (d) along with the distance threshold

Relationship between neighborhood constraint and prevalence constriant

Features	Events
A	A.1, A.2
B	B.1, B.2
C	C.1, C.2

Neighborhood constraints	Near distance	Far distance
Distance thresholds	Small	Large
Neighbor relations	Solid line	Solid line and dashed line

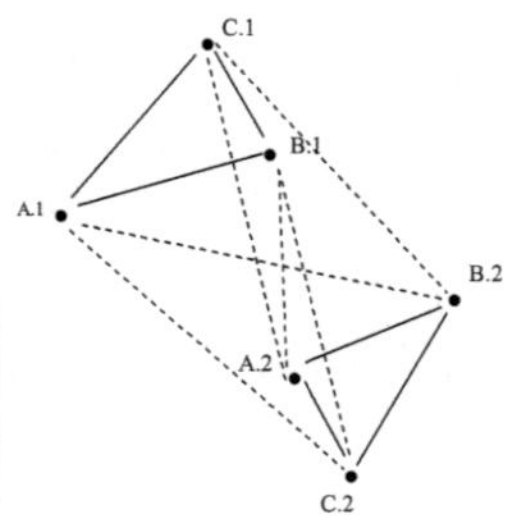

Parameters of co-locations with small distance threshold				
Co-locations	{A,B}	{A,C}	{B,C}	{A,B,C}
Clique instances	A1,B1	A1,C1	B1,C1	A1,B1,C1
	A2,B2	A2,C2	B2,C2	A2,B2,C2
Pr	2/2,2/2	2/2,2/2	2/2,2/2	2/2,2/2,2/2
Pi	1	1	1	1

Parameters of co-locations with large distance threshold				
Co-locations	{A,B}	{A,C}	{B,C}	{A,B,C}
Clique Instances (Appended instances with grey color)	A1,B1	A1,C1	B1,C1	A1,B1,C1
	A2,B2	A2,C2	B2,C2	A2,B2,C2
	A1,B2	A1,C2	B1,C2	A1,B1,C2
	A2,B1	A2,C1	B2,C1	A1,B2,C1
				A1,B2,C2
				A2,B1,C1
				A2,B1,C2
				A2,B2,C1
Pr	2/2,2/2	2/2,2/2	2/2,2/2	2/2,2/2,2/2
Pi	1	1	1	1

Fig. 2. An example to illustrate the relationship between neighborhood constraint and prevalence constraint

different types such as rail, airport, track, port and inter port. We applied the join-less algorithm [2] on the data set. The co-located relationships among these types are investigated. The geographic coordinates were transferred to projection coordinates using Universal Transverse Mercartor (UTM) projection. The detailed information about the data set is described in table 1 and table 2.

As the result of the experiments, we plotted the values of participation ratio for co-location {R,A,T} (with abbreviation described in table 2) with distance threshold (in units of miles) in figure 1(a). It is obvious that the values of participation ratio increased fast with small distance threshold. When the distance threshold reached around 20000, the increasing became slow. Figure 1(b,c) plotted the similar behavior for values of participation index along with distance threshold for different co-locations and their size 2 subsets: {R,A,T} and {A,T,P}. Moreover, {R,A,T} had a tighter bound to its subsets than that of {A,T,P}. Figure 1(d) plotted the number of clique instances with distance threshold for co-locations as figure 1(b). As can be seen from the figure, the number of clique instances for {R,A,T} increased slowly for small distance threshold compared with its subsets. But after the distance threshold reaching around 12000, the number increased faster and quickly exceeded the others.

Figure 2 simply illustrates the reason why the curves behave that way in figure 1. In the figure, each event is uniquely identified by $T.i$, where T is the feature type and i is the unique id inside each feature type. For example, A.2 represents the second event of feature A. We first describe the neighborhood constraint by a small distance threshold which confirms the neighbor relations by the solid lines. In the right-top table of figure 2, we enumerate the clique instances of the co-locations as well as the value of participation ratio and participation index. Despite of a small distance threshold, participation index reach the value of one. When it comes to a large distance threshold, the neighbor relations are confirmed by the solid lines and the dashed lines together. As can be seen from the right-bottom table in figure 2, the number of clique instances increases from 2 to 4 for size 2 co-locations and from 2 to 8 for size 3 co-locations. That is why in figure 1 (d) the long co-location run faster than the short co-locations as distance threshold increased. Because long co-locations have exponentially more combinations of clique instances than short co-locations based on the same set of events. On the contrary, the appended clique instances benefitting from the increasing of distance threshold contribute much less (nothing in the example) to the value of participation ratio as well as participation index. That is why the curves in figure 1 (a,b,c) increased slowly for large distance threshold.

Based on the analysis, we conclude the preferences for selecting qualified neighbor relations that are in favor of finding nontrivial co-locations.

1. Near distance neighbor relations are preferred.
2. Neighbor relations that benefit the value of prevalence measure are preferred.
3. Co-locations whose values of prevalence measure keep tight to that of their subsets are preferred.
4. A big gap of numbers of clique instances between co-locations and their subsets implies too much noise.

The first preference is obvious due to Tobler's first law of geography: everything is related to everything else but nearby things are more related than distance things [13]. Since the algorithms tend to find nontrivial co-locations that stand for high values of prevalence measure on early stage, the second preference promises. We prove preference 3 by comparing {R,A,T} and {A,T,P} with figure 1 (b,c). Recall the apriori approach [14] adopted by almost all of the algorithms for mining co-location patterns presently. The size k candidate co-locations are generated from their size k-1 subset co-locations as well as their instances. If these candidates are pruned a lot in the validation process or even in the earlier generated process, their values of prevalence measure leave far away from that of their size k-1 subsets, as to other shorter subsets. Hence, as can be seen from the figure 1 (b,c), domain experts may prefer {R,A,T} to {A,T,P} as a nontrivial co-location since it had a tighter bound value of prevalence measure on its subsets. The fourth observation may not be a preference but a condition the algorithm is supposed to stop its trial for those co-locations, since it imposes that the new generated clique instances benefit from far neighbor relations and they also contribute less to the value of prevalence measure.

3 Co-location Pattern Mining as an Optimization Problem

Based on the observations we made on the preferences that algorithms rely on for finding nontrivial co-locations, we can intuitively define the mining task as an optimization problem. The problem aims to discover on early stage as much as co-locations with high values of prevalence measure and near distance neighbor relations for their instances.

Specifically, we describe the task into a placement problem. Initially, the events in the spatial framework have no neighbor relation with each other. Then the neighbor relations are placed back in order to the framework. Once any neighbor relation is placed, the configuration of the framework is refreshed together with the reward of the placement. We measure the reward of the placement by the benefits of values of prevalence measure of the co-locations over the distances of neighbor relations as follows.

$$R(\alpha) = \sum_{C \subseteq F} (\theta^{-k} \cdot f(C) \cdot t(C) \cdot \sum_{d_j \in N_C} \frac{\Delta_{d_j} Pi(C)}{|d_j|^{\omega}}) \tag{1}$$

In equation 1, α is a placement and F represents the set of spatial features. Every candidate co-location C is first weighted by the step function f that is defined following (k > 2).

$$f(C_k) = \begin{cases} 1 & \text{if } Pi(C_k) \geq (\eta \cdot E[Pi(C_{k-1})]) \; where \; C_{k-1} \subset C_k \\ 0 & \text{otherwise} \end{cases} \tag{2}$$

Here, C_k is a size k co-location and $E[Pi(C_k)]$ is the average value of prevalence measure for the co-locations. The function simply prunes the size k co-locations whose values of prevalence measure leave far away from that of their size k-1 subsets. The coefficient η is used to adjust the extent of value difference which is between 0 and 1. It is obvious that algorithms only consider the benefit value as reward when the absolute value of prevalence measure is high enough. This corresponds to the preference 2 mentioned before.

In equation 1, function t weights the value difference of size k co-locations with their size k-1 subsets as described following (k > 2). It corresponds to the third preference mentioned in the last section. Due to the monotonic property of participation index [1], the value of function t is between 0 and 1.

$$t(C) = \frac{Pi(C_k)}{E[Pi(C_{k-1})]} \; where \; C_{k-1} \subset C_k \tag{3}$$

In equation 1, N_C represents the set of neighbor relations that participate in the instances of co-location C. For each neighbor relation, we calculate the incremental values of prevalence measure of co-locations that benefit from placing it to the framework. The distance of neighbor relation d_j is a penalty to the value in the equation which corresponds to the first preference. ω is a coefficient emphasizing the weight of distance to the benefit value which is usually between 0 and 1. And $\Delta_{d_j} Pi(C)$ is the incremental value of $Pi(C)$ from placing d_j into the current framework. Note, the neighbor relations are placed in selected order. Therefore the values are accumulated to evaluate the reward of placement. θ is a coefficient to adjust the user's bias on long co-locations which is between 0 and 1. In the equation, k is the size of the co-location C. The estimations for both function f and t could be skipped for size 2 co-locations since there are not comparable values of prevalence measure for them with size 1 co-locations.

As described above, the co-location mining task can be made into a optimization problem and the greedy algorithm can be applied to maximize the marginal gain when select neighbor relations for spatial framework.

$$d_i = \underset{d_i \in D \backslash \alpha_{i-1}}{\text{argmax}} \; R(d_i) = \underset{d_i \in D \backslash \alpha_{i-1}}{\text{argmax}} \; R(\alpha_{i-1} \cup d_i) - R(\alpha_{i-1}) \tag{4}$$

Here, D is the set of neighbor relations and d_i is the neighbor relation the greedy algorithm selects in the ith iteration based on *i-1*th configuration. To make the algorithm feasible, we simplify the evaluation of neighbor relations in each iteration as follows.

$$R(d_i) = \frac{1}{|d_i|^{\omega}} \cdot \sum_{C \subseteq F} (\theta^{-k} \cdot f_{i-1}(C) \cdot t_{i-1}(C) \cdot \Delta_i Pi(C)) \tag{5}$$

The deformation of function f and function t indicates that the values of the functions are calculated based on *i-1*th configuration of the spatial framework.

Recall the fourth preference presented in section 2. The ratio of the average number of clique instances of a co-location's subsets over the number of clique instances of itself could give us a signal that whether the evaluation of the co-location should be stopped. As a constraint to this optimization problem, we can simply implement this by comparing the value of ratio with a threshold κ. If the value is below κ, the evaluation of the co-location should be skipped in the iterations (k > 2) as presented following.

$$Evaluate(C_k) = \begin{cases} 1 & \text{if } \frac{E[|CI(C_{k-1})|]}{|CI(C_k)|} \geq \kappa \; where \; C_{k-1} \subset C_k \\ 0 & \text{otherwise} \end{cases} \tag{6}$$

The estimation can be easily embedded into equation 1 and equation 5 as a multiplication factor. If all the co-locations reach the line, the algorithm terminates. Other estimations can be applied in a similar way, such as defining a maximum value for the prevalence measure. Actually, as to this heuristic approach, the user can pause the estimation at any time and check the values of prevalence measure. Then, the algorithm could go on from the pausing point.

4 A Greedy Algorithm for Co-location Pattern Mining

Thus, a greedy algorithm could be carried out naturally. However, the problem is still not submodular [15] due to the possible incremental benefits for the remaining unplaced neighbor relations after refreshing the configuration when some neighbor relations were selected into the framework. Hence, the lazy evaluation approach proposed by Leskovec et al. [15] can not be adopted to speed up the algorithm.

Despite this, we propose to select a set of neighbor relations at each placing instead of a single one. We also propose to evaluate the neighbor relations in a qualified buffer instead of all the neighbor relations. The intuition comes from the thinking that we are not actually interested in maximizing the final reward of the placement but to observe as much as nontrivial co-locations. Especially on early stage, too many re-evaluations exhaust the computation, whereas a precise rank of the qualified neighbor relations is unnecessary since they will eventually take part in the early placement. Oppositely, it's also not necessary for unqualified neighbor relations to attend in early evaluations.

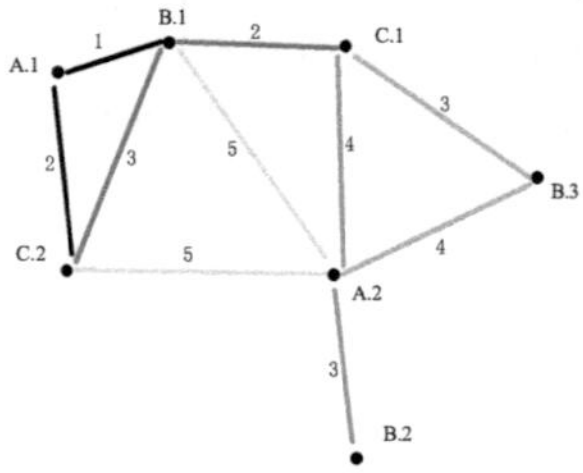

Features	Events	Coefficients	Values
A	A.1, A.2	θ	1
B	B.1, B.2, B.3	η	0
C	C.1, C.2	ω	1

Note 1:
1. Weights on the lines indicate neighbor distances;
2. Deepness of lines indicates the order of placement;
3. Other wasted neighbor relations are ignored in the figure.

Rank of neighbor relations for sequential configurations					
Rank 0	Disance	Rank 1	R(d1)	Rank 2	R(d2)
A.1,B.1	1	A.1,B.1	0.33	A.1,B.1	X
A.1,C.2	2	A.1,C.2	0.25	A.1,C.2	X
B.1,C,1	2	B.1,C.1	0.17	B.1,C.2	0.22
A.2,B.2	3	A.2,B.2	0.11	B.1,C.1	0.17
B.1,C.2	3	B.1,C.2	0.11	A.2,B.2	0.11
B.3,C.1	3	B.3,C.1	X	B.3,C.1	0.11
A.2,B.3	4	A.2,B.3	X	A.2,B.3	0.08
A.2,C.1	4	A.2,C.1	X	A.2,C.1	X
A.2,B.1	5	A.2,B.1	X	A.2,B.1	X
A.2,C.2	5	A.2,C.2	X	A.2,C.2	X

Rank of neighbor relations for sequential configurations					
Rank 3	R(d3)	Rank 4	R(d4)	Rank 5	R(d5)
A.1,B.1	X	A.1,B.1	X	A.1,B.1	X
A.1,C.2	X	A.1,C.2	X	A.1,C.2	X
B.1,C,2	X	B.1,C.2	X	B.1,C.2	X
B.1,C.1	X	B.1,C.1	X	B.1,C.1	X
A.2,C.1	0.13	A.2,C.1	X	A.2,C.1	X
A.2,B.2	0.11	A.2,B.2	X	A.2,B.2	X
B.3,C.1	0.11	B.3,C.1	0.11	B.3,C.1	X
A.2,B.3	0.08	A.2,B.3	0.08	A.2,B.3	X
A.2,B.1	0	A.2,B.1	0	A.2,B.1	0
A.2,C.2	X	A.2,C.2	0	A.2,C.2	0

Clique Instances after sequentially placing neighbor relations					
Selection	N.R.	{A,B}	{A,C}	{B,C}	{A,B,C}
1	A.1,B.1	A1,B1			
	A.1,C.2		A.1,C.2		
2	B.1,C.2			B.1,C.2	A.1,B.1,C.2
	B.1,C.1			B.1,C.1	
3	A.2,C.1		A.2,C.1		
	A.2,B.2	A.2,B.2			
4	B.3,C.1			B.3,C.1	
	A.2,B.3	A.2,B.3			A.2,B.3,C.1
5	A.2,B.1	A.2,B,1			A.2,B.1,C.1
	A.2,C.2		A.2,C.2		A.2,B.1,C.2

Rewards of sequential placements					
Placement	{A,B}	{A,C}	{B,C}	{A,B,C}	Reward
1	1/3	1/2	0	0	0.55
2	1/3	1/2	1/3	1/3	0.95
3	2/3	1	1/3	1/3	1.19
4	1	1	2/3	2/3	1.38
5	1	1	2/3	2/3	1.38

Note 2:
1. Buffer size = 5, selection size = 2;
2. Purple indicates rank buffer in reward order;
3. Green indicates neighbor relations not in the buffer but in distance order;
4. Bule indicates clique instances genereated by neighbor ralations inside a selection of placing.

Fig. 3. An example to illustrate processes of greedy algorithm for placement problem

An example is illustrated in figure 3 where we select 2 neighbor relations at each placing step and the buffer size is 5. Initially, the neighbor relations are sorted in distance order which coarsely estimates their qualifications. Then we fill the buffer with the first 5 neighbor relations in distance order. The evaluation is carried out on the buffer and the neighbor relations inside the buffer are sorted by reward order.

The rewards are calculated following the equation 5 where the values of coefficients are described in figure. Take {B.1,C.1} as an example, its reward can be calculated by dividing the benefit of *Pi*({B,C}) by its distance which is $(1/3)/2 = 0.17$. We select the first 2 neighbor relations in the ranked buffer to place into the framework. For example, after the first evaluation, the neighbor relations {A.1,B.2} and {A.1,C.2} are selected into the framework since they gain the top rewards of 0.33 and 0.25. After that, the statistics for the clique instances together with the values of prevalence measure are updated.

In the next iteration, the first 2 neighbor relations outside the buffer but in distance order are chosen to make up the buffer. In the example, they are {B.3,C.1} and {A.2,B.3}. Then the similar process is implemented in buffer. As we stated above, possible incremental benefits may be introduced for some neighbor relations. In the example, since the placed neighbor relations {A.1,B.2} and {A.1,C.2} could enhance {B.1,C.2} with potential benefit of *Pi*({A,B,C}), its reward increases and rank position arises. Therefore, iteration by iteration, sets of qualified neighbor relations are selected into the framework and the algorithm goes on.

However, it's unreasonable to calculate the final reward by simply accumulating the evaluated rewards of the selected neighbor relations. The estimated reward of a selection can be effected by two opposite cases. In the first case, some neighbor relations in the same selection may share the same benefit. For example, {B.1,C.2} and {B.1,C.1} share the same benefit of *Pi*({B,C}) with value of 1/3 in the second selection. On the contrary, in the second case, some clique instances could be generated by neighbor relations inside a selection which introduces more benefits. In the above example, {A.2,B.3,C.1} is generated by {B.3,C.1} and {A.2,B.3} inside the fourth selection. Hence, a compromise should be made to meet both the cases. We give the estimation of the reward for a selection as follows.

$$R(S) = \frac{\sum_{di \in S} R(di) \cdot |di|^{\omega}}{E[|di|^{\omega}]} \tag{7}$$

The estimation equation calculates the weighted mean for the evaluated reward of each neighbor relation. In the equation, S is the set of neighbor relations within a selection. And $E[|di|^{\omega}]$ is the average value of $|di|^{\omega}$ where di is a neighbor relation within a selection that $di \in S$. As an example in figure 3, the estimated reward of the first selection can be calculated by adding the weighted rewards of {A.1,B.1} and {A.1,C.2}, then dividing the sum by their average distance which is $(0.33 \times 1 + 0.25 \times 2)/(1.5) = 0.55$. Therefore, the reward of the placement can be calculated by accumulating the estimated rewards of the selections.

Actually, there's no need to know all the distances of neighbor relations in advance. We only need to keep a set of neighbor relations with the number of buffer size at each iteration. Hence, two operations are required: initially find the first *bsize* shortest neighbor relations for buffer where *bsize* the buffer size; at each iteration find the first *ssize* shortest neighbor relations in the rest neighbor relations where *ssize* is the selection size. The detailed algorithms are described following.

Algorithm 1. Find the first *bsize* shortest neighbor relations for buffer

```
1 for every event e in framework do
2     e' = find_distinct_nearest_neighbor(e);
3     insert_into_queue(make_neighbor_relation(e, e'));
4 sort_queue();
5 while size of buffer < bsize do
6     append_to_buffer(find_next_shortest_neighbor_relation());
7 return buffer;
```

Algorithm 1 first find the nearest neighbor for each event in the spatial framework and insert the neighbor relation into a queue. The event is called the reference event of the neighbor relation. If the found neighbor relation is duplicated with an existed one in the queue, it is discarded and the next nearest neighbor for this event is pursued until a distinct neighbor relation is found ***(Line 1-3)***. The process of finding the nearest neighbor can be implemented by plane sweeping [16]. After that, the queue is sorted in distance order and at each iteration the shortest neighbor relation outside the buffer is found through invoking method in algorithm 2 and appended to the buffer. The iteration goes on until the buffer is full ***(Line 4-6)***. Finally, the buffer is returned ***(Line 7)***.

Algorithm 2. Find the next shortest neighbor relation

1 d = pop_top_neighbor_relation_in_queue();
2 e' = find_next_distinct_nearest_neighbor(reference event e of d);
3 insert_into_queue_by_order(make_neighbor_relation(e, e'));
4 **return** d;

Algorithm 2 first pop out the top neighbor relation in the sorted queue (***Line 1***). For this neighbor relation, the next distinct nearest neighbor of its reference event is found. This neighbor relation should not be duplicated with those found before (***Line 2***). Then the neighbor relation is inserted into the queue by distance order and finally the answer is returned (***Line 3-4***). The second operation can be implemented by invoking algorithm 2 *ssize* times.

Next, we draw the greedy algorithm for mining co-location patterns following.

Algorithm 3. The greedy algorithm for mining co-location patterns

1 init_buffer(*bsize*);
2 **while** *not reach termination conditions* **do**
3   evaluate_buffer(θ, η, ω);
4   sort_buffer_by_reward_order();
5   pop_top_neighbor_relations_in_buffer_to_framework(*ssize*);
6   update_configuration();
7   make_up_buffer(*ssize*);
8 **return** co-locatins whose prevalence value above prevalence threshold;

Algorithm 3 first initializes the buffer by invoking the method of algorithm 1 (***Line 1***). Then the iteration starts by evaluating the buffer. Every neighbor relation in the buffer is evaluated following equation 5 together with a set of coefficients (***Line 3***). Then, the neighbor relations in the buffer are sorted in the evaluated reward order (***Line 4***). The top *ssize* neighbor relations are popped out and placed into the spatial framework (***Line 5***). After that, the algorithm updates the statistical information for the configuration of the spatial framework which includes the updated clique instances, the values of prevalence measure and the accumulated reward of the current placement (***Line 6***). By invoking the method of algorithm 2, the next *ssize* shortest neighbor relations are found to make up the buffer (***Line 7***). Finally, when termination conditions are met, the co-locations whose prevalence value above the prevalence threshold are returned (***Line 2,8***).

In the process of evaluating neighbor relations, the benefits of values of prevalence measure are calculated by first finding new clique instances introduced by the evaluated neighbor relation. The size 2 clique instances are naturally ready. Besides, the join approach [1] is used to generate longer clique instances. We skip the detailed explanation to save the space.

Table 3. Experimental parameters and their values in experiments

Parameters	Description	Ex1	Ex2	Ex3	Ex4	Ex5
θ	A coefficient in equation 1.	0.2				
η	A coefficient in equation 2.	0.05				
ω	A coefficient in equation 1.	0.2				
bsize	Size of buffer.	1000,1500,2000			500-2000	1000
ssize	Size of selection.	75,113,150			100	100-400
sratio	Selection ratio from buffer.	0.075			*	

5 Experimental Evaluation

In this section, we evaluate the greedy algorithm on a real world data set that is a US National Transportation Atlas Database with Intermodal Terminal Facilities (NTAD-ITF) [12]. Detailed information about the data set is described in table 1 and table 2 in section 2. The co-located relationships among the 5 types of facilities are investigated. A previous efficient algorithm (join-less algorithm [2]) based on Huang's framework [1] is also evaluated as a comparison.

The algorithms are implemented in C++ and are memory based algorithm. All the experiments were performed on an Intel Core 2 Duo 2.2GHz E4500 machine with 2G MB main memory, running Windows XP operation system. Detailed information about experiments is listed in table 3.

5.1 Capability of the Algorithms for Finding Co-location Patterns

We first evaluated the capability of the algorithms for finding co-location patterns. The capability can be specified in four ways following.

1. Detecting the high values of prevalence measure for co-locations with less neighbor relations participated in the algorithm.
2. Detecting the high values of prevalence measure for co-locations with less clique instances involved in the algorithm.
3. Finding the co-location patterns considering the distribution of the data set.

We first compared the capability of our greedy algorithm and the join-less algorithm with the number of neighbor relations and the number of clique instances. The join-less algorithm was evaluated with several repetitive executions. In each repetition, the distance threshold was incremented with a uniform value. We also compared the algorithms with the average distance of the neighbor relations which partially reflects the distribution of the neighbor relations that the algorithms selected. Detailed information is described in table 3 (Ex1-Ex3).

Capability with the number of neighbor relations. In the first experiment, we evaluated the prevalence values of the co-locations {R,A} and {R,A,T} with the number of neighbor relations as illustrated in figure 4(a) and figure 4(b) respectively. In the figure, different curves represent different algorithms. We plotted three curves of the greedy algorithm with different buffer sizes but the same

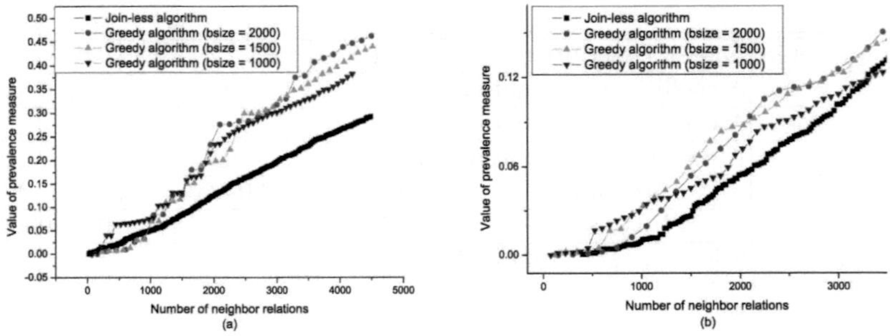

Fig. 4. The value of prevalence measure with the number of neighbor relations for co-location {R,A} (a) and co-location {R,A,T} (b)

selection ratio for the buffer. As can be seen from the figure, the greedy algorithm detected the higher value of prevalence measure with less neighbor relations involved in the calculation due to its preferences of selection. The algorithm tended to select neighbor relations in the density areas of the spatial framework on early stage, since they could benefit more values of prevalence measure. For co-location {R,A,T}, when the number of neighbor relations increased after 3000, the value difference of the two algorithms became smaller. That's because when too many neighbor relations participated in the calculation, many noises as presented in section 2 were also introduced. Then the greedy algorithm tended to select neighbor relations outside the former density areas. However on early stage, the greedy algorithm is capable of finding the co-locations with high values of prevalence measure than join-less algorithm. When buffer size became larger, the greedy algorithm had a better capability, since more qualified neighbor relations were participated in the evaluations which refined the selections.

Capability with the number of clique instances. In the second experiment, we evaluated the prevalence values of the co-locations {R,A} and {R,A,T} with the number of clique instances as illustrated in figure 5(a) and figure 5(b) respectively. As can be seen from the figure, the greedy algorithm detected the higher value of prevalence measure with very less clique instances, since the greedy algorithm selected the neighbor relations into the spatial framework that benefit the values of prevalence measure instead of many noise neighbor relations that could introduce the clique instances whose events have already participated in the calculation of the participation ratios.

Capability with the average distance of the neighbor relations. In the third experiment, we evaluated the number of neighbor relations with the average distance of the neighbor relations as illustrated in figure 6. As can be seen from the figure, the average distance of the greedy algorithm increased faster than join-less algorithm along with the number of neighbor relations. Because in the selection process of the greedy algorithm, the greedy algorithm not only considered the distances of the neighbor relations, but also took into account their quality of detecting high value of prevalence measure which potentially imposes

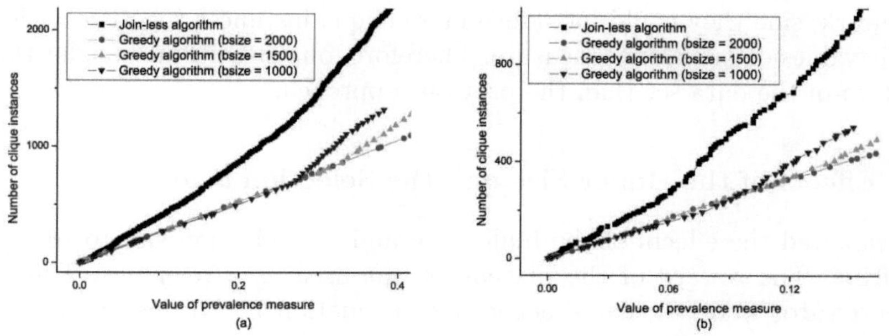

Fig. 5. The number of clique instances with the value of prevalence measure for co-location {R,A} (a) and co-location {R,A,T} (b)

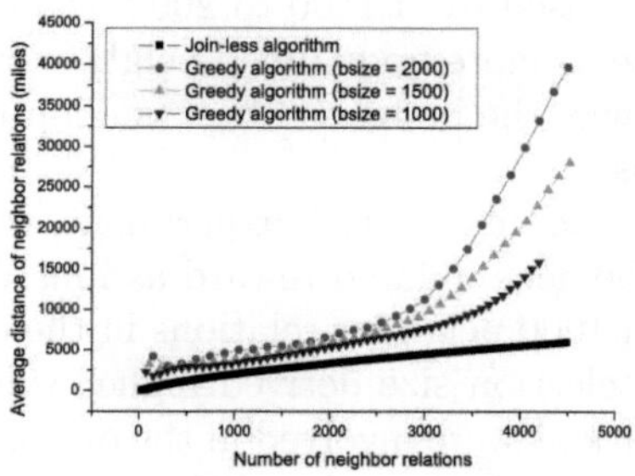

Fig. 6. The average distance of neighbor relations with number of neighbor relations

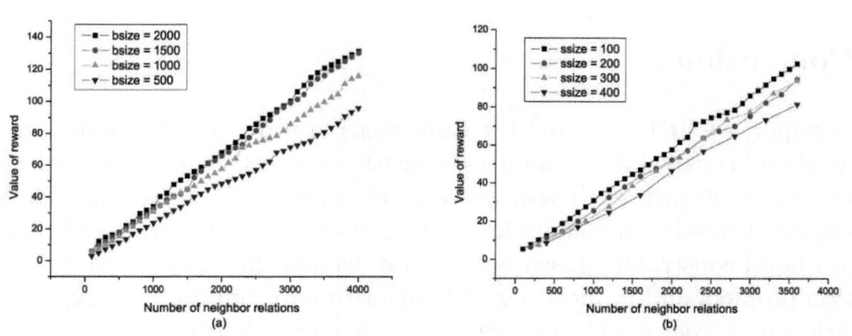

Fig. 7. The value of accumulated reward with the number of neighbor relations for different buffer sizes (a) and different selection sizes (b)

the nontrivial co-locations. Thus, from another point of view, the greedy algorithm tended to select the neighbor relations in density areas of the spatial framework, sine they could introduce more clique instances together with more benefit values of prevalence measure. Therefore, our algorithm consider the distribution of the data set than the previous approach.

5.2 Effects of the Buffer Size and the Selection Size

We evaluated the effects of the buffer size and the selection size to the greedy algorithm. The reward of the placement was used as a reference. The values of the rewards were calculated according to equation 7. Detailed information is described in table 3 (Ex4-Ex5).

Effect of the buffer size. In the fifth experiment, we evaluated the effect of the buffer size with the accumulated reward as illustrated in figure 8(a). We fixed the selection size with 100 neighbor relations in the experiment. As can be seen from the figure, algorithm with larger buffer size derived higher value of reward, since more qualified neighbor relations participated in the evaluation and rank. When the buffer size increased from 1500 to 2000, the incremental value of the reward became small due to more unqualified neighbor relations involved in the rank. That's the reason why the buffer strategy is adopted instead of evaluating all the neighbor relations.

Effect of the selection size. In the sixth experiment, we evaluated the effect of the selection size with the accumulated reward as illustrated in figure 8(b). We fixed the buffer size with 1000 neighbor relations in the experiment. The greedy algorithm with smaller selection size derived higher value of reward, since less unqualified neighbor relations were involved in the evaluation. When the number of neighbor relations increased, the value difference became larger due to the increasing of the number of unqualified neighbor relations.

We skip the discussion of the performance of the algorithms to save the space. We also skip the discussion of the effects of the coefficients (θ, η, and ω) to the greedy algorithm to save the space.

6 Conclusion

In this paper, we first proposed the drawbacks introduced by assuming a static neighborhood constraint for mining co-location patterns which adopted by previous research literatures. Then, we concluded the preferences that algorithms rely on when making decisions for mining co-location patterns with dynamic neighborhood constraint. Based on this, we defined the mining task as an optimization problem and propose a greedy algorithm for mining. The experimental evaluation on a real world data set shows that our algorithm has a better capability than previous algorithms on finding co-location patterns together with the consideration of the distribution of data set. The algorithm is also flexible for other prevalence constraints besides the participation index [1], such as the constraint with density ratio [8] and the constraint with buffer overlap [9].

References

1. Huang, Y., Shekhar, S., Xiong, H.: Discovering colocation patterns from spatial datasets: A general approach. IEEE Transactions on Knowledge and Data Engineering 16(12), 1472–1485 (2004)
2. Yoo, J., Shekhar, S.: A Joinless Approach for Mining Spatial Colocation Patterns. IEEE Transactions on Knowledge and Data Engineering 18(10), 1323–1337 (2006)
3. Yoo, J., Shekhar, S., Smith, J., Kumquat, J.: A partial join approach for mining co-location patterns. In: Proceedings of the 12th annual ACM international workshop on geographic information systems, pp. 241–249 (2004)
4. Xiao, X., Xie, X., Luo, Q., Ma, W.: Density based co-location pattern discovery. In: Proceedings of the 16th ACM SIGSPATIAL international conference on advances in geographic information systems (2008)
5. Munro, R., Chawla, S., Sun, P.: Complex spatial relationships. In: Proceedings of the 3th IEEE international conference on data mining, pp. 227–234 (2003)
6. Verhein, F., Al-Naymat, G.: Fast Mining of Complex Spatial Co-location Patterns Using GLIMIT. In: Proceedings of the 7th international conference on data mining workshop on spatial and spatio-temporal data mining, pp. 679–684 (2007)
7. Huang, Y., Pei, J., Xiong, H.: Mining Co-Location Patterns with Rare Events from Spatial Data Sets. GeoInformatica 10(3), 239–260 (2006)
8. Huang, Y., Zhang, P., Zhang, C.: On the Relationships between Clustering and Spatial Co-location Pattern Mining. International Journal on Artificial Intelligence Tools 17(1), 55 (2008)
9. Soung, X.: A framework for discovering co-location patterns in data sets with extended spatial objects. In: Proceedings of the 4th SIAM international conference on data mining, p. 78 (2004)
10. Salmenkivi, M.: Evaluating attraction in spatial point patterns with an application in the field of cultural history. In: Proceedings of the 4th IEEE international conference on data mining, pp. 511–514 (2004)
11. Sheng, C., Hsu, W., Li Lee, M., Tung, A.: Discovering Spatial Interaction Patterns. In: Haritsa, J.R., Kotagiri, R., Pudi, V. (eds.) DASFAA 2008. LNCS, vol. 4947, pp. 95–109. Springer, Heidelberg (2008)
12. Bureau of Transportation Statistics, `http://www.bts.gov/`
13. Tobler, W.: Cellular geography. Philosophy in geography, pp. 379–386 (1979)
14. Agrawal, R., Srikant, R.: Fast algorithms for mining association rules. In: Proceedings of the 20th international conference on very large data bases, vol. 1215, pp. 487–499 (1994)
15. Leskovec, J., Krause, A., Guestrin, C., Faloutsos, C., VanBriesen, J., Glance, N.: Cost-effective outbreak detection in networks. In: Proceedings of the 13th ACM SIGKDD international conference on knowledge discovery and data mining, pp. 420–429 (2007)
16. Arge, L., Procopiuc, O., Ramaswamy, S., Suel, T., Vitter, J.: Scalable sweeping-based spatial join. In: Proceedings of the 24th international conference on very large data bases, pp. 570–581 (1998)

Classifier Chains for Multi-label Classification

Jesse Read, Bernhard Pfahringer, Geoff Holmes, and Eibe Frank

Department of Computer Science
The University of Waikato
Hamilton, New Zealand
{jmr30,bernhard,geoff,eibe}@cs.waikato.ac.nz

Abstract. The widely known binary relevance method for multi-label classification, which considers each label as an independent binary problem, has been sidelined in the literature due to the perceived inadequacy of its label-independence assumption. Instead, most current methods invest considerable complexity to model interdependencies between labels. This paper shows that binary relevance-based methods have much to offer, especially in terms of scalability to large datasets. We exemplify this with a novel chaining method that can model label correlations while maintaining acceptable computational complexity. Empirical evaluation over a broad range of multi-label datasets with a variety of evaluation metrics demonstrates the competitiveness of our chaining method against related and state-of-the-art methods, both in terms of predictive performance and time complexity.

1 Introduction

The traditional data mining task of *single-label* classification, also known as *multi-class* classification, associates an instance x with a single label l from a previously known finite set of labels L. A single-label dataset D is composed of n examples $(x_1, l_1), (x_2, l_2), \cdots, (x_n, l_n)$. The *multi-label* classification task associates a *subset* of labels $S \subseteq L$ with each instance. A multi-label dataset D is therefore composed of n examples $(x_1, S_1), (x_2, S_2), \cdots, (x_n, S_n)$. The multi-label problem is receiving increased attention and is relevant to many domains such as text classification [10,2], and genomics [19,16].

A common approach to multi-label classification is by way of *problem transformation*, whereby a multi-label problem is transformed into one or more single-label problems. In this fashion, a single-label classifier can be employed to make single-label classifications, and these are then transformed back into multi-label representations. Prior problem transformation approaches have employed algorithms such as Support Vector Machines [2], Naive Bayes [5] and k Nearest Neighbor methods [19].

The alternative to problem transformation is to modify an existing single-label algorithm directly for the purpose of multi-label classification. Some well known approaches involve decision trees [16] and AdaBoost [10]. Algorithm adaption may be as simple as using a problem transformation method internally, or collecting prediction confidences and using a threshold to predict the multi-labels

W. Buntine et al. (Eds.): ECML PKDD 2009, Part II, LNAI 5782, pp. 254–269, 2009.

associated with prediction confidences that lie above the threshold. Both of these approaches can be generalised to other single-label classifiers.

By abstracting away from a specific classifier, *external* problem transformation allows greater flexibility. Any single-label classifier can be used to suit requirements. Depending on the problem context, some classifiers may demonstrate better performance than others. Moreover, external problem transformation methods can also be implemented specifically to a particular algorithm or easily integrated with meta or ensemble frameworks.

There are several families of problem transformation methods that can be found in the multi-label literature. These methods arise from one or more fundamental problem transformation approaches that either form the core of more complex frameworks or are used as modifications to other algorithms. Here we review two fundamental methods.

The most well known problem transformation method is the *binary relevance* method (BM) [13,2,19]. BM transforms any multi-label problem into one binary problem for each label. Hence this method trains $|L|$ binary classifiers $C_1, \cdots, C_{|L|}$. Each classifier C_j is responsible for predicting the 0/1 association for each corresponding label $l_j \in L$.

BM is mentioned throughout the literature but consistently sidelined on the grounds of its assumption of label independence. That is to say, during its transformation process, BM ignores label correlations that exist in the training data. The argument is that, due to this information loss, BM's predicted label sets are likely to contain either too many or too few labels, or labels that would never co-occur in practice.

We argue that BM-based methods have a lot to offer. The chaining method we present in this paper shows that the above issues can be overcome and are outweighed by the advantages of this method and any methods based closely upon it.

Another fundamental problem transformation method is the *label combination method*, or *label power-set method*, (CM), which has been the focus of several recent works [15,8]. The basis of this method is to combine entire label sets into atomic (single) labels to form a single-label problem for which the set of possible single labels represents all distinct label subsets in the original multi-label representation. Each (x, S) is transformed into (x, l) where l is the atomic label representing a distinct label subset. In this way, CM-based methods directly take into account label correlations. A disadvantage of these methods, however, is their worst-case time complexity.

The consensus view in the literature is that it is crucial to take into account label correlations during the classification process [3,2,15,8,11,18,4]. However as the size of multi-label datasets grows, most methods struggle with the exponential growth in the number of possible correlations. Consequently, these methods are able to be more accurate on small datasets, but are not as applicable to larger datasets. This necessarily restricts their usefulness as many multi-label contexts involve large numbers of examples and labels.

The paper is structured as follows. We outline the advantages of BM-based methods and present our classifier chains method CC, which overcomes disadvantages of the basic binary method. We then introduce an ensemble framework for classifier chains called ECC. Finally, we demonstrate the performance of our methods under empirical evaluation on a wide range of datasets with various evaluation measures.

The main contributions of this paper are:

- We present Classifier Chains (CC) and Ensembles of Classifier Chains (ECC)
- We introduce an evaluation metric and new datasets for multi-label classification
- We present an extensive experimental evaluation to demonstrate the effectiveness of using Classifier Chains.

2 In Defence of the Binary Method

Although BM's disadvantages are widely acknowledged, its advantages are rarely mentioned. BM is theoretically simple and intuitive. Its assumption of label independence makes it suited to contexts where new examples may not necessarily be relevant to any known labels or where label relationships may change over the test data; even the label set L may be altered dynamically – making BM ideal for active learning and data stream scenarios.

However the most important and widely relevant advantage of BM is its low computational complexity compared to other methods. Given a constant number of examples, BM scales linearly with the size of the known label set L. This set is defined in the dataset and generally limited in scope: generally $|L| < |X|$, where X is the feature space. If L is very large, or not defined prior to classification, the problem is better approached as a tag-assignment or hierarchical problem, which are beyond the scope of this paper.

CM-based methods, on the other hand, have an upper bound complexity of $\min(|D|, 2^{|L|})$, due to the exponentially expanding number of possible combinations with increasing $|L|$ (D is the training set). All methods which model all label correlations will suffer this complexity. Note that just modelling all pair-wise label correlations is $O(|L|^2)$.

Although BM involves $|L|$ single label problems, each problem only involves two classes. Depending on the dataset, CM may have to deal with thousands or tens of thousands of classes. Other methods of transformation resulting in a single problem will have to produce decisions involving at least $|L|$ classes, which may imply greater than linear complexity.

Because BM's $|L|$ binary problems are separate, under demanding circumstances it is conceivable (and desirable) to run each label problem separately, in either parallel or serial, thus only requiring $|D|$ instances in memory at any point (over $|L|$ processors, or $|L|$ iterations).

In the next section we present our new binary method, CC, which overcomes the label independence assumption of BM while maintaining acceptable computational complexity.

3 The Classifier Chain Model (CC)

The Classifier Chain model (CC) involves $|L|$ binary classifiers as in BM. Classifiers are linked along a chain where each classifier deals with the binary relevance problem associated with label $l_j \in L$. The feature space of each link in the chain is extended with the 0/1 label associations of all previous links. The training procedure is outlined in Figure 1. Recall the notation for a training example (x, S), where $S \subseteq L$ is represented by binary feature vector $(l_1, l_2, \cdots, l_{|L|}) \in \{0,1\}^{|L|}$, and x is an instance feature vector.

TRAINING($D = \{(x_1, S_1), \cdots, (x_n, S_n)\}$)
for $j \in 1 \cdots |L|$
    do ▷ single-label transformation and training
        $D' \leftarrow \{\}$
        for $(x, S) \in D$
            do $D' \leftarrow D' \cup ((x, l_1, \cdots, l_{j-1}), l_j)$
        ▷ train C_j to predict binary relevance of l_j
        $C_j : D' \rightarrow l_j \in \{0,1\}$

Fig. 1. CC's training phase for dataset D and label set L

Hence a chain $C_1, \cdots, C_{|L|}$ of binary classifiers is formed. Each classifier C_j in the chain is responsible for learning and predicting the binary association of label l_j given the feature space, augmented by all prior binary relevance predictions in the chain $l_1, \cdots, l_{j-1}$. The classification process begins at C_1 and propagates along the chain: C_1 determines $Pr(l_1|x)$ and every following classifier $C_2 \cdots C_{|L|}$ predicts $Pr(l_j|x_i, l_1, \ldots, l_{j-1})$. This classification process is outlined in Figure 2.

CLASSIFY(x)
$Y \leftarrow \{\}$
for $j \leftarrow 1$ **to** $|L|$
    do $Y \leftarrow Y \cup (l_j \leftarrow C_j : (x, l_1, \cdots, l_{j-1}))$
return (x, Y) ▷ the classified example

Fig. 2. CC's prediction phase for a test instance x

This chaining method passes label information between classifiers, allowing CC to take into account label correlations and thus overcoming the label independence problem of BM. However, CC still retains advantages of BM including low memory and runtime complexity. Although an average of $|L|/2$ features is added to each instance, because $|L|$ is invariably limited in practice, this has negligible consequences on complexity, as demonstrated in Section 6.

However in terms of computational complexity CC can be very close to BM, depending on the total number of labels and the complexity of the underlying learner. BM's complexity is $O(|L| \times f(|X|, |D|))$, where $f(|X|, |D|)$ is the complexity of the underlying learner. Using the same notation, CC's complexity is $O(|L| \times f(|X| + |L|, |D|))$, i.e. a penalty is incurred for having $|L|$ additional attributes. As demonstrated in the experiments in Section 6, in practice this penalty tends be small in many cases. For instance, assuming a linear base learner, CC's complexity simplifies to $O(|L| \times |X| \times |D| + |L| \times |L| \times |D|)$, where the first term dominates as long as $|L| < |X|$, which we expect. In this case the effective complexity of CC is then $O(|L| \times |X| \times |D|)$, which is identical to BM's complexity. CC's complexity will be worse than BM's whenever $|L| > |X|$ holds.

Also, although the chaining procedure implies that CC cannot be parallelized, it can be serialized and therefore still only requires a single binary problem in memory at any point in time – a clear advantage over other methods.

The order of the chain itself clearly has an effect on accuracy. Although there exist several possible heuristics for selecting a chain order for CC, we instead solve the issue by using an ensemble framework with a different random chain ordering for each iteration. In the next section, we present this framework.

4 Ensembles of Classifier Chains (ECC)

The classifier presented in this section is an Ensemble of Classifier Chains (ECC). Ensembles are well known for their effect of increasing overall accuracy and overcoming over-fitting, as well as allowing parallelism. They have successfully been used in various multi-label problems [10,8,15,16].

Note that binary methods are occasionally referred to as ensemble methods because they involve multiple binary models. However, none of these models is multi-label capable and therefore we use the term *ensemble* strictly in the sense of an *ensemble of multi-label methods.*

ECC trains m CC classifiers $C_1, C_2, \cdots, C_m$. Each C_k is trained with:

- a random chain ordering (of L); and
- a random subset of D.

Hence each C_k model is likely to be unique and able to give different multi-label predictions. These predictions are summed by label so that each label receives a number of votes. A threshold is used to select the most popular labels which form the final predicted multi-label set.

Each kth individual model (of m models) predicts vector $y_k = (l_1, \cdots, l_{|L|}) \in \{0,1\}^{|L|}$. The sums are stored in a vector $W = (\lambda_1, \cdots, \lambda_{|L|}) \in \mathbb{R}^{|L|}$ such that $\lambda_j = \sum_{k=1}^{m} l_j \in y_k$. Hence each $\lambda_j \in W$ represents the sum of the votes for label $l_j \in L$. We then normalise W to W^{norm}, which represents a distribution of scores for each label in $[0,1]$. A threshold is used to choose the final multi-label set Y such that $l_j \in Y$ where $\lambda_j \geq t$ for threshold t. Hence the relevant labels in Y represent the final multi-label prediction.

This is a generic voting scheme and it is straightforward to apply an ensemble of any multi-label problem transformation method. We can therefore apply BM under this same scheme to create an Ensemble of the Binary Method (EBM). This is carried out identically to ECC (except that chain ordering has no effect on BM). As far as we are aware, this scheme has not been previously evaluated in the literature.

5 Related Work

Using labels in the feature space has been approached in the past by Godbole and Sarawagi [2]. In the main contribution of their work, the authors stacked BM classification outputs along with the full original feature space into a separate meta classifier, thereby creating a two-stage classification process. Similar to CC, this process is able to take into account label correlations, but their meta classifier implies an extra iteration of both training and test data as well as internal classifications on the training data to acquire the label outputs for this meta step. In contrast, CC only requires a single training iteration like BM, and uses labels directly from the training data without any internal classification. We evaluate and compare Godbole and Sarawagi's meta-stacking (MS) in our experimental evaluation.

A method reviewed in [9] maps confidence predictions of a single-label classifier to actual label subsets (observed in the training data) using Hamming distance. The subset with the shortest Hamming distance to the predictions is chosen as the predicted set. This procedure can also be applied to the binary outputs of BM. Hence we use this Subset Mapping method (SM) in our experimental evaluation.

The work in [3] is based upon the binary approach but their algorithm adds a second part for deriving a low-dimensional shared subspace in order to model label correlations. This is computationally expensive, despite an approximation algorithm, which is reflected in their experimental setup where they randomly select 1000 data points for training – relatively small sets. Similarly [11] uses a computationally complex hypergraph method to model label correlations and, despite a proposed approximate formulation, induces high computational complexity.

ML*k*NN [19] is a nearest-neighbor based method with similar time complexity to BM. ML*k*NN performs well against BM, but in [12], it did not perform as well as RAKEL (see below), which we use in our evaluation.

A boosting algorithm is introduced by [18] that aims to reduce complexity by reducing redundancy in the learning space and sharing models between labels. Binary models are trained on subsets of the instance and feature spaces, i.e. a random forest paradigm is used. This is a good example of how the complexity of the binary approach can be significantly reduced, and supports the conclusion of this paper that binary methods have been underrated.

The binary pairwise problem has also been employed for multi-label classification. This is the *one-vs-one* approach, as opposed to the *one-vs-rest* approach used by BM, therefore requiring $|L|^2$ classifiers as opposed to $|L|$. [7] accompanies each pairwise classifier with two probabilistic models to isolate the overlapping feature space. They cite a computational bottleneck for this method for large datasets. Another pairwise approach [4] works with large datasets when used with simple and efficient perceptrons, although this method is only able to provide a ranking and not actual multi-label classification sets. A related approach is taken in [1] which can conduct classification by using a virtual label to separate relevant and irrelevant labels. This method performed well against BM in terms of ranking evaluation, but only marginally in classification.

While label rankings can in most cases be turned into a multi-label classification, the reverse is not always true. Both BM and CC, for example, cannot naturally provide prediction confidences, and therefore are unable to provide a ranking. ECC can output a ranking directly from its voting scheme, although this ranking is only coincidental as a means to achieve a classification. Ranking methods and evaluation of ranking performance itself falls outside the scope of this paper.

Several ensemble approaches have been developed based on the common problem transformation methods introduced in Section 1, particularly CM due to its inherent ability to take into account label correlations.

A good example is the RAKEL system by Tsoumakas and Vlahavas [15]. For m iterations of the training data, RAKEL draws a random subset of size k from all labels L and trains a CM classifier using these labels. A simple voting process determines the final classification set. Using appropriate values of m and k, RAKEL was shown to be better than BM and CM.

HOMER [14] is a computationally efficient multi-label classification method specifically designed for large multi-label datasets. Its efficiency is due to hierarchically splitting up the label set L using a modified k-means algorithm, and solving each subproblem individually. We note that the authors use *Naive Bayes* as their base classifier to further reduce complexity. In our experiments we use the more computationally demanding *SVM*s, known for their predictive performance, particularly on text data.

In [8] we presented EPS: an ensemble method that uses pruning to reduce the computational complexity of CM, and an instance duplication method to reduce error rate as compared to CM and other methods. This method proved to be particularly competitive in terms of efficiency.

6 Experiments

We perform an experimental comparison based on several experimental setups designed to test a variety of methods in different contexts. Initially we carry out experiments to justify the value of CC by comparing to related methods and then later compare ECC to other state-of-the-art methods.

We consider our evaluation one of the most extensive in the multi-label literature. To the best of our knowledge, our collection of multi-label datasets represents the largest one so far in multi-label evaluation, and we have used four evaluation methods. First we introduce some evaluation measures, the datasets and relevant statistics, and following this, we review our experimental method and setup, and then present the results.

6.1 Evaluation Measures

It is essential to include several evaluation measures in multi-label evaluation. Given the extra label dimension, it is otherwise possible to optimise for certain evaluation measures. We use four different evaluation measures.

Multi-label classification requires a different measure of *accuracy* from standard single-label (multi-class) classification. Recall that for each ith classified instance, Y_i is the predicted set of labels, and S_i is the actual set. It is possible to measure accuracy by example (instance i is correct if $S_i = Y_i$), or by individual label (each $l \in Y_i$ is a separate evaluation), but in practice the former tends to be overly harsh and the latter overly lenient. Instead, we use accuracy as defined in [13]:

$$Accuracy = \frac{1}{|D|} \sum_{i=1}^{|D|} \frac{|S_i \cap Y_i|}{|S_i \cup Y_i|}$$

Accuracy is micro-averaged across all examples. As a contrast we include *macro-averaged F-measure*, where the average is calculated per label and then averaged across all labels. The F-measure is the harmonic mean between precision and recall, common to information retrieval. If p_j and r_j are the precision and recall for all $l_j \in Y_i$ from $l_j \in S_i$, the macro-averaged F-measure is:

$$F1_{macro} = \frac{1}{|L|} \sum_{j=1}^{|L|} \frac{2 \times p_j \times r_j}{(p_j + r_j)}$$

We also evaluate using the average *area under the precision recall curve* ($AU(\overline{PRC})$). Instead of setting a fixed threshold, the threshold is varied on the confidence predictions W_i^{norm} (see Section 4) in steps $0.00, 0.02, \cdots, 1.00$, thus producing different precision and recall values for each label. The average across all labels is the average area under the precision recall curve. More information about this measure can be found in [16].

Finally we introduce the use of *log loss*, distinct from other measures because it punishes worse errors more harshly, rewarding conservative prediction. The error is graded by the confidence at which it was predicted: predicting false positives with low confidence induces logarithmically less penalty than predicting with high confidence. Therefore, again, we use the confidence predictions for each label $\lambda \in W_i^{norm}$ to compare to the actual value of each label $l_j \in S_i$:

$$LogLoss = \frac{1}{|D|} \sum_{i=1}^{|D|} \sum_{j=1}^{|L|} -max\Big(log(\lambda_j) l_j + log(1-\lambda_j)(1-l_j), log\frac{1}{|D|}\Big)$$

We have used a dataset-dependent maximum of $log(\frac{1}{|D|})$ to limit the magnitudes of penalty. Such a limit, as explained in [9], serves to smooth the values and prevent a small subset of poorly predicted labels from greatly distorting the overall error. Note that, as opposed to the other measures, the best possible score for the log loss is 0.0.

In the analysis of time complexity we measure train and test times in seconds.

6.2 Datasets

Table 1 displays datasets from a variety of domains and their associated statistics. *L*abel *Card*inality (*LCard*) is a standard measure of "multi-labelled-ness" introduced in [13]. It is simply the average number of labels relevant to each instance. The *P*roportion of *Dist*inct label combinations (*PDist*) is simply the number of distinct label subsets relative to the total number of examples:

$$LCard(D) = \frac{\sum_{i=1}^{|D|} |S_i|}{|D|} \qquad PDist(D) = \frac{|\{S|\exists(x,S) \in D\}|}{|D|}$$

Table 1. A collection of multi-label datasets and associated statistics; n indicates numeric attributes. D_T is the training split used in some experiments.

	$\|D\|$	$\|L\|$	$\|X\|$	$LCard(D)$	$PDist(D)$	Type	D_T	$\|D_T\| \times \|L\| \times \|X\|$
Scene	2407	6	294n	1.07	0.006	media	1211	2.14E+06
Yeast	2417	14	103n	4.24	0.082	biology	1500	2.16E+06
Medical	978	45	1449	1.25	0.096	text	652	4.25E+07
Slashdot	3782	22	1079	1.18	0.041	text	1891	4.49E+07
Enron	1702	53	1001	3.38	0.442	text	1135	6.02E+07
Reuters	6000	103	500n	1.46	0.147	text	3000	1.55E+08
OHSUMED	13929	23	1002	1.66	0.082	text	6965	1.61E+08
TMC2007	28596	22	500	2.16	0.047	text	21519	2.37E+08
MediaMill	43907	101	120n	4.38	0.149	media	30993	3.76E+08
Bibtex	7395	159	1836	2.40	0.386	text	3698	1.08E+09
IMDB	95424	28	1001	1.92	0.036	text	47712	1.34E+09
Delicious	16105	983	500	19.02	0.981	text	12920	6.35E+09

We strived to include a considerable variety and scale of multi-label datasets. In total we use 12 datasets, with dimensions ranging from 6 to 983 labels, and from less than 1,000 examples to almost 100,000. The datasets are roughly ordered by complexity ($|D_T| \times |L| \times |X|$) and divided between regular and large sizes. Included are two new real-world multi-label text collections: *Slashdot*, which we collected from `http://slashdot.org`, and *IMDB* from `http://imdb.org` (data obtained from `http://www.imdb.com/interfaces#plain`). All datasets and further information about them can be found at various sources[1].

[1] `http://www.cs.waikato.ac.nz~/jmr30/#datasets` and `http://mlkd.csd.auth.gr/multilabel.html#Datasets`

6.3 Algorithms

For easy reference, Table 2 lists all the algorithms used in the experiments, their corresponding abbreviation, and any relevant citation (where citations are absent, the method has been introduced in this paper).

Table 2. Algorithms used in the experiments, and associated citations

BM-based algorithms		Ensemble algorithms	
BM Binary Method	[13]	EBM Ensembles of Binary Method	
CM Chaining Method		ECC Ensembles of Classifier Chains	
SM Subset Mapping	[9]	EPS Ensembles of Pruned Sets	[8]
MS Meta Stacking	[2]	RAKEL RAndom K labEL subsets	[15]

6.4 Setup and Method

Our default experimental setup is as follows. We evaluate all algorithms under a WEKA-based [17] framework running under Java JDK 1.6 with the following settings. Support Vector Machines are used as the internal classifier using WEKA's SMO implementation with default parameters. Ensemble iterations are set to 10. Evaluation is done in the form of 5 × 2 fold cross validation on each dataset and the corrected paired t-test [6] determines significance under a value of 0.05. The exception to this is the experiments on large datasets where cross validation is too intensive for some methods, and a train/test split is used instead. These splits are shown in Table 1 where D_T is the training set (and therefore $(D \setminus D_T)$ the test set). Experiments are run on 64 bit machines, allowing up to 2 GB RAM *per ensemble iteration*.

The thresholds for *all* ensemble voting schemes, necessary for determining accuracy and the macro F-measure, are set as following, where D_T is the training set and a classifier H_t has made predictions for test set D_S under threshold t:

$$t = \underset{\{t \in 0.00, 0.001, \cdots, 1.00\}}{\arg\min} \left| LCard(D_T) - LCard(H_t(D_S)) \right| \quad (1)$$

This is the closest approximation of the label cardinality of the training set to the predictions made on the test set. This implies a close balance between precision and recall and therefore benefits accuracy and F-measure. It also avoids ad-hoc or arbitrary thresholds or intensive internal cross-validation.

All ensemble methods involve subsampling for the individual models. EBM,ECC, and EPS subsample the training set (we set 67% for each model), while RAKEL subsamples the label set according to its k parameter.

For RAKEL we always set parameter $k = \frac{|L|}{2}$ and for EPS we set $p = 1$ and n is set according to the $LCard(D_T)$ training set statistic. Both these algorithms allow a trade-off between predictive performance and training time costs and vice versa: using smaller k for RAKEL and higher p EPS will lead to reduced computational complexity for both algorithms. However, in these experiments we optimise for predictive performance. Results in the relevant papers [8] and [15] show that our choice of parameter values generally provides highest accuracy. One of the notable advantages of ECC is that it requires no additional parameters other than the generic ensemble parameters which we set as above.

6.5 Results

Initially we compare standalone CC to BM and BM-related methods: a reproduction of the MS method, and the SM method. CC is used as the base for determining statistical significance. Results for accuracy and macro-averaged F-measure are shown in Table 3 (the other evaluation methods are not appropriate because not all these methods can supply confidence predictions). Train times are graphed in Figure 3.

Secondly, we perform an experiment comparing ensemble implementations. We compare EBM and ECC to the state-of-the-art algorithms EPS and RAKEL. Statistical significance is taken against ECC. Results for all evaluation measures are displayed in Table 4 and train times are graphed in Figure 4.

Finally we compare ensembles separately on large datasets, for which we use train/test splits for evaluation. Results for predictive performance are displayed in Table 5; train and test times are displayed in Table 6. DNF indicates that the experiment Did Not Finish within one week under the available resources.

Table 3. Binary Methods - Predictive Performance

	Accuracy				Macro F-measure			
Dataset	CC	BM	SM	MS	CC	BM	SM	MS
Scene	**67.3**	59.1 •	63.0	61.9 •	**0.696**	0.685	0.666	0.694
Yeast	**51.5**	49.6	50.4	49.8	**0.346**	0.326	0.327	0.331
Slashdot	**46.7**	43.4 •	44.7 •	43.6 •	0.327	**0.329**	0.298	0.328
Medical	**75.1**	73.0 •	73.1	73.1 •	**0.377**	0.364	0.321 •	0.370
Enron	39.5	38.6	**40.3**	38.8	**0.198**	0.197	0.144 •	**0.198**
Reuters	**39.6**	31.9 •	33.6 •	32.4 •	**0.245**	0.224 •	0.194 •	0.229 •

⊕, • statistically significant improvement or degradation vs. CC.

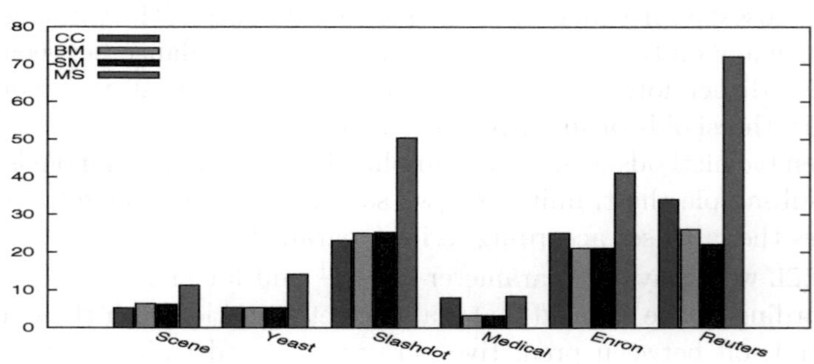

Fig. 3. Binary Methods - Train times (average seconds)

Table 4. Ensemble Methods

	Accuracy				Log Loss			
Dataset	ECC	EBM	EPS	RAK.	ECC	EBM	EPS	RAK.
Scene	70.8	68.6 •	**73.7** ⊕	72.7 ⊕	**1.32**	1.97 •	1.41	1.78 •
Yeast	53.3	52.7 •	**54.9** ⊕	54.3	9.41	11.48 •	**9.16**	10.26 •
Slashdot	51.0	50.9	50.9	**51.4**	**3.45**	3.94 •	3.81 •	4.41 •
Medical	**77.6**	76.7	75.1 •	76.2	**1.87**	1.93	2.08 •	2.19
Enron	44.6	44.2	44.5	**45.9**	**10.84**	11.00	12.15 •	12.00 •
Reuters	44.7	36.0 •	**49.6** ⊕	45.3	7.52	8.33 •	**7.09** ⊕	8.23 •
	Macro F-measure				*AU*($\overline{PRC}$)			
Dataset	ECC	EBM	EPS	RAK.	ECC	EBM	EPS	RAK.
Scene	0.742	0.729 •	**0.763**	0.750	0.778	0.706 •	**0.780**	0.736 •
Yeast	0.362	0.364	**0.420** ⊕	0.413 ⊕	**0.645**	0.618 •	0.643	0.623 •
Slashdot	0.343	0.346	0.336	**0.353**	**0.514**	0.464 •	0.498	0.443 •
Medical	**0.386**	0.382	0.324 •	0.377	**0.789**	0.782	0.752 •	0.744 •
Enron	0.201	0.201	0.155 •	**0.206**	**0.488**	0.481 •	0.440 •	0.453 •
Reuters	**0.286**	0.264 •	0.264 •	0.282	0.347	0.311 •	**0.378** ⊕	0.330 •

⊕, • statistically significant improvement or degradation vs. ECC.

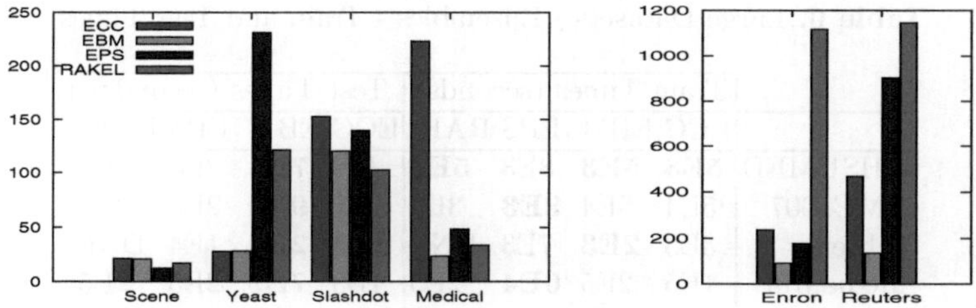

Fig. 4. Ensemble Methods - Train times (average seconds)

7 Discussion

7.1 The Value of Classifier Chains

The value of CC's chaining method can be seen in Table 3 where it is compared to related classifiers. CC improves convincingly over both the default BM method and related methods MS and SM 10 out of 12 times, and in many cases the difference is statistically significant. Hence the results justify using CC as a base method.

The training times of BM and CM (Figure 3) support the theory presented in Section 2. BM is naturally the fastest. This complexity is exceeded only marginally by CC. On smaller datasets, the effect is even negligible, and tiny variances in runtime conditions cause CC to run marginally faster in some cases - similarly to SM. SM also involves only minimal overhead over BM. Not surprisingly, because of MS's two-stage stacking process, its train times are about twice that of BM.

Table 5. Large Datasets - Ensembles - Predictive Performance

	Accuracy				Log Loss			
Dataset	ECC	EBM	EPS	RAK.	ECC	EBM	EPS	RAK.
OHSUMED	41.1	41.39	**41.98**	41.55	5.36	**5.35**	6.17	6.65
TMC2007	**53.03**	52.74	52.30	52.85	5.70	6.02	**4.87**	5.82
Bibtex	**35.50**	35.10	34.13	DNF	11.95	**11.88**	12.73	DNF
MediaMill	**40.39**	39.80	38.20	31.40	**25.37**	27.05	26.52	29.97
IMDB	**24.88**	1.92	DNF	2.06	**13.34**	21.30	DNF	20.61
Delicious	**17.93**	16.93	9.41	DNF	**119.92**	128.30	133.34	DNF
	Macro F-measure				$AU(\overline{PRC})$			
Dataset	ECC	EBM	EPS	RAK.	ECC	EBM	EPS	RAK.
OHSUMED	0.378	0.379	0.376	**0.398**	0.495	0.499	**0.506**	0.501
TMC2007	0.551	0.548	**0.561**	0.557	**0.620**	**0.620**	0.614	**0.620**
Bibtex	**0.324**	0.313	0.257	DNF	**0.437**	0.433	0.423	DNF
MediaMill	**0.395**	0.366	0.338	0.309	**0.523**	0.518	0.482	0.413
IMDB	**0.221**	0.075	DNF	0.080	**0.329**	0.023	DNF	0.025
Delicious	0.154	0.133	0.038	DNF	**0.182**	0.158	0.095	DNF

Table 6. Large Datasets - Ensembles - Train and Test Times

	Train Times (seconds)				Test Times (seconds)			
	ECC	EBM	EPS	RAK.	ECC	EBM	EPS	RAK.
OHSUMED	**5E3**	**5E3**	8E3	**5E3**	2E2	**7E1**	2E4	3E3
TMC2007	5E4	5E4	**3E3**	3E4	**9E1**	**9E1**	2E3	3E3
Bibtex	3E3	**2E3**	7E3	DNF	**2E3**	**2E3**	1E4	DNF
MediaMill	1E5	2E5	**6E4**	2E5	1E3	**7E2**	2E5	2E5
IMDB	4E5	**3E5**	DNF	3E5	2E3	**9E2**	DNF	1E4
Delicious	1E5	1E5	**3E2**	DNF	1E4	6E3	**1E2**	DNF

Shown in E notation where 5E3 $\approx$ 5000 seconds.

7.2 Ensemble Methods on Regular Datasets

In Table 4, we see that ECC competes well against the other ensemble methods. Although in some cases these methods demonstrate better performance than ECC, such gains are not paralleled under all evaluation measures. Under both log loss and $AU\overline{PRC}$, ECC improves over other methods in four out of six cases, and the improvement is for the most part statistically significant.

As expected, the timing results again weigh in favour of EBM and ECC, especially for small $|L|$. This is shown in Figure 4. An exception is the *Medical* dataset which has a very high $|L| : |D|$ ratio, spiking ECC's build time. EPS's reductions to complexity are considerable in some cases, but sporadic and not theoretically bounded as low as those of the BM-based methods, which are closely related to the dataset constants $|L|$ and $|D|$.

7.3 Ensemble Methods on Large Datasets

We presented ECC as an efficient method for multi-label classification, so the experiment on large datasets is important to justify this claim. Results are displayed in Table 5. Although the t-test could not be used on the train-test evaluation, we can still see overall superiority for ECC, even more so on regular datasets. ECC shows a clear majority of wins over the CM-based methods on all measures of predictive performance. EBM also performs particularly well in this experiment, indicating that, particularly on larger datasets, the disadvantages of BM-based methods are outweighed by the large number of examples they can train on.

Only the ECC method performs satisfactorily on the *IMDB* dataset under the experiment setup. The other methods suffer problems. EPS's pruning mechanism fails and the individual models of EBM and RAKEL predict too many empty sets. The different chain orderings prevent this effect in ECC.

Again the BM-based methods show an overall advantage in time costs (Table 6). These datasets are much larger than those typically approached in the literature. Although ECC and EBM are not always the fastest, they are the most consistent, and are the only methods to complete on every dataset under the time and memory constraints of the experiment. Again, EPS's reductions to training time are in some cases effective, but not reliable. For example, on *IMDB*, where there are few outlying label combinations to prune (indicated by a low *PDist* value), EPS's pruning mechanism is ineffective, resulting in DNF. On *Delicious*, the effect is the opposite: EPS prunes away far too much information, resulting in particularly poor predictive performance. This is arguably an improvement over RAKEL (DNF), but only about half as accurate as the binary methods.

7.4 Summary

CM-based methods and other methods that intensively model label correlations obviously have a place in multi-label classification, especially for datasets of relatively small dimensions. On larger datasets, however, not only does it become computationally challenging to model all label correlations, but there are no significant predictive advantages in doing so. These other methods work hard to model the label correlations in the training data but end up sacrificing individual label accuracy. ECC models correlations using an approach which is efficient and not prone to over-fitting, and for this reason performs strongly over a wide range of datasets and evaluation measures.

8 Conclusions

This paper presented a novel chaining method for multi-label classification. We based this method on the binary relevance method, which we argued has many advantages over more sophisticated current methods, especially in terms of time costs. By passing label correlation information along a chain of classifiers, our method counteracts the disadvantages of the binary method while maintaining acceptable computational complexity. An ensemble of classifier chains can be used to further augment predictive performance.

Using a variety of multi-label datasets and evaluation measures, we carried out empirical evaluations against a range of algorithms. Our classifier chains method proved superior to related methods, and in an ensemble scenario was able to improve on state-of-the-art methods, particularly on large datasets. Despite other methods using more complex processes to model label correlations, ensembles of classifier chains can achieve better predictive performance and are efficient enough to scale up to very large problems.

References

1. Fürnkranz, J., Hüllermeier, E., Mencía, E.L., Brinker, K.: Multilabel classification via calibrated label ranking. Machine Learning 73(2), 133–153 (2008)
2. Godbole, S., Sarawagi, S.: Discriminative methods for multi-labeled classification. In: Dai, H., Srikant, R., Zhang, C. (eds.) PAKDD 2004. LNCS (LNAI), vol. 3056, pp. 22–30. Springer, Heidelberg (2004)
3. Ji, S., Tang, L., Yu, S., Ye, J.: Extracting shared subspace for multi-label classification. In: KDD 2008: 14th ACM SIGKDD International Conference on Knowledge Discovery and Data mining, pp. 381–389. ACM, New York (2008)
4. Mencía, E.L., Fürnkranz, J.: Efficient pairwise multilabel classification for large-scale problems in the legal domain. In: Daelemans, W., Goethals, B., Morik, K. (eds.) ECML PKDD 2008, Part II. LNCS (LNAI), vol. 5212, pp. 50–65. Springer, Heidelberg (2008)
5. McCallum, A.K.: Multi-label text classification with a mixture model trained by EM. In: Association for the Advancement of Artificial Intelligence workshop on text learning (1999)
6. Nadeau, C., Bengio, Y.: Inference for the generalization error. Machine Learning 52(3), 239–281 (2003)
7. Petrovskiy, M.: Paired comparisons method for solving multi-label learning problem. In: HIS 2006: Sixth International Conference on Hybrid Intelligent Systems, p. 42. IEEE, Los Alamitos (2006)
8. Read, J., Pfahringer, B., Holmes, G.: Multi-label classification using ensembles of pruned sets. In: ICDM 2008, pp. 995–1000. IEEE, Los Alamitos (2008)
9. Schapire, R.E., Singer, Y.: Improved boosting algorithms using confidence-rated predictions. Machine Learning 37(3), 297–336 (1999)
10. Schapire, R.E., Singer, Y.: Boostexter: A boosting-based system for text categorization. Machine Learning 39(2/3), 135–168 (2000)
11. Sun, L., Ji, S., Ye, J.: Hypergraph spectral learning for multi-label classification. In: KDD 2008: 14th ACM SIGKDD International Conference on Knowledge Discovery and Data mining, pp. 668–676. ACM, New York (2008)
12. Trohidis, K., Tsoumakas, G., Kalliris, G., Vlahavas, I.: Multilabel classification of music into emotions. In: ISMIR 2008: 9th International Conference on Music Information Retrieval (2008)
13. Tsoumakas, G., Katakis, I.: Multi label classification: An overview. International Journal of Data Warehousing and Mining 3(3) (2007)
14. Tsoumakas, G., Katakis, I., Vlahavas, I.P.: Effective and efficient multilabel classification in domains with large number of labels. In: ECML/PKDD 2008 Workshop on Mining Multidimensional Data (2008)

15. Tsoumakas, G., Vlahavas, I.P.: Random k-labelsets: An ensemble method for multilabel classification. In: Kok, J.N., Koronacki, J., Lopez de Mantaras, R., Matwin, S., Mladenič, D., Skowron, A. (eds.) ECML 2007. LNCS (LNAI), vol. 4701, pp. 406–417. Springer, Heidelberg (2007)
16. Vens, C., Struyf, J., Schietgat, L., Džeroski, S., Blockeel, H.: Decision trees for hierarchical multi-label classification. Machine Learning 2(73), 185–214 (2008)
17. Witten, I.H., Frank, E.: Data Mining: Practical machine learning tools and techniques, 2nd edn. Morgan Kaufmann, San Francisco (2005)
18. Yan, R., Tesic, J., Smith, J.R.: Model-shared subspace boosting for multi-label classification. In: KDD 2007: 13th ACM SIGKDD International Conference on Knowledge Discovery and Data mining, pp. 834–843. ACM, New York (2007)
19. Zhangand, M.-L., Zhou, Z.-H.: A k-nearest neighbor based algorithm for multi-label classification. In: GnC 2005: IEEE International Conference on Granular Computing, pp. 718–721. IEEE, Los Alamitos (2005)

Dependency Tree Kernels for Relation Extraction from Natural Language Text

Frank Reichartz, Hannes Korte, and Gerhard Paass

Fraunhofer IAIS, Schloss Birlinghoven, 53754 St. Augustin, Germany

Abstract. The automatic extraction of relations from unstructured natural text is challenging but offers practical solutions for many problems like automatic text understanding and semantic retrieval. Relation extraction can be formulated as a classification problem using support vector machines and kernels for structured data that may include parse trees to account for syntactic structure. In this paper we present new tree kernels over dependency parse trees automatically generated from natural language text. Experiments on a public benchmark data set show that our kernels with richer structural features significantly outperform all published approaches for kernel-based relation extraction from dependency trees. In addition we optimize kernel computations to improve the actual runtime compared to previous solutions.

1 Introduction

Current search engines usually are not effective for complex queries, e.g. "composers born in Berlin". The retrieved documents among others contain composers who stayed some time in Berlin or have the name "Berlin". Obviously the internal representation of text in a search index as a sequence of words is insufficient to recover semantics from unstructured text. An important step towards automatic knowledge discovery is to extract semantic relations between entities.

Information extraction tackles this goal in two steps. First entity or phrase taggers detect objects of different types, such as persons, descriptions or pronouns, mentioned in the text. Some of these techniques have reached a sufficient performance level on many datasets [18]. They offer the basis for the next step: the extraction of relations that exist between the recognized entities, e.g. composer-born-in(John White, Berlin).

An early approach to relation extraction is based on patterns [6], usually expressed as regular expressions for words with wildcards. The underlying hypothesis assumes that terms sharing similar linguistic contexts are connected by similar semantic relations. Various authors follow this approach, e.g. [1] use frequent itemset mining to extract word patterns and [7] employ logic-based frequent structural patterns for relation extraction.

Syntactic parse trees provide extensive information on syntactic structure and can, for instance, represent the relation between subject, verb and object in a sentence. For feature-based methods only a limited number of structural details may be compared. On the other hand, kernel-based methods offer efficient

W. Buntine et al. (Eds.): ECML PKDD 2009, Part II, LNAI 5782, pp. 270–285, 2009.

solutions that allow to explore a much larger (often exponential, or in some cases, infinite) characteristics of trees in polynomial time, without the need to explicitly represent the features. [20] and [4] proposed kernels for dependency trees inspired by string kernels. [2] investigated a kernel that computes similarities between nodes on the shortest path of a dependency tree that connect the entities. All these kernels are used as input for a kernel classifier.

In this paper we extend current dependency tree kernels by including richer structural features. To tackle the different shortcomings of previous work we use the ordering properties as well as the labeling of nodes in dependency trees in a novel fashion to create kernels which consider most of the available information in dependency trees. To allow the usage of more substructure properties while maintaining an acceptable runtime we propose two new computation algorithms tailored for relation extraction tree kernels. Our new kernels are shown to outperform all previously published kernels in classification quality by a significant margin on a public benchmark. Our kernels reach F-measures of $77\% - 80\%$ on selected relations which is sufficient for some applications like information retrieval.

The remainder of the paper is organized as follows. In the next section we describe dependency parse trees used for relation classification. Subsequently we give a generic description of the current dependency parse trees in the literature. The following two sections outline our new kernels for relation extraction, the All-Pairs Dependency Tree Kernel as well as the Dependency Path Tree Kernel. Next we describe different versions of the algorithms optimized for efficiency. For the experiments we re-implemented existing kernels and compare them to our new kernels on a benchmark dataset. We close with a summary and conclusions.

2 Dependency Parse Trees

A dependency tree is a structured representation of the grammatical dependency between the words of a sentence by a labeled directed tree [11]. Structure is determined by the relation between a word (a head) and its dependents. The dependent in turn can be the head for other words yielding a tree structure. Each node in the tree corresponds to a word in the sentence with arrows pointing from the head to the dependents.

Dependency trees may be *typed*, specializing the "dependent" relation into many subtypes, e.g. as "auxiliary", "subject", "object", while in the untyped case there is only a "dependent" relation. In this paper we consider untyped dependency trees only, generated by the *Stanford Parser* [10] from the sentences of a text. As an example consider the two sentences $a =$ "Recently Obama became the president of the USA" and $b =$ "Ballmer is the CEO of Microsoft" which have the tree representations as shown in figure 1.

2.1 Notation

Let $w = w_1 \, w_2 \, \ldots \, w_n$ be a sequence of words, a natural language sentence with words $w_j \in \mathcal{W}$. The parser will generate a representation of the sentence w as

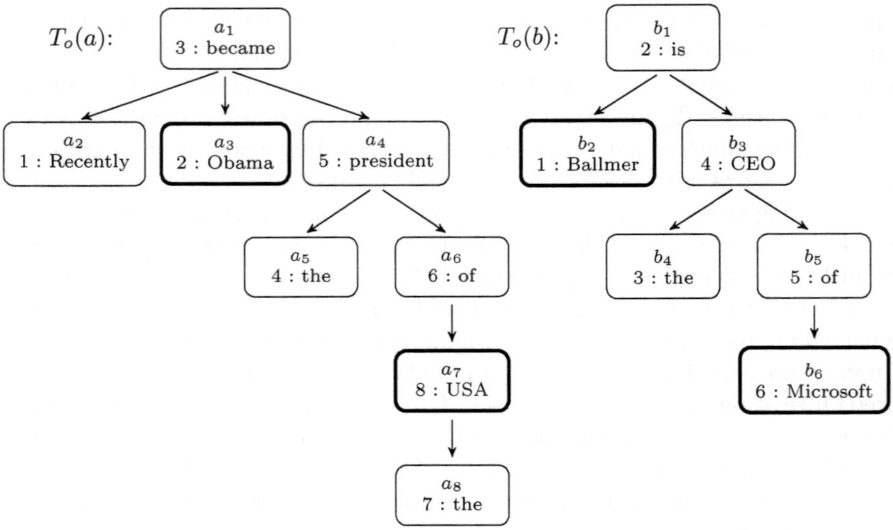

Fig. 1. The dependency trees of the two example sentences. The nodes are labeled with the position of the word in the sentence and the word itself. A thick border marks an entity mention.

a labeled rooted connected tree $T(w) = (V, E)$ with nodes $V = \{v_1, \ldots, v_n\}$ and edges in $E \subset V \times V$. Each node v_i is labeled with a corresponding word $w_{\rho(v_i)}$ in the sentence w, where $\rho(v_i)$ is a bijective function mapping a node v_i to the index j of it corresponding word $w_j \in \mathcal{W}$ in w. For example in figure 1 we have $\rho(a_4) = 5$. For each node $v \in V$ the ordered sequence of its m children $ch(v) = (u_1, \ldots, u_m)$ with $u_i \in V$ satisfies $\rho(u_i) < \rho(u_{i+1})$ for all $i \in \{1, \ldots, m-1\}$. The node $a_1 =$"became" in figure 1, for instance, has

$$ch(a_1) = (a_2 = \text{“Recently”}, a_3 = \text{“Obama”}, a_4 = \text{“president”})$$

as ordered sequence of child nodes. With the order induced by ρ we get an ordered tree $T_o(w)$.

For a sequence $s = (s_j)_{j \in [1:k]}$ of length k the set of all possible subsequence index sets can be denoted as I_k. Each ordered sequence $o(i) = \mathbf{i} = (i_1, \ldots, i_m)$ of a subset $i \subset \mathbf{I}$ of the indices $\mathbf{I}$ of a sequence $s = (s_j)_{j \in \mathbf{I}}$ implies a subsequence $(s_{i_1}, \ldots, s_{i_m})$ which we denote by $s(\mathbf{i})$, which is inline with the notation of [17]. Therefore we can write the subsequence of children referenced by $\mathbf{i}$ from a node v of an ordered tree as $ch(v, \mathbf{i})$. In figure 1, for example, $ch(a_1) = (a_2, a_3, a_4)$ has the index set $\{1, 2, 3\}$. The subset $\{1, 3\}$ in its representation as ordered sequence $(1, 3)$ defines the subsequence $ch(a_1, (1, 3)) = (a_2 = \text{“Recently”}, a_4 = \text{“president”})$ of child nodes of a_1.

We consider relations connecting entities or objects, which in this paper are collectively called entities. We assume that these entities have been extracted in a prior step, e.g. by named entity recognition tools. They are treated as a single word in the sentence w, e.g. the two words "Barack Obama" will be merged to a new word "Barack_Obama". Note that there may be different types τ of entities relevant for a relation, e.g. persons, locations or organisations which can be used as a feature associated with a entity.

Let $e(w) = (e_1(w), \ldots, e_m(w))$ be the subsequence of w of all entity mentions in w, i.e. each word $e_i(w)$ is an entity mention. Let R be a binary target relation between entities. An *instance* of the relation R in a sentence w is given by $(w, (e_i(w), e_j(w)))$, $i \neq j$, for two entity mentions $e_i(w)$ and $e_j(w)$ in $e(w)$ if the sentence semantically supports the relation between the two entity mentions, i.e. $(e_i(w), e_j(w)) \in R$. It is important to see that the same sentence w can support the same relation R between different pairs of entity mentions. The pairs $(e_i(w), e_j(w))$ of different components of $e(w)$ where the relation does not hold are negative examples of the target relation.

The example sentence a in figure 1 contains the two entities $e_1(a) = a_2 =$ "Obama" and $e_2(a) = a_8 =$ "USA" whereas sentence b contains $e_1(b) = b_1 =$ "Ballmer" and $e_2(b) = b_6 =$ "Microsoft". For the target relation *role*, where a person entity e_1 has some role in an organization entity e_2, both sentences contain a positive example e.g. $(e_1(a), e_2(a)) \in role$, which may be used as training examples.

In a rooted tree $T = (V, E)$ with root $r(T) \in V$ the *lowest common ancestor* $lca(v, w)$ of two nodes $v, w \in V$ is the lowest node in T which has both v, w as descendents. We define the set of nodes $sub(v, w)$ of two nodes v, w as the set containing all nodes of complete subtree of a tree T rooted in $lca(v, w)$. In a directed tree T we define the implied path $p(u, v)$ between two nodes u, v as the sequence $(u, \ldots, lca(u, v), \ldots, v)$ of nodes where $u, \ldots$ are the nodes on the path from the $lca(u, v)$ to u , analogous for v.

Relation extraction aims at learning a relation R from a number of positive and negative instances of the relation in natural language sentences. As a classifier we use Support Vector Machines (SVMs) [8]. They can compare complex structures, e.g. trees, by kernels, which efficiently evaluate the relevant differences. Given the kernel function, the SVM tries to find a hyperplane that separates positive from negative examples of the relation and at the same time maximizes the separation (margin) between them. This type of max-margin separator has been shown both empirically and theoretically resistant to overfitting and to provide good generalization performance on new examples.

3 Dependency Tree Kernel

The *Dependency Tree Kernel* (DTK) [4] is based on the work of [20]. A *node kernel* $\Delta(u, v)$ measures the similarity of two tree nodes u, v and its substructures. Then the tree kernel can be expressed as the node kernel of the two corresponding root nodes. Nodes may be described by different features, which are compiled in table 1. This is the feature set of [4], where *word* is the word itself, *POS* and *general-POS* are part-of-speech-tags, the *chunk tag* is the phrase the word is contained in, like noun or verb phrases. If the word describes an entity, the corresponding entity type and the mention is provided. For example "Obama" is

Table 1. List of features which may be assigned to every node in a tree. The index corresponds to k in f_v^k, hence f_v^5 denotes the "entity type" of the node v.

Feature Index k	Feature Name	Possible Feature Values
1	word	*Obama, president,...*
2	POS	NN, NNP,...
3	general-POS	noun, verb, adj,...
4	chunk tag	NP, VP, ADJP,...
5	entity type	person, location, GPE,...
6	mention type	name, nominal, pronoun
7	WordNet hypernym	corporate executive, city,...
8	relation argument	1, 2 or n/a

an entity with entity type "person" and mention type "name". If this entity is part of the current relation, the *relation argument* is set to 1 or 2 depending on the entity's role in the relation. Finally the WordNet hypernym is provided if there is one. That is a superordinate grouping word, e.g. the hypernym of "red" is "color". Unfortunately, in [4] the authors do not describe how the hypernyms are selected if there are more than one, which is a problem, because most words have different meanings, e.g. "bank" (credit institute or furniture). To tackle this problem an automatic disambiguation of words[15] is needed. So, in our experiments we exclude this feature. A specific subset of feature indices may be designated by the set f^m.

To compare the relations in two instances $X = (x, (x_1, x_2)), Y = (y, (y_1, y_2))$ [4] proposes to compare the subtrees induced by the relation arguments x_1, x_2 and y_1, y_2 in the two sentences, i.e. the least common ancestor of the relation argument nodes

$$K_{\text{DTK}}(u, v) = \Delta(lca(x_1, x_2), lca(y_1, y_2)) \tag{1}$$

In order to compute the node kernel Δ, we first define a matching function $mat(u, v) \to \{0, 1\}$ which checks whether two nodes u, v are comparable at all.

$$mat(u, v) = \begin{cases} 1 & \text{if } \forall k \in f^m : f_u^k = f_v^k \\ 0 & \text{otherwise} \end{cases}$$

where f_v^k is the k-th feature of the node v. The set f^m consists of the indices of the features used in the matching function. Based on observations in [4] this set is defined as $f^m = \{3, 5, 8\}$. Thus, two nodes u and v match, if the features *general-POS*, *entity type* and *relation argument* are equal.

The similarity function $sim(u, v) \to (0, \infty]$ counts the number of common features of the two nodes. In [4] there are 8 features (Table 1), hence $d = 8$ but in our setup $d = 7$ because the hypernym feature is not considered.

$$sim(u, v) = \sum_{k=1}^{d} comp(f_u^k, f_v^k) \qquad comp(f_u^k, f_v^k) = \begin{cases} 1 & \text{if } f_u^k = f_v^k \\ 0 & \text{otherwise} \end{cases}$$

We define the node kernel $\Delta(u, v)$ over two nodes u and v as the sum of the node similarity and their children similarity $C(ch(u), ch(v))$

$$\Delta(u,v) \begin{cases} 0 & \text{if } mat(u,v) = 0 \\ sim(u,v) + C(ch(u), ch(v)) & \text{otherwise} \end{cases} \tag{2}$$

The *child subsequence kernel* $C(s,t)$ uses a modified version of the *String Subsequence Kernel* of [17] to recursively compute the sum of node kernel values of subsequences of node sequences s and t:

$$C(s,t) = \sum_{\substack{\mathbf{i} \in I_{|s|},\ \mathbf{j} \in I_{|t|}, \\ |\mathbf{i}|=|\mathbf{j}|}} \lambda^{d(\mathbf{i})+d(\mathbf{j})} \quad \Delta'(s(\mathbf{i}), t(\mathbf{j})) \quad M(s(\mathbf{i}), t(\mathbf{j})) \tag{3}$$

where s, t are sequences of nodes and $s(\mathbf{i})$ and $t(\mathbf{j})$ are the subsequences of nodes implied by the index sets $\mathbf{i}, \mathbf{j}$. Furthermore the parameter $0 < \lambda \leq 1$ is a penalty factor for lengths and gaps of the subsequences, and $d(\mathbf{i}) = max(\mathbf{i}) - min(\mathbf{i}) + 1$ is the covered distance of the index sequence $\mathbf{i}$. The function Δ' over two node sequences of length n computes the sum over the node kernels

$$\Delta'\left((s_1, \ldots, s_n), (t_1, \ldots, t_n)\right) = \sum_{k=1}^{n} \Delta\left(s_k, t_k\right)$$

Furthermore, M represents a filter for non-matching sequences:

$$M((s_1, \ldots, s_n), (t_1, \ldots, t_n)) = \prod_{k=1}^{n} mat(s_k, t_k)$$

Less formally, the function $C(s,t)$ sums up the similarities of all subsequences in which every node matches its corresponding node. The similarity of a sequence is given by the sum of the single node similarities provided by $\Delta(u, v)$.

4 Extended Kernels

The Dependency Tree kernel forms the foundation for our proposed kernels considering a richer feature space described in the following section.

4.1 All-Pairs Dependency Tree Kernel

The dependency tree kernel [4] discards a possible relation if only one node in the subsequence does not "match" its corresponding node. We propose to use all pairs of possible combinations in the subtrees to tackle this restriction of the dependency tree kernel which can be done by generalizing the approaches of [14].

We define the All-Pairs Dependency Tree Kernel (All-Pairs-DTK), which sums up the node kernels of all possible combinations of nodes contained in the two subtrees implied by the relation argument nodes as

$$K_{\text{All-Pairs}}(X,Y) = \sum_{u \in V_x} \sum_{v \in V_y} \Delta(u,v) \tag{4}$$

where $V_x = sub(x_1, x_2)$ and $V_y = sub(y_1, y_2)$ are sets containing the nodes of the complete subtrees rooted at the respective lowest common ancestors. The consideration of all possible pairs of nodes ensure that no valuable information which resides in subtrees in which only the corresponding root does not match is discarded. However this could lead to the problem that some irrelevant subtrees are also compared which in turn leads to some noise in the kernel computation.

This Kernel is structurally comparable to the Partial Tree Kernel (PTK) described in [14]. The main difference is that the PTK does not support feature vectors linked to nodes and therefore discards important information available for our task (see table 1 for features associated with nodes). Another important difference is that our node kernel computes the sum of all node kernels of the matching subsequences, while the PTK builds a product of the node kernels.

Because Δ computes the similarity of two nodes and their substructures, Δ is likely to be called multiple times for the same nodes during the computation of $K_{\text{All-Pairs}}$. Due to its computational cost we implemented a cache to limit the number of times Δ has to be calculated to exactly $|V_x| \cdot |V_y|$ times.

It is clear that $K_{\text{All-Pairs}}$ is a kernel, because it is a sum and product of valid kernels, which has been shown to be also a valid kernel. [17]

4.2 Dependency Path Tree Kernel

In [2] the authors argue that the information expressing a relation between two entities is almost exclusively concentrated in the dependency path between them. Motivated by this they propose the Shortest Path Kernel for relation extraction. This kernel compares the nodes on the path between the two relation argument entities in the dependency tree. Our experiments revealed that this kernel performs almost as well as the Dependency Tree Kernel which shows that the path contains almost as much information as the whole subtree. However this kernel has some restrictions, like the condition that the path in both trees needs to have the same length, as otherwise the relation is discarded. Moreover the kernel computes a product of the node similarities, yielding zero similarity if only one node pair is completely dissimilar.

To avoid these constraints we propose a second new kernel combining the SPK with the DTK utilizing the main ideas of the Subsequence Kernel [3],[17]. The Dependency Path Tree Kernel (Path-DTK) not only measures the similarity of the root nodes and its descendents as in [4] or only the similarities of nodes on the path [2], it considers the similarities of all nodes (and substructures) using the node kernel on the path connecting the two relation argument entity nodes. To this end the pairwise comparison is performed using the ideas of the subsequence kernel from [17], therefore relaxing the "same length" restriction of the SPK.

In order to integrate these features into our kernel we need to define how the nodes on the dependency path contribute to the overall kernel. We use a variant of the Subsequence Kernel for the computation of the path subsequence sums of the single node kernels. In our benchmark training data (ACE 2003) the dependency paths include up to 28 nodes. So for a computationally feasible solution we need a limitation on the maximum subsequence length and hence the number of possible subsequences. We introduce an additional parameter q which acts as an upper bound on the index distance $d(\mathbf{i})$. In contrast to a limitation on the length of a sequence, this allows us to compute much longer subsequences because the number of possible combinations is very much reduced.

Besides q we define another parameter μ with $0 < \mu \leq 1$ as a factor that penalizes gaps on the path. The Dependency Path Tree Kernel is then defined as:

$$K_{\text{Path-DTK}}(X, Y) = \sum_{\substack{i \in I_{|x|},\, j \in I_{|y|}, \\ |\mathbf{i}|=|\mathbf{j}|,\, d(\mathbf{i}), d(\mathbf{j}) \leq q}} \mu^{d(\mathbf{i})+d(\mathbf{j})} \quad \Delta'(x(\mathbf{i}), y(\mathbf{j})) \quad M(x(\mathbf{i}), y(\mathbf{j})) \tag{5}$$

where x and y are the implied paths $path(x_1, x_2)$ and $path(y_1, y_2)$ between the relation arguments of the instances and $x(\mathbf{i})$ is the subsequence of x implied by the ordered sequence $\mathbf{i} = o(i)$ of the index set i analogous for j. In the examples in figure 1 this leads to the dependency paths $x = (a_3, a_1, a_4, a_6, a_7)$ and $y = (b_2, b_1, b_3, b_5, b_6)$.

The Dependency Path Tree Kernel effectively compares the nodes from paths with different lengths while maintaining the ordering information and considering the similarities of substructures. The path as well as the substructures turned out to hold valuable information in dependency trees. The Dependency Path Tree Kernel accounts for both which makes it a flexible kernel for relation extraction.

This kernel $K_{\text{Path-DTK}}$ also consists of only sums and products of a valid kernel [20], so $K_{\text{Path-DTK}}$ is also a valid kernel. [17]

5 Efficient Computation

The runtime of the kernels is highly dependent on the algorithm used to count the matching subsequences. To reduce the exponential runtime needed for naive enumeration [20] adapted the recursive algorithm of the String Subsequence Kernel [12] to dependency trees. This algorithm needs $O(mn^3)$ operations for sequence lengths n, m with $n \leq m$. Another approach to the subsequence computation for strings running in $O(mn^2)$ was proposed by [17]. We adapted this solution for the computation of $C(s, t)$ as denoted in pseudocode in algorithm 1 which leads to an $O(mn^2)$ solution for the computation of the child node subsequences $C(s, t)$.

However, because we observed in our benchmark data an average inner node child count of 1.72, another approach based on a caching strategy is useful. Let

Algorithm 1. Compute $C(s,t)$ in $O(mn^2)$ with sequence lengths $n \le m$

```
1.  k ← 0,  m ← |s|,  n ← |t|,  p ← min(m,n)
2.  for i = 1 : m do
3.    for j = 1 : n do
4.      if mat(s(i),t(j)) = 1 then
5.        DPS(i,j) ← λ^2
6.        DQS(i,j) ← λ^2 Δ(s(i),t(j))
7.        k ← k + DQS(i,j)
8.      else
9.        DPS(i,j) ← 0
10.       DQS(i,j) ← 0
11.     end if
12.   end for
13. end for
14. DP(0 : m,0) ← 0,  DQ(0 : m,0) ← 0,  DP(0,1 : n) ← 0,  DQ(0,1 : n) ← 0
15. for l = 2 : p do
16.   for i = 1 : m do
17.     for j = 1 : n do
18.       DP(i,j) ← DPS(i,j) + λ DP(i−1,j) + λ DP(i,j−1) − λ^2 DP(i−1,j−1)
19.       DQ(i,j) ← DQS(i,j) + λ DQ(i−1,j) + λ DQ(i,j−1) − λ^2 DQ(i−1,j−1)
20.       if mat(s(i),t(j)) = 1 then
21.         DPS(i,j) ← λ^2 DP(i − 1,j − 1)
22.         DQS(i,j) ← λ^2 DQ(i − 1,j − 1) + Δ(s(i),t(j)) DPS(i,j)
23.         k ← k + DQS(i,j)
24.       end if
25.     end for
26.   end for
27. end for
28. return k
```

Algorithm 2. Compute $C(s,t)$ with prepared index sequences SI

```
1.  k ← 0,  m ← |s|,  n ← |t|,  p ← min(m,n)
2.  for q = 1 : p do
3.    for i ∈ SI_{m,q} do
4.      for j ∈ SI_{n,q} do
5.        x ← 0
6.        for r = 1 : q do
7.          if mat(s(i(r)),t(j(r))) = 0 then
8.            goto 4 // jump to next index sequence j
9.          end if
10.         x ← x + Δ(s(i(r)),t(j(r)))
11.       end for
12.       k ← k + x λ^{d(i)+d(j)}
13.     end for
14.   end for
15. end for
16. return k
```

Table 2. Averaged over 5 measurements, this table shows the relative empirical runtimes of the DTK with the different subsequence algorithms

Algorithm	Theoretical runtime	Relative empirical runtime	Std. dev.
Algorithm 2	$\sum_{q=1}^{n} q \binom{m}{q} \binom{n}{q}$	1	–
Algorithm 1	$mn + mn^2$	3.2204	0.0345
Zelenko	mn^3	2.9338	0.0363

$SI_{p,q} = \{o(i) | i \in I_p \wedge |i| \leq q\}$ be the set of all ordered index subsequences of length q with highest index p. We can write the child subsequence kernel $C(s,t)$ as

$$C(s,t) = \sum_{q=1}^{min(m,n)} \sum_{\mathbf{i} \in SI_{m,q}} \sum_{\mathbf{j} \in SI_{n,q}} \lambda^{d(\mathbf{i})+d(\mathbf{j})} \Delta'(s(\mathbf{i}), t(\mathbf{j}))\ M(s(\mathbf{i}), t(\mathbf{j})) \quad (6)$$

with $m = |s|$ and $n = |t|$. This reformulation of the child subsequence kernel allows an efficient computation by utilizing a cache for the Δ function. The computation procedure is described in algorithm 2. It is easy to see that the worst case runtime (if all nodes match) of this algorithm for $n \leq m$ is

$$\mathrm{O}\left(\sum_{q=1}^{n} q \cdot \binom{n}{q} \cdot \binom{m}{q}\right)$$

However, for $n, m \leq 4$ this approach needs less operations than the $\mathrm{O}(mn^3)$ solution proposed by [20]. For $n, m \leq 3$ it also needs less operations than algorithm 1, which has an exact runtime of $\mathrm{O}(mn + mn^2)$. For 95% of the occurring calculations of $C(s,t)$ during the kernel computations on the benchmark dataset it holds that $n, m \leq 3$. Our experiments on the benchmark data set show that – though it explicitly enumerates all subsequence combinations – algorithm 2 has a better practical runtime than the two other solutions, as can be seen in table 2. Another interesting result is that the theoretically fastest approach (Algorithm 1) is even slower than the Zelenko algorithm on our dataset.

6 Related Work

There are several ways to tackle the problem of relation extraction. Prominent solution strategies have been mentioned in the introduction. Among the first who proposed kernel methods for relation extraction on shallow parse trees was [20]. The resulting kernel was then generalized to be able to handle dependency trees by [4]. [2] suggested a kernel for relation extraction which only considers the shortest path between the two entity nodes which nevertheless yielded good performance. Besides work on tree kernels for relation extraction there have been tree kernels proposed for other tasks like question classification [14] which have shown improvements over bag-of-words approaches. Considering features associable with words [5] proposed to use semantic background knowledge from various sources like WordNet, FrameNet and PropBank to enhance the DTK of [4] and showed good results.

7 Experiments

In this section we present the results of the experiments with kernel-based methods for relation extraction. Throughout this section we will compare our approaches with other state-of-the-art kernels considering their classification quality on the publicly available benchmark dataset ACE-2003 [13] which has been commonly used for the evaluation of relation extraction systems in previous work.

7.1 Technical Realization

We implemented our tree-kernels in Java and used Joachim's [9] SVM^{light} with the JNI Kernel Extension[1]. Each experiment was splitted into different tasks to allow distributed processing for better efficiency, e.g. each fold of the cross validation was computed on a different machine. Depending on the kernel and the relation this resulted in training times of minutes for the simple kernels to 12 hours for the advanced kernels. We also employed a standard grid-search to optimize the parameters of all kernels as well as the SVM-parameter C. The parameter optimization followed a strict protocol of exclusively utilizing training data to avoid a biased comparison. We only report the best configurations in the result tables. Because the implementations of the other kernels were unavailable and the experimental setup is critical for a fair comparison of the different kernels we implemented all other state of the art kernels for relation extraction, using the implementation details from the original papers. This allows for a comprehensive comparison of our approach with state-of-the-art kernels.

7.2 Benchmark Dataset

The ACE-2003 corpus consists of 519 news documents from different sources splitted in a test and training set containing 176825 words in 9256 sentences. In all documents entities and the relations between them were annotated by humans annotators. The entities are annotated by the types *named* (e.g. "Barack Obama") , *nominal* (e.g. "government") and *pronominal* (e.g. "he"). There are 5 top level relation types *role*, *part*, *near*, *social* and *at* (see table 3), which are further differentiated into 24 subtypes.

[1] Available from www.aifb.uni-karlsruhe.de/WBS/sbl/software/jnikernel/

Table 3. A list of relation types and the respective frequencies in the ACE-2003 corpus. On the left side the number of relations between named entities is counted.

Relation	Named–Named # Training	Named–Named # Test	All Entity Types # Training	All Entity Types # Test.
At	481	106	1602	389
Near	44	14	220	70
Part	265	64	749	163
Role	732	155	2927	712
Social	55	4	611	112

7.3 Experimental Setup

As currently neither nominal nor pronominal co-reference resolution can be done with sufficient quality we restricted our experiments to *named–named* relations, where named entity recognition approaches may be used to extract the arguments. Nevertheless our kernels without any modification could also applied to the all types setting as well. Throughout our experiments we conducted classification tests on the five top level relations of the dataset. For each class in each fold we trained a separate SVM following the one vs. all scheme for multi-class classification.

7.4 Evaluation Metrics

We use the standard evaluation measures for classification accuracy: precision, recall and f-measure defined as follows:

$$Prec = \frac{TP}{TP + FP} \qquad Rec = \frac{TP}{TP + FN} \qquad F = \frac{2 \cdot Prec \cdot Rec}{Prec + Rec}$$

with the number of *true positive* (TP), *false positive* (FP) and *false negative* (FN) of the classification. Because we have a multi class classification problem we report macro (per class) and micro (per instance) averaged evaluation scores as well as results for each individual class of the different kernels [19]. For each experiment we used a 5-times repeated 5-fold cross-validation. Based on this we can compute the average for all metrics and report the standard error for the F-scores. The use of a 5-times repeated CV allows to estimate the significance of the difference in classification performance of the different kernels.

7.5 Results

Table 4 gives a quick overview of the overall classification performance of the different kernels on the benchmark dataset. In this table the classification performance of the Dependency Tree Kernel (DTK)[4], Shortest Path Kernel (SPK)[2], All-Pairs Dependency Tree Kernel (All-Pairs-DTK) and Dependency Path Tree Kernel (Path-DTK) on the 5-times repeated 5-Fold CV is shown. The All-Pairs-DTK outperforms the SPK by 5.9% F_{micro} and 23.5% F_{macro} with a significance

Table 4. Table of results with 5-times repeated 5 fold cross validation. The values in parenthesis denote the standard error over the five different CV runs.

Kernel	q	μ	λ	C	Precision	Recall	F_{micro}	F_{macro}
SPK [2]	–	–	–	1	79.8%	46.4%	58.6%(0.14)	34.2%(0.14)
DTK [4]	–	–	0.65	100	71.7%	53.7%	61.4%(0.32)	44.5%(0.63)
All-Pairs-DTK	–	–	0.6	60	73.1%	57.8%	64.5%(0.26)	**57.7%**(0.54)
Path-DTK	1	0.5	0.5	10	**80.2%**	**61.2%**	**69.4%**(0.09)	57.3%(0.40)

Table 5. Table of results on the pre-specified testset

Kernel	q	μ	λ	C	Precision	Recall	F_{micro}	F_{macro}
SPK [2]	–	–	–	1	74.7%	34.4%	47.1%	25.0%
DTK [4]	–	–	0.65	100	79.5%	44.0%	56.7%	36.9%
All-Pairs-DTK	–	–	0.6	60	**80.2%**	49.6%	61.3%	40.6%
Path-DTK	1	0.5	0.5	10	76.7%	**52.8%**	**62.5%**	**41.3%**

Table 6. Table of results for the single top-level relations with 5-times repeated 5 fold cross validation

Kernel	Relation	Precision	Recall	F-score	Std. error
SPK [4]	At	78.1%	41.8%	54.5%	(0.34)
	Near	10.0%	0.5%	0.9%	(0.78)
	Part	76.8%	25.4%	38.1%	(0.76)
	Role	81.4%	62.9%	71.0%	(0.16)
	Social	54.0%	3.6%	6.8%	(0.06)
DTK [2]	At	65.2%	47.4%	54.9%	(0.52)
	Near	52.9%	29.1%	37.3%	(2.20)
	Part	71.8%	41.8%	52.8%	(0.67)
	Role	78.4%	67.1%	72.3%	(0.30)
	Social	11.5%	3.6%	5.5%	(0.73)
All-Pairs-DTK	At	65.2%	54.0%	59.1%	(0.28)
	Near	93.2%	71.8%	**80.9%**	(1.57)
	Part	70.0%	43.4%	53.6%	(0.48)
	Role	79.1%	67.8%	73.0%	(0.44)
	Social	39.6%	15.6%	**22.2%**	(2.08)
Path-DTK	At	73.4%	58.0%	**64.8%**	(0.05)
	Near	90.9%	50.5%	64.9%	(2.19)
	Part	85.5%	49.8%	**62.9%**	(0.67)
	Role	83.4%	71.9%	**77.2%**	(0.13)
	Social	42.4%	10.6%	16.8%	(1.34)

Table 7. The results of different parameters with 5-fold CV on a single seed on the training data in the optimization phase

Kernel	q	μ	λ	C	Prec	Rec	F_{micro}	F_{macro}
SPK				1	79.1%	**46.6%**	**58.7%**	**33.9%**
				10	**79.2%**	46.5%	58.6%	**33.9%**
				60	78.8%	46.6%	58.6%	**33.9%**
				100	9.9%	46.0%	16.3%	21.4%
DTK			0.65	1	**88.3%**	35.4%	50.5%	26.9%
			0.65	10	73.7%	53.1%	61.7%	43.7%
			0.65	60	72.7%	54.8%	**62.5%**	45.4%
			0.65	100	72.7%	**54.9%**	**62.5%**	**45.5%**
			0.6	1	87.8%	34.3%	49.3%	26.3%
			0.6	10	73.0%	52.9%	61.3%	43.8%
			0.6	60	72.3%	54.9%	62.4%	45.4%
			0.6	100	72.1%	55.0%	62.4%	45.4%
			0.5	1	86.1%	31.5%	46.1%	24.1%
			0.5	10	71.5%	51.5%	59.9%	41.3%
			0.5	60	69.8%	53.7%	60.7%	43.1%
			0.5	100	69.8%	54.2%	61.0%	44.2%
All-Pairs-DTK			0.65	1	89.0%	36.0%	51.2%	30.2%
			0.65	10	78.4%	54.4%	64.2%	53.1%
			0.65	60	74.4%	57.4%	64.8%	54.8%
			0.65	100	73.8%	57.6%	64.7%	54.8%
			0.6	1	89.2%	35.1%	50.3%	28.4%
			0.6	10	79.7%	53.9%	64.3%	53.3%
			0.6	60	74.3%	**58.2%**	**65.3%**	**56.7%**
			0.6	100	73.5%	58.1%	64.9%	56.5%
			0.5	1	**89.4%**	33.5%	48.8%	25.2%
			0.5	10	81.3%	51.9%	63.4%	54.3%
			0.5	60	73.6%	57.1%	64.3%	57.2%
			0.5	100	71.8%	58.1%	64.2%	58.3%
Path-DTK	1	0.5	0.65	1	86.8%	40.3%	55.1%	29.2%
	1	0.5	0.65	10	79.8%	60.6%	68.9%	55.7%
	1	0.5	0.65	60	78.4%	61.1%	68.7%	56.4%
	1	0.5	0.65	100	78.2%	61.1%	68.6%	56.3%
	1	0.5	0.6	1	**86.9%**	39.7%	54.5%	28.7%
	1	0.5	0.6	10	80.3%	61.3%	69.5%	55.6%
	1	0.5	0.6	60	77.9%	61.8%	68.9%	56.6%
	1	0.5	0.6	100	77.9%	61.8%	68.9%	56.6%
	1	0.5	0.5	1	86.8%	38.4%	53.3%	27.1%
	1	0.5	0.5	10	80.6%	61.4%	**69.7%**	56.1%
	1	0.5	0.5	100	77.4%	61.8%	68.7%	57.4%
	1	0.5	0.5	60	77.4%	61.6%	68.6%	57.3%
	2	0.5	0.5	10	79.4%	61.1%	69.1%	52.8%
	3	0.5	0.5	10	79.7%	61.3%	69.3%	52.7%
	4	0.5	0.5	10	80.1%	61.6%	69.6%	52.8%

of 99%. The All-Pairs-DTK also beats the DTK by 3.1% F_{micro} and 13.2% F_{macro} again with a significance of 99%. It has also the best F_{macro} score of all kernels although the difference is not significant. The Path-DTK exceeds all other kernels by at least 4.9% F_{micro} measure with a significance of 99%. It also beats DTK and SPK by at least 8.0% F_{micro} measure and 12.8% F_{macro} again with a significance of 99%. This experiment clearly shows that our kernels are superior to the previously published kernels by a large margin with a high significance. In table 6 we present the results of the different kernels on each of the 5 top level relations. We see that the All-Pairs-DTK outperforms all other kernels on the "Near" relation by 16.0% F-score and on the "Social" relation at least by about 5.4% F-score. This is an interesting result because for these two relations the least training data is available. This indicates that All-Pairs-DTK might need less training data than the other kernels. The Path-DTK outperforms all other kernels on the "At", "Part" and "Role" relations with a significance of 99%. These results allow the conclusion that the Path-DTK performs best on relations where many training examples are available. However it is not clear from our experiments on the benchmark data set which of our proposed kernels is suited best for an arbitrary relation. It is clear that more experiments on different data sets are needed to make such a judgement. Table 7 shows the results for all different parameter settings. Because the benchmark data set consists of a separate test set we also conducted experiments where we trained the classifier only on the training set. We show these results in table 5. The Path-DTK performs best on the test set outperforming the All-Pairs-DTK by 1.2% F_{micro} and the other published kernels by at least 5.8% F_{micro}. This further supports the conclusion that the Path-DTK is a good general kernel for relation extraction.

8 Conclusion and Future Work

In this paper we presented two new tree kernels for relation extraction using dependency parse trees. They use richer structural features and yield significantly better results than previosly published approaches. We optimized kernel computations for dependency trees reducing the runtime considerably. For some relations the accuracy level of our methods seem to be good enough for practical application, e.g. for information retrieval. An examination on how those kernels can be combined with other kernel is meaningful [16]. Another promising direction for future work is the usage of more sophisticated features which aim at capturing the semantics of words. Here word sense disambiguation approaches [15] might be employed or topic modeling algorithms may be used to assign broader semantic information to the words relevant for a relation.

Acknowledgement

The work presented here was funded by the German Federal Ministry of Economy and Technology (BMWi) under the THESEUS project. We would like to thank the reviewers for their detailed comments.

References

1. Blohm, S., Cimiano, P.: Scaling up pattern induction for web relation extraction through frequent itemset mining. In: Proc. KI 2008 WS on Ontology-Based IE Systems (2008)
2. Bunescu, R.C., Mooney, R.J.: A shortest path dependency kernel for relation extraction. In: Proc. EMNLP 2005 (2005)
3. Bunescu, R.C., Mooney, R.J.: Subsequence kernels for relation extraction. In: Proc. Neural Information Processing Systems, NIPS 2005 (2005)
4. Culotta, A., Sorensen, J.: Dependency tree kernels for relation extraction. In: Proc. ACL 2004 (2004)
5. Harabagiu, S., Bejan, C.A., Morarescu, P.: Shallow semantics for relation extraction. In: Proc. IJCAI 2005 (2005)
6. Hearst, M.: Automatic acquisition of hyponyms from large text corpora. In: Proc. COLING 1992 (1992)
7. Horvath, T., Paass, G., Reichartz, F., Wrobel, S.: A logic-based approach to relation extraction from texts. In: ILP 2009 (2009)
8. Joachims, T.: Text categorization with support vector machines: learning with many relevant features. In: Nédellec, C., Rouveirol, C. (eds.) ECML 1998. LNCS, vol. 1398. Springer, Heidelberg (1998)
9. Joachims, T.: Making large-scale SVM learning practical. In: Advances in Kernel Methods - Support Vector Learning. MIT Press, Cambridge (1999)
10. Klein, D., Manning, C.D.: Accurate unlexicalized parsing. In: Proc. ACL 2003 (2003)
11. Klein, D., Manning, C.D.: Corpus-based induction of syntactic structure: Models of dependency and constituency. In: Proc. ACL 2004 (2004)
12. Lodhi, H., Saunders, C., Shawe-Taylor, J., Cristianini, N., Watkins, C.: Text classification using string kernels. JMLR (2), 419–444 (2002)
13. Mitchell, A.: ACE-2 Version 1.0; corpus LDC2003T11. Linguistic Data Consortium, Philadelphia (2003), `http://www.ldc.upenn.edu`
14. Moschitti, A.: Efficient convolution kernels for dependency and constituent syntactic trees. In: Fürnkranz, J., Scheffer, T., Spiliopoulou, M. (eds.) ECML 2006. LNCS (LNAI), vol. 4212, pp. 318–329. Springer, Heidelberg (2006)
15. Paaß, G., Reichartz, F.: Exploiting semantic constraints for estimating supersenses with crfs. In: Proc. SDM 2009 (2009)
16. Reichartz, F., Korte, H., Paass, G.: Composite kernels for relation extraction. In: Proc. ACL 2009 (2009)
17. Shawe-Taylor, J., Cristianini, N.: Kernel Methods for Pattern Analysis. Cambridge University Press, Cambridge (2004)
18. Tjong, E.F., Sang, K., De Meulder, F.: Language-independent named entity recognition. CoRR cs.CL/0306050 (2003)
19. Yang, Y.: An evaluation of statistical approaches to text categorization. Information Retrieval 1(1-2), 69–90 (1999)
20. Zelenko, D., Aone, C., Richardella, A.: Kernel methods for relation extraction. J. Mach. Learn. Res. 3, 1083–1106 (2003)

Statistical Relational Learning with Formal Ontologies

Achim Rettinger[1], Matthias Nickles[2], and Volker Tresp[3]

[1] Technische Universität München, Germany
achim.rettinger@cs.tum.edu
[2] University of Bath, United Kingdom
M.L.Nickles@cs.bath.ac.uk
[3] Siemens AG, CT, IC, Learning Systems, Germany
volker.tresp@siemens.com

Abstract. We propose a learning approach for integrating formal knowledge into statistical inference by exploiting ontologies as a semantically rich and fully formal representation of prior knowledge. The logical constraints deduced from ontologies can be utilized to enhance and control the learning task by enforcing description logic satisfiability in a latent multi-relational graphical model. To demonstrate the feasibility of our approach we provide experiments using real world social network data in form of a $\mathcal{SHOIN}(D)$ ontology. The results illustrate two main practical advancements: First, entities and entity relationships can be analyzed via the latent model structure. Second, enforcing the ontological constraints guarantees that the learned model does not predict inconsistent relations. In our experiments, this leads to an improved predictive performance.

1 Introduction

This paper focuses on the combination of statistical machine learning with ontologies specified by formal logics. In contrast to existing approaches to the use of constraints in machine learning (ML) and data mining, we exploit a semantically rich and fully formal representation of hard constraints which govern and support the stochastic learning task. Technically, this is achieved by combining the *Infinite Hidden Relational Model* (IHRM) approach to Statistical Relational Learning (SRL) with inference guided by the constraints implied by a *Description Logic* (DL) ontology used on the Semantic Web (SW). In this way, our approach supports a tight integration of formal background knowledge resulting in an *Infinite Hidden Semantic Models* (IHSM). The term *Semantic* in IHSM stands for this integration of "meaningful", symbolic knowledge which enables the use of deductive reasoning.

Benefits of the presented approach are (1) the analysis of known entity classes of individuals by means of clustering and (2) the completion of the knowledge base (KB) with uncertain predictions about unknown relations while considering constraints as background knowledge for the machine learning process. Thus, it is guaranteed that the learned model does not violate ontological constraints and

W. Buntine et al. (Eds.): ECML PKDD 2009, Part II, LNAI 5782, pp. 286–301, 2009.

the predictive performance can be improved . While there is some research on data mining for the SW, like instance-based learning and classification of individuals, considering "hard" constraints specified in the ontology during machine learning has hardly been tried so far or only in quite restricted and semi-formal ways (see Sec. 6 for related work). Even though we use a social network OWL DL ontology and settle on SRL as an apparently natural counterpart for logical constraints, our general approach is in no way restricted to DL or SRL and could be easily adapted to other formal and learning frameworks.

To provide an intuitive understanding of the presented approach we will use a simple example throughout this paper to illustrate the application of constraints in our learning setting: Consider a social network where, amongst others, the age of persons and the schools they are attending is partially known. In addition, an ontology designer specified that persons under the age of 5 are not allowed to attend a school. All this prior knowledge is provided in a formal ontology and the ultimate task is to predict unknown elements of this network.

The remainder of this paper is structured as follows: In Sec. 2 we specify an ontology in OWL DL that defines the taxonomy, relational structure and constraints. Next, we show how to infer a relational model from the ontology and transfer the relational model into an IHSM (Sec. 3). Then, we learn the parameters of this infinite model in an unsupervised manner while taking the constraints into account (Sec. 4). In Sec. 5 the IHSM is evaluated empirically using a complex dataset from the Semantic Web. Finally, we discuss related work in Sec. 6 and conclude in Sec. 7.

2 Formal Framework

Our approach requires the specification of formal background knowledge and formal constraints for the learning process. We do so by letting the user of the proposed machine learning algorithm specify a formal ontology or use an existing ontology e.g. from the SW. In computer science, an ontology is the formal representation of the concepts of a certain domain and their relations. In the context of the (Semantic) Web and thus also in our approach, such an ontology is typically given as a so-called *TBox* and a *ABox*, each of which consists of a number of description logic formulas. The TBox comprises conceptual knowledge (i.e., knowledge about classes and their relations), whereas the ABox comprises knowledge about the instances of these classes. In our context, the given ontology and also all knowledge which is a logical consequence from it constitutes the "hard" knowledge for the learning process (i.e., knowledge which cannot be overwritten during learning), as described later.

However, our approach is not restricted to ontologies, but works in principle with all sorts of formal knowledge bases. We are using ontologies mainly because there is an obvious relatedness of clustering and ontological classification, and because formal languages, reasoners, editors and other tools and frameworks for ontologies on the SW are standardized and widely available. Consequently, we use a description logic for our examples. This is not only because ontologies and other formal knowledge on the Web (which is our application here) are usually represented using DLs, but also because the standard DL which we use is a decidable fragment of first-order logic (FOL) for which highly optimized reasoners exist.

We settle on the $\mathcal{SHOIN}(D)$ [1] description logic, because entailment in the current Semantic Web standard ontology language OWL DL can be reduced to $\mathcal{SHOIN}(D)$ knowledge base satisfiability. We could likewise work with OWL DL syntax directly, but that wouldn't have any technical advantages and would just reduce the readability of our examples. Our approach requires that the satisfiability or consistency of ontologies can be checked, which is a standard operation of most automated reasoning software for the SW. Allowing to check the satisfiability means that the reasoner is able to check whether a given KB (ontology) has a model. On the syntactic level, satisfiability corresponds to consistency, i.e., there are no sentences in the given ontology which contradict each other. The following specifies the syntax of $\mathcal{SHOIN}(D)$. Due to lack of space, please refer to [1] for a detailed account of this language.

$$\begin{aligned} C \rightarrow A|\neg C|C_1 \sqcap C_2|C_1 \sqcup C_2|\exists R.C|\forall R.C \qquad & | \\ \geq nS| \leq nS|\{a_1, ..., a_n\}| \geq nT| \leq nT & | \\ \exists T_1, ..., T_n.D|\forall T_1, ..., T_n.D|D \rightarrow d|\{c_1, ..., c_n\} & \end{aligned}$$

Here, C denote *concepts*, A denote *atomic concepts*, R denote *abstract roles* or *inverse roles* of abstract roles (R^-), S denote abstract *simple roles*, the T_i denote *concrete roles*, d denotes a concrete *domain predicate*, and the a_i / c_i denote abstract / concrete *individuals.*

A $\mathcal{SHOIN}(D)$ *ontology* or *knowledge base* is then a non-empty, finite set of TBox axioms and ABox assertions $C_1 \sqsubseteq C_2$ (inclusion of concepts), $Trans(R)$ (transitivity), $R_1 \sqsubseteq R_2$, $T_1 \sqsubseteq T_2$ (role inclusion for abstract respectively concrete roles), $C(a)$ (concept assertion), $R(a, b)$ (role assertion), $a = b$ (equality of individuals), and $a \neq b$ (inequality of individuals). Concept equality is denoted as $C_1 \equiv C_2$ which is just an abbreviation for mutual inclusion, i.e., $C_1 \sqsubseteq C_2, C_2 \sqsubseteq C_1$. Defining a semantics of $\mathcal{SHOIN}(D)$ is not required within the scope of this work, the canonical semantics which we assume in this work can be found, e.g., in [1].

2.1 Constraints

Constraints in the sense of this work are actually just formal statements. Our approach is expected to work with all kinds of logical frameworks which allow for satisfiability (or consistency) checks over some given set of logical sentences, for example an ontology. This set of given statements is denoted as the KB in the further course of this paper. Formally, we define a set of constraints C to be the deductive closure $\Theta(KB)$ of a given knowledge base KB, with $\Theta(KB) = \{c|KB \models c\}$. The deductive closure contains not only explicitly given knowledge (the knowledge base KB), but also all logical sentences which can be derived from the KB via deductive reasoning. E.g., if the KB would contain the sentences $\neg a$ and $\neg a \rightarrow b$, the deductive closure would also contain b.

The application-specific constraint set which we use as an OWL DL ontology is similar to the well-known *Friend-Of-A-Friend* (FOAF) social network schema, together with additional constraints which will be introduced later. The following ontology SN comprises only a fragment of the full FOAF-like ontology we have used (with DOB meaning "date of birth" and $hasBD$ meaning "has birthday".).

$Person \sqsubseteq Agent$	$knows^{-} \sqsubseteq knows$	$\exists knows.\top \sqsubseteq Person$
$\top \sqsubseteq \forall knows.Person$	$\exists hasBD.\top \sqsubseteq Person$	$\top \sqsubseteq \forall hasBD.DOB$
$\top \sqsubseteq \leq 1\ hasBD$	$\top \sqsubseteq \geq 1\ hasBD$	$\exists yearValue.\top \sqsubseteq DOB$
$\top \sqsubseteq \forall yearValue.gYear$	$\top \sqsubseteq \leq 1\ yearValue$	$\top \sqsubseteq \forall attends.School$

These axioms mainly express certain properties of binary relations (so-called *roles*) between classes. For example, $\top \sqsubseteq \forall attends.School$ specifies that in our example ontology the range (target set) of role *attends* is *School*.

In addition to these, we provide the machine learning algorithm with an ABox which models an incomplete social network. The later machine learning task consists essentially in a (uncertain) completion of this given network fragment. An example for such additional individuals-governing constraints A: $tim : Person, tina : Person, tom : Person$; $(tina, tim) : knows, (tina, tom) : knows$.

Note, that these relationships among persons cannot be weakened or overwritten by the learning process, even if they contradict observed data. They need to be provided manually by the KB engineer. As further constraints, we assume some specific properties G of the analyzed social network. The following set of axioms expresses that no one who is younger than six years goes to school. At this, $UnderSixYearsOld$ is the class which contains persons with an age less than six years (calculated from the given dates of birth):

$$Pupil \sqsubseteq Person \;\Big|\; Pupil \sqsubseteq \neg UnderSixOld \;\Big|\; Pupil \sqsubseteq \exists\ attendsSchool$$

The complete set of given formal and definite knowledge for our running example is then $C = \Theta(SN \sqcup A \sqcup G)$.

Example Data: The set of data used as examples for the learning tasks takes the form of ABox assertions. But in contrast to the ABox knowledge in set A above, an example here might turn out to be wrong. We also do not demand that examples are mutually consistent, or consistent with the ontology. In order to maintain compatibility with the expected input format for relational learning, we restrict the syntax of examples to the following two description logic formula patterns:

$$instance : category$$
$$(instance_a, instance_b) : role$$

At this, roles correspond to binary relations. The set of all example data given as logical formulas is denoted as D.

3 Infinite Hidden Semantic Models

The proposed *Infinite Hidden Semantic Model* (IHSM) is a machine learning algorithm from the area of SRL [2]. The novelty is its additional ability to exploit formal ontologies as prior knowledge given as a set of logical formulas. In our case, the constraints are provided as a $\mathcal{SHOIN}(D)$ ontology with a TBox and an ABox as just described in the previous section. In traditional ML, prior knowledge is just specified by the likelihood model and the prior distributions, parameters of the learning algorithm or selection of features.

In this section, we first show how the ontology from Sec. 2 defines a *Relational Model* (RM) which is the basis for an *Infinite Hidden Relational Model* (IHRM). Then, the IHSM is generated by constraining the IHRM appropriately.

3.1 Relational Models

First an abstract RM of concepts and roles defined in our social network ontology is created. Based on the TBox axioms given by the ontology we can create a simple sociogram as depicted in Fig. 1. A sociogram consists of three different elements: concept individuals (individuals that are instances of a concept (e.g. tim : $Person$)), attribute instances (relations between a concept and a literal (e.g. $tina$: $hasImage$)), role instances (relations between concepts (e.g. $(tina, tim)$: $knows$)). Please note that many TBox elements first need to be deduced from the ontology, so that all individuals can be assigned to its most specific concepts. This process is known as *realization* in DL reasoning. Fig. 2 shows the full RM we use for experiments in Sec. 5.

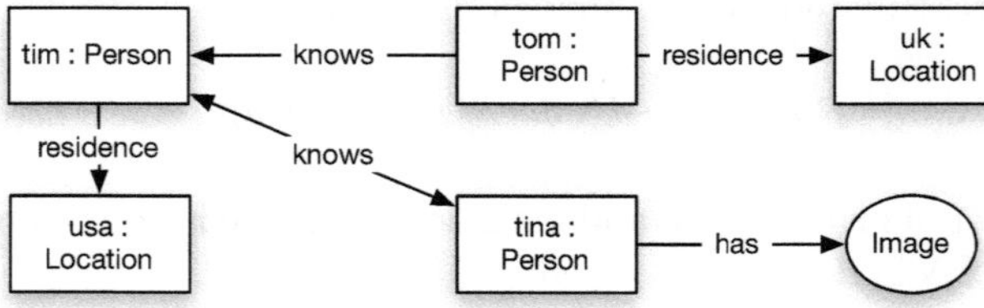

Fig. 1. Partial sociogram of the LJ-FOAF-domain

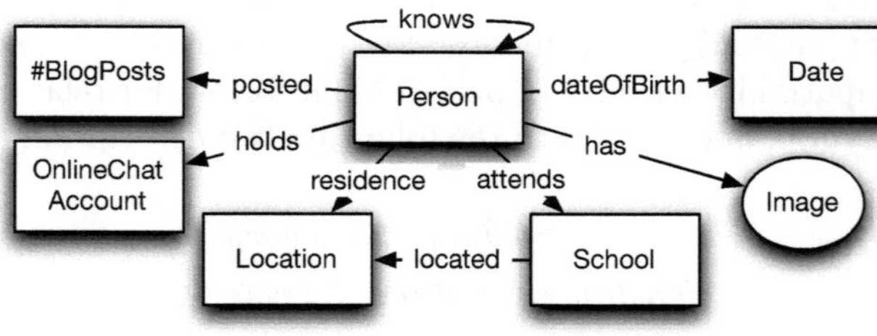

Fig. 2. Relational Model of the LJ-FOAF-domain

3.2 Infinite Hidden Relational Models

Following [3] and [4] we extend the RM to a Hidden Relational Model (HRM) by assigning a hidden variable denoted as Z_i^c to each individual i of concept c with current state k. Given that the hidden variables have discrete probability distributions they can be intuitively interpreted as clusters Z where similar individuals of the same concept c (in our case similar persons, locations, schools,...) are grouped in one specific component k. These assignments of latent states specify the component one individual is assigned to.

The resulting HRM of the sociogram shown in Fig. 1 is depicted in Fig. 3. Following the idea of hidden variables in *Hidden Markov Models* (HMMs) or *Markov Random Fields*, those additional variables can be thought of as unknown properties (roles or attributes) of the attached concept. We assume that all attributes of a concept only depend on its hidden variable and roles depend on two hidden variables of the two concepts involved. This implies that if the hidden variables were known, attributes and roles can be predicted independently. In addition, the hidden variables in the IHSM incorporate restrictions in the form of constraints imposed by the ontology (see Sec. 3.3).

Considering the HRM model shown in Fig. 3, information can now propagate via those interconnected hidden variables Z. E.g. if we want to predict whether *tom* with hidden state Z_3^1 might know *tina* (Z_2^1) we need to consider a new relationship $R_{3,2}$. Intuitively, the probability is computed based on (i) the attributes A_3^1 and A_1^1 of the latent states of immediately related persons Z_3^1 and Z_2^1; (ii) the known relations associated with the persons of interest, namely the role *knows* and *residence* $R_{2,1}$, $R_{3,1}$ and $R_{3,2}$; (iii) higher-order information indirectly transferred via hidden variables Z_3^1 and Z_2^1. In summary, by introducing hidden variables, information can globally distribute in the HRM. This reduces the need for extensive structural learning, which is known to be difficult.

Critical parameters in the HRM are the number of states in the various latent variables, which might have to be tuned as part of a complex optimization routine. A solution here offers the IHRM, that was introduced by [4] and [3]. In the IHRM, a hidden variable has a potentially infinite number of states and an estimate of the optimal number of states is determined as part of the inference process.

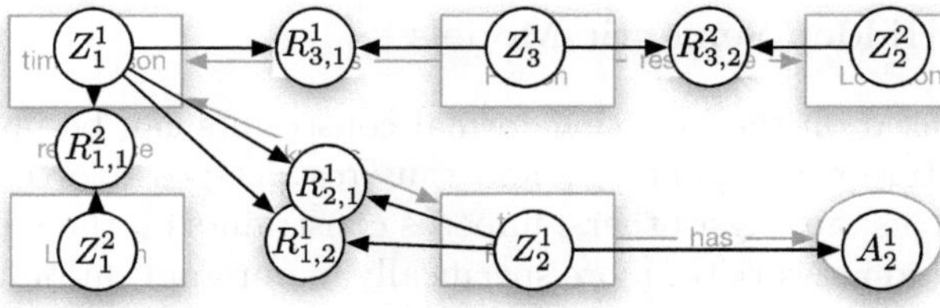

Fig. 3. Hidden relational model of the sociogram defined in Fig. 1

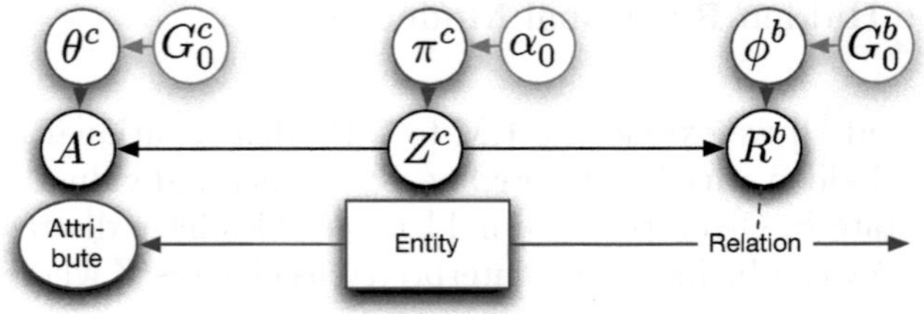

Fig. 4. Parameters of an IHRM

Finally, we need to define the remaining variables, their probability distributions and model parameters[1]. The most important parameters in our case are shown in Fig. 4. The state k of Z_i^c specifies the cluster assignment of the concept (aka entity class) c. K denotes the number of clusters in Z. Z is sampled from a multinomial distribution with parameter vector $\pi = (\pi_1, \ldots, \pi_K)$, which specifies the probability of a concept belonging to a component, i.e. $P(Z_i = k) = \pi_k$. π is referred to as mixing weights, and is drawn according to a truncated stick breaking construction with a hyperparameter α_0. α_0 is referred to as a *concentration parameter* in Dirichlet Process (DP) mixture modeling and acts as a tuning parameter that influences K. K is also limited by a truncation parameter that specifies the maximum number of components per cluster for each entity class.

Attributes A^c are generated from a Bernoulli distribution with parameters θ_k. For each component, there is an infinite number of mixture components θ_k. Each person in the component k inherits the mixture component, thus we have: $P(G_i = s|Z_i = k, \Theta) = \theta_{k,s}$. These mixture components are independently drawn from a prior G_0. The base distributions G_0^c and G_0^b are conjugated priors with hyperparameters β^c and β^b.

The truth values for the role $R_{i,j}$ involving two persons (i and j) are sampled from a binomial distribution with parameter $\phi_{k,\ell}$, where k and ℓ denote cluster assignments of the person i and the person j, respectively. $\phi_{k,\ell}^b$ is the correlation mixture component indexed by potentially infinite hidden states k for c_i and ℓ for c_j, where c_i and c_j are indexes of the individuals involved in the relationship class b. Again, G_0^b is the Dirichlet Process base distribution of a role b. If an individual i is assigned to a component k, i.e. $Z_i = k$, the person inherits not only θ_k, but also $\phi_{k,\ell}, \ell = \{1, \ldots, K\}$.

3.3 Infinite Hidden Semantic Models

The IHSM is based on the idea that formal constraints can be imposed on the correlation mixture component $\phi_{k,\ell}$ and thus restrict possible truth values for the roles $R_{i,j}$. This, amongst others, imposes constraints on the structure of the underlying ground network or, more specifically in our application, the structure

[1] Please note that we cannot focus on the technical details of an IHRM and need to refere the reader to [4] and [3] for a more detailed introduction.

of the sociogram. Recap the simple example from Sec. 1: According to this a person i known to be younger than 5 years old should not be attending any school j. The IHSM will extract this information from the ontology and set the correlation mixture component $\phi_{k,\ell}$ at entries representing according relations from *Person* component k to *School* component ℓ to 0. Here, k and ℓ denote the components the person i and the school j are assigned to. This eliminates inconsistent structural connection from the underlying ground network. More generally, all connections $R_{i,j}$ between two components k and ℓ where inconsistent individuals i and j are (partial) member of are considered void.

However, this redirection of relations by the latent variables allows IHSM not only to restrict possible connections in the ground network but makes this restriction influence the likelihood model itself. By restricting ϕ, π is affected as well. Ultimately, cluster assignments Z are influenced and information can globally propagate through the network and influence all ϕ, π and θ (see Sec. 3.2).

While this section (3.3) focused on a conceptual description of IHSM the algorithm will be specify in detail in the next section (4) before Sec. 5 presents experimental results.

4 Learning, Constraining and Predictions

The key inferential problem in the IHSM is to compute the joint posterior distribution of unobservable variables given the data. In addition, we need to avoid inconsistent correlation mixture components ϕ during learning. As computation of the joint posterior is analytically intractable, approximate inference methods need to be considered to solve the problem. We use the blocked Gibbs sampling (GS) with truncated stick breaking representation [5] a Markov chain Monte Carlo method to approximate the posterior.

Let D be the set of all available observations (observed example data, each represented as a logical formula as defined in 2.1), and let $Agents = Agent^I$ be the set of all instances of category *Agent* under interpretation I - that is informally, all persons which contribute to the social network. At each iteration, we first update the hidden variables conditioned on the parameters sampled in the last iteration, and then update the parameters conditioned on the hidden variables. So, for each entity class

1. Update hidden variable Z_i^c for each e_i^c: Assign to component with probability proportional to:

$$\pi_k^{c(t)} P(A_i^c | Z_i^{c(t+1)} = k, \theta^{c(t)}) \times \prod_{b'} \prod_{j'} P(R_{i,j'}^{b'} | Z_i^{c(t+1)} = k, Z_{j'}^{c_{j'}(t)}, \phi^{b'(t)})$$

2. Update $\pi^{c(t+1)}$ as follows:

   (a) Sample $v_k^{c(t+1)}$ from

Beta$(\lambda_{k,1}^{c(t+1)}, \lambda_{k,2}^{c(t+1)})$ for $k = \{1, \ldots, K^c - 1\}$ with

$$\lambda_{k,1}^{c(t+1)} = 1 + \sum_{i=1}^{N^c} \delta_k(Z_i^{c(t+1)}),$$

$$\lambda_{k,2}^{c(t+1)} = \alpha_0^c + \sum_{k'=k+1}^{K^c} \sum_{i=1}^{N^c} \delta_{k'}(Z_i^{c(t+1)}),$$

and set $v_{K^c}^{c(t+1)} = 1$. $\delta_k(Z_i^{c(t+1)})$ equals to 1 if $Z_i^{c(t+1)} = k$ and 0 otherwise.

(b) Compute $\pi^{c(t+1)}$ as: $\pi_1^{c(t+1)} = v_1^{c(t+1)}$ and

$$\pi_k^{c(t+1)} = v_k^{c(t+1)} \prod_{k'=1}^{k-1} (1 - v_{k'}^{c(t+1)}), \quad k > 1.$$

3. Update θ:

$$\theta_k^{c(t+1)} \sim P(\cdot|A^c, Z^{c(t+1)}, G_0^c)$$

4. Constrain ϕ to satisfiable relations:
For entity cluster k, let $F_{ext}^k = F^k \cap \{(\mathbf{e_m}, \mathbf{e_n}) : \mathbf{r} | e_m, e_n \in \mathit{Agents}, r \in R, m \neq n\}$ be the set of those logical formulas in the example data set which represent some relation ("role") r between two different individuals (persons) e_m and e_n where person e_m is assigned to component k already and e_n is assigned to a component ℓ. To keep the notation compact, we spell out role instances $(\mathbf{e_1}, \mathbf{e_2}) : \mathbf{r}$ only asymmetrically (i.e., we omit $(\mathbf{e_2}, \mathbf{e_1}) : \mathbf{r}$ if we have covered the case $(\mathbf{e_1}, \mathbf{e_2}) : \mathbf{r}$). Let $F_k \subseteq D$ be the set of *all* example formulas which have already been used to learn component k so far, that is, the subset of the data D which has been used for forming that cluster until now. Let furthermore $\vartheta(e, D)$ be the set of all sampled formulas in D where the person e appears, i.e., $f \in \vartheta(e, D)$ iff $f \in D \wedge (f \equiv \mathbf{e} : \mathbf{c} \vee f \equiv (\mathbf{e}, \mathbf{e_x}) : \mathbf{r}$ for some $c \in C$, $e_x \in \mathit{Agents}$ and $r \in R$. We use $\rho(e, j)$ to express that a certain entity e has already been assigned to a certain component j. The following steps are now used in order to check whether component k is usable w.r.t. the given set of logical constraints C:

(a) Identify the largest subset F_{clean}^k of formulas within F_{ext}^k which is consistent with C and the set of example data about person e_i^c:

$$F_{clean}^k \subseteq 2^{F_{ext}^k}, \exists\, \mathcal{I}, \mathcal{I} \models F_{clean}^k \cup \vartheta(e_i^c, D) \cup C,$$
$$\forall F \subseteq 2^{F_{ext}^k}, \exists\, \mathcal{I}, \mathcal{I} \models F \cup \vartheta(e_i^c, D) \cup C : F \subseteq F_{clean}^k$$

($\mathcal{I} \models X$ expresses that the set of logical sentences X is satisfiable, $\mathcal{I}$ being an interpretation).

(b) Verify whether F^k_{clean}, the formulas which have been used to learn "related" other clusters, $\vartheta(e^c_i, D)$ and the constraints are consistent in sum <u>if</u> we replace in F^k_{clean} the names of all persons which are assigned to components other than k with the name of person e^c_i.

Let $F^k_{upd} = \{(\mathbf{e^c_i}, \mathbf{e_m}) : \mathbf{r} | (\mathbf{e_n}, \mathbf{e_m}) : \mathbf{r} \in F^k_{ext}\}$ be the latter set of formulas. Furthermore, let $F^k_{rel} = \bigcup_{j \neq k, \rho(e_m,k), (\mathbf{e_m},\mathbf{e_n}):\mathbf{r} \in F^j} F^j$ be the set of all formulas in all other components than k which "relate" to component k using role formulas. The overall consistency check for component k yields a positive result *iff*

$$\exists\, \mathcal{I}, \mathcal{I} \models \vartheta(e^c_i, D) \cup F^k_{upd} \cup F^k_{rel} \cup C \wedge F^k_{clean} \neq \emptyset$$

Where the consistency check described above yielded a positive result:

$$\phi^{b(t+1)}_{k,\ell} \sim P(\cdot | R^b, Z^{(t+1)}, G^b_0).$$

After the GS procedure reaches stationarity the role of interest is approximated by looking at the sampled values. Here, we only mention the simple case where the predictive distribution of the existence of a relation $R_{i,j}$ between to known individuals i, j is approximated by $\phi^b_{i',j'}$ where i' and j' denote the cluster assignments of the objects i and j, respectively.

5 Experiments

The increasing popularity of social networking services like MySpace and Facebook has fostered research on social network analysis in the last years. The immense number of user profiles demands for automated and intelligent data management capabilities, e.g. formal ontologies. While data mining techniques can handle large amounts of simple facts, little effort has been made to exploit the semantic information inherent in social networks and user profiles. There is almost no work on statistical relational learning with formal ontologies in general and with SW data in particular. The lack of experiments on large and complex real world ontologies is not only due to the absence of algorithms but also due to missing suitable datasets. In this section we will present both, a large and complex SW dataset and the methodology of how to apply IHSM in practice. Ultimately, we evaluate our approach by presenting results of an empirical comparison of IHSM and IHRM in this domain.

5.1 Data and Methodology

As mentioned before our core ontology is based on Friend of a Friend (FOAF) data. The purpose of the FOAF project is to create a web of machine-readable pages describing people, the links between them and the things they create and do. The FOAF ontology is defined using OWL DL/RDF(S) and formally specified in the FOAF Vocabulary Specification 0.91[2]. In addition, we make use of

[2] http://xmlns.com/foaf/spec/

Table 1. No. of individuals, no. of instantiated roles and final number of components

Concept	#Indivi.	Role	#Inst.	#C. IHRM	#C. IHSM
Location	200	*residence*	514	18	17
School	747	*attends*	963	**36**	**48**
OnlineChatAccount	5	*holdsAccount*	427	4	4
Person	638	*knows*	8069	**38**	**45**
		hasImage	574		
Date	4	*dateOfBirth*	194	**4**	**2**
#BlogPosts	5	*posted*	629	4	4

further concepts and roles which are available in the data (see Sec. 2.1). We gathered our FOAF dataset from user profiles of the community website LiveJournal.com[3] (This specific ontology will be called LJ-FOAF from now on).

All extracted concepts and roles are shown in Fig. 2. Tab. 1 lists the number of different individuals (left column) and their known instantiated roles (middle column). Please note that *Date* and *#BlogPosts* are reduced to a small number of discrete states. As expected for a social networks *knows* is the primary source of information. This real world data set offers both, a sufficiently large set of individuals for inductive learning and a formal ontology specified in RDFS and OWL. However, while LJ-FOAF offers a taxonomy there are no complex constraints given. Thus, to demonstrate the full potential of IHSM, we additionally added constraints that are not given in the original ontology (see Sec. 2.1).

To implement all features of IHSM we made use of additional open source software packages: The Semantic Web framework Jena[4] is used to load, store and query the ontology and Pellet[5] provides the OWL DL reasoning capabilities. This outlines the workflow: First, the TBox axioms are designed and loaded into Jena. Next, all ABox assertions are added and loaded into Jena. Then, by using the taxonomy information from the ontology and the ABox assertions we extract the RM as described in Sec. 3.1. This RM is transferred into a IHSM by adding hidden variables and parameters, accordingly. Finally, the parameters are learned from the data, while constraints are constantly checked as shown in Sec. 4.

In our experiments the standard setting for the truncation parameter were *#Individuals*/10 for entity classes with over 100 instances and *#Individuals* for entity classes with less individuals. The standard iterations of the Gibbs sampler are 100. We did not engage in extensive parameter tuning because the purpose of this evaluation is to examine the influence of the constraints and not optimal predictive performance. Thus, we fixed $\alpha_0 = 5$ for every entity class and $\beta_0 = 20$ for every relationship class.

5.2 Results

We will now report our results on learning and constraining with the LJ-FOAF data set.

[3] http://www.livejournal.com/bots/

[4] http://jena.sourceforge.net/

[5] http://pellet.owldl.com/

Computational Complexity: The additional consistency check for every individual per iteration made training slower by approximately a factor of 6 if performed with Jena and Pellet. After implementing a non-generic constraining module optimized for the simple example introduced in Sec. 1 we could reduce the additional computation considerably. A comparison between IHSM and IHRM for different truncation parameter settings is given in Fig. 5. Obviously, there is almost no computational overhead in the latter case.

Evaluating the convergence of the cluster sizes is another interesting aspect in the comparison of IHSM and IHRM. Fig. 6 shows the number of individuals for the two largest components of the entity cluster Z^{Person} plotted over Gibbs sampler iterations for one exemplary training run. Apparently, the constraining does not affect the convergence speed which is desirable.

Cluster Analysis: An interesting outcome of the comparison of IHRM and IHSM is the number of components per hidden variable after convergence (see Table 1 right column). In both cases, if compared to the initialization, Gibbs sampling converged to a much smaller number of components. Most of the individuals were assigned to a few distinct components leaving most of the remaining components almost empty. There is a noticeable difference between IHRM and IHSM concerning the concepts *School* and *Person* which needed more components after training with IHSM (see bold numbers in Table 1). A closer analysis of the components revealed that IHSM generated additional components for inconsistent individuals, because both concepts are affected by constraints. However, the last concept affected by the constraints (*Date*) has fewer components. Here, IHSM divided more generally into age groups "too young" and "old enough" which also reflects the constraints. This demonstrates that the restriction of roles does influence the states of the latent variables.

Fig.7 compares the learned parameter ϕ^{attend} of IHRM to the one learned by IHSM. A brighter cell indicates stronger relations between two components. Although hard to generalize, a cell with 50% gray might indicate that no

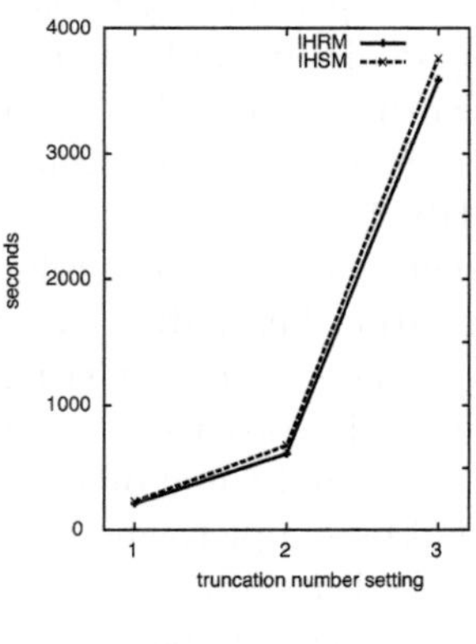

Fig. 5. Running time

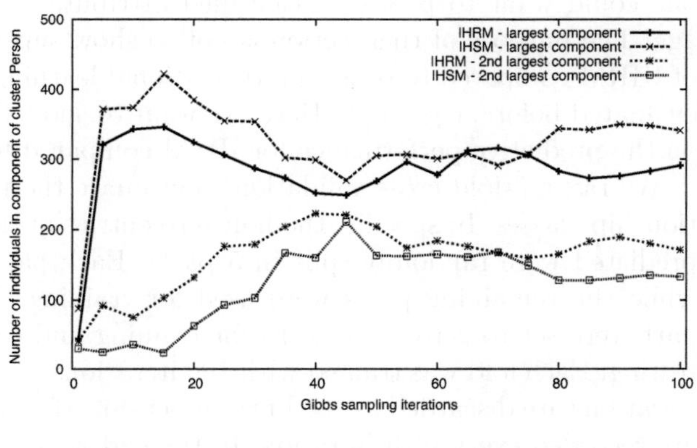

Fig. 6. Convergence of the two largest components

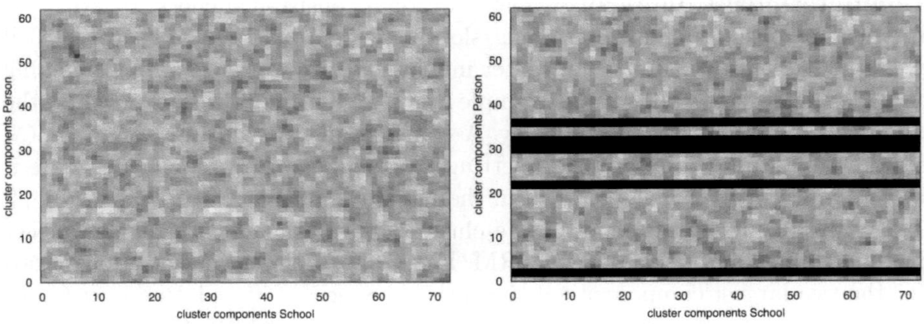

Fig. 7. Correlation mixture component ϕ^{attend} for each combination of components Z^{Person} and Z^{School}. Left: without constraining (IHRM). Right: with constraining (IHSM).

significant probabilistic dependencies for individuals in this component are found in the data. The most obvious results are the rows with black cells only which represent *Person* components that have no relation to any school. In fact, all of those cells contained at least one persons that conflicted with the ontology by having specified an age under 5. This proves that one of the main goals of IHSM is achieved, namely the exploitation of constraints provided by the ontology.

Note that the learned clusters can also be used to extract symbolic and uncertain knowledge and feed it back to the ontology. This is a promising direction of future research.

Predictive Performance: Given LJ-FOAF data for social network analysis one could for instance want to predict "who knows who" in case either this information is unknown or the systems wants to recommend new friendships. Other relations that could be interesting to predict in case they are unknown are the school someone attends/attended or the place he lives/lived. Furthermore one could want to predict unspecified attributes of certain persons, like their age. The purpose of this section is not to show superior predictive performance of IHRM compared to other multi-relational learning algorithms. This has been evaluated before, e.g. in [3]. Here, we want to show the influence of constraining on the predictive performance for IHSM compared to IHRM.

We ran a 5-fold cross validation to evaluate the predictions of different relationship classes. In specific, the non-zero entries of the relationship matrix to be predicted were randomly split in 5 parts. Each part was once used for testing while the remaining parts were used for training. The entries of each testing part were set to zero (unknown) for training and to their actual value of 1 for testing. Each fold was trained with 100 iterations of the Gibbs sampler, where 50 iterations are discarded as the burn-in period. After this, the learned parameters are recorded every fifth iteration. In the end we use the 10 recorded parameter sets to predict the unknown relationship values, average over them and calculate

Table 2. Predictive performance for different LJ-FOAF roles: AUC and 95% confidence intervals

Role	attends	dateOfBirth	knows
IHRM	0.577 (±0.013)	0.548 (±0.018)	0.813 (±0.005)
IHSM	**0.608** (±0.017)	**0.561** (±0.011)	**0.824** (±0.002)

the area under the ROC curve (AUC) as our evaluation measure[6]. Finally we average over the 5 folds and calculate the 95% confidence interval.

The obvious roles to evaluate are *attends* and *dateOfBirth*. Both are constrained by the ontology, so IHSM should have an advantage over IHRM because it cannot predict any false positives. The results in Table 2 confirm this observation. In both cases IHSM did outperform IHRM. A less obvious outcome can be examined from the influence of the constraining on a relationship that is not directly constrained by the ontology like *knows*. Still, in our experiments IHSM showed a slight advantage over IHRM. Thus, there seems to be a positive influence of the background knowledge, although a lot of users specify an incorrect age. However, there is the potential that the opposite may occur likewise. If the given constraints are conflicting with the empirical evidence there could even be a decrease in predictive performance. It is the ontology designers choice to decide whether to enforce a constraint that conflicts with the observed evidence.

Considering the numerous ongoing efforts concerning ontology learning for the Semantic Web more data sets with complex ontologies should become available in the near future. Thus, we expect to achieve more definite results of IHSM in those domains.

6 Related Work

Very generally speaking, our proposed method aims at combining machine learning with formal logic. So far, machine learning has been mainly approached either with statistical methods, or with approaches which aim at the inductive learning of formal knowledge from examples which are also provided using formal logic. The most important direction in this respect is *Inductive Logic Programming* (ILP) [6]. *Probabilistic-* and *Stochastic Logic Programming* (e.g., [7]) (SLP) are a family of ILP-based approaches which are capable of learning stochastically weighted logical formulas (the weights of formulas, respectively). In contrast to that, our approach learns probability distributions with the help of a given, formal theory which acts as a set of hard constraints. To the best of our knowledge, this direction is new. What (S)ILP and our approach have in common is that our method also uses examples formalized in a logic language as data. There

[6] Please note that SW data has no negative samples, because zero entries do not represent negative relations but unknown ones (open world assumption). Still, the AUC is appropriate because it has been shown to be a useful measure for probabilistic predictions of binary classification on imbalanced data sets.

are also some approaches which build upon other types of "uncertain logic", for example [8]. Although (S)ILP and SRL are conceptually very closely related and often subsumed under the general term *relational learning*, SRL still is rarely integrated with formal logic or ontologies as prior knowledge. [9] use ontologies in an similar model but only use taxonomic information as additional "soft" knowledge (i.e., knowledge which can be overwritten during the learning process) in the form of features for learning. They do not restrict their results using formal hard constraints. One exception are *Markov Logic Networks* [10] which combine First Order Logic and *Markov Networks* and learn weights of formulas.

Surprisingly there are also hardly any applications of (pure) SRL algorithms to (SW) ontologies. The few examples, e.g. [11], [12], do not consider formal constraints. There are various approaches to the learning of categories in formal ontologies from given instance data and/or similar categories (e.g., [13]). However, these approaches do not allow for the statistical learning of relations in the sense of SRL and their aims are all in all more related to those of ILP than to our learning goals. Although there are applications of SRL to social networks, such as [14], none of those approaches uses a formal ontology or any other kind of formal knowledge. Furthermore, the social networks examined in this work are mostly significantly less complex in regard of the underlying relation model.

The use of *hard constraints* for clustering tasks in purely statistical approaches to learning, as opposed to the ubiquitous use of "soft" prior knowledge, has been approached in, e.g., [15]. A common characteristic of these approaches is that they work with a relatively narrow, semi-formal notion of constraints and do not relate constraints to relational learning. In contrast to these efforts, our approach allows for rich constraints which take the form of a OWL DL knowledge base (with much higher expressivity). The notion of forbidden pairings of data points (*cannot-link* constraints [15]) is replaced with the more general notion of logical (un-)satisfiability w.r.t. formal background knowledge.

7 Conclusions and Future Work

In the presented approach, we explored the integration of formal ontological prior knowledge into machine learning tasks. We introduced IHSM and provided empirical evidence that hard constraints cannot only improve predictive performance of unknown roles, which are directly affected by the constraints, but also unconstraint roles via IHSMs latent variables.

In general we are hope to see more work on inductive learning with SW ontologies and on the other hand complex Semantic Web ontologies that can be supplemented with uncertain evidence. For the IHSM in particular, future work will concern a detailed theoretical analysis of the effect of constraining on clusters. Refining the ontology by extracting formal knowledge from the latent model structure is another promising research direction. As mentioned before we intend to obtain additional experimental evidence concerning computational complexity and predictive performance as soon as more suitable ontologies become available. We expect that the increased research on semantic technologies will soon result in those suitable formal ontologies that contain both, complex consistency reasoning tasks and large sets of instances.

References

1. Horrocks, I., Patel-Schneider, P.F.: Reducing owl entailment to description logic satisfiability. Journal of Web Semantics, 17–29 (2003)
2. Getoor, L., Taskar, B. (eds.): Introduction to Statistical Relational Learning. The MIT Press, Cambridge (2007)
3. Xu, Z., Tresp, V., Yu, K., Kriegel, H.P.: Infinite hidden relational models. In: Proceedings of the 22nd International Conference on Uncertainity in Artificial Intelligence, UAI 2006 (2006)
4. Kemp, C., Tenenbaum, J.B., Griffiths, T.L., Yamada, T., Ueda, N.: Learning systems of concepts with an infinite relational model. In: Proc. 21st Conference on Artificial Intelligence (2006)
5. Ishwaran, H., James, L.: Gibbs sampling methods for stick breaking priors. Journal of the American Statistical Association 96(453), 161–173 (2001)
6. Lisi, F.A., Esposito, F.: On Ontologies as Prior Conceptual Knowledge in Inductive Logic Programming. In: ECML PKDD 2008 Workshop: Prior Conceptual Knowledge in Machine Learning and Knowledge Discovery PriCKL 2007 (2007)
7. Raedt, L.D., Kersting, K.: Probabilistic logic learning. SIGKDD Explor. Newsl. 5(1), 31–48 (2003)
8. Carbonetto, P., Kisynski, J., de Freitas, N., Poole, D.: Nonparametric bayesian logic. In: Proc. 21st UAI (2005)
9. Reckow, S., Tresp, V.: Integrating Ontological Prior Knowledge into Relational Learning. In: NIPS 2008 Workshop: Structured Input - Structured Output (2008)
10. Richardson, M., Domingos, P.: Markov logic networks. Journal of Machine Learning Research 62, 107–136 (2006)
11. Kiefer, C., Bernstein, A., Locher, A.: Adding Data Mining Support to SPARQL via Statistical Relational Learning Methods. In: Bechhofer, S., Hauswirth, M., Hoffmann, J., Koubarakis, M. (eds.) ESWC 2008. LNCS, vol. 5021, pp. 478–492. Springer, Heidelberg (2008)
12. Fanizzi, N., d'Amato, C., Esposito, F.: Induction of classifiers through non-parametric methods for approximate classification and retrieval with ontologies. International Journal of Semantic Computing 2(3), 403–423 (2008)
13. Fanizzi, N., D'Amato, C., Esposito, F.: A multi-relational hierarchical clustering method for datalog knowledge bases. In: An, A., Matwin, S., Raś, Z.W., Ślęzak, D. (eds.) Foundations of Intelligent Systems. LNCS (LNAI), vol. 4994, pp. 137–142. Springer, Heidelberg (2008)
14. Xu, Z., Tresp, V., Rettinger, A., Kersting, K.: Social network mining with nonparametric relational models. In: Advances in Social Network Mining and Analysis - the Second SNA-KDD Workshop at KDD 2008 (2008)
15. Davidson, I., Ravi, S.S.: The complexity of non-hierarchical clustering with instance and cluster level constraints. Data Min. Knowl. Discov. 14(1), 25–61 (2007)

Boosting Active Learning to Optimality: A Tractable Monte-Carlo, Billiard-Based Algorithm

Philippe Rolet, Michèle Sebag, and Olivier Teytaud

TAO, CNRS – INRIA – Univ. Paris-Sud
rolet@lri.fr, sebag@lri.fr, olivier.teytaud@inria.fr

Abstract. This paper focuses on Active Learning with a limited number of queries; in application domains such as Numerical Engineering, the size of the training set might be limited to a few dozen or hundred examples due to computational constraints. Active Learning under bounded resources is formalized as a finite horizon Reinforcement Learning problem, where the sampling strategy aims at minimizing the expectation of the generalization error. A tractable approximation of the optimal (intractable) policy is presented, the *Bandit-based Active Learner* ($BAAL$) algorithm. Viewing Active Learning as a single-player game, $BAAL$ combines UCT, the tree structured multi-armed bandit algorithm proposed by Kocsis and Szepesvári (2006), and billiard algorithms. A proof of principle of the approach demonstrates its good empirical convergence toward an optimal policy and its ability to incorporate prior AL criteria. Its hybridization with the Query-by-Committee approach is found to improve on both stand-alone $BAAL$ and stand-alone QbC.

1 Introduction

Active Learning (AL), a most active topic in supervised Machine Learning (ML) [1,2,3,4,5,6,7], aims at accurately approximating a target concept with a limited number of *queries* to an *oracle*; each query asks the oracle to label a point of the problem domain (*instance*) according to the target concept. Through a judicious selection of instances to query, the hope is to learn with a significantly smaller number of queries than in the standard ML setting using iid labeled instances. Prominent AL approaches (section 2) rely on the properties of the hypothesis space (VC dimension or covering numbers) and/or propose various criteria estimating the additional information provided by an instance; they mostly proceed by iteratively selecting the optimal instance in the sense of the considered criterion.

This paper presents a new perspective on AL, formalized as a finite time horizon reinforcement learning problem. Choosing an instance, i.e. making a query is viewed as an action taken by the learner. The reward associated to a sequence of actions is the generalization error of the hypothesis learned from the training set gathered from this sequence of actions. Under mild assumptions

W. Buntine et al. (Eds.): ECML PKDD 2009, Part II, LNAI 5782, pp. 302–317, 2009.

(Bayesian realizable setting) detailed in section 3, this paper offers a provably optimal AL strategy, parameterized by the initial training set and the finite horizon T, i.e. the maximum number of queries allowed.

As could have been expected, the formal derivation of this optimal AL strategy is intractable. A tractable approximation thereof is given by the *Bandit-based Active Learner* algorithm (*BAAL*, section 4). This tractable approximation involves two main ingredients. Firstly, the exploration of the AL search space relies on UCT, the tree-structured multi-armed bandit algorithm originally introduced by Kocsis and Szepesvari [8]. The use of UCT is motivated as AL can be formulated as a single player game (section 4) and UCT has been shown quite efficient in learning game strategies [9]. Secondly, an unbiased and frugal sampler of the hypothesis and instance spaces, based on billiard-based mechanisms [10,11,12] supports the *Monte-Carlo simulations* in UCT. A proof of principle of the *BAAL* algorithm is presented, showing its practical robustness for AL under bounded resource constraints. This bounded resource constraint is motivated by targeted applications, aimed at simplified models in Numerical Engineering: in this context, training sets are limited to a few dozen or hundred labeled instances due to the high computational cost of instance labeling.

The paper is organized as follows. Section 2 briefly introduces the notations used throughout the paper and reviews the state of the art. Section 3 formalizes AL as a reinforcement learning problem, and shows how to tackle it as a single-player game. The online learning algorithm *BAAL*, implementing this single-player game using the UCT algorithm, is described in section 4; it takes as input a training set and the available computational budget, and finds the best instance to label next. Section 5 details two extensions brought to UCT in order to match the AL context. Firstly, as UCT only addresses finite action/state spaces, it needs to be extended to explore the continuous instance space; the proposed approach relies on *progressive widening* [13,14,15]. A second extension regards the hybridization of *BAAL* with prior knowledge, such as existing AL criteria. The proposed hybridization will be illustrated by embedding a variant of Query-by-Committee (QbC, [16,17]) within the progressive widening heuristics. An experimental validation of the approach is presented in section 6, using passive learning and QbC as baselines, and the paper concludes with some perspectives for further research.

2 Background and State of the Art

Notations and definitions used in the paper are as follows. Let $s_t = \{(x_i, y_i), x_i \in X, y_i \in Y, i = 1 \dots t\}$ denote a t-size training set, with X the instance space and Y the label space; only the binary classification case ($Y = \{0, 1\}$) will be considered in the paper.

From s_t, a learning algorithm $\mathcal{A}$ extracts some hypothesis h in hypothesis space $\mathcal{H}$, mapping X onto Y. The learning performance most usually refers to the expectation of the loss $\ell(h(x), h^*(x))$ incurred by h over the *target concept* h^* a.k.a. oracle, where the expectation is taken over the joint distribution P_{XY} on the problem domain. Whereas the standard supervised learning setting assumes that training examples (x_i, y_i) are independent and identically distributed (i.i.d.) after P_{XY}, AL selects at time step i some instance x_i in the instance space X (or in a pool of instances drawn from X), the label y_i of which is determined by the oracle.

A sampler S is a mapping from $\bigcup_{t\in\mathbb{N}}(X \times Y)^t$ to X, also referred to as *policy* or *strategy*, selecting a new instance x to be labeled depending on training set s_t. A learner $\mathcal{A}$ is a mapping from $\bigcup_{t\in\mathbb{N}}(X \times Y)^t$ to $\mathcal{H}$ associating a hypothesis h to any training set s_t. The *Version Space* (VS) associated to a training set s_t, noted $\mathcal{H}(s_t)$, is the set of hypotheses consistent with s_t, i.e. such that $h(x_i) = y_i$ for $1 \leq i \leq t$.

A first research direction focuses on the *uncertainty region* (set of examples where VS hypotheses disagree). Cohn et al. [2] reduce the VS by characterizing the uncertainty region via neural nets, and selecting new instances in this region. Well-known methods such as Query-by-Committee (QbC) algorithms [16,17] directly reduce the VS volume; Freund et al. [17] proved that these methods can lead to an exponentially smaller number of queries than the random querying strategy (passive learning), at least in the case of perceptrons. Dasgupta [4] has established the quasi-optimality of these methods in the realizable classification case (i.e. when h^* belongs to $\mathcal{H}$) for a finite instance pool. More generally, when there exists a probability measure P_H on $\mathcal{H}$, usual AL strategies are concerned with reducing either the measure of the Version Space, or the variance of the VS hypothesis labels. Otherwise, the reduction of the Version Space can be expressed in terms of its diameter, that is the measure of points where hypotheses in the VS differ.

A related research direction focuses on *error reduction*, meant as the expected generalization error improvement brought by an instance. Many criteria reflecting various measures of the expected error reduction have been proposed [18,19,20,21]; corresponding AL strategies proceed by greedily selecting the optimal instances in the sense of the considered criterion. Other approaches exploit prior, learner-dependent knowledge about what makes an instance informative, such as its margin [3,22,23]. Hoi et al. [6] consider *batch* active learning, querying a subset of instances that results in the largest reduction of the Fisher information. Some of these approaches however happen to face learning instabilities, which might require to mix the AL procedure with a uniform instance selection [24]. Such instabilities suggest that in some cases an optimally efficient AL system can hardly be based on a greedy selection strategy, at least using the criteria considered so far.

On the theoretical side, significant results have been obtained in terms of lower and upper bounds on the reduction of the sample complexity brought by AL, e.g. depending on the complexity of the hypothesis search space measured through covering numbers or Kolmogorov complexity [1,25,5]; the appropriateness of hypothesis spaces to AL has been studied [26,27,7] and an “almost” optimal (though intractable) algorithm has proposed by [27] in the realizable setting for finite VC-dimension.

3 An Optimal Active Learning Strategy: AL as a Markov Decision Process

This section presents a theoretically optimal strategy for Active Learning in *finite time horizon* T, where T corresponds to the total number of instances to be labeled along the AL process. As in [17,4], the proposed approach is built on a Bayesian setting [29,28]; prior knowledge about the target concept is accounted for by a probability distribution P_H on the hypothesis space[1] $\mathcal{H}$. Further, the approach relies on the *realizable* assumption, i.e. the target concept h^* to be learned is assumed to be deterministic and to belong to $\mathcal{H}$ (how to relax this assumption will be discussed in section 7).

Let us formalize AL as a Markov Decision Process (MDP). MDPs are classically described in terms of states, actions, reward, policy and transition functions [30]. In the AL context, the state space $\mathcal{S}$ consists of all possible training sets s_t. An action corresponds to the selection of a new instance to be labeled; the set of actions noted A thus coincides with the instance space $\mathcal{X}$ or a subset thereof.

The reward function associated to state s_t corresponds to the generalization error of the hypothesis $\mathcal{A}(s_t)$ learned from s_t by learner $\mathcal{A}$. In many finite time horizon settings, and particularly so for the targeted applications, the reward at the final state of the learning is the only one that matters. Accordingly, the reward function will be defined for *horizon* states only, and assimilated to the *value function* of those states (see below).

The transition function $p : \mathcal{S} \times A \times \mathcal{S} \rightarrow \mathbb{R}_+$ defines the probability of arriving at some state s_{t+1} by selecting action x in state s_t (see below). Lastly, in the AL context a MDP policy is a sampler S, mapping a state s_t onto an action, i.e. a new instance x_{t+1}.

Considering horizon T, let $S_T(h)$ denote the training set built by applying T times policy S when learning some target concept h, and let $\mathbf{Err}(\mathcal{A}(S_T(h)), h)$ denote the generalization error of the hypothesis learned from $S_T(h)$. It comes naturally that an optimal AL strategy is one minimizing the expectation of $\mathbf{Err}(\mathcal{A}(S_T(h)), h)$ when h ranges in $\mathcal{H}$:

$$S_T^* = \arg\min_S \mathbb{E}_{h \sim \mathcal{H}} \mathbf{Err}(\mathcal{A}(S_T(h)), h). \tag{1}$$

The main motivation behind the presented MDP framework is to build a policy with optimal behavior in the sense of Eq. (1). In order to do so, it remains to find the appropriate reward and transition probability functions, and such that the latter accounts for the actual behavior of the AL process.

Regarding the reward function, as the goal is to minimize the generalization error of the hypothesis learned from s_T, the reward is only known at horizon T. After the realizable assumption, the target concept h is bound to be in the Version Space of s_T, noted $\mathcal{H}(s_T)$. The reward function $V(s_T)$ (value at horizon T) therefore is the expectation of the generalization error of the learner $\mathcal{A}$, taken over $\mathcal{H}(s_T)$:

$$V(s_T) = \mathbb{E}_{h \sim \mathcal{H}(s_T)} \mathbf{Err}(\mathcal{A}(s_T), h). \tag{2}$$

[1] P_H is set to the uniform distribution on $\mathcal{H}$ in the absence of prior knowledge.

By construction, the transition function $p(s_t, x, s_{t+1})$ is such that p is zero for all s_{t+1} except the ones satisfying $s_{t+1} = (s_t, (x_{t+1} = x, y_{t+1}))$ for some y_{t+1}. In the latter case, p reflects the probability for the label of x to be y_{t+1}. $p(s_t, x, s_{t+1})$ is thus expressed as a function of the probability of the label of x, conditionally to the fact that the target hypothesis belongs to $\mathcal{H}(s_t)$:

$$p(s_t, x, (s_t, (x_{t+1} = x, y_{t+1}))) = p(h(x) = y_{t+1} | h \in \mathcal{H}(s_t)) \quad (3)$$

The above transition and value functions guarantee that the optimal MDP strategy achieves the AL goal and minimizes the expected generalization error (Eq. (1)):

Theorem 1 (Optimal AL policy). *Let $\mathbb{E}$ denote the expectation operator defined after Eq. (3). Let value function $V^{*,T}$ be recursively defined as follows, where $|s|$ denotes the number of examples in training set s:*

$$V^{*,T}(s) = \begin{cases} \mathbb{E}_{h \sim \mathcal{H}(s)} \mathbf{Err}(\mathcal{A}(s), h) & \text{if } |s| = T \\ \inf_{x \in X} \mathbb{E}_{s' \sim p(s'|x,s_t)} V^{*,T}(s') & \text{otherwise} \end{cases} \quad (4)$$

Define $S^{*,T}$ as $S^{*,T}(s) = \arg\inf_{x \in X} \mathbb{E}_{s' \sim p(s'|x,s_t)} V^{*,T}(s')$ (Bellman optimal strategy). Then $S^{*,T}$ is optimal in the sense of Eq. (1).

The proof is given in [31] due to space limitations. It essentially shows that the value function of a policy at the initial state coincides with the criterion in Eq. (1). The rest of the proof is a direct application of Bellman's optimality principle [32].

After this MDP formulation, AL can be seen as a one-player game. The active learner plays against the (unknown) target hypothesis h, belonging to the version space $\mathcal{H}(\mathbf{s}_0)$ of the initial training set[2] s_0. Upon each move (selection of some instance x), oracle h provides the label $y = h(x)$. At the end of the game – that is, after T examples have been picked, defining training set s_T – the reward is the generalization error of the hypothesis $\mathcal{A}(s_T)$ learnt from s_T. The *real* learning game is indeed played against oracle h.

It is however possible to train the AL player, and therefore devise a good AL policy beforehand, by mimicking the above game and playing against a "surrogate" oracle, made of a hypothesis h uniformly selected in the VS. The reward of such a game is computed as in the real game: it is the generalization error of the learned hypothesis w.r.t. the (surrogate) oracle. This reward implements a uniform draw of the random variable $\mathbf{Err}(\mathcal{A}(s), h)$ for h ranging in $\mathcal{H}$.

By construction, the average empirical reward collected by the AL player after many such games asymptotically converges toward the true expectation of this random variable, that is, the desired reward function (Eq. (2)).

4 Tractable Approximations of Optimal AL Strategy

This section presents a consistent and tractable approximate resolution of the above MDP, the *BAAL* algorithm. *BAAL* relies on two main ingredients. Firstly

[2] For the sake of simplicity and when no confusion is to fear, the initial training set $\mathbf{s}_0$ is omitted and $\mathcal{H}$ is used instead of $\mathcal{H}(\mathbf{s}_0)$.

the tree-structured multi-armed bandit UCT [8] is extended to the one-player game of AL. Secondly, a fair and frugal billiard-based algorithm [10,11] is used to sample the instance and hypothesis spaces.

4.1 Bandit-Based Active Learning

UCT (Upper Confidence Tree) is a Monte-Carlo tree-search algorithm [8,13], where the selection of a child node is cast into a multi-armed bandit problem [33]. UCT notably became famous in the domain of strategic games as it inspired the computer-Go program MoGo, first to ever win over professional human players [9].

The UCT-based game strategy provides the basis for *BAAL* (Fig. 1). Let $\mathcal{T}$ denote the "game" tree of AL; its root node is the initial training set $\mathbf{s}_0$. It is worth noting that *BAAL* actually explores a directed graph (a given node can be reached along different paths) although it is presented as a tree-search algorithm for simplicity.

Each tree-walk (aka game or simulation) is indexed by a surrogate hypothesis h (see below) and proceeds as follows. Each node corresponding to a training set s_t is called a *state* node. Its child nodes, referred to as *decision* nodes, stand for the actions of the AL player, that is, the possible instances to be selected. A decision node thus is made of the training set s_t (parent node), and the chosen instance x_{t+1}.

Every decision node has two child nodes (the possible labels of the instance x_{t+1}). The one selected during the tree-walk indexed by h obviously corresponds to $h(x_{t+1})$, thus leading to the next state node $s_{t+1} = s_t \cup (x_{t+1}, h(x_{t+1}))$.

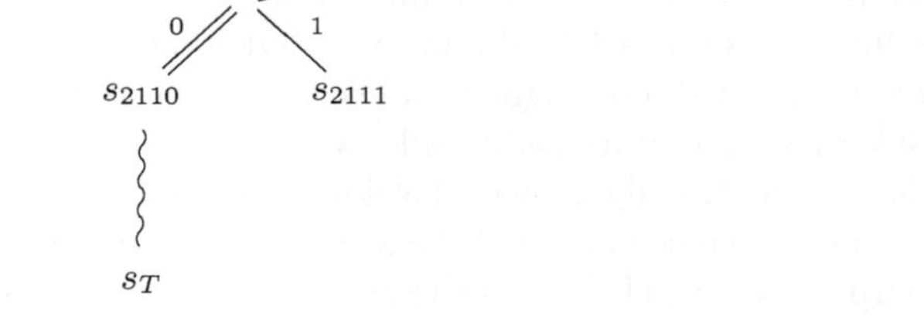

Fig. 1. Search Tree developed by *BAAL*, the Bandit Based Active Learner, in the binary classification case. The root of the tree is the initial training set s_0. Each tree walk is based on selecting some hypothesis h in the Version Space of s_0. The tree walk proceeds by iteratively selecting an instance x, whose label is set to $h(x)$. Ultimately, the tree walk is assessed from the generalization error $Err(\mathcal{A}(s_T), h)$ of the learned hypothesis w.r.t. h.

More precisely, each tree-walk proceeds as follows (Fig. 1; *BAAL* pseudo-code is given in Alg. 1):

- Firstly, a surrogate hypothesis h is uniformly drawn in the version space $\mathcal{H}(s_0)$ of the initial training set s_0;
- In each state node (i.e. some training set s_t), *BAAL* uses the famed Upper Confidence Bounds (UCB) criterion (Eq. (5), see below) to select some decision node, i.e. some instance x_{t+1} to be labeled. The associated label is set to $h(x_{t+1})$, where h is the surrogate hypothesis. The next state node thus is $s_{t+1} = s_t \cup (x_{t+1}, h(x_{t+1}))$.
- The tree-walk proceeds until arriving in a tree leaf, i.e. a state node s_{t_0} not yet visited. At this point, $T - t_0$ additional instances are uniformly selected, labeled after h, and added to the training set s_{t_0}, forming a T-size training set s_T.
- From training set s_T, a hypothesis $\hat{h}$ is learned by *BAAL*[3].
- The reward of the tree walk is the generalization error of $\hat{h}$ with respect to the surrogate hypothesis h: this reward is used to update the value of every relevant node.

The value $\hat{\mu}(s_t)$ of state node s_t is computed by averaging all rewards gathered for nodes s_T such that $s_t \subseteq s_T$. These values are exploited by the UCB criterion [33], determining which arm (instance aka decision node) should be selected. The arm selected is the one maximizing the sum of the empirical reward $\hat{\mu}_i$ (exploitation term), plus an exploration term depending on the number n_i of times arm i has been selected. The exploration vs exploitation tradeoff is adjusted using some tuned constant C:

$$\arg \max_{i \in \text{child nodes}} \hat{\mu}_i + C\sqrt{\frac{\log(\sum_{j \in \text{child nodes}} n_j)}{n_i}} \tag{5}$$

In the multi-armed bandit setting, this formula guarantees a provably optimal convergence towards the arms with maximal value, under the assumption that arm rewards are independent random variables. (Clearly this assumption does not hold in the AL game, no more than in the game of Go [34]).

The memory-wise computational tractability of *BAAL* is ensured by gradually developing the tree, initially made of the root node and its child nodes only. After each tree-walk, the first randomly selected node is stored in memory with its child nodes: the current leaf node will no longer be a leaf node, and the next time it is encountered, the selection among its child nodes will rely on the UCB formula. Thereby the tree is asymmetrically grown, developing more the subtrees where nodes have better values – since those will be selected more often.

BAAL (Alg. 1) implements the UCT scheme with two specific ingredients. The `DrawHypothesis` function selects the surrogate hypothesis h attached to each tree-walk using billiard algorithms (see below). The `ArmSet` function (section 5.1) extends UCT to deal with an infinite set of actions (the continuous instance space), using the so-called progressive widening heuristics.

[3] The learning algorithm actually is a parameter of *BAAL*. In the experiments, we used a uniform sampler of the version space of s_T.

Algorithm 1. The ***BAAL*** algorithm:
Input: measure P_H on hypothesis space $\mathcal{H}$; initial training set $\mathbf{s}_0$; time horizon T; number N of allowed tree-walks;
Output: an instance x to be labelled by the oracle.

```
BAAL(P_H, s_0, T, N)
for i=1 to N do
  h = DrawHypothesis(P_H, s_0)
  Tree-Walk(s_0, T, h)
end for
Return x = argmax_{x'∈X} {n(s ⋃ {x'})}

Tree-Walk(s, t, h)
Increment n(s)
if t==0 then
  Compute r = Err(A(s), h)
else
  X(s) = ArmSet(s, n(s))
  Select x* = UCB(s, X(s)))
  r = Tree-Walk(s ⋃ {(x*, h(x*))}, t − 1, h)
end if
r(s) ← (1 − 1/n(s)) r(s) + 1/n(s) r
Return r
```

4.2 Billiard Algorithms

As already mentioned, each tree-walk in *BAAL* is indexed by a hypothesis h uniformly sampled in version space $\mathcal{H}(s_0)$. The most straightforward sampling algorithm is based on *rejection*: hypotheses are uniformly drawn in $\mathcal{H}$ and rejected if they do not belong to the VS (if they are inconsistent with the training set s_0). Unfortunately, the rejection algorithm, although sound, is hardly tractable. Alternative algorithms such as Gibbs sampling or more generally Monte-Carlo Markov Chains (MCMC) methods, involve quite a few free parameters and might scale poorly with respect to the dimensionality of the search space and the size of the training set s_0. We thus used a billiard (a.k.a. ray-tracing) algorithm inspired from [10,35,11], assuming that the hypothesis space $\mathcal{H}$ can be parameterized by $\mathbb{R}^d$. Actually, the experimental validation (section 6) considers linear hypotheses; how to go beyond the linear case will be discussed in section 7.

Let Ω be a connected subset of $\mathbb{R}^d$, defined by a set of constraints $g_1, \ldots, g_n$: $\Omega = \{x \in \mathbb{R}^d \ s.t. \ g_i(x) \geq 0, i = 1 \ldots n\}$. The billiard algorithm considers a point $z \in \mathbb{R}^d$ (not necessarily in Ω) and a direction $\boldsymbol{v}$ ($\boldsymbol{v} \in \mathbb{R}^d, ||\boldsymbol{v}|| = 1$). The trajectory followed by z is such that: i) the set of constraints satisfied by z does not decrease; ii) z "bounces" when it meets an active constraint g, i.e. its direction v is changed to point inside the domain (see below); iii) the trajectory is stopped when the computational resources are exhausted, that is when its total length reaches some user-defined parameter L, and the final point is returned. Under a few regularity conditions on the constraints, a billiard trajectory is ergodic, i.e.

it covers the whole domain when L goes to infinity [36]; the distribution of the final trajectory point converges toward the uniform distribution on Ω. Billiard algorithms have been successfully used in Machine Learning, e.g. to estimate the Bayes classifier in a (high-dimensional) kernel feature space [10,35].

Rebound policy. Let g_i be the saturated constraint and z_i the rebound point. The rebound policy sets the current direction $\boldsymbol{v}$ to a unit vector randomly drawn in the half sphere centered on z_i and defined from the hyperplane tangent to g_i in z_i. Experimentally, this policy efficiently approximates the Knudsen law advocated by Comets et al. [36], which is more expensive to compute. This matter is however outside the scope of this work.

5 UCT with Continuous Domain and AL Criteria

This section is devoted to specific adaptations of UCT involved in BAAL: how to deal with a continuous action set (the instances) and how to take advantage of existing AL criteria, such as Query-by-Committee [16].

5.1 Progressive Widening

UCB originally requires each arm to be selected at least once for Eq. (5) to be computed. When the number of arms is large with respect to the number N of simulations, UCB thus tends to degenerate into pure exploration. UCT faces the same limitation, and even more so since many arms need be considered at each node of the tree, strongly biasing the search towards exploration.

The *progressive widening* (PW) technique has been proposed by [13] to handle such cases. Let n_s denote the number of times node s has been visited so far; then the number of arms that can be considered from s is limited to a fraction m of n_s. Empirical and theoretical studies suggest $m = O(n_s^{1/4})$ [13,14,15]. In practice, the `ArmSet` procedure implementing PW in *BAAL* (Algorithm 1), considers a finite set of options for each node (see below), and a new option is added to this set whenever $\lfloor n_s^{1/4} \rfloor$ is incremented by one. Typically, a single arm will be explored from the root node in the first fifteen tree-walks; an additional arm is considered in the sixteenth random walk; UCB will be used to select among both arms during random walks 16 to 80; a third arm is considered in random walk 81, etc. The rationale behind progressive widening is that the better (i.e. the more visited) s, the more careful the investigation of its subtrees should be.

5.2 Integrating AL Criteria in *BAAL*

The selection of the actions (instances) considered at a given time step by `ArmSet` offers some room for the use of prior knowledge. The simplest selection procedure indeed is based on the uniform sampling of the instance space. This procedure reflects an agnostic viewpoint, making no assumption about the utility of instances. It must be noted however that *BAAL* convergence might be delayed (like other UCT-based algorithms) if the optimal options are considered late during the search: instances introduced later on are clearly disadvantaged compared to the earlier, more investigated, instances.

Furthermore, the AL literature suggests that some selection criteria – although not optimal – offer generally sound indications in order to select informative instances (section 2). Such criteria can seamlessly be integrated in *BAAL* through the progressive widening procedure. Formally, a new instance is added to `ArmSet` when it satisfies the considered criterion, instead of being uniformly selected in the pool of instances[4].

5.3 QbC-*BAAL*: *BAAL* with Maximal Uncertainty

Maximal uncertainty (MU) (section 2) is a criterion stemming from Query-by-committee (QbC) algorithms. QbC works by randomly sampling hypotheses in the Version Space, and choosing a given instance only if enough hypotheses disagree about its label. An analysis of this strategy with a committee of size 2 is for instance presented in [17]. In the case of large-sized committees, one proceeds by ranking instances based on the committee disagreement: the top-ranked instances indeed correspond to those with maximal uncertainty.

The QbC approach is integrated within BAAL as follows. At each node (training set s_t) a committee of hypotheses is built by uniformly sampling the Version Space of s_t. Whenever a new instance is to be added to ArmSet, one selects the one maximizing the committee disagreement.

MU is an *aggressive* criterion, thus holding a higher AL potential than random progressive widening. In counterpart it leads to a less diversified sample; whenever the criterion is under-optimal, the limited exploration will yield poorer performance. It is also more demanding computationally.

6 Experimental Validation

This section reports on the experimental study of *BAAL*. After describing the experimental goal and setting, empirical results are discussed comparatively to random active learning, referred to as passive learning, and the Query-by-Committee (QbC) approach [16,17].

6.1 Goal of Experiments

The main questions investigated in the experiments regard the theoretical and computational performance of *BAAL*. Specifically:
`Question 1` (Optimality): Does the stand-alone *BAAL*, where the progressive widening heuristics involves a uniform instance selection, converge to an optimal strategy; does it match QbC results when these are known to be quasi-optimal

[4] Let us emphasize that hybridizing *BAAL* with any such criterion does not depend on the true instance labels.

after Dasgupta [4] ?
Question 2 (Flexibility): As shown in section 5.1, *BAAL* can embed existing AL criteria, e.g. QbC, within the progressive widening heuristics. Does the use of QbC within *BAAL* improve i) on stand-alone *BAAL* ? ii) on stand-alone QbC ?

6.2 Experimental Setting

Following [17,21,23], the instance space $\mathcal{X}$ considered in these experiments is the unit sphere of $\mathbb{R}^d$. The hypothesis space $\mathcal{H}$ is restricted to the linear classifiers, a.k.a. separating hyperplanes. The choice of this search space is motivated as it has been thoroughly studied from a theoretical perspective [17,21,23] although these results did not lead to experimental studies to our best knowledge. Accordingly, some upper and lower bounds on the optimal performance are available, and will be used to assess *BAAL* performance. The goal of this experimental study is to provide a proof of principle regarding the validity of this new AL approach; still, it must be emphasized that *BAAL* is not limited to linear hypothesis spaces (a kernelized extension will be discussed in section 7).

Two variants of *BAAL* will be considered. *BAAL* stand-alone (or *BAAL* for short), uses a uniform selection of instances within progressive widening, whereas QbC-*BAAL* uses a committee-based selection of instances. More precisely, in each node (training set s_t), a committee of 100 hypotheses uniformly selected in $\mathcal{H}(s_t)$ is built, and a set of 10,000 instances uniformly drawn from $\mathcal{X}$ is sampled and ordered after the committee disagreement. Whenever a new action is to be considered ($\lfloor n(s_t)^{1/4} \rfloor$ increased by one), the first instance not yet considered in the ordered set is returned by `ArmSet`.

A hypothesis h is characterized as a unit vector w_h ($w_h \in \mathbb{R}^d$, $||w_h||_2 = 1$), where $h(x)$ is positive if and only if the scalar product $< w_h, x >$ is positive. The lack of prior knowledge is accounted for by setting distribution P_H to the uniform distribution on the unit sphere of $\mathbb{R}^d$.

A *BAAL* run proceeds as follows:

1. The target concept h^* is uniformly selected in $\mathcal{H}$.
2. For $t = 0$ to T, where T is the horizon time and the initial training set s_0 is empty:
   (a) The *BAAL* tree rooted on s_t is constructed using N tree-walks, respectively using a randomly (a QbC-) ordered instance pool in the progressive widening heuristics for *BAAL* (resp. QbC-*BAAL*);
   (b) The best instance, i.e. the most visited one at the first level of the s_t rooted tree, noted x_{t+1}, is selected and labeled after the target concept (oracle) h^*;
   (c) $s_{t+1} = s_t \cup \{(x_{t+1}, h^*(x_{t+1})\}$.
3. From s_T, the T-size training set built by *BAAL* (QbC-*BAAL*) and labeled after the current target concept h^*, a hypothesis h_T is learned (uniformly sampled in the Version Space $\mathcal{H}(s_T)$).
4. The performance of the run, that is, the generalization error $\mathbf{Err}(h, h^*)$ defined as $P_{x \sim \mathcal{X}}(h^*(x) \neq h(x))$, is computed as the dot product of the associated vectors w_h and w_{h^*}.

BAAL and QbC-*BAAL* performances are averaged over 400 independent runs for each 3-uple {d,T,N} (dimension, horizon, simulation number). The number N of simulations, controlling the computational cost, ranges in $\{1, 2, 4, \ldots N = 2^{12}\}$. The reported experiments consider $(d = 4, T = 15)$ and $(d = 8, T = 20)$ to assess the scalability of the approach. The performance is plot against the number N of tree-walks, Fig. 2, indicating the average performance (plain line) and the standard deviation (vertical bars). The performance of *BAAL* stand-alone (respectively QbC-*BAAL*) is assessed comparatively to the passive learning (resp. the QbC greedy AL) baseline. The baseline performances correspond to those obtained for $N = 1$. Actually, *BAAL* can be viewed as an algorithm providing an "educated" sampling strategy, where the "educated sampler" is based on a computational budget of N simulations; the baseline, non-educated, sampler accordingly corresponds to $N = 1$.

6.3 Performance and Scalability

Fig. 2.(a) and (c) display the overall performance of stand-alone *BAAL* and QbC-*BAAL* versus the computational budget N (in log scale).

The competence of *BAAL* in terms of Active Learning is visible as stand-alone *BAAL* strongly outperforms the passive learning baseline ($N = 1$): the generalization error significantly decreases as N increases. For a given computational budget N, the improvement on passive learning is higher in small dimension, as could have been expected. Nevertheless, the improvement is significant both in dimensions 4 and 8 even for a small computational budget ($N > 2^6$).

The performance of stand-alone *BAAL* is further assessed by comparison with that of a stand-alone QbC using a large-sized committee. After [4,17], the Maximum Uncertainty heuristics (approximated by a large-sized QbC selection) is almost optimal (up to logarithmic terms) in the linear setting. It is thus satisfactory to see that stand-alone *BAAL* steadily approaches the QbC reference performance (the fact that it can even outperform the QbC reference in dimension 8 is discussed below). These results suggest a positive answer to the *Optimality* question (section 6.1): stand-alone *BAAL* is a competent, criterion-agnostic Active Learner, which might work well in the absence of prior knowledge about the instance selection, and which matches the known optimal performance in a simple hypothesis space.

Another picture is provided by the performance of QbC-*BAAL* (Fig. 2.(c) and (d)), suggesting that QbC-*BAAL* can get the "best of both worlds". In dimensions 4 and 8, QbC-*BAAL* outperforms the QbC baseline in a statistically significant way; it overcomes the theoretical limitations of QbC with regards to optimality (i.e. multiplicative factors w.r.t. optimality logarithmic in the precision and linear in the dimension). These results thus suggest a positive answer to the *Flexibility* question (section 6.1): *BAAL* can be hybridized with the QbC criterion, and this hybridization can improve on both stand-alone *BAAL* and stand-alone QbC. Although *BAAL* flexibility remains to be confirmed by considering other AL criteria, QbC is among the most widely used criteria in the AL literature.

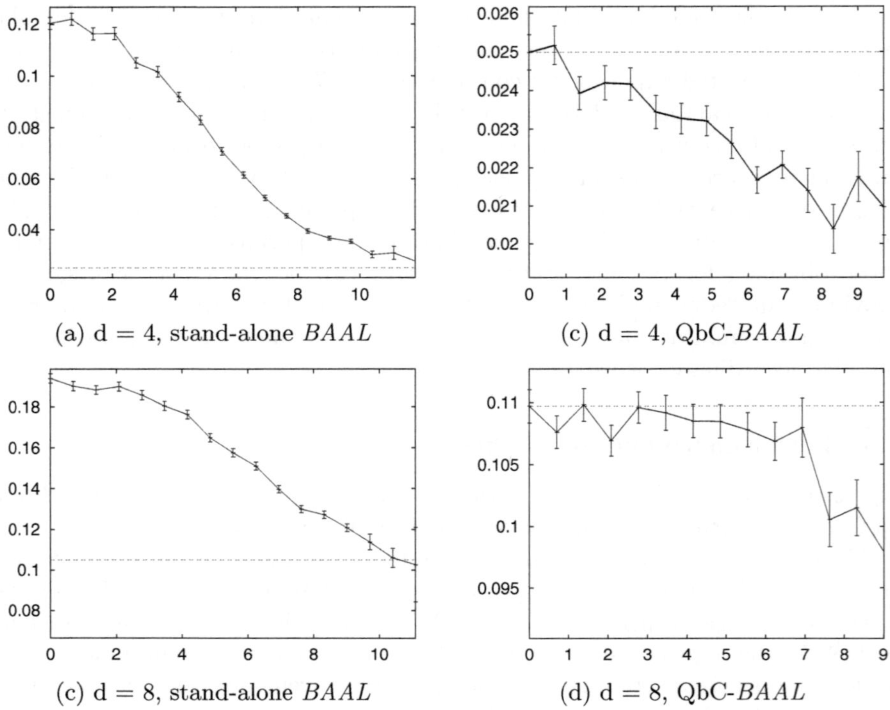

Fig. 2. *BAAL* performances: Generalization error vs the number of simulations N in log scale. Top rows report the results obtained for instance space dimension $d = 4$ and time horizon $T = 15$; bottom rows report the results for $d = 8, T = 20$. Left rows report the performance of the stand-alone *BAAL*. Right rows report the performance of QbC-*BAAL*. The performance of the baseline correspond to the performance obtained for $N = 1$ (at the origin of the curve). The horizontal line reports the results obtained for the stand-alone QbC strategy.

Due to space limitations, the tractability of *BAAL* is detailed in [31]. Overall, billiard algorithms allow for a time complexity lower by several orders of magnitude than the rejection algorithm, with an outstanding scalability w.r.t. horizon and dimension of the instance space. For $(d = 8, T = 20, N = 16,000)$, the computational time is 17 seconds on Pentium PC vs > 3 hours for the rejection-based *BAAL*.

7 Conclusion and Perspectives

This paper focuses on Active Learning with a limited number of queries, motivated by application domains such as Numerical Engineering. In such domains, computing the response of an example might require several days of computation, limiting the training set size to a few dozen or hundred examples.

Within this bounded resource setting, Active Learning was tackled as a Reinforcement Learning problem: find the sampling strategy aimed at minimizing the overall generalization error for a finite time horizon. This ideal and utterly intractable formalization leads to the proposed *BAAL* algorithm: an approximation of the optimal sampling strategy is learned within a one-player game setting. *BAAL* is inspired from an earlier work devoted to Computer Go [9], building upon the tree-structured bandit-based search algorithm UCT [37] and the progressive widening heuristics to deal with a continuous search space [13,14,15].

The experimental validation of *BAAL* investigates three main questions: the convergence towards the optimal performance when it is known; the ability to take advantage of competent AL criteria, such as Maximum Uncertainty [38], and improve on the greedy use of these criteria; the computational tractability of the overall AL scheme. A proof of principle, the experiments discussed in the paper show that the answer to all three questions is positive in the simple case of a realizable setting with linear hypotheses.

Further research aims at extending the proposed method beyond this simple setting. Getting rid of the realizable setting (considering label noise, or the case where the target concept does not belong to the hypothesis space $\mathcal{H}$) will be investigated through relaxing the billiard-based exploration of the Version Space. The billiard-based sampling of the hypotheses will likewise be extended to accommodate the priors on the hypothesis space (set to the uniform density in the present paper).

Another perspective for further study is to extend *BAAL* to non-linear hypothesis spaces, thanks to the famed kernel trick. It must be noted that billiard-based algorithms have been investigated for kernel spaces [11,10,35], featuring good theoretical and computational results. The extension of *BAAL* to Active Learning with kernels thus defines a new and appealing objective.

Acknowledgments

This work was supported in part by the Région Ile de France and Digiteo, and the PASCAL network of excellence. The authors acknowledge fruitful discussions with Jean-Marc Martinez, DM2S CEA, regarding applications of this work to Numerical Engineering, and Benoit Kloeckner, from the Mathematics Department of the University of Grenoble, regarding the theoretical properties of billiards.

References

1. Kulkarni, S.R., Mitter, S.K., Tsitsiklis, J.N.: Active learning using arbitrary binary valued queries. Mach. Learn. 11(1), 23–35 (1993)
2. Cohn, D., Atlas, L., Ladner, R.: Improving generalization with active learning. Mach. Learn. 15(2), 201–221 (1994)
3. Schohn, G., Cohn, D.: Less is more: Active learning with support vector machines. Int. Conf. on Machine Learning 282, 285–286 (2000)

4. Dasgupta, S.: Analysis of a greedy active learning strategy. In: NIPS 17, pp. 337–344. MIT Press, Cambridge (2005)
5. Castro, R., Willett, R., Nowak, R.: Faster rates in regression via active learning. In: NIPS 18, pp. 179–186. MIT Press, Cambridge (2006)
6. Hoi, S.C.H., Jin, R., Zhu, J., Lyu, M.R.: Batch mode active learning and its application to medical image classification. In: Int. Conf. on Machine Learning, pp. 417–424. ACM, New York (2006)
7. Hanneke, S.: A bound on the label complexity of agnostic active learning. In: Int. Conf. on Machine Learning, pp. 353–360. ACM, New York (2007)
8. Kocsis, L., Szepesvari, C.: Bandit-based monte-carlo planning. In: Fürnkranz, J., Scheffer, T., Spiliopoulou, M. (eds.) ECML 2006. LNCS (LNAI), vol. 4212, pp. 282–293. Springer, Heidelberg (2006)
9. Gelly, S., Silver, D.: Combining online and offline knowledge in UCT. In: Int. Conf. on Machine Learning, pp. 273–280. ACM, New York (2007)
10. Ruján, P.: Playing billiards in version space. Neural Computation 9(1), 99–122 (1997)
11. Herbrich, R., Graepel, T., Campbell, C.: Bayes point machines. Journal of Machine Learning Research 1, 245–279 (2001)
12. Warmuth, M.K., Liao, J., Rätsch, G., Mathieson, M., Putta, S., Lemmen, C.: Support vector machines for active learning in the drug discovery process. Journal of Chemical Information Sciences 43, 667–673 (2003)
13. Coulom, R.: Efficient selectivity and backup operators in Monte-Carlo tree search. In: Ciancarini, P., van den Herik, H.J. (eds.) CG 2006. LNCS, vol. 4630, pp. 72–83. Springer, Heidelberg (2007)
14. Chaslot, G., Winands, M., Uiterwijk, J., van den Herik, H., Bouzy, B.: Progressive strategies for Monte-Carlo tree search. In: Wang, P., et al. (eds.) Proc. of the 10th Joint Conf. on Information Sciences, pp. 655–661. World Scientific Publishing, Singapore (2007)
15. Wang, Y., Audibert, J.Y., Munos, R.: Algorithms for infinitely many-armed bandits. In: NIPS 21, pp. 1729–1736 (2009)
16. Seung, H.S., Opper, M., Sompolinsky, H.: Query by committee. In: COLT 1992, pp. 287–294. ACM, New York (1992)
17. Freund, Y., Seung, H.S., Shamir, E., Tishby, N.: Selective sampling using the query by committee algorithm. Mach. Learn. 28(2-3), 133–168 (1997)
18. Cohn, D., Ghahramani, Z., Jordan, M.: Active Learning with Statistical Models. Journal of Artificial Intelligence Research 4, 129–145 (1996)
19. Roy, N., McCallum, A.: Toward optimal active learning through sampling estimation of error reduction. In: Int. Conf. on Machine Learning, pp. 441–448. Morgan Kaufmann, San Francisco (2001)
20. Lindenbaum, M., Markovitch, S., Rusakov, D.: Selective sampling for nearest neighbor classifiers. Machine Learning 54, 125–152 (2004)
21. Dasgupta, S., Kalai, A.T., Monteleoni, C.: Analysis of perceptron-based active learning. In: Auer, P., Meir, R. (eds.) COLT 2005. LNCS (LNAI), vol. 3559, pp. 249–263. Springer, Heidelberg (2005)
22. Cesa-Bianchi, N., Conconi, A., Gentile, C.: Learning probabilistic linear-threshold classifiers via selective sampling. In: Schölkopf, B., Warmuth, M.K. (eds.) COLT/Kernel 2003. LNCS (LNAI), vol. 2777, pp. 373–387. Springer, Heidelberg (2003)
23. Florina Balcan, M., Broder, A., Zhang, T.: Margin based active learning. In: Bshouty, N.H., Gentile, C. (eds.) COLT. LNCS (LNAI), vol. 4539, pp. 35–50. Springer, Heidelberg (2007)

24. Xiao, G., Southey, F., Holte, R.C., Wilkinson, D.: Software testing by active learning for commercial games. In: AAAI 2005, pp. 609–616 (2005)
25. Vidyasagar, M.: A Theory of Learning and Generalization, with Applications to Neural Networks and Control Systems. Springer, Heidelberg (1997)
26. Hegedüs, T.: Generalized teaching dimensions and the query complexity of learning. In: COLT 1995, pp. 108–117. ACM, New York (1995)
27. Dasgupta, S.: Coarse sample complexity bounds for active learning. In: NIPS 18, pp. 235–242. MIT Press, Cambridge (2006)
28. Haussler, D., Kearns, M., Schapire, R.E.: Bounds on the sample complexity of bayesian learning using information theory and the VC dimension. Mach. Learn. 14(1), 83–113 (1994)
29. Mackay, D.J.C.: Bayesian interpolation. Neural Computation 4, 415–447 (1992)
30. Sutton, R., Barto, A.: Reinforcement learning: An introduction. MIT Press, Cambridge (1998)
31. Rolet, P., Sebag, M., Teytaud, O.: Boosting active learning to optimality: some results on a tractable Monte-Carlo, billiard-based algorithm. Technical report, Laboratoire de Recherche en Informatique, Univ. Paris Sud. (2009)
32. Bellman, R.: Dynamic Programming. Princeton Univ. Press, Princeton (1957)
33. Auer, P.: Using confidence bounds for exploitation-exploration trade-offs. The Journal of Machine Learning Research 3, 397–422 (2003)
34. Wang, Y., Gelly, S.: Modifications of UCT and sequence-like simulations for Monte-Carlo Go. In: IEEE Symposium on Computational Intelligence and Games, Honolulu, Hawaii, pp. 175–182 (2007)
35. Ruján, P., Marchand, M.: Computing the bayes kernel classifier (1999)
36. Comets, F., Popov, S., Schütz, G.M., Vachkovskaia, M.: Billiards in a General Domain with Random Reflections. Archive for Rational Mechanics and Analysis 191, 497–537 (2009)
37. Kocsis, L., Szepesvari, C.: Bandit Based Monte-Carlo Planning. In: Fürnkranz, J., Scheffer, T., Spiliopoulou, M. (eds.) ECML 2006. LNCS (LNAI), vol. 4212, pp. 282–293. Springer, Heidelberg (2006)
38. Freund, Y., Schapire, R.: Large margin classification using the perceptron algorithm. In: COLT 1998. Morgan Kaufmann, San Francisco (1998)

Capacity Control for Partially Ordered Feature Sets

Ulrich Rückert

International Computer Science Institute,
1947 Center Street, Suite 600, Berkeley, CA 94704
rueckert@icsi.berkeley.edu

Abstract. Partially ordered feature sets appear naturally in many classification settings with structured input instances, for example, when the data instances are graphs and a feature tests whether a specific substructure occurs in the instance. Since such features are partially ordered according to an "is substructure of" relation, the information in those datasets is stored in an intrinsically redundant form. We investigate how this redundancy affects the capacity control behavior of linear classification methods. From a theoretical perspective, it can be shown that the capacity of this hypothesis class does not decrease for worst case distributions. However, if the data generating distribution assigns lower probabilities to instances in the lower levels of the hierarchy induced by the partial order, the capacity of the hypothesis class can be bounded by a smaller term. For itemset, subsequence and subtree features in particular, the capacity is finite even when an infinite number of features is present. We validate these results empirically on three graph datasets and show that the limited capacity of linear classifiers on such data makes underfitting rather than overfitting the more prominent capacity control problem. To avoid underfitting, we propose using more general substructure classes with "elastic edges" and we demonstrate how such broad feature classes can be used with large datasets.

Keywords: capacity control, partially ordered features, graph mining, QSAR.

1 Introduction

In this paper we investigate classification with feature sets that are partially ordered. Such feature sets often appear naturally in learning settings with structured input objects. For instance, consider the task of learning whether or not a particular chemical compound inhibits tumor growth [3,8,2,10]. A popular approach to this setting is to represent the compounds by molecular graphs and to generate Boolean features that test for the occurrence of certain substructures (e.g. subgraphs) in the molecular graph. In this setting each compound can be represented by a bit vector, where each bit indicates whether or not the corresponding subgraph appears in the compound's molecular graph. Such a representation can then be used with traditional linear classifiers such as the

W. Buntine et al. (Eds.): ECML PKDD 2009, Part II, LNAI 5782, pp. 318–333, 2009.

ones output by support vector machines. The difference to most other classification settings, of course, is that the space of subgraphs is *partially ordered* via the "is subgraph of" relation. Consider, for instance, a data instance where the feature representing an aromatic ring is set to *true*. This means that all features that represent linear sequences of up to six carbon atoms connected with aromatic bonds must also be set to *true*. Consequently, the information provided by partially ordered feature sets is redundant by design.

The main question dealt with in this paper is how this redundancy affects the empirical risk minimization and capacity control behavior of a learning algorithm. In the first part of the paper we address these questions from a theoretical point of view. We show that the capacity of the class of linear classifiers (as measured by the VC-dimension) does not change, even when the features in the training data are totally ordered. However, if the underlying data distribution puts lower probabilities on data instances that are ordered in the later levels of the hierarchy induced by the partial order, one can find smaller upper bounds for the capacity of the class of linear classifiers. We show that distributions where the probability of observing an instance declines exponentially with the instance's level in the hierarchy can lead to settings, where the class of linear classifiers has finite capacity even in the presence of an infinite amount of features.

On the practical side we validate the theoretical results empirically on three datasets from quantitative structure-activity relationships. We show that adding more features does indeed not lead to overfitting for subsequence, subtree and subgraph features. Instead, we show that extending the feature set with more expressive substructures can improve predictive accuracy. This indicates that underfitting rather than overfitting is the prominent problem on datasets with partially ordered features. Finally, we investigate how an expressive and therefore large substructure feature class can be efficiently applied on large datasets.

2 Background

Before we can delve into the details, we need to introduce the used concepts and definitions. We assume an instance space $\mathcal{X}$ of possible objects and a binary output space $\mathcal{Y} := \{-1, 1\}$. We are given a set $\mathcal{F} = \{f_0, \ldots, f_m\}$ of $m+1$ substructure features, which are ordered by a (possibly partial) "is substructure of" relation $R \subset \mathcal{F} \times \mathcal{F}$ so that $(f_i, f_j) \in R$ whenever f_i is a substructure of f_j. We write $f_i(x) = 1$ or $f_i(x) = 0$ to express that the substructure for feature f_i is contained ($f_i(x) = 1$) or not contained ($f_i(x) = 0$) in object x. We also assume that the first feature f_0 represents the empty substructure, so that $f_0(x) = 1$ for all $x \in \mathcal{X}$. With this, each object $x \in \mathcal{X}$ can be represented by a $m+1$-dimensional binary vector $\boldsymbol{x} \in \{0,1\}^{m+1}$. In the following, we will not distinguish between x as an object and $\boldsymbol{x}$ as a bit vector if the meaning is clear from the context. We write $f_i \preceq f_j$ to denote that $(f_i, f_j) \in R$. Naturally, the "is substructure of" relation R is transitive so that $f_i \preceq f_k$ whenever there is a f_j with $f_i \preceq f_j$ and $f_j \preceq f_k$. Also, whenever a substructure feature f_i is more general than a feature f_j (i.e. $f_i \preceq f_j$), all objects $x \in \mathcal{X}$ with $f_j(x) = 1$ also have $f_i(x) = 1$. The relation R limits the number of bit vectors that can be used to represent the examples in $\mathcal{X}$. We denote the space of possible bit vectors by $\mathcal{X}_R := \{x \in \{0,1\}^{m+1} | \forall (f, f') \in R : f'(x) = 1 \rightarrow f(x) = 1\}$.

We follow the usual learning setting, where a data set $(X, Y) \in (\mathcal{X}_R \times \mathcal{Y})^n$ of n instances $(x_1, y_1), \ldots, (x_n, y_n)$ is drawn i.i.d. from a fixed but unknown distribution P. The task of the learning system is to find a linear classifier $w \in \mathbb{R}^{m+1}$, which minimizes the true error $\varepsilon_w := \mathbf{E}_{(x,y)\sim P}\, l(w^T x, y)$, where the loss $l(y, y') \rightarrow \mathbb{R}$ assigns some loss to each misclassification. Since P is unknown, a classifier's true error is unknown and practical learning algorithms deal with the empirical error $\hat{\varepsilon}_w = \frac{1}{n}\sum_{i=1}^{n} l(w^T x_i, y_i)$ as a computable substitute. Since the first feature f_0 represents the empty substructure and is always set to 1, the first component w_0 of the linear classifier is essentially a bias term, which controls the distance of the hyperplane induced by w to the origin.

3 Ordered Feature Sets

We are now in the position to formulate the main theoretical results. We first show that the capacity of the class of linear classifiers is the same as the capacity in the unrestricted case without a partial order. Thus, the introduction of a partial order on the features does not increase or decrease the capacity of linear classifiers for worst case distributions. In the second part we give upper bounds of the capacity for distributions where the probability of observing an instance declines with its level in the hierarchy induced by the partial order. We also show that an exponential decay in this probability can lead to linear classifiers having finite capacity, even though the number of features is infinite. This is the case for instance for subsequence and subtree features, but not for subgraph features.

3.1 Distribution-Independent Capacity

One of the main contributions of computational learning theory deals with estimates for the capacity of hypothesis classes. In the general case (i.e. without partially ordered features), it is well known that the hypothesis space of linear classifiers of size m has VC dimension m (see e.g. corollary 13.1 in [4]), giving rise to bounds of the form

$$\mathbf{Pr}\left[\varepsilon_w \leq \hat{\varepsilon}_w + O\left(\sqrt{\frac{m + \ln\frac{1}{\delta}}{n}}\right)\right] \geq 1 - \delta.$$

Thus, the "overfitting penalty" introduced by a hypothesis space of linear classifiers scales as $O(\sqrt{m/n})$ in the number of features. These bounds are tight up to the order of magnitude, that is, there are also lower bounds that scale with $O(\sqrt{m/n})$. Let us now consider the hypothesis space of linear classifiers on the *restricted* instance space $\mathcal{X}_R$, whose instances meet the constraints induced by the partial order R. Observe that the VC-dimension of the class of linear classifiers decreases, if the data instances are chosen only from a d-dimensional subspace of $\mathcal{X}$ with $d < m$. Since $\mathcal{X}_R \subset \mathcal{X}$, the class of linear classifiers on $\mathcal{X}_R$ is smaller in the sense that its hypotheses needs to distinguish between a smaller number of instances. One could thus hope that the VC-dimension of linear classifiers on $\mathcal{X}_R$ decreases in a similar way, if the constraints imposed by R are strict enough. Unfortunately, this turns out to be not the case. The following theorem states this more formally:

Theorem 1. *Let R be an arbitrary partial order on the features $\{f_0, \ldots, f_m\}$, let $\mathcal{X}_R \subset \{0,1\}^{m+1}$ be the space of instances which are consistent with the order R and let $\mathcal{H}_R$ denote the hypothesis space of linear classifiers over $\mathcal{X}_R$. Then, $\mathcal{H}_R$ has VC-dimension $m+1$.*

Proof. It is sufficient to show that there is no dataset of size $m+2$, which can be shattered and that there is a dataset of size $n \leq m+1$, which can be shattered by a linear classifier. The first statement follows directly from the fact that the VC-dimension of linear classifiers in $\mathbb{R}^m$ is also $m+1$. For the second statement, assume without loss of generality that the features $f_0, f_1, \ldots, f_m$ are ordered according to R. Then, select the set of examples $\{x_0, x_1, \ldots, x_m\}$, where $f_j(x_i) = 1$, if $f_j \preceq f_i$ and $f_j(x_i) = 0$ otherwise. The $(m+1) \times (m+1)$ training matrix X for this data set has the lower triangle set to zero and the diagonal set to one, that is, it is in upper diagonal form. A straightforward application of Gaussian elimination shows that the matrix has full rank. This means that there is a linear classifier for all 2^{m+1} possible target value assignments.

This means the VC bounds for the hypothesis space of linear classifiers with partially ordered features are essentially the same as the ones for unordered features. The lower bounds in chapter 14 of [4] ensure that there is no way to get significantly better guarantees than $O(\sqrt{m/n})$. This is quite remarkable, because we did not impose any restrictions on R. In particular, if R is a total order so that $f_i \preceq f_{i+1}$ for all $0 \leq i < m$, the instance space $\mathcal{X}_R$ contains only $m+1$ instances. Thus, the VC dimension $\mathcal{H}_R$ remains constant for any order R, regardless of whether R is a total order (and $|\mathcal{X}_R| = m+1$) or the empty order (and $|\mathcal{X}_R| = 2^{m+1}$). Even though the capacity of the hypothesis class remains constant, the best obtainable empirical risk does depend heavily on R. In particular, if R is a total order and the Bayes error is zero, it is easy to see that empirical risk minimization will find a w with $\hat{\varepsilon}_w = 0$. This is not true for non-total orders.

3.2 Distribution-Dependent Capacity

The results in the preceding section indicate that there are worst-case distributions where the partial ordered feature sets do not decrease the capacity of the class of linear classifiers. However, there may very well be distributions that lead to smaller capacity estimates. In the following we show that this is indeed the case. More specifically, we introduce a distribution-based quantity that can be used to upper-bound the capacity of the class of linear classifiers with partially ordered feature sets.

We begin with a few definitions. Since the features are partially ordered, they can be categorized by level. More formally, for a given feature f_i let the *level* $\lambda(f_i)$ denote the largest $k \in \mathbb{N}$ so that there is a sequence $i_1, i_2, \dots, i_k$ of size k with $f_{i_1} \prec \dots \prec f_{i_k}$. Similarly, the level $\lambda(x) := \max_{f \in \mathcal{F}} \{\lambda(f) | f(x) = 1\}$ of an instance x is the largest level of the features that are set to one by x. The probability $\mathbf{Pr}[f(x) = 1] = \mathbf{E}[f(x)]$ that a feature f is set to one decreases with its level. In fact, if $f_i \preceq f_j$ and $\mathbf{E}[f_i(x)] = \mathbf{E}[f_j(x)]$, then we know that the two features f_i and f_j are equivalent, because $f_i(x) = f_j(x)$ for all instances $x \in \mathcal{X}$. This means that we can remove a feature f_i from the training data whenever it is more general than another feature f_j and $\mathbf{E}[f_i(x)] = \mathbf{E}[f_j(x)]$. Removing the feature does not affect the capacity of the hypothesis class of linear classifiers, because for every classifier w that assigns a non-zero weight to f_i there is an equivalent classifier w', which simply transfers the weight from f_i to the equivalent feature f_j. Thus, we can assume without loss of generality that $\mathbf{E}[f_i(x)] > \mathbf{E}[f_j(x)]$ whenever $\lambda(f_i) < \lambda(f_j)$. With this, let $\lambda_i := \max_{x \in \mathcal{X}_R} \{\mathbf{Pr}[X = x] | \lambda(x) \geq i\}$ denote the maximum probability of obtaining an example of level at least i and let $d_i := |\{f \in \mathcal{F} | \lambda(f) = i\}|$ denote the number of features of level i. We can now state the following two results. The first one gives an upper bound of the capacity of the class of linear classifiers for the zero-one loss, whereas the second one deals with loss functions that are Lipschitz with Lipschitz constant L.

Theorem 2. *Let R be a partial order on a feature space $\mathcal{F}$, let $\mathcal{X}_R := \{x \in \{0,1\}^m | \forall (f, f') \in R : f'(x) = 1 \rightarrow f(x) = 1\}$ be an partially ordered instance space, let P be a probability distribution on $\mathcal{X} \times \mathcal{Y}$ and let ε_w and $\hat{\varepsilon}_w$ be based on the zero-one loss. Define*

$$D_R := \sum_{i=1}^{n} \log \left[\sum_{j=1}^{k} \lambda_j \exp\left(\frac{k}{i} d_j \right) \right]$$

Then it holds for all linear classifiers $w \in \mathbb{R}^m$ that

$$\Pr\left[\varepsilon_w \leq \hat{\varepsilon}_w + \sqrt{\frac{2D_R + \log \frac{1}{\delta}}{n}} + \sqrt{\frac{\log \frac{1}{\delta}}{2n}} \right] \geq 1 - \delta$$

A more straightforward bound can be achieved, if the loss function is continuous and Lipschitz. Let $\mathcal{B}_2^m := \{w \in \mathbb{R}^m | \|w\|_2 \leq 1\}$ denote the m-dimensional unit ball for the 2-norm.

Theorem 3. *Let R, $\mathcal{X}_R$, P be as above, but let ε_w and $\hat{\varepsilon}_w$ be based on a Lipschitz loss $l_L : \mathbb{R} \rightarrow [0,1]$ with Lipschitz constant L. Then it holds for all linear classifiers $w \in \mathcal{B}_2^m$ that*

$$\Pr\left[\varepsilon_w \leq \hat{\varepsilon}_w + 2L\sqrt{\frac{\sum_{i=1}^{k} d_i \lambda_i}{n}} + \sqrt{\frac{8 \log \frac{2}{\delta}}{n}} \right] \geq 1 - \delta$$

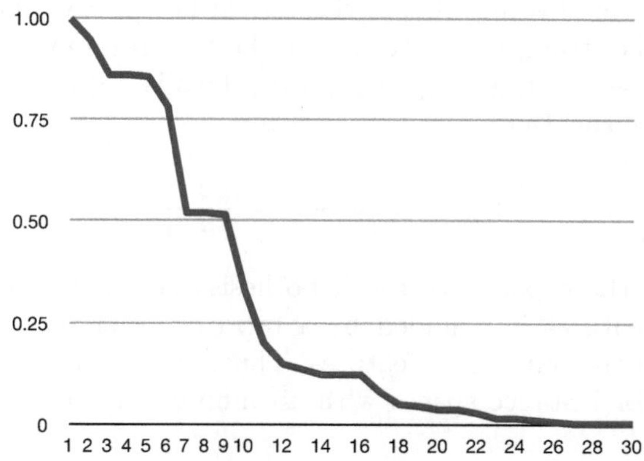

Fig. 1. The maximum probability of observing an instance for each level on the NCTRER dataset

The proofs are in the Appendix. The results essentially state that the capacity of the class of linear classifiers can be upper-bounded depending on how λ_i and d_i scale for increasing levels. The bound is small, either if there are only a limited number of features of higher levels (i.e. d_i is small for large i), or if the probability of encountering instances of higher levels is small (i.e. λ_i is small for higher levels). As explained above, the sequence $\lambda_1, \lambda_2, \ldots$ is strictly decreasing, but the extent of the decay depends on the distribution. In practice, applications with structured examples and substructure features often lead to distributions where the level probabilities features exponential decay. For instance, we plot the level sequence for the NCTRER dataset in figure 1. The figure shows that the probability decreases approximately exponentially.

The number d_i of features per level, though, grows usually exponentially. If a learning system makes use of all possible substructure features for each level, an exponential decay in the level probabilities does therefore not automatically guarantee a small or even finite learning capacity. In the following we give capacity estimates for the case where the decay in the level probabilities is exponential. In particular we assume that the level probabilities can be upper-bounded by a decay that is exponential in a constant α:

$$\lambda_i \leq \alpha^i$$

3.3 Capacity Estimates for Various Partial Orders

Given the upper bound in Theorem 3 we can consider $C := \sum_{i=1}^{k} d_i \lambda_i$ as a capacity measure for the class of linear classifiers on partially ordered feature sets. This enables us to investigate the capacity estimates for the partial orders induced by some popular substructure classes. Let us begin with the total order

$R_t := \{(f_i, f_j)|i \leq j\}$. While this order will be rarely encountered in practical applications, it is interesting from a theoretical perspective, because it is the order that puts the strongest constraints on the features. Obviously, the level of feature f_i is simply i so that $C_{R_t} = \sum_{i=1}^{k} \alpha^i$. This is a geometric series, and a basic analysis confirms that

$$C_{R_t} = \frac{1-\alpha^{k+1}}{1-\alpha} \leq \frac{1}{1-\alpha}$$

This means that the capacity of the hypothesis class of linear classifiers with totally ordered features is bounded by a term of order $O(1/(1-\alpha))$, which is independent of the number of features. Thus, empirical risk minimization is consistent even for instance spaces with an infinite amount of totally ordered features.

As a slightly more complicated case, we consider the setting where the examples are sets of items. Here, we have a set $O = \{o_1, \ldots, o_k\}$ of items and the instance space $\mathcal{X}$ is the power set of O, so that the instances and features are represented by subsets of O. A feature assigns the label 1 to an example, whenever the item set associated with the feature is a subset of the item set associated with the example. It is easy to see that the level of a feature is simply the number of items in its associated item set. Since there are $\binom{k}{i}$ possible item sets with i items, the decay capacity can be computed as:

$$C_{R_I} = \sum_{i=0}^{k} \binom{k}{i} \alpha^i = (1+\alpha)^k \leq e^{k\alpha}$$

Depending on the size of α, this is an exponential improvement over the VC-bound, which upper-bounds the capacity of the class of linear classifiers by $O(2^k)$, because there are 2^k possible features. This result is also applicable to SVM-classification with polynomial kernels on binary data. If one selects a polynomial kernel of degree t for SVM training on data with k binary (i.e. zero-one-valued) features, the kernel-induced feature space is equivalent to a feature space containing all itemset features of size at most t. In this case the decay capacity is $\sum_{i=0}^{t} \binom{k}{i} \alpha^i$.

In the next step, we handle the case where the examples are strings over an alphabet $\mathcal{A} = \{a_1, \ldots, a_h\}$ containing h characters. Here, it is a natural choice to use substring features, which are partially ordered by the "is substring of" order R_S. More precisely, we assume that each feature is associated with a string and the feature assigns the value 1 to an instance, if this string is a substring of the instance. Even though there is a potentially infinite number of instances and features, the analysis is particularly easy. It is clear that the level of a string is just its length. Also, there are h^l different strings of length l. That means that the decay capacity for linear classifiers with partial order R is:

$$C_{R_S} = \sum_{i=1}^{\infty} (h\alpha)^i$$

This is again a geometric series. If $\alpha < \frac{1}{h}$, the series converges and the capacity of the class of linear classifiers can be bounded by $\frac{1}{1-h\alpha}$. Since this quantity does not depend on the number of features, one can have finite capacity even with an infinite amount of features.

As a more complicated partial order, we consider the classification setting where the features are represented by labeled rooted trees, where the labels are taken from label set $\mathcal{L}$ of size l. Here, the features are ordered by some form of subtree isomorphism. Let T_i denote the number of rooted unlabeled trees with i vertices. It is well known that the fraction $\frac{T_i}{T_{i-1}}$ converges as $i \to \infty$ and that it converges to the limit $c_T := \lim_{i\to\infty} \frac{T_i}{T_{i-1}} = 2.955765\ldots$ from below [7]. Thus, we can use c_T^i as a crude upper bound for the number of unlabeled rooted trees with i vertices. Since there are l^i ways to assign labels, we get the following upper bound for the "is subtree of" order R_T:

$$C_{R_T} \leq \sum_{i=1}^{\infty} (lc_T\alpha)^i$$

If $\alpha < \frac{1}{c_T l} \approx \frac{0.3383\ldots}{l}$, the decay complexity of labeled rooted tree classifiers with the subtree order R_T can be upper-bounded as follows:

$$C_{R_T} \leq \frac{1}{1 - lc_T\alpha}$$

It is remarkable that this upper bound differs only in a comparably modest constant from the one for strings.

Finally, let us investigate the setting, where examples and features are connected graphs and the features are partially ordered according to the "is subgraph of" order R_G. Here, a graph with i edges has level i. The number of graphs with i edges and at most l different node labels can be upper-bounded by $li!(l+1)^i$. If there are at most k levels, the capacity can thus be upper-bounded by

$$C_{R_G} \leq \sum_{i=1}^{k} i!l\,((l+1)\alpha)^i$$

Unlike the previous substructure classes, this bound does not converge for $k \to \infty$. In fact it is easy to see that the number of graphs with i edges grows at least with $o((i/c)!)$ for some constant c. That means that an exponential decay of the level probability is not sufficient to enforce a finite capacity bound in the limit.

4 Experiments

Theoretical results are interesting for determining worst case capacity estimates and investigating the asymptotical consistency of learning systems with partially ordered feature sets. However, in practical applications one is much more interested in finding average case capacity estimates, which can be used to avoid over- and underfitting. If the capacity of the learning system's hypothesis class is too large, it might overfit on the training data. In this case, the training error will be near zero, but the validation error is worse than necessary. If, on the other hand, the learner's capacity is too small, the system might induce classifiers with high training and validation error. The theoretical results in the preceding section indicate that the capacity of the space of linear classifiers can be considerably smaller than it is the case with non-ordered features. In the following we investigate the overfitting behavior in quantitative structure-activity relationships. Here, the learning system is given a training set containing the molecular structure of compounds as labeled graphs. The task is to induce a model that can predict some biological or chemical endpoint such as tumor growth inhibition or a compound's ability to pass the blood-brain barrier. We used the three datasets from [8]. The NCTRER dataset [5] deals with the prediction of binding activity of small molecules at the estrogen receptor. It contains 232 molecules. The Yoshida dataset [12] consists of 265 molecules classified according to their bio-availability. The third dataset classifies 415 molecules according to the degree to which they can cross the blood-brain barrier (BBB) [6].

4.1 Overfitting

For the first experiment we followed the standard substructure feature generation methodology (see e.g. [8,2]) and implemented a frequent subgraph mining tool similar to gSpan [11]. The system recursively generates all subgraphs occurring in at least one graph of the training database. However, in contrast to gSpan, it discards all substructures whose instantiation vector is a duplicate of an existing subgraph, that is, a substructure which occurs in exactly the same graphs as an already generated subgraph. We used the tool to generate all possible subsequences, subtrees, and subgraphs for the three datasets. The NCTRER dataset contains 463 subsequences, 1822 subtrees and 1897 subgraphs. We sort the features by level and plot training accuracy and predictive accuracy of a support vector machine (with $C = 1$) induced on an increasing subset of the features in figure 2. The plot shows no significant overfitting; even though the training accuracy reaches 100%, the predictive accuracy as measured by tenfold cross-validation does not decrease very much. Similar plots can be generated for the yoshida and BBB datasets. Obviously, overfitting is less of an issue as compared to many other datasets. This is remarkable when one considers that the substructure features lead to training data that has many more features than examples.

4.2 Elastic Subgraph Feature Generation

As overfitting is apparently not a big problem, one might suspect that the SVM is actually underfitting. To investigate this question we need a new feature generation mechanism that leads to feature sets with higher capacity than the existing ones. The theorems in section 3.2 indicate that one should look for feature sets whose λ_is do not decrease too fast. This means we would like to use substructure occurrence tests, where even large substructures are still likely to appear in

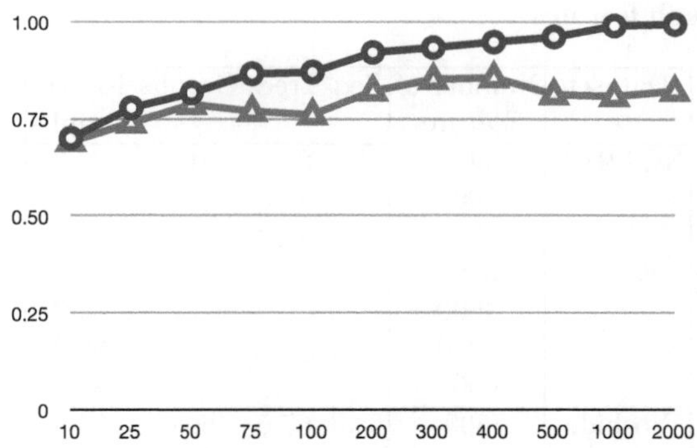

Fig. 2. Training accuracy (circles) and predictive accuracy (triangles) for the NCTRER dataset with an increasing number of features

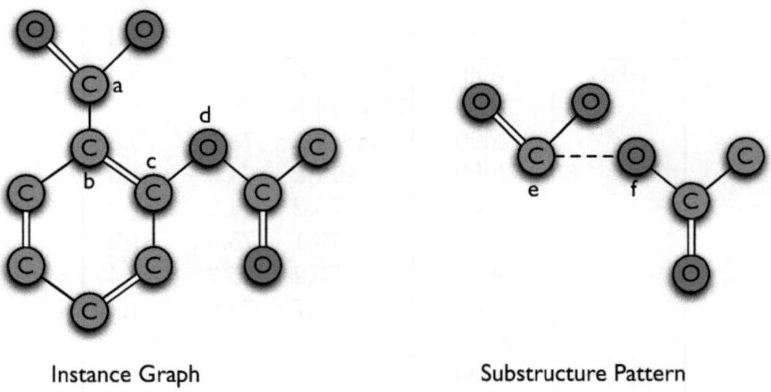

Fig. 3. The dashed edge in the substructure pattern is an elastic edge, which matches the path from vertex a over b and c to vertex d in the instance graph

many instances. In order to do so, we extend the class of subgraph features by allowing for *elastic edges*. More precisely, we introduce a new "elastic" edge label in each subgraph's edge label set. Whenever a subgraph pattern with such an elastic edge is tested for occurrence in a graph in the database, the elastic edge matches with any path containing only edges, which are not already matched with another edge in the subgraph pattern. Figure 3 illustrates this concept. Here, the substructure pattern on the right contains an elastic edge between the vertices e and f. This pattern matches with the instance graph on the left, because the elastic edge can be matched to the path from vertex a over b and c to vertex d. On the other hand, consider the case where one extends the pattern with another vertex, which is connected to the vertex e and has label C. The resulting pattern does not occur anymore in the graph. This is because the new edge can only be matched to the edge between a and b, but the elastic edge is

Table 1. Predictive accuracy of a SVM on an increasing number of subsequence, subtree and subgraph features

Dataset	Number of Features	Seq	Trees	Graphs	Elastic Graphs
NCTRER	500	81.0	81.9	81.5	79.7
	1000		82.3	81.0	82.3
	2000		82.3	82.3	83.6
	3000				85.8
	4000				84.9
	5000				86.2
	6000				86.2
Yoshida	500	63.8	67.5	66.0	64.9
	1000	64.5	69.8	69.1	68.7
	2000		67.5	66.8	64.2
	3000		69.1	67.9	63.4
	4000		67.9	69.1	64.5
	5000		67.5	67.9	63.4
	6000				67.9
	7000				69.1
	8000				68.3
BBB	500	76.1	74.0	77.6	75.4
	1000	75.7	74.5	76.1	78.1
	2000	76.6	74.5	73.7	79.3
	3000		74.2	74.0	81.0
	4000		75.2	74.2	81.2
	5000		76.1	75.9	81.2
	6000		75.2	75.7	81.4
	7000		73.7	74.0	81.7
	8000				81.9
	9000				83.4
	10000				81.9
	11000				81.7

not allowed to match with the path from a to d, if one of the edges on the path is already used in a different match.

We extended the subgraph mining tool to also generate subgraph structures which contain a limited number of elastic edges. While it is feasible to compute all non-duplicate subtree or subgraph features for the three datasets, the number of non-duplicate subgraph patterns with elastic edges is way too large to generate all of them. We therefore restricted the maximum size of the elastic subgraph patterns to eight for the NCTRER dataset, and five for the BBB and yoshida datasets. Table 1 shows the predictive accuracies of a SVM (with C=1) as estimated by tenfold cross-validation for subsequence, subtree, subgraph and subgraph with one elastic edge patterns depending on the number of used features. Elastic subgraph patterns outperform the other pattern languages by approximately four percent on the NCTRER dataset and over five percent on the BBB dataset. On the yoshida dataset, trees, graphs and elastic graph feature give approximately the same predictive accuracy. All three datasets show better accuracies than the best ones reported in [8]. These results indicate that underfitting was indeed a problem on two of the three datasets.

4.3 Feature Generation for Large Datasets

For the third experiment, the goal was to investigate how learning linear classifiers with a broad class of substructure features can be made efficient on a large dataset. The main problem here is that the considerations in section 4.1 indicate that one should use broad substructure classes with many general features to avoid underfitting. Unfortunately, the number of substructures in such classes is way too large to enumerate them exhaustively as it was possible in the preceding experiments. Consider the NCI DTP Human Tumor Cell Line Screen dataset [9]. The dataset contains 34748 compounds, which were tested for their ability to inhibit tumor growth. Mining for all non-duplicate subsequences of only up to four edges leads to over 10,000 features. Clearly, the database is too large to allow for exhaustive enumeration of all existing subsequence features and working with all subtree or subgraph features is clearly not feasible. To avoid the generation of all substructure features, we resort to a heuristic feature search algorithm inspired by the feature generation method presented in [8]. Instead of using a combinatorial search approach with an index structure (which would be too large for the NCI dataset), we perform a simple beam search. The algorithm starts with substructures of size one (i.e. single vertices) and iteratively extends the most promising candidate in the current beam with a new edge. The search heuristic is based on the *class-correlated dispersion score* as described in [8], but features an exponential rather than a quadratic penalty for features with high similarity to an existing feature:

$$h(s') := \sum_{i=1}^{m} \exp[\frac{c}{n} s_i^T s'] - m \exp[\frac{c}{n} t^T s']$$

Here, s' is the $-1/+1$-valued n-dimensional instantiation vector of the new feature candidate, the s_i are the instantiation vectors of the m existing features, and t is the target class vector. We generated one hundred substructure features for subsequences, subgraphs and subgraphs with one elastic edge. We then learned a linear classifier from a training set consisting of two thirds of the dataset and evaluated the classifier on the remaining third. The feature generation, learning and evaluation took 28 minutes on a 1 GHz Athlon 5200 computer for the sequences features, 68 minutes for the subgraphs feature set and 129 minutes for subgraphs with one elastic edge. The classifier achieved a predictive accuracy of 64.4% with subtree features, 64.1% with subgraph features and 63.3% with subgraphs with one elastic edge, so underfitting seems not to be a big issue here. It is unclear whether this is a limitation of the feature generation method or a fundamental property of the data generation process.

5 Conclusion

In the preceding sections we investigated classification with linear classifiers and partially ordered feature sets. Learning with partially ordered feature sets differs from other settings in that the partial order induces redundancy in the training and test data. From a theoretical point of view, this does not necessarily affect the over- or underfitting behavior of a learning system, because the VC-dimension of the class of linear classifiers remains the same for worst-case data distributions. However, if the data distribution features a sufficiently steep decline in the probability of observing features of higher level, the capacity of the learning system can be upper-bounded by a smaller term. This means that overfitting is less of an issue for linear classifiers on those distributions. We evaluated this theoretical result on three datasets and found that subsequence, subtree and subgraph features did indeed not show typical overfitting behavior. Instead, we were able to extend the class of subgraph features towards subgraphs with elastic edges. These patterns are more likely to occur in higher levels and thus increase the capacity estimate. Practical experiments confirmed that this extended class of subgraph features avoids underfitting and increases predictive accuracy on two of the three datasets. Finally, we showed how classification with such large substructure feature classes can be implemented efficiently on large datasets.

The work raises a couple of interesting questions. On the theoretical side one could look for lower bounds that quantify to which degree the presented upper bounds are tight and investigate how the results apply, if one uses support vector machines with non-linear kernels. On the practical side, it would be interesting to obtain more insights on the actual under- or overfitting behavior of common data (for instance with regard to the study in [2]) and how the present results apply to other partially ordered feature sets, for example in natural language processing.

References

1. Bartlett, P.L., Mendelson, S.: Rademacher and gaussian complexities: risk bounds and structural results. J. Mach. Learn. Res. 3, 463–482 (2003)
2. Bringmann, B., Zimmermann, A., De Raedt, L., Nijssen, S.: Don't be afraid of simpler patterns. In: Fürnkranz, J., Scheffer, T., Spiliopoulou, M. (eds.) PKDD 2006. LNCS (LNAI), vol. 4213, pp. 55–66. Springer, Heidelberg (2006)
3. Deshpande, M., Kuramochi, M., Karypis, G.: Frequent sub-structure-based approaches for classifying chemical compounds. In: IEEE International Conference on Data Mining, p. 35 (2003)
4. Devroye, L., Györfi, L., Lugosi, G.: A Probabilistic Theory of Pattern Recognition (Stochastic Modelling and Applied Probability). Springer, New York (1996)
5. Fang, H., Tong, W., Shi, L.M., Blair, R., Perkins, R., Branham, W., Hass, B.S., Xie, Q., Dial, S.L., Moland, C.L., Sheehan, D.M.: Structure-activity relationships for a large diverse set of natural, synthetic, and environmental estrogens. Chemical Research in Toxicology 14(3), 280–294 (2001)

6. Li, H., Yap, C.W., Ung, C.Y., Xue, Y., Cao, Z.W., Chen, Y.Z.: Effect of selection of molecular descriptors on the prediction of blood-brain barrier penetrating and nonpenetrating agents by statistical learning methods. Journal of Chemical Information and Modeling 45(5), 1376–1384 (2005)
7. Otter, R.: The number of trees. The Annals of Mathematics 49(3), 583–599 (1948)
8. Rückert, U., Kramer, S.: Optimizing feature sets for structured data. In: Kok, J.N., Koronacki, J., Lopez de Mántaras, R., Matwin, S., Mladenič, D., Skowron, A. (eds.) ECML 2007. LNCS (LNAI), vol. 4701, pp. 716–723. Springer, Heidelberg (2007)
9. Teicher, A.B. (ed.): The NCI Human Tumor Cell Line (60-Cell) Screen, 2nd edn., pp. 41–62. Humana Press, Totowa (1997)
10. Wale, N., Watson, I.A., Karypis, G.: Comparison of descriptor spaces for chemical compound retrieval and classification. Knowl. Inf. Syst. 14(3), 347–375 (2008)
11. Yan, X., Han, J.: gspan: Graph-based substructure pattern mining. In: ICDM 2002: Proceedings of the 2002 IEEE International Conference on Data Mining (ICDM 2002), Washington, DC, USA, p. 721. IEEE Computer Society, Los Alamitos (2002)
12. Yoshida, F., Topliss, J.: QSAR model for drug human oral bioavailability. J. Med. Chem. 43, 2575–2585 (2000)

6 Appendix

6.1 Proof of Theorem 2

Proof. For a data sample $S = (X_1, \dots, x_n)$ and a vector of Rademacher variables $\sigma = (\sigma_1, \dots, \sigma_n)^T$ (where σ_i has value $+1$ or -1 with probability 0.5) define $V(S, \sigma) = \sup_w [\frac{1}{n} \sum_{i=1}^n \sigma_i \operatorname{sgn}(w^T x_i)]$. First of all, we consider the conditional expectation $\mathbf{E}[V(S, \sigma)|S]$. Let $v := (\operatorname{sgn}(w^T x_1), \dots, \operatorname{sgn}(w^T x_n))^T$ denote the vector of predictions for a fixed training sample S and a fixed linear classifier w. Changing the value of a Rademacher variable σ_i changes the value of $\frac{1}{n} v^T \sigma$ by at most $\frac{2}{n}$. Thus, for a fixed data set S, one can apply McDiarmid's inequality to bound the probability that a random Rademacher vector disagrees with v by more than fraction a fixed $r > 0$:

$$\mathbf{Pr}\left[\frac{1}{n}\sum_{i=1}^{n} \sigma_i \operatorname{sgn}(w^T x_i) \geq r \,\middle|\, S\right] \leq e^{-\frac{1}{2} r^2 n} \tag{1}$$

Now, let $\lambda(x) := \max\{\lambda(f)|f(x) = 1\}$ denote the level of example x. We can assume without loss of generality that the instances in S are sorted by level. Define $S_i := \{x|\lambda(x) = i\}$ and $n_i := |S_i|$, so that $\sum_{i=1}^k n_i = n$. Sauer's lemma states that there are at most $\sum_{j=0}^{d_i} \binom{n_i}{j} \leq (n_i+1)^{d_i}$ ways, in which the different w can assign class labels to the instances in S_i. This means the number of possible class vectors induced by the w is at most $\prod_{i=1}^k (n_i+1)^{d_i}$. Taking the union bound over all possible class label assignments in (1) yields:

$$
\begin{aligned}
\mathbf{Pr}[V(S,\sigma) \geq r|S] &= \mathbf{Pr}\left[\sup_{w}\left[\frac{1}{n}\sum_{i=1}^{n}\sigma_i v_i\right] \geq r \middle| S\right] \\
&\leq e^{-\frac{1}{2}r^2 n}\prod_{i=1}^{k}(n_i+1)^{d_i} \\
&\leq e^{-\frac{1}{2}r^2 n}\exp\left[\sum_{i=1}^{k} d_i \log(n_i+1)\right] \\
&\leq e^{-\frac{1}{2}r^2 n}\exp\left[\sum_{i=1}^{k} d_i \sum_{j=1}^{n_i}\frac{1}{j}\right] \\
&\leq e^{-\frac{1}{2}r^2 n}\exp\left[\sum_{i=1}^{n} d_{\lambda(x_i)}\frac{k}{i}\right] \\
&\leq e^{-\frac{1}{2}r^2 n}\prod_{i=1}^{n} e^{\frac{k}{i}d_{\lambda(x_i)}}
\end{aligned}
$$

Taking the expectation on both sides yields:

$$
\begin{aligned}
\mathbf{Pr}[V(S) \geq r] &= e^{-\frac{1}{2}r^2 n}\prod_{i=1}^{n}\mathbf{E}\left[e^{\frac{k}{i}d_{\lambda(x_i)}}\right] \\
&\leq e^{-\frac{1}{2}r^2 n}\prod_{i=1}^{n}\sum_{j=1}^{k}\lambda_j e^{\frac{k}{i}d_j}
\end{aligned}
$$

Setting

$$
r := \sqrt{\frac{2\sum_{i=1}^{n}\log\left[\sum_{j=1}^{k}\lambda_j \exp\left(\frac{k}{i}d_j\right)\right] + \log\frac{1}{\delta}}{n}}
$$

yields that with probability larger than $1-\delta$ it holds that

$$
V(S) \leq \sqrt{\frac{2\sum_{i=1}^{n}\log\left[\sum_{j=1}^{k}\lambda_j \exp\left(\frac{k}{i}d_j\right)\right] + \log\frac{1}{\delta}}{n}}
$$

Taking the union bound with theorem 5 (b) in [1] yields the result.

6.2 Proof of Theorem 3

Proof. For a sequence of n Rademacher variables $\sigma_1, \ldots, \sigma_n$ define

$$
R_n(\mathcal{X}) := \mathbf{E}\left[\sup_{w \in \mathcal{B}_2^m}\left|\frac{2}{n}\sum_{i=1}^{n}\sigma_i l\left(y_i, w^T x_i\right)\right|\right]
$$

The result follows from theorem 8 in [1] and the fact that $R_n(\mathcal{X}) \leq 2L\sqrt{\frac{\sum_{i=1}^{k} d_i \lambda_i}{n}}$. To see this, observe that

$$
\begin{aligned}
R_n(\mathcal{X}) &\leq \mathbf{E}\left[\sup_{w \in \mathcal{H}} \frac{2}{n} \sum_{i=1}^{n} \sigma_i l(y_i, w^T x_i)\right] \\
&\leq 2L\,\mathbf{E}\left[\sup_{w \in \mathcal{H}} \frac{1}{n} \sum_{i=1}^{n} \sigma_i w^T x_i\right] && (2) \\
&\leq 2L\,\mathbf{E}\left[\sup_{w \in \mathcal{H}} \|w\|_2 \left\| \frac{1}{n} \sum_{i=1}^{n} \sigma_i x_i \right\|_2\right] && (3) \\
&\leq 2L\,\mathbf{E}\left[\sqrt{\frac{1}{n^2} \sum_{k=1}^{m} \sum_{i,j=1}^{n} \sigma_i \sigma_j f_k(x_i) f_k(x_j)}\right] \\
&\leq 2L\sqrt{\frac{1}{n^2}\mathbf{E}\left[\sum_{k=1}^{m} \sum_{i=1}^{n} f_k(x_i)^2\right]} && (4) \\
&\leq 2L\sqrt{\frac{1}{n} \sum_{k=1}^{m} \mathbf{E}\left[f_k(x)\right]} \\
&\leq 2L\sqrt{\frac{\sum_{i=1}^{k} d_i \lambda_i}{n}}
\end{aligned}
$$

Here, (2) is due to theorem 12 in [1], (3) is an application of Hölder's inequality, while (4) follows from the concavity of the square root and the independence of the Rademacher variables.

Reconstructing Data Perturbed by Random Projections When the Mixing Matrix Is Known*

Yingpeng Sang[1], Hong Shen[1], and Hui Tian[2]

[1] School of Computer Science, The University of Adelaide, SA 5005, Australia
[2] School of Mathematical Science, The University of Adelaide, SA 5005, Australia
{yingpeng.sang,hong.shen,hui.tian}@adelaide.edu.au

Abstract. Random Projection ($\mathcal{RP}$) has drawn great interest from the research of privacy-preserving data mining due to its high efficiency and security. It was proposed in [27] where the original data set composed of m attributes, is multiplied with a mixing matrix of dimensions $k \times m$ ($m > k$) which is random and orthogonal on expectation, and then the k series of perturbed data are released for mining purposes. To our knowledge little work has been done from the view of the attacker, to reconstruct the original data to get some sensitive information, given the data perturbed by $\mathcal{RP}$ and some priori knowledge, e.g. the mixing matrix, the means and variances of the original data. In the case that the attributes of the original data are mutually independent and sparse, the reconstruction can be treated as a problem of Underdetermined Independent Component Analysis (UICA), but UICA has some permutation and scaling ambiguities. In this paper we propose a reconstruction framework based on UICA and also some techniques to reduce the ambiguities. The cases that the attributes of the original data are correlated and not sparse are also common in data mining. We also propose a reconstruction method for the typical case of Multivariate Gaussian Distribution, based on the method of Maximum A Posterior (MAP). Our experiments show that our reconstructions can achieve high recovery rates, and outperform the reconstructions based on Principle Component Analysis (PCA).

Keywords: Privacy-preserving Data Mining, Data Perturbation, Data Reconstruction, Underdetermined Independent Component Analysis, Maximum A Posteriori, Principle Component Analysis.

1 Introduction

Privacy-preserving Data Mining (PPDM) concerns the problems of completing data mining tasks without any direct access to the original data sets, because the providers claim privacy on their data. The general mining tasks include classification, clustering and association rule mining. PPDM can be treated as a subset within the problems of Secure Multi-party Computation (SMC) ([19], [20], [26], [40], etc). The cryptographic techniques from SMC provide solutions which always demand high computation cost,

* This work is partially supported by Australian Research Council Discovery Project grant #DP0985063.

W. Buntine et al. (Eds.): ECML PKDD 2009, Part II, LNAI 5782, pp. 334–349, 2009.

especially on processing volumes of data in data mining applications. Alternative approaches are based on data perturbation techniques which aim to be much more efficient than techniques of SMC.

Additive data perturbation, i.e. adding random data to the original data, was used in [3] to build decision tree classifiers, but in [17] and [22] random additive noise was questioned and pointed out that it can be easily filtered out, and thus lead to compromising of privacy. Multiplicative perturbation was used in [32] where the original data of each data provider is multiplied with the same matrix which is random and orthogonal before released, while in [28] this kind of perturbation is easily reconstructed by methods such as Principle Component Analysis (PCA), i.e. recovering the original data by analyzing the covariance matrix of the perturbed data.

In [27], an improved multiplicative data perturbation was proposed, in which the original data set X with m attributes is multiplied with a $k \times m$ $(m > 2k - 1)$ matrix R, each entry of which is an independently and identically distributed (i.i.d.) random number with the zero means. We name this method *Random Projection* ($\mathcal{RP}$) following [27] to avoid confusions with the method in [32]. The security claim of RX in $\mathcal{RP}$ is based on the structure of R and the fact that there does not exist a matrix T such that the product TR is a partition matrix and TRX is a separation of some attributes of X. However in the research field of Independent Component Analysis (ICA), the separation of m series of data $\hat{X}$ from k $(m > k)$ series of linearly mixed data RX is treated as the problem of Underdetermined ICA (UICA). Plenty of methods have proposed for UICA ([31]), and most of them are not seeking for the partition matrix T, but the possible values of X with maximum probability given only RX, and they have been successful in the case that the m original sources in X are mutually independent and sparse, except some permutation and scaling ambiguities.

In [8] and [16] reconstructions based on ICA were employed to attack the perturbation method of [32]. To our knowledge the only work on attacking the $\mathcal{RP}$ of [27] was proposed in [29] or [2] (Chapter 15), which was based on Maximum A Posterior (MAP). This attack only assumed the original data are uniformly distributed, but in practice many data properties can be assumed as normally (or approximately normally) distributed (e.g. personal heights, weights, financial variables), or are sparse enough to be modeled by the Laplace distribution (e.g. the voice or image data, financial data). It is not difficult for an attacker to obtain these priori knowledge on the original data, such as whether they are sparse, or whether they are normally distributed, given enough samples extracted from the same pool where the original data are extracted. It is also possible that the attacker may know the mixing matrix R by colluding with one data provider. Little work has been done to address these considerations.

In this paper, we will propose some attacking techniques on the $\mathcal{RP}$ method of [27] under some practical scenarios. We name the recovery of the original data by the attacker who is given the perturbed data as "*reconstruction*" following [3] and [17]. We also assume the attacker has a collusion with one of the data providers from which he can know the mixing matrix in $\mathcal{RP}$, the attacker has also obtained enough samples with identical distribution with the original data set, and thus some necessary priori knowledge on the original data, including whether the data attributes are mutually independent, whether they are sparse, their means and covariance matrix. Based on these assumptions we propose the following reconstruction methods from the view of the attacker:

1) If the attributes are mutually independent and sparse, we propose *Underdetermined Independent Component Analysis (UICA) based reconstruction* for the case that the attacker knows the mixing matrix, which outperforms the reconstruction based on PCA.
2) If the attributes are not mutually independent, where the ICA-based reconstruction will not be effective, we propose *Maximum A Posterior (MAP) based Reconstruction* for the case that the attacker knows the mixing matrix, and the original data following the Multivariate Gaussian Distribution. Our reconstruction outperforms the reconstruction based on PCA.

The organization of this paper is as following. In Section 2 we briefly review the related work. In Section 3 we give formal definitions on the problems of Data Perturbation and Data Reconstruction. In Section 4 we talk about how to obtain the necessary priori knowledge. In Section 5 and Section 6 we propose the ICA-based and MAP-based Reconstructions respectively. In Section 7 we conduct some experiments to evaluate our reconstruction methods, and compare them with reconstructions based on PCA. Section 8 concludes the paper.

2 Related Work

2.1 Data Perturbation

According to the taxonomy of [1], two families of approaches, *query restriction* and *data perturbation*, are usually used to provide statistical information (sum, count, average, etc) without compromising sensitive information about individuals. The query restriction family includes restricting the size of query result, controlling the overlap amongst successive queries, suppression of data cells, clustering entities into mutually exclusive atomic populations, etc. The data perturbation family includes the methods, loosely speaking, which perturb the original value of the data, X, into a random value Y.

The perturbation methods employed until now consist of data replacement ([1], [25], [23]), data swapping ([10], [13]), additive value distortion ([3], [22], [17]), and multiplicative value distortion ([32], [27]), random perturbation on categorical or boolean data ([12],[33], [4]), etc. k-anonymity ([35], [30], [24]) and sensitive rule hiding ([5], [34], [39]) have also been employed in PPDM. Details on these methods can be found in the given references. In this paper we only focus on the multiplicative value distortion (or multiplicative data perturbation).

Multiplicative Data Perturbation of [32]. Multiplicative data perturbation was used in [32] in which the original data set $X = (x_1, ..., x_m)'$ is multiplied by a random and orthogonal matrix R of dimensions $m \times m$, and perturbed into the set $U = RX$. Given $U_1 = RX_1$ and $U_2 = RX_2$, obviously $U_2' \cdot U_1 = X_2' \cdot X_1$, i.e. the inner products $X_2' \cdot X_1$ are the same as the inner products $U_2' \cdot U_1$. The distance-related metrics such as

Euclidean distance between X_1 and X_2 can be computed based on the inner products $U_2' \cdot U_1, U_1' \cdot U_1, U_2' \cdot U_2$, and data mining tasks can continue by these metrics without knowing the private X_1 and X_2.

Random Projection of [27]. In [27], a similar R is used in their multiplicative perturbation, but is different from [32] in that R is rectangle, which is $k \times m (m \geq 2k-1, m \geq 2)$ with i.i.d entries from $N(0, \sigma_r^2)$. Two data owners respectively compute $U = \frac{1}{\sqrt{k}\sigma_r} RX$ and $V = \frac{1}{\sqrt{k}\sigma_r} RY$, and then send them to the miner. Because $U'V = \frac{1}{k\sigma_r^2} X'R'RY$, and $E(R'R) = k\sigma_r^2 I$ (by Lemma 5.2 of [27]), then $E(U'V) = X'Y$. By this statistical result, the inner product $x \cdot y$ $(\forall x \in X, \forall y \in Y)$ can be computed without knowing X and Y.

The security of this $\mathcal{RP}$ method is based on the fact that if R is a rectangle $(k \times m)$ matrix and $m \geq 2k - 1$, there does not exist a matrix T such that the product TR becomes a partition matrix which has at most one nonzero element in each column, and separates any single independent signal in TRX. Details on the partition matrix and signal separation can be referred to [7].

2.2 Reconstructions on Multiplicative Data Perturbation

Without the transformation matrix R, the recovery of X from U is infeasible given only the linear system $RX = U$. If in the $k \times m$ matrix R $k < m$, given R and $RX = U$, X can not also be determined from solving the linear system. However, the solutions can be sought from clues in some priori knowledge on X.

Reconstruction on the Perturbation of [32]. According to [28], if some input-output pairs "(x_i, u_i)" (i.e. $x_i \in X$, $u_i \in U$, and $u_i = Rx_i$), or some samples $x_i \in X$, are known apriori by an attacker, he may approximately recover X. Given some input-output pairs, the attacker can uniformly select an R to meet some criterion function, but this kind of attack can be prevented by deleting these pairs from the data owners, or de-identifying the output set U by randomly mixing the records $(u_1, ..., u_i, ...)$, before they are released for mining.

Another reconstruction of [28] is based on PCA, and does not requires the input-output pairs, but some general samples from the same pool where X is originated. By analyzing these samples, the attacker can estimate the covariance matrix of X, i.e. Σ_X. By analyzing the covariance matrix of the perturbation U, i.e. Σ_U, the attacker can obtain the information on R since $\Sigma_U = R\Sigma_X R'$. Specifically, suppose the eigenvalue decompositions of Σ_X and Σ_U are $Q_X E_X Q_X'$ and $Q_U E_U Q_U'$, then the diagonal matrices $E_X = E_U$, and the orthogonal matrices $Q_U = RQ_X$. When R is know in this way, the attacker can then have an estimate on X.

This attack based on PCA will not be effective on reconstructing the $\mathcal{RP}$ perturbation of [27], since in this method R is rectangle $(k < m)$, Σ_U is very similar as a diagonal matrix as proved by [11], and it may be difficult to make an eigenvalue decomposition on Σ_U. What's more, given Σ_U and Σ_X, it is difficult to obtain the information on the rectangle R from $\Sigma_U = R\Sigma_X R'$.

Attacks based on ICA were proposed in [2] (Chapter 15), [8] and [16], under various assumptions on the attacker's priori knowledge. They assumed an $m \times m$ mixing matrix R, thus were not suitable to attack the $\mathcal{RP}$ in [27].

Underdetermined Independent Component Analysis (UICA). Underdetermined (or overcomplete) independent component analysis (or blind source separation) has been under years of research in the field of signal processing, which addresses the problem that, X composed of m sources is linearly mixed by a $k \times m$ matrix R, and given only the mixed data RX, without knowing R, the sources are required to be separated out. A survey on the research can be referred to [31]. The methods of UICA have been mostly successful in the case that the m sources are mutually independent and sparse. They generally require two steps: 1) recovering the mixing matrix R, and 2) given R, recovering the sources. For Step 1) a lot of improvement work have been continuously done (e.g. [6], [24], [37] and [41]). For Step 2) L_1-norm minimization is the standard method that has been widely used.

L_1-norm minimization comes from the general approach of Maximum A Posterior (MAP). Considering $u = Rx$ and neglecting any additional noise, the probability of observing a vector x given R and a vector u is $p(x|R,u)$, and by Bayes Theorem

$$p(x|R,u) = \frac{p(u|R,x)p(x)}{p(u)}$$

By MAP x will be the vector in the Euclidean space of $\mathbb{R}^m$ that maximize the above equation. In the searching of this vector, $p(u)$ is a constant, $p(u|R,x)$ can be viewed as a constraint $Rx = u$, a sparse source x_i $(i = 1, ..., m)$ can be modeled by the Laplace distribution, i.e. $p(x_i) \propto e^{-|x_i|}$ (assuming they have zero means and identical variances). Then x will be the solution of the following constrained linear programming problem:

$$\begin{aligned} x &= \arg \max_{Rx=u} p(x) \\ &= \arg \max_{Rx=u} e^{-|x_1|-...-|x_m|} \\ &= \arg \min_{Rx=u} \sum_{i=1}^{m} |x_i| \end{aligned} \tag{1}$$

In Eq. (1) $\sum_{i=1}^{m} |x_i|$ is the L_1-norm of the vector $x = (x_1, ..., x_m)'$, therefore x will be the solution of the constrained L_1-norm minimization problem. There have been many methods to solve this problem and a survey of them can be referred to [9].

When the methods of UICA are used for reconstructions on the $\mathcal{RP}$-perturbed data, there is no additional noise N such as $U = RS + N$ and no need to reduce N, but they still have the following limitations:

1) When R is not known, these methods have permutation and scaling ambiguities. For each estimate of R, e.g. $\hat{R}$, there is infinite equivalent matrices $\tilde{R} = \hat{R}PL$ in which P is a permutation matrix of dimensions $m \times m$, L is a nonsingular diagonal matrix of dimensions $m \times m$ (scaling matrix), and $\tilde{R}$ is also an estimate of R. Thus the recovering of X will also have permutation and scaling ambiguities.

2) When R is known, there is no permutation ambiguity (the detailed reason is postponed to Section 5), but the means and variances of x_i should not be neglected in the constrained L_1-norm minimization of Eq. (1), since in practical scenarios their means may not be zero, and their variances may be not identical.
3) These methods generally require the original sources are mutually independent and sparse. In many scenarios of data mining, the attributes of the original data are correlated and not sparse. One typical model of these data is the Gaussian Mixture Model (GMM).

2.3 Disclosure Risk of the Distances

The risks of disclosing the mutual distances between data objects were investigated in [38]. They proposed two reconstruction methods based on the mutual distances of the data objects. The first one needs some known samples in their original forms and perturbed forms, so this method will not be effective when all perturbed data objects are mixed arbitrarily and de-identified before released, and an attacker can not find the corresponding perturbed data of the known samples. The second one needs no known sample, but makes a PCA on the perturbed data. However, this PCA will not be effective on analyzing the data perturbed by the method of [27], because R is rectangle, the covariance matrix of $U = RX$ is very similar as a diagonal matrix by [11], and it will be difficult to make a desired eigenvalue decomposition of the covariance matrix.

3 Problem Statement

3.1 Database Model

For convenience we consider a two-party case in which Alice and Bob share a distributed database. The reconstructions under the cases of more than two parties are similar as the two-party case, since in all the cases the parties use the same mixing matrix R for $\mathcal{RP}$. Suppose Alice and Bob have the data set X and Y respectively. Suppose the database has m attributes. If the database is horizontally distributed on the two parties, Alice has n_1 records, Bob has n_2 records, then X is an $m \times n_1$ matrix $[[x_{i,j}]_{i=1}^{m}]_{j=1}^{n_1}$, Y is an $m \times n_2$ matrix $[[y_{i,j}]_{i=1}^{m}]_{j=1}^{n_1}$.

If the database is vertically distributed and the database has n records, Alice has m_1 attributes, Bob has m_2 attributes, then X is an $n \times m_1$ matrix $[[x_{j,i}]_{j=1}^{n}]_{i=1}^{m_1}$, Y is an $n \times m_2$ matrix $[[y_{j,i}]_{j=1}^{n}]_{i=1}^{m_2}$. In this paper we only focus on the horizontally distributed database. In the $\mathcal{RP}$ of [27] the vertically distributed databases are perturbed as similar as the horizontally distributed databases are perturbed, so our proposed reconstruction methods can be easily extended to the vertical cases.

3.2 Network Model

We consider two kinds of network models in this paper:

1) *Centralized model*: There is an independent miner who receives perturbed data from the data owners Alice and Bob (as in Fig. 1(a)). Scenarios are some companies under the investigation of a governmental organization. The companies are data providers, and the governmental organization wants to mine their private data.

2) *Distributed model*: There is no independent miner. All the data owners, Alice and Bob, act as miners on the perturbed data of their own and those received from the other party (as in Fig. 1(b)). Scenarios are some companies which want to share their data with each other to complete the data mining tasks. Each company is simultaneously a data provider and miner.

(a) Centralized Model (b) Distributed Model

Fig. 1. Two Network Models for PPDM

3.3 Adversary Model

Adversary models have been theoretically defined in SMC ([14]), and also extensively used in PPDM. Depending on whether the participants merely gather information, or take active steps to disrupt the execution of the protocol, there are usually two types of adversaries:

1) *Semi-honest* participants, which are assumed to execute the solution exactly as what is prescribed, but may collude and analyze all the intermediate computations.
2) *Malicious* participants, which may arbitrarily deviate from the specified solution, e.g. generate arbitrary inputs, substitute the intermediate computations, or prematurely quit.

3.4 Problem Definition

Definition 1 - Privacy-preserving Data Mining based on Data Perturbation: *A database is distributed on two parties, Alice and Bob. The two providers respectively perturb their data matrices X and Y into U and V, and publish U, V to the miner. The miner may be an independent party, or replaced by Alice and Bob. All of them may be semi-honest or malicious. Privacy-preserving data ming on the miner should satisfy the following two requirements:*

1) *Privacy Requirement: No sensitive information on X and Y should be inferred from U and V by the miner.*
2) *Accuracy Requirement: The mining tasks including classification, clustering, etc, on U and V, should have statistically the same results as directly mining X and Y.*

It is worthy to note that in Definition 1 we do not specify the method of perturbing X and Y into U and V. Different perturbation methods possess different properties on privacy and accuracy, so the definition is made inclusive so as to cover as many perturbation methods as possible. In addition, to achieve the accuracy requirement, an accurate computation on the inner product of two vectors, such as $x'y(\forall x \in X, \forall y \in Y)$, is enough. Distance-related metrics like Euclidean distance, required in both the horizontally and vertically partitioned data mining, can be computed based on those inner products, the details of which can be referred to [27].

Definition 2 - Data Reconstruction: *An attacker obtains the perturbed data U and V from the data providers Alice and Bob. He wants to recover as many as possible the entries of X and Y. The attacker and any of the providers may be semi-honest or malicious.*

In the centralized model, if the miner and any of the providers are semi-honest, they may collude to reconstruct the data of the other providers. If the miner is malicious, he may not communicate the correct data mining results with the data providers.

In the distributed model, if one of the provider is semi-honest, he may reconstruct the data of another provider using R. If he is malicious, he may arbitrarily substitute his original data and publish them to another provider. Malicious attackers are not the focus of this paper.

Figure 2 shows the two mutually inverse processes, data perturbation and reconstruction.

(a) Data Perturbation by the owner Alice (b) Data Reconstruction by an adversarial miner

Fig. 2. Data Perturbation and Reconstruction for PPDM

Definition 3 - Recovery Rate: *Suppose $\hat{X}$ is a reconstruction of the original data X, $\hat{X} = [[\hat{x}_{i,j}]_{i=1}^{m}]_{j=1}^{n_1}$, and $X = [[x_{i,j}]_{i=1}^{m}]_{j=1}^{n_1}$. The Recovery Rate, $r(\hat{X}, \epsilon)$ with a given threshold ϵ, is the percentage of reconstructed entries whose relative errors are within ϵ, i.e.*

$$r(\hat{X}, \epsilon) = \frac{\#\{\hat{x}_{i,j} : |\frac{x_{i,j} - \hat{x}_{i,j}}{x_{i,j}}| \leq \epsilon, i = 1, ..., m, j = 1, ..., n_1\}}{m * n_1} \quad (2)$$

We will use the recovery rate in this definition to evaluate the performance of our reconstruction methods.

4 Obtaining the Priori Knowledge

In this section we discuss how the attacker can obtain the necessary priori knowledge on the original data, including their mean values, covariance matrix, whether they are mutually independent, under the condition that he has got enough samples, i.e. m-dimension vectors like $v = (v_1, ..., v_m)$, which are identically and independently selected from the multivariate distribution of the original data X, i.e. $p(X) = p(x_1, ..., x_m)$.

Given enough samples, the attacker can compute the *sample means* $\overline{X} = (\overline{x}_1, ..., \overline{x}_m)$ in which $\overline{x}_i$ is an estimate of x_i's mean u_i, and he can also compute the *sample covariance matrix* $\Sigma_X = [[cov(x_i, x_j)]_{i=1}^m]_{j=1}^m$ in which $cov(x_i, x_j)$ is an estimate of the covariance $E[(x_i - u_i)(x_j - u_j)]$.

When the estimated covariance matrix is diagonal, $x_1, ..., x_m$ are uncorrelated with each other. Uncorrelation is only a necessary condition of independence. In order to know the independence, the attacker should do a further test of *mutual independence* of the m attributes $(x_1, ..., x_m)$. One test method is to compute the mutual information I among the m attributes, i.e. $I(x_1, ..., x_m) = \sum_{i=1}^m H(x_i) - H(x_1, ..., x_m)$, in which $H(x_i)$ is the entropy of the i-th attribute x_i, $H(x_1, ..., x_m)$ is the joint entropy of the m attributes. I is zero if and only if the m attributes are statistically independent. There are also characteristic function-based and kernel-density based methods, which can be referred to [15] and [21].

When the attributes are mutually independent, the attacker can use statistical test, e.g. Kolmogorov-Smirnov Test, to check whether the values are following the Laplace distribution (i.e. sparse enough). When the attributes are not mutually independent, the attacker can employ some multivariate statistical test, e.g. the method of [36], to check whether the attributes are following the Multivariate Gaussian Distributions. As we discuss in Section 2.2, practically the original data may be in a Gaussian Mixture Model in which there are multiple clusters, but the attacker can easily identify some clusters given the perturbed data, and target his attacks on the original data belonging to these clusters.

In the later sections we assume the attacker has obtain all the necessary priori knowledge about whether the m attributes are sparse, whether they are in the Multivariate Gaussian Distribution, the means and covariance matrix. The reconstructions made by the attacker are summarized in Table 1.

Table 1. Types of Reconstructions

Priori Knowledge	*Reconstruction*
$I(x_1, ..., x_m) = 0$ & $p(x_i) = Laplace(\mu_i, \sigma_i)$	UICA-based
$I(x_1, ..., x_m) > 0$ & $p(X) = N(\mu, \Sigma)$	MAP-based

5 UICA-Based Reconstruction

As we have discussed in Section 2.2, UICA has permutation and scaling ambiguities in recovering X. However, in $\mathcal{RP}$ of [27], the data owner can not arbitrarily permute the rows in R and the resulting $U = RX$, before U is released, otherwise suppose P_1 and P_2 are two different permutation matrices of Alice and Bob respectively, $\mathcal{U} = P_1 RX$, $\mathcal{V} = P_2 RY$, then $\mathcal{V}'\mathcal{U}$ will not equal $Y'X$ on expectations. Therefore, when an attacker knows R and U, he will not have permutation ambiguity in the reconstruction of X.

The attacker still needs to reduce the scaling ambiguity with the priori knowledge on the mean μ_i and variance σ_i^2 of the i-th attribute x_i. In our reconstruction, we will use the same optimization function as Eq. (1), which assumes all x_i have zero means and identical variances. In order to do this, we firstly remove the means of all x_i before the use of Eq. (1), and afterwards add them again. We also change R to RL in which L is a scaling matrix whose diagonal entries are the variances, thus we can use Eq. (1) to obtain solutions with identical variances, and afterwards the solutions will be multiplied with the corresponding variances.

Specifically our UICA-based reconstruction includes the following steps:

1) *Remove the means:* Let μ_{u_i} is the sample mean of u_i, the i-th row of U ($i = 1, ..., k$). $\mu_U = (\mu_{u_1}, ..., \mu_{u_k})'$. $\Theta = (1, ..., 1)_{n_1}$, then $\tilde{U} = U - \mu_U \Theta$.
2) *Change R to $\tilde{R}$:* $\tilde{R} = RL$, in which

$$L = \begin{pmatrix} \sigma_1 & 0 & ... & 0 \\ 0 & \sigma_2 & ... & 0 \\ ... \\ 0 & 0 & ... & \sigma_m \end{pmatrix}$$

3) *L_1-norm Minimization:* By the optimization function in Eq. (1), substitute $\tilde{R}$ into R, and each column of $\tilde{U}$ into u, search for the solution x of the function. Let $\tilde{X}$ be an $m \times n_1$ matrix, each column of it is the solution vector of the function corresponding to each column of $\tilde{U}$.
4) *Reduce the scaling ambiguity:* Let $\mu = (\mu_1, ..., \mu_m)'$ in which $\mu_i (i = 1, ..., m)$ is the sample mean of x_i, then the reconstruction is $\hat{X} = L\tilde{X} + \mu\Theta$.

6 MAP-Based Reconstruction

The UICA-based reconstruction is effective when the original data $x_1, ..., x_m$ are mutually independent and non-Gaussian. For the case that the m attributes are following the Multivariate Gaussian Distribution, we use the method of Maximum A Posterior (MAP) to estimate them. The basic idea of our MAP is similar as the constrained linear programming problem in Section 2.2, but the probability density function of the original data are different, and thus our MAP becomes a constrained *quadratic* programming problem.

6.1 Priori Knowledge

As same as UICA-based reconstruction, MAP method also requires sufficient samples from the multivariate distribution $p(x_1, ..., x_m)$, from which the means $\mu = (\mu_1, ..., \mu_m)'$, and the covariance matrix (i.e. Σ_X) of X, can be successfully estimated. By the definition of Multivariate Gaussian Distribution, Σ_X is positive definite, i.e. its eigenvalues are all positive.

6.2 Reconstruction under Collusion

Given $\mathbf{u} = (\mathbf{u}_1, ..., \mathbf{u}_k)'$ which is one column of U, the attacker can search a vector $\hat{\mathbf{x}} = (\hat{\mathbf{x}}_1, ..., \hat{\mathbf{x}}_m)'$ in $\mathbb{R}^m$ to maximize the posterior probability $p(\hat{\mathbf{x}}|\mathbf{u})$. Since $p(\hat{\mathbf{x}}|\mathbf{u}) = p(\hat{\mathbf{x}})p(\mathbf{u}|\hat{\mathbf{x}})/p(\mathbf{u})$, and under collusion the attacker will know R, then

$$p(\hat{\mathbf{x}}|\mathbf{u}) = \begin{cases} \frac{p(\hat{\mathbf{x}})}{p(\mathbf{u})}, & if\ R\hat{\mathbf{x}} = \mathbf{u}, \\ 0, & if\ R\hat{\mathbf{x}} \neq \mathbf{u}, \end{cases} \tag{3}$$

$p(\mathbf{u})$ can be treated as a constant in the search of $\hat{\mathbf{x}}$, then the maximization of $p(\hat{\mathbf{x}}|\mathbf{u})$ is equivalent to the following constrained optimization problem:

$$MAX_{\hat{x}}\ p(\hat{\mathbf{x}}),\ Subject\ to\ R\hat{\mathbf{x}} = \mathbf{u} \tag{4}$$

We assume $X \sim N(\mu, \Sigma_X)$, i.e. given a vector $\mathbf{x} = (\mathbf{x}_1, ..., \mathbf{x}_m)' \in X$,

$$p(\mathbf{x}) = \frac{1}{(2\pi)^{m/2}|\Sigma_X|^{1/2}} e^{-\frac{1}{2}(\mathbf{x}-\mu)'\Sigma_X^{-1}(\mathbf{x}-\mu)}$$

Since the exponential function is a monotone one-to-one function, the problem in Eq. (4) is equivalent to the following Quadratic Programming (QP) problem:

$$MIN_{\hat{x}}\ f(\hat{\mathbf{x}}) = \frac{1}{2}(\hat{\mathbf{x}} - \mu)'\Sigma_X^{-1}(\hat{\mathbf{x}} - \mu),\ Subject\ to\ R\hat{\mathbf{x}} = \mathbf{u} \tag{5}$$

We have assumed the Σ_X is positive definite, so is Σ_X^{-1}, then $f(\hat{\mathbf{x}})$ is convex and has the unique global minimizer, which can be computed by the gradient of the Lagrange function:

$$L(\hat{\mathbf{x}}, \Lambda) = \frac{1}{2}(\hat{\mathbf{x}} - \mu)'\Sigma_X^{-1}(\hat{\mathbf{x}} - \mu) + \Lambda'(R\hat{\mathbf{x}} - \mathbf{u}) \tag{6}$$

in which $\Lambda = (\lambda_1, ..., \lambda_k)'$, λ_i $(i = 1, ..., k)$ are Lagrange multipliers.

By Eq. (6),

$$\frac{\partial L}{\partial \hat{\mathbf{x}}} = \left(\tfrac{\partial L}{\partial \hat{\mathbf{x}}_1}, ..., \tfrac{\partial L}{\partial \hat{\mathbf{x}}_m}\right)' = \Sigma_X^{-1}(\hat{\mathbf{x}} - \mu) + R'\Lambda = 0, \tag{7a}$$

$$\frac{\partial L}{\partial \Lambda} = \left(\tfrac{\partial L}{\partial \lambda_1}, ..., \tfrac{\partial L}{\partial \lambda_k}\right)' = R\hat{\mathbf{x}} - \mathbf{u} = 0 \tag{7b}$$

The $m + k$ equations in Eq. (7) can be treated as a linear system with $m + k$ variables $(\lambda_1, ..., \lambda_k, \hat{\mathbf{x}}_1, ..., \hat{\mathbf{x}}_m)$:

$$R'\Lambda + \Sigma_X^{-1}\hat{\mathbf{x}} = \Sigma_X^{-1}\mu, \tag{8a}$$

$$0 \cdot \Lambda + R\hat{\mathbf{x}} = \mathbf{u} \tag{8b}$$

Let Θ_1, Θ_2 be $m \times k$ and $k \times k$ zero matrices, I be an $m \times m$ identity matrix, then by solving the above linear system,

$$\hat{\mathbf{x}} = (\Theta_1, I)\begin{pmatrix} \Lambda \\ \hat{\mathbf{x}} \end{pmatrix} = (\Theta_1, I)\,\Omega^{-1}\begin{pmatrix} \Sigma_X^{-1}\mu \\ \mathbf{u} \end{pmatrix},\ \Omega = \begin{pmatrix} R' & \Sigma_X^{-1} \\ \Theta_2 & R \end{pmatrix}. \tag{9}$$

Lemma 1. *Ω in Eq. (9) is nonsingular with high probability.*

Proof. By the Leibniz formula,

$$\det(\Omega) = \det(\Sigma_X^{-1})\det(\Theta_2 - R\Sigma_X R') = (-1)^k \det(\Sigma_X^{-1})\det(R\Sigma_X R').$$

$R\Sigma_X R' = \Sigma_U$, which is the covariance matrix of $U = RX$.

By [11], when $\mathbf{x}$ is fixed, R is a $k \times m$ matrix each entry of which is an i.i.d random number, then $\mathbf{u} = R\mathbf{x}$ is approximately Gaussian, following the distribution $N(R\mu, ||\mathbf{x}||^2 I_k)$ in which I_k is a $k \times k$ identity matrix. Therefore $\Sigma_U \approx ||\mathbf{x}||^2 I_k$, which means when $\mathbf{x}$ is not a zero vector, Σ_U will be nonsingular with high probability. Since $\det(\Sigma_X^{-1}) \neq 0$, then $\det(\Omega) \neq 0$, i.e. Ω is nonsingular. When Σ_U is singular, it is most possible that $\mathbf{x}$ is a zero vector. □

In sum, the MAP-based reconstruction includes the following steps:

1) estimate Σ_X and μ by enough samples from the same distribution as X;
2) compute $\hat{\mathbf{x}}_i$ by Eq. (9) for each column $\mathbf{u}_i$ of U, $i = 1, ..., n_1$. Let $\Theta_3 = (1, ..., 1)_{n_1}$, the reconstructed $\hat{X} = (\hat{\mathbf{x}}_1, ..., \hat{\mathbf{x}}_{n_1})$ can be written as:

$$\hat{X} = (\Theta_1, I)\begin{pmatrix} R' & \Sigma_X^{-1} \\ \Theta_2 & R \end{pmatrix}^{-1}\begin{pmatrix} \Gamma \\ U \end{pmatrix}, \ \Gamma = \Sigma_X^{-1}\mu\Theta_3. \tag{10}$$

7 Experiments and Comparisons

7.1 Reconstruction Based on Principle Component Analysis

As we sum in Section 2.2 the PCA-based attack of [28] is not suitable for the $\mathcal{RP}$ of [27], and to our knowledge, there is no PCA-based attack proposed for the $\mathcal{RP}$. For comparison purposes, we use the pre-whitening phase of ICA ([18]) as a PCA-based attack, which includes the following steps:

1) The attacker removes the mean of each row $u_i (i = 1, ..., k)$ of U.
2) The attacker computes the covariance matrix of $U = RX$ as $\Sigma_U = \mathbf{E}(UU')$, and makes an eigenvalue decomposition of it. Let $\Sigma_U = QDQ'$, in which Q is an orthogonal matrix, D is a diagonal matrix each entry of which is an eigenvalue of Σ_U.
3) The attacker computes $\tilde{X} = QD^{-1/2}Q'U$, in which $D^{-1/2}$ is a diagonal matrix each diagonal entry of which is the inverse of the square root of the corresponding entry of D. Let $A = QD^{-1/2}Q'$, then

$$\Sigma_{\tilde{X}} = A\Sigma_U A' = (QD^{-1/2}Q')(QDQ')(QD^{-1/2}Q') = I.$$

4) Suppose $\tilde{x}_j$ is the j-th row of $\tilde{X}$. For $i = 1, ..., m$, and $j = 1, ..., k$ the attacker computes:

$$\hat{x}_j = \sigma_i \tilde{x}_j + \mu_i, \tag{11}$$

and makes a statistical test $G(\hat{x}_j, p(x_i))$ in which $p(x_i)$ is the p.d.f of x_i, e.g. using the Two-sample K-S Test. If G outputs 1, $\hat{x}_j$ has a similar distribution to x_i, and the attacker treats $\hat{x}_j$ as an estimate of x_i.

In this method, $\Sigma_{\tilde{X}}$ is an identity matrix, so the k rows of $\tilde{X}$ will be uncorrelated. This reconstruction can be an approximate recovery when the m attributes of the original data are mutually independent, or they are not mutually independent and not having high correlations.

One major limitation of this method is that it can only recover k components, so it is essential for the attacker to use the priori knowledge to reduce the permutation and scaling ambiguities, as in Step 4). Another limitation is that Σ_U may be diagonal by [11], then in Step 2) Q will be an identity matrix, and in Step 3) the reconstruction result $\tilde{X}$ is simply $D^{-1/2}U$, i.e. some scaling of U.

7.2 Experiments and Comparisons for UICA-Based Reconstruction

We use the Laplace distribution to simulate 3 series of independent and sparse data, as the original data X. We generate a random R with dimensions 2×3 following the $\mathcal{RP}$ method of [27]. The means of X are $(0.3, 0.5, 0.8)$, the variances of X are changed to get different recovery rates measured by Definition 3 in Section 3.4. ϵ for the recovery rates are set to be 0.2. To search the solutions of the L_1-norm minimization problems, we use the *fmincon* function in the Optimization Toolbox of MATLAB.

In Fig. 3(a) the x-axis is the variance σ_1 of the first row x_1 of X, we make $\sigma_2 = 0.8\sigma_1$, $\sigma_3 = 0.3\sigma_1$. From Fig. 3(a) our reconstruction based on UICA achieves higher recovery rates than PCA.

3 series of financial data from the UCI Machine Learning Repository, including Attribute 1 of the Japanese Credit Screening Data Set, Attribute 1 of the Australian Credit Approval Data Set, Attribute 5 of the German Credit Data Set, are used as the original data, and perturbed by a random 2×3 matrix. They are treated as sparse data since they have high kurtosis (subtract 3), respectively 11.1, 1.17, 3.84. With different ϵ Fig. 3(b) shows the recovery rates of UICA-based and PCA-based reconstructions, and in comparison our UICA-based reconstruction performs much better than PCA.

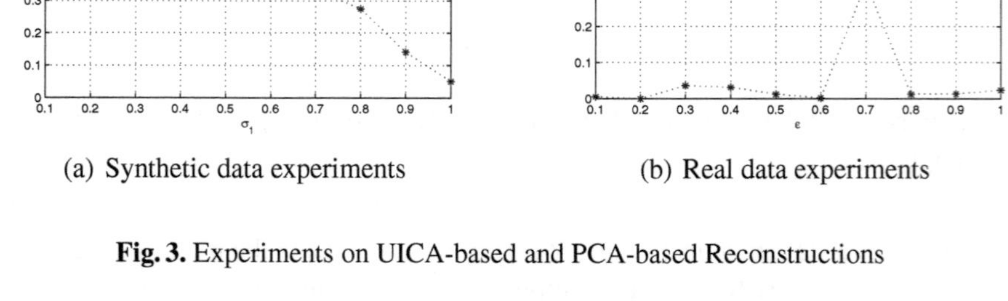

(a) Synthetic data experiments (b) Real data experiments

Fig. 3. Experiments on UICA-based and PCA-based Reconstructions

7.3 Experiments and Comparisons for MAP-Based Reconstruction

For the synthetic data experiments we assume $m = 10$, $\mu_i = 0.1 * i$ for $i = 1, ..., 10$. It is a non-trivial problem to generate Σ_X for the experiments which require variations on the structure of the covariance matrices. We use $\Sigma_X = AA'$ in which A is 10×10 matrix with i.i.d entries uniformly sampled from $[0, b]$, and b is changed (from 0.1 to 1) to get different Σ_X. We use the $mvnrnd$ function in MATLAB to generate X composed of 10 synthetic data attributes. R is a 4×10 matrix each entry of which follows $N(0, 1)$. Fig. 4(a) gives the recovery rates of our MAP-based reconstruction with different b, in comparisons with the recovery rates of PCA ($\epsilon = 0.2$). The figure shows that our method achieves higher recovery rates than PCA.

It is difficult to find real data strictly following the Multivariate Gaussian Distribution. We take the Attribute 2 ("Duration in month"), 13 ("Age in years"), 16 ("Number of existing credits at this bank") of the German Credit Data Set from the UCI Machine Learning Repository. They have $\mu = (20.9, 35.5, 1.4)$, $\Sigma_X = (145.4, -4.96, -0.08;$ $-4.96, 129.4, 0.98; -0.08, 0.98, 0.33)$. They are perturbed by a 2×3 random R. Fig. 4(b) shows the recovery rates with different ϵ, and our MAP-based reconstruction performs much better than PCA.

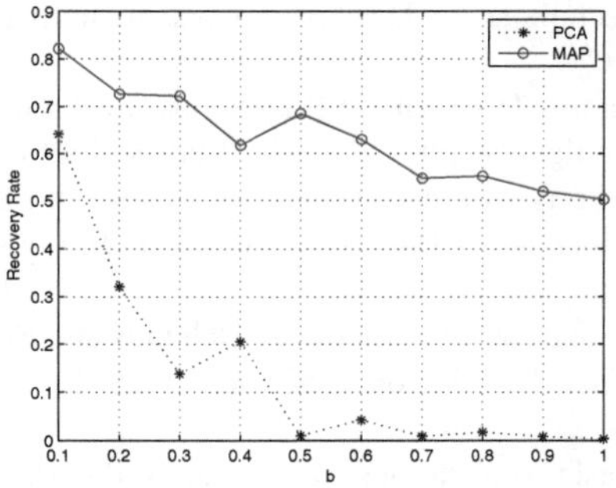

(a) Synthetic data experiments

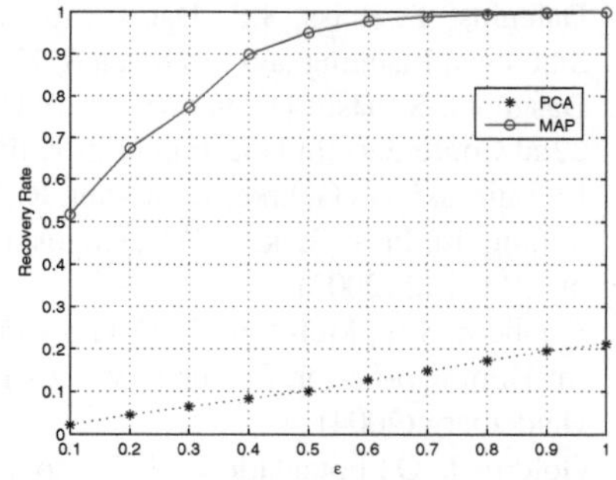

(b) Real data experiments

Fig. 4. Experiments on MAP-based and PCA-based Reconstructions

8 Conclusions

In this paper we propose two types of methods to reconstruct the original data from the data perturbed by Random Projection in [27]. Our reconstructions consider the case that the original data are mutually independent and sparse, and the case that the original data are not mutually independent and not sparse. Experiments show that our methods outperform the reconstructions based on PCA, and achieve higher recovery rates on the perturbed data. In the future work we will consider more reconstruction methods when R is not known, towards an improved perturbation method which is secure under these reconstructions.

References

1. Adam, N., Worthmann, J.: Security-control methods for statistical databases: a comparative study. ACM Computing Surveys 21(4), 515–556 (1989)
2. Aggarwal, C., Yu, P.S. (eds.): Privacy-Preserving Data Mining: Models and Algorithms. Springer, Heidelberg (2008)
3. Agrawal, R., Srikant, R.: Privacy-preserving data mining. In: Proc. of the 2000 ACM SIGMOD Conference on Management of Data, pp. 439–450. ACM, New York (2000)
4. Agrawal, S., Haritsa, J.R.: A Framework for High-Accuracy Privacy-Preserving Mining. In: Proc. 21st Int'l Conf. Data Eng. (ICDE 2005), pp. 193–204 (2005)
5. Atallah, M., Bertino, E., Elmagarmid, A., Ibrahim, M., Verykios, V.: Disclosure Limitation of Sensitive Rules. In: Proc. of IEEE Knowledge and Data Engineering Workshop, pp. C45–C52 (1999)
6. Bofill, P., Zibulevsky, M.: Underdetermined blind source separation using sparse representations. Signal Processing 81(11), 2353–2362 (2001)
7. Cao, X., Liu, R.: General Approach to Blind Source Separation. IEEE Transactions on Signal Processing 44(3), 562–571 (1996)
8. Chen, K., Sun, G., Liu, L.: Towards Attack-resilient Geometric Data Perturbation. In: Proceedings of the 2007 SIAM International Conference on Data Mining (SDM 2007), Minneapolis, MN (April 2007)
9. Chen, S.S., Donoho, D.L., Saunders, M.A.: Atomic Decomposition by Basis Pursuit. SIAM Review 43(1), 129–159 (2001)
10. Dalenius, T., Reiss, S.P.: Data-swapping: A Technique for Disclosure Control. Journal of Statistical Planning and Inference 6, 73–85 (1982)
11. Dasgupta, S., Hsu, D., Verma, N.: A Concentration Theorem for Projections. In: Proc. the 22nd Conference in Uncertainty in Artificial Intelligence, pp. 1–17. AUAI Press (2006)
12. Evfimievski, A., Gehrke, J., Srikant, R.: Limiting privacy breaches in privacy preserving data mining. In: Proc. 22nd ACM Symposium on Principles of Database Systems (PODS 2003), pp. 211–222 (2003)
13. Fienberg, S.E., McIntyre, J.: Data Swapping: Variations on a Theme by Dalenius and Reiss. In: Domingo-Ferrer, J., Torra, V. (eds.) PSD 2004. LNCS, vol. 3050, pp. 14–29. Springer, Heidelberg (2004)
14. Goldreich, O.: Foundations of Cryptography: Basic Applications, vol. 2. Cambridge University Press, Cambridge (2004)
15. Gretton, A., Fukumizu, K., Teo, C., Song, L., Scholkopf, B., Smola, A.: A Kernel Statistical Test of Independence. In: Proc. Advances in Neural Information Processing Systems (NIPS 2007), pp. 585–592. MIT Press, Cambridge (2007)
16. Guo, S., Wu, X.: Deriving private information from arbitrarily projected data. In: Zhou, Z.-H., Li, H., Yang, Q. (eds.) PAKDD 2007. LNCS (LNAI), vol. 4426, pp. 84–95. Springer, Heidelberg (2007)
17. Huang, Z., Du, W., Chen, B.: Deriving Private Information from Randomized Data. In: SIGMOD 2005, pp. 37–48. ACM, New York (2005)
18. Hyvärinen, A., Oja, E.: Independent Component Analysis: Algorithms and Applications. Neural Networks 13, 411–430 (2000)
19. Jha, S., Kruger, L., McDaniel, P.: Privacy Preserving Clustering. In: de di Vimercati, S.C., Syverson, P.F., Gollmann, D. (eds.) ESORICS 2005. LNCS, vol. 3679, pp. 397–417. Springer, Heidelberg (2005)
20. Kantarcioglu, M., Clifton, C.: Privacy-Preserving Distributed Mining of Association Rules on Horizontally Partitioned Data. IEEE Transactions on Knowledge and Data Engineering 16(9), 1026–1037 (2004)

21. Kankainen, A., Ushakov, N.: A consistent modification of a test for independence based on the empirical characteristic function. Journal of Mathematical Sciences, 1–10 (1998)
22. Kargupta, H., Datta, S., Wang, Q., Sivakumar, K.: On the privacy preserving properties of random data perturbation techniques. In: Proc. 3rd IEEE International Conference on Data Mining (ICDM 2003), p. 99 (2003)
23. Lefons, E., Silvestri, A., Tangorra, F.: An analytic approach to statistical databases. In: Proceedings of the 9th VLDB Conference (1983)
24. Li, N., Li, T., Venkatasubramanian, S.: t-closeness: Privacy beyond k-anonymity and l-diversity. In: Proc. ICDE 2007, pp. 106–115 (2007)
25. Liew, C.K., Choi, U.J., Liew, C.J.: A data distortion by probability distribution. ACM Transactions on Database Systems 10(3), 395–411 (1985)
26. Lindell, Y., Pinkas, B.: Privacy Preserving Data Mining. In: Bellare, M. (ed.) CRYPTO 2000. LNCS, vol. 1880, pp. 36–54. Springer, Heidelberg (2000)
27. Liu, K., Kargupta, H., Ryan, J.: Random projection-based multiplicative data perturbation for privacy preserving distributed data mining. IEEE Transactions on Knowledge and Data Engineering 18(1), 92–106 (2006)
28. Liu, K., Giannella, C., Kargupta, H.: An Attacker's View of Distance Preserving Maps for Privacy Preserving Data Mining. In: Fürnkranz, J., Scheffer, T., Spiliopoulou, M. (eds.) PKDD 2006. LNCS (LNAI), vol. 4213, pp. 297–308. Springer, Heidelberg (2006)
29. Liu, K.: Multiplicative Data Perturbation for Privacy Preserving Data Mining., PhD thesis, University of Maryland, Baltimore County, Baltimore, MD (January 2007)
30. Machanavajjhala, A., Gehrke, J., Kifer, D., Venkitasubramaniam, M.: l-Diversity: Privacy Beyond k-Anonymity. In: Proc. of ICDE 2006, p. 24 (2006)
31. O'Grady, P.D., Pearlmutter, B.A., Rickard, S.T.: Survey of Sparse and Non-Sparse Methods in Source Separation. International Journal of Imaging Systems and Technology 15(1), 18–33 (2005)
32. Oliveira, S.R.M., Zaïane, O.R.: A privacy-preserving clustering approach toward secure and effective data analysis for business collaboration. Computers & Security 26(1), 81–93 (2007)
33. Rizvi, S., Haritsa, J.: Maintaining Data Privacy in Association Rule Mining. In: Proc. of 28th Intl. Conf. on Very Large Databases (VLDB) (August 2002)
34. Saygin, Y., Verykios, V.S., Clifton, C.: Using unknowns to prevent discovery of association rules. ACM SIGMOD Record 30(4), 45–54 (2001)
35. Sweeney, L.: k-anonymity: a model for protecting privacy. International Journal on Uncertainty, Fuzziness and Knowledge-based Systems 10(5), 557–570 (2002)
36. Szekely, G.J., Rizzo, M.L.: Testing for Equal Distributions in High Dimension, InterStat, November (5)
37. Theis, F.J., Lang, E.W., Puntonet, C.G.: A Geometric Algorithm for Overcomplete Linear ICA. Neurocomputing 56, 381–398 (2004)
38. Turgay, E.O., Pedersen, T.B., Saygin, Y., Savas, E., Levi, A.: Disclosure Risks of Distance Preserving Data Transformations. In: Ludäscher, B., Mamoulis, N. (eds.) SSDBM 2008. LNCS, vol. 5069, pp. 79–94. Springer, Heidelberg (2008)
39. Verykios, V., Elmagarmid, A., Elisa, B., Elena, D., Saygin, Y., Dasseni, E.: Association Rule Hiding. IEEE Transactions on Knowledge and Data Engineering 16(4), 434–447 (2004)
40. Yang, Z., Zhong, S., Wright, R.N.: Privacy-Preserving Classification of Customer Data without Loss of Accuracy. In: Proc. of the 2005 SIAM International Conference on Data Mining, SDM (2005)
41. Zibulevsky, M., Pearlmutter, B.A.: Blind Source Separation by Sparse Decomposition in a Signal Dictionary. Neural Computation 13(4), 863–882 (2001)

Identifying the Original Contribution of a Document via Language Modeling

Benyah Shaparenko and Thorsten Joachims

Department of Computer Science, Cornell University,
Ithaca NY 14853, USA

Abstract. One major goal of text mining is to provide automatic methods to help humans grasp the key ideas in ever-increasing text corpora. To this effect, we propose a statistically well-founded method for identifying the original ideas that a document contributes to a corpus, focusing on self-referential diachronic corpora such as research publications, blogs, email, and news articles. Our statistical model of passage impact defines (interesting) original content through a combination of impact and novelty, and the model is used to identify each document's most original passages. Unlike heuristic approaches, the statistical model is extensible and open to analysis. We evaluate the approach both on synthetic data and on real data in the domains of research publications and news, showing that the passage impact model outperforms a heuristic baseline method.

1 Introduction

With the rapid proliferation of large text corpora, it is especially relevant to provide automatic methods to support users in understanding global aspects of a corpus without requiring them to read it in full. In diachronic corpora that grow over time, one such global aspect is the dependency structure between ideas and documents throughout the corpus. In particular, what is the original contribution that a given document makes, and how does this idea further "flow" through the corpus? In this paper, we focus on the first half of this question and develop methods that automatically identify the original ideas that a document contributes. Our methods leverage the diachronic nature of many text corpora, where ideas originate in some documents and get discussed and refined in later documents. Such corpora include research publications, email, news articles, Wikipedia content, discussion boards, and blogs. Useful applications include the visualization of corpora, the detection of important developments in news corpora, or the attribution of ideas in blogs or email discussions.

When identifying the original ideas expressed in a document, we are most interested in ideas that ultimately had impact. Anybody can write some spam on a discussion board, which would likely be novel to the discussion (at least the first time), but not particularly interesting. In addition to novelty, measuring the impact of an idea lets us focus on those ideas that are important, or that

W. Buntine et al. (Eds.): ECML PKDD 2009, Part II, LNAI 5782, pp. 350–365, 2009.

at least are interesting to a large number of people. Therefore, our operational definition of an original contribution combines both novelty and impact.

Unlike methods that rely on explicit citations that must be localizable in each document [1], our methods require only the text of the documents. This makes them more broadly applicable than citation-based measures (e.g., for email, news). Furthermore, unlike novelty detection methods [2] (e.g., based on TFIDF-style measures), our methods combine novelty with impact, which provides a way of measuring the importance of novel ideas. The originality-detection methods we propose are derived from a probabilistic language model of diachronic corpora – called the Passage Impact Model (PIM), which makes them theoretically well-founded and more extensible than heuristic approaches. The method is evaluated on a corpus of Slashdot discussions, as well as through a blind experiment with human judges on a collection of NIPS research articles. In both experiments, the language modeling approach was found to outperform a heuristic that focuses on novelty detection alone.

2 Related Work

The task of succinctly describing the original contribution of a document relates to several existing research areas, including document summarization, topic detection, topic modeling, and language modeling.

The largest body of related work is in document summarization (see e.g. [3]). Document summarization methods provide the user with a summary of the entire document, including both original and existing ideas, without explicitly making a distinction. The difference between summarization and originality detection is most apparent for documents that do not necessarily contain original content (e.g., textbooks, review articles). While such documents have a summary, their original contribution can be quite different or even non-existent.

Another area of related work lies in novelty detection for Topic Detection and Tracking [4,5] in news streams. There, the task is to identify new topics and events as they appear in the news. One major difference is that the Passage Impact Model segments the document to identify a single passage that best describes that document's original contribution. Thus the inference method can actually find a text description within the document, instead of just marking that the document contains a novel topic. A second difference is that the Passage Impact Model combines novelty with impact, focusing on ideas that not only are novel but also affect the rest of the corpus. The TREC Novelty track [2] solves a different problem, combining novelty and relevance, not novelty and impact.

One previous paper has tackled the problem of making "impact-based summaries" [1]. Their method is based on citation contexts for explicit citations to a document d. The task is to select the sentence s in document d that best describes the contribution of d that had impact in these citation contexts. That work followed a KL-divergence-based information retrieval framework where the document d stands for the corpus, the sentences s stand for the documents to be retrieved, and the citation context is descriptive of the "query." The Passage

Impact Model is quite different in model and inference, since it does not require citations. Instead, our method is based on an extensible generative and unsupervised language-modeling framework. We start from a generative model of the corpus and derive an inference method to identify the most densely-concentrated original contribution in the document d. We do not need to use a citation context, as the method is completely text-based.

On a higher level, topic models and other language models also provide generative models of corpora. In topic models, however, the focus is on discovering underlying topics, without any explicit notion of originality or impact. Typically, topics are inferred by fitting graphical models with topics as the latent variables. Latent Dirichlet Allocation (LDA) [6,7] and its extensions [8,9] are the most well-known, but there is much other work in topic modeling [10,11,12,13,14,15,16,17,18,19,20]. In this sense, topic models describe the relationship between topics and documents, but not the relationships between individual documents. Our Passage Impact Model directly models relationships between documents via a copy process. In this sense it builds on the models in [21,15], extending them to recognizing document substructure. We use simple unigram language models in the PIM, but one could also use more complex language models [22,10,6,23,24,25,26].

3 Methods

We take a language modeling approach and define a generative model for diachronic corpora. An author writes a new document using a mixture of novel ideas and ideas "copied" from earlier documents. An idea has impact if it is copied (i.e., discussed, elaborated on) by future documents. This picture is one of idea flows, originating in documents with impact and "flowing" to documents based on idea development. We directly model idea flows between documents, without an extra level of the topic as in topic models [6]. Identifying the original contribution of a document means separating novel ideas from old ideas, and simultaneously assessing impact. We assume that documents generally contain a key paragraph or sentence(s) that succinctly describe the new idea, and we aim to identify this piece of original text. The following gives more detail on our probabilistic model and inference method.

3.1 Passage Impact Model

We propose a generative model of a diachronic corpus that extends the model in [21] with respect to modeling originality. We model a document $D^{(i)}$ containing n_i words as a vector of n_i random variables $W^{(i)} = (W_1^{(i)} ... W_{n_i}^{(i)})'$, one per word. Considering the process by which authors write documents, the text can be split into several types: original content that will have impact on following documents, novel content that will not have impact, and content "copied" from already-existing ideas in the corpus. The location of the original content in $D^{(i)}$ is denoted by $Z^{(i)}$, where $Z^{(i)} \subseteq \{1...n_i\}$. More concretely, the random variables

$W^{(i)}$ are partitioned into two sets: $Z^{(i)} \subseteq \{1...n_i\}$ for the indices of the words of $D^{(i)}$ that are original and have impact, while $\bar{Z}^{(i)} = \{1...n_i\} - Z^{(i)}$ contains the rest of $D^{(i)}$ (i.e., the copied content and the novel content without impact). With these definitions, the document is described by the tuple

$$D^{(i)} = (W^{(i)}, Z^{(i)}) \tag{1}$$

and we will now define a probabilistic model of a document $P(D^{(i)}|D^{(1)}...D^{(i-1)})$. Each document $D^{(i)}$ can draw on the ideas already expressed in the existing documents $D^{(1)}...D^{(i-1)}$ in the corpus. The probability of an entire corpus $\mathcal{C}$ consisting of documents $D^{(1)}...D^{(n)}$, can be decomposed as

$$P(\mathcal{C}) = \prod_{i=1}^{n} P(D^{(i)}|D^{(1)}...D^{(i-1)}). \tag{2}$$

We decompose the probability for a single document $D^{(i)}$ into

$$\begin{aligned} P(D^{(i)}|D^{(1)}...D^{(i-1)}) &= P(W^{(i)}, Z^{(i)}|D^{(1)}...D^{(i-1)}) \\ &= P(W^{(i)}|Z^{(i)}, D^{(1)}...D^{(i-1)})P(Z^{(i)}) \end{aligned}$$

since the document text $W^{(i)}$ depends on the previous documents, but the author's selection of placement of original content is independent of previous documents. Prior information about the placement of $Z^{(i)}$ in the document can be encoded in $P(Z^{(i)})$. Furthermore, in the inference described below, the quantity $P(Z^{(i)})$ can be used to encode constraints on the form of original content summary that is desirable (e.g., a single sentence or a single paragraph).

Words in the original portion $Z^{(i)}$ are generated from a unigram language model with word probabilities $\theta^{(i)}$. The rest of the document (i.e. the words indexed by $\bar{Z}^{(i)}$) comes from a mixture of existing ideas and text that is novel but without impact. That is, the words indexed by $\bar{Z}^{(i)}$ are drawn from a mixture of a novel unigram model $\bar{\theta}^{(i)}$ (new but without impact) and words copied from the original sections of prior documents. Words are drawn uniformly and independently in this copy process so that it can also be described by a unigram model with parameters $\hat{\theta}^{(k)}$ for each prior document $D^{(k)}$. The document-specific mixing weights $\pi^{(i)}$ are $(\pi_n^{(i)}, \pi_k^{(i)})$ for $\bar{\theta}^{(i)}$ and $\hat{\theta}^{(k)}$, respectively.

With the assumption that text is generated from these unigram multinomial language models, the generative model of the text given $Z^{(i)}$ and the existing corpus at time i is

$$P(W^{(i)}|Z^{(i)}, D^{(1)}...D^{(i-1)}) = \prod_{j \in z^{(i)}} \left(\theta^{(i)}_{w_j^{(i)}}\right) \prod_{j \in \bar{z}^{(i)}} \left(\pi_n^{(i)} \bar{\theta}^{(i)}_{w_j^{(i)}} + \sum_{k=1}^{i-1} \pi_k^{(i)} \hat{\theta}^{(k)}_{w_j^{(i)}}\right).$$

Figure 1 illustrates the generative process at document $d^{(i)}$, showing how $d^{(i)}$ copies content from the original part $Z^{(k)}$ of earlier documents $d^{(k)}$ and showing how terms indexed by $Z^{(i)}$ are copied by later documents $d^{(l)}$. We summarize this generative process of a diachronic corpus in the Passage Impact Model.

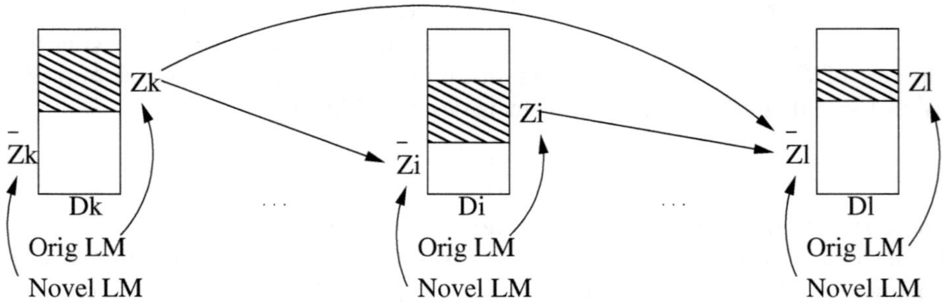

Fig. 1. The generative process for a corpus. Document $d^{(i)}$ is the current document, while $d^{(k)}$ precede $d^{(i)}$ in time and $d^{(l)}$ follow $d^{(i)}$. The shaded boxes are original content $Z^{(\cdot)}$, while the rest of the documents form $\bar{Z}^{(\cdot)}$. The arrows depict the copy process.

Model 1. (Passage Impact Model)
A corpus $\mathcal{C} = (D^{(1)}...D^{(n)})$ *of temporally-sorted documents* $D^{(i)} = (W^{(i)}, Z^{(i)})$, *each having parameters* $(\theta^{(i)}, \bar{\theta}^{(i)}, \pi^{(i)})$, *has probability* $P(\mathcal{C}) = \prod_{i=1}^{n} P(D^{(i)}|D^{(1)}...D^{(i-1)})$ *where*

$$P(D^{(i)}|D^{(1)}...D^{(i-1)}) = \prod_{j \in z^{(i)}} \left(\theta^{(i)}_{w_j^{(i)}}\right) \prod_{j \in \bar{z}^{(i)}} \left(\pi_n^{(i)} \bar{\theta}^{(i)}_{w_j^{(i)}} + \sum_{k=1}^{i-1} \pi_k^{(i)} \hat{\theta}^{(k)}_{w_j^{(i)}}\right) P(Z^{(i)})$$

and where $\hat{\theta}_w^{(k)}$ *is the probability of uniformly drawing word* w *from the words in the original section* $z^{(k)}$ *of document* $D^{(k)}$. *Note that* $\pi_n^{(i)} + \sum_k \pi_k^{(i)} = 1$, $\sum_j \theta_j^{(i)} = 1$, *and* $\sum_j \bar{\theta}_j^{(i)} = 1$.

3.2 Inference

Using the Passage Impact Model, we are primarily interested in inferring the subset $Z^{(i)}$ of words in $D^{(i)}$ where the original contribution is most succinctly contained. The only observed quantity is the text $w^{(1)}...w^{(n)}$ of all documents. We use maximum-likelihood inference based on Model 1 for inferring $Z^{(1)}...Z^{(n)}$ by maximizing $P(D^{(1)}...D^{(n)})$ given $w^{(1)}...w^{(n)}$ w.r.t. $Z^{(i)}$, $\theta^{(i)}$, $\bar{\theta}^{(i)}$, and $\pi^{(i)}$. Applying Bayes rule and independence assumptions involving the placement of original content $Z^{(\cdot)}$ in different documents $D^{(i)}...D^{(n)}$, the inferred original content $Z^{(i)^*}$ is given by the following:

$$\begin{aligned}(Z^{(1)^*}...Z^{(n)^*}) &= \operatorname*{argmax}_{Z^{(1)}...Z^{(n)}} \max_{(\theta,\bar{\theta},\pi)} P(w^{(1)}...w^{(n)}|Z^{(1)}...Z^{(n)})P(Z^{(1)}...Z^{(n)}) \\ &= \operatorname*{argmax}_{Z^{(1)}...Z^{(n)}} \max_{(\theta,\bar{\theta},\pi)} P(w^{(1)}...w^{(n)}|Z^{(1)}...Z^{(n)})P(Z^{(1)})...P(Z^{(n)})\end{aligned}$$

Note that we do not explicitly include the parameters θ, $\bar{\theta}$, and π in the notation for improved readability, since their dependence is straightforward. To avoid the intractable simultaneous maximization over all $(Z^{(i)}...Z^{(n)})$, we introduce some simplifying assumptions that allow independent optimization for each $Z^{(i)}$. First, we assume that for all prior documents $d^{(1)}...d^{(i-1)}$, the copy probabilities $\hat{\theta}^{(1)}...\hat{\theta}^{(i-1)}$ can be approximately estimated from the full set of words $w^{(1)}...w^{(i-1)}$, respectively, not merely the words indexed by the original markers $z^{(1)}...z^{(i-1)}$. In practice, this assumption can be expected to have only minor impact[1], and it can be removed if $z^{(1)}...z^{(i-1)}$ are already known. With this assumption, we have that for any i

$$
\begin{aligned}
(Z^{(i)*}...Z^{(n)*}) &= \operatorname*{argmax}_{Z^{(i)}..Z^{(n)}} \max_{Z^{(1)}..Z^{(i-1)}} \max_{(\theta,\bar{\theta},\pi)} P(w^{(1)}..w^{(n)}|Z^{(1)}..Z^{(n)})P(Z^{(1)})...P(Z^{(n)}) \\
&= \operatorname*{argmax}_{Z^{(i)}..Z^{(n)}} \max_{(\theta,\bar{\theta},\pi)} P(w^{(i)}..w^{(n)}|Z^{(i)}..Z^{(n)},\hat{\theta}^{(1)}..\hat{\theta}^{(i-1)})P(Z^{(i)})...P(Z^{(n)})
\end{aligned}
$$

Second, we introduce a simplified model for the future documents $D^{(i+1)}...D^{(n)}$ so that one can maximize over $Z^{(i)}$ independently. When inferring $Z^{(i)}$, modeling exactly how future documents $D^{(l)}$, $l > i$, had impact on each other is of minor importance, so that we do not model their $Z^{(l)}$. Instead, we assume that the original and novel content of future documents comes from a multinomial mixture, which can be captured by a single multinomial language model $\bar{\theta}^{(l)}$. Thus, each $D^{(l)}$ depends only on the documents $D^{(1)}...D^{(i)}$, and

$$
P(w^{(i+1)}...w^{(n)}|Z^{(i)}...Z^{(n)},\hat{\theta}^{(1)}...\hat{\theta}^{(i-1)}) = \prod_{l=i+1}^{n} P(w^{(l)}|Z^{(i)},w^{(i)},\hat{\theta}^{(1)}...\hat{\theta}^{(i-1)})
$$

Putting all of these assumptions together, we can rewrite the objective function as the likelihood of the documents in the corpus starting from $D^{(i)}$, given all the documents that precede $D^{(i)}$, which is $P(D^{(i)}...D^{(n)}|D^{(1)}...D^{(i-1)})$. We express this likelihood using the parameters $(\theta^{(i)}, \bar{\theta}^{(i)}, \pi^{(i)})$ as follows:

$$
\begin{aligned}
Z^{(i)*} &= \operatorname*{argmax}_{Z^{(i)}} \max_{(\theta,\bar{\theta},\pi)} P(Z^{(i)})P(w^{(i)}|Z^{(i)},\hat{\theta}^{(1)}..\hat{\theta}^{(i-1)}) \prod_{l=i+1}^{n} P(w^{(l)}|Z^{(i)},w^{(i)},\hat{\theta}^{(1)}..\hat{\theta}^{(i-1)}) \\
&= \operatorname*{argmax}_{Z^{(i)}} \max_{(\theta,\bar{\theta},\pi)} \left[P(Z^{(i)}) \prod_{j\in z^{(i)}} \left(\theta^{(i)}_{w^{(i)}_j}\right) \prod_{j\in \bar{z}^{(i)}} \left(\pi^{(i)}_n \bar{\theta}^{(i)}_{w^{(i)}_j} + \sum_{k=1}^{i-1} \pi^{(i)}_k \hat{\theta}^{(k)}_{w^{(i)}_j}\right) \right. \\
&\qquad \left. \prod_{l=i+1}^{n} \prod_{j=1}^{n_l} \left(\pi^{(l)}_n \bar{\theta}^{(l)}_{w^{(l)}_j} + \sum_{k=1}^{i} \pi^{(l)}_k \hat{\theta}^{(k)}_{w^{(l)}_j}\right)\right] \qquad (3)
\end{aligned}
$$

Note that the various $\pi^{(\cdot)}$ and $\bar{\theta}^{(\cdot)}$, as well as $\theta^{(i)}$, are linearly constrained to form proper probability distributions, and that $\hat{\theta}^{(i)}$ can be computed in closed

[1] Since each document can be a mixture of original content and previous content, when estimating $\hat{\theta}^{(\cdot)}$ from the entire document, it is equal to the true $\hat{\theta}^{(\cdot)}$, mixed with some previous content that would have come from the $\hat{\theta}^{(\cdot)}$ of even earlier documents in the corpus. This assumption means that the $\hat{\theta}^{(\cdot)}$ also could include some content from $\bar{\theta}^{(i)}$. However, if this portion's mixture component is relatively small, the $\hat{\theta}^{(i)}$ will still be quite faithful to the Passage Impact Model's definition.

form for a given $z^{(i)}$. For a fixed $z^{(i)}$, the above optimization problem is convex and has no local optima. The prior $P(Z^{(i)})$ can be used to enforce a particular form of original content description (e.g., that the algorithm has to select a whole paragraph or a single sentence).

3.3 Implementation Details

When solving the optimization problem, the method can efficiently find the maximum likelihood if given a specific $z^{(i)}$. In the following, we therefore give non-zero prior $P(Z^{(i)})$ only to a fairly small number of $z^{(i)}$ that can be enumerated explicitly. This allows us to find the globally optimal solution of Eq. 3. In particular, we break documents into consecutive passages of equal length, which we denote $s^1...s^K$. We set $P(Z^{(i)} = s^k)$ to be uniform for each $k = 1...K$, with all other $P(z^{(i)}) = 0$. One could also define a non-uniform prior over the candidate passages $z^{(i)}$ to encode additional knowledge (e.g., bias toward the beginning or end of the document). With this particular assumption on $z^{(i)}$, the entire likelihood maximization can now be reduced to a sequence of convex problems, one per s^k. The solution to this sequence of optimizations is the global maximum likelihood across the passages. We use the general software optimization tool MOSEK to solve these convex optimizations [27].

While the individual problems are convex, for efficiency reasons, we have to consider the number of parameters in the Passage Impact Model. Therefore, when performing inference on document $d^{(i)}$, instead of using the full set of previous documents $\{d^{(1)}...d^{(i-1)}\}$, we choose the set of k_P nearest neighbors from these documents according to cosine similarity. The document indices for these k_P documents are given in the set $\mathcal{P}$. Besides $d^{(i)}$, the optimization also uses the likelihood of generating the documents $d^{(i+1)}...d^{(n)}$. Each of these "future" documents $d^{(l)}$ has its own set of mixing weights and set of previous documents, again chosen from the documents $\{d^{(1)}...d^{(i-1)}\}$ nearest to $d^{(l)}$ by cosine similarity. While we do not use the following strategies for improving efficiency, one could further reduce the size of the optimization problem. For example, it is possible to consider a Passage Aggregated Impact Model, wherein all future text is "lumped" together into one single "document" for inference. Then, there would only be a single set of future document parameters. Equivalently, we could constrain all future documents to have the same mixing weights and choose the set of previous neighbors as those most similar to the concatenation of all future documents. There is a tradeoff between using more information in more future documents vs. using more parameters for a specific set of interesting previous documents.

4 Experiments

We conducted experiments to test the Passage Impact Model on both synthetic and real data from research publications and news articles.

4.1 Experiment 1: Synthetic Data

We use synthetic data to explore the range of problems and parameters under which the methods work effectively and robustly. The synthetic data is generated with underlying language models from documents in the full-text proceedings of the Neural Information Processing Systems (NIPS) conference [28] between 1987-2000. NIPS has 1955 documents with text obtained by OCR, resulting in 74731 unique words (multi-character alphabetic strings), except without stopwords.

To generate a document $d^{(i)}$, we selected a NIPS document d randomly and set the original language model $\theta^{(i)}$ for $d^{(i)}$ to be the distribution of words in d. The words indexed in $Z^{(i)}$ are then generated according to $\theta^{(i)}$. For $\bar{Z}^{(i)}$, we set the novel language models $\bar{\theta}^{(i)}$ and each $\bar{\theta}^{(l)}$ similarly, with each document selected for $\bar{\theta}^{(l)}$ following NIPS document d in time. The mixing weights $\pi_k^{(i)}$ are selected uniformly at random, except for explicitly exploring $\pi_i^{(l)}$, $l > i$, (how much future documents $d^{(l)}$ copy from $d^{(i)}$) and $\pi_n^{(i)}$ (how much novel but not original content $d^{(i)}$ has) according to the values they might take in practice.

The structure of $Z^{(i)}$ and $\bar{Z}^{(i)}$ takes the form of $K = 20$ passages with L words per passage. In the simplest case, $Z^{(i)}$ marks exactly one passage as original. In addition, we test scenarios where the original content is more diffused through the document, which poses a challenge in inference. One crucial assumption of our method is that the prior $P(Z^{(i)})$ used during inference matches the data-generating process. However, the inference procedure as implemented above aims to find a single passage containing all the original content, while the true $Z^{(i)}$ might diffuse it over other passages. To test the robustness of inference w.r.t. the degree of diffusion, we include a fraction δ of original content in the (mostly) non-original passages in data generation, but not during inference.

Evaluation on the synthetic data uses the percentage of (mostly) non-original passages with a greater likelihood than the original passage likelihood. Random performance would be that half of the non-original passages are misranked, resulting in a score of 50%. The error values show one standard error.

Impact Is Critical. In the first experiment, we explore the difference between pure novelty detection vs. the additional use of impact when identifying $Z^{(i)}$. When not using any future documents, our method might still be able to identify $Z^{(i)}$ merely by fitting the mixture model and detecting that $Z^{(i)}$ cannot be expressed as a mixture of previous documents. In this setting, our method becomes a pure novelty detection method. However, Table 1 shows that the signal from novelty alone is much weaker than novelty combined with impact. While the performance is better than random when no future documents are used ($k_F = 0$), detection accuracy substantially improves when future documents and impact are considered by the method. The table shows that two future documents that copy 5% of their content from $d^{(i)}$ already provide a robust signal.

More Information in Longer Passages. We would like to determine the size of the original passage for which the Passage Impact Model can perform well. Users may be interested in descriptions anywhere from one or more sentences

Table 1. Percentage of misranked non-original passages. Passage length $L = 100$, $\delta = 0.2$, $\pi_n^{(i)} = 0.5$, $\pi_i^{(l)} = 0.05$, and $\pi_n^{(l)} = 0.6$. 10 future documents $d^{(l)}$ were generated, and inference used the k_F documents $d^{(l)}$ most (cosine) similar to $d^{(i)}$.

k_F	% Err ± One Std Err
0	37.89 ± 3.23
1	2.95 ± 0.78
2	0.26 ± 0.16
5	0.00 ± 0.00
10	0.16 ± 0.16

Table 2. Percentage of misranked non-original passages with $k_F = 2$ future documents. The data was generated with $\delta = 0.2$, $\pi_n^{(i)} = 0.5$, $\pi_i^{(l)} = 0.05$, and $\pi_n^{(l)} = 0.6$.

Length	% Err ± One Std Err
25	8.16 ± 1.61
50	2.26 ± 0.65
100	1.00 ± 0.99
400	2.26 ± 0.74

to paragraphs. Table 2 shows that, in general, when performing inference on longer passages, the method is able to perform more accurately. The method performs very well for passages as short as 50 words. However, for very short passages of length 25 words, there is some drop in accuracy. Longer passages – and therefore longer documents – provide more observations, and it is less likely that the method will overfit to a few random draws.

Diffusiveness of Original Content in $d^{(i)}$. The inference method searches for a single passage that contains the original contribution, but realistic documents will have original content spread throughout all passages. How much original content in other passages can our inference method tolerate? Table 3 shows that the method is very robust towards small to moderate diffusion. Even as δ increases to 0.3 (i.e. 30% of each of the other passages is original content), the method is still quite accurate. After that, performance degrades rather quickly, at least when only two future documents are used.

How Much Copying Is Necessary? As shown above, the Passage Impact Model relies on future documents copying the ideas expressed in the original contribution of $d^{(i)}$. How much must each future document copy to provide a sufficient signal? Table 4 shows that the method performs with minimal errors for many values of $\pi_i^{(l)}$, even in the situation where future documents copy only 5% of their content (i.e. 100 words) from $d^{(i)}$. At lower values for copying, the percentage of correctly ranked passages smoothly decreases. As $\pi_i^{(l)}$ approaches 0, the method becomes essentially equivalent to a novelty detection method that does not using any future documents.

Table 3. Percentage of misranked non-original passages. $k_F = 2$ future documents, passages length $L = 100$ words, $\pi_n^{(i)} = 0.5$, $\pi_i^{(l)} = 0.05$, and $\pi_n^{(l)} = 0.6$.

δ	% Err ± One Std Err
0.1	0.00 ± 0.00
0.2	0.00 ± 0.00
0.3	4.74 ± 1.21
0.4	24.89 ± 2.80
0.5	45.26 ± 3.38

Table 4. Percentage of misranked non-original passages. $k_F = 2$ future documents, passage length $L = 100$ words, $\delta = 0.2$, $\pi_n^{(i)} = 0.5$, and $\pi_n^{(l)} = 0.6$.

$\pi_i^{(l)}$	% Err ± One Std Err
0.005	34.37 ± 3.16
0.01	28.58 ± 2.97
0.02	9.16 ± 1.41
0.05	0.11 ± 0.07
0.1	0.00 ± 0.00
0.2	0.00 ± 0.00

4.2 Experiment 2: Slashdot

Besides synthetic data, we also evaluate on the real world dataset of news articles linked to on Slashdot under the Games topic. When users post an entry, they often link to some article on the Web, and sometimes quote directly from it. Then other users read and respond to these postings in a discussion board format. We collect linked-to web documents and discussions from the Games topic where the original poster directly quotes from a linked-to document. We regard the sentences in the human-selected direct quotations as the label for the original content $z^{(i)}$ of the web document $d^{(i)}$.

We collected a set of 61 documents from the Games topic of Slashdot. These are the entries posted from August 2008 through February 2009, inclusive, where the initial entry quotes a portion of the referenced article. The documents are the referenced articles. In addition, we collect the first page of the user discussion on this topic, as selected by Slashdot. Figure 2 shows a screenshot of Slashdot that depicts the data we collected.

Experiment Setup. To do inference on Slashdot data, we sort the fulltext, linked-to news articles by their posting date. For each article, we use the Passage Impact Method to rank all the sentences in the linked-to web content $d^{(i)}$ by their likelihood under the model. The previous documents $d^{(1)}...d^{(i-1)}$ in this setting are the web content that have been linked to in earlier discussions. The future content $d^{(i+1)}$ in this experiment is the user discussion on this posting, except that any direct quotations from the fulltext article have been removed. The user discussion may not contain all the comments, but only those that have

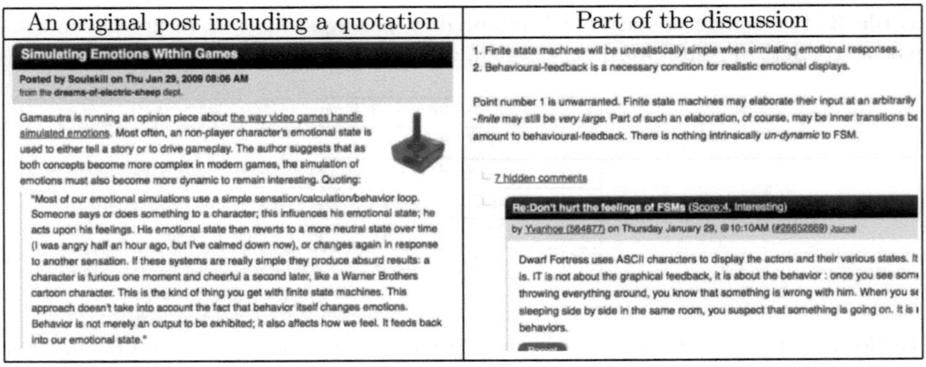

Fig. 2. Left: A post that quotes from article $d^{(i)}$ by the link "the way video games handle simulated emotions." The label for the original content $z^{(i)}$ in $d^{(i)}$ is the quotation text. Right: Part of the discussion to be used as the future document $d^{(l)}$.

been voted up enough to be selected to appear with the posting. We collected seven months (August 2008 to February 2009, inclusive) of articles that satisfy these criteria from the Games subtopic of Slashdot, which netted a corpus of 61 web documents with their associated discussions.

Evaluation Method. For evaluation, we rank the sentences in the fulltext article in decreasing order of likelihood. The user quotations typically contain no more than a handful of sentences, but often more than one. Thus, this implementation differs from the model where we assume that there is a single original contribution marked in the passage $Z^{(i)}$. As a baseline, we compare against a simple heuristic that identifies novelty. In particular, we rank the sentences by a TFIDF score given by the sum of each sentence term's IDF value. Then, since we have the labels of the true original sentences, we evaluate using the standard metrics of precision and recall at certain points in the ranking. Precision at a point in a ranking is defined to be the number of original sentences at that position in the ranking divided by the total number of sentences up to that point. For a point near the top of the ranking, precision measures whether the sentences that the method most confidently predicts as original are indeed original. Thus we report results for Prec@2. Recall at a point in the ranking is defined to be the number of original sentences at that position in the ranking divided by the total number of original sentences in the document. Recall measures how well the method can find all the original content in the document. Since each labeled quotation typically contains several sentences, we report results for Rec@10.

Results. The Prec@2 results in Table 5 show that the Passage Impact Model outperforms the TFIDF heuristic baseline for predicting the human-selected sentences at the very top of the ranking. For the task of finding a description consisting of a few good sentences that succinctly describe the original content of a news article, the Passage Impact Model is better than the baseline. The PIM also

Table 5. Prec@2 and Rec@10 are based on the predicted ranking of sentences by likelihood and TFIDF sum. Original sentences are the ones quoted word-for-word from the article. Results are for $\pi_n^{(i)} = 0.2$ and $\pi_n^{(l)} = 0.001$.

	Prec@2 ± One Std Err	Rec@10 ± One Std Err
PIM	22.13 ± 3.38	36.09 ± 3.61
TFIDF	9.84 ± 3.03	25.01 ± 4.04
RAND	10.63 ± 1.10	23.92 ± 2.27

Table 6. Comparing the PIM with future documents, and PIM as a novelty detection method (without future documents). Results are for $\pi_n^{(i)} = 0.2$ and $\pi_n^{(l)} = 0.001$.

	Prec@2 ± One Std Err	Rec@10 ± One Std Err
PIM Impact	22.13 ± 3.38	36.09 ± 3.61
PIM Novelty	9.84 ± 3.03	28.04 ± 4.24

Table 7. Prec@2 and Rec@10 for various amounts of assumed novel content $\pi_n^{(i)}$ in $d^{(i)}$. Sentences are marked as original if they appear word-for-word as in the linked article. Results are for $\pi_n^{(l)} = 0.001$.

$\pi_n^{(i)}$	Prec@2 ± One Std Err	Rec@10 ± One Std Err
0.01	18.85 ± 3.10	35.51 ± 3.55
0.05	20.49 ± 3.15	36.03 ± 3.57
0.2	22.13 ± 3.38	36.09 ± 3.61
0.8	22.95 ± 3.39	36.45 ± 3.63
0.9	22.95 ± 3.39	36.45 ± 3.63

significantly outperforms the baseline when trying to find most of the original content, as measured by Rec@10.

Importance of Impact Component. Similar to the experiment with synthetic data, the use of impact substantially improves the performance over pure novelty detection. Table 6 compares the results when using the discussion for detecting impact with the results when no future documents are used. Using the discussion significantly improves the precision of the method.

Robustness with respect to amount of novel content in $d^{(i)}$. During inference, the method needs to assume a mixture weight for the novel content in the non-original text $\bar{Z}^{(i)}$. How sensitive is the method to the selection of this parameter? Table 7 shows that the method is robust and provides good results for a wide range of values for $\pi_n^{(i)}$.

Minor Effect of Novel Language Model in Future Documents. Similarly, since Slashdot discussions are somewhat notorious for getting off topic at times, we evaluated whether changing the amount of novel content in the "future document," i.e., the discussion makes a difference. As it turns out, Table 8 shows

Table 8. Prec@2 and Rec@10 for various mixing weights $\pi_n^{(l)}$ for the noise model in fitting future documents. Sentences are marked as original if they appear word-for-word as in the linked article. The results are reported for $\pi_n^{(i)} = 0.2$.

$\pi_n^{(l)}$	Prec@2 ± One Std Err	Rec@10 ± One Std Err
0.0001	20.49 ± 3.15	36.77 ± 3.58
0.001	22.13 ± 3.38	36.09 ± 3.61
0.01	16.39 ± 3.42	34.55 ± 3.73
0.1	18.03 ± 3.29	30.34 ± 3.47
0.5	20.49 ± 3.73	31.04 ± 3.47

that for a wide range of novel content mixing weights $\pi_n^{(l)}$, the method is quite robust. The model is able to focus on the portions that the discussion derives from the underlying linked article.

4.3 Experiment 3: Evaluation Based on Human Judgments

While the Slashdot data provided a reasonable mechanism for inferring ground-truth labels, the most direct evaluation is by explicit human judgment. Therefore, we conducted an experiment with human judges to evaluate the Passage Impact Model on a corpus containing all 1955 papers from the NIPS conference [28] between 1987-2000. In a blind experiment, we asked judges to compare passages extracted by the PIM to those extracted by the TFIDF heuristic regarding how well they summarize the original contribution of a NIPS paper.

Experiment Setup. Since breaking documents into paragraphs is non-trivial, especially when they are OCR-ed and have many math equations, we arbitrarily defined passages as consecutive blocks of text of length $L = 100$ (non-stopword) words. On average, there are 14 passages per document.

For inference using the Passage Impact Model, we constrained the novel $\bar{\theta}^{(i)}$ and original $\theta^{(i)}$ language models to be equal because research publications typically discuss original contributions at length. Ideally, the identified passage should list the paper's contributions or conclusions. (Although the abstract has original content, it mostly focuses on placing the paper with the context of existing ideas.) The future document novelty mixing weight of $\pi_n^{(l)} = 0.01$ is small to force the model to "explain" the content of future documents $d^{(l)}$ by identifying copied ideas. For efficiency, we used $k_F = 5$ future documents. We compare against the TFIDF heuristic baseline. Each paper's passages predicted by the PIM and the baseline were highlighted, and three judges selected which passage better summarized the paper's original contribution. The annotators are machine learning graduate students familiar with the corpus and do not include the authors of this paper.

Since the judgment process is time-consuming, we selected a subset of NIPS publications for evaluation. We ranked all NIPS publications by their number of intra-corpus citations and selected the top 50 most-cited documents. The first publication is "Optimal Brain Damage" by Le Cun, Denker, and Solla, with 27 citations. The entire set of 50 documents includes documents down to those with only 5 intra-1987-to-2000 NIPS citations. The PIM and the baseline selected the same passage on two documents, so we use the remaining 48 for evaluation.

Results. On these 48 documents, the human judges preferred the Passage Impact Method over the baseline 58.33% of the time, with one standard error of 3.54%. Thus the judges significantly prefer the PIM over the baseline. To analyze the results more closely, we separated the 48 evaluation documents into two sets. On 20 documents, all three annotators (independently) agreed on a single passage. For these, they preferred the PIM 70% of the time. On the other 28 documents, two annotators preferred one passage, while the third annotator preferred the other passage. Here, the preferences for PIM and baseline were exactly 50%. This suggests that sometimes identifying a passage that summarizes the original contribution is quite difficult. When this is not the case, however, the PIM outperforms the baseline quite substantially with 70% preference.

5 Discussion and Future Work

While the Passage Impact Model provides a generative model of diachronic corpora and the relationships between individual documents, the model is still quite simple. For example, it is based on unigram models of text production. In modeling the probability of $W^{(i)}$, one could instead use a more sophisticated sequence model, or at least n-gram language models. Such information may help to identify coherent original ideas. Another limitation is that the model is constrained to evaluate only a small number of candidate $Z^{(i)}$ for efficiency reasons. Developing pruning criteria is a promising direction for substantially increasing the scope of $Z^{(i)}$ in hopes of finding better descriptions of original contributions.

Other information available for some corpora could be integrated into the model as well. For example, if citation information is available, it could provide additional constraints on the parameters during inference. Citations could be used as priors for mixing weights, modeling that documents copy primarily from those documents they cite. This could improve the accuracy of the model, and it could improve efficiency of the optimization since many mixing weights could be fixed at zero.

A more general direction for further work lies in the integration of originality detection with models for idea flow. The goal is to have a unified probabilistic model that identifies the dependency structure of the corpus, with ideas originating in some documents and then flowing through the corpus. Treating these inference problems separately seems suboptimal.

6 Conclusions

We have proposed an unsupervised generative model for diachronic text corpora that provides a formal structure for the process by which authors form new ideas and build on existing ideas. The model captures both novelty and impact, defining an (important) original contribution as a combination of both. For this Passage Impact Model, we have proposed an inference procedure to identify the most original passage of a document. Under reasonable approximations, the inference procedure reduces to multiple convex programs that can be solved efficiently. The method is evaluated on synthetic and real data, and it is shown to significantly outperform a heuristic baseline for selecting a passage describing the original contribution in the domains of online discussions and research articles.

Acknowledgments

We acknowledge Adam Siepel, Art Munson, Yisong Yue, Nikos Karampatziakis, and the ML Discussion Group for helpful discussions. This work was supported in part by NSF Grant IIS-0812091.

References

1. Mei, Q., Zhai, C.: Generating impact-based summaries for scientific literature. In: Proceedings of the Association for Computational Linguistics (ACL), pp. 816–824 (2008)
2. Soboroff, I., Harman, D.: Overview of the TREC 2003 novelty track. In: Proceedings of the Text Retrieval Conference, TREC (2003)
3. NIST: Document Understanding Conferences (DUC), `http://duc.nist.gov/`
4. Allan, J., Carbonell, J., Doddington, G., Yamron, J., Yang, Y.: Topic detection and tracking pilot study: Final report. In: Proceedings of the DARPA Broadcast News Transcription and Understanding Workshop (1998)
5. Allan, J., Papka, R., Lavrenko, V.: On-line new event detection and tracking. In: Proceedings of the SIGIR Conference on Research and Development in Information Retrieval, pp. 37–45 (1998)
6. Blei, D., Ng, A., Jordan, M.: Latent dirichlet allocation. Journal of Machine Learning Research (JMLR) 3(5), 993–1022 (2003)
7. Blei, D., Griffiths, T., Jordan, M., Tenenbaum, J.: Hierarchical topic models and the nested chinese restaurant process. In: Proceedings of the Conference on Advances in Neural Information Processing Systems, NIPS (2003)
8. Blei, D., Lafferty, J.: Correlated topic models. In: Proceedings of the Conference on Advances in Neural Information Processing Systems, NIPS (2005)
9. Blei, D.M., Lafferty, J.D.: Dynamic topic models. In: Proceedings of the International Conference on Machine Learning (ICML), pp. 113–120 (2006)
10. Hofmann, T.: Probabilistic latent semantic analysis. In: Proceedings of the Conference on Uncertainty in Artificial Inteligence, UAI (1999)
11. Mann, G., Mimno, D., McCallum, A.: Bibliometric impact measures leveraging topic analysis. In: Proceedings of the Joint Conference on Digital Libraries, JCDL (2006)
12. Wang, X., McCallum, A.: Topics over time: A non-markov continuous-time model of topical trends. In: Proceedings of the Conference on Knowledge Discovery and Data Mining (KDD), pp. 424–433 (2006)

13. Steyvers, M., Smyth, P., Rosen-Zvi, M., Griffiths, T.: Probabilistic author-topic models for information discovery. In: Proceedings of the Conference on Knowledge Discovery and Data Mining (KDD), pp. 306–315 (2004)
14. Griffiths, T., Steyvers, M.: A probabilistic approach to semantic representation. In: Proceedings of the Annual Conference of the Cognitive Science Society (2002)
15. Dietz, L., Bickel, S., Scheffer, T.: Unsupervised prediction of citation influences. In: Proceedings of the International Conference on Machine Learning (ICML), pp. 233–240 (2007)
16. McCallum, A., Corrada-Emanuel, A., Wang, X.: Topic and role discovery in social networks. In: Proceedings of International Joint Conference on Artificial Intelligence, IJCAI (2005)
17. Mei, Q., Ling, X., Wondra, M., Su, H., Zhai, C.: Topic sentiment mixture: Modeling facets and opinions in weblogs. In: Proceedings of the World Wide Web Conference (WWW), pp. 171–180 (2007)
18. Li, W., McCallum, A.: Pachinko allocation: Dag-structured mixture models of topic correlations. In: Proceedings of the International Conference on Machine Learning, ICML (2006)
19. Wang, X., Li, W., McCallum, A.: A continuous-time model of topic co-occurrence trends. In: AAAI Workshop on Event Detection (2006)
20. Griffiths, T., Steyvers, M., Blei, D., Tenenbaum, J.: Integrating topics and syntax. In: Proceedings of the Conference on Advances in Neural Information Processing Systems, NIPS (2004)
21. Shaparenko, B., Joachims, T.: Information genealogy: Uncovering the flow of ideas in non-hyperlinked document databases. In: Proceedings of the Conference on Knowledge Discovery and Data Mining (KDD), pp. 619–628 (2007)
22. Manning, C.D., Schuetze, H.: Foundations of Statistical Natural Language Processing. MIT Press, Cambridge (1999)
23. Jelinek, F.: Basic Language Modeling. In: Statistical Methods for Speech Recognition, pp. 57–78. MIT Press, Cambridge (1998)
24. Zhai, C.: Risk Minimization and Language Modeling in Information Retrieval. PhD thesis, Carnegie Mellon University (2002)
25. Kurland, O., Lee, L.: Corpus structure, language models, and ad hoc information retrieval. In: Proceedings of the SIGIR Conference on Research and Development in Information Retrieval, pp. 194–201 (2004)
26. Kurland, O., Lee, L.: Respect my authority! hits without hyperlinks, utilizing cluster-based language models. In: Proceedings of the SIGIR Conference on Research and Development in Information Retrieval, pp. 83–90 (2006)
27. MOSEK: `http://www.mosek.com/index.html`
28. NIPS Online: The Text Repository, `http://nips.djvuzone.org/txt.html`

Relaxed Transfer of Different Classes via Spectral Partition

Xiaoxiao Shi[1,⋆], Wei Fan[2], Qiang Yang[3], and Jiangtao Ren[4]

[1] Department of Computer Science
University of Illinois, Chicago, USA
xshi9@uic.edu

[2] IBM T.J.Watson Research, USA
weifan@us.ibm.com

[3] Department of Computer Science
Hong Kong University of Science and Technology
qyang@cse.ust.hk

[4] Department of Computer Science
Sun Yat-sen University, Guangzhou, China
issrjt@mail.sysu.edu.cn

Abstract. Most existing transfer learning techniques are limited to problems of knowledge transfer across tasks sharing the same set of class labels. In this paper, however, we relax this constraint and propose a spectral-based solution that aims at unveiling the intrinsic structure of the data and generating a partition of the target data, by transferring the eigenspace that well separates the source data. Furthermore, a clustering-based KL divergence is proposed to automatically adjust how much to transfer. We evaluate the proposed model on text and image datasets where class categories of the source and target data are explicitly different, e.g., 3-classes transfer to 2-classes, and show that the proposed approach improves other baselines by an average of 10% in accuracy. The source code and datasets are available from the authors.

1 Introduction

Traditional supervised and semi-supervised learning work well under the strict assumption that the labeled training data and unlabeled test data are drawn from the same distribution and have shared feature and category spaces. In many real world applications, however, this assumption may be violated. In fact, we often encounter the situations where we do not have sufficient labeled training examples in the target learning task. Examples include spam filtering, biological sequence annotation, web searching, and the like. To acquire more labels could usually be expensive or infeasible. For example, in the field of computational biology, many expensive and time-consuming experiments are needed to provide the labels for even a small number of examples. As an alternative solution, transfer learning was proposed to help extract some supervisory knowledge from

⋆ Part of the work was done when the author was a visiting student at HKUST.

W. Buntine et al. (Eds.): ECML PKDD 2009, Part II, LNAI 5782, pp. 366–381, 2009.

related source data to help learn the target task (e.g., [3,4,16,22]). Existing transfer learning techniques implicitly assume that there is sufficient overlap between the source data and the target data, and categories of class labels are the same, in order to allow the transfer of knowledge. However, this can significantly limit the applicability of transfer learning, as it is not always possible to find labeled data satisfying these constraints. In order to improve its applicability, we study how to transfer knowledge across tasks having different class categories.

For example, can the text documents labeled in "wikipedia" help classify those documents in "ODP[1]" even though they have different index systems? Can the labeled source image data in Fig 1(a) help classify the target data in Fig 1(b) given that they are images of different objects? The problem formulation is to partition an unlabeled target data, by the supervision from a labeled source data that has different class categories. To solve the problem, two issues need to be addressed:

1. What and how to transfer? Since the source and target data have different class labels, we can not directly take advantage of the class conditional density $p(\mathbf{x}|y)$ or posterior $p(y|\mathbf{x})$ to construct the model, and thus most of the previous transfer learning methods do not work. A new transfer learning strategy independent of class labels is needed.
2. How to avoid "negative transfer"? Given that the source and target data do not share class labels, they may come from significantly different domains. Thus, it is also necessary to avoid negative transfer (or accuracy worse than no transfer) when the source and target data are unrelated.

We propose a spectral-based solution that uses eigenspace to unveil the intrinsic structure similarities between source and target data. The key idea is that, regardless of their class category naming, if the source and target data are similar in distribution, the eigenspace constructed from the source data should be also helpful to reflect the intrinsic structure of the target data. We illustrate the intuition in Fig 1. Although the target data (Fig 1(b)) and source data (Fig 1(a)) have totally different class labels, the eigenspace Fig 1(c) constructed with the supervision from Fig 1(a) still helps group the target data. On the one hand, the images about homer-simpson are similar to the images about cartman because they are all cartoon characters; on the other hand, the shape of real bear is similar to teddybear, and the background of real bear may contain plants similar to palm tree. Thus, the eigenspace that well separates Fig 1(a) also helps separate Fig 1(b) even though their class labels are different.

To be specific, the proposed model finds an eigenspace through a combination of two optimization objectives. The first is to find the eigenspace that well separates the source data: the labeled data with the same class categories will be grouped together. The second objective is to maximize the marginal separation of the unlabeled target data. Moreover, to avoid negative transfer, we also derive a clustering-based Kullback-Leibler divergence to measure the difference in distribution between two finite datasets more effectively (see Lemma 1). We then

[1] "Open Directory Projects" (http://www.dmoz.org/). Both "wikipedia" and "ODP" are systems categorizing large amount of documents.

(a) Labeled Source Dataset (b) Unlabeled Target Dataset

(c) Solution in eigenspace

Fig. 1. An example on two image datasets with totally different class labels. Fig 1(c) is the eigenspace constructed with the supervision from the source dataset, and we plot a sub set of examples to illustrate the intuition. In this example, the eigenspace learnt from source dataset (Fig 1(a)) can also reflect the structure of the target dataset (Fig 1(b)) though their class labels are different.

make use of the measure to define "transfer risk" to regulate the effects of the two objectives. When the two datasets are very different, the effect of the first objective (or the supervision from source task) automatically decreases to minimize the risk of negative transfer. We provide a PAC bound for the proposed method in Theorem 1. In addition, the proposed algorithm is tested in several datasets where target and source data have very different class categories. For instance, in one of the experiments, we apply the 4-classes document sets "Graphics vs. Hardware vs. Politics.mid vs. Religion.misc" to supervise the partition of the binary document datasets "Comp vs. Rec". The proposed model achieves an accuracy 98%, while the accuracy of the baseline that does not apply transfer learning is only 74%.

2 Problem Formulation

We consider the problem to find a good partition of the unlabeled target data, possibly with the supervision from the labeled source data having different class labels, only when the supervision is helpful. We denote the source data as $\mathcal{L} = \{\mathbf{l}_1, \mathbf{l}_2, ..., \mathbf{l}_s\}$, and the target data as $\mathcal{U} = \{\mathbf{u}_1, \mathbf{u}_2, ..., \mathbf{u}_t\}$, where s and t are the sample size for the source and target data, respectively. Both $\mathcal{L}$ and $\mathcal{U}$ are drawn from the same feature space R^d under different distributions $\mathcal{L} \sim P_{\mathcal{L}}$ and $\mathcal{U} \sim P_{\mathcal{U}}$. We also denote that the class labels of $\mathcal{L}$ are $\mathcal{Y} = \{y_1, y_2, ..., y_r\}$ where r is the number of categories of $\mathcal{L}$, while the number of class labels of $\mathcal{U}$ is k. And we do not assume the class labels of source and target data are the same.

No Transfer: To generate a partition $C = \{c_1, c_2, ..., c_k\}$ of the target data $\mathcal{U}$, we find a clustering decision function $\mathcal{N} : \mathcal{U} \mapsto C$ to minimize $\varepsilon[\mathcal{N}(\mathcal{U})]$, where $\varepsilon[*]$ is a cost function to measure the quality of the partition. Example of such a function is Normalized Cut [19].

Transfer Learning: We learn the partition $\mathcal{U} \mapsto C$ by making optional use of the knowledge from the source data. There are many interpretations of transfer learning strategies, e.g., [3,4] reviewed in Section 5. One straightforward strategy is formulated as follows. We first learn a decision function $\mathcal{F} : \mathcal{L} \mapsto \mathcal{Y}$ on the feature space R^d that correctly reflects the true label of the source data $\mathcal{L}$. A simple transfer learning strategy can be $\mathcal{T}(\mathcal{U}) = \mathcal{M}(\mathcal{F}(\mathcal{U}))$, where $\mathcal{M} : \mathcal{Y} \mapsto C$.

Negative Transfer: Since the source and target data have different class labels, they may be from very different domains. Thus, one of the main challenges is how to avoid negative transfer when the source and target data are too different. Formally, $\mathcal{L}$ and $\mathcal{U}$ may be very different so that the performance after transfer learning is worse than no transfer. Formally,

$$\varepsilon[\mathcal{N}(\mathcal{U})] < \varepsilon[\mathcal{T}(\mathcal{U})] \tag{1}$$

where $\varepsilon[*]$ is the cost function to measure the partition quality. There can be other criteria such as error rates in classification problem. When Eq (1) holds, it is said to have negative transfer [17].

3 Risk-Sensitive Spectral Partition

We propose an improved spectral partition model to transfer the eigenspace that well separates the source data, in order to generate a good partition of the target data. Importantly, source and target data have different class labels. Since negative transfer may happen, we use the "risk of negative transfer" to automatically decide when and how much to transfer.

3.1 Divergence Measure and Transfer Risk

We first derive a new formula of the KL-divergence and propose a clustering-based approach to calculate it, in order to quantify the difference between source and target data more effectively. Normally, given two probability distributions P and Q, the KL divergence $\mathbf{KL}(P||Q)$ is defined as follows[2]:

$$\mathbf{KL}(P||Q) = \sum_x P(x)(\mathbf{log}P(x) - \mathbf{log}Q(x)) \tag{2}$$

[2] KL can be also written in terms of probability density in a continuous form, which is difficult to estimate without prior knowledge [1].

On finite datasets, one usually calculates $P(x)$ and $Q(x)$ for every value of x via Eq (3).

$$P(x=a) = \frac{|\{x|x=a \wedge x \in \mathbb{D}_P\}|}{|\{x|x \in \mathbb{D}_P\}|} \\ Q(x=a) = \frac{|\{x|x=a \wedge x \in \mathbb{D}_Q\}|}{|\{x|x \in \mathbb{D}_Q\}|} \tag{3}$$

where $\mathbb{D}_P \sim P$ and $\mathbb{D}_Q \sim Q$ are the datasets generated from the distributions P and Q, respectively. However, small changes to the datasets, such as those sparse ones typically found in text mining, can result in significant changes by the above approximations, making it difficult to distinguish different distributions [1]. To resolve this problem, we derive another format of the KL divergence and propose a clustering-based approach to calculate it. We first perform a clustering on the combined dataset $\mathbb{D}_P \cup \mathbb{D}_Q$. We then directly employ some basic statistics of the clustering results as shown in Lemma 1.

Lemma 1. *Given two distributions P and Q, the KL divergence can be rewritten with a new formula as*[3]*:*

$$\begin{aligned} &\mathbf{KL_c}(P||Q) \\ =&\frac{1}{\mathbb{E}(P)}\Big(\sum_C (P'(C)S(P',C)\mathbf{log}\frac{S(P',C)}{S(Q',C)}) \\ &+\sum_C (P'(C)S(P',C)\mathbf{log}\frac{P'(C)}{Q'(C)})\Big) + \mathbf{log}\frac{\mathbb{E}(Q)}{\mathbb{E}(P)} \end{aligned} \tag{4}$$

where C is the cluster generated from the combined dataset, and

$$S(P',C) = \frac{|\mathbb{D}_P \cap C|}{|C|}, P'(c) = \frac{|\mathbb{D}_P \cap C|}{|\mathbb{D}_P \cup \mathbb{D}_Q|}, \\ \mathbb{E}(P) = \frac{|\mathbb{D}_P|}{|\mathbb{D}_P \cup \mathbb{D}_Q|} \tag{5}$$

Likewise are the definitions of $S(Q',c)$, $Q'(c)$ *and* $\mathbb{E}(Q)$ *(by replacing* $\mathbb{D}_P$ *with* $\mathbb{D}_Q$ *in the nominator).*

Proof. Define $P'(x=a) = \frac{|\{x|x=a \wedge x \in \mathbb{D}_P\}|}{|\{x|x \in \mathbb{D}_P \vee x \in \mathbb{D}_Q\}|}$, $\mathbb{E}(P) = \frac{|\mathbb{D}_P|}{|\mathbb{D}_P \vee \mathbb{D}_Q|}$. We then have

$$P(x) = P'(x)/\mathbb{E}(P)$$

$Q'(x=a)$ and $\mathbb{E}(Q)$ are defined in a similar way. Note that the first step of the proposed calculation is to perform clustering on the combined dataset. We expect a reasonable clustering approach can guarantee that the instances with the same value are assigned to the same cluster. In other words, $\{x|x=a \wedge x \in$

[3] To distinguish the new formula with the original formula of KL, we denote the new version as $\mathbf{KL_c}$ (Clustering-based KL). Their difference is explained after the proof.

$\mathbb{D}_P \wedge x \in c\} = \{x | x = a \wedge x \in \mathbb{D}_P\}$ where c is a cluster and $x \in c$. This property can be valid for many clustering approaches, such as K-means. We then have

$$P'(x{=}a, c) = \frac{|\{x | x = a \wedge x \in \mathbb{D}_P \wedge x \in c\}|}{|\{x | x \in \mathbb{D}_P \vee x \in \mathbb{D}_Q\}|} = \frac{|\{x | x = a \wedge x \in \mathbb{D}_P\}|}{|\{x | x \in \mathbb{D}_P \vee x \in \mathbb{D}_Q\}|} = P'(x{=}a)$$

With these equations, the KL divergence in Eq (2) becomes:

$$\begin{aligned}
&\mathbf{KL_c}(P(x)||Q(x)) \\
&= \sum_x P(x)\mathbf{log}\frac{P(x)}{Q(x)} = \sum_x \frac{P'(x)}{\mathbb{E}(P)}\mathbf{log}\frac{P'(x)\mathbb{E}(Q)}{Q'(x)\mathbb{E}(P)} \\
&= \sum_x \frac{P'(x)}{\mathbb{E}(P)}\mathbf{log}\frac{P'(x)}{Q'(x)} \\
&\quad + \mathbf{log}\frac{\mathbb{E}(Q)}{\mathbb{E}(P)}\sum_x \frac{P'(x)}{\mathbb{E}(P)} \sum_c \sum_{x \in c} P(x,c)\mathbf{log}\frac{P(x,c)}{Q(x,c)} \\
&= \frac{1}{\mathbb{E}(P)}\Big(\sum_c \sum_{x \in c} (P'(x|c)P'(c)\mathbf{log}\frac{P'(x|c)P'(c)}{Q'(x|c)Q'(c)})\Big) + \mathbf{log}\frac{\mathbb{E}(Q)}{\mathbb{E}(P)} \\
&= \frac{1}{\mathbb{E}(P)}\Big(\sum_c (P'(c)\sum_{x \in c} P'(x|c)\mathbf{log}\frac{P'(x|c)}{Q'(x|c)}) \\
&\quad + \sum_c (P'(c)\mathbf{log}\frac{P'(c)}{Q'(c)}\sum_{x \in c} P'(x|c))\Big) + \mathbf{log}\frac{\mathbb{E}(Q)}{\mathbb{E}(P)}
\end{aligned}$$

Recall that the instances assigned to the same cluster are very similar to each other. We can then assume that in the same cluster, the expectation of instances from distribution P is the same as expectation of instances from Q:

$$\begin{aligned}
&\mathbb{E}_{x \in \mathbb{D}_P, x \in c}[x] = \mathbb{E}_{x \in \mathbb{D}_Q, x \in c}[x] \\
&\Rightarrow \sum_{x \in c} x \frac{P'(x|c)}{\sum_{x \in c} P'(x|c)} = \sum_{x \in c} x \frac{Q'(x|c)}{\sum_{x \in c} Q'(x|c)} \\
&\Rightarrow \sum_{x \in c} x \Big(\frac{P'(x|c)}{\sum_{x \in c} P'(x|c)} - \frac{Q'(x|c)}{\sum_{x \in c} Q'(x|c)}\Big) = 0
\end{aligned}$$

Note that this property can be guaranteed to be satisfied by applying clustering techniques such as bisecting k-means [18], with $|\mathbb{E}_{x \in P, x \in c}[x] - \mathbb{E}_{x \in Q, x \in c}[x]| < \theta$ as the termination condition, where θ set close to 0. In other words, if the condition does not satisfy in one of the clusters, a binary clustering procedure can be performed to divide the cluster smaller, until each cluster satisfies the condition. This process also adaptively decides the number of clusters. Since x can take any value, to validate the above equation, we let

$$\frac{P'(x|c)}{\sum_{x \in c} P'(x|c)} = \frac{Q'(x|c)}{\sum_{x \in c} Q'(x|c)}$$

Eq (2) can then be rewritten as

$$\begin{aligned}
&\mathbf{KL_c}(P(x)||Q(x)) \\
=&\frac{1}{\mathbb{E}(P)}\Big(\sum_c (P'(c)\sum_{x\in c} P'(x|c)\mathbf{log}\frac{\sum_{x\in c} P'(x|c)}{\sum_{x\in c} Q'(x|c)}) \\
&+\sum_c (P'(c)\mathbf{log}\frac{P'(c)}{Q'(c)}\sum_{x\in c} P'(x|c))\Big) + \mathbf{log}\frac{\mathbb{E}(Q)}{\mathbb{E}(P)} \\
=&\frac{1}{\mathbb{E}(P)}\Big(\sum_c (P'(c)S(P',c)\mathbf{log}\frac{S(P',c)}{S(Q',c)}) \\
&+\sum_c (P'(c)S(P',c)\mathbf{log}\frac{P'(c)}{Q'(c)})\Big) + \mathbf{log}\frac{\mathbb{E}(Q)}{\mathbb{E}(P)}
\end{aligned}$$

□

The main difference between the original version of the KL divergence in Eq (2) and its new formula $\mathrm{KL}_\mathbf{c}$ in Lemma 1 is that they are calculated in different ways in practice. The original version of KL in Eq (2) is usually calculated by Eq (3) because its formula requires to know every specific values of each variable x. However, with the new formula in Lemma 1, we can calculate the KL divergence by the clustering result on the whole dataset with several advantages. First, the clustering-based KL divergence in Lemma 1 can be computed efficiently and easily, because it only uses some basic statistics of the clustering. For example, $S(P',C)$ in Eq (5) represents the proportion of examples in the cluster C originally generated from the distribution P. Second, we do not explicitly calculate the marginal distribution $P(x)$, which is normally difficult to approximate with a limited number of instances. Third, "high-level structures" (clusters) of the datasets are applied as a bridge to learn their differences, which are normally a more effective way to reflect the divergence. Other than the proof, we also empirically study the proposed version of KL in the experiment.

It is important to note that the KL divergence is asymmetric. In other words, $\mathbf{KL_c}(P_\mathcal{L}||P_\mathcal{U})$ is not necessarily equal to $\mathbf{KL_c}(P_\mathcal{U}||P_\mathcal{L})$, where $\mathcal{L} \sim P_\mathcal{L}$, $\mathcal{U} \sim P_\mathcal{U}$. However, we keep this property because we are only interested in the "risk" of learning the unlabeled data $\mathcal{U}$ based on the concept learnt from the labeled data $\mathcal{L}$. In other words, we use the risk of coding $\mathcal{U}$ based on the encoding from $\mathcal{L}$, as reflected by $\mathbf{KL_c}(P_\mathcal{U}||P_\mathcal{L})$. With Lemma 1, we define the "transfer risk" $\Re(\mathcal{L};\mathcal{U})$ in the logistic form, to regularize it into $[0,1]$ and it is consistent with the known form of probability distribution:

$$\Re(\mathcal{L};\mathcal{U}) = \big(1 + \mathbf{exp}(\lambda - \mathbf{KL_c}(P_\mathcal{U}||\mathcal{P}_\mathcal{L}))\big)^{-1} \tag{6}$$

where $\mathcal{L} \sim P_\mathcal{L}$, $\mathcal{U} \sim P_\mathcal{U}$, and $\lambda = e^2$ is a deviation to make the minimum value of $\Re(\mathcal{L};\mathcal{U})$ close to 0. We then incorporate the transfer risk $\Re(\mathcal{L};\mathcal{U})$ into the proposed optimization function to automatically regulate the objectives to avoid negative transfer.

3.2 Objective Function

To generate a partition of the target data, the proposed algorithm finds an eigenspace where the target data can be clearly separated through a combination of two objectives. The first is to ensure the labeled data with the same class labels will be grouped together, and the second is to adapt the feature space to cater to the target data. Importantly, the transfer risk $\Re(\mathcal{L};\mathcal{U})$ (Section 3.1) automatically regulates the two goals.

Formally, the proposed optimization function, based on graph partition, can be written as

$$\min_{Y} \mathcal{J}(\mathcal{L},\mathcal{U}) = Cut(G_{\mathcal{L}\cup\mathcal{U}}, Y) + \beta\Big((1-\Re(\mathcal{L};\mathcal{U}))T_{\mathcal{L}} + \Re(\mathcal{L};\mathcal{U})T_{\mathcal{U}}\Big) \quad (7)$$

where $Cut(G_{\mathcal{L}\cup\mathcal{U}}, Y)$ is a cost function of the partition Y on a graph $G_{\mathcal{L}\cup\mathcal{U}}$ generated from the combined data $\mathcal{L}\cup\mathcal{U}$. Examples of such cost functions are Normalized Cut [19], MinMax Cut [6], and so on. Note that $T_{\mathcal{L}}$ and $T_{\mathcal{U}}$ are the two objectives formulated as partition constraints; β is a parameter to control the overall effect of the constraints. On one hand, $T_{\mathcal{L}}$ is directly derived from the "must-link" constraint [21] to find a subspace where the instances are close to each other if they have the same class labels. On the other hand, $T_{\mathcal{U}}$ is a partition constraint defined on the pre-clustering result of $\mathcal{U}$ to reflect its natural separation. To construct $T_{\mathcal{U}}$, we first perform unsupervised spectral clustering on the target data $\mathcal{U}$ individually by Ncut [19]. The proposed algorithm then prefers to find a subspace to "gather" the instances closer if they are in the same pre-cluster. This constraint is defined to "reinforce" the natural manifold structure of $\mathcal{U}$ by maximizing its marginal separation.

We describe a partition constraint as $T_{\mathcal{L}}$ or $T_{\mathcal{U}}$ in Eq (7). To do this, we construct a constraint matrix $\mathbb{M}$ as follows:

$$\mathbb{M} = [\mathbf{m}_1, \mathbf{m}_2, ..., \mathbf{m}_r]^T \quad (8)$$

where each $\mathbf{m}_d$ is a $(s+t)\times 1$ matrix ($s+t$ is the total size of the combined dataset $\mathcal{L}\cup\mathcal{U}$). Each $\mathbf{m}_d$ represents a constraint on the dataset. For example, if $\mathbf{m}_1$ has an entry of $+1$ in the ith row, -1 in the jth row and the rest are all zero, it represents data i and data j are constrained to be close to each other. There are a total of r constraints on the dataset. Then, let $\mathbb{M}_{\mathcal{L}}$ and $\mathbb{M}_{\mathcal{U}}$ denote the constraint matrix of $T_{\mathcal{L}}$ and $T_{\mathcal{U}}$, respectively. We have

$$\begin{aligned} T_{\mathcal{L}} &= \|\mathbb{M}_{\mathcal{L}}Y\|^2 \\ T_{\mathcal{U}} &= \|\mathbb{M}_{\mathcal{U}}Y\|^2 \end{aligned} \quad (9)$$

where Y is the partition indicator. Now consider normalized cut [19] as the graph partition cost function, the proposed optimization function in Eq (7) becomes:

$$\min_{Y} \mathcal{J}(\mathcal{L},\mathcal{U}) = \frac{Y^T(D-W)Y}{Y^TDY} + \beta\Big((1-\Re(\mathcal{L};\mathcal{U}))\|\mathbb{M}_{\mathcal{L}}Y\|^2 + \Re(\mathcal{L};\mathcal{U})\|\mathbb{M}_{\mathcal{U}}Y\|^2\Big) \quad (10)$$

where Y is the partition indicator, W is the similarity matrix of the combined dataset $\mathcal{L} \cup \mathcal{U}$, and $D = \mathbf{diag}(W{\cdot}e)$ ($\mathbb{e}$ is a vector with all coordinates as 1). In Eq (10), the first term reflects the partition quality derived from normalized cut, and the second term consists of two constraints ($\|\mathbb{M}_{\mathcal{L}}Y\|^2$ and $\|\mathbb{M}_{\mathcal{U}}Y\|^2$), representing the two objectives; β is the parameter to control the overall effect of the constraints. The transfer risk $\Re(\mathcal{L};\mathcal{U})$ serves as the pivotal component to balance the two constraints.

Then, if the transfer is too risky, or $\Re(\mathcal{L};\mathcal{U})$ is large, the effect of the first constraint decreases, and the optimization step prefers to satisfy the second constraint more in order to maintain the natural manifold structure of the target data and to avoid negative transfer. It is also important to emphasize that the parameter β is not the essential component to avoid negative transfer. When negative transfer is likely to happen, the constraint will mainly come from the target data regulated by $\Re(\mathcal{L};\mathcal{U})$. In this case, the partition constraint does not include supervision from source data, and thus negative transfer is avoided regardless of the value of β. The effect of β is also studied in the experiment.

3.3 Optimization

We introduce a key step to solve the proposed optimization function Eq (10). First, we denote

$$A = D - W + \beta\Big(\big(1 - \Re(\mathcal{L};\mathcal{U})\big)\mathbb{M}_{\mathcal{L}}^T\mathbb{M}_{\mathcal{L}} + \Re(\mathcal{L};\mathcal{U})\mathbb{M}_{\mathcal{U}}^T\mathbb{M}_{\mathcal{U}}\Big) \tag{11}$$

and $Z = D^{\frac{1}{2}}Y/\|D^{\frac{1}{2}}Y\|$. Then we have:

$$\begin{aligned} \mathcal{J}(\mathcal{L},\mathcal{U}) =& Z^T D^{-\frac{1}{2}}(D-W)D^{-\frac{1}{2}}Z \\ &+ \beta\Big(\big(1 - \Re(\mathcal{L};\mathcal{U})\big)\|\mathbb{M}_{\mathcal{L}}D^{-\frac{1}{2}}Z\|^2 \\ &+ \Re(\mathcal{L};\mathcal{U})\|\mathbb{M}_{\mathcal{U}}D^{-\frac{1}{2}}Z\|^2\Big) \\ =& Z^T D^{-\frac{1}{2}}AD^{-\frac{1}{2}}Z = \frac{Y^T D^{-\frac{1}{2}}AD^{-\frac{1}{2}}Y}{Y^TY} \end{aligned} \tag{12}$$

It is easy to prove that A is symmetric, because D, W, $\mathbb{M}_{\mathcal{L}}^T\mathbb{M}_{\mathcal{L}}$ and $\mathbb{M}_{\mathcal{U}}^T\mathbb{M}_{\mathcal{U}}$ are all symmetric while A is a liner combination of these symmetric matrices. When we relax Y to take the real values similar to other spectral clustering methods [10], we can use the k smallest orthogonal eigenvectors of $D^{-\frac{1}{2}}AD^{-\frac{1}{2}}$ to generate the partition.

The proposed algorithm is described in Fig 2. The algorithm first prepares the partition constraint matrices $\mathbb{M}_{\mathcal{L}}$ and $\mathbb{M}_{\mathcal{U}}$ as Eq (8), and the transfer risk $\Re(\mathcal{L};\mathcal{U})$ according to Eq (6). Moreover, we also construct the similarity matrix W by a distance function like cosine distance, and then construct the corresponding diagonal matrix D. With these terms, we can get a matrix A according to Eq (11). Then, we use the k smallest eigenvectors of $D^{-\frac{1}{2}}AD^{-\frac{1}{2}}$ to generate the eigenspace. Finally, we can perform clustering on the projected target data

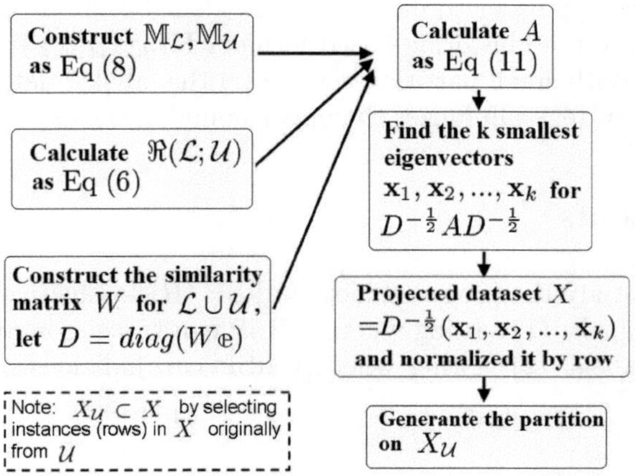

Fig. 2. Risk-sensitive Spectral Partition. Input: labeled dataset $\mathcal{L}$, target dataset $\mathcal{U}$ (number of clusters k); constraint parameter β.

$X_\mathcal{U}$. Note that although clustering is a straightforward approach to generate the partition, we can also use classifiers, such as KNN, to generate the partition if the class labels of source and target data are the same. The target data marked with the same class labels are assigned to the same cluster.

3.4 PAC Bound

The PAC-Bayes error bound in [14] is adopted with new terms to explain the behavior of the proposed model.

Theorem 1. *Let $P_\mathcal{U}$ and $P_\mathcal{L}$ be the distributions of target and source data respectively, and s be the sample size of source data. Let $\delta \in (0,1)$ be given. Then, with probability at least $1-\delta$, the following bound holds,*

$$\varepsilon_\mathcal{U}(\mathcal{N}) \leq \varepsilon_\mathcal{L}(\mathcal{N}) + \sqrt{\frac{\mathbf{KL}(P_\mathcal{U}||P_\mathcal{L}) - ln\pi(\mathcal{N}) - \eta - ln\delta}{2s}}$$

where $\mathcal{N}$ is the partition function, and $\pi(\mathcal{N})$ is a prior distribution of $\mathcal{N}$ that is usually based on domain knowledge, and η is a normalization constant[14].

In the proposed algorithm, we apply semi-supervised spectral clustering to generate the partition. The goal is to minimize the expected partition cost $\varepsilon_\mathcal{U}(\mathcal{N})$, similar to the expected error in classification. Like other PAC methods, we can minimize the empirical partition cost $\varepsilon_\mathcal{L}(\mathcal{N})$ on the given source data $\mathcal{L}$. In our case, we apply the "must-link" constraint to achieve this goal, by encouraging the labeled instances with the same class categories grouped together. However, observed from the second term of the right hand side, the bound also depends on the divergence of the two distributions $P_\mathcal{U}$ and $P_\mathcal{L}$. Thus, we apply the supervisory knowledge of source data $\mathcal{L}$ only when their divergence $\mathbf{KL}(P_\mathcal{U}||P_\mathcal{L})$

is small. Moreover, in order to distinguish the distribution divergence more effectively, we apply the clustering-based formula $\mathbf{KL_c}(P_{\mathcal{U}}||P_{\mathcal{L}})$ to calculate the KL divergence. With small partition cost $\varepsilon_{\mathcal{L}}$ on the labeled data, and small KL divergence, the strategy minimizes the upper bound of the expected cost $\varepsilon_{\mathcal{U}}(\mathcal{N})$.

4 Experiments

We empirically study the proposed method RSP (**R**isk-sensitive **S**pectral **P**artition) with two goals: (1) testing whether RSP can transfer across tasks having different class labels; (2) testing whether RSP can judiciously avoid negative transfer.

4.1 Experiment Setup

Datasets: We first conduct experiments on the text datasets as shown in Table 1. They are generated from 20-newsgroup and Reuters-21578 as in [5]. Each set of experiments contains one labeled source dataset and a corresponding target dataset. In addition, both the target dataset and the source dataset may come from different categories of documents, or even different document corpus. For example, the target dataset is from Reuters-21578, and the source dataset may be from 20-newsgroup. Thus, the class categories and data distributions of the two datasets may be significantly different. Each category contains around 1,500 documents. To speed up the optimization process, we first perform clustering on the target and source dataset respectively by Cluto [24] to generate 100 clusters each. We then choose the center of each cluster as the new data point. Finally, we label the whole cluster by its center.

Table 1. Text Datasets

Target	$Comp_1$ VS. Rec_1	Target	Org_1 VS. $People_1$
Source	2 classes: $Comp_2$ VS. Rec_2 4 classes: Graphics, Hardware, Politics.mid, Religion.misc 3 classes: Sci.crypt, Sci.med, Politics.guns	Source	2 classes: Org_2, $People_2$ 3 classes: $Place_2$, $People_2$, Org_2 3 classes: Sci.crypt, Sci.med, Politics.guns

Note: $Comp_1$ and $Comp_2$ are different datasets with different distributions [5], likewise the other dataset with different superscripts.

Table 2. Image Datasets

Target	Homer-simpson VS. Real-bear	Target	Cartman VS. Palmtree
Source	2 classes: Superman, Teddybear 3 classes: Cartman, Palmtree, Teddybear 4 classes: Laptop, Pram, Keyboard, Soccer	Source	2 classes: Superman, Bonsai 3 classes: Homer, Bonsai, Rear Bear 4 classes: Laptop, Pram, Keyboard, Umbrella

In addition, we conduct experiments on image datasets in Table 2. Similar to the setting of text data, we also generate 6 sets of experiments, where each set contains one labeled dataset and one unlabeled target dataset. All the data are generated from the Caltech-256 image corpus [9] as shown in Table 1. Each category contains around 100 image instances.

Baseline Methods: To verify the effectiveness of the proposed model, an unsupervised spectral partition approach using normalized cut [19] is set as the first baseline method abbreviated as "No-T". This baseline directly generates the partition of the target data without transfer learning. Furthermore, we design another baseline method, abbreviated as "Full-T", by setting the transfer risk $\Re(\mathcal{L};\mathcal{U}) = 0$ in the optimization function in Eq (10) to fully apply the knowledge transferred from labeled data to learn the target data. This model does not include any strategy to avoid negative transfer. In the baseline methods and the proposed model RSP, the parameter β (Eq 10) is set to be 0.6. The effect of this parameter is studied in another set of experiment.

Evaluation Criteria: Note that the outputs of the proposed model and the baseline methods are actually clusters. Thus, to compare the models, we define their accuracy by the purity of each cluster similar to [11]. The purity of a cluster c can be defined as $\max P(y_i|c)$, and $P(y_i|c) = \frac{|\{x|x \in c, y(x)=y_i\}|}{|\{x|x \in c\}|}$, where y_i is a class label, and $y(x)$ denotes the true label of x. The purity can be regarded as the accuracy when we label the whole cluster by its majority label. As a result, the accuracy is defined to be the weighted sum of the purity in all clusters; that is $\sum_c \frac{|\{x|x \in c\}|}{|x|} \max P(y_i|c)$.

4.2 Empirical Analysis

Tables 3 and 4 show the performance given by the baseline methods and the proposed model RSP in average accuracy on ten runs. We answer the following questions using three results:

(1) **Can RSP transfer knowledge across tasks having different class labels?** From the experimental result, RSP can achieve a higher accuracy than the strategy of "No-Transfer" especially when the source and target data are detected to be similar in distribution. For example, when the target dataset is "Org_1 VS. People_1" and the source dataset is a 3-classes document sets "Place_2, etc", the clustering-based formula of KL divergence is 0.51, implying that the target and source data are similar in distribution. In this case, RSP achieves an accuracy of 78% while the accuracy of "No-Transfer" is only 65%. It is clear that transfer learning helps improve the accuracy. More specifically, we plot Fig 3 to illustrate that the final eigenspace transferred from the 3-classes datasets also helps separate the binary target data.

(2) **Can the proposed model avoid negative transfer?** When the source dataset is the 3-classes document sets "Sci.crypt, etc" and the target dataset is "Comp_1 VS. Rec_1", the accuracy of full transfer (Full-T) is only 51%, close to random guessing. With the same setting, the accuracy of no transfer (No-T) is 74%. It is clear that negative transfer happens because the accuracy of the

Table 3. Experiment Result on Text Datasets

Target	Source	KL_t	KL_c	Full-T	No-T	RSP
$Comp_1$	$Comp_2$ VS. Rec_2	0.21	0.37	0.99	0.74	**0.99**±0.00
VS.	4 classes: Graphics, etc	0.01	1.17	0.94	0.74	**0.98**±0.01
Rec_1	3 classes: Sci.crypt, etc	0.05	**21.4**	0.51	0.74	**0.74**±0.03
Org_1	Org_2 VS. $People_2$	0.11	0.24	0.80	0.65	**0.80**±0.00
VS.	3 classes: Places, etc	0.05	0.51	0.73	0.65	**0.78**±0.02
$People_1$	3 classes: Sci.crypt, etc	0.21	**26.5**	0.56	0.65	**0.65**±0.06

Note: "KL_t" is the traditional calculation of KL by Eq (3); "KL_c" is the KL calculated by clustering according to Lemma 1. "Full-T" denotes the method applied transfer learning without considering the divergence between domains, while "No-T" denotes the traditional normalized cut without the strategy of transfer learning.

Table 4. Experiment Result on Image Datasets

Target	Source	KL_t	KL_c	Full-T	No-T	RSP
Homer	Superman VS. Teddy	0.62	0.17	0.85	0.72	0.85±0.02
VS.	3 classes: Cartman, etc	0.88	0.29	0.81	0.72	**0.81**±0.01
Real-bear	4 classes: Laptop, etc	0.11	**10.3**	0.53	0.72	0.72±0.01
Cartman	Superman VS. Bonsai	0.12	0.07	0.87	0.55	**0.87**±0.00
VS.	3 classes: Homer, etc	0.43	0.55	0.92	0.55	**0.92**±0.01
Fern	4 classes: Laptop, etc	0.54	1.58	0.61	0.55	**0.68**±0.01

transfer learning models are worse than no transfer. In the same situation, the proposed model RSP can judiciously avoid negative transfer and still obtains an accuracy of 74%.

(3) **How the proposed model avoids negative transfer?** In the above example, we observe that the KL divergence calculated by "Lemma 1" is 21.4, implying that the transfer is very risky according to Eq (6). In this case, the proposed model automatically decreases the effect of the first objective to avoid negative transfer. From the experimental result, it is also important to note that the clustering-based version KL_c is a more effective KL to reflect distribution divergence. It is also one of the reasons the proposed model RSP outperforms the baseline models.

Parameter Sensitivity: We plot Fig 4 to study the effect of the parameter β on the performance of the proposed model RSP (Fig 2). It is important to emphasize again that β is not the essential component to avoid negative transfer. Instead, it is the transfer risk $\Re(\mathcal{L};\mathcal{U})$ that decides where the partition constraint comes from (from source data or target data). Thus, if negative transfer may happen, the partition constraint will mainly come from the target data regulated by the transfer risk $\Re(\mathcal{L};\mathcal{U})$, and negative transfer is avoided regardless of the value of β. In Fig 4, for each unlabeled target dataset in Table 1 and Table 2, the first source dataset is selected to report the result. The best performance appears at around $\beta = 0.6$. In real world practice, there are various ways to select the best value for β. For instance, partition cost functions, such as normalized cut [19], can be directly applied to evaluate the partition quality, by which one can choose the value of β with the best performance.

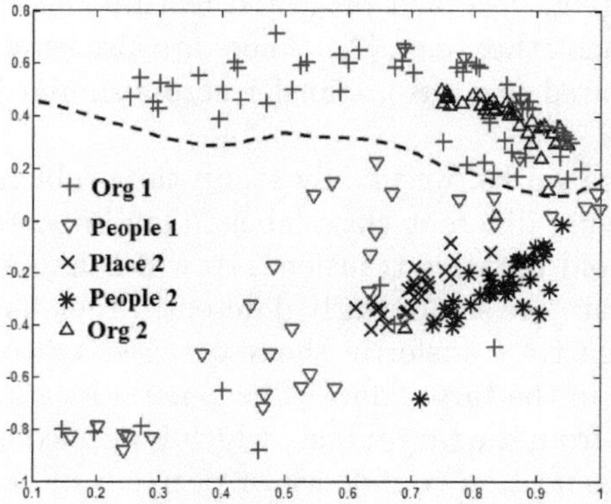

Fig. 3. Projection on the eigenspace

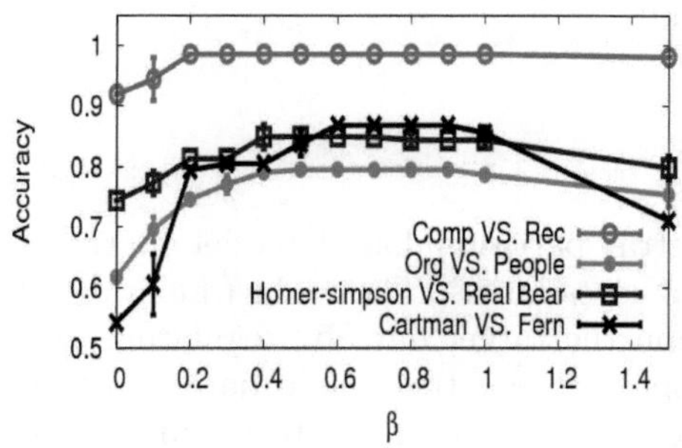

Fig. 4. Parameter Sensitivity of RSP

5 Related Work

Spectral Method. Various unsupervised spectral clustering algorithms have been proved effective in applications such as image segmentation (e.g., [6,12,19]), and the like. Moreover, several works about supervised spectral methods have been proposed to apply the labeled examples to help find the eigenspace of the target data drawn from the same or very similar distributions, such as [11,15,23]. Unlike most of these works, in this paper, we generate a partition of the unlabeled data by transferring knowledge from the given labeled data that may have very different distributions and class categories with the target data.

Transfer Learning. Transfer learning is proposed to extract knowledge from source data to help learn the target data. One of the main issues in transfer learning is how to transfer knowledge across different data distributions. A general approach is based on re-sampling (e.g., [3]), where the motivation of it is to "emphasize" the knowledge among "similar" and discriminating instances. Another line of work is to transfer knowledge based on the common features found in a subspace (e.g., [5]) or a projected feature space where the different tasks are similar to each other (e.g., [2]). There are also some other solutions like model-combination based (e.g., [8]), transfer across similar learning parameters (e.g., [13]), and so on.

Different from these works, we mainly study the problem to transfer knowledge across tasks having different class labels. One important sub-issue of the problem is how to avoid negative transfer [17], which happens when the source data and the target data are significantly different. Previous works like [7,20] are proposed to solve negative transfer in the *supervised setting* where there are a few labeled examples in the target data. The general idea is to build a classifier with the labeled data from the target task, which is applied to identify the harmful knowledge by classification confidence or decrease of accuracy. However, in our problem to transfer knowledge over different class labels, we can not directly apply statics dependant on class labels (e.g., posterior) to select those harmful knowledge. Thus, different from these works, we solve the negative transfer problem in the *unsupervised setting* where the target data does not have any labeled examples at all.

6 Conclusions

We proposed a spectral partition based model to transfer knowledge across tasks having different class labels. The main framework is to find the optimal eigenspace to partition the target data by regulating two objectives. The first is to find the eigenspace where the source data of the same class labels will be close to each other, and the second is to maximize the marginal separation of the unlabeled target data. Importantly, a transfer risk term, as defined on the basis of an effective clustering-based KL divergence, is applied to regulate these two objectives to avoid negative transfer. These two objectives are formulated as partition constraints to construct a symmetric matrix, similar to graph Laplacian, to find the optimal solution given the objective function. The most important advantage of the proposed model is that it can automatically avoid negative transfer when the source data is very different from the target data, while still benefiting from transfer learning even when the source and target data have totally different class labels.

We evaluated the proposed model on text datasets and image datasets. For example, in one of the experiments, a 3-classes image dataset was used to supervise the partition of a binary-class dataset. Even though the two datasets have totally different class labels, the proposed method still achieved an accuracy of 81%, while the baseline model that does not apply transfer learning has accuracy of only 72%.

Acknowledgement. We thank the anonymous reviewers for their greatly helpful comments. Qiang Yang thanks the support of Hong Kong CERG Project 621307. Jiangtao Ren is supported by the National Natural Science Foundation of China under Grant No. 60703110.

References

1. Batu, T., Fortnow, L., Rubinfeld, R., Smith, W.D., White, P.: Testing that distributions are close. In: Proc. 41st FOCS (2000)
2. Ben-David, S., Blitzer, J., Crammer, K., Pereira, F.: Analysis of representations for domain adaptation. In: NIPS (2007)
3. Bickel, S., Bogojeska, J., Lengauer, T., Scheffer, T.: Multi-task learning for hiv therapy screening. In: ICML (2008)
4. Caruana, R.: Multitask learning. Machine Learning 28(1), 41–75 (1997)
5. Dai, W., Xue, G.R., Yang, Q., Yu, Y.: Co-clustering based classification for out-of-domain documents. In: KDD (2007)
6. Ding, C., He, X., Zha, H., Gu, M., Simon, H.: Spectral min-max cut for graph partitioning and data clustering. In: ICDM (2001)
7. Eaton, E., desJardins, M., Lane, T.: Modeling transfer relationships between learning tasks for improved inductive transfer. In: Daelemans, W., Goethals, B., Morik, K. (eds.) ECML PKDD 2008, Part I. LNCS (LNAI), vol. 5211, pp. 317–332. Springer, Heidelberg (2008)
8. Gao, J., Fan, W., Jiang, J., Han, J.: Knowledge transfer via multiple model local structure mapping. In: KDD (2008)
9. Griffin, G., Holub, A., Perona, P.: Caltech-256 object category dataset. Technical Report 7694, California Institute of Technology (2007)
10. Golub, G., Loan, C.V.: Matrix Computation. The Johns Hopkins University Press, Baltimore (1996)
11. Ji, X., Xu, W., Zhu, S.: Document clustering with prior knowledge. In: SIGIR (2006)
12. Joachims, T.: Transductive learning via spectral graph partitioning. In: ICML (2003)
13. Lawrence, N.D., Platt, J.C.: Learning to learn with the informative vector machine. In: ICML (2004)
14. Li, X., Bilmes, J.: A divergence prior for adaptive learning. In: NIPS Workshop on Learning When Test and Training Inputs Have Different Distributions (2006)
15. Ling, X., Dai, W., Xue, G.R., Yang, Q., Yu, Y.: Spectral domain-transfer learning. In: KDD (2008)
16. Raina, R., Ng, A.Y., Koller, D.: Constructing informative priors using transfer learning. In: ICML (2006)
17. Rosenstein, M.T., Marx, Z., Kaelbling, L.P., Dietterich, T.G.: To transfer or not to transfer. In: NIPS (2005)
18. Savaresi, S.M., Boley, D.L.: On the performance of bisecting K-means and PDDP. In: SDM (2001)
19. Shi, J., Malik, J.: Normalized cuts and image segmentation. IEEE Transactions on Pattern Analysis and Machine Intelligence 22(8), 888–905 (2000)
20. Shi, X., Fan, W., Ren, J.: Actively transfer domain knowledge. In: Daelemans, W., Goethals, B., Morik, K. (eds.) ECML PKDD 2008, Part II. LNCS (LNAI), vol. 5212, pp. 342–357. Springer, Heidelberg (2008)
21. Wagstaff, K., Rogers, S., Schroedl, S.: Constrained k-means clustering with background knowledge. In: ICML (2001)
22. Wu, P., Dietterich, T.: Improving svm accuracy by training on auxiliary data sources. In: ICML (2004)
23. Yu, Stella X., Shi, Jianbo.: Grouping with Bias. newblock In: NIPS (2001)
24. Zhao, Y., Karypis, G.: Evaluation of hierarchical clustering algorithms for document datasets. In: ICKM (2002)

Mining Databases to Mine Queries Faster

Arno Siebes and Diyah Puspitaningrum

Department Of Information and Computing Sciences
Universiteit Utrecht, Utrecht, The Netherlands
{arno,diyah}@cs.uu.nl

Abstract. Inductive databases are databases in which models and patterns are first class citizens. Having models and patterns in the database raises the question: do the models and patterns that are stored help in computing new models and patterns? For example, let C be a classifier on database DB and let Q be a query. Does knowing C speed up the induction of a new classifier on the result of Q?

In this paper we answer this problem positively for the code tables induced by our KRIMP algorithm. More in particular, assume we have the code tables for all tables in the database. Then we can approximate the code table induced by KRIMP on the result of a query, using only these global code tables as candidates. That is, we do not have to mine for frequent item sets on the query result.

1 Introduction

The problem investigated in this paper can informally be phrased as follows. Let M_{DB} be the model we induced from database DB and let Q be a query on DB. Does knowing M_{DB} help in inducing a model M_Q on $Q(DB)$, i.e., on the result of Q when applied to DB. For example, if M_{DB} is a classifier and Q selects a subset of DB, does knowing M_{DB} speed-up the induction of a new classifier M_Q on the subset $Q(DB)$?

There are at least two contexts in which this question is relevant. Firstly in the context of inductive databases. Ever since their introduction in the seminal paper by Imielinski and Mannila [11], they have been a constant theme in data mining research. There is no formal definition of an inductive database, in fact, it may be too early for such a definition [14]. However, consensus is that models and patterns should be first class citizens in such a database. That is, e.g., one should be able to query for patterns. Having models and patterns in the database naturally raises the question: do the models and patterns that are stored help in computing new models and patterns?

The second context in which the problem is relevant is in every day data mining practice. In the data mining literature, the usual assumption is that we are given some database that has to be mined. In practice, however, this assumption is usually not met. Rather, the construction of the mining database is often one of the hardest parts of the KDD process. The data often resides in a data warehouse or in multiple databases, and the mining database is constructed from these underlying databases.

W. Buntine et al. (Eds.): ECML PKDD 2009, Part II, LNAI 5782, pp. 382–397, 2009.

From most perspectives, it is not very interesting to know whether one mines a specially constructed database or an original database. For example, if the goal is to build the best possible classifier on that data set, the origins of the database are of no importance whatsoever.

It makes a difference, however, if the underlying databases have already been modelled. Then, like with inductive databases, one would hope that knowing such models would help in modelling the specially constructed 'mining database. For example, if we have constructed a classifier on a database of customers, one would hope that this would help in developing a classifier for the female customers only.

So, the problem is relevant, but isn't it trivial? After all, if M_{DB} is a good model on DB, it is almost always also a good model on a random subset of DB; almost always, because a random subset may be highly untypical. The problem is, however, *not* trivial because queries in general do *not* compute a random subset. Rather, queries construct a very specific result.

For the usual "project-select-join" queries, there is not even a natural way in which the query-result can be seen as subset of the original database. Even if Q is just a "select"-query, the result is usually not random and M_{DB} can even be highly misleading on $Q(DB)$. This is nicely illustrated by the well-known example of *Simpson's Paradox*, viz., Berkeley's admission data [2]. Overall, 44% of the male applicants were admitted, while only 35% of the females were admitted. Four of the six departments, however, have a bias that is in favour of female applicants. While the overall model may be adequate for certain purposes, it is woefully inadequate for a query that selects a single department.

Solving the problem for all model classes and algorithms is a rather daunting task. Rather, in this paper we study the problem for one specific class of models, viz., the code tables induced by our KRIMP algorithm [15]. Given all frequent item sets on a table, KRIMP selects a small subset of these frequent item sets. The reason why we focus on KRIMP is threefold. Firstly, because together the selected item sets describe the underlying data distribution of the complete database very well, see, e.g., [16,17]. Secondly, because the code table consists of *local* patterns. Such a local pattern can be seen as a selection query on the database (for the transactions in its support), hence, one would expect KRIMP to do well on selection queries. Thirdly, from earlier research on KRIMP in a multi-relational setting, we noticed as a side-result that KRIMP is probably easily transformed for joins [13]; this is investigated further in this paper.

More in particular, we show that if we know the code tables for all tables in the database, then we can approximate the code table induced by KRIMP on the result of a query, using *only* the item sets in these global code tables as candidates. Since KRIMP is linear in the number of candidates and KRIMP reduces the set of frequent item sets by many orders of magnitude, this means that we can now speed up the induction of code tables on query results by many orders of magnitude.

This speed-up results in a slightly less good code table, but it approximates the optimal solution within a few percent. We will formalise "approximation" in terms of MDL [10]. Hence, the data miner has a choice: either a quick, good approximation, or the optimal result taking longer time to compute.

2 Problem Statement

This section starts with some preliminaries and assumptions. Then we introduce the problem informally. To formalise it we use MDL, which is briefly discussed.

2.1 Preliminaries and Assumptions

We assume that our data resides in relational databases. In fact, note that the union of two relational databases is, again, a relational database. Hence, we assume, without loss of generality, that our data resides in *one* relational database DB. As query language we will use the standard relational algebra. More precisely, we focus on the usual "select-project-join" queries. That is, on the selection operator σ, the projection operator π, and the (equi-)join operator $\bowtie$; see [5]. Note that, as usual in the database literature, we use *bag* semantics. That is, we do allow duplicates tuples in tables and query results.

As mentioned in the introduction, the mining database is constructed from DB using queries. Given the compositionality of the relational algebra, we may assume, again without loss of generality, that the analysis database is constructed using one query Q. That is, the analysis database is $Q(DB)$, for some relational algebra expression Q. Since DB is fixed, we will often simply write Q for $Q(DB)$; that is, we will use Q to denote both the query and its result.

2.2 The Problem Informally

In the introduction we stated that knowing a model on DB should help in inducing a model on Q. To make this more precise, let $\mathcal{A}$ be our data mining algorithm. At this point, $\mathcal{A}$ can be any algorithm, it may, e.g., compute a decision tree, all frequent item sets or a neural network.

Let $\mathcal{M}_{DB}$ denote the model induced by $\mathcal{A}$ from DB, i.e, $\mathcal{M}_{DB} = \mathcal{A}(DB)$. Similarly, let $\mathcal{M}_Q = \mathcal{A}(Q)$. We want to transform $\mathcal{A}$ into an algorithm $\mathcal{A}^*$ that takes at least two inputs, i.e, both Q and $\mathcal{M}_{DB}$, such that:

1. $\mathcal{A}^*$ gives a reasonable approximation of $\mathcal{A}$ when applied to Q, i.e.,

$$\mathcal{A}^*(Q, \mathcal{M}_{DB}) \approx \mathcal{M}_Q$$

2. $\mathcal{A}^*(Q, \mathcal{M}_{DB})$ is simpler to compute than $\mathcal{M}_Q$.

The second criterion is easy to formalise: the runtime of $\mathcal{A}^*$ should be shorter than that of $\mathcal{A}$. The first one is harder. What do we mean that one model is an approximation of another? Moreover, what does it mean that it is a *reasonable* approximation?

2.3 Model Approximation

The answer to the question how to formalise that one model approximates another depends very much on the goal. If $\mathcal{A}$ induces classifiers, approximation should probably be defined in terms of prediction accuracy, e.g., on the Area Under the ROC-curve (AUC).

KRIMP computes code tables. Hence, the quick approximating algorithm we are looking for, KRIMP* in the notation used above, also has to compute code tables. So, one way to define the notion of approximation is by comparing the resulting code tables. Let CT_{KRIMP} be the code table computed by KRIMP and similarly, let CT_{KRIMP^*} denote the code table computed by KRIMP* on the same data set. The more similar CT_{KRIMP^*} is to CT_{KRIMP}, the better KRIMP* approximates KRIMP.

While this is intuitively a good way to proceed, it is far from obvious how to compare two code tables. Fortunately, we do not need to compare code tables directly. KRIMP is based on the Minimum Description Length principle (MDL) [10], and MDL offers another way to compare models, viz., by their *compression-rate*. Note that using MDL to define "approximation" has the advantage that we can formalise our problem for a larger class of algorithms than just KRIMP. It is formalised for all algorithms that are based on MDL. MDL is quickly becoming a popular formalism in data mining research, see, e.g., [8] for an overview of other applications of MDL in data mining.

2.4 Minimum Description Length

MDL embraces the slogan *Induction by Compression.* It can be roughly described as follows.

Given a set of models[1] $\mathcal{H}$, the best model $H \in \mathcal{H}$ is the one that minimises

$$L(H) + L(D|H)$$

in which

- $L(H)$ is the length, in bits, of the description of H, and
- $L(D|H)$ is the length, in bits, of the description of the data when encoded with H.

One can paraphrase this by: the smaller $L(H) + L(D|H)$, the better H models D.

What we are interested in is comparing two algorithms on the same data set, viz., on $Q(DB)$. Slightly abusing notation, we will write $\mathcal{L}(\mathcal{A}(Q))$ for $L(\mathcal{A}(Q)) + L(Q(DB)|\mathcal{A}(Q))$, similarly, we will write $\mathcal{L}(\mathcal{A}^*(Q, \mathcal{M}_{DB}))$. Then, we are interested in comparing $\mathcal{L}(\mathcal{A}^*(Q, \mathcal{M}_{DB}))$ to $\mathcal{L}(\mathcal{A}(Q))$. The closer the former is to the latter, the better the approximation is.

[1] MDL-theorists tend to talk about *hypothesis* in this context, hence the $\mathcal{H}$; see [10] for the details.

Just taking the difference of the two, however, can be quite misleading. Take, e.g., two databases db_1 and db_2 sampled from the same underlying distribution, such that db_1 is far bigger than db_2. Moreover, fix a model H. Then necessarily $L(db_1|H)$ is bigger than $L(db_2|H)$. In other words, big absolute numbers do not necessarily mean very much. We have to *normalise* the difference to get a feeling for how good the approximation is. Therefore we define the asymmetric dissimilarity measure (ADM) as follows.

Definition 1. *Let H_1 and H_2 be two models for a dataset D. The asymmetric dissimilarity measure $ADM(H_1, H_2)$ is defined by:*

$$ADM(H_1, H_2) = \frac{|\mathcal{L}(H_1) - \mathcal{L}(H_2)|}{\mathcal{L}(H_2)}$$

Note that this dissimilarity measure is related to the Normalised Compression Distance [4]. The reason why we use this asymmetric version is that we have a "gold standard". We want to know how far our approximate result $\mathcal{A}^*(Q, \mathcal{M}_{DB})$ deviates from the optimal result $\mathcal{A}(Q)$.

Clearly, $ADM(\mathcal{A}^*(Q, \mathcal{M}_{DB}), \mathcal{A}(Q))$ does not only depend on $\mathcal{A}^*$ and on $\mathcal{A}$, but also very much on Q. We do not seek a low ADM on one particular Q, rather we want to have a reasonable approximation on all possible queries. Requiring that the ADM is equally small on all possible queries seems to strong a requirement. Some queries might result in a very untypical subset of DB, the ADM is probably higher on the result of such queries than it is on queries that result in more typical subsets. Hence, it is more reasonable to require that the ADM is small most of the time. This is formalised through the notion of an (ϵ, δ)-approximation.

Definition 2. *Let DB be a database and let Q be a random query on DB. Moreover, let $\mathcal{A}_1$ and $\mathcal{A}_2$ be two data mining algorithms on DB. $\mathcal{A}_1$ is an (ϵ, δ)-approximation of $\mathcal{A}_2$ iff*

$$P(ADM(\mathcal{A}_1(Q), \mathcal{A}_2(Q)) > \epsilon) < \delta$$

2.5 The Problem

Using the notation introduced above, we formalise the problem as follows.

Problem Statement
For a given data mining algorithm $\mathcal{A}$, devise an algorithm $\mathcal{A}^*$, such that for a random database DB:

1. $\mathcal{A}^*$ is an (ϵ, δ)-approximation of $\mathcal{A}$ for reasonable values for ϵ and δ.
2. Computing $\mathcal{A}^*(Q, \mathcal{M}_{DB})$ is faster than computing $\mathcal{A}(Q)$ for a random query Q on DB.

What reasonable values for ϵ and δ are depends very much on the application. While $\epsilon = 0.5$ for $\delta = 0.9$ might be acceptable for one application, these values may be unacceptable for others.

The ultimate solution to the problem as stated here would be an algorithm that transforms any data mining algorithm $\mathcal{A}$ in an algorithm $\mathcal{A}^*$ with the requested properties. This is a rather ambitious, ill-defined (what is the class of all data mining algorithms?), and, probably, not attainable goal. Hence, in this paper we take a more modest approach: we transform one algorithm only, our KRIMP algorithm.

3 Introducing KRIMP

For the convenience of the reader we provide a brief introduction to KRIMP in this section, it was originally introduced in [15] (although not by that name) and the reader is referred to that paper for more details.

Since KRIMP selects a small set of representative item sets from the set of all frequent item sets, we first recall the basic notions of frequent item set mining [1].

3.1 Preliminaries

Let $\mathcal{I} = \{I_1, \ldots, I_n\}$ be a set of binary (0/1 valued) attributes. That is, the domain D_i of item I_i is $\{0, 1\}$. A transaction (or tuple) over $\mathcal{I}$ is an element of $\prod_{i \in \{1,\ldots,n\}} D_i$. A database DB over $\mathcal{I}$ is a bag of tuples over $\mathcal{I}$. This bag is indexed in the sense that we can talk about the i-th transaction.

An item set J is, as usual, a subset of $\mathcal{I}$, i.e., $J \subseteq \mathcal{I}$. The item set J *occurs* in a transaction $t \in DB$ if $\forall I \in J : \pi_I(t) = 1$. The *support* of item set J in database DB is the number of transactions in DB in which J occurs. That is, $supp_{DB}(J) = |\{t \in DB|\ J \text{ occurs in } t\}|$. An item set is called *frequent* if its support is larger than some user-defined threshold called the *minimal support* or *min-sup*. Given the A Priori property,

$$\forall I, J \in \mathcal{P}(\mathcal{I}) : I \subset J \rightarrow supp_{DB}(J) \leq supp_{DB}(I)$$

frequent item sets can be mined efficiently level wise, see [1] for more details.

Note that while we restrict ourself to binary databases in the description of our problem and algorithms, there is a trivial generalisation to categorical databases. In the experiments, we use such categorical databases.

3.2 KRIMP

The key idea of the KRIMP algorithm is the code table. A code table is a two-column table that has item sets on the left-hand side and a code for each item set on its right-hand side. The item sets in the code table are ordered descending on 1) item set length and 2) support size and 3) lexicographically. The actual codes on the right-hand side are of no importance but their lengths are. To explain how these lengths are computed, the coding algorithm needs to be introduced.

A transaction t is encoded by KRIMP by searching for the first item set I in the code table for which $I \subseteq t$. The code for I becomes part of the encoding of t. If $t \setminus I \neq \emptyset$, the algorithm continues to encode $t \setminus I$. Since it is insisted that each code table contains at least all singleton item sets, this algorithm gives a unique encoding to each (possible) transaction over $\mathcal{I}$.

The set of item sets used to encode a transaction is called its *cover*. Note that the coding algorithm implies that a cover consists of non-overlapping item sets.

The length of the code of an item in a code table CT depends on the database we want to compress; the more often a code is used, the shorter it should be. To compute this code length, we encode each transaction in the database DB. The *frequency* of an item set $I \in CT$, denoted by $freq(I)$ is the number of transactions $t \in DB$ which have I in their cover. That is,

$$freq(I) = |\{t \in DB | I \in cover(t)\}|$$

The relative frequency of $I \in CT$ is the probability that I is used to encode an arbitrary $t \in DB$, i.e.

$$P(I|DB) = \frac{freq(I)}{\sum_{J \in CT} freq(J)}$$

For optimal compression of DB, the higher $\mathrm{P}(c)$, the shorter its code should be. Given that we also need a prefix code for unambiguous decoding, we use the well-known optimal Shannon code [7]:

$$l(I|CT) = -\log(P(I|DB)) = -\log\left(\frac{freq(I)}{\sum_{J \in CT} freq(J)}\right)$$

The length of the encoding of a transaction is now simply the sum of the code lengths of the item sets in its cover. Therefore the encoded size of a transaction $t \in DB$ compressed using a specified code table CT is calculated as follows:

$$L(t|CT) = \sum_{I \in cover(t,CT)} l(I|CT)$$

The size of the encoded database is the sum of the sizes of the encoded transactions, but can also be computed from the frequencies of each of the elements in the code table:

$$\begin{aligned} L(DB|CT) &= \sum_{t \in DB} L(t|CT) \\ &= -\sum_{I \in CT} freq(I) \log\left(\frac{freq(I)}{\sum_{J \in CT} freq(J)}\right) \end{aligned}$$

To find the optimal code table using MDL, we need to take into account both the compressed database size, as described above, as well as the size of the code table. For the size of the code table, we only count those item sets that have a non-zero frequency. The size of the right-hand side column is obvious; it is simply the sum of all the different code lengths. For the size of the left-hand side column, note that the simplest valid code table consists only of the singleton item sets. This is the *standard encoding (ST)*, of which we use the codes to compute the size of the item sets in the left-hand side column. Hence, the size of code table CT is given by:

$$L(CT|DB) = \sum_{I \in CT: freq(I) \neq 0} l(I|ST) + l(I|CT)$$

In [15] we defined the optimal set of (frequent) item sets as that one whose associated code table minimises the total compressed size:

$$L(CT, DB) = L(CT|DB) + L(DB|CT)$$

As before, this minimal compressed size of DB is denoted by $\mathcal{L}(DB)$. KRIMP starts with a valid code table (only the collection of singletons) and a sorted list of candidates (frequent item sets). These candidates are assumed to be sorted descending on 1) support size, 2) item set length and 3) lexicographically. Each candidate item set is considered by inserting it at the right position in CT and calculating the new total compressed size. A candidate is only kept in the code table iff the resulting total size is smaller than it was before adding the candidate. If it is kept, all other elements of CT are reconsidered to see if they still positively contribute to compression. The whole process is illustrated in Figure 1; see [15].

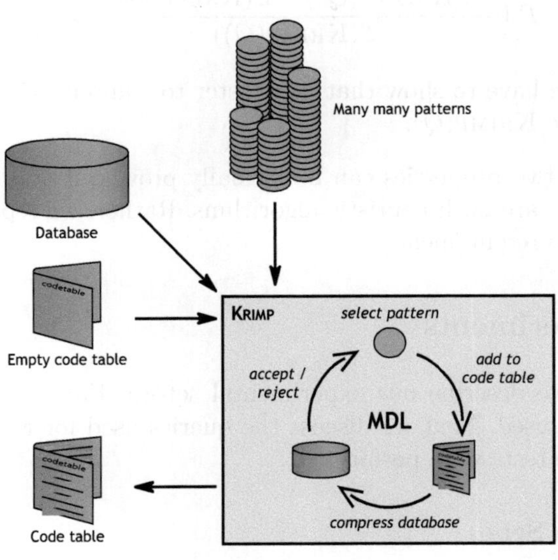

Fig. 1. KRIMP in action

4 The Problem for KRIMP

If we assume a fixed minimum support threshold for a database, KRIMP has only one essential parameter: the database. For, given the database and the (fixed) minimum support threshold, the candidate list is also specified. Hence, we will simply write CT_{DB} and KRIMP(DB), to denote the code table induced by KRIMP from DB. Similarly CT_Q and KRIMP(Q) denote the code table induced by KRIMP from the result of applying query Q to DB.

Given that KRIMP results in a code table, there is only one sensible way in which KRIMP(DB) can be re-used to compute KRIMP(Q): provide KRIMP only with the item sets in CT_{DB} as candidates. While we change nothing to the algorithm, we'll use the notation KRIMP* to indicate that KRIMP got only code table elements as candidates. So, e.g., KRIMP*(Q) is the code table that KRIMP induces from $Q(DB)$ using the item sets in CT_{DB} only.

Given our general problem statement, we now have to show that KRIMP* satisfies our two requirements for a transformed algorithm. That is, we have to show for a random database DB:

- For reasonable values for ϵ and δ, KRIMP* is an (ϵ, δ)-approximation of KRIMP, i.e, for a random query Q on DB:

$$P(ADM(\text{KRIMP}^*(Q), \text{KRIMP}(Q)) > \epsilon) < \delta$$

  Or in MDL-terminology:

$$P\left(\frac{|\mathcal{L}(\text{KRIMP}^*(Q)) - \mathcal{L}(\text{KRIMP}(Q))|}{\mathcal{L}(\text{KRIMP}(Q))} > \epsilon\right) < \delta$$

- Moreover, we have to show that it is faster to compute KRIMP*(Q) than it is to compute KRIMP(Q).

Neither of these two properties can be formally proven, if only because KRIMP and thus KRIMP* are both heuristic algorithms. Rather, we report on extensive tests of these two requirements.

5 The Experiments

In this section we describe our experimental set-up. First we briefly describe the data sets we used. Next we discuss the queries used for testing. Finally we describe how the tests were performed.

5.1 The Data Sets

To test our hypothesis that KRIMP* is a good and fast approximation of KRIMP, we have performed extensive tests mostly on 6 well-known UCI [6] data sets and one data set from the KDDcup 2004.

More in particular, we have used the data sets *connect, adult, chessBig, letRecog, PenDigits* and *mushroom* from UCI. These data sets were chosen because they are well suited for KRIMP. Some of the other data sets in the UCI repository are simply too small for KRIMP to perform well. MDL needs a reasonable amount of data to be able to function. Some other data sets are very dense. While

KRIMP performs well on these data sets, choosing them would have turned our extensive testing prohibitively time-consuming.

Since all these data sets are single table data sets, they do not allow testing with queries involving joins. To test such queries, we used tables from the "Hepatitis Medical Analysis"[2] of the KDDcup 2004. From this relational database we selected the tables *bio* and *hemat*. The former contains biopsy results, while the latter contains results on hematological analysis. The original tables have been converted to item set data and rows with missing data have been removed.

5.2 The Queries

To test our hypothesis, we need to consider randomly generated queries. On first sight this appears a daunting task. Firstly, because the set of all possible queries is very large. How do we determine a representative set of queries? Secondly, many of the generated queries will have no or very few results. If the query has no results, the hypothesis is vacuously true. If the result is very small, MDL (and KRIMP) doesn't perform very well.

To overcome these problems, we restrict ourselves to queries that are build using selections (σ), projections (π), and joins ($\bowtie$) only. The rationale for this choice is twofold. Firstly, the well-known "project-select-join" queries are among the most used queries in practice. This is witnessed by the important role they play in benchmarks for DBMSs such as the TPC family of benchmarks. Secondly, simple queries will have, in general, larger results than more complex queries.

5.3 The Experiments

The experiments preformed for each of the queries on each of the data sets were generated as follows.

Projection: The projection queries were generated by randomly choosing a set X of n attributes, for $n \in \{1, 3, 5, 7, 9\}$. The generated query is then $\pi_{\overline{X}}$. That is, the elements of X are projected out of each of the transactions. For example, $\pi_{\overline{\{I_1, I_3\}}}(\{I_1, I_2, I_3\}) = \{I_2\}$. For this case, the code table elements generated on the complete data set were projected in the same way. For each value of n, 10 random sets X were generated on each data set.

As an aside, note that the rationale for limiting X to maximally 9 elements is that for larger values too many result sets became too small for meaningful results.

Selection: The random selection queries were again generated by randomly choosing a set X of n attributes, with $n \in \{1, 2, 3, 4\}$. Next for each random attribute A_i a random value v_i in its domain D_i was chosen. Finally, for each A_i in X a random $\theta_i \in \{=, \neq\}$ was chosen The generated query is thus $\sigma(\bigwedge_{A_i \in X} A_i \theta_i v_i)$. As in the previous case, we performed 10 random experiments on each of the data sets for each of the values of n.

[2] http://lisp.vse.cz/challenge/

Project-Select: The random project-select queries generated, are essentially combinations of the simple projection and selection queries as explained above. The only difference is that we used $n \in \{1, 3\}$ for the projection and $n \in \{1, 2\}$ for the selections. That is we select on 1 or 2 attributes and we project away either 1 or 3 attributes. The size of the results is, of course, again the rationale for this choice. For each of the four combinations, we performed 100 random experiments on each of the data sets: first we chose randomly the selection (10 times for each selection), for each such selection we performed 10 random projections.

Project-Select-Join: Since we only use one "multi-relational" data set and there is only one possible way to join the *bio* and *hemat* tables, we could not do random tests for the join operator. However, in combination with projections and selections, we can perform random tests. These tests consist of randomly generated project-select queries on the join of *bio* and *hemat*. In this two-table case, KRIMP* got as input all pairs $(\mathcal{I}_1, \mathcal{I}_2)$ in which $\mathcal{I}_1$ is an item set in the code table of *bio*, and $\mathcal{I}_2$ is an item set in the code table of *hemat*. Again we select on 1 or 2 attributes and we project away either 1 or 3 attributes. And, again, we performed again 100 random experiments on the database for each of the four combinations; as above.

6 The Results

In this section we give an overview of the results of the experiments described in the previous section. Each test query is briefly discussed in its own subsection.

6.1 Projection Queries

In Figure 2 the results of the random projection queries on the *letRecog* data set are visualised. The marks in the picture denote the averages over the 10 experiments, while the error bars denote the standard deviation. Note that, while not statistically significant, the average ADM grows with the number of attributes projected away. This makes sense, since the more attributes are projected away, the smaller the result set becomes. On the other data sets, KRIMP* performs similarly. Since this is also clear from the project-select query results, we do not provide all details here. This will become clear when we report on the project-select queries.

6.2 Selection Queries

The results of the random selection queries on the *penDigits* data set are visualised in figure 3. For the same reason as above, it makes sense that the average ADM grows with the number of attributes selected on. Note, however, that the ADM averages for selection queries seem much larger than those for projection queries. These numbers are, however, not representative for the results on the other data sets. It turned out that *penDigits* is actually too small and sparse to

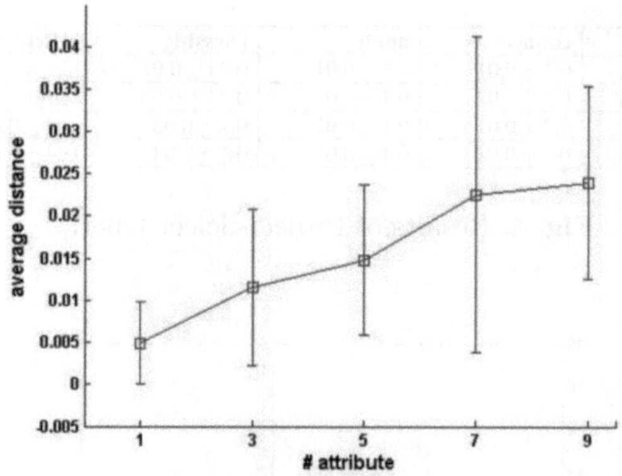

Fig. 2. Projection results on *letRecog*

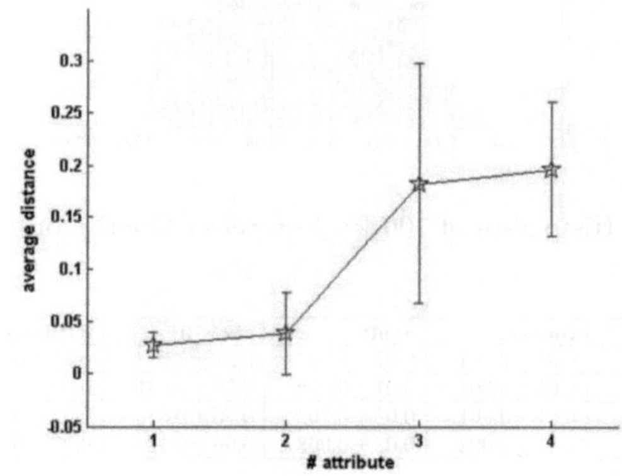

Fig. 3. Selection results on *penDigits*

test KRIMP* seriously. In the remainder of our results section, we do not report further results on *penDigits*. The reason why we report on it here is to illustrate that even on rather small and sparse data sets KRIMP* still performs reasonably well. On all other data sets KRIMP* performs far better, as will become clear when we report on the project-select queries.

6.3 Project-Select Queries

The results of the projection-select queries are given in the table in Figure 4. All numbers are the average ADM score ± the standard deviation for the 100 random experiments. All the ADM numbers are rather small, only for mushroom do they get above 0.2. Two important observations can be made from this table. Firstly, as for the projection and selection queries reported on above, the ADM scores get only slightly worse when the query results get smaller: "Select 2, Project out 3" has slightly worse ADM scores than "Select 1, Project out 1". Secondly,

ADM±STD		connect	adult	chessBig	letRecog	mushroom
Select 1	Project out 1	0.1 ± 0.01	0.1 ± 0.01	0.04 ± 0.01	0.1 ± 0.01	0.3 ± 0.02
	Project out 3	0.1 ± 0.02	0.1 ± 0.01	0.04 ± 0.03	0.1 ± 0.01	0.3 ± 0.16
Select 2	Project out 1	0.2 ± 0.01	0.1 ± 0.01	0.1 ± 0.03	0.04 ± 0.01	0.2 ± 0.04
	Project out 3	0.2 ± 0.02	0.1 ± 0.01	0.1 ± 0.03	0.04 ± 0.01	0.2 ± 0.05

Fig. 4. Results of Project-Select Queries

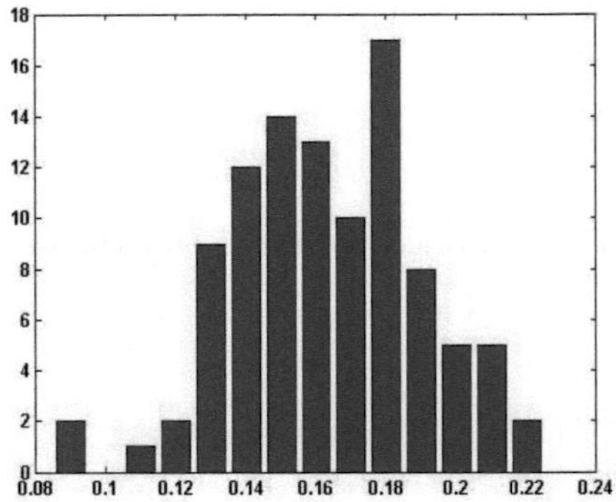

Fig. 5. Histogram of 100 Project-Select Queries on *connect*

Relative #candidates		connect	adult	chessBig	letRecog	mushroom
Select 1	Project out 1	0.01 ± 0.001	0.01 ± 0.002	0.21 ± 0.012	0.01 ± 0.001	0.01 ± 0.001
	Project out 3	0.01 ± 0.001	0.01 ± 0.004	0.26 ± 0.031	0.02 ± 0.004	0.01 ± 0.001
Select 2	Project out 1	0.01 ± 0.001	0.03 ± 0.003	0.76 ± 0.056	0.02 ± 0.002	0.03 ± 0.002
	Project out 3	0.01 ± 0.002	0.03 ± 0.008	0.96 ± 0.125	0.02 ± 0.004	0.03 ± 0.003

Fig. 6. Relative number of candidates for KRIMP*

even more importantly, combining algebra operators only degrades the ADM scores slightly. This can be seen if we compare the results for "Project out 3" on *letRecog* in Figure 2 with the "Select 1, Project out 3" and "Select 2, Project out 3" queries in Figure 4 on the same data set. These results are very comparable, the combination effect is small and mostly due to the smaller result sets. While not shown here, the same observation holds for the other data sets.

To give insight in the distribution of the ADM scores of the "Select 2, Project out 3" queries on the *connect* data set are given in Figure 5. From this figure we see that if we choose $\epsilon = 0.2$, $\delta = 0.08$. In other words, KRIMP* is a pretty good approximation of KRIMP. Almost always the approximation is less than 20% worse than the optimal result. The remaining question is, of course, how much faster is KRIMP*? This is illustrated in the table in Figure 6 This table gives the average number of candidates KRIMP* has to consider relative to those that the full KRIMP run has to consider. Since, both KRIMP* and KRIMP are linear in the number of candidates, this table shows that the speed-up is considerable; a factor of 100 is often attained; except for *chessBig* were the query results get small and, thus, have few frequent item sets. The experiments are those that are reported on in Figure 4.

6.4 Select-Project-Join Queries

The results for the select-project-join queries are very much in line with the results reported on above. In fact, they are even better. Since the join leads to rather large results, the ADM score is almost always zero: in only 15 of the 400 experiments the score is non-zero (average of non-zero values is 1%). The speed-up is also in line with the numbers reported above, a factor of 100 is again often attained.

7 Discussion

As noted in the previous section, the speed-up of KRIMP* is easily seen. The number of candidates that KRIMP* has to consider is often a factor 100 smaller than those that the full KRIMP run has to consider. Given that the algorithm is linear in the number of candidates, this means a speed-up by a factor 100. In fact, one should also note that for KRIMP*, we do not have to run a frequent item set miner. In other words, in practice, using KRIMP* is even faster than suggested by the Speed-up scores.

But, how about the other goal: how good is the approximation? That is, how should one interpret ADM scores? Except for some outliers, ADM scores are below 0.2. That is, a full-fledged KRIMP run compresses the data set 20% better than KRIMP*. Is that good?

In a previous paper [17], we took two random samples from data sets, say D_1 and D_2. Code tables CT_1 and CT_2 were induced from D_1 and D_2 respectively. Next we tested how well CT_i compressed D_j. For the four data sets also used in this paper, *Iris, Led7, Pima* and, *PageBlocks*, the "other" code table compressed 16% to 18% worse than the "own" code table; the figures for other data sets are in the same ball-park. In other words, an ADM score on these data sets below 0.2 is on the level of "natural variations" of the data distribution. Hence, given that the average ADM scores are often much lower we conclude that the approximation by KRIMP* is good.

In other words, the experiments verify our hypothesis: KRIMP* gives a fast and good approximation of KRIMP. The experiments show this for simple "project-select-join" queries, but as noticed with the results of the "project-select" queries, the effect of combining algebra operators is small. If the result set is large enough, the approximation is good.

8 Related Work

While there are, as far as the authors know, no other papers that study the same problem, the topic of this paper falls in the broad class of data mining with background knowledge. For, the model on the database, $\mathcal{M}_{DB}$, is used as background knowledge in computing $\mathcal{M}_Q$. While a survey of this area is beyond the scope of this paper, we point out some papers that are related to one of the two aspects we are interested in, viz., speed-up and approximation.

A popular area of research in using background knowledge is that of constraints. Rather than trying to speed up the mining, the goal is often to produce models that adhere to the background knowledge. Examples are the use of constraints in frequent pattern mining, e.g. [3], and monotonicity constraints [9]. Note, however, that for frequent pattern mining the computation can be speeded up considerably if the the constraints can be pushed into the mining algorithm [3]. So, speed-up is certainly a concern in this area. However, as far as we know approximation plays no role. The goal is still to find all patterns that satisfy the constraints.

Another use of background knowledge is to find unexpected patterns. In [12], e.g., Bayesian Networks of the data are used to estimate how surprising a frequent pattern is. In other words, the (automatically induced) background knowledge is used filter the output. In other words, speed-up is of no concern in this approach. Approximation clearly is, albeit in the opposite direction of ours: the more a pattern deviates from the global model, the more interesting it becomes. Whereas we would like that all patterns in the query result are covered by our approximate answer.

9 Conclusions

In this paper we introduce a new problem: given that we have a model induced from a database DB, does that help us in inducing a model on the result of a query Q on DB? We formalise the problem for algorithms based on MDL and solve it for a particular algorithm, viz., our KRIMP algorithm. More in particular we introduce KRIMP*. This is actually the same as KRIMP, but it gets a restricted input. The code tables computed by KRIMP on DB are used as input, and thus as background knowledge, for KRIMP* on $Q(DB)$.

Extensive experiments with select-project-join queries show that KRIMP* approximates the results of KRIMP very well while it computes these results upto hundreds of times faster. Hence, the data analyst has a real choice: either get good result fast, or get optimal results slower.

References

1. Agrawal, R., Mannila, H., Srikant, R., Toivonen, H., Inkeri Verkamo, A.: Fast discovery of association rules. In: Advances in Knowledge Discovery and Data Mining, pp. 307–328. AAAI, Menlo Park (1996)
2. Bickel, P.J., Hammel, E.A., O'Connell, J.W.: Sex bias in graduate admissions: Data from berkeley. Science 187(4175), 398–404 (1975)
3. Boulicaut, J.-F., Bykowski, A.: Frequent closures as a concise representation for binary data mining. In: Terano, T., Chen, A.L.P. (eds.) PAKDD 2000. LNCS, vol. 1805, pp. 62–73. Springer, Heidelberg (2000)

4. Cilibrasi, R., Vitanyi, P.: Automatic meaning discovery using google. IEEE Transactions on Knowledge and Data Engineering 19, 370–383 (2007)
5. Codd, E.F.: A relational model of data for large shared data banks. Communications of the ACM 13(6), 377–387 (1970)
6. Coenen, F.: The LUCS-KDD discretised/normalised ARM and CARM data library (2003), `http://www.csc.liv.ac.uk/~frans/KDD/Software/LUCS_KDD_DN/`
7. Cover, T.M., Thomas, J.A.: Elements of Information Theory, 2nd edn. John Wiley and Sons, Chichester (2006)
8. Faloutsos, C., Megalooikonomou, V.: On data mining, compression and kolmogorov complexity. In: Data Mining and Knowledge Discovery, vol. 15, pp. 3–20. Springer, Heidelberg (2007)
9. Feelders, A.J., van der Gaag, L.C.: Learning bayesian network parameters under order constraints. Int. J. Approx. Reasoning 42(1-2), 37–53 (2006)
10. Grünwald, P.D.: Minimum description length tutorial. In: Grünwald, P.D., Myung, I.J. (eds.) Advances in Minimum Description Length. MIT Press, Cambridge (2005)
11. Imielinski, T., Mannila, H.: A database perspective on knowledge discovery. Communications of the ACM 39(11), 58–64 (1996)
12. Jaroszewicz, S., Simovici, D.A.: Interestingness of frequent itemsets using bayesian networks as background knowledge. In: Proceedings KDD, pp. 178–186 (2004)
13. Koopman, A., Siebes, A.: Discovering relational item sets efficiently. In: Proceedings SDM 2008, pp. 585–592 (2008)
14. De Raedt, L.: A perspective on inductive databases. SIGKDD Explorations 4(2), 69–77 (2000)
15. Siebes, A., Vreeken, J., van Leeuwen, M.: Item sets that compress. In: Proceedings of the SIAM Conference on Data Mining, pp. 393–404 (2006)
16. van Leeuwen, M., Vreeken, J., Siebes, A.: Compression picks item sets that matter. In: Fürnkranz, J., Scheffer, T., Spiliopoulou, M. (eds.) PKDD 2006. LNCS (LNAI), vol. 4213, pp. 585–592. Springer, Heidelberg (2006)
17. Vreeken, J., van Leeuwen, M., Siebes, A.: Preserving privacy through data generation. In: Proceedings of the IEEE International Conference on Data Mining, pp. 685–690 (2007)

MACs: Multi-Attribute Co-clusters with High Correlation Information

Kelvin Sim[1,2], Vivekanand Gopalkrishnan[2], Hon Nian Chua[1], and See-Kiong Ng[1]

[1] Institute for Infocomm Research, A*STAR, Singapore
[2] School of Computer Engineering, Nanyang Technological University, Singapore

Abstract. In many real-world applications that analyze correlations between two groups of diverse entities, each group of entities can be characterized by multiple attributes. As such, there is a need to co-cluster *multiple* attributes' values into pairs of highly correlated clusters. We denote this co-clustering problem as the *multi-attribute co-clustering* problem. In this paper, we introduce a generalization of the mutual information between two attributes into mutual information between two attribute sets. The generalized formula enables us to use *correlation information* to discover *multi-attribute co-clusters (MACs)*. We develop a novel algorithm MACminer to mine MACs with high correlation information from datasets. We demonstrate the mining efficiency of MACminer in datasets with multiple attributes, and show that MACs with high correlation information have higher classification and predictive power, as compared to MACs generated by alternative high-dimensional data clustering and pattern mining techniques.

1 Introduction

Co-clustering values of two attributes (also known as pairwise co-clustering) is a well-established research area with many successful applications, ranging from clustering words and documents [7], to clustering video shots and video features [17]. In pairwise co-clustering, the values of two attributes are partitioned into clusters such that pairwise pairings of these clusters formed co-clusters. More recently, star-structured co-clustering [8] was proposed to handle higher dimensional data. In essence, a star-structured co-cluster is a set of pairwise co-clusters, with the constraint that each pairwise co-cluster involves the center attribute and a non-center attribute. Figure 1(a) shows some examples of pairwise co-clusters.

Such pairwise co-cluster structures are not applicable in many real-world applications that involve co-clustering multiple attributes' values into pairs of clusters that correlate. We call such problems *multi-attribute co-clustering*. Figure 1(b) depicts the difference between the multi-attribute co-clusters (MACs) and the pairwise star-structured co-clusters (Figure 1(a)). Here are some real-world examples:

W. Buntine et al. (Eds.): ECML PKDD 2009, Part II, LNAI 5782, pp. 398–413, 2009.

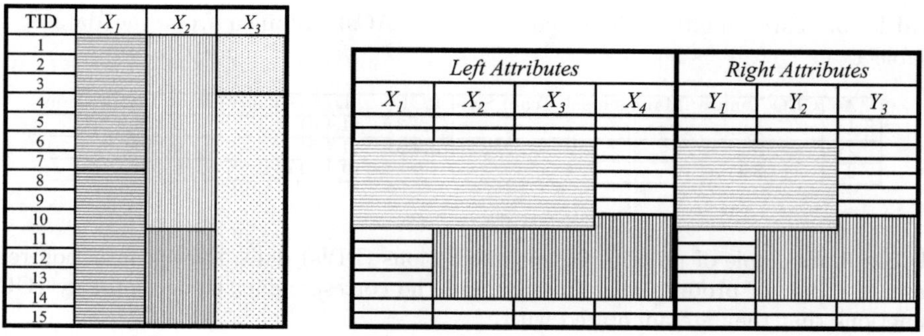

(a) (b)

Fig. 1. (a) Pairwise co-clustering. The combination of clusters with horizontal and vertical lines form pairwise co-clusters. Combination of clusters with horizontal lines, vertical lines and dots form star-structured co-clusters. (b) Multi-attribute co-clustering. The pair of clusters with horizontal lines form a multi-attribute co-cluster, and so do the pair of clusters with vertical lines.

Table 1. Dataset of company management attributes and performance indicators

	Management		Performance		
Company	CEO Tenure	Management Team Size	ROE	D/E	Correlation Information
1	1	2	3	4	0.0322
2	1	2	3.1	4.1	0.0322
3	1	2	3.2	4.2	0.0322
4	1	2	3.3	4.3	0.0322
5	5	6	3	4	0.0322
6	5	6	3.1	4.1	0.0322
7	5	6	3.2	4.2	0.0322
8	5	6	3.3	4.3	0.0322
9	7	4	9	10	0.332
10	7.1	4	9	10.1	0.332

Example 1. In finance, a key research challenge is to investigate the correlation between management and the performance of the companies [6, 14]. In a study by Murray and Goyal [14], management attributes were shown to affect the performance of the companies. Table 1 shows an example dataset of companies. The first two attributes are management attributes: CEO Tenure and Management Team Size. The next two attributes reflect the performance of the companies, as measured by their efficiency indicator (ROE ratio) and debt indicator (D/E ratio). The problem of understanding how management attributes affect the company performance is a multi-attribute co-clustering problem. This concept is illustrated in Table 2, where each row represents a MAC mined from Table 1. We can see that each MAC contains two clusters of attributes' values that are highly correlated. In this example, the last MAC has the highest correlation information as all companies with CEO Tenure(7.1,7) and Management Team Size(4) will also have ROE(9) and D/E(10,10.1).

Table 2. Three multi-attribute co-clusters (MACs) obtained from the dataset in Table 1

MAC	CEO Tenure	Management Team Size	ROE	D/E	Correlation Information
1	1	2	(3,3.3)	(4,4.3)	0.129
2	5	6	(3,3.3)	(4,4.3)	0.129
3	(7,7.1)	4	9	(10,10.1)	0.464

Table 3. Example of protein-protein interactions (PPIs) data. Each transaction represents a pair of proteins that interact and the corresponding Gene Ontology (GO) functions that they are annotated with.

	Gene Ontology (GO) functions							
Protein-Protein Interactions	F_1^1	F_2^1	...	F_n^1	F_1^2	F_2^2	...	F_n^2
P1-P7	1	1	...	0	1	1	...	0
P3-P6	1	0	...	0	0	0	...	1
P2-P5	0	1	...	1	1	1	...	1
P4-P9	1	0	...	0	0	1	...	0
P1-P6	1	1	...	0	0	0	...	1

Example 2. In biology, proteins can be profiled using biologically relevant features such as functional annotations. A relevant problem is to discover *pattern pairs* in the biological profiles of proteins that may be associated with the proteins' propensity to interact [11]. More specifically, given a set of known protein-protein interactions (PPIs), the objective is to find *correlated* pattern pairs from the biological profiles of interacting proteins. These pattern pairs can then be used to identify unknown interactions. Table 3 shows an example of PPI data. The first column of the table refers to protein-pairs that are known to be involved in an interaction. Each row of the table represents the Gene Ontology (GO) functions that the two interacting proteins are annotated with. $F_1^1, \ldots, F_n^1$ and $F_1^2, \ldots, F_n^2$ represent the two sets of GO functions associated with the first and second protein respectively. An entry of '1' in the column F_i^x denotes that the xth protein in the interaction is annotated with function i, and an entry of '0' indicates otherwise. By treating each GO function as a binary attribute, we can mine MACs from the data, where a MAC consists of two clusters of GO functions that are correlated in the presence of PPI.

Conventional pairwise co-clustering algorithms typically use information theory to partition the values of two attributes into correlated clusters. The idea that two clusters are co-clustered if they are correlated, that is, attributes' values in both clusters occur frequently together and not by chance, can be quantified by mutual information $I(X_i; Y_j) = \sum_{x_i, y_j} p(x_i, y_j) \log \frac{p(x_i, y_j)}{p(x_i)p(y_j)}$, with X_i and Y_j as two attributes of a dataset. Intuitively, we can see that the first part of this formula measures how frequent values x_i, y_j occur together and the second part measures if their occurrence together is by chance.

In this work, we adopt the information theory principles of existing (pairwise) co-clustering works [7, 8, 17] to obtain our desired MACs. Existing co-clustering techniques are limited to co-clustering pairwise attributes due to mutual information calculation being restricted to a pair of attributes. We overcome this limitation by generalizing the mutual information $I(X_i; Y_j)$ to mutual information $I(X_1, \ldots, X_n; Y_1, \ldots, Y_m)$ of two sets of attributes $X_1, \ldots, X_n$ and $Y_1, \ldots, Y_m$, which is non-trivial.

For a MAC, the correlation between its two clusters of attributes' values can then be calculated by the contribution of these attributes' values to $I(X_1, \ldots, X_n; Y_1, \ldots, Y_m)$. We denote this contribution as the *correlation information* of the MAC — high correlation information means the pair of clusters forming the MAC is highly correlated. To validate our claim, we show that these MACs have good clustering quality, as well as high predictive and classification power in our experiments.

We develop an algorithm MACminer which mines k MACs with high correlation information, so that users will not be overwhelmed by a large amount of results. To overcome the curse of dimensionality problem faced when clustering data with multiple attributes, we adopt the subspace clustering approach [15]. We also remove redundant co-clusters, keep maximal co-clusters and merge highly overlapping co-clusters, to keep the results succinct but informative. As correlation information has no anti-monotone property [9], traversing the search space of the dataset to mine co-clusters can be computationally expensive. Hence, we introduce several heuristic techniques to prune the search space.

One may think that correlation information can be calculated by considering both sets of attributes as two meta-attributes and the pairwise mutual information is used instead. However, this will create an unnatural constraint that a value from each attribute must be in a MAC. It is possible that values from a subset of attributes form a MAC with high correlation and values from the excluded attributes are noise to the MAC. An alternative way to solve multi-attribute co-clustering is to use existing high-dimensional data clustering [12,13] or pattern mining [9,10] techniques and adapt them to mine MACs. However, each of these techniques has its own set of criteria to define its results, and these criteria are not catered to mine our MACs, which are pairs of clusters that are highly correlated. In Section 5, we show that our MACs have much higher predictive and classification power than MACs mined by existing high-dimensional data clustering or pattern mining techniques [9,10,12].

In summary, we address the multi-attribute co-clustering problem with the following contributions:

- we generalize the mutual information $I(X_i; Y_j)$ between two attributes X_i and Y_j, to the mutual information $I(X_1, \ldots, X_n; Y_1, \ldots, Y_m)$ of two *sets* of attributes $X_1, \ldots, X_n$ and $Y_1, \ldots, Y_m$. (Section 3)
- we propose using *correlation information* to measure the correlation between the two clusters of attributes' values of a MAC. Correlation information is based on $I(X_1, \ldots, X_n; Y_1, \ldots, Y_m)$. (Section 3)
- we develop a novel algorithm MACminer, which efficiently mines k MACs with high correlation information. (Section 4)
- we conduct experiments to show the efficiency of MACminer and to show that MACs discovered by MACminer have good clustering quality, high predictive and classification power. (Section 5)

Table 4. Definitions of symbols

Symbol	Definition
X_i, Y_j	A left attribute and a right attribute
X_1^n	A set of left attributes $\{X_1, X_2, \ldots, X_n\}$
Y_1^m	A set of right attributes $\{Y_1, Y_2, \ldots, Y_m\}$
$x_i \in X_i$	A value from the left attribute X_i
$y_j \in Y_j$	A value from the right attribute Y_j
x_1^n	A tuple of values $(x_1, x_2, \ldots, x_n)$ from the attributes $\{X_1, X_2, \ldots, X_n\}$
$\tilde{x_i} \subseteq X_i$	A *set of values* from the attribute X_i
$\tilde{x}_1^n$	A tuple of *sets of values* $(\tilde{x_1}, \ldots, \tilde{x_n})$ from the attributes $\{X_1, X_2, \ldots, X_n\}$
$\mathcal{C} = (\tilde{x}_a^b, \tilde{y}_c^d)$	A MAC (multi-attribute co-cluster)
$\mathcal{A} = \{X_a^b, Y_c^d\}$	An attribute set. $1 \le a < b \le n$, $1 \le c < d \le m$
D	A dataset

2 Preliminaries

Let $\{X_1, X_2, \ldots, X_n\}$ and $\{Y_1, Y_2, \ldots, Y_m\}$ be two sets of attributes which we will refer to as the left and right attributes respectively. Let dataset D be a set of transactions. A transaction is a tuple $(x_1, \ldots, x_n, y_1, \ldots, y_m)$ of values from the attributes, with each $x_i \in X_i$ and each $y_j \in Y_j$, for $1 \le i \le n$ and $1 \le j \le m$. We denote $\tilde{x_i}$ as a *set of values* from the left attribute X_i, that is, $\tilde{x_i} \subseteq X_i$, and $\tilde{y_j}$ as a *set of values* from the right attribute Y_j, that is, $\tilde{y_j} \subseteq Y_j$. For brevity, we denote a set of attributes $\{X_1, \ldots, X_n\}$ as X_1^n, a tuple of values $(x_1, \ldots, x_n)$ as x_1^n, and a tuple of *sets of values* $(\tilde{x_1}, \ldots, \tilde{x_n})$ as $\tilde{x}_1^n$.

Definition 1 (MAC (multi-attribute co-cluster)). *Assume that we have a tuple of sets of values* $\tilde{x}_a^b = \{\tilde{x_i} \subseteq X_i : 1 \le a \le i \le b \le n\}$ *from a set of left attributes* X_a^b*, and a tuple of sets of values* $\tilde{y}_c^d = \{\tilde{y_i} \subseteq Y_i : 1 \le c \le i \le d \le m\}$ *from a set of right attributes* Y_c^d*. We denote a MAC as* $\mathcal{C} = (\tilde{x}_a^b, \tilde{y}_c^d)$.

Definition 2 (Attribute set of the MAC). *Given a MAC* $\mathcal{C} = (\tilde{x}_a^b, \tilde{y}_c^d)$*, we denote* $\mathcal{A} = \{X_a^b, Y_c^d\}$ *as the attribute set of* $\mathcal{C}$.

For example, the first row of Table 2 is a MAC $\mathcal{C} = ((1, 2), ((3, 3.3), (4, 4.3)))$. The attribute set of $\mathcal{C}$ is $\mathcal{A} = \{\{$CEO Tenure, Management Team Size$\}, \{$ROE, D/E$\}\}$.

Let us assume that the degree of correlation between $\tilde{x}_a^b$ and $\tilde{y}_c^d$ of a MAC is measured by some metric θ — the higher θ is, the higher is the degree of correlation. We are interested in mining the top-k MACs that have the highest θ. However, as mining the top-k MACs is a NP-complete problem [16], we propose to mine the approximate top-k MACs instead.

Problem Statement. Given a dataset D that can contain quantitative and categorical attributes, we propose to mine k MACs with high θ, sorted in descending order of their θ values.

3 Correlation Information

We propose to use correlation information as the metric θ to measure the correlation of MACs. An attribute X_i can be considered as a random variable with probability mass function $p(x_i) = Pr\{X_i = x_i\} = \frac{occ(x_i)}{|D|}$, where $occ(x_i)$ is the number of times value x_i occurs in dataset D. The conditional probability of value x_i occurring in D, given the occurrence of value y_j in D, is $p(x_i|y_j) = Pr\{X_i = x_i|Y_j = y_j\} = \frac{occ(x_i,y_j)}{occ(y_j)}$, where $occ(x_i, y_j)$ is the number of times values x_i, y_j occur together in the transactions of dataset D.

The *mutual information* $I(X_i;Y_j) = \sum_{x_i,y_j} p(x_i, y_j) \log \frac{p(x_i,y_j)}{p(x_i)p(y_j)}$ is the reduction in the uncertainty of the right attribute Y_j due to the knowledge of the left attribute X_i. In other words, it shows the amount of information about Y_j that can be described by X_i.

We are interested to know specifically *which* value x_i of the attribute X_i is correlated to *which* value y_j of the attribute Y_j. We quantify this information as the *correlation information* between values x_i and y_j. The correlation information between a pair of values x_i and y_j is defined as $ci(((x_i), (y_j))) = p(x_i, y_j) \log \frac{p(x_i,y_j)}{p(x_i)p(y_j)}$.

We are also interested in *group* phenomenon — how a group of values is correlated to another group of values. To find these kind of MACs, we must calculate the mutual information between their left attributes and right attributes, which is the mutual information of their attribute set.

For simplicity, let us consider a scenario where the attribute set contains only one left attribute and several right attributes, $\{\{X_i\}, Y_1^m\}$. To calculate the mutual information of this attribute set, we use the *chain rule for mutual information*, defined as

$$I(X_i; Y_1^m) = \sum_{j=1}^{m} I(X_i; Y_j|Y_1^{j-1})$$

The equation above shows that the chain rule for mutual information is a summation of *conditional mutual information*. The conditional mutual information of X_i and Y_j given Y_k is

$$I(X_i; Y_j|Y_k) = \sum_{x_i,y_j,y_k} p(x_i, y_k, y_j) \log \frac{p(x_i, y_j|y_k)}{p(x_i|y_k)p(y_j|y_k)}$$

We first make the assumption that the *generalized* conditional mutual information, which is the conditional mutual information of X_i and Y_j given X_1^{i-1}, Y_1^{j-1}, is

$$I(X_i; Y_j|X_1^{i-1}, Y_1^{j-1}) = \sum_{x_1^i,y_1^j} p(x_1^i, y_1^j) \log \frac{p(x_i, y_j|x_1^{i-1}, y_1^{j-1})}{p(x_i|x_1^{i-1}, y_1^{j-1})p(y_j|x_1^{i-1}, y_1^{j-1})}$$

We then use the generalized conditional mutual information to obtain the mutual information of an attribute set.

Definition 3 (Mutual information of an attribute set). *The mutual information of the attribute set* $\mathcal{A} = \{X_1^n, Y_1^m\}$ *is*

$$I(X_1^n; Y_1^m) = \sum_{i=1}^{n}\sum_{j=1}^{m} I(X_i; Y_j | X_1^{i-1}, Y_1^{j-1})$$
$$= \sum_{i=1}^{n}\sum_{j=1}^{m}\sum_{x_1^i, y_1^j} p(x_1^i, y_1^j) \log \frac{p(x_i, y_j | x_1^{i-1}, y_1^{j-1})}{p(x_i | x_1^{i-1}, y_1^{j-1}) p(y_j | x_1^{i-1}, y_1^{j-1})}$$

For the interested reader, the derivation for the equation above is given in [16]. The mutual information of an attribute set is the generalized chain rule for mutual information, thus it is not order dependent.

The correlation information of a MAC $\mathcal{C}$ is derived from the mutual information of the attribute set $\mathcal{A}$ of $\mathcal{C}$. Let the probability mass function of a *set of values* $\tilde{x}_i \subseteq X_i$ be $p(\tilde{x}_i) = \sum_{x_i \in \tilde{x}_i} Pr\{X_i = x_i\}$.

Definition 4 (Correlation information of a MAC). *The correlation information of* $\mathcal{C}$ *is*

$$ci(\mathcal{C}) = \sum_{i=a}^{b}\sum_{j=c}^{d} p(\tilde{x}_a^i, \tilde{y}_c^j) \log \frac{p(x_i, y_j | \tilde{x}_a^{i-1}, \tilde{y}_c^{j-1})}{p(x_i | \tilde{x}_a^{i-1}, \tilde{y}_c^{j-1}) p(y_j | \tilde{x}_a^{i-1}, \tilde{y}_c^{j-1})}$$

We can see that the correlation information of $\mathcal{C}$ is the total contributions of all subsets of $\tilde{x}_a^b$ and $\tilde{y}_c^d$ to the mutual information of attribute set $\mathcal{A}$. Since mutual information of attribute set $\mathcal{A}$ shows the amount of information about Y_c^d that can be described by X_a^b, correlation information shows the amount of information about $\tilde{y}_c^d$ that can be described by $\tilde{x}_a^b$. It is possible that contributions of subsets of $\tilde{x}_a^b$ and $\tilde{y}_c^d$ to the mutual information of $\mathcal{A}$ are negative, so correlation information is not biased towards large MACs.

We use Table 1 to show how correlation information of MACs is calculated. We focus on the first and third MACs $\mathcal{C} = ((1,2), ([3,3.3],[4,4.3]))$ and $\mathcal{C}' = (([7,7.1],4),(9,[10,10.1]))$. Their correlation information are $ci(\mathcal{C}) = p(1,[3,3.3]) \log \frac{p(1,[3,3.3])}{p(1)p([3,3.3])} + p(1,[3,3.3],[4,4.3]) \log \frac{p(1,[4,4.3]|[3,3.3])}{p(1|[3,3.3])p([4,4.3]|[3,3.3])} + p(1,2,[3,3.3]) \log \frac{p(2,[3,3.3]|1)}{p(2|1)p([3,3.3]|1)} + p(1,2,[3,3.3],[4,4.3]) \log \frac{p(2,[4,4.3]|1,[3,3.3])}{p(2|1,[3,3.3])p([4,4.3]|1,[3,3.3])} = 0.129$ and $ci(\mathcal{C}') = 0.464$. An intuitive reason why $ci(\mathcal{C}')$ is high is because the occurrence of $([7,7.1],4)$ in the dataset always results in the occurrence of $(9,[10,10.1])$. $ci(\mathcal{C})$ is low because $([3,3.3],[4,4.3])$ occurs not only together with $(1,2)$, but also together with $(5,6)$. Thus, $\mathcal{C}$ is less informative.

4 MACminer

We propose a novel algorithm, MACminer to mine k MACs with high correlation information, sorted in descending order of their correlation information. MACminer consists of three main parts:

1. Traversing the search space of D to obtain attribute sets
2. Obtaining MACs from each attribute set
3. Post-processing of the results by merging highly overlapping MACs

4.1 Traversing the Search Space

Given a dataset D with left attributes X_1^n and right attributes Y_1^m, MACminer obtains attribute sets $\mathcal{A}$ from the set enumeration tree of X_1^n and Y_1^m in depth-first order. Enumerating all $\mathcal{A}$ has a complexity of $O(2^{n+m})$, so pruning the search space is essential. Given that correlation information has no anti-monotone property, we have to implement heuristic pruning techniques to prune the search space.

Let us describe the pruning techniques by assuming that MACminer is in the search space of D that has attribute set $\mathcal{A} = \{X_a^b, Y_c^d\}$, and we are extending the attribute set with $X_k \in X_1^n - X_a^b$. These pruning techniques are also applicable when we extend the attribute set with $Y_k \in Y_1^m - Y_c^d$.

Pruning Technique 1: If $I(X_a^b \cup X_k; Y_c^d) = I(X_a^b; Y_c^d)$, we do not extend $\mathcal{A}$ with X_k. This is because no new information is gained between the left and right attributes of the attribute set when we add new attribute X_k to it.

Pruning Technique 2: If $I(X_a^b \cup X_k; Y_c^d) \leq \sum_{X_i \in X_a^b \cup X_k, Y_j \in Y_c^d} I(X_i; Y_j)$, we do not extend $\mathcal{A}$ with X_k. If the mutual information of the attribute set is less than the sum of pairwise mutual information of the attributes in the attribute set, then synergy between its left and right attributes does not exist.

Pruning Technique 3: If $\exists Y_j \in Y_c^d$ where $I(X_k; Y_j) > I(X_a^b \cup X_k; Y_c^d) - I(X_a^b; Y_c^d)$, we do not extend $\mathcal{A}$ with X_k. If the mutual information of X_k and Y_j is more than what X_k can contribute to the mutual information of the attribute set $\mathcal{A}$, then the attribute set $\{\{X_k\}, \{Y_j\}\}$ is more informative than $\mathcal{A}$ extended with X_k.

Pruning Technique 4: Let $f(i, j)$ be the highest mutual information that can be achieved, when the attribute set has i left attributes and j right attributes. If $I(X_a^b \cup X_k; Y_c^d) < f(|X_a^b \cup X_k|, |Y_c^d|)$, we do not extend $\mathcal{A}$ with X_k. So at a level of the set enumeration tree of X_1^n and Y_1^m, only the attribute set that has the highest mutual information at that level can be extended. We denote parameters l and r as the size of the left and right attribute sets the pruning technique is to be used.

4.2 Obtaining MACs from an Attribute Set

After obtaining an attribute set $\mathcal{A}$, MACminer takes the values of X_a^b and Y_c^d in each transaction t of D to be a MAC $\mathcal{C}$.

$\mathcal{C}$ is then used to update *list*, a list of MACs sorted in descending order of their correlation information. When updating the list, we also conduct the following two checks:

Non-redundant MACs. Given a MAC $\mathcal{C}$, we check if *list* contains a MAC $\mathcal{C}'$ such that $\mathcal{C}'$ is a proper superset of $\mathcal{C}$ and correlation information of $\mathcal{C}$ is equal to the correlation information of $\mathcal{C}'$. If so, we replace $\mathcal{C}'$ by $\mathcal{C}$ in *list*. Both $\mathcal{C}$ and $\mathcal{C}'$ giving the same amount of correlation information means that there are some redundant values in $\mathcal{C}'$ which do not increase the correlation information of $\mathcal{C}'$.

Maximal MACs. We also check in *list* if there exists any MAC $\mathcal{C}'$ such that $\mathcal{C}'$ is a proper subset of $\mathcal{C}$ and correlation information of $\mathcal{C}$ is larger than the correlation information of $\mathcal{C}'$. If so, we replace $\mathcal{C}'$ with $\mathcal{C}$ in *list*, as we prefer maximal MACs that have higher correlation information than their subsets that have lower correlation information.

4.3 Merging MACs

After mining of MACs is completed, we can merge those that are highly overlapping. This is an effective way of handling noisy datasets— the noise may be the cause of the small differences in highly similar MACs. In addition, this step of merging MACs is particularly useful for MACs that contain categorical values. Since MACminer does not partition categorical values, a MAC that only contains categorical values becomes a pair of itemsets, with each itemset containing a value from each attribute of the attribute set of the MAC. For example, if the attributes are words and documents, then a MAC can contain only a word and a document.

MACminer merges two MACs if they satisfy the merging condition (Definition 5), which is adopted from [19]. MACminer iteratively merges two MACs that satisfy the merging condition. The merging stops when no more MACs satisfy the merging condition.

Definition 5 (Merging condition). *Given two MACs* $(\tilde{x}_a^b, \tilde{y}_c^d)$ *and* $(\tilde{x'}_a^b, \tilde{y'}_c^d)$, *if* $\sum_{i=a}^{b} |\tilde{x}_i \cap \tilde{x}'_i| + \sum_{j=c}^{d} |\tilde{y}_j \cap \tilde{y}'_j| \geq \delta(\sum_{i=a}^{b} |\tilde{x}_i \cup \tilde{x}'_i| + \sum_{j=c}^{d} |\tilde{y}_j \cup \tilde{y}'_j|)$, *then we merge them to become MAC* $(\tilde{x''}_a^b, \tilde{y''}_c^d)$, *where* $\tilde{x''}_a^b = (\tilde{x}_a \cup \tilde{x'}_a, \ldots, \tilde{x}_b \cup \tilde{x'}_b)$ *and* $\tilde{y''}_c^d = (\tilde{y}_c \cup \tilde{y'}_c, \ldots, \tilde{y}_d \cup \tilde{y'}_d)$. δ *is a parameter controlling the strictness of the merging.*

5 Experimentation Results

MACminer was coded in C++ and the algorithms [7, 9, 10, 12] that we used for comparison were kindly provided by the respective authors. Except for the QCoMine [10] and *Co-cluster* [7] algorithm, all of our experiments were conducted in a Win XP environment with a 3.6Ghz Pentinum 4 CPU and 2GB memory. As QCoMine and *Co-cluster* can only be compiled in Linux, we ran them in Linux environment with a 4-way Intel Xeon based server and 8GB memory.

We used three real world in our experiments. The protein-protein interaction (PPI) dataset was downloaded from Gene Ontology [2] and BioGRID [5] databases, the 20 Newsgroup subsets dataset was downloaded from UCD Machine Learning Group [1], and the Insurance dataset was downloaded from UCI Machine Learning Repository [3].

PPI dataset. Proteins were profiled using their functional annotation from Gene Ontology (GO) [2]. GO annotations were organized as a Direct Acyclic Graph, with more specific terms at lower levels and more general terms at higher levels of the graph. To avoid overly general terms, we ignored terms in the top two levels of the graph. Existing PPIs were obtained from the BioGRID [2] database. The data consists of a total of 38,555 physical interactions between proteins. We transformed both sets of data into a set of transactions, which contains 38,555 transactions with 3851 left and 3851 right attributes. Each transaction was labeled with two proteins that have interactions, and the attributes of the transaction are the two sets of GO functions that the two proteins are annotated with respectively. A toy example is shown in Table 3.

20 Newsgroup subset datasets. Each transaction of the datasets indicates a word, a document and the occurrence number of the word in the document. The left attribute is *words* and the right attribute is *documents*. Stop-word removal and stemming were applied on these datasets by the original authors. ob-2-1 is a dataset which contains two topics and has 80,009 transactions, while ob-8-1 is a dataset which contains eight topics and has 316,867 transactions. Both datasets have balanced clusters containing 500 documents each, and they were clustered according to their topics. In these datasets, the words and documents may belong to more than one topic. For example, the word "earth" belongs to topics *Space* and *Graphics*.

Insurance dataset. This dataset contains 5,822 customer transactions with 85 left attributes relating to the customer's profile and 1 right (class) attribute indicating if the customer buys a caravan policy. This dataset contains 4 quantitative attributes, and they are discretized using a parameter-free, mutual information based technique. The interested reader may refer to [16] for the details of this discretization technique.

5.1 Performance of MACminer

As there is no existing work on mining MACs with high correlation information, we show the efficiency of MACminer by assessing the efficiency of its pruning techniques. We used the Insurance and PPI datasets, and we set $k = 100$ in this experiment. The newsgroup dataset is not used since there is no need to prune its search space of two attributes. We calculated how much times faster MACminer is with each pruning technique (denoted as speedup) and Figure 2 presents the results. Techniques 1, 2 are more effective on the PPI dataset than the Insurance dataset, while pruning techniques 3, 4 are very effective on both the PPI and Insurance datasets. In fact, for the Insurance dataset, MACminer could not complete mining after 24 hours without pruning technique 3. In general, the pruning techniques can improve the mining time of MACminer on different types of datasets.

5.2 Multi-Attribute Co-clustering: Predicting PPIs

We mined MACs from the PPI dataset, where each MAC is a pattern pairs of GO functions that are associated with proteins that interact. We then evaluated these MACs by using them to predict PPIs in the PPI dataset and the accuracy of the predictions was validated by a five-fold cross-validation on the PPI dataset. The proteins in the PPI dataset were randomly divided into five equal-sized groups.

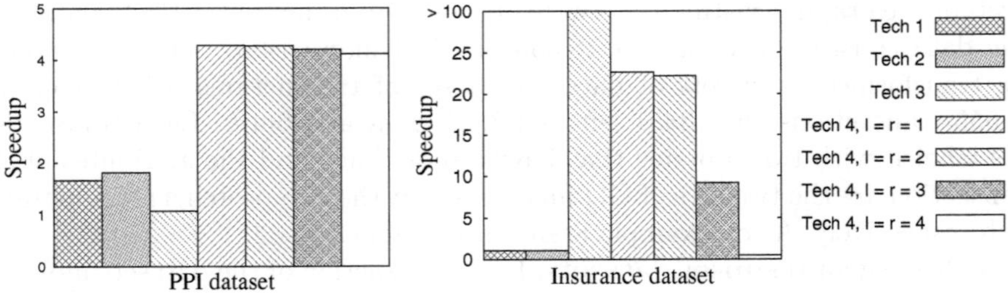

Fig. 2. The speedup based on different pruning techniques (denoted as Tech)

Table 5. AUCs of the PPI predictions by different types of patterns

Pattern Type			AUC
	minsup	avg num	
Closed itemsets	5000	24	60.31
	10000	127	59.76
	15000	434	54
	20000	2182	56.21
	25000	24631	56.5
	k		
MACs	100		74.46
	200		72.61
	300		72.24
	400		72
	500		72.3

Table 6. Performance results of the different classifiers

Classifier	tp	fp	Precision	Recall	F-measure
SVM	0	0	0	0	0
C4.5	0	238	0	0	0
Linear	17	221	0.227	0.071	0.109
k-nearest neighbor	29	209	0.155	0.122	0.136
Naïve bayes	97	141	0.135	0.408	0.203
MACs, k=100	81	679	0.119	0.34	0.177

In each cross-validation run, the proteins in one fold were isolated for evaluation, while those from the remaining four folds were used for mining pattern pairs. The pattern pairs were then used to predict PPIs from the isolated fold. Given a pattern pair, if a protein from the isolated fold is annotated with the GO functions of one of the pattern and another protein from the isolated fold is annotated with the GO functions of the other pattern, then we predict these two proteins interact, and we assign a prediction score to this prediction, which is based on the correlation information of the pattern pair.

The predictions made for all five folds were evaluated against the known protein interactions, and we computed the Receiver Operating Characteristic (ROC) curve (a graphical plot of the sensitivity versus (1-specificity)) of the predictions results. We then calculated the area under the ROC curve (AUC). An AUC of 1 means the classifier is perfect, while an AUC of 0.5 means the classifier is similar to a random guesser.

Since there are no existing algorithms that perform the same task, we used the frequent closed itemset mining approach [9] with post-processing as our baseline. Frequent closed itemsets were mined from the PPI dataset, and in the post-processing, frequent closed itemsets that do not correspond to pattern pairs were pruned. We then used the confidence of these frequent closed itemsets to calculate their prediction score.

The AUCs of MACs and frequent closed itemsets are presented in Table 5. For the frequent closed itemsets, we set minimum support ms from 5000 to 25000, and avg num in Table 5 indicates the average number of pattern pairs mined from frequent closed itemsets across the 5 folds. For MACs, we varied the number of predictors k from 100 to 500, and we disabled pruning techniques 1-3 and set $l = k = 1$. From Table 5, we can see that our approach achieves an average AUC score from 72% to 74.46%, which indicates that the pattern pairs discovered by MACs are meaningful and relevant. On the other hand, the average AUC score achieved by the baseline method is significantly lower. A possible explanation of why frequent closed itemsets are poor predictors is that some functional annotations are more general than the others and may be annotated to many proteins. Hence high co-occurrences of such functions in interacting proteins may not be associated with biologically meaningful annotations patterns. Our method, on the other hand, takes into account the correlation between two sets of pattern, and thus is able to detect dependency relationships in pattern pairs.

5.3 Pairwise Co-clustering: Co-clustering Words and Documents

We performed pairwise co-clustering of words and documents in the 20 Newsgroup subsets using MACs and *Co-cluster* [7], which was developed specifically for this task. We assessed how accurate the co-clusters mined by MACs and *Co-cluster* are in clustering documents of the same topic together.

Since the 20 Newsgroup subsets used in [7] were unavailable, and to have an unbias comparison, we used the 20 Newsgroup subsets that was prepared by a neutral party [1]. For the parameter settings of *Co-cluster* algorithm, we adjusted its two main parameters, the desired number of document clusters and the desired number of word clusters, and kept the rest as default settings. We set the number of document clusters to the number of topics for the dataset and we varied the number of word clusters from 8 to 128 (128 word clusters were shown to get good co-clustering results in [7]). Hence, the number of co-clusters mined varies from 16 to 1024. For MACminer, we merged co-clusters during mining to prevent them from being too specific, with $\delta = 0.3$. We mined 100 co-clusters without any pruning technique, since there are only two attributes, words and documents. Co-clusters mined by *Co-cluster* and MACminer that contain only one document were removed.

In the document cluster of each co-cluster, we checked the topic of each document and the *dominant topic* is the topic that the majority number of documents in the cluster belong to. So in a document cluster, documents belonging to the dominant topic c are deemed to be assigned correctly. We used the evaluation measure in [7], micro-averaged-precision $P(d)$, to measure the accuracy of clustering documents of the same topic together. $P(d) = \sum_c \alpha(c,d) / \sum_c (\alpha(c,d) + \beta(c,d))$, where $\alpha(c,d)$ is the number of documents correctly assigned to topic c, $\beta(c,d)$ is the number of documents incorrectly assigned to c and d is denoted as the set of documents in the dataset.

The micro-averaged precision of the co-clusters mined by MACminer and *Co-cluster* across different parameter settings was calculated and the results are

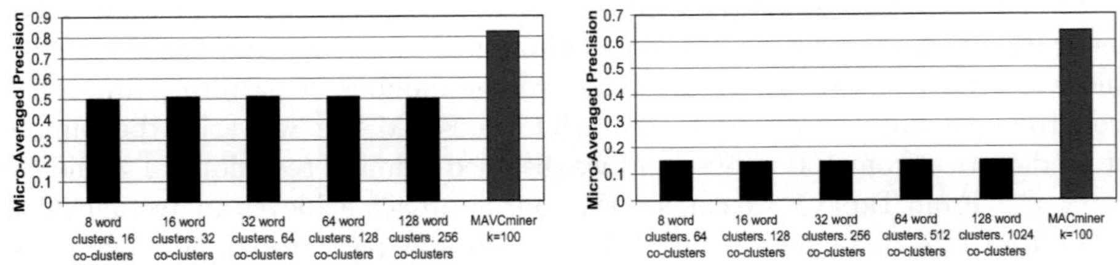

(a) Dataset: 20 Newsgroup ob-2-1 (b) Dataset: 20 Newsgroup ob-8-1

Fig. 3. Micro-averaged precision of the co-clusters mined by *Co-cluster* and MACminer

presented in Figure 3. The first five columns show the micro-averaged precision of the co-clusters mined by *Co-cluster* and the last column show the micro-averaged precision of the co-clusters mined by MACminer. In ob-2-1 dataset that contains two topics, Figure 3(a) shows that the micro-averaged precision of co-clusters mined by MACminer is much better than those by *Co-cluster*.

On the more complex ob-8-1 dataset that has 8 topics, the micro-averaged precision of the co-clusters mined by MACminer is again much better than those by *Co-cluster*, as shown in Figure 3(b). This demonstrates that MACminer is able to cluster documents of similar topics effectively, regardless of the number of topics. In these datasets, the words and documents can belong to more than one topic, so it is natural that the co-clusters of words and documents should overlap. Existing co-clustering frameworks do not allow overlapping co-clusters, thus they perform poorly in these datasets.

5.4 Multi-Attribute to One Attribute Co-clustering: Using MACs to Build a Simple Classifier

In this experiment, we used MACs as rules to build a simple classifier. The training dataset is the Insurance dataset and the testing dataset has the same attributes as the training dataset but it contains 4000 other customers. The objective is to predict the number of customers that will buy a caravan policy from the testing dataset. This dataset is challenging because only 238 of the customers bought the policy. 100 MACs were mined from the training dataset without pruning technique 4 and they were used as rules for our classifier. As these MACs were used for predicting customers buying, in each MAC, its left cluster contains values from the left attributes (customer's profile attributes) and its right cluster contains the value 'buy' of the class attribute. If a customer from the testing dataset has attributes' values that are in any of the 100 MACs, then our simple classifier predicts that this customer will buy.

For comparison, we tried using quantitative correlated patterns [10] and subspace clusters [12] for comparison. We mined and post-processed them, with the requirement that each of them contains both left and right attributes' values, so that they can be used as rules for a classifier. However, even at their lowest parameter settings, QCoMine [10] mined 0 quantitative correlated patterns from the training dataset and LCM-nCluster [12] mined 121 subspace clusters from the training dataset but none contains the value 'buy' of the class attribute. Since we cannot use related works to build classifiers, we compared our simple classifier with the major classifiers C4.5, SVM, Linear, k-nearest neighbors and Naïve Bayes [18].

To measure the performance of the classifier, we used the standard precision, recall and F-measure, where precision $P = \frac{tp}{tp+fp}$, recall $R = \frac{tp}{tp+fn}$ and F-measure $F = \frac{2(P.R)}{P+R}$. In this experiment context, tp is the number of correct predictions in predicting that the customer will buy, fp is the number of wrong predictions, and fn is the number of customers who buy but they are not predicted by the classifier. Table 6 summarizes the results of the different classifiers. Our simple classifier using 100 MACs and Naïve Bayes are the top performers, based on their F-measure. Although sophisticated techniques were not used to build our classifier, it is still able to perform better than the other major classifiers. A possible explanation of why SVM and C4.5 performed poorly is because the data is highly skewed as only a small percentage of customers bought the policy. The performance of SVM and C4.5 performed well in predicting customers that do not buy, but this classification result is not meaningful.

6 Related Work

For co-clustering that involves only two attributes such as words and documents [7], each word/document value is assigned to a word/document cluster, and a pair of word cluster and document cluster form a co-cluster. Since a word/document value can only be assigned to a word/document cluster, no overlapping of co-clusters is allowed. Our proposed framework allows a word/document value to be in multiple word/document clusters, thereby allowing overlapping of co-clusters. Gao *et al.* [8] extend the traditional pairwise co-cluster into star structured co-cluster. In a star structured co-cluster, a cluster of values of the center attribute is in the center, and the clusters of values of other attributes are linked to it. To obtain this structure, pairwise co-clusterings are performed between values of the center attribute and values of each non-center attribute. It is therefore a series of pairwise co-clusters, with the constraint that the clustering result of the values of the center attribute are the same in each pairwise co-cluster. This is in contrast to MACs, where each cluster of a MAC contains values of multiple attributes.

Biclustering algorithms [13] have been developed for co-clustering genes and experimental conditions. In a bicluster, one cluster contains genes and the other cluster contains the expression values of different experimental conditions. This in essence, is clustering a group of attributes' values, where the experimental conditions are the attributes. The clustering structure of subspace clustering [12] and correlation clustering [4] are also similar to biclustering. Thus, their clustering structures are different from MACs, which are a pair of clusters of attributes' values. Biclustering [13] and subspace clustering [12] aim to cluster attributes' values that satisfy certain homogeneity criteria, such as constancy, similarity or coherency (scaling or shifting), whereas correlation clustering aims to cluster attributes' values that exhibit linear dependency [4]. These are different from our clustering aim; we aim to find a cluster of attributes' values that is correlated to another cluster of attributes' values.

Ke *et al.* [10] mine quantitative correlated patterns based on mutual information. Quantitative correlated pattern is a set of attributes' values, with the requirement that the pairwise mutual information between the attributes exceed a threshold and the all-confidence of the pattern exceed another threshold. Hence, we are mining different things and our clustering aims are different.

7 Conclusions

In this paper, we have introduced the problem of *multi-attribute co-clustering* for data mining applications that involve co-clustering two highly correlated clusters of attributes' values (known as MAC). We proposed using correlation information to measure the correlation in a MAC. Correlation information is based on the mutual information of two sets of attributes, which we derived by generalizing the mutual information for two attributes. We developed an algorithm MACminer, which mines MACs that have high correlation information from datasets. MACminer adopts the subspace clustering approach to overcome the curse of dimensionality problem when clustering multiple attributes dataset. MACminer also used heuristic techniques to aggressively prune the search space of the dataset during mining. Our experimental results showed that MACs produced better clustering, classification and prediction results than other algorithms.

Acknowledgment

We would like to thank Zeyar Aung, Clifton Phua, Ardian Kristanto Poernomo and Ghim-Eng Yap for their constructive suggestions. We would also like to thank the authors for providing us the executables or source codes of their algorithms.

References

1. UCD Machine Learning Group (2008), `http://mlg.ucd.ie/content/view/22/`
2. Ashburner, M., et al.: Gene ontology: tool for the unification of biology. The gene ontology consortium. Nature Genetics 25(1), 25–29 (2000)
3. Asuncion, A., Newman, D.: UCI Machine Learning Repository (2007), `http://www.ics.uci.edu/~mlearn/MLRepository.html`
4. Böhm, C., Kailing, K., Kröger, P., Zimek, A.: Computing clusters of correlation connected objects. In: SIGMOD, pp. 455–466 (2004)
5. Breitkreutz, B.-J., Stark, C., Tyers, M.: The grid: The general repository for interaction datasets. Genome Biology 3, R23 (2002)

6. Denis, D.J., Denis, D.K.: Performance changes following top management dismissals. Journal of Finance 50(4), 1029–1057 (1995)
7. Dhillon, I.S., Mallela, S., Modha, D.S.: Information-theoretic co-clustering. In: KDD, pp. 89–98 (2003)
8. Gao, B., Liu, T.-Y., Ma, W.-Y.: Star-structured high-order heterogeneous data co-clustering based on consistent information theory. In: ICDM, pp. 880–884 (2006)
9. Grahne, G., Zhu, J.: Efficiently using prefix-trees in mining frequent itemsets. In: ICDM Workshop on FIMI (2003)
10. Ke, Y., Cheng, J., Ng, W.: Mining quantitative correlated patterns using an information-theoretic approach. In: KDD, pp. 227–236 (2006)
11. Li, X., Tan, S.H., Ng, S.-K.: Improving domain based protein interaction prediction using biologically significant negative dataset. International Journal of Data Mining and Bioinformatics 1(2), 138–149 (2006)
12. Liu, G., Li, J., Sim, K., Wong, L.: Distance based subspace clustering with flexible dimension partitioning. In: ICDE, pp. 1250–1254 (2007)
13. Madeira, S.C., Oliveira, A.L.: Biclustering algorithms for biological data analysis: A survey. IEEE/ACM Transactions on Computational Biology and Bioinformatics 1(1), 24–45 (2004)
14. Murray, F., Goyal, V.K.: Corporate leverage: how much do managers really matter? In: Social Science Research Network (2007)
15. Parsons, L., Haque, E., Liu, H.: Subspace clustering for high dimensional data: a review. ACM SIGKDD Explorations Newsletter 6(1), 90–105 (2004)
16. Sim, K., Gopalkrishnan, V., Chua, H.N., Ng, S.-K.: MACs: Multi-attribute co-clusters with high correlation information (2009), `http://www.ntu.edu.sg/home/asvivek/pubs/TR0902.pdf`
17. Wang, P., Cai, R., Yang, S.-Q.: Improving classification of video shots using information-theoretic co-clustering. In: ISCAS (2), pp. 964–967 (2005)
18. Witten, I.H., Frank, E.: Data Mining: Practical machine learning tools and techniques. Morgan Kaufmann, San Francisco (2005)
19. Zhao, L., Zaki, M.J.: TRICLUSTER: an effective algorithm for mining coherent clusters in 3d microarray data. In: SIGMOD 2005, pp. 694–705 (2005)

Bi-directional Joint Inference for Entity Resolution and Segmentation Using Imperatively-Defined Factor Graphs

Sameer Singh*, Karl Schultz*, and Andrew McCallum

Department of Computer Science
University of Massachusetts, Amherst MA 01002 USA
{sameer,kschultz,mccallum}@cs.umass.edu

Abstract. There has been growing interest in using joint inference across multiple subtasks as a mechanism for avoiding the cascading accumulation of errors in traditional pipelines. Several recent papers demonstrate joint inference between the segmentation of entity mentions and their de-duplication, however, they have various weaknesses: inference information flows only in one direction, the number of uncertain hypotheses is severely limited, or the subtasks are only loosely coupled. This paper presents a highly-coupled, bi-directional approach to joint inference based on efficient Markov chain Monte Carlo sampling in a relational conditional random field. The model is specified with our new probabilistic programming language that leverages imperative constructs to define factor graph structure and operation. Experimental results show that our approach provides a dramatic reduction in error while also running faster than the previous state-of-the-art system.

1 Introduction

Advances in machine learning have enabled the research community to build fairly accurate models for individual components of an information extraction and integration system, such as field segmentation and classification, relation extraction, entity resolution, canonicalization, and schema alignment. However, there has been significantly less success combining these components together into a high-accuracy end-to-end system that successfully builds large, useful knowledge bases from unstructured text.

The simplest possible combination is to run each component independently such that they each produce their output using only provided input data. We call this an *isolated* approach. For example, if asked to build a de-duplicated database of bibliographic records given citation strings, one could build field extraction and citation coreference models that operate independently of each other. However, the accuracy of coreference would almost certainly be improved if it could access the output of segmentation, and thus be able to separately compare individual bibliographic fields such as authors, titles, and venues.

* First two authors contributed equally.

W. Buntine et al. (Eds.): ECML PKDD 2009, Part II, LNAI 5782, pp. 414–429, 2009.

For this reason, the most common approach to component combination is a *pipeline*, in which components are run in some order, and later components have access to the output of already-completed earlier components. Here segmentation could be run first, and then coreference would have fruitful access to field boundaries. Nevertheless, pipeline systems also typically fail to produce high-accuracy final output. This is because errors cascade and compound in a pipeline. When an earlier component makes a mistake, later stages consuming its output are also very likely to produce errorful output. Every stage is an opportunity for compounding error—for example, six components each having 90% accuracy may result in about 50% accuracy for the final stage when pipelined.

In order to avoid this brittle accumulation of errors there has been increasing interest in probabilistic models that perform *joint* inference across multiple components of an information processing pipeline, *e.g.* [1]. Here the system does not commit to a single answer in some stage before considering other stages; rather, multiple hypotheses and uncertainty information are exchanged throughout, such that some components can correct the misconceptions of others. The need for joint inference appears not only in extraction and integration, but also in natural language processing, computer vision, robotics, and elsewhere.

Joint inference aims to predict many variables at once, and thus usually leads to complex graphical model structure with large tree-width, making exact inference intractable. Several approximate methods for joint inference have been explored.

One, described by Finkel *et al.* [2], runs a pipeline, but *samples* the output for each component rather than selecting the single most likely output; then the pipeline is run repeatedly so that different combinations of output throughout the pipeline are evaluated. This feed-forward approach to inference is a classic method in Bayesian networks, but has the drawback that it only allows information to flow in one direction. For example, correct coreference of a messy citation with a clean citation provides the opportunity for an alignment between these two citations to help the model correctly segment the messy one. However, feed-forward inference does not support this backward flow of information.

In another approach, pursued by Wellner *et al.* [3], each component produces a weighted N-best list of hypotheses for consumption by other components, and components are re-visited in both directions—both forward and backward through the pipeline. This approach allows information to flow in both directions, however an N-best list is a restrictive approximation for the full distribution of large-output components.

Yet another approach, proposed by Poon and Domingos [4], creates a single Markov random field with factor template structure specified via first-order logic—called a Markov logic network, MLN—and performs randomized approximate inference by MC-SAT [5], a variant of WalkSAT [6]. MC-SAT repeatedly reconsiders variables and factors in the model in an order independent of the pipeline. However, as described below, limitations of first-order logic make it difficult to specify a coreference factor that uses the uncertain output of segmentation. For this reason, in Poon and Domingos [4], citation coreference compatibility is measured using features of the *un*-segmented citation (see Sect. 3.2), and inference is not strongly bi-directional.

This paper presents a strong bi-directional approach to inference for joint segmentation and coreference. Like Poon and Domingos [4] we use a conditionally-trained factor graph, with randomized approximate inference that roams freely in both directions. However, rather than employing first-order logic, we leverage our new work in probabilistic programming that uses imperative procedures to define the factor template structure. We term this approach *imperatively-defined factor graphs* (IDFs); they are factor graphs whose template structure, as well as aspects of parameterization and inference, are specified with the power of a Turing-complete language. This added flexibility enables us to include more sophisticated factors between segmentation and coreference that allow information to flow bi-directionally. In spite of being more expressive, IDF's flexibility also allows our model to be more efficient.

We evaluate our approach on segmentation and coreference of the citation strings from the Cora data set. In comparison with Poon and Domingos [4] we reduce coreference error and segmentation error by $\sim 20-25\%$ while also running $3-15$ times faster, providing a new state-of-the-art result. We also analyze the nature of bi-directional inference with separate diagnostic experiments.

2 Background

2.1 Factor Graphs

A factor graph [7] is a bipartite graph over factors and variables. Let factor graph G define a probability distribution over a set of output variables $\mathbf{y}$ conditioned on input variables $\mathbf{x}$. A factor Ψ_i computes a scalar value over the subset of variables $\mathbf{x}_i$ and $\mathbf{y}_i$ that are neighbors of Ψ_i in the graph. Often this real-valued function is defined as the exponential of an inner product over sufficient statistics $\{f_{ik}(\mathbf{x}_i, \mathbf{y}_i)\}$ and parameters $\{\theta_{ik}\}$, where $k \in [1, K_i]$ and K_i is the number of parameters for factor Ψ_i. Let $Z(\mathbf{x})$ be the data-dependent partition function used for normalization. The probability distribution can be written as:

$$p(\mathbf{y}|\mathbf{x}) = \frac{1}{Z(\mathbf{x})} \prod_{\Psi_i \in G} \exp \left[\sum_{k=1}^{K_i} \theta_{ik} f_{ik}(\mathbf{x}_i, \mathbf{y}_i) \right].$$

In practice factor graphs often use the same parameters for several factors; this is termed *parameter tying*. A *factor template* T_j consists of parameters $\{\theta_{jk}\}$, sufficient statistic functions $\{f_{jk}\}$, and a description of an arbitrary relationship between variables, yielding a set of satisfying tuples $\{(\mathbf{x}_j, \mathbf{y}_j)\}$. For each of these variable tuples $(\mathbf{x}_i, \mathbf{y}_i)$ that fulfills the relationship, the factor template instantiates a factor that shares $\{\theta_{jk}\}$ and $\{f_{jk}\}$ with all other instantiations of T_j. Let $\mathcal{T}$ be the set of factor templates. In this case the probability distribution becomes:

$$p(\mathbf{y}|\mathbf{x}) = \frac{1}{Z(\mathbf{x})} \prod_{T_j \in \mathcal{T}} \prod_{(\mathbf{x}_i, \mathbf{y}_i) \in T_j} \exp\left[\sum_{k=1}^{K_j} \theta_{jk} f_{jk}(\mathbf{x}_i, \mathbf{y}_i)\right].$$

The process of instantiating individual factors from their templates is termed *unrolling*. For a factor Ψ_i that is an instantiation of factor template T_j, the inner product of $\{f_{jk}(\mathbf{x}_i, \mathbf{y}_i)\}$ and parameters $\{\theta_{jk}\}$ is termed the *score* of the factor.

In a pipeline approach to solving multiple tasks the influence between stages is uni-directional down the pipeline, since inference is performed separately for each task in a sequential order. To enable the bi-directional flow of information we add factors connecting variables of different tasks and perform inference simultaneously for the entire model. A factor template defined over variables of different tasks is termed a *joint factor template*, and an instantiation of a joint factor template is called a *joint factor*.

Introducing joint factors usually increases the complexity of the graph because it tends to create many more cycles and a larger tree width. These complex graphs often cause inference to become intractable, which we address by using imperatively-defined factor graphs, as described next.

2.2 Imperatively-Defined Factor Graphs

Imperatively-defined factor graphs (IDFs) are an approach to probabilistic programming that preserves the *declarative* semantics of factor graphs, while leveraging *imperative* constructs (pieces of procedural programming) to greatly aid both efficiency and natural intuition in specifying model structure, inference, and learning. Rather than using declarative languages, such as SQL or first-order logic, model designers have access to a Turing complete language when writing their model specification. A model written as an IDF *is* a factor graph, with all the traditional semantics of factors, variables, possible worlds, scores, and partition functions; IDFs simply provide an extremely flexible language for their succinct specification that also enables efficient inference and learning. A side benefit is that IDFs provide a mechanism whereby model designers can inject both declarative and procedural domain knowledge.

We have developed IDFs in the context of Markov chain Monte Carlo (MCMC) inference, which is a common approach to achieve efficiency in complex factor graphs [8,9,10] where variable-factor connectivity structure changes during inference. In such situations, fully unrolling the graph (creating the factors that would be necessary to score all possible worlds, as required for belief propagation) would often result in an exponential or super-exponential number of factors. However, in MCMC, we only represent a single possible world at a time. MCMC performs inference by stochastically proposing some change to the current possible world, and then accepting that change with a probability that depends on the ratio of post- and pre-proposal model scores. Calculating these acceptance probabilities is quite efficient because not only do the partition functions, $Z(\mathbf{x})$, cancel, but the contributions of all factors not touching changed variables also cancel; in fact, in our implementation they are not even created. This allows us to avoid the need to unroll and score the entire graph to evaluate a change, resulting in quite efficient inference.

We summarize four key imperative constructs in IDFs, and argue that they provide a natural interface to central operations in factor graph construction and inference. For more details see [11].

1. *Imperative structure definition* determines the connectivity between factors and their variable arguments. The key operation in efficient MCMC is finding all variable arguments of a factor template given a relevant changed variable. IDFs make this a primitive operation, with the opportunity to define its behavior in a Turing-complete language—dynamically finding all neighboring variables of an instantiated factor given one of its neighbors. For example, this approach allows arbitrary graph-search algorithms to define the arguments of a factor.
2. *Imperative constraint preservation* keeps all explored possible worlds in the feasible region by embedding the necessary control in the procedurally-defined MCMC proposal function. For example, we can avoid the large expense of having factors that aim to enforce transitivity in coreference by instead: (a) initializing to a possible world that obeys transitivity, and (b) implementing a proposal function that is guaranteed to preserve the transitivity constraint.
3. *Imperative variable coordination* also preserves constraints or encourages more fruitful proposals by including in the variable-value setting method of one variable a procedural "hook" that automatically sets other related variable(s) to a compatible value.
4. *Imperative variable-sufficient mapping* allows the data to be represented in natural, convenient variables, and then later to be functionally mapped into the sufficient statistics required for our desired parameterization.

We have implemented IDFs in the FACTORIE toolkit[1] [11]. Typically, IDF programming consists of four distinct stages: (1) defining the data representation, (2) defining the factors for scoring, (3) optionally providing domain knowledge to aid inference (including the flexibility to write the entirety of a proposal function), (4) reading in the data, learning parameters, testing, and evaluating.

For parameter estimation in this paper we use Sample-Rank [12]. This method embeds opportunities for approximate gradient ascent into each MCMC proposal step by performing perceptron-like updates whenever possible worlds, pre- and post-proposal, are ranked differently by their model scores versus their distance to the labeled true world. Sample-Rank's proof of convergence [12] is similar to the proof for perceptron. In this paper, the final values of each parameter are obtained by averaging over the learning period. To evaluate a trained model we search for the MAP configuration by running MCMC with a low temperature applied to the ratio of model scores in the proposal acceptance probability.

[1] Available at http://factorie.cs.umass.edu

3 Bi-directional Joint Inference for Segmentation and Entity Resolution

In some information extraction tasks mentions of entities must be discovered by segmenting them from background text. In other cases the mention strings are provided, but they have internal structure requiring segmentation. Here we address the latter case, in which we jointly segment the contents of many mentions, while simultaneously performing coreference on the mentions.

Consider the task of citation matching in which we are given a large collection of citation strings from the "References" section of research papers. They have different citation styles, different abbreviations, and typographical errors. Many of the citations refer to the same underlying papers. Our job is to find the citations referring to the same paper (*coreference* or *entity resolution*) and also identify the *author*, *title*, and *venue* fields of each citation (*segmentation*).

The tasks of segmentation and entity resolution are often solved in isolation, without access to each other's predictions [13,14], however, using the results of the other subtask often helps reduce errors. For example, coreference compatibility between two citations can be assessed more accurately if we can compare segmented *title* fields and *venue* fields separately, with different distance measures for each. Also, segmentation accuracy can be improved by accounting for field similarity among multiple coreferent citations. These interdependencies between the two tasks have led others to explore *joint* models of segmentation and coreference of citations with generative models [15], with conditionally-trained models performing bi-directional N-best message-passing between tasks [3], and with conditional models whose bi-directional factors mainly leverage coreference for performing segmentation [4].

Next we present a highly-coupled, bi-directional approach to joint inference for citation segmentation and coreference. We use imperatively-defined factor graphs (IDFs) to specify a single undirected graphical model that performs both tasks. It includes multiple factors that simultaneously examine variables of segmentation and coreference. In the following sections we describe the variables and factors of this model, as well as the MCMC proposal function used.

3.1 Variables

Segmentation Variables. Observed and predicted data for segmentation are represented by three types of variables: `Token`, `Label`, and `Field`. Each observed citation string consists of a sequence of words, each represented by a `Token` whose value is the word string itself. Each `Token` is associated with a corresponding unobserved `Label` variable, whose value is any of the field types, or "None." In addition, consecutive `Tokens` whose `Labels` have the same value are also represented as a `Field` variable, whose value is the string formed by the concatenation of its words. `Fields` are set in coordination with `Labels` through *imperative variable coordination*, and facilitate joint factors between coreference and segmentation. These variables are summarized in Table 1.

Table 1. Variable Types for Segmentation and Coreference: Variable types that are used in the joint model of segmentation and entity resolution

	Segmentation Variable Types
`Token:`	Observed variable that represents a word in the mention string
`Label:`	Variable that can take any of the field types as a value (or "None")
`Field:`	Indices of consecutive `Tokens` that belong to a particular field
	Coreference Variable Types
`Mention:`	Variable that takes a single `Entity` as its value
`Entity:`	Set of `Mentions` that are coreferent

Coreference Variables. There are two variable types for coreference: `Entity` and `Mention`. Each `Entity` embodies an underlying paper, to which there may be many citations. Each citation is represented by a `Mention` whose coreference involves an assignment to its corresponding `Entity`. Although each `Mention` object *contains* the citation string as a member instance variable, the `Mention`'s *value* (as an IDF random variable) is the `Entity` to which it has been assigned. The `Entity` is accordingly a *set-valued* variable—its value is the set of `Mentions` that have this `Entity` as their value. The values of `Mentions` and `Entities` are synchronized through *imperative variable coordination*. Note that this representation eliminates the need for factors enforcing mutual-exclusion or transitivity since each `Mention` can only have one `Entity` as its value. These variables are also summarized in Table 1.

Connections between Coreference and Segmentation Variables. To support joint factors between segmentation and coreference, the above variables contain references to each other. We take advantage of the fact that our random variables are objects in an object-oriented programming language, and represent arbitrary relations as member instance variables. Each `Mention` contains an array of its `Tokens`, as well as a list of its constituent `Fields`. Furthermore, each `Field` has a reference to its enclosing `Mention`, comprising `Tokens`, and adjacent `Fields`. These object-oriented cross-references are used to efficiently find the neighboring variables of a factor (*imperative structure definition*) and to access related variables for calculating sufficient statistics (*imperative variable-sufficient mapping*).

3.2 Factors Templates

Now we define the factor templates that are used to score possible worlds; they are summarized in Table 2. A small example consisting of three mentions and three fields, showing their instantiated factors and neighboring variables, is depicted in Figure 1. Given the neighboring variables of a factor, most of our factor templates employ additional computation to calculate sufficient statistics (features) from these neighbors—leveraging the flexible separation of data representation and parameterization (*imperative variable-sufficient mapping*).

Table 2. Factor Templates of Segmentation and Coreference: Factor templates that are used in the joint model of segmentation and entity resolution

	Segmentation Factors
`LabelToken:`	Factor between every token and its field type
`LabelNextToken:`	Factor between every label and the next token
`LabelPrevToken:`	Factor between every label and the previous token
`FieldFactor:`	Factor created for every `Field` to allow field-wise features
	Coreference Factors
`Affinity:`	Factor created between coreferent pairs of `Mentions`
`Repulsion:`	Factor created between pairs of `Mentions` that are not coreferent
	Joint Factors
`JntInfBased:`	Factor between `Mention` and `Label` based on joint factors in [4]
`JointAffinity:`	Factor between `Fields` of the same type of coreferent `Mentions`
`JointRepulsion:`	Factor between `Fields` of the same type of non-coref. `Mentions`

Segmentation. The segmentation factor templates express preferences about the `Tokens`' `Label` values and their segmentation into `Fields`. We define three factor templates traditional in linear-chain CRFs: factors between each `Label` and its corresponding `Token` (`LabelToken`), plus factors between the `Label` and adjacent `Tokens` on either side (`LabelPrevToken` and `LabelNextToken`).

In addition we define a factor template that examines a `Field` (and therefore has simultaneous access to all its constituent `Tokens`). This allows us to implement features similar to the rest of the segmentation rules of Poon and Domingos' "Isolated" MLN [4]. These include various first-order formulae over the `Tokens` of a `Field`. We also employ features testing the existence of punctuation at the beginning, the end, and within the `Field`. Finally we include features that take the position of the `Field` within the mention string into account. All of these features are taken in conjunction with the field type (`Label`).

Coreference. We have two coreference factor templates, which express preferences about partitioning `Mentions` into `Entities`. One measures pairwise affinity between `Mentions` in the same `Entity`; the other measures pairwise repulsion between `Mentions` in different `Entities`. Both factor templates have two `Mentions` as neighbors, and they both share the same *imperative variable-sufficient mapping* function. The number of these factors for a fully-unrolled graph is $O(m^2)$ (where m is the number of the citations). As described above, IDFs do not unroll the graph fully, and only need evaluate factors neighboring *moved* `Mentions`, which is $O(k)$ (where k is the average `Entity` size).

`Affinity` and `Repulsion` factors are scored using the same features (*i.e.* sufficient statistics function), but different parameters. These sufficient statistics are calculated from the un-segmented citation strings. We use the *SimilarTitle* and *SimilarVenue* features as defined in [4]. Furthermore, we add a *SimilarDate* feature that is true when the same year `Token` appears in both `Mentions`, as well as a *DissimilarDate* feature that is true when unequal year `Tokens` appear.

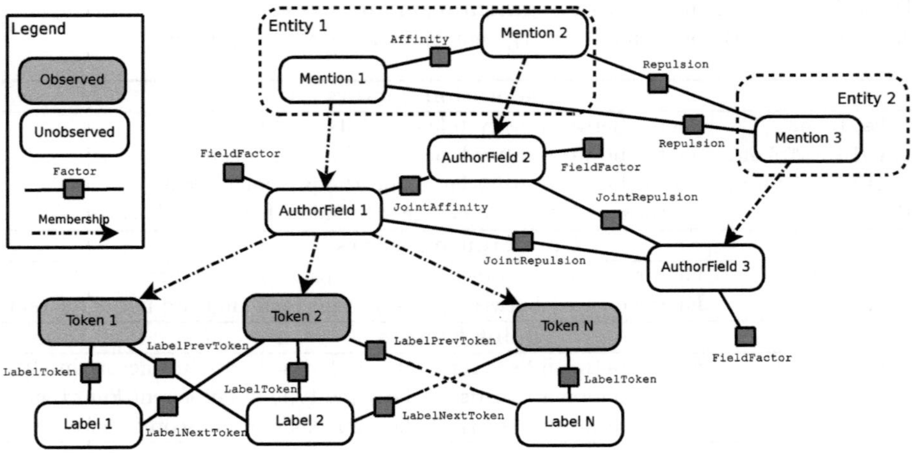

Fig. 1. Model: Variables and factors for joint segmentation and entity resolution shown on a toy example containing two entities and three mentions with a single field in each. Segmentation factors are only shown for one field. `JntInfBased` factors have been omitted for clarity.

Bi-directional. Joint factor templates express preferences about both segmentation and coreference simultaneously. In Poon and Domingos [4] all of the interaction between these tasks is captured by four similar rules that use their "JntInfCandidate" predicate. JntInfCandidate(m, i, m') is true if the `Token` trigram starting at position i in `Mention` m also appears in `Mention` m', and the trigram in m is not preceded by punctuation whereas the trigram in m' is. The trigram also must not meet certain "title exclusion" rules described in [4]. Note that JntInfCandidate is pre-calculated from the observed data, independent of segmentation and coreference predictions.

Even though Poon and Domingos' rules containing JntInfCandidate are scored on changes to both coreference and segmentation decisions, there are two reasons it forms only a weak interaction between the tasks. First, these templates only examine pairs of consecutive labels, not whole fields—failing to use information from predicted field range and non-consecutive words in the field. Second, the frequency with which the JntInfCandidate feature appears is quite data-dependent. In the Cora corpus, it occurs only 4973 times—representing an average of < 4 possible coreference candidates per `Mention`, whereas the average `Entity` size is ~ 10. On the other hand, if the feature occurs too often, it can be harmful for coreference. We hypothesize these reasons explain the empirical results in [4], in which the joint factors do not help entity resolution. We further explore this issue in Sect. 5.2.

To achieve stronger interaction between the tasks, we add factor templates that examine predicted `Fields` jointly with coreference decisions (`Mentions`). Our `JointAffinity` factor template defines factors that measure the compatibility of corresponding `Fields` in coreferent `Mentions`. Similarly `JointRepulsion` factor templates compare the corresponding `Fields` of non-coreferent `Mentions`. Hence the features (sufficient statistics) of these factor templates are able to compare full extracted field strings, and include StringMatch, SubStringMatch, PrefixMatch, SuffixMatch, AnyNTokenMatch, TokenIntersectionSize, etc.

One reason such strongly-joint factor templates have not been used in related work is due to the large number of factor instantiations in the many possible worlds. Such a factor could be created between any two pairs of `Mentions`, and any pair of their possible `Field` segmentations. This leads to $O(m^2n^4)$ factors, where m is the number of `Mentions`, and n is the maximum number of `Tokens` in a `Mention` string. The number of factors then becomes too large to be pre-processed and stored for big datasets, making such factor templates intractable for methods that fully unroll the factor graph or that pre-calculate features. This problem is common in joint inference because factors that represent dependencies between tasks often blow-up in the cross-product of the two hypothesis spaces. Since IDFs never unroll the entire factor graph and allow on-the-fly feature calculation they can efficiently use such factors.

3.3 Proposal Function

Our MCMC proposal function can make changes either to coreference or segmentation. A coreference proposal selects a random `Mention`, then with probability 0.8 moves it to another randomly selected `Entity`, or with probability 0.2 makes it a singleton in a new `Entity`. A segmentation proposal selects a random `Field` and identifies the minimum and maximum amount by which the `Field` can shrink or grow (which depends on the neighboring `Fields`). A new range for the `Field` is selected randomly based on that potential range. When the range of a `Field` is changed, the corresponding `Labels` are automatically adjusted via *imperative variable coordination*. The order of fields within a mention string is fixed: *author*, *title*, then *venue*.

4 Experimental Setup

We use the Cora dataset[2] [16] to evaluate our joint entity resolution and segmentation model. The dataset contains a total of 1,295 citations that refer to 134 ground truth entities. Each citation has three fields (*author*, *title*, and *venue*), with a total of 36,487 tokens. The dataset is divided into the same three folds used by [4]. These folds were not entirely random since they ensure no clusters are split across different folds. Ten runs of three-fold cross-validation are performed on the dataset, unless otherwise specified. Segmentation is evaluated on token-wise precision, recall, and F1. Pairwise coreference decisions are evaluated to obtain the precision, recall, and F1. Cluster recall, defined as the fraction of clusters that are correctly predicted, is calculated to compare with earlier results.

As a baseline we run isolated coreference and segmentation experiments. These isolated models use only the segmentation and coreference factors, as described

[2] The cleaned version available at http://alchemy.cs.washington.edu/papers/poon07

in Table 2, respectively (not the "joint factors"). Coreference is initialized to an all-singleton configuration, while segmentation is initialized to an equal-sized three way split of the mention string; we term these "default" configurations. Training consists of 5 loops of 100,000 proposals each. At the beginning of every loop we initialize either to the ground truth or the default configuration, selected randomly. Test-time inference consists of 300,000 proposals (for each task), starting with the default configuration. The temperatures for annealing during training and testing are set to 1.0. During training and inference for baseline isolated tasks only the respective proposal function is used.

Three different types of joint experiments are run to examine several aspects of the bi-directional nature of the joint model. For all of the joint experiments training is performed for 5 loops of 250,000 proposals each. Each task is initialized to either the default or the ground truth configuration, selected randomly, at the beginning of every training loop. Test-time inference consists of a total of 750,000 proposals across both tasks. The temperatures for training and testing are set to 3.0 and 1.0, respectively. During joint inference the proposal function randomly chooses between selecting a coreference or a segmentation proposal.

5 Results

The first joint experiment evaluates the full model including all factor templates and features, and compares these results to the isolated baseline models. The second joint experiment separately evaluates the gains resulting from the `JointInfBased` factors and the fully bi-directional factors described in Sect. 3.2. The third joint experiment examines the behavior of passing predictions between coreference and segmentation in an iterative fashion.

The experiments run very quickly, which can be attributed to *imperative variable coordination* and *imperative structure definition*, as described earlier. Training and inference of the isolated tasks finish within 3 minutes while the joint task takes approximately 18 minutes to run. By comparison, MC-SAT in [4], which does not enforce transitivity constraints for coreference, takes $50 - 90$ minutes. Adding transitivity constraints to the MLN severely increases running time further, as shown in [17].

5.1 Overall Joint Inference

Our results on the Cora dataset are shown in Table 3 and Table 4, demonstrating the benefits of a bi-directional approach to joint inference, with significant improvements on both segmentation and coreference. We also compare with the Fellegi-Sunter coreference model [14]. All improvements of our joint model over our isolated models are statistically significant at 1% using the T-test.

Table 3 shows that our isolated coreference model outperforms the previously published results in [4] on both metrics. Our joint model, which concurrently solves the segmentation task, outperforms our isolated coreference model, with a 13% error reduction compared to our isolated IDF. It also provides an overall

Table 3. Cora Coreference: Pairwise precision/recall, F1, and cluster recall, for the coreference task on the *Cora* dataset

Method	Prec/Recall	F1	Cluster Rec.
Fellegi-Sunter	78.0/97.7	86.7	62.7
Joint MLN	94.3/97.0	95.6	78.1
Isolated IDF	97.09/95.42	**96.22**	**86.01**
Joint IDF	95.34/98.25	**96.71**	**94.62**

Table 4. Cora Segmentation: Token-wise F1 for each field of the segmentation task on the *Cora* dataset

Method	Author	Title	Venue	Total
Isolated MLN	99.3	97.3	98.2	98.2
Joint MLN	99.5	97.6	98.3	98.4
Isolated IDF	99.35	97.63	98.58	**98.51**
Joint IDF	99.42	97.99	98.78	**98.72**

25.2% error reduction in pairwise coreference F1 in comparison to the joint MLN. In addition, Table 3 shows that the joint approach allows cluster recall to improve substantially, resulting in a 75.4% error reduction compared to the joint MLN, and a 61.5% error reduction compared to our isolated IDF.

Table 4 shows similar improvements on the segmentation task. Our isolated segmentation model significantly outperforms all earlier results, and the joint model uses the coreference predictions to improve segmentation further. In comparison to the joint MLN we provide an overall error reduction in token-wise segmentation F1 of 20.0%. Compared to our isolated IDF the reduction is 14.1%.

5.2 Bi-directionality

We also examine the performance as joint factors are added to our isolated models. The results are shown in Fig. 2. The isolated models produces the lowest scores amongst our models. "Semi-Joint" refers to the model containing the `JointInfBased` factors in addition to the isolated factors. They lead to a larger improvement in segmentation than in coreference, confirming their weaker effect on coreference proposals. When the fully bi-directional factors are also added ("Fully-Joint") both segmentation and coreference scores improve. However, the improvement for coreference is much higher. Recall that the factors added for the "Fully-Joint" model are prohibitively expensive to pre-calculate (as described in section 3.2), which demonstrates the benefit of using an IDF for bi-directional joint inference.

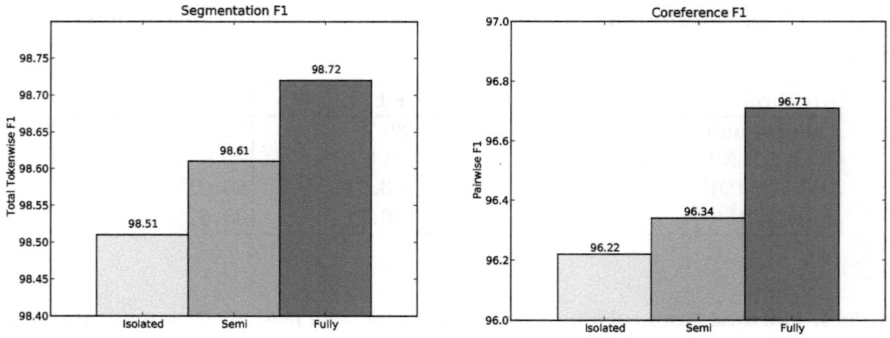

(a) Segmentation F1 (b) Coreference F1

Fig. 2. Adding Joint Factors: F1 of the joint model as different types of factors are added, starting with the base model containing only isolated segmentation and coreference factors. "Semi-Joint" refers to the model containing *weakly* joint factors while the "Fully-Joint" model consists of bi-directional highly-coupled factors.

5.3 Iterated Pipeline Comparison

Conventionally, multiple information extraction tasks are solved using a pipeline architecture in which the output predictions of one task are used as input to the next. To minimize error caused due to cascading, multiple iterations of a pipeline can be carried out, such that the output predictions of the last stage of the pipeline feed back to the first stage.

As described earlier, we switch between segmentation and coreference proposals randomly. From the iterated pipeline perspective our method of performing joint inference is similar to repeating a pipeline in which each stage consists of a single proposal. To compare the performance of the fully joint proposal function against an iterated pipeline, we vary the number of pipeline iterations—by changing the total number of stages—while keeping the total number of training and testing proposals constant.

Our results are shown in Fig. 3. Each experiment involves 20 runs of three-fold cross validation using 5 loops of 250,000 training proposals and 750,000 testing proposals. The proposals are evenly divided across the stages, for example, the 2-stage experiment consists of 125,000 training proposals in segmentation followed by 125,000 training proposals in coreference for each of the 5 loops. In comparison, the 10-stage experiment consists of 5 pipelines, in which each pipeline has a segmentation stage of 12,500 proposals followed by a coreference stage of 12,500 proposals. Thus a higher number of total stages leads to a smaller number of proposals per stage. The first stage is always segmentation, to be consistent with earlier work in citation matching [15].

For both tasks our experiments show that the fully joint model gives higher F1 than any of the iterated pipeline results. Notice that the segmentation F1 rises as the number of stages in the pipeline increases. It is possible that segmentation

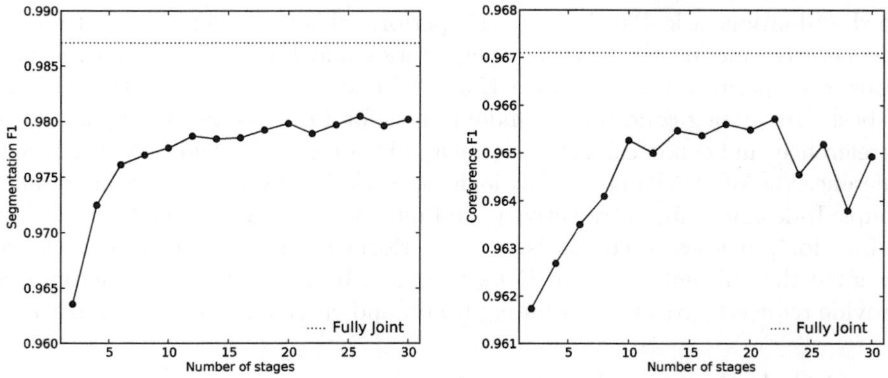

(a) Segmentation F1 (b) Coreference F1

Fig. 3. Iterated Pipeline: Performance of segmentation and coreference as the number of iterations of the pipeline is varied for a fixed number of total sampling steps over all iterations. The dotted line denotes the model that performs inference jointly, randomly switching between segmentation and coreference proposals.

F1 for a specific number of stages may be better than the fully joint results. However, finding the optimal number of stages can be expensive and we feel that a fully joint model is likely to perform competitively and generalize to a withheld validation set.

6 Related Work

Many researchers have explored issues of joint inference in text processing. McCallum and Jensen [1] present a position paper motivating the need for joint inference, and propose unified undirected graphical models for joint information extraction and data mining, describing several examples of conditional random fields. Many other papers present experimental results with various methods of inference. Sometimes joint inference can be done with standard dynamic-programming methods, for example, joint named entity recognition and parsing [18,19] via CYK, with parse non-terminal symbols augmented to include named entity information. Another common alternative is feedforward N-best lists, *e.g.* [20,21]. The feedforward probability distribution can be better approximated by sampling [2], but this flow of information is still uni-directional. Others have passed N-best list information bi-directionally between two tasks [3,22]. Multi-directional passing of full probability distributions corresponds to loopy belief propagation, which has been used for skip-chains [23]. If joint factors can be expressed as linear constraints, one can employ efficient software packages for integer linear programming (ILP) [24]. When the structure of the model is changing during inference, MCMC provides significant efficiencies [8,9,10].

Joint citation segmentation and coreference has become somewhat of a standard evaluation task. Pasula *et al.* [15] perform this task using BLOG to define a generative model of research paper entities and their noisily-rendered citations; they perform inference by MCMC. Wellner *et al.* [3] bi-directionally pass N-best lists among conditional random fields for 14-field citation segmentation, coreference, and canonicalization. Poon and Domingos [4] avoid N-best lists with inference via MC-SAT in a Markov logic network. However, the tasks are weakly-coupled, do not enforce transitivity, and only segment into three fields. In this paper, for purposes of comparison, we perform the same three-field task. We leverage the efficient power of IDFs to define bi-directional joint factors that provide reduced error, faster running times, and enforce coreference transitivity.

7 Conclusions and Future Work

In this paper we presented a highly-coupled, bi-directional model for joint inference in citation segmentation and coreference, yielding new state-of-the-art accuracy. We incorporate factors that use coreference to aid segmentation and vice-versa, and do so efficiently using imperatively-defined factor graphs (IDFs). Compared to other joint models for the same tasks, our method results in an error reduction of $20 - 25\%$, providing a new state-of-the-art result, while also running $3 - 15$ times faster. In future work we will explore similar methods for joint inference in newswire named entity extraction and coreference, in which, unlike citations, the number of mentions must be inferred.

Acknowledgements

This work was supported in part by the Center for Intelligent Information Retrieval, in part by NSF grant #CNS-0551597, in part by Army prime contract number W911NF-07-1-0216 and University of Pennsylvania subaward number 103-548106, in part by The Central Intelligence Agency, the National Security Agency and National Science Foundation under NSF grant #IIS-0326249, and in part by UPenn NSF medium IIS-0803847. Any opinions, findings and conclusions or recommendations expressed in this material are the authors' and do not necessarily reflect those of the sponsor.

References

1. McCallum, A., Jensen, D.: A note on the unification of information extraction and data mining using conditional-probability, relational models. In: IJCAI Workshop on Learning Statistical Models from Relational Data (2003)
2. Finkel, J.R., Manning, C.D., Ng, A.Y.: Solving the problem of cascading errors: Approximate bayesian inference for linguistic annotation pipelines. In: Conference on Empirical Methods on Natural Language Processing, EMNLP (2006)
3. Wellner, B., McCallum, A., Peng, F., Hay, M.: An integrated, conditional model of information extraction and coreference with application to citation matching. In: Uncertainty in Artificial Intelligence (UAI), pp. 593–601 (2004)

4. Poon, H., Domingos, P.: Joint inference in information extraction. In: AAAI Conference on Artificial Intelligence, pp. 913–918 (2007)
5. Poon, H., Domingos, P.: Sound and efficient inference with probabilistic and deterministic dependencies. In: AAAI Conference on Artificial Intelligence (2006)
6. Selman, B., Kautz, H., Cohen, B.: Local search strategies for satisfiability testing. Discrete Mathematics and Theoretical Computer Science (DIMACS) 26 (1996)
7. Kschischang, F., Frey, B., Loeliger, H.A.: Factor graphs and the sum-product algorithm. IEEE Trans on Information Theory 47(2), 498–519 (2001)
8. Culotta, A., McCallum, A.: Tractable learning and inference with high-order representations. In: International Conference on Machine Learning (ICML) Workshop on Open Problems in Statistical Relational Learning (2006)
9. Richardson, M., Domingos, P.: Markov logic networks. Machine Learning 62(1-2), 107–136 (2006)
10. Milch, B., Marthi, B., Russell, S.: BLOG: Relational Modeling with Unknown Objects. PhD thesis, University of California, Berkeley (2006)
11. McCallum, A., Rohanimanesh, K., Wick, M., Schultz, K., Singh, S.: FACTORIE: Efficient probabilistic programming via imperative declarations of structure, inference and learning. In: NIPS Workshop on Probabilistic Programming (2008)
12. Rohanimanesh, K., Wick, M., McCallum, A.: Inference and learning in large factor graphs with a rank based objective. Technical Report UM-CS-2009-08, University of Massachusetts, Amherst (2009)
13. Bilenko, M., Mooney, R.J.: Adaptive duplicate detection using learnable string similarity measures. In: ACM SIGKDD International Conference on Knowledge Discovery and Data Mining, KDD (2003)
14. Singla, P., Domingos, P.: Entity resolution with Markov logic. In: International Conference on Data Mining (ICDM), pp. 572–582 (2006)
15. Pasula, H., Marthi, B., Milch, B., Russell, S., Shpitser, I.: Identity uncertainty and citation matching. In: Neural Information Processing Systems, NIPS (2003)
16. McCallum, A., Nigam, K., Rennie, J., Seymore, K.: A machine learning approach to building domain-specific search engines. In: IJCAI, pp. 661–667 (1999)
17. Poon, H., Domingos, P., Sumner, M.: A general method for reducing the complexity of relational inference and its application to MCMC. In: AAAI (2008)
18. Miller, S., Fox, H., Ramshaw, L., Weischedel, R.: A novel use of statistical parsing to extract information from text. In: Applied Natural Language Processing Conference, pp. 226–233 (2000)
19. Finkel, J.R., Manning, C.D.: Joint parsing and named entity recognition. In: North American Association of Computational Linguistics, NAACL (2009)
20. Gildea, D., Jurafsky, D.: Automatic labeling of semantic roles. Computational Linguistics 28, 245–288 (2002)
21. Sutton, C., McCallum, A.: Joint parsing and semantic role labeling. In: Conference on Computational Natural Language Learning, CoNLL (2005)
22. Hollingshead, K., Roark, B.: Pipeline iteration. In: Annual Meeting of the Association of Computational Linguistics (ACL), pp. 952–959 (2007)
23. Sutton, C., McCallum, A.: Collective segmentation and labeling of distant entities in information extraction. In: ICML Workshop on Statistical Relational Learning and Its Connections to Other Fields (2004)
24. Roth, D., Yih, W.: Global inference for entity and relation identification via a linear programming formulation. In: Getoor, L., Taskar, B. (eds.) Introduction to Statistical Relational Learning. MIT Press, Cambridge (2007)

Latent Dirichlet Allocation for Automatic Document Categorization*

István Bíró and Jácint Szabó

Data Mining and
Web Search Research Group,
Computer and Automation
Research Institute
of the Hungarian Academy of Sciences
Budapest, Hungary
{ibiro,jacint}@ilab.sztaki.hu

Abstract. In this paper we introduce and evaluate a technique for applying latent Dirichlet allocation to supervised semantic categorization of documents. In our setup, for every category an own collection of topics is assigned, and for a labeled training document only topics from its category are sampled. Thus, compared to the classical LDA that processes the entire corpus in one, we essentially build separate LDA models for each category with the category-specific topics, and then these topic collections are put together to form a unified LDA model. For an unseen document the inferred topic distribution gives an estimation how much the document fits into the category.

We use this method for Web document classification. Our key results are 46% decrease in 1-AUC value in classification accuracy over tf.idf with SVM and 43% over the plain LDA baseline with SVM. Using a careful vocabulary selection method and a heuristic which handles the effect that similar topics may arise in distinct categories the improvement is 83% over tf.idf with SVM and 82% over LDA with SVM in 1-AUC.

1 Introduction

Generative topic models [1,2,3] have a wide range of applications in the fields of language processing, text mining and information retrieval, including categorization, keyword extraction, similarity search and statistical language modeling.

One of the most successful generative topic models is latent Dirichlet allocation (LDA) developed by Blei, Ng and Jordan [3]. LDA models every topic as a distribution over the terms of the vocabulary, and every document as a distribution over the topics. These distributions are sampled from Dirichlet distributions. LDA is an intensively studied model, and the experiments are really impressive compared to other known information retrieval techniques. The applications of

* Supported by the EU FP7 project LiWA - Living Web Archives and by grants OTKA NK 72845, *ASTOR* NKFP 2/004/05.

W. Buntine et al. (Eds.): ECML PKDD 2009, Part II, LNAI 5782, pp. 430–441, 2009.

LDA include entity resolution [4], fraud detection in telecommunication systems [5], image processing [6,7,8] and ad-hoc retrieval [9].

Another important and widely studied area of language processing is supervised text categorization (for a survey we refer to [10]). LDA, in its original form [3], cannot be used for supervised text categorization as it is an unsupervised latent model rather than an explicit topic model. This issue, to our best knowledge, remained mainly unexplored, and the goal of the present paper is to address this question. Although LDA can be applied for dimensionality reduction prior to supervised classification as in LSA [1], we show that this baseline method is not competitive with our modified LDA model.

In this paper we introduce **multi-corpus LDA** (MLDA), a modification of LDA, which incorporates explicit topic labels into LDA making it applicable for text categorization. MLDA is essentially a hierarchical method with two levels, category and topics. Assume we have a supervised document categorization task with m semantic categories. Every document is assigned exactly one category, and this assignment is known only for the training corpus. For every category we assign an own collection of topics, and the union of these collections forms the topic collection of LDA. In LDA, for every document, a Dirichlet parameter vector α is chosen such that the assigned topics to the document's words are drawn from a fixed multinomial distribution drawn from $\text{Dir}(\alpha)$. In MLDA, for every training document we require that this α Dirichlet parameter has component zero for all topics outside the document's category, in order to achieve that only topics from the document's category are sampled to the document's words. This is tantamount to building separate LDA models for every category with category-specific topics. Then for an unseen document d the fraction of topics in the topic distribution of d that belong to a given category measures how well d fits into that category. As a Dirichlet distribution allows only positive parameters, we will extend the notion of Dirichlet distribution in a natural way by allowing zeros. Although there exist hierarchical latent topic models [11,12] to tackle more than one layers as we have, the advantage of MLDA is that it is built up from plain LDA's and no complicated hierarchical models should be developed. For a more detailed description of MLDA, see Subsection 2.2.

We apply MLDA for a corpus of 12k documents from the DMOZ library, divided into $m = 8$ categories. We carry out a careful term selection method, based on the entropy of the normalized tf-vectors over the categories, resulting in a vocabulary consisting of terms with high coverage and discriminability. We also try out a heuristic, ϑ-smoothing, which tries to compensate the effect that similar topics may arise in distinct categories, and thus unseen inference may put very skewed weights on these two topics.

We test MLDA in combination with SVM over LDA and over tf.idf. The improvement is 43% decrease in 1-AUC value over LDA with SVM. Careful choice of term selection results in a further 37% decrease, while ϑ-smoothing gives a further 2% decrease over LDA in 1-AUC, summing up to 82%. MLDA with the best term selection and ϑ-smoothing results in a 83% decrease in 1-AUC over tf.idf with SVM. For a detailed explanation, see Section 3.

The MLDA technique was applied with success to Web spam filtering in the Web Spam Challenge 2008 competition [13].

The rest of the paper is organized as follows. Section 2 explains LDA and MLDA. Section 3 describes the experimental setup and Section 4 the results. Finally, Section 5 summarizes our work and envisions future research.

2 Multi-corpus LDA

2.1 The Classical LDA

We shortly describe latent Dirichlet allocation (Blei, Ng, Jordan [3]), for a detailed elaboration, we refer to Heinrich [14]. We have a vocabulary V consisting of terms, a set T of k topics and n documents of arbitrary length. For every topic z a distribution φ_z on V is sampled from $\mathrm{Dir}(\beta)$, where $\beta \in \mathbb{R}_+^V$ is a smoothing parameter. Similarly, for every document d a distribution ϑ_d on T is sampled from $\mathrm{Dir}(\alpha)$, where $\alpha \in \mathbb{R}_+^T$ is a smoothing parameter.

The words of the documents are drawn as follows: for every word-position of document d a topic z is drawn from ϑ_d, and then a term is drawn from φ_z and filled into the position.

LDA can be thought of as a Bayesian network, see Figure 1.

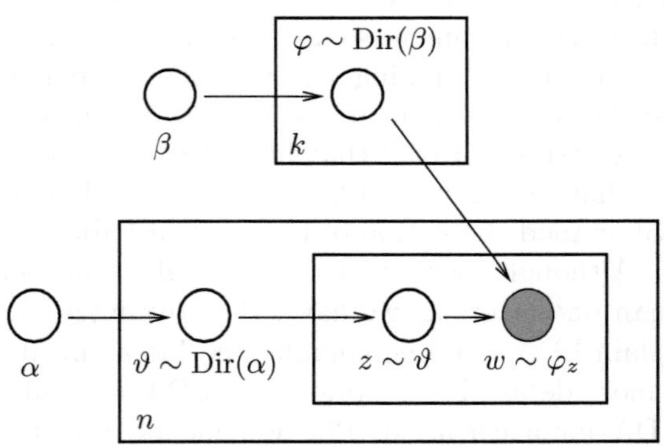

Fig. 1. LDA as a Bayesian network

One method for finding the LDA model by inference is via Gibbs sampling [15]. (Additional methods are variational expectation maximization [3], and expectation propagation [16]). Gibbs sampling is a Monte Carlo Markov-chain algorithm for sampling from a joint distribution $p(x)$, $x \in \mathbb{R}^n$, if all conditional distributions $p(x_i|x_{-i})$ are known ($x_{-i} = (x_1, \ldots, x_{i-1}, x_{i+1}, \ldots, x_n)$). In LDA the goal is to estimate the distribution $p(z|w)$ for $z \in T^P$, $w \in V^P$ where P denotes the set of word-positions in the documents. Thus in Gibbs sampling one has to calculate for all $i \in P$ and topics z' the probability $p(z_i = z'|z_{-i}, w)$. This has an efficiently computable closed form, as deduced for example in Heinrich

[14]. Before describing the formula, we introduce the usual notation. We let d be a document and w_i its word at position i. We also let count N_{dz} be the number of words in d with topic assignment z, N_{zw} be the number of words w in the whole corpus with topic assignment z, N_d be the length of document d and N_z be the number of all words in the corpus with topic assignment z. A superscript N^{-i} denotes that position i is excluded from the corpus when computing the corresponding count. Now the Gibbs sampling formula becomes [14]

$$p(z_i = z'|z_{-i}, w) \propto \frac{N^{-i}_{z'w_i} + \beta(w_i)}{N^{-i}_{z'} + \sum_{w \in V} \beta(w)} \cdot \frac{N^{-i}_{dz'} + \alpha(z')}{N^{-i}_d + \sum_{z \in T} \alpha(z)}. \tag{1}$$

After a sufficient number of iterations we arrive at a topic assignment sample z. Knowing z, the variables φ and ϑ are estimated as

$$\varphi_{z,w} = \frac{N_{zw} + \beta_w}{N_z + \sum_{w \in V} \beta_w} \tag{2}$$

and

$$\vartheta_{d,z} = \frac{N_{dz} + \alpha_z}{n_d + \sum_{z \in T} \alpha_z}. \tag{3}$$

We call the above method *model inference*. After the model (that is, φ) is built, we make *unseen inference* for every new, unseen document d. The ϑ topic-distribution of d can be estimated exactly as in (3) once we have a sample from its word-topic assignment z. Sampling z is usually performed with a similar method as before, but now only for the positions i in d:

$$p(z_i = z'|z_{-i}, w) \propto \varphi_{z',w_i} \cdot \frac{N^{-i}_{dz'} + \alpha(z')}{N^{-i}_d + \sum_{z \in T} \alpha(z)}. \tag{4}$$

To verify (4), note that the first factor in Equation (1) is approximately equal to φ_{z',w_i}, and φ is already known during unseen inference.

2.2 Multi-corpus LDA

All what is modified in LDA in order to adapt it to supervised semantic categorization is that we first divide the topics among the categories and then make sure that for a training document only topics from its own category are sampled during the training phase. To achieve this we have to extend the notion of a Dirichlet distribution in a natural way. If $\gamma = (\gamma_1, \ldots, \gamma_l, 0, \ldots, 0) \in \mathbb{R}^n$ where $\gamma_i > 0$ for $1 \leq i \leq l$, then let the distribution $\mathrm{Dir}(\gamma)$ be concentrated on the subset $\{x \in \mathbb{R}^n : x_i = 0 \ \forall i > l, \sum_{1 \leq i \leq n} x_i = 1\}$, with distribution $\mathrm{Dir}(\gamma_1, \ldots, \gamma_l)$. Thus for $p \sim \mathrm{Dir}(\gamma)$ we have that $p_i = 0$ for $i > l$ with probability 1, and $(p_1, \ldots, p_l)$ is of distribution $\mathrm{Dir}(\gamma_1, \ldots, \gamma_l)$. It can be checked in the deduction of [14] that the only property used in the calculus of Subsection 2.1 is that the Dirichlet distribution is conjugate to the multinomial distribution, which is kept for our extension, by construction. Indeed, if $x \sim \chi$ where

$\chi \sim \mathrm{Dir}(\gamma_1, \ldots, \gamma_l, 0, \ldots, 0)$ with $\gamma_i > 0$ for $1 \le i \le l$, then for $i > l$ we have that $\chi_i = 0$ and thus $x_i = 0$ with probability 1. So the maximum a posteriori estimation of χ_i is

$$\frac{\gamma_i + x_i}{\sum_{1 \le j \le n} \gamma_j + x_j},$$

because the same holds for the classical case. To conclude, every calculation of the previous subsection still holds.

As $p(z_i = z' | z_{-i}, w) = 0$ in Equation 1 if z' has 0 Dirichlet prior, that is if it does not belong to the category of the document, the model inference procedure breaks down into making separate model inferences, one for every category. In other words, if we denote by C_i, $1 \le i \le m$, the collection of those training documents which were assigned category i, then model inference in MLDA is essentially building m separate LDA models, one for every C_i, with an appropriate choice of the topic number k_i. After all model inferences have been done, we have term-distributions for all $k = \sum\{k_i : 1 \le i \le m\}$ topics.

Unseen inference is the same as for LDA. For an unseen document d, we perform Gibbs sampling as in Equation (4), and after a sufficient number of iterations, we calculate ϑ_d as in (3). We define for every category $1 \le i \le m$

$$\xi_i = \sum\{\vartheta_{d,z} : z \text{ is a topic from category } i\}. \tag{5}$$

As ξ_i estimates how relevant category i is to the document, ξ_i is a classification itself. We call this **direct classification**, and measure its accuracy in terms of the AUC value, see Section 3. It is an appealing property of MLDA that right after unseen inference the resulting topic distribution directly gives rise to a classification. This is in contrast to, say, using plain LDA for categorization, where the topic-distribution of the documents serve as features for a further advanced classifier.

MLDA also outperforms LDA in its running time. If there are k_i topics and p_i word-positions in category i, then MLDA model inference runs in time $O(I \cdot \sum_{i=1}^{m} k_i p_i)$, where I is the number of iterations. On the contrary, LDA model inference runs in time $O(I \cdot kp)$ where $k = \sum_{i=1}^{m} k_i$ and $p = \sum_{i=1}^{m} p_i$ (running times of LDA and MLDA do not depend on the number of documents). The more categories we have, the more is the gain. In addition, model inference in MLDA can be run in parallel. For measured running times see Subsection 4.3.

2.3 ϑ-Smoothing with Personalized Page Rank

There is a possibility that MLDA infers two very similar topics in two distinct categories. In such a case it can happen that unseen inference for an unseen document puts very skewed weights on these two topics, endangering the classification's performance. To avoid such a situation, we apply a modified 1-step Personalized Page Rank on the topic space, taking topic-similarity into account.

Fix a document d. After unseen inference, its topic-distribution is ϑ_d. We smooth ϑ_d by distributing the components of ϑ_d among themselves, as follows.

For all topics j replace $\vartheta_{d,j}$ by

$$S \cdot \vartheta_{d,j} + \sum_{h \text{ topic}, \, h \neq j} c_h \cdot \frac{\vartheta_{d,h}}{\mathrm{JSD}(\varphi_j, \varphi_h) + \varepsilon}.$$

The constants c_h are chosen in such a way that

$$c_h \cdot \sum_{j \text{ topic}, \, j \neq h} \frac{1}{\mathrm{JSD}(\varphi_j, \varphi_h) + \varepsilon} = 1 - S$$

for all topics h, making sure that the new ϑ_d is indeed a distribution. JSD is the Jensen-Shannon divergence, a symmetric distance function between distributions with range $[0, 1]$. Thus $\mathrm{JSD}(\varphi_j, \varphi_h)$ is a measure of similarity of topics j and h. ε is to avoid dividing with zero, we chose it to $\varepsilon = 0.001$. S is a smoothing constant. We tried four values for it, $S = 1$, 0.85, 0.75 and 0.5. Note that $S = 1$ corresponds to no ϑ-smoothing.

The experiments on ϑ-smoothing in Subsection 4.2 show slight improvement in accuracy if the vocabulary has small discriminating power among the categories. This is perhaps because it is more probable that similar topics are inferred in two categories if there are more words in the vocabulary with high occurrence in both. We mention that ϑ-smoothing can clearly be applied to the classical LDA as well.

3 Experimental Setup

We have 290k documents from the DMOZ web directory[1], divided into 8 categories: Arts, Business, Computers, Health, Science, Shopping, Society, Sports. In our experiments a document consists only of the text of the html page. After dropping those documents whose length is smaller than 2000, we are left with 12k documents with total length of 64M.

For every category, we randomly split the collection of pages assigned with that category into training (80%) and test (20%) collections, and we denote by C_i the training corpus of the ith category. We learn the MLDA model on the train corpus, that is, effectively, we build separate LDA models, one for every category. Then we carry out unseen inference on the test corpus (that is the union of the test collections). For every unseen document d we define two aggregations of the inferred topic distribution ϑ_d, we use ϑ_d itself as the feature set, and also the category-wise ξ sums, as defined in (5). As we already noted, the category-wise sum ξ gives an estimation on the relevancy of category i for document d, thus we use it in itself as a direct classification, and an AUC value is calculated.

Similarly, we learn an LDA model with the same number of topics on the training corpus (without using the category labels), and then take the inferred ϑ values as feature on the test corpus.

[1] `http://www.dmoz.org/`

We make experiments on how advanced classifiers perform on the collection of these aggregated features. For every category we do a supervised classification as follows. We take the documents of the test corpus with the feature-set and the Boolean label whether the document belongs to the category or not. We run binary classifications (linear SVM, C4.5 and Bayes-net) using 10-fold cross-validation to get the AUC value. What we report is the average of these AUC values over the 8 categories. This is carried out for both feature-sets ϑ and ξ.

Every run (MLDA model build, unseen inference and classification) is repeated 10 times to get variance of the AUC classification performance.

The calculations were performed with the machine learning toolkit Weka [17] for classification and a home developed C++ code for LDA[2].

The computations were run on a machine of 20GB RAM and 1.8GHz Dual Core AMD Opteron 865 processor with 1MB cache. The OS was Debian Linux.

3.1 Term Selection

Although the importance of term selection in information retrieval and text mining has been proved crucial by several results, most papers on LDA-based models do not put strong emphasis on the choice of the vocabulary. In this work we perform a careful term selection in order to find terms with high coverage and discriminability. There are several results published on term selection methods for text categorization tasks [18,19]. However, here we do not directly apply these, as our setup is different in that the features put into the classifier come from discovered latent topics, and are not derived directly from terms.

First we keep only terms consisting of alphanumeric characters, the hyphen, and the apostrophe, then we delete all stop-words enumerated in the Onix list[3], and then the text is run through a tree-tagger software for lemmatization[4].

Then

1. for every training corpus C_i we take the **top.tf** terms with top tf values (calculated w.r.t C_i) (the resulting set of terms is denoted by W_i),
2. we unify these term collections over the categories, that is, let $W = \bigcup\{W_i : 1 \le i \le m\}$,
3. then we drop from W those terms w for which the entropy of the normalized tf-vector over the categories exceeds a threshold **ent.thr**, that is, for which

$$H(\mathrm{tf}(w)) \ge \textbf{ent.thr}.$$

Here $\mathrm{tf}(w) \in \mathbb{R}^m$ is the vector with i^{th} component the tf value of w in the training corpus C_i, normalized to 1 to be a distribution.

Term selection has two important aspects, coverage and discriminability. Note that step 1. takes care of the first, and step 3. of the second.

[2] `http://www.ilab.sztaki.hu/~ibiro/linkedLDA/`

[3] `http://www.lextek.com/manuals/onix/stopwords1.html`

[4] `http://www.ims.uni-stuttgart.de/projekte/corplex/TreeTagger/`

Table 1. The size of the vocabulary and the average document length after filtering for different thresholds (**top.tf/ent.thr**)

vocabulary	size of vocab	avg doc length
unfiltered	21862	2602
30000 / 1.8	89299	1329
30000 / 1.5	74828	637
15000 / 1.8	35996	1207

We also made experiments with the unfiltered vocabulary (stop-wording and stemming the top 30k terms in tf). For the thresholds set in our experiments the size of the vocabulary and the average document length after filtering is shown in Table 1.

We mention that the running time of LDA is insensitive to the size of the vocabulary, so this selection is exclusively for enhancing performance.

3.2 LDA Inference

The number k of topics is chosen in three ways, see Table 2:

1. (const) $k_i = 50$ for all categories,
2. (sub) k_i is the number of subcategories of the category,
3. (sub-sub) k_i is the number of sub-sub-categories of the category.

Table 2. Three choice for topic-numbers k

category	const	sub	sub-sub
Arts	50	15	41
Business	50	23	71
Computers	50	17	44
Health	50	15	39
Science	50	12	58
Shopping	50	22	82
Society	50	20	58
Sports	50	14	28
sum, $k =$	400	138	421

The Dirichlet parameter β was chosen to be constant 0.1 throughout. For a training document in category i we chose α to be $50/k_i$ on topics belonging to category i and zero elsewhere. During unseen inference, the α smoothing parameter was defined accordingly, that is, for every topic z we have $\alpha_z = 50/k_i$ if z belongs to category i. The tests are run with and without ϑ-smoothing, with $b = 0.5, 0.75, 0.85$ and $\varepsilon = 0.001$ (Subsection 2.3).

We apply Gibbs sampling for inference with 1000 iterations throughout. We use a home developed C++ code [5] to run LDA and MLDA.

4 Results

4.1 Plain Multi-corpus LDA vs LDA

As a justification for MLDA, we compared plain MLDA (no vocabulary filtering and ϑ-smoothing) with the classical LDA [3] (as described in Subsection 2.1). The vocabulary is chosen to be the unfiltered one in both cases (see Subsection 3.1). For MLDA we tested all three variations for topic-numbers (see Table 2). We show only the ξ aggregation, as with ϑ features the AUC values were about 5% worse. The classifiers were linear SVM, Bayes network and C4.5, as implemented in Weka, together with the direct classification (defined in (5)). For LDA the number of topics was $k = 138$, which is equal to the total number of topics in MLDA 'sub', and for a test document the corresponding 138 topic probabilities served as features for the binary classifiers in the test corpus. Another baseline classifier is SVM over tf.idf, run on the whole corpus with 10-fold cross validation. The AUC values are averaged over the 8 categories. The results are shown in Table 3.

Table 3. Comparing plain MLDA with LDA (avg-AUC)

	SVM	Bayes	C4.5	direct
MLDA (const)	0.812	0.812	0.605	0.866
MLDA (sub)	0.820	0.826	0.635	**0.867**
MLDA (sub-sub)	0.803	0.816	0.639	0.866
LDA ($k = 138$)	0.765	0.791	0.640	–
SVM over tf.idf	0.755	–	–	–

The direct classification of MLDA strongly outperforms the baselines and the advanced classification methods on MLDA based ϑ features. Even the smallest improvement, for SVM over MLDA ϑ features, is 35% in 1-AUC. Table 3 indicates that MLDA is quite robust to the parameter of topic-numbers. However, as topic-number choice 'sub' was the best, in later tests we used this one.

4.2 Vocabularies and ϑ-Smoothing

We made experiments on MLDA to fine tune the vocabulary selection thresholds and to test performance of the ϑ-smoothing heuristic by a parameter sweep. Note that $S = 1$ in ϑ-smoothing corresponds to doing no ϑ-smoothing. We fixed seven kinds of vocabularies (with different choices of **top.tf** and **ent.thr**) and the topic-number was chosen to be 'sub' (see Table 2). We evaluated the direct classification, see Table 4.

[5] `http://www.ilab.sztaki.hu/~ibiro/linkedLDA/`

Table 4. Testing the performance of ϑ-smoothing and the vocabulary parameters (**top.tf/ent.thr**) in avg-AUC

vocabulary	$S = 1$	0.85	0.75	0.5
30000 / 1.8	0.954	0.955	0.956	**0.958**
30000 / 1.5	0.948	0.948	0.948	0.947
30000 / 1.0	0.937	0.937	0.936	0.934
15000 / 1.8	0.946	0.947	0.948	0.952
15000 / 1.2	0.937	0.937	0.937	0.936
10000 / 1.5	0.942	0.942	0.943	0.943
unfiltered	0.867	0.866	0.861	0.830

It is apparent that our term selection methods result in a big improvement in accuracy. This improvement is more accurate if the entropy parameter **ent.thr** and the tf parameter **top.tf** are larger. As both result in larger vocabularies, term selection should be conducted carefully to keep the size of the vocabulary big enough. Note that the more the entropy parameter **ent.thr** is the more ϑ-smoothing improves performance. This is perhaps because of the fact that large **ent.thr** results in a vocabulary consisting of words with low discriminability among the categories, and thus topics in distinct categories may have similar word-distributions.

Every run (MLDA model build, unseen inference and classification) was repeated 10 times to get variance of the AUC measure. Somewhat interestingly, these were at most 0.01 throughout, so we decided not to quote them individually.

4.3 Running Times

We enumerate the running times of some experiments. If the filtering parameters of the vocabulary are chosen to be **top.tf**=15000 and **ent.thr**=1.8, and the topic number is 'sub' then model inference took 90min for the biggest category Society (4.3M word positions), and 5min for the smallest category Sports (0.3M word positions). Unseen inference took 339min, with the same settings.

4.4 An Example

To illustrate MLDA's performance, we show what categories MLDA inferred for the site `http://www.order-yours-now.com/`. As of July 2008, this site advertises a tool for creating music contracts, and it has DMOZ categorization Computers: Software: Industry-Specific: Entertainment Industry.

The 8 category-wise ξ features (defined in (5)) of MLDA measure the relevance of the categories, see Table 5. We feel that MLDA's categorization is at par or perhaps better than that of DMOZ. The top DMOZ category is Computers, perhaps because the product is sold as a computer program. On the contrary, MLDA suggests that the site mostly belongs to Arts and Shopping, which we feel appropriate as it offers service for musicians for a profit. MLDA also detects

Table 5. Relevance of categories for site `http://www.order-yours-now.com/`, found by MLDA

category	ξ
Arts	0.246
Shopping	0.208
Business	0.107
Health	0.103
Society	0.096
Computers	0.094
Sports	0.076
Science	0.071

the Business concept of the site, however, Shopping is given more relevance than Business because of the informal style of the site.

The parameters for MLDA were set as follows: **top.tf**=15000, **ent.thr**=1.8, $S = 0.5$ for ϑ-smoothing, 'sub' as topic-numbers.

5 Conclusion and Future Work

In this paper we have described a way to apply LDA for supervised text categorization by viewing it as a hierarchical topic model. This is called multi-corpus LDA (MLDA). Essentially, separate LDA models are built for each category with category-specific topics, then these models are unified, and inference is made for an unseen document w.r.t. this unified model. As a key observation, the topic unification method significantly boosted performance by avoiding overfitting to a large number of topics, requiring lower running times.

In further research we will investigate possible modifications of MLDA for hierarchical categorization for example into the full DMOZ topic tree as well as for multiple category assignments frequently appearing in Wikipedia.

Acknowledgment. We would like to thank András Benczúr for fruitful discussions, Ana Maguitman for providing us the DMOZ corpus that we used for testing and also to Dávid Siklósi.

References

1. Deerwester, S.C., Dumais, S.T., Landauer, T.K., Furnas, G.W., Harshman, R.A.: Indexing by latent semantic analysis. Journal of the American Society of Information Science 41(6), 391–407 (1990)
2. Hofmann, T.: Unsupervised Learning by Probabilistic Latent Semantic Analysis. Machine Learning 42(1), 177–196 (2001)
3. Blei, D., Ng, A., Jordan, M.: Latent Dirichlet allocation. Journal of Machine Learning Research 3(5), 993–1022 (2003)

4. Bhattacharya, I., Getoor, L.: A latent dirichlet model for unsupervised entity resolution. In: SIAM International Conference on Data Mining (2006)
5. Xing, D., Girolami, M.: Employing Latent Dirichlet Allocation for fraud detection in telecommunications. Pattern Recognition Letters 28(13), 1727–1734 (2007)
6. Elango, P., Jayaraman, K.: Clustering Images Using the Latent Dirichlet Allocation Model (2005), `http://www.cs.wisc.edu/~pradheep/`
7. Fei-Fei, L., Perona, P.: A Bayesian hierarchical model for learning natural scene categories. In: IEEE Computer Society Conference on Computer Vision and Pattern Recognition, 2005. CVPR 2005, vol. 2 (2005)
8. Sivic, J., Russell, B., Efros, A., Zisserman, A., Freeman, W.: Discovering Objects and their Localization in Images. In: Tenth IEEE International Conference on Computer Vision, ICCV 2005, vol. 1 (2005)
9. Wei, X., Croft, W.: LDA-based document models for ad-hoc retrieval. In: Proceedings of the 29th annual international ACM SIGIR conference on Research and development in information retrieval, pp. 178–185 (2006)
10. Sebastiani, F.: Machine learning in automated text categorization. ACM Comput. Surv. 34(1), 1–47 (2002)
11. Blei, D., Griffiths, T., Jordan, M., Tenenbaum, J.: Hierarchical topic models and the nested Chinese restaurant process. In: Advances in Neural Information Processing Systems 16: Proceedings of the 2003 Conference, p. 17. Bradford Book (2004)
12. Teh, Y., Jordan, M., Beal, M., Blei, D.: Hierarchical dirichlet processes. Journal of the American Statistical Association 101(476), 1566–1581 (2006)
13. Biró, I., Szabó, J., Benczúr, A.: Latent Dirichlet Allocation in Web Spam Filtering. In: Proc. 4th AIRWeb (2008)
14. Heinrich, G.: Parameter estimation for text analysis. Technical report (2004)
15. Griffiths, T., Steyvers, M.: Finding scientific topics. Proceedings of the National Academy of Sciences 101(suppl. 1), 5228–5235 (2004)
16. Minka, T., Lafferty, J.: Expectation-propagation for the generative aspect model. In: Uncertainty in Artificial Intelligence, UAI (2002)
17. Witten, I.H., Frank, E.: Data Mining: Practical Machine Learning Tools and Techniques, 2nd edn. Morgan Kaufmann Series in Data Management Systems. Morgan Kaufmann, San Francisco (2005)
18. Forman, G., Guyon, I., Elisseeff, A.: An Extensive Empirical Study of Feature Selection Metrics for Text Classification. Journal of Machine Learning Research 3(7-8), 1289–1305 (2003)
19. Li, J., Sun, M.: Scalable Term Selection for Text Categorization. In: Proceedings of the 2007 Joint Conference on Empirical Methods in Natural Language Processing and Computational Natural Language Learning (EMNLP-CoNLL), pp. 774–782 (2007)

New Regularized Algorithms for Transductive Learning

Partha Pratim Talukdar and Koby Crammer

Computer & Information Science Department
University of Pennsylvania
Philadelphia, PA 19104
{partha,crammer}@cis.upenn.edu

Abstract. We propose a new graph-based label propagation algorithm for transductive learning. Each example is associated with a vertex in an undirected graph and a weighted edge between two vertices represents similarity between the two corresponding example. We build on Adsorption, a recently proposed algorithm and analyze its properties. We then state our learning algorithm as a convex optimization problem over multi-label assignments and derive an efficient algorithm to solve this problem. We state the conditions under which our algorithm is guaranteed to converge. We provide experimental evidence on various real-world datasets demonstrating the effectiveness of our algorithm over other algorithms for such problems. We also show that our algorithm can be extended to incorporate additional prior information, and demonstrate it with classifying data where the labels are not mutually exclusive.

Keywords: label propagation, transductive learning, graph based semi-supervised learning.

1 Introduction

Supervised machine learning methods have achieved considerable success in a wide variety of domains ranging from Natural Language Processing, Speech Recognition to Bioinformatics. Unfortunately, preparing labeled data for such methods is often expensive and time consuming, while unlabeled data are widely available in many cases. This was the major motivation that led to the development of semi-supervised algorithms which learn from limited amounts of labeled data and vast amounts of freely available unannotated data.

Recently, graph based semi-supervised algorithms have achieved considerable attention [2,7,11,14,17]. Such methods represent instances as vertices in a graph with edges between vertices encoding similarities between them. Graph-based semi-supervised algorithms often propagate the label information from the few labeled vertices to the entire graph. Most of the algorithms tradeoff between *accuracy* (initially labeled nodes should retain those labels, relaxations allowed by some methods) with *smoothness* (adjacent vertices in the graph should be assigned similar labels). Most algorithms only output label information to the unlabeled data in a transductive setting, while some algorithms are designed for the semi-supervised framework and build a classification model which can be applied to out-of-sample examples.

W. Buntine et al. (Eds.): ECML PKDD 2009, Part II, LNAI 5782, pp. 442–457, 2009.

Adsorption [1] is one such recently proposed graph based semi-supervised algorithm which has been successfully used for different tasks, such as recommending YouTube videos to users [1] and large scale assignment of semantic classes to entities within Information Extraction [13]. Adsorption has many desirable properties: it can perform multiclass classification, it can be parallelized and hence can be scaled to handle large data sets which is of particular importance for semi-supervised algorithms. Even though Adsorption works well in practice, to the best of our knowledge it has never been analyzed before and hence our understanding of it is limited. Hoping to fill this gap, we make the following contributions in this paper:

- We analyze the Adsorption algorithm [1] and show that there does not exist an objective function whose local optimization would be the output of the Adsorption algorithm.
- Motivated by this negative result, we propose a new graph based semi-supervised algorithm (Modified Adsorption, MAD), which shares Adsorption's desirable properties, yet with some important differences.
- We state the learning problem as an optimization problem and develop efficient (iterative) methods to solve it. We also list the conditions under which the optimization algorithm – MAD – is guaranteed to converge.
- The transition to an optimization based learning algorithm provides a flexible and general framework that enables us to specify a variety requirements. We demonstrate this framework using data with non-mutually exclusive labels, resulting in the Modified Adsorption for Dependent Labels (MADDL, pronounced *medal*) algorithm.
- We provide experimental evidence demonstrating the effectiveness of our proposed algorithm on various real world datasets.

2 Adsorption Algorithm

Adsorption [1] is a general algorithmic framework for transductive learning where the learner is often given a small set of labeled examples and a very large set of unlabeled examples. The goal is to label all the unlabeled examples, and possibly under the assumption of label-noise, also to relabel the labeled examples.

As many other related algorithms [17,12,5], Adsorption assumes that the learning problem is given in a *graph* form, where examples or instances are represented as nodes or vertices and edges code *similarity* between examples. Some of the nodes are associated with a pre-specified label, which is correct in the noise-free case, or can be subject to label-noise. Additional information can be given in the form of *weights* over the labels. Adsorption propagates label-information from the labeled examples to the entire set of vertices via the edges. The labeling is represented using a non-negative score for each label, with high score for some label indicating high-association of a vertex (or its corresponding instance) with that label. If the scores are additively normalized they can be thought of as a conditional distribution over the labels given the node (or example) identity.

More formally, Adsorption is given an undirected graph $G = (V, E, W)$, where a node $v \in V$ corresponds to an example, an edge $e = (a, b) \in V \times V$ indicates that the label of the two vertices $a, b \in V$ should be similar and the weight $W_{ab} \in \mathbb{R}_+$ reflects the strength of this similarity.

We denote the total number of examples or vertices by $n = |V|$, by n_l the number of examples for which we have prior knowledge of their label and by n_u the number of unlabeled examples to be labeled. Clearly $n_l + n_u = n$. Let $\mathcal{L}$ be the set of possible labels, their total number is denoted by $m = |\mathcal{L}|$ and without loss of generality we assume that the possible labels are $\mathcal{L} = \{1 \ldots m\}$. Each instance $v \in V$ is associated with two row-vectors $\mathbf{Y}_v, \hat{\mathbf{Y}}_v \in \mathbb{R}_+^m$. The lth element of the vector $\mathbf{Y}_v$ encodes the prior knowledge for vertex v. The higher the value of $\mathbf{Y}_{vl}$ the stronger we a-priori believe that the label of v should be $l \in \mathcal{L}$ and a value of zero $\mathbf{Y}_{vl} = 0$ indicates no prior about the label l for vertex v. Unlabeled examples have all their elements set to zero, that is $\mathbf{Y}_{vl} = 0$ for $l = 1 \ldots m$. The second vector $\hat{\mathbf{Y}}_v \in \mathbb{R}_+^m$ is the output of the algorithm, using similar semantics as $\mathbf{Y}_v$. For example, a high value of $\hat{\mathbf{Y}}_{vl}$ indicates that the algorithm believes that the vertex v should have the label l. We denote by $\mathbf{Y}, \hat{\mathbf{Y}} \in \mathbb{R}_+^{n \times m}$ the matrices whose rows are $\mathbf{Y}_v$ and $\hat{\mathbf{Y}}_v$ respectively. Finally, we denote by $\mathbf{0}_d$ the all-zeros row vector of dimension d.

2.1 Random-Walk View

The Adsorption algorithm can be viewed as a controlled random walk over the graph G. The control is formalized via three possible actions: `inject`, `continue` and `abandon` (denoted by *inj, cont, abnd*) with pre-defined probabilities $p_v^{inj}, p_v^{cont}, p_v^{abnd} \geq 0$ per vertex $v \in V$. Clearly their sum is unit: $p_v^{inj} + p_v^{cont} + p_v^{abnd} = 1$. To label any vertex $v \in V$ (either labeled or unlabeled) we initiate a random-walk starting at v facing three options: with probability p_v^{inj} the random-walk stops and return (i.e. *inject*) the pre-defined vector information $\mathbf{Y}_v$. We constrain $p_v^{inj} = 0$ for unlabeled vertices v. Second, with probability p_v^{abnd} the random-walk *abandons* the labeling process and return the all-zeros vector $\mathbf{0}_m$. Third, with probability p_v^{cont} the random-walk *continues* to one of v's neighbors v' with probability proportional to $W_{v'v} \geq 0$. Note that by definition $W_{v'v} = 0$ if $(v, v') \notin E$. We summarize the above process with the following set of equations. The transition probabilities are,

$$\Pr\left[v'|v\right] = \begin{cases} \dfrac{W_{v'v}}{\sum\limits_{u\,:\,(u,v)\in E} W_{uv}} & (v',v) \in E \\ 0 & \text{otherwise} \end{cases} . \tag{1}$$

The (expected) score $\hat{\mathbf{Y}}_v$ for node $v \in V$ is given by,

$$\hat{\mathbf{Y}}_v = p_v^{inj} \times \mathbf{Y}_v + p_v^{cont} \times \sum_{v'\,:\,(v',v)\in E} \Pr\left[v'|v\right] \hat{\mathbf{Y}}_{v'} + p_v^{abnd} \times \mathbf{0}_m . \tag{2}$$

2.2 Averaging View

For this view we add a designated symbol called the dummy label denoted by $\nu \notin \mathcal{L}$. This additional label explicitly encodes ignorance about the correct label and it means

Algorithm 1. Adsorption Algorithm

Input:
- **Graph:** $G = (V, E, W)$
- **Prior labeling:** $\mathbf{Y}_v \in \mathbb{R}^{m+1}$ for $v \in V$
- **Probabilities:** $p_v^{inj}, p_v^{cont}, p_v^{abnd}$ for $v \in V$

Output:
- **Label Scores:** $\hat{\mathbf{Y}}_v$ for $v \in V$

```
Ŷ_v ← Y_v for v ∈ V {Initialization}

repeat
    D_v ← (Σ_u W_uv Ŷ_u) / (Σ_u W_uv) for v ∈ V
    for all v ∈ V do
        Ŷ_v ← p_v^inj × Y_v + p_v^cont × D_v + p_v^abnd × r
    end for
until convergence
```

$$D_v \leftarrow \frac{\sum_u W_{uv}\hat{\mathbf{Y}}_u}{\sum_u W_{uv}} \text{ for } v \in V$$

$$\hat{\mathbf{Y}}_v \leftarrow p_v^{inj} \times \mathbf{Y}_v + p_v^{cont} \times D_v + p_v^{abnd} \times \mathbf{r}$$

that a dummy label can be used instead. Explicitly, we add an additional column to all the vectors defined above, and have that $\mathbf{Y}_v, \hat{\mathbf{Y}}_v \in \mathbb{R}_+^{m+1}$ and $\mathbf{Y}, \hat{\mathbf{Y}} \in \mathbb{R}_+^{n\times(m+1)}$. We set $\mathbf{Y}_{v\nu} = 0$, that is, a-priori no vertex is associated with the dummy label, and replace the zero vector $\mathbf{0}_m$ with the vector $\mathbf{r} \in \mathbb{R}_+^{m+1}$ where $\mathbf{r}_l = 0$ for $l \neq \nu$ and $\mathbf{r}_\nu = 1$. In words, if the random-walk is abandoned, then the corresponding labeling vector is zero for all true labels in $\mathcal{L}$, and an arbitrary value of unit for the dummy label ν. This way, there is always positive score for at least one label, the ones in $\mathcal{L}$ or the dummy label.

The averaging view then defines a set of *fixed-point* equations to update the predicted labels. A summary of the equations appears in Algorithm 1. The algorithm is run until convergence which is achieved when the label distribution on each node ceases to change within some tolerance value. Since Adsorption is memoryless, it scales to tens of millions of nodes with dense edges and can be easily parallelized [1].

Baluja et. al. [1] show that up to the additional dummy label, these two views are equivalent. It remains to specify the values of p_v^{inj}, p_v^{cont} and p_v^{abnd}. For the experiments reported in Section 6, we set their value using the following heuristics (adapted from Baluja et. al. [1]) which depends on a parameter β which we set to $\beta = 2$. For each node v we define two quantities: c_v and d_v and define

$$p_v^{cont} \propto c_v \quad ; \quad p_v^{inj} \propto d_v \; .$$

The first quantity $c_v \in [0, 1]$ is monotonically decreasing with the number of neighbors for node v in the graph G. Intuitively, the higher the value of c_v, the lower the number of neighbors of vertex v and higher the information they contain about the labeling of v. The other quantity $d_v \geq 0$ is monotonically increasing with the entropy (for labeled vertices), and in this case we prefer to use the prior-information rather than the computed quantities from the neighbors.

Specifically we first compute the entropy of the transition probabilities for each node,

$$H[v] = -\sum_u \Pr[u|v] \log \Pr[u|v] \ ,$$

and then pass it through the following monotonically decreasing function,

$$f(x) = \frac{\log \beta}{\log(\beta \ + \ e^x))} \ .$$

Note that $f(0) = log(\beta)/\log(\beta + 1)$ and that $f(x)$ goes to zero, as x goes to infinity. We define,

$$c_v = f\left(H[v]\right) \ .$$

Next we define,

$$d_v = \begin{cases} (1 - c_v) \times \sqrt{H[v]} & \text{the vertex v is labeled} \\ 0 & \text{the vertex v is unlabled} \end{cases}$$

Finally, to ensure proper normalization of p_v^{cont}, p_v^{inj} and p_v^{abnd}, we define,

$$z_v = \max(c_v + d_v, 1) \ ,$$

and

$$p_v^{cont} = \frac{c_v}{z_v} \quad ; \quad p_v^{inj} = \frac{d_v}{z_v} \quad ; \quad p_v^{abnd} = 1 - p_v^{cont} - p_v^{inj} \ .$$

Thus, abandonment occurs only when the continuation and injection probabilities are low enough. This is most likely to happen at unlabeled nodes with high degree. Once the random walk reaches such a node (v), the walk is terminated with probability p_v^{abnd}. This, in effect, prevents the Adsorption algorithm from propagating information through high degree nodes. We note that the probabilities p_v^{inj}, p_v^{cont} and p_v^{abnd} for node v may be set with heuristics other than the fan-out entropy heuristics shown above to suit specific application contexts.

3 Analysis of the Adsorption Algorithm

Our next goal is to find an objective function that the Adsorption algorithm minimizes. Our starting point is line 6 of Algorithm 1. We note that when the algorithm converges, both sides of the assignment operator equal each other before the assignment takes place. Thus when the algorithm terminates, we have for all $v \in V$:

$$\hat{\mathbf{Y}}_v = p_v^{inj} \times \mathbf{Y}_v + p_v^{cont} \times \frac{1}{N_v} \sum_u W_{uv} \hat{\mathbf{Y}}_u + p_i^{abnd} \times \mathbf{r} \ ,$$

where

$$N_v = \sum_{v'} W_{v'v} \ .$$

The last set of equalities is equivalent to,

$$G_v\left(\{\hat{\mathbf{Y}}_u\}_{u\in V}\right) = 0 \text{ for } v \in V\ , \tag{3}$$

where we define,

$$G_v\left(\{\hat{\mathbf{Y}}_u\}_{u\in V}\right) = p_v^{inj} \times \mathbf{Y}_v + p_v^{cont} \times \frac{1}{N_v}\sum_u W_{uv}\hat{\mathbf{Y}}_u + p_i^{abnd} \times \mathbf{r} - \hat{\mathbf{Y}}_v\ .$$

Now, if the Adsorption algorithm was minimizing some objective function (denoted by $\mathcal{Q}\left(\{\hat{\mathbf{Y}}_u\}_{u\in V}\right)$), the termination condition of Eq. (3) was in fact a condition on the vector of its partial derivatives where we would identify

$$G_v = \frac{\partial}{\partial \hat{\mathbf{Y}}_v}\mathcal{Q}\ . \tag{4}$$

Since the functions G_v are linear (and thus has continuous derivatives), necessary conditions for the existence of a function $\mathcal{Q}$ such that (4) holds is that the derivatives of G_v are symmetric [8], that is,

$$\frac{\partial}{\partial \hat{\mathbf{Y}}_v}G_u = \frac{\partial}{\partial \hat{\mathbf{Y}}_u}G_v\ .$$

Computing and comparing the derivatives we get,

$$\frac{\partial}{\partial \hat{\mathbf{Y}}_u}G_v = p_v^{cont}\left(\frac{W_{uv}}{N_v} - \delta_{u,v}\right) \neq p_u^{cont}\left(\frac{W_{vu}}{N_u} - \delta_{u,v}\right) = \frac{\partial}{\partial \hat{\mathbf{Y}}_v}G_u\ ,$$

which is true since in general $N_u \neq N_v$ and $p_v^{cont} \neq p_u^{cont}$. We conclude:

Theorem 1. *There does not exists a function $\mathcal{Q}$ with continuous second partial derivatives such that the Adsorption algorithm convergences when gradient of $\mathcal{Q}$ are equal to zero.*

In other words, we searched for a (well-behaved) function $\mathcal{Q}$ such that its local optimal would be the output of the Adsorption algorithm, and showed that this search will always fail. We use this negative results to define a new algorithm, which builds on the Adsorption algorithm and is optimizing a function of the unknowns $\hat{\mathbf{Y}}_v$ for $v \in V$.

4 New Algorithm: Modified Adsorption (MAD)

Our starting point is Sec. 2.2 where we assume to have been given a weighted-graph $G = (V, E, W)$ and a matrix $\mathbf{Y} \in \mathbb{R}_+^{n\times(m+1)}$ and are seeking for a labeling-matrix $\hat{\mathbf{Y}} \in \mathbb{R}_+^{n\times(m+1)}$. In this section it is more convenient to decompose the matrices $\mathbf{Y}$ and $\hat{\mathbf{Y}}$ into their columns, rather than rows. Specifically, we denote by $\mathbf{Y}_l \in \mathbb{R}_+^n$ the lth column of $\mathbf{Y}$ and similarly by $\hat{\mathbf{Y}}_l \in \mathbb{R}_+^n$ the lth column of $\hat{\mathbf{Y}}$. We distinguish the rows and columns of the matrices $\mathbf{Y}$ and $\hat{\mathbf{Y}}$ using their indices, the columns are indexed with the label index l, while the rows are indexed are with a vertex index v (or u).

We build on previous research [3,17,11] and construct an objective that reflects three requirements as follows. First, for the labeled vertices we like the output of the algorithm to be close to the a-priori given labels, that is $\mathbf{Y}_v \approx \hat{\mathbf{Y}}_v$. Second, for pair of vertices that are close according to the input graph, we would like their labeling to be close, that is $\hat{\mathbf{Y}}_u \approx \hat{\mathbf{Y}}_v$ if W_{uv} is large. Third, we want the output to be as uninformative as possible, this serves as additional regularization, that is $\hat{\mathbf{Y}}_v \approx \mathbf{r}$. We now further develop the objective in light of the three requirements.

We use the Euclidian distance to measure discrepancy between two quantities, and start with the first requirement above,

$$\begin{aligned}\sum_v p_v^{inj} \sum_l \left(\mathbf{Y}_{vl} - \hat{\mathbf{Y}}_{vl}\right)^2 &= \sum_l \sum_v p_v^{inj} \left(\mathbf{Y}_{vl} - \hat{\mathbf{Y}}_{vl}\right)^2 \\ &= \sum_l \left(\mathbf{Y}_l - \hat{\mathbf{Y}}_l\right)^\top \mathbf{S} \left(\mathbf{Y}_l - \hat{\mathbf{Y}}_l\right) ,\end{aligned}$$

where we define the diagonal matrix $\mathbf{S} \in \mathbb{R}^{n \times n}$ and $\mathbf{S}_{vv} = p_v^{inj}$ if vertex v is labeled and $\mathbf{S}_{vv} = 0$ otherwise. The matrix $\mathbf{S}$ captures the intuition that for different vertices we enforce the labeling of the algorithm to match the a-priori labeling with different extent.

Next, we modify the similarity weight between vertices to take into account the difference in degree of various vertices. In particular we define $\mathbf{W}'_{vu} = p_v^{cont} \times \mathbf{W}_{vu}$. Thus, a vertex u will *not* be similar to a vertex v if either the input weights $\mathbf{W}_{vu}$ are low or the vertex v has a large-degree (p_v^{cont} is low). We write the second requirement as,

$$\begin{aligned}&\sum_{v,u} \mathbf{W}'_{vu} \left\|\hat{\mathbf{Y}}_v - \hat{\mathbf{Y}}_u\right\|^2 = \sum_{v,u} \mathbf{W}'_{vu} \sum_l \left(\hat{\mathbf{Y}}_{vl} - \hat{\mathbf{Y}}_{ul}\right)^2 = \sum_l \sum_{v,u} \mathbf{W}'_{vu} \left(\hat{\mathbf{Y}}_{vl} - \hat{\mathbf{Y}}_{ul}\right)^2 \\ &= \sum_l \sum_v \left(\sum_u \mathbf{W}'_{vu}\right) \|\hat{\mathbf{Y}}_{vl}\|^2 + \sum_l \sum_u \left(\sum_v \mathbf{W}'_{vu}\right) \|\hat{\mathbf{Y}}_{ul}\|^2 - 2 \sum_l \sum_{u,v} \mathbf{W}'_{vu} \hat{\mathbf{Y}}_{ul} \hat{\mathbf{Y}}_{vl} \\ &= \sum_l \hat{\mathbf{Y}}_l^\top \mathbf{L} \hat{\mathbf{Y}}_l ,\end{aligned}$$

where,

$$\mathbf{L} = \mathbf{D} + \bar{\mathbf{D}} - \mathbf{W}' - \mathbf{W}'^\top ,$$

and $\mathbf{D}, \bar{\mathbf{D}}$ are $n \times n$ diagonal matrices with

$$\mathbf{D}_{vv} = \sum_u \mathbf{W}'_{uv} \quad , \quad \bar{\mathbf{D}}_{vv} = \sum_u \mathbf{W}'_{vu} .$$

Finally we define the matrix $\mathbf{R} \in \mathbb{R}_+^{n \times (m+1)}$ where the vth row of $\mathbf{R}$ equals $p_v^{abnd} \times \mathbf{r}$ (we define $\mathbf{r}$ in Sec 2.2). In other words the first m columns of $\mathbf{R}$ equal zero, and the last (m+1th column) equal the elements of p_v^{abnd}. The third requirement above is thus written as,

$$\sum_{vl} \left(\hat{\mathbf{Y}}_{vl} - \mathbf{R}_{vl}\right)^2 = \sum_l \left\|\hat{\mathbf{Y}}_l - \mathbf{R}_l\right\|^2 .$$

We combine the three terms above into a single objective (which we would like to minimize), giving to each term a different importance using the weights μ_1, μ_2, μ_3.

$$C(\hat{\mathbf{Y}}) = \sum_l \left[\mu_1 \left(\mathbf{Y}_l - \hat{\mathbf{Y}}_l \right)^\top \mathbf{S} \left(\mathbf{Y}_l - \hat{\mathbf{Y}}_l \right) + \mu_2 \hat{\mathbf{Y}}_l^T \mathbf{L} \hat{\mathbf{Y}}_l + \mu_3 \left\| \hat{\mathbf{Y}}_l - \mathbf{R}_l \right\|_2^2 \right] . \quad (5)$$

The objective in Equation 5 is similar to the Quadratic Cost Criteria [3], with the exception that the matrices $\mathbf{S}$ and $\mathbf{L}$ have different constructions. We remind the reader that $\hat{\mathbf{Y}}_l, \mathbf{Y}_l, \mathbf{R}_l$ are the lth columns (each of size $n \times 1$) of the matrices $\hat{\mathbf{Y}}, \mathbf{Y}$ and $\mathbf{R}$ respectively.

4.1 Solving the Optimization Problem

We now develop an algorithm to optimize (5) similar to the quadratic cost criteria [3]. Differentiating Equation 5 w.r.t. $\hat{\mathbf{Y}}_l$ we get,

$$\begin{aligned} \frac{1}{2} \frac{\delta C(\hat{\mathbf{Y}})}{\delta \hat{\mathbf{Y}}_l} &= \mu_1 \mathbf{S}(\hat{\mathbf{Y}}_l - \mathbf{Y}_l) + \mu_2 \mathbf{L} \hat{\mathbf{Y}}_l + \mu_3 (\hat{\mathbf{Y}}_l - \mathbf{R}_l) \\ &= (\mu_1 \mathbf{S} + \mu_2 \mathbf{L} + \mu_3 \mathbf{I}) \hat{\mathbf{Y}}_l - (\mu_1 \mathbf{S} \mathbf{Y}_l + \mu_3 \mathbf{R}_l) . \end{aligned} \quad (6)$$

Differentiating once more we get,

$$\frac{1}{2} \frac{\delta C(\hat{\mathbf{Y}})}{\delta \hat{\mathbf{Y}}_l \delta \hat{\mathbf{Y}}_l} = \mu_1 \mathbf{S} + \mu_2 \mathbf{L} + \mu_3 \mathbf{I} ,$$

and since both $\mathbf{S}$ and $\mathbf{L}$ are symmetric and positive semidefinite matrices (PSD), we get that the Hessian is PSD as well. Hence, the optimal minima is obtained by setting the first derivative (i.e. Equation (6)) to 0 as follows,

$$(\mu_1 \mathbf{S} + \mu_2 \mathbf{L} + \mu_3 \mathbf{I}) \hat{\mathbf{Y}}_l = (\mu_1 \mathbf{S} \mathbf{Y}_l + \mu_3 \mathbf{R}_l) .$$

Hence, the new labels ($\hat{\mathbf{Y}}$) can be obtained by a matrix inversion followed by matrix multiplication. However, this can be quite expensive when large matrices are involved. A more efficient way to obtain the new label scores is to solve a set of linear equations using Jacobi iteration which we now describe.

4.2 Jacobi Method

Given the following linear system (in x)

$$\mathbf{M} x = b$$

the Jacobi iterative algorithm defines the approximate solution at the (t+1)th iteration given the solution at tth iteration as follows,

$$x_i^{(t+1)} = \frac{1}{\mathbf{M}_{ii}} \left(b_i - \sum_{j \neq i} \mathbf{M}_{ij} x_j^{(t)} \right) . \quad (7)$$

We apply the iterative algorithm to our problem by substituting $x = \hat{\mathbf{Y}}_l$, $\mathbf{M} = \mu_1\mathbf{S} + \mu_2\mathbf{L} + \mu_3\mathbf{I}$ and $b = \mu_1\mathbf{S}\mathbf{Y}_l + \mu_3\mathbf{R}_l$ in (7),

$$\hat{\mathbf{Y}}_{vl}^{(t+1)} = \frac{1}{\mathbf{M}_{vv}}\left(\mu_1(\mathbf{S}\mathbf{Y}_l)_v + \mu_3\mathbf{R}_{vl} - \sum_{u\neq v}\mathbf{M}_{vu}\hat{\mathbf{Y}}_{ul}^{(t)}\right) \tag{8}$$

Let us compute the values of $(\mathbf{S}\mathbf{Y}_l)_i$, $\mathbf{M}_{ij(j\neq i)}$ and $\mathbf{M}_{ii}$. First,

$$\mathbf{M}_{vu(v\neq u)} = \mu_1\mathbf{S}_{vu} + \mu_2\mathbf{L}_{vu} + \mu_3\mathbf{I}_{vu} .$$

Note that since $\mathbf{S}$ and $\mathbf{I}$ are diagonal, we have that $\mathbf{S}_{vu} = 0$ and $\mathbf{I}_{vu} = 0$ for $u \neq v$. Substituting the value of $\mathbf{L}$ we get,

$$\mathbf{M}_{vu(v\neq u)} = \mu_2\mathbf{L}_{vu} = \mu_2\left(\mathbf{D}_{vu} + \bar{\mathbf{D}}_{vu} - \mathbf{W}'_{vu} - \mathbf{W}'_{uv}\right) ,$$

and as before the matrices $\mathbf{D}$ and $\bar{\mathbf{D}}$ are diagonal and thus $\mathbf{D}_{vu} + \bar{\mathbf{D}}_{vu} = 0$. Finally, substituting the values of $\mathbf{W}'_{vu}$ and $\mathbf{W}'_{uv}$ we get,

$$\mathbf{M}_{vu(v\neq u)} = -\mu_2 \times (p_v^{cont} \times \mathbf{W}_{vu} + p_u^{cont} \times \mathbf{W}_{uv}) . \tag{9}$$

We now compute the second quantity,

$$(\mathbf{S}\mathbf{Y}_l)_{vu} = \mathbf{S}_{vv}\mathbf{Y}_{vv} + \sum_{t\neq v}\mathbf{S}_{vt}\mathbf{Y}_{tu} = p_v^{inj} \times \mathbf{Y}_{vv} ,$$

where the second term equals zero since $\mathbf{S}$ is diagonal. Finally, the third term,

$$\begin{aligned}\mathbf{M}_{vv} &= \mu_1\mathbf{S}_{vv} + \mu_2\mathbf{L}_{vv} + \mu_3\mathbf{I}_{vv}\\ &= \mu_1 \times p_v^{inj} + \mu_2(\mathbf{D}_{vv} + \bar{\mathbf{D}}_{vv} - \mathbf{W}'_{vv} - \mathbf{W}'_{vv}) + \mu_3\\ &= \mu_1 \times p_v^{inj} + \mu_2\sum_{u\neq v}(p_v^{cont}\mathbf{W}_{vu} + p_u^{cont}\mathbf{W}_{uv}) + \mu_3 .\end{aligned}$$

Plugging the above equations into (8) and using the fact that the diagonal elements of $\mathbf{W}$ are zero, we get,

$$\hat{\mathbf{Y}}_v^{(t+1)} = \frac{1}{\mathbf{M}_{vv}}\left(\mu_1 p_v^{inj}\,\mathbf{Y}_v + \mu_2\sum_u\left(p_v^{cont}\mathbf{W}_{vu} + p_u^{cont}\mathbf{W}_{uv}\right)\hat{\mathbf{Y}}_u^{(t)} + \mu_3\, p_v^{abnd}\,\mathbf{r}\right) . \tag{10}$$

We call the new algorithm MAD for Modified-Adsorption and it is summarized in Algorithm 2. Note that for graphs G that are invariant to permutations of the vertices, and setting $\mu_1 = 2 \times \mu_2 = \mu_3 = 1$, MAD reduces to the Adsorption algorithm.

4.3 Convergence

A sufficient condition for the iterative process of Equation (7) to converge is that $\mathbf{M}$ is strictly diagonally dominant [10], that is if,

$$|\mathbf{M}_{vv}| > \sum_{u\neq v}|\mathbf{M}_{vu}| \quad \text{for all values of } v$$

Algorithm 2. Modified Adsorption (MAD) Algorithm

Input:
- **Graph:** $G = (V, E, W)$
- **Prior labeling:** $\mathbf{Y}_v \in \mathbb{R}^{m+1}$ for $v \in V$
- **Probabilities:** $p_v^{inj}, p_v^{cont}, p_v^{abnd}$ for $v \in V$

Output:
- **Label Scores:** $\hat{\mathbf{Y}}_v$ for $v \in V$

1: $\hat{\mathbf{Y}}_v \leftarrow \mathbf{Y}_v$ for $v \in V$ {Initialization}
2: $\mathbf{M}_{vv} \leftarrow \mu_1 \times p_v^{inj} + \mu_2 \sum_{u \neq v} (p_v^{cont} \mathbf{W}_{vu} + p_u^{cont} \mathbf{W}_{uv}) + \mu_3$
3: **repeat**
4: $\quad D_v \leftarrow \sum_u \left(p_v^{cont} \mathbf{W}_{vu} + p_u^{cont} \mathbf{W}_{uv}\right) \hat{\mathbf{Y}}_u$
5: $\quad$ **for all** $v \in V$ **do**
6: $\quad\quad \hat{\mathbf{Y}}_v \leftarrow \frac{1}{\mathbf{M}_{vv}} \left(\mu_1 \times p_v^{inj} \times \mathbf{Y}_v + \mu_2 \times D_v + \mu_3 \times p_v^{abnd} \times \mathbf{r}\right)$
7: $\quad$ **end for**
8: **until** convergence

We have,

$$
\begin{aligned}
|\mathbf{M}_{vv}| - \sum_{u \neq v} |\mathbf{M}_{vu}| &= \mu_1 \times p_v^{inj} + \mu_2 \times \sum_{u \neq v} \left(p_v^{cont} \times \mathbf{W}_{vu} + p_u^{cont} \times \mathbf{W}_{uv}\right) + \mu_3 - \\
&\qquad \mu_2 \times \sum_{u \neq v} \left(p_v^{cont} \times \mathbf{W}_{vu} + p_u^{cont} \times \mathbf{W}_{uv}\right) \\
&= \mu_1 \times p_v^{inj} + \mu_3
\end{aligned} \tag{11}
$$

Note that $p_v^{inj} \geq 0$ for all v and that μ_3 is a free parameter in (11). Thus we can guarantee a strict diagonal dominance (and hence convergence) by setting $\mu_3 > 0$.

5 Extensions: Non-mutually Exclusive Labels

In many learning settings, labels are not mutually exclusive. For example, in hierarchical classification, labels are organized in a tree. In this section, we extend the MAD algorithm to handle dependence among labels. This can be easily done using our new formulation which is based on objective optimization. Specifically, we shall add additional terms to the objective for each pair of dependent labels. Let C be a $m \times m$ matrix where m is the number of labels (excluding the dummy label) as before. Each entry, $C_{ll'}$, of this matrix C represents the dependence or similarity among the labels l and l'. By encoding dependence in this pairwise fashion, we can capture dependencies among labels represented as arbitrary graphs. The extended objective is shown in Equation 12.

$$
\begin{aligned}
C(\hat{\mathbf{Y}}) = \sum_l \Bigg[& \mu_1 \left(\mathbf{Y}_l - \hat{\mathbf{Y}}_l\right)^\top \mathbf{S} \left(\mathbf{Y}_l - \hat{\mathbf{Y}}_l\right) + \mu_2 \hat{\mathbf{Y}}_l^T \mathbf{L} \hat{\mathbf{Y}}_l + \mu_3 \left\|\hat{\mathbf{Y}}_l - \mathbf{R}_l\right\|_2^2 \\
& + \mu_4 \sum_i \sum_{l,l'} C_{ll'} (\hat{\mathbf{Y}}_{il} - \hat{\mathbf{Y}}_{il'})^2 \Bigg]
\end{aligned} \tag{12}
$$

The last term in Equation 12 penalizes the algorithm if similar labels (as determined by the matrix C) are assigned different scores, with severity of the penalty controlled by μ_4. Now, analyzing the objective in Equation 12 in the manner outlined in Section 4, we arrive at the update rule shown in Equation 13.

$$\hat{\mathbf{Y}}_{vl}^{(t+1)} = \frac{1}{\mathbf{M}_{vv}^{l}} \left(\mu_1 p_v^{inj}\, \mathbf{Y}_{vl} + \mu_2 \sum_u \left(p_v^{cont}\mathbf{W}_{vu} + p_u^{cont}\mathbf{W}_{uv}\right) \hat{\mathbf{Y}}_{ul}^{(t)} + \right.$$
$$\left. \mu_3\, p_v^{abnd}\, \mathbf{r}_l + \mu_4 \sum_{l'} C_{ll'} \hat{\mathbf{Y}}_{il'} \right) \qquad (13)$$

where,

$$\mathbf{M}_{vv}^{l} = \mu_1 \times p_v^{inj} + \mu_2 \times \sum_{u \neq v} (p_v^{cont}\mathbf{W}_{vu} + p_u^{cont}\mathbf{W}_{uv}) + \mu_3 + \mu_4 \sum_{l'} C_{ll'}$$

Replacing Line 6 in MAD (Algorithm 2) with Equation 13, we end up with a new algorithm: Modified Adsorption for Dependent Labels (MADDL). In Section 6.4, we shall use MADDL to obtain smooth ranking for sentiment classification.

6 Experimental Results

We compare MAD with various state-of-the-art learning algorithms on two tasks, text classification (Sec. 6.1) and sentiment analysis (Sec. 6.2), and demonstrate its effectiveness. In Sec. 6.3, we also provide experimental evidence showing that MAD is quite insensitive to wide variation of values of its hyper-parameters. In Sec. 6.4, we present evidence showing how MADDL can be used to obtain smooth ranking for sentiment prediction, a particular instantiation of classification with non-mutually exclusive labels. For the experiments reported in this section involving Adsorption, MAD and MADDL, the a-priori label matrix $\mathbf{Y}$ was column-normalized so that all labels have equal overall injection score. Also, the dummy label was ignored during evaluation as its main role is to add regularization during learning phase only.

6.1 Text Classification

World Wide Knowledge Base (WebKB) is a text classification dataset widely used for evaluating transductive learning algorithms. Most recently, the dataset was used by Subramanya and Bilmes [11], who kindly shared their preprocessed complete WebKB graph with us. There are a total of $4,204$ vertices in the graph, with the nodes labeled with one of four categories: *course, faculty, project, student*. A K-NN graph is created from this complete graph by retaining only top K neighbors of each node, where the value of K is treated as a hyper-parameter.

We follow the experimental protocol in [11]. The dataset was randomly partitioned into four sets. A transduction set was generated by first selecting one of the four splits at random and then sampling n_l documents from it; the remaining three sets are used as the test set for evaluation. This process was repeated 21 times to generate as many training-test splits. The first split was used to tune the hyper-parameters, with search over the

Table 1. PRBEP for the WebKB data set with n_l = 48 training and 3148 testing instances. All results are averages over 20 randomly generated transduction sets. The last row is the macro-average over all the classes. MAD is the proposed approach. Results for SVM, TSVM, SGT, LP and AM are reproduced from Table 2 of [11].

Class	SVM	TSVM	SGT	LP	AM	Adsorption	MAD
course	46.5	43.9	29.9	45.0	**67.6**	61.1	67.5
faculty	14.5	31.2	42.9	40.3	42.5	**52.8**	42.2
project	15.8	17.2	17.5	27.8	42.3	**52.6**	45.5
student	15.0	24.5	56.6	51.8	55.0	39.8	**59.6**
average	23.0	29.2	36.8	41.2	51.9	51.6	**53.7**

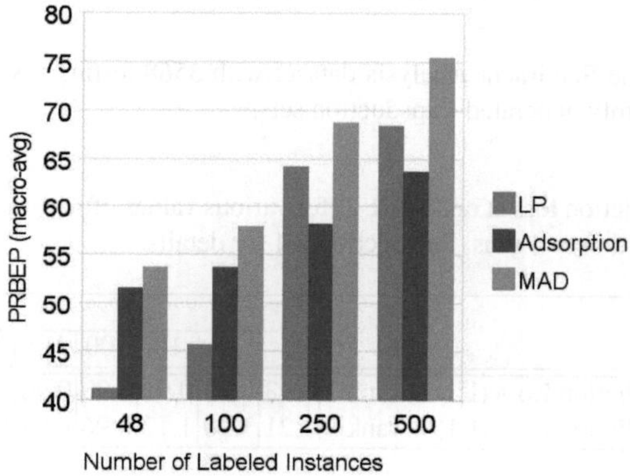

Fig. 1. PRBEP (macro-averaged) for the WebKB dataset with 3148 testing instances. All results are averages over 20 randomly generated transduction sets.

following: $K \in \{10, 50, 100, 500, 1000, 2000, 4204\}, \mu_2, \mu_3 \in \{1e{-}8, 1e{-}4, 1e{-}2, 1, 10, 1e2, 1e3\}$. The value of μ_1 was set to 1 for this experiment. Both for Adsorption and MAD, the optimal value of K was $1,000$. Furthermore, the optimal value for the other parameters were found to be $\mu_2 = \mu_3 = 1$. As in previous work [11], we use Precision-Recall Break Even Point (PRBEP) [9] as the evaluation metric. Same evaluation measure, dataset and the same experimental protocol makes the results reported here directly comparable to those reported previously [11]. For easier readability, the results from Table 2 of Subramanya and Bilmes [11] are cited in Table 1 of this paper, comparing performance of Adsorption based methods (Adsorption and MAD) to many previously proposed approaches: SVM [6], Transductive-SVM [6], Spectral Graph Transduction (SGT) [7], Label Propagation (LP) [16] and Alternating Minimization (AM) [11]. The first four rows in Table 1 shows PRBEP for individual categories, with the last line showing the macro-averaged PRBEP across all categories. The MAD algorithm achieves the best performance overall (for n_l = 48).

Performance comparison of MAD and Adsorption for increasing n_l are shown in Figure 1. Comparing these results against Fig. 2 in Subramanya and Bilmes [11], it seems that MAD outperforms all other methods compared (except AM [11]) for all

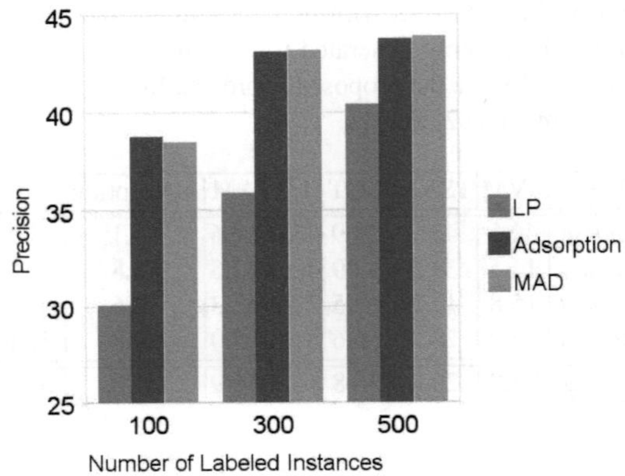

Fig. 2. Precision for the Sentiment Analysis dataset with 3568 testing instances. All results are averages over 4 randomly generated transduction sets.

Table 2. Average prediction loss at ranks 1 & 2 (for various values of μ_4) for sentiment prediction. All results are averaged over 4 runs. See Section 6.4 for details.

	μ_4					
	0	1	10	100	1e3	1e4
Prediction Loss (L1) at rank 1	0.93	0.93	0.92	0.90	0.90	0.90
Prediction Loss (L1) at rank 2	1.21	1.20	1.12	0.96	0.97	0.97

values of n_l. MAD performs better than AM for n_l = 48, but achieves second best solution for the other three values of n_l. We are currently investigating why MAD is best for settings with fewer labeled examples.

6.2 Sentiment Analysis

The goal of sentiment analysis is to automatically assign polarity scores to text collections, with a high score reflecting positive sentiment (user likes) and a low score reflecting negative sentiment (user dislikes). In this section, we report results on sentiment classification in the transductive setting. From Section 6.1 and [11], we observe that Label Propagation (LP) [16] is one of the best performing L2-norm based transductive learning algorithm. Hence, we compare the performance of MAD against Adsorption and LP.

For the experiments in this section, we use a set of $4,768$ user reviews from the electronics domain [4]. Each review is assigned one of the four scores: 1 (worst), 2, 3, 4 (best). We create a K-NN graph from these reviews by using cosine similarity as the measure of similarity between reviews. We created 5 training-test splits from this data using the process described in Section 6.1. One split was used to tune the hyper-parameters while the rest were used for training and evaluation. Hyper-parameter search was carried over the following ranges: $K \in \{10, 100, 500\}$, $\mu_1 \in \{1, 100\}$,

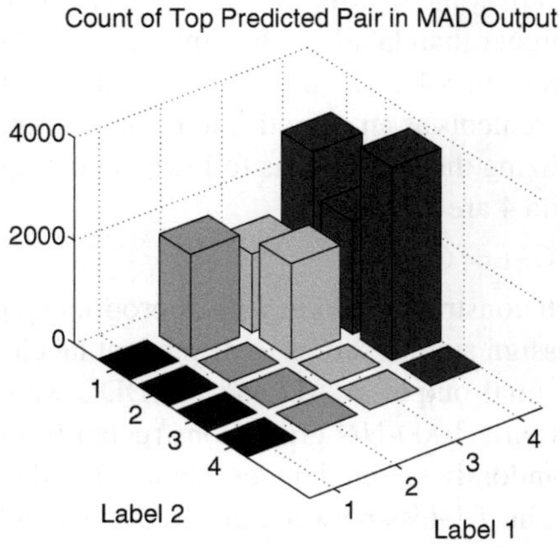

Fig. 3. Plot of counts of top predicted label pairs (order ignored) in MAD's predictions with $\mu_1 = \mu_2 = 1$, $\mu_3 = 100$

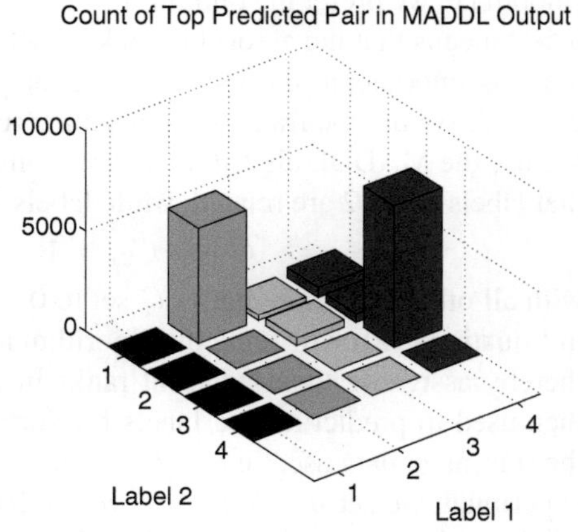

Fig. 4. Plot of counts of top label pairs (order ignored) in MADDL's predictions (Section 5), with $\mu_1 = \mu_2 = 1$, $\mu_3 = 100$, $\mu_4 = 1e3$

$\mu_2 \in \{1e{-}4, 1, 10\}$, $\mu_3 \in \{1e{-}8, 1, 100, 1e3\}$. Precision is used as the evaluation metric. Comparison of different algorithms for varying number of labeled instances are shown in Figure 2. From this, we note that MAD and Adsorption outperform LP, while Adsorption and MAD are competitive.

6.3 Parameter Sensitivity

We evaluated the sensitivity of MAD to variations of its μ_2 and μ_3 hyper-parameters, with all other hyper-parameters fixed. We used a 2000-NN graph constructed from the WebKB dataset and a 500-NN graph constructed from the Sentiment dataset. In both cases, 100 nodes were labeled. We tried three values each for μ_2 and μ_3, ranging in at least 3 order of magnitude. For the WebKB, the PRBEP varied between 43.1−49.9 and for the sentiment data, the precision varied in the range 31.4−36.4 with $\mu_2 \leq \mu_3$ while precision dropped to 25 with $\mu_2 > \mu_3$. This underscores the need for regularization in these models, which is enforced with high μ_3. We note that in both cases the algorithm is less sensitive to the value of μ_2 than the value of μ_3. In general, we have found that setting μ_3 to one or two order magnitude more than μ_2 is a reasonable choice. We have also found that the MAD algorithm is quite insensitive to variations in μ_1. For example on the sentiment dataset, we tried two values for μ_1 ranging two order of magnitude, with other hyper-parameters fixed. In this case, precision varied in the range 36.2 - 36.3.

6.4 Smooth Ranking for Sentiment Analysis

We revisit the sentiment prediction problem in Section 6.2, but with the additional requirement that ranking of the labels (1, 2, 3, 4) generated by the algorithm should be smooth i.e. we prefer the ranking $1 > 2 > 3 > 4$ over the ranking $1 > 4 > 3 > 2$, where $3 > 2$ means that the algorithm ranks label 3 higher than label 2. The ranking $1 > 2 > 3 > 4$ is smoother as it doesn't involve rough transition $1 > 4$ which is present in $1 > 4 > 3 > 2$. We use the framework of stating requirements as an objective to be optimized. We use the MADDL algorithm of Sec. 5 initializing the matrix C as follows (assuming that labels 1 and 2 are related, while labels 3 and 4 are related):

$$C_{12} = C_{21} = 1 \quad , \quad C_{34} = C_{43} = 1$$

with all other entries in matrix C set to 0. Such constraints (along with appropriate μ_4 in Equation (12)) will force the algorithm to assign similar scores to dependent labels, thereby assigning them adjacent ranks in the final output. MAD and MADDL were then used to predict ranked labels for vertices on a 1000-NN graph constructed from the sentiment data used in Sec. 6.2, with 100 randomly selected nodes labeled. For this experiment we set $\mu_1 = \mu_2 = 1$, $\mu_3 = 100$. The $L1$-loss between the gold label and labels predicted at ranks $r = 1, 2$ for increasing values of μ_4 are given in Table 2. Note that, MADDL with $\mu_4 = 0$ corresponds to MAD. From Table 2 we observe that with increasing μ_4, MADDL is ranking at $r = 2$ a label which is related (as per C) to the top ranked label at $r = 1$, but at the same time maintain the quality of prediction at $r = 1$ (first row of Table 2), thereby ensuring a smoother ranking. From Table 2, we also observe that MADDL is insensitive to variations of μ_4 beyond a certain range. This suggests that μ_4 may be set to a (high) value and that tuning it may not be necessary.

Another view of the same phenomenon is shown in Fig. 3 and Fig. 4. In these figures, we plot the counts of top predicted label pair (order of prediction is ignored for better readability) generated by the MAD and MADDL algorithms. By comparing these two figures we observe that label pairs (e.g. (2,1) and (4,3)) favored by C (above) are more frequent in MADDL's predictions than in MAD's. At the same time, non-smooth predictions (e.g. (4, 1)) are virtually absent in MADDL's predictions while they are quite frequent in MAD's. These clearly demonstrate MADDL's ability to generate smooth predictions in a principled way, and more generally the ability to handle data with non-mutually exclusive or dependent labels.

7 Related Work

LP [16] is one of the first graph based semi-supervised algorithms. Even though there are several similarities between LP and MAD, there are important differences: (1) LP doesn't allow the labels on seeded nodes to change (while MAD does). As was pointed out previously [3], this can be problematic in case of noisy seeds. (2) There is no way for LP to express label uncertainty about a node. MAD can accomplish this by assigning high score to the dummy label. More recently, a KL minimization based algorithm was presented in [11]. Further investigation is necessary to determine the merits of each approach. For a general introduction to the area of graph-based semi-supervised learning, the reader is referred to a survey by Zhu [15].

8 Conclusion

In this paper we have analyzed the Adsorption algorithm [1] and proposed a new graph based semi-supervised learning algorithm, MAD. We have developed efficient (iterative) solution to solve our convex optimization based learning problem. We have also listed the conditions under which the algorithm is guaranteed to converge. Transition to an optimization based learning algorithm allows us to easily extend the algorithm to handle data with non-mutually exclusive labels, resulting in the MADDL algorithm. We have provided experimental evidence demonstrating effectiveness of our proposed methods. As part of future work, we plan to evaluate the proposed methods further and apply the MADDL method in problems with dependent labels (e.g. Information Extraction).

Acknowledgment

This research is partially supported by NSF grant #IIS-0513778. The authors would like to thank F. Pereira and D. Sivakumar for useful discussions.

References

1. Baluja, S., Seth, R., Sivakumar, D., Jing, Y., Yagnik, J., Kumar, S., Ravichandran, D., Aly, M.: Video suggestion and discovery for youtube: taking random walks through the view graph. In: WWW (2008)
2. Belkin, M., Niyogi, P., Sindhwani, V.: Manifold regularization: A geometric framework for learning from labeled and unlabeled examples. JMLR 7, 2399–2434 (2006)
3. Bengio, Y., Delalleau, O., Roux, N.: Label Propogation and Quadratic Criterion. In: Semi-Supervised Learning (2007)
4. Blitzer, J., Dredze, M., Pereira, F.: Biographies, bollywood, boom-boxes and blenders: Domain adaptation for sentiment classification. In: ACL (2007)
5. Indyk, P., Matousek, J.: Low-distortion embeddings of finite metric spaces. In: Handbook of Discrete and Computational Geometry (2004)
6. Joachims, T.: Transductive inference for text classification using support vector machines. In: ICML (1999)
7. Joachims, T.: Transductive learning via spectral graph partitioning. In: ICML (2003)
8. Katz, V.J.: The history of stokes' theorem. Mathematics Magazine 52(3), 146–156 (1979)
9. Raghavan, V., Bollmann, P., Jung, G.: A critical investigation of recall and precision as measures of retrieval system performance. ACM TOIS 7(3), 205–229 (1989)
10. Saad, Y.: Iterative Methods for Sparse Linear Systems. Society for Industrial Math. (2003)
11. Subramanya, A., Bilmes, J.: Soft-Supervised Learning for Text Classification. In: EMNLP (2008)
12. Szummer, M., Jaakkola, T.: Partially labeled classification with markov random walks. In: NIPS (2002)
13. Talukdar, P.P., Reisinger, J., Pasca, M., Ravichandran, D., Bhagat, R., Pereira, F.: Weakly supervised acquisition of labeled class instances using graph random walks. In: EMNLP (2008)
14. Wang, J., Jebara, T., Chang, S.: Graph transduction via alternating minimization. In: ICML (2008)
15. Zhu, X.: Semi-supervised learning literature survey (2005)
16. Zhu, X., Ghahramani, Z.: Learning from labeled and unlabeled data with label propagation. Technical report, CMU CALD tech report (2002)
17. Zhu, X., Ghahramani, Z., Lafferty, J.: Semi-supervised learning using gaussian fields and harmonic functions. In: ICML (2003)

Enhancing the Performance of Centroid Classifier by ECOC and Model Refinement

Songbo Tan, Gaowei Wu, and Xueqi Cheng

Key Laboratory of Network,
Institute of Computing Technology,
Beijing, China
tansongbo@{software.ict.ac.cn,gmail.com}, {wgw,cxq}@ict.ac.cn

Abstract. With the aim of improving the performance of centroid text classifier, we attempt to make use of the advantages of Error-Correcting Output Codes (ECOC) strategy. The framework is to decompose one multi-class problem into multiple binary problems and then learn the individual binary classification problems by centroid classifier. However, this kind of decomposition incurs considerable bias for centroid classifier, which results in noticeable degradation of performance for centroid classifier. In order to address this issue, we use Model-Refinement strategy to adjust this so-called bias. The basic idea is to take advantage of misclassified examples in the training data to iteratively refine and adjust the centroids of text data. The experimental results reveal that Model-Refinement strategy can dramatically decrease the bias introduced by ECOC, and the combined classifier is comparable to or even better than SVM classifier in performance.

1 Introduction

With the advent of the Web and the enormous growth of digital content in Internet, databases, and archives, text categorization has received more and more attention in information retrieval and natural language processing community. To this date, numerous machine-learning approaches have been introduced to deal with text classification [1-6, 11-12, 19-24].

In recent years, ECOC has been applied to boost naïve bayes, decision tree and SVM classifier for text data [7-10]. Following this research direction, in this work, we explore the use of ECOC to enhance the performance of centroid classifier. The framework we adopted is to decompose one multi-class problem into multiple binary problems and then use centroid classifier to learn the individual binary classification problems.

However, this kind of decomposition incurs considerable bias [11-13] for centroid classifier. In substance, centroid classifier [2, 21] relies on a simple decision rule that a given document should be assigned a particular class if the similarity (or distance) of this document to the centroid of the class is the largest (or smallest). This decision rule is based on a straightforward assumption that the documents in one category should share some similarities with each other. However, this hypothesis is often broken by ECOC on the grounds that it ignores the similarities of original classes when disassembling one multi-class problem into multiple binary problems.

W. Buntine et al. (Eds.): ECML PKDD 2009, Part II, LNAI 5782, pp. 458–472, 2009.

In order to attack this problem, we use Model-Refinement strategy [11-12] to reduce this so-called bias. The basic idea is to take advantage of misclassified examples in the training data to iteratively refine and adjust the centroids. This technique is very flexible, which only needs one classification method and there is no change to the method in any way. The empirical evaluation shows that Model-Refinement strategy can dramatically reduce the bias and boost the performance of centroid classifier. From the perspective of mathematics, we justified that with respect to a linearly separable problem, the Model-Refinement strategy converges to the optimal solution after finite online updates.

To examine the performance of proposed method, we conduct an extensive experiment on two commonly used datasets, i.e., Newsgroup and Industry Sector. The results indicate that Model-Refinement strategy can dramatically decrease the bias introduce by ECOC, and the resulted classifier is comparable to or even better than SVM classifier in performance.

The rest of this paper is constructed as follows: Next section presents related work on applying ECOC to text classification. ECOC algorithm is described in section 3. In section 4, we present the proposed method. Experimental results are given in section 5. Finally section 6 concludes this paper.

2 Related Work

In this section, we present the related work on applying ECOC to text classification. ECOC has been applied to boost Naïve Bayes, Decision Tree, SVM and Co-Training [7-10].

Berger [7] made the first attempt to explore the application of ECOC to Naïve Bayes and Decision Tree for text categorization. He conducted his experiments on four datasets: Newsgroup, WebKB, Yahoo Science and Yahoo Health. His experiment showed that with sufficiently high bit n, combining a Decision Tree (Naïve Bayes) to an ECOC classifier can improve the performance over the one-vs.-rest Decision Tree (Naïve Bayes) approach. He also gave some theoretical evidences for the use of random codes rather than error-correcting codes.

Following the spirit of Berger, Ghani [8] explored the use of different kinds of codes, namely Error-Correcting Codes, Random Codes, Domain and Data-specific codes. Experiments conducted on Industry Sector show a reduction in classification error by up to 66% over Naive Bayes Classifier. Further more, he also gave empirical evidence for using error-correcting codes rather than random codes.

In 2002, Ghani [9] continued his research in this direction. He developed a framework to incorporate unlabeled data in ECOC setup by first decomposing multi-class problems into multiple binary problems and then using Co-Training to learn the individual binary classification problems. Experiments show that this strategy is especially useful for text classification tasks with a large number of categories and outperforms other semi-supervised learning techniques such as EM and Co-Training. In addition to being highly accurate, this method utilizes the Hamming distance from ECOC to provide high-precision results. He also present results with algorithms other than Co-Training in this framework and show that Co-Training is uniquely suited to work well within ECOC.

In 2002, Rennie [10] compared Naive Bayes and Support Vector Machines using ECOC on the task of multi-class text classification. The experimental results show that the Support Vector Machine can perform multi-class text classification very effectively when used as part of an ECOC scheme. Rennie argues that its improved ability to perform binary classification gives it much lower error scores than Naive Bayes.

3 Error-Correcting Output Codes

Error-Correcting Output Codes (ECOC) is a form of combination of multiple classifiers [8]. The ECOC method is borrowed from data transmitting task in communication. Its main idea is to add redundancy to the data being learned (transmitted) so that even if some errors occur due to the biases (noises) in the learning process (channel), the data can be correctly classified (received) in prediction stage (at the other end). It works by converting a multi-class supervised learning problem into a large number (L) of two-class supervised learning problems [8]. Any learning algorithm that can handle two-class learning problems, such as Naïve Bayes [3], can then be applied to learn each of these L problems. L can then be thought of as the length of the codewords with one bit in each codeword for each classifier. The ECOC algorithm is outlined in Figure 1.

TRAINING

1. Load training data and parameters, i.e., the length of code L and training class K.

2. Create a L-bit code for the K classes using a kind of coding algorithm.

3. For each bit, train the base classifier using the binary class (0 and 1) over the total training data.

TESTING

1. Apply each of the L classifiers to the test example.

2. Assign the test example the class with the largest votes.

Fig. 1. Outline of ECOC

Different from the use of ECOC in communication tasks, the use in classification tasks requires not only the rows of a code to be well-separated, but also the columns to be well-separated as well. The reason behind the rows being well-separated is obvious, since we want codewords or classes to be maximally far apart from each other, but the column separation is necessary because the functions being learned by the learner for each bit should be uncorrelated so that the errors in each bit are independent of each other [8].

There are three commonly used coding strategies: code theory based method [16], random method [7], and data-specific method. Ghani's work [8] indicates that code theory based method, such as BCH, performs the best. Further more, BCH can insure the code to be well-separated in row. As a result, we only use BCH in this work.

4 Methodology

This section presents the rationale of why we combine ECOC with Model-Refinement strategy. First, we illustrate why ECOC brings bias for centroid classifier. Then explain why Model-Refinement strategy can modify this kind of bias. Finally we outline the combined ECOC algorithm.

4.1 The Bias Incurred by ECOC for Centroid Classifier

Centroid classifier is a linear, simple and yet efficient method for text categorization. The basic idea of centroid classifier is to construct a centroid C_i for each class c_i using formula (1) where d denotes one document vector and $|z|$ indicates the cardinality of set z. In substance, centroid classifier makes a simple decision rule (formula (2)) that a given document should be assigned a particular class if the similarity (or distance) of this document to the centroid of the class is the largest (or smallest). This rule is based on a straightforward assumption: the documents in one category should share some similarities with each other.

$$C_i = \frac{1}{|c_i|} \sum_{d \in c_i} d \,. \tag{1}$$

$$c = \arg\max_{c_i} \left(\frac{d \cdot C_i}{\|d\|_2 \|C_i\|_2} \right). \tag{2}$$

For example, the single-topic documents involved with "sport" or "education" can meet with the presumption; while the hybrid documents involved with "sport" as well as "education" break this supposition.

As such, ECOC based centroid classifier also breaks this hypothesis. This is because ECOC ignores the similarities of original classes when producing binary problems. In this scenario, many different classes are often merged into one category. For example, the class "sport" and "education" may be assembled into one class. As a result, the assumption will inevitably be broken.

Let's take a simple multi-class classification task with 12 classes. After coding the original classes, we obtain the dataset as in Figure 2. Class 0 consists of 6 original categories, and class 1 contains another 6 categories. Then we calculate the centroids

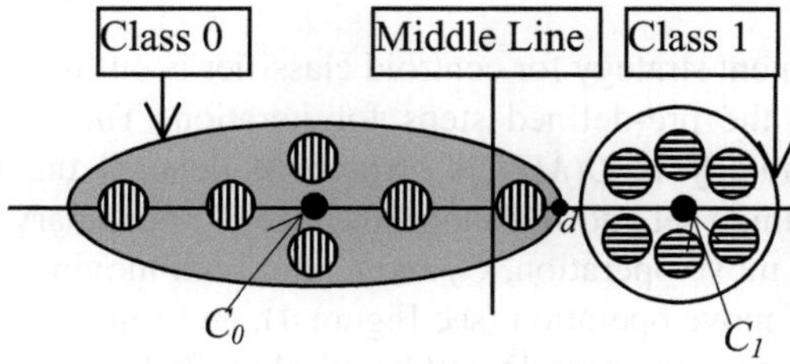

Fig. 2. Original Centroids of Merged Class 0 and Class 1

of merged class 0 and merged class 1 using formula (1), and draw a Middle Line that is the perpendicular bisector of the line between the two centroids.

According to the decision rule (formula (2)) of centroid classifier, the examples of class 0 on the right of the Middle Line will be misclassified into class 1. This is the mechanism why ECOC can bring bias for centroid classifier. In other words, the ECOC method conflicts with the assumption of centroid classifier to some degree.

4.2 Why Model-Refinement Strategy Can Reduce This Bias?

In order to decrease this kind of bias, we employ the Model-Refinement strategy to adjust the class representatives, i.e., the centroids. The basic idea of Model-Refinement strategy is to make use of training errors to adjust class centroids so that the biases can be reduced gradually, and then the training-set error rate can also be reduced gradually.

For example, if document *d* of class 0 is misclassified into class 1, both centroid C_0 and C_1 should be moved right by the following formulas (3-4) respectively,

$$C_0^* = C_0 + \eta \cdot d \, . \tag{3}$$

$$C_1^* = C_1 - \eta \cdot d \, . \tag{4}$$

where η ($0<\eta<1$) is the *Learning Rate* which controls the step-size of updating operation. The former formula (3) is called as "drag" formula and the latter (4) is called as "push" formula.

```
1. Load training data and parameters;
2. Calculate centroid for each class;
3. For iter=1 to MaxIteration Do
  3.1 For each document d in training set Do
        3.1.1 Classify d labeled "A1" into class "A2";
        3.1.2 If (A1!=A2) Do
          Drag centroid of class A1 to d using formula (3);
          Push centroid of class A2 against d using formula (4);
```

Fig. 3. Outline of Model-Refinement Strategy

The Model-Refinement strategy for centroid classifier is outlined in Figure 3 where *MaxIteration* denotes the pre-defined steps for iteration. The time requirement of Model-Refinement strategy is *O*(*MTKW*) where *M* denotes the iteration steps, *T* denotes the size of training set and *W* denotes the size of vocabulary.

With this so-called move operation, C_0 and C_1 are both moving right gradually. At the end of this kind of move operation (see Figure 4), no example of class 0 locates at the right of Middle Line so no example will be misclassified.

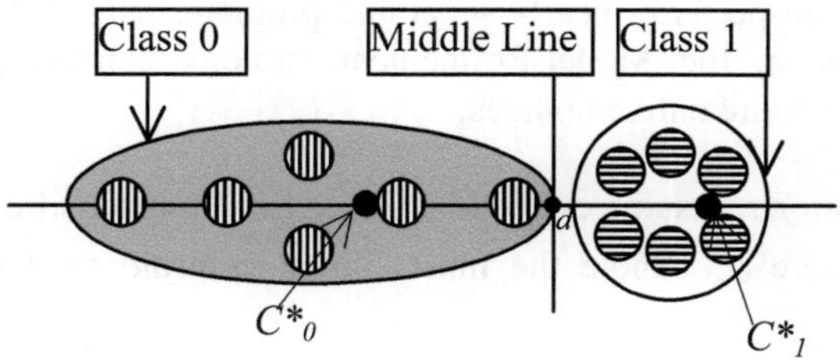

Fig. 4. Refined Centroids of Merged Class 0 and Class 1

4.3 The Combination of ECOC and Model-Refinement Strategy for Centroid Classifier

In this subsection, we present the outline (Figure 5) of combining ECOC with Model-Refinement strategy for centroid classifier. In substance, the improved ECOC combines the strengths of ECOC and Model-Refinement strategy. ECOC research in ensemble learning techniques has shown that it is well suited for classification tasks with a large number of categories. On the other hand, Model-Refinement strategy has proved to be an effective approach to reduce the bias of base classifier, that is to say, it can dramatically boost the performance of the base classifier.

TRAINING

1. Load training data and parameters, i.e., the length of code L and training class K.

2. Create a L-bit code for the K classes using a kind of coding algorithm.

3. For each bit, train centroid classifier using the binary class (0 and 1) over the total training data.

4. Use Model-Refinement approach to adjust centroids.

TESTING

1. Apply each of the L classifiers to the test example.

2. Assign the test example the class with the largest votes.

Fig. 5. Outline of combining ECOC with Model-Refinement strategy

4.4 The Convergence Analysis of Model-Refinement Strategy

Given a training set $S = \bigcup_{i=1}^{K} S_i$, where *K* denotes the number of training classes, and S_i denotes the training examples of class *i*. In the following analysis, we suppose the data is 2-norm bounded, that is, $\forall d \in S, \|d\|_2 \le R(R > 0)$. Since the size of training set is finite, this assumption always holds.

Definition 1. We said a training set *S* is a linearly separable problem if there exits $\{C_1^{opt}, C_2^{opt}, \ldots, C_K^{opt}\}$ so that $\forall i \in [1, K]$ satisfies,

$$C_i^{opt} d - C_j^{opt} d \ge \gamma(\gamma > 0) \text{ where } d \in S_i, j \ne i .$$

Theorem 1. With respect to linearly separable problem, if we select an appropriate learning parameter η, the Model-Refinement strategy converges to the optimal solution $\{C_i^{opt}\}$ after finite online updates.

Proof: In the iteration *t*, assume example $d(d \in S_A)$ is a misclassified example, that is, $C_A^t d - C_B^t d < 0$, where C_B^t denote the most similar centroid to *d* with the different label. Then,

$$\sum_{i=1}^{K}\left\|C_i^{t+1} - C_i^{opt}\right\|^2 = \sum_{i \neq A,B}\left\|C_i^t - C_i^{opt}\right\|^2$$

$$+\left\|C_A^t + \eta d - C_A^{opt}\right\|^2 + \left\|C_B^t - \eta d - C_B^{opt}\right\|^2$$

$$= \sum_{i=1}^{K}\left\|C_i^t - C_i^{opt}\right\|^2 + 2\eta^2\|d\|^2 + 2\eta d\left(C_A^t - C_A^{opt}\right) - 2\eta d\left(C_B^t - C_B^{opt}\right)$$

$$= \sum_{i=1}^{K}\left\|C_i^t - C_i^{opt}\right\|^2 + 2\eta^2 R^2 + 2\eta d\left(C_A^t - C_B^t\right) - 2\eta d\left(C_A^{opt} - C_B^{opt}\right)$$

$$\leq \sum_{i=1}^{K}\left\|C_i^t - C_i^{opt}\right\|^2 + 2\eta^2 R^2 - 2\eta\gamma$$

In this time, as long as select $\eta < \gamma / R^2$, we can guarantee that $\sum_{i=1}^{K}\left\|C_i^{t+1} - C_i^{opt}\right\|^2 < \sum_{i=1}^{K}\left\|C_i^t - C_i^{opt}\right\|^2$. In other words, after each update, class centroid C_i^t approaches optimal centroid C_i^{opt}.

Furthermore, if select an appropriate ρ so as to $0 < \rho < \left(\eta\gamma - \eta^2 R^2\right)$, then,

$$\sum_{i=1}^{K}\left\|C_i^t - C_i^{opt}\right\|^2 < \sum_{i=1}^{K}\left\|C_i^{t-1} - C_i^{opt}\right\|^2 - 2\rho = \sum_{i=1}^{K}\left\|C_i^0 - C_i^{opt}\right\|^2 - 2t\rho.$$

Obviously, $\sum_{i=1}^{K}\left\|C_i^t - C_i^{opt}\right\|^2 \geq 0$, let $\zeta = \sum_{i=1}^{K}\left\|C_i^0 - C_i^{opt}\right\|^2$, then,

$$\zeta - 2t\rho > 0\text{, that is, } t < \frac{\zeta}{2\rho}.$$ □

Lemma 1. $\left(\sum_{i=}^{K} a_i\right)^2 \leq K\sum_{i=}^{K} a_i^2$ when $a_i \geq 0$.

Proof:
$$K\sum_{i=}^{K} a_i^2 - \left(\sum_{i=}^{K} a_i\right)^2 = (K-1)\sum_{i=}^{K} a_i^2 - 2\sum_{i \neq j} a_i a_j$$
$$= \sum_{i \neq j}\left(a_i - a_j\right)^2 \geq 0$$ □

Theorem 2. With respect to a linearly separable problem, the proposed method converges after finite online updates using any learning parameter $\eta(\eta > 0)$.

Proof: In the iteration *t*, assume example $d(d \in S_A)$ is a misclassified example or small-margin example, that is, $C_A^t d - C_B^t d < \delta(0 < \delta < \gamma)$, where C_B^t denote the most similar centroid to *d* with the different label. Then,

$$\begin{aligned}\sum_{i=1}^{K} C_i^{t+1} C_i^{opt} &= \sum_{i \neq A,B} C_i^t C_i^{opt} + (C_A^t + \eta d) A_A^{opt} + (C_B^t - \eta d) C_B^{opt} \\ &= \sum_{i=1}^{K} C_i^t C_i^{opt} + \eta d (C_A^{opt} - C_B^{opt}) \\ &\geq \sum_{i=1}^{K} C_i^t C_i^{opt} + \eta \gamma,\end{aligned}$$

which indicates,

$$\sum_{i=1}^{K} C_i^t C_i^{opt} \geq \sum_{i=1}^{K} C_i^0 C_i^{opt} + t\eta\gamma \tag{5}$$

In the same way,

$$\begin{aligned}\sum_{i=1}^{K} \left\|C_i^{t+1}\right\|^2 &= \sum_{i \neq A,B} \left\|C_i^t\right\|^2 + \left\|C_A^t + \eta d\right\|^2 + \left\|C_B^t - \eta d\right\|^2 \\ &= \sum_{i=1}^{K} \left\|C_i^t\right\|^2 + 2\eta^2 \|d\|^2 + 2\eta d (C_A^t - C_B^t) \\ &\leq \sum_{i=1}^{K} \left\|C_i^t\right\|^2 + 2\eta^2 R^2 + 2\eta\delta,\end{aligned}$$

therefore,

$$\sum_{i=1}^{K} \left\|C_i^t\right\|^2 \leq \sum_{i=1}^{K} \left\|C_i^0\right\|^2 + 2t(\eta^2 R^2 + \eta\delta)$$

let $\tau = \max_i \left\|C_i^{opt}\right\|$, then,

$$\sum_{i=1}^{K} C_i^t C_i^{opt} \leq \sum_{i=1}^{K} \left\|C_i^t\right\| \cdot \left\|C_i^{opt}\right\| \leq \tau \sum_{i=1}^{K} \left\|C_i^t\right\|.$$

According to **lemma 1**,

$$\begin{aligned}\sum_{i=1}^{K} C_i^t C_i^{opt} &\leq \tau \sum_{i=1}^{K} \left\|C_i^t\right\| \\ &\leq \tau \left(K \sum_{i=1}^{K} \left\|C_i^t\right\|^2 \right)^{1/2} \\ &\leq \tau \sqrt{K} \left(\sum_{i=1}^{K} \left\|C_i^0\right\|^2 + 2t(\eta^2 R^2 + \eta\delta) \right)^{1/2} \\ &\leq \tau \sqrt{K} \left(\sum_{i=1}^{K} \left\|C_i^0\right\|^2 \right)^{1/2} + \tau \sqrt{2K(\eta^2 R^2 + \eta\delta) t}\end{aligned} \tag{6}$$

According to (5) and (6), we obtain,

$$\tau\sqrt{K}\left(\sum_{i=1}^{K}\left\|C_i^0\right\|^2\right)^{1/2}+\tau\sqrt{2K\left(\eta^2R^2+\eta\delta\right)t}\geq\sum_{i=1}^{K}C_i^0C_i^{opt}+t\eta\gamma\cdot$$

Obviously, if above inequality holds, t must be finite. That is to say, the Model-Refinement strategy converges after finite online updates. □

5 Experiment Results

In this section, we introduce the experimental data set, evaluation metrics, experiment settings and present the experimental results.

5.1 Datasets

In our experiment, we use two corpora: NewsGroup[1], and Industry Sector[2].

20NewsGroup. The 20Newsgroup (20NG) dataset contains approximately 20,000 articles evenly divided among 20 Usenet newsgroups. We use a subset consisting of total categories and 19,446 documents.

Industry Sector. The Industry Section dataset is based on the data made available by Market Guide, Inc. (www.marketguide.com). The set consists of company homepages that are categorized in a hierarchy of industry sectors, but we disregard the hierarchy. There were 9,637 documents in the dataset, which were divided into 105 classes. We use a subset called as Sector-48 consisting of 48 categories and in all 4,581 documents.

5.2 The Performance Measure

To evaluate a text classification system, we use the F1 measure introduced by van Rijsbergen [15]. This measure combines recall and precision in the following way:

$$Recall=\frac{number\ of\ correct\ \ positive\ predictions}{number\ of\ positive\ examples}$$

$$Precision=\frac{number\ of\ correct\ \ positive\ \ predictions}{number\ of\ positive\ \ predictions}$$

$$F1=\frac{2\times Recall\times Precision}{(Recall+Precision)}$$

For ease of comparison, we summarize the F1 scores over the different categories using the Micro- and Macro-averages of F1 scores [17]:

$$Micro\text{-}F1\ =\ F1\ over\ categories\ \ and\ documents$$

$$Macro\text{-}F1\ =\ average\ of\ within\text{-}category\ F1\ values$$

[1] www-2.cs.cmu.edu/afs/cs/project/theo-11/www/wwkb

[2] www-2.cs.cmu.edu/afs/cs.cmu.edu/project/theo-20/www/data/

The MicroF1 and MacroF1 emphasize the performance of the system on common and rare categories respectively. Using these averages, we can observe the effect of different kinds of data on a classification system [18].

5.3 Experimental Design

We evenly split the each dataset into two parts. Then we use one part for training and the remaining second for test. We perform the train-test procedure two times and use the average of the two performances as final result. This is so called two-fold cross validation.

In order to remove redundant features and save running time, we employ Information Gain as feature selection method because it consistently performs well in most cases [14].

We employ TFIDF as input features. The formula for calculating the TFIDF can be written as follows:

$$W(t,d) = \frac{tf(t,d) \times \log(N/n_t)}{\sqrt{\sum_{t \in \vec{d}} [tf(t,d) \times \log(N/n_t)]^2}} \tag{7}$$

where N is the total number of training documents, and n_t is the number of documents containing the word t. $tf(t,d)$ indicates the occurrences of word t in document d.

For experiments involving SVM we employed SVMTorch, which uses one-versus-the-rest decomposition and can directly deal with multi-class classification problems. (www.idiap.ch/~bengio/projects/SVMTorch.html). Particularly, it has been specifically tailored for large-scale problems.

5.4 Comparison and Analysis

Table 1 and table 2 show the performance comparison of different methods on two datasets when using 10,000 features. For ECOC, we use 63-bit BCH coding; for Model-Refinement strategy, we fix its *MaxIteration* and *LearningRate* as 8 and 0.01 respectively. For brevity, we use MR to denote Model-Refinement strategy.

From the two tables, we can observe that ECOC indeed brings significant bias for centroid classifier, which results in considerable decrease in accuracy. Especially on sector-48, the bias reduces the MicroF1 of centroid classifier from 0.7985 to 0.6422.

Table 1. The MicroF1 of different methods

Method / Dataset	Centroid	MR +Centroid	ECOC +Centroid	ECOC + MR +Centroid	SVM
Sector-48	0.7985	0.8671	0.6422	**0.9122**	0.8948
NewsGroup	0.8371	0.8697	0.8085	**0.8788**	0.8777

Table 2. The MacroF1 of different methods

Method / Dataset	Centroid	MR +Centroid	ECOC +Centroid	ECOC + MR +Centroid	SVM
Sector-48	0.8097	0.8701	0.6559	**0.9138**	0.8970
NewsGroup	0.8331	0.8661	0.7936	0.8757	0.8759

On the other hand, the combination of ECOC and Model-Refinement strategy makes a considerable performance improvement over centroid classifier. On Newsgroup, it beats centroid classifier by 4 percents; on Sector-48, it beats centroid classifier by 11 percents. More encouragingly, it yields better performance than SVM classifier on Sector-48. This improvement also indicates that Model-Refinement strategy can effectively reduce the bias incurred by ECOC.

Figure 6 displays the MicroF1 curves of different methods vs. the number of features. For ECOC, we use 63-bit BCH coding; for Model-Refinement strategy, we fix its *MaxIteration* and *LearningRate* as 8 and 0.01 respectively. From this figure, we can observe that the combination of ECOC and Model-Refinement strategy delivers consistent top-notch performance on two datasets, especially when the number of features is larger than 3000.

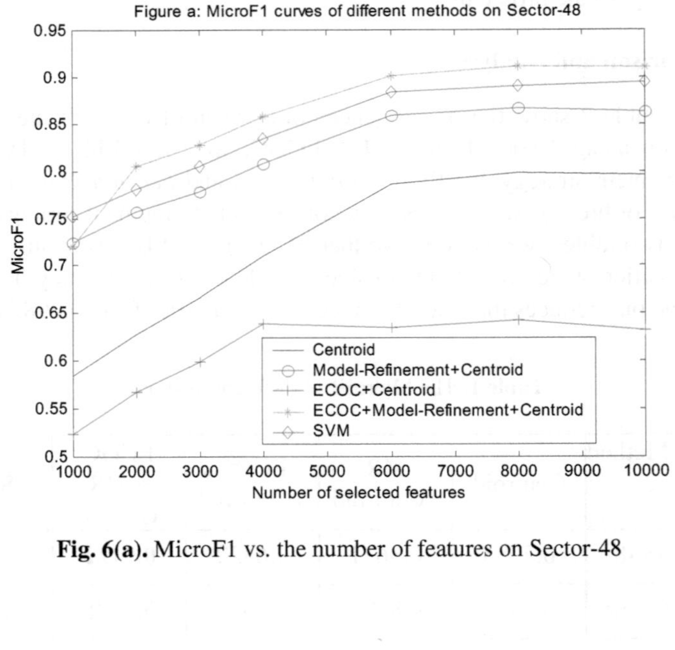

Fig. 6(a). MicroF1 vs. the number of features on Sector-48

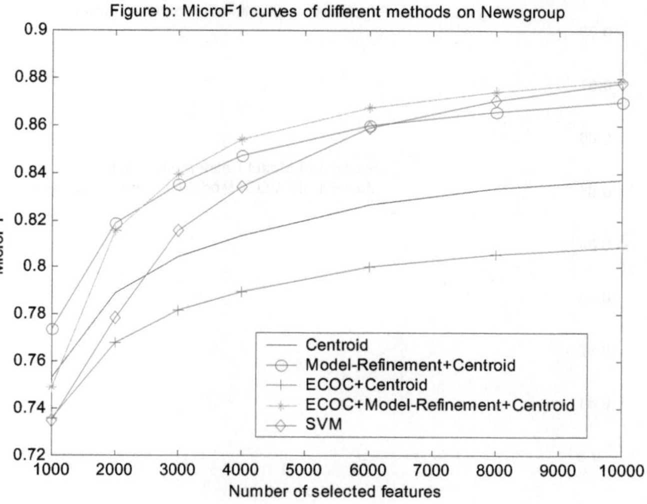

Fig. 6(b). MicroF1 vs. the number of features on Newsgroup

Figure 7 presents the MicroF1 curves of different methods vs. iteration. For ECOC, we use 63-bit BCH coding; for Model-Refinement strategy, we fix its *LearningRate* as 0.01; the number of feature is fixed as 10,000.

As we increase the iteration for Model-Refinement strategy, the combination of ECOC and Model-Refinement strategy shows an improved performance that is to be expected. This result verifies the fact that Model-Refinement strategy can dramatically reduce the bias that ECOC brings to centroid classifier.

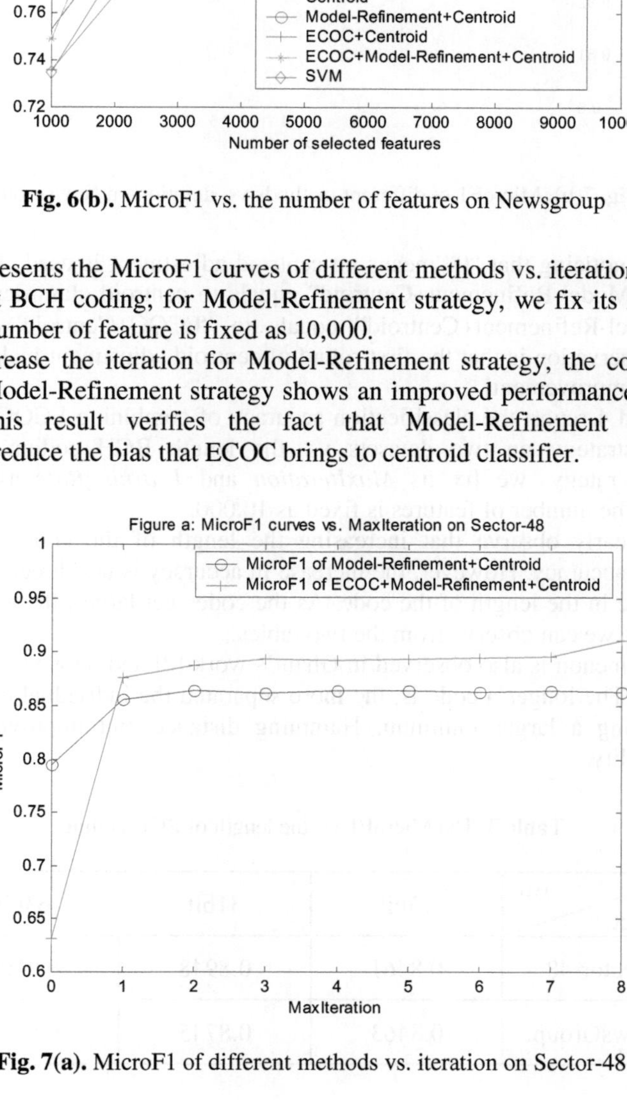

Fig. 7(a). MicroF1 of different methods vs. iteration on Sector-48

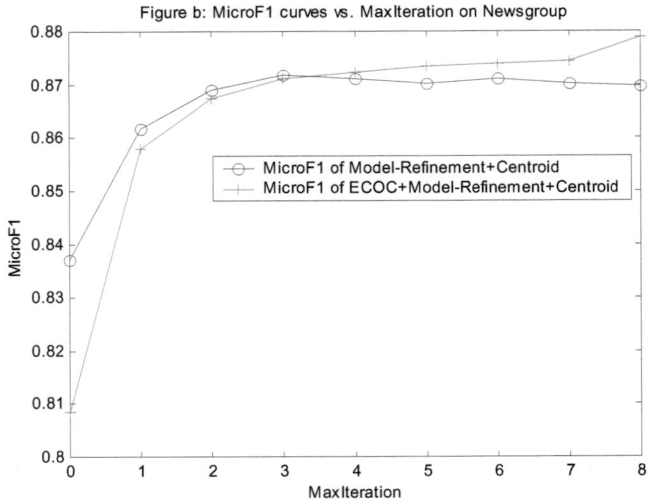

Fig. 7(b). MicroF1 of different methods vs. iteration on Newsgroup

It is worth noticing that "0" means no centroid adjustment is used. That is to say, the method "Model-Refinement+Centroid" equals to centroid classifier; the method "ECOC+Model-Refinement+Centroid" equals to "ECOC+Centroid" classifier. A noticeable observation is that the first round of centroid adjustment makes the largest performance improvement.

Table 3 and 4 report the classification accuracy of combining ECOC with Model-Refinement strategy on two datasets vs. the length BCH coding. For Model-Refinement strategy, we fix its *MaxIteration* and *LearningRate* as 8 and 0.01 respectively; the number of features is fixed as 10,000.

We can clearly observe that increasing the length of the codes increases the classification accuracy. However, the increase in accuracy is not directly proportional to the increase in the length of the code. As the codes get larger, the accuracies start leveling off as we can observe from the two tables.

This phenomenon is also observed in Ghani's work [8] and he gave an explanation for this case: The longer a code is, the more separated the individual codewords can be, thus having a larger minimum Hamming distance and improving the error-correcting ability.

Table 3. The MicroF1 vs. the length of BCH coding

Dataset \ Bit	15bit	31bit	63bit
Sector-48	0.8461	0.8948	0.9105
NewsGroup	0.8463	0.8745	0.8788

Table 4. The MacroF1 vs. the length of BCH coding

Dataset \ Bit	15bit	31bit	63bit
Sector-48	0.8459	0.8961	0.9122
NewsGroup	0.8430	0.8714	0.8757

6 Conclusion Remarks

In this work, we examine the use of ECOC for improving centroid text classifier. The implementation framework is to decompose one multi-class problem into multiple binary problems and then learn the individual binary classification problems by centroid classifier. Meanwhile, Model-Refinement strategy is employed to reduce the bias incurred by ECOC. Furthermore, we present the theoretical justification and analysis for Model-Refinement strategy.

In order to investigate the effectiveness and robustness of proposed method, we conduct an extensive experiment on two commonly used corpora, i.e., Industry Sector and Newsgroup. The experimental results indicate that the combination of ECOC with Model-Refinement strategy makes a considerable performance improvement over traditional centroid classifier, and even performs comparably with SVM classifier.

The results reported here are not necessarily the best that can be achieved. Our future effort is to seek new techniques to enhance the performance of ECOC for centroid text classifier. Additionally, we will investigate the effectiveness of proposed method on multi-label text classification problems.

Acknowledgments

This work was mainly supported by two funds, i.e., 0704021000 and 60803085, and one another project, i.e., 2004CB318109.

References

1. Yang, Y., Lin, X.: A re-examination of text categorization methods. In: SIGIR, pp. 42–49 (1999)
2. Han, E., Karypis, G.: Centroid-Based Document Classification Analysis & Experimental Result. In: Zighed, D.A., Komorowski, J., Żytkow, J.M. (eds.) PKDD 2000. LNCS (LNAI), vol. 1910, pp. 424–431. Springer, Heidelberg (2000)
3. McCallum, A., Nigam, K.: A Comparison of Event Models for Naive Bayes Text Classification. In: AAAI/ICML 1998 Workshop on Learning for Text Categorization, pp. 41–48. AAAI Press, Menlo Park (1998)
4. van Mun, P.P.T.M.: Text Classification in Information Retrieval using Winnow, `http://citeseer.ist.psu.edu/cs`

5. Tan, S.: Neighbor-weighted K-nearest neighbor for unbalanced text corpus. Expert Systems With Applications 28(4), 667–671 (2005)
6. Joachims, T.: Text categorization with support vector machines: learning with many relevant features. In: Nédellec, C., Rouveirol, C. (eds.) ECML 1998. LNCS, vol. 1398, pp. 137–142. Springer, Heidelberg (1998)
7. Berger, A.: Error-Correcting Output Coding for text classification. In: IJCAI (1999)
8. Ghani, R.: Using error-correcting codes for text classification. In: ICML (2000)
9. Ghani, R.: Combining labeled and unlabeled data for multiclass text categorization. In: ICML (2002)
10. Rennie, J., Rifkin, R.: Improving multiclass text classification with the support vector machine. In: AI Memo AIM-2001-026. MIT, Cambridge (2001)
11. Tan, S., Cheng, X., Ghanem, M., Wang, B., Xu, H.: A novel refinement approach for text categorization. In: CIKM 2005, pp. 469–476 (2005)
12. Tan, S.: An Effective Refinement Strategy for KNN Text Classifier. Expert Systems With Applications 30(2), 290–298 (2006)
13. Wu, H., Phang, T.H., Liu, B., Li, X.: A Refinement Approach to Handling Model Misfit in Text Categorization. In: SIGKDD, pp. 207–216 (2002)
14. Yang, Y., Pedersen, J.O.: A Comparative Study on Feature Selection in Text Categorization. In: ICML, pp. 412–420 (1997)
15. van Rijsbergen, C.: Information Retrieval. Butterworths, London (1979)
16. Peterson, W., Weldon, E.: Error-Correcting Codes. MIT Press, Cambridge (1972)
17. Lewis, D., Schapire, R., Callan, J., Papka, R.: Training algorithms for linear text classifiers. In: SIGIR, pp. 298–306 (1996)
18. Chai, K., Ng, H., Chieu, H.: Bayesian online classifiers for text classification and filtering. In: SIGIR, pp. 97–104 (2002)
19. Shankar, S., Karypis, G.: Weight adjustment schemes for a centroid-based classifier. In: TextMining Workshop, KDD (2000)
20. Godbole, S., Sarawagi, S., Chakrabarti, S.: Scaling multi-class support vector machine using inter-class confusion. In: SIGKDD, pp. 513–518 (2002)
21. Vapnik, V.: Statistical Learning Theory. John Wiley and Sons, New York (1998)
22. Apte, C., Damerau, F., Weiss, S.: Text mining with decision rules and decision trees. In: Proceedings of the Workshop with Conference on Automated Learning and Discovery: Learning from text and the Web (1998)
23. Schapire, R., Singer, Y.: Boostexter: A boosting-based system for text categorization. Machine Learning 39, 135–168 (2000)
24. Schapire, R., Freund, Y., Bartlett, P., Lee, W.: Boosting the Margin: A New Explanation for the Effectiveness of Voting Method. The Annals of Statistics 26(5), 1651–1686 (1998)

Optimal Online Learning Procedures for Model-Free Policy Evaluation

Tsuyoshi Ueno[1], Shin-ichi Maeda[1], Motoaki Kawanabe[2], and Shin Ishii[1]

[1] Graduate School of Informatics, Kyoto University
{tsuyos-u,ichi,ishii}@sys.i.kyoto-u.ac.jp
[2] Fraunhofer FIRST and Berlin Institute of Technology, Germany
motoaki.kawanabe@first.fraunhofer.de

Abstract. In this study, we extend the framework of semiparametric statistical inference introduced recently to reinforcement learning [1] to online learning procedures for policy evaluation. This generalization enables us to investigate statistical properties of value function estimators both by batch and online procedures in a unified way in terms of estimating functions. Furthermore, we propose a novel online learning algorithm with optimal estimating functions which achieve the minimum estimation error. Our theoretical developments are confirmed using a simple chain walk problem.

1 Introduction

Reinforcement learning is a class of machine learning based on reward-related interactions with environments, and has successfully been applied to various control problems [2]. In order to find out optimal strategies, it is important, in particular in model-free approaches, to estimate the value function which denotes goodness of the current policy, from a given sample trajectory. There are two major ways in value function estimation. The temporal difference (TD) learning [2] updates the current estimator step-by-step whose step uses a relatively small number of samples (online procedure). On the other hand, the least squares temporal difference (LSTD) learning [3,4] obtains an estimator in one shot by using all samples in the given trajectory (batch procedure). Other algorithms proposed so far are also categorized into one of these two groups.

Recently, [1] introduced a novel framework of semiparametric statistical inference to model-free policy evaluation. The semiparametric statistical models include not only parameters of interest but also additional nuisance parameters which may have infinite degrees of freedom [5,6,7]. For estimating the parameters of interest in such models, estimating functions provide a well-established toolbox: they give consistent estimators (M-estimators) without knowing the nuisance parameters [5,8]. Applying this technique to Markov decision processes (MDP), they discussed asymptotic properties of LSTD-like learning procedures and proposed the generalized LSTD (gLSTD) based on the optimal estimating function that achieved the minimum error. Although the framework by [1] has potential to bring new insights to reinforcement learning, their theory could only

W. Buntine et al. (Eds.): ECML PKDD 2009, Part II, LNAI 5782, pp. 473–488, 2009.

deal with batch procedures and a bunch of online algorithms such as TD were excluded.

In this article, we extend their semiparametric statistical techniques to be applicable to online learning procedures as to follow the existing analysis of online learning [9]. This extension leads to a general class of online learning procedures for model-free policy evaluation derived from estimating functions, which includes many popular algorithms [2,10] such as TD learning [2] and least squares policy evaluation (LSPE) [11]. This generalization also allows us to examine the convergence of statistical error and hence to see that online algorithms can achieve the same asymptotic performance as their batch counterparts if a matrix factor is properly tuned (Theorem 4). Based on this fact, we can accelerate TD learning (Section 5.4). Furthermore, we can derive the optimal choice of the estimating function and construct a novel online learning algorithm which achieves the *minimum estimation error* asymptotically (Algorithm 1).

This article is organized as follows. In Section 2, a semiparametric setting of Markov reward processes (MRPs) is presented. We explain the concept of estimating functions in Section 3, before going into those for MRPs in Section 4. Then, in Section 5, we discuss online learning procedures derived from estimating functions. Convergence theorems for such algorithms will be presented, followed by a novel algorithm with the optimal estimating function. In Section 6, the performance of the proposed algorithms are compared to a couple of well-established algorithms using a simple chain walk problem.

2 Markov Reward Process

Following the literature of policy evaluation [12], we consider Markov Reward Processes (MRPs) in this study. However, extension to Markov Decision Processes (MDPs) is straightforward as long as focusing on policy evaluation (hence the policy is fixed).

An MRP is defined by the initial state probability $p(s_0)$, the state transition probability $p(s_{t+1}|s_t)$ and the reward probability $p(r_{t+1}|s_t, s_{t+1})$. The state variable s is an element of a finite set S and the reward variable $r \in R$ can be either discrete or continuous, but a finite value.

The joint distribution of a sample trajectory $Z_T := \{s_0, s_1, r_1 \cdots, s_T, r_T\}$ of the MRP is described as

$$p(Z_T) = p(s_0) \prod_{t=0}^{T-1} p(r_{t+1}|s_t, s_{t+1}) p(s_{t+1}|s_t). \quad (1)$$

We also impose the following assumptions on MRPs.

Assumption 1. *Under $p(s_{t+1}|s_t)$, MRP has a unique invariant stationary distribution $\mu(s)$.*

Assumption 2. *For any time t, the state s_t and the reward r_t are uniformly bounded.*

Here, we introduce a statistical framework by confirming that the value function estimation can be interpreted as the estimation of certain statistics of MRP (1).

Proposition 1. *[10] Consider a conditional probability of* $\{r_{t+1}, s_{t+1}\}$ *given* s_t,

$$p(r_{t+1}, s_{t+1}|s_t) = p(r_{t+1}|s_t, s_{t+1})p(s_{t+1}|s_t).$$

Then, there is such a function V *that*

$$\mathbb{E}\left[r_{t+1}|s_t\right] = V(s_t) - \gamma\mathbb{E}[V(s_{t+1})|s_t] \tag{2}$$

holds for any state s_t*. Here,* $\mathbb{E}\left[\cdot|s\right]$ *denotes the conditional expectation for a given state* s*. The function* V *that satisfies eq. (2) is unique and found to be a value function;*

$$V(s) := \lim_{T\to\infty} \mathbb{E}\left[\sum_{t=0}^{T} \gamma^t r_{t+1} \middle| s_0 = s\right], \tag{3}$$

where $\gamma \in [0, 1)$ *is a constant called the discount factor.*

We assume throughout this article that the value function can be represented by a certain parametric function, including a nonlinear function with respect to the parameter.

Assumption 3. *The value function given by eq. (3) is represented by a parametric function* $g(s, \boldsymbol{\theta})$*;*

$$V(s) = g(s, \boldsymbol{\theta}),$$

where $g : S \to \mathbb{R}$, $\boldsymbol{\theta} \in \mathbb{R}^m$ *is a parameter. Moreover,* $g(s, \boldsymbol{\theta})$ *is assumed to be twice-differentiable with respect to* $\boldsymbol{\theta}$*, and* $g(s, \boldsymbol{\theta}) < \infty$ *for any* $s \in S$ *and* $\boldsymbol{\theta}$*.*

Under Assumption 3, $p(r_{t+1}|s_t)$ is partially parameterized by $\boldsymbol{\theta}$, through its conditional mean

$$\mathbb{E}[r_{t+1}|s_t] = g(s_t, \boldsymbol{\theta}) - \gamma\mathbb{E}[g(s_{t+1}, \boldsymbol{\theta})|s_t]. \tag{4}$$

Our goal is to find out such a value of the parameter $\boldsymbol{\theta}$ that the function $g(s, \boldsymbol{\theta})$ satisfies eq. (4), that is, it coincides with the true value function.

In order to specify the probabilistic model (4) completely, we need usually extra parameters other than $\boldsymbol{\theta}$. Let $\boldsymbol{\xi}_0$ and $\boldsymbol{\xi}_s$ be such extra parameters that initial distribution $p(s_0, \boldsymbol{\xi}_0)$ and transition distribution $p(r, s|s; \boldsymbol{\theta}, \boldsymbol{\xi}_s)$ are completely identified, respectively. In such a case, the joint distribution of the trajectory Z_T is expressed as

$$p(Z_T; \boldsymbol{\theta}, \boldsymbol{\xi}) = p(s_0; \boldsymbol{\xi}_0) \prod_{t=0}^{T-1} p(r_{t+1}, s_{t+1}|s_t; \boldsymbol{\theta}, \boldsymbol{\xi}_s), \tag{5}$$

where $\boldsymbol{\xi} = (\boldsymbol{\xi}_0, \boldsymbol{\xi}_s)$. Since there is no way to know the complexity of the target system, we attempt to estimate the parameter $\boldsymbol{\theta}$ without estimating the extra $\boldsymbol{\xi}$, which may have innumerable degrees of freedom. Statistical models which contain such (possibly infinite-dimensional) nuisance parameters ($\boldsymbol{\xi}$) in addition to the parameter of interest ($\boldsymbol{\theta}$) are said semiparametric [6]. We emphasize that the nuisance parameters are necessary only for theoretical discussions. In actual estimation of the parameters, same as in other model-free policy evaluation algorithms, we neither define them concretely, nor estimate them. This can be achieved by usage of estimating functions which is a well-established technique to obtain a consistent estimator of the parameter without estimating the nuisance parameter [5,7]. The advantages of considering such semiparametric models behind model-free approaches are:

(a) we can characterize all possible model-free algorithms,
(b) we can discuss asymptotic properties of the estimators in a unified way and obtain the optimal one with the asymptotically *minimum estimation error*.

We will summarize the estimating function method in the next section.

3 Estimating Functions in Semiparametric Models

We begin with a short overview of the estimating function theory in the i.i.d. case and then discuss the MRP case in the next section. We consider a general semiparametric model $p(\boldsymbol{x}; \boldsymbol{\theta}, \boldsymbol{\xi})$, where $\boldsymbol{\theta}$ is an m-dimensional parameter of interest and $\boldsymbol{\xi}$ is a nuisance parameter which can have infinite degrees of freedom. An m-dimensional vector function $\boldsymbol{f}$ is called an *estimating function* when it satisfies the following conditions for any $\boldsymbol{\theta}$ and $\boldsymbol{\xi}$;

$$\mathbb{E}_{\boldsymbol{\theta},\boldsymbol{\xi}}[\boldsymbol{f}(\boldsymbol{x}, \boldsymbol{\theta})] = \mathbf{0} \tag{6}$$

$$\det|\mathbf{A}| \neq 0, \quad \text{where } \mathbf{A} = \mathbb{E}_{\boldsymbol{\theta},\boldsymbol{\xi}}\left[\partial_{\boldsymbol{\theta}} \boldsymbol{f}(\boldsymbol{x}, \boldsymbol{\theta})\right] \tag{7}$$

$$\mathbb{E}_{\boldsymbol{\theta},\boldsymbol{\xi}}\left[||\boldsymbol{f}(\boldsymbol{x}, \boldsymbol{\theta})||^2\right] < \infty, \tag{8}$$

where $\partial_{\boldsymbol{\theta}} = \partial/\partial\boldsymbol{\theta}$ is the partial derivative with respect to $\boldsymbol{\theta}$, and $\det|\cdot|$ and $||\cdot||$ denote the determinant and the Euclidean norm, respectively. Here $\mathbb{E}_{\boldsymbol{\theta},\boldsymbol{\xi}}[\cdot]$ means the expectation over $\boldsymbol{x}$ with $p(\boldsymbol{x}; \boldsymbol{\theta}, \boldsymbol{\xi})$ and we further remark that the parameter $\boldsymbol{\theta}$ in $\boldsymbol{f}(\boldsymbol{x}, \boldsymbol{\theta})$ and $\mathbb{E}_{\boldsymbol{\theta},\boldsymbol{\xi}}[\cdot]$ must be the same.

Suppose i.i.d. samples $\{\boldsymbol{x}_1, \cdots, \boldsymbol{x}_N\}$ are generated from the model $p(\boldsymbol{x}; \boldsymbol{\theta}^*, \boldsymbol{\xi}^*)$. If there is an estimating function $\boldsymbol{f}(\boldsymbol{x}, \boldsymbol{\theta})$, we can obtain an estimator $\hat{\boldsymbol{\theta}}$ which has good asymptotic properties, by solving the following estimating equation;

$$\sum_{i=1}^{N} \boldsymbol{f}(\boldsymbol{x}_i, \hat{\boldsymbol{\theta}}) = \mathbf{0}. \tag{9}$$

A solution of the estimating equation (9) is called an *M-estimator* in statistics [5]. The M-estimator is consistent, that is, it converges to the true value *regardless*

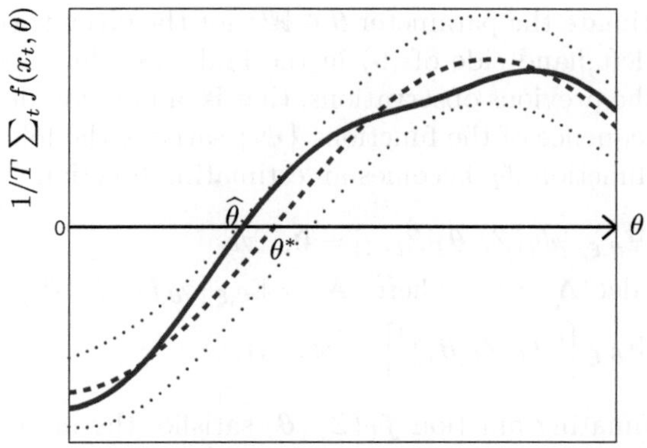

Fig. 1. An illustrative plot of $1/T \sum_t f(x_t, \theta)$ as a function of θ (the solid line). Due to the effect of finite samples, the function is slightly apart from its expectation $\mathbb{E}_{\theta^*,\xi^*}[f(x, \theta)]$ (the dashed line) which takes 0 at $\theta = \theta^*$ because of the condition (6). The condition (8) means that the expectation (the dashed line) has a non-zero slope around θ^*, which ensures the local uniqueness of the zero crossing point. On the other hand, the condition (7) guarantees that its standard deviation shown by the two dotted lines shrinks in the order of $1/\sqrt{T}$, thus we can expect to find asymptotically at least one solution $\hat{\theta}$ of the estimating equation (9) near the true value θ^*. This situation holds regardless of that the true nuisance parameter ξ^* takes any possible value.

of the nuisance parameter $\boldsymbol{\xi}^*$. Moreover, it is normally distributed, that is, $\hat{\boldsymbol{\theta}} \sim \mathcal{N}(\boldsymbol{\theta}^*, \mathrm{Av})$ when the sample size N approaches infinity. The matrix Av, which is called the asymptotic variance, can be calculated by

$$\mathrm{Av} := \mathrm{Av}(\hat{\boldsymbol{\theta}}) = \frac{1}{N}\mathbf{A}^{-1}\mathbb{E}_{\boldsymbol{\theta}^*,\boldsymbol{\xi}^*}\left[\boldsymbol{f}(\boldsymbol{x}, \boldsymbol{\theta}^*)\boldsymbol{f}(\boldsymbol{x}, \boldsymbol{\theta}^*)^\top\right](\mathbf{A}^\top)^{-1},$$

where $\mathbf{A} = \mathbb{E}_{\boldsymbol{\theta}^*,\boldsymbol{\xi}^*}[\partial_{\boldsymbol{\theta}}\boldsymbol{f}(\boldsymbol{x}, \boldsymbol{\theta}^*)]$, and the symbol $\top$ denotes the matrix transpose. Note that Av depends on $(\boldsymbol{\theta}^*, \boldsymbol{\xi}^*)$, but not on the samples $\{\boldsymbol{x}_1, \cdots, \boldsymbol{x}_N\}$. We illustrate in Fig. 3 the left hand side of the estimating equation (9) in order to explain the reason why an M-estimator has nice properties and the meaning of conditions (6)-(8).

4 Estimating Functions in the MRP Model

The notion of estimating function has been extended to be applicable to Markov time-series [13,14]. To make it applicable to MRPs, we need similar extension. For convenience, we write the triplet at time t as $z_t := \{s_{t-1}, s_t, r_t\}$. and the trajectory up to time t as $Z_t := \{s_0, s_1, r_1, \ldots, s_t, r_t\}$.

We consider an m-dimensional vector valued function of the form:

$$\boldsymbol{f}_T(Z_T, \boldsymbol{\theta}) = \sum_{i=1}^{T} \boldsymbol{\psi}_t(Z_t, \boldsymbol{\theta}),$$

we attempt to estimate the parameter $\boldsymbol{\theta} \in \mathbb{R}^m$ for the given trajectory Z_T. This is similar to the left hand side of (9) in the i.i.d. case, but now each term $\boldsymbol{\psi}_t$ depends also on the previous observations, that is, a function of the sequence up to time t. If the sequence of the functions $\{\boldsymbol{\psi}_t\}$ satisfies the following properties for any $\boldsymbol{\theta}$ and $\boldsymbol{\xi}$, function $\boldsymbol{f}_T$ becomes an estimating function.

$$\mathbb{E}_{\boldsymbol{\theta},\boldsymbol{\xi}_s}\left[\boldsymbol{\psi}_t(Z_t,\boldsymbol{\theta})|Z_{t-1}\right] = \mathbf{0}, \quad \forall t \tag{10}$$

$$\det|\mathbf{A}| \neq 0, \quad \text{where} \quad \mathbf{A} := \mathbb{E}_{\boldsymbol{\theta},\boldsymbol{\xi}}\left[\partial_{\boldsymbol{\theta}} \boldsymbol{f}_T(Z_T,\boldsymbol{\theta})\right] \tag{11}$$

$$\mathbb{E}_{\boldsymbol{\theta},\boldsymbol{\xi}}\left[\|\boldsymbol{\psi}_t(Z_t,\boldsymbol{\theta})\|^2\right] < \infty, \quad \forall t. \tag{12}$$

Note that the estimating function $\boldsymbol{f}_T(Z_T,\boldsymbol{\theta})$ satisfies the martingale properties because of the condition (10). Therefore, it is called a *martingale estimating function* in literature [5]. Although time-series estimating functions can be defined in a more general form, the above definition is enough for our theoretical consideration.

4.1 Characterizing the Class of Estimating Functions

In this section, we characterize possible estimating functions in MRPs. Let ϵ_{t+1} be the TD error, that is,

$$\epsilon_{t+1} := \epsilon(z_{t+1},\boldsymbol{\theta}) := g(s_t,\boldsymbol{\theta}) - \gamma g(s_{t+1},\boldsymbol{\theta}) - r_{t+1}.$$

From (4), its conditional expectation $\mathbb{E}_{\boldsymbol{\theta},\boldsymbol{\xi}_s}\left[\epsilon_{t+1}|s_t\right]$ is equal to 0 for any state s_t. Furthermore, this zero-mean property holds even when multiplied by any weight function $\boldsymbol{w}_t := \boldsymbol{w}_t(Z_t)$ which depends only on the past observations, that is,

$$\mathbb{E}_{\boldsymbol{\theta},\boldsymbol{\xi}_s}\left[\boldsymbol{w}_t(Z_t)\epsilon_{t+1}|s_t\right] = \boldsymbol{w}_t(Z_t)\mathbb{E}_{\boldsymbol{\theta},\boldsymbol{\xi}_s}\left[\epsilon_{t+1}|s_t\right] = \mathbf{0},$$

for any s_t. From this observation, we can obtain a class of estimating functions $\boldsymbol{f}_T(Z_T,\boldsymbol{\theta})$ in MRPs.

Lemma 1. *Suppose that the random sequence Z_T is generated from the distribution of the semiparametric model $\{p(Z_T;\boldsymbol{\theta},\boldsymbol{\xi})\,|\,\boldsymbol{\theta},\,\boldsymbol{\xi}\}$ defined by (5). If the matrix $\mathbb{E}_{\boldsymbol{\theta},\boldsymbol{\xi}}\left[\sum_{t=1}^{T} \boldsymbol{w}_{t-1}(Z_{t-1})\left\{\partial_{\boldsymbol{\theta}}\epsilon(z_t,\boldsymbol{\theta})\right\}^\top\right]$ is nonsingular for any $\boldsymbol{\theta}$ and $\boldsymbol{\xi}$, then*

$$\boldsymbol{f}_T(Z_T,\boldsymbol{\theta}) = \sum_{t=1}^{T} \boldsymbol{\psi}_t(Z_t,\boldsymbol{\theta}) := \sum_{t=1}^{T} \boldsymbol{w}_{t-1}(Z_{t-1})\epsilon(z_t,\boldsymbol{\theta}) \tag{13}$$

becomes an estimating function.

From Lemma 1, we can obtain an M-estimator $\hat{\boldsymbol{\theta}}$ by solving the estimating equation

$$\sum_{t=1}^{T} \boldsymbol{\psi}_t(Z_t,\hat{\boldsymbol{\theta}}) = \mathbf{0}. \tag{14}$$

In general, estimating equations can be nonlinear with respect to the parameter $\boldsymbol{\theta}$. Therefore, in order to obtain a solution we need to employ iterative procedures, for example, online learning procedures as will be discussed in Section 5. The estimator derived from the estimating equation (14) has such an asymptotic variance that described by the following lemma.

Lemma 2. *Suppose that the random sequence $\{Z_T\}$ is generated from the distribution $p(Z_T;\boldsymbol{\theta}^*,\boldsymbol{\xi}^*)$ and $\boldsymbol{w}_t$ is a function of $\{s_{0:t}, r_{1:t}\}$ satisfying the condition of Lemma 1. Then, the M-estimator derived from eq. (14) has the asymptotic variance*

$$\mathrm{Av} = \mathrm{Av}(\hat{\boldsymbol{\theta}}) = \frac{1}{T}\mathbf{A}^{-1}\boldsymbol{\Sigma}\left(\mathbf{A}^\top\right)^{-1},$$

where $\mathbf{A} = \mathbf{A}(\boldsymbol{\theta}^*,\boldsymbol{\xi}^*) = \lim_{t\to\infty}\mathbb{E}_{\boldsymbol{\theta}^*,\boldsymbol{\xi}^*}\left[\boldsymbol{w}_{t-1}\left\{\partial_{\boldsymbol{\theta}}\epsilon(z_t,\boldsymbol{\theta}^*)\right\}^\top\right]$,
$\boldsymbol{\Sigma} = \boldsymbol{\Sigma}(\boldsymbol{\theta}^*,\boldsymbol{\xi}^*) = \lim_{t\to\infty}\mathbb{E}_{\boldsymbol{\theta}^*,\boldsymbol{\xi}^*}\left[(\epsilon_t^*)^2\boldsymbol{w}_{t-1}\boldsymbol{w}_{t-1}^\top\right]$ *and* $\epsilon_t^* := \epsilon(z_t,\boldsymbol{\theta}^*)$ *denotes the TD error with the optimal parameter* $\boldsymbol{\theta}^*$.

Interestingly, the converse of Lemma 1 can also be shown; any martingale estimating functions for MRP take the form (13).

Theorem 1. *Any martingale estimating functions in the semiparametric model $\{p(Z_T;\boldsymbol{\theta},\boldsymbol{\xi})\,|\,\boldsymbol{\theta},\,\boldsymbol{\xi}\}$ of MRP can be expressed as*

$$\boldsymbol{f}_T(Z_T,\boldsymbol{\theta}) = \sum_{t=1}^{T}\boldsymbol{\psi}_t(Z_t,\boldsymbol{\theta}) = \sum_{t=1}^{T}\boldsymbol{w}_{t-1}(Z_{t-1})\epsilon(z_t,\boldsymbol{\theta}). \tag{15}$$

Proof. Due to space limitation, we just sketch the proof here. From the martingale property, for any t, we have

$$\mathbb{E}_{\boldsymbol{\theta},\boldsymbol{\xi}_s}\left[\boldsymbol{f}_{t+1}(Z_{t+1},\boldsymbol{\theta}) - \boldsymbol{f}_t(Z_t,\boldsymbol{\theta})|s_t\right] = \mathbf{0},$$

which should hold for any nuisance parameter $\boldsymbol{\xi}$. It can be shown that the TD error ϵ_{t+1} is the unique one that satisfies $\mathbb{E}_{\boldsymbol{\theta},\boldsymbol{\xi}}\left[\epsilon_{t+1}|s_t\right] = 0$ for any s_t and $\boldsymbol{\xi}$. This implies $\boldsymbol{f}_{t+1}(Z_{t+1},\boldsymbol{\theta}) - \boldsymbol{f}_t(Z_t,\boldsymbol{\theta}) = \boldsymbol{w}_t(Z_t)\epsilon(z_{t+1},\boldsymbol{\theta})$. By induction, we see that $\boldsymbol{f}_T(Z_T,\boldsymbol{\theta})$ must have the form (15). □

4.2 Optimal Estimating Function

Since Theorem 1 has specified the set of all martingale estimating functions, we can now discuss the optimal estimating function among them which gives an M-estimator with *minimum asymptotic variance.* Because of the same reason as described in [1], it is suffice to consider the estimating function (15) with the weight $\boldsymbol{w}_t = \boldsymbol{w}_t(s_t)$ which depends only on the current state s_t. Furthermore, by the calculus of variations, we can obtain the optimal estimating function as stated by the following theorem.

Theorem 2. *When the random sequence Z_T is generated from the distribution $p(Z_T; \boldsymbol{\theta}^*, \boldsymbol{\xi}^*)$, the optimal estimating function is given by*

$$\boldsymbol{f}_T^*(Z_T, \boldsymbol{\theta}) = \sum_{t=1}^{T} \boldsymbol{\psi}^*(z_t, \boldsymbol{\theta}) := \sum_{t=1}^{T} \boldsymbol{w}_{t-1}^*(s_{t-1})\epsilon(z_t, \boldsymbol{\theta}), \tag{16}$$

$$\text{where } \boldsymbol{w}_t^*(s_t) := \mathbb{E}_{\boldsymbol{\theta}^*,\boldsymbol{\xi}_s^*}[\epsilon(z_{t+1}, \boldsymbol{\theta}^*)^2|s_t]^{-1}\mathbb{E}_{\boldsymbol{\theta}^*,\boldsymbol{\xi}_s^*}[\partial_{\boldsymbol{\theta}}\epsilon(z_{t+1}, \boldsymbol{\theta}^*)|s_t].$$

Note that the optimal weighting function $\boldsymbol{w}_t^*$ depends on the true parameter $\boldsymbol{\theta}^*$ (but unknown) and needs the expectation with respect to $p(r_{t+1}, s_{t+1}|s_t; \boldsymbol{\theta}^*, \boldsymbol{\xi}_s^*)$, which is also unknown. Therefore, we need to substitute initial estimators for them as we will explain later. It is noted, however, that there is no need to estimate the nuisance parameter $\boldsymbol{\xi}$ itself and that consistency is always guaranteed, even if the initial estimators are based on rough approximation.

The minimum asymptotic variance can be obtained from Lemma 2 and Theorem 2.

Corollary 1. *The minimum asymptotic variance is given by*

$$\mathrm{Av}[\hat{\boldsymbol{\theta}}] = \frac{1}{T}\mathbf{Q}^{-1},$$

where $\mathbf{Q} = \lim_{t\to\infty} \mathbb{E}_{\boldsymbol{\theta}^*,\boldsymbol{\xi}^*}\left[\partial_{\boldsymbol{\theta}}\boldsymbol{\psi}^*(z_t, \boldsymbol{\theta}^*)\right] = \lim_{t\to\infty} \mathbb{E}_{\boldsymbol{\theta}^*,\boldsymbol{\xi}^*}\left[\boldsymbol{\psi}^*(z_t, \boldsymbol{\theta}^*)\boldsymbol{\psi}^*(z_t, \boldsymbol{\theta}^*)^\top\right].$

We remark that the positive definite matrix $\mathbf{Q}$ measures information of the optimal estimating function. In general, the information associated with this matrix $\mathbf{Q}$ is smaller than Fisher information, since we trade efficiency for robustness against the nuisance parameter [7].

5 Learning Algorithms

This section describes the learning algorithm of the parameter $\boldsymbol{\theta}$. In reinforcement learning, online learning is often preferred to batch learning because of its computational efficiency and adaptability to even time-variant situations. Estimating functions provide not only batch algorithms via estimating equations, but also online ones as follows. An online estimator of $\boldsymbol{\theta}$ at time t is denoted as $\hat{\boldsymbol{\theta}}_t$. Suppose that the sequence $\{\boldsymbol{\psi}_1(Z_1, \boldsymbol{\theta}), \ldots, \boldsymbol{\psi}_T(Z_T, \boldsymbol{\theta})\}$ forms a martingale estimating function for MRP. Then, an online update rule can be given by

$$\hat{\boldsymbol{\theta}}_t = \hat{\boldsymbol{\theta}}_{t-1} - \eta_t\boldsymbol{\psi}_t(Z_t, \hat{\boldsymbol{\theta}}_{t-1}), \tag{17}$$

where η_t denotes a nonnegative scalar stepsize. In fact, there exist other online update rules derived from the same estimating function
$\boldsymbol{f}_t(Z_t, \boldsymbol{\theta}) = \sum_{i=1}^{t} \boldsymbol{\psi}_i(Z_i, \boldsymbol{\theta})$ as,

$$\hat{\boldsymbol{\theta}}_t = \hat{\boldsymbol{\theta}}_{t-1} - \eta_t\mathbf{R}(\hat{\boldsymbol{\theta}}_{t-1})\boldsymbol{\psi}_t(Z_t, \hat{\boldsymbol{\theta}}_{t-1}), \tag{18}$$

where $\mathbf{R}(\boldsymbol{\theta})$ denotes an $m \times m$ nonsingular matrix depending only on $\boldsymbol{\theta}$ [15]. These variations come from the fact that $\mathbf{R}(\boldsymbol{\theta}) \sum_{i=1}^{t} \psi_i(Z_i, \boldsymbol{\theta})$ gives the same roots as its original for any $\mathbf{R}(\boldsymbol{\theta})$. This equivalence guarantees that both learning procedures, (17) and (18), have the same stable point, while their dynamics may be different; that is, even if the plain algorithm (17) is unstable, it can be stabilized by introducing an appropriate $\mathbf{R}(\boldsymbol{\theta})$ as (18).

In the next two sections, we will discuss convergence of the online learning algorithm (18).

5.1 Convergence to the True Value

Here, we give sufficient conditions to guarantee the convergence of the online learning (18) to the true parameter $\boldsymbol{\theta}^*$. For the sake of simplicity, we focus on the final convergence phase: $\hat{\boldsymbol{\theta}}_t$ are confined in a neighborhood of $\boldsymbol{\theta}^*$. Now we introduce the following conditions for the convergence.

Condition 1

(a) For any t, $(\hat{\boldsymbol{\theta}}_t - \boldsymbol{\theta}^*)^\top \mathbf{R}(\hat{\boldsymbol{\theta}}_t) \mathbb{E}_{\boldsymbol{\theta}^*, \boldsymbol{\xi}_s^*} \left[\psi_{t+1}(Z_{t+1}, \hat{\boldsymbol{\theta}}_t) \middle| s_t \right]$ is nonnegative .

(b) For any t, there exist such nonnegative constants c_1 and c_2 that $\|\mathbf{R}(\hat{\boldsymbol{\theta}}_t)\mathbb{E}_{\boldsymbol{\theta}^*, \boldsymbol{\xi}_s^*}[\psi_{t+1}(Z_{t+1}, \hat{\boldsymbol{\theta}}_t)|s_t]\|^2 \leq c_1 + c_2 \|\hat{\boldsymbol{\theta}}_t - \boldsymbol{\theta}^*\|$.

Then, the following theorem guarantees the (local) convergence of $\hat{\boldsymbol{\theta}}_t$ to $\boldsymbol{\theta}^*$.

Theorem 3. *Suppose that Condition 1 holds. If the stepsizes $\{\eta_t\}$ are all positive and satisfy $\sum_{t=1}^{\infty} \eta_t = \infty$ and $\sum_{t=1}^{\infty} \eta_t^2 < \infty$, then the online algorithm (18) converges to the true parameter $\boldsymbol{\theta}^*$ almost surely.*

Proof. The proof is given in Appendix A.

Theorem 3 ensures that an online algorithm of the form (18) is consistent, if we can find such a matrix $\mathbf{R}(\boldsymbol{\theta})$ that satisfies Condition 1.

5.2 Convergence Rate

In general, the convergence rate of an online algorithm is slow when compared to a batch algorithm that tries to obtain the solution of the estimating equation using all available samples. However, if we choose an appropriate matrix $\mathbf{R}(\boldsymbol{\theta})$ and adjust the stepsizes $\{\eta_t\}$ appropriately, then it is possible to achieve the same convergence rate with the batch algorithm [9]. First, we characterize the learning process of the batch algorithm.

Lemma 3. *Let $\tilde{\boldsymbol{\theta}}_t$ and $\tilde{\boldsymbol{\theta}}_{t-1}$ be solutions of the estimating equations $1/t \sum_{i=1}^{t} \psi_i(Z_i, \tilde{\boldsymbol{\theta}}_t) = \mathbf{0}$ and $1/(t-1) \sum_{i=1}^{t-1} \psi_i(Z_i, \tilde{\boldsymbol{\theta}}_{t-1}) = \mathbf{0}$, respectively. Then, we have*

$$\tilde{\boldsymbol{\theta}}_t = \tilde{\boldsymbol{\theta}}_{t-1} - \frac{1}{t}\hat{\mathbf{R}}_t^{-1}(\tilde{\boldsymbol{\theta}}_{t-1})\boldsymbol{\psi}_t(Z_t, \tilde{\boldsymbol{\theta}}_{t-1}) + \mathcal{O}\left(\frac{1}{t^2}\right), \tag{19}$$

where $\hat{\mathbf{R}}_t^{-1}(\tilde{\boldsymbol{\theta}}_{t-1}) = \{1/t\sum_{i=1}^t \partial_{\boldsymbol{\theta}}\boldsymbol{\psi}_i(Z_i, \tilde{\boldsymbol{\theta}}_{t-1})\}^{-1}$.

Note that (19) defines the sequence of $\tilde{\boldsymbol{\theta}}_t$ as a recursive stochastic process that is essentially same as the online learning (18) for the same $\mathbf{R}$. In other words, Lemma 3 implies that online algorithms can converge with the same convergence rate as batch counterparts by an appropriate choice of the matrix $\mathbf{R}$. Finally, the following theorem addresses the convergence rate of the (stochastic) learning process such as (19).

Theorem 4. *Consider the following learning process*

$$\hat{\boldsymbol{\theta}}_t = \hat{\boldsymbol{\theta}}_{t-1} - \frac{1}{t}\hat{\mathbf{R}}_t^{-1}\boldsymbol{\psi}_t(Z_t, \hat{\boldsymbol{\theta}}_{t-1}) + \mathcal{O}\left(\frac{1}{t^2}\right), \tag{20}$$

where $\hat{\mathbf{R}}_t = \{1/t\sum_{i=1}^t \partial_{\boldsymbol{\theta}}\boldsymbol{\psi}_i(Z_i, \hat{\boldsymbol{\theta}}_{i-1})\}$.
Assume that:

(a) $\hat{\mathbf{R}}_t^{-1}$ *can be written as* $\hat{\mathbf{R}}_t^{-1} = \mathbb{E}_{\boldsymbol{\theta}^*,\boldsymbol{\xi}_s^*}[\hat{\mathbf{R}}_t^{-1}|s_{t-1}] + o\left(t^{-1}\right)$.
(b) For any t, $\hat{\mathbf{R}}_t$ *is a nonsingular matrix.*

If the learning process (20) converges to the true parameter almost surely, then the convergence rate is given as

$$\mathbb{E}_{\boldsymbol{\theta}^*,\boldsymbol{\xi}^*}\left[\|\hat{\boldsymbol{\theta}}_t - \boldsymbol{\theta}^*\|^2\right] = \frac{1}{t}\mathrm{Tr}\left[\mathbf{A}^{-1}\boldsymbol{\Sigma}(\mathbf{A}^{-1})^\top\right] + o\left(\frac{1}{t}\right), \tag{21}$$

where $\mathbf{A} = \lim_{t\to\infty}\mathbb{E}_{\boldsymbol{\theta}^*,\boldsymbol{\xi}^*}[\boldsymbol{w}_{t-1}\{\partial_{\boldsymbol{\theta}}\epsilon(z_t, \boldsymbol{\theta}^*)\}^\top]$ *and*
$\boldsymbol{\Sigma} = \lim_{t\to\infty}\mathbb{E}_{\boldsymbol{\theta}^*,\boldsymbol{\xi}^*}\left[\epsilon(z_t, \boldsymbol{\theta}^*)^2\boldsymbol{w}_{t-1}\boldsymbol{w}_{t-1}^\top\right]$.

Theorem 4 applies to both the online and batch sequences. Note that this convergence rate (21) is neither affected by the third term of (20) nor by small variations on the matrix $\hat{\mathbf{R}}_t^{-1}$.

5.3 Implementation of Online Algorithm with Optimal Estimating Function

We now construct an optimal online learning which yields the minimum estimation error. Roughly speaking, this is given by the optimal estimating function in Theorem 2 with the best (i.e., with the fastest convergence) choice of the nonsingular matrix in Theorem 4;

$$\hat{\boldsymbol{\theta}}_t = \hat{\boldsymbol{\theta}}_{t-1} - \frac{1}{t}\hat{\mathbf{Q}}_t^{-1}\boldsymbol{\psi}^*(z_t, \hat{\boldsymbol{\theta}}_{t-1}), \tag{22}$$

where $\hat{\mathbf{Q}}_t^{-1} = \{1/t \sum_{i=1}^{t} \partial_{\boldsymbol{\theta}} \boldsymbol{\psi}^*(z_i, \hat{\boldsymbol{\theta}}_{i-1})\}^{-1}$ and $\boldsymbol{\psi}^*(z_t, \boldsymbol{\theta})$ is defined by eq. (16). If the learning equation (22) satisfies Condition 1 and Theorem 4, then it converges to the true parameter with the minimum estimation error, $(1/t)\mathbf{Q}^{-1}$.

However, the learning rule (22) still contains unknown parameters and quantities, so is impractical. For practical implementation, it is necessary to evaluate $\mathbb{E}_{\boldsymbol{\theta}^*, \boldsymbol{\xi}_s^*}[\epsilon(z_{t+1}, \boldsymbol{\theta}^*)^2 | s_t]$ and $\mathbb{E}_{\boldsymbol{\theta}^*, \boldsymbol{\xi}_s^*}[\partial_{\boldsymbol{\theta}} \epsilon(z_{t+1}, \boldsymbol{\theta}^*) | s_t]$ appearing in the optimal estimating function. Therefore, we apply the online function approximation for them. Let $\zeta(s_t, \boldsymbol{\alpha}_t)$ and $\boldsymbol{\varphi}(s_t, \boldsymbol{\beta}_t)$ be the approximations of $\mathbb{E}_{\boldsymbol{\theta}^*, \boldsymbol{\xi}_s^*}[\epsilon(z_{t+1}, \boldsymbol{\theta}_t)^2 | s_t]$ and $\mathbb{E}_{\boldsymbol{\theta}^*, \boldsymbol{\xi}_s^*}[\partial_{\boldsymbol{\theta}} \epsilon(z_{t+1}, \boldsymbol{\theta}_t) | s_t]$, respectively:

$$
\begin{aligned}
\zeta(s_t, \boldsymbol{\alpha}_t) &\approx \mathbb{E}_{\boldsymbol{\theta}^*, \boldsymbol{\xi}_s^*}[\epsilon(z_{t+1}, \hat{\boldsymbol{\theta}}_t)^2 | s_t] \\
\boldsymbol{\varphi}(s_t, \boldsymbol{\beta}_t) &\approx \mathbb{E}_{\boldsymbol{\theta}^*, \boldsymbol{\xi}_s^*}[\partial_{\boldsymbol{\theta}} \epsilon(z_{t+1}, \hat{\boldsymbol{\theta}}_t) | s_t],
\end{aligned}
$$

where $\boldsymbol{\alpha}_t$ and $\boldsymbol{\beta}_t$ are adjustable parameters. $\boldsymbol{\alpha}_t$ and $\boldsymbol{\beta}_t$ are adjusted in an online manner;

$$
\begin{aligned}
\hat{\boldsymbol{\alpha}}_t &= \hat{\boldsymbol{\alpha}}_{t-1} - \eta_t^{\boldsymbol{\alpha}} \partial_{\boldsymbol{\alpha}} \zeta(s_{t-1}, \hat{\boldsymbol{\alpha}}_{t-1}) \left(\zeta(s_{t-1}, \hat{\boldsymbol{\alpha}}_{t-1}) - \epsilon(z_t, \hat{\boldsymbol{\theta}}_{t-1})^2 \right) \\
\hat{\boldsymbol{\beta}}_t &= \hat{\boldsymbol{\beta}}_{t-1} - \eta_t^{\boldsymbol{\beta}} \partial_{\boldsymbol{\beta}} \boldsymbol{\varphi}(s_{t-1}, \hat{\boldsymbol{\beta}}_{t-1}) \left(\boldsymbol{\varphi}(s_{t-1}, \hat{\boldsymbol{\beta}}_{t-1}) - \partial_{\boldsymbol{\theta}} \epsilon(z_t, \hat{\boldsymbol{\theta}}_{t-1}) \right),
\end{aligned}
$$

where $\eta_t^{\boldsymbol{\alpha}}$ and $\eta_t^{\boldsymbol{\beta}}$ are stepsizes. By using these parameterized functions, we can replace $\boldsymbol{\psi}_t^*(z_t, \hat{\boldsymbol{\theta}}_{t-1})$ and $\hat{\mathbf{Q}}_t^{-1}$ by

$$
\begin{aligned}
\boldsymbol{\psi}_t^*(z_t, \hat{\boldsymbol{\theta}}_{t-1}) &= \zeta(s_{t-1}, \hat{\boldsymbol{\alpha}}_{t-1})^{-1} \boldsymbol{\varphi}(s_{t-1}, \hat{\boldsymbol{\beta}}_{t-1}) \epsilon(z_t, \hat{\boldsymbol{\theta}}_t) \\
\hat{\mathbf{Q}}_t^{-1} &= \left(\frac{1}{t} \sum_{i=1}^{t} \zeta(s_{i-1}, \hat{\boldsymbol{\alpha}}_{i-1})^{-1} \boldsymbol{\varphi}(s_{i-1}, \hat{\boldsymbol{\beta}}_{i-1}) \partial_{\boldsymbol{\theta}} \epsilon(z_i, \hat{\boldsymbol{\theta}}_{i-1})^\top \right)^{-1} . \qquad (23)
\end{aligned}
$$

Note that the update (23) can be done in an online manner by applying the well-known matrix inversion lemma [16]. We summarize our implementation of the optimal online learning algorithm in Algorithm 1. The empirical results of this algorithm will be shown in Section 6.

5.4 Acceleration of TD Learning

TD learning is a traditional online approach to model-free policy evaluation and has been as one of the most important algorithms in reinforcement learning. Although the TD learning is widely used due to its simplicity, it is known to converge rather slowly. In this section, we discuss the TD learning from the viewpoint of the estimating function method and propose a new online algorithm which can achieve faster convergence than the usual TD learning.

To simplify the following discussions, let $g(s, \boldsymbol{\theta})$ be a linear function of features:

$$
V(s_t) := \boldsymbol{\phi}(s_t)^\top \boldsymbol{\theta} := \boldsymbol{\phi}_t^\top \boldsymbol{\theta},
$$

Algorithm 1. The proposed online learning algorithm

Initialize $\hat{\boldsymbol{\alpha}}_0$, $\hat{\boldsymbol{\beta}}_0$, $\hat{\boldsymbol{\theta}}_0$, $\hat{\mathbf{Q}}_0^{-1} = \epsilon\mathbf{I}$, a_1, a_2
{ϵ and $\mathbf{I}$ denote a small constant and an $m \times m$ identical matrix, respectively. }

for $t = 1, 2, \cdots$ **do**
Obtain a new sample $z_t = \{s_{t-1}, s_t, r_t\}$
Compute the optimal weight function $\boldsymbol{w}^*_{t-1}$
$\hat{\boldsymbol{\alpha}}_t \leftarrow \hat{\boldsymbol{\alpha}}_{t-1} - \eta_t^{\alpha}\partial_{\boldsymbol{\alpha}}\zeta(s_{t-1}, \hat{\boldsymbol{\alpha}}_{t-1})\{\zeta(s_{t-1}, \hat{\boldsymbol{\alpha}}_{t-1}) - \epsilon(z_t, \hat{\boldsymbol{\theta}}_{t-1})^2\}$
$\hat{\boldsymbol{\beta}}_t \leftarrow \hat{\boldsymbol{\beta}}_{t-1} - \eta_t^{\beta}\partial_{\boldsymbol{\beta}}\boldsymbol{\varphi}(s_{t-1}, \hat{\boldsymbol{\beta}}_{t-1})\left(\boldsymbol{\varphi}(s_{t-1}, \hat{\boldsymbol{\beta}}_{t-1}) - \partial_{\boldsymbol{\theta}}\epsilon(z_t, \hat{\boldsymbol{\theta}}_{t-1})\right)$
$\boldsymbol{w}^*_{t-1} \leftarrow \zeta(s_{t-1}, \hat{\boldsymbol{\alpha}}_{t-1})^{-1}\boldsymbol{\varphi}(s_{t-1}, \hat{\boldsymbol{\beta}}_{t-1})$
Update $\hat{\mathbf{Q}}_t^{-1}$ using matrix inversion lemma
$\hat{\mathbf{Q}}_t^{-1} \leftarrow \frac{1}{t-1}\hat{\mathbf{Q}}_{t-1}^{-1} - \frac{1}{t}\frac{\hat{\mathbf{Q}}_{t-1}^{-1}\boldsymbol{w}^*_{t-1}\partial_{\boldsymbol{\theta}}\epsilon(z_t, \hat{\boldsymbol{\theta}}_{t-1})^\top\hat{\mathbf{Q}}_{t-1}^{-1}}{1+\partial_{\boldsymbol{\theta}}\epsilon(z_t, \hat{\boldsymbol{\theta}}_{t-1})^\top\hat{\mathbf{Q}}_{t-1}^{-1}\boldsymbol{w}^*_{t-1}}$
Update the parameter
$\tau \leftarrow \max(a_1, t - a_2)$
$\hat{\boldsymbol{\theta}}_t \leftarrow \hat{\boldsymbol{\theta}}_{t-1} - \frac{1}{\tau}\hat{\mathbf{Q}}_t^{-1}\boldsymbol{w}^*_{t-1}\epsilon(z_t, \hat{\boldsymbol{\theta}}_{t-1})$
end for

where $\boldsymbol{\phi}(s) : S \rightarrow \mathbb{R}^m$ is a feature vector and $\boldsymbol{\theta} \in \mathbb{R}^m$ is a parameter vector. In this case, we have two ways to solve the linear estimating equation; one is a batch procedure:

$$\hat{\boldsymbol{\theta}} = \{\sum_{t=1}^{T}\boldsymbol{w}_{t-1}(\boldsymbol{\phi}_{t-1} - \gamma\boldsymbol{\phi}_t)^\top\}^{-1}\{\sum_{t=1}^{T}\boldsymbol{w}_{t-1}r_t\}$$

and the other is an online procedure:

$$\hat{\boldsymbol{\theta}}_t = \hat{\boldsymbol{\theta}}_{t-1} - \eta_t\boldsymbol{w}_{t-1}\epsilon(z_t, \hat{\boldsymbol{\theta}}_{t-1}).$$

When the weight function $\boldsymbol{w}_t$ is set to $\boldsymbol{\phi}_t$, the online procedure and batch procedure correspond to the TD learning and LSTD algorithm, respectively. Note that both TD and LSTD share the same estimating function. Therefore, from Lemma 3 and Theorem 4, we can in principle construct an accelerated TD learning which converges at the same speed as the LSTD algorithm.

Here, we consider the following learning equation;

$$\hat{\boldsymbol{\theta}}_t = \hat{\boldsymbol{\theta}}_{t-1} - \frac{1}{t}\hat{\mathbf{R}}_t^{-1}\boldsymbol{\phi}_{t-1}\epsilon(z_t, \hat{\boldsymbol{\theta}}_{t-1}), \tag{24}$$

where $\hat{\mathbf{R}}_t^{-1} = \{1/t\sum_{i=1}^{t}\boldsymbol{\phi}_{i-1}(\boldsymbol{\phi}_{i-1} - \gamma\boldsymbol{\phi}_i)^\top\}^{-1}$. Since $\hat{\mathbf{R}}_t^{-1}$ converges to $\mathbf{A}^{-1} = \lim_{t\to\infty}\mathbb{E}_{\boldsymbol{\theta}^*,\boldsymbol{\xi}^*}[\boldsymbol{\phi}_{t-1}(\boldsymbol{\phi}_{t-1} - \gamma\boldsymbol{\phi}_t)^\top]^{-1}$ and $\mathbf{A}^{-1}$ must be a positive definite matrix (see Lemma 6.4 in [10]), the online algorithm (24) also converges to the true parameter almost surely. Then, if $\hat{\mathbf{R}}_t$ satisfies the condition in Theorem 4, it can achieve same convergence rate as LSTD. We call this procedure the *accelerated-TD learning*.

In both the optimal online learning and the accelerated-TD learning, it is necessary to maintain the inverse of the scaling matrix $\hat{\mathbf{R}}_t$. Since this matrix inversion operation costs $\mathcal{O}(m^2)$ in each step, maintaining the inverse matrix becomes expensive when the dimensionality of parameters increases. The computational cost can be dramatically reduced by maintaining a coarse approximation of the scaling matrix (e.g. diagonal, block diagonal, etc.). An appropriate setting ensures the convergence rate remains $\mathcal{O}(1/t)$ without spoiling computational efficiency.

6 Simulation Experiments

In order to validate our theoretical developments, we compared the performance (statistical error) of the proposed online algorithms (accelerated-TD algorithm and the optimal online learning algorithm) with those of the baselines: TD algorithm [2] (online), LSTD algorithm [3] (batch), and gLSTD algorithm [1] (batch) in a toy problem. An MRP trajectory was generated from a simple Markov random walk on a chain with ten states ($s = 1, \cdots, 10$) as depicted in Fig. 2. At each time t, the state changes to either of its left (-1) or right ($+1$) with equal probability of 0.5. A reward was given by the deterministic function $r = \exp(-0.5(s-5)^2/3^2)$, and the discount factor was set to 0.95. The value function was approximated by a linear function with three-dimensional basis functions, that is, $V(s) \approx \sum_{n=1}^{3} \theta_n \phi_n(s)$. The basis functions $\phi_n(s)$ were generated according to a diffusion model [17]. This approximation was not faithful; i.e. there remained tiny bias.

We generated $M = 200$ trajectories (episodes) each of which consisted of $T = 200$ random walk steps. The value function was estimated for each episode. We evaluated the "mean squared error" (MSE) of the value function, that is, $\frac{1}{M}\frac{1}{10}\sum_{k=1}^{M}\sum_{i\in\{1,\cdots,10\}} \|\boldsymbol{\phi}_i^\top \hat{\boldsymbol{\theta}}_k - V^*(i)\|^2$ where V^* denotes the true value function.

As is often done in online procedures, we utilized some batch procedures to obtain initial estimates of the parameter. More specifically, the first 20 steps in each episode were used to obtain an initial estimator in a batch manner and the online algorithm started after 20 steps. In this random walk problem, owing to the linear approximation, the parameter by the batch algorithm can be obtained analytically. In general situations, on the other hand, an online algorithm has

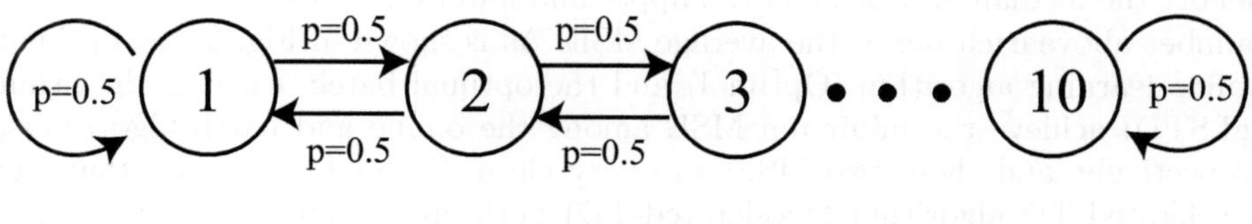

Fig. 2. A ten-states MRP

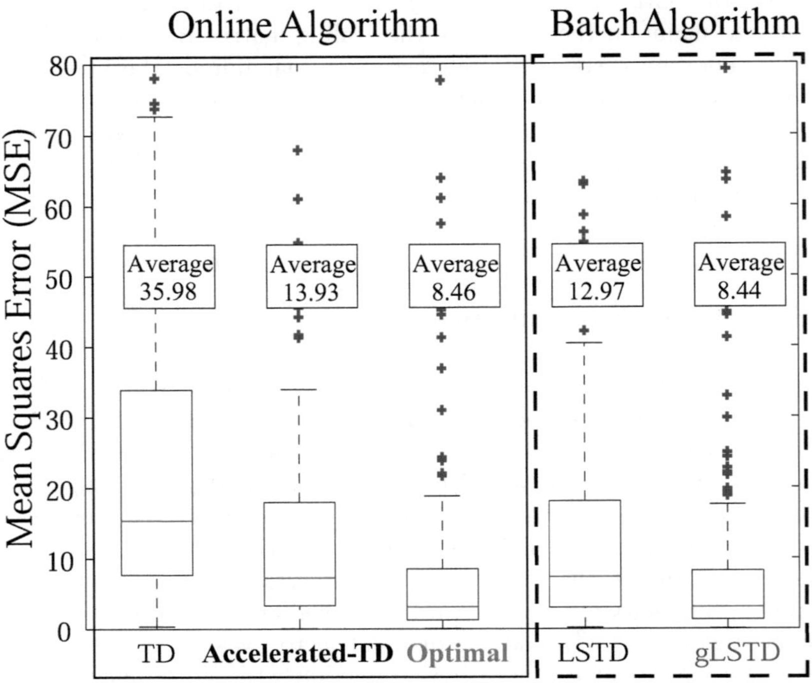

Fig. 3. Simulation results

a merit, because online procedures require less memory and are computationally more efficient. They perform only a single update at each time, while the batch algorithms must keep all trajectories and need to iterate computation until convergence which is serious when employing nonlinear estimating equations.

In the proposed online algorithms, the stepsizes were decreased as simple as $1/t$. On the other hand, the convergence of TD learning was too slow in simple $1/t$ setting due to fast decay of the stepsizes but also in certain well-chosen constant stepsize. Therefore, we adopt an ad-hoc adjustment for the stepsizes as $1/\tau$, where $\tau = \max(10, t - 100)$.

Fig. 3 shows the MSEs of the value functions estimated by our proposed algorithms and the existing algorithms, in which the MSEs of all 200 episodes are shown by box-plots; the center line, and the upper and lower sides of each box denote the median of MSE, and the upper and lower quartiles, respectively. The number above each box is the average MSE. As is shown in Fig. 3, the optimal online learning algorithm (Optimal) and the optimal batch learning algorithm (gLSTD) achieve the minimum MSE among the online and batch algorithms, respectively, and these two MSEs are very close. It should be noted that the accelerated-TD algorithm (accelerated-TD) performs significantly better than the ordinary TD algorithm showing the matrix $\mathbf{R}$ was effective for accelerating the convergence as expected by our theoretical analysis.

7 Conclusion

In this study, we extended the framework of semiparametric statistics inference for value function estimation to be applicable to online learning procedures. Based on this extension, we derived the general form of estimating functions for the model-free value function estimation in MRPs, which provides the statistical basis to many existing batch and online learning algorithms. Moreover, we found the optimal estimating function, which yields the minimum asymptotic estimation variance amongst the general class, and presented a new online learning algorithm (optimal algorithm) based on it. Using a simple MRP problem, we confirmed the validity of our analysis, that is, the optimal algorithm achieves the minimum MSE of the value function estimation and converges with almost the same speed with the batch algorithm gLSTD.

Throughout this article, we assumed that the function approximation is faithful, that is, there is no model misspecification for the value function, and analyzed only its asymptotic variance. Even in misspecified cases, the asymptotic variance can be correctly evaluated [14]. Therefore, if we can reduce the bias term to be much smaller than the variance, our optimal and accelerated-TD procedures could also improve significantly the existing algorithms. Moreover, it is an important future issue to find out a good parametric function g or set of basis functions ϕ_n for linearly approximating the value function.

References

1. Ueno, T., Kawanabe, M., Mori, T., Maeda, S., Ishii, S.: A semiparametric statistical apparoach to model-free policy evaluation. In: Proceedings of the 25th International Conference on Machine Learning, pp. 1072–1079 (2008)
2. Sutton, R., Barto, A.: Reinforcement Learning: An Introduction. MIT Press, Cambridge (1998)
3. Bradtke, S., Barto, A.: Linear least-squares algorithms for temporal difference learning. Machine Learning 22(1), 33–57 (1996)
4. Boyan, J.: Technical update: Least-squares temporal difference learning. Machine Learning 49(2), 233–246 (2002)
5. Godambe, V. (ed.): Estimating Functions. Oxford Science, Oxford (1991)
6. Bickel, D., Ritov, D., Klaassen, C., Wellner, J.: Efficient and Adaptive Estimation for Semiparametric Models. Springer, Heidelberg (1998)
7. Amari, S., Kawanabe, M.: Information geometry of estimating functions in semiparametric statistical models. Bernoulli 3(1), 29–54 (1997)
8. van der Vaart, A.: Asymptotic Statistics. Cambridge University Press, Cambridge (1998)
9. Bottou, L., LeCun, Y.: On-line learning for very large datasets. Applied Stochastic Models in Business and Industry 21(2), 137–151 (2005)
10. Bertsekas, D., Tsitsiklis, J.: Neuro-Dynamic Programming. Athena Scientific, Belmont (1996)
11. Nedić, A., Bertsekas, D.: Least squares policy evaluation algorithms with linear function approximation. Discrete Event Dynamic Systems 13(1), 79–110 (2003)

12. Mannor, S., Simester, D., Sun, P., Tsitsiklis, J.: Bias and variance in value function estimation. In: Proceedings of the twenty-first international conference on Machine learning. ACM, New York (2004)
13. Godambe, V.: The foundations of finite sample estimation in stochastic processes. Biometrika 72(2), 419–428 (1985)
14. Sørensen, M.: On asymptotics of estimating functions. Brazilian Journal of Probability and Statistics 13(2), 419–428 (1999)
15. Amari, S.: Natural gradient works efficiently in learning. Neural Computation 10(2), 251–276 (1998)
16. Horn, R., Johnson, C.: Matrix analysis. Cambridge University Press, Cambridge (1985)
17. Mahadevan, S., Maggioni, M.: Proto-value Functions: A Laplacian Framework for Learning Representation and Control in Markov Decision Processes. The Journal of Machine Learning Research 8, 2169–2231 (2007)

A Proof: Theorem 3

To simplify the following proof, we assume the true parameter is located on the origin without loss of generality: $\boldsymbol{\theta}^* = \mathbf{0}$. Let h_t be $\|\hat{\boldsymbol{\theta}}_t\|^2$. The conditional expectation of variation of h_t can be derived as

$$\begin{aligned}\mathbb{E}_{\boldsymbol{\theta}^*,\boldsymbol{\xi}_s^*}\left[h_{t+1} - h_t|s_t\right] = &-2\eta_{t+1}\hat{\boldsymbol{\theta}}_t^\top \mathbf{R}(\hat{\boldsymbol{\theta}}_t)\mathbb{E}_{\boldsymbol{\theta}^*,\boldsymbol{\xi}_s^*}\left[\boldsymbol{\psi}_{t+1}(Z_{t+1},\hat{\boldsymbol{\theta}}_t)|s_t\right]\\ &+\eta_{t+1}^2\mathbb{E}_{\boldsymbol{\theta}^*,\boldsymbol{\xi}_s^*}\left[\|\mathbf{R}(\hat{\boldsymbol{\theta}}_t)\boldsymbol{\psi}_{t+1}(Z_{t+1},\hat{\boldsymbol{\theta}}_t)\|^2|s_t\right].\end{aligned}$$

From Condition 1, the second term of this equation is bounded by the second moment, thus we obtain

$$\begin{aligned}&\mathbb{E}_{\boldsymbol{\theta}^*,\boldsymbol{\xi}_s^*}\left[h_{t+1} - (1+\eta_{t+1}^2 c_2)h_t|s_t\right]\\ &\le -2\eta_{t+1}\hat{\boldsymbol{\theta}}_t^\top \mathbf{R}(\hat{\boldsymbol{\theta}}_t)\mathbb{E}_{\boldsymbol{\theta}^*,\boldsymbol{\xi}_s^*}\left[\boldsymbol{\psi}_{t+1}(Z_{t+1},\hat{\boldsymbol{\theta}}_t)|s_t\right] + \eta_{t+1}^2 c_1.\end{aligned} \tag{25}$$

Now, let $\chi_t = \prod_{k=0}^{t-1}\frac{1}{1+\eta_{k+1}^2 c_2}$ and $h_t' = \chi_t h_t$. From the assumption $\sum_{t=1}^\infty \eta_t^2 < \infty$, we easily verify that $0 < \chi_t < 1$. Multiplying both sides of eq. (25) by χ_{t+1}, we obtain

$$\begin{aligned}&\mathbb{E}_{\boldsymbol{\theta}^*,\boldsymbol{\xi}^*}\left[h_{t+1}' - h_t'|\mathcal{P}_t\right]\\ &\le -2\eta_{t+1}\chi_{t+1}\hat{\boldsymbol{\theta}}_t^\top \mathbf{R}(\hat{\boldsymbol{\theta}}_t)\mathbb{E}_{\boldsymbol{\theta}^*,\boldsymbol{\xi}_s^*}\left[\boldsymbol{\psi}_{t+1}(Z_{t+1},\hat{\boldsymbol{\theta}}_t)|s_t\right] + \eta_{t+1}^2\chi_{t+1} c_1.\end{aligned}$$

The first term of this upper bound is negative because of Condition 1, and the second term is nonnegative because η_t, χ_{t+1}, and c_1 are nonnegative, and the sum of the second terms $\sum_{t=1}^\infty \eta_t^2\chi_{t+1}c_1$ is finite. Then, the supermartingale convergence theorem [10] guarantees that h_t' converges to a nonnegative random variable almost surely, and $\sum_{t=1}^\infty \eta_{t+1}\chi_{t+1}\hat{\boldsymbol{\theta}}_t^\top \mathbf{R}_t(\hat{\boldsymbol{\theta}}_t)\mathbb{E}_{\boldsymbol{\theta}^*,\boldsymbol{\xi}_s^*}\left[\boldsymbol{\psi}_{t+1}(Z_{t+1},\hat{\boldsymbol{\theta}}_t)|s_t\right] < \infty$. Since $\sum_{t=1}^\infty \eta_t = \infty$ and $\lim_{t\to\infty}\chi_t = \chi_\infty > 0$, we have $\hat{\boldsymbol{\theta}}_t^\top \mathbf{R}(\hat{\boldsymbol{\theta}}_t)\mathbb{E}_{\boldsymbol{\theta}^*,\boldsymbol{\xi}_s^*}\left[\boldsymbol{\psi}_{t+1}(Z_{t+1},\hat{\boldsymbol{\theta}}_t)|s_t\right] \xrightarrow{a.s.} \mathbf{0}$, where $\xrightarrow{a.s.}$ denotes the almost sure convergence. This result suggests the conclusion that the online learning algorithm converges almost surely: $\hat{\boldsymbol{\theta}}_t \xrightarrow{a.s.} \boldsymbol{\theta}^* = \mathbf{0}$.

Kernels for Periodic Time Series Arising in Astronomy

Gabriel Wachman[1], Roni Khardon[1],
Pavlos Protopapas[2,3], and Charles R. Alcock[2]

[1] Tufts University, Medford, MA USA
{gwachm01,roni}@cs.tufts.edu
[2] Harvard-Smithsonian Center for Astrophysics, Cambridge, MA USA
{pprotopapas,calcock}@cfa.harvard.edu
[3] Harvard Initiative in Innovative Computing, Cambridge, MA USA

Abstract. We present a method for applying machine learning algorithms to the automatic classification of astronomy star surveys using time series of star brightness. Currently such classification requires a large amount of domain expert time. We show that a combination of phase invariant similarity and explicit features extracted from the time series provide domain expert level classification. To facilitate this application, we investigate the cross-correlation as a general phase invariant similarity function for time series. We establish several theoretical properties of cross-correlation showing that it is intuitively appealing and algorithmically tractable, but not positive semidefinite, and therefore not generally applicable with kernel methods. As a solution we introduce a positive semidefinite similarity function with the same intuitive appeal as cross-correlation. An experimental evaluation in the astronomy domain as well as several other data sets demonstrates the performance of the kernel and related similarity functions.

1 Introduction

The concrete application motivating this research is the classification of stars into meaningful categories from astronomy literature. A major effort in astronomy research is devoted to sky surveys, where measurements of stars' or other celestial objects' brightness are taken over a period of time. Classification as well as other analyses of stars lead to insights into the nature of our universe, yet the rate at which data are being collected by these surveys far outpaces current methods to classify them. For example, microlensing surveys, such as MACHO [1] and OGLE [2] followed millions of stars for a decade taking one observation per night. The next generation panoramic surveys, such as Pan-STARRS [3] and LSST [4], will begin in 2009 and 2013, respectively, and will collect data on the order of hundreds of billions of stars. It is unreasonable to attempt manual analysis of this data, and there is an immediate need for robust, automatic classification methods.

It is this need that we address directly with our first contribution: the foundation of an automatic methodology for classifying *periodic variable stars* where

W. Buntine et al. (Eds.): ECML PKDD 2009, Part II, LNAI 5782, pp. 489–505, 2009.

a star is variable if its brightness varies over time, and periodic if the variance in brightness is periodic over time. In the data sets taken from star surveys, each example is represented by a time series of brightness measurements, and different types of stars have different periodic patterns. Fig. 1 shows several examples of such time series generated from the three major types of periodic variable stars: Cepheid, RR Lyrae, and Eclipsing Binary. In our experiments only stars of the types in Fig. 1 are present in the data, and the period of each star is given. A complete solution will automatically process an entire survey, of which a small percentage will be periodic variable stars. We are actively working on automatic methods for filtering out non-periodic variables and for identifying period, however these are outside the scope of this paper. We use the existing OGLEII periodic variable star catalog [5] to show that our classification method achieves $> 99\%$ accuracy once such processing and filtering has been done.

As our second contribution we present several insights into the use of the cross-correlation function as a similarity function for time series. Cross-correlation provides an intuitive mathematical analog of what it means for two time series to look alike: we seek the best phase alignment of the time series, where the notion of alignment can be captured by a simple Euclidean distance or inner product. We show that cross-correlation is "almost" a kernel in that it satisfies the Cauchy-Schwartz inequality and induces a distance function satisfying the triangle inequality. Therefore, fast indexing methods can be used with cross-correlation for example with the k-Nearest Neighbor algorithm [6]. We further show that although every 3×3 similarity matrix is positive semidefinite, some 4×4 matrices are not and therefore cross-correlation is not a kernel and not generally applicable with kernel methods.

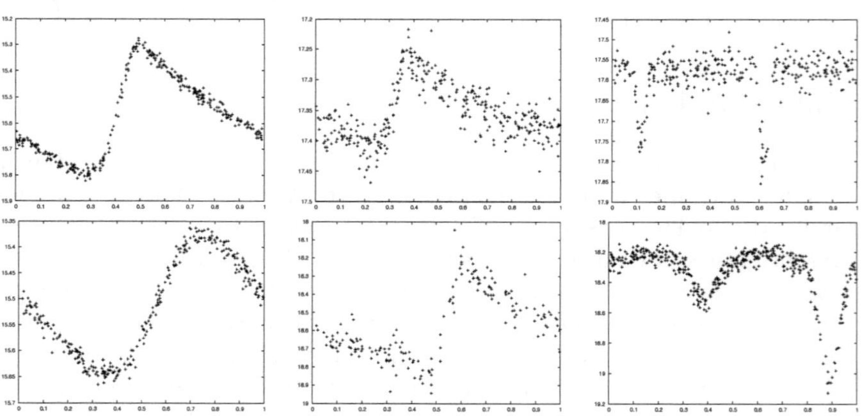

Fig. 1. Examples of light curves of periodic variable stars. Each column shows two stars of the same type. Left: Cepheid, middle: RR Lyrae, right: eclipsing binary. Examples of the same class have similar shapes but are not phase aligned. Examples are a result of folding a long sequence of observations leading to a noisy sample of one period of the light curve. The y-axis labels represent brightness in magnitude units, which is an inverse logarithmic scale (this is the convention in astronomy).

As our final contribution we introduce a positive semidefinite similarity function that has the same intuitive appeal as cross-correlation. We investigate the performance of our kernel on other data sets, both real and artificial, showing excellent performance. We show instances where the kernel outperforms all other methods as well as instances where a simple universal phasing algorithm performs comparably. Our investigation reveals that our kernel performs better than cross-correlation and that the ability to use Support Vector Machines (SVM) [7] with our kernel can provide a significant increase in performance.

The remainder of the paper is organized as follows. Section 2 investigates properties of cross-correlation, and Sect. 3 introduces the new kernel function. Related work is discussed in Sect. 4. We present our experiments and discuss results in Sect. 5. Finally, the concluding section puts this work in the larger context of fully automatic processing of sky surveys.

2 Cross-Correlation

Our examples are vectors in $\mathbb{R}^n$ but they represent an arbitrary shifts of periodic time series. We use the following notation: y_{+s} refers to the vector y shifted by s positions, where positions are shifted modulo n. We then use the standard inner product between shifted examples

$$\langle x, y_{+s} \rangle = \sum_{i=1}^{n} x_i (y_{+s})_i.$$

We define the *cross-correlation* between $x, y \in \mathbb{R}^n$ as

$$C(x, y) = \max_s \langle x, y_{+s} \rangle.$$

In the context of time series, computing the cross-correlation corresponds to aligning two time series such that their inner product, or similarity, is maximized.

2.1 Properties of Cross-Correlation

We first show that cross-correlation has some nice properties making it suitable as a similarity function:

Theorem 1

(P1) $C(x, x) = \langle x, x \rangle \geq 0$.
(P2) $C(x, y) = C(y, x)$.
(P3) The Cauchy-Schwartz Inequality holds, i.e. $\forall x, y, \quad C(x, y) \leq \sqrt{C(x, x)C(y, y)}$.
(P4) If we use the cross-correlation function to give a distance measure d such that

$$d(x, y)^2 = C(x, x) + C(y, y) - 2C(x, y) = \min_s \|x - (y_{+s})\|^2$$

then d satisfies the Triangle Inequality.

In other words cross-correlation has properties similar to an inner product, and can be used intuitively as a similarity function. In particular, we can use metric trees and other methods based only on the triangle inequality [8,6] to speed up distance based algorithms using cross-correlation.

Proof. For (P1) note that by definition $C(x,x) \geq \langle x,x \rangle$. On the other hand, $C(x,x) = \sum x_i x_{i+s}$, and by the Cauchy-Schwartz inequality,

$$\sum x_i x_{i+s} \leq \sqrt{\sum x_i^2}\sqrt{\sum x_{i+s}^2} = \sqrt{\sum x_i^2}\sqrt{\sum x_i^2} = \langle x,x \rangle. \tag{1}$$

Which means $\langle x,x \rangle \geq C(x,x) \geq \langle x,x \rangle$ or $C(x,x) = \langle x,x \rangle \geq 0$.

To prove (P2) observe that since $\langle x, y_{+s} \rangle = \langle x_{-s}, y \rangle = \langle x_{+(n-s)}, y \rangle$ maximizing over the shift for y is the same as maximizing over the shift for x.

(P3) follows from K1 of Theorem 2 below (see Proposition 2.7 of [9]) but we give a direct argument here. Let $C(x,y) = \langle x, y_{+s} \rangle = \langle x, z \rangle$, where s is the shift maximizing the correlation and where we denote $z = y_{+s}$. Then by (P1), $\sqrt{C(x,x)C(y,y)} = \sqrt{\langle x,x \rangle \langle y,y \rangle} = \|x\|\|y\|$. Therefore the claim is equivalent to $\|x\|\|y\| \geq \langle x,z \rangle$, and since the norm does not change under shifting the claim is equivalent to $\|x\|\|z\| \geq \langle x,z \rangle = C(x,y)$. The last inequality holds by the Cauchy-Schwartz inequality for normal inner products.

Finally, for (P4) let $x,y,z \in \mathbb{R}^n$. Let τ_{ab} be the shift that minimizes $d(a,b)$.

$$\begin{aligned}
d(x,y) + d(y,z) &= \|(x_{+\tau_{xy}}) - y\| + \|(y_{+\tau_{yz}}) - z\| && (2)\\
&= \|(x_{+\tau_{xy}+\tau_{yz}}) - (y_{+\tau_{yz}})\| + \|(y_{+\tau_{yz}}) - z)\| && (3)\\
&\geq \|(x_{+\tau_{xy}+\tau_{yz}}) - (y_{+\tau_{yz}}) + (y_{+\tau_{yz}}) - z\| && (4)\\
&= \|(x_{+\tau_{xy}+\tau_{yz}}) - z\| && (5)\\
&\geq \|(x_{+\tau_{xz}}) - z\| = d(x,z) && (6)
\end{aligned}$$

Where (3) holds because shifting x and y by the same amount does not change the value of $\|x - y\|$, (4) holds because of the triangle inequality, and (6) holds because by definition τ_{xz} minimizes the distance between x and z. □

Since cross-correlation shares many properties with inner products it is natural to ask whether it is indeed a kernel function. We show that, although every 3x3 similarity matrix is positive semidefinite, the answer is negative.

Theorem 2

(K1) Any 3×3 *Gram matrix of the cross-correlation is positive semidefinite.*
(K2) The cross-correlation function is not positive semidefinite.

Proof. Let $x_1, x_2, x_3 \in \mathbb{R}$, G a 3×3 matrix such that $G_{ij} = C(x_i, x_j)$, $c_1, c_2, c_3 \in \mathbb{R}$. We prove K1 by showing $Q = \sum_{i=1}^{3} \sum_{j=1}^{3} c_i c_j G_{ij} \geq 0$.

At least one of the products c_1c_2, c_1c_3, c_2c_3 is non-negative. Assume WLOG that $c_2c_3 \geq 0$ and shift x_2 and x_3 so that they obtain the maximum alignment with x_1, calling the shifted versions $\tilde{x_1}, \tilde{x_2}, \tilde{x_3}$ noting that $\tilde{x_1} = x_1$. Now $C(x_i, x_j) = \langle \tilde{x_i}, \tilde{x_j} \rangle$ except possibly when $(i, j) = (2, 3)$, so

$$\sum_{i=1}^{3}\sum_{j=1}^{3} c_i c_j G_{ij} = \sum_{i=1}^{3}\sum_{j=1}^{3} c_i c_j \langle \tilde{x_i}, \tilde{x_j} \rangle + 2c_2c_3(C(\tilde{x_2}, \tilde{x_3}) - \langle \tilde{x_2}, \tilde{x_3} \rangle)$$

$$\geq \sum_{i=1}^{3}\sum_{j=1}^{3} c_i c_j \langle x_i, x_j \rangle \geq 0$$

since $c_2c_3 \geq 0$ and $C(\tilde{x_2}, \tilde{x_3}) \geq \langle \tilde{x_2}, \tilde{x_3} \rangle$ by definition.

The negative result, K2, is proved is by giving a counter example. Consider the matrix A and the row-normalized A'

$$A = \begin{pmatrix} 0 & 1 & 2 \\ 1 & 0 & 0 \\ 2 & 1 & 2 \\ 0 & 2 & 1 \end{pmatrix} \quad A' = \begin{pmatrix} 0 & 0.4472 & 0.8944 \\ 1 & 0 & 0 \\ 0.6667 & 0.3333 & 0.6667 \\ 0 & 0.8944 & 0.4472 \end{pmatrix}$$

where each row is a vector of 3 dimensions. This illustrates a case where we have 4 time series, each with 3 samples and the time series are normalized. Using the cross-correlation function on A', we would get the following Gram matrix

$$G = \begin{pmatrix} 1 & 0.8944 & 0.8944 & 0.8 \\ 0.8944 & 1 & 0.6667 & 0.8944 \\ 0.8944 & 0.6667 & 1 & 0.8944 \\ 0.8 & 0.8944 & 0.8944 & 1 \end{pmatrix}$$

G has a negative eigenvalue of -0.0568 corresponding to the eigenvector $c = (-0.4906, 0.5092, 0.5092, -0.4906)$ and therefore G is not positive semidefinite. In other words $cGc' = \sum_{i=1}^{4}\sum_{j=1}^{4} c_i c_j G_{ij} = -0.0568$. □

3 A Kernel for Periodic Time Series

Since the cross-correlation function is not positive semidefinite, we propose an alternative kernel function that can be used in place of the cross-correlation function with kernel methods. To motivate our choice consider first the kernel

$$K(x, y) = \sum_{i=1}^{n}\sum_{j=1}^{n} \langle x_{+i}, y_{+j} \rangle.$$

Note that here K iterates over all possible shifts, so that we no longer choose the best alignment but instead aggregate the contribution of all possible alignments.

This seems to lose the basic intuition behind cross-correlation and it is indeed not a good choice. On closer inspection we can see that

$$K(x,y) = (x_{+1} + x_{+2} + \ldots + x_{+n})y_{+1} + \ldots + (x_{+1} + x_{+2} + \ldots + x_{+n})y_{+n}$$
$$= (\sum_{i=1}^{n} x_{+i})(\sum_{j=1}^{n} y_{+j}).$$

So K just calculates the product of the sums of the shifted vectors. In particular, if the data is normalized as mentioned above then this is identically zero.

Instead our kernel weights each shift with exponential function so that shifts with high correlation are highly weighted and shifts with low correlation have smaller effect.

Definition 1. *The kernel function* $K : \mathbb{R}^n \times \mathbb{R}^n \to \mathbb{R}$ *is defined as*

$$K(x,y) = \sum_{i=1}^{n} e^{\gamma \langle x, y_{+i} \rangle} \tag{7}$$

where $\gamma \geq 0$ *is a constant.*

Thus like cross-correlation the value of the kernel will be dominated by the maximizing alignment although the number of "good alignments" is also important. In this way we get positive semidefinite kernel while having the same guiding intuition as cross-correlation. Exponential weighting of various alignments of time series has been proposed previously in [10]. Despite the similarity in the construction, the proof of positive semidefiniteness in [10] does not cover our case as their set of alignments is all possible time warpings under a fixed phase and does not allow for circular shifting. Similar ideas to weight different matches exponentially have also been explored in kernels for multi-instance problems [11].

Theorem 3. *K is a positive semidefinite kernel.*

Proof. Consider the following function

$$K'(x,y) = \sum_{i=1}^{n} \sum_{j=1}^{n} e^{\gamma \langle x_{+i}, y_{+j} \rangle}.$$

By [12], $K'(x,y)$ is a convolution kernel. This can be directly shown as follows. First rewrite K' as

$$K'(x,y) = \sum_{a \in R^{-1}(x)} \sum_{b \in R^{-1}(y)} e^{\gamma \langle a, b \rangle} \tag{8}$$

where $R^{-1}(x)$ gives all shifts of x. It is well known that the exponential function $e^{\gamma \langle x, y \rangle}$ is a kernel [9]. Let $\Phi(x)$ be the underlying vector representation of the this kernel so that $e^{\gamma \langle x, y \rangle} = \langle \Phi(x), \Phi(y) \rangle$. Then

$$K'(x,y) = \sum_{a \in R^{-1}(x)} \sum_{b \in R^{-1}(y)} \langle \Phi(a), \Phi(b) \rangle = \langle (\sum_{a \in R^{-1}(x)} \Phi(a)), (\sum_{b \in R^{-1}(y)} \Phi(b)) \rangle \tag{9}$$

Thus K' is an inner product in the same vector space captured by Φ with the map being the aggregate of all elements in $R^{-1}(x)$.

Note that $K'(,)$ iterates over all shifts of both x and y, hence effectively counting each shift n times. For example, observe that for the identity shift, we have $\langle x, y\rangle = \langle x_{+1}, y_{+1}\rangle = \ldots = \langle x_{+(n-1)}, y_{+(n-1)}\rangle$. Hence we need to scale K' by $1/n$ in order to count each shift exactly once. This gives us

$$K(x,y) = \frac{1}{n}\sum_{i=1}^{n}\sum_{j=1}^{n} e^{\gamma\langle x_{+i}, y_{+j}\rangle}.$$

Since scaling a kernel (i.e. K') is also a kernel, K is a kernel. □

Previous work [13] has shown that cross-correlation can be calculated in time $O(n \log n)$ where n is the length of the time series. In particular they show that $\langle x, y_{+s}\rangle = \mathcal{F}^{-1}(\mathcal{X} \cdot \hat{\mathcal{Y}})[s]$ where $\cdot$ indicates point-wise multiplication, $\mathcal{X}$ is the discrete Fourier transform of x, and $\hat{\mathcal{Y}}$ is the complex conjugate of the discrete Fourier transform of y. Therefore cross-correlation can be calculated as $C(x,y) = \max_s \mathcal{F}^{-1}(\mathcal{X} \cdot \hat{\mathcal{Y}})[s]$ and using the fast Fourier transform we get the claimed time bound. This easily extends to our kernel by calculating $K(x,y) = \sum_s e^{\mathcal{F}^{-1}(\mathcal{X}\cdot\hat{\mathcal{Y}})[s]}$ implying:

Proposition 1. *$K(x,y)$ can be calculated in time $O(n \log n)$.*

Note that we need take the Fourier transform of each example only once. This gives a significant practical speedup over the naive quadratic time implementation.

4 Related Work

The current discoveries from the microlensing surveys such as OGLE and MACHO are predominantly transient objects such as gravitational microlensing, supernovae etc., and some periodic variable stars [14,15]. Recent work on star surveys introduced the application of semi-automatic classification techniques for periodic variable stars based on simple selection criteria over the parameter space indexed by average brightness, average difference in brightness between two spectral regions, and period, e.g [16,17]. We refer to these three parameters as *explicit features*. The semi-automatic methods require significant human intervention and hence pose an imperfect solution for a survey of even tens of millions of stars. An automatic approach has been proposed in [18]. This approach extracts explicit features from the light curves and applies machine learning methods in the resulting parameter space. Despite the similarity in terms of automation, our approach is unique in that we use the shape of the periodic time series to derive a similarity measure. Furthermore our approach is not astronomy-specific and is applicable across a range of domains.

There are many existing approaches for processing and classifying time series. A classical approach is to extract features of the time series, such as the Fourier basis, wavelets, or Hermite basis representation, and then work directly in the resulting vector space,e.g. [19]. Another major approach models the time series using a generative probabilistic model, such as Hidden Markov Models (HMM), and classifies examples using maximum likelihood or MAP estimates [20]. Our work falls into a third category: using similarity functions or distance measures for time series data [21,22]. Various similarity functions for time series have been proposed. Notably, Dynamic Time Warping (DTW) has been shown to be very effective across a large number of applications [21,23]. Such similarity functions are not phase invariant, hence they rely on a good universal phasing of the data.

Cross-correlation has been proposed precisely as an effective phase-invariant similarity function for astronomy and has been used for anomaly detection [13]. It is faster in runtime, $O(n \log n)$, than other methods that compute a maximum phase-invariant alignment. The notion of phase-invariance similarity has also been explored in the context of time series classification, specifically for time series generated from 2-d shape contours. For example, [23] present a method for applying any distance measure in a phase-invariant context. This allows for the application of Dynamic Time Warping, for instance, to data that is phase-invariant. While in general the run-time ($O(n^3)$) is as bad as brute-force methods such as in [24], they give experimental evidence that their heuristics lead to much faster, run-times in practice. We extend the work in [13] by investigating theoretical properties of cross-correlation and proposing a positive semidefinite alternative.

Several alternative approaches for working with non-positive semidefinite similarity measures exist in the literature. The simplest approach is just to use the (non-PSD) similarity function with SVM and hope for good results. Our experiments in the next section show that this does not always yield the desired performance. Another common alternative is to add a diagonal term λI to the gram matrix in order to render it positive semidefinite. More recent approaches reformulate the SVM optimization to account for the potential non-PSD kernel [25,26]. Finally, [27] show that a similarity function that meets some general requirements can be used to project examples into an explicit feature space indexed by their similarity to a fixed set of examples, and that this preserves some useful learnability properties. Unlike these generic methods, our work gives an explicit kernel construction that is useful for the time series domain.

There is significant overlap between the domain of time series classification and 2-d shape matching [23]. This is in part because a popular method for representing 2-d shapes is to create a time series from the contour of the shape. Shape classification has its own domain-specific approaches and it is beyond the scope of this paper to examine them. Nevertheless we note that shape matching is an example of a phase-invariant time series classification problem, and in fact we will present experiments from this domain.

The general issue of "maximizing alignment" appears repeatedly in work on kernels for structured objects. Dynamic Time Warping is a classic (non-positive semidefinite) example where we maximize alignment under legal potential warping of the time axes. A general treatment of such alignments, characterizing when the result is a kernel, is developed by [28]. Their results do not cover the case of cross-correlation, however. A similar alignment idea has been used for graph kernels in the application of classifying molecules, where each molecule can be seen as a graph of atoms and their bonds [29,30,31]. Here a base kernel is introduced between pairs of nodes in the two graphs. Then one can define a convolution kernel between the graphs using an equation similar to (8) where the sum ranges over all nodes in the graph [29,30]. Note that this approach does not maximize alignments, but sums over all possible alignments. A non-positive semidefinite alternative is to maximally align the two molecules by pairing their atoms in a one-to-one manner [31]. A major question is whether one could define an efficiently computable exponentially weighted version of such a (non-maximizing but PSD) graph kernel. One can show that this problem is closely related to calculating the permanent, a problem well known to be computationally hard [32,33]. As it is a special case of the permanent problem, however, where edge weights are related through the kernel function, it may be possible to calculate efficiently.

5 Experiments

In the following sets of experiments we demonstrate the performance of cross-correlation and our kernel in the context of phase-invariant time series classification.

For real-world data we use time series from astronomy surveys, and time series generated from contours of 2-d images. For artificial data, we generate examples that highlight the importance of phase invariance in an intuitive fashion. We use the same pre-processing for all time series, unless otherwise noted. The time series are smoothed as in [13,34], linearly-interpolated to 1024 evenly spaced points, and normalized to have mean of 0 and standard deviation of 1.

In all experiments we use the LIBSVM [35] implementation of SVM [7] and k-Nearest Neighbors (k-NN) to perform classification. For LIBSVM, we choose the "one-versus-one" multiclass setting, and we do not optimize the soft-margin parameter, instead using the default setting. For k-NN, we choose $k = 1$ following [23], who have published results on the shape data used in this paper[1]. When we use explicit features, we use a linear kernel. When we use cross-correlation or our kernel in addition to explicit features, we simply add the result of the inner product of the explicit features to the value of the cross-correlation or kernel[2].

We use five different similarity functions in our experiments: Euclidean Distance (ED) returns the inner product of two time series. The Universal Phasing (UP) similarity measure uses the method from [13] to phase each time series

[1] We reproduce their experiments as opposed to reporting their results in order to account for the different splits when cross-validating; our results do not differ significantly from those reported in [23].

[2] Another approach would be to perform multiple kernel learning [36] with one kernel being the cross-correlation and the other the inner product of the explicit features. However, this issue is orthogonal to the topic of the paper so we use the simple weighting.

according to the sliding window on the time series with the maximum mean, and then behaves exactly like Euclidean Distance. We use a sliding window size of 5% of the number of original points; the phasing takes place after the preprocessing explained above. In all experiments where we use K as in Equation 7, we do parameter selection by performing 10-fold cross-validation on the training set for each value of γ in $(1, 5, 10, 15, 25, 50, 80)$, then re-train using the value of γ that gave best average accuracy on the training set. When we use Dynamic Time Warping (DTW), we use the standard algorithm and do not restrict the warping window [21]. Finally we note that although cross-correlation is not positive semidefinite, we can in practice use it on some data sets with SVM.

In the first set of experiments we run on the OGLEII data set [5]. This data set consists of 14087 time series (light curves) taken from the OGLE astronomical survey. Each light curve is from one of three kinds of *periodic variable star*: Cepheid, RR Lyrae (RRL), or Eclipsing Binary (EB). We run 10-fold cross-validation over the entire data set, using the cross-correlation (CC), our kernel (K), and Universal Phasing (UP). The results, shown in the left top three rows of Tab. 1, illustrate the potential of the different similarities in this application. We see significant improvements for both cross-correlation and the kernel over Universal Phasing. We also see that the possibility to run SVM with our kernel leads to significant improvement over cross-correlation.

While the results reported so far on OGLEII are good, they are not sufficient for the domain of periodic variable star classification. Thus we turn next to improvements that are specific to the astronomy domain. In particular, the astronomy literature identifies three aggregate features that are helpful in variable star classification: the average brightness of the star, the *color* of the star which is the difference in average brightness between two different spectra, and the *period* of the star, i.e. the length of time to complete one period of brightness variation [16,17]. The right side of Tab. 1 gives the results when these features are added to the corresponding similarities. The features on their own yield very high accuracy, but there is a significant improvement in performance when we combine the features with cross-correlation or the kernel. Interestingly, while Universal Phasing on its own is not strong enough, it provides improvement over the features similar to our kernel and cross-correlation. Notice that a performance gain of 2% is particularly significant in the domain of astronomy where our goal is to publish such star catalogs with no errors or very few errors. The left confusion matrix in Tab. 2 (for SVM with our kernel plus features) shows that we can get very close to this goal on the OGLEII data. To our knowledge this is the first such demonstration of the potential of applying a shape matching similarity measure in order to automatically publish clean star catalogs from survey data. In addition, based on our domain knowledge, some of the errors reported in the left of Tab. 2 appear to be either mis-labeled or borderline cases whose label is difficult to determine.

In addition to classification, we show in Tab. 2 that the confidences produced by the classifier are well ordered. Here we do not perform any calibration and

Table 1. Accuracies with standard deviation reported from 10-fold cross-validation on OGLEII using various kernels and the cross-correlation

	1-NN	SVM		1-NN	SVM
CC	0.844 ± 0.011	0.680 ± 0.011	features + CC	0.991 ± 0.002	0.998 ± 0.001
K	0.901 ± 0.008	0.947 ± 0.005	features + K	0.992 ± 0.002	0.998 ± 0.001
UP	0.827 ± 0.010	0.851 ± 0.006	features + UP	0.991 ± 0.002	0.997 ± 0.001
			features	0.938 ± 0.006	0.974 ± 0.004

Table 2. Three confusion matrices for OGLEII, using SVM with K and features. From left to right we reject none, then the lowest 1%, 1.5% and 2%.

	Ceph	EB	RRL	Ceph	EB	RRL	Ceph	EB	RRL	Ceph	EB	RRL
Cepheid	3416	1	13	3382	1	3	3363	1	3	3352	1	0
EB	0	3389	0	0	3364	0	0	3342	0	0	3312	0
RRL	9	0	7259	1	0	7195	0	0	7166	0	0	7138

simply take the raw output of each of the 3 hyperplanes learned by the SVM[3]. To calculate the confidence in label 1, we add the raw output of the $1v2$ (the classifier separating class 1 from class 2) and $1v3$ classifiers. To calculate the confidence in label 2 we add the negative output of the $1v2$ hyperplane and the output of the $2v3$ hyperplane, etc. We can then reject the examples that received the lowest confidences and set them aside for review. When we reject the lowest 1%, for example, we reject all but 5 errors, showing that almost all of our errors have low confidences. We now have reason to believe that, when we classify a new catalog, we can reliably reject a certain percentage of the predictions that are most likely to be errors. The rejected examples can either be ignored or set aside for human review.

In the next set of experiments we use five shape data sets: Butterfly, Arrowhead, Fish, Seashells introduced in [23], as well as the SwedishLeaf data set [38][4]. These data sets were created by taking pictures of objects and creating a time series by plotting the radius of a line anchored in the center of the object as it rotates around the image [23]. As all of the pictures have aligned each object more or less along a certain orientation, we randomly permute each time series prior to classification in order to eliminate any bias of the orientation. The identification of objects from various orientations is now cast as a phase-invariant time series problem.

A natural and relatively easy problem is to use a classifier to separate the different image types from each other. In this case we attempt to separate butterflies, arrowheads, seashells, and fish. We refer to this data set as *Intershape*[5].

[3] While there are several methods in the literature to produce class-membership probabilities from SVM output [37], exploratory experiments could not confirm the reliability of these probabilities for our data. Therefore we chose to use the simple method based on raw SVM output.

[4] Detailed Information available via `www.cs.ucr.edu/~eamonn/shape/shape.htm`

[5] We treat the SwedishLeaf set differently because it has a different resolution and is not part of the same overall shape data set.

Table 3. Number of examples in each data set. For those data sets that were filtered to include 20 examples of each class, the number of examples post-filtering appears after the '/'.

	Num Examples	Num Classes	Majority Class
Arrowhead	558/474	9	0.19
Butterfly	754/312	5	0.39
Intershape	2511	4	0.30
SwedishLeaf	1125	15	0.07

We also investigate the potential to separate sub-classes of each shape type. The SwedishLeaf data has already been labeled before, and hence the sub-classes are already identified. For the other data sets that have not been explicitly labeled by class before, we generate labels as follows: for the Butterfly and Fish data set, we consider two examples to be the same class if they are in the same genus. For the Arrowhead data set, we consider two arrowheads to be the same type if they share the same type name, such as "Agate Basin", or "Cobbs." In order to make the results more statistically robust, we eliminate sub-types for which there exist fewer than 20 examples. Seashells and Fish have too few examples when processed in this way and are therefore only used in the Intershape data set. A summary of the data sets, including number of examples and majority class probability (that can be seen as a baseline) are given in Tab. 3.

For these experiments we calculate no explicit features. We run 10-fold cross-validation using k-NN with cross-correlation (1-NN CC), the kernel (1-NN K), Dynamic Time Warping (1-NN DTW), Universal Phasing (1-NN UP) and SVM with the kernel (SVM K), Universal Phasing (SVM UP), and Euclidean distance (SVM ED). The results are given in Tab. 4. We also tried using 1-NN with Euclidean Distance, but the performance was not competitive with any of the other methods so we do not include it in the comparison.

The results demonstrate that both cross-correlation and the kernel provide a significant performance advantage. It is not surprising that DTW does not do well since it only considers the one given random phasing of the data. Rather, it is surprising that it does not perform worse on this data. The only way it can expect to perform well with k-NN is if, by chance, for each example there is another example of the same class that happens to share roughly the same phase. In a large enough data set, this can happen, and this may explain why DTW does much better than random guessing. It is interesting that SVM does not always dominate k-NN and does very poorly on SwedishLeaf. It may be that the data are linearly inseparable but there are are enough examples such that virtual duplicates appear in the data allowing 1-NN to do well.

Another interesting observation is that while Universal Phasing never outperforms all methods it does reasonably well across the domains. Recall that this method phases the time series according to the maximum average brightness of a sliding window. This finds a "maximum landmark" in the data for alignment

Table 4. Performance on various shape data sets. All results are cross-validated. Data set names: A = arrowhead, B = butterfly, I = intershape, S = Swedish.

	1-NN CC	1-NN K	1-NN DTW	1-NN UP	SVM ED	SVM UP	SVM K
A	0.54 ± 0.06	0.54 ± 0.08	0.33 ± 0.06	0.49 ± 0.05	0.2 ± 0.05	0.41 ± 0.05	0.63 ± 0.04
B	0.73 ± 0.04	0.73 ± 0.04	0.59 ± 0.08	0.70 ± 0.07	0.4 ± 0.1	0.65 ± 0.08	0.76 ± 0.08
I	0.98 ± 0.01	0.98 ± 0.01	0.84 ± 0.03	0.97 ± 0.02	0.47 ± 0.03	0.8 ± 0.02	0.91 ± 0.02
S	0.84 ± 0.03	0.82 ± 0.03	0.48 ± 0.06	0.78 ± 0.04	0.08 ± 0.03	0.18 ± 0.03	0.33 ± 0.04

and is obviously not guaranteed to be informative of the class in every case. Nevertheless, it works well on the Butterfly and Intershape data sets showing that this type of landmark is useful for them.

As we show in the next set of experiments with artificial data, it is easy to construct examples where Universal Phasing will fail. We generate two classes of time series. Each example contains 1024 points. Class 1 is a multi-step function with one set of four steps beginning at time 0, as well as one spike placed randomly. Class 2 is also a multi-step function but with two sets of two steps, the first at time 0 and the second at time 665 (roughly 65% of the entire time series) and one random spike exactly as in class 1. We show two examples of each class in Fig. 2. We generate 10 disjoint training sets containing 70 examples and test sets containing 30 examples for cross-validation. We keep the training set small to avoid clobbering the results by having near-identical examples. In these experiments we normalize as above, however we do not perform smoothing.

For this type of data the random spike will always be in the center of the largest magnitude sliding-window, and hence Universal Phasing will phase each time series according to the random location of the spike. In a real world setting, the random spike could be sufficiently wide noise period in the signal, or any irrelevant feature of the time series. This is key to understanding the strength of our method: if it is easy to find a global shifting such that each example is maximally correlated with every other, our method performs identically to Euclidean Distance. On the other hand, when a global shift is not trivial to find, our method succeeds where a Universal Phasing algorithm fails. To illustrate further the performance potential of the kernel we create a second version of the data where we add noise to the labels by flipping the label of each example with probability of 0.1. When the data are completely or nearly separable, both k-NN and SVM should attain close to 100% accuracy. The noise changes the domain to make it harder to get this level of performance.

The results are shown in Tab. 5. As we expected, Universal Phasing does quite poorly in this setting. With no noise, 1-NN with cross-correlation, 1-NN with our kernel, and SVM with our kernel attain almost 100% accuracy. The results with noisy data show that SVM with our kernel is more robust to noise than 1-NN with cross-correlation or our kernel.

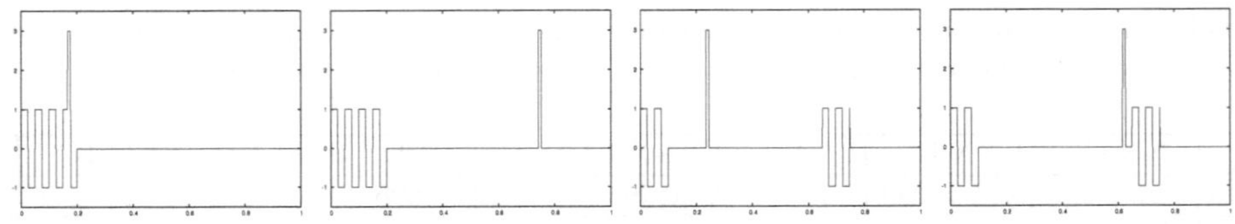

Fig. 2. Examples of artificial data. The left two examples are from class 1, the right two example are from class 2.

Table 5. Results on artificial data

	1-NN CC	1-NN K	1-NN UP	SVM UP	SVM K
Artificial	0.99 ± 0.02	1.00 ± 0.00	0.65 ± 0.04	0.50 ± 0.07	0.997 ± 0.001
Artificial w/ Noise	0.84 ± 0.14	0.84 ± 0.12	0.61 ± 0.09	0.53 ± 0.12	0.90 ± 0.05

6 Conclusion

On the OGLEII data set we have shown a basis for a completely automatic star classification algorithm. We have shown that cross-correlation is an effective similarity measure for phase-invariant time series, we proved that cross-correlation is not positive semidefinite, and we gave a positive semidefinite alternative, justifying its use in an experimental setting.

The work we have presented in the astronomy domain is a portion of our continuing effort to build an "end-to-end" automatic classification system for astronomy events. In particular we have used the work presented here to classify other star surveys such as MACHO. A complete star classification system requires several modules in addition to the classification method we have demonstrated here. For instance, the raw data from a survey contains no information about the star other than its brightness measured at specific times. To classify the star, we must first determine if it is variable, determine if it is periodic, and find its period; only then can we finally classify it. What we have shown in this paper represents the bulk of the classification portion. A manuscript detailing a methodology for the complete task and results for the MACHO catalog is currently in preparation.

As discussed in Sect. 4, using a computationally tractable approximation to the maximum alignment has potential applications in the domain of classifying graphs and other structured data. The main question is whether we can efficiently calculate an exponential weighted approximation to a maximum alignment and whether this would prove useful in an experimental setting.

Acknowledgments

This research was partly supported by NSF grant IIS-080340. The experiments in this paper were performed on the Odyssey cluster supported by the FAS

Research Computing Group at Harvard. We gratefully acknowledge Xiaoyue (Elaine) Wang, Lexiang Ye, Chotirat Ratanamahatana and Eamonn Keogh for creating the UCR Time Series Classification data repository, and for providing us with the shape data sets. We would also like to thank the Harvard Initiative in Innovative Computing for research space and computing facilities.

References

1. Alcock, C., et al.: The MACHO Project - a Search for the Dark Matter in the Milky-Way. In: Soifer, B.T. (ed.) Sky Surveys. Protostars to Protogalaxies. Astronomical Society of the Pacific Conference Series, vol. 43, p. 291 (1993)
2. Udalski, A., Szymanski, M., Kubiak, M., Pietrzynski, G., Wozniak, P., Zebrun, Z.: Optical gravitational lensing experiment. Photometry of the macho-smc-1 microlensing candidate. Acta Astronomica 47(431) (1997)
3. Hodapp, K.W., et al.: Design of the Pan-STARRS telescopes. Astronomische Nachrichten 325, 636–642 (2004)
4. Starr, B.M., et al.: LSST Instrument Concept. In: Tyson, J.A., Wolff, S. (eds.) Society of Photo-Optical Instrumentation Engineers (SPIE) Conference Series, vol. 4836, pp. 228–239 (2002)
5. Soszynski, I., Udalski, A., Szymanski, M., Kubiak, M., Pietrzynski, G., Wozniak, P., Zebrun, K., Szewczyk, O., Wyrzykowski, L.: The Optical Gravitational Lensing Experiment. Catalog of RR Lyr Stars in the Large Magellanic Cloud. Acta Astronomica 53, 93–116 (2003)
6. Elkan, C.: Using the triangle inequality to accelerate k-means. In: Fawcett, T., Mishra, N. (eds.) International Conference on Machine Learning, pp. 147–153 (2003)
7. Boser, B.E., Guyon, I., Vapnik, V.: A training algorithm for optimal margin classifiers. In: Computational Learning Theory, pp. 144–152 (1992)
8. Moore, A.W.: The anchors hierarchy: Using the triangle inequality to survive high dimensional data. In: Boutilier, C., Goldszmidt, M. (eds.) Uncertainty in Artificial Intelligence, pp. 397–405. Morgan Kaufmann, San Francisco (2000)
9. Schölkopf, B., Smola, A.J.: Learning with Kernels. The MIT Press, Cambridge (2002)
10. Cuturi, M., Vert, J.P., Birkenes, O., Matsui, T.: A kernel for time series based on global alignments. In: IEEE International Conference on Acoustics Speech and Signal Processing (ICASSP) 2007, vol. 2, pp. 413–416 (2007)
11. Gärtner, T., Flach, P.A., Kowalczyk, A., Smola, A.J.: Multi-instance kernels. In: International Conference on Machine Learning, pp. 179–186 (2002)
12. Haussler, D.: Convolution kernels for discrete structures. Technical Report UCSC-CRL-99-10, Department of Computer Science, University of California at Santa Cruz (1999)
13. Protopapas, P., Giammarco, J.M., Faccioli, L., Struble, M.F., Dave, R., Alcock, C.: Finding outlier light curves in catalogues of periodic variable stars. Monthly Notices of the Royal Astronomical Society 369, 677–696 (2006)
14. Faccioli, L., Alcock, C., Cook, K., Prochter, G.E., Protopapas, P., Syphers, D.: Eclipsing Binary Stars in the Large and Small Magellanic Clouds from the MACHO Project: The Sample. Astronomy Journal 134, 1963–1993 (2007)

15. Alcock, C., et al.: The MACHO project LMC variable star inventory. 1: Beat Cepheids-conclusive evidence for the excitation of the second overtone in classical Cepheids. Astronomy Journal 109, 1653 (1995)
16. Geha, M., et al.: Variability-selected Quasars in MACHO Project Magellanic Cloud Fields. Astronomy Journal 125, 1–12 (2003)
17. Howell, D.A., et al.: Gemini Spectroscopy of Supernovae from the Supernova Legacy Survey: Improving High-Redshift Supernova Selection and Classification. Astrophysical Journal 634, 1190–1201 (2005)
18. Debosscher, J., Sarro, L.M., Aerts, C., Cuypers, J., Vandenbussche, B., Garrido, R., Solano, E.: Automated supervised classification of variable stars. i. methodology. Astronomy and Astrophysics 475, 1159–1183 (2007)
19. Vlachos, M., Vagena, Z., Yu, P.S., Athitsos, V.: Rotation invariant indexing of shapes and line drawings. In: Herzog, O., Schek, H.J., Fuhr, N., Chowdhury, A., Teiken, W. (eds.) Conference on Information and Knowledge Management, pp. 131–138. ACM, New York (2005)
20. Ge, X., Smyth, P.: Deformable markov model templates for time-series pattern matching. In: Knowledge Discovery and Data Mining, pp. 81–90 (2000)
21. Berndt, D.J., Clifford, J.: Using dynamic time warping to find patterns in time series. In: Knowledge Discovery and Data Mining, pp. 359–370 (1994)
22. Lu, Z., Leen, T.K., Huang, Y., Erdogmus, D.: A reproducing kernel hilbert space framework for pairwise time series distances. In: Cohen, W.W., McCallum, A., Roweis, S.T. (eds.) International Conference on Machine Learning, pp. 624–631 (2008)
23. Keogh, E.J., Wei, L., Xi, X., Lee, S.H., Vlachos, M.: Lb keogh supports exact indexing of shapes under rotation invariance with arbitrary representations and distance measures. In: Dayal, U., et al. (eds.) International Conference on Very Large Databases, pp. 882–893. ACM, New York (2006)
24. Adamek, T., O'Connor, N.E.: A multiscale representation method for nonrigid shapes with a single closed contour. IEEE Trans. Circuits Syst. Video Techn. 14(5), 742–753 (2004)
25. Luss, R., d'Aspremont, A.: Support vector machine classification with indefinite kernels. In: Platt, J.C., Koller, D., Singer, Y., Roweis, S.T. (eds.) Neural Information Processing Systems. MIT Press, Cambridge (2007)
26. Ong, C.S., Mary, X., Canu, S., Smola, A.J.: Learning with non-positive kernels. In: Brodley, C.E. (ed.) International Conference on Machine Learning (2004)
27. Balcan, M.F., Blum, A., Srebro, N.: A theory of learning with similarity functions. Machine Learning 72(1-2), 89–112 (2008)
28. Shin, K., Kuboyama, T.: A generalization of Haussler's convolution kernel: mapping kernel. In: Cohen, W.W., McCallum, A., Roweis, S.T. (eds.) International Conference on Machine Learning, pp. 944–951 (2008)
29. Gärtner, T., Flach, P., Wrobel, S.: On graph kernels: Hardness results and efficient alternatives. In: Schölkopf, B., Warmuth, M.K. (eds.) COLT/Kernel 2003. LNCS (LNAI), vol. 2777, pp. 129–143. Springer, Heidelberg (2003)
30. Wachman, G., Khardon, R.: Learning from interpretations: a rooted kernel for ordered hypergraphs. In: Ghahramani, Z. (ed.) International Conference on Machine Learning, pp. 943–950 (2007)
31. Fröhlich, H., Wegner, J., Sieker, F., Zell, A.: Optimal assignment kernels for attributed molecular graphs. In: International Conference on Machine Learning, pp. 225–232 (2005)
32. Valiant, L.G.: The complexity of computing the permanent. Theor. Comput. Sci. 8, 189–201 (1979)

33. Papadimitriou, C.H.: Computational Complexity. Addison Wesley, Reading (1993)
34. Gorry, P.A.: General least-squares smoothing and differentiation by the convolution (savitzky-golay) method. Analytical Chemistry 62(6), 570–573 (1990)
35. Chang, C.C., Lin, C.J.: LIBSVM: a library for support vector machines (2001), `http://www.csie.ntu.edu.tw/~cjlin/libsvm`
36. Sonnenburg, S., Rätsch, G., Schäfer, C., Schölkopf, B.: Large scale multiple kernel learning. Journal of Machine Learning Research 7, 1531–1565 (2006)
37. Huang, T.K., Weng, R.C., Lin, C.J.: Generalized bradley-terry models and multi-class probability estimates. Journal of Machine Learning Research 7, 85–115 (2006)
38. Söderkvist, O.J.O.: Computer vision classification of leaves from Swedish trees. Master's thesis, Linköping University, SE-581 83 Linköping, Sweden (September 2001)

K-Subspace Clustering

Dingding Wang[1], Chris Ding[2], and Tao Li[1]

[1] School of Computer Science, Florida International Univ., Miami, FL 33199, USA
[2] CSE Department, University of Texas, Arlington, Arlington, TX 76019, USA

Abstract. The widely used K-means clustering deals with ball-shaped (spherical Gaussian) clusters. In this paper, we extend the K-means clustering to accommodate extended clusters in subspaces, such as line-shaped clusters, plane-shaped clusters, and ball-shaped clusters. The algorithm retains much of the K-means clustering flavors: easy to implement and fast to converge. A model selection procedure is incorporated to determine the cluster shape. As a result, our algorithm can recognize a wide range of subspace clusters studied in various literatures, and also the global ball-shaped clusters (living in all dimensions). We carry extensive experiments on both synthetic and real-world datasets, and the results demonstrate the effectiveness of our algorithm.

1 Introduction

Data clustering is useful to discover patterns in a large dataset by grouping similar data points into clusters. Traditionally a cluster is viewed as a spherical (ball-shaped) data distribution and therefore is often modeled with mixture of Gaussians. For high dimensional data, sometimes an extended cluster may live in a subspace with much smaller dimension, i.e., it deviates away from a spherical cluster very significantly (Figures 1-5 in Section 5 illustrate various subspace clusters). This type of subspace clusters is difficult to discover using traditional clustering algorithms (e.g., K-means).

Here, we begin with an observation on the key difficulty of subspace clustering. We believe the main difficulty with subspace clustering is the exact definition of clusters. If a cluster lives in a subspace, but is not extended significantly, this type of clusters can be handled by traditional algorithms such as K-means. However, if a cluster has an extended size in limited dimensions, (e.g., an extended plane in 5-dimensional space), it becomes a subspace cluster. Thus the difficulty is how to model an extended cluster.

In this paper, we propose a new clustering algorithm to handle clusters with extended shapes. Our approach is a distance-minimization based approach, much like the standard K-means clustering algorithm. This approach can be cast in the framework of mixture clustering with Expectation and Maximization steps.

In the following, we discuss the related work in Section 2. In section 3 we propose several distance minimization based models to describe extended cluster objects and derive the optimal solutions to these cluster models (Sections 3.2-3.5). We discuss the model selection issue in section 3.6. Extensive illustrative examples are given in Section 4.

W. Buntine et al. (Eds.): ECML PKDD 2009, Part II, LNAI 5782, pp. 506–521, 2009.

One strong feature of our K-subspace algorithm is that it also accommodates ball-shaped clusters which live in all dimensions, in addition to the subspace clusters. In other words, our K-subspace algorithm is a comprehensive clustering algorithm, in contrast to many previous subspace clustering algorithms which are specifically tailored to find subspace clusters.

Clusters come in all shapes and sizes. Consider an elongated ball cluster which is in-between a line-shaped cluster and a ball-shaped cluster. Our algorithm tries to model it in both ways and automatically determine the best way to model it. An algorithm specifically designed for seeking subspace clusters would have to incorporate certain threshold to define whether it meets the criteria for subspace. This specific subspace clustering search algorithm goes in contradiction to the unsupervised nature of clustering.

Our K-subspace algorithm, on the other hand, retains much of the usual unsupervised learning. It is an extension of K-means algorithm for incorporating the capability to handle subspace clusters. For these reasons, we run our algorithm on an extensive list of datasets and compare with many existing clustering algorithms. Results are presented in Section 5. The new algorithm outperforms many existing algorithms, due to its capability to model a larger number of different cluster shapes. Conclusions are given in Section 6.

2 Related Work

Current study of subspace clustering mainly follows two general approaches [20].: (1) bottom-up search methods such as Clique [3], enclus [6], mafia [14], cltree [18], and DOC [21]; and (2) top-down search methods including COSA [13], Proclus [1], ORCLUS [2], Findit [25], and δ-clusters [27]. The bottom-up approaches use the monotonicity of density to prune subspaces by selecting those subspaces with densities above a given threshold, and they often cover clusters in various shapes and sizes. However, the clustering results are much dependent on the density threshold parameter, which can be very difficult to set properly. The top-down approaches integrate clustering and subspace selection by multiple iterations of selection and feature evaluation. The required iterations of clustering are very expensive and often cause slow performance. In addition to the poor efficiency, the top-down approaches are bad at finding hyper-spherical clusters due to the nature of using cluster centers to represent similar instances. Other recent works include GPCA [24], Oriented K-windows [23], Weighted K-means [30], adaptive subspace clustering [17] etc.

Adaptive dimension reduction, which attempts to adaptively find the subspace where clusters are most well-separated or well-defined, has also been proposed. Recent work in [11] combines linear discriminant analysis (LDA) and K-means clustering in a coherent way where K-means is used to generate cluster labels and LDA is used to select subspaces. Thus the adaptive subspace selection is integrated with the clustering process. Spectral clustering (e.g., Normalized Cut algorithm) and nonnegative matrix factorization (NMF) are also able to discover arbitrarily shaped clusters. It has been shown that spectral clustering can be viewed as a weighted version of Kernel K-means, and NMF is proved to be equivalent to spectral clustering [10].

In our work, we extend K-means algorithm to K-subspace algorithm, which handles not only clusters living in all dimensions but also subspace clusters, while retaining the simple and fast implementation of K-means clustering.

3 K-Subspace Clustering Model

3.1 The Distance Minimization Clustering Model

The distance minimization clustering model can be represented as:

$$\min_{H,\theta} \sum_{i=1}^{n} \left[\min_{1 \le k \le K} Dist(\mathbf{x}_i, C_k) \right] \tag{1}$$

where H is the cluster indicator matrix, θ is a parameter in the optimization, $\mathbf{x}_i$ is a data point, C_k is the $k-th$ cluster, and

$$Dist(\mathbf{x}_i, C_k) = -\log Prob(\mathbf{x}_i; C_k).$$

Note $Prob(\mathbf{x}_i; C_k) \le 1$ and a common choice of the individual probability is the full Gaussian distribution using Mahalanobis distance.

This minimization can be equivalently written as

$$\min_{\{C_k\}} J = \sum_{k=1}^{K} \sum_{i \in C_k} Dist(\mathbf{x}_i, C_k) \tag{2}$$

Clearly for K-means clustering,

$$Dist(\mathbf{x}_i, C_k) = ||\mathbf{x}_i - \mathbf{c}_k||^2 \tag{3}$$

The K-subspace clustering model utilizes the distance minimization model. In this paper we consider only linear subspaces: 1D line-shaped (rod-like) subspace, 2D plane-shaped (slab-like) subspace, etc. The model also accommodates the ball-shaped clusters. Our main contribution here is to show that these subspace cluster models can be rigorously defined and efficiently computed. To our knowledge, this approach is novel and this type of analysis and results have not been previously investigated and obtained.

3.2 Line-Shaped Clusters

How to represent a line-shaped cluster C_k? A line is defined by a point in space and a unit direction. Let the point be $\mathbf{c}_k$ and the unit direction be $\mathbf{a}_k$. A data point $\mathbf{x}$ can be decomposed as a parallel component

$$\mathbf{x}^{\parallel} = \mathbf{a}_k[\mathbf{a}_k^T(\mathbf{x} - \mathbf{c}_k)]$$

and a perpendicular component

$$\mathbf{x}^{\perp} = (\mathbf{x} - \mathbf{c}_k) - \mathbf{x}^{\parallel}.$$

We define the distance between the data point $\mathbf{x}$ and cluster C_k as the perpendicular distance between the point and the line:

$$Dist(\mathbf{x}, C_k) = ||\mathbf{x}^{\perp}||^2 = ||\mathbf{x} - \mathbf{c}_k - \alpha \mathbf{a}_k||^2 \tag{4}$$

where

$$\alpha = (\mathbf{x} - \mathbf{c}_k)^T \mathbf{a}_k. \tag{5}$$

Computing model parameters. Given a set of data points in the cluster, and assuming they are described by a line, how do we compute the model parameters $(\mathbf{c}_k, \mathbf{a}_k)$?

We minimize the total distance (dispersion) of the cluster:

$$\min_{\mathbf{c}_k, \mathbf{a}_k} \sum_{i \in C_k} Dist(\mathbf{x}_i, C_k) = \sum_{i \in C_k} ||\mathbf{x}_i - \mathbf{c}_k - \alpha \mathbf{a}_k||^2 \tag{6}$$

It turns out that the model parameters can be computed in a surprisingly simple way:

Theorem 1. Let

$$X^{(k)} = [\mathbf{x}_1^{(k)}, \mathbf{x}_2^{(k)}, ..., \mathbf{x}_{n_k}^{(k)}] \tag{7}$$

be the data points in C_k. Compute the principal component analysis (PCA) of them, and let $\mathbf{u}_1$ be the first principal direction. Let the centroid of C_k be

$$\bar{\mathbf{x}}^{(k)} = \frac{1}{|C_k|} \sum_{i \in C_k} \mathbf{x}_i. \tag{8}$$

Then the global optimal solution to the minimization problem of Eq.(6) is given by

$$\mathbf{c} = \bar{\mathbf{x}}^{(k)}, \ \mathbf{a} = \mathbf{u}_1. \tag{9}$$

Proof. We wish to minimize Eq.(6) which is

$$\min_{\mathbf{c}_k, \mathbf{a}_k} J_1 = \sum_{i \in C_k} ||(I - \mathbf{a}_k \mathbf{a}_k^T)(\mathbf{x}_i - \mathbf{c}_k)||^2 \tag{10}$$

by substituting Eq.(5) into Eq.(6).

Using $||A||^2 = \mathrm{Tr}(A^T A)$, we have

$$J_1 = \mathrm{Tr} \sum_{i \in C_k} [(\mathbf{x}_i - \mathbf{c}_k)^T (I - \mathbf{a}_k \mathbf{a}_k^T)^T (I - \mathbf{a}_k \mathbf{a}_k^T)(\mathbf{x}_i - \mathbf{c}_k)] \tag{11}$$

Because $(I - \mathbf{a}_k \mathbf{a}_k^T)^T (I - \mathbf{a}_k \mathbf{a}_k^T) \equiv M$ is a semi-positive definite matrix, J_1 is a convex function of $\mathbf{c}_k$. Its global minima is given by the zero- gradient condition:

$$0 = \frac{\partial J_1}{\partial \mathbf{c}_k} \sum_i 2(M\mathbf{c}_k - M\mathbf{x}_i)$$

Thus $M\mathbf{c}_k = M\bar{\mathbf{x}}^{(k)}$. Clearly,$\mathbf{c}_k = \bar{\mathbf{x}}^{(k)}$ is the solution. In general, M is non-deficient and$\mathbf{c}_k = \bar{\mathbf{x}}^{(k)}$ is the unique solution. When M is deficient, $\mathbf{c}_k = \bar{\mathbf{x}}^{(k)}$ is still a correct solution, but it is not the unique solution. Because $\mathbf{a}_k$ is a unit vector,$(I - \mathbf{a}_k\mathbf{a}_k^T)^T(I - \mathbf{a}_k\mathbf{a}_k^T) = (I - \mathbf{a}_k\mathbf{a}_k^T)$. Thus

$$J_1 = \mathrm{Tr}(I - \mathbf{a}_k\mathbf{a}_k^T)\Sigma,$$

where

$$\Sigma = [\sum_{i \in C_k} (\mathbf{x}_i - \mathbf{c}_k)(\mathbf{x}_i - \mathbf{c}_k)^T]$$

is the sample covariance matrix. Now minimization for $\mathbf{a}_k$ becomes

$$\min_{\mathbf{a}_k} J_1 = \mathrm{Tr}\ (I - \mathbf{a}_k\mathbf{a}_k^T)\Sigma = const - \mathbf{a}_k^T\Sigma\mathbf{a}_k \quad (12)$$

This is equivalent to $\max_{\mathbf{a}_k} \mathbf{a}_k^T\Sigma\mathbf{a}_k$, whose solution is the first principle direction, because Σ is the covariance matrix. □

3.3 Plane-Shaped Clusters

A 2D plane is defined by a point in space and two unit directions. Let $\mathbf{c}_k$ be the point, and $\mathbf{a}_k, \mathbf{b}_k$ be the two unit directions. A data point $\mathbf{x}$ can be decomposed as a parallel component (lying inside the plane)

$$\mathbf{x}^{\parallel} = \mathbf{a}_k[\mathbf{a}_k^T(\mathbf{x} - \mathbf{c}_k)] + \mathbf{b}_k[\mathbf{b}_k^T(\mathbf{x} - \mathbf{c}_k)]$$

assuming the two unit directions are orthogonal: $\mathbf{a}_k^T\mathbf{b}_k = 0$. And the perpendicular component

$$\mathbf{x}^{\perp} = (\mathbf{x} - \mathbf{c}_k) - \mathbf{x}^{\parallel}.$$

We define the distance between the data point $\mathbf{x}$ and cluster C_k as the perpendicular distance between the point and the plane:

$$Dist(\mathbf{x}, C_k) = ||\mathbf{x}^{\perp}||^2 = ||\mathbf{x} - \mathbf{c}_k - \alpha\mathbf{a}_k - \beta\mathbf{b}_k||^2 \quad (13)$$

where

$$\alpha = (\mathbf{x} - \mathbf{c}_k)^T\mathbf{a}_k,\ \ \beta = (\mathbf{x} - \mathbf{c}_k)^T\mathbf{b}_k. \quad (14)$$

Clearly, we can extend this to 3D and higher dimensional subspace clusters.

Computing model parameters. Given a set of data points in the cluster. Assuming they are described by a plane, how do we compute the model parameters $(\mathbf{c}_k, \mathbf{a}_k, \mathbf{b}_k)$?

We minimize the dispersion (total distance) of the cluster:

$$\min_{\mathbf{c}_k, \mathbf{a}_k, \mathbf{b}_k} \sum_{i \in C_k} Dist(\mathbf{x}_i, C_k) = \sum_{i \in C_k} ||\mathbf{x}_i - \mathbf{c}_k - \alpha\mathbf{a}_k - \beta\mathbf{b}_k||^2 \quad (15)$$

As one might have expected, the solution is given by PCA.

Theorem 2. Compute the (PCA) of $X^{(k)}$ of Eq.(7) to obtain the first two principal directions $(\mathbf{u}_1, \mathbf{u}_1)$. Compute the centroid as in Eq.(8). The optimal solution to Eq.(15) is given by

$$\mathbf{c} = \bar{\mathbf{x}}^{(k)},\ \mathbf{a} = \mathbf{u}_1,\ \mathbf{b} = \mathbf{u}_2. \quad (16)$$

Proof. We outline the proof here by pointing out the main difference from the proof for Theorem 1.

We wish to minimize Eq.(15) which is

$$\min_{\mathbf{c}_k,\mathbf{a}_k,\mathbf{b}_k} J_2 = \sum_{i \in C_k} ||(I - \mathbf{a}_k\mathbf{a}_k^T - \mathbf{b}_k\mathbf{b}_k^T)(\mathbf{x}_i - \mathbf{c}_k)||^2 \tag{17}$$

by substituting Eq.(5) into Eq.(17). J_2 can be written as

$$J_2 = \sum_{i \in C_k} \text{Tr } [(\mathbf{x}_i - \mathbf{c}_k)^T B^T B(\mathbf{x}_i - \mathbf{c}_k)] \tag{18}$$

where

$$B = (I - \mathbf{a}_k\mathbf{a}_k^T - \mathbf{b}_k\mathbf{b}_k^T)$$

Because $B^T B$ is a semi-positive definite matrix, J_2 is a convex function of $\mathbf{c}_k$. Its global minima is given by$\mathbf{c}_k = \bar{\mathbf{x}}^{(k)}$.

Now focusing on $(\mathbf{a}_k, \mathbf{b}_k)$. Since we restrict $\mathbf{a}_k^T\mathbf{b}_k = 0$,and $B^T B = B = (I - \mathbf{a}_k\mathbf{a}_k^T - \mathbf{b}_k\mathbf{b}_k^T)$, the minimization over $(\mathbf{a}_k, \mathbf{b}_k)$ becomes

$$\min_{\mathbf{a}_k,\mathbf{b}_k} J_2 = \text{Tr } (I - \mathbf{a}_k\mathbf{a}_k^T - \mathbf{a}_k\mathbf{b}_k^T)\Sigma = const - \mathbf{a}_k^T\Sigma\mathbf{a}_k - \mathbf{b}_k^T\Sigma\mathbf{b}_k \tag{19}$$

The solution for $\mathbf{a}_k$ is given by the first principal direction $\mathbf{u}_1$. Since $\mathbf{a}_k^T\mathbf{b}_k = 0$, the solution for $\mathbf{b}_k$ is given by the second principal direction $\mathbf{u}_2$. □

3.4 3D and Higher Dimensional Planes

Clearly, we can model a subspace cluster using 3D and higher dimensional planes. The same definitions using the perpendicular distances can be adopted and the same computational algorithms using PCA can be easily generalized from 1D and 2D cases.

We can use this in the reverse direction. Suppose we are given a set of data points, and the question is which dimensions of the subspace cluster we should choose?

According to Theorem 1 and Theorem 2, we compute PCA of the dispersion and obtain eigenvalues and eigenvectors. If there is only one significant eigenvalue, this implies that the cluster is a 1D (line-shaped) subspace cluster, then we set $\mathbf{a}_k = \mathbf{u}_1$. If there are only two significant eigenvalues, it implies that the cluster is a 2D (plane-shaped) subspace cluster, then we set $\mathbf{a}_k = \mathbf{u}_1$ and $b_k = \mathbf{u}_2$.

In general, if there are k significant eigenvalues, then the cluster is likely a k-dimensional (kD plane) subspace cluster. For this cluster, we need k vectors (kp parameters, where $p =$ dimension of data space) to describe it.

However, to keep the simplicity of cluster modeling to prevent overfitting with too many parameters, we restrict the cluster dimensions to 2. For higher dimensions, we use a spherical cluster to describe the cluster, instead of using the high-dimensional planes which have much more parameters.

3.5 Spherical Clusters

Although Gaussian mixtures using the EM algorithm can describe more complicated clusters, in practice very often the much simpler K-means clustering algorithm provides results as good as the mixture models.

This implies two fundamental facts in statistics: (S1) complex functions do not necessarily perform better (mainly due to overfitting) and (S2) most clusters are reasonably modeled by the spherical cluster model. In fact, clusters of any shape can be modeled as spherical clusters if they are reasonably separated.

From (S1), we should not use too many parameters in describing a cluster. From (S2), we should make use of spherical clusters as much as possible. These two facts motivate us to consider spherical clusters more carefully, rather than using 3D or higher dimensional plane-shaped clusters.

A spherical (ball-shaped) data cluster is a global cluster in the sense that it exists in all dimensions. We adopt it as our K-subspace clustering because not all clusters in a dataset are subspace clusters.

Then a natural question is: what is the distance between a data point and a spherical cluster? We could use the distance to the cluster centroid

$$Dist(\mathbf{x}_i, C_k) = ||\mathbf{x}_i - \mathbf{c}_k||^2 \tag{20}$$

as in K-means clustering. If all clusters in the dataset are spherical clusters, the problem is reduced to the usual K-means clustering.

However, we wish to measure the distance between a data point to the *cluster*, not to the cluster *center*. In the case of the line-shaped cluster, the point-cluster distance is the perpendicular distance, i.e., data point $\mathbf{x}$ to the closest data points on the line. In the case of the plane-shaped cluster, the point-cluster distance is the perpendicular distance, i.e., data point $\mathbf{x}$ to the closest data points on the plane.

For these reasons, we define the point-cluster distance for spherical clusters to be the distance between the data point and the closest data point on the sphere. For this we introduce

$$Dist(\mathbf{x}_i, C_k) = \max\left(0, ||\mathbf{x}_i - \mathbf{c}_k||^2 - \eta\sigma^2\right) \tag{21}$$

where $\eta \geq 0$ is an input parameter. This distance assumes that the data points inside the cluster are located in the spherical surface of radius $\sqrt{\eta}\sigma_k$. Thus $Dist(\mathbf{x}_i, C_k)$ is the distance of a outside data point to this spherical surface. Setting $\eta = 1$, assuming Gaussian distribution, there will be about 69% data points inside this surface. Setting $\eta = 0$, the distance metric is reduced to the distance to the cluster center. In practice, we set $\eta \simeq 0.2 \sim 0.5$.

Computing model parameters. To compute the model parameter $\mathbf{c}_k$, we optimize the dispersion of C_k

$$\min_{\mathbf{c}_k} \sum_{i \in C_k} \max\left(0, ||\mathbf{x}_i - \mathbf{c}_k||^2 - \eta\sigma_k^2\right) \tag{22}$$

which, by substituting the variance, is

$$\min_{\mathbf{c}_k} \sum_{i \in C_k} \max\left(0, ||\mathbf{x}_i - \mathbf{c}_k||^2 - \eta \sum_{j \in C_k} ||\mathbf{x}_j - \mathbf{c}_k||^2\right) \tag{23}$$

The max(·) is not differentiable and thus difficult to handle analytically. We use the following continuous function to handle it:

$$\max(u, v) = \frac{1}{\tau} \lim_{\tau \to \infty} \log(e^{\tau u} + e^{\tau v}) \tag{24}$$

Thus we minimize

$$J_3(\mathbf{c}_k) = \sum_{i \in C_k} \frac{1}{\tau} \lim_{\tau \to \infty} \log f(\tau, \mathbf{x}_i, \mathbf{c}_k) \tag{25}$$

where

$$f(\tau, \mathbf{x}_i, \mathbf{c}_k) = 1 + \exp\left(\tau ||\mathbf{x}_i - \mathbf{c}_k||^2 - \tau \frac{\eta}{n} \sum_{j \in C_k} ||\mathbf{x}_j - \mathbf{c}_k||^2\right) \tag{26}$$

Taking the derivative w.r.t. $\mathbf{c}_k$, we have

$$\frac{\partial J_3(\mathbf{c}_k)}{\partial \mathbf{c}_k} = \sum_{i \in C_k} \lim_{\tau \to \infty} \frac{2e^f}{1 + e^f} \left(\mathbf{c}_k - \mathbf{x}_i - \frac{\eta}{n} \sum_{j \in C_k} (\mathbf{c}_k - \mathbf{x}_j)\right) \tag{27}$$

Note

$$\lim_{\tau \to \infty} \frac{2e^{f(\cdot)}}{1 + e^{f(\cdot)}} = 2 \quad \text{if} \quad ||\mathbf{x}_i - \mathbf{c}_k||^2 > \frac{\eta}{n} \sum_{j \in C_k} ||\mathbf{x}_j - \mathbf{c}_k||^2 \tag{28}$$

$$\lim_{\tau \to \infty} \frac{2e^{f(\cdot)}}{1 + e^{f(\cdot)}} = 0 \quad \text{if} \quad ||\mathbf{x}_i - \mathbf{c}_k||^2 < \frac{\eta}{n} \sum_{j \in C_k} ||\mathbf{x}_j - \mathbf{c}_k||^2 \tag{29}$$

Thus we obtain

$$\frac{\partial J_3(\mathbf{c}_k)}{\partial \mathbf{c}_k} = \sum_{\mathbf{x}_i^>} \left(\mathbf{c}_k - \mathbf{x}_i - \frac{\eta}{n} \sum_{j \in C_k} (\mathbf{c}_k - \mathbf{x}_j)\right) \tag{30}$$

where $\mathbf{x}_i^>$ denotes the points outside the sphere, i.e., they satisfy Eq.(28). Let the center for averaging outside points be

$$\bar{\mathbf{x}}_>^k = \left(\sum_{\mathbf{x}_i^>} \mathbf{x}_i\right) \Big/ \left(\sum_{\mathbf{x}_i^>} 1\right). \tag{31}$$

The final solution to the optimization problem is

$$\mathbf{c}_k = \frac{\bar{\mathbf{x}}_>^k - \eta\, \bar{\mathbf{x}}^k}{1 - \eta} \tag{32}$$

where $\bar{\mathbf{x}}^k$ is the centroid of C_k. Clearly, as $\eta \to 0$, $\bar{\mathbf{x}}_>^k \to \bar{\mathbf{x}}^k$. Thus this result is consistent.

3.6 Model Selection and Computation

In our distance-minimization based clustering model, because we allow each cluster to be modeled by different models, there is a model selection issue. Given the data points in C_k as in Eq.(7), which model describes the data best?

We define the model dispersion as

$$Dispersion(ball) = \sum_{i \in C_k} \max(0, ||\mathbf{x}_i - \mathbf{c}_k||^2 - \eta\sigma_k^2),$$

$$Dispersion(line) = \sum_{i \in C_k} ||\mathbf{x}_i - \mathbf{c}_k - \alpha_i \mathbf{a}_k||^2,$$

$$Dispersion(plane) = \sum_{i \in C_k} ||\mathbf{x}_i - \mathbf{c}_k - \alpha_i \mathbf{a}_k - \beta_i b_k||^2,$$

$$\alpha_i = (\mathbf{x}_i - \mathbf{c}_k)^T \mathbf{a}_k, \beta_i = (\mathbf{x}_i - \mathbf{c}_k)^T b_k;$$

We choose the model with the smallest dispersion.

Then we minimize the total distance via the fundamental EM algorithmic approach. The two steps are the Cluster Assignment step (E-step) and Model Estimation step (M-step). The *Cluster Assignment* step is estimating the posterior probability for all data points belonging to different clusters. In our distance minimization model, for every $\mathbf{x}_i$, we assign $\mathbf{x}_i$ to the closest cluster C_k:

$$k = \arg\min_{1 \le k \le K} Dist(\mathbf{x}_i, C_k) \tag{33}$$

With new assignments of $\{\mathbf{x}_i\}$ to each cluster, in the model estimation step we re-estimate the model parameters using the computational procedure described in §2.1, §2.2 and §2.4.

4 Illustrative Examples

To demonstrate the effectiveness of the K-subspace clustering in extended clusters, we give the following examples. In the following figures, points with different colors belong to different clusters.

Two lines in 3-D space: 50 points along two straight lines are generated randomly in 3-D space as shown in Figure 1(a).

A line and a 2-D plane in 3-D space: 50 points along a straight line and 300 points in a 2-D plane are generated randomly in the 3-D space as shown in Figure 2(a).

Two 2-D planes in 3-D space: 300 points in each 2-D plane are generated randomly in the 3-D space as shown in Figure 3(a).

A Line, a 2-D plane and a sphere in 3-D space: 50 points along a straight line, 300 points in a 2-D plane, and 400 points in the surface of a sphere are generated randomly in the 3-D space as shown in Figure 4a.

Four clusters in 3-D space: Points with four clusters, each of which exists in just two dimensions with the third dimension being noise [20]. Points from two clusters can be very close together, which confuses many traditional clustering

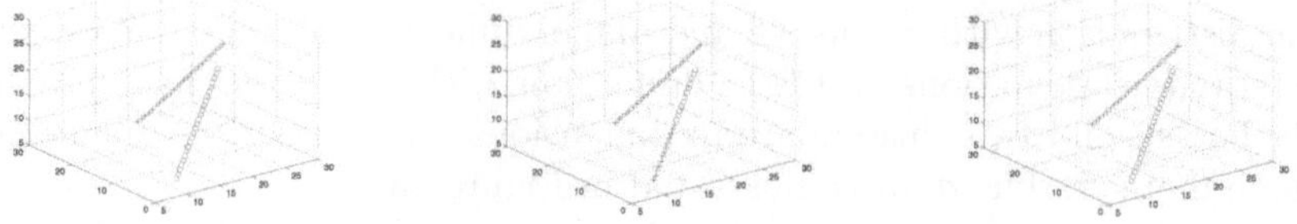

(a) Original Data (b) K-means Results (c) K-subspace Results

Fig. 1. Example (I): Two Lines in 3-D Space

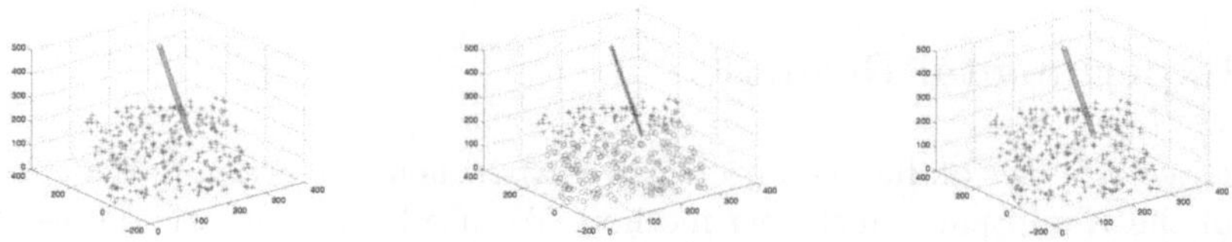

(a) Original Data (b) K-means Results (c) K-subspace Results

Fig. 2. Example (II): Line-Plane in 3-D Space

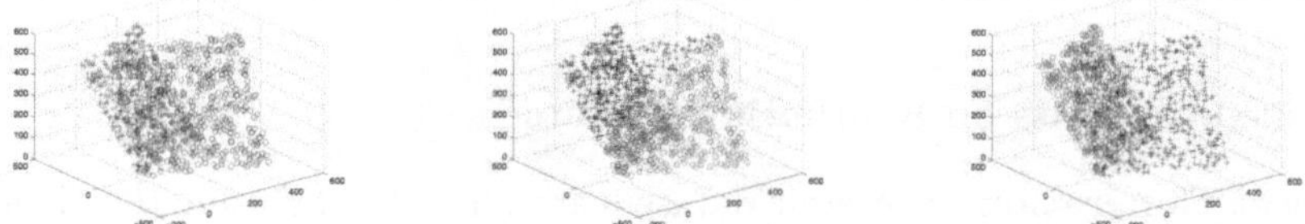

(a) Original Data (b) K-means Results (c) K-subspace Results

Fig. 3. Example (III): Plane-Plane in 3-D Space

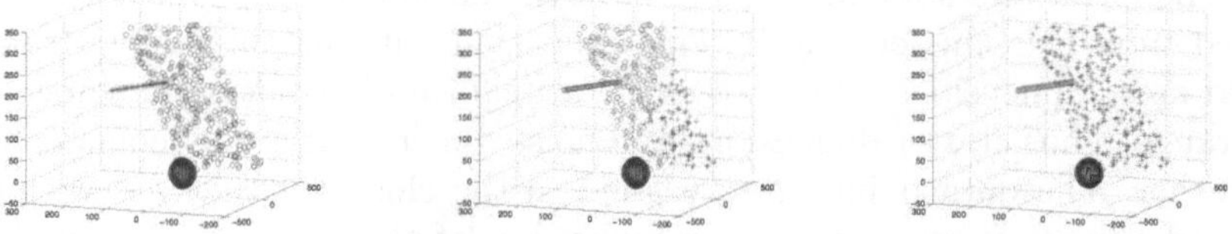

(a) Original Data (b) K-means Results (c) K-subspace Results

Fig. 4. Example (IV): Line-Plane-Ball in 3-D Space

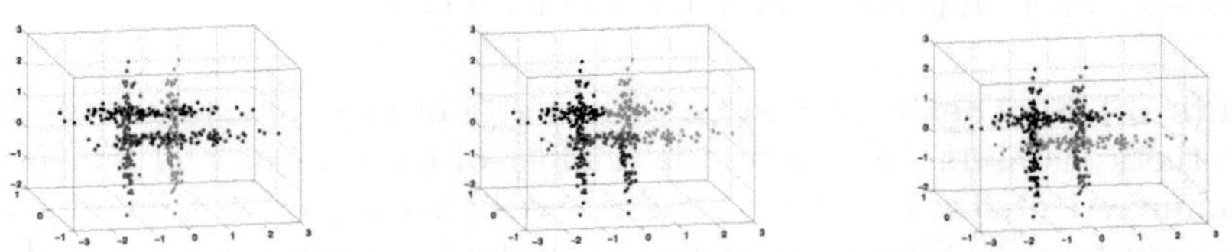

(a) Original Data (b) K-means Results (c) K-subspace Results

Fig. 5. Example (V): Four clusters in 3-D Space

algorithms. The first two clusters exist in dimensions x and y. The data forms a normal distribution with means 0.6 and -0.6 in dimension x and 0.5 in dimension y, and standard deviations of 0.1. In dimension z, these clusters have $\mu = 0$ and $\sigma = 1$. The second two clusters are in dimensions y and z and are generated in the same manner. The data is illustrated in Figure 5a.

As you can observe from these figures, K-means algorithm has difficulties in dealing with subspace clusters. However, our K-subspace algorithm is a comprehensive clustering algorithm, which can recognize a wide range of subspace clusters and is also able to discover the global sphere clusters.

5 Experimental Results

In this section, we conduct comprehensive experiments to evaluate the effectiveness of the K-subspace clustering method. We use both synthetic datasets and real-world datasets. The purpose of using synthetic data is that the detailed cluster structures are known, hence we can evaluate K-subspace with different factors such as cluster structures and sizes systematically. And we believe that those cluster structures do exist in real world applications. We use accuracy [12] and normalized mutual information (NMI) [22] as performance measures.

5.1 Experiments on Synthetic Datasets

Dataset Generation. Synthetic datasets are generated by using the algorithm proposed by Milligan [15]. The clusters are non-overlapping and truncated multivariate normal mixtures. Subspace structures are embedded in the synthetic datasets. Basically the dimension with clustering structures (called cluster dimensions) is created based on a cluster length chosen from a uniform distribution. The mean and standard deviation of the cluster on this dimension can be derived from this cluster length. The dimension without clustering structures (called noise dimensions) is created from a uniform distribution in some generated range. Thus cluster dimensions and noise dimensions are independent from each other. The data points are assigned to the cluster if they are within 1.5 standard deviation of every cluster dimension of this cluster. The number of points that are assigned to each cluster are determined as following: From the range $[a, a+d]$ where $a, d > 0$, we choose k numbers $\{P_1, \cdots, P_k\}$ uniformly, and k is the number of clusters. Then, the number of points belonging to the cluster i is $n\frac{P_i}{\sum_j P_j}$, where n is the total number of data points.

Results on Synthetic Datasets. Three sets of experiments are conducted on synthetic datasets: (1) We fix the number of data points and dimensionality and investigate the clustering performance of k-subspace as a function of the number clusters. (2) We fix the number of clusters and dimensionality and investigate the clustering performance of K-subspace as a function of the number of points. (3) We fix the number of clusters and points and investigate the clustering performance of K-subspace as a function of the number of dimensionality.

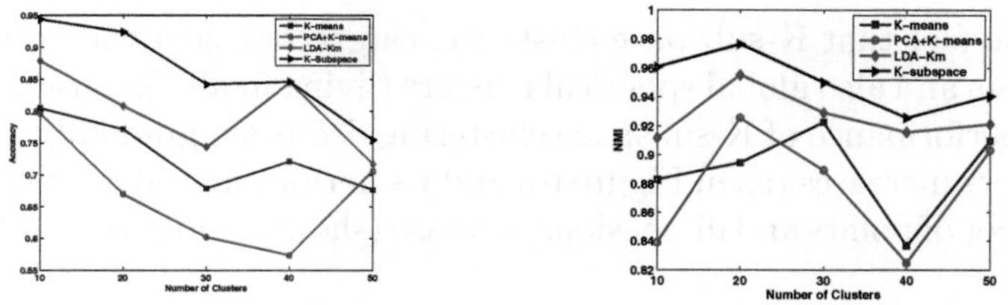

(a) Accuracy (b) NMI

Fig. 6. Clustering performance comparison on synthetic dataset (I). Remark: the number of data points is 1000; the dimensionality of the data is 500; and the number of clusters changes from 10 to 50.

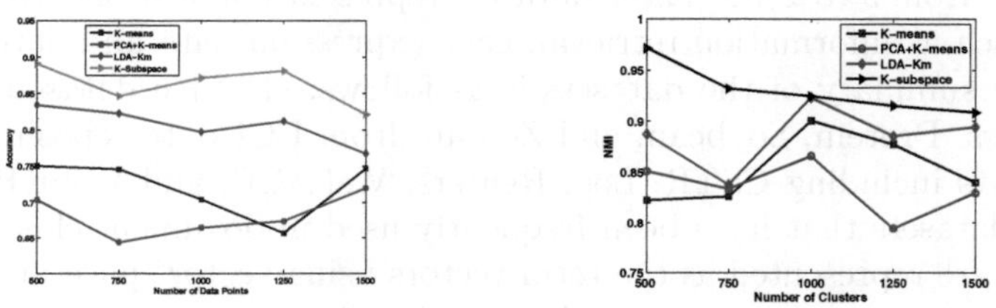

(a) Accuracy (b) NMI

Fig. 7. Clustering performance comparison on synthetic dataset (II). Remark: the number of clusters is 30; the dimensionality of the data is 1000; and the number of data points changes from 500 to 1500.

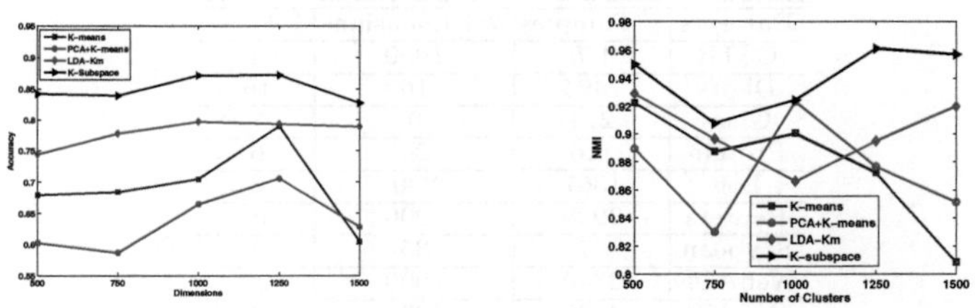

(a) Accuracy (b) NMI

Fig. 8. Clustering performance comparison on synthetic dataset (III). Remark: the number of clusters is 30; the number of data points is 1000; and the dimensionality of the data changes from 500 to 1500.

We compare the K-subspace clustering algorithm with three other algorithms: (1) Standard K-means algorithm; (2) LDA-Km algorithm [11]: an adaptive subspace clustering algorithm by integrating linear discriminant analysis (LDA) and K-means clustering into a coherent process. (3) PCA+K-means clustering algorithm: PCA is firstly applied to reduce the data dimension followed by K-means clustering. The results are shown in Figure 6, Figure 7, and Figure 8, respectively. We observe that: (1) LDA-Km outperforms K-means and PCA+K-means since it integrates adaptive subspace selection with clustering process. (2) K-subspace outperforms LDA-Km and achieves the best results among all the methods. This is due to the fact that K-subspace clustering can recognize a wide range of subspace clusters and also global spherical clusters (living in all dimensions). (3) The clustering performance of K-subspace clustering decreases gradually as the number of clusters increases (as in Figure 6), and its performance does not vary much as the number of points and dimensions increase (shown in Figure 7 and Figure 8).

5.2 Experiments on Real-World Datasets

Dataset Description. We use a wide range of real-world datasets in our experiments as summarized in Table 1. The number of classes ranges from 4 to 20, the number of samples ranges from 47 to 7494, and the number of dimensions ranges from 9 to 1000. These datasets represent applications from different domains such as information retrieval, gene expression data and pattern recognition. The summary of the datasets is as follows: (1) Five datasets including Digits, Glass, Protein, Soybean, and Zoo are from UCI data repository [5]. (2) Five datasets including CSTR, Log, Reuters, WebACE, and WebKB are standard text datasets that have been frequently used in document clustering. The documents are represented as the term vectors using vector space model. These document datasets are pre-processed (removing the stop words and unnecessary tags and headers) using rainbow package [19]. The dataset descriptions can be found in [12].

Table 1. Dataset Descriptions

Datasets	# Samples	# Dimensions	# Class
CSTR	475	1000	4
Digits	7494	16	10
Glass	214	9	7
Protein	116	20	6
Log	1367	200	8
Reuters	2900	1000	10
Soybean	47	35	4
WebACE	2340	1000	20
WebKB	4199	1000	4
Zoo	101	18	7

Results on Real Datasets. On the five datasets from UCI data repository, we compare the K-subspace clustering algorithm with standard K-means algorithm, LDA-Km algorithm, PCA+K-means clustering algorithm, and Spectral Clustering with Normalized Cuts (Ncut) [28]. We use normalized cut since it has been shown that weighted Kernel K-means is equivalent to the normalized cut [8]. The implementation of Ncut is based on [28] and the variance of the Gaussian similarity is determined using local scaling as in [29]. The results are shown in Figure 9(a) and Figure 9(b). On the text datasets, besides the above three alternative clustering methods, we also compare K-subspace algorithm with the following algorithms: (i) Non-negative Matrix Factorization (NMF) method [16]; (ii) Tri-Factorization Method (TNMF) [12]; (iii) Euclidean co-clustering (ECC) and minimum squared residue co-clustering (MSRC) algorithms [7]. Note that NMF

Table 2. Clustering accuracy comparison on text datasets

	K-means	PCA+K-means	LDA-Km	K-Subspace	ECC	MSRC	NMF	TNMF	Ncut
WebACE	0.4081	0.4432	0.4774	**0.5158**	0.4081	0.5021	0.4803	0.4996	0.4513
Log	0.6979	0.6562	0.7198	**0.7696**	0.7228	0.5655	0.7608	0.7527	0.7574
Reuters	0.436	0.3925	0.5142	**0.6426**	0.4968	0.4516	0.4047	0.4682	0.4890
CSTR	0.5210	0.5630	0.5630	**0.6555**	0.5945	0.6513	0.5945	0.6008	0.5435
WebKB	0.5951	0.6354	0.6468	**0.8583**	0.5210	0.5165	0.4568	0.6094	0.6040

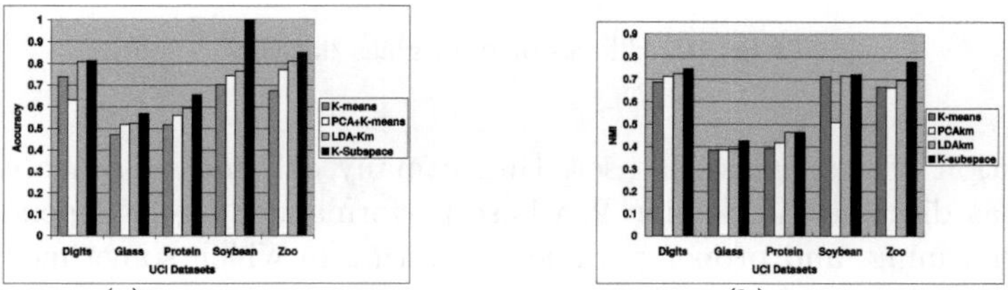

(a) Accuracy (b) NMI

Fig. 9. Clustering comparison on UCI data

has been shown to be effective in document clustering [26] and Tri-Factorization (TNMF) is an extension of NMF [12]. Both ECC and MSRC are document co-clustering algorithms that are able to find blocks in a rectangle document-term matrix. Co-clustering algorithms generally perform implicit dimension reduction during clustering process [9]. The comparison results on text datasets are shown in Table 2.

From the comparisons, we observe that: (1) On most datasets, PCA+K-means outperforms K-means due to the pre-processing by PCA. (2) Co-clustering algorithms (ECC and MSRC) generally outperform K-means since they are performing implicit dimension reduction during the clustering process. (3) NMF outperforms K-means on most datasets since NMF can model widely varying data distributions due to the flexibility of matrix factorization as compared to the rigid spherical clusters that the K-means clustering objective function attempts to capture [10]. When the data distribution is far from a spherical clustering, NMF has advantages over K-means. (4) TNMF provides a good framework for simultaneously clustering the rows and columns of the input documents. Hence TNMF generally outperforms NMF on text datasets. (5) The results of spectral clustering (Ncut) is better than K-means. Note that spectral clustering can be viewed as a weighted version of Kernel K-means and hence it is able to discover arbitrarily shaped clusters. The experimental results of Ncut is similar to those of NMF and TNMF. Note that it has also been shown that NMF is equivalent to spectral clustering [10]. (6) K-subspace clustering outperforms the other methods on all datasets. The improvements on Soybean dataset, Reuters dataset, and WebKB dataset are evident. The comparisons demonstrate the effectiveness of K-subspace algorithm in modeling different subspaces in real-life datasets.

Besides the above experimental comparisons, we also test the sensitivity of the parameter η introduced in Eq.(21). Figure 10 shows the clustering performance

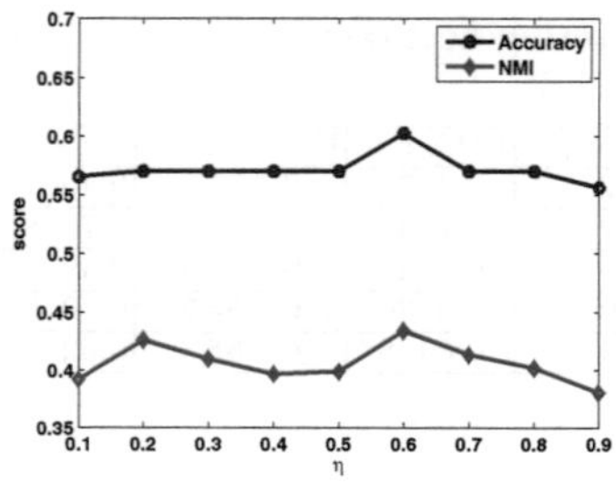

Fig. 10. Effects of η on glass dataset

as a function of η on glass dataset. Unlike many existing subspace clustering methods as discussed in Section 2, whose performance largely depends on the parameter tuning, and proper parameters setting in which sometimes are particularly difficult, we observe that the clustering performance of our K-subspace algorithm is not very sensitive to the choice of the parameters.

6 Conclusion

We extend the K-means clustering algorithm to accommodate subspace clusters in addition to the usual ball-shaped clusters. The optimal solutions to subspace models can be efficiently computed using PCA. We demonstrate that the model can capture most (if not all) of subspace clusters investigated so far. Running on synthetic datasets and 10 real life datasets, the K-subspace algorithm outperforms existing methods due to its increased capability of modeling clusters of different shapes.

Acknowledgement. The work of D. Wang and T. Li is partially supported by NSF grants IIS-0546280, CCF-0830659, and DMS-0844513. The work of C. Ding is partially supported by NSF grants DMS-0844497 and CCF-0830780.

References

1. Aggarwal, C.C., Wolf, J.L., Yu, P.S., Procopiuc, C., Park, J.S.: Fast algorithms for projected clustering. In: SIGMOD 1999 (1999)
2. Aggarwal, C.C., Yu, P.S.: Finding generalized projected clusters in high dimensional spaces. In: SIGMOD 2000 (2000)
3. Agrawal, R., Gehrke, J., Gunopulos, D., Raghavan, P.: Automatic subspace clustering of high dimensional data for data mining applications. In: SIGMOD 1998 (1998)
4. Bishop, C.M.: Pattern Recognition and Machine Learning. Springer, Heidelberg (2006)
5. Blake, C.L., Merz, C.J.: UCI repository of machine learning databases (1998)
6. Cheng, C.-H., Fu, A.W., Zhang, Y.: Entropy-based subspace clustering for mining numerical data. In: SIGKDD 1999 (1999)

7. Cho, H., Dhillon, I., Guan, Y., Sra, S.: Minimum sum squared residue co-clustering of gene expression data. In: SDM 2004 (2004)
8. Dhillon, I.S., Guan, Y., Kulis, B.: Kernel k-means: spectral clustering and normalized cuts. In: SIGKDD 2004 (2004)
9. Dhillon, I.S., Mallela, S., Modha, D.S.: Information-theoretical co-clustering. In: SIGKDD 2003 (2003)
10. Ding, C., He, X., Simon, H.: On the equivalence of nonnegative matrix factorization and spectral clustering. In: SDM 2005 (2005)
11. Ding, C., Li, T.: Adaptive dimension reduction using discriminant analysis and k-means c lustering. In: ICML 2007 (2007)
12. Ding, C., Li, T., Peng, W., Park, H.: Orthogonal nonnegative matrix tri-factorizations for clustering. In: SIGKDD 2006 (2006)
13. Friedman, J.H., Meulman, J.J.: Clustering objects on subsets of attributes. Journal of the Royal Statistical Society: Series B (Statistical Methodology) (2004)
14. Goil, S., Nagesh, H., Choudhary, A.: Efficient and scalable subspace clustering for very large data sets. Technical Report CPDC-TR-9906-010, Northwestern Univ. (1999)
15. Milligan, G.W.: An algorithm for generating artificial test clusters. Psychometrika 50, 123–127 (1985)
16. Lee, D., Seung, H.: Algorithms for non-negative matrix factorization. In: NIPS 2001 (2001)
17. Li, T., Ma, S., Ogihara, M.: Document Clustering via Adaptive Subspace Clsutering. In: SIGIR 2004 (2004)
18. Liu, B., Xia, Y., Yu, P.S.: Clustering through decision tree construction. In: CIKM 2000 (2000)
19. McCallum, A.K.: Bow: A toolkit for statistical language modeling, text retrieval, classification and clustering (1996), http://www.cs.cmu.edu/~mccallum/bow
20. Parsons, L., Haque, E., Liu, H.: Subspace clustering for high dimensional data: A review. In: SIGKDD Explorations 2004 (2004)
21. Procopiuc, C.M., Jones, M., Agarwal, P.K., Murali, T.M.: A monte carlo algorithm for fast projective clustering. In: SIGMOD 2002 (2002)
22. Strehl, A., Ghosh, J.: Cluster ensembles - a knowledge reuse framework for combining multiple partitions. The Journal of Machine Learning Research (2003)
23. Tasoulis, D.K., Zeimpekis, D., Gallopoulos, E., Vrahatis, M.N.: Oriented -windows: A pca driven clustering method. In: Advances in Web Intelligence and Data Mining (2006)
24. Vidal, R., Ma, Y., Sastry, S.: Generalized principal component analysis (GPCA). In: CVPR 2003 (2003)
25. Woo, K.-G., Lee, J.-H., Kim, M.-H., Lee, Y.-J.: Findit: a fast and intelligent subspace clustering algorithm using dimension voting. Information and Software Technology (2004)
26. Xu, W., Liu, X., Gong, Y.: Document clustering based on non-negative matrix factorization. In: SIGIR 2003 (2003)
27. Yang, J., Wang, W., Wang, H., Yu, P.: d-clusters: Capturing subspace correlation in a large data set. In: ICDE 2002 (2002)
28. Yu, S.X., Shi, J.: Multiclass spectral clustering. In: ICCV 2003 (2003)
29. Zelnik-manor, L., Perona, P.: Self-tuning spectral clustering. In: NIPS 2005 (2005)
30. Zhang, Q., Liu, J., Wang, W.: Incremental subspace clustering over multiple data streams. In: ICDM 2007 (2007)

Latent Dirichlet Bayesian Co-Clustering

Pu Wang[1], Carlotta Domeniconi[1], and Kathryn Blackmond Laskey[2]

[1] Department of Computer Science
[2] Department of Systems Engineering and Operations Research
George Mason University
4400 University Drive, Fairfax, VA 22030 USA

Abstract. Co-clustering has emerged as an important technique for mining contingency data matrices. However, almost all existing co-clustering algorithms are hard partitioning, assigning each row and column of the data matrix to one cluster. Recently a Bayesian co-clustering approach has been proposed which allows a probability distribution membership in row and column clusters. The approach uses variational inference for parameter estimation. In this work, we modify the Bayesian co-clustering model, and use collapsed Gibbs sampling and collapsed variational inference for parameter estimation. Our empirical evaluation on real data sets shows that both collapsed Gibbs sampling and collapsed variational inference are able to find more accurate likelihood estimates than the standard variational Bayesian co-clustering approach.

Keywords: Co-Clustering, Graph Learning, Dirichlet Distribution.

1 Introduction

Co-clustering [2] has emerged as an important approach for mining dyadic and relational data. Often, data can be organized in a matrix, where rows and columns present a symmetrical relation. For example, documents can be represented as a matrix, where rows are indexed by the documents, and columns by words. Co-clustering allows documents and words to be grouped simultaneously: documents are clustered based on the contained words, and words are grouped based on the documents they appear in. The two clustering processes are inter-dependent.

Some researchers have proposed a hard-partition version [3], others a soft-partition version [1] of co-clustering. In the hard-partition case, each row (column) is assigned to exactly one row (column) cluster. In the soft-partition case, each row (column) has a probability of belonging to each row (column) cluster.

The Bayesian Co-Clustering (BCC) model proposed in [1] is a kind of generative model. BCC maintains separate Dirichlet priors for the distribution of row- and column-clusters given rows and columns. To generate each entry in the data matrix, the model first generates the row and column clusters of the current entry according to this Dirichlet distribution. The value of the current entry is then generated according to the corresponding row-cluster and column-cluster. The advantage of a generative model is that it can be used to predict unseen

W. Buntine et al. (Eds.): ECML PKDD 2009, Part II, LNAI 5782, pp. 522–537, 2009.

data. Like the original *Latent Dirichlet Allocation* (LDA) [5] model, though, BCC assumes uniform priors for the entry value distributions given row- and column-clusters. The authors in [1] proposed a variational Bayesian algorithm to perform inference and estimate the BCC model. A lower bound of the likelihood function is learned and used to estimate model parameters.

In this work, we extend the BCC model and propose a collapsed Gibbs sampling and a collapsed variational Bayesian algorithm for it. Following [5], first we smooth the BCC model, by introducing priors for the entry value distributions given row- and column-clusters. Following [5], we call our approach *Latent Dirichlet Bayesian Co-Clustering* (LDCC), since it assumes Dirichlet priors for row- and column-clusters, which are unobserved in the data contingency matrix. The collapsed Gibbs sampling and collapsed variational Bayesian algorithms we propose can learn more accurate likelihood functions than the standard variational Bayesian algorithm [1]. This result is derived analytically for the collapsed variational Bayesian algorithm. More accurate likelihood estimates can lead to higher predictive performance, as corroborated by our experimental results.

The rest of the paper is organized as follows. In Section 2, we discuss related work. Section 3 introduces the LDCC model and the variational Bayesian algorithm. We then discuss the collapsed Gibbs sampling and the collapsed variational Bayesian algorithms. Section 4 demonstrates our empirical evaluation of the three methods. Finally, Section 5 summarizes the paper.

2 Related Work

Our work is closely related to [1], which we discuss in Section 3.1. Dhillon et al. proposed an information-theoretic co-clustering approach (hard-partition) in [3]. Shafiei et al. proposed a soft-partition co-clustering, called "Latent Dirichlet Co-clustering" in [4]. The proposed model, though, does not cluster rows and columns simultaneously. It first defines word-topics, i.e., groups of words, and then defines document-topics, i.e., groups of word-topics. Documents are modeled as mixtures of such document-topics. Thus, the resulting model is similar to a hierarchical extension of the "Latent Dirichlet Allocation" [5] model, since the defined document-topics are not groups of documents, but groups of word-topics. Our LDCC model and BCC [1] model assume independence between row-clusters and column-clusters, which is the same assumption as in [3].

Blei et al. proposed "Latent Dirichlet Allocation" (LDA) [5], which assumes that topics are mixtures of words, and documents are mixtures of topics. A standard variational Bayesian algorithm [5] is used to estimate the posterior distribution of model parameters given the model evidence. Griffiths et al. used a collapsed Gibbs sampling method to learn the posterior distribution of parameters for the LDA model [9]. Recently, Teh et al. proposed a collapsed variational Bayesian algorithm to perform model inference for LDA and "Hierarchical Dirichlet Processing" [7,10].

3 Latent Dirichlet Co-Clustering

In this section, we first introduce the LDCC model. We then discuss three different learning methods: variational Bayesian, collapsed Gibbs sampling, and collapsed variational Bayesian. Table 1 gives a summary of the notation used.

Symbol	Description
X	data matrix
u	index for row
v	index for column
$[u, v]$	entry of the matrix at row u and column v
x_{uv}	value for matrix entry at row u and column v
$\mathcal{X}$	entry value set
i	index for row clusters
j	index for column clusters
N_1	number of rows
N_2	number of columns
$\boldsymbol{z}_1$	row clusters
$\boldsymbol{z}_2$	column clusters
K_1	number of row clusters
K_2	number of column clusters
α_1	Dirichlet prior hyperparameter for for rows
α_2	Dirichlet prior hyperparameter for columns
β	Dirichlet prior hyperparameter for the probabilities of each entry value given a row- and a column-clusters
$\boldsymbol{\pi}_1$	The probabilities of each row-cluster given each row
$\boldsymbol{\pi}_2$	The probabilities of each column-cluster given each column
$\theta_{ijx_{uv}}$	probability of entry value x_{uv} give $z_1 = i$ and $z_2 = j$
$n_{ijx_{uv}}$	number of entries with value x_{uv} assigned to row cluster i and column cluster j
n_{ui}	number of entries in row u assigned to row cluster i
n_{vj}	number of entries in column v assigned to column cluster j
n_{ij}	number of entries in matrix assigned to row cluster i and column cluster j
n_u	number of entries in row u
n_v	number of entries in column v

Fig. 1. Notation used in this paper

Given an $N_1 \times N_2$ data matrix X, the values x_{uv} of each entry $[u, v]$, $u = 1, \ldots, N_1, v = 1, \ldots, N_2$ are defined in a value set, $x_{uv} \in \mathcal{X}$. For co-clustering, we assume there are K_1 row clusters z_1, and K_2 column clusters z_2. LDCC assumes two Dirichlet priors[1] $Dir(\alpha_1)$ and $Dir(\alpha_2)$ for rows and columns respectively, $\alpha_1 =< \alpha_{1_u} | u = 1, \ldots, N_1 >$, $\alpha_2 =< \alpha_{2_v} | v = 1, \cdots, N_2 >$, from which the probabilities of each row-cluster z_1 and column-cluster z_2 given each row u and each column v are generated, denoted as π_{1_u} and π_{2_v} respectively. Row clusters for entries in row u and column clusters for entries in column v are sampled from multinomial distributions $p(z_1|\pi_{1_u})$ and $p(z_2|\pi_{2_v})$ respectively. We denote $\boldsymbol{\pi}_1 =< \pi_{1_u} | u = 1, \ldots, N_1 >$, $\boldsymbol{\pi}_2 =< \pi_{2_v} | v = 1, \ldots, N_2 >$, $\boldsymbol{z}_1 =< z_{1_{uv}} | u = 1, \ldots, N_1, v = 1, \ldots, N_2 >$ and $\boldsymbol{z}_2 =< z_{2_{uv}} | u = 1, \ldots, N_1, v = 1, \ldots, N_2 >$, where $\boldsymbol{z}_1$ and $\boldsymbol{z}_2$ are row- and column-cluster assignment for all entries in the data matrix X. A row cluster $z_1 = i$ and a column cluster $z_2 = j$ together

[1] In the rest of the paper, we assume symmetric Dirichlet priors, which means α_1 and α_2 do not depend on u or v.

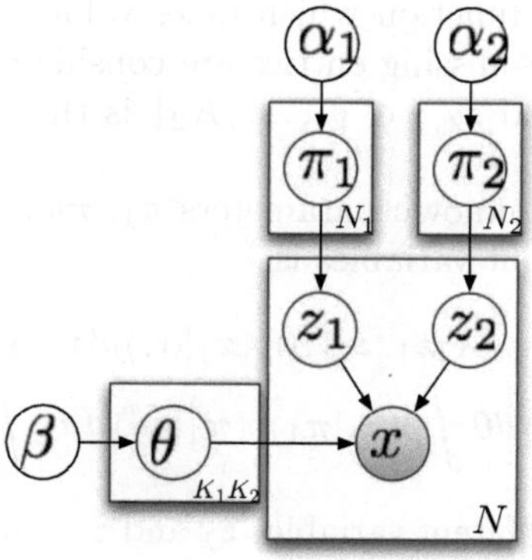

Fig. 2. Latent Dirichlet Bayesian Co-clustering Model

decide a co-cluster $(z_1, z_2) = (i, j)$, and entries in the matrix are also sampled from a multinomial distribution $p(x|\theta_{z_1,z_2})$ given a co-cluster $(z_1, z_2) = (i, j)$. We denote $\boldsymbol{\theta} =< \theta_{z_1=i,z_2=j}|i = 1, \ldots, K_1, j = 1, \ldots, K_2 >$. Here, $\boldsymbol{z}_1$ and $\boldsymbol{z}_2$ are latent variables, while $\boldsymbol{\pi}_1$, $\boldsymbol{\pi}_2$ and $\boldsymbol{\theta}$ are unknown parameters. The generative process for the whole data matrix is as follows (see Figure 2):

– For each row u, choose $\pi_{1_u} \sim Dir(\alpha_1)$
– For each column v, choose $\pi_{2_v} \sim Dir(\alpha_2)$
– To generate the entry of row u and column v:
  • choose $z_{1_{uv}} \sim p(z_1|\pi_{1_u})$, $z_{2_{uv}} \sim p(z_2|\pi_{2_v})$
  • choose $\theta_{z_{1_{uv}} z_{2_{uv}}} \sim Dir(\beta)$
  • choose $x_{uv} \sim p(x|z_{1_{uv}}, z_{2_{uv}}, \theta_{z_{1_{uv}}, z_{2_{uv}}})$.

The LDCC model proposed here departs from the BCC model [1] by introducing a prior β for $\theta_{z_1 z_2}$. Thus, LDCC can assign a probability to an unseen entry value according to $p(\theta_{z_1,z_2}|\beta)$.

The marginal probability of an entry x in the data matrix X is given by:

$$p(x|\alpha_1, \alpha_2, \beta) = \int_{\pi_1}\int_{\pi_2}\int_{\theta} p(\pi_1|\alpha_1)p(\pi_2|\alpha_2)p(\theta_{z_1 z_2}|\beta) \cdot \sum_{z_1}\sum_{z_2} p(z_1|\pi_1)p(z_2|\pi_2)p(x|\theta_{z_1 z_2})\, d\pi_1 d\pi_2 d\theta_{z_1 z_2} \quad (1)$$

Note that the entries in the same row/column are generated from the same π_{1_u} or π_{2_v}, so the entries in the same row/column are related. Therefore, the model introduces a coupling between observations in the same row/column [1].

The overall joint distribution over X, $\boldsymbol{\pi}_1$, $\boldsymbol{\pi}_2$, $\boldsymbol{z}_1$, $\boldsymbol{z}_2$ and $\boldsymbol{\theta}$ is given by:

$$p(X, \boldsymbol{\pi}_1, \boldsymbol{\pi}_2, \boldsymbol{z}_1, \boldsymbol{z}_2, \boldsymbol{\theta}|\alpha_1, \alpha_2, \beta) = \prod_u p(\pi_{1_u}|\alpha_1) \prod_v p(\pi_{2_v}|\alpha_2) \cdot \prod_{K_1}\prod_{K_2} p(\theta_{z_1,z_2}|\beta) \prod_{u,v} p(z_{1_{uv}}|\pi_{1_u})p(z_{2_{uv}}|\pi_{2_v})p(x_{uv}|\theta_{z_1,z_2}, z_{1_{uv}}, z_{2_{uv}})^{\delta_{uv}} \quad (2)$$

where δ_{uv} is an indicator function which takes value 0 when x_{uv} is empty, and 1 otherwise (only the non-missing entries are considered); $z_{1_{uv}} \in \{1, \ldots, K_1\}$ is the latent row cluster, and $z_{2_{uv}} \in \{1, \ldots, K_2\}$ is the latent column cluster for observation x_{uv}.

Marginalizing out all unknown parameters $\boldsymbol{\pi}_1$, $\boldsymbol{\pi}_2$ and $\boldsymbol{\theta}$, the marginal likelihood of observed and latent variables is:

$$p(X, \boldsymbol{z}_1, \boldsymbol{z}_2|\alpha_1, \alpha_2, \beta) = p(X|\boldsymbol{z}_1, \boldsymbol{z}_2, \beta)p(\boldsymbol{z}_1|\alpha_1)p(\boldsymbol{z}_2|\alpha_2) = \tag{3}$$
$$\int p(X|\boldsymbol{\theta}, \boldsymbol{z}_1, \boldsymbol{z}_2)p(\boldsymbol{\theta}|\beta)\, d\boldsymbol{\theta} \int p(\boldsymbol{z}_1|\boldsymbol{\pi}_1)p(\boldsymbol{\pi}_1|\alpha_1)\, d\boldsymbol{\pi}_1 \int p(\boldsymbol{z}_2|\boldsymbol{\pi}_2)p(\boldsymbol{\pi}_2|\alpha_2)\, d\boldsymbol{\pi}_2$$

Summing over all possible latent variables $\boldsymbol{z}_1$ and $\boldsymbol{z}_2$, the probability of observing the entire matrix X is:

$$p(X|\alpha_1, \alpha_2, \beta) = \int_{\boldsymbol{\pi}_1}\int_{\boldsymbol{\pi}_2}\int_{\boldsymbol{\theta}} \left(\prod_u p(\pi_{1_u}|\alpha_1)\right)\left(\prod_v p(\pi_{2_v}|\alpha_2)\right)\left(\prod_{z_1, z_2} p(\theta_{z_1, z_2}|\beta)\right)$$
$$\cdot \left(\prod_{u,v}\sum_{z_{1_{uv}}}\sum_{z_{2_{uv}}} p(z_{1_{uv}}|\pi_{1_u})p(z_{2_{uv}}|\pi_{2_v})p(x_{uv}|\theta_{z_{1_{uv}}, z_{2_{uv}}})^{\delta_{uv}}\right) d\boldsymbol{\pi}_1 d\boldsymbol{\pi}_2 d\boldsymbol{\theta} \tag{4}$$

3.1 Variational Bayesian Algorithm for BCC

In this section, we briefly describe the variational Bayesian algorithm for the original BCC model (see Appendix). The BCC model assumes uniform priors for $\boldsymbol{\theta}$ and assumes that $\boldsymbol{\theta}$ has a Gaussian distribution. The authors in [1] derived a variational algorithm for their model. In this paper, we assume that the values for each entry in the data matrix are discrete[2]. We do so for mathematical convenience of the derivation of the collapsed Gibbs sampling and the collapsed variational Bayesian algorithms. Thus, unlike [1], we do not assume that $\boldsymbol{\theta}$ has a Gaussian distribution.

The variational Bayesian algorithm introduces $q(\boldsymbol{z}_1, \boldsymbol{z}_2, \boldsymbol{\pi}_1, \boldsymbol{\pi}_2|\boldsymbol{\gamma}_1, \boldsymbol{\gamma}_2, \boldsymbol{\phi}_1, \boldsymbol{\phi}_2)$ as an approximation of the actual distribution $p(\boldsymbol{z}_1, \boldsymbol{z}_2, \boldsymbol{\pi}_1, \boldsymbol{\pi}_2|X, \alpha_1, \alpha_2, \boldsymbol{\theta})$, where $\boldsymbol{\gamma}_1$, $\boldsymbol{\gamma}_2$, $\boldsymbol{\phi}_1$ and $\boldsymbol{\phi}_2$ are called variational variables, $\boldsymbol{\gamma}_1 =< \gamma_{1_u}|u = 1, \cdots, N_1 >$, $\boldsymbol{\gamma}_2 =< \gamma_{2_v}|v = 1, \cdots, N_2 >$, $\boldsymbol{\phi}_1 =< \phi_{1_u}|u = 1, \cdots, N_1 >$, $\boldsymbol{\phi}_2 =< \phi_{2_v}|v = 1, \cdots, N_2 >$, γ_{1_u} and γ_{2_v} are variational Dirichlet distribution parameters with K_1 and K_2 dimensions respectively for rows and columns, ϕ_{1_u} and ϕ_{2_v} are multinomial parameters with K_1 and K_2 dimensions for rows and columns. It is assumed that $q(\boldsymbol{z}_1, \boldsymbol{z}_2, \boldsymbol{\pi}_1, \boldsymbol{\pi}_2|\boldsymbol{\gamma}_1, \boldsymbol{\gamma}_2, \boldsymbol{\phi}_1, \boldsymbol{\phi}_2)$ can be fully factorized as:

$$q(\boldsymbol{z}_1, \boldsymbol{z}_2, \boldsymbol{\pi}_1, \boldsymbol{\pi}_2|\boldsymbol{\gamma}_1, \boldsymbol{\gamma}_2, \boldsymbol{\phi}_1, \boldsymbol{\phi}_2) =$$
$$\left(\prod_{u=1}^{N_1} q(\pi_{1_u}|\gamma_{1_u})\right)\left(\prod_{v=1}^{N_2} q(\pi_{2_v}|\gamma_{2_v})\right)\left(\prod_{u=1}^{N_1}\prod_{v=1}^{N_2} q(z_{1_{uv}}|\phi_{1_u})q(z_{2_{uv}}|\phi_{2_v})\right) \tag{5}$$

[2] Technically, our theory applies to any exponential family distribution for data matrix.

The factorization assumption of $q(\boldsymbol{z}_1, \boldsymbol{z}_2, \boldsymbol{\pi}_1, \boldsymbol{\pi}_2|\gamma_1, \gamma_2, \boldsymbol{\phi}_1, \boldsymbol{\phi}_2)$ means that parameters and latent variables are independent, and the assignment of $z_{1_{uv}}$ and $z_{2_{uv}}$ for the current entry $[u, v]$ is independent of the assignments for other entries.

The variational Bayesian algorithm can find a lower bound of the true log-likelihood:

$$\log p(X|\alpha_1, \alpha_2, \boldsymbol{\theta}) \geq \quad (6)$$
$$E_q[\log p(X, \boldsymbol{z}_1, \boldsymbol{z}_2, \boldsymbol{\pi}_1, \boldsymbol{\pi}_2|\alpha_1, \alpha_2, \boldsymbol{\theta})] - E_q[\log q(\boldsymbol{z}_1, \boldsymbol{z}_2, \boldsymbol{\pi}_1, \boldsymbol{\pi}_2|\gamma_1, \gamma_2, \boldsymbol{\phi}_1, \boldsymbol{\phi}_2)]$$

and we denote the lower bound as $L(\gamma_1, \gamma_2, \boldsymbol{\phi}_1, \boldsymbol{\phi}_2, \alpha_1, \alpha_2, \boldsymbol{\theta})$.

The variational Bayesian algorithm is an EM-style method: the E-step estimates the values for γ_1, γ_2, $\boldsymbol{\phi}_1$ and $\boldsymbol{\phi}_2$ that maximize the lower bound of the log-likelihood based on α_1, α_2 and $\boldsymbol{\theta}$; the M-step estimates α_1, α_2 and $\boldsymbol{\theta}$ according to the log-likelihood lower bound based on γ_1, γ_2, $\boldsymbol{\phi}_1$ and $\boldsymbol{\phi}_2$ learned during the previous E-step. Thus, in the E-step, in order to maximize $L(\gamma_1, \gamma_2, \boldsymbol{\phi}_1, \boldsymbol{\phi}_2, \alpha_1, \alpha_2, \boldsymbol{\theta})$, one takes the derivative of L w.r.t γ_1, γ_2, $\boldsymbol{\phi}_1$ and $\boldsymbol{\phi}_2$ respectively, and sets it to zero. We get:

$$\phi_{1_{ui}} \propto exp\left(\Psi(\gamma_{1_{ui}}) + \frac{\sum_{u=1}^{N_1}\sum_{i=1}^{K_1}\delta_{uv}\phi_{2_{vj}}\log\theta_{ijx_{uv}}}{n_u}\right) \quad (7)$$

$$\phi_{2_{vj}} \propto exp\left(\Psi(\gamma_{2_{vj}}) + \frac{\sum_{v=1}^{N_2}\sum_{j=1}^{K_2}\delta_{uv}\phi_{1_{vi}}\log\theta_{ijx_{uv}}}{n_v}\right) \quad (8)$$

$$\gamma_{1_{ui}} \propto \alpha_{1_i} + n_u\phi_{1_{u_i}} \quad (9)$$

$$\gamma_{2_{vj}} \propto \alpha_{2_j} + n_v\phi_{2_{v_j}} \quad (10)$$

where n_u and n_v are the number of entries in row u and column v respectively, and $\Psi(\cdot)$ is the digamma function, the first derivative of $\log\Gamma(\cdot)$, the log Gamma function. In the M-step, to estimate the Dirichlet parameters α_1 and α_2, one can use Newton method, as shown in [5] for LDA, to estimate $\boldsymbol{\theta}$, one takes the derivative of L w.r.t $\boldsymbol{\theta}$ and setting it to zero. We get:

$$\theta_{ijx_{uv}} \propto \sum_{u'=1}^{N_1}\sum_{v'=1}^{N_2}\delta_{u'v'}(x_{uv})\phi_{1_{u'i}}\phi_{2_{v'j}} \quad (11)$$

where $\delta_{u'v'}(x_{uv})$ is an indicator function, which equals 1 if the value of the entry at row u' and column v' equals to x_{uv}, 0 otherwise. The variational Bayesian method iterates through the E-step and the M-step until convergence.

Although efficient and easy to implement, the variational Bayesian algorithm can potentially lead to inaccurate results. The latent variables $\boldsymbol{z}_1$, $\boldsymbol{z}_2$ and the parameters $\boldsymbol{\pi}_1$, $\boldsymbol{\pi}_2$, $\boldsymbol{\theta}$ can have a strong inter-dependence in the true posterior $p(X, \boldsymbol{z}_1, \boldsymbol{z}_2, \boldsymbol{\pi}_1, \boldsymbol{\pi}_2|\alpha_1, \alpha_2, \boldsymbol{\theta})$. This dependence is ignored in the variational Bayesian algorithm which assumes independence between the latent variables and the parameters. As a result, the lower bound learned for the log marginal likelihood can be very loose, leading to inaccurate estimates of the posterior.

3.2 Collapsed Gibbs Sampling for LDCC

Standard Gibbs sampling [8], which iteratively samples the latent variables $\boldsymbol{z}_1$ and $\boldsymbol{z}_2$, and the parameters $\boldsymbol{\pi}_1$, $\boldsymbol{\pi}_2$ and $\boldsymbol{\theta}$, may converge very slowly due to the strong dependencies between the latent variables and the parameters. Collapsed Gibbs sampling improves upon Gibbs sampling by marginalizing out the parameters $\boldsymbol{\pi}_1$, $\boldsymbol{\pi}_2$ and $\boldsymbol{\theta}$, and then sampling the latent variables $\boldsymbol{z}_1$ and $\boldsymbol{z}_2$ only, over the so called collapsed space. Consider a model in which each matrix element can have a discrete value from a value set $\mathcal{X}$, with $|\mathcal{X}| = N$. Using a symmetric Dirichlet prior, the marginal likelihood over X, $\boldsymbol{z}_1$ and $\boldsymbol{z}_2$, (Equation (3)), can be rewritten as:

$$p(X, \boldsymbol{z}_1, \boldsymbol{z}_2|\alpha_1, \alpha_2, \beta) = \prod_{u=1}^{N_1} \left(\frac{\Gamma(K_1\alpha_1)}{\Gamma(K_1\alpha_1 + n_u)} \prod_{i=1}^{K_1} \frac{\Gamma(\alpha_1 + n_{ui})}{\Gamma(\alpha_1)} \right) \tag{12}$$

$$\prod_{v=1}^{N_2} \left(\frac{\Gamma(K_2\alpha_2)}{\Gamma(K_2\alpha_2 + n_v)} \prod_{j=1}^{K_2} \frac{\Gamma(\alpha_2 + n_{vj})}{\Gamma(\alpha_2)} \right) \prod_{i=1}^{K_1} \prod_{j=1}^{K_2} \left(\frac{\Gamma(N\beta)}{\Gamma(N\beta + n_{ij})} \prod_{x=1}^{N} \frac{\Gamma(\beta + n_{ijx})}{\Gamma(\beta)} \right)$$

Given all the latent variables but the ones for entry $[u, v]$, the conditional probability of $z_{1_{uv}} = i$ and $z_{2_{uv}} = j$ is:

$$p(z_{1_{uv}} = i, z_{2_{uv}} = j|X, \boldsymbol{z}_1^{\neg uv}, \boldsymbol{z}_2^{\neg uv}, \alpha_1, \alpha_2, \beta) = \frac{(\alpha_1 + n_{ui}^{\neg uv})(\alpha_2 + n_{vj}^{\neg uv})(\beta + n_{ijx_{uv}}^{\neg uv})}{(K_1\alpha_1 + n_u^{\neg uv})(K_2\alpha_2 + n_v^{\neg uv})(N\beta + n_{ij}^{\neg uv})} \tag{13}$$

where $\neg uv$ denotes the corresponding count with x_{uv}, $z_{1_{uv}}$ and $z_{2_{uv}}$ excluded. The derivation can be found in the Appendix. The conditional probability can be rewritten as:

$$p(z_{1_{uv}} = i, z_{2_{uv}} = j|X, \boldsymbol{z}_1^{\neg uv}, \boldsymbol{z}_2^{\neg uv}, \alpha_1, \alpha_2, \beta) = \frac{(\alpha_1 + n_{ui}^{\neg uv})(\alpha_2 + n_{vj}^{\neg uv})(\beta + n_{ijx_{uv}}^{\neg uv})(N\beta + n_{ij}^{\neg uv})^{-1}}{\sum_{i'=1}^{K_1} \sum_{j'}^{K_2} (\alpha_1 + n_{ui'}^{\neg uv})(\alpha_2 + n_{vj'}^{\neg uv})(\beta + n_{i'j'x_{uv}}^{\neg uv})(N\beta + n_{i'j'}^{\neg uv})^{-1}} \tag{14}$$

where the numerator covers the factors specific to $z_{1_{uv}} = i$ and $z_{2_{uv}} = j$, and the denominator serves as a normalization factor by summing over all combination of z_1 and z_2 for the current entry $[u, v]$.

Note that since collapsed Gibbs sampling marginalizes out the parameters $\boldsymbol{\pi}_1$, $\boldsymbol{\pi}_2$ and $\boldsymbol{\theta}$, it induces new dependencies between the latent variables $z_{1_{uv}}, z_{2_{uv}}$ (which are conditionally independent given the parameters) [7]. Equation (14) shows that $z_{1_{uv}}$ and $z_{2_{uv}}$ depend on $\boldsymbol{z}_1^{\neg uv}$, $\boldsymbol{z}_2^{\neg uv}$ only through the counts $n_{ui'}^{\neg uv}$, $n_{vj'}^{\neg uv}$ and $n_{i'j'}^{\neg uv}$, which is to say that the dependence of $z_{1_{uv}} = i$ and $z_{2_{uv}} = j$ on any other variable $z_{1_{uv}} = i'$, $z_{2_{uv}} = j'$ is very small, especially for large datasets. This is precisely the right setting for a mean field (i.e., fully factorized variational) approximation: a particular variable interacts with the remaining variables only through a summary statistics called the field, and the impact of any single variable on the field is very small [7]. On the contrary, this is not true in the joint space of parameters and latent variables because fluctuations in parameters can have a signicant impact on latent variables. As a consequence, the mean field assumption fits better the collapsed space of latent variables than the joint space of latent variables and parameters.

Given Equation (13) or (14), Gibbs sampling can generate row- and column-cluster probabilities for the current entry conditioned on row- and column-clusters for the other entries. One can calculate the following stationary distributions:

$$p(x_{uv}|z_{1_{uv}} = i, z_{2_{uv}} = j) = \frac{n_{ijx_{uv}} + \beta}{n_{ij} + N\beta} \tag{15}$$

$$p(z_{1_{uv}} = i|u) = \frac{n_{ui} + \alpha_1}{n_u + K_1\alpha_1} \tag{16}$$

$$p(z_{2_{uv}} = j|v) = \frac{n_{vj} + \alpha_2}{n_v + K_2\alpha_2} \tag{17}$$

which correspond to $\theta_{z_1=i,z_2=j}$, π_{1_u} and π_{2_v}.

Although Gibbs sampling leads to unbiased estimators, it also has some drawbacks: one needs to assess convergence of the Markov chain and to have some idea of mixing times to estimate the number of samples to collect, and to identify coherent topics across multiple samples. In practice, one often ignores these issues and collects as many samples as is computationally feasible, while the question of topic identication is often sidestepped by using just one sample. Hence, there still is a need for more efficient, accurate and deterministic inference procedures.

3.3 Collapsed Variational Bayesian Algorithm for LDCC

The collapsed variational Bayesian algorithm for LDCC is similar to the standard variational Bayesian one, except for the optimization of the lower bound of the log-likelihood in the collapsed space, which is inspired by collapsed Gibbs sampling. There are two ways to derive the collapsed variational Bayesian algorithm for LDCC, either in the collapsed space or in the original joint space of latent variables and parameters.

We start from the collapsed space with parameters marginalized out. We introduce $q(\boldsymbol{z}_1, \boldsymbol{z}_2|\boldsymbol{\gamma})$ to approximate $p(\boldsymbol{z}_1, \boldsymbol{z}_2|X, \alpha_1, \alpha_2, \beta)$, where $\boldsymbol{\gamma} =< \boldsymbol{\gamma}_{uv}|u = 1, \cdots, N_1, v = 1, \cdots, N_2 >$, and $\boldsymbol{\gamma}_{uv} =< \gamma_{uvij}|i = 1, \cdots, K_1, j = 1, \cdots, K_2 >$. Assume that $q(\boldsymbol{z}_1, \boldsymbol{z}_2|\boldsymbol{\gamma})$ can be factorized as:

$$q(\boldsymbol{z}_1, \boldsymbol{z}_2|\boldsymbol{\gamma}) = \prod_{u=1}^{N_1}\prod_{v=1}^{N_2} q(z_{1_{uv}}, z_{2_{uv}}|\boldsymbol{\gamma}_{uv}) \tag{18}$$

where $q(z_{1_{uv}}, z_{2_{uv}}|\boldsymbol{\gamma}_{uv})$ is a multinomial with parameters $\boldsymbol{\gamma}_{uv}$.

The lower bound of the log-likelihood is:

$$\begin{aligned}&\log p(X|\alpha_1, \alpha_2, \beta) \geq \\ &E_{q(\boldsymbol{z}_1,\boldsymbol{z}_2|\boldsymbol{\gamma})}[\log p(X, \boldsymbol{z}_1, \boldsymbol{z}_2|\alpha_1, \alpha_2, \beta)] - E_{q(\boldsymbol{z}_1,\boldsymbol{z}_2|\boldsymbol{\gamma})}[\log q(\boldsymbol{z}_1, \boldsymbol{z}_2|\boldsymbol{\gamma})]\end{aligned} \tag{19}$$

denoted as $L(\boldsymbol{\gamma}, \alpha_1, \alpha_2, \beta)$.

When using the original joint latent variables and parameters space, we introduce $q(\boldsymbol{z}_1, \boldsymbol{z}_2, \boldsymbol{\pi}_1, \boldsymbol{\pi}_2, \boldsymbol{\theta}|\gamma)$ to approximate $p(\boldsymbol{z}_1, \boldsymbol{z}_2, \boldsymbol{\pi}_1, \boldsymbol{\pi}_2, \boldsymbol{\theta}|X, \alpha_1, \alpha_2, \beta)$, where we assume a factorization different from Equation (5):

$$q(\boldsymbol{z}_1, \boldsymbol{z}_2, \boldsymbol{\pi}_1, \boldsymbol{\pi}_2, \boldsymbol{\theta}|\gamma) = q(\boldsymbol{\pi}_1, \boldsymbol{\pi}_2, \boldsymbol{\theta}|\boldsymbol{z}_1, \boldsymbol{z}_2) \prod_{u=1}^{N_1} \prod_{v=1}^{N_2} q(z_{1_{uv}}, z_{2_{uv}}|\gamma_{uv}) \quad (20)$$

where we model the conditional distribution of parameters $\boldsymbol{\pi}_1$, $\boldsymbol{\pi}_2$, and $\boldsymbol{\theta}$ given latent variables $\boldsymbol{z}_1$ and $\boldsymbol{z}_2$ without any assumptions on their form. By doing so, we drop the assumption made in Equation (5) that the parameters and the latent variables are independent. Furthermore, from Equations (18) and (20), we can see that we make the same assumption on $\boldsymbol{z}_1$, $\boldsymbol{z}_2$, that is the assignment of $z_{1_{uv}}$ and $z_{2_{uv}}$ to the current entry $[u, v]$ is independent w.r.t the assignments of the other entries.

The lower bound of the log-likelihood is:

$$\begin{aligned}
&\log p(X|\alpha_1, \alpha_2, \beta) \geq \quad (21)\\
&E_{q(\boldsymbol{z}_1, \boldsymbol{z}_2, \boldsymbol{\pi}_1, \boldsymbol{\pi}_2, \boldsymbol{\theta}|\gamma)}[\log p(X, \boldsymbol{z}_1, \boldsymbol{z}_2, \boldsymbol{\pi}_1, \boldsymbol{\pi}_2, \boldsymbol{\theta}|\alpha_1, \alpha_2, \beta)] -\\
&E_{q(\boldsymbol{z}_1, \boldsymbol{z}_2, \boldsymbol{\pi}_1, \boldsymbol{\pi}_2, \boldsymbol{\theta}|\gamma)}[\log q(\boldsymbol{z}_1, \boldsymbol{z}_2, \boldsymbol{\pi}_1, \boldsymbol{\pi}_2, \boldsymbol{\theta}|\gamma)] =\\
&E_{q(\boldsymbol{\pi}_1, \boldsymbol{\pi}_2, \boldsymbol{\theta}|\boldsymbol{z}_1, \boldsymbol{z}_2)q(\boldsymbol{z}_1, \boldsymbol{z}_2|\gamma)}[\log p(X, \boldsymbol{z}_1, \boldsymbol{z}_2, \boldsymbol{\pi}_1, \boldsymbol{\pi}_2, \boldsymbol{\theta}|\alpha_1, \alpha_2, \beta)] -\\
&E_{q(\boldsymbol{\pi}_1, \boldsymbol{\pi}_2, \boldsymbol{\theta}|\boldsymbol{z}_1, \boldsymbol{z}_2)q(\boldsymbol{z}_1, \boldsymbol{z}_2|\gamma)}[\log(q(\boldsymbol{\pi}_1, \boldsymbol{\pi}_2, \boldsymbol{\theta}|\boldsymbol{z}_1, \boldsymbol{z}_2)q(\boldsymbol{z}_1, \boldsymbol{z}_2|\gamma))] =\\
&E_{q(\boldsymbol{z}_1, \boldsymbol{z}_2|\gamma)}[E_{q(\boldsymbol{\pi}_1, \boldsymbol{\pi}_2, \boldsymbol{\theta}|\boldsymbol{z}_1, \boldsymbol{z}_2)}[\log(p(\boldsymbol{\pi}_1, \boldsymbol{\pi}_2, \boldsymbol{\theta}|X, \boldsymbol{z}_1, \boldsymbol{z}_2)p(X, \boldsymbol{z}_1, \boldsymbol{z}_2|\alpha_1, \alpha_2, \beta))]] -\\
&E_{q(\boldsymbol{z}_1, \boldsymbol{z}_2|\gamma)}[E_{q(\boldsymbol{\pi}_1, \boldsymbol{\pi}_2, \boldsymbol{\theta}|\boldsymbol{z}_1, \boldsymbol{z}_2)}[\log q(\boldsymbol{\pi}_1, \boldsymbol{\pi}_2, \boldsymbol{\theta}|\boldsymbol{z}_1, \boldsymbol{z}_2)]] - E_{q(\boldsymbol{z}_1, \boldsymbol{z}_2|\gamma)}[\log q(\boldsymbol{z}_1, \boldsymbol{z}_2|\gamma)]
\end{aligned}$$

Since we do not assume any specific form for $q(\boldsymbol{\pi}_1, \boldsymbol{\pi}_2, \boldsymbol{\theta}|\boldsymbol{z}_1, \boldsymbol{z}_2)$, the lower bound will reach at the true posterior $p(\boldsymbol{\pi}_1, \boldsymbol{\pi}_2, \boldsymbol{\theta}|X, \boldsymbol{z}_1, \boldsymbol{z}_2)$. Therefore, the lower bound can be rewritten as:

$$\begin{aligned}
&\log p(X|\alpha_1, \alpha_2, \beta) \geq\\
&\quad E_{q(\boldsymbol{z}_1, \boldsymbol{z}_2|\gamma)}[\log p(X, \boldsymbol{z}_1, \boldsymbol{z}_2|\alpha_1, \alpha_2, \beta)] - E_{q(\boldsymbol{z}_1, \boldsymbol{z}_2|\gamma)}[\log q(\boldsymbol{z}_1, \boldsymbol{z}_2|\gamma)] \quad (22)
\end{aligned}$$

which is the same as $L(\gamma, \alpha_1, \alpha_2, \beta)$. Thus, both approaches derive the same lower bound of the log-likelihood.

Since the collapsed variational Bayesian algorithm makes a strictly weaker assumption on the variational posterior than the standard variational Bayesian algorithm, the collapsed approach can find a tighter lower bound, i.e. $L(\gamma, \alpha_1, \alpha_2, \beta) \leq L(\gamma_1, \gamma_2, \phi_1, \phi_2, \alpha_1, \alpha_2, \boldsymbol{\theta})$.

Maximizing Equation (22) w.r.t γ_{uvij} and setting it to zero, we obtain:

$$\begin{aligned}
&\gamma_{uvij} = q(z_{1_{uv}} = i, z_{1_{uv}} = j|\gamma_{uv}) = \quad (23)\\
&\frac{\exp(E_{q(\boldsymbol{z}_1^{\neg uv}, \boldsymbol{z}_2^{\neg uv})}[\log p(X, \boldsymbol{z}_1^{\neg uv}, \boldsymbol{z}_2^{\neg uv}, z_{1_{uv}} = i, z_{1_{uv}} = j|\alpha_1, \alpha_2, \beta)])}{\sum_{i'=1}^{K_1}\sum_{j'=1}^{K_2} \exp(E_{q(\boldsymbol{z}_1^{\neg uv}, \boldsymbol{z}_2^{\neg uv})}[\log p(X, \boldsymbol{z}_1^{\neg uv}, \boldsymbol{z}_2^{\neg uv}, z_{1_{uv}} = i', z_{1_{uv}} = j'|\alpha_1, \alpha_2, \beta)])}
\end{aligned}$$

Substituting Equation (3), and setting

$$\begin{aligned}
&f(u, v, i, j) =\\
&\exp(E_{q(\boldsymbol{z}_1^{\neg uv}, \boldsymbol{z}_2^{\neg uv})}[\log(\alpha_1 + n_{ui}^{\neg uv}) + \log(\alpha_2 + n_{vj}^{\neg uv}) + \log(\beta + n_{ijx_{uv}}^{\neg uv}) - \log(N\beta + n_{ij}^{\neg uv})])
\end{aligned}$$

we have:

$$\gamma_{uvij} = q(z_{1_{uv}} = i, z_{1_{uv}} = j|\gamma_{uv}) = \frac{f(u, v, i, j)}{\sum_{i'=1}^{K_1} \sum_{j'=1}^{K_2} f(u, v, i', j')} \tag{24}$$

The derivation of Equations (23) and (24) can be found in the Appendix.

Following [7], we also apply a Gaussian approximation to Equation (24). Here we just illustrate how to calculate $E_{q(\boldsymbol{z}_1^{\neg uv}, \boldsymbol{z}_2^{\neg uv})}[\log(\alpha_1 + n_{ui}^{\neg uv})]$. The calculation of the other three expectations is similar. Suppose $n_u \gg 0$, and note that $n_{ui}^{\neg uv} = \sum_{v'=1, v' \neq v}^{N_1} \sum_{j'=1, j' \neq j}^{K_2} \mathbf{1}(z_{1_{uv'}} = i, z_{1_{uv'}} = j')$ is a sum of a large number of independent Bernoulli variables $\mathbf{1}(z_{1_{uv'}} = i, z_{1_{uv'}} = j')$, each with mean parameter $\gamma_{uv'ij'}$; thus, it can be accurately approximated by a Gaussian. The mean and variance are given by the sum of the means and the variances of the individual Bernoulli variables:

$$E_{q(\boldsymbol{z}_1^{\neg uv}, \boldsymbol{z}_2^{\neg uv})}[n_{ui}^{\neg uv}] = \sum_{v'=1, v' \neq v}^{N_1} \sum_{j'=1, j' \neq j}^{K_2} \gamma_{uv'ij'} \tag{25}$$

$$Var_{q(\boldsymbol{z}_1^{\neg uv}, \boldsymbol{z}_2^{\neg uv})}[n_{ui}^{\neg uv}] = \sum_{v'=1, v' \neq v}^{N_1} \sum_{j'=1, j' \neq j}^{K_2} \gamma_{uv'ij'}(1 - \gamma_{uv'ij'}) \tag{26}$$

We further approximate $\log(\alpha_1 + n_{ui}^{\neg uv})$ using a second-order Taylor expansion, and evaluate its expectation under the Gaussian approximation:

$$E_{q(\boldsymbol{z}_1^{\neg uv}, \boldsymbol{z}_2^{\neg uv})}[\log(\alpha_1 + n_{ui}^{\neg uv})] \approx \tag{27}$$

$$\log(\alpha_1 + E_{q(\boldsymbol{z}_1^{\neg uv}, \boldsymbol{z}_2^{\neg uv})}[n_{ui}^{\neg uv}]) - \frac{Var_{q(\boldsymbol{z}_1^{\neg uv}, \boldsymbol{z}_2^{\neg uv})}[n_{ui}^{\neg uv}]}{2(\alpha_1 + E_{q(\boldsymbol{z}_1^{\neg uv}, \boldsymbol{z}_2^{\neg uv})}[n_{ui}^{\neg uv}])^2} \tag{28}$$

As discussed in [7], the Gaussian approximation will be accurate. Finally, plugging Equation (27) into (24), we have:

$$\begin{aligned} &\gamma_{uvij} \propto \\ &\left(\alpha_1 + E_{q(\boldsymbol{z}_1^{\neg uv}, \boldsymbol{z}_2^{\neg uv})}[n_{ui}^{\neg uv}]\right) \left(\alpha_2 + E_{q(\boldsymbol{z}_1^{\neg uv}, \boldsymbol{z}_2^{\neg uv})}[n_{vj}^{\neg uv}]\right) \\ &\left(\beta + E_{q(\boldsymbol{z}_1^{\neg uv}, \boldsymbol{z}_2^{\neg uv})}[n_{ijx_{uv}}^{\neg uv}]\right) \left(N\beta + E_{q(\boldsymbol{z}_1^{\neg uv}, \boldsymbol{z}_2^{\neg uv})}[n_{ij}^{\neg uv}]\right)^{-1} \\ &\exp\Bigg(-\frac{Var_{q(\boldsymbol{z}_1^{\neg uv}, \boldsymbol{z}_2^{\neg uv})}[n_{ui}^{\neg uv}]}{2(\beta + E_{q(\boldsymbol{z}_1^{\neg uv}, \boldsymbol{z}_2^{\neg uv})}[n_{ui}^{\neg uv}])^2} - \frac{Var_{q(\boldsymbol{z}_1^{\neg uv}, \boldsymbol{z}_2^{\neg uv})}[n_{vj}^{\neg uv}]}{2(\beta + E_{q(\boldsymbol{z}_1^{\neg uv}, \boldsymbol{z}_2^{\neg uv})}[n_{vj}^{\neg uv}])^2} - \\ &\frac{Var_{q(\boldsymbol{z}_1^{\neg uv}, \boldsymbol{z}_2^{\neg uv})}[n_{ijx_{uv}}^{\neg uv}]}{2(\beta + E_{q(\boldsymbol{z}_1^{\neg uv}, \boldsymbol{z}_2^{\neg uv})}[n_{ijx_{uv}}^{\neg uv}])^2} + \frac{Var_{q(\boldsymbol{z}_1^{\neg uv}, \boldsymbol{z}_2^{\neg uv})}[n_{ij}^{\neg uv}]}{2(N\beta + E_{q(\boldsymbol{z}_1^{\neg uv}, \boldsymbol{z}_2^{\neg uv})}[n_{ij}^{\neg uv}])^2}\Bigg) \end{aligned} \tag{29}$$

An EM-style iterative algorithm can be applied to estimate the γ_{uvij}'s by defining Equation (29) as the recursion equation, we can compute every γ_{uvij} for $u \in 1, \cdots, N_1$, $v \in 1, \cdots, N_2$, $i \in 1, \cdots, K_1$, $j \in 1, \cdots, K_2$, until the change of γ_{uvij} between two consecutive iterations is less than a certain threshold, which we consider as converged.

4 Experiments

4.1 Datasets

Two real datasets are used in our experiments: (a) MovieLens[3]: MovieLens is a movie recommendation dataset created by the Grouplens Research Project. It contains 100,000 ratings in a sparse data matrix for 1682 movies rated by 943 users. The ratings are ranged from 1 to 5, with 5 being the highest score. We use 5-fold cross-validation for training and testing. (b) Jester[4]: Jester is a joke rating dataset. The original dataset contains 4.1 million continuous ratings of 100 jokes from 73,421 users. The ratings are ranged from -10 to 10, with 10 being the highest. Following [1], we pick 1000 users who rate all 100 jokes and use this dense data matrix in our experiment, and binarize the dataset such that the non-negative entries become 1 and the negative entries become 0. We held out 1/4 data to do prediction.

4.2 Methodology

We train the LDCC model using the three methods discussed in Section 3, and make prediction on the test data using the learned model parameters. For prediction, we report the perplexity [1], which is defined as:

$$perp(X) = \exp\left(\frac{-\log p(X)}{N}\right)$$

where N is the number of non-missing entries in X. Perplexity monotonically decreases as the log-likelihood increases. Thus, a lower perplexity value is an indication of a better model. In fact, a higher log-likelihood on the training set means that the model fits the data better, and a higher log-likelihood on the test set implies that the model can explain the data better.

The variational Bayesian algorithm can find local optima of α_1, α_2, and $\boldsymbol{\theta}$ for the training data, given a random initialization of these parameters. If the change in log-likelihood between two consecutive iterations is less than 1.0e-6, we stop the process. For collapsed Gibbs sampling and collapsed variational Bayesian algorithms, we use uniform priors to initialize the model parameters α_1, α_2, and β. We set to 5000 the maximum number of iterations for Gibbs sampling, the first 2000 as burn-in, and 500 sample lag. Again, if the maximum change between the model parameters γ of two consecutive iterations is less than 1.0e-6, we assume that the algorithm has converged, and stop the process.

4.3 Experimental Results

In this section, we present two experimental results: perplexity comparison among the three methods, and the likelihood v.s. number of iterations comparison among the three methods.

[3] http://www.grouplens.org/node/73

[4] http://goldberg.berkeley.edu/jester-data/

Table 1. Perplexity Values

	Gibbs	CVB	VB
MovieLens	3.247	4.553	5.849
Binarized Jester	2.954	3.216	4.023

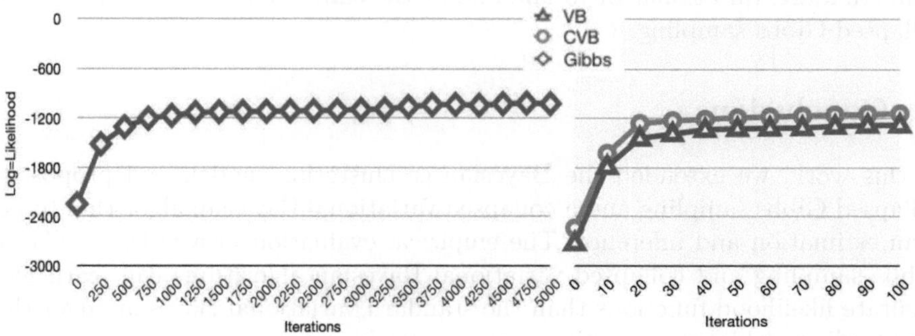

Fig. 3. Log-likelihood v.s. Number of Iterations

Following [1], for the MovieLens dataset, we set $K_1 = 20$ and $K_2 = 19$, which are the numbers of user-clusters and movie-clusters; for the Jester dataset, we set $K_1 = 20$ and $K_2 = 5$, which are the numbers of user-clusters and joke-clusters; the matrices of both datasets roughly have 100,000 entries. Table 1 shows the perplexity values of the three methods on the test data. For the MovieLens dataset, we report the average perplexity of five-fold cross-validation for all the three methods. Doing prediction for the MovieLens dataset is harder than for the binarized Jester dataset. In fact, the binarized Jester data have only two rating states, while the MovieLens has 5. For this reason the perplexity values for the MovieLens are smaller than that for the binarized Jester data. From the table, we can see that collapsed Gibbs sampling achieves the best perplexity on both datasets, followed by collapsed variational Bayesian (CVB). The worst performer is the standard variational Bayesian (VB) approach. These results corroborate our theoretical analysis: collapsed Gibbs sampling and collapsed variational Bayesian can learn more accurate likelihood functions than the standard variational Bayesian algorithm, thus leading to higher predicting performance.

Figure 3 shows the log-likelihood as a function of the number of iterations for the three methods on the binarized Jester dataset. As expected, the collapsed Gibbs sampling algorithm provides higher log-likelihood values, but needs a larger number of iterations, 5000 in our case. Collapsed variational Bayesian provides better log-likelihood values than the standard variational Bayesian, but worse than collapsed Gibbs sampling. Collapsed and standard variational Bayesian algorithms have similar numbers of iterations at convergence (100).

Although collapsed Gibbs sampling is an unbiased estimator and can find the true likelihood function, it takes a long time to achieve the stationary distribution. Standard variational Bayesian suffers from the strong assumption of independence between model parameters and latent variables. As a consequence it finds a loose lower bound of the true likelihood function. Collapsed variational Bayesian, however, can find a tighter lower bound of the likelihood function than standard variational Bayesian, and at the same time it's much faster than collapsed Gibbs sampling.

5 Conclusions

In this work, we extended the Bayesian co-clustering model, and proposed a collapsed Gibbs sampling and a collapsed variational Bayesian algorithm to perform estimation and inference. The empirical evaluation proved that collapsed Gibbs sampling and collapsed variational Bayesian algorithms can learn more accurate likelihood functions than the standard variational Bayesian algorithm, thus leading to higher predicting performance in general.

Acknowledgement

This work was in part supported by NSF CAREER Award IIS-0447814.

References

1. Shan, H., Banerjee, A.: Bayesian co-clustering. In: IEEE International Conference on Data Mining (2008)
2. Hartigan, J.A.: Direct Clustering of a Data Matrix. Journal of the American Statistical Association 337, 123–129 (1972)
3. Dhillon, I.S., Mallela, S., Modha, D.S.: Information-Theoretic Co-Clustering. In: ACM SIGKDD international conference on Knowledge discovery and data mining, pp. 89–98 (2003)
4. Shafiei, M.M., Milios, E.E.: Latent Dirichlet Co-Clustering. In: International Conference on Data Mining, pp. 542–551 (2006)
5. Blei, D., Ng, A., Jordan, M.: Latent Dirichlet Allocation. Journal of Machine Learning Research 3, 993–1022 (2003)
6. Beal, M.J.: Variational Algorithms for Approximate Bayesian Inference. PhD thesis, Gatsby Computational Neuroscience Unit, University College London (2003)
7. Teh, Y.W., Newman, D., Welling, M.: A Collapsed Variational Bayesian Inference Algorithm for Latent Dirichlet Allocation. In: Advances in Neural Information Processing Systems, vol. 19 (2007)
8. Neal, R.M.: Probabilistic Inference Using Markov Chain Monte Carlo Methods, Technical Report CRG-TR-93-1, Dept. of Computer Science, University of Toronto (1993)
9. Griffiths, T.L., Steyvers, M.: Finding Scientific Topics. National Academy of Science 101, 5228–5235 (2004)
10. Teh, Y.W., Kurihara, K., Welling, M.: Collapsed Variational Inference for HDP. In: Advances in Neural Information Processing Systems, vol. 20 (2008)

Appendix

BCC Model

For the BCC model, the marginal probability of an entry x in the data matrix X is:

$$p(x|\alpha_1,\alpha_2,\boldsymbol{\theta}) = \int_{\pi_1}\int_{\pi_2} p(\pi_1|\alpha_1)p(\pi_2|\alpha_2)\sum_{z_1}\sum_{z_2} p(z_1|\pi_1)p(z_2|\pi_2)p(x|\theta_{z_1 z_2})\, d\pi_1 d\pi_2$$

The overall joint distribution over all observable and latent variables is given by:

$$p(X,\boldsymbol{\pi}_1,\boldsymbol{\pi}_2,\boldsymbol{z}_1,\boldsymbol{z}_2|\alpha_1,\alpha_2,\boldsymbol{\theta}) = \prod_u p(\pi_{1_u}|\alpha_1)\prod_v p(\pi_{2_v}|\alpha_2)$$
$$\cdot \prod_{u,v} p(z_{1_{uv}}|\pi_{1_u})p(z_{2_{uv}}|\pi_{2_v})p(x_{uv}|\theta_{z_{1_{uv}},z_{2_{uv}}})^{\delta_{uv}}$$

The probability of observing the entire matrix X is:

$$p(X|\alpha_1,\alpha_2,\boldsymbol{\theta}) = \int_{\boldsymbol{\pi}_1}\int_{\boldsymbol{\pi}_2}\int_{\boldsymbol{\theta}} \left(\prod_u p(\pi_{1_u}|\alpha_1)\right)\left(\prod_v p(\pi_{2_v}|\alpha_2)\right)$$
$$\cdot\left(\prod_{u,v}\sum_{z_{1_{uv}}}\sum_{z_{2_{uv}}} p(z_{1_{uv}}|\pi_{1_u})p(z_{2_{uv}}|\pi_{2_v})p(x_{uv}|\theta_{z_{1_{uv}},z_{2_{uv}}})^{\delta_{uv}}\right) d\boldsymbol{\pi}_1 d\boldsymbol{\pi}_2$$

Derivation of Equation (13)

$$p(X,z_{1_{uv}}=i,z_{2_{uv}}=j,\boldsymbol{z}_1^{\neg uv},\boldsymbol{z}_2^{\neg uv}|\alpha_1,\alpha_2,\beta) = \tag{30}$$
$$\prod_{u'=1}^{N_1}\left(\frac{\Gamma(K_1\alpha_1)}{\Gamma(K_1\alpha_1+n_{u'}^{\neg uv}+\delta_{u'=u})}\prod_{i'=1}^{K_1}\frac{\Gamma(\alpha_1+n_{u'i'}^{\neg uv}+\delta_{u'=u}^{i'=i})}{\Gamma(\alpha_1)}\right)$$
$$\prod_{v'=1}^{N_2}\left(\frac{\Gamma(K_2\alpha_2)}{\Gamma(K_2\alpha_2+n_{v'}^{\neg uv}+\delta_{v'=v})}\prod_{j'=1}^{K_2}\frac{\Gamma(\alpha_2+n_{v'j'}^{\neg uv}+\delta_{v'=v}^{j'=j})}{\Gamma(\alpha_2)}\right)$$
$$\prod_{i'=1}^{K_1}\prod_{j'=1}^{K_2}\left(\frac{\Gamma(N\beta)}{\Gamma(N\beta+n_{i'j'}^{\neg uv}+\delta_{i'=i}^{j'=j})}\prod_{x'=1}^{N}\frac{\Gamma(\beta+n_{i'j'x'}^{\neg uv}+\delta_{i'=i,j'=j}^{x'=x_{uv}})}{\Gamma(\beta)}\right)$$

$$p(X,\boldsymbol{z}_1^{\neg uv},\boldsymbol{z}_2^{\neg uv}|\alpha_1,\alpha_2,\beta) = \tag{31}$$
$$\prod_{u'=1}^{N_1}\left(\frac{\Gamma(K_1\alpha_1)}{\Gamma(K_1\alpha_1+n_{u'}^{\neg uv}+\delta_{u'=u})}\prod_{i'=1}^{K_1}\frac{\Gamma(\alpha_1+n_{u'i'}^{\neg uv}+\delta_{u'=u}^{i'=i})}{\Gamma(\alpha_1)}\right)$$
$$\prod_{v'=1}^{N_2}\left(\frac{\Gamma(K_2\alpha_2)}{\Gamma(K_2\alpha_2+n_{v'}^{\neg uv}+\delta_{v'=v})}\prod_{j'=1}^{K_2}\frac{\Gamma(\alpha_2+n_{v'j'}^{\neg uv}+\delta_{v'=v}^{j'=j})}{\Gamma(\alpha_2)}\right)$$
$$\prod_{i'=1}^{K_1}\prod_{j'=1}^{K_2}\left(\frac{\Gamma(N\beta)}{\Gamma(N\beta+n_{i'j'}^{\neg uv}+\delta_{i'=i}^{j'=j})}\prod_{x'=1}^{N}\frac{\Gamma(\beta+n_{i'j'x'}^{\neg uv}+\delta_{i'=i,j'=j}^{x'=x_{uv}})}{\Gamma(\beta)}\right)$$

where $\delta_{(\cdot)}^{(\cdot)}$ is an indicator function: if all input equations are true, it takes value 1, else 0. Note that if $u' \neq u$, $n_{u'}^{\neg uv} = n_{u'}$. The same holds for the other counting variables. Thus, Equation (13) can be derived by taking the ratio of Equations (30) and (31).

Derivation of Equations (23) and (24)

$$L(\gamma,\alpha_1,\alpha_2,\beta) =$$
$$\int \prod_{u=1}^{N_1}\prod_{v=1}^{N_2} q(z_{1_{uv}}, z_{2_{uv}}|\gamma_{uv}) \log p(X, \boldsymbol{z}_1, \boldsymbol{z}_2|\alpha_1,\alpha_2,\beta)\, dq(\boldsymbol{z}_1,\boldsymbol{z}_2|\gamma) -$$
$$\int \prod_{u=1}^{N_1}\prod_{v=1}^{N_2} q(z_{1_{uv}}, z_{2_{uv}}|\gamma_{uv}) \log \prod_{u=1}^{N_1}\prod_{v=1}^{N_2} q(z_{1_{uv}}, z_{2_{uv}}|\gamma_{uv})\, dq(\boldsymbol{z}_1,\boldsymbol{z}_2|\gamma)$$

Taking the derivative of $L(\gamma,\alpha_1,\alpha_2,\beta)$ w.r.t $q(z_{1_{uv}}, z_{1_{uv}}|\gamma_{uv})$, we get:

$$\frac{\partial L(\gamma,\alpha_1,\alpha_2,\beta)}{\partial q(z_{1_{uv}}, z_{1_{uv}}|\gamma_{uv})} =$$
$$\int \prod_{u'=1,u'\neq u}^{N_1}\prod_{v'=1,v'\neq v}^{N_2} q(z_{1_{u'v'}}, z_{2_{u'v'}}|\gamma_{u'v'}) \log p(X, \boldsymbol{z}_1, \boldsymbol{z}_2|\alpha_1,\alpha_2,\beta)\, dq(\boldsymbol{z}_1^{\neg uv},\boldsymbol{z}_2^{\neg uv}|\gamma) -$$
$$\int \prod_{u'=1,u'\neq u}^{N_1}\prod_{v'=1,v'\neq v}^{N_2} q(z_{1_{u'v'}}, z_{2_{u'v'}}|\gamma_{u'v'}) \log \prod_{u'=1}^{N_1}\prod_{v'=1}^{N_2} q(z_{1_{u'v'}}, z_{2_{u'v'}}|\gamma_{u'v'})\, dq(\boldsymbol{z}_1^{\neg uv},\boldsymbol{z}_2^{\neg uv}|\gamma) -$$
$$\prod_{u'=1,u'\neq u}^{N_1}\prod_{v'=1,v'\neq v}^{N_2} q(z_{1_{u'v'}}, z_{2_{u'v'}}|\gamma_{u'v'}) =$$
$$\int \prod_{u'=1,u'\neq u}^{N_1}\prod_{v'=1,v'\neq v}^{N_2} q(z_{1_{u'v'}}, z_{2_{u'v'}}|\gamma_{u'v'}) \log p(X, \boldsymbol{z}_1, \boldsymbol{z}_2|\alpha_1,\alpha_2,\beta)\, dq(\boldsymbol{z}_1^{\neg uv},\boldsymbol{z}_2^{\neg uv}|\gamma) -$$
$$\log q(z_{1_{uv}}, z_{2_{uv}}|\gamma_{uv}) \int \prod_{u'=1,u'\neq u}^{N_1}\prod_{v'=1,v'\neq v}^{N_2} q(z_{1_{u'v'}}, z_{2_{u'v'}}|\gamma_{u'v'})\, dq(\boldsymbol{z}_1^{\neg uv},\boldsymbol{z}_2^{\neg uv}|\gamma) -$$
$$\int \prod_{u'=1,u'\neq u}^{N_1}\prod_{v'=1,v'\neq v}^{N_2} q(z_{1_{u'v'}}, z_{2_{u'v'}}|\gamma_{u'v'}) \log \prod_{u'=1,u'\neq u}^{N_1}\prod_{v'=1,v'\neq v}^{N_2} q(z_{1_{u'v'}}, z_{2_{u'v'}}|\gamma_{u'v'})\, dq(\boldsymbol{z}_1^{\neg uv},\boldsymbol{z}_2^{\neg uv}|\gamma) -$$
$$\prod_{u'=1,u'\neq u}^{N_1}\prod_{v'=1,v'\neq v}^{N_2} q(z_{1_{u'v'}}, z_{2_{u'v'}}|\gamma_{u'v'})$$

Setting the derivative to zero, it's clear that:

$$q(z_{1_{uv}}, z_{2_{uv}}|\gamma_{uv}) \propto \exp(E_{q(\boldsymbol{z}_1^{\neg uv},\boldsymbol{z}_2^{\neg uv}|\gamma)}[\log p(X, \boldsymbol{z}_1, \boldsymbol{z}_2|\alpha_1,\alpha_2,\beta)])$$

from which we derive Equation (23). From Equation (30), we can see that:

$$\log p(X, z_{1_{uv}} = i, z_{2_{uv}} = j, \boldsymbol{z}_1^{\neg uv}, \boldsymbol{z}_2^{\neg uv}|\alpha_1,\alpha_2,\beta) =$$
$$\sum_{u'=1}^{N_1}\left(\log\Gamma(K_1\alpha_1) - \log\Gamma(K_1\alpha_1 + n_{u'}^{\neg uv} + \delta_{u'=u}) + \sum_{i'=1}^{K_1}(\log\Gamma(\alpha_1 + n_{u'i'}^{\neg uv} + \delta_{u'=u}^{i'=i}) - \log\Gamma(\alpha_1))\right) +$$
$$\sum_{v'=1}^{N_2}\left(\log\Gamma(K_2\alpha_2) - \log\Gamma(K_2\alpha_2 + n_{v'}^{\neg uv} + \delta_{v'=v}) + \sum_{j'=1}^{K_2}(\log\Gamma(\alpha_2 + n_{v'j'}^{\neg uv} + \delta_{v'=v}^{j'=j}) - \log\Gamma(\alpha_2))\right) +$$
$$\sum_{i'=1}^{K_1}\sum_{j'=1}^{K_2}\left(\log\Gamma(N\beta) - \log\Gamma(N\beta + n_{i'j'}^{\neg uv} + \delta_{i'=i}^{j'=j}) + \sum_{x'=1}^{N}(\log\Gamma(\beta + n_{i'j'x'}^{\neg uv} + \delta_{i'=i,j'=j}^{x'=x_{uv}}) - \log\Gamma(\beta))\right)$$

Note that if $u' = u$, $\Gamma(K_1\alpha_1 + n_{u'}^{\neg uv} + \delta_{u'=u}) = (K_1\alpha_1 + n_{u'}^{\neg uv})\Gamma(K_1\alpha_1 + n_{u'}^{\neg uv})$, as for other Gamma functions. Then, we have:

$$\log p(X, z_{1uv} = i, z_{2uv} = j, \boldsymbol{z}_1^{\neg uv}, \boldsymbol{z}_2^{\neg uv} | \alpha_1, \alpha_2, \beta) = -\log(K_1\alpha_1 + n_u^{\neg uv}) + \log(\alpha_1 + n_{ui}^{\neg uv}) - \log(K_2\alpha_2 + n_v^{\neg uv}) + \log(\alpha_2 + n_{v'j'}^{\neg uv}) - \log(N\beta + n_{i'j'}^{\neg uv}) + \log(\beta + n_{i'j'x'}^{\neg uv}) +$$
$$\sum_{u'=1}^{N_1}\left(\log\Gamma(K_1\alpha_1) - \log\Gamma(K_1\alpha_1 + n_{u'}^{\neg uv}) + \sum_{i'=1}^{K_1}(\log\Gamma(\alpha_1 + n_{u'i'}^{\neg uv}) - \log\Gamma(\alpha_1))\right) +$$
$$\sum_{v'=1}^{N_2}\left(\log\Gamma(K_2\alpha_2) - \log\Gamma(K_2\alpha_2 + n_{v'}^{\neg uv}) + \sum_{j'=1}^{K_2}(\log\Gamma(\alpha_2 + n_{v'j'}^{\neg uv}) - \log\Gamma(\alpha_2))\right) +$$
$$\sum_{i'=1}^{K_1}\sum_{j'=1}^{K_2}\left(\log\Gamma(N\beta) - \log\Gamma(N\beta + n_{i'j'}^{\neg uv}) + \sum_{x'=1}^{N}(\log\Gamma(\beta + n_{i'j'x'}^{\neg uv}) - \log\Gamma(\beta))\right) \quad (32)$$

where for a chosen entry $[u, v]$, no matter what z_{1uv} and z_{2uv} are, $\log(K_1\alpha_1 + n_u^{\neg uv})$, $\log(K_2\alpha_2 + n_v^{\neg uv})$, and the summations in Equation (5) are the same. So it's clear that:

$$q(z_{1uv} = i, z_{1uv} = j | \gamma_{uv}) \propto \exp(E_{q(\boldsymbol{z}_1^{\neg uv}, \boldsymbol{z}_2^{\neg uv})}[\log(\alpha_1 + n_{ui}^{\neg uv}) + \log(\alpha_2 + n_{vj}^{\neg uv}) + \log(\beta + n_{ijx_{uv}}^{\neg uv}) - \log(N\beta + n_{ij}^{\neg uv})])$$

Thus, we derive Equation (24).

Variational Graph Embedding for Globally and Locally Consistent Feature Extraction

Shuang-Hong Yang[1,3], Hongyuan Zha[1], S. Kevin Zhou[2], and Bao-Gang Hu[3]

[1] College of Computing, Georgia Institute of Technology, USA
[2] Integrate Data Systems, Siemens Corporate Research, USA
[3] NLPR & LIAMA, Chinese Academy of Sciences, China
shyang@gatech.edu, zha@cc.gatech.edu
shaohua.zhou@siemens.com, hubg@nlpr.ia.ac.cn

Abstract. Existing feature extraction methods explore either global statistical or local geometric information underlying the data. In this paper, we propose a general framework to learn features that account for both types of information based on variational optimization of nonparametric learning criteria. Using mutual information and Bayes error rate as example criteria, we show that high-quality features can be learned from a variational graph embedding procedure, which is solved through an iterative EM-style algorithm where the E-Step learns a variational affinity graph and the M-Step in turn embeds this graph by spectral analysis. The resulting feature learner has several appealing properties such as *maximum discrimination*, *maximum-relevance-minimum-redundancy* and *locality-preserving*. Experiments on benchmark face recognition data sets confirm the effectiveness of our proposed algorithms.

1 Introduction

Feature extraction, the preprocessing step aimed at learning a small set of highly predictive features out of a large amount of possibly noisy or redundant raw input variables, plays a fundamental role in the success of many learning tasks where high dimensionality arises as a big challenge [8,5].

Over the past decades, a large number of algorithms have been proposed, mostly based on using either *global statistical* or *local geometric* structures underlying the data. Classical techniques, such as the Principal Component Analysis (PCA) and the Fisher Discriminant Analysis (FDA), make use of some well-defined statistical measures (e.g., variance, entropy, linear correlation, cross-correlogram, Fisher information, etc.) to evaluate the usefulness of features. Since these measures are usually defined based on the overall properties of the data set, hence, such statistical approaches are often powerful to retain the global structures of the data space, but usually perform poorly when their underlying assumptions (i.e., the optimal condition of statistical measures) are violated. For example, FDA performs poorly for multi-modal data, because Fisher's criterion is optimal only if the data in each class are sampled from a Gaussian distribution[7].

W. Buntine et al. (Eds.): ECML PKDD 2009, Part II, LNAI 5782, pp. 538–553, 2009.

In contrast, a host of algorithms were recently established by using the local geometric information to restore the submanifold structure from which the data are sampled. Examples include ISOMAP [21], Locally Linear Embedding [17], Laplacian Eigenmap [1], Locality Preserving Projection [9]. Such geometric methods, although are powerful to retain the geometric structure revealed by the training data, neglect the overall properties of the data that are fundamental to the success of extracting predictive features. For example, most geometric methods neglect the supervision (label) information of the data and therefore lack of discriminant power.

In this paper, we show that, by optimizing certain nonparametric learning criteria, both the global (statistical) and local (geometric) structures revealed by the training data can be used to build high-quality features. As case studies, we mainly consider two learning criteria, Mutual Information (MI) and Bayes Error Rate (BER). Both MI and BER are theoretically optimal for feature learning but computationally intractable in practice. Our approach, however, is able to encode these criteria into well-defined data graphs and in turn reduce the task into variational graph-embedding problem. The proposed approach is an iterative EM-style algorithm where the E-Step learns a variational affinity graph and the M-Step in turn embeds this graph by spectral analysis. More importantly, the learned graphs are capable to simultaneously capture the supervision information, the global statistical property and the local geometric structure of the data, leading to feature extractors sharing the advantages of both local geometric methods and global statistical methods while mitigating their drawbacks.

1.1 Related Work

MI is a popular criterion in machine learning. Unlike other statistical measures, such as variance, correlation or Fisher's criterion, which only account for up to second order moments, MI can capture the complete dependence between features and the target concept (i.e., label) [14,16]. A practical prohibition of using MI is the computational cost of entropy estimation which involves numerical integration of high dimensional data. Conventional approaches usually resort to histogram or discretization based methods to obtain estimations of MI [2,26], which is computationally intensive when we are dealing with high-dimensional data. In contrast, our approach encodes maximization of MI as variational graph embedding, a much easier problem to solve. In addition, our approach is significantly different from the existing MI-based feature extraction algorithms such as [12,22,10], all of which are formalized as nonlinear non-convex optimization problems and do not account for the local properties of the data, as a result, cannot capture the geometric structure of the underlying manifold, which is, however, fundamental to feature learning as being demonstrated by recent researches [21,1].

BER has also been employed to learn features. However, due to the unavailability of the underlying generative distribution, almost all the existing algorithms optimize BER indirectly [18,4], e.g., by maximizing the *Average Pairwise Divergence* or minimizing the *Union Bhattacharyya Error Bounds* or *Bhattacharyya Distance*. And Gaussian assumption usually has to be made to make these approach tractable, which strongly limits the applicability and performance of those methods. Again, our approach shows advantages over this class of methods in computational efficiency as well as the capability to use local information to restore geometric structures.

The importance of utilizing both global and local information for better feature learning has been increasingly recognized. Algorithms that are globally-and-locally consistent were reported to significantly outperform both global algorithms and local algorithms. For example, Weinberger et al [23] proposed an approach to learn a kernel matrix for nonlinear feature projections by maximizing a global statistic measure (i.e. variance of the projected data) subject to local geometric constraints (e.g., preserving the angles and distances between nearest neighbors). Sugiyama [20] proposed a localized FDA algorithm by optimizing a combination of the *Fisher's criterion* [7] and the *locality preserving cost* [9]. In this paper, we propose a principled way to derive such globally-and-locally consistent approaches for feature learning.

The last few years have witnessed a surge of interests in graph-based learning (GBL). Typically, a GBL learner is established by: (1) conveying basic assumptions and heuristical intuitions into pairwise similarity of and/or constraints over the training instances; (2) constructing a affinity graph based on the defined similarity; and (3) building a learner by spectral analysis of the graph. For instance, in dimensionality reduction, Yan et al [24] and Zhao & Liu [29] presented generalized formalization frameworks for graph-based feature extraction and attribute selection respectively. Usually, the spectral graph theory [6] is employed to provide justifications for GBL. However, it provides no guidance on how to construct the graph, which is of central importance to the success of the GBL algorithms since the performance is extremely sensitive to both graph structures and edge weight settings. As a consequence, one has to resort to heuristics to establish graph for GBL. In contrast, we show in this paper that affinity graphs can be learned by optimizing theoretically sound learning measures. In particular, we show that some nonparametric criteria lead to graphs which naturally encode both the global statistical and the local geometric structures of the data.

1.2 Our Contribution

The main contribution of this paper are three folds:

- Firstly, we propose two effective feature extraction algorithms and test them on real-world tasks.
- Secondly, the graphs learned by our method, which encodes both global (statistical) and local (geometric) structures of the data, can be used in a wide variety of graph-based learning tasks, e.g., semi-supervised learning, metric learning, etc.
- Finally, the approach we use to learn graphs, that is, *nonparametric learning measure estimation* and *variational approximation of kernel terms*, can

be applied to other learning criteria to learn predictive graphs/kernels/ similarity functions.

2 Optimal Feature Learning Criteria

Suppose we are given a set of input vectors $\{\mathbf{x}_n\}_{n=1}^N$ along with the corresponding labels $\{y_n\}_{n=1}^N$ drawn *i.i.d* from an unknown distribution $p(\mathbf{x},y)$, where $\mathbf{x}_n \in \mathcal{X} \subset \mathbb{R}^D$ is a training instance and $y_n \in \mathcal{Y}=\{1,\ldots,C\}$ is its label, N, D and C denote the training set size, the input space dimensionality and the total number of categories, respectively. The goal of feature extraction is to construct a set of M ($M \ll D$) most predictive features, i.e., to find a preprocessing of data $\mathbf{z} = \tau(\mathbf{x})$, where $\mathbf{z} \in \mathcal{Z} \subset \mathbb{R}^M$, $\tau : \mathcal{X} \to \mathcal{Z}$ and $\tau \in \mathcal{H}$ with $\mathcal{H}$ being a hypothesis space. Let the goodness of τ be measured by the feature evaluation criterion $\mathfrak{J}(\cdot)$. Then the problem of feature extraction can be formalized as:

$$\tau = \arg\max \mathfrak{J}(\tau).$$

Theoretically, two optimal criteria can be identified for feature learning. The first one [13] is based on information theory, which attempts to minimize the amount of information loss incurred in the process of dimensionality reduction, i.e.: $\min_{\tau\in\mathcal{H}} KL\{p(y|\mathbf{z})||p(y|\mathbf{x})\}$. Ideally, if the KL-divergence between these two posteriors reaches zero, we can recover the optimal Bayes classifier based on the data in the reduced dimensional space $\mathcal{Z}$. This criterion is equivalent to maximizing the mutual information, i.e., the expected information gain about y from observing $\mathbf{z}$:

$$\max_{\tau\in\mathcal{H}} I(\mathbf{z}, y) = \int_{\mathbf{z},y} p(\mathbf{z}, y) \log \frac{p(\mathbf{z}, y)}{p(\mathbf{z})p(y)} d\mathbf{z}dy, \quad (1)$$

The second optimal criterion [25] considers classification directly and naturally reflects the Bayes error rate Υ in the reduced dimensional space $\mathcal{Z}$, i.e.:

$$\min_{\tau\in\mathcal{H}} \Upsilon(\tau) = \inf_h E_{\mathbf{x}}[err(h|\mathbf{z})] = E_{\mathbf{z}}[1 - \max_{c=1,\ldots,C} P(c|\mathbf{z})], \quad (2)$$

where $E_{\mathbf{x}}\{err(h|\mathbf{z})\}$ is the generalization error of a decision rule h in the reduced-dimensional space.

However, a critical issue of learning features by directly optimizing such optimal learning criteria is that they involve unknown generative distributions. In this paper, we attempt to establish feature extractors by efficiently optimizing these criteria. We achieve this goal by two steps: (1) By nonparametric estimation of the criteria, we reduce the task to kernel-based optimization problems; (2) Based on variational approximation of each kernel term, the task can be compactly formalized as variational graph-embedding problems, which can be solved by graph spectral analysis.

We will mainly focus on MI in the next section and discuss BER in Section 4 since the derivation procedures are quite similar.

3 Graph-Based Feature Learning by Maximizing Mutual Information

In this section, we establish feature extractors by maximize the mutual information between the features and the class label. Compared with other criteria that only account for up to second-order statistics, an appealing property of MI is that it makes use of higher-order statistics and is able to capture the complete nonlinear dependence between the features and the class label [14,16]. However, the computation of MI involves integration of the unknown generative distributions. Conventional approaches usually resort to histogram or discretization based methods to obtain estimations of MI [2,26]. Such approaches are not only highly ill-formed but also computationally intractable when dealing with extremely high-dimensional data. In this section, we will show that MaxMI-features can be learned efficiently by variational graph embedding.

3.1 Nonparametric Quadratic MI

We use a nonparametric method to estimate MI. One of the advantages of using nonparametric estimators is that they can effectively capture the properties (e.g., multimodality) of the underlying distribution. We adopt an efficient estimator that was proposed by Principe et al [16] and exploited recently by Torkkola [22] to learn features. First, instead of using standard MI, we use the quadratic MI:

$$I_2(\mathbf{z}, y) = \int_{\mathbf{z},y} \left(p(\mathbf{z}, y) - p(\mathbf{z})p(y)\right)^2 d\mathbf{z}dy, \tag{3}$$

which has been justified both theoretically and experimentally by many previous works, e.g., [16,22,10].

A kernel density estimator is then employed to estimate the density function involved in Eq.(3). Particularly, consider the isotropic Gaussian kernel:

$$\mathsf{k}(\mathbf{x}, \mathbf{x}') = g(\mathbf{x} - \mathbf{x}', \sigma^2 U),$$

where σ is the standard deviation, U denotes the unit matrix, and $g(\boldsymbol{\mu},\Sigma)$ is the Gaussian distribution function with mean $\boldsymbol{\mu}$ and covariance Σ. An interesting property of this kernel is:

$$\langle \mathsf{k}(\mathbf{x}, \mathbf{x}'), \mathsf{k}(\mathbf{x}, \mathbf{x}'') \rangle = \int_{\mathbf{x}} \mathsf{k}(\mathbf{x}, \mathbf{x}')\mathsf{k}(\mathbf{x}, \mathbf{x}'')d\mathbf{x} = \mathsf{k}(\mathbf{x}', \mathbf{x}'').$$

It turns out that, by using this property and the quadratic form of Eq.(3), we are able to eliminate the integration in MI.

Given the training data, $p(\mathbf{x})$ can be estimated as $\hat{p}(\mathbf{x}) = \frac{1}{N}\sum_{n=1}^{N} \mathsf{k}(\mathbf{x}, \mathbf{x}_n)$. Plugging this into the objective $\mathfrak{J}(\mathbf{z}) = I_2(\mathbf{z}, y)$, we obtain,

$$\begin{aligned}
\mathfrak{J}(\mathbf{z}) &= \hat{I}_2(\mathbf{z}, y) = \mathfrak{J}_{\mathbf{z}y} + \mathfrak{J}_{\mathbf{z}} \times \mathfrak{J}_y, \\
\mathfrak{J}_{\mathbf{z}y} &= \frac{1}{N^2} \sum_{c=1}^{C} \sum_{i \in \mathcal{Y}_c} \Big(\sum_{j \in \mathcal{Y}_c} \mathsf{k}(\mathbf{z}_i, \mathbf{z}_j) - \frac{2N_c}{N} \sum_{n=1}^{N} \mathsf{k}(\mathbf{z}_i, \mathbf{z}_n) \Big), \\
\mathfrak{J}_{\mathbf{z}} &= \frac{1}{N^2} \sum_{i=1}^{N} \sum_{j=1}^{N} \mathsf{k}(\mathbf{z}_i, \mathbf{z}_j), \\
\mathfrak{J}_y &= \sum_{c=1}^{C} \frac{N_c^2}{N^2},
\end{aligned} \tag{4}$$

where $\mathcal{Y}_c$ denotes the index set of examples in class c and N_c the number of instances in it, $\sum_{c=1}^{C} N_c = N$.

3.2 MaxMI by Variational Graph Embedding

In this section, we show that maximization of MI can be formalized compactly as a variational graph embedding problem with a generic graph learning procedure being derived. We begin with rewriting the objective Eq.(4).

Theorem 1. *Maximizing the nonparametric quadratic MI is equivalent to the following optimization problem:*

$$\begin{aligned}
&\max \frac{1}{N^2} \sum_{i=1}^{N} \sum_{j=1}^{N} \gamma_{ij} \mathsf{k}(\mathbf{z}_i, \mathbf{z}_j), \\
&\gamma_{ij} = I(y_i = y_j) + \sum_{c=1}^{C} \frac{N_c^2}{N^2} - \hat{P}_{y_i} - \hat{P}_{y_j},
\end{aligned} \tag{5}$$

where $I(\cdot)$ denotes the indicator function, and $\hat{P}_{y_i} = \hat{P}(c = y_i) = \frac{N_c}{N}$ is the proportion of instances sharing the same label with y_i.

Proof. Let $I_{ic} = I(y_i = c)$, $\sum_c I_{ic} I_{jc} = I_{ij} = I(y_i = y_j)$, we have:

$$\begin{aligned}
&\mathfrak{J}(\mathbf{z}) = \mathfrak{J}_{\mathbf{z}y} + \mathfrak{J}_{\mathbf{z}} \times \mathfrak{J}_y \\
=&\frac{1}{N^2} \sum_{c=1}^{C} \sum_{i=1}^{N} \Big(\sum_{j=1}^{N} (I_{ic} I_{jc} - I_{ic} \hat{P}_{y_i} - I_{jc} \hat{P}_{y_j}) \mathsf{k}(\mathbf{z}_i, \mathbf{z}_j) \Big) + \frac{1}{N^2} \sum_{c=1}^{C} \frac{N_c^2}{N^2} \sum_{i=1}^{N} \sum_{j=1}^{N} \mathsf{k}(\mathbf{z}_i, \mathbf{z}_j) \\
=&\frac{1}{N^2} \sum_{i=1}^{N} \sum_{j=1}^{N} \Big(I_{ij} + \sum_{c} \frac{N_c^2}{N^2} - \hat{P}_{y_i} - \hat{P}_{y_i} \Big) \mathsf{k}(\mathbf{z}_i, \mathbf{z}_j) \\
=&\frac{1}{N^2} \sum_{i=1}^{N} \sum_{j=1}^{N} \gamma_{ij} \mathsf{k}(\mathbf{z}_i, \mathbf{z}_j).
\end{aligned}$$

□

The optimization in Eq.(5) is still computationally difficult due to its expression as a big summation of the nonconvex Gaussian density function $\mathsf{k}(\cdot,\cdot)$. To address this problem, we adopt the variational optimization method [11] to approximate each kernel term with its variational lower bound. Since exponential function is convex, a simple lower bound can be easily obtained by making first-order Taylor expansion:

$$\exp(\frac{||\mathbf{z}_i - \mathbf{z}_j||^2}{-\delta^2}) \geqslant \lambda_{ij}||\mathbf{z}_i - \mathbf{z}_j||^2/\delta^2 - \lambda_{ij} + \lambda_{ij}\log(-\lambda_{ij}), \quad (6)$$

where we have introduced variational parameters λ_{ij}. Then for given λ_{ij}, integrate Eq.(6) into the objective Eq.(5), the task is reduced to a linear graph embedding problem [24]:

$$\min \sum_{i=1}^{N}\sum_{j=1}^{N} w_{ij}||\mathbf{z}_i - \mathbf{z}_j||^2, \quad (7)$$

where $W = (w_{ij}) \in \mathbb{R}^{N\times N}$ is the adjacency matrix of the derived data graph with edge wights $w_{ij} = -\gamma_{ij}\lambda_{ij}$.

However, since the variational parameters are coupled with the feature extractors, the variational optimization is a EM-style algorithm, where the E-step corresponds to learning a graph W by optimizing the variational parameters in Eq.(6), and the M-step in turn learns feature extractors by solving the graph embedding problem Eq.(7). These two steps are repeated alternatively until convergence.

The E-Step has an analytical solution because the variational lower bound Eq.(6) is exact iff $\lambda_{ij} = -\exp(-||\mathbf{z}_i - \mathbf{z}_j||^2/\delta^2)$. In the following, we will mainly discuss the M-Step.

Let $\mathcal{G} = \{\mathbf{X}, W\}$ be the undirected weighted graph with $\mathbf{X} = \{\mathbf{x}_i\}_{i=1}^N$ being the vertex set and W being the adjacency matrix. $D = \mathrm{diag}[W\mathbf{1}]$ denotes the degree matrix of $\mathcal{G}$, $\mathbf{1} = [1, 1, \ldots, 1]^\top$, $L = D - W$ denote the Laplacian matrix of $\mathcal{G}$. The M-Step, i.e., graph-embedding, learns a set of features by maximizing their consistency with the graph $\mathcal{G}$:

$$\min_{\mathcal{T}\in\mathcal{H}} \sum_{i=1}^{N}\sum_{j=1}^{N} w_{ij}||\mathbf{z}_i - \mathbf{z}_j||^2 = Z^\top L Z, \quad (8)$$

where $Z = [\mathbf{z}_1, \mathbf{z}_2, \ldots, \mathbf{z}_n]^\top$. To obtain practical feature extractors, we consider two special forms.

Linear Case. Assume each feature is obtained as a linear projection of the input attributes, i.e., $\mathbf{z} = T^\top\mathbf{x}$, $T = [\mathbf{t}_1, \mathbf{t}_2, \ldots, \mathbf{t}_M] \in \mathbb{R}^{D\times M}$ is a transformation matrix, we have

$$\min \sum_{i=1}^{N}\sum_{j=1}^{N} w_{ij}||T^\top \mathbf{x}_i - T^\top \mathbf{x}_j||^2 = tr(T^\top X^\top LXT),$$
$$s.t.: \ \mathbf{t}_m^\top \mathbf{t}_m = 1, \ \ \forall\, m \in \{1, 2, \ldots, M\}, \tag{9}$$

where $X = [\mathbf{x}_1, \mathbf{x}_2, \ldots, \mathbf{x}_n]^\top$ is the input data matrix, $tr(\cdot)$ denotes the trace of a matrix. This is a constrained quadratic program problem. By using the Lagrangian technique, we can easily prove that the optimal projection vectors are given by the eigenvectors of $X^\top LX$ corresponding to the bottom M eigenvalues[1]. Since W is symmetric, the resulting features are naturally orthogonal projections, i.e., $T^\top T = U$. The algorithm is referred to as Mutual Information Embedding (MIE).

Nonlinear Case. The linear MIE might fail to discover nonlinear structures underlying the data. We now investigate nonlinear feature extractors. For simplicity, we only consider a special case, where the hypothesis space $\mathcal{H}$ is restricted to a reproducing kernel Hilbert space (RKHS) induced by a Mercer kernel $k(\mathbf{x}, \mathbf{x}')$, where $k(\cdot, \cdot)$ is a symmetric positive semi-definite function, $K = (k_{ij}) \in \mathbb{R}^{N\times N}$ is the kernel matrix, $k_{ij} = k(\mathbf{x}_i, \mathbf{x}_j)$ denotes its entry, $i, j = 1, 2, \ldots, N$. Based on the property of the Mercer kernel, we can assume that the nonlinearly projected features lie in the space spanned by the kernel bases, i.e., $K\times\boldsymbol{\alpha}$, where $\boldsymbol{\alpha}_i = [\alpha_{i1}, \alpha_{i2}, \ldots, \alpha_{iN}]^\top$ is a projection vector in the kernel space. We have:

$$\begin{aligned} A^* &= \arg\min \ tr(A^\top K^\top LKA), \\ s.t.: \boldsymbol{\alpha}_m^\top K\boldsymbol{\alpha}_m &= 1, m = 1, 2, \ldots, M, \end{aligned} \tag{10}$$

where $A = [\boldsymbol{\alpha}_1, \boldsymbol{\alpha}_2, \ldots, \boldsymbol{\alpha}_M]$ is the kernel space projection matrix. Similarly, the optimal projection vectors are given by the bottom M eigenvectors of $K^\top LK$. Note that the Euclidean distance in $\mathcal{H}$ is given by:

$$d_{\mathcal{H}}(\mathbf{x}, \mathbf{x}') = \sqrt{k(\mathbf{x}, \mathbf{x}) + k(\mathbf{x}', \mathbf{x}') - 2k(\mathbf{x}, \mathbf{x}')}. \tag{11}$$

For a new input instance $\mathbf{x}$, its projection in the optimal reduced kernel space is given as:

$$\begin{aligned} \mathbf{z} &= \tau(\mathbf{x}) = A^\top \boldsymbol{\kappa}, \\ \boldsymbol{\kappa} &= [k(\mathbf{x}_1, \mathbf{x}), k(\mathbf{x}_2, \mathbf{x}), \ldots, k(\mathbf{x}_N, \mathbf{x})]^\top. \end{aligned} \tag{12}$$

This algorithm is referred to as kernel MIE (kMIE).

3.3 Initial Graph and Efficient Approximate Solution

The variational graph embedding algorithm requires initialization of the variational parameters λ_{ij} or equivalently the graph w_{ij}. In this section, we construct

[1] For fast implementation, please refer to[3,24,19] and the references therein.

an initial graph, which is in the near neighborhood of the optimal one and hence makes the convergence of the variational algorithm very fast.

Again, we approximate each kernel term by its first-order Taylor expansion. Since expanding $\exp(-v)$ at v_0 leads to $\exp(-v) \approx \exp(-v_0) - \exp(-v_0)(v - v_0)$, let $v = \frac{||\mathbf{z}_i - \mathbf{z}_j||^2}{\delta^2}$, and $v_0 = \frac{||\mathbf{x}_i - \mathbf{x}_j||^2}{\delta^2}$, we have:

$$\exp(\frac{||\mathbf{z}_i - \mathbf{z}_j||^2}{-\delta^2}) \approx -\exp(\frac{||\mathbf{x}_i - \mathbf{x}_j||^2}{-\delta^2})||\mathbf{z}_i - \mathbf{z}_j||^2/\delta^2 + \text{const.}$$

Integrating this into the objective, we get an initial data graph with edge weights $w_{ij} = \gamma_{ij} e^{-||\mathbf{x}_i - \mathbf{x}_j||^2/\delta^2}$, where $\delta^2 = 2\sigma^2$. That is, the feature space distance $||\mathbf{z}_i - \mathbf{z}_j||$ is replaced with the original space distance$||\mathbf{x}_i - \mathbf{x}_j||$.

Besides being used to initialize the graph of the variational embedding algorithm, this graph could also be used as an off-the-shelf tool to build other graph-based learners. For instance, solely embedding this graph leads to an efficient non-iterative alternative to MIE (refer to as initial MIE or MIE_0).

3.4 Justifications

3.4.1 Max-Relevance-Min-Redundancy

In feature extraction, our goal is to learn a set of features that are *useful* for building the predictor. Hence, merely extracting features that are most *relevant* to the target concept is suboptimal since the learned features might be *redundant* such that adding them will gain no additional information [8]. Therefore, a good feature extractor should achieve a reasonable balance between maximizing relevance and minimizing redundancy.

It turns out that our proposed algorithm simultaneously (1)maximizes the relevance of the learned features $\mathbf{z}$ to the class label y; (2) minimizes the redundancy within $\mathbf{z}$; and (3) employs a natural tradeoff parameter for perfect balance. This can be seen clearly from the objective function Eq.(4), which consists of two terms: $\mathfrak{J}(\mathbf{z}) = \mathfrak{J}_{\mathbf{z}y} + \eta\mathfrak{J}_{\mathbf{z}}$, where $\mathfrak{J}_{\mathbf{z}y}$, depending on both the features $\mathbf{z}$ and the class label y, is a measure of relevance of $\mathbf{z}$ to y; $\mathfrak{J}_{\mathbf{z}}$, depending solely on features $\mathbf{z}$, measures the degree of redundancy within $\mathbf{z}$; and $\eta = \mathfrak{J}_y = \sum_c N_c^2/N^2$, which is invariant to $\mathbf{z}$, provides a natural compromise between relevance and redundancy.

3.4.2 Max-Discrimination

Another observation is that the proposed algorithms are supervised methods, i.e., they take advantage of the label information to maximize the discriminative ability of the learned features.

Theorem 2. *The weights $\gamma_{i,j}$ have the following properties:*

1. for $\forall\ i, j : y_i = y_j$, $\gamma_{i,j} > 0$;
2. for $\forall\ i, j : y_i \neq y_j$, $E\{\gamma_{i,j}\} < 0$.

Proof. Without loss of generality, consider a nontrivial case, i.e., $\forall\, c = 1, \ldots, C$, $C \geqslant 2$, $P_c > 0$ and $N_c \geqslant 1$.

1. if $y_i = y_j$:

$$\gamma_{ij} = 1 + \frac{1}{N^2}\sum_{c=1}^{C} N_c^2 - 2\frac{N_i}{N} = \frac{1}{N^2}(\sum_{c\neq i} N_c^2 + (\sum_{k\neq i} N_k)^2) > 0,$$

2. if $y_i \neq y_j$:

$$E\{\gamma_{ij}\} = E\{\sum_{c=1}^{C} \frac{N_c^2}{N^2} - \frac{N_i}{N} - \frac{N_j}{N}\} = -\sum_{i,j=1}^{C} (P_i - P_j)^2 - (C-1)\sum_{i=1}^{C} P_i^2 < 0,$$

completing the proof. □

From Theorem 2, we can see that instance pairs that are from the same class ($y_i = y_j$) are always assigned positive weights, while pairs from different classes ($y_i \neq y_j$) are expected to get negative weights. As a consequence, the optimization Eq.(7) actually minimizes the distances between patterns with the same label while keeping patterns with different labels as far apart as possible, i.e., maximizing the margin between within-class and between-class patterns.

3.4.3 Locality Preserving

Another appealing properties of the proposed algorithm is that, besides the statistical information used to model the global property underlying the data (e.g., relevance, redundance and discrimination), it also makes use of local geometric information to restore the submanifold structure from which the data is sampled, or in other words, it is locality preserving. Particularly, the learned graph $\mathcal{G}$ naturally includes a weight term $e^{-||\mathbf{z}_i - \mathbf{z}_j||^2/\delta^2}$ (in the initial graph: $e^{-||\mathbf{x}_i - \mathbf{x}_j||^2/\delta^2}$), which is actually the Gaussian heat weight [19] usually employed to retain the local consistency of the data. This term turns to assign small weights to the instance pairs that are far apart from each other, making sure that the results are mainly affected by neighboring instance pairs and hence sufficiently smooth with respect to the intrinsic structure revealed by the training data.

3.4.4 Connection with LPP and FDA

We provide some theoretical analysis of the connection between our proposed algorithm and two popular dimensionality reduction methods, i.e., the global statistical method FDA and the local geometric method LPP [9]. We show that even the initial solution of MIE (i.e., embedding the initial graph) shares the advantages of both FDA and LPP and mitigates their drawbacks at the same time.

Theorem 3. *Maximizing $\hat{\mathfrak{J}}_{\mathbf{z}}$ is equivalent to LPP.*

Proof. We have

$$\tau = \arg\max(\hat{\mathfrak{J}}_{\mathbf{z}} = -\sum_{i=1}^{N}\sum_{j=1}^{N} w_{ij}^{(0)}||\mathbf{z}_i - \mathbf{z}_j||^2)$$
$$= \arg\min \sum_{m=1}^{M} \mathbf{t}_m^\top X^\top L^{(0)} X \mathbf{t}_m,$$

where $w_{ij}^{(0)} = e^{-||\mathbf{x}_i - \mathbf{x}_j||^2/\delta^2}$ is the adjacency matrix of a heat graph, $L^{(0)}$ is its corresponding Laplacian matrix. Recall that LPP minimizes exactly the same objective function except that it uses a different normalization constraint (equivalent to using normalized Laplacian). □

Theorem 4. *If the classes are balanced, i.e.,* $\frac{N_c}{N} = \frac{1}{C}$, *maximizing* $\hat{\mathfrak{J}}_{\mathbf{z}y}$ *is reduced to a localized version of FDA (LFDA,[20]).*

Proof. We have

$$\tau = \arg\max(\hat{\mathfrak{J}}_{\mathbf{z}y} \propto -\sum_{i=1}^{N}\sum_{j=1}^{N}(w_{ij}^{(+)} - w_{ij}^{(-)})||\mathbf{z}_i - \mathbf{z}_j||^2)$$
$$= \arg\min \sum_{m=1}^{M} \mathbf{t}_m^\top X^\top (L^{(+)} - L^{(-)}) X \mathbf{t}_m, \quad (13)$$

where $w_{ij}^{(+)} = I(y_i = y_j)e^{-||\mathbf{x}_i - \mathbf{x}_j||^2/\delta^2})$ and $w_{ij}^{(-)} = (\hat{p}_{y_i} + \hat{p}_{y_j})e^{-||\mathbf{x}_i - \mathbf{x}_j||^2/\delta^2}$ defines two graphs, $L^{(+)}$ and $L^{(-)}$ represents their Laplacian matrices respectively. Assume the classes are balanced, $\forall i, j, \hat{p}_{y_i} = \hat{p}_{y_j} = 1/C$, the Fisher criterion can be rewritten as:

$$\max \sum_{m=1}^{M} \frac{\mathbf{t}_m S_B \mathbf{t}_m}{\mathbf{t}_m S_W \mathbf{t}_m} \propto \sum_{m=1}^{M} \frac{\mathbf{t}_m^\top X^\top L^{(-)} X \mathbf{t}_m}{\mathbf{t}_m^\top X^\top L^{(+)} X \mathbf{t}_m} + \text{const.} \quad (14)$$

The equivalence between Eq.(13) and Eq.(14) follows directly from the equivalence of trace ratio and trace difference [7]. □

To summarize, both LPP and FDA can be viewed as degraded forms of the initial MIE, They maximize solely either $\mathfrak{J}_{\mathbf{z}}$ or $\mathfrak{J}_{\mathbf{z}y}$. In contrast, MIE simultaneously optimizes both $\mathfrak{J}_{\mathbf{z}}$ and $\mathfrak{J}_{\mathbf{z}y}$. In addition, instead of combining these two goals in an ad-hoc way, MIE uses a natural parameter $\eta = \mathfrak{J}_y = \sum_c \frac{N_c^2}{N^2}$ that is derived from a theoretically optimal criterion to strive for a reasonable balance between these two goals.

As an empirical validation, we test FDA, LPP and the initial MIE on a synthetic 2-D data set [20]. The results are shown in Fig.1. We can see that in an

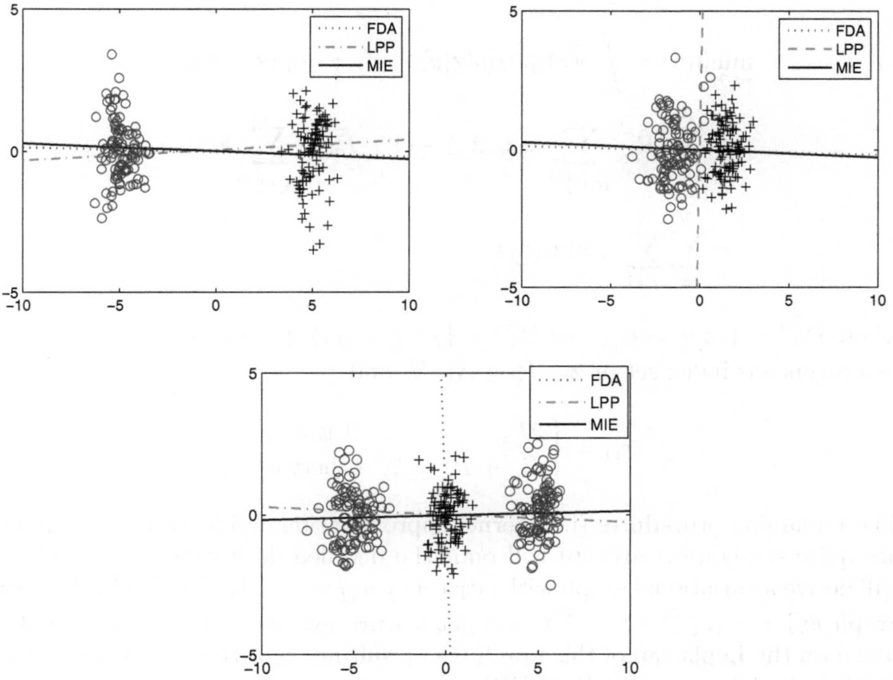

Fig. 1. Typical behaviors of global method (FDA), local method (LLP) and globally-locally consistent method (initial solution of MIE)

ideal case as depicted in the leftmost subfigure, LPP, FDA and MIE are all able to find an excellent projection for the data. In the middle subfigure, while the MIE and FDA can still find a good projection, the geometric and unsupervised method LPP gets a very poor projection such that the data from both classes are totally mixed up. In the rightmost subfigure where the data has multiple modes, the methods that account for locality geometric information (i.e., MIE and LPP) successfully obtain a good feature. However, FDA totally fails, which is not surprising since it is based on Gaussian distribution assumption.

4 Using Bayes Error Rate

We now apply the procedures to the Bayes Error Rate (BER) criterion. From Eq.(2), we have:

$$\Upsilon \propto \int_{\mathbf{z}} p(\mathbf{z})(p(y)p(\mathbf{z}|y) - p(c \neq y)p(\mathbf{z}|c \neq y))d\mathbf{z} + \text{const.}$$

Using nonparametric estimator, we have:

$$\min_{\tau \in \mathcal{H}} \mathfrak{J}(\mathbf{z}) = \int_{\mathbf{z}} p(\mathbf{z})(p(y)p(\mathbf{z}|y) - p(c \neq y)p(\mathbf{z}|c \neq y))d\mathbf{z}.$$
$$\approx \sum_{n=1}^{N} [\hat{P}_{y_n} \sum_{i \in \Omega_n^{(o)}} \mathsf{k}(\mathbf{z}_i, \mathbf{z}_n) - (1 - \hat{P}_{y_n}) \sum_{j \in \Omega_n^{(e)}} \mathsf{k}(\mathbf{z}_j, \mathbf{z}_n)]$$
$$= \sum_{i=1}^{N} \sum_{j=1}^{N} r_{ij} \mathsf{k}(\mathbf{z}_i, \mathbf{z}_j),$$

where $\Omega_n^{(o)} = \{i : y_i = y_n\}$ and $\Omega_n^{(e)} = \{j : y_j \neq y_n\}$ denote the homogenous and heterogeneous index set of $\mathbf{z}_n$, $\hat{P}_c = N_c/N$, and

$$r_{ij} = \begin{cases} 2\hat{P}_{y_i} & \text{if } y_i = y_j \\ \hat{P}_{y_i} + \hat{P}_{y_j} - 2 & \text{otherwise} \end{cases}$$

The remaining procedures (i.e., kernel approximation and feature construction) are quite straightforward; we will omit the detailed derivation. Eventually, we will derive a variational graph with adjacency $w_{ij} = -r_{ij}\lambda_{ij}$ (similarly, the initial graph $w_{ij} = r_{ij}e^{-||\mathbf{x}_i - \mathbf{x}_j||^2/\delta^2}$), and get feature extractors by spectral analysis based on the Laplacian of this graph (the resulting algorithms are referred to as BER Embedding, or BERE/kBERE). It can be easily proved that BERE also has the Max-Discrimination and Locality-Preserving properties.

5 Experiment

We test our proposed algorithms on real-world face recognition tasks. For comparison, PCA, FDA, LPP, MFA (Margin Fisher Analysis,[24]) and their kernel counterparts are selected as baselines, among which PCA and LPP are unsupervised, FDA and MFA are supervised, and LPP and MFA account for local geometric structures.

Three benchmark facial image sets are selected: (1) the `Yale`[2] data set, which contains 165 facial images of 15 persons, 11 images for each individual; (2)the `ORL`[3] data set, which consists of 400 facial images of 40 persons; and (3) the `CMU Pie`[4] data set, which contains 41368 facial images of 68 individuals. All the three sets of images were taken in different environments, at different times, with different poses, facial expressions and details. In our experiment, all the raw images are normalized to 32×32. For each data set, we randomly select ν images of each person as training data (referred to as νtrain), and leave others for testing. Only the training data are used to learn features. To evaluate the effectiveness of different methods, the classification accuracy of a k-NN classifier on testing data is used as the evaluation metric.

---
[2] http://cvc.yale.edu/projects/yalefaces/yalefaces.html
[3] http://www.uk.research.att.com/facedatabase.html
[4] http://www.ri.cmu.edu/projects/project_418.html

Table 1. Comparison of feature extraction algorithms on face recognition tasks. The results of our proposed algorithms that are better than the performances of baseline methods are highlighted in bold.

Method	Yale			ORL			CMU Pie		
	2train	3train	4train	2train	3train	4train	5train	10train	20train
PCA	.44±.022	.52±.014	.55±.017	.54±.015	.63±.012	.69±.014	.47±.017	.55±.016	.65±.015
FDA	.46±.016	.57±.013	.68±.012	.76±.021	.82±.017	.90±.016	.61±.022	.78±.015	.85±.008
LPP	.50±.019	.61±.018	.67±.015	.70±.018	.78±.017	.86±.014	.62±.022	.75±.016	.84±.013
MFA	.49±.023	.64±.019	.77±.014	.72±.021	.84±.017	.89±.017	.64±.016	.78±.011	**.91**±.013
MIE_0	**.51**±.019	**.67**±.019	**.83**±.016	**.77**±.021	**.85**±.018	**.94**±.013	.64±.017	**.81**±.016	.89±.014
MIE	.50±.016	**.67**±.018	**.85**±.018	**.75**±.019	**.88**±.018	**.92**±.017	**.65**±.019	**.83**±.018	.93±.015
$BERE_0$	**.53**±.020	**.66**±.017	**.81**±.016	**.80**±.018	**.88**±.017	**.91**±.013	**.70**±.019	**.81**±.016	**.94**±.015
BERE	**.55**±.022	**.69**±.019	**.84**±.015	**.84**±.018	**.92**±.018	**.94**±.016	**.69**±.019	**.87**±.017	**.94**±.018
KPCA	.48±.025	.55±.019	.60±.016	.63±.022	.74±.017	.78±.013	.50±.018	.57±.017	.69±.016
KDA	.49±.023	.63±.021	.69±.018	.79±.021	.89±.019	.92±.016	.60±.022	.79±.013	.91±.010
KMFA	.52±.024	.65±.024	.78±.020	.75±.024	.84±.017	.92±.015	.62±.021	.83±.017	.91±.016
$kMIE_0$	**.55**±.026	**.70**±.021	**.84**±.018	**.86**±.022	**.91**±.017	**.94**±.015	**.69**±.018	.83±.016	**.93**±.017
kMIE	.52±.021	**.73**±.022	**.87**±.017	**.88**±.020	**.91**±.017	.92±.019	**.67**±.019	**.86**±.018	**.95**±.015
$kBERE_0$	**.54**±.018	**.73**±.019	**.86**±.016	**.81**±.021	**.93**±.015	.92±.017	**.71**±.015	**.86**±.016	**.95**±.016
kBERE	**.57**±.021	**.72**±.019	**.89**±.018	**.83**±.022	**.95**±.019	**.95**±.016	**.74**±.019	**.89**±.016	**.94**±.018

The only parameter of our algorithms is the bandwidth parameter δ in the initial graph. In our experiments, we adopt the local scaling scheme used in [27]. All the other hyper-parameters (e.g., the number of neighbors k used in kNN and local scaling) are tuned by 5-fold cross validation. The experiments are averaged over 10 random runs. The results are given in Table 1, where each entry represents the mean testing accuracy ± standard deviation.

From Table 1, we can see that for almost all the entries, our proposed algorithms significantly outperform other baseline methods. Even the algorithms based on the initial graphs (i.e., MIE_0 and $BERE_0$) perform significantly better than the baselines. Note that the computation complexity of the initial algorithms are of the same order as the baseline methods. The improvements are quite evident. On average, MIE is 36% over PCA, 8% over FDA, and 4% over MFA; BERE is 39% over PCA, 11% over FDA and 6% over MFA. The improvements are even more significant in the kernel case: kMIE (31%,9%,7%) and kBERE(33%,10%,8%) over (KPCA, KDA, KMFA). For most entries in the table, we got p-values less than 0.005. We also observe in our experiment that, when using the initial graph for initialization, the variational graph embedding algorithms (i.e., MIE, BERE) usually converge within 5 iteration steps.

6 Conclusion

In this paper, we have established graph-based feature extraction algorithms based on variational optimization of nonparametric learning criteria. As case studies, we employed two theoretically optimal but computationally intractable feature learning criteria, i.e., Mutual Information and Bayes Error Rate. By nonparametric criteria estimation and kernel term approximation, we reduced the optimization of these criteria to variational graph-embedding problems, which can be solved by an iterative EM-style procedure where the E-Step learns a variational affinity graph and the M-Step in turn embeds this graph by spectral analysis. The resulting feature learner has several appealing properties such as *maximum discrimination*, *maximum-relevance-minimum-redundancy* and *locality-preserving*. Experiments on benchmark face recognition data sets confirm the effectiveness of our proposed algorithms.

We finally note that our derived graphs (e.g., the initial graphs derived in Section 3 and 4) as well as the approach we used to derive graphs (i.e., *non-parametric learning measure estimation* and *variational approximation of kernel terms*) are not confined to feature extraction scenarios. They might also be useful in a variety of graph-based learning tasks, e.g., semi-supervise learning, relational learning, metric learning. We shall leave such investigations for future research. Sparseness is a desirable property for feature learning, especially for kernel based methods since both the memory for storing the kernel matrix and the training and testing time are typically proportional to the degree of sparsity of the feature extractor. In the future, we would also like to investigate sparse feature learning in the proposed framework.

Acknowledgement. The authors would like to thank the anonymous reviewers for their helpful comments. This work was supported in part by the NSF "Computational methods for nonlinear dimensionality reduction" project (under grant #DMS-0736328), the NSF China grant #60275025 and the MOST of China grant #2007DFC10740.

References

1. Belkin, M., Niyogi, P.: Laplacian Eigenmaps and Spectral Techniques for Embedding and Clustering. In: Proceeding of Advances in Neural Information Processing Systems 14 (NIPS 2001), pp. 585–591 (2001)
2. Bollacker, K.D., Ghosh, J.: Linear Feature Extractors Based on Mutual Information. In: Proceeding of the 13th International Conference on Pattern Recognition (ICPR 1996) (1996)
3. Cai, D., He, X., Han, J.: SRDA: An Efficient Algorithm for Large-Scale Discriminant Analysis. IEEE Transactions on Knowledge and Data Engineering 20(1), 1–12 (2008)
4. Choi, E.: Feature Extraction Based on the Bhattacharyya Distance. Pattern Recognition 36, 1703–1709 (2003)
5. Fan, J., Li, R.: Statistical Challenges with High Dimensionality: Feature Selection in Knowledge Discovery. In: Proceeding of Regional Conference in Mathematics (AMS 1996), vol. 3, pp. 595–622 (1996)
6. Chung, F.R.K.: Spectral Graph Theory. In: Proceeding of Regional Conference in Mathematics (AMS 1992), vol. 92 (1997)
7. Fukunnaga, K.: Introduction to Statistical Pattern Recognition, 2nd edn. Academic Press, London (1991)
8. Guyon, I., Elissee, A.: An introduction to variable and feature selection. Journal of Machine Learning Research 3, 1157–1182 (2003)
9. He, X., Niyogi, P.: Locality Preserving Projections. In: Proceeding of Advances in Neural Information Processing Systems 16 (NIPS 2003) (2003)

10. Hild II, K.E., Erdogmus, D., Torkkola, K., Principe, C.: Feature Extraction Using Information-Theoretic Learning. IEEE Transactions on Pattern Analysis and Machine Intelligence 28(9), 1385–1392 (2006)
11. Jaakkola, T.: Tutorial on Variational Approximation Methods. In: Opper, M., Saad, D. (eds.) Advanced Mean Field Methods: Theory and Practice, pp. 129–159. MIT Press, Cambridge (2000)
12. Kaski, S., Peltonen, J.: Informative Discriminant Analysis. In: Proceeding of the 20th Annual International Conference on Machine Learning (ICML 2003), pp. 329–336 (2003)
13. Koller, D., Sahami, M.: Toward Optimal Feature Selection. In: Proceeding of the 13th International Conference on Machine Learning (ICML 1996), pp. 284–292 (1996)
14. Paninski, L.: Estimation of Entropy and Mutual Information. Neural Computation 15, 1191–1253 (2003)
15. Peng, H., Long, F., Ding, C.: Feature Selection Based on Mutual Information: Criteria of Max-Dependency, Max-Relevance, and Min-Redundancy. IEEE Transactions on Pattern Analysis and Machine Intelligence 27(8), 1226–1238 (2005)
16. Principe, J.C., Fisher III, J.W., Xu, D.: Information Theoretic Learning. In: Haykin, S. (ed.) Unsupervised Adaptive Filtering. Wiley, Chichester (2000)
17. Roweis, S., Saul, L.K.: Nonlinear Dimensionality Reduction by Locally Linear Embedding. Science 290(22), 2323–2326 (2000)
18. Saon, G., Padmanabhan, M.: Minimum Bayes Error Feature Selection for Continuous Speech Recognition. In: Proceeding of the16th Annual Conference on Neural Information Processing Systems (NIPS 2002), pp. 800–806 (2002)
19. Saul, L.K., Weinberger, K.Q., Ham, J.H., Sha, F., Lee, D.D.: Spectral Methods for Dimensionality Reduction. In: Chapelle, O., et al. (eds.) Semisupervised Learning, MIT Press, Cambridge (2006)
20. Sugiyama, M.: Dimensionality Reduction of Multimodal Labeled Data by Local Fisher Discriminant Analysis. Journal of Machine Learning Research 8, 1027–1061 (2007)
21. Tenenbaum, J.B., Silva, V., Langford, J.C.: A Global Geometric Framework for Nonlinear Dimensionality Reduction. Science 290(22), 2319–2323 (2000)
22. Torkkola, K.: Feature Extraction by Nonparametric Mutual Information Maximization. Journal of Machine Learning Research 3, 1415–1438 (2003)
23. Weinberger, K.Q., Sha, F., Saul, L.K.: Learning a kernel matrix for nonlinear dimensionality reduction. In: Proceedings of the 21st Annual International Conference on Machine Learning (ICML 2004), pp. 839–846 (2004)
24. Yan, S.C., Xu, D., Zhang, B.Y., Zhang, H.J., Yang, Q., Lin, S.: Graph Embedding and Extensions: A General Framework for Dimensionality Reduction. IEEE Transactions on Pattern Analysis and Machine Intelligence 29(1), 40–51 (2007)
25. Yang, S.H., Hu, B.G.: Discriminative Feature Selection by Nonparametric Bayes Error Minimization. In: Knowledge and Information Systems (KAIS) (to appear)
26. Yang, H., Moody, J.: Data Visualization and Feature Selection: New Algorithms for Nongaussian Data. In: Proceeding of the 14th Annual Conference on Neural Information Processing Systems (NIPS 2000), pp. 687–693 (2000)
27. Zelnik-Manor, L., Perona, P.: Self-tuning Spectral Clustering. In: Proceeding of the 18th Neural Information Processing Systems (NIPS 2004), pp. 1601–1608 (2004)
28. Zhu, X.: Semi-Supervised Learning Literature Survey. Techincal Report 1530, Department of Computer Sciences, University of Wisconsin-Madison (2005)
29. Zhao, Z., Liu, H.: Spectral feature selection for supervised and unsupervised learning. In: Proceeding of the 24th Annual International Conference on Machine Learning (ICML 2007), pp. 1151–1157 (2007)

Protein Identification from Tandem Mass Spectra with Probabilistic Language Modeling

Yiming Yang, Abhay Harpale, and Subramaniam Ganapathy

Carnegie Mellon University,
Pittsburgh, PA – 15217, United States

Abstract. This paper presents an interdisciplinary investigation of statistical information retrieval (IR) techniques for protein identification from tandem mass spectra, a challenging problem in proteomic data analysis. We formulate the task as an IR problem, by constructing a "query vector" whose elements are system-predicted peptides with confidence scores based on spectrum analysis of the input sample, and by defining the vector space of "documents" with protein profiles, each of which is constructed based on the theoretical spectrum of a protein. This formulation establishes a new connection from the protein identification problem to a probabilistic language modeling approach as well as the vector space models in IR, and enables us to compare fundamental differences in the IR models and common approaches in protein identification. Our experiments on benchmark spectrometry query sets and large protein databases demonstrate that the IR models significantly outperform well-established methods in protein identification, by enhancing precision in high-recall regions in particular, where the conventional approaches are weak.

Keywords: Proteomics, Information Retrieval.

1 Introduction

Statistical pattern matching technologies have been successfully applied to many real-world problems. Among those, text-based information retrieval (IR) is perhaps one of the most intensively studied and highly successful areas. Computational biology is another important area where pattern matching plays a key role in various forms of data analysis. This paper presents an interdisciplinary investigation, focusing on how to generalize good ideas and successful technologies in one domain (IR) into new insights and novel solutions in another (computational proteomics). Specifically, we focus on the problem of protein identification from detected peptides in tandem mass spectra.

Protein identification is important for discovering biomarkers linked to diseases, therapeutic outcomes and individualized drug toxicity. Tandem mass (MS/MS) spectra, generated by a chemical process over complex biological samples such as tissues or blood, contain rich information about proteins and peptides which are constituents of proteins. Protein identification from MS/MS data is typically carried out in two steps. Step 1 is to predict peptides based on observed empirical spectra, and step 2 is to predict proteins based on the predicted peptides. Many technical solutions have been developed for the peptide identification step in the past two decades,

W. Buntine et al. (Eds.): ECML PKDD 2009, Part II, LNAI 5782, pp. 554–569, 2009.

including commercially available software [1], [2], [4], [5], [6], [7], [8]. However, for the second step, the current literature is relatively sparse. Particularly, few interdisciplinary investigations were conducted for exploring the potential of advanced IR technologies in solving the mapping from predicted peptides to proteins. Addressing this research gap is the primary motivation of this paper, and we focus on the second step in particular, i.e., the mapping from system-predicted peptides to the true proteins in large protein databases.

Why would it be beneficial to bridge the technical and methodological gaps between the fields of protein identification and text retrieval? At first glance, the two tasks look totally different. However, at a higher level of abstraction, the two tasks and related technical solutions have important properties in common. If we consider peptides as words, proteins as documents, and peptide identification from spectra as a query generation process, then the mapping from predicted peptides to proteins in a protein database is just like ad-hoc retrieval in a vector space, albeit a particularly high-dimensional one. A database with tens of thousands of proteins would contain tens of millions of unique peptides. Some common peptides could be considered in analogous to stop-words in text, while the majority of peptides are much rarer, form a highly skewed distribution over proteins. This means that the rich body of research findings in text retrieval would provide meaningful insights into how to weight peptides in proteins, how to combine peptide-level evidence into predictions of proteins, and how to leverage state-of-the-art IR methods directly or with adaptation, including efficient inverted indexing, effective term weighting schemes, smoothing and dimensionality reduction techniques, choices of similarity measure in retrieval models, well-understood evaluation metrics, and standardized software toolkits like Lemur and Indri [9][10]. In order to leverage those potentials we need a good understanding of the background knowledge and related literature, including how biological samples, MS/MS spectra, peptides and protein sequences are related to each other, what kinds of technical solutions have been developed for peptide identification from spectra and for protein identification from predicted peptides, how the current solutions have been evaluated and compared, what the strengths and weaknesses of those methods, and how can we apply or adapt retrieval techniques to create better solutions. Achieving such an understanding is the primary contribution we target in this paper. Specifically, our main contributions can be listed as:

1) A new formulation of the protein-prediction task that enables probabilistic modeling with joint use of well-established peptide identification techniques and domain knowledge about protein sequences, as well as the rich developments in IR on language modeling and vector space models.
2) A comparative theoretical analysis of two probabilistic models for combining peptide-level evidence in the prediction of in-sample proteins: one is the well-establish method by Nesvizhskii et al., which scores candidate proteins based on estimated probability of a Boolean OR function, and the other is a language-modeling method that we propose, which uses the estimated probability of a Boolean AND instead. We show that the former has a weakness in discriminating true positives from false positives in the high-recall region of the system's predictions, and that the latter addresses such a weakness in a principled way.

3) A thorough evaluation on standard MS/MS datasets and large protein databases for comparison of methods, including probabilistic OR, probabilistic AND, cosine-similarity with a robust TF-IDF term weighting scheme, and the commonly used X!Tandem commercial software in proteomic applications. We observed significant performance enhancement by the IR models tested over the conventional methods in the current literature of protein identification from tandem mass spectra.

The rest of the paper is organized as follows. Section 2 outlines related background and representative approaches in protein identification from tandem mass spectra. Section 3 defines our new framework and discusses its connection to well-established techniques in text retrieval. Section 4 introduces three benchmark query sets (spectrometry samples) and the corresponding large protein databases for empirical evaluation. Section 5 reports the experiments and the results. Section 6 summarizes the main findings and conclusions.

2 Background and Related Work

Statistical approaches for data analysis with tandem mass spectra is an important and fast growing area in recent computational biology research. Analogous and complementary to micro-array data which are highly informative for analyzing gene-level activities under various conditions, MS/MS spectra contain rich information about proteins which are potentially responsible for diseases, therapeutic responses and drug toxicity [3]. MS/MS spectra are generated using liquid chromatography where a sampled tissue or a blood drop is digested into peptides which are segments of protein sequences. The peptides are further separated into ionized fragments and analyzed to produce MS/MS spectra. Each spectrum is a list of spikes: the location of each spike is the mass/charge (m/z) ratio of an ionized fragment, and the magnitude of the spike is the abundance or intensity of the fragment. An input sample is typically a mixture of multiple proteins but the exact number of proteins is unknown in advance. Standard MS/MS datasets for benchmark evaluations were typically constructed for sample mixtures that contain a dozen or a few dozens of proteins [19]. The numbers of MS/MS spectra obtained from those samples are in the range of a few thousands. The task of protein identification is to find a mapping from the few thousands of observed MS/MS spectra to the true proteins in the input sample. It is typically accomplished in two steps: first, identify the peptides based on observed spectra; second, predict proteins based by system-predicted peptides.

In peptide identification research, database search techniques have been commonly used to select a candidate set of peptides based on the degree of matching between the "theoretical" (expected) mass spectra of candidate peptides in a protein database and the empirical spectra in the input sample [1],[2],[4],[5],[6], [7], [8]. The theoretical spectrum of each peptide can be automatically derived by rules from the amino acid sequences of proteins. Each known protein has a unique amino acid sequence, which can be segmented by rules into peptide-level subsequences. The theoretical spectrum of each peptide can also be automatically generated based on existing knowledge about lower-level chemical properties of amino acid letters. The number of unique peptides in a protein database can be very large. For example, we found roughly 5 million unique peptides in a sample of 50,000 proteins as typical. By comparing each theoretical spectrum against the empirical spectra in an input sample, a system obtains a confidence score of each candidate peptide. Further, applying some threshold to the confidence scores yields peptide assignments by the system. The similarity measures differ from system to system. SEQUEST, for example, one of the most commonly used commercial programs in practical applications as well as in comparative evaluations of peptide identification methods on benchmark datasets, employs a Fourier Transform cross-correlation strategy [4]. X!Tandem is another popular open-source software for peptide/protein identification and has been commonly used as a baseline in comparative evaluations. It uses aggressive thresholds to reduce false-alarms and to enhance computational efficiency, and produces a ranked list of proteins for each input sample [21].

In protein identification based on system-predicted peptides from MS/MS spectra, the ProteinProphet system by Nesvizhskii et al [12] is among the most commonly used in comparative evaluations of methods on benchmark datasets. This system uses SEQUEST-predicted peptides as the input, and converts the non-probabilistic confidence scores by SEQUEST to the probabilistic scores for peptide assignments. Specifically, they used the Expectation-Maximization (EM) algorithm to obtain a mixture model for true positives and false positives in system-predicted peptides. Some empirical evaluations [6] showed performance improvement by the score refinement method over that of the original SEQUEST. ProteinProphet estimates the probability of each protein being present in the input sample using the probability that at least one of the constituent peptides in the protein is a corrected assignment to the sample. To be explicit, suppose $q_j \in [0,1]$ is the estimated probability of the presence of peptide j in the input sample. The probability that a protein i is present in the input sample is calculated in ProteinProphet as

$$p_i = 1 - \prod_{j=1}^{J} (1 - q_j) \ .$$

This formula calculates the estimated probability of the Boolean-OR function over the peptide-level evidence, assuming that the occurrence (being present or not) of each peptide is an *identically independently distributed* (i.i.d.) random event with $q_j \in [0,1], \forall j$. If any constituent peptide of a protein is predicted as present by the system, we have $q_j = 1, \exists j$ and $p_i = 1$ as the consequence. Clearly, the protein scoring function in ProteinProphet is the estimated probability for Boolean OR logic. We will refer to this method as prob-OR in the rest of the paper. A refined version of this method is also supported by the system, i.e., an EM algorithm is used to find hidden groups of proteins, and the peptide probabilities are estimated conditioned on the hidden groups of proteins instead of individual proteins.

Other work of a similar nature in protein identification includes that by MacCoss et al. [11] who used a modified version of SEQUEST to generate peptide assignments with normalized scores, and performed protein-level predictions with a prob-OR equivalent operation. Moore et al. [13] pursued a different but heuristic approach: after aggressive removal of low-scoring candidate peptides, the product of the scores of the remaining peptides that constitute a protein sequence is used to estimate the quality of the match for the protein. Theoretical comparison of their method with the probabilistic models (including Nesvizhskii et al., and others) is difficult because their scoring functions are heuristically or procedurally defined, not explicitly probabilistic; empirical comparison was not reported on the other hand. The recent work by Li et. al. [22] presents another interesting alternative which predicts proteins by modeling the input sample as a multi-protein mixture and finding the Maximum-a-Posteriori (MAP) solution for the mixture weights. They used Gibbs sampling as an approximation method because solving MAP exactly is computationally intractable. Although no theoretical upper/lower bound is guaranteed by the approximation, an empirical evaluation on a new (their own) dataset shows improved results over that of Nesvizhskii's method (prob-OR). However, repeating this comparative evaluation has been difficult as the dataset is not publicly available, and no sufficient details were published about how to reconstruct the dataset from publicly available protein databases. Other indirectly related work includes CHOMPER [14], INTERACT [15] and DTASelect [16], which focus on visualization and filtering tools for manual interaction in protein identification, and Mascot [3] and Sonar [17] which focus on commercial tool development.

3 Methods

The desiderata for a new approach are: 1) a theoretically justified function (or family of functions) for combining peptide-level evidence, and 2) higher performance in standard metrics such as average precision, compared to the best results reported in the MS/MS literature. To address these objectives we turn to modern IR approaches for mapping predicted peptides to proteins.

Notice that the commonly used prob-OR type of functions in protein scoring has a potential weakness. That is, it has the tendency to produce many false alarms due to an overly simplistic assumption because if any constituent peptide of a protein is detected, then the protein is assumed as a correct assignment. As an alternative, we propose a mapping with stronger constraints, i.e. using a probabilistic AND (prob-AND) function to combine evidence in predicted peptides. More precisely, we propose to

1) translate the predicted peptides into an empirical in-sample distribution of peptides as observed in the MS/MS spectra;
2) use the relative frequencies of peptides in the amino acid sequence of each protein as the protein profile; and
3) measure the Kullback-Leibler (KL) divergence of each protein-specific distribution of peptides from the sample distribution of peptides.

These steps together accomplish a prob-AND mapping from the predicted peptides to candidate proteins with probabilistic scores.

3.1 Data Representations

The input to our protein identification system is a set of peptides with confidence scores which are produced by a well-established method for peptide identification from a sample of MS/MS spectra [6]. We present the scored peptides using a vector $\vec{q} = (q_1, q_2, \cdots, q_J)$ whose elements $q_j \in [0,1]$ are normalized so that they sum to one, and J is the number of total unique peptides being identified. For convenience, we call vector $\vec{q}$ the "query" for protein search. Notice that a peptide identification method may not generate normalized scores. In that case, we translate scores as following:

$$a = \min\{q_1, q_2, \cdots, q_J\}, \quad b = \max\{q_1, q_2, \cdots, q_J\},$$

$$q_j' = \frac{q_j - a}{b - a}, \qquad q_j'' = \frac{q_j'}{\sum_{t=1}^{J} q_t'}.$$

We also define a normalized vector (profile) for each protein in the target database (DB) as:

$$\vec{p}_i = (p_{i1}, p_{i2}, \cdots, p_{iJ}), \quad p_{ij} = \frac{n_{ij}}{\sum_{j=1}^{J} n_{ij}},$$

where n_{ij} is the count of peptide j in protein i. Notice that query normalization is generally not required in text retrieval methods because it does not effect the ranking of documents given a query. Similarly, in our mapping from a "bag" of system-predicted peptides to protein profiles, query normalization does not affect the ranking of proteins given a query. However, with explicit normalization of both the query vector and protein profiles we can intuitively interpret the mapping criterion based the KL-divergence between the two types of vectors (Section 3.2).

We smooth the peptide probabilities using a Dirichlet prior [18], modifying the elements as

$$p_{ij} = \frac{n_{ij} + \mu \pi_j}{\sum_{j=1}^{J} n_{ij} + \mu}$$

Parameter μ controls the degree of smoothing, and $\pi_j \in [0,1]$ is calculated as:

$$\pi_j = \frac{\sum_{i \in DB} n_{ij}}{\sum_{t=1}^{J} \sum_{i \in DB} n_{it}}.$$

Smoothing is a crucial step in the formulation of protein profiles. As discussed earlier, the peptide identification step identifies peptides in the sample which are the result of the protein cleavage, i.e. breaking of protein into its constituent peptides upon the reaction with a chemical cleaving agent (e.g. Trypsin). It is not guaranteed that each protein breaks at every peptide boundary (phenomenon known as miscleavage), and consequently, not every constituent peptide is necessarily observed and not every observed component is necessarily a valid peptide. Smoothing therefore is necessary for assigning non-zero weights to unobserved peptides, just as the out-of-vocabulary words need to be handled in language modeling for document retrieval. To ensure that all the observed components (both valid peptides and peptide concatenations) are taken into account, we simulated miscleavages in creation of protein profiles.

Why do we construct protein profiles in the above way? Because we want to leverage the domain knowledge about amino acid sequences and establish the mapping from peptides to proteins accordingly. Ideally, we would like to have a large training set of MS/MS spectra with explicitly labeled correspondences to positively and negatively related proteins in a target database, which would enable supervised learning of the conditional distribution of peptides given a protein in MS/MS samples. However, such a large training set would be very expensive to produce and is not currently available in open-source benchmark datasets for protein identification evaluations. The only knowledge we have for relating predicted peptides for an input sample to the proteins in a target database are 1) the peptide occurrences in amino acid sequences of proteins, and 2) the expected (theoretical) spectrum of each valid peptide. Thus, we stay with the unsupervised setting for the mapping problem, i.e., by constructing a peptide-based profile for each protein, and by conducting proximate-search over protein profiles given a synthetic query. The normalization of both the query vector and the profile vectors of proteins enables probabilistic interpretation for the mapping criterion, and avoids an unjustified bias of favoring longer proteins (i.e. with a larger number of constituent peptides) over shorter ones, as present in the prob-OR approaches. As for the need of smoothing, it is well understood in statistical learning theory and practice that model smoothing is particularly important when the feature (input variable) space is very large and the observed data is highly sparse. In our problem, the feature space consists of a large number of peptides, with a skewed distribution over a modest number of protein sequences. For example, the PPK benchmark dataset (Section 4) contains 4,534 proteins and 325,812 unique peptides. This means that most protein profiles are both high-dimensional and extremely sparse, and that appropriate smoothing is necessary for successful mapping from a query to candidate proteins.

3.2 Protein Scoring Based on Prob-AND

The choice of scoring criterion is obviously crucial for successful ranking of proteins given a query. We may consider the presence or absence of a peptide in the predicted list as a random variable, where the randomness comes from both the sampled protein mixture, and the noisy process of generating MS/MS spectra from the protein mixture. Consequently, we may view vector $\vec{q}$ as the empirically observed in-sample distribution of peptides in an unknown protein mixture. Similarly, we may view vector $\vec{p}_i$ as the "theoretical" peptide distribution in a specific protein, derived based on the amino acid sequences of proteins in a target database, and existing knowledge (rules) about how protein sequences decompose to peptides. We use the cross entropy to score each candidate protein with respect to the query. The cross entropy of the two distributions is defined as:

$$H(\vec{q} \parallel \vec{p}_i) \equiv -\Sigma_{j=1}^{J} q_j \log p_{ij} = -\Sigma_{j=1}^{J} q_j \log\left(q_j \frac{p_{ij}}{q_j} \right)$$

$$= -\Sigma_{j=1}^{J} q_j \log q_j + \Sigma_{j=1}^{J} q_j \log\left(\frac{p_{ij}}{q_j}\right)$$

$$= H(\vec{q}) + D(\vec{q} \parallel \vec{p}_i)$$

The cross entropy decouples into two terms as shown in the last line above: the first term $H(\vec{q})$ is the entropy of the query, which is the same for every protein; the second term $D(\vec{q} \parallel \vec{p}_i)$ is the Kullback-Leibler (KL) divergence that determines the relative ranking of proteins with respect to the query. A smaller KL divergence means a better matched protein for the query.

We use prob-AND as the abbreviation of the proposed method. KL divergence has been commonly used in language modeling (LM) approaches for ad-hoc text retrieval with probabilistic ranking of documents given a query. It is a function proportional to the log-likelihood of the query conditioned on the document model under the term independence assumption in a multinomial process. Let $\vec{x} = (x_1, x_2, \cdots, x_J)$ be the vector whose elements are the within-query term frequencies. The log-likelihood is estimated as:

$$\log \Pr(\vec{x} \parallel \vec{p}_i) = \log \prod_{j=1}^{J} p_{ij}^{n_{ij}} = \Sigma_{j=1}^{J} x_j \log p_{ij}$$

$$= N_x \Sigma_{j=1}^{J} \frac{x_j}{N_x} \log p_{ij}$$

where the scaling factor $N_x = x_1 + x_2 + \cdots + x_J$ is the total count of term occurrences in the query. Except for the scaling factor (which is constant given a query), the log-likelihood and KL divergence are identical. Therefore, using the multinomial probabilistic model to rank documents for a query makes no difference compared to using the negation of the KL divergence as the metric. With respect to ranking proteins given a set of predicted peptides, the only difference is that the within-query term frequencies are not directly observed but are predicted instead. Nevertheless, the connection between the log-likelihood function and KL divergence shows clearly that the logic being used for assembling partial evidence (from individual terms or peptides) is probabilistic AND, not probabilistic OR. In other words, KL divergence imposes stronger constraints in the mapping from predicted peptides to proteins. Probabilistic AND and KL divergence have not been studied for protein identification in the current literature, to our knowledge.

3.3 Connections to Other Vector Space Models for Text Retrieval

How are prob-AND and prob-OR related to conventional retrieval Vector Space Models (VSMs) in IR? In fact, they are closely related. Let $\vec{d}_i = (d_{i1}, d_{i2}, \cdots, d_{iJ})$ be a document vector and define within-document term weighting as

$$d_{ij} = \log p_{ij} \equiv \log \Pr(\text{ term } j \mid \text{doc } i),$$

The dot-product similarity in a standard VSM is calculated as:

$$sim(\vec{q} \cdot \vec{d}_i) = \vec{q} \cdot \vec{d}_i = \sum_{j=1}^{J} q_j \log p_{ij} \, .$$

This is exactly the same formula in the prob-AND model, i.e., scoring function based on the cross entropy. On the other hand, if we choose $d_{ij} = p_{ij}$ as the term weighting scheme, the dot-product similarity becomes:

$$sim(\vec{q} \cdot \vec{d}_i) = \vec{q} \cdot \vec{d}_i = \sum_{j=1}^{J} q_j p_{ij} = \sum_{j \in \text{query}} q_j p_{ij}$$

This is a variant of soft OR. That is, a document (or protein) receives a positive weight where as long as any of its terms (or peptides) is found in the query. In a further extreme setting of $d_{ij} = I_{ij}$ which is the indicator function with $I(i,j) = 1$ if peptide j is a constituent of protein i and $I(i,j) = 0$ otherwise, we have:

$$score(\vec{d}_i \mid \vec{q}) = \sum_{j \in \text{query}} q_j I(i,j)$$

It also mimics the Boolean OR logic in a soft manner, obviously. There soft-OR scoring functions are closely related to the prob-OR metric in ProteinProphet which we analyzed in Section 2.

The connections from prob-OR and prob-AND to conventional VSMs invites a question: are they better choices than other variants of VSM, e.g., the commonly used cosine similarity with TF-IDF term weighting scheme? Since the latter is not a probabilistic scoring function, direct theoretical comparison on the basis of probabilistic modeling is impossible. However, an empirical comparison between these VSM variants would be highly informative and practically important for a thorough investigation on the applicability and effectiveness of advanced IR techniques in solving the protein identification problem. Hence, we report such a comparative evaluation in Section 5.

4 Datasets

For evaluation and benchmarking of protein identification algorithms, we use standard proteomic mixtures whose MS/MS spectra are publicly available. Purvine et al in 2003 introduced a standardized proteomics dataset to support comparative evaluation which consists of a query set of MS/MS spectra from a mixture of 12 proteins and 23 peptides[1] and a search database consisting of 4534 proteins [19]. The dataset was designed to mimic the complexity of large scale proteomics experiments and to serve as a standard in proteomics research. We refer to this dataset as PPK, after the authors Purvine S, Picone AF and Kolker E [19].

We also created two more datasets, called Mark12+50000 and Sigma49+50000, respectively. The Mark12+50000 dataset consists of a query set of MS/MS spectra from a 12-protein mixture (from Invitrogen, Carlsbad CA) called the 'Mark12

[1] The query set was generated from 12 proteins and 23 peptides. Each of the peptides was treated as a single-peptide protein in evaluation yielding a total of 35 proteins.

Electrophoresis Standard', and a target protein database which we name as M50000. The Sigma49+50000 dataset consists of the query set of MS/MS spectra from a 49 protein mixture (from Sigma-Andrich, St. Louis MO) and a target protein database which we name as S50000. Both query sets were provided by the Mass Spectrometry Research Center at Vanderbilt University and have been used as standard benchmarks in proteomics research. The target databases were generated by us by drawing two random samples from the SwissProt[2] protein database, which contains over 280,000 protein sequences, and then adding Mark12 query-set proteins to one sample and Sigma49 query-set proteins to the other sample. We chose the size (50,000) of the target protein databases to be comparable to those used in actual proteomic analyses. Tables 1 and 2 summarize the datasets[3].

Table 1. Query set statistics

Query Set	#spectra	#proteins	#peptides
PPK (queries)	2995	35	1596
Mark12	9380	12	1944
Sigma49	12498	49	4560

Table 2. Protein database statistics

Protein DB	#proteins	#peptides	#relevant proteins
PPK (protein DB)	4534	325,812	35
M50000	50012	5,149,302	12
S50000	50049	2,571,642	49

5 Experiments

We conducted a comparative evaluation with controlled experiments for three models: prob-OR, prob-AND, and a standard VSM model (supported by the Lemur) which uses TF-IDF ("ltc") for within-document term weighting and cosine similarity for the scoring function. We name the last method "TFIDF-cosine". We also used the popular X!Tandem software (available online) to generate an alternative baseline.

5.1 Experimental Settings

To ensure a controlled setting, all the four methods share the same query generation process. We used the publicly available software of SEQUEST [4] and the

---

[2] http://expasy.org/sprot/

[3] Datasets will be made publicly available to support comparative evaluation and benchmarking at the following URL: http://nyc.lti.cs.cmu.edu/clair/datasets.htm

PeptideProphet[4] pipeline to predict peptides from MS/MS data, producing the queries shared by all the methods except X!Tandem for retrieving proteins. For the experiment with X!Tandem, we use the inbuilt peptide and protein identification tools in the open-source software package. When evaluating a method on one dataset, we used the remaining two datasets as the validation sets for tuning parameters. For example, when PPK is used as the test set, we tuned the smoothing method and μ (smoothing parameter) in prob-AND on Mark12+50000 and Sigma49+50000 as the validation datasets[5]. Based on the results, we chose the Dirichlet prior over Laplace as the smoothing method and μ=5000 as the smoothing parameter.

5.2 Metrics

The output of each method is a ranked list of predicted proteins for a pre-specified MS/MS dataset and a protein database. Applying a threshold to the ranked list of each method yielded binary decisions and shifting the threshold enables us to calculate precision values at different levels of recall. Using TP (true positives), FP (false positives), FN (false negatives) and TN (true negatives) to denote the counts of predictions in the four corresponding categories, the performance at a fixed threshold is measured as:

$$\text{Recall} = TP / (TP + FN),$$

$$\text{Precision} = TP / (TP + FP)$$

To evaluate the ranking ability of each method, we computed its average precision (over all recall levels) per query, and then the mean over all queries. This produces the standard MAP score for each method.

5.3 Main Results

The performance of the four methods in average precision is summarized in Table 3. The main observations are the following:

- Prob-OR had a relatively weak performance, with the MAP score significantly below the levels of all the other methods except X!Tandem. This observation supports our theoretical analysis (Sections 7) on the weakness of the protein scoring functions based on Boolean-OR in assembling peptide-level evidence – they are not sufficiently powerful for discriminating true positives from false positives.
- Prob-AND is among the two best methods (the other is TFIDF cosine) on average, with a MAP score of 0.71. It outperformed the prob-OR method significantly on all the datasets, successfully addressing the main weakness of the latter.

---

[4] PeptideProphet is a part of the TransProteomicPipeline, a publicly available software toolkit for protein identification at http://tools.proteomecenter.org/software.php

[5] Note that the 50000 proteins in Sigma49+50000 and Mark12+50000 are different independent samples drawn from Swissprot.

Table 3. Results summary in average precision (bold case indicates best performance)

Dataset	prob-AND	prob-OR	TFIDF cosine	X!Tandem
PPK	**0.87**	0.8	0.84	0.43
Mark12	0.77	0.66	**0.81**	0.41
Sigma49	0.48	0.44	**0.49**	0.241
MAP	**0.71**	0.63	**0.71**	0.36

- The TFIDF-cosine method performed equally well as Prob-AND. This is not surprising from the view point of text retrieval model analysis. It has been well-understood that the conventional vector space model (VSM) using cosine and TFIDF term weighting is a good approximation of language modeling with a multinomial assumption and the Dirichlet prior of corpus-level term distribution [18]. And the latter is the foundation of our prob-AND approach. On the other hand, it is the first time that the conventional VSM is examined in protein identification and compared with prob-AND. We are pleased to see both methods worked equally well on average, and both superior to prob-OR as a strong baseline in the computational proteomics literature.
- X!Tandem, one of the most popular publicly available protein identification program that is commonly used as a comparative baseline algorithm, performed inferior to the other methods on all three datasets in our experiments. It has been reported in the peptide/protein identification literature that X!Tandem differs from SEQUEST significantly in the identified peptides (and proteins). X!Tandem usually suffers with poor recall (sensitivity) in the peptide identification step as compared to SEQUEST based approaches [20], as a result of an aggressive thresholding strategy for computational efficiency and for reducing false alarms in the peptide identification step. Our results of X!Tandem agree with the previously reported findings in this sense.

5.4 Performance in High-Recall Regions

While average precision or MAP is well-accepted in evaluations of IR models, they may not be sufficiently informative for judging how much the protein identification systems would help biologists in reality. Notice that for biologists to verify the validity of the system-predicted proteins, wet-lab experiments would be needed and the cost would be much higher than what is required for a user to check through a ranked list of documents. In other words, dealing with a large number of false alarms would be too costly and hence impractical in proteomic data analysis. With this concern, we further analyze the performance of the methods in the high-recall (80%, 90% and 100%).

Table 4 shows the average numbers of false positives (FP) for each method at fixed levels of recall; the average is computed over the three datasets.

Table 4. Results summary in false positive counts (averaged over the 3 datasets) at fixed levels of recall

	Average Number of False Positives		
Recall	prob-AND	prob-OR	TFIDF-cosine
80%	28	52	28
90%	74	1002	96
100%	17746	16631	16586

It can be observed that all the methods achieved 80% recall with a relative small number of FP, which is quite encouraging. However, to achieve 90% recall, the FP number of prob-OR increased from 92 (at 80% recall) to 1002 which is unacceptably high, while prob-AND and TFIDF-cosine retain their low-FP behavior. At the 100% recall level, all the methods produced a large number of FP, which is not too surprising. X!Tandem did not reach any of the recall levels higher than 60% on all the 3 datasets, thus it is not included in the table.

5.5 Statistical Significance Tests

We conducted one-sample proportion tests for comparing the average precision scores of the protein identification methods. Table 5 summarizes the results.

Table 5. Significance test summary: each element in the matrix indicates the number of datasets (out of 3) on which System A significantly outperforms System B with a p-value < 0.01

A / B	prob-AND	prob-OR	TFIDF-cosine	X!Tandem
prob-AND		3	1	3
prob-OR	0		0	3
TFIDF-cosine	1	3		3
X!Tandem	0	0	0	

Comparing the two strongest methods, i.e., prob-AND and TFIDF-cosine, each of them significantly outperformed the other on one of the three datasets, and performed equally well on the remaining dataset. Comparing prob-OR with all the others, it significantly underperformed prob-AND and TFIDF-cosine on all three datasets. X!Tandem performance was inferior to all other approaches on all the datasets.

6 Conclusion and Future Work

In this paper, we present the first interdisciplinary investigation on how to leverage the rich research insights and successful techniques in IR to better solve the challenging problem of protein identification from tandem mass spectra. We formulated the problem (the mapping from system-predicted peptides to proteins) as an ad-hoc retrieval task, proposed a prob-AND model for combining peptide-level evidence in protein retrieval, and conducted a thorough evaluation of these models in comparison with a well-established method (prob-OR by Keller et al.) and a common baseline method (X!Tandem) in the field of protein-identification and a successful vector space model (TFIDF-cosine) in IR. The results are highly encouraging: we obtained significant performance improvements by the prob-AND models and the VSM model over the representative baseline methods. We hope this investigation provides useful information and insights for future research in adapting IR techniques to proteomic applications, and invites new ideas for further improvements from both the IR community and the computational proteomics community.

Several extensions of the presented work are possible, including modeling the queries as a mixture of proteins. Such approaches are likely to rely on sampling and greedy approximation strategies as explicitly modeling mixtures of thousands of proteins is computationally intractable. One such approach by Li et. al. [22] uses the Gibbs Sampling strategy to overcome the computational limitations. It might also be possible to reduce the search space of mixtures is by grouping proteins based on co-occurrences and modeling queries as mixture of such protein groups. We would like to explore such approaches in the future..Other important extensions of the presented work include addressing the issues caused by incorrect cleaving of protein sequences into peptides, leveraging n-gram peptides in extended protein profiles, and applying supervised or semi-supervised classification and functional analysis to predicted proteins in different types of MS/MS data samples, e.g., cancerous vs. normal. Also, Nesvizhskii et al. have found that using Expectation Maximization (EM) as an additional step for finding hidden groups of proteins and for dealing with degenerate peptides can improve the performance of the prob-OR method. That suggests a potential way to further improve prob-AND and the other methods similarly by deploying the additional EM step, which is an interesting topic for future research.

Acknowledgments

We thank the Mass Spectrometry Research Center at the Vanderbilt University for providing the Mark12 and Sigma49 query sets, and many useful suggestions on processing the data as well as insightful analysis on related literature. This work is supported in parts by the National Science Foundation (NSF) under grants EIA-0225656 and IIS-0704689. Any opinions, findings and conclusions or recommendations expressed in this material are those of the authors and do not necessarily reflect the views of the sponsors.

References

[1] Bafna, V., Edwards, N.: SCOPE: a probabilistic model for scoring tandem mass spectra against a peptide database. Bioinformatics 17(suppl. 1), S13–S21 (2001)

[2] Craig, R., Beavis, R.C.: TANDEM: matching proteins with tandem mass spectra. Bioinformatics 20(9), 1466–1467 (2004)

[3] Perkins, D.N., Pappin, D.J.C., Creasy, D.M., Cottrell, J.S.: Probability based protein identification by searching sequence databases using mass spectrometry data. Electrophoresis 20, 3551–3567 (1999)
[4] Eng, J.K., McCormack, A.L., Yates III, J.R.: An approach to correlate tandem mass spectral data of peptides with amino acid sequences in a protein database. J. Am. Soc. Mass Spectrum 5, 976–989 (1994)
[5] Friedman, T., Razumovskaya, J., Verberkmoes, N., Hurst, G., Protopopescu, V., Xu, Y.: The probability distribution for a random match between an experimental-theoretical spectral pair in tandem mass spectrometry. J. Bioinformatics and Computational Biology 3(2), 455–476 (2005)
[6] Keller, A., Nesvizhskii, A.I., Kolker, E., Aebersold, R.: Empirical statistical model to estimate the accuracy of peptide identifications made by MS/MS and database search. Analytical Chemistry 74, 5383–5392 (2002)
[7] Sadygov, R., Yates III, J.: A Hypergeometric probability model for protein identification and validation using tandem mass spectral data and protein sequence databases. Anal. Chem. 75, 3792–3798 (2003)
[8] Zhang, N., Li, X.J., Ye, M., Pan, S., Schwikowski, B., Aebersold, R.: ProbIDtree: an automated software program capable of identifying multiple peptides from a single collision-induced dissociation spectrum collected by a tandem mass spectrometer. Proteomics 5(16), 4096–4106 (2005)
[9] `http://www.lemurproject.org/indri/`
[10] `http://www.lemurproject.org/`
[11] MacCoss, M.J., Wu, C.C., Yates III, J.R.: Probability-based validation of protein identifications using a modified SEQUEST algorithm. Analytical Chemistry 74, 5593–5599 (2002)
[12] Nesvizhskii, A.I., Keller, A., Kolker, E., Aebersold, R.: A statistical model for identifying proteins by Tandem mass spectrometry. Analytical Chemistry 75, 4646–4658 (2003)
[13] Moore, R.E., Young, M.K., Lee, T.D.: QScore: An algorithm for evaluating SEQUEST database search results. Journal of the American Society for Mass Spectrometry 13(4), 378–386 (2002)
[14] Eddes, J.S., Kapp, E.A., Frecklington, D.F., Connolly, L.M., Layton, M.J., Moritz, R.L., Simpson, R.: CHOMPER: a bio-informatics tool for rapid validation of tandem mass spectrometry search results associated with high-throughput proteomic strategies. Proteomics 2(9), 1097–1103 (2002)
[15] Han, D.K., Eng, J., Zhou, H., Aebersold, R.: Quantitative profiling of differentiation induced microsomal proteins using isotope-coded affinity tags and mass spectrometry. Nature Biotechnology 19(10), 946–951 (2001)
[16] Tabb, D.L., Hayes MacDonald, W., Yates III, J.R.: DTASelect and Contrast: Tools for assembling and comparing protein identifications from shotgun proteomics. Journal of Proteome Research 1(1), 21–26 (2002)
[17] Field, H.I., Fenyo, D., Beavis, R.C.: RADARS, a bio-informatics solution that automates proteome mass spectral analysis, optimizes protein identification, and archives data in a relational database. Proteomics, 36–47 (2002)
[18] Zhai, C., Lafferty, J.: A study of smoothing methods for language models applied to ad hoc information retrieval. In: Proceedings of ACM SIGIR 2001, pp. 334–342 (2001)
[19] Purvine, S., Picone, A.F., Kolker, E.: Standard Mixtures for Proteome Studies. OMICS 1(1), 79–92 (2004)

[20] Kapp, E.A., Schütz, F., Connolly, L.M., Chakel, J.A., Meza, J.E., Miller, C.A., Fenyo, D., Eng, J.K., Adkins, J.N., Omenn, G.S., Simpson, R.J.: An evaluation, comparison, and accurate benchmarking of several publicly available MS/MS search algorithms: Sensitivity and specificity analysis. Proteomics 5(13), 3475–3490
[21] Craig, R., Beavis, R.C.: Tandem: Matching Proteins with mass spectra. Bioinformatics 20, 1466–1467 (2004)
[22] Li, Y.F., Arnold, R.J., Li, Y., Radivojac, P., Sheng, Q., Tang, H.: A Bayesian Approach to Protein Inference Problem in Shotgun Proteomics. In: Vingron, M., Wong, L. (eds.) RECOMB 2008. LNCS (LNBI), vol. 4955, pp. 167–180. Springer, Heidelberg (2008)

Causality Discovery with Additive Disturbances: An Information-Theoretical Perspective

Kun Zhang[1] and Aapo Hyvärinen[1,2]

[1] Dept of Computer Science & HIIT, University of Helsinki, Finland
[2] Dept of Mathematics and Statistics, University of Helsinki, Finland

Abstract. We consider causally sufficient acyclic causal models in which the relationship among the variables is nonlinear while disturbances have linear effects, and show that three principles, namely, the causal Markov condition (together with the independence between each disturbance and the corresponding parents), minimum disturbance entropy, and mutual independence of the disturbances, are equivalent. This motivates new and more efficient methods for some causal discovery problems. In particular, we propose to use multichannel blind deconvolution, an extension of independent component analysis, to do Granger causality analysis with instantaneous effects. This approach gives more accurate estimates of the parameters and can easily incorporate sparsity constraints. For additive disturbance-based nonlinear causal discovery, we first make use of the conditional independence relationships to obtain the equivalence class; undetermined causal directions are then found by nonlinear regression and pairwise independence tests. This avoids the brute-force search and greatly reduces the computational load.

1 Introduction

Given some observed variables, scientists, engineers, and policy-makers often wish to find their causal relations, as well as to understand how to control a particular variable by manipulating others. Discovering causal relations from non-experimental data has attracted the interests of researchers in many areas, such as philosophy, psychology, machine learning, etc [13,19]. In this paper we focus on the causally sufficient acyclic causal models [13] of continuous variables. That is, we assume that there are no confounders nor any feedback in the causal relations.

There are some frequently-used models for acyclic causal discovery, and traditionally they are estimated with different principles. For example, for Gaussian variables with linear relations, conditional independence between the variables allows one to find a set of acyclic causal models which are in the d-separation equivalence class [13]. Generally speaking, with more specific information about the disturbance distribution or the causal structure, one can find the underlying causal model more accurately. Based on the independent component analysis (ICA [8]) technique, a class of linear, non-Gaussian, and acyclic models (LiNGAM) can be estimated very efficiently [18]. Moreover, in economics,

W. Buntine et al. (Eds.): ECML PKDD 2009, Part II, LNAI 5782, pp. 570–585, 2009.

Granger causality analysis [4] is a popular way to examine the causal relations between times series. It exploits the temporal constraint that causes must precede effects and uses the vector auto-regression (VAR) for parameter estimation.

In this paper, we consider a large class of acyclic causal models in which the causal relations among observed variables are nonlinear but the effect of disturbances is linear, as extensions of the linear models mentioned above. We show that for such causal models, mutual independence of the disturbances is equivalent to conditional independence of observed variables (as well as the independence between the disturbance and the parents affecting the same variable). Furthermore, they are achieved if and only if the total entropy of the disturbances is minimized. The three criteria above, namely, conditional independence of variables (together with the independence between each disturbance and the corresponding parents), mutual independence of disturbances, and minimum disturbance entropy, can all be exploited to estimate such causal models. In practice, which one should be chosen depends on the problem at hand.

We then consider two causal discovery problems, and show how our results help solve them efficiently. One is discovery of Granger causality with instantaneous effects. Previous methods consist of two separate steps: the first step performs ordinary Granger causality analysis by using VAR's, and the second step finds the instantaneous causal relations [16,10]. Although these methods are consistent in large samples, they are not efficient, because the Gaussianity assumption for the innovations made in the first step is usually not true. We propose a more efficient approach to estimate this model by making the disturbances mutually independent, as achieved by multichannel blind deconvolution (MBD) [1], an extension of ICA.

The second problem on which we apply our theory is nonlinear causal discovery with additive disturbances in the case of more than two variables. The existing approach requires an exhaustive search over all possible causal structures and testing if the disturbances are mutually independent [6]. It becomes impractical when we have more than three or four variables. The proposed approach, which can easily solve this problem with tens of variables, consists of two stages. First, using nonlinear regression and statistical independence tests, one can find the d-separation equivalence class. Next, among all possible causal models in the equivalence class, the one consistent with the data can be found by examining if the disturbance is independent of the parents for each variable.

2 Equivalence of Three Estimation Principles for Acyclic Causal Models

In this section we consider a kind of acyclic data generating processes in which each variable is generated by a nonlinear function of its parents plus the disturbance. Such processes can be represented graphically by a directed acyclic graph (DAG). Mathematically, each of the observed variables x_i, $i = 1, \cdots, n$, is written as

$$x_i = f_i(pa_i) + e_i, \tag{1}$$

where pa_i denotes the parents of x_i, and the disturbances e_i are independent from each other.

A well-known approach to identify the acyclic causal relations is based on a test called d-separation, which examines conditional independence between variables [13,19]. In the following theorem, we show that mutual independence of the disturbances and conditional independence between observed variables (together with the independence between e_i and pa_i) are equivalent. Furthermore, they are achieved if and only if the total entropy of the disturbances is minimized.

Theorem 1. *Assume that the data $x_1, \cdots, x_n$ are causally sufficient, and were generated by the acyclic causal model in Eq. 1. Then, when fitting the model Eq. 1 with the causal structure represented by a DAG to $x_1, \cdots, x_n$, the following three properties are equivalent:*

(i) The causal Markov condition holds (i.e., each variable is independent of its non-descendants in the DAG conditional on its parents), and in addition, the disturbance in x_i is independent from the parents of x_i[1].
(ii) The total entropy of the disturbances, i.e., $\sum_i H(e_i)$, is minimized, with the minimum $H(x_1, \cdots, x_n)$.
(iii) The disturbances e_i are mutually independent.

See Appendix for a proof. From this theorem, one can see that $\sum_{i=1}^n H(e_i)$ can be used as a measure to compare the quality of different acyclic causal model. Generally, we prefer the model with the smallest $\sum_{i=1}^n H(e_i)$. This is intuitively appealing from a physics viewpoint. In physics, it is often claimed that causality increases entropy [3]. Therefore, the disturbances e_i, which are pure causes of the observed variables x_i, should have the least total entropy, compared to any observed variables they generate. Another interesting point is that although the causal relations among x_i increases the total entropy of the observed variables, i.e., $\sum_i H(x_i)$, the joint entropy of the x_i remains the same, with the value $\sum_i H(e_i)$. In addition, the property (*ii*) for the acyclic causal model relates causality and predictability from an information theory viewpoint. The causal relations give the best prediction all variables in the system in an acyclic manner, since the uncertainty (measured by entropy) in all errors e_i is minimized.

All the three criteria in Theorem 1 can be used for causal discovery. However, for a given problem, they may result in different techniques and involve different computational loads. It should be noted that when using the minimum disturbance entropy criterion, one needs to constrain the causal models to be acyclic; otherwise, minimization of this criterion tends to introduce cyclic relations into the model, to reduce the magnitude of the disturbances.

It should be noted that the theorem does not state when the solution of the problem is unique. It is well-known that for Gaussian variables, there may be several solutions providing independent disturbances, and thus minimum entropy.

---

[1] The property that the disturbance in x_i is independent from the parents of x_i is trivial in the linear Gaussian case, since using linear regression, the noise (disturbance) is uncorrelated from the explanatory variables (parents), and uncorrelatedness is equivalent to independence for jointly Gaussian variables.

Finding exact conditions for uniqueness is an important problem for future research. For example, non-Gaussianity, or temporal order, have been shown to make the solution unique in the linear case [18,4].

3 On Granger Causality with Instantaneous Effects

Granger causality [4] exploits the temporal information that the cause occurs before the effect, such that the directions of the possible causal effects are known. Consequently, it can be determined uniquely. A stationary process $X_1 : \{x_{1t}\}$ is said to Granger cause the process $X_2 : \{x_{2t}\}$ if it contains information about the predictability for $x_{2,t+1}$ contained nowhere else in some large information set, which includes $x_{1,t-k}, k \geq 0$ [4]. Using the language of conditional independence, this means that X_1 is not conditionally independent of $x_{2,t+1}$ given the large information set. In this sense, Granger causality is a kind of conditional independence-based causality combined with the constraint that effects must follow causes. Here we would like to treat $x_{i,t}$ as random variables, and use a DAG to represent the possible causal relations among them. If there exist significant causal relations $x_{i,t-k} \rightarrow x_{j,t}$ ($k > 0$, and $i \neq j$), then the process X_i Granger causes X_j.

As the causal relations among $x_{i,t}$ are linear and acyclic, the three properties in Theorem 1 can all be used to estimate this model. Conventionally, all of $e_{i,t}$ are treated as Gaussian, and minimization of the disturbance entropy is reduced to minimizing the prediction error (in the mean square error sense), as achieved by VAR's. It should be noted that when e_{it} are not Gaussian, although the estimate given by VAR's is consistent in large samples, it is not efficient. With Granger causality analysis, it is sometimes observed that there is significant contemporaneous dependency between the innovations e_i. This means that there are some instantaneous relations among x_{it}, which cannot be captured by traditional Granger causality.

3.1 Granger Causality with Instantaneous Effects

Mathematically, Granger causality analysis of $x_{1t}, \cdots, x_{nt}$ with instantaneous effects can be formulated as

$$\mathbf{x}_t = \sum_{\tau=0}^{p} \mathbf{B}_\tau \mathbf{x}_{t-\tau} + \mathbf{e}_t, \tag{2}$$

where $\mathbf{x}_t = (x_{1t}, \cdots, x_{nt})^T$, $\mathbf{e}_t = (e_{1t}, \cdots, e_{nt})^T$, and $\mathbf{B}_\tau$ are $n \times n$ matrix of coefficients. Here we have assumed that all involved random variables have been made zero-mean. We also assume that the instantaneous causal relations, which are implied in $\mathbf{B}_0$, are acyclic. That is, $\mathbf{B}_0$ can be transformed to a strictly lower-triangular matrix by simultaneous equal row and column permutations [18]. Equivalently, the representation Eq. 2 can be written as

$$(\mathbf{I} - \mathbf{B}_0)\mathbf{x}_t = \sum_{\tau=1}^{p} \mathbf{B}_\tau \mathbf{x}_{t-\tau} + \mathbf{e}_t. \tag{3}$$

3.2 Existing Methods

Existing methods for Granger causality analysis with instantaneous effects consist of two steps [10,16]. Multiplying both sides of Eq. 3 by $(\mathbf{I}-\mathbf{B}_0)^{-1}$ from the left, one can get

$$\mathbf{x}_t = \sum_{\tau=1}^{p} (\mathbf{I}-\mathbf{B}_0)^{-1} \cdot \mathbf{B}_\tau \cdot \mathbf{x}_{t-\tau} + (\mathbf{I}-\mathbf{B}_0)^{-1} \cdot \mathbf{e}_t. \tag{4}$$

This is exactly the Granger causality model without instantaneous effects with the errors $(\mathbf{I}-\mathbf{B}_0)^{-1} \cdot \mathbf{e}_t$. Therefore, one can first find $(\mathbf{I}-\mathbf{B}_0)^{-1} \cdot \mathbf{B}_\tau$, $\tau = 1, \cdots, p$, and $(\mathbf{I}-\mathbf{B}_0)^{-1} \cdot \mathbf{e}_t$, by ordinary VAR analysis. In the second step, one needs to estimate $\mathbf{B}_0$ by examining (the estimate of) the errors $(\mathbf{I}-\mathbf{B}_0)^{-1} \cdot \mathbf{e}_t$. One way is based on conditional independence graphs [16]. It is a combination of VAR for lagged causality and conditional independence-based method [13] for instantaneous causality. Due to the Gaussianity assumption, this method produces a distribution-equivalence class of causal models. Another way, recently proposed in [10], resorts to the ICA-based LiNGAM analysis [18]. The method is very easy to implement, and also consistent in large samples. But when at most one of the disturbance sequence is Gaussian, it is not efficient due to the wrong assumption of Gaussianity of the disturbances in the first step and the error accumulation of the two-step method.

3.3 Estimation by Multichannel Blind Deconvolution

According to the causal model Eq. 2, the causal relations among random variables x_{it} are linear and acyclic. According to Theorem 1, such a causal model can be estimated by making the disturbances e_{it} mutually independent for different i and different t. That is, we need to make e_{it}, which are a mixed and filtered version of $\mathbf{x}_t$, both spatially and temporally independent. Estimation of the model Eq. 3 (or equivalently, Eq. 2) is then closely related to the multichannel blind deconvolution (MBD) problem with causal finite impulse response (FIR) filters [1,8]. MBD, as a direct extension of ICA [8], assumes that the observed signals are convolutive mixtures of some spatially and independently and identically distributed (i.i.d.) sources. Under the assumption that at most one of the sources is Gaussian, by making the estimated sources spatially and temporally independent, MBD can recover the mixing system (here corresponding to e_{it} and $\mathbf{B}_\tau$) up to some scaling, permutation, and time shift indeterminacies [11]. This implies that Granger causality with instantaneous effects is identifiable if at most one of the disturbances e_i is Gaussian.

In Eq. 2, the observed variables x_{it} can be considered as convolutive mixtures of the disturbances e_{it}. We aim to find the estimate of $\mathbf{B}_\tau$, as well as e_{it}, in Eq. 2, by MBD with the filter matrix $\mathbf{W}(z) = \sum_{\tau=0}^{p} \mathbf{W}_\tau z^{-\tau}$ ($\mathbf{W}_\tau$ are $n \times n$ matrices):

$$\hat{\mathbf{e}}_t = \sum_{\tau=0}^{p} \mathbf{W}_\tau \mathbf{x}_{t-\tau}. \tag{5}$$

There exist several well-developed algorithms for MBD. For example, one may adopt the one based on natural gradient [1]. Comparing Eq. 5 and Eq. 3, one can see that the estimate of $\mathbf{B}_\tau$ ($\tau \geq 0$) can be constructed by analyzing $\mathbf{W}_\tau$: by extending the LiNGAM analysis procedure [18], we can find the estimate of $\mathbf{B}_\tau$ in the following three steps, based on the MBD estimates of $\mathbf{W}_\tau$.

1. Find the permutation of rows of $\mathbf{W}_0$ which yields a matrix $\widetilde{\mathbf{W}}_0$ without any insignificant entries on the main diagonal. Note that here we also need to apply the same permutations to rows of $\mathbf{W}_\tau$ ($\tau > 0$) to produce $\widetilde{\mathbf{W}}_\tau$.
2. Divide each row of $\widetilde{\mathbf{W}}_0$ and $\widetilde{\mathbf{W}}_\tau$ ($\tau > 0$) by the corresponding diagonal entry in $\widetilde{\mathbf{W}}_0$. This gives $\widetilde{\mathbf{W}}'_0$ and $\widetilde{\mathbf{W}}'_\tau$, respectively. The estimate of $\mathbf{B}_0$ and $\mathbf{B}_\tau$ ($\tau > 0$) can be computed as $\widehat{\mathbf{B}}_0 = \mathbf{I} - \widetilde{\mathbf{W}}'_0$ and $\widehat{\mathbf{B}}_\tau = -\widetilde{\mathbf{W}}'_\tau$, respectively.
3. To obtain the causal order in the instantaneous effects, find the permutation matrix $\mathbf{P}$ (applied equally to both rows and columns) of $\widehat{\mathbf{B}}_0$ which makes $\widetilde{\mathbf{B}}_0 = \mathbf{P}\widehat{\mathbf{B}}_0\mathbf{P}^T$ as close as possible to strictly lower triangular.

3.4 Sparsification of the Causal Relations

For the interpretation or generalization purpose, we need to do model selection on the causal structure (i.e., to set insignificant entries of $\hat{\mathbf{B}}_\tau$ to zero, and to determine p, if needed). This is difficult to do in the two-step methods [16,10], but is easily achieved in our method. Analogously to the development of ICA with sparse connections [20], we can incorporate the adaptive L_1 penalties into the likelihood of the MBD model to achieve fast model selection. More importantly, the model selection result obtained by this approach is consistent with that by traditional information criteria, such as BIC [17][2]. To make $\mathbf{W}_\tau$ in Eq. 5 as sparse as possible, we maximize the penalized likelihood

$$pl(\{\mathbf{W}_\tau\}) = l(\{\mathbf{W}_\tau\}) - \lambda \sum_{i,j,\tau} |w_{i,j,\tau}|/|\hat{w}_{i,j,\tau}|, \tag{6}$$

where $l(\{\mathbf{W}_\tau\})$ is the likelihood, $w_{i,j,\tau}$ the (i,j)th entry of $\mathbf{W}_\tau$, and $\hat{w}_{i,j,\tau}$ a consistent estimate of $w_{i,j,\tau}$, such as the maximum likelihood estimate. To achieve BIC-like model selection, one can set $\lambda = \log T$, where T is the sample size.

3.5 Simulation

To investigate the performance of our method, we conducted a series of simulations. We set $p = 1$ lag and the dimensionality $n = 5$. We randomly constructed the strictly lower-triangular matrix $\mathbf{B}_0$ and matrix $\mathbf{B}_1$. About 60% of the entries

[2] Note that traditional model selection based on the information criteria involves a combinatorial optimization problem, whose complexity increases exponentially in the dimensionality of the parameter space. In the MBD problem, it is not practical to set insignificant entries of $\hat{\mathbf{B}}_\tau$ to zero by directly minimizing the information criteria, as the number of parameters is too large.

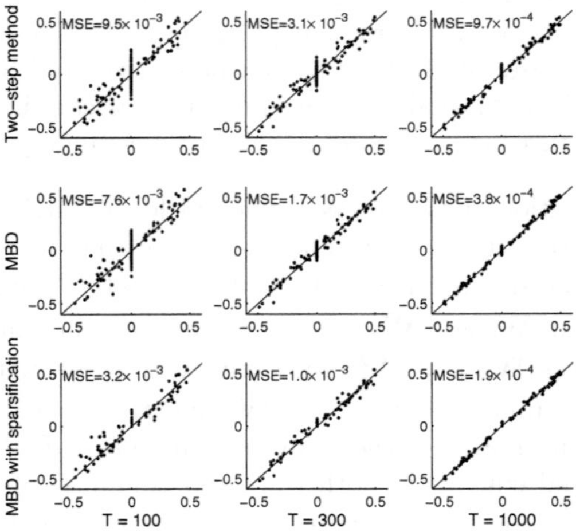

Fig. 1. Scatter plots of the estimated coefficients (y axis) versus the true ones (x axis) for different sample sizes and different methods

in these matrices were set to zero, while the magnitude of the others is uniformly distributed between 0.05 and 0.5 and the sign is random. The disturbances e_{it} were generated by passing i.i.d. Gaussian samples through a power nonlinearity with exponent between 1.5 and 2.0 (the original sign was kept). The observations $\mathbf{x}_t$ were then generated according to Eq. 4. Various sample sizes ($T = 100$, 300, and 1000) were tested. We compared the performance of the two-step method proposed in [10], the method by MBD (Section 3.3) and the MBD-based method with the sparsity constraint (Section 3.4). In the last method, we set the penalization parameter in Eq. 6 as $\lambda = \log T$ to make its results consistent with those obtained by BIC. In each case, we repeated the experiments for 5 replications.

Fig. 1 shows the scatter plots of the estimated parameters (including the strictly lower triangular part of $\mathbf{B_0}$ and all entries of $\mathbf{B}_1$) versus the true ones. Different subplots correspond to different sample sizes or different methods. The mean square error (MSE) of the estimated parameters is also given in each subplot. One can see that as the sample sizes increases, all methods give better results. For each sample size, the method based on MBD is always better than the two-step method, showing that the estimate by the MBD-based method is more efficient. Furthermore, due to the prior knowledge that many parameters are zero, the MBD-based method with the sparsity constraint behaves best.

3.6 Application in Finance

In this section we aim at using Granger causality analysis with instantaneous effects to find the causal relations among several world stock indices. The chosen indices are Dow Jones Industrial Average (DJI) in USA, Nikkei 225 (N225) in Japan, Hang Seng Index (HSI) in Hong Kong, and the Shanghai Stock Exchange

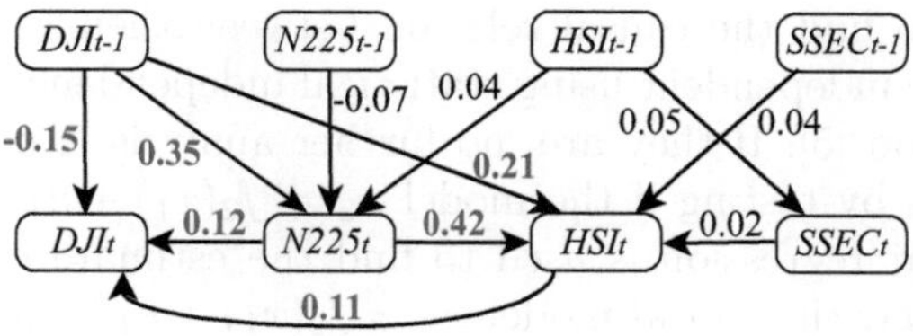

Fig. 2. Result of applying Granger causality analysis with instantaneous effects to daily returns of the stock indices DJI, N225, HSI, and SSEC, with $p = 1$ lag. Large coefficients (greater than 0.1) are shown in bold.

Composite Index (SSEC) in China. We used the daily dividend/split adjusted closing prices from Dec. 4, 2001 to Jul. 11, 2006, obtained from the Yahoo finance database. For the few days when the price is not available, we use simple linear interpolation to estimate the price. Denoting the closing price of the ith index on day t by P_{it}, the corresponding return is calculated by $x_{it} = \frac{P_{it} - P_{i,t-1}}{P_{i,t-1}}$. The data for analysis are $\mathbf{x}_t = [x_{1t}, ..., x_{4,t}]^T$, with 1200 samples.

We applied the MBD-based method with the sparsity constraint to $\mathbf{x}_t$. The kurtoses of the estimated disturbances $\hat{e}_{it}$ are 3.9, 8.6, 4.1, and 7.6, respectively, implying that the disturbances are non-Gaussian. We found that more than half of the coefficients in the estimated $\mathbf{W}_0$ and $\mathbf{W}_1$ are zero. $\widehat{\mathbf{B}}_0$ and $\widehat{\mathbf{B}}_0$ were constructed based on $\mathbf{W}_0$ and $\mathbf{W}_1$, using the procedure given in Section 3.3. It was found that $\widehat{\mathbf{B}}_0$ can be permuted to a strictly lower-triangular matrix, meaning that the instantaneous effects follow a linear acyclic causal model. Finally, based on $\widehat{\mathbf{B}}_0$ and $\widehat{\mathbf{B}}_1$, one can plot the causal diagram, as shown in Fig. 2.

Fig. 2 reveals some interesting findings. First, DJI_{t-1} has significant impacts on N225_t and HSI_t, which is a well-known fact in the stock market. Second, the causal relations $\text{DJI}_{t-1} \to \text{N225}_t \to \text{DJI}_t$ and $\text{DJI}_{t-1} \to \text{HSI}_t \to \text{DJI}_t$ are consistent with the time difference between Asian and USA. That is, the causal effects from N225_t and HSI_t to DJI_t, although seeming to be instantaneous, are actually mainly caused by the time difference. Third, unlike SSEC, HSI is very sensitive to others; it is even strongly influenced by N225, another Asian index. Fourth, it may be surprising that there is a significant negative effect from DJI_{t-1} to DJI_t; however, it is not necessary for DJI_t to have significant negative autocorrelations, due to the positive effect from DJI_{t-1} to DJI_t going through N225_t and HSI_t.

4 Additive Disturbance-Based Nonlinear Causal Discovery with More Than Two Variables

The additive disturbance causal model Eq. 1 has been proposed for nonlinear causal discovery very recently by Hoyer et al. [6]. They mainly focused on the two-variable case and showed that the model is able to distinguish the cause from effect for some real-world data sets. Suppose we have two variables x_1 and

x_2. The method to find the causal relation between them is as follows. First, examine if they are independent using statistical independence tests, such as the kernel-based method [5]. If they are, no further analysis is needed. Otherwise, one then continues by testing if the model $x_2 = f_2(x_1) + e_2$ is consistent with the data. Nonlinear regression is used to find the estimate of f_2 and e_2. If $\hat{e}_2$ is independent of x_1, the causal model $x_2 = f_2(x_1) + e_2$ is accepted; otherwise it is rejected. One then needs to test if the reverse model $x_1 = f_1(x_2) + e_1$ is acceptable. Finally, if one model is accepted and the other is rejected, the additive disturbance nonlinear causal model is uniquely found. If both models hold, one can conclude that in this situation the additive disturbance causal model cannot distinguish the cause from effect. If neither of the models holds, one can conclude that the model cannot explain the data well, possibly due to the existence of hidden variables or a different data generating process. When performing nonlinear regression, one should carefully avoid over-fitting.

4.1 Existing Methods

It is much more difficult to find the causal relations implied by the causal model Eq. 1 when we have more than two observed variables. In this case, a brute-force search was exploited in [5]; for each possible acyclic causal structure, represented by a DAG, one performs nonlinear regression of each variable on its parents, and tests if the residuals are *mutually* independent with statistical independence tests. The simplest causal model which gives mutually independent disturbances is preferred. Clearly this approach may encounter two difficulties. One is that the test of mutual independence is difficult to do when we have many variables. The other is that the search space of all possible DAG's increases too rapidly with the variable number. In fact, it is well-known that the total number of all possible DAG's is super-exponential in the number of variables. Consequently, this approach involves high computational load, and is not suitable when we have more than three or four variables.

4.2 A More Practical Approach

We propose an approach which is suitable for identifying the nonlinear causal model Eq. 1 with moderate-sized variables, say, with tens of variables. According to Theorem 1, the nonlinear causal model Eq. 1 can be identified by enforcing the causal Markov property and the independence between the disturbance and the parents associated with the same variable. This motivates a two-stage approach to identify the whole causal model. One can first use conditional independence-based methods to find the d-separation equivalence class. Next, the nonlinear causal model Eq. 1 is used to identify the causal relations that cannot be determined in the first step: for each possible causal model contained in the equivalence class, we estimate the disturbances, and determine if this model is plausible, by examining if the disturbance in each variable x_i is independent of the parents of x_i[3]. In this way, one avoids the exhaustive search over all possible causal structures and statistical tests of mutual independence of more than two variables.

In the first stage of the proposed approach, we need to find an efficient way to derive the conditional independence relationships and to construct the equivalence class. This brings up two issues. One is how to reliably test conditional independence for variables with nonlinear relations. Traditionally, in the implementation of most conditional independence-based causal discovery algorithms, such as PC [19], it is assumed that the variables are either discrete or Gaussian with linear causal relations. This assumption greatly simplifies the difficulty in conditional independent tests. But here we need to capture the nonlinear causal effects. Some methods, such as the probabilistic non-parametric test proposed in [12], have been developed for this task. However, they may be unreliable when the conditional set contains many variables, due to the curse of dimensionality. Alternatively, one may simplify the conditional independence test procedure by making use of the particular structure of the nonlinear causal model Eq. 1. This can be done by extending the partial correlation concept to the nonlinear case.

Partial correlation measures the degree of linear association between two variables, with the effect of a set of conditional variables removed. In particular, the partial correlation between X and Y given a set of variables $\mathbf{Z}$, denoted by $\rho_{XY \cdot \mathbf{Z}}$, is the correlation between the residuals R_X and R_Y resulting from the linear regression of X with $\mathbf{Z}$ and of Y with $\mathbf{Z}$, respectively. Here we assume that the data follow the nonlinear generating process Eq. 1. Due to the additive disturbance structure, one can examine if X and Y are conditionally independent given the variable set $\mathbf{Z}$ by performing independent tests on the residuals $R_X^{\mathcal{N}}$ and $R_Y^{\mathcal{N}}$, which are obtained by *nonlinear* regression of X with $\mathbf{Z}$ and of Y with $\mathbf{Z}$, respectively. In our implementation, Gaussian process regression with a Gaussian kernel [15] is adopted for nonlinear regression, and the involved hyperparameters are learned by maximizing the marginal likelihood. The kernel-based independence test [5] with the significance level 0.01 is then used to test if the residuals are independent.

The other issue is how to construct the independence-based equivalence class with as few conditional independence tests as possible. We adopt the total conditioning scheme discussed in [14], which was shown to be very efficient provided the underlying graph is sparse enough. It first finds the Markov blanket of each variable and builds the moral graph. The Markov blanket of the variable X is the set of parents, children, and children's parents (spouses) of X. Let $\mathbf{V}$ be the set of all variables. The variable Y is in the Markov blanket of X if X and Y are not conditionally independent given $\mathbf{V} \setminus \{X, Y\}$. In particular, in our case we use nonlinear regression combined with the kernel-based independence test to perform the conditional independence test, as discussed above. Next, it removes the possible spouse links between linked variables X and Y by looking

[3] According to Theorem 1, one can use the total entropy of the disturbances as the criterion to find the "best" causal model in the equivalence class. However, this approach does not easily provide a criterion for testing model validity.

for a d-separating set around X and Y. When spouse links are removed, the V-structures can be oriented using collider sets. One can then find the equivalence class by propagating orientation constraints. For details of the total conditioning scheme for causal discovery based on Markov blankets, see [14]. In order to construct the moral graph, one needs to perform nonlinear regressions for $n(n-1)$ times and independence tests for $\frac{n(n-1)}{2}$ times, where n is the number of observed variables. To find the possible spouse links, one further needs to do nonlinear regressions for $2^{\alpha+1}$ times and corresponding independence tests for 2^{α} times, where $\alpha = \max_{X,Y} |\mathbf{Tri}(X-Y)|$, with $\mathbf{Tri}(X-Y)$ denoting the set of variables forming a triangle with X and Y.

Finally, for each DAG in the equivalence class, we use nonlinear regression to estimate the disturbances and then test if the disturbance and parents are independent for each variable. Those that make each disturbance independent of the parents associated with the same variable are valid models.

4.3 Simulation

In this section we investigate how the proposed approach behaves with a simulation study. The data generating process is given in Fig. 3. It consists of seven variables with both linear and strongly nonlinear causal relations, and the disturbances are Gaussian, uniform, or super-Gaussian. The sample size is 1000.

We first used nonlinear regression and independence test to construct the moral graph, with the result shown in Fig. 4(a). One can see that it is exactly the moral graph corresponding to the causal model generating the data. The edge $x_2 - x_4$ was then found to be a spouse link, since they are (unconditionally) independent. Consequently, x_3 and x_7 are colliders and thus common children of x_2 and x_4. That is, we have $x_2 \rightarrow x_3 \leftarrow x_4$ and $x_2 \rightarrow x_7 \leftarrow x_4$. Furthermore, since x_5 (x_6) is not connected to x_4 in the moral graph, one can find the orientation $x_3 \rightarrow x_5$ ($x_3 \rightarrow x_6$). To avoid cyclicity, the causal direction between x_2 and x_5 must be $x_2 \rightarrow x_5$. The resulting equivalence class is shown in Fig. 4(b), with only the causal direction between x_1 and x_2 and that between x_3 and x_7 are

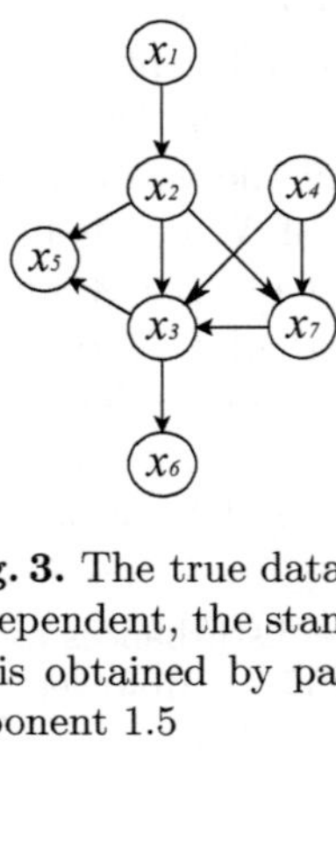

Data-generating process:
$x_1 = v_1$, with normal v_1;
$x_2 = x_1/2 + x_1^3/10 + v_2$, with normal v_2;
$x_4 = v_4$, with normal v_4;
$x_3 = x_2 + x_4 + v_3$, with super-Gaussian v_3;
$x_5 = x_2 x_3 + v_5$, with uniform v_5;
$x_6 = x_3/2 + x_3^3/10 + v_6$, with normal v_6;
$x_7 = |x_3|^{1.3}(1 + x_2 + \tanh(x_4)) + v_7$, with normal v_7.

Fig. 3. The true data-generating model used in the simulation, where v_i are mutually independent, the standard deviation of v_i is a random number between 0.2 and 1, and v_3 is obtained by passing a Gaussian variable through the power nonlinearity with exponent 1.5

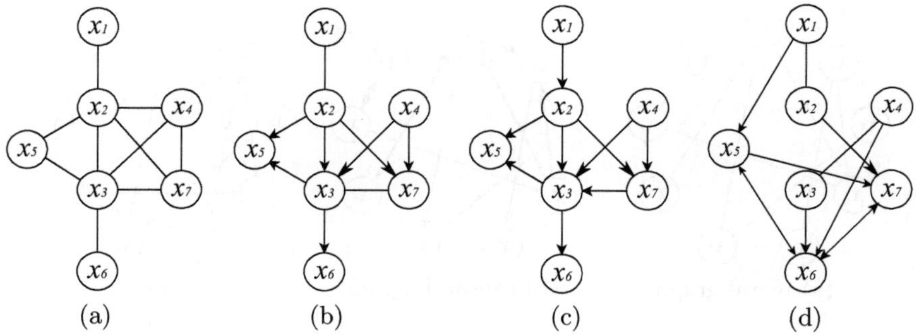

Fig. 4. Simulation results of nonlinear causal discovery with additive disturbances. (a) The moral graph obtained by our approach (the significance level for independence tests is 0.01). (b) The graph after removing spouse link ($x_2 - x_4$), orienting V-structures, and propagating orientation constraints. (c) The final result. (d) For comparison, the result obtained by the PC algorithm, with the significance level 0.01.

not determined. Under the hypothesis $x_3 \leftarrow x_7$, we found that the disturbance estimated by nonlinear regression is independent of the assumed parents (x_2, x_4, and x_7), while the disturbance is not independent of the parents for the variable x_7 under the hypothesis $x_3 \rightarrow x_7$, so we obtained $x_3 \leftarrow x_7$. Similarly, one can find the causal direction $x_1 \rightarrow x_2$. The obtained causal model is given in Fig. 4(c), which turns out to be the same as the one generating the data (Fig. 3). For comparison, we also show the equivalence class obtained by the PC algorithm [19] implemented in Tetrad[4] with the significance level 0.01. One can see that since the linearity assumption in PC is violated in this case, the resulting equivalence class is significantly different from the true causal model; in fact, half of the edges are spurious.

4.4 Application in Causal Discovery of MEG Data

As an illustration of the applicability of the method on real data, we applied it on magnetoencephalography (MEG), i.e., measurements of the electric activity in the brain. The raw data consisted of the 306 MEG channels measured by the Vectorview helmet-shaped neuromagnetometer (Neuromag Ltd., Helsinki, Finland) in a magnetically shielded room at the Brain Research Unit, Low Temperature Laboratory, Helsinki University of Technology. The measurements consisted of 300 seconds of resting state brain activity earlier used in [9]. The subject was sitting with eyes closed, and did not perform any specific task nor was there any specific sensory stimulation.

As pre-processing, we performed a blind separation of sources using the method called Fourier-ICA [9]. This gave nine sources of oscillatory brain activity. Our goal was to analyze the causal relations between the powers of the source, so we divided the data into windows of length of one second (half overlapping, i.e., the initial

4 Available at http : //www.phil.cmu.edu/projects/tetrad/

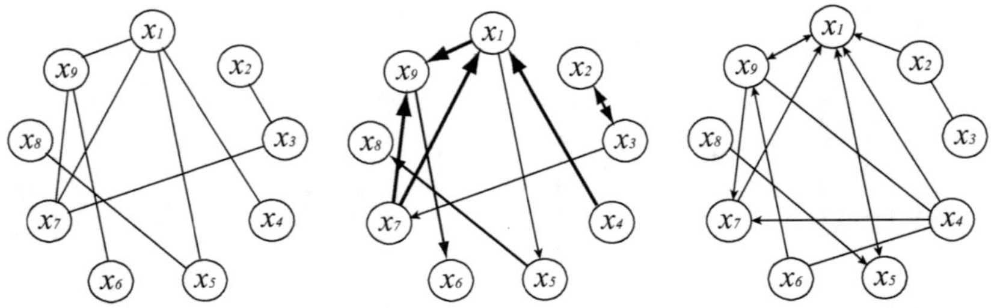

(a) moral graph (b) causal diagram (c) by PC

Fig. 5. Experimental results of the MEG data. (a) The moral graph obtained by conditional independence tests. Here no spouse link exists. (b) The final causal model. The thickness of the lines indicates the strength of the causal effects, as measured by the contributed variance. (c) For comparison, the result by the PC algorithm [19] is given.

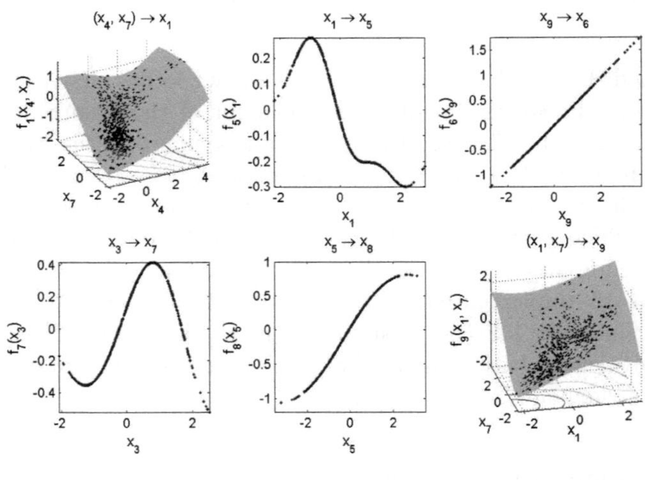

Fig. 6. The effect of the causes on each target variable in the MEG data. For clarity, we also give the fitted surface in 3D plots.

points were at a distance of 0.5 seconds each) and computed the logarithm of the local standard deviation in each window. This gave a total of 604 observations of a nine-dimensional random vector, on which we applied our method.

We first tested if the obtained variables have linear causal relations. If the variables have linear acyclic causal relations and at most one of the disturbances is Gaussian, the de-mixing matrix obtained by ICA can be permuted to lower-triangularity, and the causal model can be uniquely found [18]. We applied FastICA [7], a widely-used linear ICA algorithm, to the data, and found that the de-mixing matrix is far from a permuted lower-triangular matrix. Furthermore, some independent components are not even truly independent, as verified by the kernel-based independence test [5], which gave the p-value 7×10^{-3}. These findings imply that a linear acyclic causal model does not fit the data well.

We then applied the proposed approach to do nonlinear causal analysis. The results, including the moral graph and the final causal diagram, are shown in Fig. 5. Note that there is a bidirected edge $x_2 \leftrightarrow x_3$ in the final result (Fig. 5(b)); in fact, neither of the causal relations $x_2 \rightarrow x_3$ and $x_2 \leftarrow x_3$ could make the disturbance independent of the parents. This means that the causal relation between x_2 and x_3 could not be represented by Eq. 1, or that there exists some confounder. For comparison, the result by the PC algorithm is given in Fig. 5(c). It contains clearly more edges, with two bidirected and three undirected.

For illustration, Fig.6 plots the effect of the causes for each variable which has parents. One can see that three of the causal relations ($x_9 \rightarrow x_6$, $x_5 \rightarrow x_8$, and $(x_1, x_7) \rightarrow x_9$) are close to linear, while others are clearly nonlinear. The obtained causal connections are something completely new in neuroscience. Their interpretation will require a lot of work from domain experts.

5 Conclusion

In this paper we focused on the acyclic causality discovery problem with an additive disturbance model. For the acyclic causal models in which the causal relations among observed variables are nonlinear while disturbances have linear effects, we have shown that the following criteria are equivalent: 1. mutual independence of the disturbances, 2. causal Markov property of the causal model, as well as the independence between the disturbance and the parents associated with the same variable, and 3. minimum disturbance entropy. From this viewpoint, conventional conditional independence-based methods, non-Gaussianity-based linear methods, and the Granger causality analysis could be unified.

The criterion of mutual independence of disturbances then inspires us to exploit multichannel blind deconvolution, a well-developed extension of ICA, to estimate Granger causality with instantaneous effects. Compared to other methods, this approach is more efficient (in the statistical sense), and it admits simple ways for model selection of the causal structure by incorporating suitable penalties on the coefficients. Finally, we showed that nonlinear causal discovery with additive disturbances can be achieved by enforcing the causal Markov condition and the independence between the disturbance and parents of the same variable. The resulting approach is suitable for moderate-sized problems. Simulations and real-world applications showed the usefulness of the proposed approaches.

Acknowledgement

We are very grateful to Pavan Ramkumar and Lauri Parkkonen for providing the MEG data. We thank Patrik Hoyer for helpful discussions.

References

1. Cichocki, A., Amari, S.: Adaptive Blind Signal and Image Processing: Learning Algorithms and Applications. John Wiley & Sons, UK (2003) (corrected and revisited edition)
2. Cover, T.M., Thomas, J.A.: Elements of Information Theory. Wiley, Chichester (1991)
3. Gibbs, P.: Event-Symmetric Space-Time. Weburbia Press, Great Britain (1998)
4. Granger, C.: Testing for causality: A personal viewpoint. Journal of Economic Dynamics and Control 2 (1980)
5. Gretton, A., Fukumizu, K., Teo, C.H., Song, L., Schölkopf, B., Smola, A.J.: A kernel statistical test of independence. In: NIPS 20, pp. 585–592. MIT Press, Cambridge (2008)
6. Hoyer, P.O., Janzing, D., Mooji, J., Peters, J., Schölkopf, B.: Nonlinear causal discovery with additive noise models. In: NIPS 21, Vancouver, B.C., Canada (2009)
7. Hyvärinen, A.: Fast and robust fixed-point algorithms for independent component analysis. IEEE Transactions on Neural Networks 10(3), 626–634 (1999)
8. Hyvärinen, A., Karhunen, J., Oja, E.: Independent Component Analysis. John Wiley & Sons, Inc., Chichester (2001)
9. Hyvärinen, A., Ramkumar, P., Parkkonen, L., Hari, R.: Independent component analysis of short-time Fourier transforms for spontaneous EEG/MEG analysis (2008) (submitted manuscript)
10. Hyvärinen, A., Shimizu, S., Hoyer, P.O.: Causal modelling combining instantaneous and lagged effects: an identifiable model based on non-gaussianity. In: ICML 2008, Helsinki, Finland, pp. 424–431 (2008)
11. Liu, R.W., Luo, H.: Direct blind separation of independent non-Gaussian signals with dynamic channels. In: Proc. Fifth IEEE Workshop on Cellular Neural Networks and their Applications, London, England, April 1998, pp. 34–38 (1998)
12. Margaritis, D.: Distribution-free learning of bayesian network structure in continuous domains. In: Proceedings of the 20th Conference on Artificial Intelligence (AAAI 2005), Pittsburgh, PA, July 2005, pp. 825–830 (2005)
13. Pearl, J.: Causality: Models, Reasoning, and Inference. Cambridge University Press, Cambridge (2000)
14. Pellet, J.P., Elisseeff, A.: Using markov blankets for causal structure learning. Journal of Machine Learning Research 9, 1295–1342 (2008)
15. Rasmussen, C.E., Williams, C.K.I.: Gaussian Processes for Machine Learning. MIT Press, Cambridge (2006)
16. Reale, M., Tunnicliffe Wilson, G.: Identification of vector ar models with recursive structural errors using conditional independence graphs. Statistical Methods and Applications 10(1-3), 49–65 (2001)
17. Schwarz, G.: Estimating the dimension of a model. The Annals of Statistics 6, C461–C464 (1978)
18. Shimizu, S., Hoyer, P.O., Hyvärinen, A., Kerminen, A.J.: A linear non-Gaussian acyclic model for causal discovery. Journal of Machine Learning Research 7, 2003–2030 (2006)
19. Spirtes, P., Glymour, C., Scheines, R.: Causation, Prediction, and Search, 2nd edn. MIT Press, Cambridge (2001)
20. Zhang, K., Peng, H., Chan, L., Hyvärinen, A.: ICA with sparse connections: Revisisted. In: Proc. 8rd Int. Conf. on Independent Component Analysis and Blind Signal Separation (ICA 2009), Paraty, Brazil, pp. 195–202 (2009)

Appendix: Proof of Theorem 1

Implication $(iii) \Rightarrow (i)$ is obviously, as shown below. Suppose x_k is neither x_i nor any descendant of x_i. Apparently x_k is a function of the disturbances which do not include e_i, and hence x_k (as well as any parent of x_i) is independent of e_i. According to Eq. 1, we have $p(x_i|pa_i, x_k) = p(e_i|pa_i, x_k) = p(e_i|pa_i) = p(x_i|pa_i)$, meaning x_i and x_k are conditionally independent given pa_i. Below we shall prove $(i) \Rightarrow (ii)$ and $(ii) \Rightarrow (iii)$.

As the causal relations of x_i are acyclic, x_i can be arranged in an order such that no later variable causes any earlier one. Let $k(i)$ denote such an order. Let $\tilde{\mathbf{x}}_i$ be the vector of $\mathbf{x}$ with the order $k(i)$, i.e., $\tilde{x}_{k(i)} = x_i$, and denote by $\tilde{e}_i$ and $\tilde{pa}_i$ the disturbance in $\tilde{x}_i$ and the parents of $\tilde{x}_i$, respectively.

According to the properties of conditional entropy, for any $1 \leq i \leq n$, we have

$$H(\tilde{e}_i) \geq H(\tilde{e}_i|\tilde{pa}_i) \tag{7}$$

$$= H(\tilde{x}_i|\tilde{pa}_i) \geq H(\tilde{x}_i|\tilde{x}_1, \cdots, \tilde{x}_{i-1}), \tag{8}$$

where the equality in 7 holds if and only if e_i is independent of $\tilde{pa}_i$, and the equality in 8 holds if and only if $\tilde{x}_j$ ($j < i$, and $\tilde{x}_j \notin \tilde{pa}_i$) are conditionally independent of $\tilde{e}_i$ given $\tilde{pa}_i$ [2]. Summation of the above inequality over all i yields

$$\sum_i H(e_i) = \sum_i H(\tilde{e}_i) \geq H(\tilde{x}_1) + H(\tilde{x}_2|\tilde{x}_1) + \cdots + H(\tilde{x}_n|\tilde{x}_1, \cdots, \tilde{x}_{n-1}) \tag{9}$$

$$= H(\tilde{x}_1, \cdots, \tilde{x}_n) = H(x_1, \cdots, x_n),$$

where the equality in Eq. 9 holds when (i) is true. This implies $(i) \Rightarrow (ii)$.

Now let us suppose (ii) is true. Denote by $\mathcal{G}$ the transformation from $(\tilde{x}_1, \cdots, \tilde{x}_n)$ to $(\tilde{e}_1, \cdots, \tilde{e}_n)$. As $\tilde{e}_i$ only depends on $\tilde{x}_i$ and its parents, the Jacobian matrix of $\mathcal{G}$, denoted by $\mathbf{J}_{\mathcal{G}}$, is a lower-triangular matrix with 1 on its diagonal. This gives $|\mathbf{J}_{\mathcal{G}}| = 1$, and hence $H(\tilde{e}_1, \cdots, \tilde{e}_n) = -E\{\log p_{\tilde{\mathbf{e}}}(\tilde{e}_1, \cdots, \tilde{e}_n)\} = -E\{\log[p_{\tilde{\mathbf{x}}}(\tilde{x}_1, \cdots, \tilde{x}_n)/|\mathbf{J}_{\mathcal{G}}|]\} = H(x_1, \cdots, x_n)$. Consequently, the mutual information $I(e_1, \cdots, e_n) = \sum_i H(e_i) - H(\tilde{e}_1, \cdots, \tilde{e}_n) = H(x_1, \cdots, x_n) - H(x_1, \cdots, x_n) = 0$, meaning that e_i are mutually independent. We then have $(ii) \Rightarrow (iii)$. Therefore, (i), (ii), and (iii) are equivalent. □

Subspace Regularization: A New Semi-supervised Learning Method

Yan-Ming Zhang, Xinwen Hou, Shiming Xiang, and Cheng-Lin Liu

National Laboratory of Pattern Recognition,
Institute of Automation, Chinese Academy of Sciences,
95 Zhongguancun East Road, Beijing 100190, P.R. China
{ymzhang,xwhou,smxiang,liucl}@nlpr.ia.ac.cn

Abstract. Most existing semi-supervised learning methods are based on the smoothness assumption that data points in the same high density region should have the same label. This assumption, though works well in many cases, has some limitations. To overcome this problems, we introduce into semi-supervised learning the classic low-dimensionality embedding assumption, stating that most geometric information of high dimensional data is embedded in a low dimensional manifold. Based on this, we formulate the problem of semi-supervised learning as a task of finding a subspace and a decision function on the subspace such that the projected data are well separated and the original geometric information is preserved as much as possible. Under this framework, the optimal subspace and decision function are iteratively found via a projection pursuit procedure. The low computational complexity of the proposed method lends it to applications on large scale data sets. Experimental comparison with some previous semi-supervised learning methods demonstrates the effectiveness of our method.

1 Introduction

We consider the general problem of learning from labeled and unlabeled data. Given an input data set $\{x_1, \ldots, x_l, x_{l+1}, \ldots, x_n\}$, the first l points have labels $\{y_1, \ldots, y_l\} \in \{-1, +1\}$ and the remaining points are unlabeled. The goal is to learn a prediction function which has low classification error on test points.

To make use of unlabeled data, assumption of the relationship between the marginal distribution $p(x)$ and the conditional distribution $p(y|x)$ should be made. This prior assumption plays an essential role in semi-supervised learning [1,2]. Most of the semi-supervised learning methods proposed by far are based on the smoothness assumption that "two points in the same high-density region are likely of the same label" [1]. The effectiveness and generality of this assumption has made the smoothness-based methods very successful, and in fact most of the state-of-the-art semi-supervised learning methods are based on this assumption.

Although the smoothness assumption is effective and methods based on it have obtained good performance on some problems, there are two main limitations of it. First, according to Mann and McCallum [3], most of the current

W. Buntine et al. (Eds.): ECML PKDD 2009, Part II, LNAI 5782, pp. 586–601, 2009.

semi-supervised methods lack in scalability. Typical semi-supervised learning methods, such as label propagation [4,5,2], Manifold Regularization [6], Transductive Support Vector Machines (TSVM) [7], require learning time of order $O(n^3)$ where n is the size of data set. Second, when data of different classes overlap heavily, the ideal decision boundary should cross the overlapping area which is a high density region. Thus, smoothness-based methods may fail since they always avoid a decision boundary that crosses high density region to satisfy the smoothness assumption [3,8].

In this paper, we turn to consider the low-dimensionality embedding assumption in the semi-supervised learning setting. This assumption can be roughly described as most information of high dimensional data is embedded in a low dimensional manifold. It has been traditionally used in dimension reduction methods to map the high dimensional data to a low-dimensional representation while preserving most original information [9].

In the typical setting of semi-supervised learning, there is very limited number of labeled data in a high-dimensional input space. According to statistical learning theory, because of the sparseness of data it is impossible to learn a good classifier for a general problem [10]. Based on the low-dimensional embedding assumption, we naturally hope to find a low-dimensional representation of data so as to make the labeled data more dense and therefore much easier for training. To this end, both labeled and unlabeled data can be used to explore the low-dimensional structure in data. Specifically, we propose a combined criterion to evaluate the candidate subspace and the classifier simultaneously: the data-fitting term evaluates how well the labeled data points of different classes are separated in the subspace, and the regularization term evaluates how much information has been lost by mapping the whole (labeled and unlabeled) data to the subspace. As the regularization term tends to find the subspace that preserves most interesting information, we call this framework as subspace regularization.

Within the above general framework, we instantiate a specific algorithm called PCA-based Least Square (PCA-LS), where the regularization term aims at reducing the reconstruction error of input points just like Principal Component Analysis (PCA) [11]. We also kernelize the PCA-LS to extend our method to nonlinear cases. The method we use to solve the optimal problem turns out to be a special case of the classic projection pursuit procedure which constructs the subspace and the decision function defined on this subspace in an incremental manner.

Compared to the smoothness-based methods, subspace regularization has two remarkable advantages. First, our methods still work when data from different classes overlap heavily, while smoothness-based methods may fail. Roughly speaking, after the subspace is fixed, subspace regularization looks for a decision boundary that can optimally classify the projected labeled data. Thus, although the data still overlap in the subspace, the decision boundary will not be affected by the data density directly and avoids the problem that fails smoothness-based methods. Second, our method has very low computational complexity which implies that it can be applied in large-scale applications. For linear PCA-LS, the computational complexity is linear in the number of data points and dimension of the input space. As a result, for a data set of 80,000 points, our method is about 60 times faster than typical smoothness-based methods in training. Beside these, the method is rather robust to hyperparameters and still suitable when l/n is large. We will examine these in much more detail in the experiment section.

The remainder of this paper is organized as follows. In section 2, we briefly review the existing semi-supervised learning methods. Section 3 introduces the subspace regularization framework and the details of algorithms. Experimental results are reported in section 4. We conclude the paper in section 5.

2 Related Work

Over the past decade, there are many methods have been proposed to handle the semi-supervised problem. Based on the prior assumptions they use, these methods can be roughly divided into three categories.

The methods from the first category are generative models in which they make a prior assumption on the form of the input data distribution, for example Gaussian mixture models or naive Bayes models [8]. Then, models are trained by Expectation Maximization algorithm using both labeled and unlabeled data. Nigam et al. applied this method to text classification problem and improved the performance dramatically [12]. However, this prior assumption on data distribution is too strict for general problems, and when the model is wrong unlabeled data may hurt accuracy.

The methods from the second category are based on the smoothness assumption as stated in the previous section. Based on this assumption, two families of methods have been developed.

The methods in the first family, namely, graph-based methods, have been developed to satisfy the smoothness assumption by penalizing the variance of decision function in high density region of data. Specifically, using both labeled and unlabeled data, an adjacency graph is constructed to explore the intrinsic geometric structure of the data. The decision function is then found by minimizing the training error on labeled data and the variance on the adjacency graph. Many famous semi-supervised learning methods, like label propagation [4,5,2], spectral methods [13,14], manifold regularization [6], belong to this family.

The methods in the second family, namely, low-density-separation-based methods, implement the smoothness assumption based on an equivalent assumption that "the best decision boundary should be located in low-density region". The aim of this kind of methods is to find a decision boundary which can correctly classify the labeled data and meanwhile is far away from the high density region of unlabeled data points. The distance from one point to the decision boundary is evaluated by the absolute value of the decision function or the value of posterior probability on the point. Methods of this family include the TSVM [7], semi-supervised Gaussian processes [15], entropy regularization [16], information regularization [17], low density separation [18], etc.

The methods from the third category make use of the prior knowledge of label distribution. They force the model predictions on unlabeled data to match the prior label distribution. For example, Zhu et al. used class mean normalization (CMN) as a post-processing step of Gaussian Random Fields (GRF) method to adjust the classification results [4]. Mann et al. proposed a method named Expectation Regularization that directly augments the objective function by adding a regularization term which is defined as the KL-divergence between prior label distribution and empirical label distribution predicted by model [3]. One important advantage of the methods of this type is their high efficiency in computation. However, as they do not explicitly explore the underling structure of data, these methods can not utilize the information of unlabeled data sufficiently.

3 Subspace Regularization

In this section, we first present the subspace regularization framework. Then a specific algorithm is given to learn linear decision function, and kernelized to tackle the nonlinear case. Finally, we analyze the computational complexity of our method.

3.1 Objective Function

Given a set of labeled and unlabeled data, we denote the input data by matrix $X = [x_1, \ldots, x_n]$, and the output by vector $Y = [y_1, \ldots, y_l]^T$. Without confusion, we use W to denote the subspace $W = span\{w_1, \ldots, w_p | w_i \perp w_j, i \neq j\}$ and the matrix $W = [w_1, \ldots, w_p]$ depending on the context.

From the above discussion, we aim to find a low-dimensional subspace $W = span\{w_1, \ldots, w_p | w_i \perp w_j, i \neq j\}$ and a decision function g defined on the W such that the following objective is minimized:

$$L(X, Y, W, g) = \sum_{i=1}^{l} L_F(y_i, g(x_i^T W)) + \lambda L_R(X, X^W), \tag{1}$$

where L_F and L_R are loss functions, and $X^W = [x_1^W, \ldots, x_n^W]$ in which x_i^W is the projection of x_i onto the subspace W. The first term evaluates how well the projected labeled data can be separated by g in the subspace W, and the second term evaluates how much information is lost by projecting data onto the subspace.

Specifically, we choose L_F as the least square error, L_R as the reconstruction error, and let g be an arbitrary linear function defined on W. Then, the objective function can be rewritten as

$$L(X, Y, W, g) = \sum_{i=1}^{l} (y_i - \sum_{t=1}^{p} \alpha_t x_i^T w_t)^2 + \lambda \sum_{i=1}^{n} \|x_i - x_i^W\|^2. \tag{2}$$

The parameters $\alpha = [\alpha_1, ..., \alpha_p]$ and W are estimated by minimizing (2). The dimension of subspace p and the regularization factor λ are hyperparameters which can be fixed by cross-validations. In experiment section, we will show our method is surprisingly robust to the λ, while the p should be carefully chosen. The hyperparameter λ is introduced to trade off between terms of data-fitting error and reconstruction error. When λ becomes large, the optimal subspace approximates the PCA subspace. Thus, we name our algorithm as PCA-based least square (PCA-LS).

3.2 PCA-Based Least Square Algorithm

We employ the traditional projection pursuit procedure [19,20] to incrementally construct the optimal subspace W and decision function g in problem (2). More specifically, we use an iterative procedure to minimize the objective function. In each iteration, based on the current model, we select one projection direction to add into the subspace and choose the coefficient in g for the selected direction such that the objective function has a maximum reduction.

one iteration in projection pursuit

Suppose that, at t^{th} iteration, we have $W = span\{w_1, \ldots, w_{t-1} | w_i \perp w_j, i \neq j\}$ and $g(v) = \sum_{j=1}^{t-1} \alpha_j v_j$. Then the residual r_i of decision response y_i is $r_i = y_i - \sum_{j=1}^{t-1} \alpha_j x_i^T w_j$, and the residual R_i of data point x_i is $R_i = x_i - x_i^W = x_i - \sum_{j=1}^{t-1} \beta_i^j w_j$. Note that R_i is orthogonal to the subspace W. Our goal in the tth iteration is to optimize the following problem:

$$\begin{aligned}
\min_{\alpha,\beta,w} I(\alpha, \beta, w) &= \sum_{i=1}^{l} (y_i - \sum_{j=1}^{t-1} \alpha_j x_i^T w_j - \alpha x_i^T w)^2 + \sum_{i=1}^{n} \|x_i - \sum_{j=1}^{t-1} \beta_i^j w_j - \beta_i w\|^2 \\
&= \sum_{i=1}^{l} (r_i - \alpha x_i^T w)^2 + \lambda \sum_{i=1}^{n} \|R_i - \beta_i w\|^2 \\
&= \|r - \alpha {X^L}^T w\|^2 + \lambda \sum_{i=1}^{n} \|R_i - \beta_i w\|^2, \\
s.t. \quad & w \perp w_j \quad \forall j = 1, \ldots, t-1, \qquad (3)
\end{aligned}$$

where X^L is the first l columns of X, α is a scalar, $\beta = [\beta_1, \ldots, \beta_n]^T$ and $r = [r_1, \ldots, r_l]^T$. After w is solved from problem (3), we denote it by w_t and add it to W. In this way, we finish one iteration of projection pursuit.

The problem (3) is difficult to optimize due to the high order of variables and the orthogonal constraints. To eliminate the constraints, we limit the searching scope of direction to be the subspace spanned by the residuals, which means $w = \sum_{i=1}^{n} \eta_i R_i = R\eta$ in which $R = [R_1, \ldots, R_n]$ and $\eta = [\eta_1, \ldots, \eta_n]^T$. As $R_i \perp W \quad \forall i = 1, \ldots, n$, the orthogonal constraints are automatically met. It thus results in the following unconstrained minimization problem:

$$H(\alpha, \beta, \eta) = \|r - \alpha {X^L}^T R\eta\|^2 + \lambda \sum_{i=1}^{n} \|R_i - \beta_i R\eta\|^2. \qquad (4)$$

Fortunately, it is guaranteed that the optimal w^* of problem (3) is indeed a linear combination of R_i:

Proposition 1: The minimum point of problem (3) can be represented as a linear combination of R_i.

Proof: We decompose $\mathbb{R}^d$ as $\mathbb{R}^d = X^{\parallel} \oplus X^{\perp}$, where $X^{\parallel}$ is the subspace spanned by $x_i, 1 \leq i \leq n$ and $X^{\perp}$ is the orthogonal complement space of $X^{\parallel}$. By construction, $X^{\parallel}$ can be further decomposed as $X^{\parallel} = W \oplus R$, where W is the subspace spanned by $w_j, 1 \leq j \leq t-1$, and R is the subspace spanned by $R_i, 1 \leq i \leq n$. As the optimal solution w^* of (3) should be perpendicular to the subspace W, thus $w^* \in X^{\perp} \oplus R$. Assume $w^* = w^{\perp} + w^R$, where $w^{\perp} \in X^{\perp}$ and $w^R \in R$. So,

$$\begin{aligned} I(\alpha, \beta, w^*) &= \|r - \alpha X^{L^T} w^*\|^2 + \lambda \sum_{i=1}^{n} \|R_i - \beta_i w^*\|^2, \\ &= \|r - \alpha X^{L^T} (w^{\perp} + w^R)\|^2 + \lambda \sum_{i=1}^{n} \|R_i - \beta_i (w^{\perp} + w^R)\|^2, \\ &= \|r - \alpha X^{L^T} w^R\|^2 + \lambda \sum_{i=1}^{n} (\|R_i - \beta_i w^R\|^2 + \beta_i^2 \|w^{\perp}\|^2). \end{aligned} \quad (5)$$

The third equation follows from the fact that $X^{L^T} w^{\perp} = 0$ and $R_i^T w^{\perp} = 0$. Since $I(\alpha, \beta, w^*) \geq I(\alpha, \beta, w^R)$ and w^* minimize $I(\alpha, \beta, w)$, we have $w^* = w^R$. Thus, w^* is in the subspace R, and can be represented as a linear combination of $R_i, 1 \leq i \leq n$. ■

To minimize the objective function (4), the iterative coordinate decent method is used. Briefly speaking, in each step, we optimize α, β for fixed η , and then optimize η for fixed α, β.

For a fixed η, the optimal α and β can be obtained by setting the partial derivatives $\frac{\partial H(\alpha,\beta,\eta)}{\partial \alpha}$, $\frac{\partial H(\alpha,\beta,\eta)}{\partial \beta}$ to zeros, and are given by:

$$\alpha = \frac{\langle r, X^{L^T} R\eta \rangle}{\langle X^{L^T} R\eta, X^{L^T} R\eta \rangle}, \qquad \beta_i = \frac{\langle R_i, R\eta \rangle}{\langle R\eta, R\eta \rangle}. \quad (6)$$

For fixed α and β, gradient decent is used to update η. The partial derivative of $H(\alpha, \beta, \eta)$ with respect to η is given by

$$\frac{\partial H(\alpha, \beta, \eta)}{\partial \eta} = -2\alpha R^T X^L r + 2\alpha^2 R^T X^L X^{L^T} R\eta + 2\lambda R^T R((\sum_{i=1}^{n} \beta_i^2)\eta - \beta). \quad (7)$$

After the iterative coordinate decent method converges, we get the optimal solution $\alpha^*, \beta^*, \eta^*$ for the problem (4). The new projection direction $w_t = R\eta^*$ is then added into $\{w_1, \ldots, w_{t-1}\}$ to form the new basis of subspace W. The residual of the response and the residual of inputs are updated by $r \leftarrow r - \alpha^* X^{L^T} R\eta^*$ and $R_i \leftarrow R_i - \beta_i^* R\eta^*$. Note that the new residual R_i preserves the property that it is orthogonal to the new subspace $W = span\{w_1, \ldots, w_t\}$. This fact follows the observation that the method we use to update the residual of input data is exactly the Gram-Schmidt orthogonalization.

After p times greedy search, we get the p dimensional subspace W and a decision function $g(v) = \sum_{t=1}^{p} \alpha_t v_t$ defined on W. If only classification is concerned, these two things can be combined to get the final decision function defined on the input space $f(x) = \sum_{t=1}^{p} \alpha_t x^T w_t = x^T W \overline{\alpha}$.

The whole procedure is summarized in Algorithm 1.

Algorithm 1. PCA-based Least Square (PCA-LS)

Init: $r = [r_1, \ldots, r_l]^T$; $R_i = x_i \quad i = 1, \ldots, n$
for $t = 1$ **to** p **do**
  repeat
    1. Compute α, β using (6)
    2. Compute $\frac{\partial H(\alpha,\beta,\eta)}{\partial \eta}$ using (7)
    3. $\eta = \eta - StepSize * \frac{\partial H(\alpha,\beta,\eta)}{\partial \eta}$
  until η is convergent
  $w_t = R\eta$
  $\alpha_t = \alpha$
  $r = r - \alpha_t X^L w_t$
  $R_i = R_i - \beta_i w_t \quad i = 1, \ldots, n$
end for
Output:
$f(x) = \sum_{t=1}^{p} \alpha_t x^T w_t = x^T W \overline{\alpha}$
$\overline{\alpha} = [\alpha_1, \ldots, \alpha_p]^T$

3.3 Kernel PCA-Based Least Square

When the data set has highly nonlinear structure, the PCA-based least square may fail. One common technique to tackle the nonlinear problem in machine learning is the kernel trick, which can be briefly described as follows: with a feature mapping $\phi : x \rightarrow \phi(x)$, the input data is mapped to a feature space. For linear learning in this feature space, the inner product of two mapped data is defined as the kernel function: $k(x, y) = \langle \phi(x), \phi(y) \rangle$, and the matrix K with $(K)_{ij} = k(x_i, x_j)$ is the Gram matrix.

At the tth iteration, suppose that the residual R_i of $\phi(x_i)$ can be expressed as $R_i = \sum_{j=1}^{n} M_i^j \phi(x_j) = \phi(X) M_i$, where $\phi(X) = [\phi(x_1), \ldots, \phi(x_n)]$ and $M_i = [M_i^1, \ldots, M_i^n]^T$. Thus, $R = [R_1, \ldots, R_n] = \phi(X)[M_1, \ldots, M_n] = \phi(X)M$, where M is a $n \times n$ matrix. In parallel with the linear case, we constrain the projection direction w to be a linear combination of residuals: $w = \sum_{i=1}^{n} \eta_i R_i = R\eta = \phi(X)M\eta$. Now we get the objective function $H(\alpha, \beta, \eta)$ of the kernel method similar to the linear case:

$$H(\alpha, \beta, \eta) = \|r - \alpha \phi(X^L)^T R\eta\|^2 + \lambda \sum_{i=1}^{n} \|R_i - \beta_i R\eta\|^2$$

$$= \|r - \alpha\phi(X^L)^T\phi(X)M\eta\|^2 + \lambda\sum_{i=1}^{n}\|R_i - \beta_i R\eta\|^2$$

$$= \|r - \alpha K^{L^T} M\eta\|^2 + \lambda\sum_{i=1}^{n}\|R_i - \beta_i R\eta\|^2, \tag{8}$$

where K^L is the first l columns of the Gram matrix K.

As before, we employ the iterative coordinate decent method to minimize the objective function (8). In each step, α, β are given by

$$\alpha = \frac{r^T K^{L^T} M\eta}{\eta^T M^T K^L K^{L^T} M\eta}, \qquad \beta_i = \frac{M_i K M\eta}{\eta^T M^T K M\eta}, \tag{9}$$

and the partial derivative of $H(\alpha, \beta, \eta)$ with respect to η is given by

$$\frac{\partial H(\alpha,\beta,\eta)}{\partial\eta} = -2\alpha M^T K^L r + 2\alpha^2 M^T K^L K^{L^T} M\eta + 2\lambda M^T K M((\sum_{i=1}^{n}\beta_i^2)\eta - \beta). \tag{10}$$

After the iterative coordinate decent method converges, we get the optimal solution $\alpha^*, \beta^*, \eta^*$ for the problem (8). Then, the direction $w_t = \phi(X)M\eta^*$ is added to $\{w_1, \ldots, w_{t-1}\}$ to form the new basis of subspace W. The residual of response and residual of inputs are updated by: $r \leftarrow r - \alpha^*\phi(X^L)^T R\eta^* = r - \alpha^* K^{L^T}\eta^*$ and $M_i \leftarrow M_i - \beta_i^* M\eta^*$. Again the new residual R_i is orthogonal to the new subspace $W = span\{w_1, \ldots, w_t\}$.

The whole process is summarized in Algorithm 2. Different from the linear case, we can not express the projection direction w_t explicitly. Instead, the coefficient vector s_t of w_t's linear representation by $\{\phi(x_1), \ldots, \phi(x_n)\}$ is stored.

3.4 Computational Complexity

As discussed above, large-scale problem is extremely important for semi-supervised learning. The linear PCA-LS algorithm consists of p times greedy search iterations, where p is the dimensionality of the subspace W. The complexity of every iteration is dominated by computing the gradient of η which scales as $O(nd)$ where d is the dimensionality of the input space. Thus, the computational complexity of the linear PCA-LS is $O(pnd)$. By a similar analysis, the computational complexity of the kernel PCA-LS algorithm scales as $O(pn^2)$.

4 Experiments

In this section, we first conduct experiments on synthetic data sets to show the ability of subspace regularization methods in handling overlapping data and manifold data. Then comparison experiments are given on several real data sets. Finally, we analyze the robustness of our methods to hyperparameters.

Algorithm 2. Kernel PCA-based Least Square (Kernel PCA-LS)

Init: $r = [r_1, \ldots, r_l]^T$; $M = I$
for $t = 1$ **to** p **do**
  repeat
    1. Compute α, β using (9)
    2. Compute $\frac{\partial H(\alpha,\beta,\eta)}{\partial \eta}$ using (10)
    3. $\eta = \eta - StepSize * \frac{\partial H(\alpha,\beta,\eta)}{\partial \eta}$
  until η is convergent
  $s_t = M\eta$
  $\alpha_t = \alpha$
  $r = r - \alpha_t {K^L}^T s_t$
  $M_i = M_i - \beta_i s_t \quad i = 1, \ldots, n$
end for
Output:
$f(x) = k_n(x)^T S\overline{\alpha}$
$k_n(x) = [k(x, x_1), \ldots, k(x, x_n)]^T$
$S = [s_1, \ldots, s_p]$
$\overline{\alpha} = [\alpha_1, \ldots, \alpha_p]^T$

4.1 Overlapping Data

When data of different classes overlap heavily, the optimal decision boundary or Bayesian decision boundary may cross the overlapping area which is a high density region. In this case, the smoothness-based methods tend to fail as they prefer a decision boundary located in low density area [3,8]. With following experiments, we demonstrate that subspace regularization methods can effectively avoid this problem.

Two dimension case. 200 data points were drawn from each of two unit-variance Gaussians, the centers of which are $(0, 0)$ and $(1.5, 1.5)$. Only one point

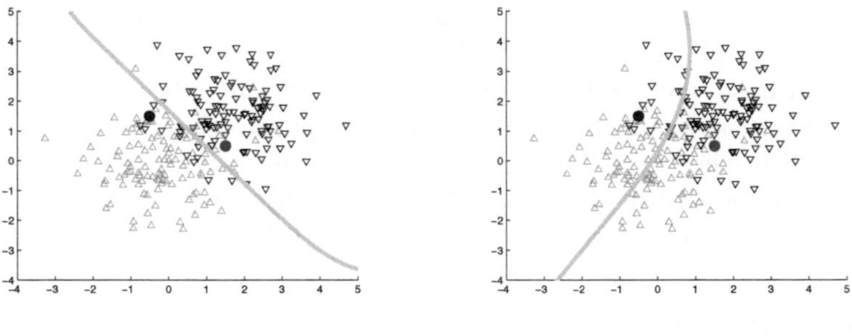

(a) (b)

Fig. 1. Two Gaussians Data Set: Decision boundary using RBF kernel for KPCA-LS (a) and KLapRLS (b) was labeled in each class and all the rest points were treated as unlabeled. The data set and the best decision boundary across a range of hyperparameters of kernel PCA-LS (KPCA-LS) and kernel LapRLS (KLapRLS) [6] are shown in Figure 1. Just as PCA, KPCA-LS finds the first principal component direction as the subspace to project on. In Figure 1 (a), this direction is parallel to the line passing through the centers of two Gaussians. As a result, the decision boundary is a line that is perpendicular to the selected subspace and can correctly classify the labeled data. However, for KLapRLS, the smoothness regularization term makes the decision boundary avoid the overlapping area as crossing such a high density region would lead to a big penalty. This makes the KLapRLS fail to find a satisfactory solution.

High dimension case. G241c and G241d are commonly-used data sets in semi-supervised learning which are constructed by Chapelle et al. [1]. Each of them is composed of 1500 points with dimension of 241. For G241c, two classes data come from two unit-variance Gaussians respectively, and centers of the two Gaussians have a distance of 2.5. For G241d, the data of the first class come from the two unit-variance Gaussians, the centers of which have a distance of 6; and the data of second class come from another two Gaussians which are fixed by moving each of the former centers a distance of 2.5. By the construction of G241c and G241d, we see that there exists overlapping between different class.

We compare subspace regularization methods with 3 smoothness-based methods, including Gaussian Random Field (GRF) [5], Learning with Local and Global Consistency (LLGC) [2] and manifold regularization (linear LapRLS and kernel LapRLS) [6]. 50 points are randomly sampled as labeled data, and the rest are left as unlabeled data. Data set is split 10 times, and the reported results are averaged classification accuracy over these 10 splits. 5-fold cross-validation is used to select hyperparameters, and the detailed setting for every method is introduced in following section. Experiments results is summarized in Table 1.

From the results, we can conclude that, when data of different classes overlap heavily, the subspace regularization methods outperform smoothness-based methods dramatically.

4.2 Manifold Data

This experiment was conducted on the two moons data set which was designed to satisfy the smoothness assumption and have a strong geometric structure. 200 data points were drawn from each of two moons, and three points drawn from

Table 1. Averaged classification accuracy (%) for 50 labeled data

	GRF	LLGC	LapRLS	KLapRLS	PCA-LS	KPCA-LS
G241c	48.28	48.28	71.66	68.36	**84.62**	83.59
G241d	65.62	63.53	67.90	66.28	**83.92**	74.01

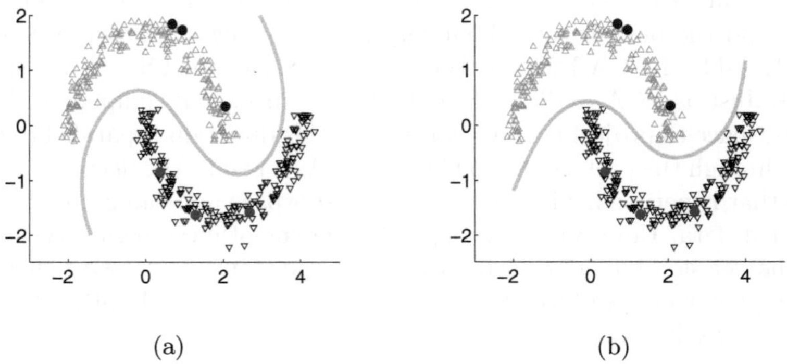

Fig. 2. Two Moons Data Set: Decision boundary using RBF kernel for KPCA-LS (a) and KLapRLS (b)

each class were labeled. The data set and the best decision boundary across a range of hyperparameters of kernel PCA-LS and kernel LapRLS are shown in Figure 2. It is well-known that the kernel LapRLS algorithm can work perfectly well on such problems. As can be seen from Figure 2, the kernel PCA-LS also learns a satisfactory decision function that correctly classify all the data points.

4.3 Real Data Sets

To evaluate our methods, We perform experiments on seven data sets of binary classification problem that come from the benchmark in Chapelle's book [1] and UCI machine learning repository. These data set originate from areas like image, text, biometrics etc., with size from 277 to 83,679 and dimension from 9 to 11,960. The characteristics of the data sets are shown in Table 2.

We compare subspace regularization methods with 3 smoothness-based semi-supervised learning methods, including GRF [5], LLGC [2], manifold regularization (linear LapRLS and kernel LapRLS) [6]. We also use both labeled and unlabeled data to perform dimension reduction with PCA, and then use regularized least square (RLS) to learn classifier on labeled data. For convenience, we name this baseline algorithm as PCA+RLS. We details the experimental setting for each method in the following:

Table 2. Description of the data sets

	Breast-cancer	BCI	USPS	Text	Image	Waveform	SecStr
size	277	400	1500	1500	2310	5000	83,679
dimension	9	117	241	11,960	18	21	315

Table 3. Averaged classification accuracy (%)

	PCA+RLS	GRF	LLGC	LapRLS	KLapRLS	PCA-LS	KPCA-LS
Breast-cancer	67.84	70.83	70.22	67.36	70.53	**73.44**	71.98
BCI	57.77	50.34	50.00	**62.29**	61.69	**62.29**	58.94
USPS	86.18	89.81	**94.5**	84.61	87.48	84.10	86.18
Text	**73.25**	52.68	62.63	71.84	68.62	72.25	71.32
Image	78.00	49.53	73.17	76.90	**83.60**	76.36	80.13
Waveform	78.99	70.84	79.69	78.34	-	85.33	**86.18**
SecStr	60.26	-	-	60.41	-	**61.28**	-

- PCA-LS: the dimension of subspace is chosen from $\{2; 2^2; 2^3; 2^4; 2^5\}$, and the regularization factor λ is simply fixed to 1.
- KPCA-LS: the dimension of subspace is chosen from $\{2; 2^2; 2^3; 2^4; 2^5\}$, the regularization factor λ is simply fixed to 100. RBF kernel is selected as kernel function, and the σ_k is chosen from $\sigma_0 * \{2; 1; 2^{-1}; 2^{-2}; 2^{-3}; 2^{-4}\}$ where σ_0 is the averaged distance between every two points in data set.
- GRF and LLGC methods: we use RBF kernel as similarity function to construct the graph, and the hyperparameter σ_g is chosen from $\sigma_0 * \{2; 1; 2^{-1}; 2^{-2}; 2^{-3}; 2^{-4}\}$ where σ_0 is defined as above.
- LapRLS: γ_A and γ_I are chosen from $\{10^4; 10^2; 1; 10^{-2}; 10^{-4}\}$. Linear kernel is used as kernel function. For all data sets except SecStr, RBF kernel is used as similarity function to construct the graph, and the hyperparameter σ_g is chosen from $\sigma_0 * \{2; 1; 2^{-1}; 2^{-2}; 2^{-3}; 2^{-4}\}$ where σ_0 is defined as above. For SecStr, the graph entry $G_{ij} = 1$ if x_i is among x_j's k nearest neighbors, else $G_{ij} = 0$. k is fixed to 5 which is the value used in [1].
- KLapRLS: γ_A, γ_I and σ_g is selected as in linear LapRLS. RBF kernel is used as kernel function, and the σ_k is chosen from $\sigma_0 * \{2; 1; 2^{-1}; 2^{-2}; 2^{-3}; 2^{-4}\}$.
- PCA+RLS: the dimension of subspace is chosen from $\{2; 2^2; 2^3; 2^4; 2^5\}$, and the regularization factor λ is chosen from $\{10^4; 10^2; 1; 10^{-2}; 10^{-4}\}$.

As some graph-based methods like GRF and LLGC can not directly predict the out-of-sample points, we adopt the transductive setting in the experiments which means all the data points are available before training. But we emphasize that the subspace regularization method is inductive.

For all data sets except SecStr, 50 points are randomly sampled as labeled data, and the rest are left unlabeled. We split each data set 10 times, and report the averaged classification accuracy over these 10 splits. 5-fold cross-validation is used to select hyperparameters. For SecStr, 500 points are randomly sampled as labeled data in which 400 points are used for training and 100 points are used to select the hyperparameters. The reported results are averaged accuracy on 10 splits.

Table 3 summarizes the experiments results. There are blank entries in Table 3 as some algorithms require too many memory, or the results can not be achieved within 3 days.

For all data sets except image (Image and USPS), subspace regularization method works as well as or better than smoothness-based methods. It demonstrates that subspace regularization is a powerful semi-supervised learning method. We will analyze this result in more detail in the next section.

For SecStr data set, it costs PCA-LS less than 5 minutes to learn a decision function and a subspace of dimension $p = 8$ which is the dimensionality selected by most splits. However, for graph-based methods, only to construct a neighborhood graph will take more than 2 hours without saying to inverse a matrix of same size with neighborhood graph. All experiments are performed on a PC with 3.0GHz CPU, 2GB memory, using Matlab.

4.4 Robustness Analysis

Recall that in all of the above experiments the regularization factor λ for subspace regularization methods is set to a fixed number, and here we will examine this robustness of our methods to hyperparameters in more details.

To evaluate the PCA-LS's robustness to the regularization factor λ, we fix the dimension p of subspace to 16 and report the classification accuracy for $\lambda = \{10^{-4}, 10^{-2}, 10^{0}, 10^{2}, 10^{4}\}$. For SecStr, 500 points are randomly labeled, and for other data sets 50 data are randomly labeled. The result is averaged over 10 splits which is shown in the Figure 3. We can see that PCA-LS is rather robust to the variation of λ, so no careful tuning of λ is needed.

To evaluate the PCA-LS's robustness to the dimension p of subspace, we fix the λ to 1 and report the classification accuracy for $p = \{1, 2, 2^2, 2^3, 2^4, 2^5, 2^6\}$. For SecStr, 500 points are randomly labeled, and for other data sets 50 data are randomly labeled. The result shown in the Figure 4 is averaged over 10 splits.

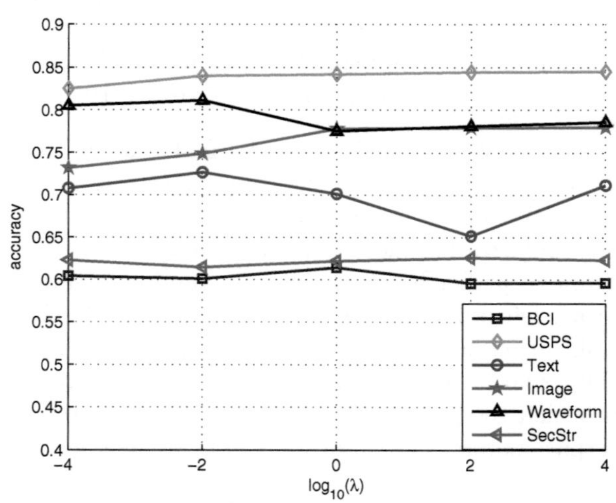

Fig. 3. PCA-LS's classification accuracy with different λ ($p = 16$)

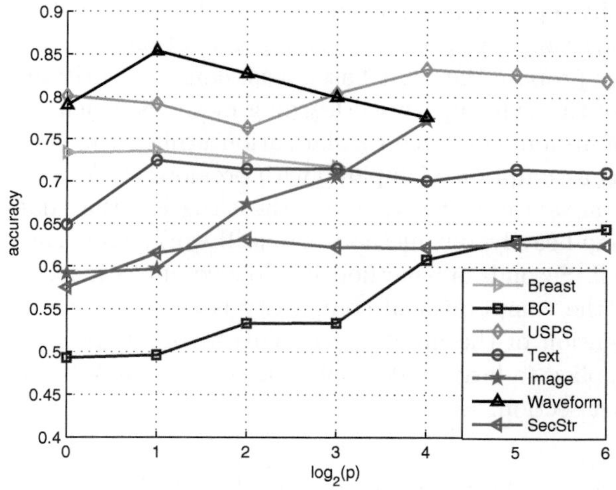

Fig. 4. PCA-LS's classification accuracy with different p ($\lambda = 1$)

Unlike λ, the dimension p of the subspace is very important to the success of subspace regularization methods, and needs to be chosen carefully. According to the performance of PCA-LS, these seven data sets can be clearly divided into two categories. For Breast-cancer, waveform, Text and SecStr, the accuracy peaks at a very small p and then decrease gradually, which means most of the information valuable for classification is actually embedding in a low dimensional subspace. For Image, USPS and BCI, however, the accuracy increases with the dimension of subspace monotonically which means the discriminative information is highly nonlinear and can not be captured by a low dimensional subspace. It is very interesting to see that it is exactly on the first class of data sets that the subspace regularization methods beat all the smoothness-based methods, while on the second class of data sets smoothness-based methods work better. Thus the conclusion is that, for very limited labeled data finding a general semi-supervised learning method that works well for most problem may be extremely hard, and we need to choose method carefully according to the characteristics of problem at hands. For manifold or highly nonlinear data set, smoothness-based methods are good choices. For those linear or approximately linear problems, subspace regularization methods are more effective.

5 Conclusions

This paper has presented subspace regularization, a new method for semi-supervised learning. The motivation of our work is to find a low-dimensional representation of data to avoid the curse of dimensionality and reduce the complexity of the problem. Specifically, we hope to find a subspace and a decision function defined on this subspace such that the projection of labeled data onto this subspace can be easily separated and meanwhile the data information does not loss too much by projection. By specifying the regularization term as reconstruction error as PCA, we propose the PCA-based least square algorithm.

Unlike most of the semi-supervised learning methods which are based on the smoothness assumption, subspace regularization utilizes the classic low dimensional embedding assumption. Compared with previous works, our methods have two remarkable advantages. First, under the situation that data from different classes overlap heavily, our methods can still work, while smoothness-based methods may fail. Second, our method has low computational complexity. For linear PCA-LS, the computational complexity is linear in the number of data points and dimension of the input space. This favorable property enables our method to be applied to large-scale applications which have been demonstrated in the experiment section.

Acknowledgments. This work was supported by the National Natural Science Foundation of China (NSFC) under Grant nos. 60825301 and 60723005. The authors thank the anonymous reviewers for their valuable comments.

References

1. Chapelle, O., Schölkopf, B., Zien, A., NetLibrary, I.: Semi-supervised learning. MIT Press, Cambridge (2006)
2. Zhou, D., Bousquet, O., Lal, T., Weston, J., Scholkopf, B.: Learning with local and global consistency. In: Advances in Neural Information Processing Systems 16, pp. 321–328 (2004)
3. Mann, G., McCallum, A.: Simple, robust, scalable semi-supervised learning via expectation regularization. In: Proceedings of the International Conference on Machine Learning, pp. 593–600 (2007)
4. Zhu, X., Ghahramani, Z.: Learning from labeled and unlabeled data with label propagation. Technical Report CMU-CALD-02, Computer Science, University of Wisconsin-Madison (2002)
5. Zhu, X., Ghahramani, Z., Lafferty, J.: Semi-supervised learning using Gaussian fields and Harmonic functions. In: Proceedings of the International Conference on Machine Learning, pp. 912–919 (2003)
6. Belkin, M., Niyogi, P., Sindhwani, V.: Manifold regularization: a geometric framework for learning from labeled and unlabeled examples. The Journal of Machine Learning Research 7, 2399–2434 (2006)
7. Joachims, T.: Transductive inference for text classification using support vector machines. In: Proceedings of the International Conference on Machine Learning, pp. 200–209 (1999)
8. Zhu, X.: Semi-supervised learning literature survey. Technical report, Computer Science, University of Wisconsin-Madison (2005)
9. Burges, C.: Geometric methods for feature extraction and dimensional reduction: A guided tour. In: The Data Mining and Knowledge Discovery Handbook, pp. 59–92 (2005)
10. Vapnik, V.: The Nature of Statistical Learning Theory. Springer, Heidelberg (1995)

11. Jolliffe, I.: Principal component analysis. Springer, New York (2002)
12. Nigam, K., McCallum, A., Thrun, S., Mitchell, T.: Text classification from labeled and unlabeled documents using EM. Machine learning 39(2), 103–134 (2000)
13. Joachims, T.: Transductive learning via spectral graph partitioning. In: Proceedings of the International Conference on Machine Learning, pp. 290–297 (2003)
14. Belkin, M., Niyogi, P.: Semi-supervised learning on Riemannian manifolds. Machine Learning 56(1), 209–239 (2004)
15. Lawrence, N.D., Jordan, M.I.: Semi-supervised learning via Gaussian processes. In: Saul, L.K., Weiss, Y., Bottou, L. (eds.) Advances in Neural Information Processing Systems 17, pp. 753–760 (2005)
16. Grandvalet, Y., Bengio, Y.: Semi-supervised learning by entropy minimization. In: Advances in Neural Information Processing Systems 17, pp. 529–536 (2005)
17. Szummer, M., Jaakkola, T.: Information regularization with partially labeled data. In: Becker, S., Thrun, S., Obermayer, K. (eds.) Advances in Neural Information Processing Systems 15 (2003)
18. Chapelle, O., Zien, A.: Semi-supervised classification by low density separation. In: Cowell, R.G., Ghahramani, Z. (eds.) Society for Artificial Intelligence and Statistics, pp. 57–64 (2005)
19. Friedman, J., Stuetzle, W.: Projection pursuit regression. Journal of the American Statistical Association 76(376), 817–823 (1981)
20. Huber, P.: Projection pursuit. The Annals of Statistics 13(2), 435–475 (1985)

Heteroscedastic Probabilistic Linear Discriminant Analysis with Semi-supervised Extension

Yu Zhang and Dit-Yan Yeung

Hong Kong University of Science and Technology
{zhangyu,dyyeung}@cse.ust.hk

Abstract. *Linear discriminant analysis* (LDA) is a commonly used method for dimensionality reduction. Despite its successes, it has limitations under some situations, including the small sample size problem, the homoscedasticity assumption that different classes have the same Gaussian distribution, and its inability to produce probabilistic output and handle missing data. In this paper, we propose a semi-supervised and heteroscedastic extension of probabilistic LDA, called S^2HPLDA, which aims at overcoming all these limitations under a common principled framework. Moreover, we apply automatic relevance determination to determine the required dimensionality of the low-dimensional space for dimensionality reduction. We empirically compare our method with several related probabilistic subspace methods on some face and object databases. Very promising results are obtained from the experiments showing the effectiveness of our proposed method.

1 Introduction

The need for dimensionality reduction is pervasive in many applications of pattern recognition and machine learning due to the high dimensionality of the data involved. Dimensionality reduction techniques seek to project high-dimensional data either linearly or nonlinearly into a lower-dimensional space according to some criterion so as to facilitate subsequent processing, such as classification. Classical linear dimensionality reduction methods include principal component analysis (PCA) [1] and *linear discriminant analysis* (LDA) [2], with the former being an unsupervised technique while the latter a supervised one that exploits the label information in the labeled data. For classification applications, LDA generally outperforms PCA because label information is usually useful for finding a projection to improve class separability in the lower-dimensional space.

Although LDA is widely used in many applications, the method in its original form does have limitations under some situations. One of them is a well-known limitation often referred to as the small sample size (SSS) problem [3], which arises in applications when the sample size is much smaller than the feature dimensionality and hence the within-class scatter matrix is singular. A number of methods have been proposed to address this problem, e.g., PseudoLDA [4], PCA+LDA [5], LDA/QR [6], NullLDA [3], DCV [7], DualLDA [8] and 2DLDA [9]. The main idea underlying these methods is to

W. Buntine et al. (Eds.): ECML PKDD 2009, Part II, LNAI 5782, pp. 602–616, 2009.

seek a space or subspace in which the within-class scatter matrix is nonsingular and then perform LDA or its variants there without suffering from the singularity problem. More recently, another approach has been pursued by some researchers [10,11,12] to alleviate the SSS problem via *semi-supervised learning* [13], by utilizing unlabeled data in performing dimensionality reduction in addition to labeled data. Another limitation of LDA arises from the fact that the solution it gives is optimal only when the classes are homoscedastic with the same Gaussian distribution. However, this requirement is too rigid in practice and hence it does not hold in many real-world applications. To overcome this limitation, mixture discriminant analysis [14] and a maximum likelihood approach [15] have been proposed. Recently, Loog and Duin [16] proposed a heteroscedastic extension to LDA based on the Chernoff criterion, with a kernel extension proposed later in [17]. The third limitation of LDA comes from its non-probabilistic nature. As such, it cannot produce probabilistic output and handle missing data in a principled manner. While producing probabilistic output can help the subsequent decision-making process in incorporating uncertainty under a probabilistic framework, the missing data problem is so commonly encountered in applications that being able to deal with it is very essential to the success of pattern recognition tools for practical applications. Some probabilistic LDA models have been proposed, e.g., [18,19,20]. A by-product of most probabilistic LDA models except the one in [18] is that it imposes no restriction on the maximum number of reduced dimensions, but the original LDA model can only project data into at most $C - 1$ dimensions where C is the number of classes. Nevertheless, previous research in probabilistic LDA [19,20] did not pay much attention to the issue of how to determine the reduced dimensionality needed.

While various attempts were made previously to address the above limitations individually, mostly one or at most two at a time, we are more aggressive here in trying to address all of them within a common principled framework. Specifically, in this paper, we will go through a two-step process in our presentation. First, we propose a *heteroscedastic probabilistic LDA* (HPLDA) model which relaxes the homoscedasticity assumption in LDA. However, in HPLDA, the parameters for each class can only be estimated using labeled data from that class. This may lead to poor performance when labeled data are scarce. Motivated by previous attempts that applied semi-supervised learning to alleviate the SSS problem, we then extend HPLDA to *semi-supervised heteroscedastic probabilistic LDA* (S^2HPLDA) by making use of (usually large quantities of) unlabeled data in the learning process. In S^2HPLDA, each class can have a different class covariance matrix and unlabeled data are modeled by a Gaussian mixture model in which each mixture component corresponds to one class. We also use automatic relevance determination (ARD) [21] to determine the required dimensionality of the lower-dimensional space which can be different for different classes and hence is fairly flexible.

The remainder of this paper is organized as follows. In Section 2, we first briefly review some previous work on probabilistic LDA. We then present HPLDA in Section 3 and S^2HPLDA in Section 4. Section 5 reports some experimental results based on face and object databases to demonstrate the effectiveness of our proposed method. Finally, Section 6 concludes the paper.

2 Related Work

To the best of our knowledge, three variants of probabilistic LDA [18,19,20] were proposed before.

In [18], each class is modeled by a Gaussian distribution with a common covariance matrix shared by all classes and the mean vectors of different classes are modeled by another Gaussian distribution whose covariance matrix is similar to the between-class scatter matrix in LDA. The solution of this probabilistic LDA model is so similar to that of LDA that it, unfortunately, also inherits some limitations of LDA. For example, it needs probabilistic PCA (PPCA) to perform (unsupervised) dimensionality reduction first to alleviate the SSS problem and it can only project data to $(C-1)$ dimensions.

Yu et al. [19] proposed a supervised extension of probabilistic PCA (PPCA) [22] called SPPCA. This approach can be viewed as first concatenating each data point with its class indicator vector and then applying PPCA to this extended form. From the analysis of [23], the maximum likelihood solution of this approach is identical to that of LDA. Yu et al. [19] also proposed a semi-supervised extension of SPPCA, called S^2PPCA, which can utilize unlabeled data as well.

The model in [20] is slightly different from others. It directly models the between-class and within-class variances. So each data point can be described as the aggregation of three parts: the common mean which is the mean of the whole dataset, the between-class variance which describes the characteristics of different classes, and the within-class variance which describes the characteristics of each data point. Prince and Elder [20] also gave some extensions of this model useful for face recognition.

3 HPLDA: Heteroscedastic Probabilistic Linear Discriminant Analysis

Suppose the whole dataset contains l labeled data points $\{(\mathbf{x}_i, y_i)\}_{i=1}^{l}$ from C classes $\boldsymbol{\Pi}_k$ $(k=1,\ldots,C)$, where $\mathbf{x}_i \in \mathbb{R}^D$ with its label $y_i \in \{1,\ldots,C\}$ and class $\boldsymbol{\Pi}_k$ contains n_k examples. Moreover, all data points $\{\mathbf{x}_i\}_{i=1}^{l}$ are independent and identically distributed.

HPLDA is a latent variable model. It can be defined as follows:

$$\begin{aligned}
\mathbf{x}_i &= \mathbf{W}_{y_i}\mathbf{t}_i + \boldsymbol{\mu}_{y_i} + \boldsymbol{\varepsilon}_i \\
\mathbf{t}_i &\sim \mathcal{N}(0, \mathbf{I}_d) \\
\boldsymbol{\varepsilon}_i &\sim \mathcal{N}(0, \tau_{y_i}^{-1}\mathbf{I}_D),
\end{aligned} \tag{1}$$

where τ_i specifies the noise level of the ith class, $\mathbf{t}_i \in \mathbb{R}^d$ with $d < D$, $\mathbf{I}_D$ is the $D \times D$ identity matrix and $\mathcal{N}(\mathbf{m}, \boldsymbol{\Sigma})$ denotes a multivariate Gaussian distribution with mean $\mathbf{m}$ and covariance matrix $\boldsymbol{\Sigma}$. So for each class $\boldsymbol{\Pi}_k$, we have a different $\mathbf{W}_k$. This is different from the models proposed in [18,19,20] in which different classes share the same matrix $\mathbf{W}$. The graphical model for HPLDA is shown in Figure 1. From (1), we can get

$$P(\mathbf{x}_i|\mathbf{t}_i) = \mathcal{N}(\mathbf{W}_{y_i}\mathbf{t}_i + \boldsymbol{\mu}_{y_i}, \tau_{y_i}^{-1}\mathbf{I}_D)$$

and

$$P(\mathbf{x}_i) = \mathcal{N}(\boldsymbol{\mu}_{y_i}, \boldsymbol{\Phi}_{y_i}),$$

where $\boldsymbol{\Phi}_k = \mathbf{W}_k\mathbf{W}_k^T + \tau_k^{-1}\mathbf{I}_D$. So the log-likelihood L of the data set can be calculated as

$$L = -\frac{1}{2}\sum_{k=1}^{C}\sum_{y_i=k}\left[(\mathbf{x}_i - \boldsymbol{\mu}_k)^T\boldsymbol{\Phi}_k^{-1}(\mathbf{x}_i - \boldsymbol{\mu}_k) + D\ln 2\pi + \ln|\boldsymbol{\Phi}_k|\right], \quad (2)$$

where $|\mathbf{A}|$ denotes the determinant of a square matrix $\mathbf{A}$. We set the derivative of L with respect to $\boldsymbol{\mu}_k$ to 0 to obtain the maximum likelihood estimate of $\boldsymbol{\mu}_k$ as

$$\boldsymbol{\mu}_k = \bar{\mathbf{m}}_k \equiv \frac{1}{n_k}\sum_{y_i=k}\mathbf{x}_i. \quad (3)$$

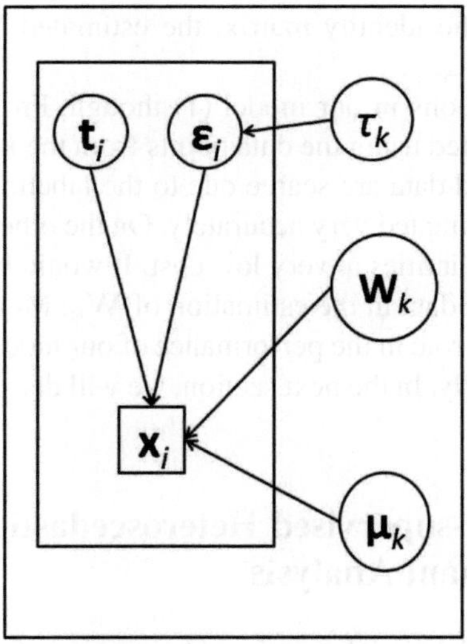

Fig. 1. Graphical model for HPLDA

Plugging Eq. (3) into (2), the log-likelihood can be simplified as

$$L = -\frac{1}{2}\sum_{k=1}^{C} n_k\left[\mathrm{tr}(\boldsymbol{\Phi}_k^{-1}\mathbf{S}_k) + D\ln 2\pi + \ln|\boldsymbol{\Phi}_k|\right], \quad (4)$$

where $\mathbf{S}_k = \frac{1}{n_k}\sum_{y_i=k}(\mathbf{x}_i - \bar{\mathbf{m}}_k)(\mathbf{x}_i - \bar{\mathbf{m}}_k)^T$ is the estimated covariance matrix for the kth class. Since $\mathbf{W}_k$ for different classes are independent, we can estimate each $\mathbf{W}_k$ from the following expression:

$$L_k = -\frac{1}{2}n_k\left[\mathrm{tr}(\boldsymbol{\Phi}_k^{-1}\mathbf{S}_k) + D\ln 2\pi + \ln|\boldsymbol{\Phi}_k|\right], \quad (5)$$

which is similar to the log-likelihood in PPCA. So, following the analysis in [22], we can obtain the maximum likelihood estimate of $\mathbf{W}_k$ as the eigenvectors of $\mathbf{S}_k$ corresponding to the largest eigenvalues and τ_k^{-1} is equal to the mean of the discarded eigenvalues.

3.1 Discussion

If all $\mathbf{W}_k$ and τ_k in (1) are the same, denoted by $\mathbf{W}$ and τ, then, from Eq. (4), the log-likelihood can be expressed as

$$L = -\frac{l}{2}\Big[\mathrm{tr}(\boldsymbol{\Phi}^{-1}\mathbf{S_w}) + D\ln 2\pi + \ln|\boldsymbol{\Phi}|\Big], \tag{6}$$

where $\mathbf{S_w} = \frac{1}{l}\sum_{k=1}^{C}\sum_{y_i=k}(\mathbf{x}_i - \bar{\mathbf{m}}_k)(\mathbf{x}_i - \bar{\mathbf{m}}_k)^T$ is the within-class scatter matrix in LDA and $\boldsymbol{\Phi} = \mathbf{W}\mathbf{W}^T + \tau^{-1}\mathbf{I}_D$. So, also following the analysis in [22], $\mathbf{W}$ consists of the top eigenvectors of $\mathbf{S_w}$ and τ^{-1} is equal to the mean of the discarded eigenvalues. Then if the data points are whitened by the total scatter matrix, i.e., the total scatter matrix of the dataset is the identity matrix, the estimated $\mathbf{W}$ is just the solution in traditional LDA.

There are some limitations in our model (1) though. From the above analysis, we can see that $\mathbf{W}_k$ is estimated using the data points from the kth class only. However, in many applications, labeled data are scarce due to the labeling effort required. So, as a result, $\mathbf{W}_k$ may not be estimated very accurately. On the other hand, unlabeled data are often available in large quantities at very low cost. It would be desirable if we can also make use of the unlabeled data in the estimation of $\mathbf{W}_k$. Moreover, the dimensionality of $\mathbf{W}_k$ plays an important role in the performance of our model and it should preferably be determined automatically. In the next section, we will discuss how to solve these two problems together.

4 S²HPLDA: Semi-supervised Heteroscedastic Probabilistic Linear Discriminant Analysis

As in HPLDA, there are l labeled data points $\{(\mathbf{x}_i, y_i)\}_{i=1}^{l}$ from C classes. In addition, there are u unlabeled data points $\{\mathbf{x}_{l+1}, \ldots, \mathbf{x}_{l+u}\}$, with $n = l + u$. Each class $\boldsymbol{\Pi}_k$ contains n_k labeled examples. For the labeled data points, we still use (1) to model them. For the unlabeled data points, we model them using a mixture model in which each mixture component follows (1) with prior probability $p(\boldsymbol{\Pi}_k) = \pi_k$. Thus the new model can be defined as:

$$\begin{aligned}
\mathbf{x}_i &= \mathbf{W}_{y_i}\mathbf{t}_i + \boldsymbol{\mu}_{y_i} + \boldsymbol{\varepsilon}_i, \text{ for } i \le l \\
\mathbf{t}_i &\sim \mathcal{N}(0, \mathbf{I}_d) \\
\boldsymbol{\varepsilon}_i &\sim \mathcal{N}(0, \tau_{y_i}^{-1}\mathbf{I}_D) \\
p(\mathbf{x}_i) &= \sum_{k=1}^{C}\pi_k p(\mathbf{x}_i|\boldsymbol{\Pi}_k), \text{ for } i > l,
\end{aligned} \tag{7}$$

where $\mathbf{t}_i \in \mathbb{R}^d$. Moreover, we use the ARD method [21] to determinate the dimensionality of $\mathbf{W}_k$ by introducing a data-dependent prior distribution

$$p(\mathbf{W}_{k,j}) \sim \mathcal{N}(0, \nu_{kj}^{-1}\mathbf{X}\mathbf{L}\mathbf{X}^T),$$

where $\mathbf{W}_{k,j}$ is the jth column of $\mathbf{W}_k$, $\mathbf{X} \in \mathbb{R}^{D\times n}$ is the total data matrix including both labeled and unlabeled data, and $\mathbf{L}$, whose construction will be described later, is the graph Laplacian matrix defined on $\mathbf{X}$. The graphical model is shown in Figure 2. Using the data-dependent prior on $\mathbf{W}_{k,j}$, we are essentially adopting the manifold assumption, which has been widely used in dimensionality reduction [24] and semi-supervised learning [25]. More specifically, if two points are close with respect to the intrinsic geometry of the underlying manifold, they should remain close in the embedding space after dimensionality reduction. The parameter ν_{kj} can be viewed as an indicator of the importance of the corresponding dimension of $\mathbf{W}_k$ to determine whether that dimension should be kept.

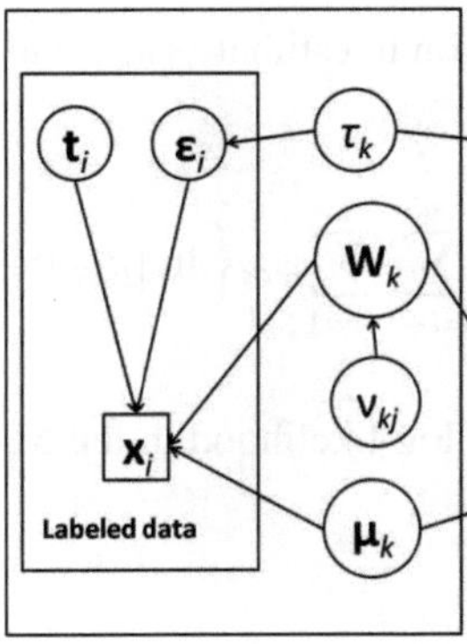

Fig. 2. Graphical model for S^{2}HPLDA

We now describe the construction of $\mathbf{L}$. Given the dataset $\mathcal{D} = \{\mathbf{x}_1, \ldots, \mathbf{x}_n\}$, we first construct a K nearest neighbor graph $G = (V, E)$, with the vertex set $V = \{1, \ldots, n\}$ corresponding to the labeled and unlabeled data points and the edge set $E \subseteq V \times V$ representing the relationships between data points. Each edge is assigned a weight r_{ij} which reflects the similarity between points $\mathbf{x}_i$ and $\mathbf{x}_j$:

$$r_{ij} = \begin{cases} \exp\left(-\frac{\|\mathbf{x}_i - \mathbf{x}_j\|^2}{\sigma_i \sigma_j}\right) & \text{if } \mathbf{x}_i \in N_K(\mathbf{x}_j) \text{ or } \mathbf{x}_j \in N_K(\mathbf{x}_i) \\ 0 & \text{otherwise} \end{cases}$$

where $N_K(\mathbf{x}_i)$ denotes the neighborhood set of the K-nearest neighbors of $\mathbf{x}_i$, σ_i the distance between $\mathbf{x}_i$ and its Kth nearest neighbor, and σ_j the distance between $\mathbf{x}_j$ and its Kth nearest neighbor. This way of constructing the nearest neighbor graph is called *local scaling* [26]. Then $\mathbf{G}$ is the similarity graph with its (i, j)th element being r_{ij}, $\mathbf{D}$ is a diagonal matrix whose entries are the column sums of $\mathbf{G}$, and $\mathbf{L} = \mathbf{D} - \mathbf{G}$.

Model (7) has parameters $\{\boldsymbol{\mu}_k\}$, $\{\tau_k\}$, $\{\pi_k\}$, $\{\nu_{kj}\}$, $\{\mathbf{W}_k\}$. We use the expectation maximization (EM) algorithm [27] to estimate them from data. Here we introduce $\mathbf{z}_i$ as a hidden indicator vector for each unlabeled data point $\mathbf{x}_i$, with z_{ik} being 1 if $\mathbf{x}_i$ belongs to the kth class. Since the number of parameters in this model is quite large, we apply two-fold EM [28] here to speed up convergence. In the E-step of the outer-fold EM, $\{\mathbf{z}_i\}$ are the hidden variables. We estimate $p(z_{ik} = 1)$ as:

$$p(z_{ik} = 1) = p(\boldsymbol{\Pi}_k|\mathbf{x}_i) = \frac{\pi_k p(\mathbf{x}_i|\boldsymbol{\Pi}_k)}{\sum_{j=1}^{C} \pi_j p(\mathbf{x}_i|\boldsymbol{\Pi}_j)}$$

where

$$\begin{aligned} p(\mathbf{x}_i|\boldsymbol{\Pi}_k) &= \int p(\mathbf{x}_i|\mathbf{t}_i, \mathbf{W}_k, \boldsymbol{\mu}_k, \tau_k) p(\mathbf{t}_i) d\mathbf{t}_i \\ &= \mathcal{N}(\mathbf{x}_i|\boldsymbol{\mu}_k, \mathbf{W}_k\mathbf{W}_k^T + \tau_k^{-1}\mathbf{I}_D). \end{aligned}$$

In the M-step of the outer-fold EM, we aim to estimate $\{\pi_k\}$ and $\{\boldsymbol{\mu}_k\}$. The complete-data log-likelihood is defined as

$$L_C = \sum_{i=1}^{l} \ln p(\mathbf{x}_i|\boldsymbol{\Pi}_{y_i}) + \sum_{i=l+1}^{n} \sum_{k=1}^{C} z_{ik} \Big\{ \ln \big[\pi_k p(\mathbf{x}_i|\boldsymbol{\Pi}_k) \big] \Big\}.$$

So the expectation of the complete-data log-likelihood in the M-step of the outer-fold EM can be calculated as

$$\begin{aligned} &\langle L_C \rangle = \\ &\sum_{k=1}^{C} \sum_{y_i=k} \left\{ -\frac{1}{2}(\mathbf{x}_i - \boldsymbol{\mu}_k)^T \boldsymbol{\Phi}_k^{-1} (\mathbf{x}_i - \boldsymbol{\mu}_k) - \frac{1}{2} \ln |\boldsymbol{\Phi}_k| - \frac{D}{2} \ln 2\pi \right\} + \\ &\sum_{k=1}^{C} \sum_{i=l+1}^{n} \langle z_{ik} \rangle \left\{ -\frac{1}{2}(\mathbf{x}_i - \boldsymbol{\mu}_k)^T \boldsymbol{\Phi}_k^{-1} (\mathbf{x}_i - \boldsymbol{\mu}_k) - \frac{1}{2} \ln |\boldsymbol{\Phi}_k| + \ln \pi_k - \frac{D}{2} \ln 2\pi \right\}, \end{aligned}$$

where $\boldsymbol{\Phi}_k = \mathbf{W}_k\mathbf{W}_k^T + \tau_k^{-1}\mathbf{I}_D$. We maximize the expectation of the complete-data log-likelihood with respect to $\{\pi_i\}$ and $\{\boldsymbol{\mu}_i\}$. The update rules are given by

$$\begin{aligned} \tilde{\pi}_k &= \frac{\sum_{i=l+1}^{n} \langle z_{ik} \rangle}{\sum_{i=l+1}^{n} \sum_{k=1}^{C} \langle z_{ik} \rangle} = \frac{1}{u} \sum_{i=l+1}^{n} \langle z_{ik} \rangle \\ \tilde{\boldsymbol{\mu}}_k &= \frac{\sum_{y_i=k} \mathbf{x}_i + \sum_{i=l+1}^{n} \langle z_{ik} \rangle \mathbf{x}_i}{n_k + \sum_{i=l+1}^{n} \langle z_{ik} \rangle}, \end{aligned}$$

where $\langle \cdot \rangle$ denotes the expectation of a variable.

In the E-step of the inner-fold EM, $\{\mathbf{t}_i\}$ are the hidden variables. We estimate

$$\begin{aligned} p(\mathbf{t}_i|\mathbf{x}_i, \boldsymbol{\Pi}_k) &= \mathcal{N}(\mathbf{t}_i|\boldsymbol{\Sigma}_k^{-1}\mathbf{W}_k^T(\mathbf{x}_i - \tilde{\boldsymbol{\mu}}_k), \tau_k^{-1}\boldsymbol{\Sigma}_k^{-1}) \\ \langle \mathbf{t}_i|\boldsymbol{\Pi}_k\rangle &= \boldsymbol{\Sigma}_k^{-1}\mathbf{W}_k^T(\mathbf{x}_i - \tilde{\boldsymbol{\mu}}_k) \\ \langle \mathbf{t}_i\mathbf{t}_i^T|\boldsymbol{\Pi}_k\rangle &= \tau_k^{-1}\boldsymbol{\Sigma}_k^{-1} + \langle \mathbf{t}_i|\boldsymbol{\Pi}_k\rangle\langle \mathbf{t}_i|\boldsymbol{\Pi}_k\rangle^T, \end{aligned}$$

with $\boldsymbol{\Sigma}_k = \tau_k^{-1}\mathbf{I}_d + \mathbf{W}_k^T\mathbf{W}_k$.

In the M-step of the inner-fold EM, we aim to estimate $\{\mathbf{W}_k\}$, $\{\nu_{kj}\}$ and $\{\tau_k\}$. The complete-data log-likelihood can be calculated as

$$\tilde{L}_C = \sum_{i=1}^{l} \ln p(\mathbf{x}_i, \mathbf{t}_i|\boldsymbol{\Pi}_{y_i}) + \sum_{k=1}^{C}\sum_{j=1}^{d} p(\mathbf{W}_{k,j}) + \sum_{i=l+1}^{n}\sum_{k=1}^{C} \langle z_{ik}\rangle \ln\{\pi_k p(\mathbf{x}_i, \mathbf{t}_i|\boldsymbol{\Pi}_k)\}.$$

The expectation of the complete-data log-likelihood can be calculated as

$$\begin{aligned} \langle \tilde{L}_C\rangle = \\ &\sum_{i=1}^{C}\sum_{y_j=i}\Big\{\frac{D}{2}\ln\tau_i - \frac{1}{2}\mathrm{tr}(\langle \mathbf{t}_j\mathbf{t}_j^T|C_i\rangle) - \frac{\tau_i}{2}\|\mathbf{x}_j - \tilde{\boldsymbol{\mu}}_i\|^2 + \\ &\qquad \tau_i\langle \mathbf{t}_j|C_i\rangle^T\mathbf{W}_i^T(\mathbf{x}_j - \tilde{\boldsymbol{\mu}}_i) - \frac{\tau_i}{2}\mathrm{tr}(\mathbf{W}_i^T\mathbf{W}_i\langle \mathbf{t}_j\mathbf{t}_j^T|C_i\rangle)\Big\} + \\ &\sum_{i=1}^{C}\sum_{j=l+1}^{n}\langle z_{ji}\rangle\Big\{\frac{D}{2}\ln\tau_i - \frac{1}{2}\mathrm{tr}(\langle \mathbf{t}_j\mathbf{t}_j^T|C_i\rangle) - \frac{\tau_i}{2}\|\mathbf{x}_j - \tilde{\boldsymbol{\mu}}_i\|^2 + \\ &\qquad \tau_i\langle \mathbf{t}_j|C_i\rangle^T\mathbf{W}_i^T(\mathbf{x}_j - \tilde{\boldsymbol{\mu}}_i) - \frac{\tau_i}{2}\mathrm{tr}(\mathbf{W}_i^T\mathbf{W}_i\langle \mathbf{t}_j\mathbf{t}_j^T|C_i\rangle) + \ln\tilde{\pi}_i\Big\} + \\ &\sum_{i=1}^{C}\sum_{j=1}^{d}\Big\{\frac{D}{2}\ln\nu_{ij} - \frac{1}{2}\nu_{ij}\mathbf{W}_{i,j}^T\mathbf{L}_\star^{-1}\mathbf{W}_{i,j}\Big\}, \end{aligned}$$

where $\mathbf{L}_\star = \mathbf{X}\mathbf{L}\mathbf{X}^T$. Maximization of the expected complete-data log-likelihood with respect to $\mathbf{W}_k$, τ_k and ν_{kj} gives the following update rules:

$$\begin{aligned} \tilde{\mathbf{W}}_k &= (\tau_k\mathbf{S}_k\mathbf{W}_k - \mathbf{L}_\star^{-1}\mathbf{W}_k\boldsymbol{\Lambda}_k\boldsymbol{\Sigma}_k)(\tilde{n}_k\mathbf{I}_d + \tau_k\boldsymbol{\Sigma}_k^{-1}\mathbf{W}_k^T\mathbf{S}_k\mathbf{W}_k)^{-1} \\ \tilde{\tau}_k &= \frac{D\tilde{n}_k}{\mathrm{tr}\left\{(\mathbf{I}_D - \mathbf{W}_k\boldsymbol{\Sigma}_k^{-1}\mathbf{W}_k^T)^2\mathbf{S}_k + \tilde{n}_k\tau_k^{-1}\mathbf{W}_k\boldsymbol{\Sigma}_k^{-1}\mathbf{W}_k^T\right\}} \\ \tilde{\nu}_{kj} &= \frac{D}{\tilde{\mathbf{W}}_{k,j}^T\mathbf{L}_\star^{-1}\tilde{\mathbf{W}}_{k,j}}, \end{aligned}$$

where $\mathbf{L}_\star = \mathbf{X}\mathbf{L}\mathbf{X}^T$, $\boldsymbol{\Lambda}_k = \mathrm{diag}(\nu_{k1}, \ldots, \nu_{kM})$ is a diagonal matrix with the (j, j)th element being ν_{kj}, $\tilde{n}_k = n_k + u\tilde{\pi}_k$, and $\mathbf{S}_k = \sum_{y_i=k}(\mathbf{x}_i - \tilde{\boldsymbol{\mu}}_k)(\mathbf{x}_i - \tilde{\boldsymbol{\mu}}_k)^T + \sum_{i=l+1}^{n}\langle z_{ik}\rangle(\mathbf{x}_i - \tilde{\boldsymbol{\mu}}_k)(\mathbf{x}_i - \tilde{\boldsymbol{\mu}}_k)^T$.

After estimating the parameters, we can use ν_{kj} to determinate the dimensionality of $\mathbf{W}_k$. We can set a threshold η and discard the $\mathbf{W}_{k,j}$ whose corresponding ν_{kj} is larger than η. In our experiments, we set η to be 10000.

For a test data point $\mathbf{x}_{\text{test}}$, we classify it to class $\boldsymbol{\Pi}_k$ where $k = \arg\max_j p(\boldsymbol{\Pi}_j|\mathbf{x}_{\text{test}})$.

4.1 Discussion

Our S^2HPLDA model has advantages over existing probabilistic subspace methods. In our method, each class is modeled by a Gaussian distribution with a possibly different covariance matrix, giving our model higher expressive power than existing methods. Moreover, our model, being a semi-supervised method, can utilize unlabeled data but most other probabilistic LDA models cannot, except S^2PPCA.

There exist several variants of LDA [10,11,12] which also utilize unlabeled data to alleviate the SSS problem. Cai et al. [10] and Zhang and Yeung [11] used unlabeled data to define a regularization term to incorporate the manifold and cluster assumptions, which are two widely adopted assumptions in semi-supervised learning. Zhang and Yeung [12] used unlabeled data to maximize the criterion of LDA and estimate the labels simultaneously, in a way similar to the idea behind transductive SVM (TSVM) [29,30]. Unlike these methods, our method works in a different way. We use a Gaussian mixture model to model the unlabeled data with each component corresponding to one class. From previous research in semi-supervised learning, unlabeled data are more suitable for generative models since unlabeled data can help to estimate the data density [13] and our method also follows this strategy.

According to [31], integrating out all parameters is better than performing point estimation in terms of the generalization performance. In our future research, we plan to propose a fully Bayesian extension of S^2HPLDA by placing priors on the parameters of S^2HPLDA. For example, we can add a Dirichlet prior to $(\pi_1, \ldots, \pi_C)$, a Gaussian prior to $\boldsymbol{\mu}_k$, and Gamma priors to τ_k and ν_{kj}:

$$\begin{aligned}(\pi_1, \ldots, \pi_C) &\sim \mathrm{Dir}(\alpha_0, \ldots, \alpha_0) \\ \boldsymbol{\mu}_k &\sim \mathcal{N}(\boldsymbol{\mu}_0, \beta_0 \mathbf{I}_D) \\ \tau_k &\sim \mathrm{Gamma}(a_0, b_0) \\ \nu_{kj} &\sim \mathrm{Gamma}(c_0, d_0).\end{aligned}$$

Since direct inference is intractable, we may resort to the variational approximation approach [32].

5 Experiments

In this section, we report experimental results based on two face databases and one object database to evaluate the performance of our method and compare it with some related probabilistic subspace methods.

5.1 Experimental Setup

Subspace methods are widely used in face recognition and object recognition applications. Previous research found that face and object images usually lie in a low-dimensional subspace of the ambient image space. Eigenface [33] (based on PCA) and Fisherface [5] (based on LDA) are two representative subspace methods. Many variants have also been proposed in recent years. These subspace methods use different dimensionality reduction techniques to obtain a low-dimensional subspace and then perform classification in the subspace using some classifier. Some researchers also proposed probabilistic versions of these subspace methods, with PPCA [22] and SPPCA [19] being two popular ones. From the analysis in [22], the maximum likelihood solution to PPCA is identical to that to PCA. Since the models proposed in [19] and [23] are identical, then from the analysis in [23], the maximum likelihood solution to SPPCA is also the same as that to LDA. Moreover, PPCA and SPPCA can deal with missing data using the EM algorithm, but PCA and LDA cannot. In our experiments, we study our method empirically and compare it with several probabilistic subspace methods, including PLDA [20], SPPCA [19] and S^2PPCA [19]. Note that PLDA and SPPCA are supervised, but S^2PPCA and our method S^2HPLDA are semi-supervised in nature. For SPPCA and S^2PPCA, we use a simple nearest-neighbor classifier to perform classification after dimensionality reduction.

5.2 Face Recognition

We use the ORL face database [5] for the first experiment. The ORL face database contains 400 face images of 40 persons, each having 10 images. These face images contain significant variations in pose and scale. Some images from the database are shown in Figure 3. We randomly select seven images for each person to form the training set and the rest for the test set. Of the seven images for each person, $p \in \{2, 3\}$ images are randomly selected and labeled while the other images remain unlabeled. We perform 10 random splits and report the average results across the 10 trials. Table 1 reports the error rates of different methods evaluated on the unlabeled training data and the test data separately. For each setting, the lowest classification error is shown in bold. Since S^2PPCA exploits the structure of unlabeled data, we can see that its performance is better than PLDA and SPPCA. Moreover, S^2HPLDA relaxes the homoscedasticity assumption and so it achieves better performance than its homoscedastic counterpart S^2PPCA in our settings. From Table 1, we can see that the performance of PLDA is very bad, probably because it gets trapped in an unsatisfactory local optimum when running the EM algorithm.

The PIE database [34] is used in our second experiment. This database contains 41,368 face images from 68 individuals and these images have large variations in pose, illumination and expression conditions. For our experiments, we select the frontal pose

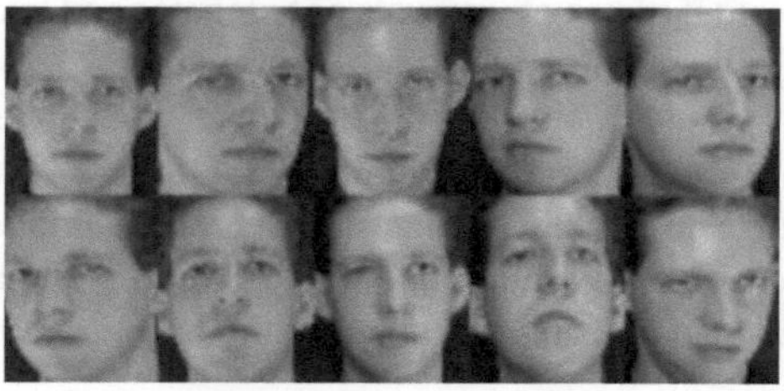

Fig. 3. Some images for one person in the ORL database

Table 1. Recognition error rates (in mean±std-dev) on ORL for two different p values. 1ST TABLE: $p = 2$; 2ND TABLE: $p = 3$.

Method	Error rate (unlabeled)	Error rate (test)
PLDA	0.7141±0.0803	0.7016±0.0640
SPPCA	0.4562±0.1219	0.4578±0.0710
S^2PPCA	0.2703±0.0332	0.2422±0.0366
S^2HPLDA	**0.1406±0.0231**	**0.1781±0.0308**

Method	Error rate (unlabeled)	Error rate (test)
PLDA	0.5156±0.0744	0.5042±0.0603
SPPCA	0.4359±0.0713	0.4604±0.0322
S^2PPCA	0.2625±0.0595	0.2000±0.0245
S^2HPLDA	**0.1375±0.0135**	**0.1562±0.0336**

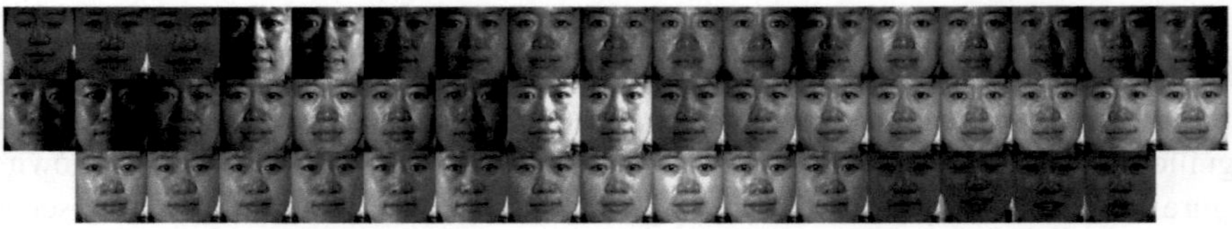

Fig. 4. Some images for one person in the PIE database

(C27)[1] with varying lighting and illumination conditions and there are about 49 images for each subject. Some images from the database are shown in Figure 4. The experimental setting is almost the same as that of the first experiment. The only difference is that we use 22 images to form the training set. Of these 22 images, we randomly select $p \in \{3, 4, 5, 6\}$ images and label them, leaving the remaining images unlabeled. Each setting is also repeated 10 times. Table 2 reports the average results over the 10 trials. From the results, we can see that our method again gives the best performance.

5.3 Object Recognition

We use the COIL database [35] for our object recognition experiment. This database contains 1,440 grayscale images with black background for 20 objects. For each object, the camera moves around it in pan at intervals of 5 degrees and takes a total of 72 different images. These objects exhibit a wide variety of complex geometric and reflectance characteristics. Some sample images for the 20 objects are shown in Figure 5. We use 22 images from each object to form the training set. Of the 22 images, $p \in \{3, 4, 5, 6\}$ images are randomly selected as labeled data and the rest as unlabeled data. We perform 10 random splits on each configuration and Table 3 reports the average results. From the results, our method also outperforms other methods under all four settings.

[1] This face database can be downloaded from
`http://www.cs.uiuc.edu/homes/dengcai2/Data/FaceData.html`

Table 2. Recognition error rates (in mean±std-dev) on PIE for four different p values. 1ST TABLE: $p = 3$; 2ND TABLE: $p = 4$; 3RD TABLE: $p = 5$; 4TH TABLE: $p = 6$.

Method	Error rate (unlabeled)	Error rate (test)
PLDA	0.8421±0.0142	0.8492±0.0212
SPPCA	0.4509±0.0487	0.4798±0.0590
S^2PPCA	0.3367±0.0088	0.3639±0.0139
S^2HPLDA	**0.3066±0.0131**	**0.3109±0.0397**
Method	Error rate (unlabeled)	Error rate (test)
PLDA	0.7549±0.0451	0.7469±0.0532
SPPCA	0.2741±0.0202	0.2654±0.0073
S^2PPCA	0.2545±0.0110	0.2520±0.0046
S^2HPLDA	**0.2096±0.0324**	**0.2225±0.0066**
Method	Error rate (unlabeled)	Error rate (test)
PLDA	0.7029±0.0018	0.7201±0.0154
SPPCA	0.2080±0.0153	0.2409±0.0120
S^2PPCA	0.2011±0.0055	0.2330±0.0046
S^2HPLDA	**0.1743±0.0177**	**0.1933±0.0108**
Method	Error rate (unlabeled)	Error rate (test)
PLDA	0.7096±0.0351	0.7215±0.0420
SPPCA	0.1875±0.0104	0.2119±0.0143
S^2PPCA	0.1590±0.0390	0.1724±0.0347
S^2HPLDA	**0.1220±0.0149**	**0.1450±0.0204**

Fig. 5. Some images for different objects in the COIL database

Table 3. Recognition error rates (in mean±std-dev) on COIL for four different p values. 1ST TABLE: $p = 3$; 2ND TABLE: $p = 4$; 3RD TABLE: $p = 5$; 4TH TABLE: $p = 6$.

Method	Error rate (unlabeled)	Error rate (test)
PLDA	0.4026±0.0112	0.4000±0.0311
SPPCA	0.7303±0.1172	0.7195±0.1393
S^2PPCA	0.3303±0.0428	0.3270±0.0410
S^2HPLDA	**0.3145±0.0651**	**0.3015±0.0474**

Method	Error rate (unlabeled)	Error rate (test)
PLDA	0.3694±0.0118	0.3850±0.0156
SPPCA	0.6958±0.0727	0.7075±0.0658
S^2PPCA	0.3500±0.0039	0.3195±0.0021
S^2HPLDA	**0.3167±0.0314**	**0.3005±0.0375**

Method	Error rate (unlabeled)	Error rate (test)
PLDA	0.3471±0.0208	0.3290±0.0792
SPPCA	0.7691±0.0769	0.7815±0.0884
S^2PPCA	0.3221±0.0062	0.2865±0.0346
S^2HPLDA	**0.2438±0.0265**	**0.2670±0.0566**

Method	Error rate (unlabeled)	Error rate (test)
PLDA	0.3156±0.0707	0.3085±0.0559
SPPCA	0.7844±0.0398	0.7840±0.0226
S^2PPCA	0.3391±0.0420	0.3270±0.0028
S^2HPLDA	**0.2250±0.0312**	**0.2200±0.0354**

6 Conclusion

In this paper, we have presented a new probabilistic LDA model. This semi-supervised, heteroscedastic extension allows it to overcome some serious limitations of LDA. As said earlier in the paper, one natural extension is a fully Bayesian extension to boost the generalization performance of the probabilistic model. Another possibility is to apply the kernel trick to introduce nonlinearity into the model using techniques such as that in [36].

Acknowledgments

This research has been supported by General Research Fund 621407 from the Research Grants Council of the Hong Kong Special Administrative Region, China. We would like to thank Shipeng Yu for providing the source code of SPPCA and S^2PPCA.

References

1. Jolliffe, I.T.: Principal Component Analysis, 2nd edn. Springer, New York (2002)
2. Fukunaga, K.: Introduction to Statistical Pattern Recognition. Academic Press, New York (1991)

3. Chen, L., Liao, H., Ko, M., Lin, J., Yu, G.: A new LDA-based face recognition system which can solve the small sample size problem. Pattern Recognition 33(10), 1713–1726 (2000)
4. Krzanowski, W.J., Jonathan, P., McCarthy, W.V., Thomas, M.R.: Discriminant analysis with singular covariance matrices: methods and applications to spectroscopic data. Applied Statistics 44(1), 101–115 (1995)
5. Belhumeur, P.N., Hespanha, J.P., Kriegman, D.J.: Eigenfaces vs. Fisherfaces: Recognition using class specific linear projection. IEEE Transactions on Pattern Analysis and Machine Intelligence 19(7), 711–720 (1997)
6. Ye, J., Li, Q.: A two-stage linear discriminant analysis via QR-decomposition. IEEE Transactions on Pattern Analysis and Machine Intelligence 27(6), 929–941 (2005)
7. Cevikalp, H., Neamtu, M., Wilkes, M., Barkana, A.: Discriminative common vectors for face recognition. IEEE Transactions on Pattern Analysis and Machine Intelligence 27(1), 4–13 (2005)
8. Wang, X., Tang, X.: Dual-space linear discriminant analysis for face recognition. In: Proceedings of the IEEE Computer Society Conference on Computer Vision and Pattern Recognition, Washington, DC, pp. 564–569 (2004)
9. Ye, J., Janardan, R., Li, Q.: Two-dimensional linear discriminant analysis. In: Advances in Neural Information Processing Systems 17, Vancouver, British Columbia, Canada, pp. 1529–1536 (2005)
10. Cai, D., He, X., Han, J.: Semi-supervised discriminant analysis. In: Proceedings of the IEEE International Conference on Computer Vision, Rio de Janeiro, Brazil (2007)
11. Zhang, Y., Yeung, D.Y.: Semi-supervised discriminant analysis using robust path-based similarity. In: Proceedings of the IEEE Computer Society Conference on Computer Vision and Pattern Recognition, Anchorage, Alaska (2008)
12. Zhang, Y., Yeung, D.Y.: Semi-supervised discriminant analysis via CCCP. In: Proceedings of the 19th European Conference on Machine Learning and Principles and Practice of Knowledge Discovery in Databases, Antwerp, Belgium, pp. 644–659 (2008)
13. Chapelle, O., Zien, A., Schölkopf, B. (eds.): Semi-Supervised Learning. MIT Press, Boston (2006)
14. Hastie, T., Tibshirani, R.: Discriminant analysis by Gaussian mixture. Journal of the Royal Statistical Society, Series B 58(1), 155–176 (1996)
15. Kumar, N., Andreou, A.G.: Heteroscedastic discriminant analysis and reduced rank HMMS for improved speech recognition. Speech Communication 26(4), 283–297 (1998)
16. Loog, M., Duin, R.P.W.: Linear dimensionality reduction via a heteroscedastic extension of LDA: the Chernoff criterion. IEEE Transactions on Pattern Analysis and Machine Intelligence 26(6), 732–739 (2004)
17. Dai, G., Yeung, D.Y., Chang, H.: Extending kernel fisher discriminant analysis with the weighted pairwise Chernoff criterion. In: Leonardis, A., Bischof, H., Pinz, A. (eds.) ECCV 2006. LNCS, vol. 3954, pp. 308–320. Springer, Heidelberg (2006)
18. Ioffe, S.: Probabilistic linear discriminant analysis. In: Leonardis, A., Bischof, H., Pinz, A. (eds.) ECCV 2006. LNCS, vol. 3954, pp. 531–542. Springer, Heidelberg (2006)
19. Yu, S., Yu, K., Tresp, V., Kriegel, H.P., Wu, M.: Supervised probabilistic principal component analysis. In: Proceedings of the 12th ACM SIGKDD International Conference on Knowledge Discovery and Data Mining, Philadelphia, PA, USA, pp. 464–473 (2006)
20. Prince, S.J.D., Elder, J.H.: Probabilistic linear discriminant analysis for inferences about identity. In: Proceedings of the 11th IEEE International Conference on Computer Vision, Rio de Janeiro, Brazil, pp. 1–8 (2007)
21. Neal, R.M.: Bayesian Learning for Neural Network. Springer, New York (1996)
22. Tipping, M.E., Bishop, C.M.: Probabilistic principal component analysis. Journal of the Royal Statistic Society, B 61(3), 611–622 (1999)

23. Bach, F.R., Jordan, M.I.: A probabilistic interpretation of canonical correlation analysis. Technical Report 688, Department of Statistics, University of California, Berkeley (2005)
24. He, X., Yan, S., Hu, Y., Niyogi, P., Zhang, H.J.: Face recognition using Laplacianfaces. IEEE Transactions on Pattern Analysis and Machine Intelligence 27(3), 328–340 (2005)
25. Belkin, M., Niyogi, P., Sindhwani, V.: Manifold regularization: A geometric framework for learning from labeled and unlabeled examples. Journal of Machine Learning Research 7, 2399–2434 (2006)
26. Zelnik-Manor, L., Perona, P.: Self-tuning spectral clustering. In: Advances in Neural Information Processing Systems 17, Vancouver, British Columbia, Canada, pp. 1601–1608 (2005)
27. Dempster, A.P., Laird, N.M., Rubin, D.B.: Maximum likelihood from incomplete data via the EM algorithm. Journal of the Royal Statistic Society, B 39(1), 1–38 (1977)
28. Tipping, M.E., Bishop, C.M.: Mixtures of probabilistic principal component analysers. Neural Computation 11(2), 443–482 (1999)
29. Joachims, T.: Transductive inference for text classification using support vector machines. In: Proceedings of the Sixteenth International Conference on Machine Learning, San Francisco, CA, USA, pp. 200–209 (1999)
30. Bennett, K., Demiriz, A.: Semi-supervised support vector machines. In: Advances in Neural Information Processing Systems 11, Vancouver, British Columbia, Canada, pp. 368–374 (1998)
31. MacKay, D.J.C.: Comparison of approximate methods for handling hyperparameters. Neural Computation 11(5), 1035–1068 (1999)
32. Bishop, C.M.: Pattern Recognition and Machine Learning. Springer, New York (2006)
33. Turk, M., Pentland, A.: Eigenfaces for recognition. Journal of Cognitive Neuroscience 3(1), 71–86 (1991)
34. Sim, T., Baker, S., Bsat, M.: The CMU pose, illumination and expression database. IEEE Transactions on Pattern Analysis and Machine Intelligence 25(12), 1615–1618 (2003)
35. Nene, S.A., Nayar, S.K., Murase, H.: Columbia object image library (COIL-20). Technical Report 005, CUCS (1996)
36. Lawrence, N.D.: Probabilistic non-linear principal component analysis with Gaussian process latent variable models. Journal of Machine Learning Research 6, 1783–1816 (2005)

Semi-Supervised Multi-Task Regression

Yu Zhang and Dit-Yan Yeung

Hong Kong University of Science and Technology
{zhangyu,dyyeung}@cse.ust.hk

Abstract. Labeled data are needed for many machine learning applications but the amount available in some applications is scarce. Semi-supervised learning and multi-task learning are two of the approaches that have been proposed to alleviate this problem. In this paper, we seek to integrate these two approaches for regression applications. We first propose a new supervised multi-task regression method called SMTR, which is based on Gaussian processes (GP) with the assumption that the kernel parameters for all tasks share a common prior. We then incorporate unlabeled data into SMTR by changing the kernel function of the GP prior to a data-dependent kernel function, resulting in a semi-supervised extension of SMTR, called SSMTR. Moreover, we incorporate pairwise information into SSMTR to further boost the learning performance for applications in which such information is available. Experiments conducted on two commonly used data sets for multi-task regression demonstrate the effectiveness of our methods.

1 Introduction

Many machine learning applications require that labeled data be available for model training. Unfortunately the amount of labeled data available in some applications is very scarce because labeling the data points manually is very tedious and costly. As a consequence, the model thus learned is often not satisfactory in performance. To alleviate this problem, machine learning researchers have investigated various approaches, with *semi-supervised learning* and *multi-task learning* being two popular ones.

Semi-supervised learning [1] can be seen as an extension of the conventional supervised learning paradigm by augmenting the (labeled) training data set with unlabeled data so as to exploit the useful information in the unlabeled data to boost learning performance. Early semi-supervised learning methods include co-training [2], which builds two learning models based on two different views of the data and then uses each learning model to select confident unlabeled data for the other, and transductive SVM [3,4], which uses both labeled and unlabeled data to maximize the margin of a support vector machine (SVM). More recent development includes many graph-based methods [5,6,7], which model the geometric relationship between all data points in the form of a graph and then propagate the label information from the labeled data points to the unlabeled data points throughout the graph. In order for the unlabeled data to be useful for semi-supervised learning, some assumptions about the data have to be satisfied. Two widely used assumptions are the cluster assumption and manifold assumption. The cluster assumption simply means that if two points are in the same cluster, they are more likely to belong to the same class. Equivalently, this means that the class decision boundary only

W. Buntine et al. (Eds.): ECML PKDD 2009, Part II, LNAI 5782, pp. 617–631, 2009.

goes through low-density regions. Transductive SVM is one popular method based on the cluster assumption. As for the manifold assumption, it means that the data points in some high-dimensional space span a low-dimensional manifold. If two points are close to each other with respect to some metric on the manifold, their outputs are likely to be similar. As a result, if we want to preserve the manifold structure when performing some projection, then two points that are close in the manifold should remain close after projection. The manifold assumption is the underlying model assumption of many semi-supervised learning methods, particularly graph-based methods.

On the other hand, multi-task learning [8,9,10] seeks to improve the learning performance of one task with the help of some related tasks. This approach has been inspired by psychological observations that humans can often benefit from previous learning experience when learning a new but related task, sometimes referred to as *transfer of learning*. Many multi-task learning methods have been proposed over the past decade. For example, multi-task feature learning [11] learns a common representation for all tasks under the regularization framework, regularized multi-task SVM [12] extends SVM by requiring that the SVM parameters for all tasks be close to each other, task clustering methods [13,14] group the tasks into multiple clusters and then learn a similar or common representation for all tasks within a cluster, and GP-based multi-task learning methods [15,16,17,18] utilize Gaussian processes (GP) as the base model for multi-task learning. For multi-task learning, many existing methods assume that all the tasks are related to each other and hence similar or identical data features or model parameters are shared by all tasks or subsets of tasks, for example, all tasks in the same cluster. Methods based on neural networks and multi-task feature learning all assume that the data features are shared by all tasks. On the other hand, regularized multi-task SVM and the methods in [14] assume that similar model parameters are shared by all tasks or tasks in the same cluster. Moreover, some methods incorporate both assumptions in their models, e.g., [13].

Since semi-supervised learning and multi-task learning share the common objective of seeking to improve the learning performance of the original supervised learning task by exploiting some auxiliary data available (unlabeled data for the current task or labeled data for other related tasks), it makes sense to combine them in an attempt to get the best of both worlds. Indeed, some such attempts have been made recently. The method proposed by Ando and Zhang [19] bears some relationship with these two learning paradigms even though its objective is mainly to improve the performance of semi-supervised learning. There exists only a single task (target task) to start with as well as some unlabeled data. The unlabeled data are then utilized to create more tasks to help the learning of the target task. Moreover, Liu et al. [20] proposed a semi-supervised learning method called parameterized neighborhood-based classification (PNBC), which applies random walk to logistic regression and uses a task clustering method for multi-task learning. However, these two methods only consider the classification problem and cannot be extended readily to the regression problem. Indeed, there exist some applications that can be modeled as the combination of semi-supervised regression and multi-task regression, for example, personalized pose estimation. In personalized pose estimation, each task corresponds to the pose estimation for one person. In this application, there exist large amount of images with unknown pose information for each person.

To the best of our knowledge, there does not exist any work in the literature that integrates semi-supervised regression and multi-task regression. In this paper, we want to fill the gap by proposing a scheme for such integration. We first propose a new supervised multi-task regression method called SMTR, which is based on GP with the assumption that the kernel parameters for all tasks share a common Gaussian prior. We then incorporate unlabeled data into SMTR by changing the kernel function of the GP prior to a data-dependent kernel function, resulting in a semi-supervised extension of SMTR, called SSMTR. Moreover, as in [21], we incorporate pairwise information into SSMTR to further boost the learning performance for applications in which such information is available.

We first present SMTR in Section 2. SSMTR, our semi-supervised extension of SMTR, and the incorporation of pairwise information into SSMTR are then presented in Sections 3 and 4, respectively. Section 5 reports some experimental results to provide empirical evaluation of our proposed methods.

2 Supervised Multi-Task Regression

Let there be m related regression tasks $T_1, \ldots, T_m$. For task T_i, the training set D_i consists of n_i labeled data points $\{(\mathbf{x}_j^i, y_j^i)\}_{j=1}^{n_i}$ with the jth point $\mathbf{x}_j^i \in \mathbb{R}^d$ and its output $y_j^i \in \mathbb{R}$.

For each task, we use a GP [22] as the base regressor. For task T_i, we define a latent variable f_j^i for each data point $\mathbf{x}_j^i$. The prior of $\mathbf{f}^i$ is defined as

$$\mathbf{f}^i \mid \mathbf{X}^i \sim \mathcal{N}(\mathbf{0}_{n_i}, \mathbf{K}_{\boldsymbol{\theta}_i}), \tag{1}$$

where $\mathbf{f}^i = (f_1^i, \ldots, f_{n_i}^i)^T$, $\mathbf{X}^i = (\mathbf{x}_1^i, \ldots, \mathbf{x}_{n_i}^i)$, $\mathcal{N}(\mathbf{m}, \boldsymbol{\Sigma})$ denotes a multivariate Gaussian distribution with mean $\mathbf{m}$ and covariance matrix $\boldsymbol{\Sigma}$, $\mathbf{0}_{n_i}$ denotes an $n_i \times 1$ zero vector, and $\mathbf{K}_{\boldsymbol{\theta}_i}$ denotes the kernel matrix defined on $\mathbf{X}^i$ where the kernel function is parameterized by $\boldsymbol{\theta}_i$.

The likelihood for each task T_i is defined based on the Gaussian noise model:

$$\mathbf{y}^i \mid \mathbf{f}^i \sim \mathcal{N}(\mathbf{f}^i, \sigma^2 \mathbf{I}_{n_i}), \tag{2}$$

where $\mathbf{y}^i = (y_1^i, \ldots, y_{n_i}^i)^T$, σ^2 denotes the noise level, and $\mathbf{I}_{n_i}$ is the $n_i \times n_i$ identity matrix.

Since all tasks are assumed to be related, we impose a common prior on the kernel parameters $\{\boldsymbol{\theta}_i\}_{i=1}^m$ for all m tasks:

$$\boldsymbol{\theta}_i \sim \mathcal{N}(\mathbf{m}_\theta, \boldsymbol{\Sigma}_\theta). \tag{3}$$

The graphical model for SMTR is depicted in Figure 1.

In some formulation of GP regression, the noise level σ^2 can also be regarded as one element of the kernel parameters $\boldsymbol{\theta}_i$ since GP regression has an analytical form for $p(y_j^i \mid \mathbf{x}_j^i)$. So the noise levels for different tasks can also share a common prior as

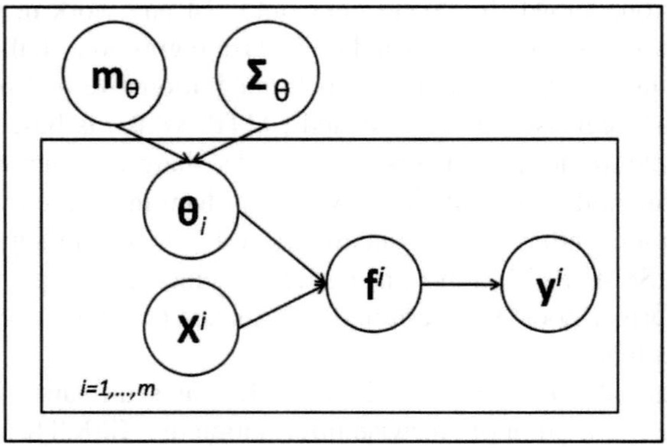

Fig. 1. Graphical model for Supervised Multi-Task Regression

in Eq. (3) but they are not identical. Note that the noise level can be estimated from the labeled data. Since the number of labeled data points in semi-supervised learning is typically not very large, it may not be possible to obtain an accurate estimate of the noise level if estimation is done independently for each task based on limited labeled data. For this reason, we assume in this paper that all tasks have the same noise level. The more general case that allows different noise levels for different tasks will be studied in the future.

There exist some GP-based multi-task regression models [15,16,17,18]. Lawrence and Platt [15] proposed a multi-task regression model in which the kernel parameters are shared by all tasks. This assumption becomes unreasonable when there exist outlier tasks. This problem also exists in the models of [16] and [17], which later motivated the development of a robust model using t-processes [23]. Unlike the model in [15], the models in [16,17] learn the kernel matrix in a nonparametric way. This makes it difficult to perform inductive inference since there is no parametric form for the kernel function. Bonilla et al. [18] proposed a powerful multi-task GP regressor which is especially suitable for multi-output regression problems. Their method directly models the similarity between multiple tasks and is equivalent to using a matrix-variate normal distribution to model the multiple latent function values. Due to the use of the Kronecker product, this method incurs high storage and computational costs. However, these difficulties do not exist in our proposed model. In our model, the kernel parameters for different tasks just share the same prior but are not identical, making it capable of modeling outlier tasks. Our model has a parametric form for the kernel function and hence it can be used to make inductive inference directly. Even though our model does not directly characterize the relatedness between tasks, it is implicitly characterized by the kernel parameters. Moreover, since the dimensionality of $\boldsymbol{\theta}_i$ is usually not very large, the storage cost is not high. Although there exist some multi-task learning methods which also place a common prior on the model parameters of different tasks [24,25,13], to the best of our knowledge none of them is based on GP.

2.1 Learning and Inference

Since

$$p(\mathbf{y}^i|\mathbf{X}^i) = \int p(\mathbf{y}^i|\mathbf{f}^i)p(\mathbf{f}^i|\mathbf{X}^i)d\mathbf{f}^i$$
$$= \mathcal{N}(\mathbf{y}^i|\mathbf{0}_{n_i}, \mathbf{K}_{\boldsymbol{\theta}_i} + \sigma^2\mathbf{I}_{n_i}),$$

the log-likelihood of all tasks can be computed as

$$L = -\frac{1}{2}\sum_{i=1}^{m}\left[(\mathbf{y}^i)^T(\mathbf{K}_{\boldsymbol{\theta}_i} + \sigma^2\mathbf{I}_{n_i})^{-1}\mathbf{y}^i + \ln\left|\mathbf{K}_{\boldsymbol{\theta}_i} + \sigma^2\mathbf{I}_{n_i}\right|\right]$$
$$-\frac{1}{2}\sum_{i=1}^{m}\left[(\boldsymbol{\theta}_i - \mathbf{m}_\theta)^T\boldsymbol{\Sigma}_\theta^{-1}(\boldsymbol{\theta}_i - \mathbf{m}_\theta) + \ln\left|\boldsymbol{\Sigma}_\theta^{-1}\right|\right] + \text{Const},$$

where $|\mathbf{A}|$ denotes the determinant of a square matrix $\mathbf{A}$. We maximize L to estimate the optimal values of $\boldsymbol{\theta}_i$, σ, $\mathbf{m}_\theta$ and $\boldsymbol{\Sigma}_\theta$. Since the number of parameters to estimate is large, we use an alternating method to solve the problem.

In the $(t+1)$st iteration, given $\mathbf{m}_\theta^{(t)}$ and $\boldsymbol{\Sigma}_\theta^{(t)}$ as estimates of $\mathbf{m}_\theta$ and $\boldsymbol{\Sigma}_\theta$ from the tth iteration, we apply gradient ascent to maximize the log-likelihood to estimate $\boldsymbol{\theta}_i^{(t+1)}$ and $\sigma^{(t+1)}$. The form of the kernel function we adopt is $k(\mathbf{x}_1, \mathbf{x}_2) = \theta_1\mathbf{x}_1^T\mathbf{x}_2 + \theta_2\exp(-\frac{\|\mathbf{x}_1-\mathbf{x}_2\|_2^2}{2\theta_3^2})$ where $\|\cdot\|_2$ denotes the 2-norm of a vector. Since each element of $\boldsymbol{\theta}_i$ and σ is positive, we instead treat $\ln\boldsymbol{\theta}_i$ and $\ln\sigma$ as variables, where each element of $\ln\boldsymbol{\theta}_i$ is the logarithm of the corresponding element in $\boldsymbol{\theta}_i$. The gradients of the log-likelihood with respect to $\ln\boldsymbol{\theta}_i$ and $\ln\sigma$ can be computed as:

$$\frac{\partial L}{\partial \ln\sigma} = \frac{\partial L}{\partial\sigma^2}\frac{\partial\sigma^2}{\partial\ln\sigma}$$
$$= \sigma^2\sum_{i=1}^{m}\left\{(\mathbf{y}^i)^T(\mathbf{K}_{\boldsymbol{\theta}_i} + \sigma^2\mathbf{I}_{n_i})^{-2}\mathbf{y}^i - \operatorname{tr}\left[(\mathbf{K}_{\boldsymbol{\theta}_i} + \sigma^2\mathbf{I}_{n_i})^{-1}\right]\right\}$$

$$\frac{\partial L}{\partial\ln\boldsymbol{\theta}_i} = \frac{1}{2}\operatorname{diag}(\boldsymbol{\theta}_i)\left[\operatorname{Tr}\left(\mathbf{A}\frac{\partial\mathbf{K}_{\boldsymbol{\theta}_i}}{\partial\boldsymbol{\theta}_i}\right) - 2(\boldsymbol{\Sigma}_\theta^{(t)})^{-1}(\boldsymbol{\theta}_i - \mathbf{m}_\theta^{(t)})\right],$$

where $\mathbf{A} = (\mathbf{K}_{\boldsymbol{\theta}_i}+\sigma^2\mathbf{I}_{n_i})^{-1}\mathbf{y}^i(\mathbf{y}^i)^T(\mathbf{K}_{\boldsymbol{\theta}_i}+\sigma^2\mathbf{I}_{n_i})^{-1}-(\mathbf{K}_{\boldsymbol{\theta}_i}+\sigma^2\mathbf{I}_{n_i})^{-1}$, $\operatorname{tr}(\cdot)$ denotes the trace function defined on a square matrix, $\operatorname{Tr}(\mathbf{A}\frac{\partial\mathbf{K}_{\boldsymbol{\theta}_i}}{\partial\boldsymbol{\theta}_i})$ denotes a vector whose jth element is $\operatorname{tr}(\mathbf{A}\frac{\partial\mathbf{K}_{\boldsymbol{\theta}_i}}{\partial\theta_{ij}})$ where θ_{ij} is the jth element of $\boldsymbol{\theta}_i$, and $\operatorname{diag}(\boldsymbol{\theta}_i)$ denotes the diagonal matrix whose (j, j)th element is the jth element of $\boldsymbol{\theta}_i$.

After we obtain $\boldsymbol{\theta}_i^{(t+1)}$ and $\sigma^{(t+1)}$, we keep them fixed and maximize the log-likelihood with respect to $\mathbf{m}_\theta$ and $\boldsymbol{\Sigma}_\theta$. With some simple algebraic calculations, we can get

$$\mathbf{m}_\theta^{(t+1)} = \frac{1}{m}\sum_{i=1}^{m}\boldsymbol{\theta}_i$$
$$\boldsymbol{\Sigma}_\theta^{(t+1)} = \frac{1}{m}\sum_{i=1}^{m}(\boldsymbol{\theta}_i - \mathbf{m}_\theta^{(t+1)})(\boldsymbol{\theta}_i - \mathbf{m}_\theta^{(t+1)})^T.$$

These two steps are repeated until the model parameters converge.

Given a test data point $\mathbf{x}_\star^i$ of task T_i, the predictive distribution $p(y_\star^i \mid x_\star^i, \mathbf{X}^i, \mathbf{y}^i)$ is a Gaussian distribution with mean $m_\star^i$ and variance $(\sigma_\star^i)^2$ given by

$$m_\star^i = (\mathbf{k}_\star^i)^T(\mathbf{K}_{\boldsymbol{\theta}_i} + \sigma^2\mathbf{I}_{n_i})^{-1}\mathbf{y}^i$$
$$(\sigma_\star^i)^2 = k_{\boldsymbol{\theta}_i}(\mathbf{x}_\star^i, \mathbf{x}_\star^i) - (\mathbf{k}_\star^i)^T(\mathbf{K}_{\boldsymbol{\theta}_i} + \sigma^2\mathbf{I}_{n_i})^{-1}\mathbf{k}_\star^i,$$

where $k_{\boldsymbol{\theta}_i}(\cdot,\cdot)$ denotes the kernel function parameterized by $\boldsymbol{\theta}_i$ and $\mathbf{k}_\star^i = (k_{\boldsymbol{\theta}_i}(\mathbf{x}_\star^i, \mathbf{x}_1^i), \ldots, k_{\boldsymbol{\theta}_i}(\mathbf{x}_\star^i, \mathbf{x}_{n_i}^i))^T$.

The computational complexity of our model is $O\left(\sum_{i=1}^m (n_i)^3\right)$. Since the data set sizes n_i for different tasks are generally small in typical semi-supervised learning applications, our model is usually quite efficient.

2.2 Inductive Inference for New Tasks

The model presented above assumes that all tasks are given in advance for multi-task learning to take place. This setting is sometimes referred to as symmetric multi-task learning [14]. If a newly arrived task does not belong to any of the tasks in the training set, our model can still deal with this situation easily without having to retrain the whole model from scratch using an augmented training set. Instead, we can utilize the common prior in Eq. (3) as the prior of the kernel parameters for the new task and then perform maximum a posteriori (MAP) estimation to obtain the kernel parameters and maximum likelihood estimation (MLE) to obtain the noise level. Therefore, we only need to store $\mathbf{m}_\theta$ and $\boldsymbol{\Sigma}_\theta$ instead of all the training data points for all tasks.

3 Semi-Supervised Multi-Task Regression

We now extend the SMTR model to the semi-supervised setting, which is called SSMTR. For task T_i, the training set D_i consists of a set of labeled data points $\{(\mathbf{x}_j^i, y_j^i)\}_{j=1}^{l_i}$ and a set of unlabeled data points $\{\mathbf{x}_j^i\}_{j=l_i+1}^{n_i}$. Typically, we have $n_i \gg l_i$.

Like in many semi-supervised learning methods which are based on the manifold assumption as described above, the unlabeled data in our model serve to enforce the smoothness of the regression function. For each task T_i, we use *local scaling* [26] to construct the similarity graph $\mathbf{S}^i$ in which each element is defined as follows:

$$S_{jr}^i = \begin{cases} \exp\left(-\frac{\|\mathbf{x}_j^i - \mathbf{x}_r^i\|^2}{\sigma_j^i\sigma_r^i}\right) & \text{if } \mathbf{x}_j^i \in N_K(\mathbf{x}_r^i) \text{ or } \mathbf{x}_r^i \in N_K(\mathbf{x}_j^i) \\ 0 & \text{otherwise} \end{cases}$$

where $N_K(\mathbf{x}_j^i)$ denotes the neighborhood set of the K nearest neighbors of $\mathbf{x}_j^i$ in task T_i, σ_j^i is the distance between $\mathbf{x}_j^i$ and its Kth nearest neighbor, and σ_r^i is the distance between $\mathbf{x}_r^i$ and its Kth nearest neighbor.

We introduce a random variable G^i to reflect the geometric structure contained in the training set of task T_i. The prior for G is defined as

$$p(G^i \mid \mathbf{f}^i, D_i) \propto \exp\left[-\frac{\alpha_i}{2}(\mathbf{f}^i)^T \mathbf{L}^i \mathbf{f}^i\right], \tag{4}$$

where $\mathbf{f}^i = (f_1^i, \ldots, f_{n_i}^i)^T$ includes the latent variables for both labeled and unlabeled data, $\mathbf{L}^i$ is the Laplacian matrix or normalized Laplacian matrix [27] of the similarity graph $\mathbf{S}^i$ defined on the training set D_i, and α_i is a hyperparameter which needs to be estimated. So if the probability of G^i is high, it means that the data set is more likely to contain manifold structure according to the graph structure implied by $\mathbf{L}^i$.

Thus the joint prior of $\mathbf{f}^i$ conditioned on D_i and G^i can be computed based on

$$p(\mathbf{f}^i \mid D_i, G^i) \propto p(\mathbf{f}^i \mid D_i) p(G^i \mid \mathbf{f}^i, D_i)$$

and so

$$\mathbf{f}^i \mid D_i, G^i \sim \mathcal{N}\left(\mathbf{0}_n, (\mathbf{K}_{\boldsymbol{\theta}_i}^{-1} + \alpha_i \mathbf{L}^i)^{-1}\right). \tag{5}$$

This formulation is similar to that of [28]. However, [28] focused on semi-supervised classification but not the regression problem. The graphical model for SMTR is depicted in Figure 2.

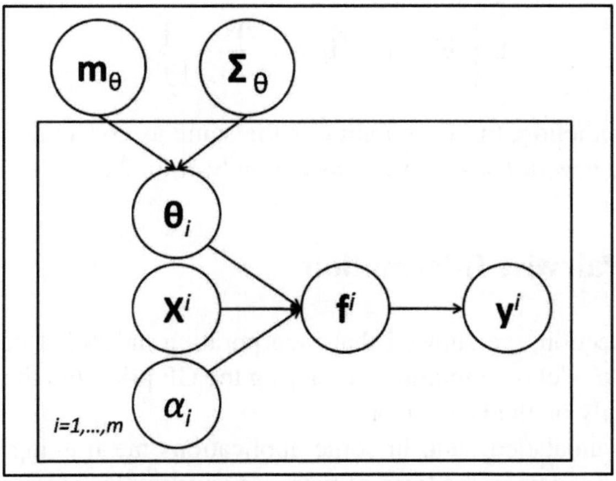

Fig. 2. Graphical model for Semi-Supervised Multi-Task Regression. Here $\mathbf{X}^i$ contains labeled and unlabeled data in the ith task.

From the joint prior defined in Eq. (5), the new kernel function for task T_i can be defined as:

$$k_i(\mathbf{x}, \mathbf{z}) = k_{\boldsymbol{\theta}_i}(\mathbf{x}, \mathbf{z}) - (\mathbf{k}_{\mathbf{x}}^i)^T \left(\alpha_i^{-1}\mathbf{I} + \mathbf{L}^i \mathbf{K}_{\boldsymbol{\theta}_i}\right)^{-1} \mathbf{L}^i \mathbf{k}_{\mathbf{z}}^i, \tag{6}$$

where $\mathbf{k}_{\mathbf{x}}^i = \left(k_{\boldsymbol{\theta}_i}(\mathbf{x}, \mathbf{x}_1^i), \ldots, k_{\boldsymbol{\theta}_i}(\mathbf{x}, \mathbf{x}_{n_i}^i)\right)^T$. The kernel function in Eq. (6) is similar to the semi-supervised kernel function defined in [29].

The hyperparameter α_i in Eq. (4) can be viewed as a measure of usefulness of the unlabeled data. We expect to automatically learn α_i from data. If the optimal α_i is very small or even 0 after learning, then the prior of $\mathbf{f}^i$ will degenerate to the Gaussian prior. This means that the unlabeled data points have negligible effect on improving the performance of GP regression. From the new kernel function in Eq. (6), we can view α_i as a parameter in the kernel function.

There exist some works on semi-supervised or transductive GP regression that work in a different way, such as [30]. The assumption of [30] is that the mean and variance of the predictive results on the unlabeled data are close to those of the labeled data. We can show that it is easy to incorporate this more restrictive assumption into our model.

3.1 Learning and Inference

Since the likelihood is only related to the labeled data, we first marginalize the joint prior with respect to f_j^i corresponding to the unlabled data. In this section, we still use $\mathbf{K}_{\boldsymbol{\theta}_i}$ to denote the kernel matrix whose elements are calculated by the modified kernel function in Eq. (6) on the labeled data of task T_i.

We still use an alternating method to maximize the log-likelihood. The update rules for $\mathbf{m}_\theta$, $\boldsymbol{\Sigma}_\theta$, $\boldsymbol{\theta}_i$ and σ are the same as those in Section 2.1 with the kernel function being the only difference. Moreover, for α_i, the gradient can be calculated as

$$\frac{\partial L}{\partial \ln \alpha_i} = \frac{\alpha_i}{2}\left\{ \mathrm{tr}\left[(\mathbf{K}_{\boldsymbol{\theta}_i} + \sigma^2 \mathbf{I}_{l_i})^{-1} \mathbf{y}^i (\mathbf{y}^i)^T (\mathbf{K}_{\boldsymbol{\theta}_i} + \sigma^2 \mathbf{I}_{l_i})^{-1} \frac{\partial \mathbf{K}_{\boldsymbol{\theta}_i}}{\partial \alpha_i} \right] - \mathrm{tr}\left[(\mathbf{K}_{\boldsymbol{\theta}_i} + \sigma^2 \mathbf{I}_{l_i})^{-1} \frac{\partial \mathbf{K}_{\boldsymbol{\theta}_i}}{\partial \alpha_i} \right] \right\}.$$

When making prediction, the formulation is the same as conventional GP. Moreover, the way to handle new tasks is the same as that in Section 2.2.

4 Utilizing Pairwise Information

In the previous section, we showed that incorporating unlabeled data into SMTR to give the SSMTR model only requires modifying the GP prior, but the likelihood is still defined based solely on the labeled data.

In addition to unlabeled data, in some applications the training set also contains some other auxiliary data in the form of pairwise constraints [21]. Let the jth pairwise constraint for task T_i take the form $(i, u(j), v(j), d_j^i)$, which means that $y_{u(j)}^i - y_{v(j)}^i \geq d_j^i$ where $y_{u(j)}^i$ and $y_{v(j)}^i$ are the true outputs of two data points $\mathbf{x}_{u(j)}^i$ and $\mathbf{x}_{v(j)}^i$ in task T_i with at least one of them being an unlabeled point. For personalized pose estimation, it is easy to add a constraint that the pose angle difference between a frontal face image and a left profile face image is not less than 45 degrees.

In semi-supervised classification or clustering applications, one may also find pairwise constraints such as 'must-link' and 'cannot-link' constraints [31], which state whether or not two data points should belong to the same class or cluster. Many methods have been proposed to incorporate such pairwise constraints into their learning models.

For semi-supevised regression, however, very little has been studied on this topic. Here we offer a preliminary study in the context of SSMTR.

The jth pairwise constraint of task T_i is denoted by ξ^i_j. The noise-free likelihood function $p_{\text{ideal}}(\xi^i_j \mid f^i_{u(j)}, f^i_{v(j)})$ is defined as

$$p_{\text{ideal}}(\xi^i_j \mid f^i_{u(j)}, f^i_{v(j)}) = \begin{cases} 1 & \text{if } f^i_{u(j)} - f^i_{v(j)} \geq d^i_j \\ 0 & \text{otherwise} \end{cases}$$

In real applications, however, the pairwise constraints are often noisy. To model this more realistic setting, we introduce a random variable δ which follows some normal distribution with zero mean and unknown variance ε^2. The variance is the same for all tasks. So the corresponding likelihood function is defined as

$$\begin{aligned} &p(\xi^i_j \mid f^i_{u(j)}, f^i_{v(j)}) \\ &= \int\int p_{\text{ideal}}(\xi^i_j \mid f^i_{u(j)} + \delta_1, f^i_{v(j)} + \delta_2) \mathcal{N}(\delta_1 \mid 0, \varepsilon^2) \mathcal{N}(\delta_2 \mid 0, \varepsilon^2)\, d\delta_1 d\delta_2 \\ &= \Phi\left(\frac{f^i_{u(j)} - f^i_{v(j)} - d^i_j}{\sqrt{2}\varepsilon}\right), \end{aligned}$$

where $\Phi(z) = \int_{-\infty}^{z} \mathcal{N}(a \mid 0, 1) da$ is the probit function.

The noise level ε^2 in the pairwise constraints has some relationship to the noise level σ^2 in the likelihood function since they both relate the latent variable f^i_j to the output y^i_j. However, it should be noted that the noise sources they represent are different. For instance, one may have noise in the pairwise constraints but not in the likelihood, or their noise levels may be different. For flexibility, we use two different parameters for the two noise sources in our model.

Although it appears that the likelihood function of our model is similar to that of [32], their differences are worth pointing out here. The model in [32] is for classification and the constraints there refer to label preference. On the other hand, our model is for semi-supervised regression with pairwise constraints as auxiliary data and the constraints specify the differences between the outputs of pairs of data points. Moreover, the likelihood function in [32] can be seen as a special case of our likelihood function when each d^i_j takes the value 0.

4.1 Learning and Inference

Since direct integration of $\mathbf{f}^i$ is intractable, we resort to the Laplace approximation [33] to approximate the posterior of $\mathbf{f}^i$. We first compute $\mathbf{f}^i_{\text{MAP}}$ by maximizing the posterior of $\mathbf{f}^i$, which is equivalent to minimizing the following function:

$$g(\mathbf{f}^i) = \frac{1}{2}(\mathbf{f}^i)^T \mathbf{K}^{-1}_{\boldsymbol{\theta}_i} \mathbf{f}^i + \frac{\sigma^{-2}}{2} \|\tilde{\mathbf{y}}^i - \tilde{\mathbf{f}}^i\|^2_2 - \sum_{j=1}^{c_i} \ln \Phi(\omega^i_j) + l_i \ln \sigma + \frac{1}{2} \ln |\mathbf{K}_{\boldsymbol{\theta}_i}|,$$

where $\tilde{\mathbf{y}}^i$ and $\tilde{\mathbf{f}}^i$ denote the subsets of $\mathbf{y}^i$ and $\mathbf{f}^i$, respectively, corresponding to the labeled data, c_i is the number of pairwise constraints available in task T_i, and $\omega^i_j = \frac{f^i_{u(j)} - f^i_{v(j)} - d^i_j}{\sqrt{2}\varepsilon}$. We want to find $\mathbf{f}^i_{\text{MAP}}$ that minimizes $g(\mathbf{f}^i)$:

$$\mathbf{f}^i_{\text{MAP}} = \arg\min_{\mathbf{f}^i} g(\mathbf{f}^i).$$

It is easy to show that $g(\mathbf{f}^i)$ is a convex function since the Hessian matrix of $g(\mathbf{f}^i)$ is $\frac{\partial^2 g(\mathbf{f}^i)}{\partial \mathbf{f}^i \partial (\mathbf{f}^i)^T} = \mathbf{K}_{\boldsymbol{\theta}_i}^{-1} + \sigma^{-2}\mathbf{I}_{n_i}^{l_i} + \boldsymbol{\Omega}^i$, which is positive definite, where $\mathbf{I}_{n_i}^{l_i}$ is the $n_i \times n_i$ zero matrix with the first l_i diagonal elements being 1, and $\boldsymbol{\Omega}^i = \frac{\partial^2 - \sum_{j=1}^{c_i} \ln \Phi(\omega_j^i)}{\partial \mathbf{f}^i \partial (\mathbf{f}^i)^T}$ is positive semidefinite. The proof is similar to that in [32] and we omit it here. So we can apply gradient descent to find the global optimum.

After obtaining $\mathbf{f}_{\text{MAP}}^i$, we can approximate the likelihood or evidence of task T_i according to the analysis in [33] as

$$p(\mathbf{y}^i) \approx \exp\{-g(\mathbf{f}_{\text{MAP}}^i)\} \frac{(2\pi)^{n_i/2}}{|\mathbf{K}_{\boldsymbol{\theta}_i}^{-1} + \sigma^{-2}\mathbf{I}_{n_i}^{l_i} + \boldsymbol{\Omega}_{\text{MAP}}^i|^{1/2}},$$

where $\boldsymbol{\Omega}_{\text{MAP}}^i$ is the value of the function $\boldsymbol{\Omega}^i$ taking on $\mathbf{f}_{\text{MAP}}^i$.

So the total negative log-likelihood of all m tasks can be approximated as

$$\begin{aligned} L = \sum_{i=1}^m \Bigg[& \frac{1}{2}(\mathbf{f}_{\text{MAP}}^i)^T \mathbf{K}_{\boldsymbol{\theta}_i}^{-1} \mathbf{f}_{\text{MAP}}^i + \frac{\sigma^{-2}}{2} \|\tilde{\mathbf{y}}^i - \tilde{\mathbf{f}}_{\text{MAP}}^i\|_2^2 + \\ & \frac{1}{2} \ln |\mathbf{K}_{\boldsymbol{\theta}_i}| + \frac{1}{2} \ln |\mathbf{K}_{\boldsymbol{\theta}_i}^{-1} + \sigma^{-2}\mathbf{I}_{n_i}^{l_i} + \boldsymbol{\Omega}_{\text{MAP}}^i| \Bigg] + \\ & \sum_{i=1}^m \frac{1}{2} \Big[(\boldsymbol{\theta}_i - \mathbf{m}_\theta)^T \boldsymbol{\Sigma}_\theta^{-1} (\boldsymbol{\theta}_i - \mathbf{m}_\theta) + \ln |\boldsymbol{\Sigma}_\theta^{-1} \Big] + \\ & \sum_{i=1}^m \Big[l_i \ln \sigma - \sum_{j=1}^{c_i} \ln \Phi(\omega_j^i) \Big] + \text{Const.} \end{aligned}$$

We still use an alternating method to minimize the negative log-likelihood. In the $(t+1)$st iteration, given $\mathbf{m}_\theta^{(t)}$ and $\boldsymbol{\Sigma}_\theta^{(t)}$, the gradient of L with respect to each variable is given by

$$\begin{aligned} \frac{\partial L}{\partial \ln \sigma} &= \sigma^2 \sum_{i=1}^m \Big\{ -\sigma^{-4} \|\tilde{\mathbf{y}}^i - \tilde{\mathbf{f}}_{\text{MAP}}^i\|_2^2 + \frac{l_i}{\sigma^2} - \sigma^{-4} \text{tr} \left[(\mathbf{K}_{\boldsymbol{\theta}_i}^{-1} + \sigma^{-2}\mathbf{I}_{n_i}^{l_i} + \boldsymbol{\Omega}_{\text{MAP}}^i)^{-1} \right] \Big\} \\ \frac{\partial L}{\partial \ln \boldsymbol{\theta}_i} &= \frac{\text{diag}(\boldsymbol{\theta}_i)}{2} \left\{ \text{Tr}(\mathbf{B} \frac{\partial \mathbf{K}_{\boldsymbol{\theta}_i}^{-1}}{\partial \boldsymbol{\theta}_i}) + 2(\boldsymbol{\Sigma}_\theta^{(t)})^{-1}(\boldsymbol{\theta}_i - \mathbf{m}_\theta^{(t)}) \right\} \\ \frac{\partial L}{\partial \ln \alpha_i} &= \frac{\alpha_i}{2} \Bigg\{ -(\mathbf{f}_{\text{MAP}}^i)^T \frac{\partial \mathbf{K}_{\boldsymbol{\theta}_i}^{-1}}{\partial \alpha_i} \mathbf{f}_{\text{MAP}}^i - \text{tr} \left(\mathbf{K}_{\boldsymbol{\theta}_i} \frac{\partial \mathbf{K}_{\boldsymbol{\theta}_i}^{-1}}{\partial \alpha_i} \right) \\ &\quad + \text{tr} \left[(\mathbf{K}_{\boldsymbol{\theta}_i}^{-1} + \sigma^{-2}\mathbf{I}_{n_i}^{l_i} + \boldsymbol{\Omega}_{\text{MAP}}^i)^{-1} \frac{\partial \mathbf{K}_{\boldsymbol{\theta}_i}^{-1}}{\partial \alpha_i} \right] \Bigg\} \\ \frac{\partial L}{\partial \ln \varepsilon} &= \varepsilon \sum_{i=1}^m \left\{ \frac{1}{2} \text{tr} \left[(\mathbf{K}_{\boldsymbol{\theta}_i}^{-1} + \sigma^{-2}\mathbf{I}_{n_i}^{l_i} + \boldsymbol{\Omega}_{\text{MAP}}^i)^{-1} \frac{\partial \boldsymbol{\Omega}_{\text{MAP}}^i}{\partial \varepsilon} \right] + \sum_{j=1}^{c_i} \frac{\omega_j^i \mathcal{N}(\omega_j^i|0,1)}{\varepsilon \Phi(\omega_j^i)} \right\}, \end{aligned}$$

where $\mathbf{B} = \mathbf{f}_{\text{MAP}}^i (\mathbf{f}_{\text{MAP}}^i)^T - \mathbf{K}_{\boldsymbol{\theta}_i} + \left(\mathbf{K}_{\boldsymbol{\theta}_i}^{-1} + \sigma^{-2}\mathbf{I}_{n_i}^{l_i} + \boldsymbol{\Omega}_{\text{MAP}}^i \right)^{-1}$.

The update rules for $\mathbf{m}_\theta$ and $\boldsymbol{\Sigma}_\theta$ are the same as those in Section 2.1.

4.2 Extension

In some applications, there exist auxiliary data given in another form as $\xi_j^i = (i, u(j), v(j), w(j), z(j))$, which means that $y^i_{u(j)} - y^i_{v(j)} \geq y^i_{w(j)} - y^i_{z(j)}$. Let us take the personalized pose estimation problem again as example. It is often easy to know that the pose angle difference between a left profile face image and a right profile face image is larger than that of two nearly frontal face images. The pairwise constraints considered before may be seen as a special case of ξ_j^i when two of the data points $\mathbf{x}^i_{u(j)}, \mathbf{x}^i_{v(j)}, \mathbf{x}^i_{w(j)}$ and $\mathbf{x}^i_{z(j)}$ are labeled. The special case with $v(j) = w(j)$ is also interesting in the pose estimation application. For example, the pose angle difference between a left profile face image and a right profile face image is larger than that between the left profile image and a frontal image. Similar to the pairwise constraints above, the noise-free likelihood function $p_{\text{ideal}}(\xi_j^i \mid f^i_{u(j)}, f^i_{v(j)}, f^i_{w(j)}, f^i_{z(j)})$ is defined as

$$p_{\text{ideal}}(\xi_j^i \mid f^i_{u(j)}, f^i_{v(j)}, f^i_{w(j)}, f^i_{z(j)}) = \begin{cases} 1 & \text{if } f^i_{u(j)} - f^i_{v(j)} \geq f^i_{w(j)} - f^i_{z(j)} \\ 0 & \text{otherwise} \end{cases}$$

For more realistic situations, we again introduce a random variable δ following a normal distribution with zero mean and unknown variance ε^2. The likelihood function is thus defined as

$$\begin{aligned} & p(\xi_j^i \mid f^i_{u(j)}, f^i_{v(j)}, f^i_{w(j)}, f^i_{z(j)}) \\ & = \int p_{\text{ideal}}(\xi_j^i \mid f^i_{u(j)} + \delta_1, f^i_{v(j)} + \delta_2, f^i_{w(j)} + \delta_3, f^i_{z(j)} + \delta_4) \\ & \mathcal{N}(\delta_1 \mid 0, \varepsilon^2)\mathcal{N}(\delta_2 \mid 0, \varepsilon^2)\mathcal{N}(\delta_3 \mid 0, \varepsilon^2)\mathcal{N}(\delta_4 \mid 0, \varepsilon^2)\, d\boldsymbol{\delta} \\ & = \Phi\left(\frac{f^i_{u(j)} - f^i_{v(j)} - f^i_{w(j)} + f^i_{z(j)}}{2\varepsilon}\right), \end{aligned}$$

where $\boldsymbol{\delta} = (\delta_1, \delta_2, \delta_3, \delta_4)^T$. It is easy to show that the posterior distribution of $\mathbf{f}^i$ is unimode, so we can also use the Laplace approximation to make inference.

5 Experiments

We compare a single-task regression method, SMTR, SSMTR and SSMTR with pairwise information in this section using two benchmark data sets, one for learning the inverse dynamics of a robot arm and another for predicting the student performance in terms of examination scores.

5.1 Learning Inverse Dynamics

This data set[1] was used in [34]. For each instance, there are 7 joint positions, 7 joint velocities and 7 joint accelerations forming 21 input attributes, together with 7 joint torques for the outputs corresponding to 7 degrees of freedom. We treat each output

[1] http://www.gaussianprocess.org/gpml/data/

Table 1. nMSE results on learning inverse dynamics (SSTR: supervised single-task regression which uses one GP for each task; SSMTRPI: semi-supervised multi-task regression with pairwise information)

Method	Transductive Error	Inductive Error
SSTR	1.0228±0.1318	1.0270±0.1450
SMTR	0.4149±0.1109	0.4368±0.1020
SSMTR	0.3810±0.1080	0.3905±0.1123
SSMTRPI	0.3500±0.1088	0.3486±0.1010

Table 2. nMSE results on predicting student performance

Method	Transductive Error	Inductive Error
SSTR	1.2914±0.3146	1.3240±0.3274
SMTR	1.1151±0.3025	1.1535±0.3128
SSMTR	1.0506±0.2804	1.0612±0.2813
SSMTRPI	0.9817±0.2809	0.9824±0.2832

(i.e., degree of freedom) as a separate learning task. To simulate a more general multi-task learning setting, we randomly select 2000 data points independently for each task so that the input data points for different tasks are different. We randomly partition the whole data set into three subsets, with 1% as labeled data, 10% as unlabeled data and the rest as test data. The kernel we use is the RBF kernel. Moreover, we randomly select 100 pairs of data points and generate the pairwise constraint using their labels. Ten random splits are performed and the mean and standard derivation of the performance measure over different splits are reported. We adopt the normalized mean squared error (nMSE), which is defined as the mean squared error divided by the variance of the test label, as the performance measure. Table 2 shows the results. From the results, we can see that the performance of our proposed SMTR is significantly better than that of supervised single-task learning which uses one GP for each task. Moreover, the performance of SSMTR and SSMTR using pairwise information is better than that of SMTR, which shows that both the unlabeled data and the pairwise information are effective in improving performance.

5.2 Predicting Student Performance

This data set[2] was used in [11] for multi-task learning. The goal is to predict the student performance in terms of examination scores. The data set consists of 15362 students from 139 secondary schools, recording their examination scores in three years (1985–87). We treat each school as a different task, so that there are 139 learning tasks in total. For each instance, the input consists of the year of the examination as well as 4 school-specific and 3 student-specific attributes. We still use nMSE as performance measure. We randomly select 2% of the data as labeled data, 20% as unlabeled data, and the rest as test data. The kernel we adopt is the RBF kernel. We also generate 100 pairwise

[2] http://www.cs.ucl.ac.uk/staff/A.Argyriou/code/

constraints just as the last experiment did. We perform 10 random splits and report the mean and standard derivation over different splits. Table 2 shows the results. Similar to the results on learning inverse dynamics, SSMTR with pairwise information gives the best performance.

6 Conclusion

In this paper, we have proposed an approach for integrating semi-supervised regression and multi-task regression under a common framework. We first propose a new supervised multi-task regression method based on GP and then extend it to incorporate unlabeled data by modifying the GP prior. In addition, if auxiliary data in the form of pairwise constraints are available, we propose a scheme to incorporate them into our semi-supervised multi-task regression framework by modifying the likelihood term. In our future research, we will investigate sparse extension of our models, possibly by using the informative vector machine [35].

Acknowledgments

This research has been supported by General Research Fund 621407 from the Research Grants Council of the Hong Kong Special Administrative Region, China.

References

1. Chapelle, O., Zien, A., Schölkopf, B. (eds.): Semi-Supervised Learning. MIT Press, Boston (2006)
2. Blum, A., Mitchell, T.: Combining labeled and unlabeled data with co-training. In: Proceedings of the Workshop on Computational Learning Theory, Madison, Wisconsin, USA, pp. 92–100 (1998)
3. Bennett, K., Demiriz, A.: Semi-supervised support vector machines. In: Advances in Neural Information Processing Systems 11, Vancouver, British Columbia, Canada, pp. 368–374 (1998)
4. Joachims, T.: Transductive inference for text classification using support vector machines. In: Proceedings of the Sixteenth International Conference on Machine Learning, San Francisco, CA, USA, pp. 200–209 (1999)
5. Zhou, D., Bousquet, O., Lal, T., Weston, J., Schölkopf, B.: Learning with local and global consistency. In: Advances in Neural Information Processing Systems 16, Vancouver, British Columbia, Canada (2003)
6. Zhu, X., Ghahramani, Z., Lafferty, J.: Semi-supervised learning using Gaussian fields and harmonic functions. In: Proceedings of the Twentieth International Conference on Machine Learning, Washington, DC, pp. 912–919 (2003)
7. Belkin, M., Niyogi, P., Sindhwani, V.: Manifold regularization: A geometric framework for learning from labeled and unlabeled examples. Journal of Machine Learning Research 7, 2399–2434 (2006)
8. Caruana, R.: Multitask learning. Machine Learning 28(1), 41–75 (1997)
9. Baxter, J.: A Bayesian/information theoretic model of learning to learn via multiple task sampling. Machine Learning 28(1), 7–39 (1997)

10. Thrun, S.: Is learning the n-th thing any easier than learning the first? In: Touretzky, D.S., Mozer, M., Hasselmo, M.E. (eds.) Advances in Neural Information Processing Systems 8, Denver, CO, pp. 640–646 (1996)
11. Argyriou, A., Evgeniou, T., Pontil, M.: Convex multi-task feature learning. Machine Learning 73(3), 243–272 (2008)
12. Evgeniou, T., Pontil, M.: Regularized multi-task learning. In: Proceedings of the Tenth ACM SIGKDD International Conference on Knowledge Discovery and Data Mining, Seattle, Washington, USA, pp. 109–117 (2004)
13. Bakker, B., Heskes, T.: Task clustering and gating for Bayesian multitask learning. Journal of Machine Learning Research 4, 83–99 (2003)
14. Xue, Y., Liao, X., Carin, L., Krishnapuram, B.: Multi-task learning for classification with Dirichlet process priors. Journal of Machine Learning Research 8, 35–63 (2007)
15. Lawrence, N.D., Platt, J.C.: Learning to learn with the informative vector machine. In: Proceedings of the Twenty-first International Conference on Machine Learning, Banff, Alberta, Canada (2004)
16. Schwaighofer, A., Tresp, V., Yu, K.: Learning Gaussian process kernels via hierarchical Bayes. In: Saul, L.K., Weiss, Y., Bottou, L. (eds.) Advances in Neural Information Processing Systems 17, Vancouver, British Columbia, Canada, pp. 1209–1216 (2005)
17. Yu, K., Tresp, V., Schwaighofer, A.: Learning Gaussian processes from multiple tasks. In: Proceedings of the Twenty-Second International Conference on Machine Learning, Bonn, Germany, pp. 1012–1019 (2005)
18. Bonilla, E., Chai, K.M.A., Williams, C.: Multi-task Gaussian process prediction. In: Platt, J., Koller, D., Singer, Y., Roweis, S. (eds.) Advances in Neural Information Processing Systems 20, Vancouver, British Columbia, Canada, pp. 153–160 (2008)
19. Ando, R.K., Zhang, T.: A framework for learning predictive structures from multiple tasks and unlabeled data. Journal of Machine Learning Research 6, 1817–1853 (2005)
20. Liu, Q., Liao, X., Carin, L.: Semi-supervised multitask learning. In: Platt, J., Koller, D., Singer, Y., Roweis, S. (eds.) Advances in Neural Information Processing Systems 20, Vancouver, British Columbia, Canada, pp. 937–944 (2008)
21. Zhu, X., Goldberg, A.: Kernel regression with order preferences. In: Proceedings of the Twenty-Second AAAI Conference on Artificial Intelligence, Vancouver, British Columbia, Canada, pp. 681–686 (2007)
22. Rasmussen, C.E., Williams, C.K.I.: Gaussian Processes for Machine Learning. The MIT Press, Cambridge (2006)
23. Yu, S., Tresp, V., Yu, K.: Robust multi-task learning with t-processes. In: Proceedings of the Twenty-Fourth International Conference on Machine Learning, Corvalis, Oregon, USA, pp. 1103–1110 (2007)
24. Heskes, T.: Solving a huge number of similar tasks: a combination of multi-task learning and a hierarchical Bayesian approach. In: Proceedings of the Fifteenth International Conference on Machine Learning, Madison, Wisconson, USA, pp. 233–241 (1998)
25. Heskes, T.: Empirical bayes for learning to learn. In: Proceedings of the Seventeenth International Conference on Machine Learning, Stanford University, Standord, CA, USA, pp. 367–374 (2000)
26. Zelnik-Manor, L., Perona, P.: Self-tuning spectral clustering. In: Advances in Neural Information Processing Systems 17, Vancouver, British Columbia, Canada, pp. 1601–1608 (2005)
27. Chung, F.R.K.: Spectral Graph Theory. American Mathematical Society, Rhode Island (1997)
28. Sindhwani, V., Chu, W., Keerthi, S.S.: Semi-supervised Gaussian process classifiers. In: Proceedings of the 20th International Joint Conference on Artificial Intelligence, Hyderabad, India, pp. 1059–1064 (2007)

29. Sindhwani, V., Niyogi, P., Belkin, M.: Beyond the point cloud: from transductive to semi-supervised learning. In: Proceedings of the Twenty-Second International Conference on Machine Learning, Bonn, Germany, pp. 824–831 (2005)
30. Le, Q.V., Smola, A.J., Gärtner, T., Altun, Y.: Transductive Gaussian process regression with automatic model selection. In: Proceedings of the 17th European Conference on Machine Learning, Berlin, Germany, pp. 306–317 (2006)
31. Bilenko, M., Basu, S., Mooney, R.J.: Integrating constraints and metric learning in semi-supervised clustering. In: Proceedings of the Twenty-first International Conference on Machine Learning, Banff, Alberta, Canada (2004)
32. Chu, W., Ghahramani, Z.: Preference learning with Gaussian processes. In: Proceedings of the Twenty-Second International Conference on Machine Learning, pp. 137–144 (2005)
33. Bishop, C.M.: Pattern Recognition and Machine Learning. Springer, New York (2006)
34. Vijayakumar, S., D'Souza, A., Schaal, S.: Incremental online learning in high dimensions. Neural Computation 17(12), 2602–2634 (2005)
35. Lawrence, N.D., Seeger, M., Herbrich, R.: Fast sparse Gaussian process methods: The informative vector machine. In: Becker, S., Thrun, S., Obermayer, K. (eds.) Advances in Neural Information Processing Systems 15, Vancouver, British Columbia, Canada, pp. 609–616 (2003)

A Flexible and Efficient Algorithm for Regularized Fisher Discriminant Analysis

Zhihua Zhang[1], Guang Dai[1], and Michael I. Jordan[2]

[1] College of Computer Science and Technology
Zhejiang University
Hangzhou, Zhejiang 310027, China
{zhzhang,daiguang116}@gmail.com

[2] Department of Statistics and Division of Computer Science
University of California, Berkeley
Berkeley, CA 94720 USA
jordan@cs.berkeley.edu

Abstract. Fisher linear discriminant analysis (LDA) and its kernel extension—kernel discriminant analysis (KDA)—are well known methods that consider dimensionality reduction and classification jointly. While widely deployed in practical problems, there are still unresolved issues surrounding their efficient implementation and their relationship with least mean squared error procedures. In this paper we address these issues within the framework of regularized estimation. Our approach leads to a flexible and efficient implementation of LDA as well as KDA. We also uncover a general relationship between regularized discriminant analysis and ridge regression. This relationship yields variations on conventional LDA based on the pseudoinverse and a direct equivalence to an ordinary least squares estimator. Experimental results on a collection of benchmark data sets demonstrate the effectiveness of our approach.

1 Introduction

In this paper we are concerned with the supervised dimensionality reduction problem, an enduring issue in data mining and machine learning. Fisher linear discriminant analysis (LDA) provides a classical example of supervised dimension reduction. LDA estimates an effective dimension reduction space defined by linear transformations by maximizing the ratio of between-class scatter to within-class scatter.

The LDA formulation reduces to the solution of a generalized eigenproblem [6] that involves the pooled between-class scatter matrix and total scatter matrix of the input vectors. To solve the generalized eigenproblem, LDA typically requires the pooled scatter matrix to be nonsingular. This can become problematic when the dimensionality is high, because the scatter matrix is likely to be singular. In applications such as information retrieval, face recognition and microarray analysis, for example, we often meet undersampled problems which are in a "small n but large p" regime; i.e., there are a small number of samples but a very large number of variables. There are two main variants of LDA in the literature that aim to deal with this issue: the *pseudoinverse* method and the *regularization* method [7,19].

W. Buntine et al. (Eds.): ECML PKDD 2009, Part II, LNAI 5782, pp. 632–647, 2009.

Another important family of methods for dealing with singularity is based on a two-stage process in which two symmetric eigenproblems are solved successively. This approach was pioneered by Kitter and Young [12]. Recently, Howland *et al.* [10] used this approach to introduce the generalized singular value decomposition (GSVD) [14] into the LDA solution by utilizing special representations of the pooled scatter matrix and between-class scatter matrix. A similar general approach has been used in the development of efficient approximate algorithms for LDA [2,21]. However, the challenge of developing an efficient general implementation methodology for LDA still remains.

It is well known that LDA is equivalent to a least mean squared error procedure in the binary classification problem [3]. It is of great interest to obtain a similar relationship in multi-class problems. A significant literature has emerged to address this issue [7,16,20]. However, the results obtained by these authors are subject to restrictive conditions. The problem of finding a general theoretical link between LDA and least mean squares is still open.

In this paper we address the issues within a regularization framework. We propose a novel algorithm for solving the regularized LDA (RLDA) problem. Our algorithm is more efficient than the GSVD-based algorithm [10], especially in the setting of "small n but large p" problems. More importantly, our algorithm leads us to an equivalence between RLDA and a ridge estimator for multivariate linear regression [8]. This equivalence is derived in a general setting and it is fully consistent with the established result in the binary problem [3].

Our algorithm is also appropriate for the pseudoinverse variant of LDA. Indeed, we establish an equivalence between the pseudoinverse variant and an ordinary least squares (OLS) estimation problem. Thus, we believe that we completely solve the open problem concerning the relationship between the multi-class LDA problem and multivariate linear estimation problems.

LDA relies on the assumption of linearity of the data manifold. In recent years, kernel methods [18] have aimed at removing such linearity assumptions. The kernel technology can circumvent the linearity assumption of LDA, because it works by nonlinearly mapping vectors in the input space to a higher-dimensional feature space and then implementing traditional versions of LDA in the feature space. Many different approaches have been proposed to extend LDA to kernel spaces in the existing literature [1,13,17].

The KDA method in [13] was developed for binary problems only, and it was solved by using the relationship between KDA and the least mean squared error procedure. A more general method, known as generalized discriminant analysis (GDA) [1], requires that the kernel matrix be nonsingular. Unfortunately, centering in the feature space will violate this requirement. Park and Park [15] argued that this might break down the theoretical justification for GDA and proposed their GSVD method to avoid this requirement for nonsingularity. The approach to LDA that we present in the current paper also handles the nonsingularity issue and extends naturally to KDA, both in its regularization and pseudoinverse forms.

The paper is organized as follows. Section 2 reviews LDA and KDA. In Section 3 we propose a new algorithm for LDA as well as KDA. An equivalence between LDA and multivariate linear regression problems is presented in Section 4. We conduct the empirical comparisons in Section 5 and conclude in Section 6.

2 Problem Formulation

We are concerned with a multi-class classification problem. Given a set of n p-dimensional data points, $\{\mathbf{x}_1, \ldots, \mathbf{x}_n\} \in \mathcal{X} \subset \mathbb{R}^p$, we assume that the $\mathbf{x}_i$ are to be grouped into c disjoint classes and that each $\mathbf{x}_i$ belongs to one and only one class. Let $V = \{1, 2, \ldots, n\}$ denote the index set of the data points $\mathbf{x}_i$ and partition V into c disjoint subsets V_j; i.e., $V_i \cap V_j = \varnothing$ for $i \neq j$ and $\cup_{j=1}^{c} V_j = V$, where the cardinality of V_j is n_j so that $\sum_{j=1}^{c} n_j = n$. We also make use of a matrix representation for the partitions. In particular, we let $\mathbf{E} = [e_{ij}]$ be an $n \times c$ indicator matrix with $e_{ij} = 1$ if input $\mathbf{x}_i$ is in class j and $e_{ij} = 0$ otherwise.

In this section we review LDA and KDA solutions to this multi-class classification problem. We begin by presenting our notation.

2.1 Notation

Throughout this paper, $\mathbf{I}_m$ denotes the $m \times m$ identity matrix, $\mathbf{1}_m$ the $m \times 1$ vector of ones, $\mathbf{0}$ the zero vector or matrix with appropriate size, and $\mathbf{H}_m = \mathbf{I}_m - \frac{1}{m}\mathbf{1}_m\mathbf{1}_m'$ the $m \times m$ centering matrix. For an $m \times 1$ vector $\mathbf{a} = (a_1, \ldots, a_m)'$, diag($\mathbf{a}$) represents the $m \times m$ diagonal matrix with $a_1, \ldots, a_m$ as its diagonal entries. For an $m \times m$ matrix $\mathbf{A} = [a_{ij}]$, we let $\mathbf{A}^+$ be the Moore-Penrose inverse of $\mathbf{A}$, $\mathrm{tr}(\mathbf{A})$ be the trace of $\mathbf{A}$, $\mathrm{rk}(\mathbf{A})$ be the rank of $\mathbf{A}$ and $\|\mathbf{A}\|_F = \sqrt{\mathrm{tr}(\mathbf{A}'\mathbf{A})}$ be the Frobenius norm of $\mathbf{A}$.

For a matrix $\mathbf{A} \in \mathbb{R}^{p \times q}$ with $p \geq q$, we always write the the singular value decomposition (SVD) of $\mathbf{A}$ as $\mathbf{A} = \mathbf{U}\boldsymbol{\Gamma}\mathbf{V}'$ where $\mathbf{U} \in \mathbb{R}^{p \times q}$ and $\mathbf{V} \in \mathbb{R}^{q \times q}$ are orthogonal, and $\boldsymbol{\Gamma} = \mathrm{diag}(\gamma_1, \ldots, \gamma_q)$ is arrayed in descending order of $\gamma_1 \geq \gamma_2 \geq \cdots \geq \gamma_q$ (≥ 0). Let the rank of $\mathbf{A}$ be $r \leq \min\{p, q\}$ (denoted $\mathrm{rk}(\mathbf{A}) = r$). The thin SVD [6] of $\mathbf{A}$ is then $\mathbf{A} = \mathbf{U}_A\boldsymbol{\Gamma}_A\mathbf{V}_A'$ where $\mathbf{U}_A \in \mathbb{R}^{p \times r}$ and $\mathbf{V}_A \in \mathbb{R}^{q \times r}$ are orthogonal, and $\boldsymbol{\Gamma}_A = \mathrm{diag}(\gamma_1, \ldots, \gamma_r)$ satisfies $\gamma_1 \geq \gamma_2 \geq \cdots \geq \gamma_r > 0$.

Given two matrices $\boldsymbol{\Phi}$ and $\boldsymbol{\Sigma} \in \mathbb{R}^{p \times p}$, we refer to $(\boldsymbol{\Lambda}, \mathbf{B})$ where $\boldsymbol{\Lambda} = \mathrm{diag}(\lambda_1, \ldots, \lambda_q)$ and $\mathbf{B} = [\mathbf{b}_1, \ldots, \mathbf{b}_q]$ as q eigenpairs of the matrix pencil $(\boldsymbol{\Phi}, \boldsymbol{\Sigma})$ if $\boldsymbol{\Phi}\mathbf{B} = \boldsymbol{\Sigma}\mathbf{B}\boldsymbol{\Lambda}$, namely,

$$\boldsymbol{\Phi}\mathbf{b}_i = \lambda_i \boldsymbol{\Sigma}\mathbf{b}_i, \quad \text{for } i = 1, \ldots, q.$$

The problem of finding eigenpairs of $(\boldsymbol{\Phi}, \boldsymbol{\Sigma})$ is known as a *generalized eigenproblem*.

2.2 Linear Discriminant Analysis

Let $\mathbf{m} = \frac{1}{n}\sum_{i=1}^{n} \mathbf{x}_i$ be the sample mean, and $\mathbf{m}_j = \frac{1}{n_j}\sum_{i \in V_j} \mathbf{x}_i$ to the jth class mean for $j = 1, \ldots, c$. We then have the pooled scatter matrix $\mathbf{S}_t = \sum_{i=1}^{n}(\mathbf{x}_i - \mathbf{m})(\mathbf{x}_i - \mathbf{m})'$ and the pooled between-class scatter matrix $\mathbf{S}_b = \sum_{j=1}^{c} n_j(\mathbf{m}_j - \mathbf{m})(\mathbf{m}_j - \mathbf{m})'$. Conventional LDA solves the generalized eigenproblem as

$$\mathbf{S}_b\mathbf{a}_j = \lambda_j \mathbf{S}_t \mathbf{a}_j, \quad \lambda_1 \geq \lambda_2 \geq \cdots \geq \lambda_q > \lambda_{q+1} = 0 \tag{1}$$

where $q \leq \min\{p,\ c-1\}$ and refers to $\mathbf{a}_j$ as the jth discriminant direction. Note that we ignore a multiplier $1/n$ in these scatter matrices for simplicity.

Since $\mathbf{S}_t = \mathbf{S}_b + \mathbf{S}_w$ where $\mathbf{S}_w$ is the pooled within-class scatter matrix, LDA is equivalent to finding a solution to

$$\mathbf{S}_b\mathbf{a} = \lambda/(1-\lambda)\mathbf{S}_w\mathbf{a}.$$

We see that LDA involves solving the generalized eigenproblem in (1), which can be expressed in matrix form:

$$\mathbf{S}_b\mathbf{A} = \mathbf{S}_t\mathbf{A}\boldsymbol{\Lambda}, \tag{2}$$

where $\mathbf{A} = [\mathbf{a}_1, \ldots, \mathbf{a}_q]$ and $\boldsymbol{\Lambda} = \mathrm{diag}(\lambda_1, \ldots, \lambda_q)$. If $\mathbf{S}_t$ is nonsingular, it may be shown that

$$\mathbf{S}_t^{-1}\mathbf{S}_b\mathbf{A} = \mathbf{A}\boldsymbol{\Lambda}.$$

Thus, $(\lambda_j, \mathbf{a}_j)$ is eigenpair of $\mathbf{S}_t^{-1}\mathbf{S}_b$ and the eigenvectors corresponding to the largest eigenvalues of $\mathbf{S}_t^{-1}\mathbf{S}_b$ are used for the discriminant directions. Since $\mathrm{rk}(\mathbf{S}_b)$ is at most $c-1$, the projection will be onto a space of dimension at most $c-1$ (i.e., $q \le c-1$).

In applications such as information retrieval, face recognition and microarray analysis, however, we often meet a "small n but large p" problem. Thus, $\mathbf{S}_t$ is usually ill-conditioned; that is, it is either singular or close to singular. In this case, $\mathbf{S}_t^{-1}\mathbf{S}_b$ cannot be computed accurately.

Let $\boldsymbol{\Pi} = \mathrm{diag}(n_1, \ldots, n_c)$, $\boldsymbol{\Pi}^{\frac{1}{2}} = \mathrm{diag}(\sqrt{n_1}, \ldots, \sqrt{n_c})$, $\boldsymbol{\pi} = (n_1, \ldots, n_c)'$, $\sqrt{\boldsymbol{\pi}} = (\sqrt{n_1}, \ldots, \sqrt{n_c})'$ and $\mathbf{H}_\pi = \mathbf{I}_c - \frac{1}{n}\sqrt{\boldsymbol{\pi}}\sqrt{\boldsymbol{\pi}}'$. It follows that $\mathbf{1}_n'\mathbf{E} = \mathbf{1}_c'\boldsymbol{\Pi} = \boldsymbol{\pi}'$, $\mathbf{E}\mathbf{1}_c = \mathbf{1}_n$, $\mathbf{1}_c'\boldsymbol{\pi} = n$, $\mathbf{E}'\mathbf{E} = \boldsymbol{\Pi}$, $\boldsymbol{\Pi}^{-1}\boldsymbol{\pi} = \mathbf{1}_c$, and

$$\mathbf{M} = \boldsymbol{\Pi}^{-1}\mathbf{E}'\mathbf{X}, \tag{3}$$

where $\mathbf{X} = [\mathbf{x}_1, \ldots, \mathbf{x}_n]'$ and $\mathbf{M} = [\mathbf{m}_1, \ldots, \mathbf{m}_c]'$. In addition, we have

$$\mathbf{E}\boldsymbol{\Pi}^{-\frac{1}{2}}\mathbf{H}_\pi = \mathbf{H}_n\mathbf{E}\boldsymbol{\Pi}^{-\frac{1}{2}} \tag{4}$$

due to $\mathbf{E}\boldsymbol{\Pi}^{-\frac{1}{2}}\mathbf{H}_\pi = \mathbf{E}\boldsymbol{\Pi}^{-\frac{1}{2}} - \frac{1}{n}\mathbf{1}_n\sqrt{\boldsymbol{\pi}}'$ and $\mathbf{H}_n\mathbf{E}\boldsymbol{\Pi}^{-\frac{1}{2}} = \mathbf{E}\boldsymbol{\Pi}^{-\frac{1}{2}} - \frac{1}{n}\mathbf{1}_n\sqrt{\boldsymbol{\pi}}'$.

Using these results and the idempotency of $\mathbf{H}_n$, we reexpress $\mathbf{S}_t$ as

$$\mathbf{S}_t = \mathbf{X}'\mathbf{H}_n\mathbf{H}_n\mathbf{X} = \mathbf{X}'\mathbf{H}_n\mathbf{X}.$$

We also have

$$\begin{aligned}
\mathbf{S}_b &= \mathbf{M}'\left[\boldsymbol{\Pi} - \frac{1}{n}\boldsymbol{\pi}\boldsymbol{\pi}'\right]\mathbf{M} \\
&= \mathbf{M}'\left[\boldsymbol{\Pi}^{\frac{1}{2}} - \frac{1}{n}\boldsymbol{\pi}\sqrt{\boldsymbol{\pi}}'\right]\left[\boldsymbol{\Pi}^{\frac{1}{2}} - \frac{1}{n}\sqrt{\boldsymbol{\pi}}\boldsymbol{\pi}'\right]\mathbf{M} \\
&= \mathbf{X}'\mathbf{E}\boldsymbol{\Pi}^{-1}\boldsymbol{\Pi}^{\frac{1}{2}}\mathbf{H}_\pi\mathbf{H}_\pi\boldsymbol{\Pi}^{\frac{1}{2}}\boldsymbol{\Pi}^{-1}\mathbf{E}'\mathbf{X} \\
&= \mathbf{X}'\mathbf{H}_n\mathbf{E}\boldsymbol{\Pi}^{-1}\mathbf{E}'\mathbf{H}_n\mathbf{X}.
\end{aligned}$$

Utilizing the above representations of $\mathbf{S}_t$ and $\mathbf{S}_b$, Howland *et al.* [10] proved that the GSVD method can be used to solve the problem in (2).

There are also two variants of conventional LDA in the literature that aim to handle this problem [19]. The first variant involves replacing $\mathbf{S}_t^{-1}$ by $\mathbf{S}_t^{+}$ and solving the following eigenproblem:

$$\mathbf{S}_t^{+}\mathbf{S}_b\mathbf{A} = \mathbf{A}\boldsymbol{\Lambda}. \tag{5}$$

Note that $\mathbf{S}_t^{+}$ exists and is unique [6]. Moreover, $\mathbf{S}_t^{+}$ is equal to $\mathbf{S}_t^{-1}$ whenever $\mathbf{S}_t$ is nonsingular. Thus, we will use (5) when $\mathbf{S}_t$ is either nonsingular or singular.

The second variant is referred to as *regularized discriminant analysis* (RDA) [4]. It replaces $\mathbf{S}_t$ by $\mathbf{S}_t + \sigma^2\mathbf{I}_p$ and solves the following eigenproblem:

$$(\mathbf{S}_t + \sigma^2\mathbf{I}_p)^{-1}\mathbf{S}_b\mathbf{A} = \mathbf{A}\boldsymbol{\Lambda}. \tag{6}$$

It is a well known result that LDA is equivalent to a least mean squared error procedure in the binary classification problem ($c = 2$) [3]. Recently, similar relationships have been studied for multi-class ($c > 2$) problems [7,16,20]. In particular, Park and Park [16] proposed an efficient algorithm for LDA via a least mean squared error procedure in the multi-class problem.

We can see that the solution $\mathbf{A}$ for (5) or (6) is not unique. For example, if $\mathbf{A}$ is the solution, then so is $\mathbf{AD}$ whenever $\mathbf{D}$ is a $q\times q$ nonsingular diagonal matrix. Thus, constraint $\mathbf{A}'(\mathbf{S}_t + \sigma^2\mathbf{I}_p)\mathbf{A} = \mathbf{I}_q$ is typically imposed in the literature. In this paper we concentrate on the solution of (6) with or without this constraint, and investigate the connection with ridge regression problems in the multi-class setting.

2.3 Kernel Discriminant Analysis

Kernel methods [18] work in a feature space $\mathcal{F}$, which is related to the original input space $\mathcal{X} \subset \mathbb{R}^p$ by a mapping,

$$\varphi : \mathcal{X} \rightarrow \mathcal{F}.$$

That is, φ is a vector-valued function which gives a vector $\varphi(\mathbf{s})$, called a *feature vector*, corresponding to an input $\mathbf{s} \in \mathcal{X}$. In kernel methods, we are given a reproducing kernel $K : \mathcal{X} \times \mathcal{X} \rightarrow \mathbb{R}$ such that $K(\mathbf{s}, \mathbf{t}) = \varphi(\mathbf{s})'\varphi(\mathbf{t})$ for $\mathbf{s}, \mathbf{t} \in \mathcal{X}$. The mapping $\varphi(\cdot)$ itself is typically not given explicitly.

In the sequel, we use the tilde notation to denote vectors and matrices in the feature space. For example, the data vectors and mean vectors in the feature space are denoted as $\tilde{\mathbf{x}}_i$ and $\tilde{\mathbf{m}}_j$. Accordingly, $\tilde{\mathbf{X}} = [\tilde{\mathbf{x}}_1, \ldots, \tilde{\mathbf{x}}_n]'$ and $\tilde{\mathbf{M}} = [\tilde{\mathbf{m}}_1, \ldots, \tilde{\mathbf{m}}_n]'$ are the data matrix and mean matrix in the feature space.

Fisher kernel discriminant analysis (KDA) seeks to solve the following generalized eigenproblem:

$$\tilde{\mathbf{S}}_b\tilde{\mathbf{A}} = \tilde{\mathbf{S}}_t\tilde{\mathbf{A}}\tilde{\boldsymbol{\Lambda}}, \tag{7}$$

where $\tilde{\mathbf{S}}_t$ and $\tilde{\mathbf{S}}_b$ are the total scatter matrix and the between-class scatter matrix in $\mathcal{F}$, respectively:

$$\tilde{\mathbf{S}}_t = \sum_{i=1}^{n}(\tilde{\mathbf{x}}_i - \tilde{\mathbf{m}})(\tilde{\mathbf{x}}_i - \tilde{\mathbf{m}})',$$

$$\tilde{\mathbf{S}}_b = \sum_{j=1}^{c} n_j(\tilde{\mathbf{m}}_j - \tilde{\mathbf{m}})(\tilde{\mathbf{m}}_j - \tilde{\mathbf{m}})'.$$

The KDA problem is to solve (7), doing so by working solely with the kernel matrix $\mathbf{K} = \tilde{\mathbf{X}}\tilde{\mathbf{X}}'$. This is done by noting [13,15] that the eigenvectors $\tilde{\mathbf{a}}_j$ are in the space spanned by $\tilde{\mathbf{x}}_1, \ldots, \tilde{\mathbf{x}}_n$ and $\tilde{\mathbf{A}}$ can be expressed as

$$\tilde{\mathbf{A}} = \sum_{i=1}^{n} (\tilde{\mathbf{x}}_i - \tilde{\mathbf{m}})\boldsymbol{\beta}_i' = \tilde{\mathbf{X}}'\mathbf{H}_n\mathbf{B},$$

where $\mathbf{B} = [\boldsymbol{\beta}_1, \ldots, \boldsymbol{\beta}_n]$ is an $n \times q$ coefficient matrix. Hence, (7) is equivalent to

$$\tilde{\mathbf{X}}'\mathbf{H}_n\mathbf{E}\boldsymbol{\Pi}^{-1}\mathbf{E}'\mathbf{H}_n\tilde{\mathbf{X}}\tilde{\mathbf{X}}'\mathbf{H}_n\mathbf{B} = \tilde{\mathbf{X}}'\mathbf{H}_n\mathbf{H}_n\tilde{\mathbf{X}}\tilde{\mathbf{X}}'\mathbf{H}_n\mathbf{B}\tilde{\boldsymbol{\Lambda}}.$$

Pre-multiplying the equation by $\mathbf{H}_n\tilde{\mathbf{X}}$, we have a new generalized eigenvalue problem

$$\mathbf{C}\mathbf{E}\boldsymbol{\Pi}^{-1}\mathbf{E}'\mathbf{C}\mathbf{B} = \mathbf{C}\mathbf{C}\mathbf{B}\tilde{\boldsymbol{\Lambda}}, \tag{8}$$

which involves only the kernel matrix $\mathbf{K} = \tilde{\mathbf{X}}\tilde{\mathbf{X}}'$. Here $\mathbf{C} = \mathbf{H}_n\mathbf{K}\mathbf{H}_n$ is the centered kernel matrix. Moreover, given a new input vector $\mathbf{x}$, we can compute the projection $\mathbf{z}$ of its feature vector $\tilde{\mathbf{x}}$ onto $\tilde{\mathbf{A}}$ through

$$\begin{aligned}\mathbf{z} = \tilde{\mathbf{A}}'(\tilde{\mathbf{x}} - \tilde{\mathbf{m}}) &= \mathbf{B}'\mathbf{H}_n\tilde{\mathbf{X}}\Big(\tilde{\mathbf{x}} - \frac{1}{n}\tilde{\mathbf{X}}'\mathbf{1}_n\Big) \\ &= \mathbf{B}'\mathbf{H}_n\Big(\mathbf{k}_{\mathbf{x}} - \frac{1}{n}\mathbf{K}\mathbf{1}_n\Big),\end{aligned} \tag{9}$$

where $\mathbf{k}_{\mathbf{x}} = \big(K(\mathbf{x}, \mathbf{x}_1), \ldots, K(\mathbf{x}, \mathbf{x}_n)\big)'$. This shows that the kernel trick can be used for KDA.

The concern then becomes that of solving the problem (8). Although $\mathbf{K}$ can be assumed to be nonsingular, $\mathbf{C}$ is positive semidefinite but not positive definite because the centering matrix $\mathbf{H}_n$ is singular. In fact, the rank of $\mathbf{C}$ is not larger than $n{-}1$ because the rank of $\mathbf{H}_n$ is $n{-}1$. In this case, the method devised by [1] cannot be used for the problem (8). Thus, Park and Park [15] proposed a GSVD-based algorithm to solve problem (8). Running this algorithm requires the complete orthogonal decomposition [6] of matrix $[\mathbf{C}\mathbf{E}\boldsymbol{\Pi}^{-\frac{1}{2}}, \mathbf{C}]'$, which is of size $(n{+}c) \times n$. Thus, this approach is infeasible for large values of n.

Another treatment is based on the following regularized variant of (8):

$$\mathbf{C}\mathbf{E}\boldsymbol{\Pi}^{-1}\mathbf{E}'\mathbf{C}\mathbf{B} = (\mathbf{C}\mathbf{C} + \sigma^2\mathbf{I}_n)\mathbf{B}\tilde{\boldsymbol{\Lambda}}. \tag{10}$$

This RKDA problem can be equivalently expressed as

$$(\tilde{\mathbf{S}}_t + \sigma^2\mathbf{I}_d)^{-1}\tilde{\mathbf{A}} = \tilde{\mathbf{S}}_b\tilde{\mathbf{A}}\tilde{\boldsymbol{\Lambda}}, \tag{11}$$

where d is the dimension of the feature space. Although d is possibly infinite, we here assume that it is finite but not necessarily known.

3 RR-SVD Algorithms for RDA

In this section, we propose a novel approach to solving the RLDA problem in (6). We then extend this approach for the solution of the RKDA problem in (11).

3.1 The Algorithm for RLDA

We reformulate the eigenproblem in (6) as

$$\mathbf{G}\boldsymbol{\Pi}^{-\frac{1}{2}}\mathbf{E}'\mathbf{H}_n\mathbf{X}\mathbf{A} = \mathbf{A}\boldsymbol{\Lambda}, \tag{12}$$

where

$$\begin{aligned}\mathbf{G} &= (\mathbf{X}'\mathbf{H}_n\mathbf{X} + \sigma^2\mathbf{I}_p)^{-1}\mathbf{X}'\mathbf{H}_n\mathbf{E}\boldsymbol{\Pi}^{-\frac{1}{2}} \\ &= (\mathbf{X}'\mathbf{H}_n\mathbf{X} + \sigma^2\mathbf{I}_p)^{-1}\mathbf{M}'\boldsymbol{\Pi}^{\frac{1}{2}}\mathbf{H}_\pi\end{aligned} \tag{13}$$

due to (3) and (4). We also have

$$\mathbf{G} = \mathbf{X}'\mathbf{H}_n(\mathbf{H}_n\mathbf{X}\mathbf{X}'\mathbf{H}_n + \sigma^2\mathbf{I}_n)^{-1}\mathbf{E}\boldsymbol{\Pi}^{-\frac{1}{2}} \tag{14}$$

due to $(\mathbf{X}'\mathbf{H}_n\mathbf{X} + \sigma^2\mathbf{I}_p)^{-1}\mathbf{X}'\mathbf{H}_n = \mathbf{X}'\mathbf{H}_n(\mathbf{H}_n\mathbf{X}\mathbf{X}'\mathbf{H}_n + \sigma^2\mathbf{I}_n)^{-1}$. This implies that if $n < p$, we may wish to use (14) to reduce the computational cost. More importantly, we will see that (14) plays a key role in the development of an efficient algorithm for KDA to be presented shortly.

Let $\mathbf{R} = \boldsymbol{\Pi}^{-\frac{1}{2}}\mathbf{E}'\mathbf{H}_n\mathbf{X}\mathbf{G}$. Since $\mathbf{G}\boldsymbol{\Pi}^{-\frac{1}{2}}\mathbf{E}'\mathbf{H}_n\mathbf{X}$ $(p{\times}p)$ and $\mathbf{R}$ $(c{\times}c)$ have the same nonzero eigenvalues [9], the $\lambda_j, j = 1, \ldots, q$, are the nonzero eigenvalues of $\mathbf{R}$. Moreover, if $(\boldsymbol{\Lambda}, \mathbf{V})$ is the nonzero eigenpair of $\mathbf{R}$, $(\boldsymbol{\Lambda}, \mathbf{G}\mathbf{V})$ is the nonzero eigenpair of $\mathbf{G}\boldsymbol{\Pi}^{-\frac{1}{2}}\mathbf{E}'\mathbf{H}_n\mathbf{X}$. Note that

$$\begin{aligned}\mathbf{R} &= \boldsymbol{\Pi}^{-\frac{1}{2}}\mathbf{E}'\mathbf{H}_n\mathbf{X}(\mathbf{X}'\mathbf{H}_n\mathbf{X} + \sigma^2\mathbf{I}_p)^{-1}\mathbf{X}'\mathbf{H}_n\mathbf{E}\boldsymbol{\Pi}^{-\frac{1}{2}} \\ &= \mathbf{H}_\pi\boldsymbol{\Pi}^{\frac{1}{2}}\mathbf{M}\mathbf{G}.\end{aligned} \tag{15}$$

This shows that $\mathbf{R}$ is positive semidefinite. Thus, its SVD is equivalent to the eigenvalue decomposition.

We thus develop an algorithm for solving the RLDA problem in (6). This is a two-stage process, which is presented in Algorithm 1. We will prove that the first stage is equivalent to the solution to a ridge regression (RR) problem in Section 4. Thus, we refer to this two-stage process as an RR-SVD algorithm. It is easily obtained that

$$\mathbf{A}'(\mathbf{S}_t + \sigma^2\mathbf{I}_p)\mathbf{A} = \boldsymbol{\Gamma}_R \quad \text{and} \quad \mathbf{A}'\mathbf{S}_b\mathbf{A} = \boldsymbol{\Gamma}_R^2.$$

This implies that $\mathbf{A}\boldsymbol{\Gamma}_R^{-\frac{1}{2}}$ is also a solution of problem (6) such that $\boldsymbol{\Gamma}_R^{-\frac{1}{2}}\mathbf{A}'(\mathbf{S}_t + \sigma^2\mathbf{I}_p)\mathbf{A}\boldsymbol{\Gamma}_R^{-\frac{1}{2}} = \mathbf{I}_q$.

The first stage calculates $\mathbf{G}$ by either (13) or (14). The computational complexity is $O(m^3)$ where $m = \min(n, p)$. The second stage makes use of the thin SVD of $\mathbf{R}$ and the computational complexity is $O(c^3)$. If both n and p are large, we recommend to use the incomplete Cholesky decomposition of $\mathbf{H}_n\mathbf{X}\mathbf{X}'\mathbf{H}_n$ (or $\mathbf{X}'\mathbf{H}_n\mathbf{X}$) and then the Sherman-Morrison-Woodbury formula [6] for calculating $(\mathbf{H}_n\mathbf{X}\mathbf{X}'\mathbf{H}_n + \sigma^2\mathbf{I}_n)^{-1}$ (or $(\mathbf{X}'\mathbf{H}_n\mathbf{X} + \sigma^2\mathbf{I}_p)^{-1}$). Compared with the GSVD-based algorithm [15], the RR-SVD algorithm is more efficient for a "small n but large p" problem.

When $\sigma^2 = 0$, we can solve the problem in (5) by simply adjusting the first stage in the RR-SVD algorithm. In particular, we calculate $\mathbf{G}$ by

$$\mathbf{G} = (\mathbf{X}'\mathbf{H}_n\mathbf{X})^{+}\mathbf{M}'\boldsymbol{\Pi}^{\frac{1}{2}}\mathbf{H}_\pi \overset{(or)}{=} \mathbf{X}'\mathbf{H}_n(\mathbf{H}_n\mathbf{X}\mathbf{X}'\mathbf{H}_n)^{+}\mathbf{E}\boldsymbol{\Pi}^{-\frac{1}{2}}. \tag{16}$$

Algorithm 1. RR-SVD Algorithm for RLDA problem (6)

procedure RLDA($\mathbf{X}, \mathbf{E}, \boldsymbol{\Pi}, \sigma^2$)
    Calculate $\mathbf{G}$ by (13) or (14) and $\mathbf{R}$ by (15);
    Perform the thin SVD of $\mathbf{R}$ as $\mathbf{R} = \mathbf{V}_R \boldsymbol{\Gamma}_R \mathbf{V}_R'$;
    Return $\mathbf{A} = \mathbf{G}\mathbf{V}_R$ or $\mathbf{G}\mathbf{V}_R \boldsymbol{\Gamma}_R^{-\frac{1}{2}}$ as the solution of RLDA problem (6).
end procedure

3.2 The Algorithm for RKDA

We now apply Algorithm 1 to the RKDA problem in (11), which is the kernel extension of RLDA in (6).

It immediately follows from (14) that

$$\tilde{\mathbf{G}} = \tilde{\mathbf{X}}'\mathbf{H}_n(\mathbf{H}_n\tilde{\mathbf{X}}\tilde{\mathbf{X}}'\mathbf{H}_n + \sigma^2\mathbf{I}_n)^{-1}\mathbf{E}\boldsymbol{\Pi}^{-\frac{1}{2}}$$

from which, using (15), we can calculate $\tilde{\mathbf{R}}$ by

$$\tilde{\mathbf{R}} = \boldsymbol{\Pi}^{-\frac{1}{2}}\mathbf{E}'\mathbf{C}(\mathbf{C}+\sigma^2\mathbf{I}_n)^{-1}\mathbf{E}\boldsymbol{\Pi}^{-\frac{1}{2}}.$$

Moreover, given a new input vector $\mathbf{x}$, we can compute the projection $\mathbf{z}$ of its feature vector $\tilde{\mathbf{x}}$ onto $\tilde{\mathbf{A}}$ through

$$\begin{aligned}\mathbf{z} = \tilde{\mathbf{A}}'(\tilde{\mathbf{x}} - \tilde{\mathbf{m}}) &= \mathbf{V}_{\tilde{\mathbf{R}}}'\boldsymbol{\Pi}^{-\frac{1}{2}}\mathbf{E}'(\mathbf{C} + \sigma^2\mathbf{I}_n)^{-1}\mathbf{H}_n\tilde{\mathbf{X}}\Big(\tilde{\mathbf{x}} - \frac{1}{n}\tilde{\mathbf{X}}'\mathbf{1}_n\Big) \\ &= \mathbf{V}_{\tilde{\mathbf{R}}}'\boldsymbol{\Pi}^{-\frac{1}{2}}\mathbf{E}'(\mathbf{C} + \sigma^2\mathbf{I}_n)^{-1}\mathbf{H}_n\Big(\mathbf{k}_{\mathbf{x}} - \frac{1}{n}\mathbf{K}\mathbf{1}_n\Big).\end{aligned} \tag{17}$$

This shows that we can calculate $\tilde{\mathbf{R}}$ and $\mathbf{z}$ directly using $\mathbf{K}$ and $\mathbf{k}_{\mathbf{x}}$. We thus obtain a RR-SVD algorithm for RKDA, which is given in Algorithm 2. Also, when $\sigma^2 = 0$, we can calculate $\tilde{\mathbf{R}}$ by

$$\tilde{\mathbf{R}} = \boldsymbol{\Pi}^{-\frac{1}{2}}\mathbf{E}'\mathbf{C}\mathbf{C}^{+}\mathbf{E}\boldsymbol{\Pi}^{-\frac{1}{2}}$$

and exploit the RR-SVD algorithm to solve the following variant of KDA:

$$\tilde{\mathbf{S}}_t^{+}\tilde{\mathbf{S}}_b\tilde{\mathbf{A}} = \tilde{\mathbf{A}}\tilde{\boldsymbol{\Lambda}}.$$

We see that the RR-SVD algorithm is more efficient than the GSVD-based algorithm [15] for the RKDA problem in (11). Recall that problem (11) is not equivalent to that in (10). Moreover, it is not feasible to develop a GSVD-based algorithm for solving problem (11). However, we also have an RR-SVD algorithm for solving (10), by replacing $\mathbf{C}$ by $\mathbf{CC}$ in calculating $\tilde{\mathbf{R}}$ and (17) by (9) in calculating $\mathbf{z}$. The resulting algorithm may be less computationally efficient, however, because it involves more matrix computations.

4 Relationships between RFDA and Ridge Regression

It is a well known result that LDA (or KDA) is equivalent to a least mean squared error procedure in the binary classification problem ($c = 2$) [3,13]. Recently, relationships

Algorithm 2. RR-SVD Algorithm for RKDA problem (11)

procedure RKDA($\mathbf{K}, \mathbf{E}, \mathbf{k_x}, \boldsymbol{\Pi}, \sigma^2$)
    Calculate $\tilde{\mathbf{R}} = \boldsymbol{\Pi}^{-\frac{1}{2}}\mathbf{E}'\mathbf{C}(\mathbf{C}+\sigma^2\mathbf{I}_n)^{-1}\mathbf{E}\boldsymbol{\Pi}^{-\frac{1}{2}}$;
    Perform the thin SVD of $\tilde{\mathbf{R}}$ as $\tilde{\mathbf{R}} = \tilde{\mathbf{V}}_{\tilde{R}}\tilde{\boldsymbol{\Gamma}}_{\tilde{R}}\tilde{\mathbf{V}}'_{\tilde{R}}$;
    Calculate $\mathbf{z}$ by (17);
    Return $\mathbf{z}$ as the q-dimensional representation of $\mathbf{x}$.
end procedure

between LDA and a least mean squared error procedure in multi-class ($c > 2$) problems were discussed by [7,16,20].

Motivated by this line of work, we investigate a possible equivalency between RLDA and ridge regression [8]. We then go on to consider a similar relationship between RKDA and the corresponding ridge regression problem.

Let $\mathbf{Y} = [\mathbf{y}_1, \ldots, \mathbf{y}_n]' = \mathbf{E}\boldsymbol{\Pi}^{-\frac{1}{2}}\mathbf{H}_\pi$. That is, $\mathbf{y}_i = (y_{i1}, \ldots, y_{ic})$ is defined by

$$y_{ij} = \begin{cases} \frac{n-n_j}{n\sqrt{n_j}} & \text{if } i \in V_j, \\ -\frac{\sqrt{n_j}}{n} & \text{otherwise.} \end{cases}$$

Regarding $\{(\mathbf{x}_i, \mathbf{y}_i), i = 1, \ldots, n\}$ as the training samples, we fit the following multivariate linear function:

$$\mathbf{f}(\mathbf{x}) = \mathbf{w}_0 + \mathbf{W}'\mathbf{x}$$

where $\mathbf{w}_0 \in \mathbb{R}^c$ and $\mathbf{W} \in \mathbb{R}^{p\times c}$. We now find ridge estimates of $\mathbf{w}_0$ and $\mathbf{W}$. In particular, we consider the following minimization problem:

$$\min_{\mathbf{w}_0, \mathbf{W}} L(\mathbf{w}_0, \mathbf{W}) = \frac{1}{2}\|\mathbf{Y}-\mathbf{1}_n\mathbf{w}_0'-\mathbf{X}\mathbf{W}\|_F^2 + \frac{\sigma^2}{2}\mathrm{tr}(\mathbf{W}'\mathbf{W}). \tag{18}$$

The solution $\mathbf{W}^*$ for (18) is

$$\mathbf{W}^* = (\mathbf{X}'\mathbf{H}_n\mathbf{X} + \sigma^2\mathbf{I}_p)^{-1}\mathbf{M}'\boldsymbol{\Pi}^{\frac{1}{2}}\mathbf{H}_\pi. \tag{19}$$

The derivation is given in Appendix A. It then follows from (13) that $\mathbf{W}^* = \mathbf{G}$. Moreover, when $\sigma^2 = 0$, $\mathbf{W}^*$ reduces to the ordinary least squares (OLS) estimate of $\mathbf{W}$, which is the solution of the following minimization problem:

$$\min_{\mathbf{w}_0, \mathbf{W}} L(\mathbf{w}_0, \mathbf{W}) = \frac{1}{2}\|\mathbf{Y}-\mathbf{1}_n\mathbf{w}_0'-\mathbf{X}\mathbf{W}\|_F^2. \tag{20}$$

In this case, if $\mathbf{X}'\mathbf{H}_n\mathbf{X}$ is singular, a standard treatment is to use the Moore-Penrose inverse $(\mathbf{X}'\mathbf{H}_n\mathbf{X})^+$ in (19). Such a $\mathbf{W}^*$ is identical with $\mathbf{G}$ in (16).

Consequently, we have found a relationship between the ridge estimation problem in (18) and the RLDA problem in (6). This is summarized in the following theorem.

Theorem 1. *Let $\mathbf{W}^*$ be the solutions of the ridge estimation problem in (18)* (resp. the OLS estimation problem in (20)) *and $\mathbf{A}$ be defined in Algorithm 1 for the solution of the RLDA problem in (6)* (resp. the LDA problem in (5)). *Then,*

$$\mathbf{A} = \mathbf{W}^*\mathbf{V}_R$$

where the columns of $\mathbf{V}_R$ *are the eigenvectors of* $\mathbf{R}$ *associated with its* q *nonzero eigenvalues.*

Theorem 1 provides a connection between $\mathbf{A}$ and $\mathbf{W}^*$. Recall that the eigenvector matrix $\mathbf{V}_R$ $(c \times q)$ is orthogonal. This leads us to the following main result of this paper.

Theorem 2. *Under the conditions in Theorem 1, we have*

$$\mathbf{A}\mathbf{A}' = \mathbf{W}^*(\mathbf{W}^*)'.$$

Accordingly, we have

$$(\mathbf{x}_i - \mathbf{x}_j)'\mathbf{A}\mathbf{A}'(\mathbf{x}_i - \mathbf{x}_j) = (\mathbf{x}_i - \mathbf{x}_j)'\mathbf{W}^*(\mathbf{W}^*)'(\mathbf{x}_i - \mathbf{x}_j)$$

for any $\mathbf{x}_i$ *and* $\mathbf{x}_j \in \mathbb{R}^p$.

The proof of this theorem is given in Appendix A. Theorem 2 shows that when applying a distance-based classifier such as the K-nearest neighbor (KNN) in the reduced dimensional space, the classification results obtained by multi-class LDA and multivariate linear estimators are the same. Theorem 2 holds in general. Thus we obtain a complete solution to the open problem concerning the relationship between multi-class LDA problems and multivariate linear estimators.

Similar results have been obtained by [16,20], but under restrictive conditions. The key difference between our work and that of [16] revolves around a different definition for the label scores $\mathbf{Y}$. Ye [20] used the same definition of $\mathbf{Y}$ as ours, but they aimed to establish a connection of the solution $\mathbf{W}^*$ with a matrix $\mathbf{A}$ defined differently from ours.

5 Experimental Study

To evaluate the performance of the proposed algorithm for LDA and KDA, we conducted experimental comparisons with other related algorithms for LDA and KDA on several real-world data sets. In particular, the comparison was implemented on four face datasets[1], two gene datasets, the USPS dataset, the "letters" dataset and the WebKB dataset. Table 1 summarizes the benchmark datasets we used. All algorithms were implemented in Matlab on a PC configured with an Intel Dual Core 2.53GHz CPU and 2.06GB of memory. Matlab code to implement the algorithms can be obtained from the first author.

In our experiments, each dataset was randomly partitioned into disjoint training and test data sets, according to the percentage n/k listed in the last column of Table 1. Ten random partitions were obtained for each data set, and several evaluation criteria were reported, including average classification accuracy rate, standard deviation and average computational time.

[1] The YaleB(+E) dataset was collected from the YaleB database and its extension.

Table 1. Summary of the benchmark datasets: c—the number of classes; p—the dimension of the input vector; k—the size of the dataset; n—the number of the training data

Data set	c	p	k	n/k
ORL	40	1024	400	40%
Yale	15	1024	165	50%
YaleB(+E)	38	1024	2414	30%
PIE	68	1024	6800	20%
11_Tumors	11	12533	174	31%
14_Tumors	25	15009	305	41%
USPS	10	256	2007	10%
Letters	3	17	2341	5%
WebKB	4	300	4192	10%

In the linear setting, we compared our method to the LDA/GSVD method [11] and the LDA/MSE method [16]. In the kernel setting, we compared our method to the KDA/GSVD method [15] and the KDA/MSE, i.e., the kernel-based extensions of the two linear methods in above. All of the hyperparameters (such as σ^2) were selected by cross-validation [5]. In the kernel setting, the RBF kernel $K(\mathbf{x}_i, \mathbf{x}_j) = \exp(-\|\mathbf{x}_i - \mathbf{x}_j\|^2/\theta^2)$ was employed, and θ was set to the mean Euclidean distance between training data points. After having obtained the q-dimensional representations $\mathbf{z}_i$ of the $\mathbf{x}_i$ from each dimensionality reduction method, we used a simple nearest neighbor classifier to evaluate the classification accuracy.

Figure 1 presents comparative classification results on the four face datasets. We implemented our Algorithm 1 with both $\mathbf{A}$ and $\mathbf{A}\boldsymbol{\Gamma}_R^{-\frac{1}{2}}$. We found that the result with $\mathbf{A}$ was slightly better than that with $\mathbf{A}\boldsymbol{\Gamma}_R^{-\frac{1}{2}}$. Moreover, a similar result was found for kernel leaning. The results reported here were based on the setting with $\mathbf{A}$. From Figure 1, it is clear that in the linear and kernel settings, our method has better classification accuracy than that of the LDA/GSVD and LDA/MSE methods over a range of choices of number of discriminant covariates. Moreover, an appealing characteristic of our method is its effectiveness when q (i.e., the number of discriminant covariates) is small.

We also compared the computational time of the different methods in the linear and kernel settings on the four face datasets. Figure 2 shows the comparisons with respect to the training percentage n/k on the four face datasets. We can also see that our method has an overall low computational time in comparison with the other methods on the four face datasets. As the training percentage n/k increases, our method yields more efficient performance.

Finally, Tables 2 and 3 summarize the different evaluation criteria on all the data sets. As these results show, our method yields accurate and computationally efficient performance in both the linear and kernel settings. Additionally, it should be mentioned here that the data sets in our experiments range over small sample and large sample problems.

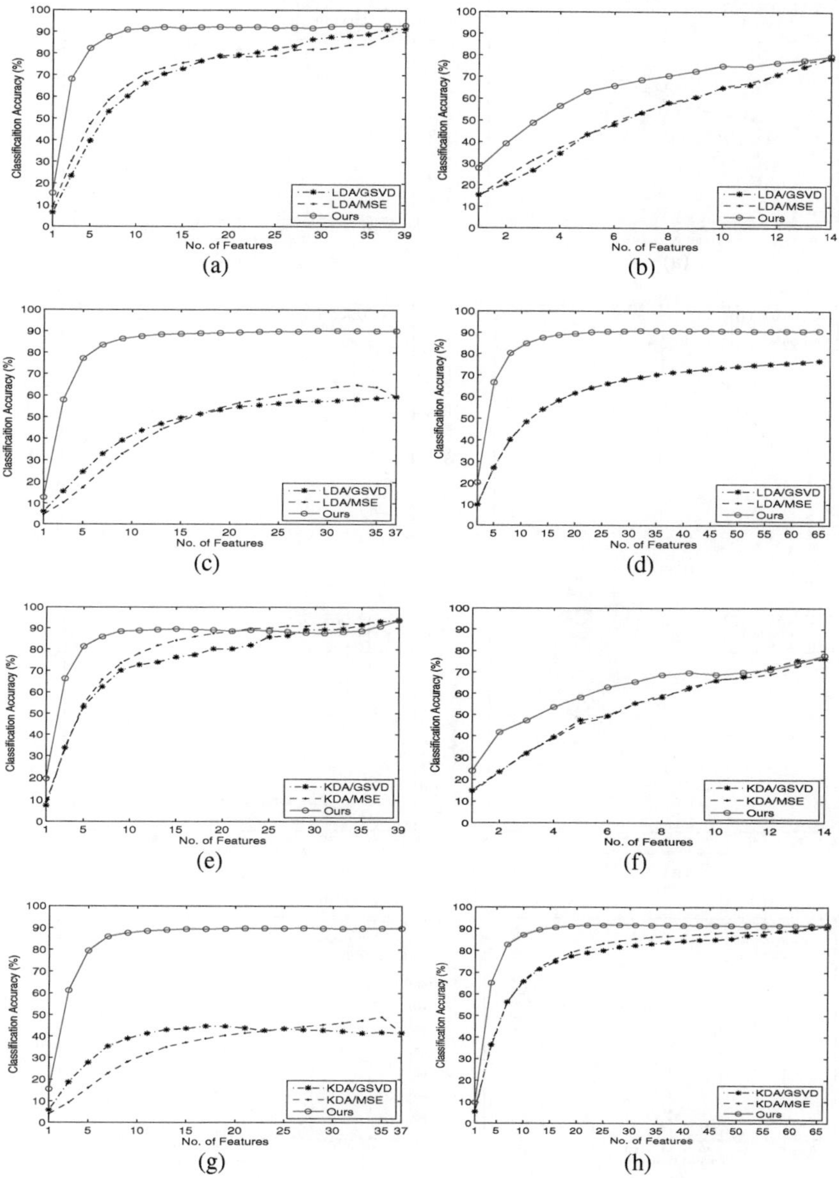

(a) (b) (c) (d) (e) (f) (g) (h)

Fig. 1. Comparison of the three different methods on the four face datasets, where (a)$\sim$(d) denote the results in the linear setting and (e)$\sim$(h) denote the results in the kernel setting: (a) ORL−linear; (b) Yale−linear; (c) YaleB(+E)−linear; (d) PIE−linear; (e) ORL−kernel; (f) Yale−kernel; (g) YaleB(+E)−kernel; (h) PIE−kernel. Here "No. of features" is equal to q.

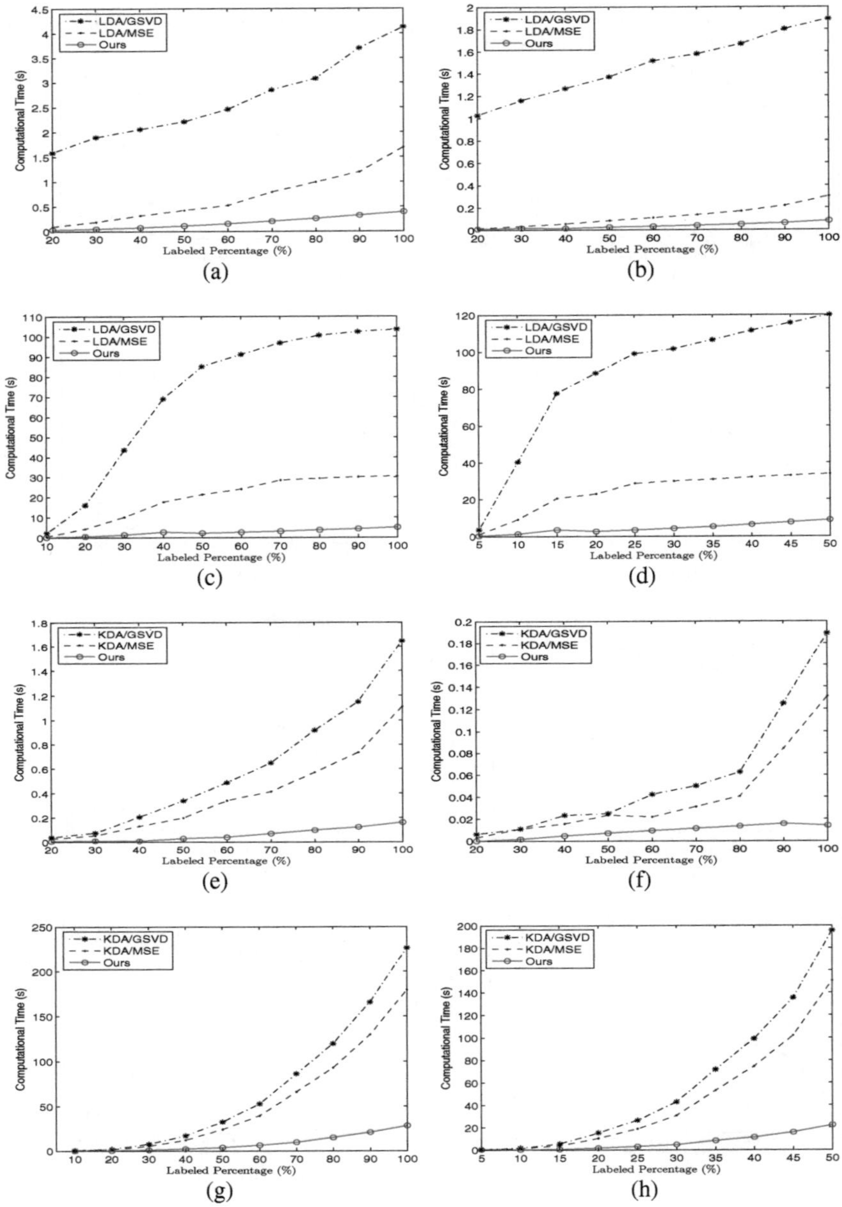

Fig. 2. Comparison of the computational times for the three different methods as the training percentage k/n increases on the four face datasets, where (a)$\sim$(d) denote the results in the linear setting and (e)$\sim$(h) denote the results in the kernel setting: (a) ORL$-$linear; (b) Yale$-$linear; (c) YaleB(+E)$-$linear; (d) PIE$-$linear; (e) ORL$-$kernel; (f) Yale$-$kernel; (g) YaleB(+E)$-$kernel; (h) PIE$-$kernel.

Table 2. Experimental results of the three methods on different datasets in the linear setting: $acc-$ the best classification accuracy percentage; $std-$ the corresponding standard deviation; $q-$ the corresponding number of discriminant covariates; $time-$ the corresponding computational time (s)

Dataset	LDA/GSVD			LDA/MSE			Ours		
	$acc\ (\pm std)$	q	$time$	$acc\ (\pm std)$	q	$time$	$acc\ (\pm std)$	q	$time$
ORL	91.54 (±1.98)	39	2.059	91.58 (±2.00)	39	0.316	93.13 (±2.00)	39	0.077
Yale	78.56 (±2.29)	14	1.370	78.44 (±2.47)	14	0.084	79.56 (±3.75)	14	0.025
YaleB(+E)	59.54 (±11.8)	37	43.62	65.19 (±8.36)	34	10.17	90.20 (±1.09)	31	1.422
PIE	77.00 (±0.81)	67	88.51	77.01 (±0.81)	67	23.01	90.91 (±0.55)	45	2.681
11_Tumors	92.35 (±1.51)	10	0.652	90.25 (±2.10)	10	0.877	92.44 (±1.43)	10	0.495
14_Tumors	66.33 (±1.82)	24	2.808	64.94 (±1.69)	24	4.499	66.39 (±1.91)	24	2.035
USPS	52.23 (±3.02)	9	0.979	52.24 (±3.02)	9	0.579	86.84 (±1.31)	9	0.114
Letters	89.16 (±0.81)	2	0.129	89.16 (±0.81)	2	0.021	89.27 (±0.86)	2	0.102
WebKB	65.92 (±1.77)	3	2.348	65.92 (±1.77)	3	1.635	81.76 (±0.87)	3	0.225

Table 3. Experimental results of the three methods on different datasets in the kernel setting: $acc-$ the best classification accuracy percentage; $std-$ the corresponding standard deviation; $q-$ the corresponding number of discriminant covariates; $time-$ the corresponding computational time (s)

Dataset	KDA/GSVD			KDA/MSE			Ours		
	$acc\ (\pm std)$	q	$time$	$acc\ (\pm std)$	q	$time$	$acc\ (\pm std)$	q	$time$
ORL	93.75 (±1.95)	39	0.339	93.75 (±1.89)	39	0.198	93.75 (±1.73)	39	0.031
Yale	76.33 (±2.67)	14	0.025	76.44 (±2.67)	14	0.023	77.78 (±2.87)	14	0.007
YaleB(+E)	44.74 (±24.6)	17	7.751	48.95 (±25.4)	35	5.477	89.80 (±1.02)	27	0.844
PIE	91.04 (±0.49)	67	46.84	91.04 (±0.49)	67	36.44	91.77 (±0.43)	26	8.811
11_Tumors	88.74 (±2.94)	10	0.031	89.16 (±2.36)	10	0.022	89.58 (±2.10)	10	0.011
14_Tumors	59.56 (±2.76)	24	0.170	60.56 (±2.73)	10	0.108	66.33 (±1.51)	24	0.035
USPS	85.13 (±3.57)	9	0.697	85.15 (±3.57)	9	0.493	89.22 (±1.42)	9	0.082
Letters	95.25 (±0.99)	2	0.120	95.24 (±0.99)	2	0.071	99.36 (±1.20)	2	0.024
WebKB	74.79 (±2.78)	3	4.285	74.79 (±2.77)	3	3.255	83.35 (±1.31)	3	0.499

6 Conclusion

In this paper we have provided an appealing solution to an open problem concerning the relationship between multi-class LDA problems and multivariate linear estimators, both in the linear setting and the kernel setting. Our theory has yielded efficient and effective algorithms for LDA and KDA within both the regularization and pseudoinverse frameworks. The favorable performance of our algorithms has been demonstrated empirically on a collection of benchmark data sets. In future work we plan to extend our algorithms to a broader class of generalized eigenproblems.

A Proof of Theorem 1

Proof. The first-order derivatives of $L(\mathbf{w}_0, \mathbf{W})$ with respect to $\mathbf{w}_0$ and $\mathbf{W}$ are given by

$$\frac{\partial L}{\partial \mathbf{w}_0} = n\mathbf{w}_0 + \mathbf{W}'\mathbf{X}'\mathbf{1}_n - \mathbf{Y}'\mathbf{1}_n,$$
$$\frac{\partial L}{\partial \mathbf{W}} = (\mathbf{X}'\mathbf{X} + \sigma^2\mathbf{I}_p)\mathbf{W} + \mathbf{X}'\mathbf{1}_n\mathbf{w}_0' - \mathbf{X}'\mathbf{Y}.$$

Letting $\frac{\partial L}{\partial \mathbf{w}_0} = \mathbf{0}$, $\frac{\partial L}{\partial \mathbf{W}} = \mathbf{0}$ and $\bar{\mathbf{x}} = \frac{1}{n}\sum_{i=1}^n \mathbf{x}_i = \frac{1}{n}\mathbf{X}'\mathbf{1}_n$, we obtain

$$\begin{cases} \mathbf{w}_0 + \mathbf{W}'\bar{\mathbf{x}} = \mathbf{0} \\ n\bar{\mathbf{x}}\mathbf{w}_0' + (\mathbf{X}'\mathbf{X} + \sigma^2\mathbf{I}_p)\mathbf{W} = \mathbf{M}'\boldsymbol{\Pi}^{\frac{1}{2}}\mathbf{H}_\pi \end{cases}$$

due to $\mathbf{Y}'\mathbf{1}_n = \mathbf{0}$ and $\mathbf{X}'\mathbf{Y} = \mathbf{M}'\boldsymbol{\Pi}^{\frac{1}{2}}\mathbf{H}_\pi$. Further, it follows that $\mathbf{w}_0 = -\mathbf{W}\bar{\mathbf{x}}$, and hence,

$$(\mathbf{X}'\mathbf{H}_n\mathbf{X} + \sigma^2\mathbf{I}_p)\mathbf{W} = \mathbf{M}'\boldsymbol{\Pi}^{\frac{1}{2}}\mathbf{H}_\pi$$

because $\mathbf{X}'\mathbf{X} - n\bar{\mathbf{x}}\bar{\mathbf{x}}' = \mathbf{X}'\mathbf{H}_n\mathbf{X}$. We thus obtain $\mathbf{W}^*$ in (19). It then follows from (13) that $\mathbf{W}^* = \mathbf{G}$. Moreover, when $\sigma^2 = 0$, $\mathbf{W}^*$ reduces to the solution of the minimization problem in (20). In this case, if $\mathbf{X}'\mathbf{H}_n\mathbf{X}$ is singular, a standard treatment is to use the Moore-Penrose inverse $(\mathbf{X}'\mathbf{H}_n\mathbf{X})^+$ in (19). Such a $\mathbf{W}^*$ is identical with $\mathbf{G}$ in (16).

Consequently, we have the relationship between the ridge estimation problem in (18) and the RLDA problem in (6). This is summarized in Theorem 1.

B Proof of Theorem 2

Proof. Since $\mathbf{V}_R$ is an $c\times q$ orthogonal matrix, there exists an $c\times(c-q)$ orthogonal matrix $\mathbf{V}_2$ such that $\mathbf{V} = [\mathbf{V}_R, \mathbf{V}_2]$ is an $c\times c$ orthogonal matrix. Noting that $\mathbf{R} = \mathbf{V}_R\boldsymbol{\Gamma}_R\mathbf{V}_R'$, we have $\mathbf{R}\mathbf{V}_2 = \mathbf{0}$ and $\mathbf{V}_2'\mathbf{R}\mathbf{V}_2 = \mathbf{0}$. Let $\mathbf{Q} = \mathbf{M}'\boldsymbol{\Pi}^{\frac{1}{2}}\mathbf{H}_\pi\mathbf{V}_2$. Then we obtain $\mathbf{Q}'(\mathbf{X}'\mathbf{H}_n\mathbf{X}+\sigma^2\mathbf{I}_p)^{-1}\mathbf{Q} = \mathbf{0}$. This implies $\mathbf{Q} = \mathbf{0}$ because $(\mathbf{X}'\mathbf{H}_n\mathbf{X}+\sigma^2\mathbf{I}_p)^{-1}$ is positive definite. Hence, $\mathbf{W}^*\mathbf{V}_2 = (\mathbf{X}'\mathbf{H}_n\mathbf{X} + \sigma^2\mathbf{I}_p)^{-1}\mathbf{Q} = \mathbf{0}$. As a result, we have

$$\begin{aligned} \mathbf{W}^*(\mathbf{W}^*)' &= \mathbf{W}^*\mathbf{V}\mathbf{V}'(\mathbf{W}^*)' \\ &= \mathbf{W}^*\mathbf{V}_R\mathbf{V}_R'(\mathbf{W}^*)' + \mathbf{W}^*\mathbf{V}_2\mathbf{V}_2'(\mathbf{W}^*)' \\ &= \mathbf{A}\mathbf{A}'. \end{aligned}$$

Note that if $\sigma^2 = 0$ and $\mathbf{X}'\mathbf{H}_n\mathbf{X}$ is nonsingular, we still have $\mathbf{W}^*(\mathbf{W}^*)' = \mathbf{A}\mathbf{A}'$. In the case that $\mathbf{X}'\mathbf{H}_n\mathbf{X}$ is singular, we have $\mathbf{Q}'(\mathbf{X}'\mathbf{H}_n\mathbf{X})^+\mathbf{Q} = \mathbf{0}$. Since $(\mathbf{X}'\mathbf{H}_n\mathbf{X})^+$ is positive semidefinite, its square root matrix (denoted $\mathbf{F}$) exists. It thus follows from $\mathbf{Q}'(\mathbf{X}'\mathbf{H}_n\mathbf{X})^+\mathbf{Q} = \mathbf{Q}\mathbf{F}\mathbf{F}\mathbf{Q}' = \mathbf{0}$ that $\mathbf{F}\mathbf{Q}' = \mathbf{0}$. This shows that $\mathbf{W}^*\mathbf{V}_2 = (\mathbf{X}'\mathbf{H}_n\mathbf{X})^+\mathbf{Q} = \mathbf{0}$. Thus, we also obtain $\mathbf{W}^*(\mathbf{W}^*)' = \mathbf{A}\mathbf{A}'$. The proof is complete.

Acknowledgements. Zhihua Zhang is supported in part by program for Changjiang Scholars and Innovative Research Team in University (IRT0652, PCSIRT), China.

References

1. Baudat, G., Anouar, F.: Generalized discriminant analysis using a kernel approach. Neural Computation 12, 2385–2404 (2000)
2. Cheng, Y.-Q., Zhuang, Y.-M., Yang, J.-Y.: Optimal Fisher discriminant analysis using the rank decomposition. Pattern Recognition 25(1), 101–111 (1992)
3. Duda, R.O., Hart, P.E., Stork, D.G.: Pattern Classification, 2nd edn. John Wiley and Sons, New York (2001)
4. Friedman, J.H.: Regularized discriminant analysis. Journal of the American Statistical Association 84(405), 165–175 (1989)
5. Golub, G.H., Heath, M., Wahba, G.: Generalized cross-validation as a method for choosing a good ridge parameter. Technometrics 21, 215–223 (1979)
6. Golub, G.H., Van Loan, C.F.: Matrix Computations. Johns Hopkins University Press, Baltimore (1996)
7. Hastie, T., Tishiran, R., Friedman, J.: The Elements of Statistical Learning: Data Mining, Inference, and Prediction. Springer, Heidelberg (2001)
8. Hoerl, A.E., Kennard, R.W.: Ridge regression. Technometrics 12, 56–67, 69–82 (1970)
9. Horn, R.A., Johnson, C.R.: Matrix Analysis. Cambridge University Press, Cambridge (1985)
10. Howland, P., Jeon, M., Park, H.: Structure preserving dimension reduction for clustered text data based on the generalized singular value decomposition. SIAM Journal on Matrix Analysis and Applications 25(1), 165–179 (2003)
11. Howland, P., Park, H.: Generalizing discriminant analysis using the generalized singular value decomposition. IEEE Transactions on Pattern Analysis and Machine Intelligence 26(8), 995–1006 (2004)
12. Kittler, J., Young, P.C.: A new approach to feature selection based on the Karhunen-Loève expansion. Pattern Recognition 5, 335–352 (1973)
13. Mika, S., Rätsch, G., Weston, J., Schölkopf, B., Smola, A., Müller, K.R.: Invariant feature extraction and classification in kernel space. In: Advances in Neural Information Processing Systems 12, vol. 12, pp. 526–532 (2000)
14. Paige, C.C., Saunders, M.A.: Towards a generalized singular value decomposition. SIAM Journal on Numerical Analysis 18(3), 398–405 (1981)
15. Park, C.H., Park, H.: Nonlinear discriminant analysis using kernel functions and the generalized singular value decomposition. SIAM Journal on Matrix Analysis and Applications 27(1), 87–102 (2005)
16. Park, C.H., Park, H.: A relationship between linear discriminant analysis and the generalized minimum squared error solution. SIAM Journal on Matrix Analysis and Applications 27(2), 474–492 (2005)
17. Roth, V., Steinhage, V.: Nonlinear discriminant analysis using kernel functions. In: Advances in Neural Information Processing Systems 12, vol. 12, pp. 568–574 (2000)
18. Shawe-Taylor, J., Cristianini, N.: Kernel Methods for Pattern Analysis. Cambridge University Press, Cambridge (2004)
19. Webb, A.R.: Statistical Pattern Recognition. John Wiley & Sons, Hoboken (2002)
20. Ye, J.: Least squares linear discriminant analysis. In: The Twenty-Fourth International Coference on Machine Learning (ICML) (2007)
21. Ye, J., Li, Q., Xiong, H., Park, H., Janardan, R., Kumar, V.: An incremental dimension reduction algorithm via QR decomposition. In: ACM SIGKDD, pp. 364–373 (2004)

Debt Detection in Social Security by Sequence Classification Using Both Positive and Negative Patterns

Yanchang Zhao[1], Huaifeng Zhang[2], Shanshan Wu[1], Jian Pei[3], Longbing Cao[1], Chengqi Zhang[1], and Hans Bohlscheid[1,2]

[1] Centre for Quantum Computation and Intelligent Systems, Faculty of Engineering & IT, University of Technology, Sydney, Australia
{yczhao,shanshan,lbcao,chengqi}@it.uts.edu.au
[2] Business Integrity Review Operations Branch, Centrelink, Australia
{huaifeng.zhang,hans.bohlscheid}@centrelink.gov.au
[3] School of Computing Science, Simon Fraser University, Canada
jpei@cs.sfu.ca

Abstract. Debt detection is important for improving payment accuracy in social security. Since debt detection from customer transactional data can be generally modelled as a fraud detection problem, a straightforward solution is to extract features from transaction sequences and build a sequence classifier for debts. The existing sequence classification methods based on sequential patterns consider only positive patterns. However, according to our experience in a large social security application, negative patterns are very useful in accurate debt detection. In this paper, we present a successful case study of debt detection in a large social security application. The central technique is building sequence classification using both positive and negative sequential patterns.

Keywords: sequence classification, negative sequential patterns.

1 Introduction and Application Background

Centrelink Australia (http://www.centrelink.gov.au) is a Commonwealth Government agency delivering a wide range of services to the Australian community. It is one of the largest data intensive applications in Australia. For example, in financial year 2004-2005 (from 1 July 2004 to 30 June 2005), Centrelink distributes approximately 63 billion dollars in social security payments to 6.4 million customers, makes 9.98 million individual entitlement payments, and records 5.2 billion electronic customer transactions [5].

Qualification for payment of an entitlement is assessed against a customer's personal circumstance. If all criteria are met, the payment to a customer continues until a change of the customer's circumstance precludes the customer from obtaining further benefit. However, for various reasons, customers on benefit payments or allowances sometimes get overpaid. The overpayments collectively lead to a large amount of debt owed to Centrelink. For instance, in financial year

W. Buntine et al. (Eds.): ECML PKDD 2009, Part II, LNAI 5782, pp. 648–663, 2009.

2004-2005, Centrelink raised over $900 million worth of customer debts (excluding Child Care Benefits and Family Tax Benefits) [5]. To achieve high payment accuracy, detection of debts is one of the most important tasks in Centrelink. Centrelink uses a number of processes to determine customers at risk of incurring a debt, however, the processes used only find a certain types of debts, for example, earnings-related debts. Another problem is that processes are applied to all customers, so a lot of time and efforts are spent on customers who are subsequently identified as being non-debtors. In this paper, we discuss our case study of debt detection in Centrelink using data mining techniques.

Debt detection can be generally modelled as a fraud detection problem. Therefore, we can adopt a classification approach. All transactions about a customer form a transaction sequence. If no debt happens to a customer, the sequence is labelled as normal (i.e., no-debt). If a debt happens to a customer, the corresponding customer sequence is labelled as debt. We can collect a training set containing both no-debt and debt sequences and learn a sequence classifier. The classifier can then be applied to new customer sequences to detect possible debts.

A classifier needs to extract features for classification. Since sequences are the data objects in debt detection, it is natural to use sequential patterns, i.e., subsequences that are frequent in customer sequences, as features. The traditional techniques for sequential pattern based classifiers consider only positive patterns, which capture a set of positively correlated events. Moreover, to detect debt at an early stage and prevent debt occurrence, a classification model is needed to predict the likelihood of debt occurrence based on the transactional activity data. Nevertheless, to the best of our knowledge, there are no techniques for building classifiers based on negative sequential patterns like $A \rightarrow \neg B$, $\neg A \rightarrow B$ and $\neg A \rightarrow \neg B$, where A and B and sequential patterns.

To tackle the above problems, based on our previous work on *negative sequential patterns* [27,28], we designed a new technique, *sequence classification using both positive and negative patterns*, to build sequence classifiers with relationship between activity sequences and debt occurrences. The contributions of this paper are:

- A new technique of *sequence classification using both positive and negative sequential patterns*; and
- An application in social security, demonstrating: 1) the effectiveness of our previous technique on *negative sequential pattern mining* to find both positive and negative sequential patterns; and 2) the effectiveness of our new technique on sequence classification using both positive and negative sequential patterns.

The rest of this paper is organized as follows. Our proposed technique of sequence classifiers using both positive and negative sequential patterns is described in Section 2. An application of the above technique in social security is presented in Section 3. Section 4 presents the related work on negative sequential pattern mining, sequence classification and existing systems and applications for fraud detection. Section 5 concludes this paper.

2 Sequence Classification Using Both Positive and Negative Sequential Patterns

From the data mining perspective, sequence classification is to build classifiers using sequential patterns. To the best of our knowledge, all of the existing sequence classification algorithms use positive sequential patterns only. However, the sequential patterns negatively correlated to debt occurrence are very important in debt detection. In this section, we first introduce negative sequential patterns and then propose a novel technique for sequence classification using both negative and positive sequential patterns.

2.1 Negative Sequential Patterns

Traditional sequential pattern mining deals with positive correlation between sequential patterns only, without considering negative relationship between them. To find negative relationship in sequences, we previously designed a notion of *negative sequential rules* [27,28] as follows.

Definition 1. *A negative sequential rule (NSR) is in the form of* $A \rightarrow \neg B$, $\neg A \rightarrow B$ *or* $\neg A \rightarrow \neg B$, *where* A *and* B *are sequential patterns.*

Based on the above definition, there are four types of sequential rules, including the tradition positive sequential rules (see Type I).

- Type I: $A \rightarrow B$, which means that pattern A is followed by pattern B;
- Type II: $A \rightarrow \neg B$, which means that pattern A is not followed by pattern B;
- Type III: $\neg A \rightarrow B$, which means that if pattern A does not appear, then pattern B will occur; and
- Type IV: $\neg A \rightarrow \neg B$, which means that if pattern A does not appear, then pattern B will not occur.

For types III and IV whose left sides are the negation of a sequence, there is no time order between the left side and the right side. Note that A and B themselves are sequential patterns, which make them different from negative association rules. The supports, confidences and lifts of the above four types of sequential

Table 1. Supports, Confidences and Lifts of Four Types of Sequential Rules

	Rules	Support	Confidence	Lift
I	$A \rightarrow B$	$P(AB)$	$\frac{P(AB)}{P(A)}$	$\frac{P(AB)}{P(A)P(B)}$
II	$A \rightarrow \neg B$	$P(A)-P(AB)$	$\frac{P(A)-P(AB)}{P(A)}$	$\frac{P(A)-P(AB)}{P(A)(1-P(B))}$
III	$\neg A \rightarrow B$	$P(B)-P(A\&B)$	$\frac{P(B)-P(A\&B)}{1-P(A)}$	$\frac{P(B)-P(A\&B)}{P(B)(1-P(A))}$
IV	$\neg A \rightarrow \neg B$	$1-P(A)-P(B)+P(A\&B)$	$\frac{1-P(A)-P(B)+P(A\&B)}{1-P(A)}$	$\frac{1-P(A)-P(B)+P(A\&B)}{(1-P(A))(1-P(B))}$

rules are shown in Table 1. In the table, $P(A\&B)$ denotes the probability of the concurrence of A and B in a sequence, no matter which one occurs first, or whether they are interwoven with each other.

2.2 Sequence Classification

Let $\mathcal{S}$ be a sequence dataset and $\mathcal{T}$ be a finite set of *class labels*. A *sequence classifier* is a function

$$\mathcal{F} : \mathcal{S} \rightarrow \mathcal{T}. \tag{1}$$

In sequence classification, a classifier $\mathcal{F}$ is built with frequent *classifiable sequential patterns* $\mathcal{P}$.

Definition 2. *A Classifiable Sequential Pattern (CSP) is in the form of* $p_a \rightarrow \tau$, *where* τ *is a class ID and* p_a *is a frequent pattern in the sequence dataset* $\mathcal{S}$.

The *support* of a sequential pattern p_a is the proportion of sequences containing p_a, and a sequential pattern is *frequent* in a dataset if its support in the dataset exceeds a user-specified minimum support threshold. Based on the mined sequential patterns, a sequence classifier can be formulized as

$$\mathcal{F} : \mathcal{S} \xrightarrow{\mathcal{P}} \mathcal{T}, \tag{2}$$

where $\mathcal{P}$ is a set of classifiable sequential patterns. That is, for each sequence $s \in \mathcal{S}$, $\mathcal{F}$ predicts the target class label of s based on the sequence classifier built using the classifiable sequential pattern set $\mathcal{P}$. A sequence instance s is said to be *covered* by a classifiable sequential pattern p ($p \in \mathcal{P}$) if s contains p_a, the antecedent of p.

2.3 Discriminative Sequential Patterns

Given a sequence dataset $\mathcal{S}$ and a set of target classes $\mathcal{T}$, a number of frequent classifiable sequential patterns need to be discovered for building a sequence classifier. The conventional algorithms use only positive sequential patterns to build classifiers. However, negative sequential patterns can also contribute to classification. To achieve better classification results, we use both negative and positive sequential patterns to build classifiers. Furthermore, instead of using the complete set of frequent patterns, we select a small set of discriminative classifiable sequential patterns according to Class Correlation Ratio (CCR) [22].

CCR measures how much a sequential pattern p_a is correlated with the target class τ compared to the negative class $\neg\tau$. Based on the contingency table (see Table 2), CCR is defined as

$$CCR(p_a \rightarrow \tau) = \frac{\hat{corr}(p_a \rightarrow \tau)}{\hat{corr}(p_a \rightarrow \neg\tau)} = \frac{a \cdot (c+d)}{c \cdot (a+b)}, \tag{3}$$

Table 2. Feature-Class Contingency Table

	p_a	$\neg p_a$	$\sum$
τ	a	b	$a+b$
$\neg\tau$	c	d	$c+d$
$\sum$	$a+c$	$b+d$	$n=a+b+c+d$

where $\hat{corr}(p_a \rightarrow \tau)$ is the correlation between p_a and the target class τ, defined as

$$\hat{corr}(p_a \rightarrow \tau) = \frac{sup(p_a \cup \tau)}{sup(p_a) \cdot sup(\tau)} = \frac{a \cdot n}{(a+c) \cdot (a+b)}. \tag{4}$$

CCR falls in $[0, +\infty)$. $CCR = 1$ means that the antecedent is independent of the target class. $CCR < 1$ indicates that the antecedent is negatively correlated with the target class, while $CCR > 1$ suggests a positive correlation between them.

In order to use the mined classifiable sequential patterns to build a classifier, we need to rank the patterns according to their capability to make correct classification. The ranking is based on a weighted score

$$W_s = \begin{cases} CCR, & \text{if } CCR \geq 1 \\ \frac{1}{CCR}, & \text{if } 0 < CCR < 1 \\ M, & \text{if } CCR = 0 \end{cases}, \tag{5}$$

where M is the maximum W_s of all rules where $CCR \neq 0$.

2.4 Building Sequence Classifiers

Our algorithm for building a sequence classifier with both positive and negative sequential patterns is composed of five steps.

1) Finding negative and positive sequential patterns using a negative sequential pattern mining algorithm, such as our previous techniques [27,28].
2) Calculating the frequency, chi-square and CCR of every classifiable sequential pattern, and only those patterns meeting *support*, *significance* (measured by chi-square) and *CCR* criteria are extracted into the classifiable sequential pattern set $\mathcal{P}$.
3) Pruning patterns in the obtained classifiable sequential pattern set with the pattern pruning algorithm in [13]. The only difference is that, in our algorithm, *CCR*, instead of confidence, is used as the measure for pruning.
4) Conducting serial coverage test by following the ideas in [15,13]. The patterns which can correctly cover one or more training samples in the test are kept for building a sequence classifier.
5) Ranking selected patterns with W_s and building the classifier as follows. Given a sequence instance s, all the classifiable sequential patterns covering s are extracted. The sum of the weighted score corresponding to each target class is computed and then s is assigned with the class label corresponding to the largest sum.

Table 3. Examples of Activity Transaction Data

Person_ID	Activity_Code	Activity_Date	Activity_Time
******002	DOC	20/08/2007	14:24:13
******002	RPT	20/08/2007	14:33:55
******002	DOC	05/09/2007	10:13:47
******002	ADD	06/09/2007	13:57:44
******002	RPR	12/09/2007	13:08:27
******002	ADV	17/09/2007	10:10:28
******002	REA	09/10/2007	07:38:48
******002	DOC	11/10/2007	08:34:36
******002	RCV	11/10/2007	09:44:39
******002	FRV	11/10/2007	10:18:46
******002	AAI	07/02/2008	15:11:54

3 A Case Study

Our technique was applied in social security to study the relationship between transactional activity patterns and debt occurrences and build sequence classifiers for debt detection.

3.1 Data

The data we used is the debt and activity transactions of 10,069 Centrelink customers from July 2007 to February 2008. In Centrelink, every single contact (e.g., because of a circumstance change) of a customer may trigger a sequence of activities running. As a result, large volumes of activity based transactions are recorded in an activity transactional database. In the original activity transactional table, each activity has 35 attributes, and we selected four of them which are related to this study. These attributes are "Person ID", "Activity Code", "Activity Date" and "Activity Time", as shown in Table 3. We sorted the activity data according to "Activity Date" and "Activity Time" to construct activity sequences. The debt data consists of "Person ID" and "Debt Transaction Date".

There are 155 different activity codes in the sequences. Different from supermarket basket analysis, every transaction in the application is composed of one activity only. The activities in four months before a debt were believed by domain experts to be related to the debt occurrence. If there were no debts for a customer during the period from July 2007 to February 2008, the activities in the first four months were taken as a sequence associated with no debts. After data cleaning and preprocessing, there are 15,931 sequences constructed with 849,831 activity records in this case study.

Table 4. Selected Positive and Negative Sequential Rules

Type	Rule	Support	Confidence	Lift
I	REA ADV ADV→DEB	0.103	0.53	2.02
	DOC DOC REA REA ANO→DEB	0.101	0.33	1.28
	RPR ANO→DEB	0.111	0.33	1.25
	RPR STM STM RPR→DEB	0.137	0.32	1.22
	MCV→DEB	0.104	0.31	1.19
	ANO→DEB	0.139	0.31	1.19
	STM PYI→DEB	0.106	0.30	1.16
II	STM PYR RPR REA RPT→ ¬DEB	0.166	0.86	1.16
	MND→ ¬DEB	0.116	0.85	1.15
	STM PYR RPR DOC RPT→ ¬DEB	0.120	0.84	1.14
	STM PYR RPR REA PLN→ ¬DEB	0.132	0.84	1.14
	REA PYR RPR RPT→ ¬DEB	0.176	0.84	1.14
	REA DOC REA CPI→ ¬DEB	0.083	0.83	1.12
	REA CRT DLY→ ¬DEB	0.091	0.83	1.12
	REA CPI→ ¬DEB	0.109	0.83	1.12
III	¬{PYR RPR REA STM}→DEB	0.169	0.33	1.26
	¬{PYR CCO}→DEB	0.165	0.32	1.24
	¬{STM RPR REA RPT}→DEB	0.184	0.29	1.13
	¬{RPT RPR REA RPT}→DEB	0.213	0.29	1.12
	¬{CCO RPT}→DEB	0.171	0.29	1.11
	¬{CCO PLN}→DEB	0.187	0.28	1.09
	¬{PLN RPT}→DEB	0.212	0.28	1.08
IV	¬{ADV REA ADV}→ ¬DEB	0.648	0.80	1.08
	¬{STM EAN}→ ¬DEB	0.651	0.79	1.07
	¬{REA EAN}→ ¬DEB	0.650	0.79	1.07
	¬{DOC FRV}→ ¬DEB	0.677	0.78	1.06
	¬{DOC DOC STM EAN}→ ¬DEB	0.673	0.78	1.06
	¬{CCO EAN}→ ¬DEB	0.681	0.78	1.05

3.2 Results of Negative Sequential Pattern Mining

Our previous technique on *negative sequential rules* [28] was used to find both positive and negative sequential patterns from the above data. By setting the minimum support to 0.05, that is, 797 out of 15,931 sequences, 2,173,691 patterns were generated and the longest pattern has 16 activities. From the patterns, 3,233,871 positive and negative rules were derived. Some selected sequential rules are given in Table 4, where "DEB" stands for debt and the other codes are activities. The rules marked by "Type I" are positive sequential rules, while others are negative ones.

3.3 Evaluation of Sequence Classification

The performance of the classifiers using both positive and negative sequential patterns were tested and compared with the classifiers using positive patterns only.

In the discovered rules shown in Table 4, generally speaking, Type I rules are positive patterns and all the other three types are negative ones. However, in the binary classification problem in our case study, $A \rightarrow \neg DEB$ can be taken as a positive rule $A \rightarrow c_2$, where c_2 denotes "no debt". Therefore, we treated Type I and Type II patterns as positive and Type III and Type IV as negative. That is, in the results shown in Tables 6–9, the traditional classifiers (labelled as "Positive") were built using both Type I and II rules, while our new classifiers (labelled as "Neg& Pos") were built using all four types of rules. However, in applications where there are multiple classes, Type II rules are negative.

By setting the minimum support to 0.05 and 0.1, respectively, we got two sets of sequential patterns, "PS05" and "PS10". The numbers of the four types of patterns are shown in Table 5. There are $775,175$ patterns in "PS10" and $3,233,871$ patterns in "PS05". It is prohibitively time consuming to do coverage test and build classifiers on so large sets of patterns. In this experiment, we ranked the patterns according to W_s. Then, we extracted the top $4,000$ and $8,000$ patterns from "PS05" and "PS10" and referred to them as "PS05-4K", "PS05-8K", "PS10-4K" and "PS10-8K", respectively.

After that, two groups of classifiers were built. The first group, labelled as "Neg& Pos", were built with both negative and positive patterns (i.e., all four types of rules), and the other group, labelled as "Positive", were built with positive patterns (i.e., Type I and II rules) only. In order to compare the two groups of classifiers, we selected various numbers of patterns from the ones passing coverage test to build the final classifiers and the results are shown in Tables 6–9. In the four tables, the first rows show the number of patterns used in the classifiers. In Tables 8 and 9, some results are not available for pattern number as 200 and 300, because there are less than 200 (or 300) patterns remaining after coverage test.

From the four tables, we can see that, if built with the same number of rules, in terms of recall, our classifiers built with both positive and negatives rules outperforms traditional classifiers with only positive rules under most conditions. It means that, with negative rules involved, our classifiers can predict more debt occurrences.

As shown by the results on "PS05-4K" in Table 6, our classifiers is superior to traditional classifiers with 80, 100 and 150 rules in recall, accuracy and precision.

From the results on "PS05-8K" shown in Table 7, we can see that our classifiers with both positive and negatives rules outperforms traditional classifiers with only positive rules in accuracy, recall and precision in most of our experiments. Again, it also shows that the recall is much improved when negative rules are involved.

As shown by Tables 8 and 9, our classifiers have higher recall with 80, 100 and 150 rules. Moreover, our best classifier is the one with 60 rules, which has accuracy=0.760, specificity=0.907 and precision=0.514. It is better in all the three measures than all traditional classifiers given in the two tables.

One interesting thing we found is that, the number of negative patterns used for building our classifiers is very small, compared with that of positive patterns (see Table 10). Especially for "PS05-4K" and "PS05-8K", the two pattern sets chosen from the mined patterns with minimum support=0.05, there are respectively only 4 and 7 negative patterns used in the classifiers. However, these

Table 5. The Number of Patterns in PS10 and PS05

	PS10 ($min\_sup = 0.1$)		PS05 ($min\_sup = 0.05$)	
	Number	Percent(%)	Number	Percent(%)
Type I	93,382	12.05	127,174	3.93
Type II	45,821	5.91	942,498	29.14
Type III	79,481	10.25	1,317,588	40.74
Type IV	556,491	71.79	846,611	26.18
Total	775,175	100	3,233,871	100

Table 6. Classification Results with Pattern Set PS05-4K

Pattern Number		40	60	80	100	150	200	300
Neg&Pos	Recall	.438	.416	**.286**	**.281**	**.422**	.492	.659
	Precision	.340	.352	.505	**.520**	**.503**	.474	.433
	Accuracy	.655	.670	**.757**	**.761**	**.757**	.742	.705
	Specificity	.726	.752	.909	.916	.865	.823	.720
Positive	Recall	.130	.124	.141	.135	.151	.400	.605
	Precision	.533	.523	.546	.472	.491	.490	.483
	Accuracy	.760	.758	.749	.752	.754	.752	.745
	Specificity	.963	.963	.946	.951	.949	.865	.790

several negative patterns do make a difference when building classifiers. Three examples of them are given as follows.

- $\neg ADV \rightarrow \neg DEB$ (CCR=1.99, conf=0.85)
- $\neg(STM, REA, DOC) \rightarrow \neg DEB$ (CCR=1.86, conf=0.84)
- $\neg(RPR, DOC) \rightarrow \neg DEB$ (CCR=1.71, conf=0.83)

Some examples of other rules used in our classifiers are

- $STM, RPR, REA, EAD \rightarrow DEB$ (CCR=18.1)
- $REA, CCO, EAD \rightarrow DEB$ (CCR=17.8)
- $CCO, MND \rightarrow \neg DEB$ (CCR=2.38)

4 Related Work

Our study is related to the previous work on negative sequential pattern mining, sequence classification and fraud/intrusion detection. In this section, we review the related work briefly.

Table 7. Classification Results with Pattern Set PS05-8K

Pattern Number		40	60	80	100	150	200	300
Neg&Pos	Recall	**.168**	**.162**	**.205**	**.162**	**.173**	**.341**	**.557**
	Precision	**.620**	**.652**	**.603**	**.625**	**.615**	**.568**	.512
	Accuracy	**.771**	**.774**	**.773**	**.771**	**.771**	**.775**	**.762**
	Specificity	.967	.972	.956	.969	.965	.916	.829
Positive	Recall	.141	.103	.092	.092	.108	.130	.314
	Precision	.542	.576	.548	.548	.488	.480	.513
	Accuracy	.761	.762	.760	.760	.754	.753	.760
	Specificity	.962	.976	.976	.976	.963	.955	.904

Table 8. Classification Results with Pattern Set PS10-4K

Pattern Number		40	60	80	100	150
Neg&Pos	Recall	0	.303	**.465**	**.535**	**.584**
	Precision	0	**.514**	.360	.352	.362
	Accuracy	.756	**.760**	.667	.646	.647
	Specificity	1	**.907**	.733	.682	.668
Positive	Recall	.373	.319	.254	.216	.319
	Precision	.451	.421	.435	.430	.492
	Accuracy	.736	.727	.737	.738	.753
	Specificity	.853	.858	.893	.907	.893

Table 9. Classification Results with Pattern Set PS10-8K

Pattern Number		40	60	80	100	150	200
Neg&Pos	Recall	0	.303	**.465**	**.535**	**.584**	N/A
	Precision	0	**.514**	.360	.352	.362	N/A
	Accuracy	.756	**.760**	.667	.646	.647	N/A
	Specificity	1	**.907**	.733	.682	.668	N/A
Positive	Recall	.459	.427	.400	.378	.281	.373
	Precision	.385	.397	.430	.438	.464	.500
	Accuracy	.688	.701	.724	.729	.745	.756
	Specificity	.762	.790	.829	.843	.895	.879

Table 10. The Number of Patterns in the Four Pattern Sets

Pattern Set	PS10-4K	PS10-8K	PS05-4K	PS05-8K
Type I	2,621	5,430	1,539	1,573
Type II	648	1,096	2,457	6,420
Type III	2	5	0	0
Type IV	729	1,469	4	7
Total	4,000	8,000	4,000	8,000

4.1 Negative Sequential Pattern Mining

Since sequential pattern mining was first proposed in [1], a few sequential methods have been developed, such as GSP (Generalized Sequential Patterns) [19], FreeSpan [8], PrefixSpan [17], SPADE [26] and SPAM [2]. Most of the sequential pattern mining algorithms focus on the patterns appearing in the sequences, i.e., the positively correlated patterns. However, the absence of some items in sequences may also be interesting in some scenarios. For example, in social welfare, the lack of follow-up examination after the address change of a customer may result in overpayment to him/her. Such kind of sequences with the non-occurrence of elements are negative sequential patterns. Only several studies look at this issue.

Sun *et al.* [20] proposed negative event-oriented patterns in the form of $\neg P \xrightarrow{T} e$, where e is a target event, P is a negative event-oriented pattern, and the occurrence of P is unexpectedly rare in T-sized intervals before target events. P is supposed to be an "existence pattern" (i.e., a frequent itemset without time order), instead of a sequential pattern, though it is claimed that the discussion can be extended to sequential patterns.

Bannai *et al.* [3] proposed a method for finding the optimal pairs of string patterns to discriminate between two sets of strings. The pairs are in the form of $p' \wedge q'$ or $p' \vee q'$, where p' is either p or $\neg p$, q' is either q or $\neg q$, and p and q are two substrings. Their concern is whether p and q appear in a string s.

Ouyang and Huang [16] proposed the notion of negative sequences as $(A, \neg B)$, $(\neg A, B)$ and $(\neg A, \neg B)$. Negative sequential patterns are derived from infrequent sequences. A drawback is that both frequent and infrequent sequences have to be found at the first stage, which demands a large amount of space.

Lin *et al.* [14] designed an algorithm NSPM (Negative Sequential Patterns Mining) for mining negative sequential patterns. In their negative patterns, only the last element can be negative, and all other elements are positive.

4.2 Sequence Classification

Classification on sequence data is an important problem. A few methods have been developed.

Wu *et al.* [23] proposed a neural network classification method for molecular sequence classification. The molecular sequences are encoded into input vectors of a neural network classifier, by either an n-gram hashing method or a SVD (Singular Value Decomposition) method.

Chuzhanova *et al.* [6] proposed to use Gamma (or near-neighbour) test to select features from l-grams over the alphabet. The method was used to classify the large subunits rRNA, and the nearest-neighbour criterion was used to estimate the classification accuracy based on the selected features.

Lesh *et al.* [11] used sequential patterns as features in classification. Sequence mining is first employed to find sequential patterns correlated with the target classes, and then the discovered patterns are used as features to build classifiers with standard classification algorithms, such as Naïve Bayes. Their experimental results show that using sequential patterns as features can improve the accuracy substantially. Compared to our work, they did not consider negative sequential patterns.

Tseng and Lee [21] designed algorithm CBS (Classify-By-Sequence) for classifying large sequence data sets. Sequential pattern mining and probabilistic induction are integrated for efficient extraction of sequential patterns and accurate classification.

Li and Sleep [12] proposed a robust approach to sequence classification, where n-grams of various lengths are used to measure the similarity between sequences, a modified LZ78 algorithm is employed for feature selection, and a Support Vector Machine (SVM) is used as the classifier.

A discriminatively trained Markov Model (MM(k-1)) for sequence classification was proposed by Yakhnenko *et al.* [25]. Their experimental results show that their classifiers are comparable in accuracy and more efficient than Support Vector Machines trained by k-gram representations of sequences.

Lei and Govindaraju [10] proposed to use an intuitive similarity measure, ER^2, for multi-dimensional sequence classification based on SVM. The measure is used to reduce classification computation and speed up the decision-making of multi-class SVM.

Exarchos *et al.* [7] proposed a two-stage methodology for sequence classification based on sequential pattern mining and optimization. In the first stage, sequential pattern mining is used and a sequence classification model is built based on the extracted sequential patterns. Then, weights are applied to both sequential patterns and classes. In the second stage, the weights are tuned with an optimization technique to achieve optimal classification accuracy.

Xing *et al.* [24] studied the problem of early prediction using sequence classifiers. The prefix of a sequence as short as possible is used to make a reasonably accurate prediction. They proposed a sequential classification rule method to mine sequential classification rules, which are then selected by an early-prediction utility measure. Based on the selected rules, a generalized sequential decision tree method is used to build a classification model with a divide-and-conquer strategy.

In all the above studies, no negative sequential patterns are considered.

4.3 Fraud/Intrusion Detection

Some applications similar to debt detection are fraud detection, terrorism detection, financial crime detection, network intrusion detection and spam detection. Different from transactional fraud detection which attempts to classify a transaction or event as being legal or fraud, our techniques try to predict the likelihood of a customer being fraud based on his past activities. It is at customer level instead of transaction level.

Bonchi *et al.* [4] proposed a classification-based methodology for planning audit strategies in fraud detection and presented a case study on illustrating how classification techniques can be used to support the task of planning audit strategies. The models are constructed by analysing historical audit data. Then, the models are used to plan effectively future audits for the detection of tax evasion. A decision tree algorithm, $C5.0$, was used in their case study. Although the target problem is similar to ours, the data used is different. We used transactional data which records activities related to customers. Because the time order in activities is important for predicting debt occurrences, sequence classifiers instead of decision trees are used in our application.

Rosset *et al.* [18] studied the fraud detection in telecommunication and presented a two-stage system based on $C4.5$ to find fraud rules. They adapted the $C4.5$ algorithm for generating rules from bi-level data, i.e., customer data and behaviour-level data. However, the behaviour data they used is the statistics in a short time frame, such as the number of international calls and total duration of all calls in a day, which is different from the sequential patterns in our techniques.

Julisch and Dacier [9] used techniques of episode rules and conceptual clustering to mine historical alarms for network intrusion detection. Their episode rules are designed to predict the occurrence of certain alarms based on other alarms. Negative sequential patterns are not taken into account in their model.

5 Conclusions and Discussion

We presented a new technique for building sequence classifiers with both positive and negative sequential patterns. We also presented an application for debt detection in the domain of social security, which shows the effectiveness of the proposed technique.

A limitation of our proposed technique is that an element in a sequence is assumed to be a single event, which is based on the transaction data in this application in social security. However, in other applications, an element may be composed of multiple items. Therefore, to extend our techniques to such general sequence data will be part of our future work.

Another limitation is that time constraints are only partly considered in our techniques. What we did is setting the time window so that a pattern is less than 4 months, based on domain experts' suggestions. Nevertheless, we have not set any other time constraints, such as the time interval between adjacent elements. In other applications, it may be interesting to find patterns with the above constraints and use them to build sequence classifiers.

A third limitation is that, in real world applications, there are different costs with correct predictions, false positives and false negatives, and it will be more fair and more useful when measuring the performance of classifiers by taking the above costs into consideration. We are currently in the progress of the above work.

In our future work, we will also use time to measure the performance of our classifiers, because it is desirable in real-world applications to predict debt occurrences as early as possible. Using time to measure the utility of negative patterns and to build sequence classifiers for early detection will be part of our future work.

Last but not least, patterns in data may keep changing as time goes on, and the learned patterns and the built classifiers may become out-of-date. New labelled data (e.g., new debts) from oncoming data can be used to improve the classifiers. Therefore, it is imperative to build an adaptive online classifier which can adapt itself to the changes in new data.

Acknowledgments

This work was supported by the Australian Research Council (ARC) Linkage Project LP0775041 and Discovery Projects DP0667060 & DP0773412, and by the Early Career Researcher Grant from University of Technology, Sydney, Australia.

We would like to thank Mr. Peter Newbigin and Mr. Brett Clark from Business Integrity Review Operations Branch, Centrelink, Australia for their support of domain knowledge and helpful suggestions.

References

1. Agrawal, R., Srikant, R.: Mining sequential patterns. In: Proc. of the 11th International Conference on Data Engineering, Taipei, Taiwan, 1995, pp. 3–14. IEEE Computer Society Press, Los Alamitos (1995)
2. Ayres, J., Flannick, J., Gehrke, J., Yiu, T.: Sequential pattern mining using a bitmap representation. In: KDD 2002: Proc. of the 8th ACM SIGKDD International Conference on Knowledge Discovery and Data Mining, pp. 429–435. ACM, New York (2002)
3. Bannai, H., Hyyro, H., Shinohara, A., Takeda, M., Nakai, K., Miyano, S.: Finding optimal pairs of patterns. In: Jonassen, I., Kim, J. (eds.) WABI 2004. LNCS (LNBI), vol. 3240, pp. 450–462. Springer, Heidelberg (2004)
4. Bonchi, F., Giannotti, F., Mainetto, G., Pedreschi, D.: A classification-based methodology for planning audit strategies in fraud detection. In: Proc. of the 5th ACM SIGKDD International Conference on Knowledge Discovery and Data Mining, San Diego, CA, USA, pp. 175–184. ACM Press, New York (1999)
5. Centrelink. Centrelink annual report 2004-2005. Technical report, Centrelink, Australia (2005)
6. Chuzhanova, N.A., Jones, A.J., Margetts, S.: Feature selection for genetic sequence classification. Bioinformatics 14(2), 139–143 (1998)
7. Exarchos, T.P., Tsipouras, M.G., Papaloukas, C., Fotiadis, D.I.: A two-stage methodology for sequence classification based on sequential pattern mining and optimization. Data and Knowledge Engineering 66(3), 467–487 (2008)
8. Han, J., Pei, J., Mortazavi-Asl, B., Chen, Q., Dayal, U., Hsu, M.-C.: Freespan: frequent pattern-projected sequential pattern mining. In: KDD 2000: Proc. of the 6th ACM SIGKDD international conference on Knowledge discovery and data mining, Boston, Massachusetts, USA, pp. 355–359. ACM, New York (2000)
9. Julisch, K., Dacier, M.: Mining intrusion detection alarms for actionable knowledge. In: Proc. of the 8th ACM SIGKDD International Conference on Knowledge Discovery and Data Mining, Edmonton, Alberta, Canada, pp. 366–375. ACM, New York (2002)

10. Lei, H., Govindaraju, V.: Similarity-driven sequence classification based on support vector machines. In: ICDAR 2005: Proc. of the 8th International Conference on Document Analysis and Recognition, Washington, DC, USA, 2005, pp. 252–261. IEEE Computer Society, Los Alamitos (2005)
11. Lesh, N., Zaki, M.J., Ogihara, M.: Mining features for sequence classification. In: KDD 1999: Proc. of the 5th ACM SIGKDD international conference on Knowledge discovery and data mining, pp. 342–346. ACM, New York (1999)
12. Li, M., Sleep, R.: A robust approach to sequence classification. In: ICTAI 2005: Proc. of the 17th IEEE International Conference on Tools with Artificial Intelligence, Washington, DC, USA, pp. 197–201. IEEE Computer Society, Los Alamitos (2005)
13. Li, W., Han, J., Pei, J.: Cmar: Accurate and efficient classification based on multiple class-association rules. In: ICDM 2001: Proc. of the 2001 IEEE International Conference on Data Mining, Washington, DC, USA, pp. 369–376. IEEE Computer Society, Los Alamitos (2001)
14. Lin, N.P., Chen, H.-J., Hao, W.-H.: Mining negative sequential patterns. In: Proc. of the 6th WSEAS International Conference on Applied Computer Science, Hangzhou, China, pp. 654–658 (2007)
15. Liu, B., Hsu, W., Ma, Y.: Integrating classification and association rule mining. In: KDD 1998: Proc. of the 4th International Conference on Knowledge Discovery and Data Mining, pp. 80–86. AAAI Press, Menlo Park (1998)
16. Ouyang, W., Huang, Q.: Mining negative sequential patterns in transaction databases. In: Proc. of 2007 International Conference on Machine Learning and Cybernetics, Hong Kong, pp. 830–834. China (2007)
17. Pei, J., Han, J., Mortazavi-Asl, B., Pinto, H., Chen, Q., Dayal, U., Hsu, M.-C.: Prefixspan: Mining sequential patterns efficiently by prefix-projected pattern growth. In: ICDE 2001: Proc. of the 17th International Conference on Data Engineering, Washington, DC, USA, pp. 215–224. IEEE Computer Society, Los Alamitos (2001)
18. Rosset, S., Murad, U., Neumann, E., Idan, Y., Pinkas, G.: Discovery of fraud rules for telecommunications - challenges and solutions. In: Proc. of the 5th ACM SIGKDD International Conference on Knowledge Discovery and Data Mining, San Diego, CA, USA, August 1999, pp. 409–413 (1999)
19. Srikant, R., Agrawal, R.: Mining sequential patterns: Generalizations and performance improvements. In: Apers, P.M.G., Bouzeghoub, M., Gardarin, G. (eds.) EDBT 1996. LNCS, vol. 1057, pp. 3–17. Springer, Heidelberg (1996)
20. Sun, X., Orlowska, M.E., Li, X.: Finding negative event-oriented patterns in long temporal sequences. In: Dai, H., Srikant, R., Zhang, C. (eds.) PAKDD 2004. LNCS (LNAI), vol. 3056, pp. 212–221. Springer, Heidelberg (2004)
21. Tseng, V.S.-M., Lee, C.-H.: Cbs: A new classification method by using sequential patterns. In: SDM 2005: Proc. of the 2005 SIAM International Data Mining Conference, Newport Beach, California, USA, pp. 596–600 (2005)
22. Verhein, F., Chawla, S.: Using significant, positively associated and relatively class correlated rules for associative classification of imbalanced datasets. In: ICDM 2007: Proc. of the 7th IEEE International Conference on Data Mining, pp. 679–684 (2007)
23. Wu, C.H., Berry, M.W., Fung, Y.-S., McLarty, J.: Neural networks for molecular sequence classification. In: Proc. of the 1st International Conference on Intelligent Systems for Molecular Biology, pp. 429–437. AAAI Press, Menlo Park (1993)
24. Xing, Z., Pei, J., Dong, G., Yu, P.: Mining sequence classifiers for early prediction. In: SDM 2008: Proc. of the 2008 SIAM international conference on data mining, Atlanta, GA, USA, April 2008, pp. 644–655 (2008)

25. Yakhnenko, O., Silvescu, A., Honavar, V.: Discriminatively trained markov model for sequence classification. In: ICDM 2005: Proc. of the 5th IEEE International Conference on Data Mining, Washington, DC, USA, pp. 498–505. IEEE Computer Society, Los Alamitos (2005)
26. Zaki, M.J.: Spade: An efficient algorithm for mining frequent sequences. Machine Learning 42(1-2), 31–60 (2001)
27. Zhao, Y., Zhang, H., Cao, L., Zhang, C., Bohlscheid, H.: Efficient mining of event-oriented negative sequential rules. In: WI 2008: Proc. of the 2008 IEEE/WIC/ACM International Conference on Web Intelligence, Sydney, Australia, December 2008, pp. 336–342 (2008)
28. Zhao, Y., Zhang, H., Cao, L., Zhang, C., Bohlscheid, H.: Mining both positive and negative impact-oriented sequential rules from transactional data. In: Proc. of the 13th Pacific-Asia Conference on Knowledge Discovery and Data Mining (PAKDD 2009), Bangkok, Thailand, April 2009, pp. 656–663 (2009)

Learning the Difference between Partially Observable Dynamical Systems

Sami Zhioua[1], Doina Precup[1], François Laviolette[2], and Josée Desharnais[2]

[1] School of Computer Science, McGill University, QC, Canada
[2] Department of Computer Science and Software Engineering, Laval University, QC, Canada

Abstract. We propose a new approach for estimating the difference between two partially observable dynamical systems. We assume that one can interact with the systems by performing actions and receiving observations. The key idea is to define a Markov Decision Process (MDP) based on the systems to be compared, in such a way that the optimal value of the MDP initial state can be interpreted as a divergence (or dissimilarity) between the systems. This dissimilarity can then be estimated by reinforcement learning methods. Moreover, the optimal policy will contain information about the actions which most distinguish the systems. Empirical results show that this approach is useful in detecting both big and small differences, as well as in comparing systems with different internal structure.

1 Introduction

When building or learning models of dynamical systems, one is often confronted with the problem of choosing between two models, or assessing whether a given model is faithful to the original system. In such scenarios, a notion of behavioral distance (or divergence) between dynamical systems can be very helpful. For example, when learning a model of a Partially Observable Markov Decision Process (POMDP), one might make assumptions about the number of internal states, then learn the parameters of the system from data. It would be very useful for the learning process to have a notion of how close the learned model is to the true system. Moreover, an algorithm that can point out what parts of the learned model are most different from the original would be very useful, as it would suggest where the learned model can be improved. The results of such an algorithm can be used to guide both the process of selecting the structure of a new model, as well as the data acquisition. An even greater need for determining differences arises if the systems under consideration are different in structure (e.g., a first and a second-order Markov model, or a POMDP and a predictive state representation). In this case, simple inspection of the models cannot determine if they behave similarly. Instead, we need a way of assessing dissimilarity based purely on data, independently of the model structure.

In this paper, we develop a novel approach to this problem, inspired by ideas from probabilistic verification [1]. The key idea is to define a Markov Decision Process (MDP) whose states and actions are built based on the processes that we want to compare. The optimal value function of this MDP will be interpreted as a divergence (or distance) between the processes. No knowledge of the original processes is needed; instead, we only need the ability to interact with them. The optimal policy in the MDP

W. Buntine et al. (Eds.): ECML PKDD 2009, Part II, LNAI 5782, pp. 664–677, 2009.

will point out the actions which distinguish the systems most. This is very useful both in learning applications, as well as in knowledge transfer, as we discuss in detail later.

The paper is structured as follows. Section 2 presents the problem and notation. Section 3 presents the core ideas of our approach. Section 4 establishes the theoretical properties of the proposed approach. We show that the MDP we define has a positive value function, with strict inequality if and only if the two dynamical systems under consideration are different. We also discuss the sample complexity of our approach. In Section 5, we illustrate the behavior of the algorithm on several standard POMDP examples from the literature. We show that the algorithm can both identify when the systems are different, as well as quantify the extent of the difference. Section 6 contains a discussion and related work. Finally, in Section 7 we conclude and present avenues for future work.

2 Background

An MDP is a tuple $\langle \mathcal{S}, \mathcal{A}, \mathcal{T}, \mathcal{R} \rangle$ where $\mathcal{S}$ is a set of states, $\mathcal{A}$ is a set of actions, $\mathcal{T} : \mathcal{S} \times \mathcal{A} \times \mathcal{S} \longrightarrow [0, 1]$ is a transition function such that $\mathcal{T}(s, a, s')$ represents the probability of making a transition from s to s' after action a, and $\mathcal{R} : \mathcal{S} \times \mathcal{A} \longrightarrow \mathbb{R}$ is a reward function such that $\mathcal{R}(s, a)$ is the reward obtained when action a is performed in state s.

We will use MDPs to establish the differences between two stochastic, partially observable systems. We make the assumption that it is possible to interact with the systems by choosing actions a_t at discrete time steps and receiving in response observations o_t. We also assume that there is a well-defined (but possibly unknown) probability distribution $Pr(o_{t+1}|a_1 o_1 \dots a_t o_t), \forall t$. The most common model for such systems in the AI literature is the Partially Observable Markov Decision Process (POMDP) (see [2] for a survey). However, this formulation is more general and it includes, for example, higher-order Markov systems, as well as systems with internal structure represented as a finite state machine.

The notion of equivalence or similarity between probabilistic processes has been studied extensively in the literature on verification of concurrent processes. In their pioneering work [3], Larsen and Skou have defined a notion of *tests*, which consist conceptually of logical formulas defined based on a given grammar. The formulas are defined over the observations produced by the systems, and give a characterization of what statements can made true in the two processes. Two processes are considered equivalent if they accept all tests with equal probabilities. Different grammars for generating tests give rise to different notions of equivalence between processes. The notion of equivalence can be relaxed by using *metrics*, which represent quantitative measures of similarity between processes. Metrics are often defined based on the complexity of the grammar needed to distinguish the processes.

In the AI literature, a similar idea is captured by predictive state representations (PSRs) [4], which characterize the "state" of a dynamical system by the set of conditional probabilities of different observation sequences that can be observed the current history, given different sequences of actions. In some cases, PSRs can provide a compact representation of the dynamical system. However, from the point of view of this paper, the main idea we want to emphasize is the use of tests, and their associated probabilities.

A *test* is a sequence of actions of the form $a_1 a_2 ... a_n$ where $a_1, a_2, ..., a_n \in \mathcal{A}$. Running a test on a dynamical system produces a sequence of observations of the form $o_1 o_2 ... o_n$ where $o_1, o_2, ..., o_n \in \mathcal{O}$. The set of all such sequences will be denoted $\mathcal{O}^*$. Under our definition of dynamical system, any test induces a well-defined probability distribution over observations sequences:

$$P(o_1 \dots o_n | a_1 \dots a_n) = P(o_1|a_1)P(o_2|a_1, o_1, a_2) \dots P(o_n|a_1, o_1, \dots a_{n-1}o_{n-1}a_n).$$

Definition 1. Two dynamical systems with probability distributions P_1 and P_2 are *trace equivalent* if and only if, for any test $a_1 \dots a_n$, and for any sequence of observations $o_1 \dots o_n$,

$$P_1(o_1 \dots o_n | a_1 \dots a_n) = P_2(o_1 \dots o_n | a_1 \dots a_n).$$

3 Quantifying the Difference between Dynamical Systems

Let P_1 and P_2 be two processes that we are interested in comparing (for ease of notation, we will denote a system by its associated probability distributions). We are interested in determining tests which maximize the difference between the two systems. Our strategy will be two-fold. First, we will break down the action sequence of a test into steps, and consider one action at a time. Second, we will establish a reward function which emphasizes the differences between processes: if the same action causes *different* observations in the two processes, it should receive a high reward, otherwise, it should receive a low reward. However, because the systems under consideration are stochastic, different observations may be obtained in the two systems even if they are identical. The likelihood of this occurrence is higher if the probability distributions associated with the systems have high entropy. In order to correctly account for this possibility, we introduce a third and a fourth processes, called respectively P_{1c} and P_{2c}, which are simply clones of the original systems P_1 and P_2 (see Fig.1). There will be a high reward if P_1 and P_2 differ for some action, but this reward can be cancelled if P_1 and P_{1c} differ and/or P_2 and P_{2c} differ. The intuition is that in this case, the difference we observed may well be due to the inherent stochasticity in the processes and not to an actual behavioral difference between them.

The intuition for our approach is inspired by the well-known Kullback-Leibler (KL) divergence [5]. A divergence between two dynamical systems could be defined as the maximum divergence between the probability distributions over observation sequences generated by all tests run in the systems. The KL divergence is a good way of measuring the similarity of two probability distributions. For two distributions P_1 and P_2, it is defined as:

$$\mathrm{KL}(P_1 \| P_2) := \mathrm{E}_{h \sim P_1} \ln \frac{1}{P_2(h)} - \mathrm{E}_{h \sim P_1} \ln \frac{1}{P_1(h)} \qquad (1)$$

Unfortunately, because of the high number of possible tests (especially large systems), computing the maximum value over all Kullback-Leibler divergences is not tractable. Nevertheless, there is an analogy between our intuition and the KL divergence. The first term in the formula can be considered as comparing the two processes of interest, P_2 and P_1. The second term accounts for the entropy of the first process, P_1, for a given test. However, the KL divergence is not symmetrical, whereas for our purpose, the two processes should be treated in the same way. Hence, we would like a divergence which is closer in spirit to the "symmetrized" KL divergence, $KL(P_1|P_2) + KL(P_2|P_1)$. Also, instead of striving to compute the KL divergence, we will compute a different divergence, which will be more efficient. We are now ready to detail our divergence computation.

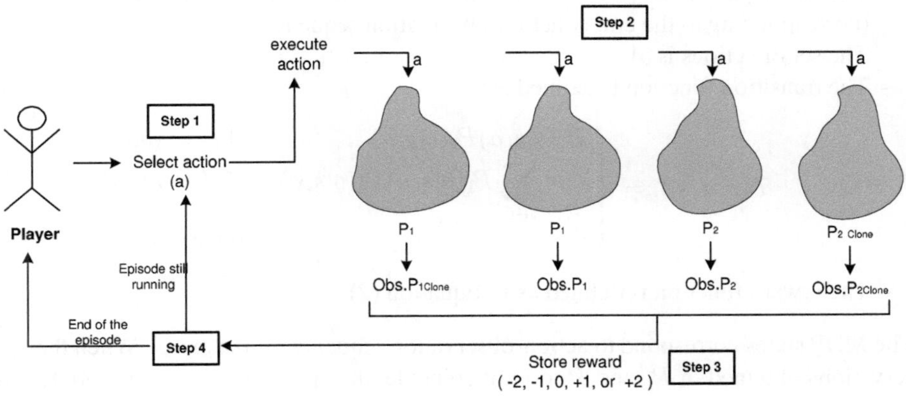

Fig. 1. Sketch of the divergence computation

Figure 1 depicts the setup that we will use. The interaction with the processes will consist of the following steps:

Initialize the four processes (P_{1c}, P_1, P_2, and P_{2c})
Repeat:

Step 1 : Choose an action a.
Step 2 : Execute a on P_{1c}, P_1, P_2. and P_{2c}.
Step 3 : Compute the reward by adding the following components:

- a reward of $(+2)$ for different observations between P_1 and P_2
- a (-1) reward for different observations between P_1 and P_{1c}
- a (-1) reward for different observations between P_2 and P_{2c}.

In other words, if $Obs.P_i$ denotes the observation obtained in system i, the immediate (stochastic) reward is given by the formula:

$$R := 2\ (Obs.P_1 \neq Obs.P_2) - (Obs.P_1 \neq Obs.P_{1c}) - \ (Obs.P_2 \neq Obs.P_{2c}) \quad (2)$$

where 0 and 1 are used as both truth values and numbers.

until either $Obs.P_1 \neq Obs.P_2$, or the episode reaches a certain horizon.

We will now show that this interaction gives rise to an MDP, whose value is 0 if and only if the two processes have the same probability distribution over observation sequences for all tests.

4 MDP Formulation and Theoretical Properties

Definition 2. Given two dynamical systems P_1 and P_2 with the same sets of actions and observations, $\mathcal{A}$ and $\mathcal{O}$, we define an MDP $\mathcal{M}$ as follows:

- The state space is the set $S := \{(\mathcal{A}\mathcal{O})^*\} \cup \{Dead\}$, with an initial state $s_0 = \epsilon$ (corresponding to the empty action-observation sequence).
- The set of actions is $\mathcal{A}$.
- The transition function is defined as:

$$\mathcal{T}(s,a,s') := \begin{cases} P_1(o|s,a)P_2(o|s,a) & \text{if } s' = sao \\ 1 - \sum_{o\in\mathcal{O}} P_1(o|s,a)P_2(o|s,a) & \text{if } s' = Dead \\ 0 & \text{otherwise} \end{cases}$$

- The reward function is defined as in Equation (2)

The MDP states correspond to action-observation sequences $a_1o_1a_2o_2\ldots$. When the observations obtained in P_1 and P_2 do not coincide, the episode stops. In the MDP, this corresponds to a transition to the $Dead$ state. Note that for a state $s = a_1o_1 \ldots a_no_n \in S$, the notation $P_1(o|s,a) = P_1(o|a_1\ldots a_n, o_1\ldots o_n, a)$ represents the probability of observing o after executing action a, following the history $a_1o_1\ldots a_no_n$.

We will now show that this MDP has optimal value function equal to 0 if and only if the two systems are identical. Moreover, the value function is never negative; therefore, it defines a divergence notion. First, we will show the following lemma:

Lemma 1. *The expected value of the reward in the MDP given by Def. 2 is:*

$$\mathcal{R}(s,a) = \sum_{o\in\mathcal{O}} (P_1(o|s,a) - P_2(o|s,a))^2$$

Proof. Suppose we have n possible observations. Let $P_1(o_i|s,a)$ and $P_2(o_i|s,a)$ be the probabilities of a designated observation o_i in the two processes. We analyze the reward by cases:

- W.p. $P_1(o_i|s,a)P_2(o_i|s,a)$, the same observation o_i is emitted in the two original processes. In this case, the MDP continues to a state sao_i. The rewards that can be obtained are:
  - 0, w.p. $P_1(o_i|s,a)P_2(o_i|s,a)P_1(o_i|s,a)P_2(o_i|s,a)$ (i.e. if both clones also produce o_i)
  - -1 w.p. $P_1(o_i|s,a)P_2(o_i|s,a)[P_1(o_i|s,a)(1 - P_2(o_i|s,a)) + P_2(o_i|s,a)(1 - P_1(o_i|s,a))]$ (i.e. one clone emits o_i, the other emits a different observation)
  - -2 w.p. $P_1(o_i|s,a)P_2(o_i|s,a)[(1 - P_1(o_i|s,a))(1 - P_2(o_i|s,a))]$ (i.e. neither clone emits o_i)

The total expected reward in this case will then be:

$$-P_1(o_i|s,a)P_2(o_i|s,a)[(1-P_1(o_i|s,a)) + (1-P_2(o_i|s,a))]$$

– W.p. $P_1(o_i|s,a)P_2(o_j|s,a)$, different observations o_i and o_j are emitted in the two original processes. In this case, the MDP goes to the *Dead* state. The rewards obtained are:
  - $+2$, w.p. $P_1(o_i|s,a)P_2(o_j|s,a)P_1(o_i|s,a)P_2(o_j|s,a)$ (the clones have identical observations with the original systems)
  - $+1$, w.p. $P_1(o_i|s,a)P_2(o_j|s,a)[(1-P_1(o_i|s,a))P_2(o_j|s,a)+)P_1(o_i|s,a)(1-P_2(o_j|s,a))]$ (one clone emits an observation identical to the one produced by the original system, the other clone emits a different observation)
  - 0, w.p. $P_1(o_i|s,a)P_2(o_j|s,a)[(1-P_1(o_i|s,a))(1-P_2(o_j|s,a))]$ (both clones have different emissions than the originals)

  The total expected reward in this case will be:

$$P_1(o_i|s,a)P_2(o_j|s,a)[P_1(o_i|s,a) + P_2(o_j|s,a)]$$

Now, we can write the immediate expected reward as:

$$\begin{aligned}
\mathcal{R}(s,a) &= \sum_i \sum_{j\neq i} P_1(o_i|s,a)P_2(o_j|s,a)[P_1(o_i|s,a) + P_2(o_j|s,a)] \\
&- \sum_i P_1(o_i|s,a)P_2(o_i|s,a)[(1-P_1(o_i|s,a)) + (1-P_2(o_i|s,a))] \\
&= \sum_i \sum_{j\neq i} P_1(o_i|s,a)P_2(o_j|s,a)[P_1(o_i|s,a) + P_2(o_j|s,a)] \\
&+ \sum_i P_1(o_i|s,a)P_2(o_i|s,a)[P_1(o_i|s,a) + P_2(o_i|s,a)] \\
&- \sum_i P_1(o_i|s,a)P_2(o_i|s,a)[P_1(o_i|s,a) + P_2(o_i|s,a)] \\
&- \sum_i P_1(o_i|s,a)P_2(o_i|s,a)[(1-P_1(o_i|s,a)) + (1-P_2(o_i|s,a))] \\
&= \sum_i \sum_j P_1(o_i|s,a)P_2(o_j|s,a)[P_1(o_i|s,a) + P_2(o_j|s,a)] \\
&- 2\sum_i P_1(o_i|s,a)P_2(o_i|s,a) \\
&= \sum_i (P_1(o_i|s,a))^2 + \sum_j (P_2(o_j|s,a))^2 - 2\sum_i P_1(o_i|s,a)P_2(o_i|s,a) \\
&= \sum_i P_1(o_i|s,a)(P_1(o_i|s,a) - P_2(o_i|s,a)) \\
&+ \sum_i P_2(o_i|s,a)(P_2(o_i|s,a) - P_1(o_i|s,a)) \\
&= \sum_{o\in\mathcal{O}} (P_1(o|s,a) - P_2(o|s,a))^2
\end{aligned}$$

which concludes the proof. □

We note that the MDP $\mathcal{M}$ has a very special form: it is a tree, with no loops. Using Lemma 1, and the Bellman equation for policy evaluation (see, e.g. [6] for a detailed description) we have:

$$\begin{aligned}
V^{\pi}(s_0) &= \sum_{a_1 \in \mathcal{A}} \pi(a_1|s_0) \sum_{o_1 \in \mathcal{O}} [(P_1(o_1|s,a_1) - P_2(o_1|s,a_1))^2 \\
&+ \gamma P_1(o_1|s,a_1) P_2(o_1|s,a_1) V^{\pi}(sa_1o_1)] \\
&= \sum_{a_1 \in \mathcal{A}} \pi(a_1|s_0) \sum_{o_1 \in \mathcal{O}} [(P_1(o_1|s,a_1) - P_2(o_1|s,a_1))^2] \\
&+ \gamma \sum_{a_1 \in \mathcal{A}} \pi(a_1|s_0) \sum_{o_1 \in \mathcal{O}} \mathcal{T}(s,a_1,sa_1o_1) \sum_{a_2 \in \mathcal{A}} \pi(a_2|sa_1o_1) \\
&\sum_{o_2 \in \mathcal{O}} (P_1(o_2|sa_1o_1,a_2) - P_2(o_2|sa_1o_1,a_2))^2 \\
&+ \gamma^2 \sum_{a_1 \in \mathcal{A}} \pi(a_1|s_0) \sum_{o_1 \in \mathcal{O}} \mathcal{T}(s,a_1,sa_1o_1) \sum_{a_2 \in \mathcal{A}} \pi(a_2|sa_1o_1) \\
&\sum_{o_2 \in \mathcal{O}} \mathcal{T}(sa_1o_1,a_2,sa_1o_1a_2o_2) V^{\pi}(sa_1o_1a_2o_2) \\
&= \ldots
\end{aligned}$$

Since all rewards are strictly positive, and 0 only if the two systems are identical, it is now intuitively clear that all policies will have value 0 if and only if P_1 and P_2 are identical, and strictly positive value otherwise. More formally:

Theorem 1. *Let $\mathcal{M}$ be the MDP induced by P_1 and P_2. If the discount factor $\gamma < 1$ or the size of MDP $|\mathcal{M}| < \infty$ then the optimal value $V^{\star}(s_0) \geq 0$, and $V^{\star}(s_0) = 0$ if, and only if, P_1 and P_2 are trace equivalent.*

The proof is in the appendix.

Note that the size of the MDP increases exponentially with the desired horizon (or depth). However, the computation of the divergence can still be done quite efficiently. In particular, in order to obtain sample complexity bounds, we can use the sparse sampling result of Kearns, Mansour and Ng [7]; they show that the optimal value function of an MDP can be approximated within a desired degree of accuracy ϵ, for all states, in an amount of time which is independent on the number of states in the MDP. The running time still depends on the number of actions; however, in many tasks of interest the number of actions is small, while the number of observations is very large. This would cause the number of states in $\mathcal{M}$ to explode, but the number of samples will still depend mainly on the desired accuracy parameters, the discount factor γ and the maximum reward in the MDP (which in our case is 2).

Moreover, if the two systems are very different, intuitively many sequences will be able to witness this discrepancy. As a result, it may become clear with much fewer samples than this theoretical bound that the systems are different. In the case when we are only interested in the number of samples needed to detect a difference, if one exists, Monte Carlo estimation, along with Hoeffding bounds, can be used to establish tighter bounds. We will now study the speed of the algorithm empirically.

5 Experimental Results

We experimented with this approach on several benchmarks from Tony Cassandra's POMDP repository [8] as well as the hand washing domain [9], which models a real-time system to assist persons with dementia during hand washing. The goal was to see whether this approach can successfully detect differences between models, and to assess the speed with which these can be detected. The experimentation consists in comparing each POMDP with itself and then with slightly modified (noisy) versions of itself. We experimented with introducing noise both in the transition dynamics as well as in the observation probabilities. In both cases, we studied two different kinds of changes: inserting one single important modification, or inserting small perturbations in several places in the model.

The single modification is generated as follows. We choose at random an action a, an internal state of the POMDP and two other internal states reachable through a; then we swap the probabilities of the two next-state transitions. A similar approach is used when perturbing the observation distributions. In the second case, for each state-action pair (s, a), we choose uniformly randomly a number $\delta \in [0, 1]$ and a state in the POMDP. Then, we increase the corresponding transition probability via a by δ. For the remaining states we decrease the transition probabilities by $\delta/(|S| - 1)$. An analogous transformation is applied when we modify the observation distribution.

For each setting, we compare the modified POMDP with the original version. We always perform 10 independent runs, each one consisting of 10^6 episodes. The other systems used for comparison are as follows:

- P_2 : a single modification in the internal dynamics
- P_3 : a single modification in the observation probabilities
- P_4 : small perturbations are inserted throughout in the internal dynamics
- P_5 : small perturbations are inserted throughout in the observation probabilities.

Figure 2 presents some representative learning curves for the value of the divergence. We picked for visualization domains which exhibit typical behavior, as we do not have

Table 1. Summary of empirical results on different POMDPs

	♯ of states	♯ of actions	♯ of observations	Divergence ratios P_2	P_3	P_4	P_5
DataSet4.3	4	4	20	1.35*	2*	2.51*	4.73*
4x3 Maze	11	4	18	1.02	1.023	1.4*	1.99*
Cheese Maze	11	4	7	1.28*	1.01	1.83*	1.72*
Tiger	2	3	6	1.01	1.64*	1.29*	1.77*
Shuttle Docking	8	3	10	1.21*	2.16*	12.26*	9.2*
Paint	4	4	6	1.17	1.16	1.78*	4.79*
MiniHall	13	3	9	4.79*	3.03*	2.74*	3.15*
HallWay	60	5	21	1.0	1.001	1.24*	1.32*
Aloha	30	9	3	1.01	1.37*	2.91*	3.06*
Hand Washing	180	6	6	1.62*	1.56*	1.66*	1.63*

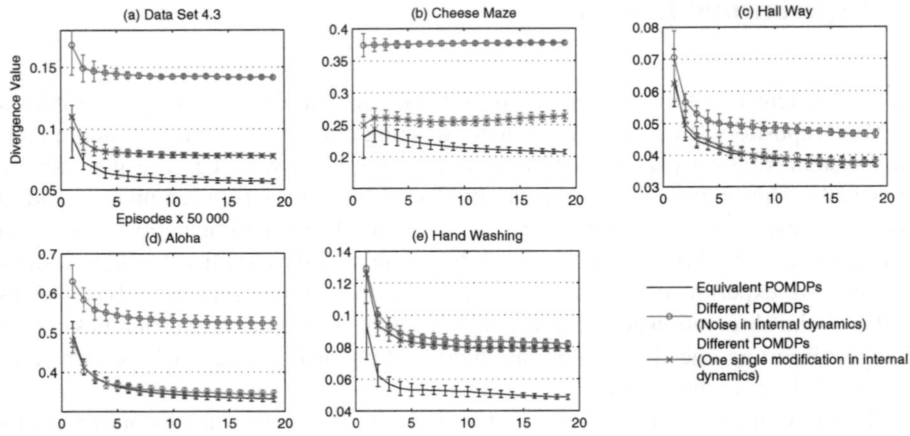

Fig. 2. Learning curves for the divergence values

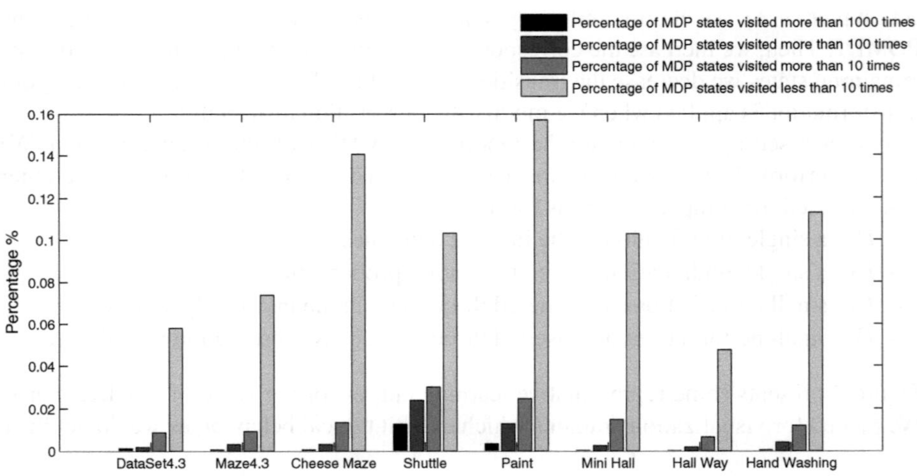

Fig. 3. Statistics on the visited MDP states

space to include the curves for all the benchmarks. As can be seen, in most cases the differences between models are detected very quickly, and the learning curves are significantly different throughout (the bars on the graphs represent standard deviations over the runs). This suggest that one would not even need to run this approach to convergence, if a significant change is detected early on. There are three problems, Hallway, Tiger and Aloha, in which the change in the internal dynamics is not detected correctly. This is because, as it turns out, this one change is not significant enough to affect the probability distribution over observations.

From this figure, one can notice that the magnitude of the divergence values differs from task to task. Hence, in order to decide if two systems are indeed identical or not, we divide the obtained divergence by the divergence that would be obtained by comparing

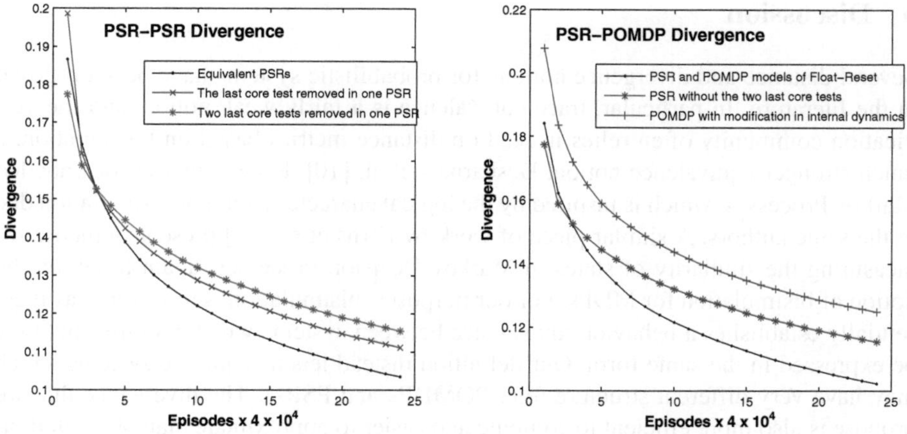

Fig. 4. Comparing (a) two PSRs and (b) a PSR and a POMDP

the system to itself. Hence, two identical systems would be expected to have a ratio of 1, while two very different systems would have a large ratio. Table 1 summarizes the results of the experimentation. The first four columns describe the names of the domains. For each benchmark, we include the number of states, the number of actions, and the number of observations. The remaining columns indicate the divergence ratios. The $*$ symbols indicate statistically significant differences (according to a bootstrap test). From the table, it is clear that almost all differences are detected successfully.

Figure 3 presents statistics on the state visitation frequencies in the MDP. One of the main concerns a priori regarding our method is that the number of states in the MDP that we use is very large (increases exponentially with the size of the horizon). However, because of the special tree structure and because of the particular way we set up the interaction, we expect that only very few states will be visited. Hence, in our implementation, we only maintain values, in a hash table, for states that are actually encountered. For each benchmark, we computed the percentages of states that are visited less than 10 times, between 10 and 100 times, between 100 and 1000 ties, and more than 1000 times. The results confirm the fact that indeed, very few states are ever visited (between 10% and 30% total). A really tiny percentage of states are visited many times. If we stopped the process as soon as significant differences are detected, these percentages would be even lower.

Finally, the last plot (Figure 4) shows the divergence measures computed for predictive state representation (PSR) models. We use the standard float-reset problem from the literature. In the left graph, we compare a PSR representation with itself, against a PSR with one less core test, and against a PSR with two less core tests. In the right graph, we compare PSR and POMDP representations of float-reset. First, we compare the PSR representation with an equivalent POMDP, then against a modified POMDP representation, and finally we compared the original POMDP representation against a PSR with two fewer core tests. The graph shows the potential of our method to detect differences even representations which are not POMDPs.

6 Discussion

Several distance and divergence notions for probabilistic systems have been proposed in the literature. In particular, trace equivalence is a fairly weak notion, and the verification community often relies instead on distance metrics based on bisimulation, a much stronger equivalence notion. Desharnais et al. [10] define a metric for Labelled Markov Processes, which is inspired by the logical characterization of bisimulation due to the same authors. A similar piece of work by Ferns et al. [11] presents a metric for measuring the similarity of states in Markov Decision Processes, again based on the notion of bisimulation for MDPs. For our purposes, bisimulation is too strong, as it essentially establishes a behavior equivalence between systems which requires that they be expressed in the same form. Our definition instead lets us compare systems which may have very different structure (e.g. POMDPs and PSRs). The divergence that we propose is also more efficient to compute and easier to approximate than bisimulation-based metrics. Note that our algorithm generalizes to a family of equivalence notions, called k-moment equivalence (see [1] for details); these equivalences are stronger than trace equivalence but weaker than bisimulation.

In his work on automatic speech recognition, Rabiner developed two distance measures for Hidden Markov Models (HMMs). One is based on a state permutation function [12] and the other is inspired by relative entropy [13]. The connection to KL divergence through relative entropy is interesting, but our setting involves actions, which are chosen actively by the learning algorithm in order to probe the differences in the systems.

The group on Foundations of Software Engineering at Microsoft Research developed a line of research which represents the problem of model-based testing as a game between a tester and the implementation under test. Blass et al. [14] model the game as a negative non-discounted MDP whose goal is to reach final states with minimum total cost. The final states are states in which shared resources are freed, i.e., where new tests can be initiated. Their algorithmic approaches are all offline: all the states are known in advance and the MDP that they set up is solved using a value iteration algorithm. Veanes et al. [15] use instead an online approach for testing through reinforcement learning. The objective of the learning is to achieve the best coverage possible within fixed resource constraints. In each episode the algorithm runs one test case until either a non-conformance between the model (specification) and the implementation under test occurs, or the maximum number of steps has been reached. The cost, or negative reward, associated to an action is proportional to the number of times that actions has been chosen from that state. In both these offline and online approaches, the goal of the learning task is to achieve the best possible coverage. In our work, however, we defined a reward model which is based on the non-conformance between the two processes we compare. We speculate that our approach focuses the computation on more useful paths, though an empirical comparison between these approaches would be useful.

The approach we propose has great potential for two important applications: learning dynamical systems from data, and knowledge transfer. In both cases, the advantage of our method is that it provides an optimal policy, in addition to the divergence. This is very useful information, focusing on the action sequence which distinguishes the systems most.

In the case of learning, it would be natural to take this action sequence and try to refine the model around it (e.g. by adding internal states, in the case of a POMDP, or by adding tests, in the case of a PSR). Alternatively, this action sequence can be played preferentially, along with exploratory paths starting from it. This would provide additional data to improve model parameter estimates, in such a way as to make this critical action path more similar (behaviorally) in the two systems.

In the case of knowledge transfer, the goal is to decide if a policy learned in one system can be applied successfully in a different system. This problem can be formalized as policy evaluation in our MDP, which is very simple and efficient to solve. Hence, the divergence we propose could be used, for example, to search a library of small dynamical systems for one whose optimal policy can be applied successfully in a larger system. We are currently investigating this line of research.

7 Conclusions and Future Work

We presented an approach for estimating the differences between partially observable dynamical systems, based on trying out actions and observing their behavior. Our algorithm is applicable even if the models of the two systems are not available, provided that we are allowed to interact with them. Moreover, the approach allows us to pinpoint the action sequences which most distinguish the systems. Our empirical results are very encouraging, but more practical experience with this approach is needed in the future. One nice aspect is that the algorithm only depends on the number of actions and observations, not on the internal structure of the dynamical system. Moreover, as illustrated above, differences are identified quickly and typically with few interactions with the environment. This is due to the fact that we use reinforcement learning methods, which quickly focus on differences that matter. An early detection mechanism will be implemented in future work.

One important direction for future work is scaling up this approach for systems with many actions and/or observations. In this case, function approximation techniques, or methods by which one expresses interest in a limited set of observations, will need to be used.

Acknowledgements

This work was supported in part by NSERC and FQRNT. The authors thank Prakash Panangaden for helpful discussions.

References

1. Desharnais, J., Laviolette, F., Zhioua, S.: Testing probabilistic equivalence through reinforcement learning. In: Arun-Kumar, S., Garg, N. (eds.) FSTTCS 2006. LNCS, vol. 4337, pp. 236–247. Springer, Heidelberg (2006)
2. Kaelbling, L.P., Littman, M.L., Cassandra, A.R.: Planning and acting in partially observable stochastic domains. Artificial Intelligence 101, 99–134 (1998)

3. Larsen, K.G., Skou, A.: Bisimulation through probabilistic testing. Inf. Comput. 94, 1–28 (1991)
4. Singh, S., James, M., Rudary, M.: Predictive state representations: a new theory for modeling dynamical systems. In: The 20th Conference on Uncertainty in Artificial Intelligence, Banff, Canada, pp. 512–519 (2004)
5. Cover, T.M., Thomas, J.A.: 12. In: Elements of Information Theory. Wiley, Chichester (1991)
6. Sutton, R.S., Barto, A.G.: Introduction to Reinforcement Learning. MIT Press, Cambridge (1998)
7. Kearns, M.J., Mansour, Y., Ng, A.Y.: A sparse sampling algorithm for near-optimal planning in large markov decision processes. Machine Learning 49, 193–208 (2002)
8. Cassandra, T.: Tony's POMDP Page (2009), `http://www.cs.brown.edu/research/ai/pomdp/`
9. Hoey, J., Bertoldi, A., Poupart, P., Mihailidis, A.: Assisting persons with dementia during handwashing using a partially observable markov decision process. In: The 5th International Conference on Computer Vision Systems, Bielefeld, March 21-24 (2007)
10. Desharnais, J., Gupta, V., Jagadeesan, R., Panangaden, P.: Metrics for labeled Markov processes. Theoretical Computer Science 318, 323–354 (2004)
11. Ferns, N., Castro, P., Panangaden, P., Precup, D.: Methods for computing state similarity in markov decision processes. In: Proceedings of the 22nd Conference on Uncertainty in Artificial intelligence, Cambridge, MA, USA, July 13 - 16 (2006)
12. Levinson, S., Rabiner, L.: An introduction to the application of the theory of probabilistic functions of a return process to automatic speech recognition. The Bell System Technical Journal 62, 1035–1074 (1983)
13. Juang, B., Rabiner, L.: A probabilistic distance measure for hidden return models. AT&T Technical Journal 62, 391–408 (1985)
14. Blass, A., Gurevich, Y., Nachmanson, L., Veanes, M.: Play to test. Technical report, Microsoft Research (2005)
15. Veanes, M., Roy, P., Campbell, C.: Online testing with reinforcement learning. In: Havelund, K., Núñez, M., Roşu, G., Wolff, B. (eds.) FATES 2006 and RV 2006. LNCS, vol. 4262, pp. 240–253. Springer, Heidelberg (2006)

Appendix: Proof of Theorem 1

We need the following three lemmas.

Lemma 2. *Let P_1 and P_2 two dynamical systems. Then the following are equivalent:*

(i) *P_1 and P_2 are trace equivalent*
(ii) $\forall a \in \mathcal{A}, s \in \mathcal{S}, \quad \mathcal{R}(s,a) = 0$
(iii) $\forall \pi, \quad V^\pi(s_0) = 0.$

Proof. $(i) \Rightarrow (ii)$. If P_1 and P_2 are trace equivalent, then

$$\forall a \in Act, o \in \mathcal{O}, s \in S, P_1(o|s,a) = P_2(o|s,a) \Rightarrow \quad \mathcal{R}(s,a) = 0 \quad \text{(by Lemma 1)}$$

$(ii) \Rightarrow (i)$. Immediate from Lemma 1. $(ii) \Rightarrow (iii)$ and $(iii) \Rightarrow (ii)$ follow from the definition of V^π given in Section 4: all policies have value greater than or equal to 0, with equality occurring only if all the squared differences in transition probabilities are 0. □

Now we need we need to link the value of the optimal policy to policies acting to finite depth.

Lemma 3. *For every policy π that prescribes actions only for states up to a maximum depth H,*

$$\exists \pi' \text{ such that } V^{\pi}(s_0) \leq V^{\pi'}(s_0),$$

where π' coincides with π up to depth H.

Proof. The result follows from Lemma 1 and the value function unrolling computation in Section 4. □

Lemma 4. *Let $\mathcal{M}$ be the MDP induced by P_1 and P_2. If the discount factor $\gamma < 1$ or the size of MDP $|\mathcal{M}| < \infty$ then $V^{\star}(s_0) \geq V^{\pi}(s_0)$ for any policy π.*

Proof. If $|\mathcal{M}| < \infty$, since $\mathcal{M}$ has a tree structure, the result is a direct consequence of Lemma 3. Otherwise, it is sufficient to show that

$$\forall \epsilon > 0 \; \forall \pi \; \exists \, \pi' \text{ such that } |V^{\pi'}(s_0) - V^{\pi}(s_0)| < \epsilon.$$

Because of Lemma 3, w.l.o.g., we may suppose n to be large enough to satisfy

$$\sum_{i=n+1}^{\infty} \gamma^i < \epsilon.$$

Since on each episode, the reward signal is between (-2) and $(+2)$, it is easy to see that any policy π' of $\mathcal{M}$ that coincides with π on $\mathcal{M}'$ will have the desired property. □

Theorem 1. *Let $\mathcal{M}$ be the MDP induced by P_1 and P_2. If the discount factor $\gamma < 1$ or the size of MDP $|\mathcal{M}| < \infty$ then the optimal value $V^{\star}(s_0) \geq 0$, and $V^{\star}(s_0) = 0$ if, and only if, P_1 and P_2 are trace equivalent.*

Proof. If P_1 and P_2 are trace equivalent, then Theorem 2 implies that $V^*(s_0) = 0$. If they are not trace equivalent, then there exists at least one policy π such that $V^{\pi}(s_0) > 0$. Finally, by Lemma 4, we can conclude that $V^*(s_0) > 0$. □

Universal Learning over Related Distributions and Adaptive Graph Transduction

Erheng Zhong[1], Wei Fan[2], Jing Peng[3],
Olivier Verscheure[2], and Jiangtao Ren[1,⋆]

[1] Sun Yat-Sen University, Guangzhou, China
{sw04zheh,issrjt}@mail2.sysu.edu.cn
[2] IBM T.J. Watson Research, USA
{weifan,ov1}@us.ibm.com
[3] Montclair State University, USA
pengj@mail.montclair.edu

Abstract. The basis assumption that "training and test data drawn from the same distribution" is often violated in reality. In this paper, we propose one common solution to cover various scenarios of learning under "different but related distributions" in a single framework. Explicit examples include (a) sample selection bias between training and testing data, (b) transfer learning or no labeled data in target domain, and (c) noisy or uncertain training data. The main motivation is that one could ideally solve as many problems as possible with a single approach. The proposed solution extends graph transduction using the maximum margin principle over unlabeled data. The error of the proposed method is bounded under reasonable assumptions even when the training and testing distributions are different. Experiment results demonstrate that the proposed method improves the traditional graph transduction by as much as 15% in accuracy and AUC in all common situations of distribution difference. Most importantly, it outperforms, by up to 10% in accuracy, several state-of-art approaches proposed to solve specific category of distribution difference, i.e, BRSD [1] for sample selection bias, CDSC [2] for transfer learning, etc. The main claim is that the adaptive graph transduction is a general and competitive method to solve distribution differences implicitly without knowing and worrying about the exact type. These at least include sample selection bias, transfer learning, uncertainty mining, as well as those alike that are still not studied yet. The source code and datasets are available from the authors.

1 Introduction

One important assumption in many learning scenarios is that training and test data are drawn from the same distribution. However, this may not be true in many applications, and the following are some examples. First, suppose we wish

⋆ The author is supported by the National Natural Science Foundation of China under Grant No. 60703110.

W. Buntine et al. (Eds.): ECML PKDD 2009, Part II, LNAI 5782, pp. 678–693, 2009.

to generate a model in clinical trial of a new drug. Since people self-select, the training data is most likely having a different distribution from the general public. Second, if we want to use a topic model to classify articles from New York Times, but the only labeled data is from Reuters, we have to transfer knowledge across these two collections. Third, for data collected over sensor networks, the feature values are likely noisy or uncertain. Obviously, the distributions of training and test data in the above problems are different but related. In this paper, we formulate this situation within a "universal learning" framework.

Given a space of instances X and labels $Y = [-1, 1]$, let $p_{tr}(\mathbf{x}, y)$ denotes the joint distribution of training data $L \subseteq X \times Y$ and $p_{te}(\mathbf{x}, y)$ denotes the test data $U \subseteq X \times Y$. Using the standard decomposition, $p(\mathbf{x}, y)$ can be represented by $p(\mathbf{x}, y) = p(\mathbf{x})p(y|\mathbf{x})$ where $p(\mathbf{x})$ and $p(y|\mathbf{x})$ are the marginal and conditional distributions. Let $h \in X \rightarrow Y$ be a function from a fixed hypothesis space $\mathcal{H}$ for X. Then, we define the distance between training and tests sets using a hypothesis class-specific distance measure. Let $A_{\mathcal{H}}$ be the subsets of X that are the support of some hypothesis in $\mathcal{H}$. In other words, for every hypothesis $h \in \mathcal{H}$, $\{\mathbf{x} : \mathbf{x} \in X, h(\mathbf{x}) = 1\} \in A_{\mathcal{H}}$. Then the distance between two distributions is:

$$d_{\mathcal{H}}(p_{tr}(\mathbf{x}), p_{te}(\mathbf{x})) = 2 * \sup_{A \in A_{\mathcal{H}}} |Pr_{p_{tr}}[A] - Pr_{p_{te}}[A]| \quad (1)$$

Using the conclusion from [3], we compute a finite-sample approximation to $d_{\mathcal{H}}$, where $\mathcal{H}$ has finite VC dimensions. Thus, $d_{\mathcal{H}}$ can be treated as the indicator to measure the relationship between training and test set. We define the following framework as "universal learning over related but different distributions":

Definition 1. *Given the training and test set, learning is universal iif* $p_{tr}(\mathbf{x}, y) \neq p_{te}(\mathbf{x}, y)$ *and* $d_{\mathcal{H}}$ *is small.*

With this definition, we study solutions for problems where $d_{\mathcal{H}}$ is reasonably small or training and testing data is related but different.

Note that this definition is unified in the sense that it covers many related problem formulations, such as sample selection bias, transfer learning, uncertainty mining and the alike that are not well studied and reported yet. For both sample selection bias and uncertainty mining, $p_{tr}(\mathbf{x}) \neq p_{te}(\mathbf{x})$ but $p_{tr}(y|\mathbf{x}) = p_{te}(y|\mathbf{x})$. For transfer learning, $p(\mathbf{x})$ and $p(y|\mathbf{x})$ of training and test data sets may both be different, but $p(y|\mathbf{x})$ are assumed to be related. As follows, we propose a generalized graph approach under this framework. One can solve as many different but similar problems as possible and avoid employing and remembering different approaches under different formulations.

1.1 Solution Based on Adaptive Graph Transduction

To solve universal learning problem, we propose an approach based on "maximum margin graph transduction". Graph transduction explores the information from both training and test set. Its most important advantage for universal learning is that it does not explicitly assume the distributions of training and test set to be the same, but only makes a weaker assumption that the decision boundary lies on the low density regions of the unlabeled data. However, the original graph transduction still suffers from the "unsmooth label problems" and these are common when learning different distributions. It may instead mislead the decision boundary to go through the high density regions, when labeled examples of training examples stand on the wrong location in the space of the testing data [4]. In margin-terms, unlabeled data with low margin are likely misclassified [5]. Clearly, if one employs graph transduction for universal learning, label information ought be regularized in order to maintain smoothness. Based on this motivation, we propose a maximal margin based graph transduction on the basis of "maximal unlabeled data margin principal."

To solve this problem, we cast the graph transduction into a joint optimization over both the classification function and the labeled data. The optimization is solved iteratively. In each iteration, sample selection is performed on labeled set to select data which can maximize the unlabeled data margin. Based on the selected sample, graph transduction is invoked to predict the labels of unlabeled data. Those closest examples will be predicted with the same class label. Then, the prediction results from each iteration are averaged to form an ensemble, in order to remove the bias from any single graph [6] and further reduce the prediction error as shown in Section 3. By the analysis of Section 3, the risk of proposed method is bounded under reasonable terms even when training and testing distributions are different.

2 Graph Transduction over Related Distributions

We first summarize the traditional graph transduction using harmonic function [7], and then present details on the proposed algorithm MarginGraph that is generalized by the maximal margin principal over unlabeled data. The notations are summarized in Table 1.

2.1 Preliminaries: Graph Transduction Using Harmonic Functions

Suppose we have ℓ training examples $L = \{(\mathbf{x}_1, y_1), \ldots, (\mathbf{x}_\ell, y_\ell)\}$, and u test examples $U = \{(\mathbf{x}_{\ell+1}, y_{\ell+1}), \ldots, (\mathbf{x}_{\ell+u}, y_{\ell+u})\}$. The labels of U are not known

Table 1. Definition of notation

Notation	Notation Description	Notation	Notation Description
X	Instance space	$maxIt$	Iteration times in algorithm
Y	Label space	G	Graph whose nodes are data
$\mathbf{x}_i$	Instance(without label), $x_i \in X$		points and the edges describe
y_i	Label of instance $\mathbf{x}_i$, $y_i \in \{-1, 1\}$		the similar between any nodes
L	Training data set	W	Weight matrix of the graph G,
ℓ	Number of instances in L		w_{ij} is the weight of the
U	Test data set		edge between node i and j
u	Number of instances in U	D	Diagonal matrix, $D = diag(d_i)$,
Δ	Laplacian matrix, $\Delta = D - W$		$d_i = \sum_j w_{ij}$
$p(\mathbf{x}, y)$	Joint distribution	$p(\mathbf{x})$	Margin distribution
$p(y\|\mathbf{x})$	Conditional distribution	$\mathcal{H}$	A fixed hypothesis space
Q	A posterior distribution over $\mathcal{H}$	$m_Q(U)$	Unlabeled data margin

apriori. In universal learning setting, L and U are drawn from the related but different distributions. We construct a connected graph $G = \{V, E\}$. Vertex V corresponds to both labeled and unlabeled data examples, and the edges $(i, j) \in E$, $i, j = 1, \ldots, \ell + u$, are weighted according to the similarity between $\mathbf{x}_i$ and $\mathbf{x}_j$. According to the results from [7], we set the weight $w_{ij} = \exp(-\frac{||\mathbf{x}_i - \mathbf{x}_j||^2}{\lambda^2})$, where λ is a bandwidth hyper-parameter. It is adaptive according to the data. Following the analysis in [7], we set $\lambda = d^0/3$, where d^0 is the minimal distance between class regions. Also, Let D be a diagonal matrix, $D = diag(d_i)$ where $d_i = \sum_j w_{ij}$. Based on the intuition that unlabeled examples that are close-by in the graph ought to have similar labels [7], a harmonic function is defined as $E(f) = \frac{1}{2}\sum_{i,j} w_{ij}(f(\mathbf{x}_i) - f(\mathbf{x}_j))^2 = f^T \Delta f$, where $f : V \to [-1, 1]$ is a real-valued mapping function to minimize the value of harmonic function and $\Delta = D - W$ is the Laplacian matrix. The harmonic solution $\Delta f = 0$ is given by:

$$f_U = \Delta_{UU}^{-1} \Delta_{UL} f_L \tag{2}$$

where f_U denotes the values on the unlabeled data, f_L represents the values on the labeled data (equal to the true label), Δ_{UU} is the sub-matrix of Δ relating the unlabeled data to unlabeled data, and Δ_{UL} relates unlabeled data to labeled data. For binary classification, we obtain a classifier based on the mapping function f and use it to predict the labels of unlabeled examples.

$$\hat{y}_i = I[f(\mathbf{x}_i)] = \begin{cases} 1, f(\mathbf{x}_i) > \theta \\ -1, \text{otherwise} \end{cases} \tag{3}$$

For balanced problem, θ is typically chosen to be 0. Based on the discussion in introduction, in universal learning over related but different distributions, the typical graph transduction suffers from label unsmooth problem that the initial labels stand on the wrong locations in the domain of testing data.

2.2 Adaptive Graph Transduction by Maximizing Unlabeled Data Margin

As follows, we focus on how to use sample selection to resolve the unsmooth problem based on maximal margin principal. Let Q be a posterior distribution over hypothesis space $\mathcal{H}$. Then the "Q-weighted majority vote Bayes classifier B_Q" is

$$B_Q(\mathbf{x}) = I[\mathbb{E}_{h \sim Q} h(\mathbf{x})] \tag{4}$$

where $I[x] = 1$ if $x > \theta$ and -1 otherwise, and h is a classifier in $\mathcal{H}$ according to the distribution Q. Thus, the unlabeled data margin is:

$$\begin{aligned} m_Q(\mathbf{x}_i) &= |\mathbb{E}_{h \sim Q}[\![h(\mathbf{x}_i) = 1]\!] - \mathbb{E}_{h \sim Q}[\![h(\mathbf{x}_i) = -1]\!]| \\ &= |1 - 2\mathbb{E}_{h \sim Q}[\![h(\mathbf{x}_i) \neq y_i]\!]| \end{aligned} \tag{5}$$

where $[\![\pi]\!]$ denotes the probability that π is true. In graph transduction,

$$m_Q(\mathbf{x}_i) = ||f(\mathbf{x}_i)| - |1 - |f(\mathbf{x}_i)||| \tag{6}$$

And the unlabeled margin on whole unlabeled data set is $m_Q(U) = \sum_{i=\ell+1}^{\ell+u} m_Q(\mathbf{x}_i)$. From the intuition in introduction (formally analyzed in Section 3), graph transduction can be generalized by maximizing $m_Q(U)$ by performing sample selection in training set. This can be formulated as

$$S_L = \arg\max_{S_L' \subseteq L} m_Q(U)_{S_L',h} \tag{7}$$

where S_L is the selected labeled subset used for building a maximal margin classifier h. To obtain S_L, we employ a greedy sequence searching procedure in the labeled data. Suppose the current mapping function is f. When we select one more labeled example $(\mathbf{x}_k, y_k) \in L$, we denote the new mapping function as f^{+k}. Then, in each iteration, we select one labeled example from L as follows:

$$k = \arg\max_{k'}\left(\sum_{i=\ell+1}^{\ell+u} ||f^{k'}(\mathbf{x}_i)| - |1 - |f^{k'}(\mathbf{x}_i)|||\right) \tag{8}$$

The above equation means that we select those data that can build a classifier to maximize the unlabeled data margin. This procedure is a joint optimization that simultaneously (1) maximizes the unlabeled data margin and (2) minimizes the harmonic function. Detailed description of MarginGraph can be found in Algorithm 1. In summary, it runs iteratively. In the beginning, the selected labeled data set S_L is empty. In each iteration, one labeled example is selected according to Eq(8), and added into the training set S_L in order to obtain the maximal unlabeled margin. Then the graph transduction is invoked on the new training set to calculate the harmonic function, as well as, to find a new mapping function f^{+k}. After the iteration, we combine all mapping functions f calculated during each iteration and obtain an averaging ensemble classifier which aims to remove the bias by any single graph and reduce the prediction error. In order to keep balanced prior class distribution, we select one positive and one negative labeled example alternatively during the iterative procedure. To practically guarantee the margin $m_Q(U)$ is not small, we propose a criterion to stop the iteration. If the following Eq(9) holds, we stop the iterative procedure.

$$m_Q(U)^* - m_Q(U)^k \geq \varepsilon m_Q(U)^* \tag{9}$$

where $m_Q(U)^k$ is the margin value after we select the k_{th} labeled data point, $m_Q(U)^*$ is the maximal margin value obtained in the procedure so far and ε is a real-value in $(0, 1]$ to control the iterations.

2.3 Implementation Details

At each iteration, we need to compute the mapping function f^{+k} after adding $(\mathbf{x}_k, y_k)$ into the selected labeled data set. We discuss an efficient implementation to retrain and reduce the computational cost. We first "suppose" data from both L and U are all "unlabeled". Thus the problem can be treated as labeling the selected data from L to maximize the margin on U. Denote the unlabeled data

Input: L,U,$maxIt$
Output: B_Q: Averaging ensemble or Bayes classifier
$S_L = \{\}$, $f_0(\mathbf{x}_i) = 0, x_i \in U$;
for $i = 1$ *to* $maxIt$ **do**
  Select one point $(\mathbf{x}_k, y_k) \in L$ using Eq(8), $S_L{=}S_L \cup (\mathbf{x}_k, y_k)$;
  Calculate f_i using Eq(2) based on S_L and new margin value $m_Q(U)^k$;
  IF Eq(9) holds **Break**. ;
  Otherwise, Update the maximal margin $m_Q(U)^* = m_Q(U)^k$;
end
return $B_Q = I[\frac{1}{maxIt}\sum_{i=1}^{maxIt} f_i]$;

Fig. 1. MarginGraph

in U and the unselected labeled data in L as $\mathcal{U}$, and the selected data from L as $\mathcal{L}$ with the labels $Y_{\mathcal{L}}$. Suppose we are labeling the k_{th} point from L, once we give the point $\mathbf{x}_k$ label y_k, we obtain: $f_{\mathcal{U}}^{+k} = f_{\mathcal{U}} + (y_k - f_k)\frac{(\Delta_{\mathcal{UU}}^{-1})_{\cdot k}}{(\Delta_{\mathcal{UU}}^{-1})_{kk}}$, where $(\Delta_{\mathcal{UU}}^{-1})_{\cdot k}$ is the k_{th} column of the inverse Laplacian on "unlabeled" data $\mathcal{U}$, and $(\Delta_{\mathcal{UU}}^{-1})_{kk}$ is the k_{th} diagonal element of the same matrix. Importantly, this means updating the values of mapping function f is a linear computation. When one "unlabeled" example in L is labeled, the corresponding row and column are removed from $\Delta_{\mathcal{UU}}^{-1}$, so $\Delta_{\mathcal{UU}}^{-1}$ should be recomputed. Instead of naively taking the inverse, there are efficient algorithms to compute in linear time [8]. In summary, the proposed approach has computation complexity of $O(\ell * (\ell + u) * maxIt)$.

3 Formal Analysis

Theorem 1 shows that the error of a classifier across different distributions is bounded. Lemma 1 shows that this bound is reduced as margin being maximized. Finally, Lemma 2 demonstrates that the averaging ensemble achieves larger margin than any single classifiers, implying that the ensemble classifier has a lower error bound.

To measure the difference between training and test distributions, we define a distance based on Eq(1): $d_{\mathcal{H}\Phi\mathcal{H}}(p_{tr}, p_{te})$, where $\mathcal{H}\Phi\mathcal{H} = \{h(\mathbf{x}) \oplus h^{'}(\mathbf{x}) : h, h^{'} \in \mathcal{H}\}$ represents the symmetric difference hypothesis space, and $\oplus$ denotes the XOR operator. A hypothesis $h \in \mathcal{H}\Phi\mathcal{H}$ assigns label $+1$ to $\mathbf{x}$ when there is a pair of hypotheses in $\mathcal{H}$ that disagree on $\mathbf{x}$. Thus, $A_{\mathcal{H}\Phi\mathcal{H}}$ is the subset of $A_{\mathcal{H}}$ for some h, $h^{'} \in \mathcal{H}$, such that $A_{\mathcal{H}\Phi\mathcal{H}} = \{\mathbf{x} | \mathbf{x} \in X, h(\mathbf{x}) \neq h^{'}(\mathbf{x})\}$. Let $\epsilon_{tr}(h, h^{'})$ be the probability that a hypothesis h disagrees with another one $h^{'}$ according to the marginal distribution $p_{tr}(\mathbf{x})$. Then

$$\epsilon_{tr}(h, h^{'}) = \mathbb{E}_{\mathbf{x} \sim p_{tr}(\mathbf{x})} |h(\mathbf{x}) - h^{'}(\mathbf{x})| \tag{10}$$

In particular, if $f^* : X \to Y$ denotes the (unknown) target (labeling) function, $\epsilon_{tr}(h) = \epsilon_{tr}(h, f_{tr}^*)$ represents the risk of the hypothesis h. Similarly, $\epsilon_{te}(h, h^{'})$

and $\epsilon_{te}(h)$ are the corresponding definitions for the test distribution. It can be shown that for any hypotheses $h, h^{'} \in \mathcal{H}$ [3],

$$|\epsilon_{tr}(h, h^{'}) - \epsilon_{te}(h, h^{'})| \leqslant \frac{1}{2} d_{\mathcal{H}\Phi\mathcal{H}}(p_{tr}, p_{te}) \tag{11}$$

Since the margin evaluates the confidence of a classifier with regard to its decision [5], we define an ideal hypothesis as the one that maximizes the margin on unlabeled data.

$$h^* = \arg\max_{h \in \mathcal{H}} (m_Q(U)_h) \tag{12}$$

where $m_Q(U)_h$ is the margin obtained by classifier h. Following these definitions, the bound for the risk of any classifiers h can be established when the distributions of training and test data are related but different. The bound is adopted from Theorem 1 of [3], but with different terms.

Theorem 1. *Let $\mathcal{H}$ be a hypothesis space of VC-dimension d_{VC} and U_{tr} and U_{te} be unlabeled samples of size m each, drawn according to $p_{tr}(\mathbf{x})$ and $p_{te}(\mathbf{x})$, respectively. Let $\hat{d}_{\mathcal{H}\Phi\mathcal{H}}$ be the empirical distance on U_{tr} and U_{te}. With probability of at least $1 - \delta$ (over the choice of the samples), for any classifiers $h \in \mathcal{H}$,*

$$\epsilon_{te}(h) \leq \epsilon_{tr}(h, h^*) + \epsilon_{te}(h^*) + \frac{1}{2}\hat{d}_{\mathcal{H}\Phi\mathcal{H}}(U_{tr}, U_{te}) + 4\sqrt{\frac{2d_{VC}\log(2m) + \log(\frac{4}{\delta})}{m}}$$

Proof. The proof uses the inequality Eq(11). Also, we assume the triangle inequality holds for classification error [3]. It implies for any functions f_1, f_2 and f_3, $\epsilon(f_1, f_2) \leq \epsilon(f_1, f_3) + \epsilon(f_2, f_3)$. Thus,

$$\begin{aligned}
\epsilon_{te}(h) &\leq \epsilon_{te}(h, h^*) + \epsilon_{te}(h^*) \\
&\leq \epsilon_{tr}(h, h^*) + \epsilon_{te}(h^*) + |\epsilon_{tr}(h, h^*) - \epsilon_{te}(h, h^*)| \\
&\leq \epsilon_{tr}(h, h^*) + \epsilon_{te}(h^*) + \frac{1}{2} d_{\mathcal{H}\Phi\mathcal{H}}(p_{tr}, p_{te}) \\
&\leq \epsilon_{tr}(h, h^*) + \epsilon_{te}(h^*) + \frac{1}{2}\hat{d}_{\mathcal{H}\Phi\mathcal{H}}(U_{tr}, U_{te}) + 4\sqrt{\frac{2d\log(2m) + \log(\frac{4}{\delta})}{m}}. \qquad \square
\end{aligned}$$

Note that this bound is constructed by three terms. The first term, $\epsilon_{tr}(h, h^*)$ represents the training error in terms of approximating the ideal hypothesis h^*. The second is the risk of h^*. Recall the definition of Eq(12), h^* just relies on the test data set and is independent from any algorithms. When the unlabeled data are given, the risk of h^* is fixed. In addition, the third term is the distance between the training and test distributions, $d_{\mathcal{H}\Phi\mathcal{H}}(p_{tr}, p_{te})$. If the training and test distributions are related, $d_{\mathcal{H}\Phi\mathcal{H}}$ can be bounded. Therefore, the bound mostly relies on $\epsilon_{tr}(h, h^*)$.

The following lemma and analysis show when training and test distributions are related, if a classifier h achieves larger margin, $\epsilon_{tr}(h, h^*)$ becomes smaller. We assume that for a given instance $\mathbf{x}$, the misclassification probabilities of h and h^* are smaller than 50%.

Lemma 1. *Let $m_Q(U)_h$ denotes the unlabeled margin obtained by h, then $\epsilon_{tr}(h, h^*)$ is related to $|m_Q(U)_{h^*} - m_Q(U)_h|$.*

Proof. By the definition of Eq(6),

$$\begin{aligned}
&|m_Q(U)_{h^*} - m_Q(U)_h| \\
&= u * \mathbb{E}_{\mathbf{x}_i \sim p_{te}}\{|1 - 2[\![h^*(\mathbf{x}_i) \neq y_i]\!]| - |1 - 2[\![h(\mathbf{x}_i) \neq y_i]\!]|\} \\
&= u * \mathbb{E}_{\mathbf{x}_i \sim p_{te}}\{(1 - 2[\![h^*(\mathbf{x}_i) \neq y_i]\!]) - (1 - 2[\![h(\mathbf{x}_i) \neq y_i]\!])\} \\
&= u * \mathbb{E}_{\mathbf{x}_i \sim p_{te}} 2 * ([\![h(\mathbf{x}_i) \neq y_i]\!] - [\![h^*(\mathbf{x}_i) \neq y_i]\!]) \\
&= u * \mathbb{E}_{\mathbf{x}_i \sim p_{te}}([\![h(\mathbf{x}_i) \neq y_i]\!] - [\![h^*(\mathbf{x}_i) \neq y_i]\!]) + ([\![h^*(\mathbf{x}_i) = y_i]\!] - [\![h(\mathbf{x}_i) = y_i]\!])
\end{aligned}$$

Obviously, when $|m_Q(U)_{h^*} - m_Q(U)_h|$ is small, h and h^* give similar classification probabilities. Thus, if the margins achieved by h^* and h are close, $\epsilon_{te}(h, h^*)$ is small. Recall the Eq(11), if the training and test distribution are related, smaller $\epsilon_{te}(h, h^*)$ induces smaller $\epsilon_{tr}(h, h^*)$. In summary, if one classifier h has larger unlabeled data margin, it will make the $\epsilon_{tr}(h, h^*)$ smaller. □

More specifically, $\epsilon_{tr}(h, h^*)$ and $\epsilon_{te}(h, h^*)$ are equivalent when the training and test distributions are the same. Under this situation, the bound becomes $\epsilon_{te}(h) \leq \epsilon_{te}(h, h^*) + \epsilon_{te}(h^*)$. That implies, when the distributions of training and test data are the same, the error bound of the maximal margin classifier is the lowest.

As follows, we analyse the idea behind the averaging ensemble.

Lemma 2. *Unlabeled data margin achieved by averaging ensemble is not smaller than any single classifiers.*

Proof. Let $h_E = \mathbb{E}_{h \sim Q} h(\mathbf{x})$ denotes the averaging ensemble where Q is the posterior distribution of selecting h. And let $d_m(h, h^*)$ denotes the difference between the margins obtained by a classifier h and the ideal hypothesis h^*

$$\begin{aligned}
d_m(h, h^*) &= \mathbb{E}_{h \sim Q, x_i \sim p_{te}}(m_Q(x_i)_h - m_Q(x_i))^2 \\
&= \mathbb{E}_{h \sim Q, x_i \sim p_{te}}(m_Q(x_i)_h^2 - m_Q(x_i)_h * m_Q(x_i) + m_Q(x_i)^2)
\end{aligned}$$

The margin difference between the ensemble h_E and the ideal hypothesis h^* is

$$\begin{aligned}
d_m(h_E, h^*) &= \mathbb{E}_{x_i \sim p_{te}}(\mathbb{E}_{h \sim Q} m_Q(x_i)_h - m_Q(x_i))^2 \\
&= \mathbb{E}_{x_i \sim p_{te}}((\mathbb{E}_{h \sim Q} m_Q(x_i)_h)^2 - \mathbb{E}_{h \sim Q} m_Q(x_i)_h * m_Q(x_i) + m_Q(x_i)^2) \\
&\leq d_m(h, h^*) \qquad as \qquad \mathbb{E}[f(x)]^2 \leq \mathbb{E}[f(x)^2]
\end{aligned}$$ □

Therefore, on average, margin distance between averaging ensemble and ideal hypothesis h^* is smaller than any single classifiers. In other words, the ensemble achieves larger margin. According to Lemma 1, the difference between the ensemble and ideal hypothesis h^* is smaller, and the bound is lower.

In summary, the error of a classifier can be bounded in universal learning and the proposed maximal margin classifier has a lower bound. Moreover, averaging ensemble further reduces the prediction error on the unlabeled data.

Table 2. Data Set Summary

Data Set	#Training	#Test	Description	Data Set	#Training	#Test	Description
Transfer learning							
O vs Pe	500	500	Documents	Sheep	61	65	Web pages
O vs Pl	500	500	from different	Biomedical	61	131	with different
Pe vs Pl	500	500	sub categories	Goats	61	70	contents
Sample Selection Bias Correction				Uncertainty Mining			
Ionosphere	34	317		ColonTumor	30	35	Training and
Diabetes	80	688	Samples with	CNS	30	30	Test set
Haberman	30	276	feature bias	Leukemia	30	42	contain different
Wdbc	30	539		ProstateCancer	30	106	Gaussian noises

4 Experiment

MarginGraph is evaluated in three different scenarios of universal learning: transfer learning, sample selection bias correction and uncertainty mining. For each scenario, several frequently used data collections are selected. Results show that MarginGraph can reduce the domain adaptation risk significantly. Both maximal margin principal and ensemble play an important role in its performance.

4.1 Experiments Setting

The proposed approach is compared against different state-of-art algorithms specifically designed for each scenario (transfer learning, etc), as well as, the original graph transduction algorithm [7]. As a fair comparison, the original graph transduction is implemented in two different ways. The first uses the entire training data set, and the second one chooses a randomly selected sample whose size is equal to the number of examples chosen by MarginGraph. For naming convenience, the first one is called Graph, and the second as RandGraph. In transfer learning, CDSC [2] is selected as the comparative method. Its main idea is to find a mapping space which optimizes over consistency measure between the out-domain supervision and in-domain intrinsic structure. In sample selection bias correction, two approaches BRSD-BK and BRSD-DB [1] are adopted. Both methods correct the bias through structural discovery and re-balancing using unlabeled data. For both CDSC and BRSD, we use Graph as their base classifiers. For the proposed method, the number of iterations is chosen to be $n_t * 2$, where n_t is the number of labeled samples of the minority class. In addition, we set $\epsilon = 0.1$ for the stop criterion. The analysis of parameters can be found in Section 4.3. Both accuracy and AUC are reported as the evaluation metrics. Due to the randomness of obtained data set on sample selection bias and uncertainty mining tasks, the results below are averaged over 10 runs. The algorithm implementations are based on Weka [9].

4.2 Experiments Procedure

The descriptions, pre-processing procedures of different scenarios and the experiment results are presented below.

Table 3. Accuracy and AUC in Transfer Learning Data Set (%)

Methods	O vs Pe	O vs Pl	Pe vs Pl	Biomedical	Goats	Sheep
	Reuters-21578			SyskillWebert		
	Accuracy					
Graph	0.682	0.672	0.698	0.687	0.586	0.708
RandGraph	0.632	0.648	0.664	0.656	0.529	0.677
CDSC	0.704	0.720	0.736	0.610	0.586	0.677
MarginGraph	**0.752**	**0.810**	**0.780**	**0.740**	**0.743**	**0.815**
	AUC					
Graph	0.773	0.723	0.677	0.600	0.562	**0.720**
RandGraph	0.719	0.715	0.652	0.561	0.510	0.587
CDSC	0.783	0.799	0.682	0.582	0.523	0.536
MarginGraph	**0.841**	**0.837**	**0.741**	**0.725**	**0.681**	0.678

Transfer Learning. Two data collections from two different domains are employed. Among them, Reuters-21578 [10] is the primary benchmark of text categorization, and SyskillWebert [10] is the standard data set used to test web page ratings. Reuters-21578 collection is formed by different news with a hierarchial structure where it contains five top categories of news wire articles, and each main category contains several sub categories. Three top categories, "orgs", "people" and "places" are selected in our study. All of the subcategories from each category are divided into two parts, one in-domain and one out-domain. They have different distributions and are approximately equal in size. Details are summarized in Table 2. The learning objective aims to classify articles into top categories. SyskillWebert collection is formed by the HTML source of web pages plus the a user rating ("hot" or "not hot") on those web pages. It contains four separate subjects belonging to different topics. In the experiment, we randomly reserve "Bands-recording artists" as out-domain and the other three as in-domain data (Table 2). The learning task is to predict the user's preferences for the given web pages.

Table 3 presents accuracy and AUC for each domain transfer data set, given by Graph, RandGraph, CDSC and the proposed algorithm MarginGraph. It is evident that MarginGraph achieves the best performance (accuracy and AUC) in 11 out of 12 runs. Due to "labeling unsmooth problem" caused by distribution difference between the training and test data, both Graph and RandGraph fail to make correct predictions most of the time. To be specific, they achieve accuracies just no more than 71% on the Reuters and SyskillWebert collections. By considering information from both domains, CDSC gets better performance in that it boosts the accuracy by 2% to 8%. However, the proposed approach, MarginGraph, has the highest accuracy in all data set, and highest AUC in 5 out of 6 data set. We notice that MarginGraph performs better than both Graph and RandGraph at least 7% in accuracy on most data set. Specifically, it achieves as high as 16% better than these baseline methods on the Goat data set. The better performance of MarginGraph than CDSC can be ascribed to both the maximal margin based sample selection and ensemble strategy. As analyzed, these criterions can give a low prediction risk. This, from the empirical perspective, provides justification to the analysis in Section 3.

Table 4. Accuracy and AUC in Sample Selection Bias Data Set (%)

Methods	Ionosphere	Diabetes	Haberman	Wdbc	Ionosphere	Diabetes	Haberman	Wdbc
	Accuracy				StDev			
Graph	0.642	0.602	0.649	0.892	**0.028**	**0.034**	0.074	**0.001**
RandGraph	0.590	0.601	0.586	0.890	0.070	0.068	**0.002**	0.001
BRSD-BK	0.699	0.643	0.631	0.890	0.038	0.043	0.004	0.002
BRSD-DB	0.649	0.624	0.627	0.887	0.010	0.018	0.059	0.887
MarginGraph	**0.817**	**0.709**	**0.717**	**0.896**	0.077	0.259	0.295	0.002
	AUC				StDev			
Graph	0.701	0.583	0.582	0.962	0.030	0.069	0.091	0.001
RandGraph	0.605	0.554	0.511	0.961	0.010	**0.115**	**0.003**	0.002
BRSD-BK	**0.726**	**0.671**	0.561	0.962	0.053	0.061	0.012	0.001
BRSD-DB	0.686	0.634	0.551	0.962	**0.008**	0.032	0.075	0.001
MarginGraph	0.654	0.650	**0.592**	**0.963**	0.259	0.063	0.276	0.001

Table 5. Accuracy and AUC in Uncertainty Mining Data Set (%)

Methods	CNS	ColonTumor	Leukemia	ProCancer	CNS	ColonTumor	Leukemia	ProCancer
	Accuracy				StDev			
Graph	0.647	0.813	0.928	0.762	0.078	0.032	0.042	0.040
RandGraph	0.566	**0.838**	0.916	0.721	0.461	0.85	0.88	0.741
MarginGraph	**0.713**	0.787	**0.944**	**0.794**	0.078	**0.026**	0.042	**0.029**
	AUC				StDev			
Graph	0.606	0.761	0.914	0.762	0.086	**0.035**	0.020	0.052
RandGraph	0.444	0.740	0.910	0.698	0.154	0.075	0.035	0.126
MarginGraph	**0.640**	**0.792**	**0.930**	**0.782**	**0.036**	0.063	**0.015**	**0.031**

Sample Selection Bias Correction. Four data sets from UCI Repository [10] are selected. "Haberman" aims to predict the survival of patients who had undergone surgery for breast cancer. "Ionosphere" is to detect which radar is "Good" or "Bad" based on the radar signals information. "Wdbc" contains digitized image with characteristics of the cell nuclei in a fine needle aspirate of breast masses. "Diabetes" records test information about diabetes of patients and the task is to figure out which patients have diabetes. To generate the sample selection bias data set, we first randomly select 50% of the features, and then we sort the data set according to each of the selected features (dictionary sort for categorical features and numerical sort for continuous). Then, we attain top instances from every sorted list as training set, with "#Training" instances, and use the remain examples as test set.

Table 4 summarizes the accuracy, AUC as well as their standard deviations of baselines: Graph, RandGraph, BRSD-BK, BRSD-DB, and the proposed algorithm MarginGraph on four biased data sets. Clearly, MarginGraph achieves higher accuracies (from 5% to 15%) for each data set than the corresponding baseline approaches. For example, on the Haberman data set, the accuracy has been improved from no more than 65% to 71%. In AUC, MarginGraph wins Graph at 3 rounds and just loses at 1 comparison. Importantly, MarginGraph outperforms BRSD-BK and BRSD-DB consistently, specifically designed for bias correction, in accuracy. Moreover, MarginGraph performs compatibly with them in AUC, 2 wins and 2 loses. These results demonstrate that MarginGraph also has good generalization in sample selection bias task. This is attributed to the ensemble strategy, which makes the classifier more robust to bias [6], and the maximal margin strategy guarantees error bound of MarginGraph is low when the test and training distributions are related as shown in Section 3.

Uncertainty Mining. Four biomedical and gene expression data sets are selected from Kent Ridge Biomedical Repository [11] for this comparison. "Colon-Tumor" contains 62 samples collected from colon-cancer patients. "Central Nervous System(CNS)" aims to detect which patients are survivors or failures. "Leukemia" is the data set about classifying subtypes of pediatric acute lymphoblastic leukemia. "ProstateCancer" is about the tumor versus normal classification. Each of them is a typical example of high dimensional, low sample size (HDLSS) problem. To generate the uncertainty, we randomly partition the data set into training and test parts first. Then, we form two f_m-dimension Gaussian noises with different means and variances where f_m is the dimensions of the data set, and add them into two parts of data set separately. Thus, we obtain four uncertain data sets where the training set and test set contain different noises (Table 2).

The accuracy and AUC of the proposed method, and original label propagation on uncertainty mining can be found in Table 5. Among the 4 tasks, MarginGraph outperforms the baseline by 3-1 in accuracy and 4-0 in AUC with smaller standard deviations. In particular, on the CNS data set, MarginGraph performs better than the baseline by as much as 6% in accuracy and 4% in AUC. The performance improvement results from the adaptation of maximal margin strategy that makes the decision boundary go through the low density region. In addition, the averaging ensemble strategy increases the resilience of the base classifier for feature noises [12].

4.3 Parameter Analysis

Three extended experiments were conducted on the SyskillWebert collection to test the parameter sensitivity and the relationship between the unlabeled data margin and prediction accuracy. As discussed in Section 2, there are two parameters to run MarginGraph, the parameter ϵ for the stopping criterion, as well as $maxIt$, the maximal number of iterations.

Figure 2(a) shows the different AUC and accuracy vs. different values of ϵ. We observe that AUC is insensitive to the value of ϵ, but the accuracy drops down when ϵ becomes large. The reason is that ϵ determines the threshold of average margin. Clearly, if the margin is too small, the prediction risk will increase and the accuracy will decrease, as analyzed in Section 3.

Figure 2(b) illustrates the relationship between the number of iterations and prediction accuracies. Because both Graph and CDSC use the entire training data, their accuracy results do not change. We observe that the accuracy of MarginGraph increases but then drops when the number of iterations is more than 40. That is because the number of labeled data is too few to build an effective classifier at the beginning. When useful labeled data are selected enough, adding more labeled data will reduce the unlabeled data margin and create unsmooth problems against the test data. This can be observed from Figure 2(c)

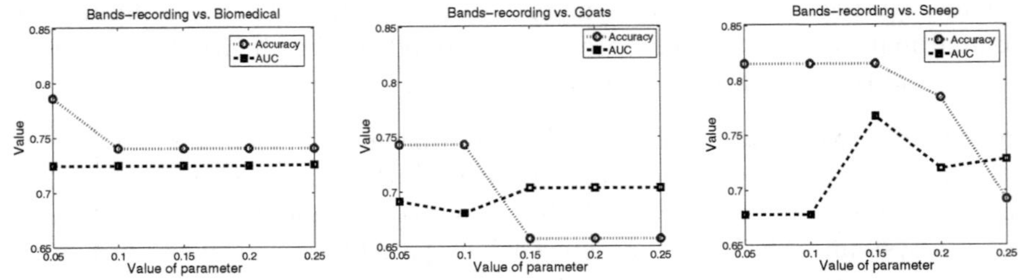

(a) Accuracy and AUC over ϵ

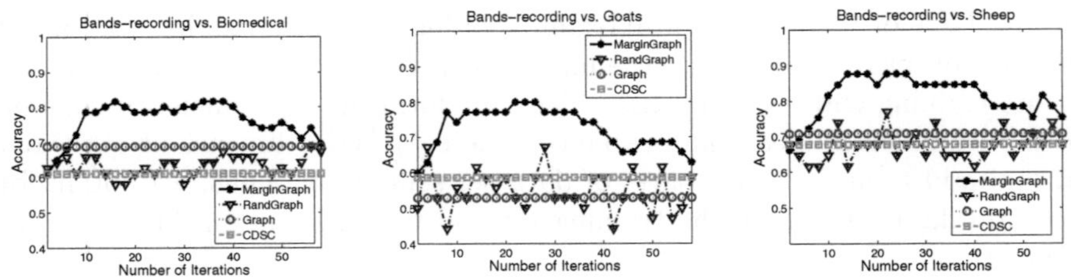

(b) Accuracy over Number of Iterations

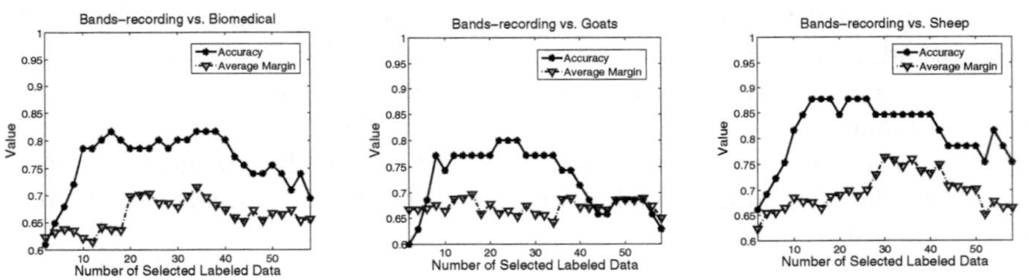

(c) Accuracy over Average Margin

Fig. 2. Parameter Analysis

that the margin also drops down when we select more than 40 labeled points. However, we still see that MarginGraph achieves the best overall performance even with a large number of iterations.

Figure 2(c) shows that the average margin and accuracy have similar trend with a function of the number of selected examples. This implies that the same set of sampled training examples that reduce the unlabeled data margin can also reduce the prediction accuracy, and vice versa. Thus, the unlabeled data margin is a good criterion to select samples for graph transduction in universal learning.

4.4 Margin Analysis

As shown in Section 3, maximal margin and ensemble are two main contributing factors to guarantee a low error bound. As follows, we perform an additional experiment to study each factor. For comparison, we adopt three other approaches, similar with MarginGraph but with slight modifications. The first "MarginBase"

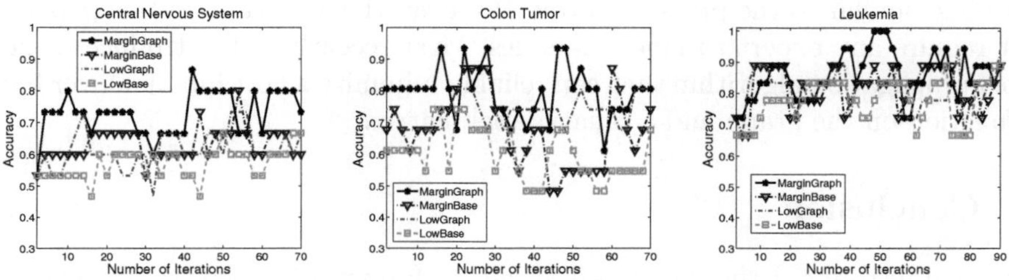

Fig. 3. Margin Analysis

is the base classifier of MarginGraph in each iteration. The second is a "minimal margin classifier" which selects samples for building a classifier with minimal unlabeled data margin, called "LowBase". The third one is the averaging ensemble of LowBase, called "LowGraph".

Figure 3 shows the relationship between the number of iterations and prediction accuracies on three uncertain data set. It is obvious that MarginGraph outperforms others in most cases. Especially, MarginGraph achieves the highest accuracy during iterations in all three data sets. In addition, LowGraph outperforms LowBase and MarginBase performs better than LowBase. That means maximal margin is better than minimal margin and ensemble is better than any single classifiers.

5 Related Works

Many solutions for transfer learning, sample selection bias correction and uncertainty mining have been proposed previously, such as but not limited to [2,13,1,14,15,16]. Among them, [2] designs a novel cost function from normalized cut that test data supervision is regularized by training data structural constraints. [13] learns a low-dimensional latent feature space where the distributions between the training data and the test data are the same or close to each other. [1] proposes to discover the natural structure of the test distribution, by which different types of sample selection biases can be evidently observed and then be reduced. [14] proposes a direct importance estimation method for sample selection bias that does not involve density estimation. [15] discusses a new method for handling error-prone and missing data with the use of density based approaches to data mining. [16] aims at minimizing the worst-case value of a loss function, over all possible realizations of the uncertainty data within given interval bounds. However, these methods are designed for each specific scenario, and are not generalized over universal learning where training and testing data are related but different. In our work, we do not distinguish transfer learning, sample selection bias, uncertainty mining and the alike. There are several significant extensions to graph transduction. For example, recently [7] introduces a semi-supervised learning framework based on Gaussian random fields and harmonic functions. Previously, [8] combines the graph transduction and active learning, similar to the proposed work. However, the algorithm in this paper do not require any expert to label examples. Most recently, [4] introduces a new label propagation algorithm that can reliably minimize a cost function over both a function on the graph and a binary label matrix.

6 Conclusion

We have introduced the "universal learning" framework to cover different formulations where the training and test set are drawn from related but different distributions. Explicit scenarios include transfer learning, sample selection bias and uncertainty mining. We have proposed an adaptive graph transduction method using unlabeled data maximum margin principle to solve universal learning tasks. Unlike prior work, the proposed framework implicitly encompasses all three problem definitions and the alike. The same proposed solution can address different scenarios. The maximum margin graph transduction works as a joint optimization to maximize the unlabeled data margin and minimize the harmonic function over unlabeled data in the same time. It is an iterative strategy that removes the bias of any single graph. Formal analysis shows that the maximum margin based sample selection strategy has good generality over testing data with related but different distribution.

Empirical studies demonstrate that with different problem formulations in universal learning, the proposed approach significantly improves the original graph transduction. For transfer learning, it outperforms the original graph transduction by as much as 16% in accuracy and 12% in AUC. For sample selection bias correction, it achieves around 10% higher in accuracy in most cases. For uncertainty mining, its performance is the highest in 7 out of 8 comparisons. Most importantly, it consistently outperforms, by as much as much 10% in accuracy, than state-of-art approaches specifically designed for transfer learning and bias correction. These base line methods include CDSC [2] for transfer learning, and BRSD-BK and BRSD-DB [1] for sample selection bias correction. The main claims are that (1) universal learning framework provides a general formulation to cover and study various real-world application scenarios where training and testing data do not follow the same distribution, and (2) unlike previously proposed methods that cover only one scenario, the proposed adaptive graph transduction provides a more accurate solution to encompass all distribution differences under universal learning, and this provides utility and ease of use.

References

1. Ren, J., Shi, X., Fan, W., Yu, P.S.: Type-independent correction of sample selection bias via structural discovery and re-balancing. In: Proceedings of the Eighth SIAM International Conference on Data Mining, SDM 2008, pp. 565–576. SIAM, Philadelphia (2008)
2. Ling, X., Dai, W., Xue, G.R., Yang, Q., Yu, Y.: Spectral domain-transfer learning. In: KDD 2008: Proceeding of the 14th ACM SIGKDD international conference on Knowledge discovery and data mining, pp. 488–496. ACM, New York (2008)

3. Blitzer, J., Crammer, K., Kulesza, A., Pereira, F., Wortman, J.: Learning bounds for domain adaptation. In: Platt, J., Koller, D., Singer, Y., Roweis, S. (eds.) Advances in Neural Information Processing Systems 20, pp. 129–136. MIT Press, Cambridge (2008)
4. Wang, J., Jebara, T., Chang, S.F.: Graph transduction via alternating minimization. In: ICML 2008: Proceedings of the 25th international conference on Machine learning, pp. 1144–1151. ACM, New York (2008)
5. Amini, M., Laviolette, F., Usunier, N.: A transductive bound for the voted classifier with an application to semi-supervised learning. In: Koller, D., Schuurmans, D., Bengio, Y., Bottou, L. (eds.) Advances in Neural Information Processing Systems 21 (2009)
6. Fan, W., Davidson, I.: On sample selection bias and its efficient correction via model averaging and unlabeled examples. In: Proceedings of the Seventh SIAM International Conference on Data Mining, SDM 2007, Minneapolis, Minnesota. SIAM, Philadelphia (2007)
7. Zhu, X.: Semi-supervised learning with graphs. PhD thesis, Carnegie Mellon University, Pittsburgh, PA, USA (2005)
8. Zhu, X., Lafferty, J.: Combining active learning and semi-supervised learning using gaussian fields and harmonic functions. In: ICML 2003 workshop on The Continuum from Labeled to Unlabeled Data in Machine Learning and Data Mining, pp. 58–65 (2003)
9. Witten, I.H., Frank, E.: Data Mining: Practical Machine Learning Tools and Techniques, 2nd edn. Morgan Kaufmann Series in Data Management Systems. Morgan Kaufmann, San Francisco (2005)
10. Asuncion, A., Newman, D.J.: UCI machine learning repository (2007), `http://www.ics.uci.edu/mlearn/MLRepository.html`
11. Liu, H., Li, J.: Kent ridge bio-medical data set repository (2005), `http://leo.ugr.es/elvira/DBCRepository/index.html`
12. Melville, P., Shah, N., Mihalkova, L., Mooney, R.J.: Experiments on ensembles with missing and noisy data. In: Roli, F., Kittler, J., Windeatt, T. (eds.) MCS 2004. LNCS, vol. 3077, pp. 293–302. Springer, Heidelberg (2004)
13. Pan, S.J., Kwok, J.T., Yang, Q.: Transfer learning via dimensionality reduction. In: Proceedings of the Twenty-Third AAAI Conference on Artificial Intelligence, AAAI 2008, Chicago, Illinois, USA, July 13-17, pp. 677–682 (2008)
14. Sugiyama, M., Nakajima, S., Kashima, H., Buenau, P.V., Kawanabe, M.: Direct importance estimation with model selection and its application to covariate shift adaptation. In: Platt, J., Koller, D., Singer, Y., Roweis, S. (eds.) Advances in Neural Information Processing Systems 20, pp. 1433–1440. MIT Press, Cambridge (2008)
15. Aggarwal, C.C.: On density based transforms for uncertain data mining. In: Proceedings of the 23rd International Conference on Data Engineering, ICDE 2007, The Marmara Hotel, Istanbul, Turkey, pp. 866–875. IEEE, Los Alamitos (2007)
16. El Ghaoui, L., Lanckriet, G.R.G., Natsoulis, G.: Robust classification with interval data. Technical Report UCB/CSD-03-1279, EECS Department, University of California, Berkeley (October 2003)

The Feature Importance Ranking Measure

Alexander Zien[1,2], Nicole Krämer[3],
Sören Sonnenburg[2], and Gunnar Rätsch[2]

[1] Fraunhofer FIRST.IDA, Kekuléstr. 7, 12489 Berlin, Germany
[2] Friedrich Miescher Laboratory, Max Planck Society,
Spemannstr. 39, 72076 Tübingen, Germany
[3] Machine Learning Group, Berlin Institute of Technology,
Franklinstr. 28/29, 10587 Berlin, Germany

Abstract. Most accurate predictions are typically obtained by learning machines with complex feature spaces (as e.g. induced by kernels). Unfortunately, such decision rules are hardly accessible to humans and cannot easily be used to gain insights about the application domain. Therefore, one often resorts to linear models in combination with variable selection, thereby sacrificing some predictive power for presumptive interpretability. Here, we introduce the *Feature Importance Ranking Measure* (FIRM), which by retrospective analysis of arbitrary learning machines allows to achieve both excellent predictive performance and superior interpretation. In contrast to standard raw feature weighting, FIRM takes the underlying correlation structure of the features into account. Thereby, it is able to discover the most relevant features, even if their appearance in the training data is entirely prevented by noise. The desirable properties of FIRM are investigated analytically and illustrated in simulations.

1 Introduction

A major goal of machine learning — beyond providing accurate predictions — is to gain understanding of the investigated problem. In particular, for researchers in application areas, it is frequently of high interest to unveil which features are indicative of certain predictions. Existing approaches to the identification of important features can be categorized according to the restrictions that they impose on the learning machines.

The most convenient access to features is granted by linear learning machines. In this work we consider methods that express their predictions via a real-valued output function $s : \mathcal{X} \to \mathbb{R}$, where $\mathcal{X}$ is the space of inputs. This includes standard models for classification, regression, and ranking. Linearity thus amounts to

$$s(\mathbf{x}) = \mathbf{w}^\top \mathbf{x} + b \ . \tag{1}$$

One popular approach to finding important dimensions of vectorial inputs ($\mathcal{X} = \mathbb{R}^d$) is *feature selection*, by which the training process is tuned to make sparse use of the available d candidate features. Examples include ℓ_1-regularized methods like Lasso [13] or ℓ_1-SVMs [1] and heuristics for non-convex ℓ_0-regularized formulations. They all find feature weightings $\mathbf{w}$ that have few non-zero components,

W. Buntine et al. (Eds.): ECML PKDD 2009, Part II, LNAI 5782, pp. 694–709, 2009.

for example by eliminating redundant dimensions. Thus, although the resulting predictors are economical in the sense of requiring few measurements, it can not be concluded that the other dimensions are unimportant: a different (possibly even disjoint) subset of features may yield the same predictive accuracy. Being selective among correlated features also predisposes feature selection methods to be unstable. Last but not least, the accuracy of a predictor is often decreased by enforcing sparsity (see e.g. [10]).

In multiple kernel learning (MKL; e.g. [5,10]) a sparse linear combination of a small set of kernels [8] is optimized concomitantly to training the kernel machine. In essence, this lifts both merits and detriments of the selection of individual features to the coarser level of feature spaces (as induced by the kernels). MKL thus fails to provide a principled solution to assessing the importance of sets of features, not to speak of individual features. It is now urban knowledge that ℓ_1-regularized MKL can even rarely sustain the accuracy of a plain uniform kernel combination [2].

Alternatively, the sparsity requirement may be dropped, and the j-th component w_j of the trained weights $\mathbf{w}$ may be taken as the importance of the j-th input dimension. This has been done, for instance, in cognitive sciences to understand the differences in human perception of pictures showing male and female faces [4]; here the resulting weight vector $\mathbf{w}$ is relatively easy to understand for humans since it can be represented as an image.

Again, this approach may be partially extended to kernel machines [8], which do not access the features explicitly. Instead, they yield a kernel expansion

$$s(\mathbf{x}) = \sum_{i=1}^{n} \alpha_i k(\mathbf{x}_i, \mathbf{x}) + b \ , \tag{2}$$

where $(\mathbf{x}_i)_{i=1,\ldots,n}$ are the inputs of the n training examples. Thus, the weighting $\boldsymbol{\alpha} \in \mathbb{R}^n$ corresponds to the training examples and cannot be used directly for the interpretation of features. It may still be viable to compute explicit weights for the features $\Phi(\mathbf{x})$ induced by the kernel via $k(\mathbf{x}, \mathbf{x}') = \langle \Phi(\mathbf{x}), \Phi(\mathbf{x}') \rangle$, provided that the kernel is benign: it must be guaranteed that only a finite and limited number of features are used by the trained machine, such that the equivalent linear formulation with

$$\mathbf{w} = \sum_{i=1}^{n} \alpha_i \Phi(\mathbf{x}_i)$$

can efficiently be deduced and represented.

A generalization of the feature weighting approach that works with general kernels has been proposed by Üstün et. al. [14]. The idea is to characterize input variables by their correlation with the weight vector $\boldsymbol{\alpha}$. For a linear machine as given by (1) this directly results in the weight vector $\mathbf{w}$; for non-linear functions s, it yields a projection of $\mathbf{w}$, the meaning of which is less clear.

A problem that all above methods share is that the weight that a feature is assigned by a learning machine is not necessarily an appropriate measure of its importance. For example, by multiplying any dimension of the inputs by a positive scalar and dividing the associated weight by the same scalar, the conjectured importance of the corresponding feature can be changed arbitrarily, although the predictions are not altered at all, i.e. the trained learning machine is unchanged. An even more practically detrimental shortcoming of the feature weighting is its failure to take into account correlations between features; this will be illustrated in a computational experiment below (Section 3).

Further, all methods discussed so far are restricted to linear scoring functions or kernel expansions. There also exists a range of customized importance measures that are used for building decision trees and random forests (see e.g. [11,12] for an overview).

In this paper, we reach for an importance measure that is "universal": it shall be applicable to any learning machine, so that we can avoid the clumsiness of assessing the relevance of features for methods that produce suboptimal predictions, and it shall work for any feature. We further demand that the importance measure be "objective", which has several aspects: it may not arbitrarily choose from correlated features as feature selection does, and it may not be prone to misguidance by feature rescaling as the weighting-based methods are. Finally, the importance measure shall be "intelligent" in that it exploits the connections between related features (this will become clearer below).

In the next section, we briefly review the state of the art with respect to these goals and in particular outline a recent proposal, which is, however, restricted to sequence data. Section 2 exhibits how we generalize that idea to continuous features and exhibits its desirable properties. The next two sections are devoted to unfolding the math for several scenarios. Finally, we present a few computational results illustrating the properties of our approach in the different settings. The relevant notation is summarized in Table 1.

Table 1. Notation

symbol	definition	reference
$\mathcal{X}$	input space	
$s(x)$	scoring function $\mathcal{X} \to \mathbb{R}$	
w	weight vector of a linear scoring function s	equation (1)
f	feature function $\mathcal{X} \to \mathbb{R}$	equation (6)
$q_f(t)$	conditional expected score $\mathbb{R} \to \mathbb{R}$	definition 1
Q_f	feature importance ranking measure (firm) $\in \mathbb{R}$	definition 2
$\mathbf{Q}$	vector $\in \mathbb{R}^d$ of firms for d features	subsection 2.4
$\mathbf{\Sigma}$, $\mathbf{\Sigma}_{j\bullet}$	covariance matrix, and its jth column	

1.1 Related Work

A few existing feature importance measures satisfy one or more of the above criteria. One popular "objective" approach is to assess the importance of a variable by measuring the decrease of accuracy when retraining the model based on a random permutation of a variable. However, it has only a narrow application range, as it is computationally expensive and confined to input variables.

Another approach is to measure the importance of a feature in terms of a sensitivity analysis [3]

$$I_j = \mathbb{E}\left[\left(\frac{\partial s}{\partial x_j}\right)^2 \mathbb{V}\mathrm{ar}\left[X_j\right]\right]^{1/2} \quad . \tag{3}$$

This is both "universal" and "objective". However, it clearly does not take the indirect effects into account: for example, the change of X_j may imply a change of some X_k (e.g. due to correlation), which may also impact s and thereby augment or diminish the net effect.

Here we follow the related but more "intelligent" idea of [17]: to assess the importance of a feature by estimating its total impact on the score of a trained predictor. While [17] proposes this for binary features that arise in the context of sequence analysis, the purpose of this paper is to generalize it to real-valued features and to theoretically investigate some properties of this approach. It turns out (proof in Section 2.2) that under normality assumptions of the input features, FIRM generalizes (3), as the latter is a first order approximation of FIRM, and because FIRM also takes the correlation structure into account.

In contrast to the above mentioned approaches, the proposed *feature importance ranking measure* (FIRM) also takes the dependency of the input features into account. Thereby it is even possible to assess the importance of features that are not observed in the training data, or of features that are not directly considered by the learning machine.

1.2 Positional Oligomer Importance Matrices [17]

In [17], a novel feature importance measure called Positional Oligomer Importance Matrices (POIMs) is proposed for substring features in string classification. Given an alphabet Σ, for example the DNA nucleotides $\Sigma = \{\mathtt{A}, \mathtt{C}, \mathtt{G}, \mathtt{T}\}$, let $\mathbf{x} \in \Sigma^L$ be a sequence of length L. The kernels considered in [17] induce a feature space that consists of one binary dimension for each possible substring $\mathbf{y}$ (up to a given maximum length) at each possible position i. The corresponding weight $w_{\mathbf{y},i}$ is added to the score if the substring $\mathbf{y}$ is incident at position i in $\mathbf{x}$. Thus we have the case of a kernel expansion that can be unfolded into a linear scoring system:

$$s(\mathbf{x}) = \sum_{\mathbf{y},i} w_{\mathbf{y},i} \mathbb{I}\left\{\mathbf{x}[i] = \mathbf{y}\right\} \quad , \tag{4}$$

where $\mathbb{I}\{\cdot\}$ is the indicator function. Now POIMs are defined by

$$Q'(\mathbf{z}, j) := \mathbb{E}\left[\, s(\mathbf{X}) \mid \mathbf{X}[j] = \mathbf{z}\,\right] - \mathbb{E}\left[s(\mathbf{X})\right] \quad , \tag{5}$$

where the expectations are taken with respect to a D-th order Markov distribution.

Intuitively, Q' measures how a feature, here the incidence of substring $\mathbf{z}$ at position j, would change the score s as compared to the average case (the unconditional expectation). Although positional sub-sequence incidences are binary features (they are either present or not), they posses a very particular correlation structure, which can dramatically aid in the identification of relevant features.

2 The Feature Importance Ranking Measure (FIRM)

As explained in the introduction, a trained learner is defined by its output or scoring function $s : \mathcal{X} \to \mathbb{R}$. The goal is to quantify how important any given feature

$$f : \mathcal{X} \to \mathbb{R} \tag{6}$$

of the input data is to the score. In the case of vectorial inputs $\mathcal{X} = \mathbb{R}^d$, examples for features are simple coordinate projections $f_j(\mathbf{x}) = x_j$, pairs $f_{jk}(\mathbf{x}) = x_j x_k$ or higher order interaction features, or step functions $f_{j,\tau}(\mathbf{x}) = \mathbb{I}\{x_j > \tau\}$ (where $\mathbb{I}\{\cdot\}$ is the indicator function).

We proceed in two steps. First, we define the expected output of the score function under the condition that the feature f attains a certain value.

Definition 1 (conditional expected score). *The conditional expected score of s for a feature f is the expected score $q_f : \mathbb{R} \to \mathbb{R}$ conditional to the feature value t of the feature f:*

$$q_f(t) = \mathbb{E}\left[\, s(X) \mid f(X) = t \,\right] \ . \tag{7}$$

We remark that this definition corresponds — up to normalization — to the marginal variable importance studied by van der Laan [15]. A flat function q_f corresponds to a feature f that has no or just random effect on the score; a variable function q_f indicates an important feature f.

Consequently, the second step of FIRM is to determine the importance of a feature f as the variability of the corresponding expected score $q_f : \mathbb{R} \to \mathbb{R}$.

Definition 2 (feature importance ranking measure). *The feature importance $Q_f \in \mathbb{R}$ of the feature f is the standard deviation of the function q_f:*

$$Q_f := \sqrt{\mathbb{V}ar\left[q_f(f(X))\right]} = \left(\int_{\mathbb{R}} \left(q_f(t) - \bar{q}_f\right)^2 \mathbb{P}r(f(X) = t)\, dt \right)^{\frac{1}{2}} \ , \tag{8}$$

where $\bar{q}_f := \mathbb{E}\left[q_f(f(X))\right] = \int_{\mathbb{R}} q_f(t) \mathbb{P}r(f(X) = t)\, dt$ is the expectation of q_f.

In case of (i) known linear dependence of the score on the feature under investigation or (ii) an ill-posed estimation problem (8) — for instance, due to scarce data —, we suggest to replace the standard deviation by the more reliably estimated slope of a linear regression. As we will show later (Section 2.3), for binary features identical feature importances are obtained by both ways anyway.

2.1 Properties of FIRM

FIRM generalizes POIMs. As we will show in Section Section 2.3, FIRM indeed contains POIMs as special case. POIMs, as defined in (5), are only meaningful for binary features. FIRM extends the core idea of POIMs to continuous features.

FIRM is "universal". Note that our feature importance ranking measure (FIRM) can be applied to a very broad family of learning machines. For instance, it works in both classification, regression and ranking settings, as long as the task is modeled via a real-valued output function over the data points. Further, it is not constrained to linear functions, as is the case for l_1-based feature selection. FIRM can be used with any feature space, be it induced by a kernel or not. The importance computation is not even confined to features that are used in the output function. For example, one may train a kernel machine with a polynomial kernel of some degree and afterwards determine the importance of polynomial features of higher degree. We illustrate the ability of FIRM to quantify the importance of unobserved features in Section 3.3.

FIRM is robust and "objective". In order to be sensible, an importance measure is required to be robust with respect to perturbations of the problem and invariant with respect to irrelevant transformations. Many successful methods for classification and regression are translation-invariant; FIRM will immediately inherit this property. Below we show that FIRM is also invariant to rescaling of the features in some analytically tractable cases (including all binary features), suggesting that FIRM is generally well-behaved in this respect. In Section 2.4 we show that FIRM is even robust with respect to the choice of the learning method. FIRM *is* sensitive to rescaling of the scoring function s. In order to compare different learning machines with respect to FIRM, s should be standardized to unit variance; this yields importances $\tilde{Q}_f = Q_f / \mathbb{V}\mathrm{ar}\left[s(X)\right]^{1/2}$ that are to scale. Note, however, that the relative importance, and thus the ranking, of all features for any single predictor remains fixed.

Computation of FIRM. It follows from the definition of FIRM that we need to assess the distribution of the input features and that we have to compute conditional distributions of nonlinear transformations (in terms of the score function s). In general, this is infeasible. While in principle one could try to estimate all quantities empirically, this leads to an estimation problem due to the limited amount of data. However, in two scenarios, this becomes feasible. First, one can impose additional assumptions. As we show below, for normally distributed inputs and linear features, FIRM can be approximated analytically, and we only need the covariance structure of the inputs. Furthermore, for linear scoring functions (1), we can compute FIRM for (a) normally distributed inputs (b) binary data with known covariance structure and (c) — as shown before in [16] — for sequence data with (higher-order) Markov distribution. Second, one can approximate the conditional expected score q_f by a linear function, and to then estimate the feature importance Q_f from its slope. As we show in Section 2.3, this approximation is exact for binary data.

2.2 Approximate FIRM for Normally Distributed Features

For general score functions s and arbitrary distributions of the input, the computation of the conditional expected score (7) and the FIRM score (8) is in general intractable, and the quantities can at best be estimated from the data. However, under the assumption of normally distributed features, we can derive an analytical approximation of FIRM in terms of first order Taylor approximations. More precisely, we use the following approximation.

Approximation. *For a normally random variable $\widetilde{X} \sim \mathcal{N}\left(\widetilde{\mu}, \widetilde{\mathbf{\Sigma}}\right)$ and a differentiable function $g : \mathbb{R}^d \to \mathbb{R}^p$, the distribution of $g(X)$ is approximated by its first order Taylor expansion:*

$$g(X) \sim \mathcal{N}\left(g(\widetilde{\mu}), J\widetilde{\mathbf{\Sigma}}J^\top\right)$$

with

$$J = \left.\frac{\partial g}{\partial \mathbf{x}}\right|_{\mathbf{x}=\widetilde{\mu}}$$

Note that if the function g is linear, the distribution is exact.

In the course of this subsection, we consider feature functions $f_j(\mathbf{x}) = \mathbf{x}_j$ (an extension to linear feature functions $f(\mathbf{x}) = \mathbf{x}^\top \mathbf{a}$ is straightforward.)

First, recall that for a normally distributed random variable $X \sim \mathcal{N}(\mathbf{0}, \mathbf{\Sigma})$, the conditional distribution of $X|X_j = t$ is again normal, with expectation

$$\mathbb{E}\left[\, X \mid X_j = t \,\right] = \frac{t}{\mathbf{\Sigma}_{jj}} \mathbf{\Sigma}_{j\bullet} =: \widetilde{\mu}_j\,.$$

Here $\mathbf{\Sigma}_{j\bullet}$ is the jth column of $\mathbf{\Sigma}$.

Now, using the above approximation, the conditional expected score is

$$q_f(t) \approx s\left(\widetilde{\mu}_j\right) = s((t/\mathbf{\Sigma}_{jj})\, \mathbf{\Sigma}_{j\bullet})$$

To obtain the FIRM score, we apply the approximation again, this time to the function $t \mapsto s(((t/\mathbf{\Sigma}_{jj})\, \mathbf{\Sigma}_{j\bullet})$. Its first derivative at the expected value $t = 0$ equals

$$J = \frac{1}{\mathbf{\Sigma}_{jj}} \mathbf{\Sigma}_{j\bullet}^\top \left.\frac{\partial s}{\partial \mathbf{x}}\right|_{\mathbf{x}=\mathbf{0}}$$

This yields

$$Q_j \approx \sqrt{\frac{1}{\mathbf{\Sigma}_{jj}} \left(\mathbf{\Sigma}_{j\bullet}^\top \left.\frac{\partial s}{\partial \mathbf{x}}\right|_{\mathbf{x}=\mathbf{0}}\right)^2} \tag{9}$$

Note the correspondence to (3) in Friedman's paper [3]: If the features are uncorrelated, (9) simplifies to

$$Q_j \approx \sqrt{\mathbf{\Sigma}_{jj} \left(\left.\frac{\partial s}{\partial \mathbf{x}_j}\right|_{\mathbf{x}_j=0}\right)^2}$$

(recall that $0 = E[X_j]$). Hence FIRM adds an additional weighting that corresponds to the dependence of the input features. These weightings are based on the true covariance structure of the predictors. In applications, the true covariance matrix is in general not known. However, it is possible to estimate it reliably even from high-dimensional data using mean-squared-error optimal shrinkage [7].

Note that the above approximation can be used to compute FIRM for the kernel based score functions (2). E.g., for Gaussian kernels

$$k_\gamma(\mathbf{x}, \mathbf{x}_i) = exp\left(-\frac{\|\mathbf{x} - \mathbf{x}_i\|^2}{\gamma^2}\right)$$

we have

$$\left.\frac{\partial k_\gamma(\mathbf{x}, \mathbf{x}_i)}{\partial \mathbf{x}}\right|_{\mathbf{x}=0} = \frac{2k(\mathbf{0}, \mathbf{x}_i)}{\gamma^2}\mathbf{x}_i^\top = \frac{2e^{-(\|\mathbf{x}_i\|^2/\gamma^2)}}{\gamma^2}\mathbf{x}_i^\top$$

and hence obtain

$$\left.\frac{\partial s}{\partial \mathbf{x}}\right|_{\mathbf{x}=\mathbf{0}} = \sum_{i=1}^{N} \alpha_i y_i \frac{2e^{-(\|\mathbf{x}_i\|^2/\gamma^2)}}{\gamma^2}\mathbf{x}_i^\top .$$

2.3 Exact FIRM for Binary Data

Binary features are both analytically simple and, due to their interpretability and versatility, practically highly relevant. Many discrete features can be adequately represented by binary features, even if they can assume more than two values. For example, a categorical feature can be cast into a sparse binary encoding with one indicator bit for each value; an ordinal feature can be encoded by bits that indicate whether the value is strictly less than each of its possibilities. Therefore we now try to understand in more depth how FIRM acts on binary variables.

For a binary feature $f : \mathcal{X} \to \{a, b\}$ with feature values $t \in \{a, b\}$, let the distribution be described by

$$p_a = \mathbb{P}\mathrm{r}\left(f(X) = a\right), p_b = 1 - p_a ,$$

and let the conditional expectations be $q_a = q_f(a)$ and $q_b = q_f(b)$. Simple algebra shows that in this case $\mathbb{V}\mathrm{ar}\left[q(f(X))\right] = p_a p_b (q_a - q_b)^2$. Thus we obtain the feature importance

$$Q_f = (q_a - q_b)\sqrt{p_a p_b} \ . \tag{10}$$

(By dropping the absolute value around $q_a - q_b$ we retain the directionality of the feature's impact on the score.) Note that we can interpret firm in terms of the slope of a linear function. If we assume that $a, b \in \mathbb{R}$, the linear regression fit

$$(w_f, c_f) = \arg\min_{w_f, c_f} \int_{\mathbb{R}} \left((w_f t + c_f) - q_f(t)\right)^2 d\mathbb{P}\mathrm{r}\,(t)$$

the slope is $w_f = \frac{q_a - q_b}{a-b}$. The variance of the feature value is $\mathbb{V}\mathrm{ar}\left[f(X)\right] = p_a p_b (a-b)^2$. (10) is recovered as the increase of the linear regression function along one standard deviation of feature value. As desired, the importance is independent of feature translation and rescaling (provided that the score remains unchanged). In the following we can thus (without loss of generality) constrain that $t \in \{-1, +1\}$.

Let us reconsider POIMS Q', which are defined in equation (5). We note that $Q'(b) := q_b - \bar{q} = p_a(q_b - q_a) = \sqrt{p_a/p_b}Q(b)$; thus $Q(\mathbf{z}, j)$ can be recovered as

$$Q(\mathbf{z}, j) = Q'(\mathbf{z}, j)\sqrt{\mathbb{P}\mathrm{r}\left(\mathbf{X}[j] \neq \mathbf{z}\right) / \mathbb{P}\mathrm{r}\left(\mathbf{X}[j] = \mathbf{z}\right)} \,.$$

Thus, while POIMs are not strictly a special case of FIRM, they differ only in a scaling factor which depends on the distribution assumption. For a uniform Markov model (as empirically is sufficient according to [17]), this factor is constant.

2.4 FIRM for Linear Scoring Functions

To understand the properties of the proposed measure, it is useful to consider it in the case of linear output functions (1).

Independently Distributed Binary Data. First, let us again consider the simplest scenario of uniform binary inputs, $X \sim unif(\{-1, +1\}^d)$; the inputs are thus pairwise independent.

First we evaluate the importance of the input variables as features, i.e. we consider projections $f_j(\mathbf{x}) = x_j$. In this case, we immediately find for the conditional expectation $q_j(t)$ of the value t of the j-th variable that $q_j(t) = tw_j + b$. Plugged into (10) this yields $Q_j = w_j$, as expected. When the features are independent, their impact on the score is completely quantified by their associated weights; no side effects have to be taken into account, as no other features are affected.

We can also compute the importances of conjunctions of two variables, i.e.

$$f_{j \wedge k}(\mathbf{x}) = \mathbb{I}\left\{x_j = +1 \wedge x_k = +1\right\} \,.$$

Here we find that $q_{j\wedge k}(1) = w_j + w_k + b$ and $q_{j\wedge k}(0) = -\frac{1}{3}(w_j + w_k) + b$, with $\mathbb{P}\mathrm{r}\left(f_{j\wedge k}(X) = 1\right) = \frac{1}{4}$. This results in the feature importance $Q_{j\wedge k} = (w_j + w_k)/\sqrt{3}$. This calculation also applies to negated variables and is easily extended to higher order conjunctions.

Another interesting type of feature derives from the xor-function. For features $f_{j\otimes k}(\mathbf{x}) = \mathbb{I}\left\{x_j \neq x_k\right\}$ the conditional expectations vanish, $q_{j\otimes k}(1) = q_{j\otimes k}(0) = 0$. Here the FIRM exposes the inability of the linear model to capture such a dependence.

Binary Data With Empirical Distribution. Here we consider the empirical distribution as given by a set $\{\, \mathbf{x}_i \mid i = 1, \ldots, n \,\}$ of n data points $\mathbf{x}_i \in \{-1, +1\}^d$: $\mathbb{P}\mathrm{r}\left(X\right) = \frac{1}{n}\sum_{i=1}^{n} \mathbb{I}\left\{X = \mathbf{x}_i\right\}$. For input features $f_j(\mathbf{x}) = x_j$, this leads to $q_j(t) =$

$\frac{1}{n_{jt}} \sum_{i:\mathbf{x}_{ij}=t} \mathbf{w}^\top \mathbf{x}_i + b$, where $n_{jt} := |\{ i \mid \mathbf{x}_{ij} = t \}|$ counts the examples showing the feature value t. With (10) we get

$$Q_j = (q_j(+1) - q_j(-1)) \sqrt{\mathbb{P}\mathrm{r}(X_j = +1)\, \mathbb{P}\mathrm{r}(X_j = -1)}$$
$$= \sum_{i=1}^{n} \frac{\mathbf{x}_{ij}}{n_{j,\mathbf{x}_{ij}}} \left(\mathbf{w}^\top \mathbf{x}_i\right) \sqrt{\frac{n_{j,+1} n_{j,-1}}{n^2}}$$

It is convenient to express the vector $\mathbf{Q} \in \mathbb{R}^d$ of all feature importances in matrix notation. Let $\mathbf{X} \in \mathbb{R}^{n \times d}$ be the data matrix with the data points $\mathbf{x}_i$ as rows. Then we can write

$$\mathbf{Q} = \mathbf{M}^\top \mathbf{X} \mathbf{w} \quad \text{with} \quad \mathbf{M} \in \mathbb{R}^{n \times d} = \mathbf{1}_{n \times d} \mathbf{D}_0 + \mathbf{X} \mathbf{D}_1$$

with diagonal matrices $\mathbf{D}_0, \mathbf{D}_1 \in \mathbb{R}^{d \times d}$ defined by

$$(\mathbf{D}_1)_{jj} = \frac{1}{2\sqrt{n_{j,+1} n_{j,-1}}} \;, \quad (\mathbf{D}_0)_{jj} = \frac{n_{j,+1} - n_{j,-1}}{2n\sqrt{n_{j,+1} n_{j,-1}}} \;. \tag{11}$$

With the empirical covariance matrix $\hat{\mathbf{\Sigma}} = \frac{1}{n} \mathbf{X}^\top \mathbf{X}$, we can thus express $\mathbf{Q}$ as $\mathbf{Q} = \mathbf{D}_0 \mathbf{1}_{d \times n} \mathbf{X} \mathbf{w} + n \mathbf{D}_1 \hat{\mathbf{\Sigma}} \mathbf{w}$. Here it becomes apparent how the FIRM, as opposed to the plain $\mathbf{w}$, takes the correlation structure of the features into account. Further, for a uniformly distributed feature j (i.e. $\mathbb{P}\mathrm{r}(X_j = t) = \frac{1}{2}$), the standard scaling is reproduced, i.e. $(\mathbf{D}_1)_{jj} = \frac{1}{n} \mathbf{I}$, and the other terms vanish, as $(\mathbf{D}_0)_{jj} = 0$.

For $\mathbf{X}$ containing each possible feature vector exactly once, corresponding to the uniform distribution and thus independent features, $\mathbf{M}^\top \mathbf{X}$ is the identity matrix (the covariance matrix), recovering the above solution of $\mathbf{Q} = \mathbf{w}$.

Continuous Data With Normal Distribution. If we consider normally distributed input features and assume a linear scoring function (1), the approximations above (Section 2.2) are exact. Hence, the expected conditional score of an input variable is

$$q_j(t) = \frac{t}{\mathbf{\Sigma}_{jj}} \mathbf{w}^\top \mathbf{\Sigma}_{j\bullet} + b \;. \tag{12}$$

With the diagonal matrix $\mathbf{D}$ of standard deviations of the features, i.e. with entries $\mathbf{D}_{jj} = \sqrt{\mathbf{\Sigma}_{jj}}$, this is summarized in

$$\mathbf{q} = b \mathbf{1}_d + t \mathbf{D}^{-2} \mathbf{\Sigma} \mathbf{w} \,.$$

Exploiting that the marginal distribution of X with respect to the j-th variable is again a zero-mean normal, $X_j \sim \mathcal{N}(0, \mathbf{\Sigma}_{jj})$, this yields $\mathbf{Q} = \mathbf{D}^{-1} \mathbf{\Sigma} \mathbf{w}$. For uncorrelated features, $\mathbf{D}$ is the square root of the diagonal covariance matrix $\mathbf{\Sigma}$, so that we get $\mathbf{Q} = \mathbf{D} \mathbf{w}$. Thus rescaling of the features is reflected by a corresponding rescaling of the importances — unlike the plain weights, FIRM cannot be manipulated this way.

As FIRM weights the scoring vector by the correlation $D^{-1}\Sigma$ between the variables, it is in general more stable and more reliable than the information obtained by the scoring vector alone. As an extreme case, let us consider a two-dimensional variable (X_1, X_2) with almost perfect correlation $\rho = cor(X_1, X_2) \approx 1$. In this situation, L1-type methods like lasso tend to select randomly only one of these variables, say $\mathbf{w} = (w_1, 0)$, while L2-regularization tends to give almost equal weights to both variables. FIRM compensates for the arbitrariness of lasso by considering the correlation structure of X: in this case $q = (w_1, \rho w_1)$, which is similar to what would be found for an equal weighting $\mathbf{w} = \frac{1}{2}(w, w)$, namely $q = (w(1+\rho)/2, w(1+\rho)/2)$.

Linear Regression. Here we assume that the scoring function s is the solution of an unregularized linear regression problem, $\min_{\mathbf{w},b} \|\mathbf{X}\mathbf{w} - \mathbf{y}\|^2$; thus $\mathbf{w} = (\mathbf{X}^\top\mathbf{X})^{-1}\mathbf{X}^\top\mathbf{y}$.

Plugging this into the expression for $\mathbf{Q}$ from above yields

$$\mathbf{Q} = \mathbf{D}^{-1}\mathbf{\Sigma}\left(n\hat{\mathbf{\Sigma}}\right)^{-1}\mathbf{X}^\top\mathbf{y}\ . \tag{13}$$

For infinite training data, $\hat{\mathbf{\Sigma}} \longrightarrow \mathbf{\Sigma}$, we thus obtain $\mathbf{Q} = \frac{1}{n}\mathbf{D}^{-1}\mathbf{X}^\top\mathbf{y}$. Here it becomes apparent how the normalization makes sense: it renders the importance independent of a rescaling of the features. When a feature is inflated by a factor, so is its standard deviation $\mathbf{D}_{jj}$, and the effect is cancelled by multiplying them.

3 Simulation Studies

We now illustrate the usefulness of FIRM in a few preliminary computational experiments on artificial data.

3.1 Binary Data

We consider the problem of learning the Boolean formula $x_1 \vee (\neg x_1 \wedge \neg x_2)$. An SVM with polynomial kernel of degree 2 is trained on all 8 samples that can be drawn from the Boolean truth table for the variables $(x_1, x_2, x_3) \in \{0,1\}^3$. Afterwards, we compute FIRM both based on the trained SVM ($\mathbf{w}$) and based on the true labelings (y). The results are displayed in Figure 1.

Note that the raw SVM $\mathbf{w}$ can assign non-zero weights only to feature space dimensions (here, input variables and their pairwise conjunctions, corresponding to the quadratic kernel); all other features, here for example pairwise disjunctions, are implicitly assigned zero. The SVM assigns the biggest weight to x_2, followed by $x_1 \wedge x_2$. In contrast, for the SVM-based FIRM the most important features are $x_1 \wedge \neg x_2$ followed by $\neg x_{1/2}$, which more closely resembles the truth. Note that, due to the low degree of the polynomial kernel, the SVM not capable of learning the function "by heart"; in other words, we have an underfitting situation. In fact, we have $s(\mathbf{x}) = 1.\bar{6}$ for $(x_1, x_2) = (0, 1)$.

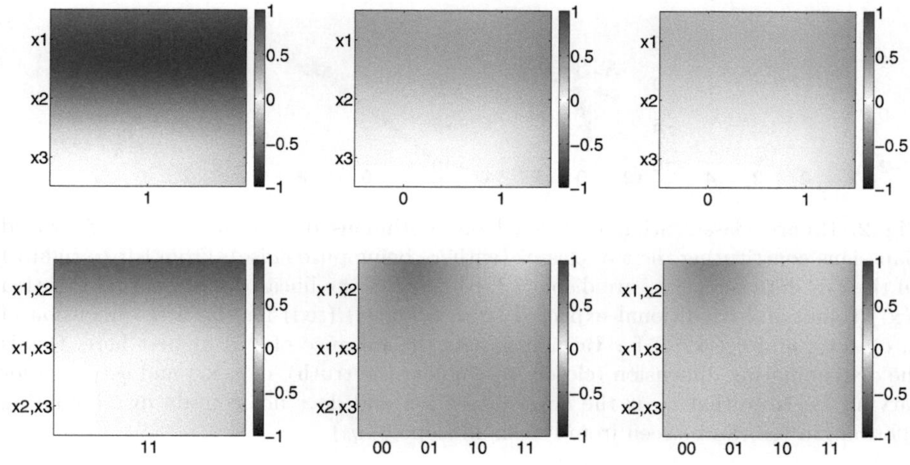

Fig. 1. FIRMs and SVM-w for the Boolean formula $x_1 \vee (\neg x_1 \wedge \neg x_2)$. The figures display heat maps of the scores, blue denotes negative label, red positive label, white is neutral. The upper row of heat maps shows the scores assigned to a single variable, the lower row shows the scores assigned to pairs of variables. The first column shows the SVM-w assigning a weight to the monomials x_1, x_2, x_3 and x_1x_2, x_1x_3, x_2x_3 respectively. The second column shows FIRMs obtained from the trained SVM classifier. The third column shows FIRMs obtained from the true labeling.

The difference in $y-$FIRM and SVM-FIRM underlines that — as intended — FIRM helps to understand the learner, rather than the problem. Nevertheless a quite good approximation to the truth is found as displayed by FIRM on the true labels, for which all seven 2-tuples that lead to true output are found (black blocks) and only $\neg x_1 \wedge x_2$ leads to a false value (stronger score). Values where $\neg x_1$ and x_2 are combined with x_3 lead to a slightly negative value.

3.2 Gaussian Data

Here, we analyze a toy example to illustrate FIRM for real valued data. We consider the case of binary classification in three real-valued dimensions. The first two dimensions carry the discriminative information (cf. Figure 2a), while the third only contains random noise. The second dimension contains most discriminative information and we can use FIRM to recover this fact. To do so, we train a linear SVM classifier to obtain a classification function $s(\mathbf{x})$. Now we use the linear regression approach to model the conditional expected scores q_i (see Figure 2b-d for the three dimensions). We observe that dimension two indeed shows the strongest slope indicating the strongest discriminative power, while the third (noise) dimension is identified as uninformative.

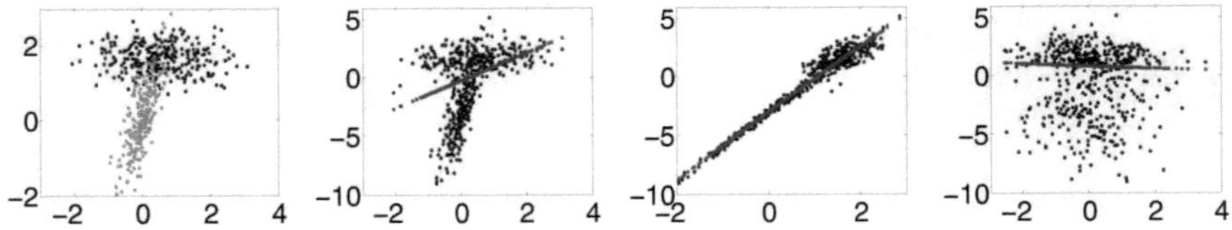

Fig. 2. Binary classification performed on continuous data that consists of two 3d Gaussians constituting the two classes (with x_3 being pure noise). From left to right a) Of the raw data set x_1, x_2 are displayed. b) Score of the linear discrimination function $s(\mathbf{x}_i)$ (blue) and conditional expected score $q_1((\mathbf{x}_i)_1)$ (red) for the first dimension of $\mathbf{x}$. c) $s(\mathbf{x}_i)$ and $q_2((\mathbf{x}_i)_2)$ for varying x_2. As the variance of q is highest here, this is the discriminating dimension (closely resembling the truth). d) $s(\mathbf{x}_i)$ and $q_3((\mathbf{x}_i)_3)$ for varying x_3. Note that x_3 is the noise dimension and does not contain discriminating information (as can be seen from the small slope of q_3).

3.3 Sequence Data

As shown above (Section 1.2), for sequence data FIRM is essentially identical to the previously published technique POIMs [17]. To illustrate its power for sequence classification, we use a toy data set from [9]: random DNA sequences are generated, and for the positive class the sub-sequence `GATTACA` is planted at a random position centered around 35 (rounded normal distribution with SD=7). As biological motifs are typically not perfectly conserved, the planted consensus sequences are also mutated: for each planted motif, a single position is randomly chosen, and the incident letter replaced by a random letter (allowing for no change for $\sim 25\%$ of cases). An SVM with WDS kernel [6] is trained on 2500 positive and as many negative examples.

Two analyses of feature importance are presented in Figure 3: one based on the feature weights $\mathbf{w}$ (left), the other on the feature importance Q (right). It is apparent that FIRM identifies the `GATTACA` feature as being most important at positions between 20 and 50, and it even attests significant importance to the strings with edit distance 1. The feature weighting $\mathbf{w}$, on the other hand, fails completely: sequences with one or two mutations receive random importance, and even the importance of the consensus `GATTACA` itself shows erratic behavior.

The reason is that the appearance of the exact consensus sequence is not a reliable feature, as is mostly occurs mutated. More useful features are substrings of the consensus, as they are less likely to be hit by a mutation. Consequently there is a large number of such features that are given high weight be the SVM. By taking into account the correlation of such short substrings with longer ones, in particular with `GATTACA`, FIRM can recover the "ideal" feature which yields the highest SVM score. Note that this "intelligent" behavior arises automatically; no more domain knowledge than the Markov distribution (and it is only 0-th order uniform!) is required. The practical value of POIMs for real world biological problems has been demonstrated in [17].

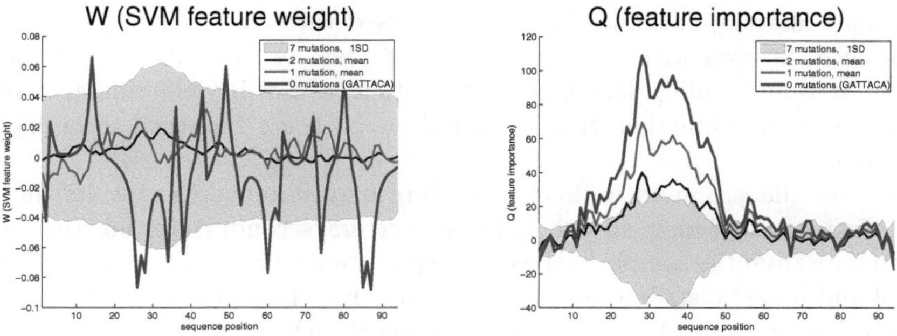

Fig. 3. Feature importance analyses based on (left) the SVM feature weighting **w** and (right) FIRM. The shaded area shows the ±1 SD range of the importance of completely irrelevant features (length 7 sequences that disagree to `GATTACA` at every position). The red lines indicate the positional importances of the exact motif `GATTACA`; the magenta and blue lines represent average importances of all length 7 sequences with edit distances 1 and 2, respectively, to `GATTACA`. While the feature weighting approach cannot distinguish the decisive motiv from random sequences, FIRM identifies it confidently.

4 Summary and Conclusions

We propose a new measure that quantifies the relevance of features. We take up the idea underlying a recent sequence analysis method (called POIMs, [17]) — to assess the importance of substrings by their impact on the expected score — and generalize it to arbitrary continuous features. The resulting *feature importance ranking measure* FIRM has invariance properties that are highly desirable for a feature ranking measure. First, it is "objective": it is invariant with respect to translation, and reasonably invariant with respect to rescaling of the features. Second, to our knowledge FIRM is the first feature ranking measure that is totally "universal", i.e. which allows for evaluating *any* feature, irrespective of the features used in the primary learning machine. It also imposes no restrictions on the learning method. Most importantly, FIRM is "intelligent": it can identify features that are not explicitly represented in the learning machine, due to the correlation structure of the feature space. This allows, for instance, to identify sequence motifs that are longer than the considered substrings, or that are not even present in a single training example.

By definition, FIRM depends on the distribution of the input features, which is in general not available. We showed that under various scenarios (e.g. binary features, normally distributed features), we can obtain approximations of FIRM that can be efficiently computed from data. In real-world scenarios, the underlying assumptions might not always be fulfilled. Nevertheless, e.g. with respect to the normal distribution, we can still interpret the derived formulas as an estimation based on first and second order statistics only.

While the quality of the computed importances does depend on the accuracy of the trained learning machine, FIRM can be used with any learning framework. It can even be used without a prior learning step, on the raw training data. Usually, feeding training labels as scores into FIRM will yield similar results as using a learned function; this is natural, as both are supposed to be highly correlated.

However, the proposed indirect procedure may improve the results due to three effects: first, it may smooth away label errors; second, it extends the set of labeled data from the sample to the entire space; and third, it allows to explicitly control and utilize distributional information, which may not be as pronounced in the training sample. A deeper understanding of such effects, and possibly their exploitation in other contexts, seems to be a rewarding field of future research.

Based on the unique combination of desirable properties of FIRM, and the empirical success of its special case for sequences, POIMs [17], we anticipate FIRM to be a valuable tool for gaining insights where alternative techniques struggle.

Acknowledgements. This work was supported in part by the FP7-ICT Programme of the European Community under the PASCAL2 Network of Excellence (ICT-216886), by the Learning and Inference Platform of the Max Planck and Fraunhofer Societies, and by the BMBF grant FKZ 01-IS07007A (ReMind). We thank Petra Philips for early phase discussion.

References

1. Bennett, K., Mangasarian, O.: Robust linear programming discrimination of two linearly inseparable sets. Optimization Methods and Software 1, 23–34 (1992)
2. Cortes, C., Gretton, A., Lanckriet, G., Mohri, M., Rostamizedeh, A.: Outcome of the NIPS 2008 workshop on kernel learning: Automatic selection of optimal kernels (2008)
3. Friedman, J.: Greedy function approximation: a gradient boosting machine. Annals of Statistics 29, 1189–1232 (2001)
4. Graf, A., Wichmann, F., Bülthoff, H.H., Schölkopf, B.: Classification of faces in man and machine. Neural Computation 18, 143–165 (2006)
5. Lanckriet, G.R.G., Cristianini, N., Ghaoui, L.E., Bartlett, P., Jordan, M.I.: Learning the kernel matrix with semidefinite programming. Journal of Machine Learning Research 5, 27–72 (2004)
6. Rätsch, G., Sonnenburg, S., Schölkopf, B.: RASE: Recognition of alternatively spliced exons in C. elegans. Bioinformatics 21(suppl. 1), i369–i377 (2005)
7. Schäfer, J., Strimmer, K.: A Shrinkage Approach to Large-Scale Covariance Matrix Estimation and Implications for Functional Genomics. Statistical Applications in Genetics and Molecular Biology 4(1), 32 (2005)
8. Schölkopf, B., Smola, A.J.: Learning with Kernels. MIT Press, Cambridge (2002)
9. Sonnenburg, S., Rätsch, G., Schäfer, C.: Learning interpretable SVMs for biological sequence classification. In: Miyano, S., Mesirov, J., Kasif, S., Istrail, S., Pevzner, P.A., Waterman, M. (eds.) RECOMB 2005. LNCS (LNBI), vol. 3500, pp. 389–407. Springer, Heidelberg (2005)

10. Sonnenburg, S., Rätsch, G., Schäfer, C., Schölkopf, B.: Large Scale Multiple Kernel Learning. Journal of Machine Learning Research 7, 1531–1565 (2006)
11. Strobl, C., Boulesteix, A., Kneib, T., Augustin, T., Zeileis, A.: Conditional variable importance for random forests. BMC Bioinformatics 9(1), 307 (2008)
12. Strobl, C., Boulesteix, A., Zeileis, A., Hothorn, T.: Bias in random forest variable importance measures: Illustrations, sources and a solution. BMC bioinformatics 8(1), 25 (2007)
13. Tibshirani, R.: Regression Shrinkage and Selection via the Lasso. Journal of the Royal Statistical Society, Series B 58(1), 267–288 (1996)
14. Üstün, B., Melssen, W.J., Buydens, L.M.: Visualisation and interpretation of support vector regression models. Analytica Chimica Acta 595(1-2), 299–309 (2007)
15. van der Laan, M.: Statistical inference for variable importance. The International Journal of Biostatistics 2(1), 1008 (2006)
16. Zien, A., Philips, P., Sonnenburg, S.: Computing Positional Oligomer Importance Matrices (POIMs). Res. Report; Electronic Publ. 2, Fraunhofer FIRST (December 2007)
17. Zien, A., Sonnenburg, S., Philips, P., Rätsch, G.: POIMs: Positional Oligomer Importance Matrices – Understanding Support Vector Machine Based Signal Detectors. In: Proceedings of the 16th International Conference on Intelligent Systems for Molecular Biology (2008)

OTTHO: On the Tip of My THOught

Pierpaolo Basile, Marco de Gemmis, Pasquale Lops, and Giovanni Semeraro

Dept. of Computer Science - University of Bari
Via E. Orabona, 4 - 70125 Bari - Italia
{basilepp,degemmis,lops,semeraro}@di.uniba.it

Abstract. This paper describes OTTHO (On the Tip of my THOught), a system designed for solving a language game called *Guillotine*. The rule of the game is simple: the player observes five words, generally unrelated to each other, and in one minute she has to provide a sixth word, semantically connected to the others. The system exploits several knowledge sources, such as a dictionary, a set of proverbs, and Wikipedia to realize a knowledge infusion process. The main motivation for designing an artificial player for Guillotine is the challenge of providing the machine with the cultural and linguistic background knowledge which makes it similar to a human being, with the ability of interpreting natural language documents and reasoning on their content. Our feeling is that the approach presented in this work has a great potential for other more practical applications besides solving a language game.

1 Background and Motivation

Words are popular features of many games, and they play a central role in many language games. A *language game* is defined as a game involving natural language in which word meanings play an important role. Language games draw their challenge and excitement from the richness and ambiguity of natural language. In this paper we present a system that tries to play the *Guillotine* game. The Guillotine is a language game played in a show on RAI, the Italian National Broadcasting Service, in which a player is given a set of five words (clues), each linked in some way to a specific word that represents the unique solution of the game. She receives one word at a time, and must choose between two different proposed words: one is correct, the other one is wrong. Each time she chooses the wrong word, the prize money is divided by half (the reason for the name *Guillotine*). The five words are generally unrelated to each other, but each of them is strongly related to the word representing the solution. Once the five clues are given, the player has one minute to provide the solution. Often the solution is not so intuitive and the player needs different knowledge sources to reason and find the correct word.

The literature classifies games related to the language in two main categories: 1) *word games* and 2) *language games*. *Word games* do not involve true language, because word meanings are not important. A typical example of word game is *Scrabble*, in which players take turn placing letters in a grid to form words.

W. Buntine et al. (Eds.): ECML PKDD 2009, Part II, LNAI 5782, pp. 710–713, 2009.

On the other side, *language games*, such as crosswords, strongly involve natural language, since word meanings play an important role. WebCrow [1] is the first solver for Italian crosswords and the first system that tackles a language game using the Web as knowledge base. OTTHO (On the Tip of my THOught) is the system we designed to solve the final stage of the *Guillotine* game. We assume that the five words are provided at the same time, neglecting the initial phase of choosing the words, that only concerns the reduction of the initial prize. With respect to other language games, our "questions" or "clues" are *single words*, and the answer is a *single word*. Co-occurrences of terms, providing the evidence of a strong relationship between words, is the key factor for finding a set of candidate words that likely contains the solution. Our system exploits different knowledge sources, such as a dictionary, an encyclopedia, a list of proverbs, etc., as described in the next section.

2 OTTHO

Guillotine is a *cultural* and *linguistic* game. Therefore, we need to define an extended knowledge base for representing the both the *cultural* and *linguistic* background knowledge of the player. After a deep analysis of the correlation between the clues and the solution, we chose to include the following knowledge sources, ranked according to the frequency with which they were helpful in finding the solution of the game:

1) Dictionary[1]: the word representing the solution is contained in the description of a lemma or in some example phrases using that lemma;
2) Encyclopedia[2]: as for the dictionary, the description of an article contains the solution, but in this case it is necessary to process a more detailed description of information;
3) Proverbs and aphorisms[3]: short pieces of text in which the solution is found very close to the clues.

In order to exploit these types of sources, it is necessary to organize data gathered from different sources of the same type, to process that information in order to extract and model relationships between words, and to define a reasoning mechanism that, given the clues, is able to select the correct solution of the game among a set of candidate words.

The above mentioned types of sources have different characteristics, therefore different heuristics should be used for building the model. Another important aspect is to define a uniform representation of that model. We decided to use a *term-term matrix* containing terms occurring in the modeled knowledge source; each cell of the matrix contains the weight representing the degree of correlation between the term on the row and the one on the column. The computation of

1 De Mauro Italian Dictionary: http://old.demauroparavia.it/
2 Italian Wikipedia: it.wikipedia.org
3 Italian proverbs: web.tiscali.it/proverbiitaliani

the weights is different for each type of knowledge source and takes into account several parameters, as described thoroughly in [2]. Natural language processing techniques adopted for modeling the knowledge sources do not depend on the language used, therefore OTTHO is language independent.

Modeling several knowledge sources realizes a sort of "knowledge infusion" process into the system, in order to create a memory of world facts and linguistic knowledge. An algorithm for retrieving the most appropriate pieces of knowledge associated with the clues is needed to replicate the cognitive mechanism of a human being in the most faithful way. We adopt a *spreading activation model* [3], which consists of a network data structure of nodes interconnected by links, that may be labeled and/or weighted and usually have directions. The processing is initiated by labeling a set of *source nodes* with activation weights and proceeds by iteratively propagating that activation to other nodes linked to the source nodes. For each iteration, a termination condition is checked in order to end the search process over the network. We decided to adopt this model as reasoning mechanism of OTTHO. In the network for Guillotine, nodes represent words, while links denote associations between words, obtained from the knowledge sources. The spreading activation is triggered by words given as clues. The activation of clues causes words with related meaning (as modeled in the sources) to become active. At the end of the weight propagation process, the most "active" words represent good candidates to be the solution of the game.

3 Into the Game

Our idea is to build a system which helps the user to solve the *Guillotine* game. OTTHO is able to provide some suggestions in order to make easier finding the correct answer. Figure 1 shows the OTTHO user interface. The clues are visualized on the left of the window, while the suggestions are shown within the text area on the bottom. A timer is displayed on the right of the window, which warns the player on time to provide the answer.

Fig. 1. OTTHO user interface

As shown in Figure 1, the clues are: *doppio (double), carta (paper), soldi (money), pasta and regalo (gift)*. The solution is **pacco (pack)**, because *carta (paper)* is a kind of pack (wad), the word pack composes a phrase with both the words *gift* and *double* (gift-wrapped package/double pack). Moreover, *un pacco di soldi* is an Italian expression which means "a lot of money", and pasta can be stored in a package. Notice that the word "pacco" appears in the list of suggestions provided by the system. The accuracy of OTTHO was measured on a dataset of $N = 50$ games attempted during the show by human players. A game is *solved* when the solution occurs in the Candidate Solution List (CSL) produced by OTTHO. Since the component that picks up the unique answer from the CSL is not yet complete, the evaluation is based on the first k words in the CSL (words are ranked according a relevance score computed by the spreading activation model). The accuracy is the percentage of games for which the solution occurs in the CSL. For $k = 10$, accuracy is 8%, while for $k = 100$, 40% of accuracy is achieved. These results are encouraging, given that the average accuracy of the human player on the dataset is about 16%.

4 Conclusions

We proposed an artificial player for a language game consisting in guessing a hidden word semantically related to five words given as clues. The system could be used also for implementing an alternative paradigm for *associative retrieval* on collections of text documents [4], in which an initial indexing phase of documents can *spread* further "hidden" terms for retrieving other related documents. The identification of hidden terms might rely on the integration of specific pieces of knowledge relevant for the domain of interest. This might represent a valuable strategy for several domains, such as search engine advertising, in which customers' search terms need to be matched with those of advertisers. Spreading activation can be also combined with document retrieval for semantic desktop search [5].

References

1. Ernandes, M., Angelini, G., Gori, M.: WebCrow: A Web-Based System for Crossword Solving. In: Veloso, M.M., Kambhampati, S. (eds.) Proc. 20th Nat. Conf. Artif. Intell., 17th Innov. Appl. Artif. Intell., pp. 1412–1417. AAAI Press/MIT Press (2005)
2. Semeraro, G., Lops, P., Basile, P., de Gemmis, M.: On the Tip of my Thought: Playing the Guillotine Game. In: Boutilier, C. (ed.) IJCAI 2009, Proc. of the 21th Int. Joint Conf. on Artificial Intelligence. Morgan Kaufmann, San Francisco (in press, 2009)
3. Collins, A.M., Loftus, E.F.: A spreading activation theory of semantic processing. Psychological Review 82(6), 407–428 (1975)
4. Crestani, F.: Application of Spreading Activation Techniques in Information Retrieval. Artificial Intelligence 11(6), 453–482 (1997)
5. Schumacher, K., Sintek, M., Sauermann, L.: Combining Fact and Document Retrieval with Spreading Activation for Semantic Desktop Search. In: Bechhofer, S., Hauswirth, M., Hoffmann, J., Koubarakis, M. (eds.) ESWC 2008. LNCS, vol. 5021, pp. 569–583. Springer, Heidelberg (2008)

Protecting Sensitive Topics in Text Documents with PROTEXTOR

Chad Cumby

Accenture Technology Labs, Chicago IL 60601, USA
chad.m.cumby@accenture.com

Abstract. This is a demonstration of a system for protecting sensitive topics present in text documents. Our system works in a privacy framework where the topic is characterized as a multiclass classification problem in a generative setting. We show how our system helps a user redact a document in a business setting to obscure what company the text pertains to, and show some experimental results on redacting the topic for a standard text classification data set.

1 Introduction

Many companies are just starting to realize that the databases that they have released to partners or shared with the public, after cursory data "masking" exercises, can in fact be linked against other public sources to recover the sensitive information that they thought had been obscured. Luckily the machine learning and data mining communities have been very active in devising means to protect individual records from discovery by attacks such as these. K-anonymity [5], L-diversity [2], and noise based methods are effective both in identifying records where possible privacy breaches might occur, and in suppressing or generalizing fields until these records are protected.

Unfortunately, general techniques to protect sensitive information in text and semi-structured data have not been developed. Text datasets are often multiple times larger than structured databases, and the risks associated with divulging sensitive information in them are no less large. For example, when AOL released an "anonymized" set of user search queries, which were linked against external data sources to discover sensitive information about the users, the release was considered to be a major privacy breach. The challenge we address with our system is not just to identify words that are correlated with the sensitive topic (see [1]), but to optimally suppress those words within a confusion set of topics (for example Ford vs GM), to control possible inferences about the topic.

2 PROTEXTOR System

We define the problem of obscuring a sensitive topic in a text document x in a generative multi-class classification framework. In our scenario, the trusted party possesses a corpus of documents X from which x is drawn, and the un-trusted

W. Buntine et al. (Eds.): ECML PKDD 2009, Part II, LNAI 5782, pp. 714–717, 2009.

party has some subset of X minus x. All the documents in X can be classified into a finite set Y of classes, with the true class of x being y. The goal of our redaction method is to perturb x in such a way as to reduce the likelihood of y and increase the likelihood of some k closest of the other classes.

To approximate a model the attacker might have of Y, we build a Naive Bayes classifier from X by creating a bag-of-words example from each document and training using the maximum likelihood method.

In our system, as shown in the screenshot below, we would then like to present to the user along with the document a suggested list of the most important words to suppress in order to make y confusable with at least k other classes of documents. We have experimented with several ordering metrics based on feature selection techniques from the text categorization domain, along with a new metric **PosMin** designed to explicitly minimize the posterior likelihood $P(y|x)$. We define this metric relative to the k closest classes to y (called K) as measured by the KL divergence $KL(P(y)||P(y'))$ for each $y' \in Y$.

We'd like to minimize $log(P(x|y))$. Notice that $log(P(y|x)) = log(P(x|y)) + log(P(y)) - log(P(x)) = log(P(x|y)) + log(P(y)) - log(\sum_{y' \in Y} P(x|y') * P(y'))$. Here we re-estimate $P(x)$ relative to the set $K \cup y$ to obtain $PosMin(x, y) = log(P(x|y)) + log(P(y)) - log(\sum_{y' \in K \cup y} P(x|y') * P(y'))$. By varying K for the same document, different sets of words will be suggested in order to vary the number of "guesses" necessary for an attacker to pick the correct class. For example, below is a table of the top 10 words picked by our metric for a document taken from the alt.atheism class from the 20 newsgroups dataset. In this dataset, the top 5 closest classes to alt.atheism in order are: soc.religion.christianity, talk.religion.misc, talk.politics.misc, talk.politics.mideast, & talk.politics.guns. The list of words to confuse the document with soc.religion.christianity is very different than the one to confuse it with the whole top 5. In particular the entire $k = 1$ list does not contain the words *jesus* or *christianity*, since these words do not distinguish alt.atheism documents from soc.religion.christianity documents.

As we show in Sec.3 in a controlled evaluation, compared to the other feature selection metrics we tested, automated redacting with our new $PosMin$ metric creates documents which are relatively unclassifiable up to k guesses for classifiers trained on the unobscured documents.

Implementation. The PROTEXTOR implementation at Accenture has been created as an add-in panel for Microsoft Office 2007. In many large companies to reuse business materials from project to project it is necessary to remove client identifiable information. In order to comply with client contracts it is insufficient to simply remove the canonical name of the client. Additionally a user must remove abbreviations of the client as well as any uniquely identifying information that an attacker could use to deduce the client. So, in the first implementation we have created topic models for 450 of Accenture's clients from documents in a 350000 document corpus.

When a document is loaded, the add-in builds the word vector of the existing text and returns the most likely client classification along with the suggested words to redact to resemble the closest client ($k = 1$). The user can adjust the

k=1	k=5
religion	faith
dogma	christian
system	dogma
encourages	system
toronto	encourages
humans	beliefs
beliefs	humans
genocide	secular
philosopher	philosopher
prison	christianity

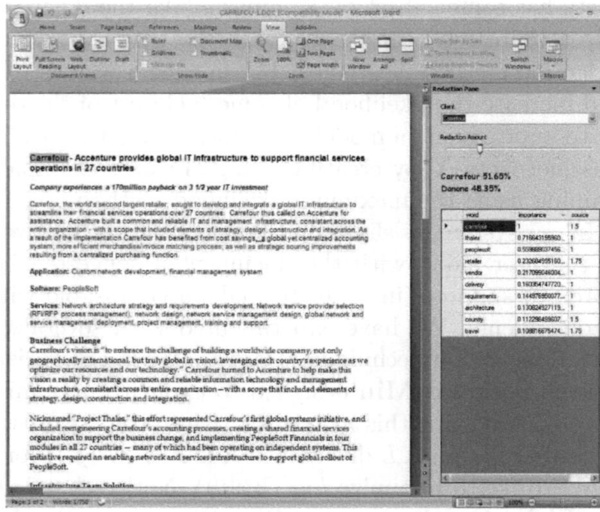

Fig. 1. (*Left*) Top 10 words to confuse alt.atheism doc with $k = 1$ and $k = 5$ other classes. (*Right*) Screenshot of PROTEXTOR system.

k parameter via a slider to change the suggested redaction list. For example if the true client is Carrefour and the next closest client model is for Danone, the list contains the word "Carrefour" and project specific terms that distinguish the document from Danone. If the user increases the k parameter to indicate that they want to obscure the document to resemble additional clients, it might suggest more general words such as "supermarket" or "france". As the user edits the text, the add-in will update the likelihoods returned for the closest client classes.

3 Evaluation

Here we describe a controlled experiment done in the context of the PROTEXTOR system. We set out to test what the effects of our redaction process with different parameter settings would be on learned classifiers. If our system can foil these classifiers, then an attacker scanning for sensitive information in a corpus of masked documents using them would be deterred. Also, the performance of learned classifiers seems to correlate with human performance in defeating our redaction (which we explore in further work).

The dataset used in the experiments was the Industry Sector dataset introduced in [3]. It contains 6440 documents corresponding to company websites in a two level hierarchy of industry classes. We trained naive Bayes classifiers for the 12 top level classes using a standard stoplist, to serve as the class models for creating redaction word lists.

In this experiment, we created redacted sets of documents using a simple greedy redaction algorithm as follows: Given an input k, for each document x of

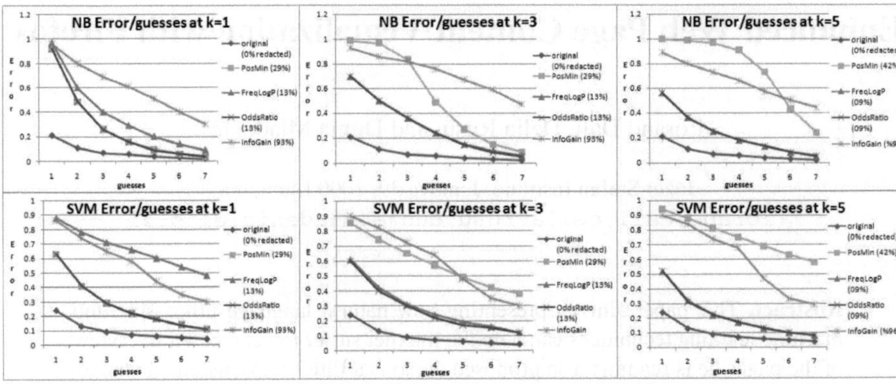

Fig. 2. All results for the Industry Sector experiment

class y create an ordered list of words to redact using the four metrics below. For each list, suppress words from x until the log-likelihood (LL) of y is less than the LL of the k closest classes. In addition to the PosMin metric as described in Sec 2, we used three standard feature selection metrics described in [4]: InfoGain, OddsRatio, and FreqLogP. Using a leave-one-out procedure we trained new naive Bayes and SVM classifiers on the corpus minus each redacted document, and tested on the redacted sets. In the results below we see the desired performance as very high error rates occur for both test classifiers up to k guesses, and then a sharp drop off as the true class becomes clear. Also our PosMin metric seems to give the best performance profile at most k settings, and exhibits the best tradeoff given the percentage of the document redacted.

Acknowledgements. Thanks to Yaron Rachlin for all his ideas and hard work.

References

1. Chow, R., Golle, P., Staddon, J.: Detecting privacy leaks using corpus-based association rules. In: Proceedings of KDD 2008 (2008)
2. Kifer, D., Gehrke, J.: l-diversity: Privacy beyond k-anonymity. In: Proceedings of ICDE 2006 (2006)
3. McCallum, A.K., Rosenfeld, R., Mitchell, T.M., Ng, A.Y.: Improving text classification by shrinkage in a hierarchy of classes. In: Proceedings of ICML 1998, pp. 359–367 (1998)
4. Mladenic, D.: Feature subset selection in text-learning. In: Nédellec, C., Rouveirol, C. (eds.) ECML 1998. LNCS, vol. 1398. Springer, Heidelberg (1998)
5. Sweeney, L.: Achieving k-anonymity privacy protection using generalization and suppression. International Journal on Uncertainty, Fuzziness and Knowledge-based Systems 10 (2002)

Enhanced Web Page Content Visualization with Firefox

Lorand Dali, Delia Rusu, and Dunja Mladenić

Jožef Stefan Institute, Jamova 39, 1000 Ljubljana
{lorand.dali,delia.rusu,dunja.mladenic}@ijs.si

Abstract. This paper aims at presenting how natural language processing and machine learning techniques can help the internet surfer to get a better overview of the pages he is reading. The proposed demo is a Firefox extension which can show a semantic graph of the text in the page that is currently loaded in the browser. The user can also get a summary of the web page she is looking at by choosing to display only the more important nodes in the semantic graph representation of the document, where importance of the nodes is obtained by machine learning techniques.

Keywords: natural language processing, document visualization, triplet ranking, Firefox extension.

1 Introduction

The popularity and size of the Internet has greatly increased lately. So has the time people spend surfing. Moreover, a significant part of the traditional media has been replaced by online content. While it is very good that information can be retrieved so quickly and easily, the increased amount of time needed to be spent online by the users to process all this information raises important health and sociological problems. We think that a tool which could render most of the text contained in web pages easier to grasp, would greatly improve the user experience, and decrease the time needed to get at least a first impression of the page content. In our opinion such a tool should have the following features:

- provide a graphical representation of text, as most people prefer pictures over text
- emphasize and display the most important content (e.g., text summary)
- the user should be able to adjust the amount of information he wants to see
- the user should be able to see the content of the summary in the order in which he would see it during a normal read of the page
- the tool should be available and easy to use for the most users on the internet, and be applicable to most of the web pages

We believe that the tool presented in this paper satisfies the abovementioned needs. It is implemented as an extension to one of the most popular browsers, Firefox, and is designed to be used with any web page which has textual content.

The screenshot in Fig. 1 shows a typical usage scenario. The user wants to read a news article online and with the help of our browser extension he can also get a graphical representation of the text in the article.

W. Buntine et al. (Eds.): ECML PKDD 2009, Part II, LNAI 5782, pp. 718–721, 2009.

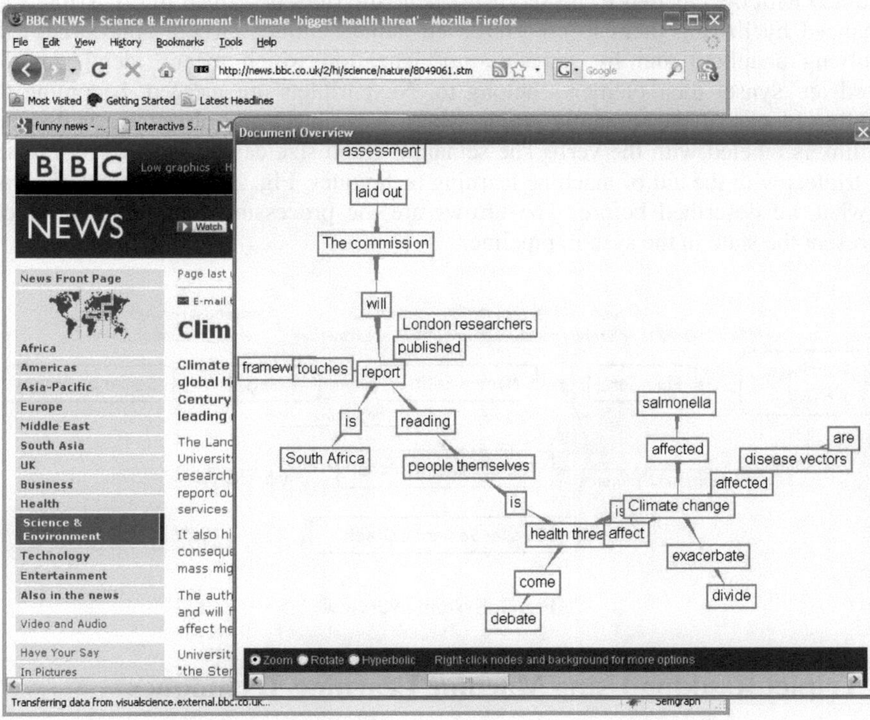

Fig. 1. Screenshot showing a web page and its visualization as a graph

The natural language processing and machine learning techniques used are exposed to the Firefox extension through RESTful web services and the visualization is done using the TouchGraph[1] graph drawing library. In the following sections we shall present the general architecture and the building blocks of the system in more detail.

2 System Architecture

In this section we describe the general architecture of our system, as a pipeline [1][2]. First the HTML code of the web page under consideration is preprocessed to obtain plain text which is split into sentences. For removing the HTML markup we use the Python HTML parsing library Beautiful Soup[2] whose main advantage is that it can handle badly formatted HTML, whereas splitting the text into sentences is done with the Natural Language Toolkit's[3] sentence detector. After that named entities and triplets are extracted from the sentences. For extracting named entities (people, places and organizations) we use GATE[4] (General Architecture for Text Engineering), while (*subject – verb – object*) triplets are extracted from the parse tree of each sentence

[1] http://sourceforge.net/projects/touchgraph
[2] http://www.crummy.com/software/BeautifulSoup/
[3] http://www.nltk.org/
[4] http://gate.ac.uk/

obtained with the OpenNLP[5] parser, using heuristic rules described in [3]. Triplets are enhanced by linking them to co-referenced named entities (where appropriate), by resolving anaphora and by semantic normalization which means identifying the WordNet[6] synset their elements belong to. Next triplets are merged, becoming the building blocks of the semantic graph, where the subject and object are nodes, and the link is labeled with the verb. The semantic graph size can be reduced by ranking the triplets with the aid of machine learning techniques. Fig. 2 shows a block diagram of what we described before. The arrows are the processing tasks and the blocks represent the state in the system pipeline.

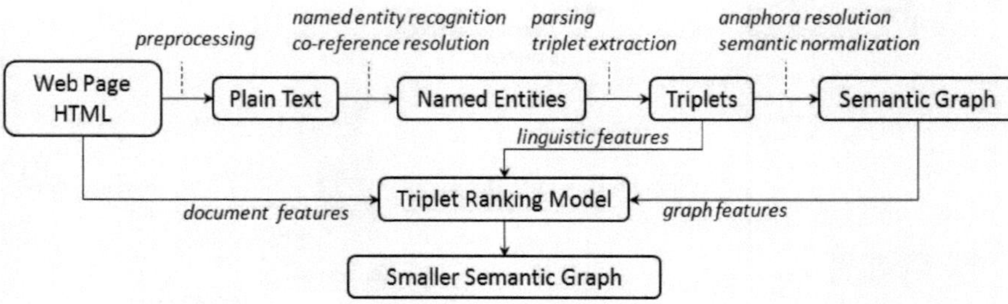

Fig. 2. System overview

3 Triplet Ranking Using Machine Learning Techniques

Ranking those triplets which compose the semantic graph, in decreasing order of importance has the goal of reducing the size of the semantic graph while retaining the most important nodes. From the list of ranked triplets a smaller semantic graph of any chosen size can be built.

The ranking method tried consists of training an SVM model for binary classification of triplets into important and not important. The 69 features used are of three kinds, as already proposed in [1]: *document features* (e.g. position of the sentence in the document, position of the triplet in the sentence, words in the triplet elements), *linguistic features* (e.g. part of speech tags, location of the triplet in the parse tree) and *graph features* (e.g. hub and authority weights, page rank, node degrees, connected components). For training the linear SVM model and for evaluating the triplet ranking, we use the DUC (Document Understanding Conferences)[7] datasets from 2002 and 2007, respectively. The DUC datasets contain news articles from various sources like Financial Times or Wall Street Journal. The 2002 dataset comprises 300 newspaper articles on 30 different topics and for each article we have a 100 word human written abstract. The DUC 2007 dataset comprises 250 articles for the update task and 1125 articles for the main task, part of the AQUAINT dataset[8]; the articles are grouped in clusters and 4 NIST assessors

[5] http://opennlp.sourceforge.net/
[6] http://wordnet.princeton.edu/
[7] http://duc.nist.gov/
[8] http://www.ldc.upenn.edu/Catalog/CatalogEntry.jsp?catalogId=LDC2008T25

manually create summaries (of 100 or 250 words) for the documents in the clusters. As training data we used the DUC 2002 articles, as well as the DUC 2007 main task articles, while the DUC 2007 update task articles were used for testing. We extracted triplets from the training and test data, and learned to classify if a triplet will appear in the summary or not. If we order the classified triplets by the confidence weights of their class we obtain a ranked list of triplets which is exactly what we need. The approach was also used to automatically create document summaries [2]. It was evaluated by comparing the results to the ones obtained by other systems competing in the DUC 2007 update task. Other competing systems that were ranked close to ours generate compressed versions of source sentences as summary candidates and use weighted features of these candidates to construct summaries 4, or learn a log-linear sentence ranking model by maximizing three metrics of sentence goodness 5 (see [2] for more details on the evaluation outcome).

4 Conclusions

We have presented a tool which is meant to make the reading of web pages easier by providing interactive graphical visualization of the read document. To achieve this task we have used natural language processing and machine learning techniques such as: named entity recognition, triplet extraction, co-reference and anaphora resolution and triplet ranking. The tool is implemented as a Firefox extension, which makes it available to a wide range of users browsing a big variety of web sites.

Acknowledgements

This work was supported by the Slovenian Research Agency and the IST Programme of the EC under NeOn (IST-4-027595-IP) and PASCAL2 (IST-NoE-216886).

References

1. Leskovec, J., Grobelnik, M., Milic-Frayling, N.: Learning Sub-structures of Document Semantic Graphs for Document Summarization. In: Workshop on Link Analysis and Group Detection (LinkKDD) at KDD 2004, Seattle, USA, August 22 – 25 (2004)
2. Rusu, D., Fortuna, B., Grobelnik, M., Mladenić, D.: Semantic Graphs Derived From Triplets With Application In Document Summarization. Informatica Journal (2009)
3. Rusu, D., Dali, L., Fortuna, B., Grobelnik, M., Mladenić, D.: Triplet Extraction from Sentences. In: Proceedings of the 10th International Multiconference "Information Society - IS 2007", Ljubljana, Slovenia, October 8 – 12, pp. 218–222 (2007)
4. Madnani, N., Zajic, D., Dorr, B., Ayan, N.F., Lin, J.: Multiple Alternative Sentence Compressions for Automatic Text Summarization. In: Proceedings of the Document Understanding Conference, DUC (2007)
5. Toutanova, K., Brockett, C., Gamon, M., Jagarlamudi, J., Suzuki, H., Vanderwende, L.: The PYTHY Summarization System: Microsoft Research at DUC 2007. In: Proceedings of the Document Understanding Conference, DUC (2007)

ClusTR: Exploring Multivariate Cluster Correlations and Topic Trends

Luigi Di Caro[1] and Alejandro Jaimes[2]

[1] Universita' di Torino, Torino, Italy
[2] Telefonica Research, Madrid, Spain
dicaro@di.unito.it, ajaimes@tid.es

Abstract. We present a demonstration of ClusTR, a highly interactive system for exploring relationships between different clusterings of a dataset and for viewing the evolution in time of topics (e.g., tags associated with objects in the dataset) within and across such clusters. In particular, ClusTR allows exploration of generic multi-dimensional, text labeled and time sensitive data.

1 Introduction

In many applications (e.g., bioinformatics, telecommunications, marketing, etc.) the focus of Data Mining is often on Knowledge Discovery (KD), largely because of the complexity of the data and the heterogeneity in terms of attributes and possible groupings of objects. The same reasons that create the needs for a focus on Knowledge Discovery, however, make it a very challenging process. On one hand, it is an activity that cannot be fully automated—by definition, the discovery process aims to gain insight from the data and therefore implies a user-in-the-loop approach. On the other hand, except in very simple cases, insight cannot be gained without often significant automatic processing intertwined in the human analysis loop. The combination of automatic techniques and human analysis is thus central to the KD process and the two could offer new potentiality in both areas: presenting information in a way that can exploit human perceptual skills, and user input to improve automatic results. Thus, a large part of the KD process focuses on discovering and understanding of not only similarities between objects, but also on how different groupings of objects, based on different attributes relate to each other. Furthermore, in cases in which time is one of the attributes and objects have textual attributes, it is also often important to discover the evolution of such textual attributes (e.g., topics or categories) over time with respect to the possible groupings of the objects.

In this paper, we address these problems with ClusTR, a highly interactive system for exploring relationships between different clusterings of a dataset and for viewing the evolution in time of topics (e.g., tags associated with objects in the dataset) within and across such clusters. In particular, ClusTR allows exploration of generic multi-dimensional, text labeled and time-sensitive data. Our system follows the goals of visual analysis tasks [9], as well as user interface design principles (e.g., overview first, zoom-and-filter, then details-on-demand and others [3]), placing strong emphasis on a highly interactive environment that combines automatic techniques with human analysis. In particular, we apply the concept of dynamic queries [3],

W. Buntine et al. (Eds.): ECML PKDD 2009, Part II, LNAI 5782, pp. 722–725, 2009.

which continuously update the data that is visualized: user actions work instantly, triggering clustering, filtering, and allowing other operations in the database.

Related work. There is a large body of work on interactive KD, the most relevant being in the areas of interactive exploration and queries, data visualization and temporal data exploration. Related systems and approaches include [2][1][10][11][4] [6][7][8], and [14].

2 ClusTR

Our system works with any kind of multi-dimensional data, but some of the functionality was implemented specifically for data that includes time and textual features. For the demonstration we use a freely available Flickr image dataset [13] containing 25,000 images where each image has the following features: textual tags, geo-tag coordinates, and camera settings (aperture, exposure, ISO, and a Flash use flag). In the current implementation we use EM for clustering, which assigns a probability distribution to each instance that indicates the probability of it belonging to each of the clusters. The value of correlation between two clusters is calculated using the Jaccard coefficient (the size of the intersection divided by the size of the union).

Interaction with the system takes place as follows. The first step is selection of the data for analysis. After data is loaded into ClusTR, exploration consists of two stages: (1) selection of features to create multiple cluster groups (Figure 1); and (2) exploration of correlations between resulting clusters and analysis of time-evolution of topics within clusters (Figure 2). We describe the two stages below.

Given k features (figure 1) in the dataset, in the first interaction screen the system creates a correlation panel of k x k squares. The user selects features by drawing sets of dots (D) and lines (L), whose combination determine the features to use in the creating cluster groups. In figure 1, for example, two cluster groups are created, one using flash, ISO and one using ISO, aperture, and exposure. The two cluster groups are created on-line when the user presses the "next" button (each cluster group contains the same data). Each dot and each line drawn by the user defines one combination of features to create a cluster group.

The drawing style of interaction used is meant to intuitively allow quick selection of features instead of cumbersome interaction with boxes or lists. ClusTR plots the value distributions for the dataset for each feature (Figure 1, right) to give the user quick

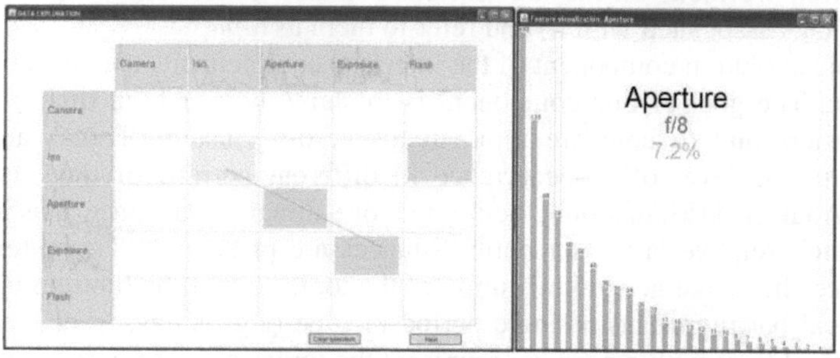

Fig. 1. The panel for selecting the features to create cluster groups

insight into the variability of the data in each feature space. The correlation layout has the advantage of allowing us to graphically represent correlations between features.

In the second stage, each of the clusters is represented by a solid circle and a label on the outside of a circle (Figure 2). The circular interface aims at showing correlations between clusters, while the bottom panel (bottom of Figure 2) shows topic trends. Each cluster group is represented by a different color. Horizontal bars on the outside of the circle associated with each cluster indicate the size of the cluster relative to its corresponding group. Lines between clusters represent correlations (thicker edge implies higher correlation). User interaction takes place in the following ways:

- Clicking anywhere on the screen, holding the button and moving the mouse up or down: the threshold for correlation is modified and shown. Lines dynamically (immediately) appear, disappear, increase or decrease thickness.
- Clicking on a cluster (solid circle) shows details of the cluster (topic evolution described next, and a tag cloud with their most frequent values).

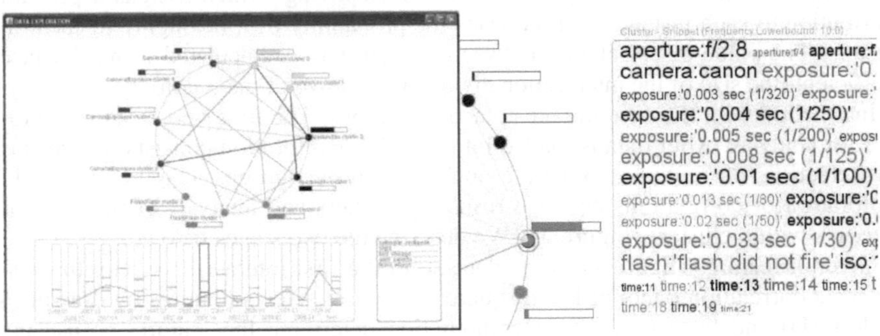

Fig. 2. Visualization of correlation strengths and evolution of topics for the clustered data. When the user selects a specific cluster, the system shows a dynamic tag cloud with the most frequent contents in the selected cluster.

Note that for clusters with the same color there's no edge/correlation given that they belong to the same clustering group so they contain distinct instances. For data that contains a textual feature (e.g., tag, free-text, etc.), it is possible to consider such values to be topics, or to extract them using a variety of techniques. For this demonstration we focus on the tag features available in the data (each image has a number of tags associated with it) and refer to them as *topics*.

The topic evolution component of the interface complements the cluster correlation exploration. The goal of this component is to show the trends in time of the topics within clusters and to compare topic trends across clusters. Topics are analyzed according to the level of co-occurrence in different periods of time using Latent Semantic Analysis [15], mapping them into a one-dimensional space **T** as in [12], in a way that their relative latent semantic distances are preserved. The bottom panel of the interface shows the general evolution of the topics over time (bottom of Figure 2). Each vertical bar represents a single period of time (e.g., a day, week, season, etc.). The horizontal lines within each bar represent the topics of items in the cluster and the vertical location of each line represents its value in **T**. The user can click over one single period of time to have an insight of the topics/labels contained in it. In addition, the system allows the user to explore the differences in topic trends for different clusters. When the user clicks on a cluster in the circle, its topics trace is projected in the topics evolution panel. In this way, the user can visually relate clusters to time, seeing how their topic evolution overlaps.

3 Conclusions and Future Work

We have presented a demonstration of ClusTR, a highly interactive system for exploring relationships between different clusterings of a dataset and for viewing the evolution of topics within and across clusters. The main features of the system are its interactive-query mechanisms, the functionality to select features for clusterings using a 2D correlation space, and allowing the dynamic exploration of topics over time for multiple clusters. Future work includes further addition to the functionality (e.g., adding several clustering methods, more visualization options, etc.), a user study, case studies with other datasets, etc.

References

1. Ahn, J.-w., Brusilovsky, P., Grady, J., He, D., Syn, S.Y.: Open user profiles for adaptive news systems: help or harm? Int. conf. WWW 2007 (2007)
2. Olsen, K.A., Korfhage, R.R., Sochats, K.M., Spring, M.B., Williams, J.G.: Visualisation of a document collection: The VIBE system. In: Inf. Proc. and Managem. (1993)
3. Shneiderman, B.: Designing the User Interface. Addison Wesley, Reading (1997)
4. Zytkow, J.M., Rauch, J.: Circle Graphs: New Visualization Tools for Text-Mining. In: Żytkow, J.M., Rauch, J. (eds.) PKDD 1999. LNCS (LNAI), vol. 1704, pp. 277–282. Springer, Heidelberg (1999)
5. Kliger, J.: Methods for Visualizing User Models. MIT Media Lab, Cambridge
6. Don, A., Zheleva, E., Gregory, M., Tarkan, S., Auvil, L., Clement, T., Shneiderman, B., Plaisant, C.: Discovering interesting usage patterns in text collections: Integrating text mining with visualization. HCIL Technical report 2007-08
7. Seo, J., Shneiderman, B.: A Rank-by-Feature Framework for Unsupervised Multidimensional Data Exploration Using Low Dimensional Projections. In: Infovis 2004 (2004)
8. Inselberg, A.: Visual Data Mining with Parallel Coordinates. Comp. Statistics (1998)
9. Nocke, T., Schumann, H.: Goals of Analysis for Visualization and Visual Data Mining Tasks. In: CODATA Workshop Information, Presentation and Design (2004)
10. Catarci, T., Costabile, M.F., Levialdi, S., Batini, C.: Visual Query Systems for Databases: A Survey. Journal of Visual Languages and Computing 8
11. Derthick, M., Kolojejchick, J., Roth, S.F.: An Interactive Visual Query Environment for Exploring Data
12. Qi, Y., Candan, K.S.: CUTS: CUrvature-Based Development Pattern Analysis and Segmentation for Blogs and other Text Streams. In: Hypertext 2006 (2006)
13. Hu., M.J., Lew, M.S.: The MIR Flickr Retrieval Evaluation
14. Berchtold, S., Jagadish, H.V., Ross, K.A.: Independence Diagrams: A Technique for Visual Data Mining. AT&T Laboratories (1998)
15. Deerwester, S., Dumais, S., Furnas, G., Harshman, R., Landauer, T., Lochbaum, K., Streeter, L.: Computer Information Retrieval using Latent Semantic Structure

Visual OntoBridge: Semi-automatic Semantic Annotation Software

Miha Grcar and Dunja Mladenic

Jozef Stefan Institute, Dept. of Knowledge Technologies, Jamova cesta 39,
1000 Ljubljana, Slovenia
{Miha.Grcar,Dunja.Mladenic}@ijs.si

Abstract. Machine learning methods have been successfully used for data labeling, also referred to as data annotation, either in semi-automatic or fully automatic way. We present a system for semi-automatic annotation of Web service schemas and other resources with the motivation to support efficient browse and search through those resources and to enable efficient composition and execution of Web services. The presented system, Visual OntoBridge (VOB), provides a graphical user interface and employs a set of machine learning algorithms to support the user in the annotation task.

Keywords: annotation, ontologies, graphs, machine learning, text mining, PageRank, Visual OntoBridge.

1 Introduction and Motivation

In order to support efficient browse and search through resources and to enable efficient composition and execution of Web services, these resources need to be semantically annotated. In our work, the semantic annotation is defined as a set of interlinked domain-ontology instances being associated with the resource being annotated. For example, let us assume that our resource is a database table. We want to annotate its fields in order to provide compatibility with databases from other systems. Further on, let us assume that this table has a field called "employee_name" that contains employee names (as given in Fig. 1, left side). On the other hand, we have a domain ontology containing knowledge and vocabulary about companies (an excerpt is given in Fig. 1, right side). In order to state that our table field in fact contains employee names, we first create an instance of the domain-ontology concept *Name* and associate it with the field. We then create an instance of *Person* and link it to the instance of *Name* via the *hasName* relation, and an instance of *Company* and link it to the instance of *Person* via the *hasEmployee* relation. Such annotation (shown in the middle in Fig. 1) indeed holds the desired semantics: the annotated field contains names of people which some company employs (i.e. names of employees). Formulating annotations in one of the ontology-description languages (e.g. Web Service Modeling Language <http://www.wsmo.org/wsml/>) is not a trivial task and requires specific expertise. Therefore, the users would benefit from a system that allows them to visualize the resource and the domain ontology (much like this is done in Fig. 1), create instances by clicking on concepts, and interlink them by

W. Buntine et al. (Eds.): ECML PKDD 2009, Part II, LNAI 5782, pp. 726–729, 2009.

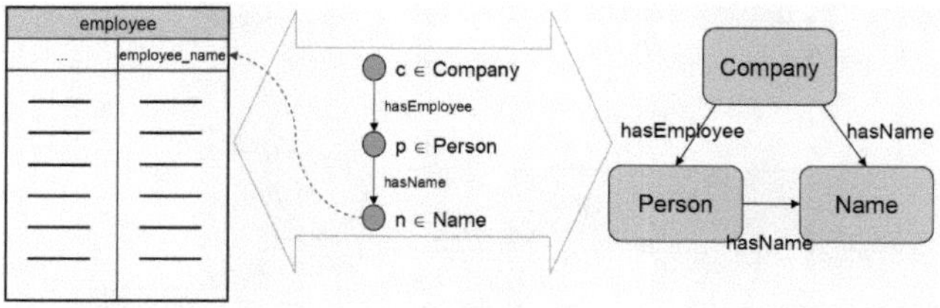

Fig. 1. Annotation as a "bridge" between the resource and the domain ontology

"drawing" relations between them. In addition, the users would benefit from non-trivial "building blocks" provided to them by the software with respect to some Google-like queries.

With this in mind, we have developed Visual OntoBridge [1], a system that provides a graphical user interface and a set of machine learning algorithms that support the user in the annotation task.

2 Software, Approach, Related Systems

When the application is launched, the domain ontology is visualized as a graph of concepts. As already said, VOB provides functionality to create several domain ontology instances and interlink them as appropriate to formulate an annotation (see the screenshot in Fig. 2). Establishing annotations manually is not a trivial task, especially if the domain ontology contains a large amount of entities and/or the user is not fully familiar with the conceptualizations in the domain ontology. VOB provides an advanced tool for querying the domain ontology with the purpose of finding the appropriate concepts and triples. The user is able to enter a set of Google-like natural-language queries and the system then provides her with two lists of ontology entities: the list of proposed concepts and the list of proposed triples. VOB employs text mining techniques, Page Rank-like algorithms, and consults a Web search engine to populate the two lists of recommended building blocks.

Let us briefly summarize the ontology-querying algorithm. The domain ontology is first represented as a graph in which concepts and triples are represented with vertices and edges represent relations. The user's natural-language query is also represented as a vertex. This is illustrated in Fig. 3. Next, each graph vertex (including the user's query) is assigned a set of documents. The documents are obtained by querying a Web search engine, the search term being either the label of the corresponding ontology entity or the user's query. The set of documents at each particular vertex is converted into TF-IDF vectors and the corresponding centroid vector is computed. The cosine similarity measure is now used to determine similarities between the centroid of the user's query and the rest of the graph. The computed similarity scores are used to weight the edges of the graph. When the graph is properly weighted, PageRank is employed to rank vertices (i.e. concepts and triples) according to the relevance to the query. The vertex representing the query is therefore used as the source vertex for PageRank. Since every concept and every triple has now been ranked by PageRank, it is possible to populate the two required lists of annotation building blocks and present these to the user. Many details and insights can be found in [1].

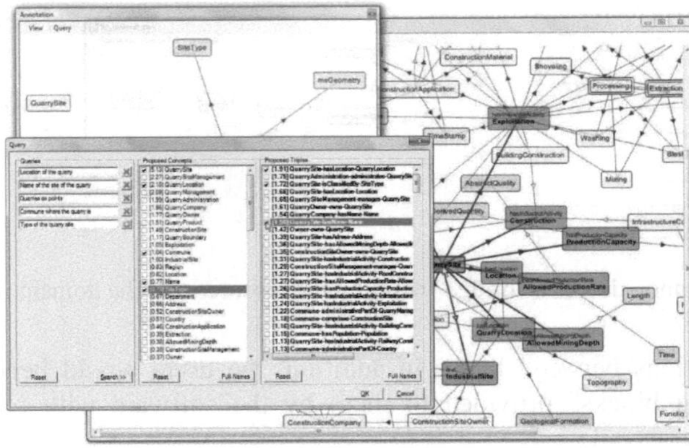

Fig. 2. Visual OntoBridge graphical user interface

Conceptually, our approach can be compared to that of QuestIO [5], a natural-language interface to ontologies. The authors of QuestIO argue that related systems—such as SemSearch, AquaLog, and Orakel—are either keyword-based or require full-blown and correctly phrased questions. Furthermore, such systems are usually domain-specific and require the user to be familiar with the domain. QuestIO does not share these drawbacks. It analyzes a question—written in a natural language—and creates a SeRQL[1] query from it. Their matching algorithm relies on morphological analysis, POS tagging, and string matching. In contrast to this, our system is not a question-answering system but rather a triple-retrieval system. It does not rely on language analysis algorithms, which makes it suitable for languages that do not yet have adequate support for natural-language processing (e.g. Slovene). Furthermore, our approach is "softer" in the sense that it will always provide answers (i.e. retrieve triples) even if there are no string similarities between query terms and ontology entity labels.

3 Discussion and Experimental Results

The presented software was developed and successfully employed in the European project SWING[2]. In SWING, we were annotating geospatial Web services (Web Feature Services or WFS's [3]). More accurately, we were annotating their schemas (i.e. capability documents) to achieve semantic interoperability of services for the purpose of discovery, composition, and execution.

During the project, several WFS were annotated manually (by formulating annotations in the WSML language). For the purpose of evaluating the techniques discussed in Section 0, we asked the domain experts at Bureau of Geological and Mining Research (BRGM, France) to provide us with natural-language queries with which they would hope to retrieve building blocks for these annotations.

[1] Sesame RDF Query Language <http://en.wikipedia.org/wiki/RDF_query_language>

[2] Semantic Web Service Interoperability for Geospatial Decision Making <http://www.swing-project.org/>

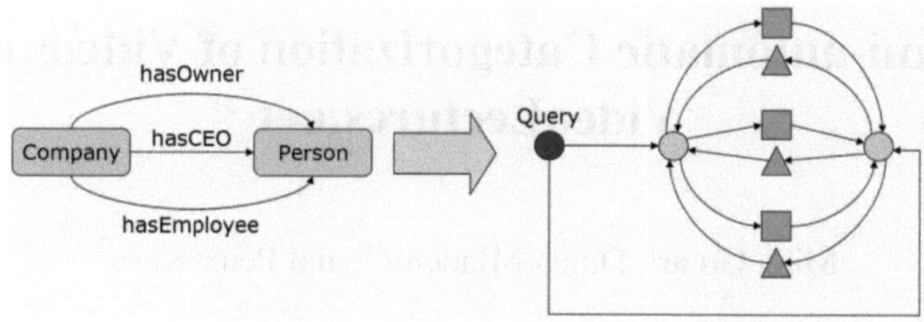

Fig. 3. Representing ontologies as graphs

We measured the Area Under the ROC Curve (AUC) to evaluate the lists of recommendations produced by the algorithm. The rewarding fact is that we managed to significantly beat the baselines which were based solely on term matching (term matching algorithm is described in [4]). We have increased the average AUC for 5.48% (from 91.46% to 96.94%) on concepts and for 3.18% on triples (from 93.16% to 96.34%). This presents a big difference from the application point of view. Roughly speaking, the graph-based algorithms are twice as good as the baseline algorithm. Also, the user is able to interact with the system and reformulate queries to achieve even better results. To support this claim, we computed the average AUC by taking, for each annotation, only the most successful annotator into account. The average AUC on the triples rose to 98.15%. The high AUC achieved in the evaluation process is also reflected in practice.

To conclude, VOB incorporates a novel approach to semi-automatic annotation of resources and represents a substantial engineering effort wrapped in an advanced visualization-based graphical user interface.

References

1. Grcar, M.: D4.5 Software Module for Semantic Annotation of a Web Service. SWING Project Deliverable (2008), `http://www.swing-project.org/deliverables/document/131`
2. Page, L., Brin, S., Motwani, R., Winograd, T.: The PageRank Citation Ranking: Bringing Order to the Web (1999)
3. Open Geospatial Consortium, OGC (2005), Web Feature Service Implementation Specification, `http://portal.opengeospatial.org/files/?artifact_id=8339`
4. Grcar, M., Klien, E., Novak, B.: Using Term-matching Algorithms for the Annotation of Geo-services (extended). In: Post-proceedings of the ECML-PKDD 2007 Workshop on "Web Mining 2.0". Springer, Heidelberg (2007)
5. Tablan, V., Damljanovic, D., Bontcheva, K. In: Proceedings of the 5th European Semantic Web Conference (ESWC 2008), Tenerife, Spain (2008)

Semi-automatic Categorization of Videos on VideoLectures.net

Miha Grcar[1], Dunja Mladenic[1], and Peter Kese[2]

[1] Jozef Stefan Institute, Dept. of Knowledge Discovery, Jamova cesta 39,
1000 Ljubljana, Slovenia
{miha.grcar,dunja.mladenic}@ijs.si
[2] Viidea Ltd., Glavni trg 25,
4000 Kranj, Slovenia
peter.kese@viidea.com

Abstract. Automatic or semi-automatic categorization of items (e.g. documents) into a taxonomy is an important and challenging machine-learning task. In this paper, we present a module for semi-automatic categorization of video-recorded lectures. Properly categorized lectures provide the user with a better browsing experience which makes her more efficient in accessing the desired content. Our categorizer combines information found in texts associated with lectures and information extracted from various links between lectures in a unified machine-learning framework. By taking not only texts but also the links into account, the classification accuracy is increased by 12–20%.

Keywords: categorization, classification, machine learning, multi-modal data mining, multimedia, video, VideoLectures.net.

1 Introduction and Motivation

In this paper, we present a module for semi-automatic categorization of video-recorded lectures. The presented categorizer is part of the software powering one of the world's largest educational Web portals, the VideoLectures video-hosting portal[1].

VideoLectures is growing fast. There are roughly 200 videos added each month. In order to provide the audience with efficient browsing facilities, each video needs to be categorized into one or more categories in the taxonomy. The taxonomy now has roughly 150 categories and is still growing. The categorization of lectures is a time-consuming task and is far from trivial.

Our categorizer combines information found in texts associated with lectures (e.g. title, description, slide texts) and information extracted from various links between lectures (e.g. lectures recorded at the same event, lectures sharing an author, lectures interlinked according to the click-stream data) in a unified machine-learning framework. The VideoLectures team provided us with their dataset of English lectures; roughly one third of them were already manually categorized and served as

[1] <http://videolectures.net>

W. Buntine et al. (Eds.): ECML PKDD 2009, Part II, LNAI 5782, pp. 730–733, 2009.

training data. Each lecture is described with a title; more than two thirds of the lectures also have a short description and/or come with slide titles. In addition to this unstructured textual data, each lecture can be seen as a vertex in several undirected weighted graphs representing the structural part of the data. These graphs are the following:

- **Same-event graph.** Two vertices are interconnected if the two corresponding lectures were recorded at the same event (e.g. a specific conference or workshop).
- **Same-author graph.** Two vertices are interconnected if the two corresponding lectures were prepared and/or given by the same author or authors. The weight of a link is the number of authors the two lectures have in common.
- **Also-watched graph.** Two vertices are interconnected if the two corresponding lectures were viewed together in the same browsing session. The weight of a link is the number of times the two lectures were viewed together.

The categorizer has already been integrated into the VideoLectures Web site. When the author or the editor decides to categorize a lecture, she is provided with a set of suggested categories (called "quick links" in the categorization panel). These suggestions were pre computed by the categorizer. The author/editor simply selects the appropriate categories in contrast to browsing the taxonomy to find them. The categorization user interface is shown in Fig. 1.

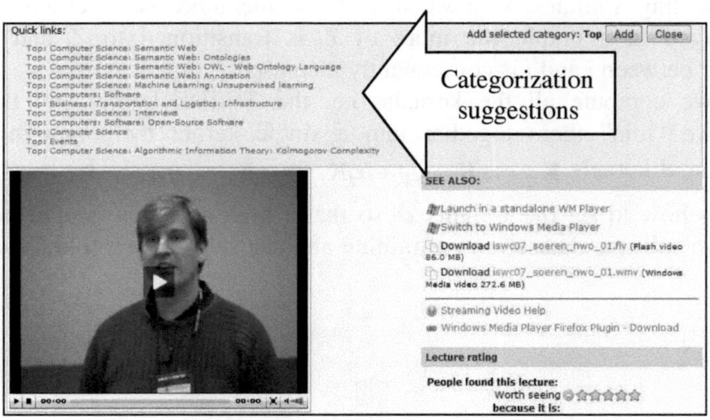

Fig. 1. Categorization user interface on VideoLectures

2 Approach and Experimental Results

Classification of documents into a taxonomy has already been addressed by several researchers, either on a flattened taxonomy [2] or by taking the structure into account [3]. We decided to flatten the taxonomy and employ Centroid Classifier as suggested in [1].

To evaluate the approach, we first represented each lecture by the associated text as TF-IDF vector as usual in text mining and performed 5-fold cross validation on the manually categorized lectures. We measured classification accuracies on top 1, 3, 5, and 10 predicted categories. We tested several classifiers: the Centroid, SVM, and k-Nearest Neighbors (k-NN) classifier. From the results, we learned that the three algorithms perform comparably well in terms of accuracy, with the Centroid Classifier standing out with respect to the accuracy on top 10 items (achieving 83.48% accuracy on average) and SVM achieving a slightly better accuracy on the topmost item (achieving 53.89% accuracy on average). Regarding the efficiency, it is important to note that SVM is slow in the training phase while k-NN is slow in the prediction phase. These two facts give even more appeal to the Centroid Classifier which was our choice for further experimentation and also final implementation.

To incorporate structural data in an attempt to boost accuracy, we decided to compute a kernel for each different "instance space" of the data, different spaces being the TF-IDF representation of the textual data on one hand and the three undirected weighted graphs (i.e. same-event, same-author, and also-watched graph) on the other.

We can look at a set of normalized TF-IDF vectors $\mathbf{f}_i$, $\| \mathbf{f}_i \|_2 = 1$, $f_{i,j} \geq 0$, as if they are rows in matrix $\mathbf{F}_{\text{TF-IDF}} = [f_{i,j}]$. If so, the corresponding kernel $\mathbf{K}_{\text{TF-IDF}}$ is obtained by multiplying $\mathbf{F}_{\text{TF-IDF}}$ with the transposed form of itself. When dealing with data in the form of an undirected weighted graph, on the other hand, we compute a diffusion kernel [4]. Let us consider attaching a random variable Z_i to each vertex i. Now let each variable send some of its value (fraction α) to each of the immediate neighbors at discrete time steps. It turns out that the covariance matrix of such random field is a kernel reflecting similarities between vertices: the more two vertices i and j are interconnected in a graph, the more of Z_i is transitioned to Z_j and vice versa. Covariance between i and j is consequently increased.

When we compute all the kernels (i.e. the TF-IDF kernel and the diffusion kernels), we "join" them together into a single kernel by computing a convex combination of kernels $\mathbf{K}_\Sigma = \alpha_1 \mathbf{K}_{\text{TF-IDF}} + \alpha_2 \mathbf{K}_{G_1} + \alpha_3 \mathbf{K}_{G_2} + \alpha_4 \mathbf{K}_{G_3}$. $\mathbf{K}_\Sigma$ is again a kernel. The task is now to set the weights α_i so that a high accuracy is achieved. For this purpose, we split the dataset into a training and test set for each fold as we did when

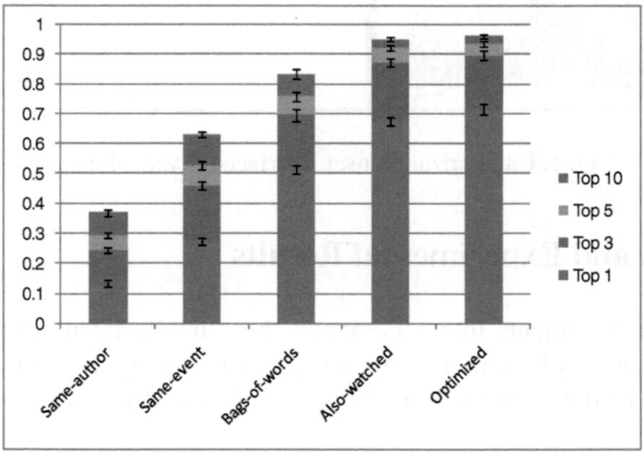

Fig. 2. Classification accuracies for each type of information in isolation as well as for the optimized combined kernels

establishing the baseline. However, we then run an optimization algorithm called Differential Evolution [5] to optimize the convex combination weights. In the optimization loop, we evaluate the intermediate weights by performing another (inner) 5-fold cross validation.

The accuracy of the optimized convex combination as well as the accuracies of the baseline and each of the graphs in isolation are shown in Fig. 2. From the chart, we can conclude the following. The same-author and same-event graphs contain the *least* relevant information for our categorization task. The same-event graph performs a bit better than same-author but is still below the baseline. By far the most relevant information is contained in the also-watched graph. This graph, in isolation, outperforms the other graphs as well as the text-based baseline. The optimized feature kernel indeed yields the best accuracies. However, the performance of the optimized data representation is close to that of the also-watched graph-based representation.

3 Conclusions

We can conclude that we were successful in our task as the categorizer is highly accurate—the proposed approach using text and graph features achieves accuracies that stretch 12–20% above the baseline obtained using text features only—and highly robust in terms of missing data. The latter means that a lecture might be missing textual annotations (such as the description and slide titles) but is still categorized correctly. Furthermore, the categorizer has been successfully integrated into the VideoLectures Web site. Categorization suggestions (termed “quick links”) are provided to the author in the categorization panel.

The current implementation of the categorizer suffers from high memory consumption as each diffusion kernel is a $n \times n$ dense matrix, where n is the number of lectures. Also, computing diffusion kernels is a time-consuming operation. We are currently working on reducing these time-space requirements to make the categorizer more scalable.

References

1. Grobelnik, M., Mladenic, D.: Simple Classification into Large Topic Ontology of Web Documents. Journal of Computing and Information Technology 13, 279–285 (2005)
2. Koller, D., Sahami, M.: Hierarchically Classifying Documents Using Very Few Words. In: Proceedings of the 14th International Conference on Machine Learning ICML 1997, pp. 170–178. Morgan Kaufmann, San Francisco (1997)
3. Mladenic, D.: Turning Yahoo into an Automatic Webpage Classifier. In: Proceedings 13th European Conference on Artificial Intelligence ECAI 1998, pp. 473–474. John Wiley & Sons, Chichester (1998)
4. Kondor, R.I., Lafferty, J.: Diffusion Kernels on Graphs and Other Discrete Structures. In: Proceedings of the ICML 2002, pp. 315–322 (2002)
5. Storn, R., Price, K.: Differential Evolution – A Simple and Efficient Heuristic for Global Optimization over Continuous Spaces. Journal of Global Optimization 11, 341–359 (1997)

Discovering Patterns in Flows: A Privacy Preserving Approach with the ACSM Prototype*

Stéphanie Jacquemont, François Jacquenet, and Marc Sebban

Université de Lyon, Université de Saint-Etienne,
Laboratoire Hubert Curien, UMR CNRS 5516,
18 rue Benoît Lauras, 42000 Saint-Etienne, France

Abstract. In this demonstration, we aim to present the ACSM prototype that deals with the discovery of frequent patterns in the context of flow management problems. One important issue while working on such problems is to ensure the preservation of private data collected from the users. The approach presented here is based on the representation of flows in the form of probabilistic automata. Resorting to efficient algebraic techniques, the ACSM prototype is able to discover from those automata sequential patterns under constraints. Contrary to standard sequential pattern techniques that may be applied in such contexts, our prototype makes no use of individuals data.

1 Introduction

Sequential pattern mining has been a very active domain of research since the mid 90's (see [1] for an overview of the domain) and it has provided various powerful tools used in many applications. Nevertheless, those researches have mainly focused on increasing the efficiency of algorithms in terms of speed and space consumption. Parallel to this line of research, there has been a great interest in privacy preserving data mining techniques [2], but not directly dedicated to the specific domain of sequence mining, except the work of Zhan et al. [3], and more recently that of Kapoor et al. [4] or Kim et al. [5]. However, in those approaches, it is supposed that we have a set of sequences on which we can apply some specific techniques in order to do some kind of anonymization, directly or indirectly on the data to be processed.

We can note that there are many situations where the database of sequences results from the study of flows. For example, we may study the flow of cars in a town and get some set of paths used by car drivers on which the sequence mining task could be achieved. We may study the flow of visits on a particular web site and get some history files of sequences of pages visited by some users. We may study the flow of IP packets in some networks and get some set of routes used by

* This work has been supported in part by the french national research agency under the Bingo2 project.

W. Buntine et al. (Eds.): ECML PKDD 2009, Part II, LNAI 5782, pp. 734–737, 2009.

IP packets sent and received by users. We may also study the flow of customers in a store and get some video recordings of customer behaviors in the store, etc. In such situations, anonymization techniques may be used on the data with more or less difficulties. However, this would result in a huge amount of pre-processing and moreover, and would require to collect non anonymous datasets that might be used by malicious agents before the end of the anonymization process. The approach we propose with the ACSM (Automata-based Constrained Sequence Mining) prototype is completely different. We assume that the underlying problem can be modeled in the form of a probabilistic automaton. This hypothesis is particularly verified in flow management problems such as those previously mentioned. In such a context, ACSM achieves a sequence mining task directly from the automaton using algebraic methods. Hence, it does not need to get the sequences of data making the flow but rather the structure of the flow itself (an automaton) and the values assigned to each path of the flow.

2 The ACSM Prototype

Due to space limitation we cannot present the theoretical aspects behind ACSM. For that purpose, the reader may refer to [6]. From an operational point of view, ACSM takes as input an XML file describing a flow graph. This file is loaded by ACSM that generates the corresponding probabilistic automaton while satisfying some statistical constraints.

ACSM can then be used to extract frequent sequential patterns given a support threshold. Our software has been implemented to satisfy statistical constraints in order to reduce the risk to extract false frequent patterns or to overlook true frequent ones. In this probabilistic framework, the user can then parameterize ACSM according to a priori fixed type I and type II errors. We dealt with the control of the false positive and false negative rates in [7] by providing a lower bound of the sample size on which the sequence mining task has to be performed. This means that we can formally decide when we can stop observing the flow, pick up the XML file and begin to mine the data.

The probabilities on the edges and vertices of the automaton are used to calculate the frequency of each discovered sequence and ACSM returns the paths (sequences), made up of not necessarily consecutive edges, whose frequency is greater than the fixed support threshold. Efficient algebraic calculus are used to optimize the extraction process, based on the LU factorization.

3 Overview of the Demonstration

To put forward the interest of ACSM, we show an application (called TrafficMiner) in the domain of car flow management that has been designed on top of it. Using ACSM to discover car flow patterns in a town, that is sequences of non necessarily consecutive streets frequently used by car drivers of that town, is a way to ensure privacy preservation of drivers' behavior. Indeed, until now, if we wanted to discover frequent routes in a town we had to install cameras in each street and then record the routes traversed by each driver. This solution is of course unrealistic because of its cost, but moreover it would be a great breach to drivers privacy and this would not be acceptable. The solution we propose consists in just counting, in each street of the part of the town we want to study, the number of cars using this street. The map of the town and the resulting counters provide a probabilistic automaton that can be processed by ACSM.

Figure 1 shows a screenshot of TrafficMiner discovering frequent routes in the city of Arlington (Virginia, USA).

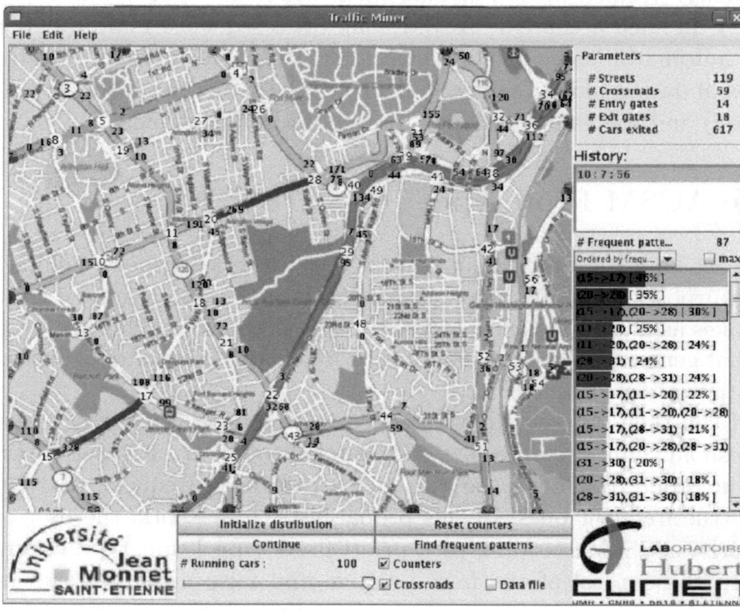

Fig. 1. TrafficMiner running on a map of Arlington

The demonstration shows the way we can design the automaton associated with the map using a specific interface and the way we can simulate some traffic on this map. Then, we look at the XML file describing the flows of cars focusing on each of its markup. Then, we run ACSM to get the frequent routes used by the virtual car drivers. At the demo, it will also be possible for users to provide specific flows that may have some interests for them. The input data will have to be provided in an XML form explained on site. Then running ACSM will provide patterns for the users.

3.1 Contribution of ACSM for the ML and DM Communities

What makes our piece of software unique and special? To our knowledge, ACSM is the first prototype able to deal with flow management problems and to extract sequential patterns in datasets completely preserving the privacy of users involved in the flows.

What are the innovative aspects or in what way/area does it represent the state of the art? The innovative aspects of our system concerns privacy preservation issues. Contrary to standard sequential pattern mining algorithms, our prototype does not need any user dependent sequential data. No anonymization step is required because the sequential data are no more needed.

For whom is it most interesting/useful? Our prototype, and the ideas behind, may be very useful for the industrial practitioners. To give an example, we may consider webmasters who might want to mine visits of users without using classical history files that lead to many known problems due to proxies, caches, etc. Instead of a tedious pre-processing step on history files, they may easily proceed by putting counters on each page and link of their site and then use ACSM on the automaton built using the structure of the website and the probabilities assigned thanks to the counters.

4 Conclusion

We think this demonstration of the ACSM system proves the interest of processing automata instead of sequences in various situations. This is the case for example if privacy concerns are crucial issues of the applications. Interoperability with ACSM is very easy as it only requires an XML file as an input, which gives ACSM the ability to be easily encapsulated in some specific applications.

References

1. Dong, G., Pei, J.: Sequence Data Mining. Springer, Heidelberg (2007)
2. Aggarwal, C., Yu, P.S. (eds.): Privacy-Preserving Data Mining: Models and Algorithms. Springer, Heidelberg (2008)
3. Zhan, J., Chang, L., Matwin, S.: Privacy-preserving collaborative sequential pattern mining. In: Proc. of the Workshop on Link Analysis, Counter-terrorism, and Privacy in conjunction with the SIAM ICDM, pp. 61–72 (2004)
4. Kapoor, V., Poncelet, P., Trousset, F., Teisseire, M.: Privacy preserving sequential pattern mining in distributed databases. In: Proc. CIKM, pp. 758–767 (2006)
5. Kim, S.W., Park, S., Won, J.I., Kim, S.W.: Privacy preserving data mining of sequential patterns for network traffic data. Information Sciences 178(3), 694–713 (2008)
6. Jacquemont, S., Jacquenet, F., Sebban, M.: Mining probabilistic automata: A statistical view of sequential pattern mining. Machine Learning Journal 75(1), 91–127 (2009)
7. Jacquemont, S., Jacquenet, F., Sebban, M.: A lower bound on the sample size needed to perform a significant frequent pattern mining task. Pattern Recognition Letters (to appear, 2009)

Using Temporal Language Models for Document Dating

Nattiya Kanhabua and Kjetil Nørvåg

Dept. of Computer Science
Norwegian University of Science and Technology
Trondheim, Norway

Abstract. In order to increase precision in searching for web pages or web documents, taking the temporal dimension into account is gaining increased interest. A particular problem for web documents found on the Internet is that in general, no trustworthy timestamp is available. This is due to its decentralized nature and the lack of standards for time and date. In previous work we have presented techniques for solving this problem. In this paper, we present a tool for determining the timestamp of a non-timestamped document (using file, URL or text as input) using temporal language models. We also outline how this tool will be demonstrated.

1 Introduction

In order to increase precision in searching for web pages or web documents, taking the temporal dimension into account is gaining increased interest. In this way, the search engine will retrieve documents according to both text and temporal criteria, i.e., *temporal text-containment search* [5].

Due to its decentralized nature and the lack of standards for time and date, it is difficult to determine an accurate and trustworthy timestamp of a web document. In a web warehouse or a web archive, there is no guarantee that the creation time and the time of retrieval by the crawler are related.

In this paper, we present a tool for determining timestamp of a non-timestamped document using temporal language models. The tool can take as input a file, contents from an URL, or text entered directly. As output it will present an estimation of possible creation time/periods, with confidence of each of the estimated time periods. Obviously, the one with highest confidence is the most probable based on the language model. An example of the interface is shown in Fig. 1(a) and example of results are shown in Fig. 1(b-e).

To build a system for dating a document, we compare document contents with word statistics and usages over time. The dating approach is based on the *temporal language model* presented in [1]. The intuition behind this approach is that, for a given document with unknown timestamp, it is possible to find the time partition that mostly overlaps in term usage with the document. For example, if the document contains the word "tsunami" and corpus statistics shows this word was very frequently used in 2004/2005, it can be assumed that this time period is a good candidate for the document timestamp. The model assigns a probability to a document according to word statistics over time. By partitioning a document corpus into time partitions, it is possible to determine the timestamp of a non-timestamped document d_i by computing a similarity score (*NLLR*)

W. Buntine et al. (Eds.): ECML PKDD 2009, Part II, LNAI 5782, pp. 738–741, 2009.

between the language model of d_i with each partition p_j. The timestamp of the document is the partition which maximizes the similarity score.

The rest of the paper is organized as follows. In Sect. 2 we outline the temporal language models used in our approach. In Sect. 3 we describe our document dating prototype. Finally, in Sect. 4 we outline our proposed demo.

2 Temporal Language Models

Timestamp estimation is based on the statistic language model presented by de Jong, Rode and Hiemstra [1]. This *temporal language model* is a variant of the time-based model in [4], based on a probabilistic model from [6]. The temporal language model assigns a probability to a time partition according to word usage or word statistics over time.

A document is modeled as $d_i = \{\{w_1, \ldots, w_n\}, (t_i, t_{i+1})\}$ where $t_i < t_{i+1}$ and (t_i, t_{i+1}) is a temporal view of document which can be represented by a time partition associated to its timestamp. A normalized log-likelihood ratio [3] is used to compute the similarity between two language models. Given a partitioned corpus, it is possible to determine the timestamp of a non-timestamped document d_i by comparing the language model of d_i with each corpus partition p_j using the following equation:

$$Score(d_i, p_j) = \sum_{w \in d_i} P(w|d_i) \times \log \frac{P(w|p_j)}{P(w|C)} \tag{1}$$

where C is the background model estimated on the entire collection and p_j is a time partition. The timestamp of the document is the partition maximizing a score according to the equation above, and the confidence *Conf* of the estimation is calculated as the logarithm of the score of the highest ranked relative to the second ranked partition.

In [2] we presented improvements to the approach of [1], the most important being temporal entropy, use of search statistics and adapted semantic-based preprocessing.

We use *temporal entropy* (TE) to weight terms differently depending on how well a term is suitable for separating time partitions among overall time partitions and also indicates how important a term is in a specific time partition. Temporal entropy of a term w_i is given as follows:

$$TE(w_i) = 1 + \frac{1}{\log N_P} \sum_{p \in \mathbf{P}} P(p|w_i) \times \log P(p|w_i) \tag{2}$$

where $P(p_j|w_i) = \frac{tf(w_i, p_j)}{\sum_{k=1}^{N_P} tf(w_i, p_k)}$, N_P is the total number of partitions in a corpus $\mathbf{P}$, and $tf(w_i, p_j)$ is the frequency of w_i in partition p_j. Modifying the score in Equation (1), each term w can be weighted with temporal entropy $TE(w)$ as follows:

$$Score_{te}(d_i, p_j) = \sum_{w \in d_i} TE(w) \times P(w|d_i) \times \log \frac{P(w|p_j)}{P(w|C)} \tag{3}$$

Search statistics provided by Google Zeitgeist (GZ) can be integrated as an additional score in order to increase the probability of a tentative time partition. GZ essentially gives statistics of trends of search terms, i.e., increasing and decreasing popularity. By analyzing search statistics, we are able to increase the probability for a particular partition which contains top-ranked queries. The higher probability the partition acquires, the more potential time candidate it becomes. *GZ* is integrated as an additional score into Equation (1) in order to increase the probability of partition p_j:

$$Score_{gz}(d_i, p_j) = \sum_{w \in d_i} \left(P(w|p_j) \times \log \frac{P(w|p_j)}{P(w|C)} + \beta GZ(p_j, w) \right) \quad (4)$$

where β is the weight for the *GZ* function (see [2] for more details on calculating GZ).

In order to further increase accuracy of the dating, we have also integrated *semantic-based techniques* into document preprocessing, i.e., part-of-speech tagging (POS), collocation extraction (COLL), word sense disambiguation (WSD), and concept extraction (CON).

3 Document Dating System

Our prototype implements the ideas from [2], and uses a web-based interface. It allows to estimate the date of different input formats (i.e., a file, an URL, or plain text) as shown by Fig. 1(a). Example inputs can be URL: "http://tsunami-thailand.blogspot.com" or text: "the president Obama". The user can select parameters: preprocessing (POS, COLL, WSD, or CON), similarity score (NLLR, GZ or TE), and time granularity (1-month, 3-months, 6-months, or 12-months). Given an input to be dated, the system computes similarity scores between a given document/text and temporal language models. The document is then associated with tentative time partitions or its likely originated timestamps. The results can be displayed in two ways. First, a rank list of partitions is shown in an descending order according to their scores. Second, each tentative time partition is drawn in a timeline with its score as a height.

4 Demo Outline

In the demo, we will present the features of our dating tool, including the impact of the variants of our temporal language approach:

Basic vs. advanced preprocessing: There is a trade-off among semantic-based preprocessing. We compare a *basic* preprocessing (POS only) to an *advanced* preprocessing (a combination of POS, COLL, WSD, and CON). As will be shown, *basic* used less time, but gains a poorer quality than the *advanced*.

How GZ enhances scores: To improve the accuracy, we compute scores by using *GZ* in addition to *NLLR*. The correct time period (2004/12 to 2005/11) is raised from the 7th rank in Fig. 1(b) to the 1st rank with higher confidence in Fig. 1(c).

TE as a trend: A term occurring in few partitions is weighted high by *TE* and it provides high scores for partitions in which the term appears. Fig. 1(d-e) display trends of the web page about "US presidential election" with and without *TE* respectively and *TE* gives higher scores for relevant periods (2000, 2004 and 2008).

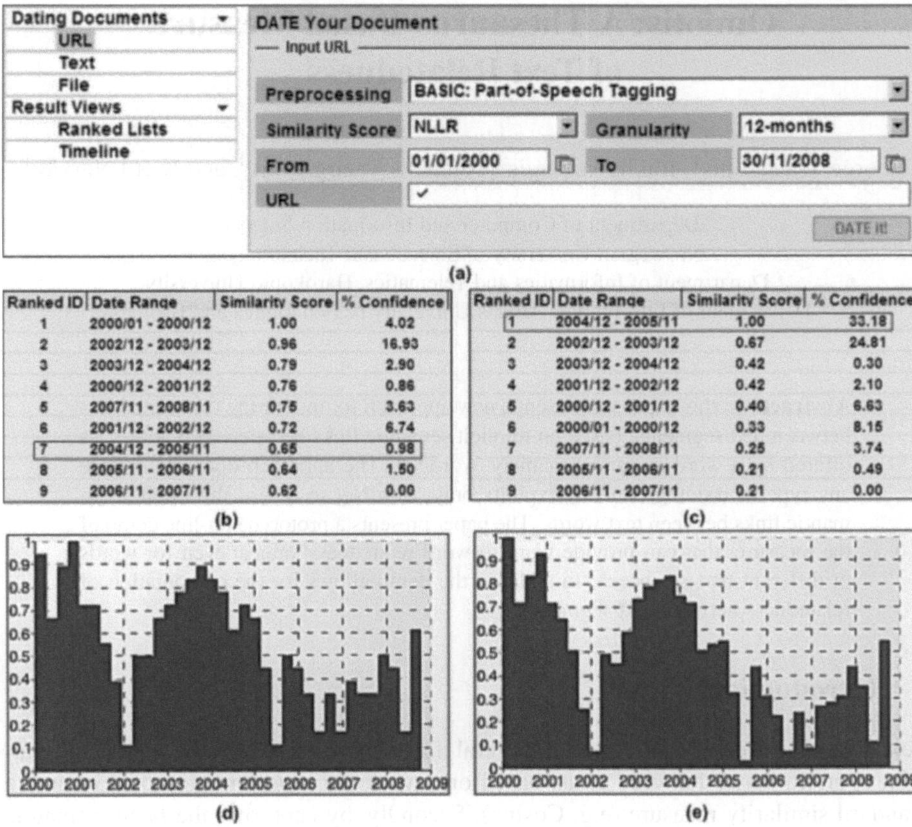

(b)

Ranked ID	Date Range	Similarity Score	% Confidence
1	2000/01 - 2000/12	1.00	4.02
2	2002/12 - 2003/12	0.96	16.93
3	2003/12 - 2004/12	0.79	2.90
4	2000/12 - 2001/12	0.76	0.86
5	2007/11 - 2008/11	0.75	3.63
6	2001/12 - 2002/12	0.72	6.74
7	2004/12 - 2005/11	0.65	0.98
8	2005/11 - 2006/11	0.64	1.50
9	2006/11 - 2007/11	0.62	0.00

(c)

Ranked ID	Date Range	Similarity Score	% Confidence
1	2004/12 - 2005/11	1.00	33.18
2	2002/12 - 2003/12	0.67	24.81
3	2003/12 - 2004/12	0.42	0.30
4	2001/12 - 2002/12	0.42	2.10
5	2000/12 - 2001/12	0.40	6.63
6	2000/01 - 2000/12	0.33	8.15
7	2007/11 - 2008/11	0.25	3.74
8	2005/11 - 2006/11	0.21	0.49
9	2006/11 - 2007/11	0.21	0.00

Fig. 1. (a) System interface, (b) Results of *basic* preprocessing and *NLLR*, (c) Results of *basic* preprocessing and *GZ*, (d-e) Trends of "US presidential election" with and without *TE*

References

1. de Jong, F., Rode, H., Hiemstra, D.: Temporal language models for the disclosure of historical text. In: Proceedings of AHC 2005 (History and Computing) (2005)
2. Kanhabua, N., Nørvåg, K.: Improving temporal language models for determining time of non-timestamped documents. In: Christensen-Dalsgaard, B., Castelli, D., Ammitzbøll Jurik, B., Lippincott, J. (eds.) ECDL 2008. LNCS, vol. 5173, pp. 358–370. Springer, Heidelberg (2008)
3. Kraaij, W.: Variations on language modeling for information retrieval. SIGIR Forum 39(1), 61 (2005)
4. Li, X., Croft, W.B.: Time-based language models. In: Proceedings of CIKM 2003 (2003)
5. Nørvåg, K.: Supporting temporal text-containment queries in temporal document databases. Journal of Data & Knowledge Engineering 49(1), 105–125 (2004)
6. Ponte, J.M., Croft, W.B.: A language modeling approach to information retrieval. In: Proceedings of SIGIR 1998 (1998)

Omiotis: A Thesaurus-Based Measure of Text Relatedness

George Tsatsaronis[1], Iraklis Varlamis[2], Michalis Vazirgiannis[3], and Kjetil Nørvåg[1]

[1] Department of Computer and Information Science,
Norwegian University of Science and Technology
[2] Department of Informatics and Telematics, Harokopio University
[3] Department of Informatics, Athens University of Economics and Business

Abstract. In this paper we present a new approach for measuring the relatedness between text segments, based on implicit semantic links between their words, as offered by a word thesaurus, namely WordNet. The approach does not require any type of training, since it exploits only WordNet to devise the implicit semantic links between text words. The paper presents a prototype on-line demo of the measure, that can provide word-to-word relatedness values, even for words of different part of speech. In addition the demo allows for the computation of relatedness between text segments.

1 Introduction

Text-relatedness can be perceived in several different ways. Primarily, as lexical relatedness or similarity between texts, based on a vectorial representation of texts and a standard similarity measure (e.g. Cosine). Secondly, by capturing the latent semantic relations between dimensions (words) in the constructed vector space model, by using techniques of latent semantic analysis [1]. Another aspect of text relatedness, probably of equal importance, is the semantic relatedness between two text segments. For example, the sentences "*The shares of the company dropped 14 cents*" and "*The business institution's stock slumped 14 cents*" have an obvious semantic relatedness, which traditional measures of text similarity fail to recognize. In this paper we present an on-line demo of a new measure of semantic relatedness between words (SR) and its expansion (Omiotis, from the Greek word for relatedness or similarity) to measure semantic relatedness between text segments. Our measure of relatedness lies in the use of WordNet, and all of its available semantic information. The contribution of this demo is twofold: (a) it can measure the semantic relatedness between words of any part of speech, and consequently between sentences, and (b) it computes relatedness values very fast, based on an index of all the pair-wise relatedness values between WordNet synsets (11 billion combinations). This is the first time, to the best of our knowledge, that such an index has been created.

2 Measuring Semantic Relatedness

The expansion of WordNet with semantic relations that cross parts of speech has widened the possibilities of semantic network construction from text. Recent approaches in

W. Buntine et al. (Eds.): ECML PKDD 2009, Part II, LNAI 5782, pp. 742–745, 2009.

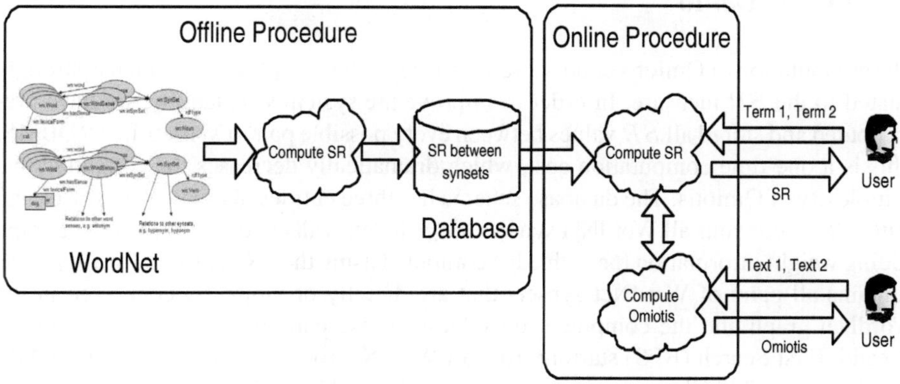

Fig. 1. System architecture

semantic network construction from word thesauri, e.g. from Navigli and Velardi [2], utilize all of the semantic relations in WordNet. The applications of these methods in areas like Word Sense Disambiguation highlight the advantages of using the full range of semantic relations that WordNet offers. In this work, we find all the semantic paths that connect two sense nodes in WordNet using all of the provided semantic relations by WordNet. To the best of our knowledge it is the first time that a measure of semantic relatedness combines three factors in tandem: (a) path length connecting concepts; (b) concepts' depth in the used thesaurus, and (c) thesaurus' edges importance. An analysis of state of the art measures that use semantic information from word thesauri can be found in [3]. Figure 1 shows the architecture of the developed system. For the computation of the semantic relatedness between two terms (**SR**), we first compute the semantic relatedness values for all pairwise combinations of their senses. The semantic relatedness between a pair of senses considers the path length, captured by compactness, and the path depth, captured by semantic path elaboration as explained in details in [4]. The values of SR range in [0,1]. In the case when only one of the terms exists in WordNet, the semantic relatedness between them is 0. If both terms are the same, and this term exists in WordNet, then it is considered 1.

Following the work we presented in [5], the measure of semantic relatedness between text segments (**Omiotis**) perceives texts as sets of terms (*bag of words*) in a vector space, and uses TF-IDF for term weighting. Omiotis value $O_{A,B}$ for a pair of texts A and B is defined as:

$$O_{A,B}=\frac{1}{2}\left[\frac{1}{|A|}\left(\sum_{i=1}^{|A|}\lambda_{i,x(i)}\cdot SR(k_i,h_{x(i)})\right)+\frac{1}{|B|}\left(\sum_{j=1}^{|B|}\lambda_{y(j),j}\cdot SR(k_{y(j)},h_j)\right)\right] \quad (1)$$

In the above equation, $SR(t_i, t_j)$ is considered as the semantic relatedness between terms t_i and t_j, as defined previously, λ_{t_i,t_j} is the sum of the terms' TF-IDF values, k_i the ith term of A, h_i the ith term of B and $x(i)$ and $y(j)$ are defined respectively:

$$x(i) = \arg\max_{j\in[1,|B|]}(\lambda_{i,j}\cdot SR(k_i,h_j)) \; and \; y(j) = \arg\max_{i\in[1,|A|]}(\lambda_{i,j}\cdot SR(k_i,h_j)) \quad (2)$$

3 On-line Demo

The computation of Omiotis entails a series of steps, the complexity of which is strongly related to the SR measure. In order to improve the system's scalability, we have pre-computed and stored all SR values between every possible pair of synsets in a RDBMS. This is a one-time computation cost, which dramatically decreases the computational complexity of Omiotis. The database schema has three entities, namely *Node*, *Edge* and *Paths*. *Node* contains all WordNet synsets. *Edge* indexes all edges of the WordNet graph adding weight information for each edge computed using the SR measure. Finally, *Paths* contains all pairs of WordNet synsets that are directly or indirectly connected in the WordNet graph and the computed relatedness. These pairs were found by running a Breadth First Search (BFS) starting from all WordNet roots for all POS. The RDBMS, which exceeds 220 Gbytes in size, has indexed in total 155.327 unique synsets, whereas the number of processed edges is 324.268. In total, the number of processed synset pairs exceeds the 11 billion combinations. The current implementation takes advantage of the database structures (indices, stored procedures etc) in order to decrease the running time of Omiotis. Because we have pre-computed the relatedness values for all synset pairs, the time required for processing 100 pairs of terms is $\simeq 1$ sec, which makes the computation of Omiotis feasible and scalable. As a proof of concept, we have developed an on-line demo version of the SR and the Omiotis measures (publicly available at http://omiotis.hua.gr), where the user can test the term-to-term and sentence-to-sentence semantic relatedness measures. Figure 2 shows a walk-through example of the demo for computing relatedness between *database* and *systems*. Similarly, the user can compute text-to-text relatedness. Once the relatedness computation takes place for the desired input, a screen showing the relatedness value appears to the user's browser.

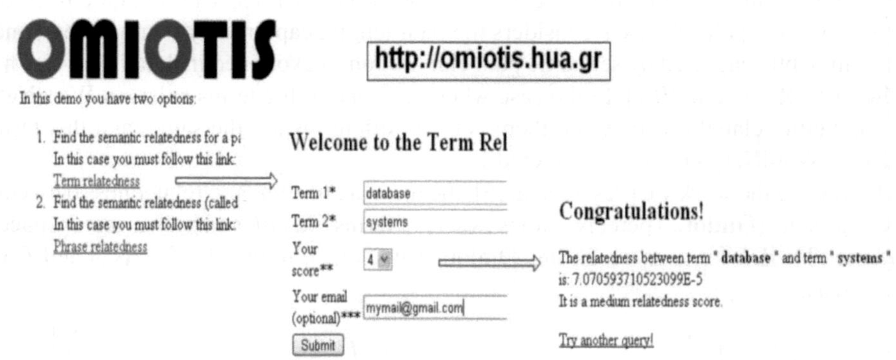

Fig. 2. Omiotis on-line demo

4 Applications

SR and Omiotis have been evaluated in several different text related tasks. Initially, SR has been evaluated as a measure for Word Sense Disambiguation in [4]. The results showed that the measure produces state of the art precision in the Senseval competitions (*all English words* task). Furthermore, SR has been evaluated in measuring word-to-word relatedness, in three widely used data sets [4], where experiments showed that it produces the highest Spearman rank order correlation coefficient compared to the human judgements, than any other dictionary-based measure [3]. In addition, Omiotis has been embedded in the text retrieval task [6], by using the implementation of the measures described in the next section, inside the TERRIER retrieval platform. Experiments in three TREC collections show that Omiotis can boost retrieval performance by even up to 2% compared to traditional retrieval models that do not take into account semantic information from text. Finally, measures of semantic relatedness have been embedded in the past in many different linguistic exercises, like for example the SAT analogy tests and the TOEFL synonym questions (consult http://www.aclweb.org/aclwiki/), as well as in paraphrase recognition [7]. From the aforementioned, it is induced that both SR and Omiotis can be embedded in a variety of applications, mainly due to the fact that they can measure relatedness for words of all parts of speech. Regarding related systems, the offered functionality of computing word-to-word relatedness by SR is also offered by the *WordNet::Similarity* software package [8], which implements a wide range of measures, but as our experimental analysis in [4] shows, SR outperforms all of these measures in capturing semantic relatedness compared to the human judgements in three data sets. The functionality offered by Omiotis, in computing text relatedness between text segments, taking into account semantic information from WordNet, is not offered, to the best of our knowledge, by another on-line system. Finally, as far as WordNet synset relatedness values is concerned, currently there is no other on-line system that has indexed all the possible WordNet synset combination relatedness values.

References

1. Landauer, T., Foltz, P., Laham, D.: Introduction to latent semantc analysis. Discourse Processes 25, 259–284 (1998)
2. Navigli, R., Velardi, P.: Structural semantic interconnections: A knowledge-based approach to word sense disambiguation. IEEE Trans. Pattern Anal. Mach. Intell. 27(7), 1075–1086 (2005)
3. Budanitsky, A., Hirst, G.: Evaluating wordnet-based measures of lexical semantic relatedness. Computational Linguistics 32(1), 13–47 (2006)
4. Tsatsaronis, G., Varlamis, I., Vazirgiannis, M.: Word sense disambiguation with semantic networks. In: Sojka, P., Horák, A., Kopeček, I., Pala, K. (eds.) TSD 2008. LNCS (LNAI), vol. 5246, pp. 219–226. Springer, Heidelberg (2008)
5. Varlamis, I., Vazirgiannis, M., Halkidi, M., Nguyen, B.: Thesus: Effective thematic selection and organization of web document collections based on link semantics. IEEE TKDE Journal 16(6), 585–600 (2004)
6. Tsatsaronis, G., Panagiotopoulou, V.: A generalized vector space model for text retrieval based on semantic relatedness. In: Proc. of the 12th EACL (SRW session), pp. 70–78 (2009)
7. Mihalcea, R., Corley, C., Strapparava, C.: Corpus-based and knowledge-based measures of text semantic similarity. In: Proc. of the 21st AAAI (2006)
8. Patwardhan, S., Banerjee, S., Pedersen, T.: Senserelate:targetword - a generalized framework for word sense disambiguation. In: Proc. of AAAI (demo session), pp. 1692–1693 (2005)

Found in Translation

Marco Turchi[1], Ilias Flaounas[2], Omar Ali[1], Tijl De Bie[1], Tristan Snowsill[1], and Nello Cristianini[1,2]

[1] Department of Engineering Mathematics, Queen's Building
[2] Department of Computer Science, Merchant Venturers Building
Bristol University, Bristol, United Kingdom
`http://patterns.enm.bris.ac.uk`

Abstract. We present a complete working system that gathers multilingual news items from the Web, translates them into English, categorises them by topic and geographic location and presents them to the final user in a uniform way. Currently, the system crawls 560 news outlets, in 22 different languages, from the 27 European Union countries. Data gathering is based on RSS crawlers, machine translation on Moses and the text categorisation on SVMs. The system also presents on a European map statistical information about the amount of attention devoted to the various topics in each of the 27 EU countries. The integration of Support Vector Machines, Statistical Machine Translation, Web Technologies and Computer Graphics delivers a complete system where modern Statistical Machine Learning is used at multiple levels and is a crucial enabling part of the resulting functionality.

1 Introduction

This paper describes a web system that makes use of Statistical Machine Learning technology at various levels, in order to process and aggregate the contents of hundreds of news outlets in dozens of languages: Found in Translation (FIT) (`http://foundintranslation.enm.bris.ac.uk/`).

News outlets produce a huge amount of daily news in different languages which are usually read only by the citizens of the country where the media that publish the story is located. Found in Translation is an online service that makes all this information available to English speakers under a common interface.

Many systems have been proposed to show different aspects of the information contained in various news outlets. NewsBlaster [1] generates multi-document summaries, while European Media Monitor [2] uses cross-language retrieval to generate news clusters without translating. The TextMap system [3] is mostly based on detecting entities.

Found in Translation takes advantage of the technologies obtained by the Statistical Machine Translation (SMT) and text categorisation communities in recent years. All of these advances have been possible only due to statistical machine learning technology. Instead of using cross-language information retrieval techniques to analyse multilingual news, FIT translates daily news from 21 European languages into English. A set of classifiers categorises each translated article

W. Buntine et al. (Eds.): ECML PKDD 2009, Part II, LNAI 5782, pp. 746–749, 2009.

into six topics, namely 'Accidents', 'Business', 'Crime', 'Environmental Issues', 'Politics' and 'War and Conflict'. All the news are geolocated and presented according to their topic on a 'heat-map' of Europe to allow an easy comparison of the topic bias present in the media in the different countries of EU.

Our approach has the advantage that an English speaker can easily read content translated into English from the different languages spoken in the EU. He or she can also easily compare the current levels of interest in the media for a specific topic, in each of the EU countries. Comparisons like this could be useful in understanding the different perspectives and biases that the media present to each country, and such an analysis could not be performed by a single human being.

From the point of view of machine learning, the problem of creating a classifier for each topic and for each language is very difficult to tackle: It is almost infeasible to gather all the multilingual labelled data that are needed to train the classifiers. The use of machine translated news allows us to focus on only one language, English, where labelled data are more easily available.

2 Methodology

FIT is based on state of the art technology both in text categorisation, using Support Vector Machines (SVMs) [4] and in SMT, using Moses [5].

We consider the popular class of SMT approaches based on a noisy channel model originally developed for speech recognition [6], where a Markovian Language Model coupled with a phrase-to-phrase translation table are at the heart. In recent years, the noisy channel model has been extended in different directions. The most fruitful has been the phrase based statistical machine translation (PBSMT) introduced by [7] that is based on the use of phrases rather than words. We use Moses, a complete phrase based translation toolkit for academic purposes. It provides all the state of the art components needed to create a PBSMT system from one language to another. For each language pair, an instance of Moses is trained using Europarl [8] data and JRC-Acquis Multilingual Parallel Corpus [9]. To train the language model, this data is complemented with the English news of the last fifteen days.

We use a set of Support Vector Machines (SVMs) classifiers, one per topic we want to detect. SVMs are kernel based approaches of supervised learning, delivering state of the art performance in text-categorisation [4]. Each classifier is trained offline, on a dataset of articles which have been pre-labelled. Linear kernels were used and the classifiers were trained with a choice of parameters aimed at achieving a low false positive rate, in other words a high precision of above 90%. This means that documents will be tagged only when we are very confident about the topic, while articles for which the confidence is lower will be left untagged.

To calculate an unbiased influence of a topic in a country, a subset of news outlets is selected. For each country, all the outlets of that country are ordered using their Alexa rank score (`http://www.alexa.com`). This score, which has been used before in the spatial analysis of news sources [3], is calculated using a combination of criteria and it estimates the traffic rank of each website. For FIT, we use as input dataset only the top ten ranked outlets in that country. For each country and each topic, we count the number of articles which referred to that topic during a sliding window, and we normalize it by the total number of articles discovered in that country in these ten outlets. This quantity is used on the web interface to colour the European map.

3 Infrastructure

Nowadays most main news outlets provide their content in Really Simple Syndication (RSS) format. We developed a web crawler infrastructure that targets these RSS feeds and collects all the advertised articles on the main page of each outlet we track. The RSS feeds provide just the summary of each article and a link to the web page of the full article. The crawler uses that link to get the full HTML page that contains the main body of the article. Finally a scraper is used to remove undesired elements from the HTML page such as images, advertisements, external links, etc., and returns the raw text of the article.

Each article we collect is stored in a database for further processing. The information stored includes a title and a summary (from the RSS feed), the full text of the article (from scraping the relevant HTML page), the publication date, the outlet name of its origin, the geographic location of the outlet and the language it is written in.

Every day, roughly 10,000 multilingual news articles are extracted and translated into English before stored in the database. Subsequently, all the English and machine-translated English articles are tagged by topic using the six SVM classifiers described previously.

4 Web Interface

FIT has a web-based user interface that presents all the translated news articles in an organised way. Articles are divided by topic and each topic is further divided by country. Two different data representations are available. The first is using colour maps of Europe, indicating topic activity. Each topic is shown on a different map as for example in Fig. 1. The second representation visualizes the same information using bar charts (see web site).

On the map, each country is coloured according to an unbiased statistic computed only on the top-ten outlets of the country as ranked by their Alexa score. There are six different maps, one map per topic. Green colours identify a weak interest for a particular topic while red colours a strong interest. On the bar chart view, the size of each bar represents the total number of discovered, translated and categorised news articles for each country for a particular topic. The list of articles we show for each country and topic is produced by using only high quality machine translated news: A filter has been introduced to hide articles that contain more than one untranslated word.

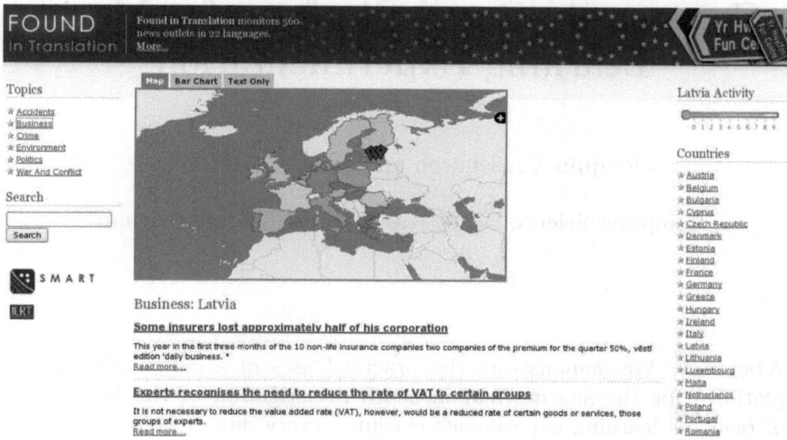

Fig. 1. Found in Translation web site

For each article the translated title and summary are reported with links to their site of origin. For United Kingdom and Ireland only English language news are shown. Multi-lingual countries, like Belgium, contain news from more than one language, all translated in English. This aspect is completely transparent to the user.

Acknowledgements. We are grateful to Simon Price and IRLT for their contribution to the web interface. This work is supported by the EU projects SMART and PASCAL2. Flaounas is supported by the A. S. Onassis Public Benefit Foundation. Cristianini is supported by a Royal Society Wolfson Merit Award.

References

1. McKeown, K., Barzilay, R., et al.: Tracking and summarizing news on a daily basis with columbia's newsblaster. In: Proceedings of HLT 2002, San Diego, USA (2002)
2. Best, C., der Goot, E.V., et al.: Europe media monitor. Technical report, JRC (2002)
3. Mehler, A., Bao, Y., et al.: Spatial analysis of news sources. IEEE Transactions on Visualization and Computer Graphics 12(5), 765–772 (2006)
4. Joachims, T.: SVM light (2002), `http://svmlight.joachims.org`
5. Koehn, P., Hoang, H., et al.: Moses: Open source toolkit for statistical machine translation. In: Proceedings ACL 2007, demonstration session (2007)
6. Brown, P.F., Pietra, S.D., et al.: The mathematic of statistical machine translation: Parameter estimation. Computational Linguistics 19(2), 263–311 (1994)
7. Koehn, P., Och, F.J., Marcu, D.: Statistical phrase-based translation. In: Proceedings NAACL 2003, Morristown, NJ, USA, pp. 48–54 (2003)
8. Koehn, P.: Europarl: A parallel corpus for statistical machine translation. In: Machine Translation Summit X, pp. 79–86 (2005)
9. Steinberger, R., Pouliquen, B., et al.: The JRC-Acquis: A multilingual aligned parallel corpus with 20+ languages. Arxiv preprint cs/0609058 (2006)

A Community-Based Platform for Machine Learning Experimentation

Joaquin Vanschoren and Hendrik Blockeel

Computer Science Dept., K.U. Leuven, Leuven, Belgium

Abstract. We demonstrate the practical uses of a community-based platform for the sharing and in-depth investigation of the thousands of machine learning experiments executed every day. It is aimed at researchers and practitioners of data mining techniques, and is publicly available at `http://expdb.cs.kuleuven.be`. The system offers standards and API's for sharing experimental results, extensive querying capabilities of the gathered results and allows easy integration in existing data mining toolboxes. We believe such a system may speed up scientific discovery and enhance the scientific rigor of machine learning research.

1 Introduction

Experimentation is the lifeblood of machine learning (ML) research. A considerable amount of effort and resources are invested in assessing the usefulness of new algorithms, finding the optimal approach for new applications or just to gain some insight into, for instance, the effect of a parameter. Yet in spite of all these efforts, experimental results are often discarded or forgotten shortly after they are obtained, or at best averaged out to be published, which again limits their future use. If we could collect all these ML experiments in a central resource and make them publicly available in an organized (searchable) fashion, the combined results would provide a highly detailed picture of the performance of algorithms on a wide range of data configurations, which can be tapped into to gain quick insights, reuse previous experiments, and increase the speed and depth of ML research.

In this paper, we demonstrate a prototype of a community-based platform designed to do just this. It allows researchers and practitioners to share their results and/or to keep an organized, detailed log of past experiments, even during early phases of algorithm development. This functionality can even be integrated seamlessly in DM toolboxes, without requiring any additional effort from the user. The biggest benefit, however, lies in the flexible querying capabilities it offers to browse and reorganize large amounts of publicly available (or locally stored) data. In the remainder of this paper we first we describe the design of this platform in Section 2. Next, Section 3 illustrates the querying capabilities and their utility in ML research. Finally, Section 4 concludes with future work.

W. Buntine et al. (Eds.): ECML PKDD 2009, Part II, LNAI 5782, pp. 750–754, 2009.

2 Anatomy of the Platform

The components of the platform are shown Fig. 1. To promote the free exchange of experiments, an extensible XML-based language for experiment exchange is introduced, dubbed ExpML, which is formally described in Vanschoren et al. [6], and adheres to an ontology of experimental concepts. It captures the basic structure of a machine learning experiment, yet allows each element (e.g. algorithms, kernels, datasets, evaluation metrics,...) to be described further to define the exact setup and cover task-specific properties. It currently supports a very wide range of classification and regression tasks, and is being extended further. It also supports the submission of new elements of any kind, verifies that experiments are reproducible, and allows each element to be 'tagged' with properties, such as dataset and algorithm characteristics. To facilitate the description of new experiments, a Java API is provided to compose experiments from scratch and submit them to the system. The same interface can also be used by existing DM tools to automatically stream new experiments to the database.

The experiments are then stored in an *experiment database* (ExpDB) [1,2]: a database designed to store all details of machine learning experiments in order to make them reproducible and clearly interpretable. It is designed to be very extensible, scale easily to large numbers of experiments and allow queries on practically any aspect of the experimental setup and outcome. As a result, its structure is very complex. Though we cannot include a complete discussion here, all the details of its design can be viewed at the project webpage. It currently contains over 600,000 experiments on 67 classification and regression algorithms, 149 different datasets and 2 data preprocessing techniques. Access is provided through web services for submitting experiments and launching queries.

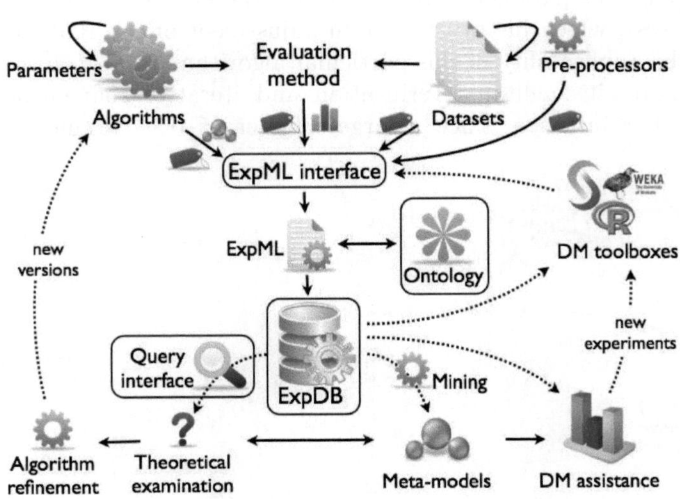

Fig. 1. The components of our platform and their use

Finally, the platform offers various ways to use the stored data. First, it offers two query interfaces to explore all the stored information. The first is an online interface, which offers an SQL query interface and basic visualizations of the returned results. The second is a stand-alone explorer tool offering more advanced querying and visualization features. Both are available at `http://expdb.cs.kuleuven.be`, including example queries and a video tutorial for using the explorer tool. Second, large sets of experiments can be downloaded to build meta-models of algorithm performance, e.g. in function of data characteristics. Finally, data can also be streamed to DM tools. For instance, experimental toolboxes can download experiments that have run before, visualization tools can download data points to provide new insights and DM assistance tools can check whether a certain approach is viable on a specific problem.

3 Benefits and Illustrations

The benefits of experiment databases have been listed systematically in previous work [5,2]. Here we simply illustrate these benefits with some example cases.

One of the biggest benefits is the reuse of previously run experiments. Collecting the experiments originally run to test various hypotheses, when combined, can be reused to test other hypotheses. As such, new experimental work can often be conducted by simply querying the database, instead of setting up new experiments; this is tremendously faster and easier. One clear example are benchmarking studies: competing algorithms probably have run before on the same datasets, and those experiments can therefore be simply downloaded instead of repeating them. The same strategy can also be used in algorithm development or refinement. For instance, we queried for the effect of the 'gamma' parameter in RBF kernels on several datasets, shown in Fig. 2. This analysis showed that the useful range of this parameter correlates with the number of attributes (shown in parentheses), which in turn led to an adjustment of the RBF function and improved the applicability of this particular algorithm [5].

Furthermore, it facilitates verification and iterative refinement of public knowledge. For instance, since a large number of bias-variance experiments

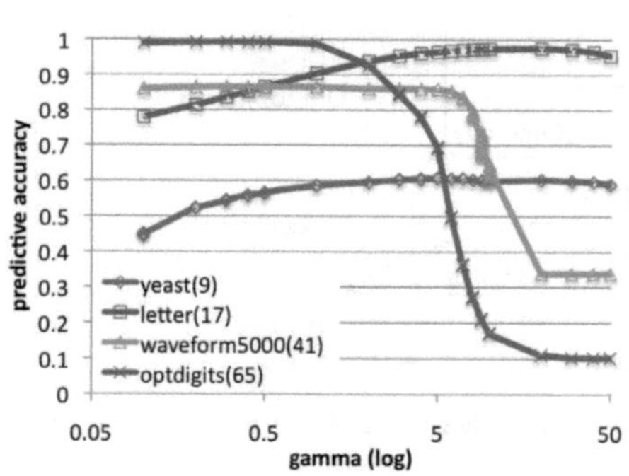

Fig. 2. The effect of kernel width in SVMs

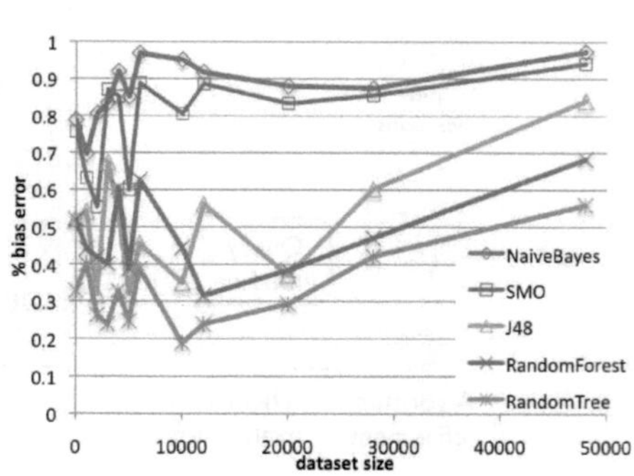

Fig. 3. Bias error ratio vs. dataset size

are stored, we could verify the claim that bias error becomes dominant on large datasets [3]. A single query reorganized the available experimental results and produced Fig. 3, showing that, at least for larger datasets, the percentage of bias error in the total error indeed increases with size. Again, not a single new experimental run was required; querying the stored experiments was sufficient.

Even when the stored experiments are not sufficient to answer a new experimental question, they may reduce the need for new experiments. New experimental runs may even be generated automatically as a query requires them (using, for instance, active learning principles), thus making the decision of which new experiments to perform in order to answer a new scientific question entirely transparent. Besides offering these practical advantages, the platform also simply serves as an ultimate reference, a 'map' of all known approaches and how they behave on certain types of datasets.

4 Conclusion and Future Work

Although standardization of experimental results has been a driving force in bio-informatics [4], our approach is, to the best of our knowledge, unique in ML and extends the latter by providing much more powerful querying abilities. Functioning as a open portal for ML experimentation, we are confident that this platform will be very useful for ML researchers, practitioners and students. Currently planned work includes a full integration in the WEKA toolbox [7] (beyond the interfaces available now), a graphical query interface, an account system and further work on the ontology used.

Acknowledgements

This research is supported by GOA 2003/08 'Inductive Knowledge Bases' and F.W.O.-Vlaanderen 'Foundations of Inductive Databases for Data Mining'.

References

1. Blockeel, H.: Experiment databases: A novel methodology for experimental research. In: Bonchi, F., Boulicaut, J.-F. (eds.) KDID 2005. LNCS, vol. 3933, pp. 72–85. Springer, Heidelberg (2006)
2. Blockeel, H., Vanschoren, J.: Experiment databases: Towards an improved experimental methodology in machine learning. In: Kok, J.N., Koronacki, J., Lopez de Mantaras, R., Matwin, S., Mladenič, D., Skowron, A. (eds.) PKDD 2007. LNCS (LNAI), vol. 4702, pp. 6–17. Springer, Heidelberg (2007)
3. Brain, D., Webb, G.: The Need for Low Bias Algorithms in Classification Learning from Large Data Sets. In: Elomaa, T., Mannila, H., Toivonen, H. (eds.) PKDD 2002. LNCS (LNAI), vol. 2431, pp. 62–73. Springer, Heidelberg (2002)

4. Stoeckert, C., Causton, H., Ball, C.: Microarray databases: standards and ontologies. Nature Genetics 32, 469–473 (2002)
5. Vanschoren, J., Pfahringer, B., Holmes, G.: Learning From The Past with Experiment Databases. In: Ho, T.-B., Zhou, Z.-H. (eds.) PRICAI 2008. LNCS (LNAI), vol. 5351, pp. 485–496. Springer, Heidelberg (2008)
6. Vanschoren, J., Blockeel, H., Pfahringer, B., Holmes, G.: Organizing the world's machine learning information. Comm. in Comp. and Inf. Science 17, 693–708 (2008)
7. Witten, I.H., Frank, E.: Data Mining: Practical Machine Learning Tools and Techniques, 2nd edn. Morgan Kaufmann, San Francisco (2005)

TeleComVis: Exploring Temporal Communities in Telecom Networks

Qi Ye, Bin Wu, Lijun Suo, Tian Zhu, Chao Han, and Bai Wang

Beijing Key Laboratory of Intelligent Telecommunications Software and Multimedia
Beijing University of Posts and Telecommunications, Beijing, China, 100876
yeqibupt@gmail.com, {wubin,wangbai}@bupt.edu.cn

Abstract. By the structure of call graphs derived from huge amounts of Call Detail Records, we can find out the social communities in the call graphs and make different market strategies for these social communities in real telecom applications. However, traditional telecom business intelligence methods are short of ways to understand the social interactions. To fill this gap, we propose a Telecom Community Visual Analysis prototype tool, called *TeleComVis*, to analyze the call graphs derived from Call Detail Records. In the demo, we will show (1) the functions of *TeleComVis*; (2) the critical techniques of finding statistically significant communities in real-world telecom applications. Using *TeleComVis*, users can both analyze the statistical properties of massive call graphs and explore the statistically significant communities and the temporal links interactively.

Keywords: Visual Analytics, Community, Call Graph, Social Network.

1 Introduction

Nowadays, researchers are increasingly interested in addressing a wide range of challenges residing in social networks. Taking a social network analysis and visual analytics approach, based on the framework of JSNVA [1], we develop a tool called *TeleComVis* in Java programming language to analyze the structure of massive call graphs derived from Call Detail Records (CDR) and the relationships among customers. Currently, many famous social network visualization tools have been proposed, such as Vizster [2]. To overcome the problem of scalability, Vizster allows users to explore massive social networks through egocentric networks. In real telecom applications, there are two scenarios that are not well addressed by current available visual analytical tools: one is how to detect the statistically significant communities accurately and interactively, and the other is how to show the temporal comparison of the call patterns.

2 Tool—*TeleComVis*

TeleComVis keeps three data structures for graphs: the *raw graph* which provides the structure of original graph, the *subgraph* which contains a subgraph in

W. Buntine et al. (Eds.): ECML PKDD 2009, Part II, LNAI 5782, pp. 755–758, 2009.

the *raw graph* and the *community graph* which is an abstract graph derived from the *raw graph* in which each vertex is a community and the edges indicate the relationships between communities. *TeleComVis* classifies the operations into the following steps: preprocessing, topological statistical analysis, community detection and visual analysis.

2.1 Preprocessing and Statistical Analysis

The 'Preprocessing' tab enables users to load network data into *TeleComVis*, as shown in Fig. 1 (a). Once the graph file has been imported, users can import the weight or temporal information of the edges. In the temporal information file, each edge has a natural number $s \in \mathbb{N}$ standing for a time step in which the edge is active. The 'Statistics' tab enables users to use different network algorithms to get the topological statistical properties of the *raw graph*. As shown in Fig. 1 (b), there are several typical network analysis methods to get the statistical properties of the whole graph, such as degree distribution, component distribution, cluster coefficient, maximal clique distribution, etc. To measure the relative importance of customers in call graphs, various centrality algorithms have also been provided. *TeleComVis* also provides several algorithms to sample the *subgraph* from the *raw graph*.

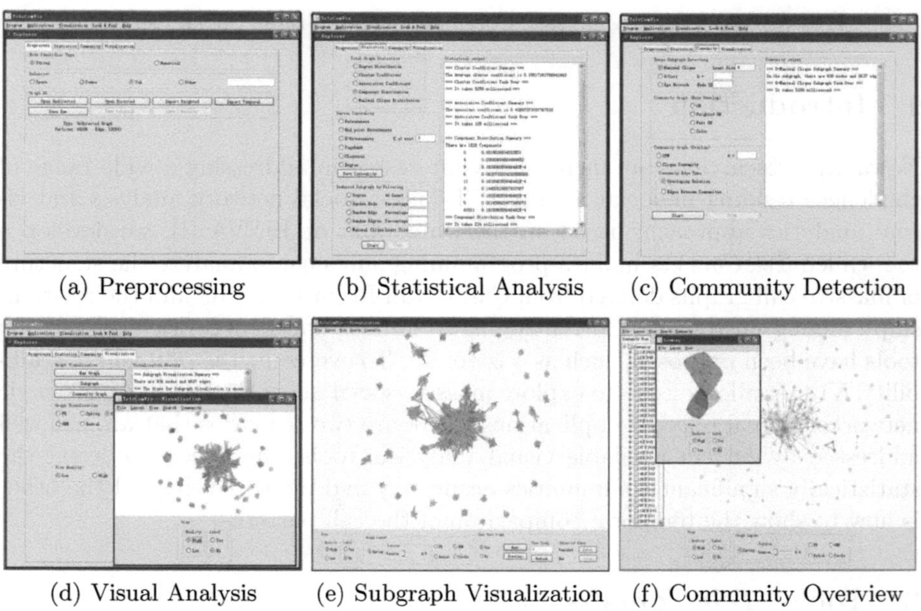

(a) Preprocessing (b) Statistical Analysis (c) Community Detection

(d) Visual Analysis (e) Subgraph Visualization (f) Community Overview

Fig. 1. *TeleComVis* user interface

2.2 Community Detection

The 'Community' tab enables users to get the dense subgraphs or communities from the original *raw graph*, as shown in Fig. 1 (c). To get an overview of massive graphs, one practical strategy is to show the cohesive subgraphs. We call the maximal cliques which contain at least k vertices as k-maximal cliques. We provide k-maximal clique enumeration algorithm and k-core algorithm to explore the dense subgraph in call graphs. *TeleComVis* can enumerate all the maximal cliques in a call graph with tens of thousands of vertices and edges in just several seconds which makes it possible to explore real-world call graphs interactively.

In *TeleComVis*, we provide two types of *community graph* to be analyzed: separated communities and overlapping communities, such as CPM [3]. There are two types of edges between overlapping communities: one shows the connection relationships and the other one shows the overlapping relationships. Fig 1 (f) shows the 8-clique-community graph of a mobile call graph with tens of thousands of vertices and hundreds of thousands of edges using the CPM algorithm.

2.3 Visual Analysis

As shown in Fig. 1 (d), the 'Visualization' tab enables users to show the *raw graph*, *subgraph* and *community graph* in new network visualization frames, respectively. As shown in Fig. 1 (d), (e) and (f), in the network visualization

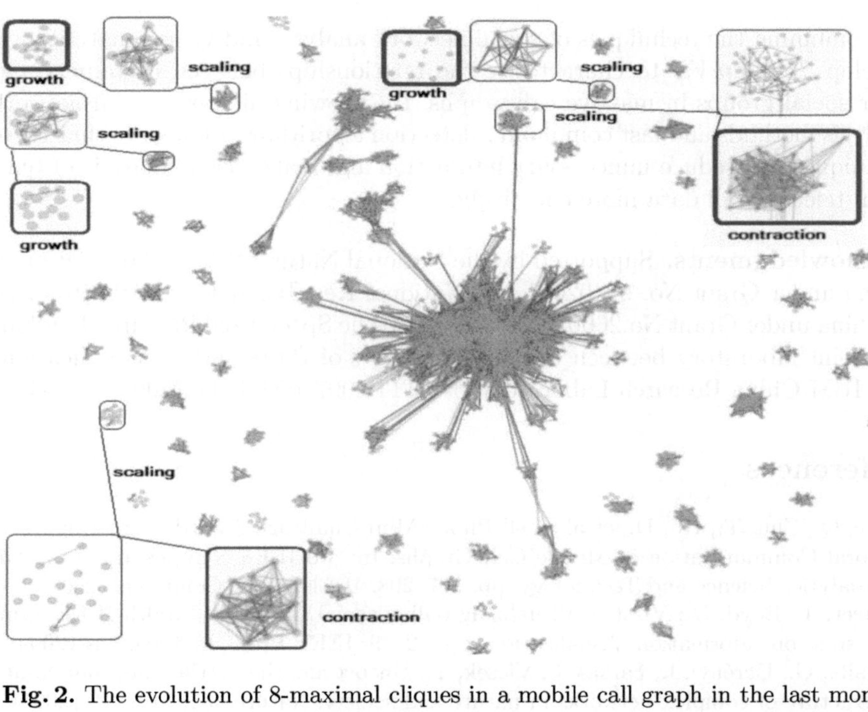

Fig. 2. The evolution of 8-maximal cliques in a mobile call graph in the last month

frame, basic interaction is done with simple mouse operations. Dragging the mouse to select a group of community vertices causes these communities to automatically expand, and users can zoom into the overlapping relationships between actors. They can also drag the vertex in the panel to change its position. It also provides better capabilities to manipulate graph display details, such as labeling, color coding, size normalization and scaling. As show in Fig. 1 (e) and (f), different graph layout algorithms in the framework JSNVA [1] are offered, such as FR layout algorithm, n-body based spring algorithm [2], radial algorithm, etc.

To show temporal links, we define 3 types of edges: the persistent edges which appear both in previous and current time steps, the vanished edges which appear in previous time step but disappear in current time step and the new born edges which do not appear in previous time step but appear in current time step. Fig. 2 shows the dynamical patterns of the subgraph made by 8-maximal cliques in a temporal mobile call graph in a city of China during 7 months which has 49035 vertices and 158543 edges. The grey edges indicate the persistent edges, and the red edges indicate the vanished edges, and the green edges indicate the new born ones. Fig. 2 shows the structure of the subgraph in the last month. To show temporal links, the link relationships in the second month are shown in the red frame.

3 Summary

By combining the techniques of social network analysis and visual analytics, we develop *TeleComVis* to characterize the relationships between customers and their social groups in massive call graphs. By allowing different social network analysis methods and fast community detection algorithms, we believe that these techniques can reduce unnecessary interaction and make visual analysis of temporal telecom call data more effectively.

Acknowledgments. Supported by the National Natural Science Foundation of China under Grant No. 60402011, the National Key Technology R&D Program of China under Grant No.2006BAH03B05, and the Specialized Research Fund for the Joint laboratory between Beijing University of Posts and Communications and IBM China Research Laboratory (No.JTP20071002-4, JTP200806014-3).

References

1. Ye, Q., Zhu, T., Hu, D., et al.: Cell Phone Mini Challenge Award: Exploring Temporal Communication in Mobile Call Graphs. In: 3rd IEEE Symposium on Visual Analytics Science and Technology, pp. 207–208. IEEE Press, Columbus (2008)
2. Heer, J., Boyd, D.: Vizster: Visualizing online social networks. In: 9th IEEE Symposium on Information Visualization, pp. 32–39. IEEE Press, Minneapolis (2005)
3. Palla, G., Derényi, I., Farkas, I., Vicsek, T.: Uncovering the overlapping community structure of complex networks in nature and society. Nature 435, 814–817 (2005)

Author Index

MIX
Papier aus verantwortungsvollen Quellen
Paper from responsible sources
FSC® C105338
FSC
www.fsc.org

If you have any concerns about our products,
you can contact us on
ProductSafety@springernature.com

In case Publisher is established outside the EU,
the EU authorized representative is:
Springer Nature Customer Service Center GmbH
Europaplatz 3, 69115 Heidelberg, Germany

Printed by Libri Plureos GmbH
in Hamburg, Germany